Dieter Meschede • Thomas Udem •
Tilman Esslinger
Editors

Exploring the World with the Laser

Dedicated to Theodor Hänsch on his 75th birthday

Springer

Editors
Dieter Meschede
Institute for Applied Physics
University of Bonn
Bonn, Germany

Thomas Udem
Max-Planck-Institute for Quantenoptik (MPQ)
Garching, Germany

Tilman Esslinger
ETH Zurich
Institute for Quantum Electronics
Zurich, Switzerland

ISBN 978-3-319-64345-8 ISBN 978-3-319-64346-5 (eBook)
https://doi.org/10.1007/978-3-319-64346-5

Library of Congress Control Number: 2017951119

Printed on acid-free paper

This Springer imprint is published by Springer Nature
The registered company is Springer International Publishing AG
The registered company address is: Gewerbestrasse 11, 6330 Cham, Switzerland

"Enlightening the World with the Laser"—Honoring T. W. Hänsch[1]

As guest editors of the T. W. Hänsch special issue of *Applied Physics B*, we are delighted to introduce you to a series of articles authored in honor of Professor Theodor W. "Ted" Hänsch, by colleagues, co-workers and (former) students of his. The occasion is Ted's 75th birthday, but the issue celebrates some 50 years of cutting-edge research that he performed first at the University of Heidelberg in the late 1960s, then at Stanford University, and since 1986 at the Max Planck Institute of Quantum Optics in Garching and the Ludwig-Maximilian University of Munich. *Applied Physics B* is a natural venue for such a Festschrift, given that Ted is serving on its Editorial Board since 1983.

How Ted Hänsch has influenced the way we think about lasers, and how we use them, is truly remarkable. His 75th birthday marks a great occasion to pause and reflect on his achievements. Ted Hänsch's passion is precision measurements and the study of the hydrogen atom. Yet, even if the Nobel Prize came for the ingenious frequency comb, "precision" and "hydrogen" do not quite cover his approach to physics. More generally, Ted keeps teaching us how we can use laser light, sometimes for rather playful applications, sometimes for fundamental break-throughs—and from time to time also for commercial devices. His unique way of doing physics, however, goes much further. In addition to leading the way with groundbreaking research on topics from precision laser spectroscopy to ultracold quantum gases, Ted has motivated an entire generation of physicists to pursue related goals. He has generated a lasting impact in several communities, not least thanks to the large number of alumni from his research group who have gone on to develop careers of their own, inspired by Ted's example.

The esteem in which Ted is held by colleagues, collaborators and friends from around the world is reflected in the large number of excellent articles in this volume. We are grateful to the many authors who contributed such interesting papers. As

[1]This article is part of the topical collection "Enlightening the World with the Laser" - Honoring T. W. Hänsch guest edited by Tilman Esslinger, Nathalie Picqué, and Thomas Udem.

you browse this issue, you will find papers by outstanding scientists who are working in the broad, interdisciplinary field of atomic, molecular and optical physics. Most contributions highlight the influence of Ted's scientific activities, current and past. Exciting new findings regarding precision spectroscopy of atoms and molecules are reported, alongside intriguing contributions in the areas of opto-mechanics, ultracold atomic and molecular quantum gases and matter-wave optics, as well as works on the development and application of novel laser sources, including frequency combs. Other articles provide insightful perspectives and reviews dedicated to various aspects of quantum and optical sciences.

We hope you will join us in celebrating Ted's achievements and enjoy this collection of papers.

Congratulations on your 75th birthday, Ted! We look forward to many more ideas on how to explore the world with laser light.

Swiss Federal Institute of Technology
Switzerland
Tilman Esslinger

Max Planck Institute of Quantum Optics
Germany
Nathalie Picqué

Max Planck Institute of Quantum Optics
Germany
Thomas Udem

Zurich and Garching
October 19, 2016

Contents

Unexpectedly Large Difference of the Electron Density at the Nucleus in the 4*p* $^2P_{1/2,3/2}$ Fine-Structure Doublet of Ca^+ 1
C. Shi, F. Gebert, C. Gorges, S. Kaufmann, W. Nörtershäuser, B.K. Sahoo, A. Surzhykov, V.A. Yerokhin, J.C. Berengut, F. Wolf, J.C. Heip, and P.O. Schmidt

New Avenues for Matter-Wave-Enhanced Spectroscopy 21
Jonas Rodewald, Philipp Haslinger, Nadine Dörre, Benjamin A. Stickler, Armin Shayeghi, Klaus Hornberger, and Markus Arndt

Yb Fiber Amplifier at 972.5 nm with Frequency Quadrupling to 243.1 nm 35
Z. Burkley, C. Rasor, S.F. Cooper, A.D. Brandt, and D.C. Yost

Optical Autler–Townes Spectroscopy in a Heteronuclear Mixture of Laser-Cooled Atoms 45
C. Bruni, F. Münchow, and A. Görlitz

From Quantum Transitions to Electronic Motions 59
Ferenc Krausz

Measuring and Imaging Nanomechanical Motion with Laser Light . . . 71
Andreas Barg, Yeghishe Tsaturyan, Erik Belhage, William H.P. Nielsen, Christoffer B. Møller, and Albert Schliesser

Quantum Regime of a Free-Electron Laser: Relativistic Approach . . . 87
Peter Kling, Roland Sauerbrey, Paul Preiss, Enno Giese, Rainer Endrich, and Wolfgang P. Schleich

Playing Tricks to Ions 117
Dietrich Leibfried

Coherence in Laser-Driven Electrons at the Surface and in the Volume of Solid Matter 129
Peter Hommelhoff

Atomic "Bomb Testing": The Elitzur–Vaidman Experiment Violates the Leggett–Garg Inequality 141
Carsten Robens, Wolfgang Alt, Clive Emary, Dieter Meschede, and Andrea Alberti

Adjacency Graphs and Long-Range Interactions of Atoms in Quasi-degenerate States: Applied Graph Theory 159
C.M. Adhikari, V. Debierre, and U.D. Jentschura

Frequency-Comb Referenced Collinear Laser Spectroscopy of Be$^+$ for Nuclear Structure Investigations and Many-Body QED Tests 175
A. Krieger, W. Nörtershäuser, Ch. Geppert, K. Blaum, M.L. Bissell, N. Frömmgen, M. Hammen, K. Kreim, M. Kowalska, J. Krämer, R. Neugart, G. Neyens, R. Sánchez, D. Tiedemann, D.T. Yordanov, and M. Zakova

High-Accuracy Deep-UV Ramsey-comb Spectroscopy in Krypton 207
Sandrine Galtier, Robert K. Altmann, Laura S. Dreissen, and Kjeld S.E. Eikema

Phase-Stable, Multi-μJ Femtosecond Pulses from a Repetition-Rate Tunable Ti:Sa-Oscillator-Seeded Yb-Fiber Amplifier 225
T. Saule, S. Holzberger, O. De Vries, M. Plötner, J. Limpert, A. Tünnermann, and I. Pupeza

Quantum Electrodynamics, High-Resolution Spectroscopy and Fundamental Constants 237
Savely G. Karshenboim and Vladimir G. Ivanov

High-Precision Spectroscopy of the HD$^+$ Molecule at the 1-p.p.b. Level 267
J. Biesheuvel, J.-Ph. Karr, L. Hilico, K.S.E. Eikema, W. Ubachs, and J.C.J. Koelemeij

Orientational Dependence of Optically Detected Magnetic Resonance Signals in Laser-Driven Atomic Magnetometers 309
Simone Colombo, Vladimir Dolgovskiy, Theo Scholtes, Zoran D. Grujić, Victor Lebedev, and Antoine Weis

All Polarization-Maintaining Fiber Laser Architecture for Robust Femtosecond Pulse Generation 331
Wolfgang Hänsel, Heinar Hoogland, Michele Giunta, Sebastian Schmid, Tilo Steinmetz, Ralf Doubek, Peter Mayer, Sven Dobner, Carsten Cleff, Marc Fischer, and Ronald Holzwarth

A >2-MJ, 10^{14}-W Laser System for DT Fusion—NIF: A Note in Celebration of the 75th Birthday of Prof. Theodore Haensch 341
John F. Holzrichter and Kenneth R. Manes

Topological Spin Models in Rydberg Lattices 351
Martin Kiffner, Edward O'Brien, and Dieter Jaksch

Ultrafast Optomechanical Pulse Picking 371
Nikolai Lilienfein, Simon Holzberger, and Ioachim Pupeza

Shifting the Phase of a Coherent Beam with a $^{174}Yb^{+}$ Ion: Influence of the Scattering Cross Section 389
Martin Fischer, Bharath Srivathsan, Lucas Alber, Markus Weber, Markus Sondermann, and Gerd Leuchs

Thirty Years of Hydrogen Spectroscopy in Paris 401
François Biraben, Lucile Julien, and François Nez

A Dense Gas of Laser-Cooled Atoms for Hybrid Atom–Ion Trapping 417
Bastian Höltkemeier, Julian Glässel, Henry López-Carrera, and Matthias Weidemüller

A Compact Echelle Spectrograph for Characterization of Astro-Combs 433
Rafael A. Probst, Tilo Steinmetz, Yuanjie Wu, Frank Grupp, Thomas Udem, and Ronald Holzwarth

Mode Locking Based on the Temporal Talbot Effect 447
Thomas Udem and Akira Ozawa

T^3-Interferometer for Atoms 457
M. Zimmermann, M.A. Efremov, A. Roura, W.P. Schleich, S.A. DeSavage, J.P. Davis, A. Srinivasan, F.A. Narducci, S.A. Werner, and E.M. Rasel

Double-Electron Ionization Driven by Inhomogeneous Fields 491
A. Chacón, L. Ortmann, F. Cucchietti, N. Suárez, J.A. Pérez-Hernández, M.F. Ciappina, A.S. Landsman, and M. Lewenstein

Parity-Nonconserving Interaction-Induced Light Shifts in the $7S_{1/2}$–$6D_{3/2}$ Transition of the Ultracold ^{210}Fr Atoms to Probe New Physics Beyond the Standard Model 509
T. Aoki, Y. Torii, B.K. Sahoo, B.P. Das, K. Harada, T. Hayamizu, K. Sakamoto, H. Kawamura, T. Inoue, A. Uchiyama, S. Ito, R. Yoshioka, K.S. Tanaka, M. Itoh, A. Hatakeyama, and Y. Sakemi

Single-Slit Focusing and Its Representations 529
Manuel R. Gonçalves, William B. Case, Ady Arie, and Wolfgang P. Schleich

Nonadiabatic Molecular Alignment of Linear Molecules Probed by Strong-Field Ionization Yields of Photoelectrons 577
G. Kaya, N. Kaya, J. Strohaber, N.A. Hart, A.A. Kolomenskii, and H.A. Schuessler

Ultracold Metastable Helium: Ramsey Fringes and Atom Interferometry 597
W. Vassen, R.P.M.J.W. Notermans, R.J. Rengelink, and R.F.H.J. van der Beek

Monolayer Graphene as Dissipative Membrane in an Optical Resonator 617
Hendrik M. Meyer, Moritz Breyer, and Michael Köhl

Cavity Ring-Up Spectroscopy for Dissipative and Dispersive Sensing in a Whispering Gallery Mode Resonator 629
Yong Yang, Ramgopal Madugani, Sho Kasumie, Jonathan M. Ward, and Síle Nic Chormaic

Sensitivity and Resolution in Frequency Comb Spectroscopy of Buffer Gas Cooled Polyatomic Molecules 647
P. Bryan Changala, Ben Spaun, David Patterson, John M. Doyle, and Jun Ye

Laser Phase Spectroscopy in Closed-Loop Multilevel Schemes 665
Ennio Arimondo

Precision Measurements and Test of Molecular Theory in Highly Excited Vibrational States of H_2 ($v = 11$) 679
T. Madhu Trivikram, M.L. Niu, P. Wcisło, W. Ubachs, and E.J. Salumbides

Ranging with Frequency-Shifted Feedback Lasers: From µm-Range Accuracy to MHz-Range Measurement Rate 701
J.I. Kim, V.V. Ogurtsov, G. Bonnet, L.P. Yatsenko, and K. Bergmann

Absorption Spectroscopy of Xenon and Ethylene–Noble Gas Mixtures at High Pressure: Towards Bose–Einstein Condensation of Vacuum Ultraviolet Photons 729
Christian Wahl, Rudolf Brausemann, Julian Schmitt, Frank Vewinger, Stavros Christopoulos, and Martin Weitz

Extraction of Enhanced, Ultrashort Laser Pulses from a Passive 10-MHz Stack-and-Dump Cavity 741
Sven Breitkopf, Stefano Wunderlich, Tino Eidam, Evgeny Shestaev, Simon Holzberger, Thomas Gottschall, Henning Carstens, Andreas Tünnermann, Ioachim Pupeza, and Jens Limpert

External Cavity Diode Laser Setup with Two Interference Filters 753
Alexander Martin, Patrick Baus, and Gerhard Birkl

Bose–Einstein Condensates in an Optical Cavity with Sub-recoil Bandwidth 765
J. Klinder, H. Keßler, Ch. Georges, J. Vargas, and A. Hemmerich

Ted's Stanford Days 791
David H. McIntyre and Allister I. Ferguson

From Laser Cooling to the Superfluid Mott Insulator 805
Tilman Esslinger

Unexpectedly Large Difference of the Electron Density at the Nucleus in the 4p $^2P_{1/2,3/2}$ Fine-Structure Doublet of Ca^+

C. Shi, F. Gebert, C. Gorges, S. Kaufmann, W. Nörtershäuser, B.K. Sahoo, A. Surzhykov, V.A. Yerokhin, J.C. Berengut, F. Wolf, J.C. Heip, and P.O. Schmidt

Abstract We measured the isotope shift in the $^2S_{1/2} \rightarrow {}^2P_{3/2}$ (D2) transition in singly ionized calcium ions using photon recoil spectroscopy. The high accuracy of the technique enables us to compare the difference between the isotope shifts of this transition to the previously measured isotopic shifts of the $^2S_{1/2} \rightarrow {}^2P_{1/2}$ (D1) line. This so-called splitting isotope shift is extracted and exhibits a clear signature of field shift contributions. From the data, we were able to extract the small difference of the field shift coefficient and mass shifts between the two transitions with high

---

This article is part of the topical collection "Enlightening the World with the Laser" - Honoring T. W. Hänsch guest edited by Tilman Esslinger, Nathalie Picqué, and Thomas Udem.

C. Shi • F. Gebert • F. Wolf • J.C. Heip
Physikalisch-Technische Bundesanstalt, 38116 Braunschweig, Germany

C. Gorges • S. Kaufmann • W. Nörtershäuser
Institut für Kernphysik, Technische Universität Darmstadt, 64289 Darmstadt, Germany

B.K. Sahoo
Atomic, Molecular and Optical Physics Division, Physical Research Laboratory, Navrangpura, Ahmedabad 380009, India

A. Surzhykov
Physikalisch-Technische Bundesanstalt, 38116 Braunschweig, Germany

Technische Universität Braunschweig, 38106 Braunschweig, Germany

V.A. Yerokhin
Physikalisch-Technische Bundesanstalt, 38116 Braunschweig, Germany

Center for Advanced Studies, Peter the Great St. Petersburg Polytechnic University, St. Petersburg, Russia, 195251

J.C. Berengut
School of Physics, University of New South Wales, New South Wales 2052, Sydney, Australia

P.O. Schmidt (✉)
Physikalisch-Technische Bundesanstalt, 38116 Braunschweig, Germany

Institut für Quantenoptik, Leibniz Universität Hannover, 30167 Hannover, Germany
e-mail: piet.schmidt@quantummetrology.de

D. Meschede et al. (eds.), *Exploring the World with the Laser*,
https://doi.org/10.1007/978-3-319-64346-5_1

accuracy. This J-dependence is of relativistic origin and can be used to benchmark atomic structure calculations. As a first step, we use several ab initio atomic structure calculation methods to provide more accurate values for the field shift constants and their ratio. Remarkably, the high-accuracy value for the ratio of the field shift constants extracted from the experimental data is larger than all available theoretical predictions.

1 Introduction

The study of isotopic shifts in atomic systems has a long history [1], and a profound understanding of the isotope shift and theoretical calculations of the atomic properties is important in many applications. These reach from the extraction of nuclear properties from atomic spectra to applications in astronomy and fundamental physics. The spectrum of atoms and ions encodes information that provides a key to the ground-state properties of nuclei [2, 3], small parity-violating effects caused by the weak interaction [4], for unitarity tests of the Cabibbo–Kobayashi–Maskawa (CKM) matrix [5], or for probing the Higgs coupling between electrons and quarks [6].

Light appearing on earth from stars at large distances is red-shifted and can provide information about the spectra of atoms in ancient times and whether there have been changes, for example, from a variation of the fine-structure constant α. However, isotopic composition can also contribute to the observed shifts since the isotopes have different resonance frequencies, but this so-called isotope shift is usually not resolved [7–9]. In this respect, isotope shift calculations became recently an important topic with the goal to either determine the influence of the isotopic abundance of the observed species on the analysis for a change in α [7, 10] or to provide information of the isotopic composition in the ancient times of the universe. Calcium is an element of considerable interest for many of the cases mentioned above. For example, the isotope shift information in the $3d\ {}^2\mathrm{D}_J \rightarrow 4p\ {}^2\mathrm{P}_J$ infrared triplet [11] led to the discovery of an anomalous isotopic composition in mercury-manganese (HgMn) stars, in which the isotopic Ca ratio in the stellar atmosphere is dominated by ^{48}Ca [12, 13]. Isotope shifts in other calcium transitions have been studied, e.g., to extract nuclear charge radii along the long chain of isotopes [14–17] or to perform ultra-trace analysis using isotope selective resonance ionization [18]. Moreover, the calcium ion is a workhorse in the field of quantum-optical applications and transition frequencies and isotope shifts of stable isotopes were measured with high accuracy [19–22] and supported on-line studies of exotic isotopes since they serve as calibration points [17, 23].

The calcium isotopic chain is quite unique since it contains two stable doubly magic isotopes ${}^{40,48}\mathrm{Ca}^+$ that have practically identical mean-square charge radii ($\delta\langle r_c^2\rangle^{40,48} = -0.0045(60)$ [22]) even though they are 20% different in mass. This has been established using a variety of techniques, i.e., elastic electron scattering [24], muonic atom spectroscopy [25] as well as optical isotope shift data, e.g. [15] (for a synopsis see, e.g. [26] and references therein). The negligible change in nuclear charge radius between the two isotopes allows for a cleaner separation of mass and field shifts than in most other multi-electron systems. The high-precision data presented here will provide important benchmarks for improved calculations of this reference system.

The extraction of nuclear parameters from atomic spectra is strongly facilitated by atomic structure calculations. Despite being a lighter system with only 19 electrons, isotope shifts in the singly ionized calcium (Ca^+) have not been studied rigorously. High-precision calculations of magnetic dipole and electric quadrupole hyperfine-structure constants have been performed in this ion using an all order relativistic many-body theory in the coupled-cluster (RCC) theory framework [27]. However, calculations of field shift and mass shift constants that are required to estimate isotope shifts have not been performed at the same level of accuracy yet.

Here we present a high-precision absolute frequency measurement of the D2 line of $^{40}Ca^+$ with 100 kHz accuracy, representing a fivefold improvement over previous results [20]. Isotope shift measurements of this transition with the same resolution are compared with a measurement of the D1 line [22]. A clear signature of field shift contributions to the splitting isotope shift is observed. To explain this finding, we have also performed several ab initio calculations of field shift constants employing a hydrogenic method, a mean-field method using Dirac-Fock (DF) equation, and state-of-the-art atomic structure calculations.

2 Experimental Setup

The isotope shift in the $^2S_{1/2} \rightarrow {}^2P_{3/2}$ (D2) transition of even calcium isotopes was measured by photon recoil spectroscopy, as described in detail in references [21, 22]. In brief, we trap a singly charged $^{25}Mg^+$ ion together with the calcium isotope under investigation in a linear Paul trap. The $^{25}Mg^+$ ion is used to sympathetically cool the axial normal mode of the two-ion crystal to the ground state [28]. To probe the transition, a series of 70 pulses of the spectroscopy laser with a pulse length of 125 ns, synchronized to one of the motional frequencies of the two-ion crystal, are applied. Each spectroscopy laser pulse is followed by 200 ns short repump pulses on the $^2D_{3/2} \rightarrow {}^2P_{1/2}$ and $^2D_{5/2} \rightarrow {}^2P_{3/2}$ transitions at 866 and 854 nm, respectively. Recoil kick upon photon absorption on the spectroscopy transition results in excitation of nearly coherent motion. This motional excitation is mapped into an electronic excitation using a stimulated rapid adiabatic passage (STIRAP) pulse on the $^{25}Mg^+$ ion [29]. The high photon sensitivity of this technique provides a large signal-to-noise ratio, resulting in a resolution of about 100 kHz in less than 15 min of averaging time. As a consequence of the smallness of systematic effects, the accuracy of absolute frequency measurements is also about 100 kHz. Isotope shift measurements benefit from further suppression of systematic effects, since most of them are common to all isotopes.

Spectroscopy is performed using a cw single-mode Ti:Sa laser (Sirah, Matisse TS) which is frequency doubled in an enhancement cavity (Spectra-Physics, WaveTrain). The frequency of the laser is locked to an erbium-fiber-laser-based frequency comb. The comb is stabilized in its offset and repetition frequencies to a H-maser frequency which is calibrated by a cesium fountain at PTB (German National Metrology Institute) as shown in Fig. 1. A beat signal between the spectroscopy laser and the nearest comb tooth of a narrow-band frequency-doubled output of the frequency comb is detected with a fast photo diode. The nearest comb tooth is identified by measuring the frequency of the spectroscopy laser with a wavemeter (High Finesse, model WS-7).

The RF signal is monitored by a spectrum analyzer (Rohde & Schwarz, FSL3) and mixed down to 10 or 35 MHz using an rf synthesizer. The band-pass-filtered signal is used as the input signal of a self-build phase frequency comparator (PFC) operating at 10 or 35 MHz, which produces an error signal that is tailored by a proportional-integral (PI) controller. The generated control signal is used to correct the frequency of the Ti:Sa laser by changing the length of the laser cavity with a fast piezo-electric actuator. By adjusting the frequency of the rf synthesizer, the frequency of the spectroscopy laser can be adjusted to the resonances of the different isotopes. Figure 2 shows the in-loop beat signal between the Ti:Sa laser and the frequency comb, indicating an upper bound for the linewidth of the spectroscopy laser below 260 kHz. The long-term frequency instability is determined by counting

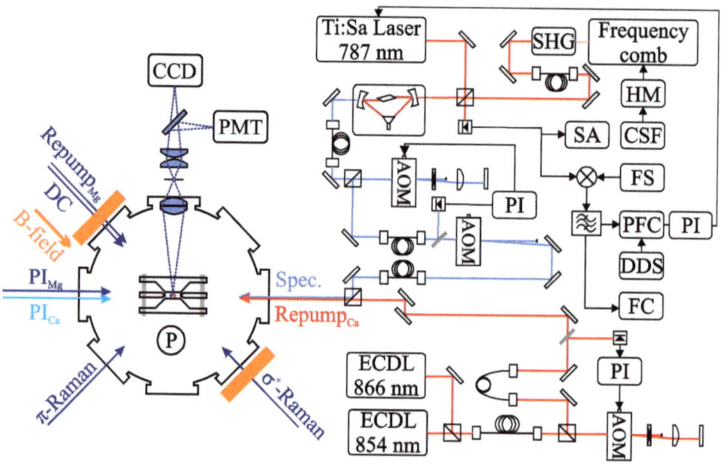

Fig. 1 Simplified experimental setup. The *left part* of this figure indicates the configuration of all optical beams and bias magnetic field with respect to the ions in the trap. The *right part* shows the optical setup for the spectroscopy beam and the calcium repump beam. For details, see text. All RF sources are referenced to a 10-MHz signal from the H-maser (HM). *AOM* acousto-optical modulator, *CCD* electron-multiplication charge-coupled device, *CSF* cesium fountain clock, *DC* Doppler cooling beam, *ECDL* external-cavity diode laser, *FC* frequency counter, *FS* frequency synthesizer, *P* Parabolic mirror, *PFC* phase and frequency comparator, *PI* proportional-integral controller, $PI_{\mathrm{Mg/Ca}}$ photo-ionization beam for Mg/Ca, *PMT* photomultiplier tube, *SA* spectrum analyzer, *SHG* second harmonic generation

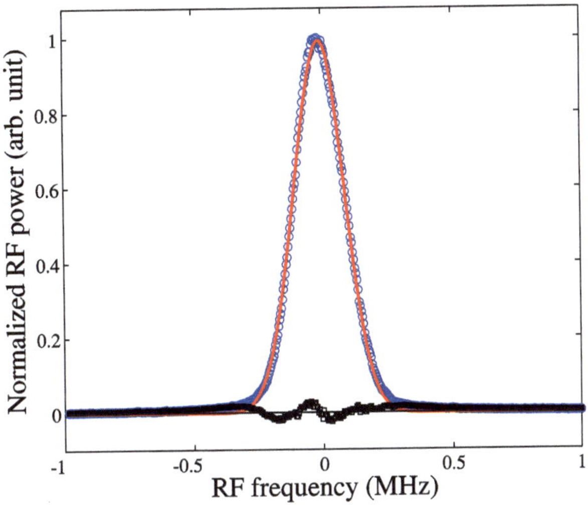

Fig. 2 In-loop beat signal of the Ti:Sa laser and the nearest frequency comb tooth. The data recorded by the spectrum analyzer with 10 kHz resolution bandwidth (*blue circles*) is fit with a Gaussian profile (*red curve*) that gives an upper bound of 260 kHz for the linewidth of the spectroscopy laser. The *black squares* present the residuals of the fit

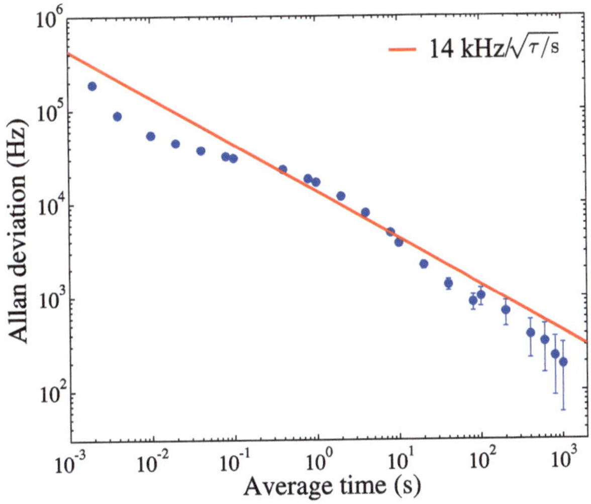

Fig. 3 Allan deviation of the Ti:Sa laser frequency as measured by the frequency counter. The long-term instability of the locked spectroscopy laser is 14 kHz $/\sqrt{\tau/s}$ for integration times longer than 1 s as derived from a fit to the data (*red line*)

the mixed-down RF signal with a frequency counter (Kramer+Klische FXE). We derive a frequency instability of about 14 kHz/$\sqrt{\tau/s}$ from an Allan deviation of these measurements as shown in Fig. 3. The linewidth and the frequency instability both fulfil the requirements for the anticipated resolution and accuracy of below 100 kHz. The spectroscopy beam is intensity stabilized and frequency scanned by an acousto-optical modulator (AOM), while another AOM is used for switching.

The 866 and 854 nm repump beams depopulating the two D-states are generated by two external-cavity diode lasers. The combined beam consisting of the two repumpers is intensity stabilized by an AOM which is also used for fast switching. The left part of Fig. 1 shows the direction of the beams together with the magnetic bias field. It also contains the Doppler cooling, Raman, and repump beams for controlling the $^{25}Mg^+$ ion as described in detail, e.g., in [28].

3 Isotope Shift

Isotope shifts in atomic transition frequencies originate from two effects: (1) a change in the size of the nucleus and a corresponding change of potential experienced by electrons having a finite probability density inside the nucleus (field shift), and (2) a change of the center of mass motion of the nucleus (mass shift). The mass effect is more important for light nuclei due to the relatively large mass differences, while the field shift is proportional to the electron density inside the nucleus, which increases roughly with Z^2 and is the dominant contribution in elements with large atomic number Z [1]. The isotope shift is defined as $\delta\nu_i^{A,A'} = \nu_i^{A'} - \nu_i^{A}$, where $\nu_i^{A'}$ and ν_i^{A} are the frequencies of transition i for isotopes with mass $m_{A'}$ and m_A, respectively. The mass shift contribution to the isotope shift of a one-electron atom can be calculated by replacing the electron mass m_e by the reduced mass of the system and leads to

$$\delta\nu_{i,\mathrm{NMS}}^{A,A'} = m_e \nu_i^A \times \frac{m_{A'} - m_A}{m_A \cdot (m_{A'} + m_e)}. \tag{1}$$

It is called the normal mass shift (NMS) and the prefactor of the mass scaling $K_{i,\mathrm{NMS}} = m_e \nu_i^A$ is the normal mass shift constant. In a multi-electron system, the nuclear recoil depends on the sum of all electron momenta and therefore the mass polarization term or specific mass shift (SMS) can have a considerable contribution with the same mass scaling. The corresponding constant $K_{i,\mathrm{SMS}}$ is notoriously difficult to calculate since it includes all electron momenta. The difference in the transition frequency, $\delta\nu_i^{A,A'}$, between isotopes with mass m_A and $m_{A'}$ can in total be expressed as

$$\delta\nu_i^{A,A'} = K_i \frac{m_{A'} - m_A}{m_A \cdot m_{A'}} + F_i\, \delta\langle r_c^2\rangle^{A,A'}, \tag{2}$$

where $K_i = K_{i,\mathrm{NMS}} + K_{i,\mathrm{SMS}}$ is the (total) mass shift constant, F_i is the field shift constant, and $\delta\langle r_c^2\rangle^{A,A'}$ is the corresponding change in the mean-square nuclear charge radius of the two isotopes. Here we simplified the mass-dependence term by the approximated dependence as it is usally used in the literature. Neglecting the additional electron mass in the denominator of Eq. (1) leads to a change only of the order of $(m_e/m_A)^2$. To extract nuclear charge radii from isotope shifts, one needs reliable numbers for the mass shift and field shift constants, which can be obtained only from semi-empirical approaches or from ab initio calculations. Reasonable agreement is usually obtained between these techniques. One of the most important procedures in this respect is the King plot [30], of which two versions exist that have different applications. Both will be used below when analyzing the data.

The general approach is to multiply both sides of Eq. (2) with the inverse mass factor $\mu = m_A \cdot m_{A'}/(m_{A'} - m_A)$ and obtain

$$\mu\,\delta\nu_i^{A,A'} = K_i + F_i\,\mu\,\delta\langle r_c^2\rangle^{A,A'}\,. \tag{3}$$

The relation allows one to eliminate the unknown nuclear charge radii if measurements are performed in two different transitions i and j for the same isotopes

$$\mu\,\delta\nu_i^{A,A'} = K_i - \frac{F_i}{F_j}K_j + \frac{F_i}{F_j}\mu\,\delta\nu_j^{A,A'}\,. \tag{4}$$

This is a linear relation between the so-called modified isotope shifts $\mu\,\delta\nu^{A,A'}$ in the two transitions i, j and can be used to extract the respective ratio of the field shift constants or the relation of the mass shift constants. Alternatively, Eq. (3) can be used with data of a single transition to directly obtain field shift and mass shift constants. This requires mean-square charge radii or their respective changes $\delta\langle r_c^2\rangle^{A,A'}$ for a subset of at least 3 isotopes from other sources. Usually, radii from elastic electron scattering or X-ray transitions in muonic atoms are used, see, e.g., Ref. [31]. By plotting the modified isotope shift versus the modified changes in the rms charge radii $\mu\,\delta\,\langle r_c^2\rangle^{A,A'}$, a linear regression delivers the field shift constant as the slope and the mass shift constant as the intercept with the y-axis. A multidimensional regression can be used if information from several transitions is available.

Applying the extracted mass shift and field shift constants, charge radii of other, especially short-lived isotopes can be obtained. However, the accuracy of the extracted nuclear charge radii with these procedures is often hampered by the insufficient accuracy of the charge radii data from external sources or the limited number of stable isotopes. Odd-Z elements, for example, have only one or two stable isotopes, which renders the usage of the King-plot procedure impossible. High-precision isotope shift data with accuracy better than 1 MHz has rarely been obtained beyond the lightest elements hydrogen [32], helium [33–38], lithium [39–44] and beryllium [45–47] where they are used to either extract nuclear charge radii of stable and short-lived isotopes or to test many-body non-relativistic quantum electrodynamics calculations (NR-QED). Similarly accurate data does so far only exist for the D1 and D2 lines in magnesium [48, 49] and the D1 line in calcium [21, 22], all obtained by laser spectroscopy in Paul traps. Small differences in mass corrected isotope shifts between transitions of the same fine-structure multiplet provide information on subtle (relativistic) effects of the electronic wavefunction. Briefly, the solution of the Dirac equation for a bound electron is a bispinor, of which the upper component has a large and the lower component has a small amplitude [50]. In the nonrelativistic limit only the large (i.e., upper) component remains. With increasing Z, however, the contribution of the small component increases. It has parity opposite to that of the large component, but the same total angular momentum. Hence, the $p_{1/2}$ state acquires through this mechanism a small contribution of an s state and therefore some small probability density within the nucleus. The $p_{3/2}$ state on the other hand has only an admixture of the corresponding d state and therefore a lower probability density at the nucleus than the $p_{1/2}$. Such a J-dependence of the field shift constant has already been reported in the $6s \rightarrow 6p$ doublet in Ba II, where a field shift difference of 2.5(3)% was observed [51]. In lighter isotopes these effects, caused by different contributions of the smaller component of the Dirac wavefunction, are expected to be much smaller and were so far not reported. Even in the most precise ab initio calculations up to Be, field shift factors are assumed to be equal for both transitions of the respective doublets, whereas a small relativistic mass-dependent change was included. In the case of Mg, the transition frequencies were measured only in two isotopes and, thus, a King-plot analysis could not be performed. Here, we report on high-precision calcium isotope shift measurements in the D2 line for ^{40,42,44,48}Ca from which we extract differences in the probability density of the electrons at the nucleus and the mass dependence of the difference between the isotope shifts in the two transitions i, j of a fine-structure doublet, the so-called splitting isotope shift (SIS) [39]

$$\delta\nu_{\mathrm{SIS}}^{A,A'} = \delta\nu_i^{A,A'} - \delta\nu_j^{A,A'}, \tag{5}$$

which can also be understood as the change of the fine structure splitting between the respective isotopes.

4 Experimental Results

4.1 Transition Frequencies

The absolute frequency of the D2 transition of $^{40}Ca^+$ is measured using the photon recoil spectroscopy technique with an accuracy of better than 100 kHz. Table 1 lists this new value together with previously measured transitions using the same technique [21, 22] and literature values. As for the previous measurements, the main systematic shifts of the D2 transition frequency include the lineshape shift, AC-Stark shift, Zeeman shift, and AOM-envelope shift. We evaluate the shifts

Table 1 Absolute frequency of the $^2S_{1/2} \rightarrow {}^2P_{3/2}$, $^2S_{1/2} \rightarrow {}^2P_{1/2}$, and $^2D_{3/2} \rightarrow {}^2P_{1/2}$ transition of $^{40}Ca^+$

Transition	Frequency (MHz)	Refs.
$^2S_{1/2}\rightarrow{}^2P_{3/2}$	761,905,012.599 (82)	This work
	761,905,012.7 (5)	[20]
$^2S_{1/2}\rightarrow{}^2P_{1/2}$	755,222,765.896 (88)	[21]
	755,222,766.2 (17)	[19]
$^2D_{3/2}\rightarrow{}^2P_{1/2}$	346,000,234.867 (96)	[22]

Table 2 Uncertainty estimation of the absolute D2 transition frequency in $^{40}Ca^+$

Systematic effect	Shift	Uncertainty
Zeeman (static magnetic field)	−8	59
AC Stark (spectroscopy laser)	60	44
Lineshape (detection scheme)	152	20
Spectroscopy pulse envelope	0	16
Spectroscopy laser lock	0	0.6
Statistics	0	27
Total	204	82

All values are in kHz

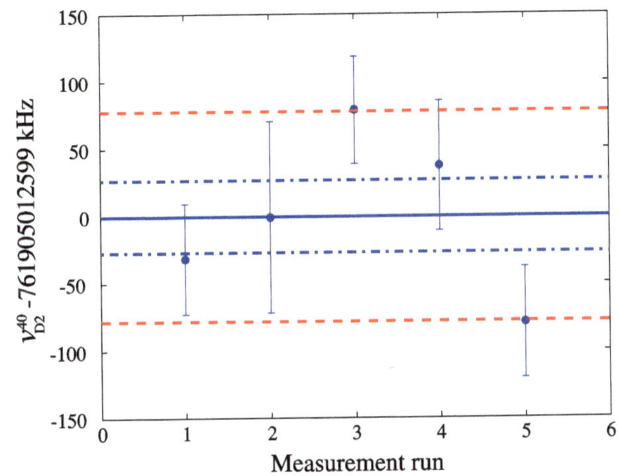

Fig. 4 Results of five independent measurements of the absolute transition frequency in the D2 line of $^{40}Ca^+$. The *blue solid line* is the weighted average frequency. The *blue dashed-dotted lines* represent the statistical uncertainty and the *red dashed lines* the systematic uncertainty. The *error bar* assigned to each measurement point indicates the statistical uncertainty. All uncertainties given correspond to a confidence interval of 68.3%

following the procedures outlined in reference [21] and adopt the results for the Zeeman and Stark shift measurements. The frequency shift estimation is listed in Table 2. The absolute frequency of the $^{40}Ca^+$ D2 transition is 761,905,012,599 (82) kHz obtained as the weighted average of the five measurements shown in Fig. 4.

4.2 *Isotope Shift*

We analyze the data to extract the different contributions to the isotope shift and to establish a self-consistent set of observables that can be used to provide benchmarks for atomic structure calculations of this 19-electron system. The isotope shifts are

Table 3 Measured isotope shifts in the D1 and D2 transitions referenced to ^{40}Ca$^+$

A'	$\delta\nu_{\mathrm{D1}}^{40,A'}$	$\delta\nu_{\mathrm{D2}}^{40,A'}$	$\delta\nu_{\mathrm{SIS}}^{40,A'}$
42	425.706 (94)	425.932 (71)	0.226 (118)
44	849.534 (74)	850.231 (65)	0.697 (98)
48	1705.389 (60)	1707.945 (67)	2.556 (90)

All values are given in MHz. The last column provides the difference between the isotope shifts in the two lines of the fine-structure doublet, which is commonly known as the splitting isotope shift $\delta\nu_{\mathrm{SIS}}^{40,A'}$. The uncertainties are dominated by statistics

evaluated with respect to the most abundant isotope ^{40}Ca$^+$. The transition frequencies of the reference and the isotope ^{A}Ca$^+$ are measured interleaved on one day to cancel all common systematic shifts. The resulting isotope shifts are listed in Table 3 together with the isotopic shift of the D1 transition obtained in a previous measurement [22], and their difference. The latter, the so-called splitting isotope shift $\delta\nu_{\mathrm{SIS}}$ will be discussed in Sect. 5.

We performed a King-plot analysis of the D1 versus the D2 transition according to Eq. (4). The result is shown in Fig. 5 and demonstrates the high quality of the data. The excellent linearity of the data points indicates that at this level of accuracy the higher even multipole moments in the nuclear shape parametrization as well as second-order mass polarization terms, having mass dependence $(1/m_A^2 - 1/m_{A'}^2)$, can still be neglected. Due to this mass dependence, the latter are expected to have contributions 10^{-4} of the linear mass polarization which is still below our measurement uncertainty unless the coefficient of the quadratic term would be enormously large. Additionally, this coefficient is usually considerably smaller than the coefficient of the linear term and there is no simple mechanism in atomic theory to invert this [52]. The slope of the line provides the ratio of the field shift constants $f := F_{\mathrm{D2}}/F_{\mathrm{D1}}$ in the two transitions while a relation between mass and field shift is obtained from the intersection with the y-axis as $k := K_{\mathrm{D2}} - f \cdot K_{\mathrm{D1}}$. We used two fitting routines that are able to take the measurement uncertainties in x and y direction into account to determine the parameters f and k with corresponding uncertainties: we employed the algorithm described by York et al. [53] and performed a Monte Carlo analysis [22]. The results of both methods are listed in Table 4 and are fully consistent. A standard linear regression performed with MATLAB gives consistent values for the fit parameters, but differs in the assigned uncertainty, since it neglects the x uncertainty of the experimental data.

From f we can clearly conclude that the field shift in the D2 transition is by 0.85 (12)% larger than in the D1 transition. The size of this difference comes as a surprise compared to a simple estimate using the hydrogenic approach of 0.52% which is expected to provide an upper bound and will be discussed in Sect. 6.

Absolute values for the field shift and mass shift constant are required to extract nuclear properties. This is not possible solely based on spectroscopic data without additional information on the finite-nuclear-size effect. Fortunately, there is plenty of data for the stable calcium isotopes not only for the mean-square charge radius

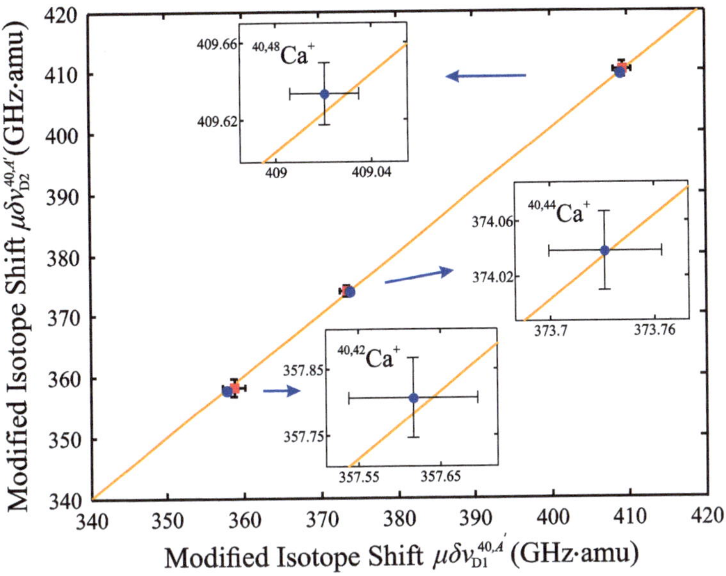

Fig. 5 King plot of the D2 versus the D1 transition. Plotted are the modified isotope shifts $\mu\,\delta\,\nu_i^{40,A'}$ in the D2 against the D1 transition. The *red dots* represent the formerly most accurate measurements [23] and the *blue dots* are the measurements presented here, with uncertainties smaller than the symbol. The *line* is the result of a linear regression of Eq. (4) taking uncertainties in both axis into account (for details see text). The *insets* show the relevant ranges around the data points enlarged by more than two orders of magnitude to illustrate the quality of the fit

Table 4 Ratio of field shift factors $f := F_{D2}/F_{D1}, f_1 := F_{DP}/F_{D1}, f_2 := F_{DP}/F_{D2}$ and difference $k := K_{D2} - f \cdot K_{D1}, k_1 := K_{DP} - f_1 \cdot K_{D1}, k_2 := K_{DP} - f_2 \cdot K_{D2}$ in GHz·amu as obtained from the King plots of the D1, D2 and $^2D_{3/2} \rightarrow {}^2P_{1/2}$ (DP) transitions

	York et al.	Monte Carlo	std. fit
f	1.0085 (11)	1.0085 (12)	1.0083 (6)
k	−2.881 (472)	−2.873 (473)	−2.787 (212)
f_1	−0.3110 (10)	−0.3114 (10)	−0.3116 (15)
k_1	−1862.9 (4)	−1862.8 (4)	−1862.7 (6)
f_2	−0.3084 (10)	−0.3088 (10)	−0.3090 (17)
k_2	−1863.8 (4)	−1863.6 (4)	−1863.6 (7)

but also for form factors and their isotopic change. The most intriguing point in the calcium isotope chain is the fact that the two doubly magic isotopes $^{40,48}Ca^{+}$ have practically identical mean-square charge radii. Form factor measurements indicate that this is due to the fact that charge is being transferred from the center and the skin of the nucleus toward the surface region where the nuclear density dropped to about half the saturation value, resulting in identical mean-square charge radii for both isotopes [24, 26, 55]. Since there is clear evidence for this from non-optical

Table 5 Parameters of a three-dimensional King plot seeded with values of $\delta\langle r_c^2\rangle^{40,A'}$ taken from [56]

Param.	Previous [22]	This work
F_{D1}	−281.8 (7.0)	−281.8 (6.9)
K_{D1}	408.73 (40)	408.73 (40)
F_{D2}		−284.7 (8.2)
K_{D2}		409.35 (42)
F_{DP}	87.7 (2.2)	87.6 (2.2)
K_{DP}	−1990.9 (1.4)	−1990.0 (1.2)
$\delta\langle r_c^2\rangle^{40,42}$	0.2160 (49)	0.2160 (49)
$\delta\langle r_c^2\rangle^{40,44}$	0.2824 (65)	0.2824 (64)
$\delta\langle r_c^2\rangle^{40,48}$	−0.0045 (60)	−0.0045 (59)

The units for the field shift constants F_i and mass shift constants K_i and the changes in mean-square nuclear charge radii $\delta\langle r_c^2\rangle^{40,A'}$ are MHz/fm^2, GHz·amu and fm^2, respectively

data and it is also confirmed with high accuracy from optical data, we use this particularity of the isotope chain to separate mass and field shift. Below, we will present the result of a full analysis with all uncertainties included. First we assume that $\delta\langle r_c^2\rangle^{40,48} = 0$ in order to explore the limits inherent in our measurement uncertainty rather than being limited by the uncertainty of the nuclear-size correction. With this assumption, the isotope shift between the doubly magic isotopes arises entirely by the mass shift, which is $K_{D1} = 409.020\,(14)[304]$ GHz·amu and $K_{D2} = 409.633\,(16)[307]$ GHz·amu for the D1 and D2 transitions, respectively. The parentheses represent the uncertainty excluding any uncertainty of $\delta\langle r_c^2\rangle^{40,48}$, while the value in square brackets indicates the change of K_i for $\delta\langle r_c^2\rangle^{40,48} = -0.0045$ fm^2 taken into account. The significant deviation of the ratio $K_{D2}/K_{D1} = 1.00150\,(5)$ from one is caused by relativistic effects. Please note that the uncertainty from $\delta\langle r_c^2\rangle^{40,48}$ given in the square brackets largely cancels in this ratio since it changes both mass shift constants by the same (small) amount.

For a full analysis, we include the newly measured transition and extend the analysis performed in [22]. From this, we extract field and mass shift constants for the D2 line and improve the uncertainties of the constants in the D1 transition. The result of this analysis is displayed in Table 5.

5 Splitting Isotope Shift

The splitting isotope shift (SIS), i.e., the change of the fine structure splitting between isotopes, has been recently the subject of investigations in He [57], Li [43, 44] and Be [47]. It is known to have in first order a mass dependence linear in $1/\mu = 1/m_A - 1/m_{A'}$. For light isotopes, the SIS is nearly independent of both, QED and nuclear volume effects and has therefore served as an important

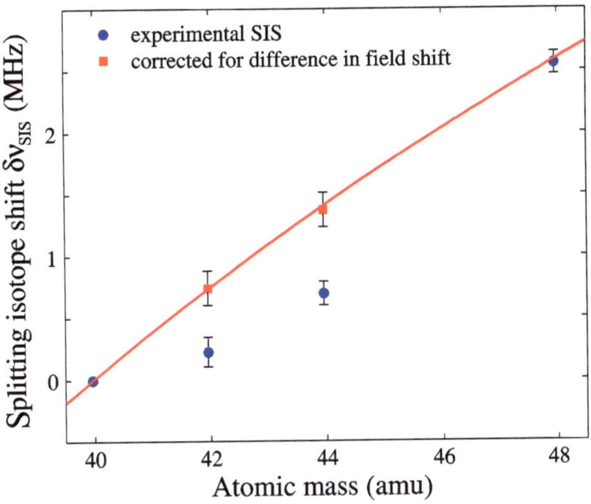

Fig. 6 Splitting isotope shift as extracted from the measurements in the D1 and the D2 transitions. The *blue circles* are the experimentally determined splitting isotope shifts. The *red line* represents the mass dependence as expected from the measured $\delta\nu_{\mathrm{SIS}}^{40,48}$ assuming that $\delta\langle r_c^2\rangle^{40,48} = 0$ and neglecting the field shift contribution. The *red squares* are the SIS of $^{42}\mathrm{Ca}^+$ and $^{44}\mathrm{Ca}^+$ after correction for the remaining field shift in the SIS

consistency check for theory and experiment [58]. From our previous analysis of the isotope shifts, we can conclude that in calcium a small contribution of the finite-nuclear-size effect is still inherent in the SIS since the field shift coefficients for both transitions are slightly different. Again, we can utilize the identical mean-square charge radii of $^{40,48}\mathrm{Ca}^+$, extract the mass polarization factor $K_{\mathrm{SIS}} = -613$ (21) MHz · amu and plot the mass dependence of the SIS (red solid line in Fig. 6). The experimental SIS values for $^{42,44}\mathrm{Ca}^+$ clearly deviate from this prediction for light isotopes. However, the typical mass dependence is restored to very high accuracy if the SIS is corrected for the difference in the field shift contribution according to

$$\delta\nu_{\mathrm{SIS,fs-corr}} = \delta\nu_{\mathrm{SIS,exp}} - \left(\frac{F_{\mathrm{D2}}}{F_{\mathrm{D1}}} - 1\right) \cdot F_{\mathrm{D1}} \cdot \delta\langle r_c^2\rangle \tag{6}$$

using the field shift ratio f as determined above, the field shift factor $F_{\mathrm{D1}} = -284.7(8.2)\ \mathrm{MHz/fm}^2$ and the known change in the mean-square charge radii according to Table 5.

The obvious discrepancy between the expected mass dependence of the first-order mass polarization term and the experimentally observed SIS for $^{42,44}\mathrm{Ca}^+$ can be perceived as the first detection of the field shift in a fine-structure transition ($^2\mathrm{P}_{1/2} \rightarrow {}^2\mathrm{P}_{3/2}$) of a light ion.

In conclusion, the experiment provided high-accuracy data that can now be used to guide and benchmark theoretical mass shift and field shift calculations. In the following section, we present improved calculations of the field shift in these transitions.

6 Theoretical Background

Theoretical analysis of the isotope shift in heavy atoms is generally a rather complicated task which requires a proper account of relativistic, many-body and even quantum electrodynamics (QED) effects. Since the detailed evaluation of these effects is beyond the scope of the present paper, we just briefly recall main theoretical approaches and their predictions here. We will focus again on the field shift constants F_i that characterize the finite-nuclear-size contribution to the difference in transition frequencies $\delta\nu_i^{A,A'}$, see Eq. (2). Our experiment has shown that the field shift in the D2 ($^2S_{1/2}\rightarrow{}^2P_{3/2}$) transition is clearly larger than that in the D1 ($^2S_{1/2}\rightarrow{}^2P_{1/2}$) channel. This J-dependence of the constant F_i can be understood from the different behavior of the Dirac wave functions of the $4p_{1/2}$ and $4p_{3/2}$ states at the coordinate origin. Namely, while $\psi_{4p_{3/2}}(r)$ vanishes at $r = 0$, the $\psi_{4p_{1/2}}(0)$ is nonzero due to the contribution of its small (relativistic) component with the orbital angular momentum $\ell = 0$. Hence the reduction of the electron density at the nucleus, being the origin of the field shift, is larger for the $^2S_{1/2} \rightarrow {}^2P_{3/2}$ transition.

Having discussed the (qualitative) reason for the difference of the D1 and D2 field shifts, we will compute now the ratio $f = F_{D2}/F_{D1}$. As a first approximation, we employ the known formulas for the finite-nuclear-size correction for the $n = 4$ *hydrogenic* states from Ref. [59] to find:

$$f_{\mathrm{hydr}} = 1 + (Z\alpha)^2\frac{15}{64} + O((Z\alpha)^2)\,. \tag{7}$$

This expression yields $f_{\mathrm{hydr}} = 1.0050$ for the nuclear charge $Z = 20$, which is in good agreement with the result $f_{\mathrm{hydr,\,num}} = 1.0051$ of a direct numerical evaluation of hydrogenic field shifts.

Within the naïve hydrogenic approach, one can also estimate the contribution of the QED effects to the ratio F_{D2}/F_{D1} of the field shift constants. Namely, by using results for nuclear-size correction to the Lamb shift of one-electron atoms, Ref. [60], we estimate:

$$\delta f_{\mathrm{hydr}}^{\mathrm{QED}} = (Z\alpha)^2\frac{15}{64}\times\frac{\alpha}{\pi}\left[1 + O(Z\alpha)\right]. \tag{8}$$

This implies that the QED effects are negligible for the f ratio in the Ca^+ ion, $\delta f_{\mathrm{hydr}}^{\mathrm{QED}} \ll f_{\mathrm{hydr}}$. Therefore, any further corrections to the f_{hydr} may arise only from the electron–electron (*e–e*) interactions which were neglected in the hydrogenic model. These interactions should *reduce* the f_{hydr} as can be expected, for example, from Eq. (7) in which we can decrease the charge Z to account for the screening of the nucleus by core electrons.

In order to describe properly the many-electron contributions to the field shift ratio f, one needs to apply theories more advanced than the "screening" hydrogenic model. Therefore, as a second approximation, we solved the DF equation and found

Table 6 Transition field shift constants and ratios obtained in the CCSD(T) and CI +MBPT calculations and comparison with experimental values

Param.	Experiment	CCSD(T)	CI+MBPT
F_{D1}	−281.8 (7.0)	−279.0 (6.0)	−288.6 (1.2)
F_{D2}	−284.7 (8.2)	−280.3 (6.0)	−289.0 (1.2)
f	1.0085 (12)	1.0048	1.0014 (4)
F_{DP}	87.7 (2.2)	103 (10)	90.3 (1.0)
f_2	−0.3088 (10)	−0.367	−0.312 (5)

We use the notation defined in Table 4. F_i is in MHz/fm^2, f dimensionless

Table 7 Theoretical predictions for the ratio $f = F_{D2}/F_{D1}$ of the field shift constants for the Ca^+ ion

Theoretical model	f	Refs.
Hydrogenic	1.0051	This work
Dirac-Fock	1.0010	This work
Dirac-Fock + Core Pol.	1.0009	This work
CCSD	1.0029	This work
CCSD(T)	1.0048	This work
MBPT	1.0011	Ref. [64]
CI+MBPT	1.0014 (4)	This work
Experimental value	1.0085 (12)	This work

The theoretical results are compared, moreover, with the present experimental value of f

the finite-nuclear-size corrections to the Ca^+ energy levels ΔE_{fns}. The field shift ratio is then simply evaluated as $f_{DF} = \Delta E_{fns}(D1)/\Delta E_{fns}(D2)$ and is 1.0010 for Ca^+. This DF result includes the first order *e–e* correlations and can be further improved by taking into account the higher-order corrections. For example, to consider (partially) the second-order *e–e* effects we employed the DF equation with the effective core-polarization potential [61–63]. As seen from the third line of the Table 7, this leads to a further slight reduction of the f ratio.

To further investigate the role of *e–e* interactions, we employ a RCC theory starting with the DF equation by considering singles and doubles excitation approximation (CCSD method) and accounting important triple excitations in the CCSD method (referred to as CCSD(T) method) as described in Refs. [65, 66]. In this calculation, we consider a Dirac-Coulomb-Breit Hamiltonian and lower order QED corrections in the approximations described in Ref. [67]. We obtain the field shift constant for the D1 and D2 lines as $F_{D1} = -284.0$ MHz/fm^2 and $F_{D2} = -284.8$ MHz/fm^2 in the CCSD method and $F_{D1} = -279.0$ MHz/fm^2 and $F_{D2} - 280.3$ MHz/fm^2 in the CCSD(T) method. This corresponds to $f_{CCSD} = 1.0029$ and $f_{CCSD(T)} = 1.0048$ in the CCSD and CCSD(T) methods, respectively. We compare the experimental results on the field shift constants and their ratios in Table 6 with the ones obtained from the CCSD(T) method. The ratio obtained in CCSD(T) is close to the hydrogenic value, but still considerably smaller than the experimental result.

As an alternative approach to the CCSD(T) method, we have examined the correlation potential method [68] and the combination of configuration interaction and many-body perturbation theory (CI+MBPT) [69, 70]. Both methods give very consistent results. In the first method, we create a correlation potential $\hat{\Sigma}$ to second order in the residual Coulomb interaction. We include this potential in the DF Hamiltonian and solve to create "Brueckner" orbitals. The field shift is obtained by varying the nuclear radius, repeating the entire calculation, and extracting the dependence of the energy eigenvalue. This 'finite-field' method is similar to that previously applied to calculations of field shift in Ca^+ [71]. It includes certain chains of diagrams to all-orders, for example so-called random-phase approximation corrections are included. Our result for the D2/D1 ratio using Brueckner orbitals is $f = 1.0010$. This value is directly comparable to the third-order MBPT results of Safronova and Johnson [64] who also obtain $f = 1.0010$.

In the second method, we perform a configuration interaction (CI) calculation allowing for single and double excitations from the 3*s* and 3*p* hole states. Correlations with the frozen core (*K* and *L* shells) are included at second-order in MBPT by modifying the radial integrals of the CI calculation. This 'particle-hole' CI+MBPT method is similar to that introduced in Hg^+ [72]. It improves upon the Brueckner calculation in that correlations with the core 3*s* and 3*p* shells are taken into account to all order (only double excitations are included, but triple excitations in this calculation are not important). Again we use the finite-field method to obtain the field shift. The final results presented in Tables 6 and 7 are the CI+MBPT method, with the difference from the Brueckner calculation taken as an indicative uncertainty. Note that QED effects are not included in these results. They can contribute up to 2 MHz/fm^2, to F_{D1} and F_{D2}, but as discussed previously they have a much smaller effect on the ratio.

In Table 7 we summarize our theoretical predictions of the field shift ratio *f*. As expected, the hydrogenic value provides the upper bound of the ratio $F_{\mathrm{D2}}/F_{\mathrm{D1}}$ while the electron correlations lead to a reduction of *f*. Moreover, one can see from the table that the experimental result for the relativistic correction to the ratio of field shift constants, i.e., $1 - f_{\mathrm{exp}}$, is 70% larger than all available calculations, including even f_{hydr}. The discrepancy between f_{exp} and the best theoretical models is about 3σ of the experimental uncertainty; its reason is still unclear and will require future study.

7 Summary

We presented absolute frequency and isotope shift measurements of the D2 line in even calcium isotopes at the 100 kHz level. The high accuracy of the measurement was enabled through photon recoil spectroscopy on trapped and sympathetically cooled ions using a spectroscopy laser referenced via an optical frequency comb to a calibrated hydrogen maser at PTB. From a multi-dimensional King-plot analysis including data from previous measurements on the D1 and ${}^2D_{3/2} \rightarrow {}^2P_{1/2}$ transitions, we derived a slightly improved set of field and mass constants for these transitions, and changes in the nuclear charge radius between isotopes. In particular, the measurements revealed for the first time field shift contributions to the splitting isotope shift in a light ion. Using the experimental data as a benchmark, we have performed theoretical calculations of the field shift constants and their ratios by using a simple hydrogenic approach, by solving the Dirac-Fock equation, also including core polarization, by relativistic coupled-cluster calculations including up to triples excitation, and by combining configuration interaction and many-body perturbation theory. The individual field shift constants for the D1 and D2 line derived from the coupled-cluster and many-body perturbation theory calculation show satisfactory agreement with the experimental data. However, we experimentally found a surprisingly large ratio of the field shifts in the D2 and D1 fine-structure doublet, $f_{\rm exp}$. The relativistic correction to this value, $1 - f_{\rm exp}$, is about 70% larger than the theoretical estimates, which is a 3 σ of experimental uncertainty. The account of electron–electron interactions leads to a further reduction of the theoretical value $f_{\rm theo}$ and, hence, to an even larger discrepancy with the experiment. For example, based on the CCSD theory we found that the difference of the D1 and D2 field shifts of 0.18% is considerably smaller than the 0.83% found in the experiment. The origin of this unexpectedly large difference must be clarified by further theoretical and experimental studies.

Acknowledgements We acknowledge financial support from the German Federal Ministry for Education and Research (BMBF) under contract 05P15RDFN1, the Helmholtz International Center for FAIR (HIC for FAIR) within the LOEWE program by the State of Hesse, the State of Lower Saxony, Hannover, Germany and DFG through grants SCHM2678/3-1 and CRC 1227 DQ-*mat*, project B05. SK received support from HGS-Hire. BKS acknowledges use of the Vikram-100 HPC cluster at the Physical Research Laboratory, Ahmedabad for performing calculations. V.A.Y acknowledges support by the Russian Federation program for organizing and carrying out scientific investigations. WN thanks G.W.F. Drake, Z.C. Yan, and R. Neugart for stimulating discussions. PS and WN thank K. Pachucki for stimulating discussions.

References

1. G. Breit, Rev. Mod. Phys. **30**, 507 (1958)
2. P. Campbell, I.D. Moore, M.R. Pearson, Prog. Part. Nucl. Phys. **86**, 127 (2016)
3. K. Blaum, J. Dilling, W. Nörtershäuser, Phys. Scr. **T152**, 014017 (2013)
4. V.A. Dzuba et al., Phys. Rev. A **72**, 022503 (2005)
5. E. Mané et al., Phys. Rev. Lett. **107**, 212502 (2011)
6. C. Delaunay et al., arXiv:1601.05087 (2016)
7. M.G. Kozlov et al., Phys. Rev. A **70**, 062108 (2004)
8. J.C. Berengut, V.A. Dzuba, V.V. Flambaum, J.A. King, M.G. Kozlov, M.T. Murphy, and J.K. Webb, *Atomic transition frequencies, isotope shifts, and sensitivity to variation of the fine structure constant for studies of quasar absorption spectra*, in *From Varying Couplings to Fundamental Physics*, edited by C. Martins and P. Molaro, Astrophysics and Space Science Proceedings, (Springer Berlin Heidelberg, 2011), pp. 9–16, arxiv preprint arXiv:1011.4136
9. M.T. Murphy, J.C. Berengut, MNRAS **438**, 388–411 (2014)
10. V.A. Korol, M.G. Kozlov, Phys. Rev. A **76**, 022103 (2007)

11. W. Nörtershäuser et al., Eur. Phys. J. D **2**, 33 (1998)
12. F. Castelli, S. Hubrig, Astron. Astrophys. **421**, L1 (2004)
13. C.R. Cowley, S. Hubrig, Astron. Astrophys. **432**, L21 (2005)
14. C.W. Palmer et al., J. Phys. B **17**, 2197 (1984)
15. A.-M. Mårtensson-Pendrill et al., Phys. Rev. A **45**, 4675 (1992)
16. L. Vermeeren et al., Phys. Rev. Lett. **68**, 1679 (1992)
17. R.F. Garcia Ruiz et al., Nat. Phys. **12**, 594 (2016)
18. Z.-T. Lu, K.D.A. Wendt, Rev. Sci. Instrum. **74**, 1169 (2003)
19. A.L. Wolf et al., Phys. Rev. A **78**, 032511 (2008)
20. A.L. Wolf et al., Phys. Rev. Lett. **102**, 223901 (2009)
21. Y. Wan et al., Nat. Commun. **5**, 4096 (2014)
22. F. Gebert et al., Phys. Rev. Lett. **115**, 053003 (2015)
23. C. Gorges et al., J. Phys. B **48**, 245008 (2015)
24. H.J. Emrich et al., Nucl. Phys. A **396**, 401 (1983)
25. H.D. Wohlfahrt et al., Phys. Rev. C **23**, 533 (1981)
26. H. Rebel et al., Proceedings of the Karlsruhe International Discussion Meeting, held at Kernforschungszentrum Karlsruhe, ed. by H. Rebel, H.J. Gils, G. Schatz, Kernforschungszentrum Karlsruhe GmbH, ISSN 0303–4003 (1979)
27. B.K. Sahoo, Phys. Rev. A **80**, 012515 (2009)
28. Y. Wan et al., Phys. Rev. A **91**, 043425 (2015)
29. F. Gebert et al., New J. Phys. **18**, 13037 (2016)
30. W.H. King, *Isotope Shifts in Atomic Spectra*, 1st edn. (Springer Verlag, New York, 1984)
31. G. Fricke, K. Heilig, *Nuclear Charge Radii, Landolt-Börnstein, Group I: Elementary Particles, Nuclei and Atoms*, vol. 20 (Springer-Verlag, Berlin Heidelberg, 2004)
32. F. Schmidt-Kaler et al., Phys. Rev. A **51**, 2789 (1995)
33. P. Zhao et al., Phys. Rev. Lett. **66**, 592 (1991)
34. F. Marin et al., Phys. Rev. A **49**, R1523 (1994)
35. D. Shiner et al., Phys. Rev. Lett. **74**, 3553 (1995)
36. L.-B. Wang et al., Phys. Rev. Lett. **93**, 142501 (2004)
37. P. Müller et al., Phys. Rev. Lett **99**, 252501 (2007)
38. P.C. Pastor et al., Phys. Rev. Lett. **108**, 143001 (2012)
39. E. Riis et al., Phys. Rev. A **49**, 207 (1994)
40. G. Ewald et al., Phys. Rev. Lett. **93**, 94 (2004)
41. R. Sánchez et al., Phys. Rev. Lett. **96**, 033002 (2006)
42. W. Nörtershäuser et al., Phys. Rev. A **83**, 012516 (2011)
43. C.J. Sansonetti et al., Phys. Rev. Lett. **107**, 023001 (2011)
44. R.C. Brown et al., Phys. Rev. A **87**, 032504 (2013)
45. W. Nörtershäuser et al., Phys. Rev. Lett. **102**, 062503 (2009)
46. A. Krieger et al., Phys. Rev. Lett. **108**, 142501 (2012)
47. W. Nörtershäuser et al., Phys. Rev. Lett. **115**, 033002 (2015)
48. M. Herrmann et al., Phys. Rev. Lett. **102**, 013006 (2009)
49. V. Batteiger et al., Phys. Rev. A **80**, 022503 (2009)
50. L.D. Landau et al., *Relativistic Quantum Mechanics* (Pergamon, Oxford, 1971)
51. K. Wendt et al., Z. Phys. A **318**, 125 (1984)
52. G. Drake, Windsor University, priv. commun
53. D. York et al., Am. J. Phys. **72**, 367 (2004)
54. G. Audi et al., Chin. Phys. C **36**, 1287 (2012)
55. E.W. Otten, Treatise Heavy Ion Sci. **8**, 517 (1989)
56. H.D. Wohlfahrt et al., Phys. Lett. **73B**, 131 (1978)
57. K. Pachucki, V. Yerokhin, Can. J. Phys. **89**, 95 (2011)
58. M. Puchalski, D. Kdziera, K. Pachucki, Phys. Rev. A **87**, 032503 (2013)
59. P.J. Mohr, B.N. Taylor, D.B. Newell, Rev. Mod. Phys. **84**, 1527 (2012)
60. V.A. Yerokhin, Phys. Rev. A **83**, 012507 (2011)

61. J. Mitroy, D.W. Norcross, Phys. Rev. A **37**, 3755 (1988)
62. D.W. Norcross, M.J. Seaton, J. Phys. B **9**, 2983 (1976)
63. V.A. Yerokhin, S.Y. Buhmann, S. Fritzsche, and A. Surzhykov, arXiv:1608.04515, (2016)
64. M.S. Safronova, W.R. Johnson, Phys. Rev. A **64**, 052501 (2001)
65. B.K. Sahoo, J. Phys. B **43**, 231001 (2010)
66. L.W. Wansbeek et al., Phys. Rev. A **86**, 015503 (2012)
67. B.K. Sahoo, Phys. Rev. A **93**, 022503 (2016)
68. V.A. Dzuba, V.V. Flambaum, P.G. Silvestrov, O.P. Sushkov, J. Phys. B **20**, 1399 (1987)
69. V.A. Dzuba, V.V. Flambaum, M.G. Kozlov, Phys. Rev. A **54**, 3948 (1996)
70. J.C. Berengut, V.V. Flambaum, M.G. Kozlov, Phys. Rev. A **73**, 012504 (2006)
71. J.C. Berengut, V.A. Dzuba, V.V. Flambaum, Phys. Rev. A **68**, 022502 (2003)
72. J.C. Berengut, Phys. Rev. A **94**, 012502 (2016)

New Avenues for Matter-Wave-Enhanced Spectroscopy

Jonas Rodewald, Philipp Haslinger, Nadine Dörre, Benjamin A. Stickler, Armin Shayeghi, Klaus Hornberger, and Markus Arndt

Abstract We present matter-wave interferometry as a tool to advance spectroscopy for a wide class of nanoparticles, clusters and molecules. The high sensitivity of de Broglie interference fringes to external perturbations enables measurements in the limit of an individual particle absorbing only a single photon on average, or even no photon at all. The method allows one to extract structural and electronic information from the loss of the interference contrast. It is minimally invasive and works even for dilute ensembles.

1 Introduction

Our contribution to this special issue is dedicated to Theodor W. Hänsch, who has inspired generations of physicists as a role model for scientific creativity, genius and passion for precision. Seeing how many methods in laser physics, atomic and molecular physics, quantum optics, and high-level spectroscopy Ted Hänsch advanced to unprecedented precision, we are reminded of a remark by Whitehead about philosophy: *The safest general characterization of the European philosophical tradition is that it consists of a series of footnotes to Plato.* [1]. In that spirit, we offer here a 'footnote' to Hänsch's work on spectroscopy and matter-wave interferometry.

In the following, we will focus on prospects for measurements in *OTIMA* [2, 3], an Optical TIme-domain near-field MAtter-wave interferometer for clusters and molecules with pulsed photo-depletion gratings. However, our arguments can be readily transferred to other interferometers for atoms, clusters, or macromolecules

---

This article is part of the topical collection "Enlightening the World with the Laser" - Honoring T. W. Hänsch guest edited by Tilman Esslinger, Nathalie Picqué, and Thomas Udem.

J. Rodewald • N. Dörre • A. Shayeghi • M. Arndt (✉)
Faculty of Physics, VCQ, University of Vienna, Boltzmanngasse 5, 1090 Vienna, Austria
e-mail: markus.arndt@univie.ac.at

P. Haslinger
University of California, Berkeley, Leconte/Birge Hall, Berkeley, CA 94720, USA

B.A. Stickler • K. Hornberger
Faculty of Physics, University of Duisburg-Essen, Lotharstraße 1-21, Duisburg, Germany

D. Meschede et al. (eds.), *Exploring the World with the Laser*,
https://doi.org/10.1007/978-3-319-64346-5_2

that use combinations of mechanical and optical gratings, operating in the matter-wave near-field or far-field, in position space or in the time domain [4, 5].

2 Matter-Wave Interferometry with Pulsed Photo-Depletion Gratings

Near-field matter-wave interferometry is based on the discovery of coherent self-imaging behind periodic structures by Talbot [6] and Lau [7]: When a transmission grating of period d is illuminated by a plane wave of wavelength λ, an image of the mask will be reproduced at multiples of the Talbot distance $L_T = d^2/\lambda$ behind the grating—without the need of any focusing optics, simply by virtue of near-field interference. The trick works even for spatially incoherent sources if another grating is inserted before the diffraction mask, again at multiples of the Talbot length. This concept was realized for light [8], X-rays [9, 10] and atoms [11, 12]—also in the time domain [13–15]. Throughout the last decade, Talbot–Lau interferometry has been extended to organic molecules, clusters and biomolecules [3, 16–18].

OTIMA, in particular, is an interferometer that utilizes three pulsed photo-depletion gratings [2, 3, 18] to prepare, diffract and detect beams of complex nanoscale particles (Fig. 1). In our experiments, the gratings are realized as retro-reflected fluorine laser beams, at a vacuum ultraviolet wavelength of $\lambda = 157.6\,\text{nm}$, yielding standing light waves with a period of $d \simeq 79$ nm. In the antinodes of the standing light waves, the molecular beam is depleted by ionization, dissociation, or any other mechanism that renders these molecules invisible to the detector further downstream. This way, the light field acts effectively as a periodic absorptive mask. The high laser photon energy of 7.9 eV allows manipulating a wide range of molecules or clusters in the same machine—largely independent of particle-specific narrow optical resonances.

Three gratings are combined to form a complete Talbot–Lau interferometer: the first grating G_1 establishes a periodic array of possible molecular locations, close to the nodes of the standing wave. The tight confinement of the wave function around these nodes then imposes a momentum uncertainty which ensures a rapid increase in transverse coherence behind the grating—even for an initially incoherent molecular beam. The second grating is positioned such that the incident molecular coherence extends at least over two nodes or antinodes of G_2. This way, the propagating molecular wave covers two or more semiclassical paths on the way to the final state at G_3, further downstream. Resonant near-field interference occurs around multiples of the Talbot time $T_T = d^2 m/h$, corresponding to a Talbot length $L_T = vT_T = d^2/\lambda_{dB}$, for particles of mass m. In time-domain interferometry, all particles within the grating area see the same pulse sequence for the same duration, independent of their own velocity v.

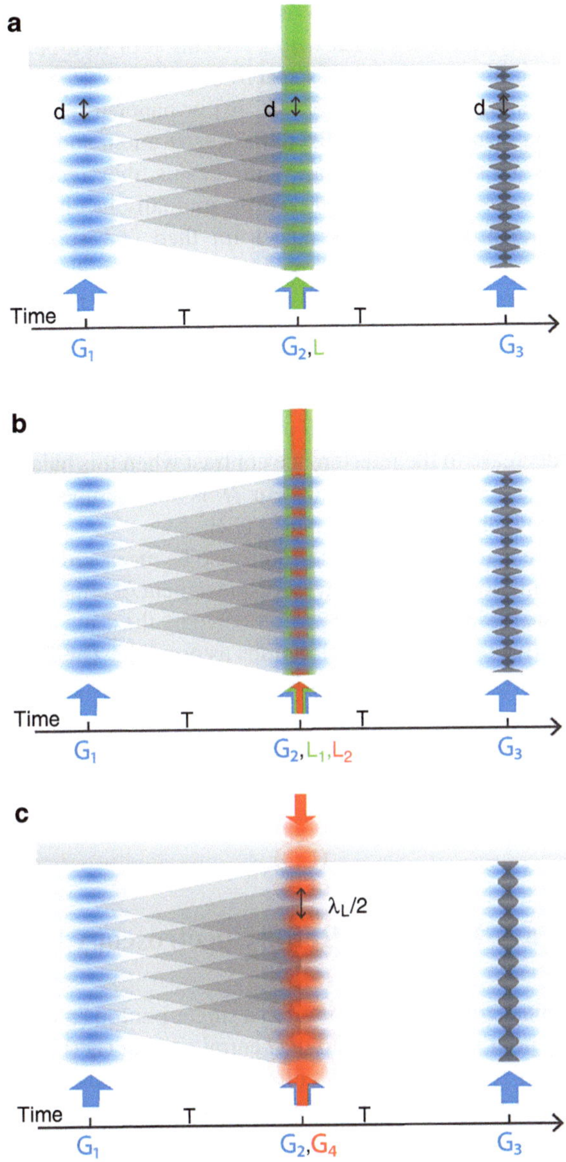

Fig. 1 **a** UV–VIS spectroscopy in OTIMA: absorption of a single photon from a running laser wave imparts a recoil to the absorbing cluster or molecule. If the wavelength of the light is comparable to the semiclassical path separation of the delocalized particle, the interference fringe pattern experiences a measurable dephasing (Sect. 3) [19, 20]. Because of the small grating period (79 nm), *single-color* visible or infrared (VIS/IR) spectroscopy requires the collective momentum transfer of several photons or operation of the matter-wave interferometer in higher Talbot orders. **b** VIS/IR spectroscopy: can also be realized by combining a single (VIS/IR) photon of laser beam L_1 (*red arrow*) with a single UV photon from beam L_2 (*green arrow*) which provides the required momentum transfer (Sects. 6 and 7). **c** Polarizability spectroscopy: is the least invasive of all three

The molecular fringe pattern can be visualized in various ways: the third grating G_3 acts as a spatially resolving mask with a resolution of well below $\lambda/2 = 79$ nm and a post-ionizing time-of-flight mass spectrometer allows recording all particles transmitted by this mask. If the clusters [3, 18] or nanoparticles [21] in the beam have a broad mass distribution with fixed mass separation, and if they all have the same velocity, as often the case in supersonic beams, they realize a 'comb' of de Broglie waves. The particles remain, however, mutually incoherent since they are distinguishable. Recording the mass spectrum then corresponds to reading an interference pattern as a function of mass m or wavelength λ_{dB}. One may also describe this phenomenon as a wave function rephasing in the time-domain [13], without reference to position and independent of the velocity distribution.

We exploit in particular the resonance in particle transmission behind grating G_3 as a function of the pulse delay between two subsequent gratings $\tau_{ij} = t(G_i) - t(G_j)$. This resonance occurs for a symmetric interferometer timing, $T = \tau_{12} = \tau_{23}$, and we find a rapid decrease in the interference contrast when this balance is skewed by more than $\Delta t = \tau_{23} - \tau_{12} \simeq \tau_{23}/N$, where N is the number of grating nodes illuminated by the incident molecular beam [22].

In principle, the matter-wave fringes could also be measured directly by plotting the particle transmission versus the lateral displacement of either grating. However, in our case, all three laser beams are retro-reflected by the same mirror to render the system as insensitive to mechanical vibrations as possible. The fringes are thus not affected by slow tilts or shifts of the mirror. Instead, in OTIMA interferometry, the interference contrast can be extracted from a comparison of the interferometer transmission for the case of resonant (symmetric) and non-resonant (slightly asymmetric) laser pulse delays [3]. For this setting, we here propose a variety of new spectroscopy tools and procedures.

3 Matter-Wave-Enhanced Recoil Spectroscopy (MERS)

A matter-wave interferometer can be used as a single-photon recoil spectrometer by adding a running laser wave L close to the central grating G_2 (Fig. 1a). Absorption of a single photon then imparts a recoil onto the molecule, without providing 'which-path information'. Subsequent spontaneous reemission of photons would introduce a random phase and decoherence [23], but most macromolecules dissipate the energy radiationless to many lower-lying electronic and vibrational states [20, 24].

Fig. 1 (continued) techniques. The off-resonant dipole interaction with the intense laser field G_4 deforms the matter-wave front—leading to a loss of fringe contrast even without any photo-absorption. This method may be particularly useful for weakly bound van der Waals clusters (Sect. 8)

Heating of the internal molecular state does not destroy the center-of-mass coherence [25, 26] as long as the internal and external degrees of freedom remain separable. Wavelets associated with the same internal state remain coherent to each other [24]. Absorption inside a matter-wave interferometer thus creates shifted and unshifted molecular fringe patterns which are correlated with heated and unheated internal states. Even if the shifted and the unshifted fringes cannot be resolved, the loss of the total fringe visibility can be used for spectroscopy with high accuracy [19, 20].

In OTIMA interferometry, the momentum imparted by each VUV grating exceeds the absorption recoil of a 0.3–100 μm spectroscopy photon by a factor up to 300. Visible (VIS) and near-infrared (NIR) spectroscopy will therefore work best in higher Talbot orders, when the grating pulse separation time amounts to about two or three Talbot times and the molecular state is delocalized over two or three periods of G_2. Probing photons with wavelengths around 270–320 nm are for instance required to study the electronic states of aromatic amino acids and nucleotides, peptides and oligonucleotides. Comparing UV spectra of biomolecules in the gas phase with molecules in solution could later provide valuable information about structural changes in these different environments [27, 28].

4 Fluorescence Recoil Spectroscopy (FRS)

If, contrary to the previous assumptions, absorption is followed by fluorescence, the emitted photon will add a recoil to the molecular motion, whose orientation varies randomly for each molecule. This leads to a reduction of the fringe contrast. One can use this loss of visibility to extract fluorescence quantum yields. When the exciting laser illuminates the molecular beam from the front, the absorption recoil does not blur the interference pattern and the timing of the laser pulse determines when and where the molecule is hit relative to the position and time of the second grating pulse. If the fluorescence wavelength distribution is known, the contrast reduction of the matter-wave interference pattern provides a measure for the product of the absorption cross section and the fluorescence yield. The absorption cross section can be extracted independently at low laser power and with the laser beam oriented parallel to the grating k-vector. When as little as 10% of all molecules are excited [20], the absorption measurement is only minimally affected by fluorescence.

5 Multi-Photon Recoil Spectroscopy (MPRS)

If the probing laser wavelength exceeds the grating period substantially, a single photon cannot provide the recoil to shift the interference pattern sufficiently far. This is for instance the case for vibrational transitions, driven by near-infrared (NIR) or far-infrared (FIR) photons with wavelengths around 3–100 μm.

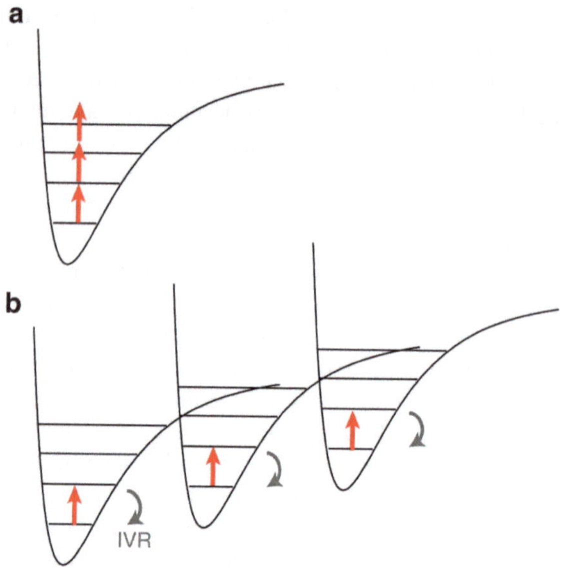

Fig. 2 **a** The absorption of multiple photons from a monochromatic source is suppressed due to the anharmonicity bottleneck. **b** Internal vibrational relaxation (IVR) to other modes dissipates the energy and enables the repeated excitation of the same IR transition until sufficient momentum recoil has been accumulated to shift the fringe pattern measurably

Multi-photon absorption can then still be a viable option if the cumulated recoil of many absorbed photons has sufficient momentum.

Multi-photon recoil spectroscopy is conceptually similar to infrared multi-photon dissociation spectroscopy (IR-MPD) [29]. The anharmonicity of molecular potentials usually prevents the subsequent absorption of many monochromatic photons within the same vibrational energy ladder (*anharmonicity bottleneck*, Fig. 2a) [30]. On the other hand, couplings between the vibrational modes can dissipate the absorbed energy (Fig. 2b). In complex particles, vibrational excitations can relax on the picosecond time scale to many vibrational states, i.e., very fast compared to the duration of the nanosecond spectroscopy pulse. Even though multi-photon absorption will lead to internal heating, this is compatible with high-contrast interference as long as it does not provide which-path information by emission of thermal radiation [31]. Sequential absorption with a Poissonian photon number distribution will lead to a biased quantum random walk in momentum. In contrast to the single-photon case, extracting an absolute absorption cross section from the visibility loss is then less direct. However, the spectral line positions and widths will remain measurable.

6 Resonance-Enhanced Multi-Photon Recoil Spectroscopy (REMPRS)

In order to avoid heating and the risk of spectroscopic shifts, conformation changes or even fragmentation, it is desirable to limit the number of photons required to retrieve information—even in the infrared regime. This challenge has been

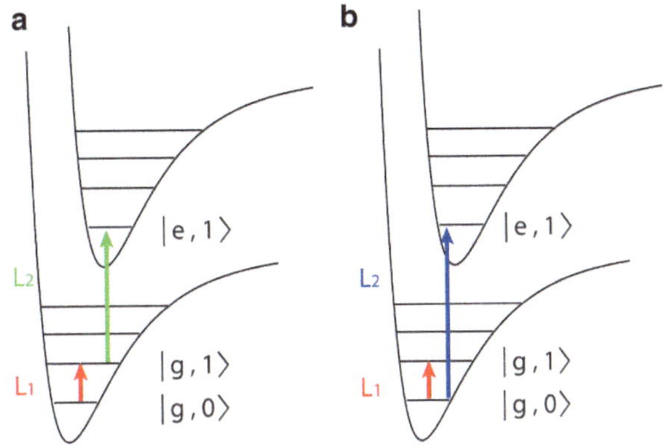

Fig. 3 **a** In IR–UV-recoil dip spectroscopy the matter-wave dephasing action of a UV photon is suppressed by emptying the ground state in a resonant IR transition. **b** In double resonant IR–UV-recoil spectroscopy the kick of the UV photon is conditioned on the prior absorption of the IR or VIS photon

addressed in physical chemistry by *action spectroscopy* where the absorption of a few photons may lead to a detectable 'action', for instance the detachment of an additional messenger atom. Action spectroscopy has been very successful in cluster physics [29]. A prominent example is the spectroscopy of impurities in helium nanodroplets where the deposition of 1 eV of energy even suffices to boil off 2000 helium atoms [32]. However, the attached messenger atom or the environment, such as a liquid helium nanodroplet, may also influence the electronic structure of the host molecule [33].

We suggest that it is possible to avoid the need for messengers and artificial environments based on a recoil analog of resonance-enhanced multi(two)-photon ionization spectroscopy (REMPI/R2PI) [35]. In matter-wave-enhanced resonant multi-photon recoil spectroscopy *(REMPRS/R2PRS)*, the spectroscopy photon from laser beam L_1 triggers the absorption of a photon of high momentum from laser beam L_2. We illustrate the idea in Figs. 1b and 3a where the first photon from laser beam L_1 excites the molecule for instance from the electronic and vibrational ground state $|g, 0\rangle$ to the higher-lying vibrational state $|g, 1\rangle$ and a photon from the more energetic laser L_2 couples this state to the upper electronic state $|e, 1\rangle$, imparting the required kick (see Fig. 3a). This method is appealing for particles where photo-ionization has been notoriously difficult and photodissociation channels are not available, as is the case for many massive biomolecules [36–38].

7 Matter-Wave-Enhanced Recoil Dip Spectroscopy (RDS)

While in our previous examples the resonant reduction of matter-wave contrast was assumed to provide the spectroscopic signal, we illustrate in Figs. 1b and 3b how *recoil dip spectroscopy* can even restore and enhance this contrast on resonance. We assume that the absorption of a single (V)UV photon from $|g, 0\rangle$ to $|e, 1\rangle$ imparts sufficient recoil to reduce the matter-wave visibility. However, we can deplete the ground state $|g, 0\rangle$ by coupling it resonantly to a neighboring vibrational state of the same electronic manifold $|g, 1\rangle$. This reduces the UV absorption and raises the fringe contrast again. Dip spectroscopy may appear counterintuitive in comparison with earlier results from atom interferometry [34] where an increase in the number of absorbed quanta led to a decrease in fringe contrast. In contrast to that, reemission is suppressed in many molecules during their transit through the interferometer. OTIMA offers a suitable frame for this scheme since the nanosecond precise timing allows depleting the ground state prior to the UV absorption and with a lead time shorter than the life time of the excited state.

IR–UV dip spectroscopy requires that the UV photon couples efficiently to one particular vibrational ground state but substantially less to the IR excited vibrational mode. In many small- and medium-sized molecules, it is possible to excite electronic transitions with vibrational resolution. In these cases, recoil dip spectroscopy (RDS) is a realistic option. Even if the UV transitions are broadened when they couple to short-lived excited states, IR dip spectroscopy should provide resolution of the vibrational ground states, as seen in the modulation of the fringe visibility.

In VIS–UV dip spectroscopy the transitions couple electronic states and absorption of a visible spectroscopy photon is followed by a UV photon with higher momentum. As before, the method requires that the ground state and the excited state of the electronic transition couple differently to the UV photon.

8 Matter-Wave-Enhanced Polarizability Spectroscopy (MEPS)

Valuable spectroscopic information can be obtained even without exchanging a single real photon: The atomic or molecular *polarizability* provides important information about the particle composition and structure as well as their van der Waals interactions with molecules or surfaces.

In atom interferometry, the optical polarizability has for instance been measured by imprinting a differential phase on two spatially separated parts of a cloud of ultracold atoms that were then recombined to interfere [39]. Even if the path separation of the matter-wave packets is smaller than the width of the spectroscopy laser beam, they accumulate state-selective phase shifts in the interference pattern, which may provide information about optical polarizabilities [40] or transition dipole matrix elements [41].

This can be generalized to high-mass particles, too. The optical polarizability of complex molecules at fixed wavelength (532 and 157 nm) can be extracted from the diffraction efficiency in the standing light wave in Kapitza–Dirac–Talbot–Lau [42] and OTIMA interferometry [43]. Here, we propose to measure it across a wide spectrum using OTIMA interferometry. By interaction with a tunable *standing light-wave grating* (G_4), close and parallel to G_2 (see Fig. 1c), the molecular matter-waves acquire a phase shift which reduces their interference contrast.

The effect of the additional grating can be understood in both a classical and a quantum picture: Quantum mechanically, the grating acts like a phase grating, whose period varies with wavelength and whose impact on the matter-wave is a function of the molecular optical polarizability. In a classical picture, the fluctuating array of dipole force microlenses in G_4 scrambles the molecular interferogram. Tuning the spectroscopy laser then allows one to modulate its fringe contrast (see below).

In contrast to the absorptive spectroscopy, which can be done already with running laser waves, we here rely on the presence of an optical grating to impose strong local dipole forces. They scale with the gradient of the dipole potential and are maximized in a standing light wave. It is favorable if the spectroscopy grating (G_4) phase is unstable since a fluctuating phase ensures that we can ignore residual effects of constructive matter-wave interference that might emerge when the spectroscopy grating G_4 and the diffraction grating G_2 have commensurate periods.

9 Theoretical Description

In order to quantify these statements, we here discuss how the fringe visibility is affected in OTIMA interferometry by the presence of a spectroscopy beam directly after the second grating, G_2. In general, the interference signal is calculated by combining the effect of each individual grating on the incoming matter wave with its free propagation between the gratings [2, 22, 44].

Exploiting that the transit through each individual laser grating can be described in the eikonal approximation [45], the interaction between the matter wave and grating G_k, $k = 1, 2, 3$, is characterized by the eikonal phase shift $\phi_0^{(k)} = 4\pi E^{(k)}\alpha(\lambda)/hc\varepsilon_0 A$, and by the mean number of absorbed photons per molecule or cluster, $n_0^{(k)} = 4E^{(k)}\lambda\sigma_{\text{abs}}(\lambda)/hcA$ [2]. Here, $E^{(k)}$ is the pulse energy, A denotes the laser spot area (flat top assumed), $\alpha(\lambda)$ and $\sigma_{\text{abs}}(\lambda)$ are the molecular polarizability and absorption cross section at the laser wavelength λ, respectively.

OTIMA contrast—In the absence of any additional laser, the sinusoidal visibility of the interferogram can be computed as a function of the laser grating pulse separation time T and all known laser parameters

$$\mathcal{V}_{\text{sin}} = \frac{2I_1(n_0^{(1)}/2)I_1(n_0^{(3)}/2)}{I_0(n_0^{(1)}/2)I_0(n_0^{(2)}/2)I_0(n_0^{(3)}/2)} \times \left| \frac{\zeta_{\text{coh}} - \zeta_{\text{dep}}}{\zeta_{\text{coh}} + \zeta_{\text{dep}}} J_2\left(\sqrt{\zeta_{\text{coh}}^2 - \zeta_{\text{dep}}^2}\right) \right|, \tag{1}$$

where J_n and I_n are Bessel functions. The parameter $\zeta_{\text{coh}} = \phi_0^{(2)} \sin(\pi T/T_{\text{T}})$ describes the coherent evolution induced by the phase grating component in G_2 and $\zeta_{\text{dep}} = n_0^{(2)} \cos(\pi T/T_{\text{T}})/2$ is related to the photo-depletion of the molecular beam in the anti-nodes of the standing light wave, also in G_2. The visibility $\mathcal{V}_{\rm sin}$ varies periodically as a function of the pulse separation T, and its period is determined by the Talbot time $T_{\rm T}$.

Recoil Spectroscopy—Absorption of photons from a pulsed running wave laser of wavelength $\lambda_{\rm L}$ in the instant after the second grating pulse will impart a recoil on the absorbing molecule [19]. In practice, one may even overlap G_2 and the spectroscopy laser on the same spot at the same time using dichroic optics. The resulting reduction of the signal visibility can then be used to extract the absolute absorption cross section of the molecule [20]. Assuming that the probability of absorbing n photons is described by a Poisson distribution with mean $n_{\rm L}(\lambda_{\rm L}) = \sigma_{\rm abs}(\lambda_{\rm L})E_{\rm L}\lambda_{\rm L}/A_{\rm L}hc$, the sinusoidal visibility $\widetilde{\mathcal{V}}_{\rm sin}$ in the presence of the spectroscopy beam can be written as

$$\frac{\widetilde{\mathcal{V}}_{\rm sin}}{\mathcal{V}_{\rm sin}} = \exp\left[-2n_{\rm L}\sin^2\left(\pi\frac{d\ T}{\lambda_{\rm L}T_{\rm T}}\right)\right]. \tag{2}$$

Thus, $\ln \widetilde{\mathcal{V}}_{\rm sin}/\mathcal{V}_{\rm sin}$ decreases linearly with the product of the total absorption cross section and the recoil laser energy, $\sigma_{\rm abs}(\lambda_{\rm L})E_{\rm L}$. One can therefore measure the molecular absorption spectrum by varying the laser power at $\lambda_{\rm L}$ and observing the fringe contrast. This idea can be extended in a straightforward way to recoil dip spectroscopy, where only the readout of the spectrum is modified.

Polarizability Spectroscopy—Replacing the running wave laser by a tunable standing light wave grating allows us to measure the molecular polarizability. In this case, the spectroscopy laser acts as a fourth grating with period $\lambda_{\rm L}/2$. It is timed such that that the free flight to the second grating is negligible. Hence, the interaction between the spectroscopy laser and the molecule is characterized by the eikonal phase $\phi_{\rm L}(\lambda_{\rm L}) = 4\pi E_{\rm L}\alpha(\lambda_{\rm L})/A_{\rm L}hc\varepsilon_0$ and the mean photon number $n_{\rm L}(\lambda_{\rm L}) = 4E_{\rm L}\lambda_{\rm L}\sigma_{\rm abs}(\lambda_{\rm L})/A_{\rm L}hc$. To avoid moiré-type effects, we propose to induce or maintain phase fluctuations between the spectroscopy grating and the three (phase stable) interferometer gratings. The signal visibility reduction is then

$$\frac{\widetilde{\mathcal{V}}_{\rm sin}}{\mathcal{V}_{\rm sin}} = \frac{\left|J_0\left[\sqrt{\phi_{\rm L}^2\sin^2\left(\pi\frac{2d\ T}{\lambda_{\rm L}\ T_{\rm T}}\right) - \frac{n_{\rm L}^2}{4}\cos^2\left(\pi\frac{2d\ T}{\lambda_{\rm L}\ T_{\rm T}}\right)}\right]\right|}{I_0(n_{\rm L}/2)}. \tag{3}$$

Varying the laser wavelength $\lambda_{\rm L}$ in a regime in which photon absorption can be neglected, $n_{\rm L} \ll 1$, the spectroscopy laser acts as a pure phase grating and the contrast reduction is

$$\frac{\widetilde{\mathcal{V}}_{\rm sin}}{\mathcal{V}_{\rm sin}} = \left|J_0\left[\phi_{\rm L}\sin\left(\pi\frac{2d\ T}{\lambda_{\rm L}\ T_{\rm T}}\right)\right]\right|. \tag{4}$$

Thus, one can directly extract the spectral molecular polarizability from measuring the contrast reduction for different pulse energies $E_{\rm L}$.

In deriving the visibility (1), we have neglected additional contrast-reducing processes such as scattering with residual gas atoms [44, 46], thermal decoherence [31] or phase averaging due to machine vibrations or internal molecular dynamics [47, 48]. Such processes would affect the signal visibility with a common pre-factor which cancels in the ratio of the visibility with and without spectroscopy laser. This renders the measurement rather robust with respect to decoherence and dephasing.

10 Conclusion

Spectroscopy is an important field of atomic, molecular and optical physics with close ties to areas as diverse as physical and biochemistry, environmental science or laboratory astrophysics. It is therefore important to explore methods which are minimally invasive in the sense that they require the scattering of very few real photons to eventually not even a single one.

Matter-wave interference offers an interesting option as it imposes a very narrow comb of molecular density fringes which serves as a nanoscale ruler, whose position can be read with a sensitivity and accuracy of 10 nm or less.

While a conceptual similarity with classical Moiré shadows is obvious [49], operating in the quantum regime allows one to prepare even narrower fringes and a substantially enhanced sensitivity to fringe displacements. Compared to classical deflectometers, which usually operate with position resolution on the order of tens of micrometers [50, 51], quantum interferometry has the potential of improving the position sensitivity by three to four orders of magnitude. However, substantial future work still needs to be invested in generating sufficiently brilliant molecular beam sources to turn this idea into a generic and universal tool.

Matter-wave-enhanced spectroscopy is promising and useful for isolated molecules and clusters in the gas phase under diverse boundary conditions. It can be beneficial when the absorbed energy is dissipated in internal conversion processes and fluorescence or action spectroscopy fails. This applies to a large class of complex biomolecules and van der Waals clusters.

Interference-assisted absorption spectroscopy is also expected to be favorable for many gas phase neutral vitamins, peptides and proteins with a low vapor pressure, forming only very dilute molecular beams. While interferometry can operate eventually even with a single molecule per shot, direct absorption using Beer's law would require beam densities many orders of magnitude higher.

Matter-wave interferometry-assisted two-photon and polarizability spectroscopy is also favored over fluorescence methods, where one would usually want to scatter many photons per particle. Multi-photon scattering may lead to excessive heating, particle dissociation or modification. This is the case for weakly bound van der Waals clusters, whose quantum wave nature has been successfully demonstrated in OTIMA interferometry [3, 18].

Acknowledgements Open access funding provided by University of Vienna. We acknowledge financial support by the European Research Council in Project (320694) and by the Austrian Science Funds (FWF) in the Projects W1201-3 and J3680. We thank Sandra Eibenberger, Lukas Mairhofer, Joseph Cotter, and Stefan Nimmrichter for many stimulating discussions on interference-assisted molecule spectroscopy.

References

1. A.N. Whitehead, *Process and Reality* (Free Press, New York, 1979), p. 39
2. S. Nimmrichter, P. Haslinger, K. Hornberger, M. Arndt, Concept of an ionizing time-domain matter-wave interferometer. New J. Phys. **13**, 075002 (2011)
3. P. Haslinger, N. Dörre, P. Geyer, J. Rodewald, S. Nimmrichter, M. Arndt, A universal matter-wave interferometer with optical ionization gratings in the time domain. Nat. Phys. **9**, 144–148 (2013)
4. K. Hornberger, S. Gerlich, P. Haslinger, S. Nimmrichter, M. Arndt, Colloquium: Quantum interference of clusters and molecules. Rev. Mod. Phys. **84**, 157–173 (2012)
5. M. Arndt, N. Dörre, S. Eibenberger, P. Haslinger, J. Rodewald, K. Hornberger, S. Nimmrichter, M. Mayor, Matter-wave interferometry with composite quantum objets. In: Tino G.N., Kasevich M. (eds.) *Atom Interferometry*, Proceedings of the International School of Physics "Enrico Fermi", vol. 188, IOS Press (2014)
6. H.F. Talbot, Facts relating to optical science. Philos. Mag. **9**, 401–407 (1836)
7. E. Lau, Beugungserscheinungen an Doppelrastern. Ann. Phys. **6**, 417 (1948)
8. K. Patorski, *Self-Imaging and Its Applications* (Elsevier, Amsterdam, 1989)
9. T. Weitkamp, A. Diaz, C. David, F. Pfeiffer, M. Stampanoni, P. Cloetens, E. Ziegler, X-ray phase imaging with a grating interferometer. Opt. Express **13**, 6296–6304 (2005)
10. F. Pfeiffer, T. Weitkamp, O. Bunk, C. David, Phase retrieval and differential phase-contrast imaging with low-brilliance X-ray sources. Nat. Phys. **2**, 258–261 (2006)
11. J.F. Clauser, S. Li, "Heisenberg microscope" decoherence atom interferometry. Phys. Rev. A **50**, 2430 (1994)
12. S. Nowak, Ch. Kurtsiefer, T. Pfau, C. David, High-order Talbot fringes for atomic matter waves. Opt. Lett. **22**, 1430–32 (1997)
13. S.B. Cahn, A. Kumarakrishnan, U. Shim, T. Sleator, P.R. Berman, B. Dubetsky, Time-domain de Broglie wave interferometry. Phys. Rev. Lett. **79**, 784–787 (1997)
14. A. Turlapov, A. Tonyushkin, T. Sleator, Talbot–Lau effect for atomic de Broglie waves manipulated with light. Phys. Rev. A **71**, 43612 (2005)
15. S. Wu, E.J. Su, M. Prentiss, Time domain de Broglie wave interferometry along a magnetic guide. Eur. Phys. J. D **35**, 111–118 (2005)
16. B. Brezger, L. Hackermüller, S. Uttenthaler, J. Petschinka, M. Arndt, A. Zeilinger, Matter-wave interferometer for large molecules. Phys. Rev. Lett. **88**, 100404 (2002)
17. S. Gerlich, L. Hackermüller, K. Hornberger, A. Stibor, H. Ulbricht, M. Gring, F. Goldfarb, T. Savas, M. Müri, M. Mayor, M. Arndt, A Kapitza–Dirac–Talbot–Lau interferometer for highly polarizable molecules. Nat. Phys. **3**, 711–715 (2007)
18. N. Dörre, J. Rodewald, P. Geyer, B. von Issendorff, P. Haslinger, M. Arndt, Photofragmentation beam splitters for matter-wave interferometry. Phys. Rev. Lett. **113**, 233001 (2014)
19. S. Nimmrichter, K. Hornberger, H. Ulbricht, M. Arndt, Absolute absorption spectroscopy based on molecule interferometry. Phys. Rev. A **78**, 063607 (2008)
20. S. Eibenberger, X. Cheng, J.P. Cotter, M. Arndt, Absolute absorption cross sections from photon recoil in a matter-wave interferometer. Phys. Rev. Lett. **112**, 250402 (2014)

21. S. Eibenberger, S. Gerlich, M. Arndt, M. Mayor, J. Tüxen, Matter-wave interference with particles selected from a molecular library with masses exceeding 10,000 amu. Phys. Chem. Chem. Phys. **15**, 14696–14700 (2013)
22. S. Nimmrichter, *Macroscopic Matter Wave Interferometry*, Springer Thesis, Springer, Heidelberg, 2014
23. M.S. Chapman, T.D. Hammond, A. Lenef, J. Schmiedmayer, R.A. Rubenstein, E. Smith, D.E. Pritchard, Photon scattering from atoms in an atom interferometer: coherence lost and regained. Phys. Rev. Lett. **75**, 3783–3787 (1995)
24. J.P. Cotter, S. Eibenberger, L. Mairhofer, X. Cheng, P. Asenbaum, M. Arndt, K. Walter, S. Nimmrichter, K. Hornberger, Coherence in the presence of absorption and heating in a molecule interferometer. Nat. Commun. **6**, 7336 (2015)
25. C. Bordé, Atomic interferometry with internal state labelling. Phys. Lett. A **140**, 10 (1989)
26. M. Kasevich, S. Chu, Atomic interferometry using stimulated raman transitions. Phys. Rev. Lett. **67**, 181–184 (1991)
27. R. Weinkauf, J.P. Schermann, M.S. de Vries, K. Kleinermanns, Molecular physics of building blocks of life under isolated or defined conditions. Eur. Phys. J. D **20**, 309–316 (2002)
28. M.S. de Vries, P. Hobza, Gas-phase spectroscopy of biomolecular building blocks. Annu. Rev. Phys. Chem. **58**, 585–612 (2007)
29. A. Fielicke, A. Kirilyuk, C. Ratsch, J. Behler, M. Scheffler, G. Von Helden, G. Meijer, Structure determination of isolated metal clusters via far-infrared spectroscopy. Phys. Rev. Lett. **93**, 023401 (2004)
30. A.M. Rijs, J. Oomens, *Gas-Phase IR Spectroscopy and Structure of Biological Molecules* (Springer, New York, 2015)
31. L. Hackermüller, K. Hornberger, B. Brezger, A. Zeilinger, M. Arndt, Decoherence of matter waves by thermal emission of radiation. Nature **427**(6976), 711–714 (2004)
32. F. Stienkemeier, K.K. Lehmann, Spectroscopy and dynamics in helium nanodroplets. J. Phys. B **39**, R127 (2006)
33. A. Shayeghi, R.L. Johnston, D.M. Rayner, R. Schäfer, A. Fielicke, The nature of bonding between argon and mixed gold–silver trimers. Angew. Chem. Int. Ed. **54**, 10675–10680 (2015)
34. D.A. Kokorowski, A.D. Cronin, T.D. Roberts, D.E. Pritchard, From single- to multiple-photon decoherence in an atom interferometer. Phys. Rev. Lett. **86**, 2191 (2001)
35. L. Zandee, R.B. Bernstein, Resonance-enhanced multiphoton ionization and fragmentation of molecular beams: NO, I2, benzene, and butadiene. J. Chem. Phys. **71**(3), 1359–1371 (1979)
36. E. Schlag, J. Grotemeyer, R. Levine, Do large molecules ionize? Chem. Phys. Lett. **190**, 521–527 (1992)
37. C.H. Becker, K.J. Wu, On the photoionization of large molecules. J. Am. Soc. Mass Spectrom. **6**, 883–888 (1995)
38. A. Akhmetov, J.F. Moore, G.L. Gasper, P.J. Koin, L. Hanley, Laser desorption postionization for imaging ms of biological material. J. Mass Spectrom. **45**(2), 137–45 (2010)
39. B. Deissler, K.J. Hughes, J.H.T. Burke, C.A. Sackett, Measurement of the ac Stark shift with a guided matter-wave interferometer. Phys. Rev. A **77**(3), 031604 (2008)
40. A. Morinaga, T. Tako, N. Ito, Sensitive measurement of phase shifts due to the ac Stark effect in a Ca optical Ramsey interferometer. Phys. Rev. A **48**, 1364–1368 (1993)
41. S. Liu, H. Knöckel, E. Tiemann, Matter wave interferometry for measuring a molecular transition dipole moment. Eur. Phys. J. D **60**, 269–277 (2010)
42. K. Hornberger, S. Gerlich, H. Ulbricht, L. Hackermüller, S. Nimmrichter, I. Goldt, O. Boltalina, M. Arndt, Theory and experimental verification of Kapitza–Dirac–Talbot–Lau interferometry. New J. Phys. **11**, 043032 (2009)
43. N. Dörre, P. Haslinger, J. Rodewald, P. Geyer, M. Arndt, Refined model for Talbot–Lau matter-wave optics with pulsed photodepletion gratings. J. Opt. Soc. B **32**, 114 (2015)
44. K. Hornberger, J.E. Sipe, M. Arndt, Theory of decoherence in a matter wave Talbot–Lau interferometer. Phys. Rev. A **70**(5), 53608 (2004)

45. S. Nimmrichter, K. Hornberger, Theory of Talbot–Lau interference beyond the eikonal approximation. Phys. Rev. A **78**, 023612 (2008)
46. K. Walter, B.A. Stickler, K. Hornberger, Collisional decoherence of polar molecules. Phys. Rev. A **93**, 063612 (2016)
47. M. Gring, S. Gerlich, S. Eibenberger, S. Nimmrichter, T. Berrada, M. Arndt, H. Ulbricht, K. Hornberger, M. Müri, M. Mayor, M. Böckmann, N. Doltsinis, Influence of conformational molecular dynamics on matter wave interferometry. Phys. Rev. A **81**, 031604 (2010)
48. B.A. Stickler, K. Hornberger, Molecular rotations in matter-wave interferometry. Phys. Rev. A **92**, 023619 (2015)
49. M.K. Oberthaler, St Bernet, E.M. Rasel, J. Schmiedmayer, A. Zeilinger, Inertial sensing with classical atomic beams. Phys. Rev. A **54**, 3165–3176 (1996)
50. R. Antoine, D. Rayane, A.R. Allouche, M. Aubert-Frëcon, E. Benichou, F.W. Dalby, Ph Dugourd, M. Broyer, Static dipole polarizability of small mixed sodium–lithium clusters. J. Chem. Phys. **110**, 5568–5577 (1999)
51. S. Heiles, R. Schaefer, *Dielectric Properties of Isolated Clusters: Beam Deflection Studies*. Springer Briefs in Molecular Science. (Springer, Heidelberg, 2014)

Yb Fiber Amplifier at 972.5 nm with Frequency Quadrupling to 243.1 nm

Z. Burkley, C. Rasor, S.F. Cooper, A.D. Brandt, and D.C. Yost

Abstract We demonstrate a continuous-wave ytterbium-doped fiber amplifier which produces 6.3 W at a wavelength of 972.5 nm. We frequency-quadruple this source in two resonant doubling stages to generate 530 mW at 243.1 nm. Radiation at this wavelength is required to excite the 1S–2S transition in atomic hydrogen and could therefore find application in experimental studies of hydrogen and anti-hydrogen.

1 Introduction

The hydrogen 1S–2S two-photon transition was first observed by Hänsch et al. [1]. Over the following four decades, the continued improvement in the spectroscopy of this transition has led to increasingly precise determinations of the Rydberg constant and proton charge radius—ultimately providing a stringent test of quantum electrodynamics [2]. The importance of the 1S–2S transition stems in part from the simplicity of hydrogen, which makes it amenable to theoretical study, and also from its narrow natural linewidth of only 1.3 Hz.

When reviewing the well-known measurements of the 1S–2S transition, one can also observe a continual refinement of the spectroscopy lasers used—first, by a transition from pulsed to continuous-wave (cw) lasers [3, 4] and then by an increase in power, coherence and robustness [5–9]. The most recent result was reported by the Hänsch group in 2011, in which they determined the transition to a fractional frequency uncertainty of 4.2×10^{-15} [10]. By that time, the UV laser source had evolved to an all solid state system that produced 13 mW of 243 nm cw radiation. This radiation was then cavity enhanced to 368 mW within the hydrogen spectrometer. In addition to the impressive intracavity power, this radiation source possessed an extremely narrow linewidth of $\approx$1 Hz which is commensurate with the hydrogen

This article is part of the topical collection "Enlightening the World with the Laser" - Honoring T. W. Hänsch guest edited by Tilman Esslinger, Nathalie Picqué, and Thomas Udem.

Z. Burkley (✉) • C. Rasor • S.F. Cooper • A.D. Brandt • D.C. Yost
Colorado State University, Fort Collins, CO 80523, USA
e-mail: zakary.burkley@colostate.edu

D. Meschede et al. (eds.), *Exploring the World with the Laser*,
https://doi.org/10.1007/978-3-319-64346-5_3

1S–2S transition width itself. More recently, in 2013, Beyer et al. [11] reported on a 243 nm laser which was capable of producing 75 mW before cavity enhancement.

Notwithstanding these accomplishments, we believe that continuing to increase the laser power at 243 nm would be very beneficial. For instance, the 1S–2S transition could be excited with laser beams of large transverse dimensions, which could decrease transit-time broadening and increase the proportion of atoms in the atomic beam that are excited. With the recent trapping of anti-hydrogen in its ground state, a larger beam would also prove beneficial in mitigating the difficulties created by the low number of trapped anti-hydrogen atoms available [12, 13]. However, we are mainly motivated to develop a power-scalable 243 nm laser in order to explore proposals to laser cool atomic hydrogen using the 1S–2S transition [14–16].

Spectroscopy of hydrogen and the recently trapped anti-hydrogen would benefit tremendously from robust laser cooling. Two-photon laser cooling could be more rapid and flexible than the more traditional approach using Lyman-alpha radiation at 121.6 nm—mostly due to the greater ease of producing radiation at 243 nm. To obtain reasonable scattering rates with such schemes requires that the 2S state be coupled to a state with short lifetime—for instance either the 2P [15] or 3P [16] states—and the average power of the cavity enhanced 243 nm radiation source should be at the $\sim$100 W level. For a beam diameter of 500 μm, this would lead to a scattering rate of $\sim$500 Hz when maximally coupling the 2S and 2P states [15]. Power enhancement within an optical cavity can reach factors of $\sim$100 with commercially available mirrors so that Watt-level 243 nm sources could be sufficient for an initial demonstration of cooling.

Here, we present a laser system that is a major step toward laser cooling hydrogen with the two-photon 1S–2S transition. The system is composed of an extended cavity laser diode (ECDL) at 972 nm followed by a tapered amplifier (TA), a ytterbium-doped double-clad fiber amplifier, and two consecutive resonant doubling stages. The ytterbium (Yb) fiber amplifier is a notable feature of this source since gain is much more readily obtained in Yb systems near 1030 nm due to the low absorption cross section at this wavelength. Gain near the emission cross-sectional peak at 976 nm is also possible but requires population inversions near 50% because the absorption cross section in that spectral region has approximately the same magnitude. Despite this difficulty, there have been demonstrations of 100 W Yb-doped fiber lasers near the emission cross-sectional peak at 976 nm [17, 18]. These lasers, however, operated at or above 976 nm. Below 976 nm, the emission cross section drops rapidly and the inversion required to obtain gain is $>$50%. For our Yb-doped fiber, the emission cross section at 972 nm is approximately one-half the peak value. To the best of our knowledge, there have been only a few Yb fiber-based laser systems which operate at wavelengths shorter than 976 nm and these produced relatively low power ($\sim$10 mW) [19, 20]. The Yb fiber amplifier we demonstrate here outputs 6.3 W of power at 972 nm which upon frequency quadrupling yields 530 mW of power at 243 nm. We believe this approach is power scalable and that we can continue to increase our UV radiation with additional power at 972 nm.

2 Seed Laser and Yb Fiber Amplifier

The master oscillator is an ECDL with an extended 10-cm-long cavity. This relatively long cavity length was chosen to increase the coherence of the oscillator [9]. The ECDL produces 30 mW of power at 972 nm and the design is shown in Fig. 1. The output of the oscillator is amplified to 3 W with a commercial tapered amplifier (DILAS). The manufacturer specified M^2 of the TA is only <1.7, and we have measured ≈2.4 W within a TEM_{00} mode at the full power of 3 W. The output from the TA is then further amplified within a double-clad Yb-doped fiber (CorActive) with a 20 μm diameter core and 128 μm cladding. The core of this fiber has a numerical aperture of 0.075 which is large enough to support the propagation of a few higher-order modes. However, we have observed the majority of the TA output power not contained within the TEM_{00} mode exists in transverse modes of significantly higher order. These modes cannot propagate through the fiber; thus, the fiber acts as a spatial filter.

The population inversion necessary to obtain gain at 972 nm requires that a high pump intensity at 915 nm be maintained along the entire length of the fiber, which results in low pump absorption. In our case, we use a short section of fiber (≈10 cm) that absorbs only 0.6 dB of the incident pump power. Due to the high population inversion within the fiber amplifier, there is also significant gain within the 1015 and 976 nm spectral regions. This is problematic as significant amplified spontaneous emission (ASE) would degrade the amplifier performance by reducing the population inversion. As shown in [21], the gain at a given wavelength within a homogeneously broadened amplifier can be written as a function of the gain or absorption at two other wavelengths and their respective absorption and emission cross sections. Using the cross-sectional data for our fiber [22], we find an expression for the gain (G_λ) at 976 nm given by

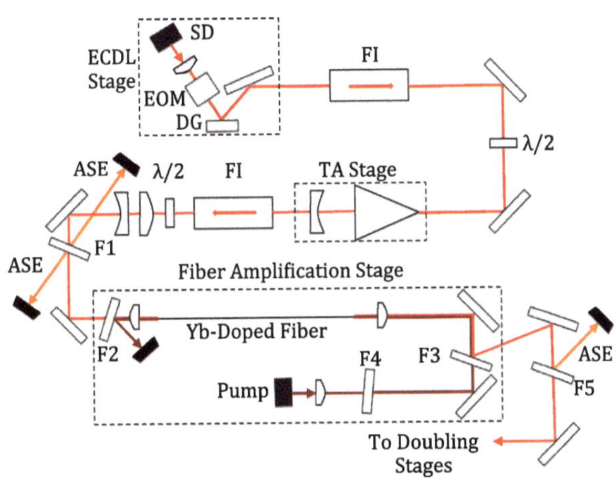

Fig. 1 Schematic of the ECDL master oscillator and amplification stages. SD: Seed diode, DG: diffraction grating, FI: Faraday isolator, F1 and F5: bandpass filters, F2: longpass filter, F3 and F4: shortpass filters. The ECDL contains an electro-optic modulator (EOM) for fast frequency control although it was not used for the studies here

$$G_{976} = 2.41 \cdot G_{972} + 1.08 \cdot \beta A_{915}. \tag{1}$$

Here A_{915} is the absorption of the pump at 915 nm, and β is the ratio of the cladding area to the core area. In our amplifier, $\beta \approx 41$ and $A_{915} \approx 0.6$ dB. With a gain of $G_{972} \approx 4$ dB (given by our experimental results), the gain at 976 nm is $G_{976} \approx 37$ dB. Similarly, the gain at 1015 nm where there is also a peak in the emission cross section of our fiber is given by

$$G_{1015} = 0.70 \cdot G_{972} + 1.64 \cdot \beta A_{915}, \tag{2}$$

which results in $G_{1015} \approx 43$ dB. As can be seen from the previous expressions, the gain at 976 and 1015 nm depends sensitively on the pump absorption due to the large value of β. As discussed extensively in [17, 18, 21], increasing the pump absorption, and therefore efficiency of the amplifier, would need to be accompanied by a decrease in β to keep the gain near 976 and 1015 nm manageable. Even in our current configuration, the large gain at 976 and 1015 nm would cause the amplifier to lase if the ends of the gain fiber were flat cleaved. To mitigate these effects, we angle-polish the ends of the gain fiber and use bandpass filters (FWHM = 4 nm) to reduce ASE originating from both the fiber amplifier and the TA [20]. The high inversion in the fiber can also lead to photodarkening—a poorly understood decrease in the optical transmission of gain fibers which degrades performance [23]. Mitigation of this effect is possible by codoping the gain fiber with phosphorous or cerium [24, 25]. We use a fiber codoped with phosphorous because it was commercially available; we have yet to observe any such degradation of the amplifier performance due to this effect.

Figure 2 shows the backward propagating radiation from the Yb fiber amplifier when pumping with 22.5 W of 915 nm radiation. Without seed power, the high inversion favors gain at 976 nm. With seed power, the peak at 976 nm is still dominant; however, this is due to ASE from the TA which is incompletely attenuated by filter F1 in Fig. 1. Back reflections from the output facet of the amplifier fiber also cause this radiation to be amplified in the reverse direction, along with a

Fig. 2 Backward propagating radiation from the Yb fiber amplifier with and without seed. Both spectra were taken with 22.5 W of pump power at 915 nm

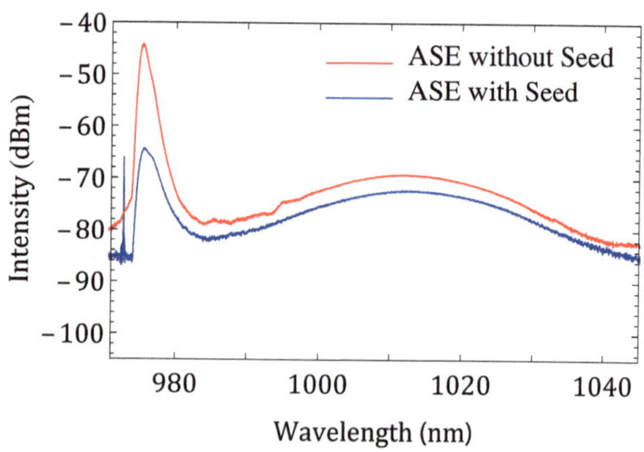

small amount of the 972 nm radiation. Without seed, the integrated power in Fig. 2 is ≈270 mW, but this drops to only ≈9 mW when seeded. The ASE is roughly linear with pump power and therefore does not significantly degrade the amplifier performance for this short length of fiber.

The 972 nm output power of the fiber amplifier as a function of 915 nm pump power is shown in Fig. 3. Amplification of the 2.4 W of seed occurs at ≈15 W of pump power. However, even with zero pump power, significant inversion of the gain medium can occur due only to the input seed radiation, so that the fiber exhibits semi-transparency before the amplification threshold. At our maximum pump power of 41.5 W, we obtain an output power of 6.3 W at 972 nm, corresponding to a gain of 4.2 dB. The roll-off at high pump power is a result of the pump wavelength shifting away from the absorption peak at 915 nm as the diode current is increased. This effect is more pronounced in our system due to the low absorption of the pump. If the pump radiation did not shift with diode current, our models indicate there would be a near linear increase in the output power as a function of pump power.

Unfortunately, several common techniques to increase the efficiency of fiber amplifiers are not possible for our system. For example, an increase in fiber length without a decrease in β will result in unmanageable gain at 976 and 1015 nm. A double pass pump configuration is also not possible as the low single pass absorption of the Yb fiber amplifier could damage our pump diode. More specific techniques such as ring-doping [21] and ultra large-mode-area rod-type photonic crystal fibers [17, 18] enabled Yb-doped fiber amplifiers with ≈60% efficiency between 976 and 980 nm. To our knowledge, the former is not commercially

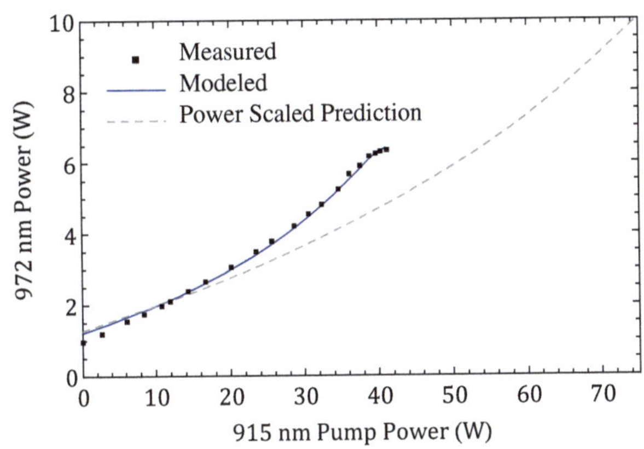

Fig. 3 Measured radiation at the output of the fiber amplifier as a function of 915 nm pump power. A linear fit to our data indicates a slope efficiency of ≈13%. The measured values agree well with a theoretical model (*solid line*) based on [18]. The nonlinear behavior is due to an increase in the pump wavelength as the diode current is increased. This effect is more pronounced in our system due to the low absorption of the pump power. Our model predicts that more power at 972 nm could be obtained with a pump diode that provides 130 W of output power (*dashed line*) and reaches the 915 nm absorption peak at a higher power [26]

available. The latter, although commercially available, have no long-term studies testing their resilience against photodarkening at high inversion. In addition, their length cannot be changed once constructed. One potentially interesting avenue would be Yb-doped fiber with smaller cladding/core ratio (β), which is also codoped with phosphorous to prevent photodarkening.

Despite the low efficiency of this amplifier, the outlook for scaling the power, which is the primary concern for laser cooling atomic hydrogen, seems promising. Modeling the performance of our amplifier using the method found in [18], we can theoretically reproduce the measured performance. As shown in Fig. 3, the model also indicates that we can continue to increase the output power with a pump diode that reaches 915 nm at higher power [26]. The same model suggests an additional amplification stage could also be an effective means to increase the output power [27].

3 Doubling Stages

As shown in Fig. 4, the output of the fiber amplifier is frequency-quadrupled to 243 nm in two consecutive resonant doubling stages. The first stage uses lithium triborate (LBO) as the nonlinear crystal (United Crystals), while for the second we tested both beta barium borate (BBO) and cesium lithium triborate (CLBO) as the nonlinear crystals (Crystals of Siberia and Altechna, respectively).

The first nonlinear doubling stage uses a standard bowtie geometry. The curved mirrors have a radius of curvature (ROC) of 200 mm, producing a 62 μm beam waist within the 25-mm-long LBO crystal. This mode size, which is ≈1.8 times the optimal waist determined from the Boyd–Kleinman focusing criteria [29], increases the robustness of the doubling stage with minimal effect on the overall conversion efficiency. We use type I non-critical phase matching in order to eliminate spatial walk-off and improve the 486 nm output beam quality, which requires the LBO be kept at a temperature of 283 °C. Alternatively, walk-off could be eliminated with

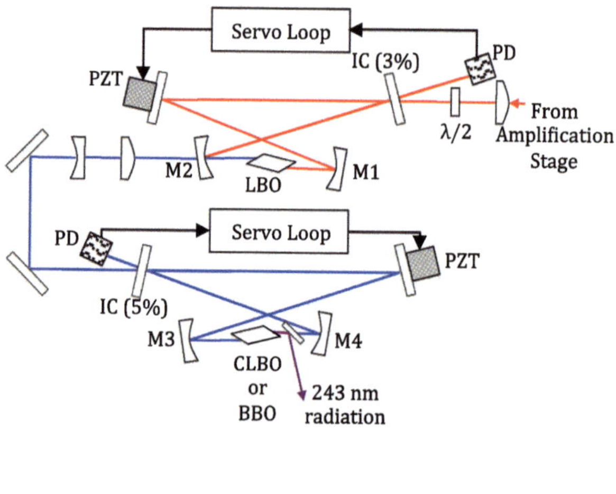

Fig. 4 Experimental setup of doubling stages. PD: Photodiode, PZT: piezoelectric transducer, IC: input coupler, M1–M4: 200 mm ROC mirrors. Since the optimal input coupler transmission depends on the power of the fundamental [28], we use input couplers with higher transmission than optimal to increase robustness and allow for power scaling

the use of periodically poled nonlinear materials. However, one of our primary concerns for this work is robustness and high damage threshold, and we believe periodically poled nonlinear crystals are less proved in this regard. The first stage uses an input coupler with a transmission of 3%, which is larger than the theoretical optimal ($T_{opt} \approx 2\%$ at the highest fundamental power) [28]. This allows for power scaling since the optimal input coupler transmission will increase with additional fundamental power.

Because the performance of typical dual wavelength anti-reflection coatings is not guaranteed at high temperatures, we use a Brewster-cut crystal to reduce the loss of the resonant 972 nm light. This leads to an 18% loss of the generated 486 nm radiation from the Fresnel reflection on the crystal output facet. The remaining 82% of the 486 radiation is coupled out of the cavity through a dichroic curved mirror with high reflectivity at 972 nm and high transmission (>90%) at 486 nm. The 486 nm output power as a function of incident fundamental power is shown in Fig. 5. We obtain 2.4 W of 486 nm radiation with 6.3 W of 972 nm fundamental power. The theoretical fit for harmonic conversion used in Fig. 5 follows the model presented in [28, 30]. The data can be fit reasonably well with a range of parameters in which the transverse mode matching is greater than 80% and the roundtrip linear losses are less than 1.2%.

To stabilize the doubling cavity to the fundamental radiation, we modulate one cavity mirror at a frequency of ≈300 kHz and detect the modulated cavity power with a photodiode (see Fig. 4). We then demodulate this signal with a double-balanced mixer to generate an error signal, which is sent to a loop filter followed by a fast (∼50 kHz bandwidth) piezoelectric transducer and a slow (∼100 Hz) transducer with greater range. A nearly identical setup is used to maintain the resonance condition within the second doubling stage.

In the second doubling stage, which produces 243 nm radiation, we tested both a BBO crystal and a CLBO crystal [31–34] in type I critical phase matching configurations. CLBO has a lower nonlinear coefficient than BBO, but also less spatial walk-off and a higher damage threshold [35]. Similar to the first stage, the second doubling stage is a bow tie design with 200 mm ROC mirrors to produce a focus in the nonlinear crystal. This produces a beam waist of 44 μm that is ≈1.9 times the

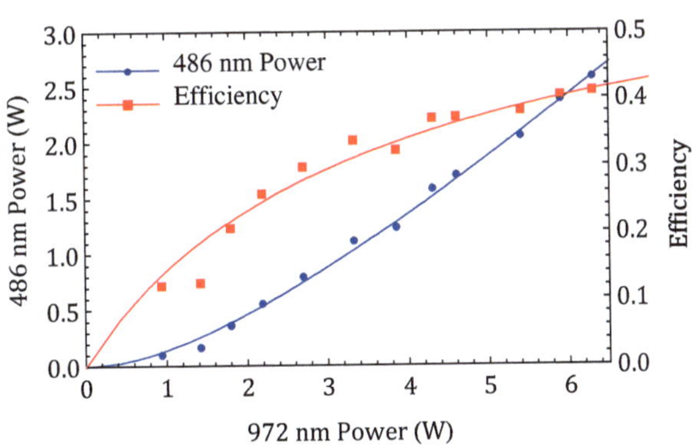

Fig. 5 Experimental results of frequency doubling fiber amplified 972 nm radiation using LBO as the nonlinear medium. With 6.3 W of power at 972 nm, up to 2.60 W of power at 486 nm is generated. The *solid lines* follow the model in [28, 30]

Boyd–Kleinman focusing criteria [29] in order to prevent damage of the nonlinear crystal at high intensities and to minimize walk-off effects. The input coupler for this cavity was 5% to increase robustness and allow for power scaling. Both crystals are Brewster cut and 10 mm long. In this case, Brewster-cut crystals are used because AR coatings are not yet well developed for CLBO crystals [32]. This introduces a 27% output coupling loss for the 243 nm light with the BBO crystal, and an 18% loss with the CLBO crystal due to the Fresnel reflection on the output facet of the crystal. The BBO crystal is cut at $\theta = 55°$ and has a double refraction angle of $\rho = 82$ mrad. For CLBO, $\rho = 18$ mrad and the crystal is cut at $\theta = 77°$. As shown in Fig. 4, a Brewster oriented dichroic mirror with high reflectivity at 243 nm and high transmission at 486 nm is used to output couple the 243 nm radiation.

The observed 243 nm output power as a function of the 486 nm input power is shown in Fig. 6 when using the BBO crystal and in Fig. 7 for the CLBO crystal. The theoretical curves again follow the model presented in [28, 30], and assume a 2% roundtrip linear loss along with 80% transverse mode matching. As can be seen from the figures, a greater efficiency was obtained with the CLBO crystal due to the smaller walk-off and smaller Fresnel loss from the Brewster-cut crystal. This also produces a 243 nm beam with less ellipticity, which will be easier to shape and couple into a 243 nm enhancement cavity.

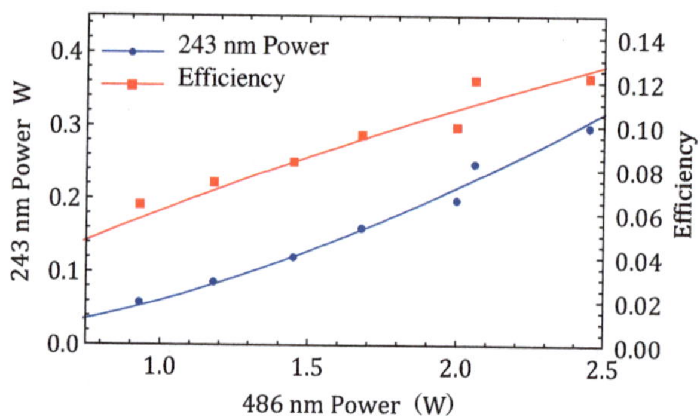

Fig. 6 Experimental results of frequency doubling 486 nm radiation to 243 nm radiation using BBO as the nonlinear medium. With 2.45 W of 486 nm radiation, up to 300 mW of power at 243 nm is generated. The *solid lines* follow the model in [28, 30]

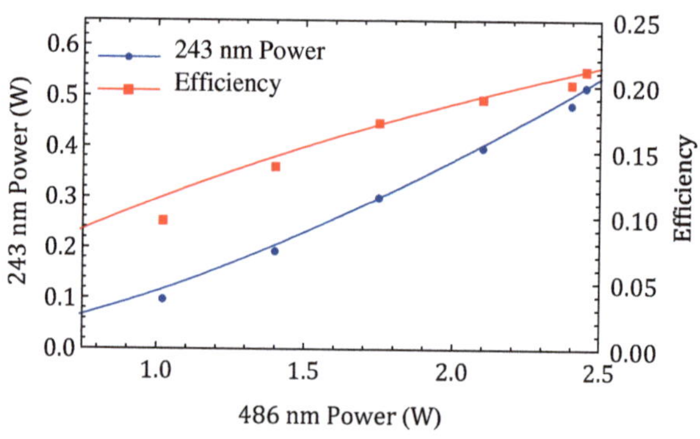

Fig. 7 Experimental results of frequency doubling 486 nm radiation to 243 nm radiation using CLBO as the nonlinear medium. With 2.45 W of 486 nm radiation, up to 530 mW of power at 243 nm is generated. The *solid lines* follow the model in [28, 30]

Due to the high UV power generated, crystal degradation is a concern. Frequency doubling studies at a similar wavelength have demonstrated 5 W of 266 nm power without damaging the CLBO crystal [32]. This corresponds to $\approx$5 times greater UV intensity within the crystal compared to the results reported here. Additional studies have shown that if degradation in CLBO occurs, it appears to be reversible [36]. This is in contrast to BBO, which shows irreversible damage caused by the formation of absorption centers [31, 36, 37]. Therefore, by utilizing CLBO we should be able to power scale our UV output as more fundamental power becomes available without crystal degradation.

CLBO is also known to be hygroscopic and some performance change has been reported as the crystal absorbs or desorbs water [38]. For this reason, the CLBO crystal is operated at a temperature of 130 °C. Over a few days at this elevated temperature, the conversion efficiency increased slightly above that shown in Fig. 7 and we observed >530 mW of 243 nm radiation over 50 min with no degradation.

4 Conclusion

We have demonstrated a fiber-based amplifier laser system capable of generating 6.3 W of power at 972 nm. Upon frequency doubling in successive resonant cavities, this laser source can generate 2.4 W at 486 nm and 530 mW at 243 nm. We are encouraged by the power scalability of our system. Simulations indicate that our fiber amplifier platform should be able to produce additional 972 nm radiation either with more powerful, commercially available pump diodes or with an additional fiber amplifier stage of similar design [27]. To use the 915 nm pump radiation more efficiently would require that we obtain fibers with a smaller cladding/core area ratio (β), such as the rod-type fibers used in [17, 18], with the addition of phosphorous codoping to increase resilience against photodarkening. The doubling stages were designed with relatively loose focusing in the crystals and high transmission input couplers. This, in conjunction with the high damage thresholds of LBO and CLBO, make us hopeful that these cavities can also be power scaled.

Although we made no in-depth studies of the linewidth of our laser source for the work described here, our seed laser copies many aspects of the low phase noise design described in [9] and we were able to couple our radiation into doubling cavities with few MHz resonance widths without any difficulty. For two-photon laser cooling of hydrogen, the transition width will be broadened to $\approx$50 MHz by coupling the 2S and 2P states. Therefore, the laser source we describe here already has the coherence necessary for that application. That being said, spectroscopy of the hydrogen and anti-hydrogen 1S–2S transition with a power scaled 243 nm system would be very beneficial but would also require the source possess an extremely narrow linewidth. Stabilizing the frequency of this source to that level will therefore be the subject of future work.

Acknowledgements We gratefully acknowledge Jacob Roberts for useful discussions and for carefully reviewing this manuscript.

References

1. T.W. Hänsch, S.A. Lee, R. Wallenstein, C. Wieman, Phys. Rev. Lett. **34**, 307 (1975)
2. P.J. Mohr, B.N. Taylor, D.B. Newell, Rev. Mod. Phys. **84**, 1527 (2012)
3. B. Couillaud, T. Hänsch, S. MacLean, Opt. Commun. **50**, 127 (1984)
4. C.J. Foot, B. Couillaud, R.G. Beausoleil, T.W. Hänsch, Phys. Rev. Lett. **54**, 1913 (1985)
5. M.G. Boshier et al., Phys. Rev. A **40**, 6169 (1989)
6. R. Kallenbach, F. Schmidt-Kaler, M. Weitz, C. Zimmermann, T.W. Hänsch, Opt. Commun. **81**, 63 (1991)
7. C. Zimmermann, V. Vuletic, A. Hemmerich, T.W. Hänsch, Appl. Phys. Lett. **66**, 2318 (1995)
8. N. Kolachevsky, J. Alnis, S.D. Bergeson, T.W. Hänsch, Phys. Rev. A. (Rapid Communications) **73**(021801(R)) (2006)
9. N. Kolachevsky, J. Alnis, C. Parthey, A. Matveev, R. Landig, T. Hänsch, Opt. Lett. **36**, 4299 (2011)
10. C. Parthey, A. Matveev, J. Alnis, B. Bernhardt, A. Beyer, R. Holzwarth, A. Maistrou, R. Pohl, K. Predehl, T. Udem, T. Wilken, N. Kolachevsky, M. Abgrall, D. Rovera, C. Salomon, P. Laurent, T. Hänsch, Phys. Rev. Lett. **107**, 203001 (2011)
11. A. Beyer, J. Alnis, K. Khabarova, A. Matveev, C. Parthey, D. Yost, R. Pohl, T. Udem, T. Hänsch, N. Kolachevsky, Ann. Phys. (Berlin) **525**, 671 (2013)
12. G. Gabrielse, R. Kalra, W.S. Kolthammer, R. McConnell, P. Richerme, D. Grzonka, W. Oelert, T. Sefzick, M. Zielinski, D.W. Fitzakerley, M.C. George, E.A. Hessels, C.H. Storry, M. Weel, A. Müllers, J. Walz, Phys. Rev. Lett. **108**, 113002 (2012)
13. G. Andersen et al., Nat. Phys. **7**, 558 (2011)
14. V. Zehnlé, J.C. Garreau, Phys. Rev. A. (Rapid Communications) **63**, 021402(R) (2001)
15. D. Kielpinski, Phys. Rev. A **73**, 063407 (2006)
16. S. Wu, R. Brown, W. Phillips, J. Porto, Phys. Rev. Lett. **106**, 213001 (2011)
17. J. Boullet, Y. Zaouter, R. Desmarchelier, M. Cazaux, F. Salin, J. Saby, R. Bello-Doua, E. Cormier, Opt. Express **16**, 17891 (2008)
18. F. Röser, C. Jauregui, J. Limpert, A. Tünnermann, Opt. Express **16**, 17310 (2008)
19. D. Hanna, R. Percival, I. Perry, R. Smart, P. Suni, A. Tropper, J. Mod. Opt. **37**, 517 (1990)
20. J. Yi, Y. Fan, S. Huang, IEEE Photonics J. **4**, 2278 (2012)
21. J. Nilsson, J. Minelly, R. Paschotta, A. Tropper, D. Hanna, Opt. Lett. **23**, 355 (1998)
22. Fiber data given to authors from CorActive upon author's request
23. J. Koponen, M. Söderlund, H. Hoffman, Opt. Express **14**, 11539 (2006)
24. M. Engholm, L. Norin, Opt. Express **16**, 1260 (2008)
25. M. Engholm, P. Jelger, F. Laurell, L. Norin, Opt. Lett. **34**, 1285–1287 (2009)
26. nLIGHT Photonics – model: e12.1300915105
27. D. Richardson, J. Nilsson, W. Clarkson, J. Opt. Soc. Am. B **27**, 63 (2010)
28. E. Polzik, H. Kimble, Opt. Lett. **16**, 1400 (1991)
29. G. Boyd, D. Kleinman, J. App. Phys. **39**, 3597 (1968)
30. R. Targat, J. Zondy, P. Lemonde, Opt. Commun. **247**, 471 (2005)
31. M. Scheid, F. Markert, J. Walz, J. Wang, M. Kirchner, T. Hänsch, Opt. Lett. **32**, 955 (2007)
32. J. Sakuma, Y. Asakawa, T. Sumiyoshi, H. Sekita, IEEE J. Sel. Top. Quantum Electron **10**, 1244 (2004)
33. J. Hu, L. Zhang, H. Liu, K. Liu, Z. Xu, Y. Feng, Opt. Express **21**, 30958 (2013)
34. Y. Kaneda, J. Yarborough, L. Li, N. Peyghambarian, L. Fan, C. Hessenius, M. Fallahi, J. Hader, J. Moloney, Y. Honda, M. Nishioka, Y. Shimizu, K. Miyazono, H. Shimatani, M. Yoshimura, Y. Mori, Y. Kitaoka, T. Sasaki, Opt. Lett. **33**, 1705 (2008)
35. Y. Mori, I. Kuroda, S. Nakajima, T. Sasaki, S. Nakai, Jpn. J. Appl. Phys. **34**, 296 (1995)
36. K. Takachiho, M. Yoshimura, Y. Takahashi, M. Imade, T. Sasaki, Y. Mori, Opt. Mat. Exp. **4**, 559 (2014)
37. K. Kondo, M. Oka, H. Wada, T. Fukui, N. Umezu, K. Tatsuki, S. Kubota, Opt. Lett. **23**, 195 (1998)
38. T. Kawamura, M. Yoshimura, Y. Honda, M. Nishioka, Y. Shimizu, Y. Kitaoka, Y. Mori, T. Sasaki, App. Opt. **48**, 1658 (2008)

Optical Autler–Townes Spectroscopy in a Heteronuclear Mixture of Laser-Cooled Atoms

C. Bruni, F. Münchow, and A. Görlitz

Abstract We report on optical Autler–Townes spectroscopy in a heteronuclear mixture of ^{87}Rb and ^{176}Yb in a continuously loaded double-species magneto-optical trap. An excited vibrational level of Rb*Yb which is energetically close to the $5^2P_{1/2}$ state of Rb is coupled by a strong laser field to a vibrational level in the ground state of RbYb and probed by a weak probe laser field. The induced Autler–Townes splittings in the photoassociation spectra allow us to determine relative Franck–Condon factors of molecular transitions in RbYb.

1 Introduction

Over the last decade, there has been tremendous progress in the emerging field of ultracold molecules and especially ultracold heteronuclear molecules. Ultracold heteronuclear molecules promise a variety of potential applications ranging from model systems for many-body physics over ultracold chemistry and precision measurements to quantum computation [1, 2]. Of particular interest for these applications are heteronuclear molecules with an unpaired electron since they possess an electronic as well as a magnetic dipole moment in the ground state, and thus, the possibilities for interaction and manipulation are enhanced. In order to make the best use of the interesting features of ultracold molecules, it is required to prepare them in the rotational, vibrational and electronic ground state.

While laser cooling is the standard starting point for the preparation of ultracold atomic gases, the complex internal structure of molecules allows for laser cooling of molecules only in very special cases [3–5] and moreover with a reduced efficiency. Therefore, the most commonly followed pathway towards ultracold heteronuclear molecules starts by creating mixtures of ultracold atoms using laser cooling. Ultracold molecules are then formed by using magnetically tunable

This article is part of the topical collection "Enlightening the World with the Laser" - Honoring T. W. Hänsch guest edited by Tilman Esslinger, Nathalie Picqué, and Thomas Udem.

C. Bruni • F. Münchow • A. Görlitz (✉)
Institut Für Experimentalphysik, Heinrich-Heine-Universität Düsseldorf, Universitätsstraße 1, 40225 Düsseldorf, Germany
e-mail: axel.goerlitz@uni-duesseldorf.de

D. Meschede et al. (eds.), *Exploring the World with the Laser*,
https://doi.org/10.1007/978-3-319-64346-5_4

Feshbach resonances [6, 7] or by photoassociation (PA) [8], and subsequently, the molecules are transferred to their internal ground state using laser STIRAP [9]. Magnetic Feshbach resonances have been used very successfully to create heteronuclear bialkali molecules in the electronic ground state such as KRb [9–11], NaK [12, 13] and RbCs [14, 15]. However, in mixtures of a closed-shell atom (such as Yb) and an alkali atom (such as Rb), Feshbach resonances of atoms in the electronic ground state have been predicted theoretically [16, 17] but not yet reported experimentally. Only magnetic-field-dependent interactions between alkali ground-state atoms and metastable Yb atoms have recently been reported [18]. Due to these complications, the more promising route to molecules with an unpaired electron seems to be light-assisted PA where unbound atoms are coupled to an excited molecular state. Molecules in the electronic ground state can then either be created by spontaneous decay as e.g. in the case of RbCs [19], LiCs [20] or homonuclear Cs_2 [21] or Sr_2 [22] or by using an additional laser which couples the excited molecular state to a molecular ground state as it was demonstrated, for example for KRb [23] or homonuclear Sr_2 [24].

Whenever light is used to prepare ultracold molecules in a controlled way or to manipulate their internal state, precise knowledge of the molecular structure is required. In our effort to create ultracold ensembles of RbYb molecules, we have previously performed one-photon spectroscopy in a magneto-optical trap (MOT) [25] and in a hybrid conservative trap [26] to determine the energetic position of weakly bound rovibrational levels of the excited ${}^2\Pi_{1/2}$ state of Rb*Yb. In addition, we have used two-photon spectroscopy to measure the binding energies of rovibrational levels close to the dissociation limit of the molecular ground state [27, 28].

A method which not only yields the binding energy of molecular levels, but also gives direct access to the coupling strengths and Franck–Condon factors of molecular bound-bound transitions is optical Autler–Townes spectroscopy, where a splitting is observed due to strong resonant coupling of two molecular states. This effect is experimentally and theoretically closely related to two-photon PA but was originally observed in the microwave range [29]. By optically coupling specific excited rovibrational levels to well-defined rovibrational levels in the ground state, a splitting of a one-photon photoassociation resonance to the excited rovibrational level can be observed. This splitting can be directly related to the Rabi frequency and thus the transition matrix element of the molecular transition.

One of the first observations of an Autler–Townes splitting in ultracold ensembles was reported in Lithium [30] which was also used in more detailed studies later on [31]. Since then the technique has been used in a wide range of studies of ultracold atoms, for example in investigations of ultracold Rydberg atoms [32] or to optimize the pathway for ultracold molecule creation [33].

In this paper, we report on optical Autler–Townes spectroscopy in an ultracold heteronuclear mixture of ^{87}Rb and ^{176}Yb. We use the observation and analysis of Autler–Townes splittings to infer information on the strengths of molecular transitions in RbYb, complimentary to our previous studies [27, 28] which were focused on investigating the molecular ground state.

2 Experimental Setup

The experiments presented in this manuscript are performed in a continuously loaded double-species MOT of Rb and Yb. The Rb atoms are confined in a forced dark-spot MOT operating on the $^2S_{1/2} \rightarrow {}^2P_{3/2}$ transition. We trap 5×10^8 atoms at a temperature of 140 μK with more than 95% of the atoms in the dark $|^2S_{1/2}, F = 1\rangle$ state. The Yb MOT is operated on the $^1S_0 \rightarrow {}^3P_1$ transition at 555.8 nm and is loaded from a Zeeman slower using the $^1S_0 \rightarrow {}^1P_1$ transition at 399 nm. The combination of these two atomic species requires a compromise regarding the experimental settings since the presence of the Rb MOT induces huge losses in the number of trapped Yb atoms. While, typically, 3×10^6 Yb atoms are trapped in a single species MOT with a $1/e^2$ radius of 400 μm at a temperature of 400 μK the number of trapped Yb atoms is reduced to $\approx 5 \times 10^4$ in the presence of the Rb MOT due to inelastic light-assisted interspecies interactions. Both atom numbers reach their final steady-state values after a loading time of a few seconds and are monitored independently by the respective MOT fluorescence.

A prerequisite for Autler–Townes spectroscopy is the exact determination of a one-photon photoassociation transition from unbound atoms to a bound rovibrational level in the electronically excited $^2\Pi_{1/2}$ of Rb*Yb. As described in detail in our previous work [25, 26], one-photon PA transitions are determined using trap-loss spectroscopy, where the double-species MOT is illuminated by a photoassociation probe laser (PA probe laser) tuned close to the Rb D_1 transition $^2S_{1/2} \rightarrow {}^2P_{1/2}$ at 795 nm (see Fig. 1). If the PA probe laser is on resonance with a transition, excited-state molecules are formed reducing the number of atoms in the MOT and thus the fluorescence. Due to a large imbalance of the atom numbers of the two species by a factor of 10^4, the PA signal is only observed in the Yb MOT fluorescence and not in the Rb MOT fluorescence. In the following, the wavenumber of the PA probe laser $\tilde{\nu}_{PA}$ is expressed relative to the wavenumber $\tilde{\nu}_{D1} = 12579.1037\,\text{cm}^{-1}$ of the Rb D_1 transition $|^2S_{1/2}, F = 1\rangle \rightarrow |^2P_{1/2}, F' = 2\rangle$ by $\Delta_{PA} = \tilde{\nu}_{PA} - \tilde{\nu}_{D1}$ which corresponds to the binding energy Δ'_{bind} of the excited rovibrational level if the PA probe laser is on resonance. In our setup, Δ_{PA} can be measured with an absolute accuracy of $\pm 5 \times 10^{-3}\,\text{cm}^{-1}$ using a home-built wavemeter which is based on a Michelson interferometer and is referenced to an atomic Rb transition. The rovibrational level of the excited molecular state is identified by the relative vibrational quantum number $\Delta v' = v' - v'_{max}$ (see Fig. 1) where v' is the *real* vibrational quantum number and v'_{max} is the vibrational quantum number of the most weakly bound state.[1]

Figure 2 a shows a typical scan of the vibrational level $\Delta v' = -11$ at $\Delta_{PA} \approx -4.9\,\text{cm}^{-1}$ of the molecular excited state. If PA spectroscopy is performed in a MOT at a temperature of a few hundred μK, a few rotational levels of the excited state can

[1]In this paper, we follow the notation in [25] with $\Delta v' = 0$ being the most weakly bound state. In contrast, in [27] the vibrational state is labelled by $v = -\Delta v' + 1$.

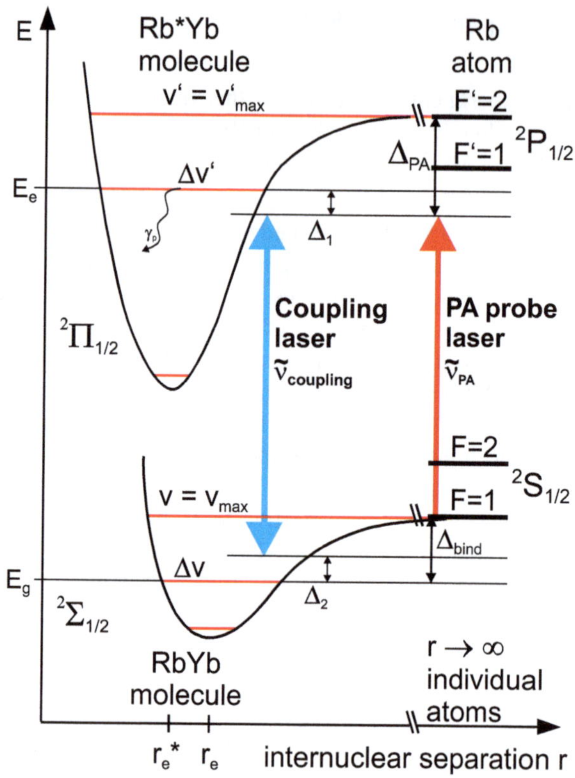

Fig. 1 Relevant level structure (not to scale) of the RbYb molecule and detuning conventions as described in the text. Both the coupling and the PA probe laser are operated near the D1-line of Rb at 795nm. For clarity sake, not all molecular potentials converging onto the ground $^2S_{1/2}$ and the excited $^2P_{1/2}$ state of Rb are depicted but only representative ones for the ground and the excited molecular state, which converge to a specific hyperfine level of the atomic ground (F) or excited (F') state

be observed (here $R' = 0$ and $R' = 1$) [25] while only the $R' = 0$ level appears if the temperature is reduced well below the p-wave threshold to a few μK in a conservative trap [26]. The rotational state belonging to $R' = 1$ shows subcomponents resulting from the coupling of R' and F' to which we assign here the quantum numbers $m'_R = -1, 0, 1$. A detailed analysis of the one-photon PA and the vibrational structure of the excited Rb*Yb molecule can be found in [25, 26].

In order to be able to address a molecular transition and perform optical Autler–Townes spectroscopy, not only the rovibrational levels of the excited molecular state but also those of the rovibrational levels of the ground molecular state have to be known. We have previously measured the binding energies of the ground-state levels by two-photon PA spectroscopy in a double-species MOT [27, 28]. In brief, for two-photon PA spectroscopy, a coupling laser (see Fig. 1) is overlapped with the PA probe laser. The PA laser frequency is fixed to a free-bound transition leading to a continuous loss of atoms from the double-species MOT and thus a reduction of the Yb fluorescence. The frequency of the coupling laser is scanned and if it is on resonance with a molecular transition to the excited state that is addressed by the PA probe laser, it perturbs the one-photon transition resulting in a reduced loss of atoms from the MOT and thus higher fluorescence (see Fig. 2 b). With the help of

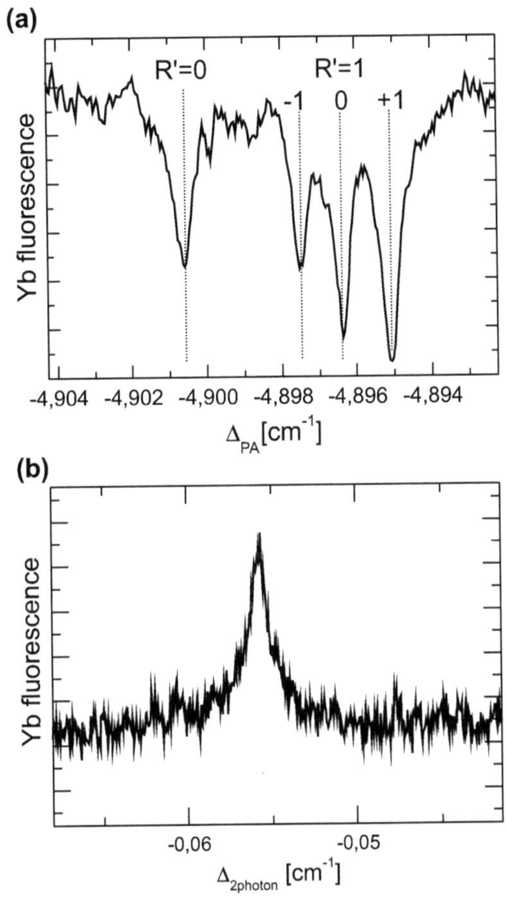

Fig. 2 **a** Typical one-photon PA spectrum of the vibrational level $|\Delta v' = -11 F' = 2\rangle$. A rotational structure with rotational quantum numbers up to $R' = 1$ is observed where the $R' = 1$ level splits into subcomponents with $m_R' = -1, 0, 1$. **b** Typical two-photon PA spectrum. Here, the PA probe laser is kept on resonance with the transition to the $|\Delta v' = -9, F' = 2, R' = 1, m_R' = +1\rangle$ state and the coupling laser is scanned over the $|\Delta v = -2, F = 1, R = 1\rangle \Rightarrow |\Delta v' = -9, F' = 2, R' = 1, m_R' = +1\rangle$ transition

this two-photon PA, we were able to detect seven vibrational levels of the molecular ground state. Similar to the notation for the excited state, the vibrational levels of the molecular ground state are identified by a relative vibrational quantum number $\Delta v = v - v_{\max}$ and the binding energy Δ_{bind}.

For the Autler–Townes spectroscopy presented in this work, the roles of the PA probe laser and the coupling laser are exchanged. Instead of fixing the frequency of the PA probe laser to a free-bound transition, the frequency of the coupling laser is fixed to a previously determined bound-bound transition. The PA probe laser is then scanned over the one-photon PA resonance which connects an unbound atom pair to the excited molecular state of the bound-bound transition. On resonance with the molecular transition, the coupling laser induces a splitting of the excited state which can be detected by the PA probe laser. The magnitude of this Autler–Townes splitting is determined by the Rabi frequency of the bound-bound transition and is hence directly related to the transition matrix element.

The PA probe laser and the coupling laser beams are both obtained from tapered amplifiers injected by diode lasers, one of which is home-built in the good old style established in the Hänsch laboratories [34]. Both laser beams are focused to a $1/e^2$ radius of about $w_0 = 450\,\mu\text{m}$ at the MOT position in order to match the size of the MOTs. Thus, for a typical output power of $P = 10\,\text{mW}$ at the position of the atoms, the peak intensity of either laser is $I_0 = 2P/(\pi w_0^2) = 31\,\text{mW/cm}^2$. In order to frequency stabilize the coupling laser, we use a transfer cavity which is referenced to the Rb D_2-line.

Typical Autler–Townes spectroscopy signals for different coupling laser intensities are depicted in Fig. 3. Here, the coupling laser frequency is fixed to the bound-bound transition $|\Delta v = -7, F = 1, R = 1\rangle \rightarrow |\Delta v' = -11, F' = 2, R' = 1, m'_R = +1\rangle$ where the binding energies of the involved vibrational levels are $\Delta_{\text{bind}} = -1.957\,\text{cm}^{-1}$ and $\Delta'_{\text{bind}} = -4.896\,\text{cm}^{-1}$, respectively.

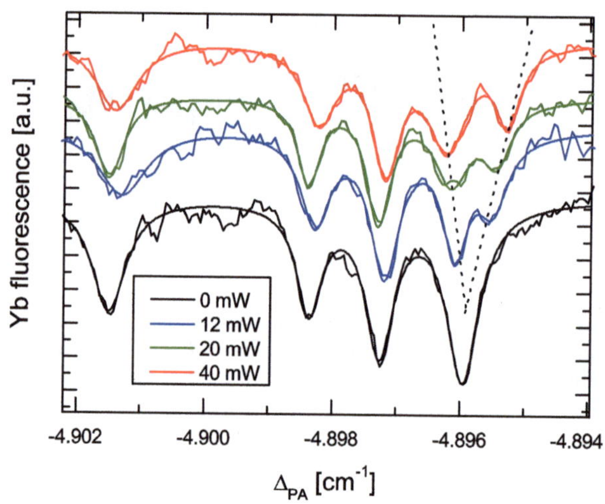

Fig. 3 Autler–Townes scans (*vertically* shifted) with the coupling laser locked to the $|\Delta v = -7, F = 1\rangle \rightarrow |\Delta v' = -11, F' = 2, R' = 1, m_R = +1\rangle$ transition of the RbYb molecule. Here, the PA probe laser has a power of 20 mW corresponding to an intensity of $63\,\text{mW/cm}^2$. The Autler–Townes splitting is observed on the $m_R = +1$ subcomponent of the $R' = 1$ rotational level as indicated by the *dashed lines*. For the other resonances, the coupling laser is too far detuned to induce an observable splitting

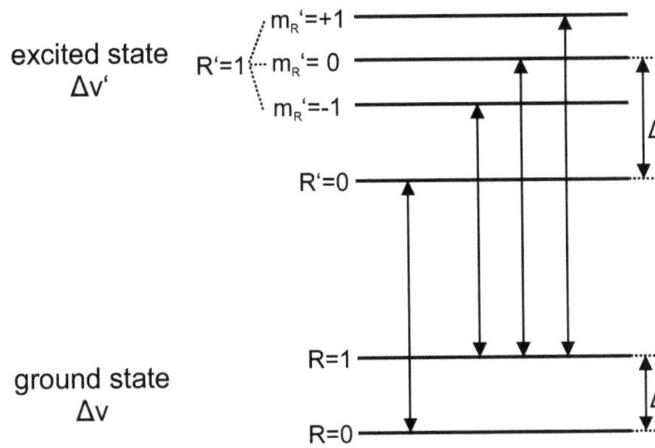

Fig. 4 Relevant molecular transitions between the vibrational ground-state level Δv and the excited-state level $\Delta v'$. Details are described in the *text*

3 Results and Analysis

In Fig. 4, the possible transitions between rovibrational levels with vibrational quantum numbers Δv and $\Delta v'$ and rotational quantum numbers $R = 0, 1$ and $R' = 0, 1$ are depicted. Here, the rotational splittings $\Delta R'$ and ΔR are $2 \cdot B_{\text{rot}}$ and $2 \cdot B'_{\text{rot}}$, respectively, where B_{rot} and B'_{rot} are the rotational constants of the ground- and excited-state vibrational levels. In addition, the splitting of the excited $|\Delta v', R' = 1\rangle$ level into three subcomponents (to which we have assigned the quantum numbers $m'_R = -1, 0, 1$) is relevant here. Based on our experimental observations in the two-photon PA spectra [28], we assume that there is no splitting of the ground-state level with $R = 1$.

The coupling laser couples only molecular states with equal rotational quantum numbers R and R'. However, for the transition from $|\Delta v, R = 1\rangle$ to $|\Delta v', R' = 1\rangle$, the splitting of the excited rotational level into subcomponents leads to three transitions with different transition frequencies. Therefore, there are overall four possible transitions between the vibrational ground-state level Δv and the excited-state level $\Delta v'$ which have to be taken into account to analyse our optical Autler–Townes spectra.

Following the derivation in [35], the scattering probability for the PA probe laser in the presence of a coupling laser for an individual transition from state $|g\rangle$ to state $|e\rangle$ is given by

$$|S_{1g}|^2 = \frac{(\epsilon - \Delta_2)^2 \gamma_1 \gamma_s / (2\pi)^2}{((\epsilon - \Delta_+)(\epsilon - \Delta_-))^2 + \left(\frac{\gamma}{4\pi}\right)^2 (\epsilon - \Delta_2)^2}. \tag{1}$$

Here $E = h\epsilon c$ is the thermal energy of the atoms, while the total width $\gamma = \gamma_1 + \gamma_s + \gamma_0$ is composed of the spontaneous emission rate γ_1 of the excited state, the stimulated emission γ_s back to the starting level and γ_0 which accounts for all other additional losses. The two lasers are detuned by $\Delta_1 = \Delta_{\text{PA}} = \tilde{\nu}_{\text{PA}} - \tilde{\nu}_{\text{D1}}$ and Δ_2

$= \Delta_{\text{bind}} - (\tilde{\nu}_{\text{PA}} - \tilde{\nu}_{\text{coupling}})$ from the one- and the two-photon PA resonance, respectively. The positions of the two peaks that appear in Autler–Townes spectroscopy are given by

$$\Delta_{\pm} = \frac{1}{2}(\Delta_1 + \Delta_2) \pm \frac{1}{2}\sqrt{(\Delta_1 - \Delta_2)^2 + \frac{1}{c^2}\left(\frac{\Omega_{12}}{2\pi}\right)^2} \tag{2}$$

with the Rabi frequency

$$\Omega_{12} = |\langle g|e\rangle|\gamma_1\sqrt{\frac{I}{2I_{\text{sat}}}} \tag{3}$$

of the bound-bound transition. Here, I is the intensity of the coupling laser and I_{sat} the saturation intensity of the transition. The overlap integral $\langle g|e\rangle$ is related to the Franck–Condon factor f_{FC} through $f_{\text{FC}} = |\langle g|e\rangle|^2$. The splitting of the two peaks can be calculated by the difference of Δ_+ and Δ_- and is hence given by

$$\Delta = \Delta_+ - \Delta_- = \sqrt{\delta^2 + \frac{1}{c^2}\left(\frac{\Omega_{12}}{2\pi}\right)^2} \tag{4}$$

with $\delta = \Delta_1 - \Delta_2$ being the total detuning of the coupling laser from the molecular bound-bound transition. According to [35] and [36], in a thermal cloud, the total loss rate can be obtained from

$$K = \frac{1}{hQ_T}\int_0^{\infty} |S_{1g}|^2 e^{-E/k_BT}\,\text{d}E \tag{5}$$

with $Q_T = \left(\frac{2\pi k_B T\mu}{h^2}\right)^{3/2}$ and the reduced mass μ.

Figure 5 compares the experimental spectrum of the transition $|\Delta v = -6, F = 1\rangle \rightarrow |\Delta v' = -11, F = 2\rangle$ to a spectrum that is modelled following Eq. 5. Here, the frequency of the coupling laser was chosen to be resonant with a transition from $|R = 1\rangle$ to $|R' = 1, m_R' = 1\rangle$. Starting from fits for each line appearing in the one-photon PA spectrum (see Fig. 2), the lineshape was modelled for each line independently (see Fig. 5 a). The red line in Fig. 5 b is the weighted sum of the four modelled individual loss rates which is in good qualitative agreement with our experimental data represented by the black line. The Autler–Townes splitting is in general only significant for small detunings and hence only clearly visible in the particular level $|R' = 1, m_R' = 1\rangle$ which is addressed by the coupling laser.

Figure 6 illustrates that, as already assumed above, the coupling laser can be used to address different m_R' subcomponents in the excited molecular state. In Fig. 6 a, the coupling laser is locked to the transition $|R = 1\rangle \rightarrow |R' = 1, m_R' = -1\rangle$ and accordingly the one-photon PA transition to the $|R' = 1, m_R' = -1\rangle$ exhibits a clearly visible Autler–Townes splitting. If the coupling laser is fixed to the $|R = 0\rangle \rightarrow |R' = 0\rangle$ transition as illustrated in Fig. 6 b, an Autler–Townes splitting is not

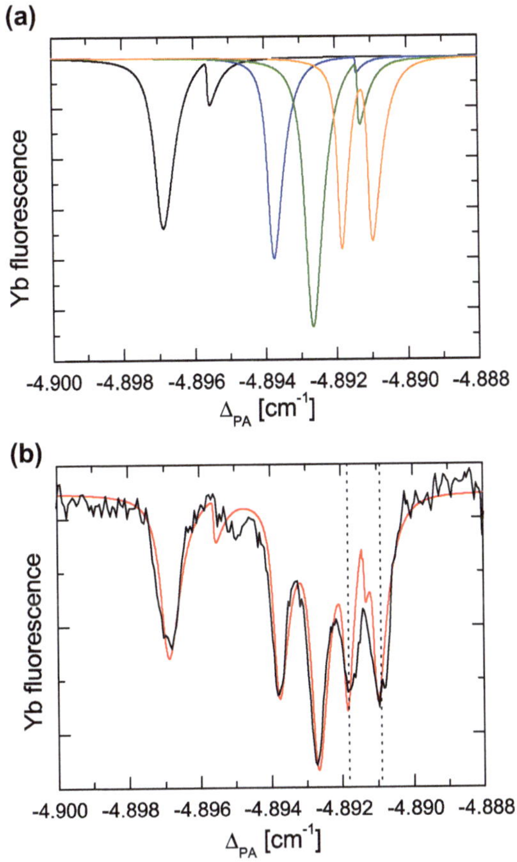

Fig. 5 Autler–Townes spectroscopy of the $|\Delta v = -6, F = 1\rangle \rightarrow |\Delta v' = -11, F' = 2\rangle$ vibrational transition. **a** Lineshapes of individual one-photon PA lines modelled by Eq. 5 for the coupling laser frequency fixed to the molecular transition from $|R = 1\rangle$ to $|R' = 1, m_R' = 1\rangle$. The *black line* corresponds to a one-photon PA transition from an unbound atom pair to the excited $|R' = 0\rangle$ level, while the *blue*, *green* and *orange lines* correspond to transitions to the $|R' = 1, m_R' = -1\rangle$, $|R' = 1, m_R' = 0\rangle$ and $|R' = 1, m_R' = 1\rangle$ levels, respectively. **b** Comparison between a weighted sum of the modelled lineshapes (*red*) and experimental data (*black*)

only observed for the one-photon PA transition to the $|R' = 0\rangle$ level but also for the one to the $|R' = 1, m_R' = 0\rangle$ level. This implies that the two bound-bound transitions $|R = 0\rangle \rightarrow |R' = 0\rangle$ and $|R = 1\rangle \rightarrow |R' = 1, m_R' = 0\rangle$ have nearly the same frequency, and hence, it can be inferred that for the specific vibrational levels $|\Delta v = -6, F = 1\rangle$ of the ground and $|\Delta v' = -11, F' = 2\rangle$ of the excited electronical state, the rotational splittings are approximately equal. In turn, this implies that the nuclear wavefunctions for the ground and excited state have a similar spatial extent and the Franck–Condon overlap is large.

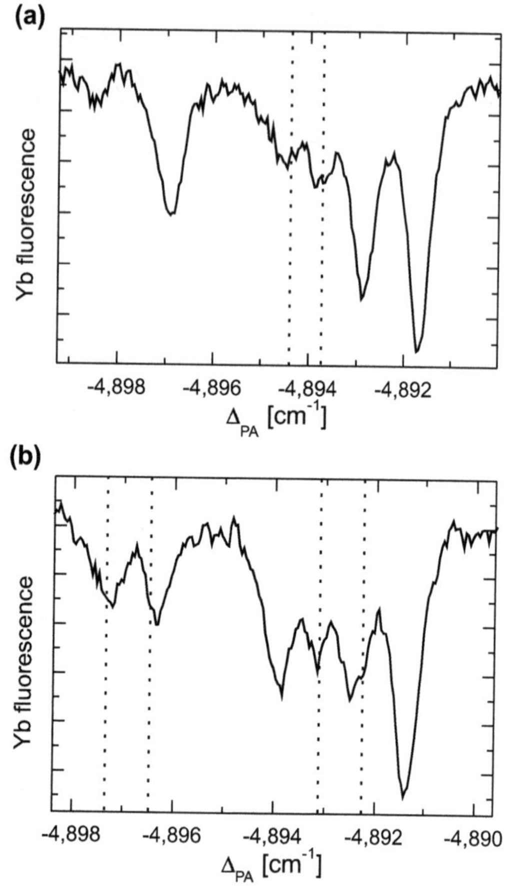

Fig. 6 Autler–Townes spectroscopy of the $|\Delta v = -6, F = 1\rangle \rightarrow |\Delta v' = -11, F' = 2\rangle$ transition with the coupling laser addressing the transition to **a** $|R' = 1, m'_R = -1\rangle$ and to **b** $|R' = 1, m'_R = 0\rangle$

Though the conditions in our experiment, where the internal state of the atoms in the MOT is not well defined, complicate the quantitative determination of a Franck–Condon factor from the Autler–Townes splitting, we can try to estimate the Franck–Condon factor of the $|\Delta v = -6, F = 1\rangle \rightarrow |\Delta v' = -11, F' = 2\rangle$ molecular transition. Generally, the Franck–Condon factor is given by

$$f_{\mathrm{FC}} = |\langle g|e\rangle|^2 = \frac{\Omega_{12}^2}{\gamma_1^2}\frac{2I_{\mathrm{sat}}}{I}\,. \tag{6}$$

To evaluate this equation for our experimental conditions, we assume that $\gamma_1 = 2\pi \cdot 6\,\mathrm{MHz}$ which is the natural linewidth of the Rb atom. The saturation intensity for Rb is $1.6\,\mathrm{mW/cm^2}$ for the $|F = 2, m_F = 2\rangle \rightarrow |F' = 3, m'_F = 3\rangle$ transition of the D_2-line. To correct for the fact that we are operating close to the D_1-line and that in the dark-spot MOT the Rb atoms are distributed over all Zeeman sublevels of the $F = 1$ hyperfine ground state, a correction factor of 18/5 for the saturation intensity is introduced which takes into account that the coupling and the probe PA laser have the same linear polarization. The Autler–Townes splitting and thus the Rabi frequency Ω_{12} obtained from Figs. 5 and 6 has a value of $\approx 30\,\text{MHz}$. With a peak intensity of the coupling laser of $30\,\text{mW/cm}^2$, we obtain a lower limit for the Franck–Condon factor of the $|\Delta v = -6, F = 1\rangle \rightarrow |\Delta v' = -11, F' = 2\rangle$ transition of $f_{\text{FC,min}}^{-6,-11} = 0.1$.

We regard the stated value for $f_{\text{FC,min}}^{-6,-11}$ as a lower limit only since systematic effects in our experiment lead to an underestimation of the Franck–Condon factor. The most important effect is that the Yb density distribution in the combined MOT, where we observe huge Yb losses due to the presence of the Rb MOT, is not well known and it is likely that most of the Yb atoms are located at the edge of the Rb cloud. In addition, the intensity profile of the coupling laser is inhomogeneous and the overlap between the Yb atomic cloud and the coupling laser, which is aligned to optimize the overlap with the Rb cloud, is not very well controlled. We believe that both effects together lead to a significant underestimation of the Franck–Condon factor since it is likely that Yb atoms are predominantly located in regions with low coupling laser intensity, while our estimate assumes that they experience the peak intensity.

Though a precise determination of absolute values of the Franck–Condon factors is not possible due to the limitations of our experimental scheme, ratios of Franck–Condon factors for different vibrational transitions can be reasonably evaluated if the data are taken under similar experimental conditions. This is indeed the case for the two data sets shown in Fig. 3 (corresponding to $|\Delta v = -7, F = 1\rangle \rightarrow |\Delta v' = -11, F' = 2\rangle$) and in Fig. 5 (corresponding to $|\Delta v = -6, F = 1\rangle \rightarrow |\Delta v' = -11, F' = 2\rangle$) which have been taken on the same day, i.e. with the same MOT alignment. In the same way as explained above, we estimate a lower limit $f_{\text{FC,min}}^{-7,-11} = 0.025$ for the Franck–Condon factor of the $|\Delta v = -7, F = 1\rangle \rightarrow |\Delta v' = -11, F' = 2\rangle$ transition leading to a ratio $f_{\text{FC,min}}^{-6,-11}/f_{\text{FC,min}}^{-7,-11} \approx 4$. This is in agreement with the assumption that the $|\Delta v = -6, F = 1\rangle \rightarrow |\Delta v' = -11, F' = 2\rangle$ transition is a strong transition with a large overlap between the ground- and excited-state nuclear wavefunction.

4 Conclusion

In conclusion, we have successfully performed Autler–Townes spectroscopy in the isotopologue $^{87}\text{Rb}^{176}\text{Yb}$ by coupling vibrational levels $\Delta v'$ of the excited molecular state to vibrational levels Δv of the ground state and monitoring the modified one-photon photoassociation spectrum. The observed Autler–Townes splittings provide information on the relative transition strengths of vibrational transitions connecting weakly bound levels of the $^2\Sigma_{1/2}$ electronic ground state to the $^2\Pi_{1/2}$ excited state of the RbYb molecule. For the two transitions investigated here, we obtain a ratio of the Franck–Condon factors of 4 and a lower bound of 0.1 for the Franck–Condon factor of the stronger $|\Delta v = -6, F = 1\rangle \rightarrow |\Delta v' = -11, F' = 2\rangle$

transition. Therefore, it is concluded that excited-state molecules which are produced by one-photon PA in the $|\Delta v' = -11, F' = 2\rangle$ level of the $^2\Pi_{1/2}$ state decay with a significant probability into the vibrational level $|\Delta v = -6, F = 1\rangle$ of the $^2\Sigma_{1/2}$ ground state. This might serve as a starting point for the creation of an ensemble of RbYb ground-state molecules in a well-defined rovibrational level which can subsequently be transferred to the rovibrational ground state using STIRAP. A prerequisite to achieve this will be that photoassociation spectroscopy is performed on a mixture of Rb and Yb atoms in well-defined internal states in a conservative trap as we have recently demonstrated [26].

Acknowledgements A. G. wants to say thanks to Ted Hänsch for introducing him to laser spectroscopy and ultracold atoms, the topics that still guide his research today. He is grateful that Ted Hänsch convinced him as a young diploma student that it would offer great perspectives to learn about lasers and atoms in his group and he was right. It is amazing to see that after almost 25-year- home-built ECDL diode lasers in the Hänsch design are as abundant in laboratories worldwide as ever including the laboratory of A. G. where the knowledge about lasers that A. G. has acquired during his diploma and Ph.D. thesis in the Hänsch laboratories is still transferred to Bachelor students.

References

1. L.D. Carr, D. DeMille, R.V. Krems, J. Ye, Cold, and ultracold molecules: science, technology and applications. New J. Phys. **11**, 055049 (2009)
2. G. Quéméner, P.S. Julienne, Ultracold molecules under control!. Chem. Rev. **112**(9), 4949–5011 (2012)
3. E.S. Shuman, J.F. Barry, D. DeMille, Laser cooling of a diatomic molecule. Nature **467**(7317), 820–823 (2010)
4. M.T. Hummon, M. Yeo, B.K. Stuhl, A.L. Collopy, Y. Xia, J. Ye, 2D magneto-optical trapping of diatomic molecules. Phys. Rev. Lett. **110**, 143001 (2013)
5. B. Hemmerling, E. Chae, A. Ravi, L. Anderegg, G.K. Drayna, N.R. Hutzler, A.L. Collopy, J. Ye, W. Ketterle, J.M. Doyle, Laser slowing of CaF molecules to near the capture velocity of a molecular MOT. J. Phys. B **49**, 174001 (2016)
6. T. Köhler, K. Góral, P.S. Julienne, Production of cold molecules via magnetically tunable Feshbach resonances. Rev. Mod. Phys. **78**, 1311–1361 (2006)
7. C. Chin, R. Grimm, P. Julienne, E. Tiesinga, Feshbach resonances in ultracold gases. Rev. Mod. Phys. **82**(2), 1225 (2010)
8. K.M. Jones, E. Tiesinga, P.D. Lett, P.S. Julienne, Ultracold photoassociation spectroscopy: long-range molecules and atomic scattering. Rev. Mod. Phys. **78**, 483–535 (2006)
9. K.-K. Ni, S. Ospelkaus, M.H.G. de Miranda, A. Pe'er, B. Neyenhuis, J.J. Zirbel, S. Kotochigova, P.S. Julienne, D.S. Jin, J. Ye, A high phase-space-density gas of polar molecules. Science **322**(5899), 231–235 (2008)
10. C. Ospelkaus, S. Ospelkaus, L. Humbert, P. Ernst, K. Sengstock, K. Bongs, Ultracold heteronuclear molecules in a 3D optical lattice. Phys. Rev. Lett. **97**, 120402 (2006)
11. S.A. Moses, J.P. Covey, M.T. Miecnikowski, B. Yan, B. Gadway, J. Ye, D.S. Jin, Creation of a low-entropy quantum gas of polar molecules in an optical lattice. Science **350**(6261), 659–662 (2015)

12. W. Cheng-Hsun, J.W. Park, P. Ahmadi, S. Will, M.W. Zwierlein, Ultracold fermionic Feshbach molecules of ^{23}Na ^{40}K. Phys. Rev. Lett. **109**, 085301 (2012)
13. J.W. Park, S.A. Will, M.W. Zwierlein, Ultracold dipolar gas of fermionic ^{23}Na ^{40}K molecules in their absolute ground state. Phys. Rev. Lett. **114**, 205302 (2015)
14. T. Takekoshi, M. Debatin, R. Rameshan, F. Ferlaino, R. Grimm, H.-C. Nägerl, C.R.L. Sueur, J.M. Hutson, P.S. Julienne, S. Kotochigova, E. Tiemann, Towards the production of ultracold ground-state RbCs molecules: Feshbach resonances, weakly bound states, and the coupled-channel model. Phys. Rev. A **85**, 032506 (2012)
15. M.P. Köppinger, D.J. McCarron, D.L. Jenkin, P.K. Molony, H.-W. Cho, S.L. Cornish, C.R.L. Sueur, C.L. Blackley, J.M. Hutson, Production of optically trapped ^{87}RbCs Feshbach molecules. Phys. Rev. A **89**, 033604 (2014)
16. P.S. Zuchowski, J. Aldegunde, J.M. Hutson, Ultracold RbSr molecules can be formed by magnetoassociation. Phys. Rev. Lett. **105**(15), 153201 (2010)
17. D.A. Brue, J.M. Hutson, Prospects of forming ultracold molecules in $^2\Sigma$ states by magnetoassociation of alkali-metal atoms with Yb. Phys. Rev. A **87**, 052709 (2013)
18. W. Dowd, R.J. Roy, R.K. Shrestha, A. Petrov, C. Makrides, S. Kotochigova, S. Gupta, Magnetic field dependent interactions in an ultracold Li–Yb(P-3(2)) mixture. New J. Phys. **17**, 055007 (2015)
19. J.M. Sage, S. Sainis, T. Bergeman, D. DeMille, Optical production of ultracold polar molecules. Phys. Rev. Lett. **94**(20), 203001 (2005)
20. J. Deiglmayr, A. Grochola, M. Repp, K. Mörtlbauer, C. Glück, J. Lange, O. Dulieu, R. Wester, M. Weidemüller, Formation of ultracold polar molecules in the rovibrational ground state. Phys. Rev. Lett. **101**(13), 133004 (2008)
21. A. Fioretti, D. Comparat, A. Crubellier, O. Dulieu, F. Masnou-Seeuws, P. Pillet, Formation of cold Cs-2 molecules through photoassociation. Phys. Rev. Lett. **80**(20), 4402–4405 (1998)
22. G. Reinaudi, C.B. Osborn, M. McDonald, S. Kotochigova, T. Zelevinsky, Optical production of stable ultracold ^{88} Sr_2 molecules. Phys. Rev. Lett. **109**, 115303 (2012)
23. K. Aikawa, D. Akamatsu, M. Hayashi, K. Oasa, J. Kobayashi, P. Naidon, T. Kishimoto, M. Ueda, S. Inouye, Coherent transfer of photoassociated molecules into the rovibrational ground state. Phys. Rev. Lett. **105**(20), 203001 (2010)
24. S. Stellmer, B. Pasquiou, R. Grimm, F. Schreck, Creation of ultracold Sr_2 molecules in the electronic ground state. Phys. Rev. Lett. **109**, 115302 (2012)
25. N. Nemitz, F. Baumer, F. Münchow, S. Tassy, A. Görlitz, Production of heteronuclear molecules in an electronically excited state by photoassociation in a mixture of ultracold Yb and Rb. Phys. Rev. A **79**(6), 061403(R) (2009)
26. C. Bruni, A. Görlitz, Observation of hyperfine interaction in photoassociation spectra of ultracold RbYb. Phys. Rev. A **94**(1), 022503 (2016)
27. M. Borkowski, P.S. Żuchowski, R. Ciuryło, P.S. Julienne, D. Kedziera, Ł. Mentel, P. Tecmer, F. Münchow, C. Bruni, A. Görlitz, Scattering lengths in isotopologues of the RbYb system. Phys. Rev. A **88**, 052708 (2013)
28. F. Münchow, C. Bruni, M. Madalinski, A. Görlitz, Two-photon photoassociation spectroscopy of heteronuclear YbRb. Phys. Chem. Chem. Phys. **13**, 18734–18737 (2011)
29. S.H. Autler, C.H. Townes, Stark effect in rapidly varying fields. Phys. Rev. **100**(2), 703–722 (1955)
30. E.R.I. Abraham, W.I. McALexander, C.A. Sackett, R.G. Hulet, Spectroscopic determination of the s-wave scattering length of lithium. Phys. Rev. Lett. **74**, 1315 (1995)
31. U. Schlöder, T. Deuschle, C. Silber, C. Zimmermann, Autler–Townes splitting in two-color photoassociation of ^{6}Li. Phys. Rev. A **68**(5), 051403 (2003)
32. B.J. DeSalvo, J.A. Aman, C. Gaul, T. Pohl, S. Yoshida, J. Burgdörfer, K.R.A. Hazzard, F.B. Dunning, T.C. Killian, Rydberg-blockade effects in Autler–Townes spectra of ultracold strontium. Phys. Rev. A **93**, 022709 (2016)
33. J.W. Park, S.A. Will, M.W. Zwierlein, Two-photon pathway to ultracold ground state molecules of (NaK)-Na-23-K-40. New J. Phys. **17**, 075016 (2015)

34. L. Ricci, M. Weidemüller, T. Esslinger, A. Hemmerich, C. Zimmermann, V. Vuletic, W. König, T.W. Hänsch, A compact grating-stabilized diode laser system for atomic physics. Opt. Commun. **117**, 541 (1995)
35. J.L. Bohn, P.S. Julienne, Semianalytic treatment of two-color photoassociation spectroscopy and control of cold atoms. Phys. Rev. A **54**(6), R4637–R4640 (1996)
36. Y.N. Martinez de Escobar, P.G. Mickelson, P. Pellegrini, S.B. Nagel, A. Traverso, M. Yan, R. Cote, T.C. Killian, Two-photon photoassociative spectroscopy of ultracold Sr-88. Phys. Rev. A **78**, 062708 (2008)

From Quantum Transitions to Electronic Motions

Ferenc Krausz

Abstract Laser spectroscopy and chronoscopy permit precision measurement of quantum transitions and captures atomic-scale dynamics, respectively. Frequency- and time-domain metrology ranks among the supreme laser disciplines in fundamental science. For decades, these fields evolved independently, without interaction and synergy between them. This has changed profoundly with controlling the position of the equidistant frequency spikes of a mode-locked laser oscillator. By the self-referencing technique invented by Theodor Hänsch, the comb can be coherently linked to microwaves and used for precision measurements of energy differences between quantum states. The resultant optical frequency synthesis has revolutionized precision spectroscopy. Locking the comb lines to the resonator round-trip frequency by the same approach has given rise to laser pulses with controlled field oscillations. This article reviews, from a personal perspective, how the bridge between frequency- and time-resolved metrology emerged on the turn of the millennium and how synthesized several-cycle laser fields have been instrumental in establishing the basic tools and techniques for attosecond science.

The frequency of microwave to optical radiation absorbed or emitted upon electronic transitions between quantum states of atoms has been the most accurately measured physical quantity to date, allowing to put the laws of quantum physics to a test, determine fundamental constants (and their possible drifts), and define the standards for time and length [1]. Precision spectroscopy underlying these applications relies on excited quantum states the energy of which (with respect to the ground state) is precisely defined. Heisenberg's uncertainty principle dictates that

Dedicated to the 75th birthday of Theodor W. Hänsch.

This article is part of the topical collection "Enlightening the World with the Laser" - Honoring T. W. Hänsch guest edited by Tilman Esslinger, Nathalie Picqué, and Thomas Udem.

F. Krausz (✉)
Max-Planck-Institut für Quantenoptik, Hans-Kopfermann-Strasse 1, 85748 Garching, Germany

Fakultät für Physik, Ludwig-Maximilians-Universität, Am Coulombwall 1, 85748 Garching, Germany
e-mail: Ferenc.Krausz@mpq.mpg.de

D. Meschede et al. (eds.), *Exploring the World with the Laser*,
https://doi.org/10.1007/978-3-319-64346-5_5

such states decay on very long timescales (by atomic standards). Accurate measurement of this energy requires highly stationary sinusoidal waves consisting of trillions to quadrillions of identical cycles.

By sharp contrast, wavepackets formed by the superposition of stationary quantum states evolve on ultrafast timescales. Nuclear wavepacket motions, in terms of which quantum mechanics describes molecular dynamics of any kind, occur on a timescale of several to several hundred femtoseconds (1 fs $= 10^{-15}$ s), see Ref. [2], whereas electronic wavepacket dynamics underlying any dynamic change of electronic structure unfolds over tens to thousands of attoseconds (1 as $= 10^{-18}$ s), see Ref. [3]. Observing and controlling dynamic changes of molecular and electronic structure in real time therefore call for femtosecond- and attosecond-duration pulses and femtosecond- and attosecond-varying forces, respectively.

From these considerations, it appears plausible that measuring quantum transitions and capturing dynamics require very different forms of radiation, with seemingly little room for any synergies. Hence, not very surprisingly, the fields of precision laser spectroscopy and ultrafast laser science evolved independently over decades with little interaction or even communication between the respective scientific communities. This state of affairs changed radically on the turn of millennium, thanks to gaining full control over the series of equidistant eigenfrequencies (henceforth, frequency comb) of a mode-locked laser by Theodor Hänsch and his coworkers, Ronald Holzwarth and Thomas Udem. In doing so, they have created a clockwork able to count optical field oscillations of more than 10^{15} cycles per second. Their method dramatically simplified and improved the instrumentation for high-precision optical spectroscopy and opened the door for the development of next-generation clocks based on optical quantum transitions [4–15].

The resultant optical frequency synthesis also led to the first fully controlled femtosecond pulses from a mode-locked laser, controlled not only in terms of their pulse envelope, but also in terms of their field oscillations. These advances have been instrumental in creating the ability to observe and control atomic-scale electronic motions in real time [3]. Here, I pay tribute to these advances from a personal perspective, without attempting a comprehensive review and by referring only to a few selected references thought to be helpful to readers seeking deeper insight. For a more complete survey of relevant literature, the reader is kindly referred to Ref. [3].

1 Time-Resolved Metrology Meets Frequency-Domain Metrology

I felt greatly honored when I was invited to give a colloquium at the Max-Planck-Institute of Quantum Optics as a young postdoc in 1995. Our group in Vienna just finished a study of the non-stationary nature of the pulses emerging from locking the longitudinal modes of a laser (in our case: Ti:sapphire laser) over a bandwidth of nearly 100 THz, implying a pulse duration of less than 10 femtosecond (<4 wave cycles). We found that the phase φ of the carrier wave with respect to the pulse envelope of the electric field,

$$\mathbf{E}_L(t) = (1.2)\mathbf{E}_{\text{peak}} a_L(t) e^{-i(\omega_L t + \varphi)} + c.c. \tag{1}$$

where $\mathbf{E}_{\text{peak}}$, $a_L(t)$, and ω_L stand for the electric field vector at the pulse peak, the normalized complex amplitude, and the (angular) carrier frequency of the laser field $\mathbf{E}_L(t)$, respectively, was slipping all the time from one pulse to the next in the output pulse train even in an apparently perfectly stationary mode of operation.

We devised a scheme that allowed us to measure this phase shift $\Delta\varphi$ occurring upon each round trip in the cavity and its dependence on the linear and nonlinear properties of the laser resonator. I concluded the presentation of these results 16 with arguing that gaining control over this phase slippage would be of tremendous value for the study of strong-field-induced electron processes, enabling their control within the optical field cycle, i.e., on an attosecond timescale, which might open the door for the extension of time-resolved science to real-time observation and control of the electrons' atomic-scale motion.

Subsequent illuminating discussions with Theodor Hänsch, Thomas Udem, and coworkers ignited a collaboration between our groups which continues to this day and culminated—on the turn of the new millennium—in the generation of strong waveform-controlled light and its use for controlling attosecond electron processes [17, 18].

2 Laser Mode Locking: In The Time and Frequency Domain

Our study in 1996 revealed that a continuous-wave mode-locked laser does not deliver identical pulses, no matter how perfectly its longitudinal modes are phase locked to each other and how perfectly they are equidistantly spaced in frequency. This can be intuitively understood by considering that a laser resonator always contains some dispersive components, in which the group and phase velocities, determining the propagation speed of the pulse envelope and the carrier wave, respectively, are (slightly) different. This gives rise to a carrier wave shifting with respect to the pulse envelope by

$$\Delta\varphi = \phi_R - \omega_L T_R \tag{2}$$

upon each round trip in the laser resonator, where ϕ_R is the round-trip phase delay (determined by the phase velocity) and T_R is the round-trip time of the laser pulse (determined by the group velocity), see Fig. 1a. A hypothetical sensor measuring the laser field at any (arbitrarily chosen) position on the optical axis observes a periodically reappearing pulse with identical $\mathbf{E}_{\text{peak}}$, $a_L(t)$, and ω_L but with a carrier-envelope (CE) phase, φ, changed by $\Delta\varphi$ as given by (2).

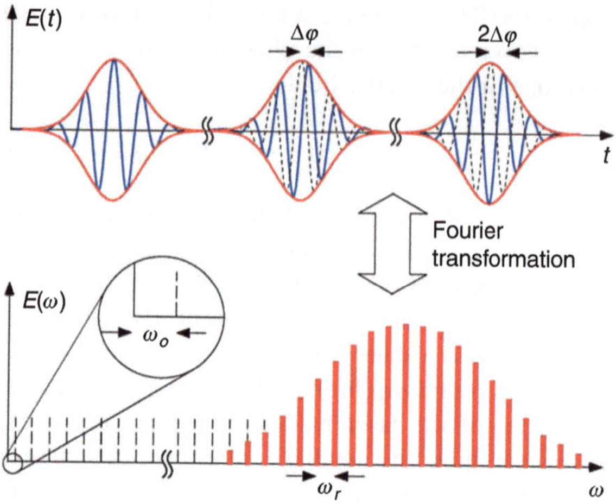

Fig. 1 Electric field versus time (*upper panel*) and its Fourier transform (*lower panel*) in a mode-locked laser, as observable at a fixed position on the resonator axis.. Courtesy of Thomas Udem

This quasi-periodic train has a frequency spectrum that consists of a series of equidistant lines, with their frequency given by

$$\omega_n = \omega_0 + n\omega_r, \tag{3}$$

where $\omega_r = 2\pi/T_R$ is the round-trip frequency and n is a large ($\sim 10^6$) integer. The offset frequency ω_0 is connected to $\Delta\varphi$ by the simple relationship [4]

$$\omega_0 = \frac{\Delta\varphi}{T_R}. \tag{4}$$

We demonstrated that the round-trip CE phase shift $\Delta\varphi$ can be adjusted by a component of variable dispersion, such as a pair of glass wedges [16]. However, this manual control of $\Delta\varphi$ is insufficient for laser pulse generation with precisely reproduced field evolution. This is because small fluctuations in $\mathbf{E}_{\text{peak}}$, $a_L(t)$, and ω_L translate into perturbations to φ. These small perturbations typically remain nearly constant over many round trips allowing their accumulation to substantial random shifts of φ over multiple round trips [16]. Therefore, control over the electric field evolution of the pulses emitted by a mode-locked laser requires active stabilization of the CE phase.

Equation (4) reveals that control over the CE phase calls for controlling the offset frequency ω_0. To this end, ω_0 has to be converted into an electronic signal. This was accomplished by the self-referencing technique invented by Theodor Hänsch. The conceptually simplest implementation of the method was first described and published in Ref. [14]. The low-frequency end of an octave-spanning mode-locked spectrum is frequency doubled and brought into interference with the high-frequency end of the same spectrum. The resultant beating occurs at a frequency of $2(n\omega_r + \omega_0) - (2n\omega_r + \omega_0) = \omega_0$ and yields the required intensity modulation at ω_0, which can be easily measured by a photodiode 10, 17. The approach has been dubbed the f-to-2f technique. Phase locking ω_0 obtained this way to a sub-harmonic of the round-trip frequency, ω_r/N, yields—according to (4)—$\Delta\varphi = 2\pi/N$ and hence a pulse train in which every Nth pulse has identical CE phase and—assuming constancy of all other pulse parameters—thereby identical electric field waveforms. This approach resulted in the first waveform-controlled few-cycle (sub-10-fs) pulses from a Ti:sapphire oscillator [17].

3 Strong, Controlled, Few-Cycle Waveforms

Few-cycle pulses from the Ti:sapphire oscillator carry only nanojoule energies, far too low to produce field strengths needed for strong-field-induced electron phenomena. The controlled waveforms from the oscillator therefore need to be amplified, and possible random shifts and drift of the CE phase introduced in the amplifier system need to be compensated. Figure 2 outlines the system developed to this end by Andrius Baltuska, Matthias Uiberacker, Michael Hentschel and co in Vienna in cooperation with Christoph Gohle, Ronald Holzwarth, and Thomas Udem from Garching, which delivered intense laser pulses with controlled field oscillations for the first time [18]. The octave-spanning spectrum of the pulses produced by the oscillator as well as the pulse compressor following the amplification allowed straightforward implementation of the f-to-2f technique both for controlling the evolution of φ in the oscillator output and for stabilizing φ of the amplified pulses compressed to a duration of 5 fs (full width at intensity half maximum).

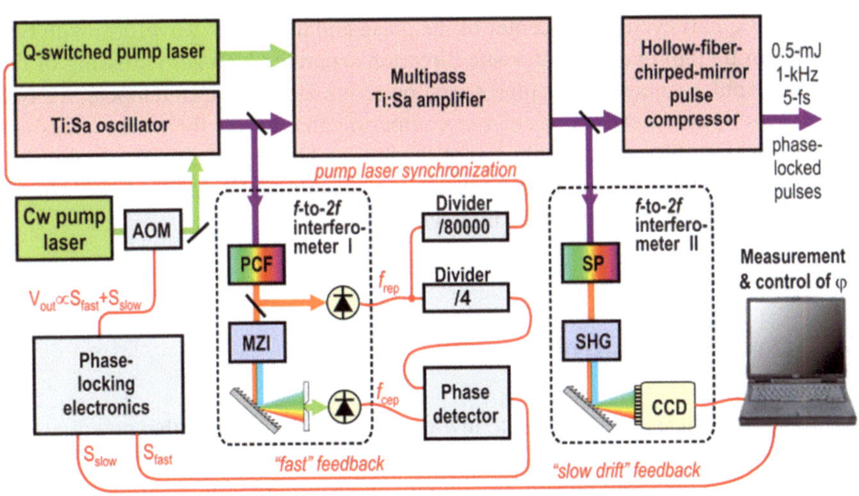

Fig. 2 Laser system producing the first intense, waveform-controlled light pulses. For more details, see text and Ref. [18]

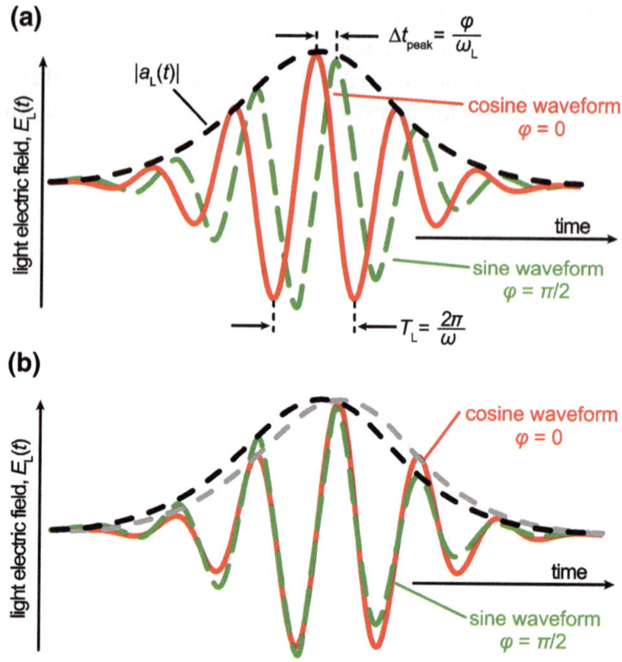

Fig. 3 Few-cycle waveforms with different CE phase, shown with the peak of their amplitude envelope (*panel a*) and the zero transitions of their field oscillations (*panel b*) coinciding

The system permits generation of this few-femtosecond duration, few-cycle pulses with arbitrary setting of the CE phase. Figure 3a depicts two representative waveforms with most distinct differences, the "cosine" waveform with a single most intense half cycle at the center of the pulse and the "sine" waveform with two equally strong half cycles of opposite direction separated by a zero crossing of the field at the pulse center. In this representation (coinciding pulse envelopes), the two waveforms appear quite different but comparing them with their zero crossings coinciding (Fig. 3b) reveals only minor deviations between their field evolutions. However, even these miniscule differences can have dramatic ramifications when it comes to *nonlinear* field–electron interactions.

4 Attosecond Control

A highly nonlinear response to a few-cycle excitation offers the potential for confining the effective light–electron interaction to a *single* wave cycle. High-order harmonic emission constitutes a textbook example for an electronic process responding to the driving electric field in an extremely nonlinear fashion [19]. As illustrated in Fig. 4, the peak intensity of a few-cycle laser pulse can be adjusted

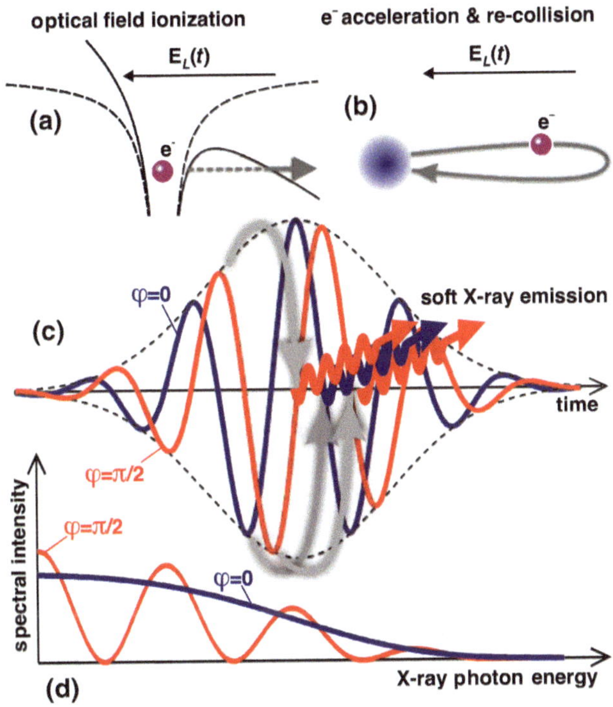

Fig. 4 XUV emission from optical field ionization (also known as high-order harmonic generation when induced by multi-cycle optical fields). **a** Liberation of most weakly bound electron by tunneling ionization. **b** Acceleration of the freed wavepacket by the linearly polarized laser field, recollision of the wavepacket with its parent ion. **c** Emission of XUV radiation upon recollision, for cosine- (*blue*) and sine (*red*)-shaped few-cycle excitation

such that the electric field is strong enough to liberate the most weakly bound electron of an atom *only* at the oscillation peak(s) of the central, most intense wave cycle with substantial probability. The liberated electron is initially pulled away from its original atomic location and—after a brief excursion in the strong laser field—pushed back to its parent ion. Here, interference of the returning wavepacket with the bound-state portion of the wavefunction of the same electron gives rise to high-frequency charge oscillations. The resultant XUV burst emitted by this tiny atomic dipole antenna is expected to last a fraction of the field half cycle, i.e., a fraction of a femtosecond.

Once this process is confined to a single wave cycle, the apparently tiny (see Fig. 3b) difference between a cosine and sine excitation waveform has striking implications. As a matter of fact, differing *temporal symmetry* results in different system response: the cosine waveform is expected to give rise to the emission of a *single* sub-femtosecond burst of XUV light, whereas the sine-shaped driver should yield—for symmetry reasons—*twin* pulses. The spectral distribution of the emitted radiation provided a first clear indication of this picture being correct. The high-

Fig. 5 Spectra of XUV emission emerging from neon atoms ionized by linearly polarized few-cycle laser pulses of different CE phase. For more details, see text and Ref. [18]

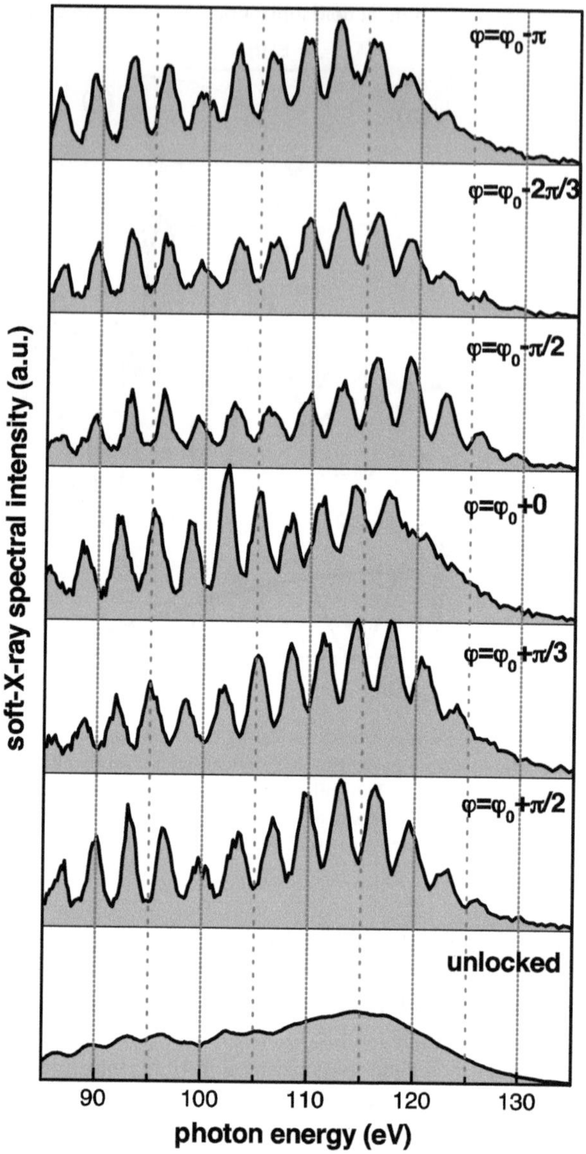

energy end (cutoff region) of the XUV spectra in Fig. 5 summarizing experimental results for different driving waveforms reveals a quasiperiodic modulation and an unmodulated continuum for sine and cosine excitation, respectively. This is exactly the behavior expected for the emission of twin and single bursts, respectively. To obtain more direct evidence about the temporal structure of the emitted high-frequency radiation, an attosecond-resolution metrology is in demand. Once again, light waveform control is the key to meet this demand.

5 Attosecond Metrology

The sub-laser-cycle XUV emission emerging from the ionization process described above is locked to the laser field oscillation driving the process. This attosecond synchronism offers the potential for using the generating laser field also for the measurement (temporal characterization) of the XUV light. Because the laser field can't directly interact with another light field, the temporal profile of the XUV light needs to be reproduced with high fidelity in some electronic process. To this end, a second set of atoms are exposed simultaneously to the XUV and laser light. The XUV photons liberate the most weakly bound electron of some of the atoms in a process that has been thought instantaneous. Meanwhile, precision attosecond metrology revealed a delay of this photoemission process. This delay is of the order of 10^{-17} s and does not affect the temporal profile of the emission significantly. Hence, the (hypothetical) temporal emission profile of the emitted photoelectrons shown in Fig. 6 mimics that of the incident XUV light.

This temporal profile can be captured by using the strong, controlled electric field of the laser field which produced the XUV pulse a few nanoseconds earlier (in another set of atoms) as a "chronoscope." The laser field changes the initial velocity of the freed electron by an amount proportional to the vector potential of the field at the instant of photoelectron ejection. For an emission process confined to a fraction of the half cycle of the laser field, the change in electron velocity varies near linearly with the emission time, projecting the temporal profile of the emission to a corresponding final velocity distribution of photoelectrons as illustrated in Fig. 6. Even if this linearity condition is not fulfilled, the temporal emission profile can be unambiguously and accurately retrieved from recording several XUV-induced and "laser-streaked" photoelectron spectra at different timing of the XUV burst with respect to the "streaking" laser field.

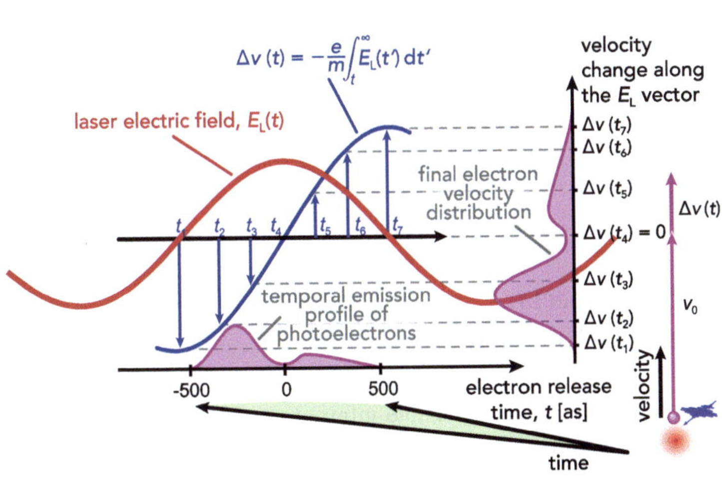

Fig. 6 Basic concept of an attosecond streak camera, see text for explanation

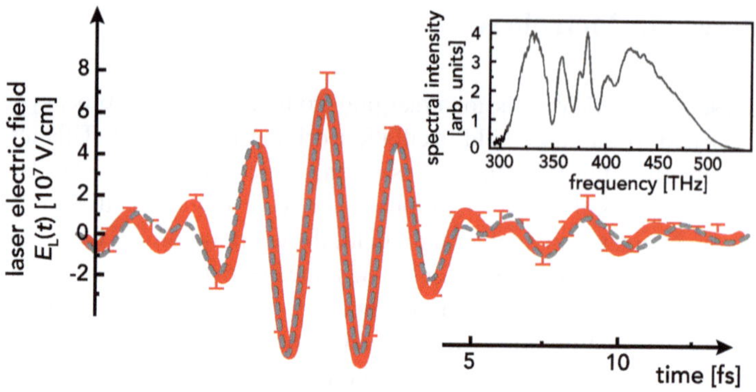

Fig. 7 First light pulse with its waveform measured. For details, see Ref. [21]

The method has been dubbed laser-field-driven (or briefly attosecond) streaking and constitutes, ever since its first demonstrations (in 2001 and 2004 without and with waveform-controlled laser light, respectively) [20, 21], the gold standard in attosecond metrology. Attosecond streaking permits much more than just measuring the duration of attosecond pulses, which can meanwhile be produced with durations shorter than 100 as. With the timing of the attosecond pulses being scanned across the entire few-cycle laser field in the light-field-driven streak camera, the evolution of the laser field can also be retrieved. Figure 7 shows the first measured light waveform, which was obtained this way [22].

With the attosecond pulse and its few-cycle driver laser field fully characterized, attosecond streaking and pump-probe approaches constitute a powerful set of approaches for interrogating ultrafast electron phenomena occurring in excited atomic systems. With the attosecond pulse as a trigger, laser-field-driven streaking can be used as an attosecond chronoscope for timing photoemission from different atomic states or measuring electron propagation delay in solids [23]. With the fully characterized oscillating laser field as probe, one can trace the nonlinear polarization response of matter to optical fields [24]. The attosecond pulse, on the other hand, can be applied to probe strong-field-induced electron phenomena via attosecond transient absorption or photoelectron spectroscopy [23]. With these tools and techniques, attosecond metrology and spectroscopy are now well established and being used in a rapidly growing number of laboratories for studying electron dynamics.

6 Future Prospects: Exploring Electronic Motions in Complex Systems

The attosecond-varying force of visible-infrared waveforms along with attosecond-duration pulses of short-wavelength light synchronized to this force has enabled real-time observation (and—to some extent—control) of atomic-scale electron dynamics for the first time. The physical observables captured with attosecond temporal resolution, such as photoelectron momentum, transient absorption, induced polarization, have provided—and continue to provide—unprecedented insight into the dynamic response of the electrons to external excitation.

However, there is another observable of crucial importance: the expectation value of the electrons' spatial coordinates. Dynamic changes in the electrons' atomic-scale distribution and the atomic-scale current densities associated with these changes play a pivotal role in chemical reactions underlying vital biological processes of living organisms as well as in the quest for ever-smaller and ever-faster electronic signal processing circuitry.

Whereas the electrons' spatially averaged excursion from their equilibrium (ground state) position has been accessed by attosecond polarization spectroscopy [23], atomic-scale rearrangements of electron charge distributions occurring on the electronic timescale remained inaccessible in complex systems, such as biological molecules and solid-state nanostructures, to this day. Making them perceivable requires attosecond-duration multi-keV photon (X-ray) or electron pulses, permitting diffraction imaging with picometer spatial and attosecond temporal resolution.

Again, controlled light forces may offer the solution. The synthesis of multi-octave, multi-terawatt visible-infrared optical transients holds promise for the generation of isolated attosecond pulses at several keV photon energies by ionizing an ensemble of helium atoms [25], whereas moderate-power, precisely controlled infrared transients may pave the way toward electron pulses with sub-femtosecond duration [26].

In conclusion, *optical field synthesis*, first enabled by the frequency-comb stabilization technique of Theodor Hänsch, has been playing a central role in establishing the basic techniques and tools for precision attosecond control and metrology and is likely to continue playing a pivotal role in its future evolution. There is plenty of room for further synergies between time- and frequency-domain metrologies, motivating the respective communities to continue their collaboration initiated by the pioneering research of Theodor Hänsch some 20 years ago.

Acknowledgement Open access funding provided by Max Planck Society. I gratefully thank Mandy Singh for her support in preparing this manuscript for publication.

References

1. T.W. Hänsch, Nobel lecture: passion for precision. Rev. Mod. Phys. **78**, 1297 (2006)
2. A.H. Zewail, Femtochemistry: atomic-scale dynamics of the chemical bond. J. Phys. Chem. A **104**, 5660 (2000). **(Nobel Lecture)**
3. F. Krausz, The birth of attosecond physics and its coming of age. Phys. Scr. **91**, 063011 (2016)
4. T. Udem, *Phasenkohärente optische Frequenzmessungen am Wasserstoffatom*. Thesis, Ludwig-Maximilians Univ (1997)
5. J. Reichert, R. Holzwarth, T. Udem, T.W. Hänsch, Measuring the frequency of light with modelocked lasers. Opt. Commun. **172**, 59 (1999)

6. Th Udem, J. Reichert, R. Holzwarth, T.W. Hänsch, Accurate measurement of large optical frequency differences with a mode-locked laser. Opt. Lett. **24**, 881 (1999)
7. Th Udem, J. Reichert, R. Holzwarth, T.W. Hänsch, Absolute optical frequency measurement of the cesium $D1$ line with a mode-locked laser. Phys. Rev. Lett. **82**, 3568 (1999)
8. H.R. Telle, G. Steinmeyer, A.E. Dunlop, D.H. Sutter, U. Keller, Carrier-envelope offset phase control: a novel concept for absolute optical frequency measurement and ultrashort pulse generation. Appl. Phys. B **69**, 327 (1999)
9. D.J. Jones et al., Carrier-envelope phase control of femtosecond mode-locked lasers and direct optical frequency synthesis. Science **288**, 635 (2000)
10. S.A. Diddams et al., Direct link between microwave and optical frequencies with a 300 THz femtosecond laser comb. Phys. Rev. Lett. **84**, 5102 (2000)
11. J. Reichert et al., Phase coherent vacuum-ultraviolet to radio frequency comparison with a modelocked laser. Phys. Rev. Lett. **84**, 3232 (2000)
12. M. Niering et al., Measurement of the hydrogen $1S$–$2S$ transition frequency by phase coherent comparison with a microwave cesium fountain clock. Phys. Rev. Lett. **84**, 5496 (2000)
13. R. Holzwarth et al., Optical frequency synthesizer for precision spectroscopy. Phys. Rev. Lett. **85**, 2264 (2000)
14. J. Ye et al., Accuracy comparison of absolute optical frequency measurement between harmonic generation synthesis and a frequency division femtosecond-comb. Phys. Rev. Lett. **85**, 3797 (2000)
15. T. Udem, R. Holzwarth, T.W. Hänsch, Optical frequency metrology. Nature **416**, 233 (2002)
16. L. Xu et al., Route to phase control of ultrashort light pulses. Opt. Lett. **21**, 2008 (1996)
17. A. Apolonski et al., Controlling the phase evolution of few-cycle light pulses. Phys. Rev. Lett. **85**, 740 (2000)
18. A. Baltuska et al., Attosecond control of electronic processes by intense light fields. Nature **421**, 611 (2003)
19. X.F. Li, A. L'Huillier, M. Ferray, L.A. Lompre, G. Mainfray, Multiple-harmonic generation in rare gases at high laser intensity. Phys. Rev. A **39**, 5751 (1989)
20. M. Hentschel et al., Attosecond metrology. Nature **414**, 509 (2001)
21. R. Kienberger et al., Atomic transient recorder. Nature **427**, 817 (2004)
22. E. Goulielmakis et al., Direct measurement of light waves. Science **305**, 1267 (2004)
23. F. Krausz, M. Ivanov, Rev. Mod. Phys. **81**, 163 (2009)
24. A. Sommer et al., Attosecond nonlinear polarization and light-matter energy transfer in solids. Nature **534**, 86 (2016)
25. H. Fattahi et al., Third-generation femtosecond technology. Optica **1**, 45 (2014)
26. C. Kealhofer et al., All-optical control and metrology of electron pulses. Science **352**, 429 (2016)

Measuring and Imaging Nanomechanical Motion with Laser Light

Andreas Barg, Yeghishe Tsaturyan, Erik Belhage, William H.P. Nielsen, Christoffer B. Møller, and Albert Schliesser

This article is dedicated to Theodor W. Hänsch on the occasion of his 75th birthday. Fortunate enough to have several chances to work with him, I could learn about his unique approach to experimental science. The first bit came right during my interview for a PhD position: When somebody claimed that all simple interesting things had already been done, he insisted that great experiments do not have to be complicated—if they are clever. It felt wise already then, now I know (better) how true it is. And I'm looking forward to seeing more clever experiments emerge from the Munich laboratories. Happy Birthday!

Abstract We discuss several techniques based on laser-driven interferometers and cavities to measure nanomechanical motion. With increasing complexity, they achieve sensitivities reaching from thermal displacement amplitudes, typically at the picometer scale, all the way to the quantum regime, in which radiation pressure induces motion correlated with the quantum fluctuations of the probing light. We show that an imaging modality is readily provided by scanning laser interferometry, reaching a sensitivity on the order of $10\,\mathrm{fm}/\mathrm{Hz}^{1/2}$, and a transverse resolution down to $2\,\mu\mathrm{m}$. We compare this approach with a less versatile, but faster (single-shot) dark-field imaging technique.

This article is part of the topical collection "Enlightening the World with the Laser" - Honoring T. W. Hänsch guest edited by Tilman Esslinger, Nathalie Picqué, and Thomas Udem.

Electronic supplementary material The online version of this chapter (doi:10.1007/978-3-319-64346-5_6) contains supplementary material, which is available to authorized users.

A. Barg • Y. Tsaturyan • E. Belhage • W.H.P. Nielsen • C.B. Møller • A. Schliesser (✉)
Niels Bohr Institute, Blegdamsvej 17, 2100 Copenhagen, Denmark
e-mail: albert.schliesser@nbi.dk

D. Meschede et al. (eds.), *Exploring the World with the Laser*,
https://doi.org/10.1007/978-3-319-64346-5_6

1 Introduction

Lasers are indispensable tools in science and technology today. They heal eyes, power the Internet, and print objects in 3D. They have also revolutionized atomic physics: Techniques such as laser cooling and optical frequency metrology have enabled the creation of new states of matter, precision tests of fundamental physical laws, and the construction of clocks more accurate than ever before. The lasers' key feature—high spatial and temporal coherence of the emitted light—is a unique asset, too, for the measurement of distance and motion. The laser interferometric gravitational wave observatory (LIGO) has provided the most recent, spectacular demonstration of this fact, with the direct detection of gravity waves [1].

While LIGO is concerned with the apparent displacement of kg-scale test masses, laser-based techniques are also an excellent choice to track the motion of micro- and nanoscale objects. Indeed, lasers have been used to measure a microcantilever's motion induced by the magnetic force of a single electron spin [2], providing only one example of the force and mass sensing capabilities of laser-transduced mechanical devices. The interaction of laser light and nanomechanical motion, which lies at the heart of any such measurement scheme, has, itself, moved to the center of attention recently. Research in the field of cavity optomechanics [3] explores the fundamental mechanisms—governed by the laws of quantum mechanics, of course—and the limitations and opportunities for mechanical measurements that they imply. Without even making an attempt at a comprehensive review of the vast activity in this field, we illustrate recent progress through a selection of our own results below.

For this research, it is often crucial to understand not only the spectral properties of the mechanical resonators, such as their eigenmodes' frequency and lifetime, but also the modes' spatial displacement patterns, as it determines the effective mass m_{eff}, and therefore the optomechanical interaction strength. The pattern can also strongly affect the modes' coherence properties. Both are particularly important for the development of new resonator systems. For example, the full knowledge of the mode shape has allowed us to design resonators with a "soft" phononic crystal clamping that enables unprecedented room-temperature quality factors $Q > 10^8$ at MHz frequencies [4]. While finite-element simulations of mechanical modes become ever more powerful and accurate, they often miss fabrication imperfections and substrate effects that can lead to broken symmetries or mode hybridization, among others. For this reason, we have developed several laser-based imaging techniques of micro- and nanomechanical devices. In this article, we provide a description of these highly useful tools.

2 Laser Interferometry and Spectroscopy

A simple two-path interferometer (Fig. 1a) constitutes the most straightforward approach to measuring mechanical displacements. One arm's path involves the reflection off the mechanical device's surface, so that its motion modulates the path

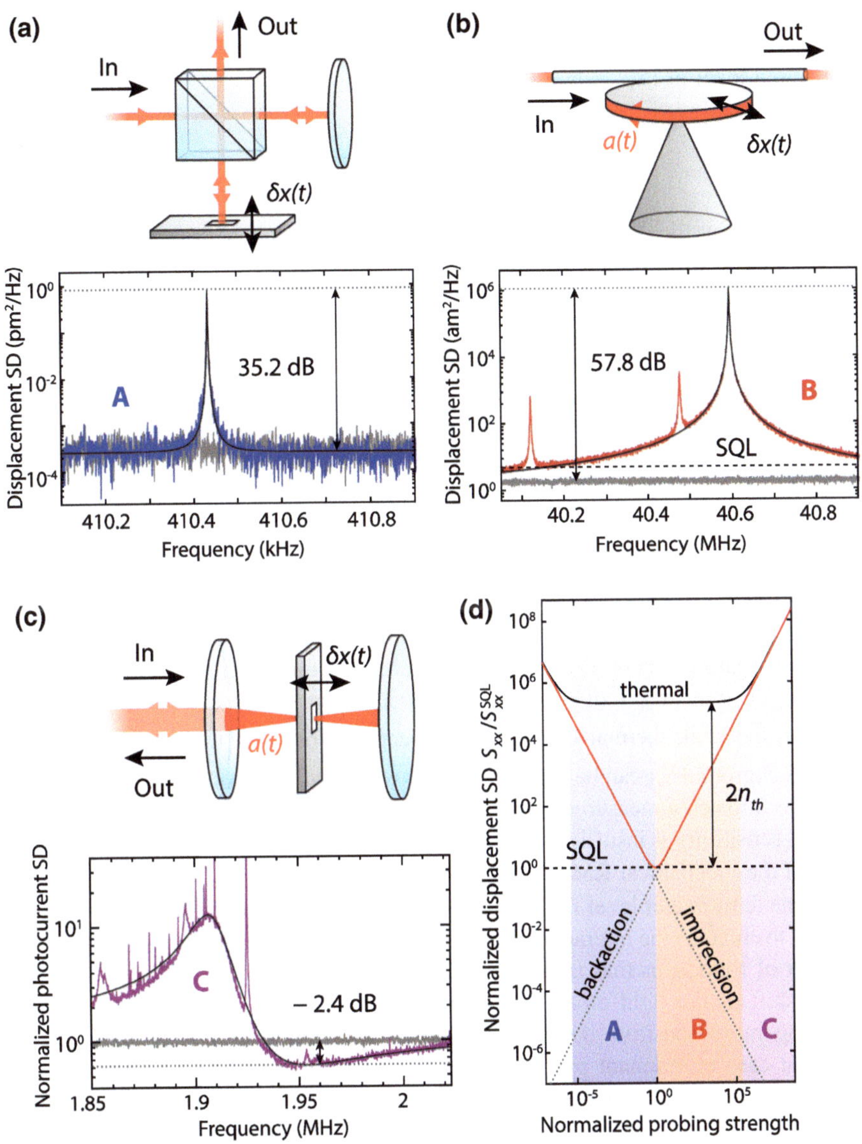

Fig. 1 Probing mechanical motion $\delta x(t)$ by laser interferometry. **a** Simple two-path interferometer, involving reflection off the mechanical device (*top*). The thermal motion (*blue trace*) of a high-Q membrane is readily resolved above the measurement imprecision background (*gray*). **b** Cavity-enhanced measurement, here of a radial-breathing mode of an optical whispering-gallery-mode resonator (*top*). Thermal motion (*red trace*) is far above the imprecision background (*gray*), which is itself below the resonant standard quantum limit (SQL) for this mechanical mode (from [10]). **c** Cavity-based measurements of highly coherent mechanical resonators, here a high-Q silicon nitride membrane placed inside a Fabry–Perot resonator (*top*). Quantum backaction starts to dominate over the thermal motion of the device, inducing correlations that lead to squeezing of the output light (*violet trace*) below the vacuum noise (*gray*), among others (from [20]). **d** Comparison of the relative levels of measurement imprecision, backaction, and thermomechanical noise in the measurement regimes depicted in the examples (**a**)–(**c**)

length difference between the two arms. If the interferometer is biased to the optimum point, it can detect displacement (double-sided) spectral densities S_{xx} down to a level of [5]

$$S_{xx}^{1/2} = \frac{\lambda}{2\pi} \frac{1}{\sqrt{\eta_\mathrm{d} P/\hbar\omega}}. \tag{1}$$

It is limited by the quantum phase uncertainty of the coherent state that the laser emits, referred to as the measurement imprecision. Here, λ, ω, and P are the wavelength, angular frequency, and power of the employed laser light, respectively. η_d is the detection efficiency, which also absorbs penalties in the sensitivity due to optical losses, insufficient interference contrast, etc. Equation (1) implies that within a bandwidth BW, the smallest displacements that can be recovered with unity signal-to-noise ratio are given by $\delta x_\mathrm{min}/\sqrt{\mathrm{BW}} = \sqrt{S_{xx}}$.

Our instrument (detailed below) employs a near-infrared laser and mW-scale probing powers and typically achieves a $S_{xx}^{1/2} \sim 10\,\mathrm{fm}/\sqrt{\mathrm{Hz}}$ displacement sensitivity, consistent with Eq. (1). This compares favorably with the picometer-scale thermal root-mean-square (RMS) displacement $\delta x_\mathrm{th} = \sqrt{k_\mathrm{B} T/m_\mathrm{eff}\Omega_\mathrm{m}^2}$ of the mechanical resonators we employ [4, 6], with nanogram mass m_eff and MHz frequency $\Omega_\mathrm{m}/2\pi$ at room temperature T. In the Fourier domain, the spectral density of the thermal motion is spread over the mechanical linewidth $\Gamma_\mathrm{m} = \Omega_\mathrm{m}/Q$. Correspondingly, a nearly four-order-of-magnitude signal-to-noise ratio $S_{xx}^\mathrm{th}(\Omega_m)/S_{xx}$ between the peak thermal displacement spectral density $S_{xx}^\mathrm{th}(\Omega_m) = \frac{\delta x_\mathrm{th}^2}{\Gamma_\mathrm{m}/2}$ and the noise background S_{xx} can be reached already with quality factors in the millions. An example for such a measurement is shown in Fig. 1a.

This sensitivity is insufficient, however, for the detection of displacements at the level of the mechanical RMS zero-point fluctuations $\delta x_\mathrm{zpf} = \sqrt{\hbar/2m_\mathrm{eff}\Omega_\mathrm{m}}$, which are at the femtometer level for the parameters discussed above. An optical cavity is needed to enhance the interaction between light and motion, recycling the light for a number of roundtrips that is commensurate with the finesse $\mathcal{F}$ of the cavity. The phase shift of the light emerging from the cavity is multiplied correspondingly, allowing more sensitive detection with the same amount of laser light. In the simplest case of resonant probing ($\omega = \omega_\mathrm{c}$, the cavity resonance frequency), the quantum imprecision noise is equivalent to displacement spectral densities of [7]

$$S_{xx}^{1/2}(\Omega) = \frac{\lambda}{16\eta_\mathrm{c}\mathcal{F}} \frac{1}{\sqrt{\eta_\mathrm{d} P/\hbar\omega}} \sqrt{1 + \left(\frac{\Omega}{\kappa/2}\right)^2}, \tag{2}$$

for a Fabry–Perot resonator with a moving end mirror (in the case of a whispering-gallery-mode resonator whose radius is measured, $\lambda \to \lambda/\pi$). Note that the sensitivity now acquires a dependence on the Fourier frequency Ω, here a simple cutoff behavior for frequencies larger than the cavity half linewidth $\kappa/2$, as well as the degree of cavity overcoupling η_c.

Figure 1b shows an example of such a measurement, in this case performed on the radial-breathing mode of a silica whispering-gallery-mode resonator [7], with the help of a polarization spectroscopy technique [8]. It resolves not only thermal motion with a large signal-to-noise ratio (here, about 58 dB), but also achieves an imprecision noise below that at the resonant standard quantum limit (SQL), $S_{xx}^{\mathrm{SQL}}(\Omega_{\mathrm{m}}) = \frac{\delta x_{\mathrm{zpf}}^2}{\Gamma_{\mathrm{m}}/2}$. Note that this coincides with the peak spectral density of ground-state fluctuations [9], for this device with $\Omega_{\mathrm{m}}/2\pi = 40.6\,\mathrm{MHz}$, $\Gamma_{\mathrm{m}} = 1.3$ kHz and $m_{\mathrm{eff}} = 10\,\mathrm{ng}$ at the level of $S_{xx}^{\mathrm{SQL}}(\Omega_{\mathrm{m}}) = (2.2\,\mathrm{am})^2/\mathrm{Hz}$ [10].

Cavity-enhanced laser interferometry has also been applied to nanomechanical resonators all the way down to the molecular scale. For example, it was shown that a fiber-based optical microcavity can resolve the thermal motion of carbon nanotubes [11]. Another successful sensing scheme consists in introducing nanomechanical resonators in the near field of optical whispering-gallery-mode resonators. It achieves imprecision well below that at the SQL of stressed silicon nitride nanostrings with picogram masses and $Q \sim 10^6$ [12–14]. It is also expected that optical cavities suppress diffraction losses through preferential scattering into the cavity mode.

To track or steer coherent dynamics of mechanical resonators at the level of their vacuum fluctuations, yet higher sensitivities are required [14]. In particular, it is necessary to resolve the ground state—which entails averaging for a time $4S_{xx}/x_{\mathrm{zpf}}^2$—before it decoheres, e.g., by heating. The latter happens at a rate $n_{\mathrm{th}}\Gamma_{\mathrm{m}}$, where $n_{\mathrm{th}} = k_{\mathrm{B}}T/\hbar\Omega_{\mathrm{m}} \gg 1$ is the mean occupation of the dominant thermal bath at temperature T. It follows from Eq. (2) that a resolution at the level of the zero-point-fluctuations is acquired at the measurement rate [9] $\Gamma_{\mathrm{opt}} = 4g^2/\kappa$, where $g = x_{\mathrm{zpf}}(\partial\omega_{\mathrm{c}}/\partial x)a$, and $|a|^2$ the number of photons in the cavity (assuming $\eta_{\mathrm{c}}\eta_{\mathrm{d}} = 1$, $\Omega \ll \kappa$). The above-mentioned requirement can then be written as $\Gamma_{\mathrm{opt}} \gtrsim n_{\mathrm{th}}\Gamma_{\mathrm{m}}$.

Interestingly, a completely new effect becomes relevant in this regime as well: the quantum fluctuations of radiation pressure linked to the quantum amplitude fluctuations of the laser light, representing the quantum backaction of this measurement [15]. And indeed the ratio of radiation pressure to thermal Langevin force fluctuations is given by $\frac{S_{FF}^{\mathrm{qba}}(\Omega_{\mathrm{m}})}{S_{FF}^{\mathrm{th}}(\Omega_{\mathrm{m}})} = \frac{\Gamma_{\mathrm{opt}}}{n_{\mathrm{th}}\Gamma_{\mathrm{m}}}$. While these force fluctuations induce random mechanical motion that can mask a signal to be measured, it is important to realize that motion and light become correlated, at the quantum level, via this mechanism. As a consequence, the mere interaction of cavity light with a nanomechanical device can induce optical phase–amplitude quantum correlations, which squeeze the optical quantum fluctuations, in a particular quadrature, below the level of the vacuum noise. This effect is referred to as ponderomotive squeezing [16–19].

An example of this phenomenon is shown in Fig. 1c [20]. A 1.928-MHz nanomechanical membrane resonator of dimensions $(544\,\mu\mathrm{m})^2 \times 60\,\mathrm{nm}$ is placed in a laser-driven high-finesse optical cavity and thereby measured at a rate of $\Gamma_{\mathrm{opt}}/2\pi = 96\,\mathrm{kHz}$. Its decoherence rate is reduced to $n\Gamma_{\mathrm{m}}/2\pi \approx 20\,\mathrm{kHz}$, by cooling it in a simple cryostat to $T = 10\,\mathrm{K}$. A slight detuning of the laser field with respect to the optical resonator ($\Delta = \omega - \omega_c = -2\pi \times 1.4\,\mathrm{MHz}$) leads to further cooling of the mechanical mode [21–24], akin to Doppler cooling of atomic gases [25]—here to a mean occupation of $n_{\mathrm{eff}} \sim 5$. It also allows direct observation of the squeezing in the amplitude fluctuations of the light emerging from the resonator: Its normalized spectral density assumes the form

$$\begin{aligned} S_{XX}^{\mathrm{out}}(\Omega) \; &\approx 1 - 2\frac{8\Delta}{\kappa}\Gamma_{\mathrm{opt}} Re\{\chi_{\mathrm{eff}}(\Omega)\} + \\ &+ \left(\tfrac{8\Delta}{\kappa}\right)^2 \Gamma_{\mathrm{opt}} |\chi_{\mathrm{eff}}(\Omega)|^2 \left(\Gamma_{\mathrm{opt}} + n_{\mathrm{th}}\Gamma_{\mathrm{m}}\right). \end{aligned} \tag{3}$$

Note that the second term represents the correlations, which can assume negative values and thus lead to noise below the vacuum level $S_{XX}^{\mathrm{out}} = 1$ (χ_{eff} is the effective mechanical susceptibility [3]). Ponderomotive squeezing down to $-2.4\,\mathrm{dB}$ has been observed, the strongest value so far, and simultaneous squeezing in a multitude of mechanical modes [20]. Schemes that exploit such quantum correlations for sub-SQL measurements of displacement and forces are subject of ongoing research [26–28].

The above examples show that laser-based measurements resolve the motion of nanomechanical oscillators all the way to the level of their vacuum fluctuations. In a simple classification (Fig. 1d), basic interferometers can readily resolve thermal motion, as required in many sensing and characterization experiments. Cavity-enhanced approaches achieve imprecision below the resonant SQL. To measure and control motion at the quantum level, displacements at the scale of the vacuum fluctuations must be resolved within the coherence time of the mechanical resonator. Then the imprecision (of an ideal setup) is more than n_{th} times below the resonant SQL, and quantum backaction exceeds thermal force fluctuations and induces quantum correlations [3, 9].

While the above-described techniques can be considered variants of laser interferometry, there are a number of techniques to characterize mechanical devices that are laser spectroscopic in nature. A prominent example is optomechanically induced transparency (OMIT), first described in Refs. [29, 30]. It consists in the observation that a laser-driven cavity containing a dispersively coupled mechanical device will have a modified transmission spectrum for a second "probe" laser beam at the frequency $\omega_{\mathrm{p}} = \omega + \Omega_{\mathrm{m}} + \Delta'$, where $|\Delta'| \lesssim \kappa$ is the two-photon detuning. The intracavity probe field,

$$a_{\mathrm{p}} \propto \frac{\sqrt{\kappa}}{(-i\Delta' + \kappa/2) + \dfrac{g^2}{-i\Delta' + \Gamma_{\mathrm{m}}/2}} \tag{4}$$

in the simplest case $-\Delta = \Omega_{\mathrm{m}} \ll \kappa$, encodes the coupling strength g. It is thus possible to derive g, for example, from probe transmission measurements [31, 32].

3 Laser-Based Imaging

As already indicated, it can be of great interest to also spatially resolve mechanical displacement patterns. With laser light, this can be accomplished in an extremely sensitive and virtually non-perturbing manner [33–37]. In the following, we present two methods that we have implemented for characterizing nano- and micromechanical resonators with micrometer transverse resolution, sufficient for resolving the spatial patterns of MHz mechanical modes.

3.1 *Scanning Laser Interferometry*

The first setup, shown in Fig. 2, is a Michelson interferometer based on a Nd: YAG laser at $\lambda = 1064$ nm. A polarizing beam splitter (PBS1) splits its output into two interferometer arms. In one arm, a single-mode fiber guides light to a probe

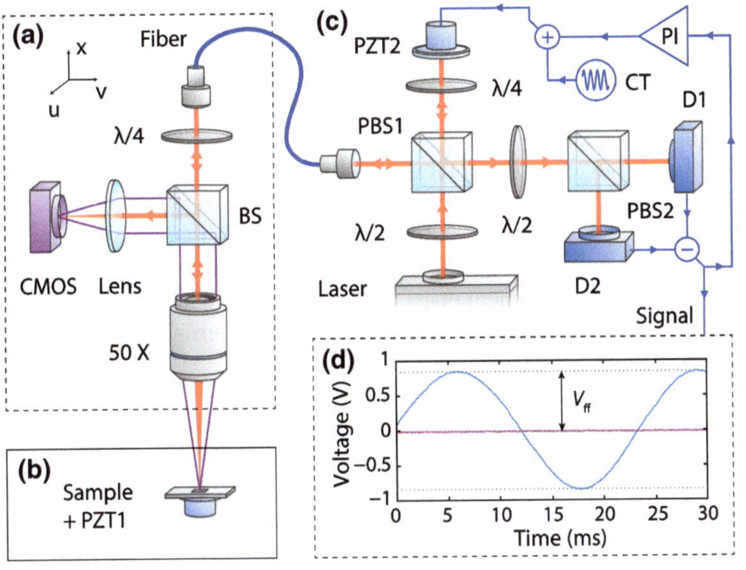

Fig. 2 Setup for interferometric imaging of mechanical motion. **a** Probe head with microscope objective mounted on a motorized 3-axis translation stage to position a focused laser spot on the sample. The sample is imaged simultaneously onto a CMOS camera via a beam splitter (BS) and a lens. **b** Sample placed on top of a piezo (PZT1) inside a high vacuum chamber. **c** Main part of the Michelson interferometer. A balanced receiver (detectors D1 and D2) measures the relative phase between the light returned from the sample and the reference arm. Electronic feedback to a piezomounted mirror (PZT2) stabilizes this phase with a low ($\lesssim 10$ kHz) bandwidth. **d** Signal from the balanced receiver as a function of time while scanning (*blue*) and actively stabilizing (*purple*) the relative phase

head mounted on a motorized 3-axis translation stage. The probe head (Fig. 2a) consists of a microscope objective focusing the laser light to a spot of diameter ~ 2 μm on the sample and a CMOS camera capturing images of the sample in real time. To reduce viscous (gas) damping of the nanomechanical motion, the sample is placed inside a high vacuum chamber at a pressure of $< 10^{-5}$ mbar. A piezoelectric shaker (PZT1) can excite mechanical eigenmodes (Fig. 2b).

Light reflected off the sample is spatially overlapped with the local oscillator from the other interferometer arm in PBS1 (Fig. 2c). Projection on a common polarization basis subsequently enforces interference in a second polarizing beam splitter (PBS2), whose outputs are monitored with a high-bandwidth (0 – 75 MHz) InGaAs-balanced receiver. This configuration ensures shot-noise-limited detection of the reflected light when a typical $\sim 800\,\mu$W beam is sent to the sample. In the correct polarization base, one obtains a receiver signal $V_{\mathrm{ff}} \cos(\phi)$, where ϕ is the relative phase between the two beams and V_{ff} is the full fringe voltage, which we check with an oscilloscope (Fig. 2d). For maximal transduction, ϕ is actively stabilized to the mid-fringe position by means of a mirror mounted on a piezoelectric transducer in the local oscillator arm (PZT2) and a proportional-integral (PI) feedback control.

In this case, small measured voltages $\delta V(t) \ll V_{\mathrm{ff}}$ convert to displacement via $\delta x(t) \approx \pm \delta V(t)\lambda/4\pi V_{\mathrm{ff}}$. Modulating PZT2 continuously with known frequency and amplitude generates a reference displacement and provides an independent calibration tone (CT) in the spectra.

As a first example, Fig. 3 shows a raster scan of a stoichiometric silicon nitride (SiN) membrane with side length $l = 1$ mm. We scan the membrane surface with the probe head using stepper motor actuation and record traces $\delta x(t)$ at each of the 22×22 positions. The traces are spectrally filtered around the peaks of several mechanical modes via digital post-processing. In this manner, we extract RMS displacements of each mechanical eigenmode in each scan pixel. Figure 3 shows the corresponding displacement maps for the modes, which are thermally excited at room temperature (PZT1 off). The measured mode patterns compare well with the hybridized eigenmodes of a square membrane:

$$w_{n,m} \propto \sin(k_n u)\sin(k_m v) + \beta \sin(k_m u)\sin(k_n v), \tag{5}$$

where $k_n = 2\pi n/l$, $k_m = 2\pi m/l$, and $n, m \geq 1$ denote the number of antinodes along in-plane coordinates u and v, respectively, and $|\beta| < 1$ quantifies the degree of hybridization between degenerate mode pairs. We find that the measured maximum RMS displacements, as calibrated by the CT, are in good agreement with the expected thermal motion (Fig. 3). Here, we have assumed a mass $m_{\mathrm{eff}} = \rho l^2 h/4 \sim 34$ ng, given the thickness $h = 50$ nm and density $\rho = 2.7$ g/cm^3 of the membrane. Note that the modes $(n, m) = (1, 2)$ and (2, 1) show hybridization with $|\beta| \sim 0.2$.

Scanning laser interferometry is particularly useful to characterize complex mode structures, such as SiN membranes patterned with phononic crystal structures [4] (Fig. 4). A scan measured on a grid of 100×100 points with a $5\,\mu$m spacing

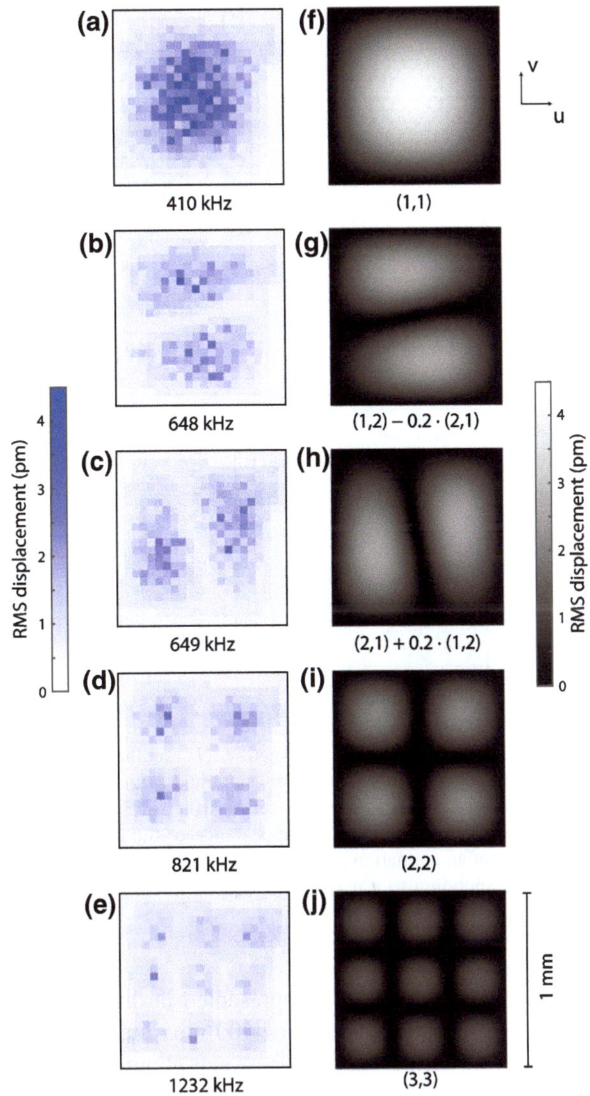

Fig. 3 Nanomechanical modes of a stoichiometric SiN membrane measured with the raster-scan interferometer. **a–e** Measurements of thermal motion on a 22 × 22 point grid. **f–j** Calculated displacement for mode numbers (n, m), accounting for hybridization between mode (1, 2) and (2, 1)

resolves also the 9.3 μm-wide tethers in between two holes, as Fig. 4b shows. At the expense of measurement time, the grid spacing could be further reduced; however, the spatial resolution of the obtained image is eventually limited to the $\sim 2\,\mu$m diameter of the laser spot. Figure 4c shows another mode of the same device imaged over a larger area. At a distance of 500 μm from the center, the mode's amplitude has decayed to the measurement noise level, illustrating the localization of the mode to the defect.

An advantage of measuring thermally excited modes is that information on all modes within the detector bandwidth is acquired simultaneously. This large set of

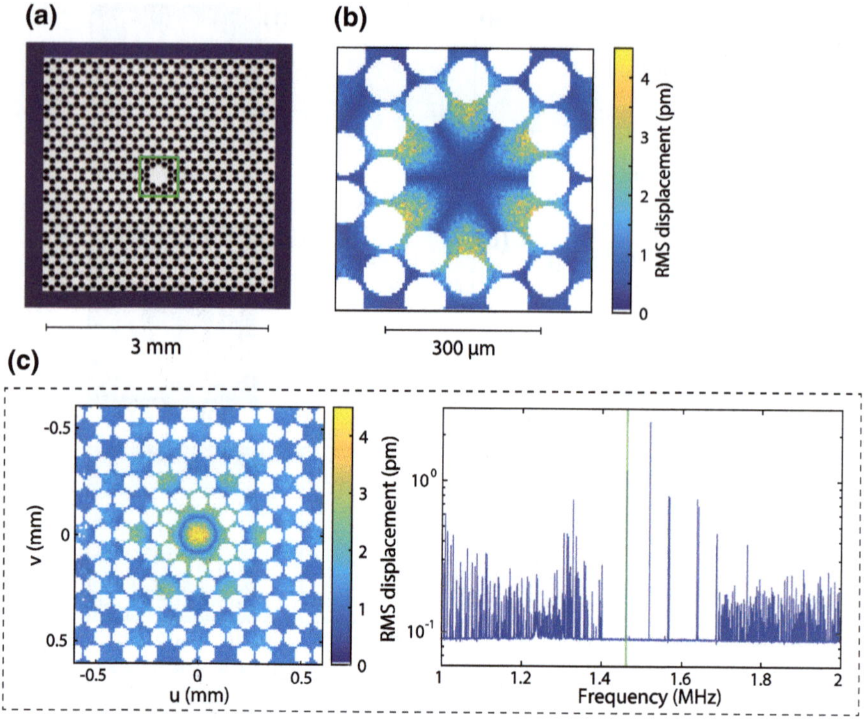

Fig. 4 Measurements of patterned SiN membrane with the raster-scan interferometer. **a** Micrograph of a SiN membrane patterned with a phononic crystal structure. **b** Localized nanomechanical mode imaged on 100 × 100 grid in the scan area indicated by a *green square* in (**a**). *Holes* are detected by disappearance of the calibration peak and shown as *white pixels*. **c** Snapshot of an animation provided as supplementary material. It shows the displacement pattern (*left*) corresponding to a particular frequency bin (*green line*) of the averaged spectrum (*right*) (from [4])

data can be processed and represented in different ways. As an example, Fig. 4c shows an average spectrum of 400 measurement points on the defect. It clearly reveals a phononic bandgap between about 1.41 and 1.68 MHz, containing five defect mode peaks, as well as the calibration peak at 1.52 MHz. The left panel shows a displacement map corresponding to a specific frequency bin of this spectrum. We can also create an animation that composes the displacement maps for each of the frequency bins in the spectrum. It is provided as electronic supplementary material to this article (see supplementary material). It delivers an instructive illustration of the effect of the phononic crystal structure, contrasting the small number of localized modes inside the bandgap with a "forest" of distributed modes at frequencies outside the bandgap.

A disadvantage of the scanning laser interferometer is its long measurement time. For instance, a high-resolution scan, such as the one shown in Fig. 4b, takes more than 8 hours. This is because for each pixel of the image we probe thermal motion during several seconds, averaging over timescales longer than Γ_{m}^{-1}. Some acceleration is possible by either artificially increasing Γ_{m}, e.g., by controlled gas damping, or by driving the modes coherently using PZT1. The latter can furthermore provide information about the mechanical phase at each position, if mechanical frequency drifts are properly accounted for.

3.2 Dark-Field Imaging

A powerful approach to single-shot characterization of mechanical modes is provided by dark-field imaging [35]. Figure 5a shows the setup which we have implemented to this end. It directly captures the squared displacement patterns of two-dimensional resonators such as membranes or cantilevers on a CCD camera. Its functional principle is described with simple Fourier optics [38]. A collimated laser beam with a wavelength $\lambda = 1064$ nm and beam diameter of 2.4 mm impinges perpendicularly on the sample, here a SiN membrane with side length $l = 1$ mm. The reflected electric field E_{r} at transverse position (u, v) is subject to a phase shift proportional to the membrane displacement $w(u, v, t)$. We assume that the incident electric field $E_0 e^{i\omega t}$ is constant across the membrane, since the incident beam diameter is 2.4 times larger than the membrane. Assuming furthermore $w(u, v, t) \ll \lambda$, the reflected electric field reads $E_{\mathrm{r}}(u, v, t) \approx rE_0 e^{i\omega t}[1 + ikw(u, v, t)]$, where r is the absolute value of the reflection coefficient, $k = 2\pi/\lambda$ and $\omega = ck$. A lens (focal length $f_1 = 75$ mm) performs an optical Fourier transform $\mathcal{F}$ with respect to the coordinates (u, v), yielding

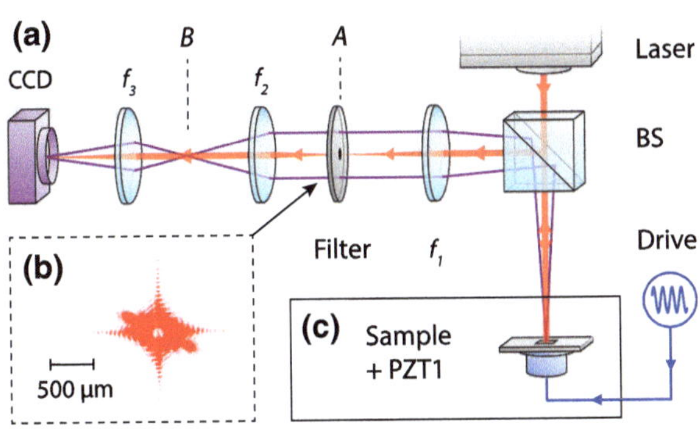

Fig. 5 Setup for dark-field imaging of mechanical motion. **a** Optical configuration with small opaque disk in the Fourier plane A, creating a dark-field image of the sample in the image plane B. A lens (f_3) projects a magnified image onto a CCD camera. A ray diagram illustrates how the image is formed (*purple lines*). **b** Image of the Fourier plane, where an opaque disk blanks out undiffracted zero-order light, when a membrane mode is excited. **c** Sample is mounted on a piezoelectric actuator (PZT1) in a high vacuum chamber

$$\mathcal{F}(E_r) = rE_0 e^{i\omega t}[\mathcal{F}(1) + \mathcal{F}(ikw(u, v, t))]. \tag{6}$$

The zero-order peak (first term in Eq. (6)) is removed from the beam by an opaque disk in the Fourier plane. This extracts the diffracted light due to the membrane displacement w. A second, subsequent lens (focal length $f_2 = 50$ mm) performs another Fourier transform on the filtered light. The time-averaged intensity pattern

$$\begin{aligned} I(u', v') &= \left\langle \left| rE_0 e^{i\omega t} \mathcal{F}(\mathcal{F}(ikw(u, v, t))) \right|^2 \right\rangle \\ &= I_0 r^2 k^2 \left\langle w(-u, -v, t)^2 \right\rangle, \end{aligned} \tag{7}$$

is then recorded by a camera, where $I_0 = |E_0 e^{i\omega t}|^2$ is the incident intensity. It directly shows an intensity pattern proportional to the squared displacement of an eigenmode.

In our setup, a third lens with focal length $f_3 = 35$ mm is placed in front of the camera to magnify the image. It also allows imaging the Fourier plane by adjusting the distance between camera and lens to f_3. Figure 5b shows a Fourier image of the membrane with diffraction patterns extending in two orthogonal directions due to the sharp edges of the membrane. Two bright spots close to the center originate from diffraction due to a driven eigenmode, a hybridization between the modes (1,2) and (2,1) at a frequency of 645 kHz. The opaque disk made of aluminum deposited on a thin piece of glass is seen as a white disk in the center. With a diameter $d = 100\,\mu$m, it blocks diffraction angles $\alpha \lesssim d/2f_1$ generated by mechanical modes with a distance between nodes of $\gtrsim \lambda/2\alpha \approx 800\,\mu$m.

A piezoelectric actuator (PZT1) successively excites the eigenmodes of the SiN membrane inside a vacuum chamber, by slowly sweeping a strong drive tone across the frequency window of interest (here $0.4 \ldots 2$ MHz). Figure 6 shows images of several modes recorded with an incident optical power of $\sim 100\,\mu$W and a typical integration time of 10 ms. Comparison with mode patterns calculated from Eq. (5) allows inferring the mode numbers (n, m), and the degree of hybridization, as seen, for example, on the 1.683-MHz mode.

While it enables much shorter measurement times than the scanning laser interferometer, the dark-field imaging setup has a relatively low displacement sensitivity. For this reason, PZT1 has to be driven with a stroke of $\gtrsim$300 pm, significantly increasing the membrane oscillation amplitude, up to a regime where mechanical nonlinearities (e.g., Duffing-type frequency shifts) can play a role. In principle, the sensitivity can be enhanced by increasing the laser intensity I_0, yet in practice it is often limited by background noise due to scattered light from optical components increasing equally with I_0. Another important limitation is that diffraction from the sample's geometry cannot be discriminated from modal displacements. In this simple implementation, the approach is thus unsuitable for devices with fine structures in their geometry, such as the patterned membranes.

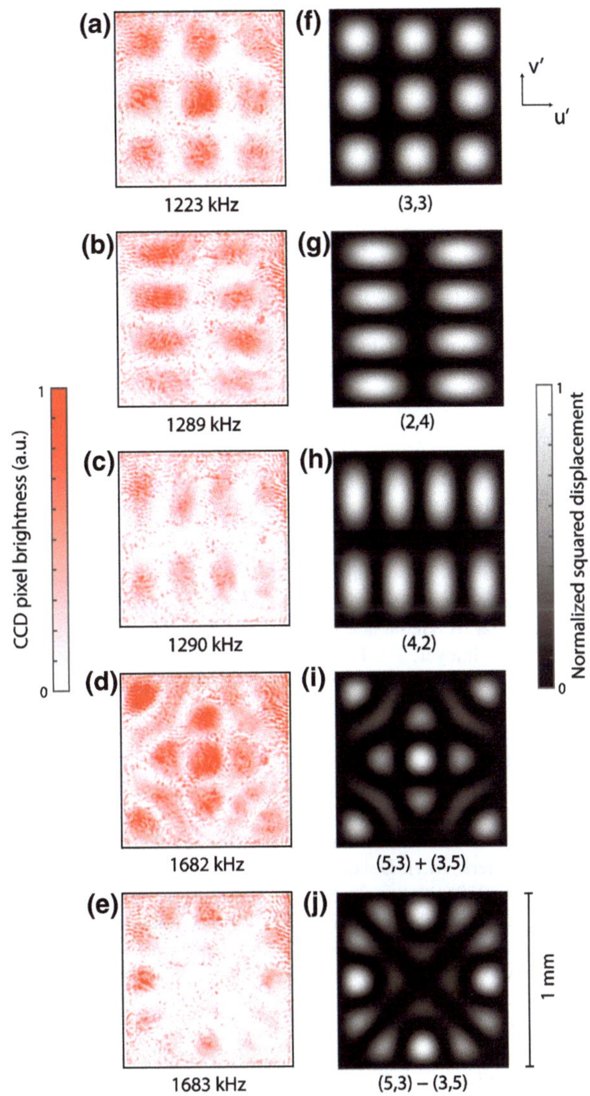

Fig. 6 Nanomechanical modes of a stoichiometric SiN membrane measured with dark-field imaging. **a–e** Single-shot (acquisition time ∼ 10 ms) measurements of mechanical modes at various frequencies while driving the membrane with PZT1. *Pixels colored in dark red* indicate high intensities and correspond to large values of squared mechanical displacement. **f–j** Calculated squared mechanical mode patterns with mode numbers (n, m). Modes (3, 5) and (5, 3) show nearly complete hybridization with $|\beta| \sim 1$

4 Conclusion

In summary, we have described several laser-based techniques to measure and image nanomechanical motion. As we show, exquisite displacement sensitivity can be reached, well into the regime in which quantum backaction and the ensuing light-motion quantum correlations dominate over thermomechanical noise. This sensitivity is rivaled only by techniques based on superconducting microwave electromechanical systems, which operate at ultra-low ($T \ll 1$ K) cryogenic temperatures [39, 40]. Interest in this quantum domain has originally been motivated by observatories such as LIGO and can now, for the first time, be explored with optical and microwave experiments [3, 9, 15, 41, 42]. In addition, laser-based techniques can provide spatial imaging of mechanical displacement patterns. They constitute not only highly useful tools to develop and characterize novel micro- and nanomechanical devices [4, 6, 35–37]. Similar techniques could also be used to address individual elements in multimode devices [20] or (opto-)mechanical arrays [4, 43]—if need be, also in combination with cavity-enhanced readout [33, 44].

Acknowledgements We would like to acknowledge our (former and present) colleagues Georg Anetsberger, Olivier Arcizet, Tobias Kippenberg, Jörg H. Müller, Eugene S. Polzik, Andreas Næsby Rasmussen, Remi Rivière, Anders Simonsen, Koji Usami, Stefan Weis, and Dalziel J. Wilson for their contributions to the work discussed here. Financial support came from the ERC starting grant Q-CEOM, a starting grant from the Danish Council for Independent Research, the EU FP7 grant iQUOEMS, and the Carlsberg Foundation.

References

1. B.P. Abbott, R. Abbott, T.D. Abbott, M.R. Abernathy, F. Acernese, K. Ackley, C. Adams, T. Adams, P. Addesso, R.X. Adhikari et al., Phys. Rev. Lett. **116**, 061102 (2016)
2. D. Rugar, R. Budakian, H.J. Mamin, B.W. Chui, Nature **430**, 329 (2004)
3. M. Aspelmeyer, T.J. Kippenberg, F. Marquardt, Rev. Mod. Phys. **86**, 1391 (2014)
4. Y. Tsaturyan, A. Barg, E.S. Polzik, A. Schliesser, arXiv:1608.00937 (2016)
5. J.W. Wagner, J.B. Spicer, J. Opt. Soc. Am. B **4**, 1316 (1987)
6. Y. Tsaturyan, A. Barg, A. Simonsen, L.G. Villanueva, S. Schmid, A. Schliesser, E.S. Polzik, Opt. Express **6**, 6810 (2013)
7. A. Schliesser, G. Anetsberger, R. Rivière, O. Arcizet, T.J. Kippenberg, New J. Phys. **10**, 095015 (2008)
8. T.W. Hänsch, B. Couillaud, Opt. Commun. **35**, 441 (1980)
9. A.A. Clerk, M.H. Devoret, S.M. Girvin, F. Marquardt, R.J. Schoelkopf, Rev. Mod. Phys. **82**, 1155 (2010)
10. A. Schliesser, T. J. Kippenberg, Cavity optomechanics with whispering-gallery modeoptical micro-resonators. in *Advances in Atomic, Molecular and Optical Physics*, vol. 58, Chap. 5, ed. by P. Berman, E. Arimondo, C. Lin (Elsevier Academic Press, 2010), pp. 207–323
11. S. Stapfner, L. Ost, D. Hunger, J. Reichel, I. Favero, E.M. Weig, Appl. Phys. Lett. **102**, 151910 (2013)
12. G. Anetsberger, O. Arcizet, Q.P. Unterreithmeier, R. Rivière, A. Schliesser, E.M. Weig, J.P. Kotthaus, T.J. Kippenberg, Nat. Phys. **5**, 909 (2009)
13. G. Anetsberger, E. Gavartin, O. Arcizet, Q.P. Unterreithmeier, E.M. Weig, M.L. Gorodetsky, J.P. Kotthaus, T.J. Kippenberg, Phys. Rev. A **82**, 061804 (2010)
14. D.J. Wilson, V. Sudhir, N. Piro, R. Schilling, A. Ghadimi, T.J. Kippenberg, Nature **524**, 325 (2015)
15. C.M. Caves, Phys. Rev. Lett. **45**, 75 (1980)
16. C. Fabre, M. Pinard, S. Bourzeix, A. Heidmann, E. Giacobino, S. Reynaud, Phys. Rev. A **49**, 1337 (1994)
17. S. Mancini, P. Tombesi, Phys. Rev. A **49**, 4055 (1994)
18. T.P. Purdy, P.-L. Yu, R.W. Peterson, N.S. Kampel, C.A. Regal, Phys. Rev. X. **3**, 031012 (2013)
19. A.H. Safavi-Naeini, S. Gröblacher, J.T. Hill, J. Chan, M. Aspelmeyer, O. Painter, Nature **500**, 185 (2013)
20. W.H.P. Nielsen, Y. Tsaturyan, C.B. Møller, E.S. Polzik, A. Schliesser, arXiv:1605.06541 (2016)

21. O. Arcizet, P.-F. Cohadon, T. Briant, M. Pinard, A. Heidmann, Nature **444**, 71 (2006)
22. S. Gigan, H.R. Böhm, M. Paternosto, F. Blaser, G. Langer, J.B. Hertzberg, K.C. Schwab, D. Bäuerle, M. Aspelmeyer, A. Zeilinger, Nature **444**, 67 (2006)
23. A. Schliesser, P. Del'Haye, N. Nooshi, K. Vahala, T. Kippenberg, Phys. Rev. Lett. **97**, 243905 (2006)
24. J.D. Thompson, B.M. Zwickl, A.M. Jayich, F. Marquardt, S.M. Girvin, J.G.E. Harris, Nature **452**, 72 (2008)
25. T.W. Hänsch, A.L. Schawlow, Opt. Commun. **13**, 68 (1975)
26. O. Arcizet, T. Briant, A. Heidmann, M. Pinard, Phys. Rev. A **73**, 033819 (2006)
27. L.F. Buchmann, S. Schreppler, J. Kohler, N. Spethmann, D.M. Stamper-Kurn, Phys. Rev. Lett. **117**, 030801 (2016)
28. N.S. Kampel, R.W. Peterson, R. Fischer, P.-L. Yu, K. Cicak, R.W. Simmonds, K.W. Lehnert, C.A. Regal, arXiv:1607.06831 (2016)
29. A. Schliesser, Cavity optomechanics and optical frequency comb generation with silica whispering-gallery-mode microresonators, Ph.D. thesis, Ludwig-Maximilians-Universität München, 2009
30. G.S. Agarwal, S. Huang, Phys. Rev. A **81**, 041803 (2010)
31. S. Weis, R. Rivière, S. Deléglise, E. Gavartin, O. Arcizet, A. Schliesser, T.J. Kippenberg, Science **330**, 1520 (2010)
32. A.H. Safavi-Naeini, T.P. Mayer, I. Alegre, J. Chan, M. Eichenfield, M. Winger, J.Q. Lin, J.T. Hill, D.E. Chang, O. Painter, Nature **472**, 69 (2011)
33. T. Briant, P.-F. Cohadon, A. Heidmann, M. Pinard, Phys. Rev. A **68**, 033823 (2003)
34. O. Arcizet, P.-F. Cohadon, T. Briant, M. Pinard, A. Heidmann, J.-M. Mackowski, C. Michel, L. Pinard, O. Francais, L. Rousseau, Phys. Rev. Lett. **97**, 133601 (2006)
35. S. Chakram, Y.S. Patil, L. Chang, M. Vengalattore, Phys. Rev. Lett. **112**, 127201 (2014)
36. Z. Wang, J. Lee, P.X.L. Feng, Nat. Commun. **5**, 5158 (2014)
37. D. Davidovikj, J.J. Slim, S.J. Cartamil-Bueno, H.S.J. van der Zant, P.G. Steeneken, W.J. Venstra, Nano Lett. **16**, 2768 (2016)
38. See Supplementary Material
39. W. Lauterborn, T. Kurz, M. Wiesenfeldt, *Coherent Optics* (Springer, Berlin, 1995)
40. J.D. Teufel, R. Donner, M.A. Castellanos-Beltran, J.W. Harlow, K.W. Lehnert, Nat. Nanotech. **4**, 820 (2009)
41. J.D. Teufel, F. Lecocq, R.W. Simmonds, Phys. Rev. Lett. **116**, 013602 (2016)
42. V.B. Braginsky, F.Y. Khalili, *Quantum Measurement* (Cambridge University Press, Cambridge, 1992)
43. I. Tittonen, G. Breitenbach, T. Kalkbrenner, T. Müller, R. Conradt, S. Schiller, E. Steinsland, N. Blanc, N.F. de Rooij, Phys. Rev. A **59**, 1038–1044 (1999)
44. G. Heinrich, M. Ludwig, J. Qian, B. Kubala, F. Marquardt, Phys. Rev. Lett. **107**, 043603 (2011)
45. M. Mader, J. Reichel, T.W. Hänsch, D. Hunger, Nat. Commun. **6**, 7249 (2015)

Quantum Regime of a Free-Electron Laser: Relativistic Approach

Peter Kling, Roland Sauerbrey, Paul Preiss, Enno Giese, Rainer Endrich, and Wolfgang P. Schleich

Abstract In the quantum regime of the free-electron laser, the dynamics of the electrons is not governed by continuous trajectories but by discrete jumps in momentum. In this article, we rederive the two crucial conditions to enter this quantum regime: (1) a large quantum mechanical recoil of the electron caused by the scattering with the laser and the wiggler field and (2) a small energy spread of the electron beam. In contrast to our recent approach based on *nonrelativistic*

This article is part of the topical collection "Enlightening the World with the Laser" - Honoring T. W. Hänsch guest edited by Tilman Esslinger, Nathalie Picqué, and Thomas Udem.

It is a great honor and pleasure for us to dedicate this article to Theodor W. Hänsch on the occasion of his 75th birthday. On first sight, there seems to be no connection between his research interests and the field of free-electron lasers (FELs). However, a closer examination reveals once more his crucial influence in shaping a new field. Indeed, in one [1] of the pioneering articles on FEL theory, W. B. Colson acknowledges 'fruitful discussions,' he had with Hänsch on this topic. Therefore, we find it appropriate to contribute an article devoted to the Quantum FEL to this Special Issue celebrating the scientific achievements of Theodor W. Hänsch. Our warmest wishes, and Happy Birthday Ted.

P. Kling (✉) • P. Preiss • R. Endrich
Helmholtz-Zentrum Dresden-Rossendorf eV, 01328 Dresden, Germany

Institut für Quantenphysik and Center for Integrated Quantum Science and Technology (IQST), Universität Ulm, Albert-Einstein-Allee 11, 89081 Ulm, Germany
e-mail: peter.kling@uni-ulm.de

R. Sauerbrey
Helmholtz-Zentrum Dresden-Rossendorf eV, 01328 Dresden, Germany

E. Giese
Institut für Quantenphysik and Center for Integrated Quantum Science and Technology (IQST), Universität Ulm, Albert-Einstein-Allee 11, 89081 Ulm, Germany

Department of Physics, University of Ottawa, 25 Templeton Street, Ottawa, ON K1N 6N5, Canada

W.P. Schleich
Institut für Quantenphysik and Center for Integrated Quantum Science and Technology (IQST), Universität Ulm, Albert-Einstein-Allee 11, 89081 Ulm, Germany

Texas A&M University Institute for Advanced Study (TIAS), Institute for Quantum Science and Engineering (IQSE), Department of Physics and Astronomy, Texas A&M University, College Station, TX 77843-4242, USA

D. Meschede et al. (eds.), *Exploring the World with the Laser*,
https://doi.org/10.1007/978-3-319-64346-5_7

quantum mechanics in a *co-moving frame* of reference, we now pursue a model in the *laboratory frame* employing *relativistic* quantum electrodynamics.

1 Introduction

An important insight gained already in the early days of the free-electron laser (FEL), especially in the articles by Colson [1] and Hopf et al. [2], was that the FEL can be completely understood in terms of classical physics [3]. However, recent years have seen a rising interest in a novel regime of FEL operation where quantum effects become crucial [4, 5]: the so-called quantum regime or Quantum FEL. For example, in Refs. [4, 5] the authors consider the Heisenberg equations of motion while in Ref. [6] as well as in Refs. [7, 8] formalisms in second quantization are discussed. Moreover, a quantum fluid model is developed in Ref. [9] and in Refs. [10, 11] the authors use the Klein–Gordon equation coupled to classical electromagnetic fields to derive a dispersion relation.

While in all these approaches the respective sets of equations were solved and the quantum regime was identified as one limit of the solution, our article [12] on the Quantum FEL takes a different route: We search *in the dynamical equations* for a regime where quantum mechanics becomes prevalent and solve these equations in an asymptotic way. To achieve this goal, we have used a nonrelativistic approach in the co-moving Bambini–Renieri frame [13, 14]. In contrast, we develop in the present article a fully relativistic model in the laboratory frame and derive within this framework the conditions to enter the quantum regime.

1.1 Conditions for Quantum FEL

In Refs. [12, 15–17], we have shown that quantum effects become important for the FEL dynamics when the discrete quantum mechanical recoil due to the scattering of the electron is the dominating momentum scale. Moreover, we have defined the quantum regime of the FEL as the limit where the electron occupies only two resonant momentum states giving rise to a system analogous to the Jaynes–Cummings model [18], where a two-level atom interacts with a quantized mode of the radiation field.

This transition to the quantum regime is quantified by two conditions: The first one relates to the 'quantum parameter' α which is defined as the ratio of the coupling strength between the electron and the fields, and a frequency connected to the quantum mechanical recoil. To be in the quantum regime, α has to be much smaller than unity.

The second requirement for the Quantum FEL demands the width Δp of the initial momentum distribution to be smaller, or of the order of the recoil.

To analyze these two conditions for a possible experimental realization, we have expressed α and Δp in the laboratory frame by transforming from the co-moving frame to the laboratory frame. With this procedure, we have obtained [12] the requirement[1]

$$\alpha \equiv \frac{\gamma^2 g \sqrt{n}}{\hbar (k_\mathrm{L} + k_\mathrm{W})^2/(2m)} \ll 1 \tag{1}$$

for the quantum parameter α where g, n, m and $\hbar$ denote the coupling constant between the electron and the fields, the number of laser photons, the mass of the electron and the reduced Planck constant, respectively, while γ is the scaled energy of the electron. The wave numbers of the laser field and the wiggler are given by k_L and k_W, respectively.

On the other hand, we have found the condition [12]

$$\frac{\Delta\gamma}{\gamma} < 4\gamma \frac{\lambda_\mathrm{C}}{\lambda_\mathrm{W}} \tag{2}$$

on the relative energy spread $\Delta\gamma/\gamma$ of the electron beam with the Compton wavelength λ_C of an electron and the wiggler wavelength $\lambda_\mathrm{W} \equiv 2\pi/k_\mathrm{W}$.

However, we could have taken a more direct route avoiding the detour to the Bambini–Renieri frame and perform the calculations in the laboratory frame. The price for this procedure is of course that we have to use the more tedious methods of relativistic quantum electrodynamics instead of the straightforward nonrelativistic computations.

Nonetheless, the physical situation in both frames of reference is the same and we should find analogous results. Hence, we formulate the goal of the present article: Find the conditions for the Quantum FEL in a fully relativistic approach, and compare them to the ones obtained in the framework of the Bambini–Renieri frame.

1.2 *Outline*

Our article is organized as follows: In Sect. 2, we review a model [19, 20] for a quantum theory of the FEL in the laboratory frame based on the Furry or bound interaction picture [21] as well as on the Volkov solution [22].

---

[1] In Ref. [12], we have rewritten α in terms of the wiggler parameter a_0, the electron density n_e, the classical electron radius r_e and the Compton wavelength λ_C. Indeed, the form

$$\alpha = \frac{1}{\gamma^3} \frac{a_0 \sqrt{r_\mathrm{e} n_\mathrm{e}}}{32\sqrt{\pi}} \frac{\lambda_\mathrm{W}^{5/2}}{\lambda_\mathrm{C}^{3/2}}$$

is more convenient to compare to experimental parameters.

With the help of this model, we derive in Sect. 3 the conditions for the Quantum FEL in the laboratory frame, which turn out to be equivalent to the ones of Ref. [12]. For our calculations, we restrict ourselves to a single-electron model which limits us to the low-gain regime of FEL operation. Moreover, we solve the FEL dynamics in a perturbative manner, which is only valid for short times. Despite these simplifications, we are able to identify the important parameters governing the transition from the classical to the quantum regime of the FEL. In Sect. 4, we summarize our results.

Appendices A and B give a more detailed description of the Furry picture and the Volkov solution, respectively. In Appendix C, we present the calculations of second-order perturbation theory to prove also in the laboratory frame that multi-photon processes are suppressed in the Quantum FEL.

2 Quantum Model of FEL in Laboratory Frame

Due to the relativistic velocities of the electrons in the laboratory frame, we have to use methods of relativistic quantum electrodynamics to establish a quantum theory of the FEL in this frame of reference. A suitable model, introduced in Refs. [19, 20], can be found in the so-called Furry or bound interaction picture [21]. In this description, the effects of the wiggler field, which is modeled as a classical and fixed quantity, are incorporated into the free dynamics of the electron while the quantized laser field acts as a perturbation of this motion.

Due to difficulties in treating multi-photon processes [23], which are crucial for the *classical* FEL [24], this relativistic model was discarded and theories were developed [25] in the co-moving Bambini–Renieri frame.[2] However, since processes involving more than one photon are strongly suppressed [12] in the Quantum FEL, we assume that the Furry picture is a good starting point for our investigations.

To introduce our model, we first consider the electromagnetic fields and then turn to the electron. We conclude by discussing the interaction between the electron and the fields, and deriving the Hamiltonian for the FEL in the laboratory frame.

2.1 Wiggler and Laser Fields

Due to its high intensity, we assume that the wiggler field can be treated as a classical and fixed electromagnetic field. In contrast, we describe the laser field in second quantization.

---

[2]Even the relativistic quantum theory of the FEL pursued in Refs. [24, 26] is effectively in the Bambini–Renieri frame. Indeed, this approach begins by using the Klein–Gordon equation in the laboratory frame, but after several transformations of variables one reproduces the same equations of motion as if one had considered the co-moving frame from the start.

We model the four-potentials

$$\hat{A}^{\mu}_{\mathrm{L}}(x) \equiv \mathcal{A}_{\mathrm{L}}\left(\hat{a}_{\mathrm{L}}\,\epsilon^{\mu}\mathrm{e}^{-ik_{\mathrm{L}}\cdot x} + \mathrm{h.c.}\right) \tag{3}$$

and

$$A^{\mu}_{\mathrm{W}}(x) \equiv \tilde{\mathcal{A}}_{\mathrm{W}}\left(\epsilon^{\mu}\mathrm{e}^{-ik_{\mathrm{W}}\cdot x} + \mathrm{c.c.}\right) \tag{4}$$

of the laser field and of the wiggler as two counterpropagating one-mode plane waves with the vacuum amplitude $\mathcal{A}_{\mathrm{L}}$ for the laser, and the amplitude $\tilde{\mathcal{A}}_{\mathrm{W}}$ for the wiggler. The four-vector $x \equiv x^{\mu} \equiv (ct, \boldsymbol{r})$ specifies time t and position $\boldsymbol{r}$ in the four-dimensional Minkowski space, while c denotes the speed of light.

The photon annihilation and creation operators $\hat{a}_{\mathrm{L}}$ and $\hat{a}^{\dagger}_{\mathrm{L}}$ fulfill the commutation relation $[\hat{a}_{\mathrm{L}}, \hat{a}^{\dagger}_{\mathrm{L}}] = 1$, while the potentials obey circular polarization with the polarization vectors $\epsilon \cdot \epsilon = \epsilon^* \cdot \epsilon^* = 0$ and $\epsilon \cdot \epsilon^* = -1$, written in a four-dimensional form.

The scalar products of the four-wave vectors k_{L} and k_{W} with themselves vanish for plane waves, i.e., $k_{\mathrm{L}} \cdot k_{\mathrm{L}} = k_{\mathrm{W}} \cdot k_{\mathrm{W}} = 0$, provided we consider the case of a laser wiggler. In the case of a magnetostatic wiggler, $k_{\mathrm{W}} \cdot k_{\mathrm{W}}$ would give a nonzero contribution, because the zeroth component which corresponds to the frequency of the field is zero in this case.

Since for an experimental operation of a Quantum FEL a laser wiggler or 'optical undulator' [27–30] was proposed [31], we concentrate in this article on this kind of undulator. However, even for a magnetostatic wiggler we find results [20] analogous to the optical ones, because in its rest frame the electron experiences the wiggler as a counterpropagating wave at approximately the speed of light [13, 14].

2.2 *Electron in Wiggler Field*

To simplify our calculations, we use the Klein–Gordon instead of the Dirac equation, which is justified since spin effects in the FEL can be neglected [32]. Indeed, according to Ref. [23] the corrections due to the spin scale with the ratio of the energy of a single laser photon and the kinetic energy of the electron in the laboratory frame. Although one expects that in the quantum regime a genuine quantum mechanical quantity such as the photon energy plays a significant role, it is still much smaller than the energy of the massive electron moving at relativistic velocities. Hence, we can safely neglect spin effects in any description of the FEL—in the classical as well as in the quantum regime.

The unperturbed dynamics of the Klein–Gordon field operator $\hat{\varphi}(x)$ in the Furry picture is determined by the Klein–Gordon equation including the interaction of the electron with the wiggler field, which is derived in Appendix A. To obtain a particle interpretation of the electron field, we first have to find a solution of the corresponding Klein–Gordon equation

$$\left[\left(i\hbar\partial^\mu - eA_{\mathrm{W}}^\mu(x)\right)^2 - m_0^2c^2\right]\varphi(x) = 0 \tag{5}$$

for the 'classical' field $\varphi(x)$. Here, m_0 and e describe the mass of the electron and the elementary charge, respectively.

The Volkov solution of Eq. (5) is derived in Appendix B and reads

$$\varphi_p(x) \equiv \frac{1}{\sqrt{2\gamma mcV}}\, \mathrm{e}^{-ip\cdot x/\hbar} \tag{6}$$

where

$$p^\mu \equiv p_{\mathrm{f}}^\mu + \frac{\mathrm{e}^2\tilde{A}_{\mathrm{W}}^2}{p_{\mathrm{f}}\cdot k_{\mathrm{W}}}\, k_{\mathrm{W}}^\mu, \tag{7}$$

denotes the effective four-momentum of the electron in the wiggler and p_{f} is the momentum of the free electron before it enters the wiggler.

The effective mass m is connected to p and to the 'free mass' m_0 via the relation

$$p\cdot p = m_0^2c^2 + 2\mathrm{e}^2\tilde{A}_{\mathrm{W}}^2 \equiv m^2c^2. \tag{8}$$

The fact that we are dealing with the effective electron momentum, p, instead of the free one, p_{f}, can be interpreted by a dressed electron model, where the effects of the external field A_{W} are already taken into account in form of the effective mass and momentum.

Other quantities that occur in the Volkov solution, Eq. (6), are the relativistic factor $\gamma \equiv p_0/mc$ and the quantization volume V.

In order to arrive at the familiar particle interpretation of the field operator $\hat{\varphi}(x)$, we expand [20] it in terms of the Volkov solutions, Eq. (6), and arrive at

$$\hat{\varphi}(x) = \sum_p \varphi_p(x)\hat{b}_p \tag{9}$$

where $\hat{b}_p$ annihilates an electron with momentum p and $\hat{b}_p^\dagger$ creates one.

One could argue that the creation and annihilation operators for a scalar field fulfill bosonic commutation relations, $[\hat{b}_p, \hat{b}_{p'}^\dagger] = \delta_{p,p'}$, instead of the required anti-commutator relations for the fermionic electrons. However, since we are describing a single-electron model, the difference in the occupations of fermionic and bosonic many-particle quantum states does not enter into our theory.

We also emphasize that the expansion, Eq. (9), considers only *particle* operators and the corresponding *positive* energy solutions. Indeed, we have discarded *anti-particle* operators with *negative* energy solutions in complete analogy to Ref. [20].

2.3 Electron in Wiggler Plus Laser Field

The Hamiltonian

$$\hat{H}(t) \equiv \mathrm{e} \int \mathrm{d}^3 r\, \hat{j}(x) \cdot \hat{A}_{\mathrm{L}}(x) \tag{10}$$

describing the interaction between the electron and the electromagnetic fields in the Furry picture is derived in Appendix A and couples the electron current $\hat{j}^{\mu}$ to the four-potential of the quantized laser field $\hat{A}^{\mu}_{\mathrm{L}}$.

The components

$$\begin{aligned} \hat{j}^{\mu}(x) &\equiv c\left[\hat{\varphi}^{\dagger}(x) i\hbar\partial^{\mu}\hat{\varphi}(x) - \left(i\hbar\partial^{\mu}\hat{\varphi}^{\dagger}(x)\right)\hat{\varphi}(x)\right] \\ &\quad + 2\mathrm{e}A^{\mu}_{\mathrm{W}}(x)\hat{\varphi}^{\dagger}(x)\hat{\varphi}(x) \end{aligned} \tag{11}$$

of the current corresponding to the electron in the wiggler field involve the field operator $\hat{\varphi}$ of the Klein–Gordon field and the four-potential A_{W} of the classical wiggler field.

We assume that the electron is injected parallel to the wiggler axis which we denote as the z-direction. Hence, all transverse components of the initial momentum vanish. Since the laser field and the wiggler field form a collinear geometry their phases also depend only on z and due to the interaction with these fields, the electron dynamics is maintained in the z-direction. Hence, we can consider an effective one-dimensional theory for the FEL dynamics.

Thus, by examining the Volkov solution, Eq. (6), and the electron current, Eq. (11), we observe that only the transverse components of the current, proportional to $A^{\mu}_{\mathrm{W}}(x)$, couple to the laser field $\hat{A}^{\mu}_{\mathrm{L}}(x)$, because in our one-dimensional model we have

$$p \cdot \hat{A}_{\mathrm{L}} = p_{\mathrm{f}} \cdot \hat{A}_{\mathrm{L}} = 0. \tag{12}$$

For our choice of circular polarization, we can derive from Eqs. (3), (4), (9), (10) and (11) the Hamiltonian

$$\hat{H}(t) = \int_V \mathrm{d}^3 r \left[\hat{\mathcal{H}}_{\mathrm{em}}(x) + \hat{\mathcal{H}}_{\mathrm{abs}}(x)\right] \tag{13}$$

where

$$\hat{\mathcal{H}}_{\mathrm{em}}(x) \equiv \frac{\hbar g}{V} \sum_{p',p''} \frac{\mathrm{e}^{i\Phi(x)}}{\sqrt{\gamma'\gamma''}}\, \hat{b}^{\dagger}_{p'} \hat{b}_{p''}\, \hat{a}^{\dagger}_{\mathrm{L}} \tag{14}$$

describes the emission of photons into the laser field with the phase

$$\Phi(x) \equiv \frac{1}{\hbar}\left[p'^{\mu} - p''^{\mu} + \hbar\left(k_{\mathrm{L}}^{\mu} - k_{\mathrm{W}}^{\mu}\right)\right] x_{\mu} \tag{15}$$

while

$$\mathcal{H}_{\mathrm{abs}}(x) = \left(\mathcal{H}_{\mathrm{em}}(x)\right)^{\dagger} \tag{16}$$

is the corresponding absorption term.

The coupling constant g is defined as

$$g \equiv \frac{\mathrm{e}^2 \mathcal{A}_{\mathrm{L}} \tilde{\mathcal{A}}_{\mathrm{W}}}{\hbar m} \tag{17}$$

in complete analogy to Ref. [12].

The dynamics of the system consisting of electron and laser field is given by the state vector $|\Psi(t)\rangle$. In the following, we write $|\Psi(t)\rangle$ in terms of photon number states $|n\rangle$ of the laser field and momentum eigenstates $|p\rangle \equiv |1_p\rangle = \hat{b}_p^{\dagger}|0\rangle$ of the electron. The vacuum $|0\rangle$ of the electron field is defined by the relation $\hat{b}_p|0\rangle = 0$, valid for all momenta p. Projecting $\langle n, p|$ on the total state $|\Psi(t)\rangle$ and taking the modulus square, that is, $|\langle n,p|\Psi(t)\rangle|^2$, gives us the joint probability to find n photons in the laser mode and the momentum p of the electron.

3 Quantum FEL in the Laboratory Frame

In this section, we analyze the FEL dynamics in the laboratory frame governed by the Hamiltonian, Eq. (13), and ultimately derive the necessary requirements to enter the quantum regime. First, we investigate the microscopic processes for the FEL from the point of view of energy–momentum conservation; then, we use first-order perturbation theory to solve the equations of motion for the system. Finally, we identify the Quantum FEL as the limit where single-photon processes dominate, while multi-photon transitions are suppressed.

3.1 Resonance Condition

For our study of the basic processes in the FEL dynamics, it is convenient to consider energy–momentum conservation in the formulation of co- and contravariant four-vectors. As described in the previous section, we use the effective four-momentum

$$p^{\mu} \equiv (p_0, 0, 0, p_z) \equiv \gamma mc(1, 0, 0, \beta) \tag{18}$$

of the electron in the wiggler instead of the free momentum

$$p_{\mathrm{f}}^{\mu} \equiv \gamma_0 m_0 c(1,0,0,\beta_0), \tag{19}$$

where we have chosen the direction of the propagation parallel to the z-axis. Moreover, we have defined the relativistic factors

$$\gamma \equiv \frac{1}{\sqrt{1-\beta^2}} \tag{20}$$

and

$$\gamma_0 \equiv \frac{1}{\sqrt{1-\beta_0^2}} \tag{21}$$

together with the scaled velocities β and β_0 such that the four-momentum fulfills the conditions $p \cdot p = m^2c^2$ and $p_{\mathrm{f}} \cdot p_{\mathrm{f}} = m_0^2c^2$.

The reason for considering the effective momentum p is its appearance in the Volkov solution, Eq. (6), and in the phase, Eq. (15), of the Hamiltonian, Eq (13). Classically it can be understood by identifying p^{μ} as an effective longitudinal momentum which contains the effects of the transverse motion of the electron in the wiggler due to the Lorentz force [33].

The microscopic processes of the FEL can be described in terms of Compton scattering, where l wiggler photons with the four-wave vector

$$k_{\mathrm{W}}^{\mu} = k_{\mathrm{W}}\,(1,0,0,\,-1) \tag{22}$$

and l laser photons with the four-wave vector

$$k_{\mathrm{L}}^{\mu} = k_{\mathrm{L}}\,(1,0,0,1) \tag{23}$$

interact with an electron, which possesses the initial momentum p^{μ}.

Due to energy–momentum conservation, the final momentum p'^{μ} of the electron reads [34]

$$p'^{\mu} = p^{\mu} \mp l\hbar\,\left(k_{\mathrm{L}}^{\mu} - k_{\mathrm{W}}^{\mu}\right), \tag{24}$$

where the minus sign occurs in the case of emission of l laser photons and absorption of l wiggler photons ('emission'), while the plus sign describes the opposite process ('absorption').

We use the relation, Eq. (8), in order to eliminate the final momentum p'^{μ} in Eq. (24). After straightforward algebra, we observe, for fixed values of the wave numbers k_{L} and k_{W}, the resonance condition

$$\beta = \frac{k_{\mathrm{L}} - k_{\mathrm{W}}}{k_{\mathrm{L}} + k_{\mathrm{W}}} \pm l\frac{2k_{\mathrm{W}}}{k_{\mathrm{L}} + k_{\mathrm{W}}}\frac{\hbar k_{\mathrm{L}}}{\gamma mc}. \tag{25}$$

For the time being, we neglect the recoil term in Eq. (25), since the photon momentum $\hbar k_{\mathrm{L}}$ is much smaller than the electron momentum γmc, as well as $k_{\mathrm{W}} \ll k_{\mathrm{L}}$. Indeed, we would need an unrealistic high number l of scattered photons to obtain noticeable corrections to the leading term.

Hence, we obtain for *classical* resonance

$$\beta \cong \frac{k_{\mathrm{L}} - k_{\mathrm{W}}}{k_{\mathrm{L}} + k_{\mathrm{W}}} \equiv \beta_{\mathrm{BR}} \tag{26}$$

where only the wave numbers of the fields occur and no term with $\hbar$ is present. As shown in Appendix A of Ref. [12], Eq. (26) gives the velocity β_{BR} of the Bambini–Renieri frame relative to the laboratory frame.

Alternatively we can derive the resonance condition for k_{L} with fixed k_{W} and γ. An analogous calculation [26] yields the expression[3]

$$k_{\mathrm{L}} \cong \frac{4\gamma_0^2 k_{\mathrm{W}}}{\left(1 + a_0^2\right)}, \tag{27}$$

where we have defined [35] the wiggler parameter $a_0 \equiv \mathrm{e}\sqrt{2}\tilde{A}_{\mathrm{W}}/(m_0 c)$. Moreover, we have made the approximations $1 + \beta \cong 2$ and $\gamma m \cong \gamma_0 m_0$, which can be easily derived from $\beta \lesssim 1$ and $p_{\mathrm{f}} \cdot k_{\mathrm{W}} = p \cdot k_{\mathrm{W}}$.

We emphasize that neglecting the recoil contributions in Eq. (25) is a reasonable procedure to estimate the resonant velocity for the electron relative to the laboratory frame. However, the recoil is the reason for gain in the FEL regardless of whether we consider the classical regime, where many photons are scattered, but the discrete nature of the jumps is washed out, or we are in the quantum regime where only few photons are scattered. Hence, we have to consider the relative velocity of the electron with respect to the resonant one, Eq. (26), in a complete theory of the FEL.

In other words, the recoil term in Eq. (25) gives a *nonrelativistic* correction to Eq. (26) which may or may not be neglected, depending on the observable of interest. The recoil corrections in Eq. (25) are much smaller than unity while the scaled velocity β_{BR} of the Bambini–Renieri frame is close to unity. For this reason we can identify the recoil terms as nonrelativistic which justifies the approach in the co-moving frame of Ref. [12].

We now further rewrite the expression in Eq. (25). With the help of the identity

$$\gamma_{\mathrm{BR}} = \frac{k_{\mathrm{L}} + k_{\mathrm{W}}}{2\sqrt{k_{\mathrm{L}} k_{\mathrm{W}}}}, \tag{28}$$

derived in Appendix A of Ref. [12], we obtain the form

[3]We note that the result for the resonant wave number k_{L} differs by a factor of 2 from the one for a magnetostatic wiggler [26].

$$\frac{2k_\mathrm{W}}{k_\mathrm{L}+k_\mathrm{W}}\frac{\hbar k_\mathrm{L}}{\gamma mc}=\frac{q}{2\gamma_\mathrm{BR}^2\gamma mc} \tag{29}$$

for the recoil term in Eq. (25). Here we have defined the quantum mechanical recoil

$$q\equiv\hbar(k_\mathrm{L}+k_\mathrm{W}) \tag{30}$$

which is the change of the electron momentum in z-direction due to a single scattering process according to Eq. (24).

Finally we cast Eq. (25) into the form

$$\beta=\beta_\mathrm{BR}\pm l\frac{\mathcal{Q}}{2} \tag{31}$$

where we have recalled the definition of β_BR, Eq. (26), and have introduced the dimensionless recoil parameter

$$\mathcal{Q}\equiv\frac{\hbar(k_\mathrm{L}+k_\mathrm{W})}{\gamma^3 mc}. \tag{32}$$

In the transition from Eq. (29) to (32), we have additionally approximated $\gamma_\mathrm{BR}\cong\gamma$ since all processes take place close to the classical resonance.

3.2 Dynamics

The dynamics of the state $|\Psi(t)\rangle$ describing the motion of the electron and the laser field during an interaction time t is given by

$$|\Psi(t)\rangle=\hat{\mathcal{U}}(t)|\Psi(0)\rangle \tag{33}$$

with the time evolution operator [36]

$$\begin{aligned}\hat{\mathcal{U}}(t)\ &\equiv\sum_{j=0}^{\infty}\left(\frac{-i}{\hbar}\right)^j\int_0^t \mathrm{d}t_j\int_0^{t_j}\mathrm{d}t_{j-1}\cdots\int_0^{t_2}\mathrm{d}t_1\\ &\times\hat{H}(t_j)\hat{H}(t_{j-1})\cdots\hat{H}(t_2)\hat{H}(t_1)\end{aligned} \tag{34}$$

originating from the time-dependent Hamiltonian $\hat{H}(t)$ given by Eq. (13).

Following Ref. [19], we extend the spatial integration in the z-direction to infinity in Eq. (13), while considering a finite integration time t. This way, the integral over space leads to a delta function for the momentum.

In analogy to Ref. [12], we use perturbation theory to solve the dynamics, i.e., we subsequently evaluate the individual terms of the sum in Eq. (34). However, before we proceed, we examine the action of the Hamiltonian, Eq. (13), on a state.

As mentioned above, the Hamiltonian density can be split into an emission $\hat{\mathcal{H}}_{\text{em}}$ and absorption term $\hat{\mathcal{H}}_{\text{abs}}$. The combination of the creation and annihilation operators of the electron and of the laser field ensures that in the time evolution Eq. (33), for the initial state $|n, p_z\rangle$ consisting of a photon number state $|n\rangle$ and a momentum eigenstate $|p_z\rangle$, only a state

$$|\Psi(t)\rangle = \sum_{l=-n}^{\infty} c_l(t)|n+l,\ p_z - lq\rangle \tag{35}$$

can be generated, where l photons are emitted into, or absorbed from the laser field while the electron looses or gains the momentum lq. The quantum mechanical recoil $q \equiv \hbar(k_{\text{L}} + k_{\text{W}})$ is defined according to Eq. (30).

This effect corresponds to momentum conservation in Eq. (24) but not to energy conservation. Due to the finite integration time, we will see that the zeroth component of the four-momentum, i.e., the energy, is not exactly conserved leading to combinations of sinc functions in the probabilities.

The expansion coefficients c_l in Eq. (35) represent probability amplitudes for a process involving l photons and the electron recoiling l times. The basis states $|n \pm l,\ p_z \mp lq\rangle$ used in this expansion are also known [37] as 'scattering basis.'

In contrast to Eq. (35), we allow in the following a distribution of the initial momenta with a nonzero width. In accordance with Ref. [12], we assume the initial condition

$$c_l(p_z; 0) = \delta_{l,0}\psi(p_z) \tag{36}$$

which describes the situation where the initial state of the system is characterized by a photon number state with n photons in the laser field and by the wave function $\psi = \psi(p_z)$ of the electron depending on the momentum p_z in z-direction.

3.3 Single-Photon Processes

Finally, we are in the position to calculate the probability densities $|c_1|^2$ and $|c_{-1}|^2$ of single-photon emission and absorption, respectively, in lowest order of perturbation theory. The resulting expressions for $|c_{\pm 1}|^2$ involve the selectivity functions $\mathcal{S}_\pm$ which contain the resonance conditions for absorption and emission discussed in Sect. 3.1. Moreover, they are also sensitive to the initial momentum distribution of the electron. Indeed, the displacement of $\mathcal{S}_+$ relative to $\mathcal{S}_-$, and its size compared to the width of the momentum distribution leads to the conditions for the Quantum FEL.

3.3.1 Selectivity Functions

We start our analysis by computing the first terms of Eq. (34), which yields

$$\left[1 - \frac{i}{\hbar}\int_0^t \mathrm{d}t_1 \int \mathrm{d}^3 r\left(\hat{\mathcal{H}}_{\text{em}}(t_1, \boldsymbol{r}) + \hat{\mathcal{H}}_{\text{abs}}(t_1, \boldsymbol{r})\right)\right]|n, p_z\rangle$$
$$= |n, p_z\rangle + c_1|n+1, p_z - q\rangle + c_{-1}|n-1, p_z + q\rangle, \tag{37}$$

when applied to the state $|n, p_z\rangle$. As expected, this state has the same form as the expansion Eq. (35), with only three nonvanishing contributions, that is, for $l = 0$ as well as for $l = 1$ and $l = -1$.

After straightforward algebra and using the initial condition Eq. (36), we arrive at the probability densities

$$|c_1|^2 = \left(\frac{gt}{\gamma}\right)^2 (n+1)\, \mathcal{S}_+^2 |\psi(p_z)|^2 \tag{38}$$

for single-photon emission, and

$$|c_{-1}|^2 = \left(\frac{gt}{\gamma}\right)^2 n\, \mathcal{S}_-^2\, |\psi(p_z)|^2 \tag{39}$$

for single-photon absorption. Here we have weighted every momentum eigenstate with the corresponding initial momentum distribution $|\psi(p_z)|^2$.

Moreover, we have neglected the small energy change in the pre-factor yielding $\sqrt{\gamma'\gamma} \cong \gamma$ and have introduced the selectivity functions

$$\mathcal{S}_\pm \equiv \operatorname{sinc}\left[\frac{ct}{2\hbar}\left(p_0(p_z \mp q) - p_0 \pm \hbar(k_{\mathrm{L}} - k_{\mathrm{W}})\right)\right] \tag{40}$$

which characterize the position and the width of the resonant energies.

To simplify the argument of $\mathcal{S}_\pm$, Eq. (40), we expand [19] the final energy

$$p_0(p_z \mp q) = \sqrt{(p_z \mp q)^2 + m^2c^2} \tag{41}$$

to second order in the recoil, that is,

$$p_0(p_z \mp q) \cong p_0 \mp \frac{p_z q}{p_0} + \frac{q^2}{2p_0}\left(1 - \frac{p_z^2}{p_0^2}\right), \tag{42}$$

while keeping in mind that q is always a small quantity in comparison with the electron momentum, characterized by $\gamma \gg 1$. Moreover, we have used the fact that p'_μ fulfills the relation $p' \cdot p' = m^2c^2$ due to Eq. (8).

We finally obtain

$$\mathcal{S}_{\pm} = \operatorname{sinc}\left[\frac{1}{2}(k_{\mathrm{L}} + k_{\mathrm{W}})ct\left(\beta - \beta_{\mathrm{BR}} \mp \frac{\mathcal{Q}}{2}\right)\right] \tag{43}$$

where we have recalled the definitions, Eqs. (26) and (32), of the scaled velocity β_{BR} of the Bambini–Renieri frame and of the recoil parameter $\mathcal{Q}$, respectively.

Hence, the resonances are centered around β_{BR} and shifted to the left for absorption or to the right for emission, by the recoil term $\mathcal{Q}/2$. This structure is already predicted by energy–momentum conservation[4] in Eq. (31).

3.3.2 Conditions

We now use the results of first-order perturbation theory to derive the requirements for realizing the Quantum FEL. Here we follow closely the arguments of Ref. [12].

Quantum parameter Therefore, we introduce the recoil frequency

$$\omega_{\mathrm{r}} \equiv \frac{c}{2}(k_{\mathrm{L}} + k_{\mathrm{W}})\frac{\mathcal{Q}}{2} = \frac{\hbar(k_{\mathrm{L}} + k_{\mathrm{W}})^2}{4\gamma^3 m} \tag{44}$$

in the laboratory frame. The connection to the corresponding quantity, introduced in Eq. (6) of Ref. [12] in the Bambini–Renieri frame, is achieved by considering the product $\omega_{\mathrm{r}}t$ and not by just transforming ω_{r}. The latter transformation yields a factor of γ^{-2}. On the other hand, the interaction time t' in the Bambini–Renieri frame has to be translated via $t' \to t/\gamma$ to t in the laboratory frame, due to time dilatation. These two steps provide us with the dependency on γ^{-3} of $\omega_{\mathrm{r}}t$ in the laboratory frame as predicted in Eq. (44).

As depicted in Fig. 1, the parameter $\omega_{\mathrm{r}}t$ is a measure for the width and the separation of the resonances. In the classical case, that is, for $\omega_{\mathrm{r}}t \ll 1$, the selectivity functions overlap. For large recoil, that is, for $\omega_{\mathrm{r}}t \gg 1$, these functions are well separated and possess a small width. This limit with two distinguishable and sharp resonances, defines the quantum regime of the FEL.

For the asymptotic expansion of the time evolution, Eq. (34), to converge, we require that the expansion parameter $gt\sqrt{n}/\gamma$, which emerges in Eqs. (38) and (39) is small, i.e., $gt\sqrt{n}/\gamma \ll 1$. We emphasize that the connection to the parameter $gt'\sqrt{n}$ of the nonrelativistic theory of Ref. [12] is again established by time dilatation $t' \to t/\gamma$.

The ratio α of the expansion parameter $gt\sqrt{n}/\gamma$ to $\omega_{\mathrm{r}}t$ can be written as

[4] In contrast to the derivation of $\mathcal{Q}$ in Eq. (31), we did not have to make the approximation $\gamma_{\mathrm{BR}} \approx \gamma$ when we derived Eq. (43). This difference stems from the fact that we have performed a Taylor expansion of the relativistic square root in Eq. (41) to linearize the argument of the selectivity functions.

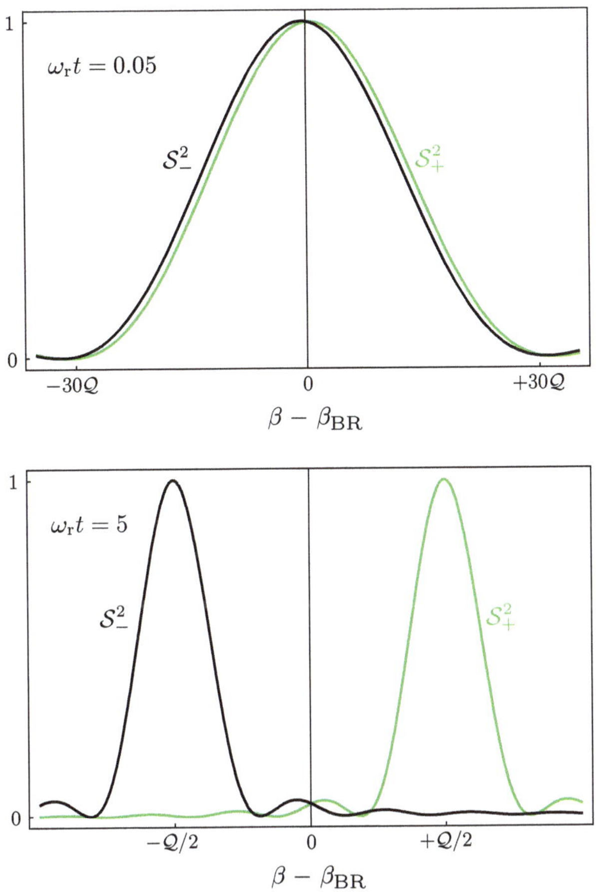

Fig. 1 Transition from the classical FEL (*top*) to the Quantum FEL (*bottom*) illustrated by the behavior of the selectivity functions $\mathcal{S}_\pm^2$, Eq. (43), for single-photon emission $\mathcal{S}_+^2$ (*green curve*) and absorption $\mathcal{S}_-^2$ (*black curve*) as functions of the scaled velocity β of the electron in the laboratory frame. While $\mathcal{S}_+^2$ and $\mathcal{S}_-^2$ overlap in the classical regime shown here for $\omega_r t = 0.05$, we obtain well-separated resonances for the Quantum FEL exemplified by $\omega_r t = 5$

$$\alpha \equiv \gamma^2 \frac{g\sqrt{n}}{q^2/(2m\hbar)} \ll 1 \tag{45}$$

where the inequality emerges for the quantum case $\omega_r t \gg 1$. That is why we call α the 'quantum parameter.'

This definition is identical to the one in Eq. (1), when we recall the expression $q \equiv \hbar(k_L + k_W)$ for the recoil. In Ref. [12], this expression was derived by rewriting the result in the Bambini–Renieri frame in terms of the laboratory frame.

We emphasize that the dependency of the quantum parameter in Eq. (45) on γ^2 does not mean that a low energy leads to a small α. In Ref. [12], we have shown that α in fact is suppressed for increasing energies when we express the recoil $q \equiv \hbar(k_L + k_W)$ solely in terms of γ and k_W and not in terms of the wave number k_L of the laser. Indeed, by inspecting Eqs. (27) and (28) and setting $\gamma_{BR} \approx \gamma$ we obtain that α is proportional[5] to γ^{-2}.

Energy spread The second requirement for a Quantum FEL derived in Ref. [12] concerns the relative energy spread $\Delta\gamma/\gamma$ of the electron beam. To obtain this condition in our present approach, we have to consider the initial momentum distribution $|\psi(p_z)|^2$ in the probability densities Eqs. (38) and (39).

A narrow momentum distribution of width $\Delta\beta$ localized at a resonance $\beta = \beta_{BR} \pm Q/2$ prefers a single process, emission for $+Q/2$ or absorption for $-Q/2$. On the other hand, for a broad distribution that covers both resonances, both emission and absorption occur nearly with the same probability, and the net gain averages out.

The separation of the resonant velocities is given by the recoil parameter Q and we demand $\Delta\beta < Q$ in order to obtain gain in the Quantum FEL. This condition can be cast into a more convenient form by approximating $\Delta\beta \approx \Delta\gamma/\gamma^3$ valid for relativistic energies, $\gamma \gg 1$. Thus, we arrive at the requirement

$$\frac{\Delta\gamma}{\gamma} < \frac{\hbar(k_L + k_W)}{\gamma mc}, \tag{46}$$

where we have recalled the definition, Eq. (32), of the recoil parameter.

The right-hand side of Eq. (46) represents the ratio of the quantum mechanical recoil to the momentum of the electron in the laboratory frame. This ratio is of course very small which makes the condition on the energy spread of the electrons the most difficult hurdle to realize the Quantum FEL in an experiment.

We can also express the relation Eq. (46) in terms of the energy scaled γ_0 of the free electron, that is,

$$\frac{\Delta\gamma_0}{\gamma_0} < \frac{4\gamma}{1 + a_0^2}\frac{\lambda_C}{\lambda_W} \tag{47}$$

with the Compton wavelength $\lambda_C \equiv h/(m_0 c)$ of the electron. In the derivation of Eq. (47), we have used Eqs. (8) and (28), set $\gamma_{BR} \cong \gamma$ and have employed the resonance condition, Eq. (27).

When we compare Eq. (47) with Eq. (2), derived in Ref. [12], we recognize that the formulas are similar with the difference that the effect of the effective momentum, visible in the factor $1 + a_0^2$, was neglected in Ref. [12]. However, for reasonable parameters, $a_0 \sim \mathcal{O}(1)$, both results differ only by a small amount.

---

[5]The proportionality of α to γ^{-3} in Eq. (37) of Ref. [12] emerges when we take into account the dependencies of the coupling constant g. We omit this calculation here and refer the reader to Ref. [12].

3.4 Higher-Order Processes

So far, we have only considered single-photon processes which is in contrast to the classical FEL where many photons are relevant. This fact implies that many momentum levels are involved in the interaction.

Therefore, to prove that processes which include more than the two resonant momenta, $\mathcal{Q}/2$ and $-\mathcal{Q}/2$, are suppressed in the quantum regime, we have to take into account higher orders of the perturbative solution of Eq. (34). This calculation, which is analogous to the one presented in Ref. [12], is explicitly performed in Appendix C.

Indeed, we find that multi-photon processes are suppressed with powers of $\alpha \ll 1$, which is equivalent to a negligible population in higher momentum states. Thus, we really can identify the Quantum FEL as a two-level system for the momentum levels of the electron.

4 Conclusion and Outlook

In conclusion, we have rederived the conditions for the Quantum FEL, Eqs. (45) and (47). Originally these constraints were introduced in Ref. [12]. We have now verified that both approaches, the relativistic one in the laboratory frame and the nonrelativistic in the Bambini–Renieri frame, yield equivalent results.

A relativistic effect which is not contained in our current model is the so-called slippage [38]. When no cavity is present, the laser field which propagates with the speed of light is slightly faster than the electrons, $\beta \lesssim 1$, and can slip over the electron bunch during the interaction. The inclusion of slippage and the development of a many-electron model, which is essential for the high-gain regime, represent necessary ingredients of a theory for a self-amplified spontaneous emission (SASE) FEL in the quantum regime. Both topics will be considered in future publications.

Acknowledgements We thank W. Becker, M. Bussmann, A. Debus, M. Knobl, K. Steiniger, S. Varró and M.S. Zubairy for many fruitful discussions. E.G. is grateful to the Center for Integrated Quantum Science and Technology (IQST) for a fellowship and the Friedrich-Alexander-Universität Erlangen-Nürnberg for the Eugen Lommel Stipend. W.P.S. is thankful to Texas A&M University for a Texas A&M University Institute for Advanced Study (TIAS) Faculty Fellowship.

Appendix A: Furry Picture for Scalar quantum Electrodynamics

In this appendix, we derive, starting from basic principles, the model introduced in Sect. 2, namely the Furry or bound interaction picture [21] for a scalar field [19]. Here we first discuss the classical Klein–Gordon and Maxwell fields, as well as their interaction. Then we perform second quantization in the Schrödinger picture, before we finally investigate the transformation to the Furry picture. We conclude by presenting the Hamiltonian in the Furry picture.

Lagrangian Formulation

We start from the classical Lagrangian density [39]

$$\mathcal{L} \equiv \mathcal{L}_{\mathrm{K-G}} + \mathcal{L}_{\mathrm{M}} + \mathcal{L}_{\mathrm{Int}} \tag{48}$$

for the Klein–Gordon field $\varphi = \varphi(x)$ interacting with an electromagnetic field described by the four-potential A^μ.

The dynamics of the free Klein–Gordon field is determined by

$$\mathcal{L}_{\mathrm{K-G}} \equiv c\left[\hbar^2 \partial_\mu \varphi^*(x) \partial^\mu \varphi(x) - m_0^2 c^2 \varphi^*(x) \varphi(x)\right] \tag{49}$$

with the electron mass m_0, the speed of light c and the reduced Planck constant $\hbar \equiv h/2\pi$.

The free Maxwell field evolves in time according to the Lagrangian density

$$\mathcal{L}_{\mathrm{M}} \equiv -\frac{1}{4\mu_0} F_{\mu\nu}(x) F^{\mu\nu}(x) \tag{50}$$

where we have introduced the field tensor

$$F^{\mu\nu}(x) \equiv \partial^\mu A^\nu(x) - \partial^\nu A^\mu(x)$$

and the vacuum permeability μ_0.

The interaction Lagrangian

$$\begin{aligned} \mathcal{L}_{\mathrm{Int}} &\equiv -\mathrm{e}cA_\mu(x)\left[\varphi^*(x) i\hbar\partial^\mu \varphi(x) - \varphi(x) i\hbar \partial^\mu \varphi^*(x)\right] \\ &\quad + \mathrm{e}^2 c\, A^2(x) \varphi^*(x) \varphi(x) \end{aligned} \tag{51}$$

is chosen such that the Euler–Lagrange equation with respect to $\varphi^*(x)$, that is,

$$\partial^\mu \frac{\partial \mathcal{L}}{\partial(\partial^\mu \varphi^*(x))} - \frac{\partial \mathcal{L}}{\partial \varphi^*(x)} = 0, \tag{52}$$

leads to the Klein–Gordon equation

$$\left[\left(i\hbar\partial^{\mu} - eA^{\mu}(x)\right)^2 - m_0^2c^2\right]\varphi(x) = 0 \tag{53}$$

with minimal coupling $\hat{p}^{\mu} \rightarrow \hat{p}^{\mu} - eA^{\mu}(x)$ between the operator $\hat{p}^{\mu} \equiv i\hbar\partial^{\mu}$ for the four-momentum and the potential $A^{\mu}(x)$.

When we compute the corresponding Euler–Lagrange equations for A^{μ}, we obtain the Maxwell equations

$$\partial^{\nu}\partial_{\nu}A^{\mu}(x) = \mu_0 e j^{\mu}(x), \tag{54}$$

where A^{μ} is driven by the four current

$$\begin{aligned} j^{\mu}(x) \quad &\equiv c[\varphi^*(x)i\hbar\partial^{\mu}\varphi(x) - \varphi(x)i\hbar\partial^{\mu}\varphi^*(x)] \\ &+ 2eA^{\mu}(x)\varphi^*(x)\varphi(x), \end{aligned} \tag{55}$$

which represents a conserved quantity for Eq. (53), in the sense that $\partial_{\mu}j^{\mu}(x) = 0$.

Hamiltonian Formulation

In order to find a quantized theory for this interaction, we change to the Hamiltonian formalism of field theory. To simplify the calculations, we choose from the beginning the gauge

$$\phi(x) \equiv \frac{1}{c}A^{(0)}(x) = 0 \tag{56}$$

and

$$\nabla \cdot \boldsymbol{A}(x) = 0 \tag{57}$$

since we are only interested in the transverse degrees of freedom. In this way, we avoid that the momentum density

$$\pi(x) \equiv \partial\mathcal{L}/\partial\dot{\varphi}(x) \tag{58}$$

of the Klein–Gordon field is coupled to ϕ. The fact that we recover the same results as if we would have retained ϕ and set it to zero at the end of the calculations [19] justifies this procedure.

With the help of Eq. (49), we obtain

$$\pi(x) = (\hbar^2/c)\dot{\varphi}^*(x) \tag{59}$$

and thus arrive at the Hamiltonian density

$$\mathcal{H} \equiv \mathcal{H}_{\text{K-G}} + \mathcal{H}_{\text{M}} + \mathcal{H}_{\text{Int}} \tag{60}$$

with

$$\begin{aligned} \mathcal{H}_{\text{K-G}} \quad &\equiv c \left[\frac{1}{\hbar^2} \pi^*(x)\pi(x) + \hbar^2 \nabla\varphi^*(x) \cdot \nabla\varphi(x) \right. \\ &\left. + m_0^2 c^2 \varphi^*(x)\varphi(x) \right], \end{aligned} \tag{61}$$

for the free Klein–Gordon field,

$$\begin{aligned} \mathcal{H}_{\text{Int}} \quad &\equiv ecA^\mu(x) \left[\varphi^*(x) i\hbar\partial_\mu\varphi(x) - \left(i\hbar\partial_\mu\varphi^*(x) \right)\varphi(x) \right] \\ &- e^2 c A^2(x)\varphi^*(x)\varphi(x) \end{aligned} \tag{62}$$

for the interaction and $\mathcal{H}_{\text{M}}$ as the Hamiltonian density [39] of the free Maxwell field.

Quantization in Schrödinger Picture

Next we perform second quantization in the Schrödinger picture where the Klein–Gordon field operator $\hat{\varphi}_{\text{S}}(\boldsymbol{r})$ and its conjugate momentum $\hat{\pi}_{\text{S}}(\boldsymbol{r})$ fulfill the commutation relations

$$\left[\hat{\pi}_{\text{S}}(\boldsymbol{r}), \hat{\varphi}_{\text{S}}(\boldsymbol{r}') \right] = \left[\hat{\pi}_{\text{S}}^\dagger(\boldsymbol{r}), \hat{\varphi}_{\text{S}}^\dagger(\boldsymbol{r}') \right] = -i\hbar\delta(\boldsymbol{r} - \boldsymbol{r}') \tag{63}$$

while all other possible combinations of commutators vanish.

We emphasize that the operators $\hat{\varphi}_{\text{S}}(\boldsymbol{r})$ and $\hat{\pi}_{\text{S}}(\boldsymbol{r})$ do not depend on time in contrast to their classical counterparts $\varphi(x) \equiv \varphi(ct, \boldsymbol{r})$ and $\pi(x) \equiv \pi(ct, \boldsymbol{r})$. In the Schrödinger picture, all dynamics is contained in the state vector $|\Psi_{\text{S}}(t)\rangle$ for the electron and the electromagnetic field and follows from the Schrödinger equation

$$i\hbar \frac{\mathrm{d}}{\mathrm{d}t} |\Psi_{\text{S}}(t)\rangle = \hat{H}_{\text{S}} |\Psi_{\text{S}}(t)\rangle. \tag{64}$$

Here we have replaced all fields and momenta by their operators in the Hamiltonian

$$\hat{H}_{\text{S}} \equiv \int \mathrm{d}^3 r \, \hat{\mathcal{H}}_{\text{S}}(\boldsymbol{r}). \tag{65}$$

The Furry picture is most convenient when the interaction with the electromagnetic field consists of two parts: one due to an external field which is treated as classical and fixed, which is its depletion is neglected, and one with a quantized field.

In our case, we identify these two components as the classical wiggler field A_{W}^μ and the quantized laser field $\hat{A}_{\text{L}}^\mu$, that is,

$$\hat{A}^{\mu}(x) = \hat{A}_{\mathrm{L}}^{\mu}(\boldsymbol{r}) + A_{\mathrm{W}}^{\mu}(x). \tag{66}$$

Here we have fixed the dynamics of the wiggler field in the time-dependent potential $A_{\mathrm{W}}(x) \equiv A_{\mathrm{W}}(ct, \boldsymbol{r})$, while the operator $\hat{A}_{\mathrm{L}}(\boldsymbol{r})$ is independent of time.

We can make a similar distinction of Eq. (66) for the Hamiltonian

$$\hat{H}_{\mathrm{S}}(t) \equiv \hat{H}_{\mathrm{W}}(t) + \hat{H}_{\mathrm{L}} \equiv \int \mathrm{d}^3 r \, \left[\hat{\mathcal{H}}_{\mathrm{W}}(t) + \hat{\mathcal{H}}_{\mathrm{L}}\right] \tag{67}$$

with

$$\begin{aligned}\hat{\mathcal{H}}_{\mathrm{W}}(t) \quad &\equiv c\left[\frac{1}{\hbar^2}\hat{\pi}_{\mathrm{S}}^{\dagger}(\boldsymbol{r})\hat{\pi}_{\mathrm{S}}(\boldsymbol{r}) + \hbar^2\nabla\hat{\varphi}_{\mathrm{S}}^{\dagger}(\boldsymbol{r})\cdot\nabla\hat{\varphi}_{\mathrm{S}}(\boldsymbol{r})\right.\\ &+ \left(m_0^2c^2 - \mathrm{e}^2A_{\mathrm{W}}^2(x)\right)\hat{\varphi}_{\mathrm{S}}^{\dagger}(\boldsymbol{r})\hat{\varphi}_{\mathrm{S}}(\boldsymbol{r}) + \hat{\mathcal{H}}_{\mathrm{M}}(\boldsymbol{r})\\ &+ \left.\mathrm{e}A_{\mathrm{W}}^{\mu}(x)\Big(\hat{\varphi}_{\mathrm{S}}^{\dagger}(\boldsymbol{r})i\hbar\partial_{\mu}\hat{\varphi}_{\mathrm{S}}(\boldsymbol{r}) - \big(i\hbar\partial_{\mu}\hat{\varphi}_{\mathrm{S}}^{\dagger}(\boldsymbol{r})\big)\hat{\varphi}_{\mathrm{S}}(\boldsymbol{r})\Big)\right]\end{aligned} \tag{68}$$

and

$$\begin{aligned}\hat{\mathcal{H}}_{\mathrm{L}} = &\mathrm{e}c\hat{A}_{\mathrm{L}}^{\mu}(\boldsymbol{r})\left[\hat{\varphi}_{\mathrm{S}}^{\dagger}(\boldsymbol{r})i\hbar\partial_{\mu}\hat{\varphi}_{\mathrm{S}}(\boldsymbol{r}) - \big(i\hbar\partial_{\mu}\hat{\varphi}_{\mathrm{S}}^{\dagger}(\boldsymbol{r})\big)\hat{\varphi}_{\mathrm{S}}(\boldsymbol{r})\right]\\ &- c\left[\mathrm{e}^2\hat{A}_{\mathrm{L}}^2(\boldsymbol{r}) + 2e^2A_{\mathrm{W}}(x)\cdot\hat{A}_{\mathrm{L}}(\boldsymbol{x})\right]\hat{\varphi}_{\mathrm{S}}^{\dagger}(\boldsymbol{r})\hat{\varphi}_{\mathrm{S}}(\boldsymbol{r}).\end{aligned} \tag{69}$$

From these expressions, we observe two important aspects: (1) Although we use the Schrödinger picture, the Hamiltonian $\hat{H}_{\mathrm{W}}$ is explicitly time dependent. This feature originates from the fact that we have used for the wiggler a time-dependent classical field instead of a field operator in the Schrödinger picture. By prescribing the dynamics of the wiggler, we have already incorporated the free motion of this field mode.

(2) We realize that $\hat{H}_{\mathrm{L}}$ still contains a term linear in A_{W}^{μ}. This linear contribution arises as the cross-term to the square A^2 in Eq. (62) of the sum in Eq. (66). Later on, we shall recognize that this term is crucial for the FEL.

Transformation to the Furry Picture

Similar to the transition into the familiar interaction picture, we obtain the Furry picture by transforming the original state $|\Psi_{\mathrm{S}}(t)\rangle$ by a unitary operator $V(t)$ such that

$$|\Psi_{\mathrm{F}}(t)\rangle = V^{-1}(t)|\Psi_{\mathrm{S}}(t)\rangle. \tag{70}$$

With the help of the Schrödinger equation, Eq. (64), the time derivative of the new state $|\Psi_{\mathrm{F}}(t)\rangle$ reads

$$i\hbar\frac{\mathrm{d}}{\mathrm{d}t}|\Psi_{\mathrm{F}}(t)\rangle = V^{-1}(t)\big[-i\hbar\dot{V}(t)V^{-1}(t) + \big(\hat{H}_{\mathrm{W}}(t)+\hat{H}_{\mathrm{L}}\big)\big]V(t)|\Psi_{\mathrm{F}}(t)\rangle. \tag{71}$$

If we demand the Schrödinger-like equation

$$i\hbar\dot{V}(t) = \hat{H}_{\mathrm{W}}(t)V(t) \tag{72}$$

for $V(t)$, we arrive at the time evolution

$$i\hbar\frac{\mathrm{d}}{\mathrm{d}t}|\Psi_{\mathrm{F}}(t)\rangle = \hat{H}_{\mathrm{F}}(t)|\Psi_{\mathrm{F}}(t)\rangle \tag{73}$$

with

$$\hat{H}_{\mathrm{F}}(t) \equiv V^{-1}(t)\hat{H}_{\mathrm{L}}V(t) \tag{74}$$

for the state $|\Psi_{\mathrm{F}}\rangle$ which is focused on the interaction with the laser field. Indeed, the Hamiltonian $\hat{H}_{\mathrm{F}}$ in this picture is a function of the transformed field operators

$$\hat{\varphi}_{\mathrm{F}}(x) \equiv V^{-1}(t)\hat{\varphi}_{\mathrm{S}}(\boldsymbol{r})V(t), \tag{75}$$

and

$$\hat{\pi}_{\mathrm{F}}(x) \equiv V^{-1}(t)\hat{\pi}_{\mathrm{S}}(\boldsymbol{r})V(t) \tag{76}$$

with

$$\hat{A}^{\mu}_{\mathrm{L}}(x) \equiv V^{-1}(t)\hat{A}^{\mu}_{\mathrm{L}}(\boldsymbol{r})V(t). \tag{77}$$

When we examine Eq. (72), we notice the difficulties to specify the explicit form of the transformation matrix V for the Furry picture, since we are dealing with the time-dependent Hamiltonian $\hat{H}_{\mathrm{W}}$. Indeed, we find the time-ordered exponential [36]

$$V(t) = \mathcal{T}\exp\left[-\frac{i}{\hbar}\int \mathrm{d}t\hat{H}_{\mathrm{W}}(t)\right] \tag{78}$$

for $V(t)$, in contrast to the simple exponential in the ordinary Dirac picture.

Only the operator of the electromagnetic field can be written in the same form as in the familiar interaction picture, because the Hamiltonian of the Maxwell field commutes with the other contributions in $\hat{H}_{\mathrm{W}}$.

Nevertheless, we can formulate the equation of motion

$$\frac{\partial}{\partial t}\hat{\varphi}_{\mathrm{F}}(x) = \frac{i}{\hbar}V^{-1}(t)\left[\hat{H}_{\mathrm{W}}(t), \hat{\varphi}_{\mathrm{S}}(\boldsymbol{r})\right]V(t) \tag{79}$$

for the field operator of the Klein–Gordon field where we have used Eq. (72) and the fact that $\hat{\varphi}_{\mathrm{S}}(\boldsymbol{r})$ is independent of time in the Schrödinger picture.

The evaluation of the commutator in Eq. (79), with the help of Eqs. (63) and (68), yields

$$\frac{\partial}{\partial t}\hat{\varphi}_{\mathrm{F}}(x) = \frac{c}{\hbar^2}\hat{\pi}^{\dagger}_{\mathrm{F}}(x). \tag{80}$$

In a similar way, we compute

$$\begin{aligned} \frac{\partial^2}{\partial t^2}\hat{\varphi}_{\mathrm{F}}(x) &= \frac{c}{\hbar^2}\frac{\partial}{\partial t}\hat{\pi}^{\dagger}_{\mathrm{F}}(x) \\ &= \frac{c}{\hbar^2}\frac{i}{\hbar}V^{-1}(t)\left[\hat{H}_{\mathrm{W}}(t), \hat{\pi}^{\dagger}_{\mathrm{S}}(\boldsymbol{r})\right]V(t), \end{aligned} \tag{81}$$

and observe that the field operator in the Furry picture fulfills the Klein–Gordon equation in the external wiggler field

$$\left[\left(i\hbar\partial^{\mu} - \mathrm{e}A^{\mu}_{\mathrm{W}}(x)\right)^2 - m_0^2c^2\right]\hat{\varphi}_{\mathrm{F}}(x) = 0. \tag{82}$$

Hence, we can expand the field operator in terms of solutions of this equation of motion. This procedure is analogous to the ordinary interaction picture, where the field operator is expanded in solutions of the free Klein–Gordon equation. The task is, of course, to find a solution of this external field problem, Eq. (82), which in case of a plane wave field is given by the Volkov solution [22], as discussed in Appendix B.

Hamiltonian in the Furry Picture

We conclude by noting that the interaction Hamiltonian defined by Eq. (74) can be written as

$$\hat{H}_{\mathrm{F}} = \mathrm{e}\int \mathrm{d}^3r\,\hat{j}_{\mathrm{F}}(x)\cdot\hat{A}_{\mathrm{L}}(x) \tag{83}$$

where the electron current $\hat{j}$ couples to the laser field $\hat{A}_{\mathrm{L}}$. Here we have neglected the term proportional to $\hat{A}^2_L(x)$ and have recalled from Eq. (55) that the current

$$\hat{j}^{\mu}_{\rm F}(x) \equiv c\Big[\hat{\varphi}^{\dagger}_{\rm F}(x) i\hbar\partial^{\mu}\hat{\varphi}_{\rm F}(x) - \big(i\hbar\partial^{\mu}\hat{\varphi}^{\dagger}_{\rm F}(x)\big)\hat{\varphi}_{\rm F}(x)\Big] + 2{\rm e}A^{\mu}_{\rm W}(x)\hat{\varphi}^{\dagger}_{\rm F}(x)\hat{\varphi}_{\rm F}(x) \tag{84}$$

is conserved.

For the sake of clarity, we omit in the main body of this article the subscript F indicating the Furry picture. Since, apart from this appendix, no quantities outside the framework of the Furry picture are considered, this procedure seems justified.

Appendix B: Volkov Solution for Klein–Gordon Equation

The term 'Volkov solution' usually refers to a solution of the Dirac equation coupled to an external electromagnetic field in the form of a plane wave and was derived in 1935 by D. M. Volkov [22]. However, in the present context 'Volkov solution' refers to a solution of the corresponding problem of the Klein–Gordon equation. In this appendix, we rederive [19] the Volkov solution in the case of an optical undulator described by the potential

$$A^{\mu}_{\rm W} = \tilde{A}_{\rm W}\left(\epsilon^{\mu}{\rm e}^{-ik_{\rm W}\cdot x} + {\rm c.c.}\right) \tag{85}$$

given by Eq. (4).

In order to solve the equation

$$\Big[\big(i\hbar\partial^{\mu} - {\rm e}A^{\mu}_{\rm W}(x)\big)^2 - m_0^2c^2\Big]\varphi(x) = 0 \tag{86}$$

we note that $A^{\mu}_{\rm W}$ depends only on the phase $\xi \equiv k_{\rm W}\cdot x$ and assume that the free field solution is only modified by a function $f = f(\xi)$. This assumption gives rise to the ansatz

$$\varphi(x) = \mathcal{N} f(\xi)\, {\rm e}^{-ip_{\rm f}\cdot x/\hbar}. \tag{87}$$

Here $p^{\mu}_{\rm f}$ describes the free four-momentum of the electron and fulfills the relation $p_{\rm f}\cdot p_{\rm f} = m_0^2c^2$, while $\mathcal{N}$ is a normalization constant.

In addition, we demand for a one-dimensional theory that before the electron enters, the wiggler its momentum is parallel to the propagation direction of the field, corresponding to the relation

$$p_{\rm f}\cdot A^{\mu}_{\rm W}(x) = p_{\rm f}\cdot \hat{A}^{\mu}_{\rm L}(x) = 0. \tag{88}$$

Since the potentials are transverse, all scalar products between the four-wave vectors and the four-potentials vanish, and in particular, we have

$$k_{\mathrm{W}} \cdot A_{\mathrm{W}} = 0 \tag{89}$$

for the wiggler.

Under these assumptions, we insert the ansatz Eq. (87) into Eq. (86) and obtain the first-order differential equation

$$f'(\xi) = -\frac{i}{\hbar}\frac{\mathrm{e}^2 \tilde{A}_{\mathrm{W}}^2}{p_{\mathrm{f}} \cdot k_{\mathrm{W}}} f(\xi) \tag{90}$$

for f with $f'(\xi) \equiv \mathrm{d}f(\xi)/\mathrm{d}\xi$. Here we have made use of the identity $A_{\mathrm{W}}^2(x) = -2\tilde{A}_{\mathrm{W}}^2$ valid for circular polarization.

Crucial for the change from a second-order differential equation, Eq. (86), to a first-order one, Eq. (90), is the fact that the only term with a second derivative, that is, the contribution $-\hbar^2 k_{\mathrm{W}} \cdot k_{\mathrm{W}} f''(\xi)$ vanishes due to the dispersion relation $k_{\mathrm{W}} \cdot k_{\mathrm{W}} = 0$ for the wiggler field.

Although the dispersion relation for the four-wave vector is correct only in the case of a laser wiggler the second derivative can still be neglected [19, 20] in the magnetostatic case. Indeed, the corrections due to the second derivative scale with a_0^2/γ_0^2, which is negligible for all reasonable values of the field strength of a wiggler.

The solution of the differential equation, Eq. (90), reads

$$f(\xi) = \exp\left[-\frac{i}{\hbar}\frac{\mathrm{e}^2 \tilde{A}_{\mathrm{W}}^2}{p_{\mathrm{f}} \cdot k_{\mathrm{W}}} \xi\right] \tag{91}$$

where we have paid attention to the constraint that f should reduce to $f(\xi) = 1$ for a vanishing wiggler field, that is, for $A_{\mathrm{W}}^{\mu}(x) = 0$.

When we combine all terms we obtain the Volkov solution [19]

$$\varphi_p(x) = \frac{1}{\sqrt{2p_0 V}}\, \mathrm{e}^{-ip\cdot x/\hbar}, \tag{92}$$

where we have defined the effective momentum

$$p^{\mu} = p_{\mathrm{f}}^{\mu} + \frac{\mathrm{e}^2 \tilde{A}_{\mathrm{W}}^2}{p_{\mathrm{f}} \cdot k_{\mathrm{W}}} k_{\mathrm{W}}^{\mu} \tag{93}$$

of the electron in the wiggler field which fulfills the modified energy–momentum relation

$$p \cdot p = p_0^2 - \boldsymbol{p}^2 = m_0^2 c^2 + 2\, \mathrm{e}^2 \tilde{A}^2 \equiv m^2 c^2 \tag{94}$$

with the effective mass m.

The normalization constant $\mathcal{N}$ follows from the condition that the integral of the density, i.e., zeroth component of the four current, Eq. (55), divided by c, over the quantization volume V yields unity, that is,

$$\int_V d^3r \left[\varphi_p^*(x) \frac{i\hbar}{c} \frac{\partial}{\partial t} \varphi_p(x) - \varphi_p(x) \frac{i\hbar}{c} \frac{\partial}{\partial t} \varphi_p^*(x) \right] = 1 \tag{95}$$

In conclusion, we observe that, similar to the free field case, the solution of Eq. (86) is a plane wave where the free momentum p_{f} is replaced by the effective momentum p which takes the effect of the wiggler field into account. Due to the linearity of the problem any other solution of Eq. (86) can be expanded in plane waves of this kind.

Appendix C: Two-Photon Processes

In this appendix, we show that higher-order photon processes are suppressed. For this purpose, we first analyze the influence of two-photon transitions within our perturbative approach and demonstrate that they scale with α^2. We conclude by briefly analyzing the general case.

The second-order terms in the expression Eq. (34) for the time evolution operator lead to the evolved state

$$|\Psi^{(2)}(t)\rangle \equiv \int_0^t dt_2 \int_0^{t_2} dt_1 \hat{H}(t_2)\hat{H}(t_1)|n, p_z\rangle. \tag{96}$$

for the initial state $|n, p_z\rangle$. Apart from the contributions with $\hat{\mathcal{H}}_{\mathrm{em}}(t_2)\hat{\mathcal{H}}_{\mathrm{em}}(t_1)$ representing two-photon emission, and with $\hat{\mathcal{H}}_{\mathrm{abs}}(t_2)\hat{\mathcal{H}}_{\mathrm{abs}}(t_1)$ representing two-photon absorption, there also emerge cross-terms with $\hat{\mathcal{H}}_{\mathrm{abs}}(t_2)\hat{\mathcal{H}}_{\mathrm{em}}(t_1)$ and $\hat{\mathcal{H}}_{\mathrm{em}}(t_2)\hat{\mathcal{H}}_{\mathrm{abs}}(t_1)$ describing processes where the number of photons remains unchanged. We omit the latter ones and discuss only the two-photon transitions.

From a calculation analogous to the single-photon case we obtain the probability density

$$|c_2|^2 = \left(\frac{gt}{\gamma}\right)^4 (n+1)(n+2)\, \mathcal{E}^{(+2)} |\psi(p_z)|^2 \tag{97}$$

for two-photon *emission*, and the probability density

$$|c_{-2}|^2 = \left(\frac{gt}{\gamma}\right)^4 n(n-1)\, \mathcal{E}^{(-2)} |\psi(p_z)|^2 \tag{98}$$

for two-photon *absorption*. Both expressions are weighted with the initial momentum distribution $|\psi(p_z)|^2$.

Moreover, we have defined the selectivity functions

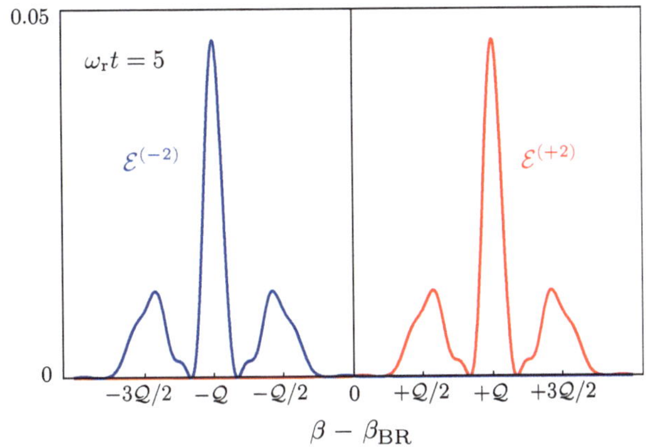

Fig. 2 Selectivity functions $\mathcal{E}^{(+2)}$ (*red curve*) or $\mathcal{E}^{(-2)}$ (*blue curve*) for two-photon emission or absorption, defined by Eq. (99), in the quantum regime for $\omega_r t = 5$, as functions of the scaled velocity β of the electron. Although minor maxima exist, the dominant ones are located at $\beta = \beta_{BR} \pm \mathcal{Q}$ as predicted by energy–momentum conservation

$$\begin{aligned}\mathcal{E}^{(\pm 2)} \equiv \frac{1}{(2\phi_{\pm 1})^2}\big[&\text{sinc}^2(2\phi_{\pm 2}) + \text{sinc}^2\phi_{\pm 3}\\ &-2\cos\phi_{\pm 1}\text{sinc}(2\phi_{\pm 2})\text{sinc}\phi_{\pm 3}\big]\end{aligned} \tag{99}$$

for these processes with the phases

$$\phi_{\pm j} \equiv \frac{(k_L + k_W)ct}{2}\left(\beta - \beta_{BR} \mp j\frac{\mathcal{Q}}{2}\right). \tag{100}$$

As shown in Fig. 2, the dominant maxima of $\mathcal{E}^{(\pm 2)}$ are located at $\beta = \beta_{BR} \pm \mathcal{Q}$ as expected from energy–momentum conservation, Eq. (31). However, there are also less pronounced maxima separated from these major ones by $\mathcal{Q}/2$.

The probability for two-photon emission, Eq. (97), is maximized for $\beta - \beta_{BR} = \mathcal{Q}$ where we can estimate

$$\mathcal{E}^{(+2)} \cong \frac{1}{(2\omega_r t)^2}. \tag{101}$$

Thus, the probability for this process scales as

$$|c_2|^2 \sim \left(\frac{gt}{\gamma}\right)^2 (n+2)\frac{\alpha^2}{4}, \tag{102}$$

where we have used the definitions Eqs. (44) and Eq. (45) of the recoil frequency and the quantum parameter α, respectively.

On the other hand, the maximum probability for *single*-photon emission, Eq. (38), scales with

$$|c_1|^2 \sim \left(\frac{gt}{\gamma}\right)^2 (n+1) \tag{103}$$

and we arrive at the ratio

$$\frac{|c_2|^2}{|c_1|^2} \sim \frac{\alpha^2}{4}, \tag{104}$$

which we have also obtained in Ref. [12] in an analogous way.

This relation demonstrates that in a Quantum FEL defined by $\alpha \ll 1$, two-photon emission is strongly suppressed in comparison with the single-photon transition. Since processes involving more than two photons occur only at even higher orders of perturbation theory, we argue that these processes are suppressed with corresponding powers of α.

References

1. W.B. Colson, Phys. Lett. A **64**, 190 (1977)
2. F.A. Hopf, P. Meystre, M.O. Scully, W.H. Louisell, Opt. Commun. **18**, 413 (1976)
3. Z. Huang, K.-J. Kim, Phys. Rev. Spec. Top. Accel. Beams **10**, 034801 (2007)
4. R. Bonifacio, N. Piovella, G.R.M. Robb, Fortschr. Phys. **57**, 1041 (2009)
5. C.B. Schroeder, C. Pellegrini, P. Chen, Phys. Rev. E **64**, 056502 (2001)
6. R. Bonifacio, N. Piovella, M.M. Cola, L. Volpe, A. Schiavi, G.R.M. Robb, Nucl Instrum Methods Phys. Res. Sect. A Accel. Spectrom. Detect. Assoc. Equip. **593**, 69 (2008)
7. H.K. Avetissian, G.F. Mkrtchian, Phys. Rev. E **65**, 046505 (2002)
8. H.K. Avetissian, G.F. Mkrtchian, Phys. Rev. Spec. Top. Accel. Beams **10**, 030703 (2007)
9. A. Serbeto, L.F. Monteiro, K.H. Tsui, J.T. Mendonça, Plasma Phys. Control. Fusion **51**, 124024 (2009)
10. B. Eliasson, P.K. Shukla, Plasma Phys. Control. Fusion **54**, 124011 (2012)
11. B. Eliasson, P.K. Shukla, Phys. Rev. E **85**, 065401 (2012)
12. P. Kling, E. Giese, R. Endrich, P. Preiss, R. Sauerbrey, W.P. Schleich, New J. Phys. **17**, 123019 (2015)
13. A. Bambini, A. Renieri, Lett. Nuovo Cimento **21**, 399 (1978)
14. A. Bambini, A. Renieri, S. Stenholm, Phys. Rev. A **19**, 2013 (1979)
15. P. Preiss, R. Sauerbrey, M. S. Zubairy, R. Endrich, E. Giese, P. Kling, M. Knobl, W. P. Schleich, in *Proceedings of the 34th Free-Electron Laser Conference, Nara, Japan*, (2012), pp. 93–96
16. R. Endrich, E. Giese, P. Kling, R. Sauerbrey, W. P. Schleich, in *Proceedings of the 36th Free-Electron Laser Conference, Basel, Switzerland*, (2014), pp. 353–357
17. P. Kling, R. Endrich, E. Giese, R. Sauerbrey, W. P. Schleich, in *Proceedings of the 36th Free-Electron Laser Conference, Basel Switzerland*, (2014), pp. 348–352
18. E.T. Jaynes, F.W. Cummings, Proc. IEEE **51**, 89 (1963)
19. W. Becker, J. McIver, Z. Phys. D Atoms Mol. Clust. **7**, 353 (1988)
20. J. Gea-Banacloche, Phys. Rev. A **31**, 1607 (1985)
21. W. H. Furry, Phys. Rev. **81**, 115 (1951)
22. D.M. Wolkow, Z. Phys. **94**, 250 (1935)
23. W. Becker, J.K. McIver, J. Phys. Colloq. **44**, C1 (1983)
24. J.K. McIver, M.V. Fedorov, J. Exp. Theor. Phys. **49**, 1012 (1979)

25. W. Becker, J. K. McIver, Phys. Rev. A **27**, 1030 (1983)
26. M.V. Fedorov, *Atomic and Free Electrons in a Strong Light Field* (World Scientific, Singapore, 1997)
27. R.R. Schlicher, M.O. Scully, H. Walther, Vorschlag für einen kompakten Freie-Elektronen Laser mit einem elektromagnetischen Undulator für den Infrarot- und weichen Röntgenbereich, Tech. rep, Max-Planck Institut für Quantenoptik (1987)
28. J. Gea-Banacloche, G.T. Moore, R.R. Schlicher, M.O. Scully, H. Walther, IEEE J. Quantum Electron. **23**, 1558 (1987)
29. P. Sprangle, B. Hafizi, J. Peñano, Phys. Rev. Spec. Top. Accel. Beams **12**, 050702 (2009)
30. K. Steiniger, M. Bussmann, R. Pausch, T. Cowan, A. Irman, A. Jochmann, R. Sauerbrey, U. Schramm, A. Debus, J. Phys. B Atomic Mol. Opt. Phys. **47**, 234011 (2014)
31. R. Bonifacio, N. Piovella, G.R.M. Robb, Nucl Instrum Methods Phys. Res. Sect. A Accel. Spectrom. Detect. Assoc. Equip. **543**, 645 (2005)
32. W. Becker, H. Mitter, Z. Phys. B Condens. Matter **35**, 399 (1979)
33. S. Varró, *Intensity Effects and Absolute Phase Effects in Nonlinear Laser-Matter Interactions* (InTech, Rijeka, 2010)
34. W. Becker, Opt. Commun. **33**, 69 (1980)
35. P. Schmüser, M. Dohlus, J. Rossbach, *Ultraviolet and Soft X-Ray Free-Electron Lasers* (Springer, Heidelberg, 2008)
36. W.P. Schleich, *Quantum Optics in Phase Space* (Wiley, Weinheim, 2001)
37. F. Ciocci, G. Dattolí, A. Renieri, A. Torre, Phys. Rep. **141**, 1 (1986)
38. R. Bonifacio, L. De Salvo, P. Pierini, N. Piovella, C. Pellegrini, Phys. Rev. Lett. **73**, 70 (1994)
39. J. D. Bjorken, S. D. Drell, *Relativistische Quantenfeldtheorie* (Bibliographisches Institut, Mannheim, 1967)

Playing Tricks to Ions

Dietrich Leibfried

Abstract Ted Hänsch's career is defined by breaking new ground in experimental physics. Curiosity, vivid imagination, deep understanding, patience and tenacity are part of the winning formula, but perhaps an equally important ingredient may be Ted's favorite past-time of exploring new tricks in his "Spiellabor" (play-lab), that often resurfaced as key ingredients in rather serious experiments later. On the occasion of Ted's 75th birthday, a few past and potential future experiments with trapped ions are playfully surveyed here. Some of these tricks are already part of the trade, some are currently emerging and a few are mostly speculation today. Maybe some of the latter will be realized and even prove useful in the future.

1 Unconventional Ion Traps

The first Ph.D. student to graduate from Ted's brand-new Munich/MPQ group, Reinald Kallenbach, joined Dick Brewer and Ralph DeVoe at IBM labs where he helped to build some ion traps with unconventional shapes [1]. When learning of this work, Ted's imagination was sparked and after a few intense days in his play-lab, he had produced a number of unconventional traps from materials such as office clips, fishing hooks and the earrings of his secretary. When driven with high voltage at line frequency, all these shapes produce points or lines of zero electric field. Charged lycopodium spores can be trapped in the pseudo-potential minima around these zeros, and align due to their mutual Coulomb-repulsion. In Ted's lab, these fleeting ion crystals moved around in rhythmical patterns driven by line frequency micro-motion as well as small air currents that Ted produced with his hands in the style of a master conductor. The Lycopodium dances were preserved for posteriority by a whimsical TWH-productions video that honored the father of

This article is part of the topical collection "Enlightening the World with the Laser" - Honoring T. W. Hänsch guest edited by Tilman Esslinger, Nathalie Picqué, and Thomas Udem.

D. Leibfried (✉)
Ion Storage Group, Time and Frequency Division, National Institute of Standards and Technology, Boulder, CO 80305, USA
e-mail: dil@boulder.nist.gov

D. Meschede et al. (eds.), *Exploring the World with the Laser*,
https://doi.org/10.1007/978-3-319-64346-5_8

RF-ion traps, Wolfgang Paul [2]. The visual spectacle was completed with a soundtrack of ballet music [3].

In the years to follow, aspects of Ted's playful explorations made their way into several more serious experimental efforts. In analogy to the cherry-endcaps that mysteriously disappear while the video proceeds to successively simpler Paul traps (Bon appetite!), non-standard "ring-only" and "end-caps only" electrode shapes were suggested as a way to increase optical access to the trapped ions [4] compared to the rather closed electrodes of traditional trap designs that were based on hyperbolical equipotential surfaces [5]. Endcap traps are incorporated in some of the most accurate ion-based frequency standards today [6, 7].

The next trap that appears in the movie is made of a simple paperclip. The geometry resembles slot traps such as those micro-fabricated in the CMOS foundry of Sandia National Laboratories [8] and used in laboratories around the world to perform experiments in quantum information processing, for example within the IARPA MQCO and LogiQ programs. All these geometries feature a point or a line where the electric field driving the trap vanishes. Even if the remainder of the field does not even remotely look like a quadrupole, the pseudo-potential in the the vicinity of such a field-zero is approximately harmonic and will confine charged particles with a charge-to-mass ratio that is compatible with the stability criteria that are familiar from Paul's work [5]. Laser cooling can ensure that the ions stay close enough to the pseudo-potential minimum to experience an essentially harmonic potential.

A geometry that is not featured in the movie has all trap electrodes in one plane and generates field zeros at a distance on the order of the electrode dimensions perpendicular to that plane [9, 10]. The planar electrode structure simplifies microfabrication, which in turn allows for industrial fabrication of trap structures with many zones and junctions, all arranged in a plane [11]. In the meantime, such traps are used by many groups and may become a crucial resource for scaling up quantum information processing with trapped ions. Single ions have been trapped, laser cooled and coherently manipulated at distances as small as 30 μm away from the nearest electrode surface [12]. Such small distances open up new possibilities, for example excitation of the ion motion with microwave near-fields [12] and precisely tailored individual potentials for ions spaced by a few 10 μm [13], where they still experience considerable mutual coupling due to their Coulomb-interaction. It is possible to expand surface trap arrays with separate wells into two dimensions and arrange the field-zeros in the center of such wells in arbitrary, pre-designed lattices [14]. Efforts are now under way to to perform quantum simulations with ions coupled across such designer lattices [15] with the goal of studying complex spin-spin interactions and phonon dynamics.

The geometry of endcap traps can be further simplified to have only one set of concentric electrodes [16]. The resulting "stylus" ion-trap has the ion(s) hovering above the tip of the center electrode, while obscuring only a very small part of the solid angle around them. This can be useful, if such a trap is inserted inside a large solid angle reflector [17] or for sensing applications [18]. More about quantum enabled sensing with single ions in stylus traps will follow in Sect. 3.

2 A Gearbox with Clutch for Quantum Logic Spectroscopy

In the hey-days of the Garching frequency chain, Ted's viewgraphs had lots of gears on them to create analogies of frequency doubling and explain how the fleetingly famous frequency divider stages are supposed to work. In no small part thanks to Ted's vision, the optical frequency comb has made frequency chains obsolete, but gears can still provide useful analogies. Here, gears can be used to illustrate a method that may streamline the operations for quantum logic spectroscopy and quantum logic clocks with trapped ions. The original approach to quantum logic spectroscopy [19] has a lot in common with the Cirac–Zoller gate that started the field of ion-based quantum information processing [20]. The Cirac–Zoller gate uses the ground state and the first excited state of a normal mode of ion motion in the trap as a helper-qubit that communicates between the two ions participating in the gate. Similarly, quantum logic spectroscopy excites the ion of interest on a so-called "blue sideband" transition [21] that simultaneously deposits a quantum in the motion shared with with a "logic ion" forming the other part of a two-ion Coulomb-crystal. If the excitation succeeds, a quantum appears in the motion which was previously cooled to its ground state. This quantum can then be converted into excitation of the internal state of the logic ion, which is subsequently detected by electron shelving [19, 21]. Effectively, by linking the ions through their common motion, one can cool otherwise inaccessible ions and detect their transitions with nearly the same ease as the corresponding operations are implemented on the well explored and much easier to access logic ions. However, any quantum logic done in this way can only be as good as the motional states mediating it.

The main advantage of more recent gate protocols for quantum computing, for example the Mølmer and Sørensen gate [22], is that their fidelity depends less on the purity of the mediating motional state. The new spectroscopy protocol extends this advantage to quantum logic spectroscopy and has some additional useful features. Since it directly entangles the logic ion with one or $N > 1$ spectroscopy ion(s), the resolution of the method ideally achieves the fundamental Heisenberg limit where the signal-to-noise increases proportional to N, as shown in [23] and and for a specific example in more detail below. Moreover, the initial state of the spectroscopy ions can be any superposition of the two levels under study, as explained below. To this end, we can set up a situation where spectroscopy and logic ions are entangled like in a Mølmer and Sørensen gate when the spectroscopy transition is resonant, while the logic ion stays in its initial state if the spectroscopy ions are off-resonant. This can be accomplished, if the transition of interest at frequency ω_s is probed by two coherent light fields at $\omega \pm (\omega_m + \delta)$ where $|\delta|$ is much smaller than the frequency of the motion ω_m. At the same time, the logic ion is driven with detuning δ from its blue and red sideband at $\omega_l \pm (\omega_m + \delta)$, where ω_l is the logic ion transition frequency. If $\omega = \omega_s$, the average frequency of the spectroscopy probe beams is on resonance with the transition in the spectroscopy ion. In this case, all four driving fields implement the interaction of the Mølmer and

Sørensen gate, which effectively couples the spins together like they were linked by a 1:1 gear. As the light fields excite and de-excites the spectroscopy ion in analogy to Rabi-flopping, the state of the logic ion is switched synchronously, thus faithfully indicating whether the state of the spectroscopy ion has flipped. Far off resonance ($\omega \neq \omega_s$), the gear disengages such that the logic ion is not affected by the driving fields and remains in its initial state. An example that shows this basic behavior is a generalized Ramsey-experiment with an odd N such that the total number of ions, $N+1$, is even. We can then use the operator identities stated as Eq. (1) in [23] so that a generalized $\pi/2$-pulse for exact resonance has the form

$$\hat{U}_{\pi/2} = \frac{e^{-i\pi/4}}{\sqrt{2}}\left(1 + (-1)^{\frac{N+1}{2}}\, i\,\hat{\sigma}_l\hat{\sigma}_1, \ldots, \hat{\sigma}_N\right), \tag{1}$$

where the $\hat{\sigma}_l$ operator of the logic ion and the $\hat{\sigma}_j$ operators ($j = \{1, 2, \ldots, N\}$) of the spectroscopy ions can be any of the Pauli operators, possibly rotated in the two-level state spaces (see [23]). Keeping in mind that $\hat{\sigma}_j^2 = 1$, it is easy to compute the effect of two generalized $\pi/2$-pulses, $\hat{U}_{\pi/2}^2 = (-1)^{(N+1)/2}\hat{\sigma}_l\hat{\sigma}_1, \ldots, \hat{\sigma}_N$, so all spins of any initial state flip like they were connected by a 1:1 gear. If all spectroscopy ions are off-resonant, the laser beams act on the logic ion only to yield $\hat{U}_{\pi/2}^2 = (e^{i(\pi/2)\hat{\sigma}_l/2})^2 = e^{-i\pi/8}$ so that any initial state only picks up a global phase that has no consequence for state detection on the logic ion. Far off resonance, the gears disengage and the detuning acts as the clutch. Intermediate detuning leads to mathematical complications that make an awkward birthday present but the behavior can be gleaned from Fig. 1 where the probability of flipping the spin of the logic ion as a function of the detuning $\delta_s = \omega - \omega_s$ is plotted for a few special cases. The lineshape of a single spectroscopy ion for no free precession time is shown in Fig. 1a with the motional mode in its ground state (red) as well as for an imperfectly cooled mode having a thermal energy distribution with one phonon on average ($\bar{n} = 1$, blue). The off-resonant features are more pronounced for imperfect cooling, but the resonance still yields a fairly clean peak while traditional quantum logic spectroscopy would completely fail for $\bar{n} = 1$. A free precession time between the two $U_{\pi/2}$-pulses acts in analogy to a conventional Ramsey experiment for a single spectroscopy ion. However, with N spectroscopy ions, any phase picked up in the free precession time is magnified by a N:1 gear, because the intermediate state is a generalized GHZ state, like the states realizing the Heisenberg limit with N-fold precession frequency investigated in [23]. This can be seen in Fig. 1b, c. The Ramsey fringes resulting from 100 time units of free precession of a single spectroscopy ion are shown in (b). The fringe spacing is approximately 1/100 frequency units. In contrast to a regular Ramsey experiment, the state of the spectroscopy ion is transduced by the 1:1 gearbox to the state of the logic ion which produces a response that is easier to detect and faithfully maps the resonance behavior of the spectroscopy ion. Fig. 1c shows the line-shape for the same precession time, but with $N = 3$ spectroscopy ions. The fringe spacing is three

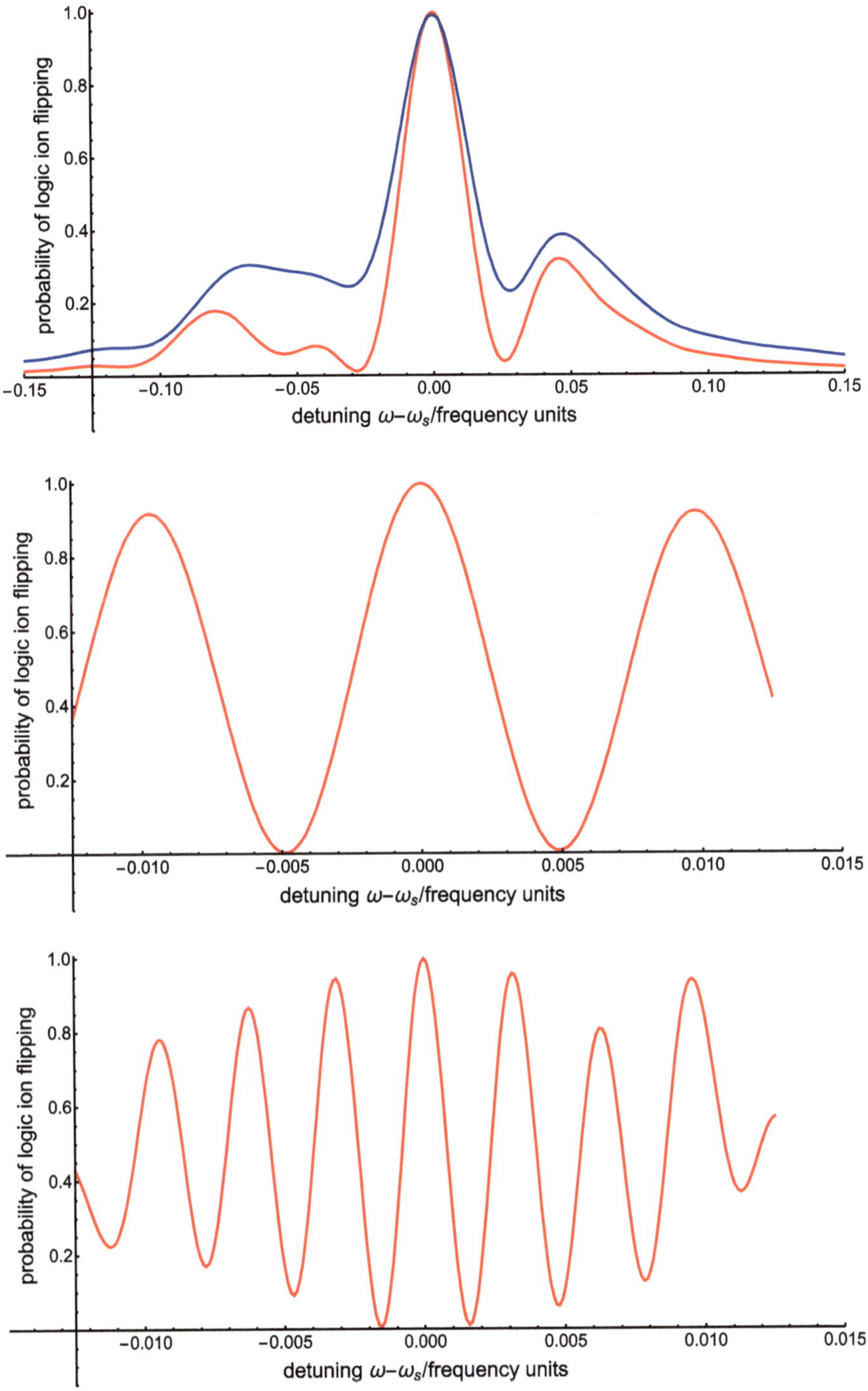

Fig. 1 Quantum logic spectroscopy resonance lines: **a** spin-flip probability of the logic ion as a function of the detuning δ_s from resonance of the spectroscopy ion for no precession time between

times narrower, in accord with a 3:1 gearbox turning the logic ion and Heisenberg-limited frequency resolution.

The on-resonance features and resilience of this method to imperfect ground state cooling have already been verified in [24] with a $^{9}Be^{+}$ ion as the spectroscopy ion and a $^{25}Mg^{+}$ ion as the logic ion. We hope to demonstrate the remaining properties in the near future. Possibly, this method will be useful for quantum logic ion clocks as well as quantum logic spectroscopy on highly charged ions or molecular ions (see Sect. 3 below).

3 Opportunities with Molecular Ions

Precise control and manipulation of single molecules is a long standing goal in chemistry and physics. Manipulation of pure quantum states of single atoms has become routine, but so far the complicated level structures of molecules has made laser cooling and quantum manipulation largely elusive, except in a few fortuitous cases, for example if the molecule can be synthesized in pure quantum states from ultra-cold atoms. Quantum logic spectroscopy as described in [19] and Sect. 2 may be combined with stimulated Raman transitions that can be driven with far-off-resonant lasers and are therefore not specific to a particular species. In this way, quantum control of a wide range of molecules with essentially the same experimental setup may become a reality [25, 26].

The molecular ion of interest is trapped together with an atomic ion that serves as the quantum logic messenger. The ions are strongly coupled by Coulomb-repulsion, therefore the atomic ion can be laser cooled such that the normal modes of both ions are near the ground state. Increases in motional energy that may be caused by a state change of the molecule can then be detected with the logic ion [25, 26]. The simultaneous state change of the molecule and the common motion can be induced by Raman transitions that are driven by continuous wave or mode-locked frequency comb lasers on either a blue sideband of the motion as in conventional quantum logic spectroscopy [19] or with two simultaneous sidebands, as sketched in Sect. 2. If a single blue sideband is used, the excitation of the motion can be subsequently transferred to the internal states of the logic ion and detected by electron shelving. Once a phonon is detected in this way, this constitutes a projection of the molecule into the electronically, vibrationally and rotationally pure final quantum state of the sideband transition that produced it and the detection with the atomic ion keeps the projected state of the molecule intact. This allows for subsequent coherent

Fig. 1 (continued) generalized $\pi/2$ pulses. The *red trace* is for the motion in the ground state, the *blue trace* for imperfect cooling with a thermal occupation with $\bar{n} = 1$. **b** With 100 time units of free precession between the generalized $\pi/2$ pulses. **c** With 100 time units of free precession and three spectroscopy ions manipulation of the molecular state, as well as spectroscopy of molecular transitions with high resolution, in principle only limited by the lifetime of the probed molecular states.

The NIST Ion Storage Group is working on a demonstration experiment with $^{40}Ca^+$ as the logic ion and $^{40}CaH^+$ as the molecule. Currently all manipulations of the molecule are induced by a single continuous wave laser at 1051 nm that is far-detuned from all molecular transitions and split into two Raman beams. The beams can be tuned relative to each other with acousto-optic modulators to address different Raman transitions between substates of different orientation quantum number m_J of the same rotational manifold with quantum number J. To also bridge transitions between states with different rotational quantum number $J' \neq J$, where transition frequencies range from hundreds of GHz to a few THz, we plan to use the frequency comb produced by a femtosecond laser [26]. So far our experiments confirm, that the basic scheme is working [27] and we believe that the same laser and methods are readily applicable to precision spectroscopy and quantum state manipulation of a wide range of molecular species, including molecules of astrophysical interest and species that have been identified as candidates for stringent limits of the electric dipole moment of the electron or for the time dependence of fundamental constants. Other exciting avenues for molecular manipulation with unprecedented precision and versatility could be controlling collisions and ultimately chemical reactions when overlapping the controlled molecular ion with reactive cold gases where the atoms are also prepared in certain quantum states and no extra energy other than the one supplied by laser photons or other targeted excitations are available. A few, possibly less obvious possibilities involving quantum control of molecular ions are listed in the following paragraphs.

It should be interesting to prepare rotational cat-states, superpositions of states with the same rotational quantum number J and extreme opposite components $m_J = \pm J$ along the quantization axis. In effect, these are superpositions where the molecule spins clockwise and counter-clockwise around the same axis at the same time. Besides the usual challenge of growing such Schrödinger kittens to the largest size possible while keeping them coherent, these superpositions might be useful for quantum enhanced sensing applications or provide a precise way of keeping and "reversing" the orientation of a molecule at the same time, for example in a search for the electron electric dipole [28].

Currently, one of the most important systematics in the best optical frequency standards are the shifts of the clock transition due to blackbody radiation. It seems conceivable to introduce a suitable molecule as the third ion in a quantum logic ion clock, which can be used as a temperature sensor a few micrometers away from the clock ion, similar to the proposal in [29], where a different technique involving vibrational transitions is considered. By measuring the occupation probability of the rotational levels one may derive the temperature of the blackbody-environment directly from the rotational quantum statistics, without using any artifacts. Moreover, if the measured statistics differ from a thermal distribution, for example due to non-thermal ambient noise fields, the deviations may reveal hints about the nature of the noise in a spectral range that is currently hard to access.

Taking these ideas even further, it could be fruitful to combine the quantum logic spectroscopy of molecular ions with the stylus ion trap sensor mentioned near the end of Sect. 1. As described in [18], one might be able to then bring atomic or molecular ions within a few tens of microns of a test surface, where they can serve as unique probes of evanescent electromagnetic fields that are crucial to the behavior of surfaces, but hardly propagate into free space. Surfaces and adsorbates are well known to have enhanced non-linearities due to the reduced symmetry at the interface. Established techniques to explore such phenomena include surface enhanced Raman spectroscopy (SERS), surface enhanced frequency doubling (SESHG), frequency summing (SESFG), and surface enhanced multi-photon ionization. In all these techniques photons or electrons are detected far away from the sample. This has precluded taking advantage of signals in the frequency range of MHz–THz that are non-linearly generated on the surface, with the notable exception of laser beat-frequency measurements using photo-multipliers, photo-diodes or other media that can produce measurable photo-currents.

Difference frequencies in the MHz–THz range are very poorly matched to the typical surface dipoles with length scales of order Angstrom to nanometers, orders of magnitude smaller than the free-space wavelength of THz-radiation. This leads to almost complete suppression of fields in this frequency range propagating away from the surface. An atomic or molecular ion confined in a stylus trap within a few 10 μm of the surface is inside the evanescent field of a THz surface wave and is sensitive to frequencies within a wide interval. The tunable, resonant motion of the atomic ions reacts sensitively to electric fields in the 100 kHz–50 MHz region. Rotational transitions of judiciously chosen molecular ions can provide sensitivity at many different frequencies in the range of 10 s of GHz–THz when prepared and interrogated with quantum logic spectroscopy as described above. Surface excitations can be sensed with a spatial resolution of the order of the ion-to-surface distance (currently several 10 μm), or even finer, if either a precisely localized excitation mechanism or sufficiently high signal-to-noise ratio permits that. Mapping of surfaces may, for example, help with understanding and tailoring the electronic structure of nano-catalysts to optimize reactivity. One can also envision measuring the lifetimes (some tens of fs) of surface resonances that participate in surface chemistry [30], using the molecular-ion surface probe to detect the evanescent field excited by a short pump-pulse in a time-resolved manner. Such a pulse can be focused to a few μm. For example, one could use two light beams at a frequency difference that can be detected by the molecule. If the surface permits a non-linear process that is proportional to the product of the two field amplitudes, an electromagnetic field at this difference frequency will be produced. This signal can be background-free as the exciting light beams are in a completely different frequency range. Due to the nonlinear character of the excitation, this may even permit resolution beyond the wavelength of the exciting light fields, an effect that has been very successfully exploited in STED microscopy [31], and related non-linear imaging techniques with propagating light fields, while here the sensor can detect fields that are evanescent on the surface under study. The resolution of such methods could potentially reach nanometers or better.

4 Hydrogen with a Handle

One of Ted's life-long dreams is to trap atomic hydrogen and keep it confined for precise spectroscopy experiments. This is hard because even the coldest hydrogen sources are poorly matched to the shallow depth of typical traps for neutral atoms and laser cooling of hydrogen would be most straightforward on the Lyman-α transition at 121 nm. Building laser-based sources for this wavelength is still a heroic effort. At first glance, it seems like ion trapping can not offer any alternative, since the electronic structure of the hydrogen atomic ion is not particularly interesting. Nevertheless, experiments involving ion traps and quantum logic spectroscopy have been proposed for precise comparisons of protons and antiprotons [32]. Another alternative is the hydrogen molecular ion H_2^+, the simplest of all molecules or its antiparticle-version. Of course I am aware that this idea contradicts the famous statement about molecules by the late Art Schawlow, who was one of Ted's best friends [33]. On the positive side, the extra proton gives H_2^+ a charge, an extra handle that ion traps with a depth of order electronvolts can hold on to. The single electron makes H_2^+ similarly accessible to theory as neutral hydrogen, allowing for precise calculations of the level structure including higher order QED corrections [34]. Precise spectroscopic measurements should be able to open a window on many fascinating questions about the limits of the standard model [35]. Variations of the fine-structure constant, the electron-to-proton mass ratio and the quark mass over time can be probed all in the same system by precisely measuring and comparing electronic, vibrational and hyperfine-transitions.

Historically, spectroscopic experiments on H_2^+ have been few and far between, because of the high level of symmetry and isolation of all ground states from the environment. For example, the lifetime of the first vibrational state $\nu = 1$ is estimated to be 10^6 s (11.6 days), while a common method of producing H_2^+, electron impact ionization, will populate a large number of vibrational states. Some access has been gained through radio-frequency-spectroscopy and state selective dissociation of large ensembles stored in ion traps [36], in beam experiments [37] and through Rydberg states of neutral H_2 [38]. Alternatively, experiments have been done in HD^+ [39] with much shorter lifetimes of the rotational-vibrational states in the ground state manifold due to a finite dipole moment and the broken ortho-para symmetry. It seems conceivable, that quantum logic spectroscopy will be able to project either isotopic composition into one of the many initial states. If this is possible, spectroscopy of the ro-vibrational and hyperfine structure with unprecedented precision could become a reality. For such experiments, resonant charge-exchange collisions with H_2 from the background gas would be highly detrimental, therefore a 4 K environment that freezes out all background hydrogen gas is preferable. Finally, if it just has to be neutral hydrogen, once H_2^+ is in a well defined internal state, it could be gently separated into a proton and a neutral hydrogen with low kinetic energy and a very well defined initial position by resonant multi-photon dissociation. At this point a quite modest magnetic trap might suffice to confine the single hydrogen atom. In this way one might be able to trap a single H atom in isolation for a long time and probe it like it is done with single trapped clock ions.

5 Anti-matters

By far the most challenging but potentially very rewarding experiment proposed here would be the combination of the molecular ions of hydrogen and antihydrogen $\bar{H}_2^-$ and a logic ion, each trapped in its own potential well. To trap each ion in its own well and still have a non-neglible Coulomb-interaction, a miniaturized multi-segmented linear trap may be used, as demonstrated in [13, 40] and proposed for anti-protons in [32]. In this arrangement, comparisons of the physics of matter and antimatter can be done more directly and for more interactions than antiprotons or anti-hydrogen would allow on their own. A molecule and anti-molecule would produce a playground for comparisons of magnetic moments, (anti-) electron to (anti-) proton mass ratios, (anti-) molecular structure and quantum electrodynamics. Getting this experiment to work is a true challenge, but I believe Ted could make this happen, if he only set his mind to it. Maybe this will become the seedling for more and different tall tales told at a future celebration. Many happy returns, Ted!

Acknowledgements Experiments, concepts and ideas are never created out of a vacuum. Most of what is described here and is not referenced as prior work was conceived and grown in the environment of the NIST Ion Storage Group, which therefore owns a great deal of the credit. However, all misconceptions and oversights are solely my own fault. I would like to acknowledge Jim Bergquist, John Bollinger, James Chou, David Hume, Wayne Itano, David Leibrandt, Andrew Wilson and Dave Wineland as well as all the post-docs, grad students and summer students and the administrative and technical staff of the Time and Frequency Division at NIST, a list of persons that is impossible to reproduce here, let alone to give them their proper share of the credit. Thanks for all your contributions, help and inspiration, I owe you big-time.

References

1. R.G. Brewer, R.G. DeVoe, R. Kallenbach, Planar ion microtraps. Phys. Rev. A **46**, R6781–R6784 (1992)
2. T.W. Hänsch, *Paul Traps*. https://www.youtube.com/watch?v=bkYXNeJ8IP0 (1991)
3. T.W. Hänsch, Ballet music seems to nicely disguise that the spores don't really move in rhythm with it. *Private Communication* (1991)
4. C.A. Schrama, E. Peik, W.W. Smith, H. Walther, Novel miniature ion traps. Opt. Commun. **101**, 32–36 (1993)
5. W. Paul, Electromagnetic traps for charged and neutral particles. Rev. Mod. Phys. **62**, 531–540 (1990)
6. H.S. Margolis et al., Hertz-level measurement of the optical clock frequency in a single $^{88}Sr^+$ ion. Science **306**, 1355–1358 (2004)
7. N. Huntemann, C. Sanner, B. Lipphardt, C. Tamm, E. Peik, Single-ion atomic clock with 3 $\times 10^{-18}$ systematic uncertainty. Phys. Rev. Lett. **116**, 063001 (2016)
8. Sandia National Laboratories, *High Optical Access Trap 2.0*. http://prod.sandia.gov/techlib/access-control.cgi/2016/160796r.pdf (2016)
9. J. Chiaverini et al., Surface electrode architecture for ion-trap quantum information porcessing. Quantum Inf. Comput. **5**, 419–439 (2005)

10. S. Seidelin et al., Microfabricated surface-electrode ion trap for scalable quantum information processing. Phys. Rev. Lett. **96**, 253003 (2006)
11. J.M. Amini et al., Toward scalable ion traps for quantum information processing. New J. Phys. **12**, 033031 (2010)
12. C. Ospelkaus et al., Microwave quantum logic gates for trapped ions. Nature **476**, 181–184 (2011)
13. A.C. Wilson et al., Tunable spinspin interactions and entanglement of ions in separate potential wells. Nature **512**, 57–60 (2014)
14. R. Schmied, J.H. Wesenberg, D. Leibfried, Optimal surface-electrode trap lattices for quantum simulation with trapped ions. Phys. Rev. Lett. **102**, 233002 (2009)
15. M. Mielenz et al., Arrays of individually controlled ions suitable for two-dimensional quantum simulations. Nat. Commun. **7**, 11839 (2016)
16. R. Maiwald et al., Stylus ion trap for enhanced access and sensing. Nat. Phys. **5**, 551–554 (2009)
17. N. Lindlein et al., A new 4π geometry optimized for focusing on an atom with a dipole-like radiation pattern. Laser Phys. **17**, 927–934 (2007)
18. C.L. Arrington et al., Micro-fabricated stylus ion trap. Rev. Sci. Instrum. **84**, 085001 (2013)
19. P.O. Schmidt et al., Spectroscopy using quantum logic. Science **309**, 749–752 (2005)
20. J.I. Cirac, P. Zoller, Quantum computations with cold trapped ions. Phys. Rev. Lett. **74**, 4091–4094 (1995)
21. D. Leibfried, R. Blatt, C. Monroe, D. Wineland, Quantum dynamics of single trapped ions. Rev. Mod. Phys. **75**, 281–324 (2003)
22. A. Sørensen, K. Mølmer, Quantum computation with ions in thermal motion. Phys. Rev. Lett. **82**, 1971–1974 (1999)
23. D. Leibfried et al., Toward Heisenberg-limited spectroscopy with multiparticle entangled states. Science **304**, 1476–1478 (2004)
24. T.R. Tan et al., Multi-element logic gates for trapped-ion qubits. Nature **528**, 380–383 (2015)
25. S. Ding, D.N. Matsukevich, Quantum logic for the control and manipulation of molecular ions using a frequency comb. New J. Phys. **14**, 023028 (2012)
26. D. Leibfried, Quantum state preparation and control of single molecular ions. New J. Phys. **14**, 023029 (2012)
27. C.W. Chou, C. Kurz, P. Plessow, D. Leibrandt, D. Leibfried (in preparation)
28. H. Loh et al., Precision spectroscopy of polarized molecules in an ion trap. Science **342**, 1220–1222 (2013)
29. J.C.J. Koelemeij, B. Roth, S. Schiller, Blackbody thermometry with cold molecular ions and application to ion-based frequency standards. Phys. Ref. A **76**, 023413 (2007)
30. J.P. Gauyacq, A.G. Borisov, M. Bauer, Excited states in the alkali/noble metal surface systems: a model system for the study of charge transfer dynamics at surfaces. Prog. Surf. Sci. **82**, 244–292 (2007). doi:10.1016/j.progsurf.2007.03.006
31. S.W. Hell, Nanoscopy with focused light (nobel lecture). Angew. Chem. Int. Ed. **54**, 8054–8066 (2015)
32. M. Niemann, A.-G. Paschke, T. Dubielzig, S. Ulmer, C. Ospelkaus, CPT test with (anti)proton magnetic moments based on quantum logic cooling and readout, ed. by V. Alan Kostolecky. Proceedings of the Sitxth Meeting on CPT and Lorentz Symmetry, World Scientific Publishing, Singapore (2014)
33. A.L. Schawlow, A diatomic molecule is a molecule with an atom too many, *Private Communication* (1955)
34. V.I. Korobov, J.C.J. Koelemeij, L. Hilico, J.-P. Karr, Theoretical hyperfine structure of the molecular hydrogen ion at the 1 ppm level. Phys. Rev. Lett. **116**, 053003 (2016)
35. W. Ubachs, J. Koelemeij, K. Eikema, E. Salumbides, Physics beyond the standard model from hydrogen spectroscopy. J. Mol. Spectrosc. **320**, 1–12 (2016)
36. K.B. Jefferts, Rotational Hfs spectra of H_2^+ molecular ions. Phys. Rev. Lett. **20**, 39–41 (1968)

37. A. Carrington, I.R. McNab, C.A. Montgomerie, Spectroscopy of the hydrogen molecular ion. J. Phys. B: At. Mol. Opt. Phys. **22**, 3551 (1989)
38. C. Haase, M. Beyer, C. Jungen, F. Merkt, The fundamental rotational interval of para-H_2^+ by MQDT-assisted Rydberg spectroscopy of H_2. J. Chem. Phys. **142**, 064310 (2015)
39. J. Biesheuvel, J.-Ph. Karr, L. Hilico, K.S.E. Eikema, W. Ubachs, J.C.J. Koelemeij, Probing QED and fundamental constants through laser spectroscopy of vibrational transitions in HD^+. Nat. Comm. **7**, 10385 (2016). doi:10.1038/ncomms10385
40. K.R. Brown et al., Coupled quantized mechanical oscillators. Nature **471**, 196–199 (2011)

Coherence in Laser-Driven Electrons at the Surface and in the Volume of Solid Matter

Peter Hommelhoff

Abstract The femtosecond frequency comb allows controlling the carrier field of ultrashort laser pulses. We show two examples on how this control over fields oscillating with a few hundred terahertz can be utilized to control electrons at the surface and in the volume of solids. After a brief discussion of strong-field physics at metal needle tips, we show how ultrafast two-color laser pulses allow quantum path interference to dramatically alter the emission current from sharp tips, with an interference visibility of 94%. With carrier-envelope phase-controlled laser pulses, we show furthermore how light-field sensitive currents can be excited in monolayer graphene via an interplay of interband and intraband electron dynamics including multiple Landau–Zener transitions.

1 Introduction

The femtosecond frequency comb allows direct control over the optical carrier field in ultrashort laser pulses by virtue of controlling the carrier-envelope phase (CEP) [1]. Hence, with the femtosecond frequency comb the optical carrier wave of ultrashort laser pulses can now be fully controlled, which was so far only possible at much smaller frequencies of the electromagnetic spectrum, such as in the microwave domain. This progress allows new control over processes that are driven by the optical carrier field rather than by the intensity envelope of laser pulses. Prominent examples include high-harmonic and attosecond pulse generation, which are based on strongly driving electrons in gas-phase atoms [2–4].

In this contribution, the recent development of phase-coherent electron control at the surface and in the volume of solids will be discussed. While spectacular control over electrons in atoms and molecules in the gas phase has been obtained in the last

This article is part of the topical collection “Enlightening the World with the Laser” - Honoring T. W. Hänsch guest edited by Tilman Esslinger, Nathalie Picqué, and Thomas Udem.

P. Hommelhoff (✉)
Department Physik, Friedrich-Alexander-Universität Erlangen-Nürnberg (FAU), Staudtstr. 1, 91058 Erlangen, EU, Germany
e-mail: peter.hommelhoff@fau.de; https://www.laser.physik.fau.de

D. Meschede et al. (eds.), *Exploring the World with the Laser*,
https://doi.org/10.1007/978-3-319-64346-5_9

one to two decades [5], the work on solids was rather limited. In his Nobel Prize Lecture [6], Theodor Hänsch refers to initial work by Nakajima and Lambropoulos on above-threshold ionization (ATI) of gas-phase atoms in this context [7]. The corresponding process in solids is above-threshold photoemission (ATP) and has been investigated by Theodor Hänsch and colleagues already in 2004 [8]. ATP describes (multi-)photon-driven emission of electrons with a (kinetic) energy larger than the minimally required energy by one and up to many photons. Even though the authors of [8] used extremely short laser pulses with a duration of only 4 fs, corresponding to 1.6 optical cycles, the carrier-envelope phase-dependent current they could detect was minuscule. This was for one obvious reason, which is spatial averaging over the Gaussian-shaped focal profile on the flat metal surface. But not only does the spatially varying intensity lead to a varying nonlinear order and spatially varying light shifts, also the focal *phase* may vary transversally for such broadband light fields [9, 10]. It turns out that working with sharp needle tips circumvents these issues: When laser pulses are focused at tips with an apex radius of curvature much smaller than the central laser wavelength, optical field enhancement takes place [11]. The optical field is preferentially enhanced at the apex, so that especially nonlinear effects such as ATP are quite dramatically enhanced at the apex and are thus limited in space to the tip apex region [12–14]. Typical tip radii lie in the range of a few nanometers to around 100 nm, so much smaller than focal spot sizes.

Observing photon orders in electron spectra implies that the coherence of the laser beam is transferred into the electron domain. While early electron spectra recorded from flat metals hinted at above-threshold photoemission [15], it remained open if the coherence of the laser beam was not washed out by a thermal process because photon orders could not be observed. It was therefore insightful that in spectrally resolved photoemission measurements from sharp needle tips above-threshold multi-photon orders were clearly visible [16]. With increasing intensity, these photon orders shifted to smaller energies, evidencing the onset of strong-field effects such as peak shifting and channel closing [16–19]. With only small DC voltages applied to the tips, the characteristic plateau in electron spectra became apparent [20–23], indicating that field-driven electron dynamics is starting to take place. With carrier-envelope phase-stable laser pulses, the laser-field-driven dynamics could be clearly proven [20, 24]. The original work, which is now several years old, has been reviewed and commented on in several occasions to which we refer the interested reader (see, for example, [25–28]).

We could show that electron emission in the energy range of the electron re-collision cutoff can be switched on and off by virtue of the carrier-envelope phase [20]. From there it is only a small step to realize that a light-field driven interface can be built from two tips that can result in an ultrafast, light-field driven switch. With two tips facing each other we have started to work in this direction. In essence, we have so far demonstrated a nanoscale version of a vacuum tube diode [29]. Yet, the demonstration of light-field driven electronic switching in a nanoscale device that allows utilizing the attosecond fast timescales is still elusive.

The physics discussed so far is based on photoemission of electrons from a tip and subsequent (strong-field) dynamics of the electrons inside of the laser field in front of the tip. For this very reason, namely that it is mostly the dynamics of the field-driven electron after it has been born into the laser field that governs its final state, we see such a great similarity between gas-phase strong-field physics spectra and the spectra obtained from solid tips. Examples include the exponentially dropping low-energy part, but most importantly the famous plateau and the similarly famous "10-U_p cutoff." For this reason we can explain the physics nicely with models well known from atomic physics [20–22, 30–33], which is also discussed in the original and review papers mentioned above. We now turn to discuss two new phenomena recently observed.

2 Two-Color Experiments at Metal Tips

Intriguing effects can arise when a harmonic of the fundamental driving field is superimposed on the fundamental of the optical pulse. A two-color or bichromatic field pulse results, representing a first step (next to CEP control) toward more complex, tailored light fields, with the goal of improving the level of control over reaction or emission pathways, extending electron energies, shaping electron pulses etc. The relative phase between the colors represents an additional control parameter. Investigations of the effects of bichromatic light fields on atomic ATI date back more than two decades [34] and have been utilized to further explore the transition from the multi-photon to the tunneling ionization regime, for example [35]. An extensive review on earlier work of the atomic (strong-field) physics in bichromatic laser fields can be found in [36]. More recently, bichromatic fields have been used to probe molecular strong-field dynamics on the 10-attosecond timescale [37, 38].

We focus the fundamental and the second harmonic of near-infrared laser pulses on a sharp tungsten tip at ambient temperature and vary the temporal delay between the two pulses. Two cases may be considered: In the strong-field regime, the carrier fields add up to yield an asymmetric optical field structure in which the carrier maxima can be enhanced by the second harmonic for one polarity, while for the other the extrema are reduced, for instance. The role of the polarity exactly changes if the second harmonic is delayed by half an optical carrier period, yielding a reversed field structure. This will have direct consequences for the strongly driven electron. On the other hand, for weaker fields, or in the perturbative regime, different electron emission quantum pathways, comprised of sequences of photons of fundamental and second harmonic (with different phases), may constructively or destructively interfere. Here we report on observations in this regime.

We frequency-double a part of the $\sim$ 70-fs output pulses of an erbium-doped fiber oscillator–amplifier running at a central wavelength of 1560 nm and at

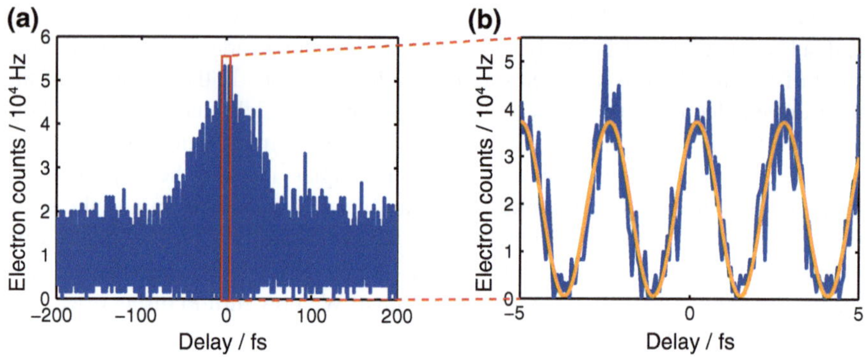

Fig. 1 Electron current emitted from a 10-nm sharp tungsten tip when irradiated by a two-color laser field as function of time delay between the fundamental and second-harmonic laser pulse. **a** A large time span between full and hardly no overlapping pulses is shown. **b** The central region of **a**, clearly revealing oscillations with the period of the second harmonic (2.6 fs). Data and a sine fit are shown. The large contrast of 94% in this central region is also directly visible. *Data points* are connected by line segments as a guide to the eye

100 MHz repetition rate. With the help of a Mach–Zehnder interferometer with dichroic beam splitters, we can continuously vary the delay (or phase) between the frequency-doubled pulses ("2ω") and the fundamental pulses ("ω"). When we focus these $\omega - 2\omega$ pulses on the tungsten tip and record the emitted electron current as function of time delay, we observe a strong modulation of the emitted current as function of delay, see Fig. 1. Notably, the modulation reaches up to 94% for near-field intensities of 330 GW / cm^2 of the fundamental and 6.6 GW/cm^2 for the second harmonic. So a weak second-harmonic admixture of $\sim 1/50$ in relative intensity magnitude suffices to induce such a strong modulation of the output current.

We explain these effects with quantum path interference as indicated in Fig. 2. For the DC-biased tungsten tip, the emission barrier at the (310)-W apex is Schottky-lowered from an unbiased 4.3–3.6 eV. Hence, four fundamental photons of 0.8 eV energy are required to lift an electron to the barrier. Alternatively, two fundamental and one frequency-doubled photon provide the same energy to an electron. We interpret our data in such a way that it is these two quantum pathways that interfere. A much more detailed discussion can be found in [39].

While the discussed two-color effects are already of fundamental interest, they may enable boosting the maximum output current without damaging the tip with the fundamental laser intensity. Furthermore, it will be even more interesting to investigate bichromatic effects at tips in the strong-field regime. With the relatively long laser pulses employed here, the strong-field regime can hardly be reached without damaging the tip, but with shorter fundamental laser pulses at around 1800 nm [40], we expect that this will be easily possible, in particular because of the longer central wavelength as compared the standard 800 nm that have been mostly used so far for strong-field studies at tips.

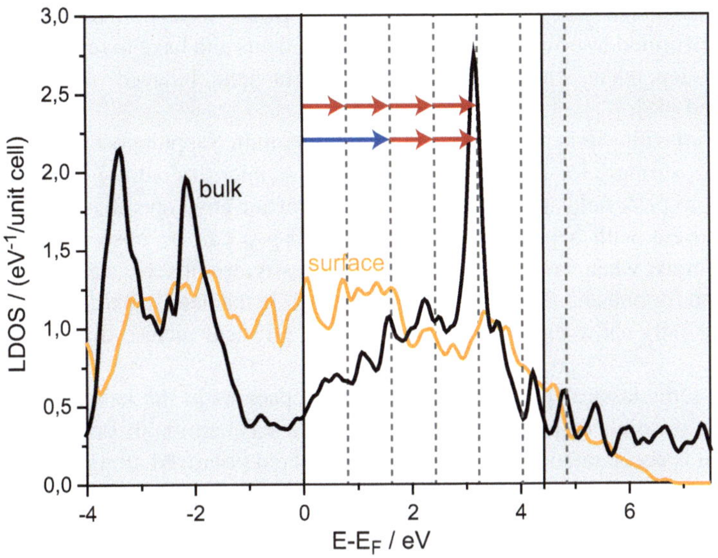

Fig. 2 Quantum path interference model we employ to explain the strong modulation of the emission current as function of delay between fundamental and second-harmonic pulses, as shown in Fig. 1. One pathway consists of four photons of the fundamental required to lift an electron from the Fermi level to a bulk state extending to the tip surface. The other pathway consists of two fundamental photons and one photon of the second harmonic (*blue*). From there the electrons can easily overcome the Schottky-lowered barrier. The full lines show the local density of states, both for the surface (*orange*) and the bulk (*black*), as discussed in [39]

3 Light-Field-Induced Currents in Graphene

The effects discussed in the previous section relate to quantum effects inside of tips, however, in the perturbative regime. In recent years, strong-field physics inside of solids has found much attention, as new effects related to the solid state nature of the samples are expected that do not show up in the atomic and molecular case. So far, large-bandgap materials and/or long-wavelength driving pulses have been used so as not to excite electrons from the valence to the conduction band easily. For example, high harmonic generation has been observed from the large-bandgap semiconductor ZnO (band gap of 3.2 eV), driven with pulses with a central wavelength of 3.2– 3.7 μm [41] or with terahertz frequencies [42]. Reversible attosecond dynamics of strongly driven fused silica glass and silicon have been investigated with the help attosecond transient absorption spectroscopy [43, 44]. The question arises if next to dielectrics and semiconductors also metals can be strongly driven by the optical field of ultrashort laser pulses.

Graphene, a (semi-) metal, is ideally suited for strong-field studies, in particular graphene on SiC. It is a robust material on a robust large-bandgap substrate with good heat conductivity, but first and foremost monolayer graphene stands out for its two-dimensional nature and its cone-shaped Dirac dispersion relation [45]. We have performed laser-induced current measurements and have investigated whether a CEP-dependent current, so again a light-field induced current, can be observed [46].

Driven with ~ 6-fs pulses derived from a Titanium:Sapphire oscillator running at 80 MHz, strongly focused down to 1.6 μm ($1/e^2$ intensity radius) on graphene and leading to peak field strengths of 0.3 V/Å, a carrier-envelope phase-(CEP)-dependent current with a peak magnitude of ~ 15 pA can be observed, see Fig. 3. Surprisingly, when we vary the peak field intensity, we observe a change in current direction for one and the same CEP setting. While the details are discussed in [46], here we only show the intriguing underlying physical picture that we assume to hold.

The semiclassical picture of electron wavepackets in the band structure environment of solids helps to gain insight. Bloch's acceleration theorem states that an electron is accelerated by an external electric field polarized along x in x-direction only: $\hbar\dot{\mathbf{k}} = -e\mathbf{E}$, where $\mathbf{k}$ represents the electron momentum, $-e$ the electron charge and $\mathbf{E}$ the externally applied electric field. Because of the cone-shaped nature of graphene's dispersion, a slice exists in the dispersion relation such that electrons can be driven diabatically across the band crossing even for vanishing field strengths (at $k_y = 0$, for example), but there also exist slices where the energy between the bands is large so that the electrons remain inside of the band and are just driven within it by the laser field (adiabatically) [47], see Fig. 4. Here, diabatic means that the electron is transferred from the valence to the conduction band, or vice versa, so an interband transition. Conversely, the adiabatic motion implies intraband motion, so an adiabatic exchange of pseudospins labeling the quantum number of the wave functions [45]. The pseudospins are conserved in the diabatic case.

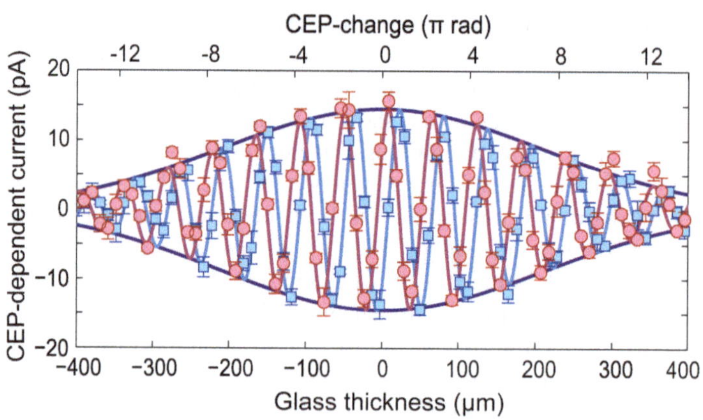

Fig. 3 CEP-dependent current as function of glass thickness in the beam path, leading to a variation of CEP and pulse duration. In-phase (*spheres*) and quadrature component (*squares*) of a lock-in measurement are shown, together with fits to the data

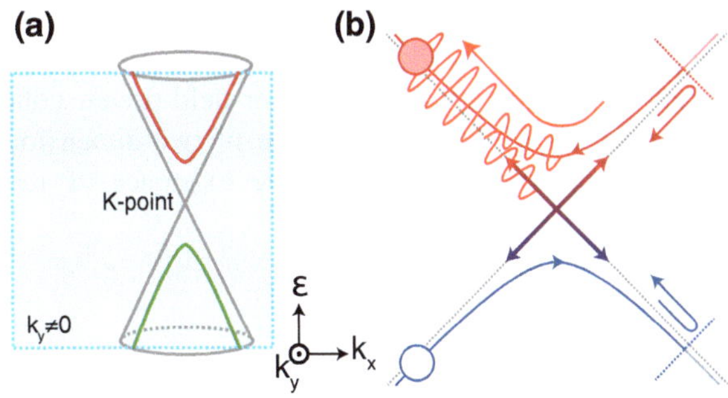

Fig. 4 **a** Graphene's Dirac-cone-shaped dispersion. Here the hyperbolic bands are highlighted that result from the intersection of a plane with a finite k_y-value with the dispersion cones; note that the K-point lies outside of that plane. Electrons with all initial k_y-values are driven by the laser field. **b** Sketch of the dynamics in the two hyperbolic bands shown in **a**. The laser field is polarized in x-direction. Because of the oscillatory nature of the laser field, the electron is driven repeatedly over the apparent band gap where it can always either undergo a diabatic transition from one band to the other, or stay in the same band (Landau–Zener dynamics). The apparent band gap thus serves as an electron beam splitter. The relative phase between the two arms determines whether the electron ends up in the conduction band (as shown here) or in the valence band. The relative phase depends on the CEP via the electric waveform of the driving field. In order to obtain the resulting CEP-dependent current, an integration over all initial k_y-values is required

Between these two extrema, a dispersion cone slice exists in which valence band electrons may become excited via Landau–Zener transitions to the conduction band *or* stay in the valence band, with equal probabilities. This represents a 50 / 50-beam splitter for electrons. It is this beam splitter that is responsible for the CEP-dependent current generation in graphene, see Fig. 4b. Because the optical pulse drives electrons more than once through the energy minimum, at least two beamsplitting events can take place. Much like in an optical interferometer, it is the relative phase between the two arms that determines if interference into a given output port, here the conduction band, for example, is constructive or destructive. By virtue of the CEP we can vary the phase, as it directly influences the propagation of the electrons inside of the interferometer arms and hence the relative interferometer phase. An in-depth discussion including numerical modeling is presented in [46].

To summarize, initial work on strong-field effects at solids showed already a high degree of coherence of the light-field-driven electrons, exemplified in the existence and the complex but well-understood behavior of above-threshold orders. Here we have discussed more recent work on electron emission from needle tips with bichromatic laser fields. We could show that coherent quantum path interference leads to a large variation of the emitted current as a function of delay between fundamental and second harmonic of the driving pulses. The visibility of up to 94% seems to be the highest observed in two-color photoemission experiments over all investigated systems, which is surprising given the metallic nature of the room temperature metal tip as compared to clean gas-phase atomic systems [48], for instance. In the second discussed experiment, we focus on the coherence of electrons inside of the special metal graphene. Here we observe a CEP-dependent current that we explain with a light-field-driven interplay of interband and intraband dynamics, representing another example of laser-field-driven coherent quantum path interference, now inside of a peculiar type of two-dimensional metal—the discovery of which is possible because of the existence of the femtosecond frequency comb.

4 Concluding Remarks

The work described here has been done long after the author had finished his PhD thesis in Professor Hänsch's group at Ludwig-Maximilians-Universität in Munich in 2002, which was already a most enjoyable time. At Stanford University in the group of Mark Kasevich, Yvan Sortais was setting up a femtosecond frequency comb for a precision spectroscopy experiment. At the other end of the optical table, the author had just succeeded in achieving field emission from a single-atom tip in early 2004 (published only much later [49]). From the atomic precision spectroscopy point of view certainly an abuse but probably in line with the Hänsch (and Schawlow) style of doing physics [50, 51], Yvan Sortais and the author decided to rather fire the femtosecond frequency comb laser pulses at the apex of a sharp metal needle tip and so realized an electron source able to generate femtosecond electron pulses from a nanometric emission area [52]—at the time without being able to observe a carrier-envelope phase-dependent current, even though they had tried right away [53]. Equipped with Max Planck Research Group funding, the author went to MPQ to again become a member of the Hänsch department in 2007. It was there where the author's small group performing *Spitzenforschung* (translate: tip research) was engulfed and fostered in Theo Hänsch's group doing *Spitzenforschung* (translate: top research): From then on until 2013, the author and his group enjoyed a most fruitful and stimulating atmosphere. The weekly Hänsch department seminars provided excellent exchange of scientific and technical ideas, as did the annual highlight of they year, the Hänsch department retreat at Ringberg Castle, resulting in research highlights discussed in the introductory part of this contribution and further including the demonstration of an on-chip Paul trap for electrons as well as the demonstration of laser acceleration of electrons at dielectric structures [54, 55]. A number of joint publications resulted from common experimental interests and requirements, for example on ultrashort pulse generation and single-pass laser amplification [56–60] and on the CEP stability of laser oscillators with various pump lasers [61, 62]. In 2013, with the author's appointment to Friedrich-Alexander-Universität Erlangen-Nürnberg, his group had to leave the Hänsch department. Theodor Hänsch has profoundly influenced the author's life and also that of the many students and postdocs who had the pleasure to interact with him. Happy birthday and ad multos annos, lieber Herr Professor Hänsch!

Acknowledgements The author acknowledges the many contributions of and is grateful to all past and current members of his group, in particular the long-time members Drs. Johannes Hoffrogge, Markus Schenk, Michael Krüger, John Breuer, Jakob Hammer, Sebastian Thomas, Michael Förster, Takuya Higuchi, Joshua McNeur and Martin Kozák. Funding from ERC (NearFieldAtto), SFB 953, SPP 1840 QUTIF and the Gordon and Betty Moore Foundation is gratefully acknowledged.

References

1. T. Udem, R. Holzwarth, T.W. Hänsch, Nature **416**, 233 (2002)
2. A. McPherson, G. Gibson, H. Jara, U. Johann, T.S. Luk, I.A. McIntyre, K. Boyer, C.K. Rhodes, J. Opt. Soc. Am. B **4**, 595 (1987)
3. M. Ferray, A. L'Huillier, X.F. Li, L.A. Lompre, G. Mainfray, C. Manus, J. Phys. B At. Mol. Opt. Phys. **21**, L31 (1988)
4. P.B. Corkum, F. Krausz, Nat. Phys. **3**, 381 (2007)
5. T. Schultz, M. Vrakking (eds.), *Attosecond and XUV Spectroscopy: Ultrafast Dynamics and Spectroscopy* (Wiley-VCH, Weinheim, 2014)
6. T.W. Hänsch, Rev. Mod. Phys. **78**, 1297 (2006)
7. T. Nakajima, P. Lambropoulos, Phys. Rev. A **50**, 595 (1994)
8. A. Apolonski, P. Dombi, G.G. Paulus, M. Kakehata, R. Holzwarth, T. Udem, C. Lemell, K. Torizuka, J. Burgdörfer, T.W. Hänsch, F. Krausz, Phys. Rev. Lett. **92**, 073902 (2004)
9. M.A. Porras, Phys. Rev. E **65**, 026606 (2002)
10. D. Hoff, M. Krüger, L. Maisenbacher, A. M. Sayler, G. G. Paulus, P. Hommelhoff, to be published (2017)
11. L. Novotny, B. Hecht, *Principles of Nano-Optics* (Cambridge University Press, Cambridge, 2012)
12. S. Thomas, M. Krüger, M. Förster, M. Schenk, P. Hommelhoff, Nano Lett. **13**, 4790 (2013)
13. M. Krüger, S. Thomas, M. Förster, P. Hommelhoff, J. Phys. B At. Mol. Opt. Phys. **47**, 124022 (2014)
14. S. Thomas, G. Wachter, C. Lemell, J. Burgdörfer, P. Hommelhoff, New J. Phys. **17**, 063010 (2015)
15. M. Aeschlimann, C.A. Schmuttenmaer, H.E. Elsayed-Ali, R.J.D. Miller, J. Cao, Y. Gao, D.A. Mantell, J. Chem. Phys. **102**, 8606 (1995)
16. M. Schenk, M. Krüger, P. Hommelhoff, Phys. Rev. Lett. **105**, 257601 (2010)
17. N.B. Delone, V.P. Krainov, *Multiphoton Processes in Atoms* (Springer, Berlin, 1994)
18. R. Bormann, M. Gulde, A. Weismann, S.V. Yalunin, C. Ropers, Phys. Rev. Lett. **105**, 147601 (2010)
19. M. Bionta, B. Chalopin, J. Champeaux, S. Faure, A. Masseboeuf, P. Moretto-Capelle, B. Chatel, J. Mod. Opt. **61**, 833 (2014)
20. M. Krüger, M. Schenk, P. Hommelhoff, Nature **475**, 78 (2011)
21. G. Wachter, C. Lemell, J. Burgdörfer, M. Schenk, M. Krüger, P. Hommelhoff, Phys. Rev. B **86**, 035402 (2012)
22. M. Krüger, M. Schenk, P. Hommelhoff, G. Wachter, C. Lemell, J. Burgdörfer, New J. Phys. **14**, 085019 (2012)
23. G. Herink, D.R. Solli, M. Gulde, C. Ropers, Nature **483**, 190 (2012)
24. B. Piglosiewicz, S. Schmidt, D.J. Park, J. Vogelsang, P. Gross, C. Manzoni, P. Farinello, G. Cerullo, C. Lienau, Nat. Photonics **8**, 37 (2014)
25. P. Dombi, A.Y. Elezzabi, in *Attosecond Nanophysics*, ed. by P. Hommelhoff, M. Kling (Wiley-VCH, Weinheim, 2015)

26. C. Lienau, M. Raschke, C. Ropers, in *Attosecond Nanophysics*, ed. by P. Hommelhoff, M. Kling (Wiley-VCH, Weinheim, 2015)
27. F. Süßmann, M.F. Kling, P. Hommelhoff, in *Attosecond Nanophysics*, ed. by P. Hommelhoff, M. Kling (Wiley-VCH, Weinheim, 2015)
28. M. Krüger, M. Schenk, M. Förster, P. Hommelhoff, J. Phys. B **45**, 074006 (2012)
29. T. Higuchi, L. Maisenbacher, A. Liehl, P. Dombi, P. Hommelhoff, Appl. Phys. Lett. **106**, 051109 (2015)
30. P.B. Corkum, Phys. Rev. Lett. **71**, 1994 (1993)
31. M. Lewenstein, P. Balcou, M.Y. Ivanov, A. L'Huiller, P.B. Corkum, Phys. Rev. A **49**, 2117 (1994)
32. S.V. Yalunin, M. Gulde, C. Ropers, Phys. Rev. B **84**, 195426 (2011)
33. S.V. Yalunin, G. Herink, D.R. Solli, M. Krüger, P. Hommelhoff, M. Diehn, A. Munk, C. Ropers, Ann. Phys. **525**, L12 (2013)
34. H.G. Muller, P.H. Bucksbaum, D.W. Schumacher, A. Zavriyev, J. Phys. B At. Mol. Opt. Phys. **23**, 2761 (1990)
35. D.W. Schumacher, F. Weihe, H.G. Muller, P.H. Bucksbaum, Phys. Rev. Lett. **73**, 1344 (1994)
36. F. Ehlotzky, Phys. Rep. **345**, 175 (2001)
37. X. Xie, S. Roither, D. Kartashov, E. Persson, D.G. Arbó, L. Zhang, S. Gräfe, M.S. Schöffler, J. Burgdörfer, A. Baltuška, M. Kitzler, Phys. Rev. Lett. **108**, 193004 (2012)
38. S. Skruszewicz, J. Tiggesbäumker, K.-H. Meiwes-Broer, M. Arbeiter, T. Fennel, D. Bauer, Phys. Rev. Lett. **115**, 043001 (2015)
39. M. Förster, T. Paschen, M. Krüger, C. Lemell, G. Wachter, F. Libisch, T. Madlener, J. Burgdörfer, P. Hommelhoff, Phys. Rev. Lett. **117**, 217601 (2016)
40. C. Homann, M. Bradler, M. Förster, P. Hommelhoff, E. Riedle, Opt. Lett. **37**, 1673 (2012)
41. S. Ghimire, A.D. DiChiara, E. Sistrunk, P. Agostini, L.F. DiMauro, D.A. Reis, Nat. Phys. **7**, 138 (2010)
42. M. Hohenleutner, F. Langer, O. Schubert, M. Knorr, U. Huttner, S.W. Koch, M. Kira, R. Huber, Nature **523**, 527 (2015)
43. M. Schultze, E.M. Bothschafter, A. Sommer, S. Holzner, W. Schweinberger, M. Fiess, M. Hofstetter, R. Kienberger, V. Apalkov, V.S. Yakovlev, M.I. Stockman, F. Krausz, Nature **493**, 75 (2013)
44. M. Schultze, K. Ramasesha, C. Pemmaraju, S. Sato, D. Whitmore, A. Gandman, J.S. Prell, L.J. Borja, D. Prendergast, K. Yabana, D.M. Neumark, S.R. Leone, Science **346**, 1348 (2014)
45. A.H. Castro Neto, F. Guinea, N.M.R. Peres, K.S. Novoselov, A.K. Geim, Rev. Mod. Phys. **81**, 109 (2009)
46. T. Higuchi, C. Heide, K. Ullmann, H. B. Weber, P. Hommelhoff, arXiv:1607.04198 (2016)
47. H.K. Kelardeh, V. Apalkov, M.I. Stockman, Phys. Rev. B **93**, 155434 (2016)
48. P. Ackermann, A. Scharf, T. Halfmann, Phys. Rev. A **89**, 063804 (2014)
49. P. Hommelhoff, C. Kealhofer, A. Aghajani-Talesh, Y.R. Sortais, S.M. Foreman, M.A. Kasevich, Ultramicroscopy **109**, 423 (2009)
50. T.W. Hänsch, Opt. Photonics News, 14 (2005)
51. Superlaser123, Arthur L. Schawlow pops a balloon, https://www.youtube.com/watch?v=WCFRYpGoC98. Accessed 11 Aug 2016
52. P. Hommelhoff, Y. Sortais, A. Aghajani-Talesh, M.A. Kasevich, Phys. Rev. Lett. **96**, 077401 (2006)
53. P. Hommelhoff, C. Kealhofer, M. A. Kasevich, in *Proceedings of the 2006 I.E. International Frequency Control Symposium and Expositions, Vols 1 and 2*, pp. 470–474 (2006)
54. J. Hoffrogge, R. Fröhlich, M.A. Kasevich, P. Hommelhoff, Phys. Rev. Lett. **106**, 193001 (2011)
55. J. Breuer, P. Hommelhoff, Phys. Rev. Lett. **111**, 134803 (2013)
56. A. Ozawa, W. Schneider, T.W. Hänsch, T. Udem, P. Hommelhoff, New J. Phys. **11**, 083029 (2009)

57. A. Ozawa, W. Schneider, F. Najafi, T.W. Hänsch, T. Udem, P. Hommelhoff, Laser Phys. **20**, 967 (2010)
58. A. Ozawa, T. Udem, U.D. Zeitner, T.W. Hänsch, P. Hommelhoff, Phys. Rev. A **82**, 033815 (2010)
59. A. Vernaleken, J. Weitenberg, T. Sartorius, P. Russbueldt, W. Schneider, S.L. Stebbings, M.F. Kling, P. Hommelhoff, H.-D. Hoffmann, R. Poprawe, F. Krausz, T.W. Hänsch, T. Udem, Opt. Lett. **36**, 3428 (2011)
60. S. Thomas, R. Holzwarth, P. Hommelhoff, Opt. Exp. **20**, 13663 (2012)
61. A. Vernaleken, B. Schmidt, M. Wolferstetter, T.W. Hänsch, R. Holzwarth, P. Hommelhoff, Opt. Exp. **20**, 18387 (2012)
62. A. Vernaleken, B. Schmidt, T.W. Hänsch, R. Holzwarth, P. Hommelhoff, Appl. Phys. B **117**, 33 (2014)

Atomic "Bomb Testing": The Elitzur–Vaidman Experiment Violates the Leggett–Garg Inequality

Carsten Robens, Wolfgang Alt, Clive Emary, Dieter Meschede, and Andrea Alberti

Abstract Elitzur and Vaidman have proposed a measurement scheme that, based on the quantum superposition principle, allows one to detect the presence of an object—in a dramatic scenario, a bomb—without interacting with it. It was pointed out by Ghirardi that this interaction-free measurement scheme can be put in direct relation with falsification tests of the macro-realistic worldview. Here we have implemented the "bomb test" with a single atom trapped in a spin-dependent optical lattice to show explicitly a violation of the Leggett–Garg inequality—a quantitative criterion fulfilled by macro-realistic physical theories. To perform interaction-free measurements, we have implemented a novel measurement method that correlates spin and position of the atom. This method, which quantum mechanically entangles spin and position, finds general application for spin measurements, thereby avoiding the shortcomings inherent in the widely used push-out technique. Allowing decoherence to dominate the evolution of our system causes a transition from quantum to classical behavior in fulfillment of the Leggett–Garg inequality.

This article is part of the topical collection "Enlightening the World with the Laser" - Honoring T. W. Hänsch guest edited by Tilman Esslinger, Nathalie Picqué, and Thomas Udem.

The original version of this article has been revised: In the editorial publishing process, the shaded band in Fig. 3 representing the theoretical model was omitted. Now, the Fig. 3 has been corrected.

C. Robens • W. Alt • A. Alberti (✉)
Institut Für Angewandte Physik, Universität Bonn, Wegelerstr. 8, 53115 Bonn, Germany
e-mail: alberti@iap.uni-bonn.de

C. Emary
Joint Quantum Centre Durham-Newcastle, Newcastle University, Newcastle upon Tyne NE1 7RU, UK

D. Meschede
Institute for Applied Physics, University of Bonn, Wegeler Str. 8, 53125 Bonn, Germany

D. Meschede et al. (eds.), *Exploring the World with the Laser*,
https://doi.org/10.1007/978-3-319-64346-5_10

1 Introduction

Measuring physical properties of an object—whether macroscopic or microscopic—is in most cases associated with an interaction. For example, scattering photons off an object allows one to detect its presence in a given region of space. However, this also produces a small perturbation of its state by direct momentum transfer. It is well known from numerous discussions on the physics of the quantum measurement process (see, e.g., Refs. [1, 2]) that a measurement in general modifies the quantum evolution unless the object is already in an eigenstate of the measurement apparatus [3]. This is even the case when the measurement yields a negative outcome, that is, when we did *not* find the particle on a certain trajectory that had originally a nonvanishing probability amplitude to be occupied. For example, in a double-slit experiment, quantum interference is suppressed as soon as a measurement detects the which-way information, regardless of whether the information is acquired by direct interaction or indirect negative inference. Quantum mechanics formalizes the loss of interference in terms of the quantum measurement process, showing that measurements are generally invasive as they entail a modification of the subsequent quantum evolution. While the quantum measurement process is still intensely debated in the literature [4], we adopt here the pragmatic view that a measurement applied to a superposition state causes a sudden reduction of the wave function to a subspace of the Hilbert space.

Ideal negative measurements, that is, measurements without direct interaction, play an important role in a physical scenario known as *macro-realism* [5–7]. This scenario advocates a classical worldview describing the state of macroscopic objects, according to which macroscopic objects are always in one of several possible macroscopically distinct states. In a macro-realistic worldview, it is plausible to assume that a negative outcome of a measurement cannot affect the evolution of a macroscopic system, meaning that negative measurements are *noninvasive* [8]. In order to rigorously test the macro-realistic point of view, Leggett and Garg have derived an inequality from the assumptions of macro-realism and noninvasive measurability, which can be violated by quantum-mechanical superposition states [9]. Relying on the implementation of an ideal negative measurement protocol—a prerequisite for any genuine test of the Leggett–Garg inequality—experimental violations of the macro-realistic worldview have been experimentally demonstrated with phosphor impurities in silicon by Knee et al. [10] and with trapped atoms by Robens et al. [11].

The definition of the degree of "macroscopic distinctness" has been a matter of discussion in the literature ever since [12], and is likely to remain as such till an experiment [13] will shed new light, for example, discovering a physical "classicalization" mechanism that causes an objective reduction of wave packets. Recently, Nimmrichter and Hornberger proposed a quantitative criterion based on a minimal macro-realistic extension of quantum mechanics to quantify the macroscopicity of an object [14]. Their objective criterion of macroscopicity allows us to experimentally test the behavior of a single trapped atom—however microscopic it is, according to our intuition—under the hypothesis of macro-realism, as we can put its macroscopicity directly in relation to that of other, ideally more massive physical objects.

It was pointed out by Ghirardi [15] that a Leggett–Garg test of macro-realism is naturally related to the notion of interaction-free measurements introduced by Elitzur and Vaidman [16]. In a rather dramatic scenario, Elitzur and Vaidman proposed a quantum device able to single out live "bombs" from a collection containing also duds without triggering them nor interacting with them. While the first realizations of the Elitzur–Vaidman experiment employed "flying" photons [17] and "flying" neutrons [18], we here implement a variation of this experiment with neutral atoms trapped in a one-dimensional optical lattice. A convenient scheme for interaction-free measurements with neutral atoms has been demonstrated by Robens et al. [11] exploiting state-dependent optical potentials. Following the idea of Ghirardi, we tested the hypothesis of macro-realism with our atomic implementation of the Elitzur–Vaidman "bomb testing" *Gedankenexperiment*. Our experiment shows explicitly that the Leggett–Garg inequality is violated by 21 σ. In addition, trapped atoms can be held for long times. By controlling the duration of a suitably chosen wait interval, it is straightforward to study the influence of decoherence and experimentally observe the gradual transition from quantum to classical behavior.

It is our pleasure to honor with these recent experimental results Theodor W. Hänsch. For many decades he has laid the foundations in laser physics and technology, without which present-day laser control of quantum particles is hardly conceivable. In this sense, the objective of this article is to experimentally demonstrate yet another example of exploring the world with lasers and laser-controlled atoms at the quantum-classical boundary.

2 Brief Review: The Elitzur–Vaidman Interaction-Free "Bomb Test"

Let us illustrate the concept of the interaction-free "bomb test" as presented in the original work by Elitzur and Vaidman [16], based on a single photon travelling on a quantum superposition of trajectories along the two paths of a Mach—Zehnder (MZ) interferometer.

The two branches A and B of a MZ interferometer can be balanced in such a way that one of the two output ports, say D2, is always bright while the other one, D1, is always dark, see Fig. 1a. If any object (e.g., a "live bomb" triggered by a single photon) intercepts the trajectory of the photon in the lower branch B, the photon field is removed from that branch and the balance of the MZ interferometer is disturbed. As shown in Fig. 1b, the photon has a 50% probability to be blocked without ever reaching the detectors. This event is of course highly invasive: the "bomb" is triggered to explode by absorbing the travelling photon.

In all other events, the photon must have followed branch A, thus avoiding interacting with the object, and is subsequently routed to the detectors D1 and D2

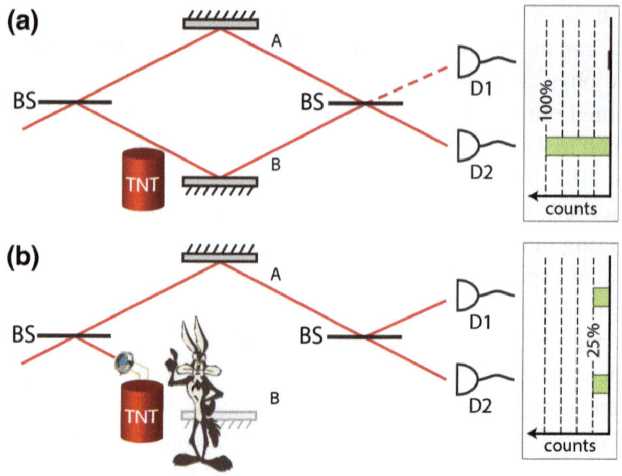

Fig. 1 Bomb testing with a Mach—Zehnder interferometer operated with a single photon at a time. The beam splitters have split ratio 50:50. **a** The phase difference between the two arms is adjusted such that all photons are directed to detector D2. The object situated close to the lower branch B (e.g., a dud bomb not equipped with the trigger) does not intercept the photons. **b** The object (e.g., a "live bomb" equipped with an avalanche photodiode which is triggered by a single photon[1]) intercepts the photons in the lower branch B

with equal probability. In case detector D2 clicks, insufficient knowledge is gained to conclude on whether the object is present, as this outcome also occurs for no object present ("dud bomb"). If the dark detector D1 lights up, however, the presence of an object in one of the two arms is signaled with certainty. Since the photon could not reach the detector if it had touched the object, finding a photon with detector D1 detects the presence of the object without touching it, therefore leaving the "bomb" intact. Following Elitzur and Vaidman, we call this measurement scheme *interaction free* [19].

We obtain further insight in the Elitzur–Vaidman "bomb test" by analyzing its outcomes from the perspective of statistical hypothesis testing, where the Elitzur–Vaidman quantum device is employed to test whether the "bomb" is live (*positive* test result). The typical figure of merit in hypothesis testing is the statistical power $1-\beta$, that is, the fraction of "live bombs" which are correctly identified (without being triggered) and rescued. In their original work [16], Elitzur and Vaidman showed that this fraction amounts to 25% (*true positives*). It is worth extending the analysis of "bomb testing" to quantum devices that, under realistic conditions, are impaired by decoherence. We assume that decoherence reduces the fringe amplitude of the MZ interferometer in an unbiased way, resulting in a contrast C less than unity. In the presence of decoherence, the statistical power of the test remains

[1]For Wile's safety, the avalanche photodiode must be cooled to near absolute zero in order to suppress dark counts.

unchanged, since this quantity describes the situation of a "live bomb," where the coherence of the photon plays no role, see Fig. 1b. We therefore introduce a second figure of merit accounting for decoherence, which is given by the statistical error of type I, that is, the probability α that we erroneously rescue the bomb even though the bomb is a dud (*false positive*). One can show that the probability of type I errors is $\alpha = (1 - C)/2$. This measure vanishes only for a decoherence-free "bomb tester" and becomes $\alpha = 1/2$ in the completely incoherent limit (equivalent to random selection). It is also worth mentioning that, by allowing for repeated measurements (this is always possible until the "bomb" has not exploded), the statistical power $1 - \beta$ can be straightforwardly increased to 1/3 for 50:50 beam splitters and to 1/2 for beam splitters with different split ratios. Furthermore, a more complex variant of this scheme based on the quantum Zeno effect has been suggested [20] and experimentally demonstrated using traveling photons [21], allowing one to approach unity efficiency ($\beta = 0$).

Let us now reconsider the situation illustrated in Fig. 1b from the perspective of a macro-realist, who conducts the same experiment but with a massive, ideally macroscopic object traveling along the two branches of the MZ interferometer. For a macro-realist, the massive particle travels either along branch A or along branch B, but not in a superposition state of the trajectories A and B. If one discards by post-selection all events where the particle's presence on one of the two trajectories has been directly detected through the absorption by the object, then only interaction-free measurements (i.e., ideal negative measurements) of the particle's position are considered. By intercepting the particle at one time in branch A and at another time in branch B, the macro-realist learns about the particle's position in the MZ interferometer avoiding any interaction with it. Therefore, to a macro-realist, ideal negative measurements must appear *noninvasive*, since the subsequent evolution of the particle could have not been influenced by the presence of an object where the particle was *not*. Noninvasive measurements constitute an important prerequisite for any rigorous test of macro-realism through a violation of the Leggett–Garg inequality [22].

3 Relation to the Leggett–Garg Inequality

Based on two assumptions, (A1) macro-realism and (A2) noninvasive measurements, the Leggett–Garg inequality bounds a linear combinations of two-time correlation measurements,

$$K = \langle Q(t_2)Q(t_1)\rangle + \langle Q(t_3)Q(t_2)\rangle - \langle Q(t_3)Q(t_1)\rangle \leq 1, \tag{1}$$

where $Q(t_i)$ denote the outcome of measurements carried out on the object at three subsequent times ($i = 1, 2, 3$). The values assigned to the individual measurement outcomes have to fulfill the condition $|Q(t_i)| \leq 1$ but can otherwise be freely chosen. While macro-realism fulfills this inequality, standard quantum mechanics allows experiments violating it.

Observing a violation of the inequality (1) allows one to refute (i.e., *falsify*) the assumptions underlying the Leggett–Garg test in the range of parameters investigated by the experiment, which in general are represented by the mass of the superposition states and their spatial split distance [14]. For simple logical reasons, a violation of Eq. (1) implies that *at least one* of the two assumptions (A1) and (A2) must not hold. In conducting a Leggett–Garg test, it is therefore crucial that assumption (A2) of noninvasive measurability cannot be simply dismissed by a macro-realist claiming that a measurement operation, due experimental inadvertence, has influenced the subsequent evolution of the particle; or else, even in case of an observed violation, no claim can be made concerning assumption (A1) of macro-realism. In the literature, this type of objection is addressed as the *clumsiness* loophole [23]. To circumvent it, Leggett and Garg suggested using ideal negative measurements, which, to say it à? la Vaidman and Elitzur, are interaction free. It should also be noted that in standard quantum theory both assumptions do not hold. In fact, standard quantum theory postulates (1) no limit on the mass and split distance of a superposition state and that (2) even a negative measurement can cause the wave packet's reduction. The latter point is indeed central to the Elitzur–Vaidman interaction-free experiment.

In the Elitzur–Vaidman experiment, the object that is put to the test of the Leggett–Garg inequality is the single photon travelling along the two paths of the MZ interferometer. In the following we define the three measurement operations performed on the photon and their assigned values $Q(t_i)$, which are employed to perform the Leggett–Garg test in Eq. (1):

$Q(t_1)$: We identify the first measurement at t_1 with the preparation of the initial state - a photon in the input of the MZ interferometer. This measurement is by definition noninvasive, as it leaves the particle in the initial state. We assign to this measurement the value $Q(t_1) = +1$.

$Q(t_2)$: This measurement is performed at time t_2 when the photon is either in branch A or B of the MZ interferometer. It detects in which branch the photon travels by removing the photon at one time from branch A, at another time from branch B. The events in which the photon was directly intercepted by the object (i.e., when the "bomb" exploded) are discarded by post-selection in order to ensure that only ideal negative measurements are performed; this measurement must thus appear to a macro-realist as noninvasive as it avoids any direct interaction with the photon itself. We assign to this measurement the constant value $Q(t_2) = +1$ regardless of which trajectory the photon has taken. It should be noted that this measurement is not performed when evaluating the correlation function $\langle Q(t_3)Q(t_1)\rangle$ of the Leggett–Garg test. In fact, from the perspective of a macro-realistic who advocates (A1) and (A2), an ideal negative measurement could not have influenced the evolution of the particle.

$Q(t_3)$: The final measurement is performed at time t_3 when the photon has reached the output of the MZ interferometer. Depending on whether detector D1 or D2 has clicked, we assign to this measurement the value

$Q(t_3) = +1$ or $Q(t_3) = -1$, respectively. Because we are not interested in the system's evolution after t_3, noninvasiveness of this measurement operation is not required.

Previous Leggett–Garg experimental tests prior to Robens et al. [11] only considered dichotomic designations of $Q(t_2)$, as opposed to the constant choice of $Q(t_2) = +1$ here. By deliberately disregarding the unnecessary, dichotomic constraint, we allow the Leggett–Garg correlation function of the Elitzur–Vaidman experiment to reach the maximum value, $K = 2$, as permitted by quantum theory for a two level system (i.e., the photon in a superposition state of the trajectories A and B). It can be shown that a violation of the inequality in Eq. (1) is also produced in the Elitzur–Vaidman experiment for a dichotomic choice of $Q(t_2)$ (see "Appendix C"). However, in this case the maximum value of K predicted by quantum theory is only 3/2 [24, 25] instead of 2.

Taking into account our specific designation of $Q(t_i)$, we can recast Eq. (1) into a simpler form. Since $Q(t_2) = 1$, the correlation function $\langle Q(t_3)Q(t_2)\rangle$ equals $\langle Q(t_3)\rangle_{\text{with } Q2}$, that is, the average value of $Q(t_3)$ conditioned on a negative result of the measurement $Q(t_2)$. Likewise, since $Q(t_1) = 1$, the correlation function $\langle Q(t_3)Q(t_1)\rangle$ simplifies to $\langle Q(t_3)\rangle_{\text{without } Q2}$, that is, the average value of $Q(t_3)$ without having measured the position of the photon at t_2. Our experiment can thus be analyzed with a simplified version of Eq. (1),

$$K = 1 + \langle Q(t_3)\rangle_{\text{with } Q2} - \langle Q(t_3)\rangle_{\text{without } Q2} \leq 1\,. \tag{2}$$

It is interesting to go one step further to provide an interpretation of the Leggett–Garg correlation function K from the point of view of quantum theory. We know that the values of $Q(t_3) = \pm 1$ are evenly distributed when the $Q(t_2)$ measurement is performed, since the latter reveals the which-way information; hence $\langle Q(t_3)\rangle_{\text{with } Q2} = 0$. Moreover, one can prove that $\langle Q(t_3)\rangle_{\text{without } Q2}$ is identical to the contrast C of the MZ interferometer (see "Appendix D"). Thus, we find that the correlation function K takes the suggestive form

$$K = 1 + C\,. \tag{3}$$

Intuitively, the function K provides a quantitative indication, say a witness, of the amount of superposition involved in the evolution of the quantum particle. Note that K can be put in relation to the first quantum witness W of superposition states introduced in Ref. [26] (also described as no-signaling in time in Ref. [27]). In fact, $W = |K - 1| = C$ as shown in Ref. [11]. This demonstrates that the figure of merit α of a partially decohered "bomb tester" (see Sec. 2) is directly related to the quantum witness W since $\alpha = (1 - W)/2$. Furthermore, our results shows that any quantum particle exhibiting a nonvanishing interference contrast should allow, according to quantum theory, for a violation of Eq. (1), provided that one can additionally show through an experiment that the which-way information acquired through interaction-free measurements yields a vanishing contrast, that is, $\langle Q(t_3)\rangle_{\text{with } Q2} = 0$.

4 Interaction-Free Measurement with Trapped Atoms

Our atomic realization of the Elitzur–Vaidman experiment employs single neutral atoms trapped in an optical potential instead of flying photons. Instead of delocalizing the particle on two distinct trajectories as in the MZ interferometer shown in Fig. 1, we let the particle evolve in a superposition of two long-lived internal states, which we denote by $|\uparrow\rangle$ and $|\downarrow\rangle$ hereafter. In our experiment, an atomic Ramsey interferometer plays the role of the optical MZ interferometer.

4.1 Experimental Apparatus

4.1.1 State-Dependent Optical Conveyor Belts

At the core of our realization of the Elitzur–Vaidman experiment with trapped atoms are polarization-synthesized (PS) optical lattices, which were recently introduced by Robens et al. [28]: two one-dimensional, periodic optical potentials can be independently shifted along their common longitudinal direction to selectively transport atoms in either one of two internal states, $|\uparrow\rangle$ and $|\downarrow\rangle$. In essence, two copropagating laser beams of opposite circular polarization interfere with a third, counterpropagating, linearly polarized beam. Their interference gives rise to two standing waves of left- and right-handed polarization, whose positions are actively controlled by means of two independent optical phase-locked loops. We obtain a residual jitter of their relative position on the order of 1 Å, which is much smaller than the longitudinal extent of the atom's wave function of $\approx$20 nm. At the so-called magic wavelength $\lambda_L = 866\,\text{nm}$ of cesium atoms, the internal state $|\uparrow\rangle = |F = 4, m_F = 4\rangle$ interacts exclusively with the σ^+-polarized component, while $|\downarrow\rangle = |F = 3, m_F = 3\rangle$ predominantly interacts with the σ^--polarized component [29]. Moreover, we choose a relatively deep lattice with a depth of $U/k_B \approx 80\,\mu\text{K}$ to prevent tunneling between different sites. Hence, atoms in the two internal states are bound to two spatially superimposed, but orthogonally polarized lattices, which can be individually shifted much like two independent optical conveyor belts: atoms in the $|\uparrow\rangle$ and $|\downarrow\rangle$ states follow, nearly rigidly, the σ^+-polarized and σ^--polarized standing waves, respectively.

State-dependent optical lattices have been pioneered first in the MPQ laboratories [30, 31], demonstrating another example of Theodor W. Hänsch's legacy as an inspiration for future generations of experiments. Compared to former realizations of state-dependent optical lattices, PS optical lattices have replaced the polarization control formerly based on an electro-optic modulator by a direct synthesis of light polarization, which enable arbitrary, state-dependent displacements of atoms. Polarization synthesis is realized through rf-control of the optical phases (0.1° RMS phase jitter) of two overlapped beams with opposite circular polarization [28].

4.1.2 Microwave Control

We employ microwave radiation at the cesium clock frequency of 9.2 GHz to induce coherent oscillations between the two atomic hyperfine states, $|\uparrow\rangle$ and $|\downarrow\rangle$ with a Rabi frequency of 55 kHz. Therefore, the application of the microwave radiation field for 4.5 μs realizes a so-called $\pi/2$ pulse, which transforms a pure internal state into an equal superposition of $|\uparrow\rangle$ and $|\downarrow\rangle$. In our realization of the Elitzur–Vaidman experiment, microwave $\pi/2$ pulses represent the atomic analogue of the beam splitters for photons, which are illustrated in Fig. 1.

4.1.3 Nondestructive Spin State Measurement

Exploiting PS optical lattices, we devised a novel measurement method to detect the internal state of the atom in the most gentle way possible. In quantum mechanics, least perturbative measurements are called nondestructive (or, equivalently, quantum nondemolition measurements): the state of the object is preserved after the measurement in the eigenstate of the measured quantity corresponding to the observed outcome [32]; a repeated measurement of the same quantity would therefore leave the state unchanged. Conversely, the widely employed push-out method, which expels atoms in one particular internal state from the trap by applying state-selective radiation pressure [33], represents a destructive measurement.

Our method is closely related to an optical Stern–Gerlach experiment [34, 35], where spin-position entanglement is created by state-dependent light fields, and is reminiscent of the nondestructive Stern–Gerlach experiment by Dehmelt [36, 37]. We realize nondestructive measurements by displacing atoms by a discrete number of lattice sites conditioned on the internal state, thereby transferring spin states to well-separated positions. The position can be detected efficiently at a later time by fluorescence imaging under molasses illumination without any atom loss [38]; we identify the correct position, and therefore the spin state, with >99% reliability. The translational invariance of the optical lattice ensures that this measurement protocol constitutes a nondestructive measurement of the internal state. Note also that it is not necessary that the position readout immediately follows the state-dependent displacement. The possibility to postpone the "destructive" fluorescence image of the atoms till the end of the evolution allows us to leave the evolution of the system minimally perturbed as required by a nondestructive spin measurement. We use this technique for the ideal negative measurement $Q(t_2)$, in which case only one spin component at a time is displaced, leaving the other unperturbed (see Sect. 4.2). It should also be noted that nondestructive measurements are not strictly needed in order to perform the ideal negative measurements that are instead required for the Elitzur–Vaidman experiment and the Leggett–Garg test [20]. This fact is illustrated in Fig. 1 where the photon removal represents indeed a destructive measurement, since the particle is destroyed if intercepted by the "bomb."

4.2 Measurement Protocols

We start each experimental sequence with, on average, 1.2 atoms loaded into the lattice. The loading procedure is stochastic and only atoms sitting at sufficiently separated lattice sites are considered. Atoms are cooled to the longitudinal ground state using first molasses cooling and then microwave sideband cooling [39]. With the quantization axis chosen along the lattice direction, optical pumping by a σ^+-polarized laser beam initializes >99% of the atoms in the state $|\uparrow\rangle$.

We outline in Fig. 2 the protocols employed to measure the two terms forming in the Leggett–Garg inequality in Eq. (2). On the left-hand side, we present the procedure with no "bomb" present, which comprises $Q(t_1)$ (preparation of the initial spin state preparation) and $Q(t_3)$ (detection of the final spin state) measurements, but not $Q(t_2)$ (the "live bomb" in one of the interferometer's branch): the spin preparation is followed by a $\pi/2$ pulse, a variable waiting time, a second $\pi/2$ pulse with adjustable microwave phase ϕ, and a nondestructive spin measurement mapping the spin state onto different positions as described in Sect. 4.1. This sequence describes a Ramsey experiment interrogating the spin coherence of a trapped atom. Note that this situation is fully equivalent to the unobstructed MZ interferometer of Fig. 1a, where the atomic internal states here take the place of the distinct trajectories of the photon. Here we adjust the microwave phase ϕ beforehand in order to ensure the highest probability to detect $|\downarrow\rangle$ at time t_3, much like in the MZ interferometer one must balance the two branches to route all photons to detector D2.

Let us turn to the application of the atomic "bomb test" illustrated on the right-hand side of Fig. 2. After the initial microwave $\pi/2$ pulse, which puts the atom in a coherent superposition state (for a macro-realist, a stochastic mixture of both states), we spatially remove atoms, in different experiments, at one time in state $|\uparrow\rangle$ and at another time in state $|\downarrow\rangle$ by transporting them apart by seven lattice sites in about 200 μs. The number of sites is chosen sufficiently large to avoid any error in the final position measurement, and therefore in the spin reconstruction. For the Leggett–Garg test, we post-select the events where an atom is indeed found at position D1 or D2, meaning that it has not been removed at time t_2, and its spin has thus been measured by an ideal negative measurement $Q(t_2)$, as was argued in Sect. 3. The excluded events instead correspond to the atoms removed from the Ramsey interferometer, and are tantamount to having triggered the "bomb" in one of the two branches of the MZ interferometer of Fig. 1b.

5 Experimental Results

In panels Fig. 3a–c, we show the raw data corresponding to our atomic realization of the Elitzur–Vaidman experiment for three different situations: (a) without the "bomb," (b) with the "bomb" removing atoms in $|\uparrow\rangle$, and (c) again with the

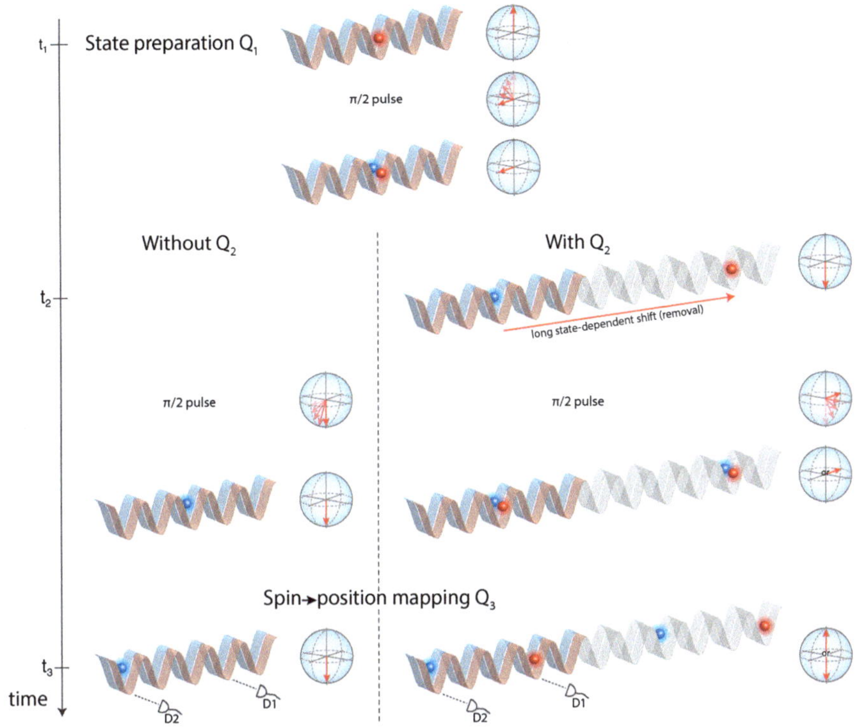

Fig. 2 Illustration of the Elitzur–Vaidman experiment using single atoms. Atoms are trapped in state-dependent optical lattices, which consist of two independently movable, periodic optical potentials for atoms in the internal states $|\uparrow\rangle$ and $|\downarrow\rangle$. The two atomic states form a spin-1/2 system, which is represented on the Bloch sphere at different moments of the time evolution; short microwave pulses allow us to rotate the spin. On the left-hand side, protocol of a Ramsey interferometer, whose pulses are configured to produce the state $|\downarrow\rangle$; this situation is equivalent to that in Fig. 1a. The spin information is eventually mapped onto two different positions on the lattice, D1 and D2, which are efficiently detected by fluorescence imaging. On the right-hand side, protocol of a Ramsey interferometer where an interaction-free measurement (i.e., an ideal negative measurement) of the spin state is performed at the intermediate time t_2. This measurement intercepts only atoms in one spin state by transporting them far apart (*grayed lattice regions*); this situation is equivalent to that in Fig. 1b

"bomb" but removing atoms in $|\downarrow\rangle$. From the dataset (a) we reconstruct the correlation function $\langle Q(t_3)\rangle_{\text{without } Q2}$, while by merging the datasets (b) and (c) we obtain the correlation function $\langle Q(t_3)\rangle_{\text{with } Q2}$. The combination of these two correlation functions produce a violation of the Leggett–Garg inequality as defined in Eq. (2). In Fig. 3d we present the recorded Leggett–Garg correlation function K for different durations of the waiting time, which separates the two microwave $\pi/2$ pulses of the Ramsey interferometer. For a minimum waiting time of 5 μs, we record a value of $K = 1.958 \pm 0.033$, which violates the Leggett–Garg

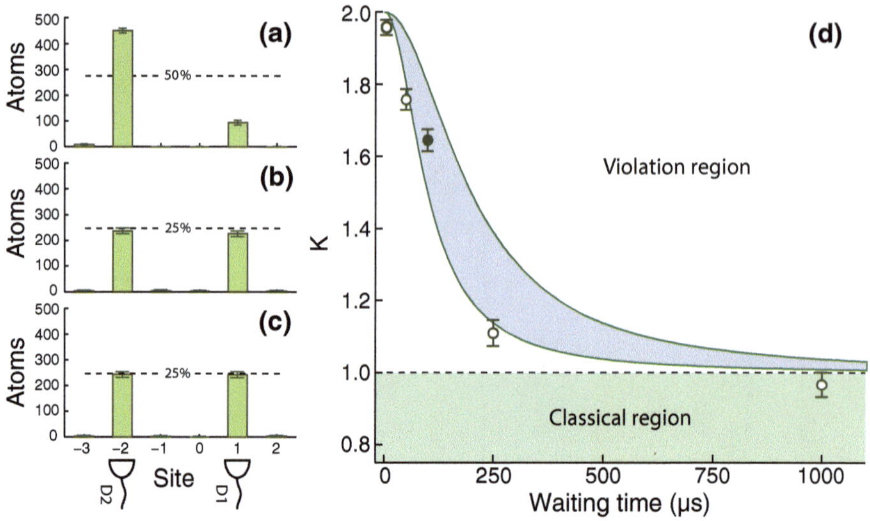

Fig. 3 Experimental violation of the Leggett–Garg inequality in the quantum-to-classical transition. From **a** to **c**, distributions at time t_3 of the detected atom at sites D1 and D2 for a waiting time of 100 μs, corresponding to the solid point in (d) for three different protocols. **a** Without the $Q(t_2)$ measurement (left-hand-side protocol in Fig. 2). **b** With the $Q(t_2)$ measurement shifting atoms in $|\uparrow\rangle$ away at time t_2 (right-hand-side protocol in Fig. 2). **c** The same but with atoms in $|\downarrow\rangle$ shifted away. **d** Values of the Leggett–Garg correlation function K of Eq. (2) for increasing waiting times between the two $\pi/2$ pulses. Decoherence gradually suppresses the quantum behavior of the atom. The *shaded band* represents the theoretical quantum-mechanical prediction for coherence times between 75 and 200 μs caused by differential scalar light shift [40]. Percentage values are referred to the total number of interrogated atoms in each dataset. The *vertical error bars* represent 1 σ statistical uncertainty

inequality by 21 σ. While this value of K lies very close to the decoherence-free prediction ($K = 2$), the recorded values of K decrease visibly for longer waiting times (i.e., increasing decoherence) till they reach the value of 1 for fully decohered spin dynamics, in fulfillment of the Leggett–Garg inequality.

In the present experiment, we attribute the main source of decoherence to scalar differential light shift [40], causing inhomogeneous spin dephasing of the atoms, which in our case are thermally distributed over more than 100 vibrational levels in the directions transverse to the lattice. Besides producing a rigorous violation of the Leggett–Garg inequality for a pseudo-spin-1/2 particle, our results show that the correlation function K can be interpreted, from the point of view quantum theory, as a quantum witness of superposition states, and employed to study decoherence—one of the most basic mechanisms affecting atoms trapped in optical potentials [41].

6 Conclusions

In this paper, we have shown that the "bomb-testing" experiment not only provides dramatic evidence of the "weirdness" of quantum mechanics, as originally intended by Elitzur und Vaidman, but also can be recast in a rigorous test of the macro-realistic worldview based on the violation of the Leggett–Garg inequality. With our realization of the Elitzur–Vaidman experiment, we can refute the macro-realistic assumptions for individual cesium atoms with a 21 σ statistical confidence. While from the point of view of "macroscopicity", the present Leggett–Garg experiment does not improve on the results obtained with quantum walks by Robens et al. [11], our atomic implementation of the Elitzur–Vaidman experiment can be directly extended in the future to superposition states involving splitting distances on the macroscopic scale of a millimeter and beyond [42, 43]. Moreover, the analysis of the experimental results allows us to understand the exact relation between the Leggett–Garg correlation function K and the interference contrast C of the corresponding Ramsey interferometer, thus providing intuition about the quantum-to-classical transition of the Leggett–Garg test.

It is also worth noting that our experiment demonstrates a nondestructive measurement technique of the spin state of the atoms, where spin-position entanglement is used to transfer information from spin to position space. This mapping technique allows us to directly read out both spin states in a Ramsey interferometer, thus avoiding the shortcomings of the widely used push-out technique, where atoms are lost after the measurement and, most importantly, the measurement outcomes must be corrected for atom losses. Further, with the nondestructive measurement technique demonstrated here we could recycle atoms multiple times. We finally anticipate that nondestructive spin measurements, preserving spatial coherence of atoms delocalized over several lattice sites, can find application in the realization of dissipative quantum-walk protocols [44].

Acknowledgements We are very grateful to Jean-Michel Raimond for insightful discussions. We acknowledge financial support from NRW-Nachwuchsforschergruppe "Quantenkontrolle auf der Nanoskala," ERC grant DQSIM, and EU Integrated Project SIQS. In addition, C.R. acknowledges support from the BCGS program and the Studienstiftung des deutschen Volkes.

Appendix A: Statistical Errors

In this work, the confidence intervals of the correlation measurements represent 1 σ statistical uncertainty, which has been computed by fitting a Gaussian profile to the bootstrapped distribution (i.e., the distribution obtained by resampling with replacement). Independently from bootstrapping, we also computed the statistical uncertainties using Monte Carlo resampling, where the statistical errors of position distributions are estimated with binomial statistics (Clopper–Pearson method). The two estimation methods lead to consistent results. While Monte Carlo analysis requires invariant statistical properties to be valid, bootstrapping analysis remains valid also in the presence of slow drifts of experimental parameters. The close agreement between the two statistical analyses indicates that each correlation measurement of K (lasting about 120 min) is performed under constant experimental conditions.

Appendix B: Systematic Errors

Provided that the experiment is performed under constant experimental conditions, systematic errors do not invalidate the result of a Leggett–Garg test. In fact, if we consider the three main mechanisms that bring about systematic fluctuations: (1) Imperfect initialization prepares $<1\%$ of the atoms in the wrong internal state. However, to derive the Leggett–Garg inequality, a statistical mixture defining the initial state is perfectly admissible. (2) Imperfect reconstruction of the atom's position can be accounted for in terms of a noisy measurement apparatus. (3) Spontaneous spin flips due to the finite T_1 time can be accounted for in terms of an additional stochastic process, which also contributes to determine the system's evolution. We estimate that each of these three mechanisms actually affects the position distribution by $<1\%$, that is less than the statistical uncertainty.

Appendix C: Dichotomic Choice

We verified that our system produces a violation also with a dichotomic definition of $Q(t_2)$. We performed a Ramsey sequence using two subsequent $\pi/3$ pulses. In this case, we set $Q(t_1) = 1$ by preparation and designated both $Q(t_2)$ and $Q(t_3)$ as $+1$ for $|\uparrow\rangle$ and as -1 for $|\downarrow\rangle$. The measured value $K = 1.503 \pm 0.051$ is consistent with the quantum mechanical expectation of $K = 3/2$.

Appendix D: Relation with Ramsey Contrast

A Ramsey interference fringe is represented by the probability $p_\uparrow$ of measuring $|\uparrow\rangle$ as a function of the Ramsey phase. Assuming the fringe is centered around the average value $\bar{p}_\uparrow = 1/2$ (this is true in case of, e.g., pure spin dephasing), the contrast can be expressed as

$$C = p_{\uparrow,\max} - p_{\uparrow,\min} = 1 - 2p_{\uparrow,\min}, \tag{4}$$

where $p_{\uparrow,\max/\min}$ is the maximum/minimum value of the fringe. Because the two Ramsey $\pi/2$ pulses in the Elitzur–Vaidman experiment are set to have the same phase (Sect. 4), the value of $p_\uparrow$ measured when evaluating $\langle Q(t_3)Q(t_1)\rangle$ corresponds to $p_{\uparrow,\mathrm{min}}$. Hence, we obtain that the correlation function reads

$$\langle Q(t_3)Q(t_1)\rangle = p_\uparrow - p_\downarrow = p_\uparrow - (1 - p_\uparrow) = -C, \tag{5}$$

which together with the definition of the LG inequality in Eq. (1) proves Eq. (3).

References

1. J.A. Wheeler, W.H. Zurek (eds.), *Quantum Theory and Measurement, Princeton Series in Physics* (Princeton University Press, Princeton, 1983)
2. A.J. Leggett, The quantum measurement problem. Science **307**, 871 (2005)
3. G. Lüders, Concerning the state-change due to the measurement process. Ann. Phys. **8**, 332 (1951)
4. A. Bassi, K. Lochan, S. Satin, T.P. Singh, H. Ulbricht, Models of wave-function collapse, underlying theories, and experimental tests. Rev. Mod. Phys. **85**, 471 (2013)
5. G. Ghirardi, A. Rimini, T. Weber, Unified dynamics for microscopic and macroscopic systems. Phys. Rev. D **34**, 470 (1986)
6. L. Diósi, Models for universal reduction of macroscopic quantum fluctuations. Phys. Rev. A **40**, 1165 (1989)
7. I. Pikovski, M. Zych, F. Costa, C. Brukner, Universal decoherence due to gravitational time dilation. Nat. Phys. **11**, 668 (2015)
8. C. Emary, N. Lambert, F. Nori, Leggett-Garg inequalities. Rep. Prog. Phys. **77**, 016001 (2013)
9. A.J. Leggett, A. Garg, Quantum mechanics versus macroscopic realism: is the flux there when nobody looks? Phys. Rev. Lett. **54**, 857 (1985)
10. G.C. Knee, S. Simmons, E.M. Gauger, J.J.L. Morton, H. Riemann, N.V. Abrosimov, P. Becker, H.-J. Pohl, K.M. Itoh, M.L.W. Thewalt, G.A.D. Briggs, S.C. Benjamin, Violation of a Leggett-Garg inequality with ideal non-invasive measurements. Nat. Commun. **3**, 606 (2012)
11. C. Robens, W. Alt, D. Meschede, C. Emary, A. Alberti, Ideal negative measurements in quantum walks disprove theories based on classical trajectories. Phys. Rev. X **5**, 011003 (2015)
12. A.J. Leggett, Testing the limits of quantum mechanics: motivation, state of play, prospects. J. Phys.: Condens. Matter **14**, 415 (2002)
13. M. Arndt, K. Hornberger, Testing the limits of quantum mechanical superpositions. Nat. Phys. **10**, 271 (2014)
14. S. Nimmrichter, K. Hornberger, Macroscopicity of mechanical quantum superposition states. Phys. Rev. Lett. **110**, 160403 (2013)
15. G. Ghirardi, *Sneaking a Look at God's Cards, Revised Edition: Unraveling the Mysteries of Quantum Mechanics* (Princeton University Press, Princeton, 2007)
16. A.C. Elitzur, L. Vaidman, Quantum mechanical interaction-free measurements. Found. Phys. **23**, 987 (1993)
17. P. Kwiat, H. Weinfurter, T. Herzog, A. Zeilinger, M.A. Kasevich, Interaction-free measurement. Phys. Rev. Lett. **74**, 4763 (1995a)
18. M. Hafner, J. Summhammer, Experiment on interaction-free measurement in neutron interferometry. Phys. Lett. A **235**, 563 (1997)
19. L. Vaidman, The meaning of the interaction-free measurements. Found. Phys. **33**, 491 (2003)

20. P. Kwiat, H. Weinfurter, T. Herzog, A. Zeilinger, M. Kasevich, Experimental Realization of "Interaction-Free" Measurements, in Fundamental Problems in Quantum Theory: A Conference Held in Honor ofProfessor John A. Wheeler, Vol. 755, edited by N. Y. A. of Sciences, New York, p. 129 (1995)
21. P.G. Kwiat, A.G. White, J.R. Mitchell, O. Nairz, G. Weihs, H. Weinfurter, A. Zeilinger, High-efficiency quantum interrogation measurements via the quantum zeno effect. Phys. Rev. Lett. **83**, 4725 (1999)
22. C. Emary, Decoherence and maximal violations of the Leggett-Garg inequality. Phys. Rev. A **87**, 032106 (2013)
23. M.M. Wilde, A. Mizel, Addressing the clumsiness loophole in a Leggett-Garg test of macrorealism. Found. Phys. **42**, 256 (2012)
24. C. Budroni, C. Emary, Temporal quantum correlations and Leggett-Garg inequalities in multi-level systems, preprint (2013). arXiv:1309.3678
25. G. Schild, C. Emary, Maximum violations of the quantum-witness equality. Phys. Rev. A **92**, 032101 (2015)
26. C.-M. Li, N. Lambert, Y.-N. Chen, G.-Y. Chen, F. Nori, Witnessing quantum coherence: from solid-state to biological systems. Sci. Rep. **2**, 885 (2012)
27. J. Kofler, C. Brukner, Condition for macroscopic realism beyond the Leggett-Garg inequalities. Phys. Rev. A **87**, 052115 (2013)
28. C. Robens, J. Zopes, W. Alt, S. Brakhane, D. Meschede, A. Alberti, Low-entropy states of neutral atoms in polarization-synthesized optical lattices. arXiv:1608.02410 [quant-ph] (2016)
29. D. Jaksch, C. Bruder, J.I. Cirac, C.W. Gardiner, P. Zoller, Cold bosonic atoms in optical lattices. Phys. Rev. Lett. **81**, 3108 (1998)
30. O. Mandel, M. Greiner, A. Widera, T. Rom, T.W. Hänsch, I. Bloch, Coherent transport of neutral atoms in spin-dependent optical lattice potentials. Phys. Rev. Lett. **91**, 010407 (2003)
31. O. Mandel, M. Greiner, A. Widera, T. Rom, T. Hänsch, I. Bloch, Controlled collisions for multi-particle entanglement of optically trapped atoms. Nature **425**, 937 (2004)
32. S. Haroche, J.M. Raimond, *Exploring the Quantum: Atoms, Cavities, and Photons* (Oxford University Press, New York, 2006)
33. S. Kuhr, W. Alt, D. Schrader, I. Dotsenko, Y. Miroshnychenko, W. Rosenfeld, M. Khudaverdyan, V. Gomer, A. Rauschenbeutel, D. Meschede, Coherence properties and quantum state transportation in an optical conveyor belt. Phys. Rev. Lett. **91**, 213002 (2003)
34. T. Sleator, T. Pfau, V. Balykin, O. Carnal, J. Mlynek, Experimental demonstration of the optical Stern-Gerlach effect. Phys. Rev. Lett. **68**, 1996 (1992)
35. C.Y. Park, J.Y. Kim, J.M. Song, D. Cho, Optical Stern-Gerlach effect from the Zeeman-like ac Stark shift. Phys. Rev. A **65**, 033410 (2002)
36. H. Dehmelt, Stored-Ion Spectroscopy, in *Advances in Laser Spectroscopy*, F. T. Arecchi, F. Strumia, H. Walther (Springer, Boston, 1983), p. 153
37. H. Dehmelt, Continuous Stern-Gerlach effect: principle and idealized apparatus. Proc. Natl. Acad. Sci. USA **83**, 2291 (1986)
38. A. Alberti, C. Robens, W. Alt, S. Brakhane, M. Karski, R. Reimann, A. Widera, D. Meschede, Super-resolution microscopy of single atoms in optical lattices. New J. Phys. **18**, 053010 (2016)
39. N. Belmechri, L. Förster, W. Alt, A. Widera, D. Meschede, A. Alberti, Microwave control of atomic motional states in a spin-dependent optical lattice. J. Phys. B: At. Mol. Opt. Phys. **46**, 4006 (2013)
40. S. Kuhr, W. Alt, D. Schrader, I. Dotsenko, Y. Miroshnychenko, A. Rauschenbeutel, D. Meschede, Analysis of dephasing mechanisms in a standing-wave dipole trap. Phys. Rev. A **72**, 023406 (2005)
41. A. Alberti, W. Alt, R. Werner, D. Meschede, Decoherence models for discrete-time quantum walks and their application to neutral atom experiments. New J. Phys. **16**, 123052 (2014)

42. T. Kovachy, P. Asenbaum, C. Overstreet, C.A. Donnelly, S.M. Dickerson, A. Sugarbaker, J.M. Hogan, M.A. Kasevich, Quantum superposition at the half-metre scale. Nature **528**, 530 (2015)
43. D.M. Stamper-Kurn, G.E. Marti, H. Müller, Verifying quantum superpositions at metre scales. Nature **537**, E1 (2016)
44. T. Rakovszky, J. Asbóth, A. Alberti, Detecting topological invariants in chiral symmetric insulators via losses (2016)

Adjacency Graphs and Long-Range Interactions of Atoms in Quasi-degenerate States: Applied Graph Theory

C.M. Adhikari, V. Debierre, and U.D. Jentschura

Abstract We analyze, in general terms, the evolution of energy levels in quantum mechanics, as a function of a coupling parameter, and demonstrate the possibility of level crossings in systems described by irreducible matrices. In long-range interactions, the coupling parameter is the interatomic distance. We demonstrate the utility of adjacency matrices and adjacency graphs in the analysis of "hidden" symmetries of a problem; these allow us to break reducible matrices into irreducible subcomponents. A possible breakdown of the no-crossing theorem for higher-dimensional irreducible matrices is indicated, and an application to the $2S$–$2S$ interaction in hydrogen is briefly described. The analysis of interatomic interactions in this system is important for further progress on optical measurements of the $2S$ hyperfine splitting.

1 Introduction

In quantum mechanical systems described by a (2×2)-matrix, no level crossings can typically occur [1, 2]. This is known as the "no level crossing theorem" and often illustrated on the basis of the simple (2×2)-model Hamiltonian matrix

$$H' = H + P = \begin{pmatrix} E_1 & 0 \\ 0 & E_2 \end{pmatrix} + \begin{pmatrix} 0 & Cg \\ Cg & 0 \end{pmatrix}, \tag{1}$$

where E_1 and E_2 are the unperturbed energy levels, C is a parameter, and g is the coupling constant. The energy levels are

---

This article is part of the topical collection "Enlightening the World with the Laser" - Honoring T. W. Hänsch guest edited by Tilman Esslinger, Nathalie Picqué, and Thomas Udem.

C.M. Adhikari • V. Debierre • U.D. Jentschura (✉)
Department of Physics, Missouri University of Science and Technology, Rolla, MO 65409-0640, USA
e-mail: ulj@mst.edu

D. Meschede et al. (eds.), *Exploring the World with the Laser*,
https://doi.org/10.1007/978-3-319-64346-5_11

$$E_{\pm} = \frac{1}{2}(E_1 + E_2) + \frac{1}{2}\sqrt{(E_1 - E_2)^2 + 4(C\,g)^2}\,. \tag{2}$$

As a function of g, one obtains two hyperbolas, with the distance of "closest approach" between the energy levels occurring for $g = 0$, with a separation $|E_+ - E_-| = |E_1 - E_2|$. For a level crossing to occur at $g = 0$, one has to have $E_1 = E_2$. The larger the perturbation, the more the energy levels "repel" each other.

However, the situation is less clear for more complex systems involving more than two energy levels. To this end, we shall analyze a higher-rank matrix which describes energy levels some of which repel each other on the basis of interlevel couplings, in a system which obviously can be broken into smaller subcomponents (i.e., the Hamiltonian is a reducible matrix having irreducible submatrices). As the levels in the irreducible subsystems evolve from the weak-coupling to the strong-coupling regime, those coming from different irreducible submatrices cross. When additional couplings are introduced between the subsystems, the matrix becomes irreducible. In this case, we shall demonstrate that some of the level crossings are avoided, but not all. Our example will be based on a (6×6)-matrix.

Another question which sometimes occurs in the analysis of interatomic interactions, and other contexts in quantum mechanics, concerns the reducibility of a matrix. Reducible tensors are usually introduced in the context of the rotation group. Under a rotation, scalars transform into scalars, vectors transform into vectors, quadrupole tensors transform into quadrupole tensors, and so on. It means that a matrix representation of the rotation would have an obvious block structure when formulated in terms of the irreducible tensor components. For example, a trivially reducible matrix is

$$H'' = \begin{pmatrix} E_1 & C\,g & 0 \\ C\,g & E_2 & 0 \\ 0 & 0 & E_3 \end{pmatrix}, \tag{3}$$

as it can obviously be broken into an upper (2×2) submatrix equal to H' and a lower (1×1) submatrix just consisting of the uncoupled energy level E_3.

The question of whether a higher-dimensional matrix is reducible can be far less trivial to analyze. For example, in a (24×24) matrix, as has been recently encountered in our analysis of the $2S$–$2S$ hyperfine-resolved interactions in hydrogen [3], entries can follow a rather irregular pattern, and the analysis then becomes far less trivial. The possibility to break up a matrix into irreducible subcomponents is equivalent to a search for "hidden" symmetries of the interaction which imply that only sublevels of specific symmetry are coupled.

After a brief look at level crossings in Sect. 2, we continue with an analysis of irreducible (sub-)matrices in Sect. 3. An application of the concepts developed to the $2S$–$2S$ hyperfine interaction in hydrogen is briefly described in Sect. 4.

2 Couplings and Level Crossings

Let us consider the (6×6)-matrix

$$H_0 = \begin{pmatrix} E_1 & C_1\, g & C_1\, g & 0 & 0 & 0 \\ C_1\, g & E_2 & C_1\, g & 0 & 0 & 0 \\ C_1\, g & C_1\, g & E_3 & 0 & 0 & 0 \\ 0 & 0 & 0 & E_4 & 0 & 0 \\ 0 & 0 & 0 & 0 & E_5 & C_1\, g \\ 0 & 0 & 0 & 0 & C_1\, g & E_6 \end{pmatrix} \tag{4}$$

This matrix consists of a mutually coupled (irreducible) upper (3×3)-block, an irreducible lower (2×2)-block, and one uncoupled state in the middle, with energy E_3. For the choice

$$E_j = j\,, \qquad C_1 = 1\,, \tag{5}$$

the evolution of the eigenenergies $E_j \to E_j(g)$ is analyzed in Fig. 1. Specifically, the level crossings occur at

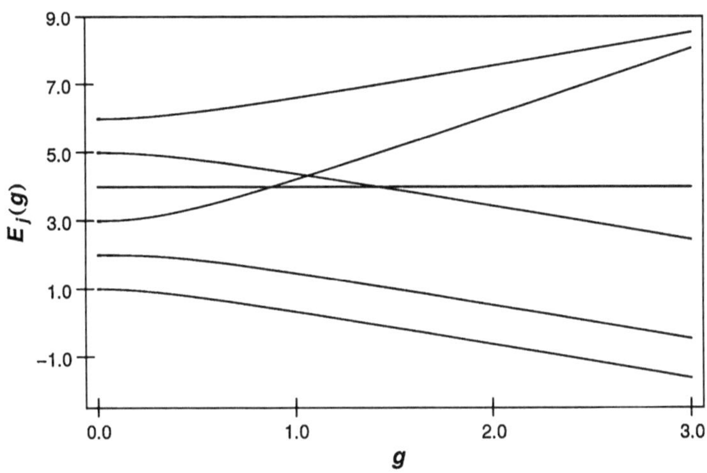

Fig. 1 Evolution of the energy levels $E_j(g)$ of the matrix H_0 given in Eq. (4), for the parameter choice given in Eq. (5). One can clearly discern the mutual "repulsion" between the lowest three energy levels $E_{1,2,3}$, stemming from the upper (3×3)-block of the matrix (4), and the same repulsion among the highest energies $E_{4,5}$, stemming from the upper (2×2)-block of the matrix (4). The level crossings occur with respect to the uncoupled level E_3, which is independent of g

$$E_3(g') = E_4(g') = 4\,, \qquad g' = 0.879\,385\,, \tag{6a}$$

$$\begin{aligned} E_3(g'') &= E_5(g'') = 4.326\,328\,, \\ g'' &= 1.061\,840\,, \end{aligned} \tag{6b}$$

$$E_4(g''') = E_5(g''') = 4, \qquad g''' = \sqrt{2}\,. \tag{6c}$$

Let us now add a further perturbation H_1,

$$H_1 = \begin{pmatrix} 0 & 0 & 0 & C_2\,g & 0 & C_2\,g \\ 0 & 0 & 0 & 0 & 0 & 0 \\ 0 & 0 & 0 & 0 & 0 & 0 \\ C_2\,g & 0 & 0 & 0 & 0 & 0 \\ 0 & 0 & 0 & 0 & 0 & 0 \\ C_2\,g & 0 & 0 & 0 & 0 & 0 \end{pmatrix}, \tag{7}$$

where C_2 is another parameter, to obtain the total Hamiltonian

$$H = H_0 + H_1 = \begin{pmatrix} E_1 & C_1\,g & C_1\,g & C_2\,g & 0 & C_2\,g \\ C_1\,g & E_2 & C_1\,g & 0 & 0 & 0 \\ C_1\,g & C_1\,g & E_3 & 0 & 0 & 0 \\ C_2\,g & 0 & 0 & E_4 & 0 & 0 \\ 0 & 0 & 0 & 0 & E_5 & C_1\,g \\ C_2\,g & 0 & 0 & 0 & C_1\,g & E_6 \end{pmatrix}. \tag{8}$$

In the total Hamiltonian H, the previously uncoupled level E_4 is now coupled to the upper (3×3) block by the term C_2, and an additional coupling between the lower (2×2) block and the upper (3×3) block is introduced in the extreme upper right and lower left corners of the matrices H_1 and H. In Fig. 2, we study the evolution of the energy levels of H for the parameter choice

$$C_2 = \frac{3}{10}. \tag{9}$$

It is clearly seen that the level crossings (6a) and (6b) now turn into avoided crossings, while the crossing (6c) is retained, but now occurs between E_3 and E_4 and not between E_4 and E_5. This difference is due to the avoided crossings.

The value of the energy at the crossing occurs at the coupling $g = g_{\rm cr}$,

$$E_{\rm cr} = E_3(g_{\rm cr}) = E_4(g_{\rm cr}) = 4\,, \qquad g_{\rm cr} = \sqrt{2}\,. \tag{10}$$

We have verified (see Fig. 3) that the crossing persists under the use of extended precision arithmetic, where the parameter $\epsilon = 10^{-128}$ (on the level of Fortran "hexadecuple precision") is employed in a numerical calculation of the eigenvalue near the crossing point, in order to ensure that the persistence of the crossing is not an artifact due to an insufficient numerical accuracy in the calculation. One might

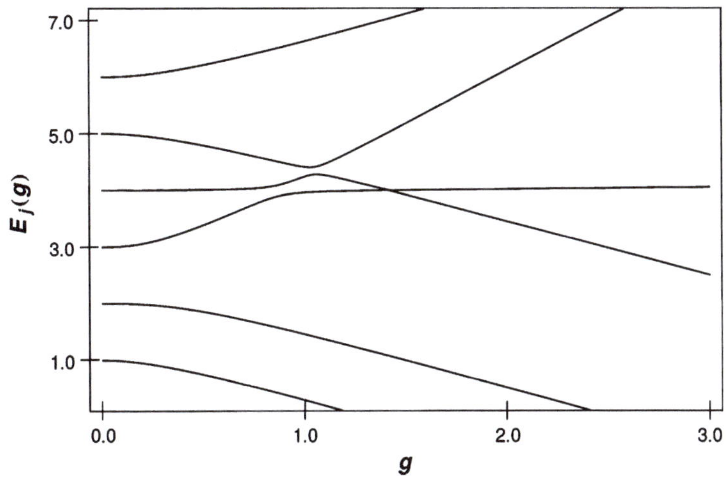

Fig. 2 Evolution of the energy levels $E_j(g)$ of the matrix H given in Eq. (8), for the parameter choices given in Eqs. (5) and (9). In comparison with Fig. 1, the ordinate axis is compressed in order to focus on the level crossings. The crossings (6a) and (6b) have turned into anticrossings, in view of the mutual level repulsion as the interlevel couplings are introduced, in accordance with the no-crossing theorem. However, the crossing (6c) is retained, with the twist that it takes place between E_3 and E_4 this time (instead of E_4 and E_5 as in the previous case). This change is due to the fact that the crossing (6a) between E_3 and E_4 and the crossing (6b) between E_3 and E_5 are now avoided

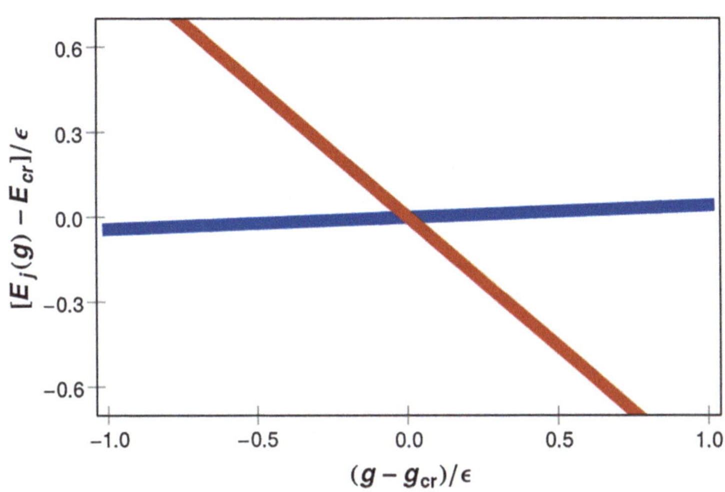

Fig. 3 Close-up of Fig. 2 in the region $E_3(g) \approx E_4(g) \approx 4$, and $g \approx g_{cr} = \sqrt{2}$, with $\epsilon = 10^{-128}$. This plot was obtained using extended precision arithmetic, using a computer algebra system [4]. The observed numerical behavior is consistent with the persistence of the level crossing for the irreducible matrix

otherwise conjecture that the "crossing" would turn into an "avoided crossing" when looking at the crossing point with finer numerical resolution.

For $C_1 = 1$, as a function of C_2, the eigenvectors at the degenerate eigenvalue $E_{\rm cr} = 4$ (where the crossing occurs) can be determined analytically; they read as

$$v_3 = \left(0, -\frac{C_2}{1+\sqrt{2}}, -\frac{\sqrt{2}\,C_2}{1+\sqrt{2}}, 0, -\sqrt{2}, 1\right), \tag{11a}$$

$$v_4 = \left(0, -\frac{C_2}{1+\sqrt{2}}, -\frac{\sqrt{2}\,C_2}{1+\sqrt{2}}, 1, 0, 0\right). \tag{11b}$$

A comparison of Figs. 1, 2 reveals that the crossing "actually" occurs between the levels 4 and 5. According to the adjacency graph in Fig. 5, the levels 4 and 5 are the most distant ones in comparison with the levels 2 and 3 which constitute the C_2-dependent "admixtures" at the crossing. According to Eq. (11), Furthermore, in the limit $C_2 \to 0$, the eigenvectors v_3 and v_4 given in Eq. (11) have contributions only from the unperturbed levels 4, 5, and 6; the latter are not directly coupled to the levels 2 and 3 in the adjacency graph in Fig. 5. Apparently, the no-crossing theorem discussed in Ref. [5] does not hold for higher-dimensional matrices, while crossings in (2×2)-matrices are strictly avoided in view of this theorem (see Chap. 79 of Ref. [6]).

A comparison of Figs. 1 and 2 reveals that the number of crossings is seen to be reduced for the case of the irreducible Hamiltonian matrix, but it is not zero.

3 Finding Irreducible Submatrices

We shall briefly discuss how to establish, by a formal, generalizable, method, that the matrix given in Eq. (4) is reducible, while the matrix (8) is irreducible.

Let us look at a general $(n \times n)$ matrix and associate it with the flight plan of a specific airline, with a nonvanishing entry, equal to unity, at position (i, j), denoting the existence of a direct flight between the cities i and j. If the matrix element (i, j) is zero, then no such direct connection exists. This matrix is known as the "adjacency matrix" U of the airline connection. A nonvanishing entry at position (i, i) could be interpreted as a "sightseeing flight" starting and ending at city i. There could be an indirect coupling between cities i and j, if not by a direct flight, then via a connection through some city k. If there is a connection with one intermediate stop, then it is obvious that the square of the adjacency matrix will have a unit entry at position (i, j). Nonzero entries in U^2 represent the cities that connect with connecting flights (one intermediate stop only). More specifically, the entries in the square of the adjacency matrix count the number of possibilities that one can fly from city i to city j with exactly one intermediate stop. If the airline serves n airports and one cannot go from city i to city j with $n - 1$ intermediate stops, then one cannot go city i to city j at all. One has exhausted the possibilities. Let U denote the adjacency matrix. It means that if the matrix

$$A = \sum_{i=1}^{n} U^i = U + U^2 + \ldots + U^n \tag{12}$$

still has a zero entry at position (i, j), then the airline must be serving at least two disconnected sets of destinations; this in turn is equivalent to showing that the adjacency matrix is reducible. The algorithm for testing the reducibility of an input matrix M is now clear. One replaces all nonzero entries in the input matrix M by unity, obtaining the adjacency matrix U. One then calculates the accumulated adjacency matrix A according to Eq. (12). If there are zero entries in A, then M must be reducible.

The adjacency matrix U_0 for H_0 given in Eq. (4) is

$$U_0 = \begin{pmatrix} 1 & 1 & 1 & 0 & 0 & 0 \\ 1 & 1 & 1 & 0 & 0 & 0 \\ 1 & 1 & 1 & 0 & 0 & 0 \\ 0 & 0 & 0 & 1 & 0 & 0 \\ 0 & 0 & 0 & 0 & 1 & 1 \\ 0 & 0 & 0 & 0 & 1 & 1 \end{pmatrix}, \tag{13}$$

resulting in

$$A_0 = \sum_{i=1}^{6} U_0^i = \begin{pmatrix} 364 & 364 & 364 & 0 & 0 & 0 \\ 364 & 364 & 364 & 0 & 0 & 0 \\ 364 & 364 & 364 & 0 & 0 & 0 \\ 0 & 0 & 0 & 6 & 0 & 0 \\ 0 & 0 & 0 & 0 & 63 & 63 \\ 0 & 0 & 0 & 0 & 63 & 63 \end{pmatrix}, \tag{14}$$

clearly displaying the reducibility and the three submatrices. The corresponding adjacency graph is given in Fig. 4. These observations only confirm the intuitive understanding gathered by inspection of H_0.

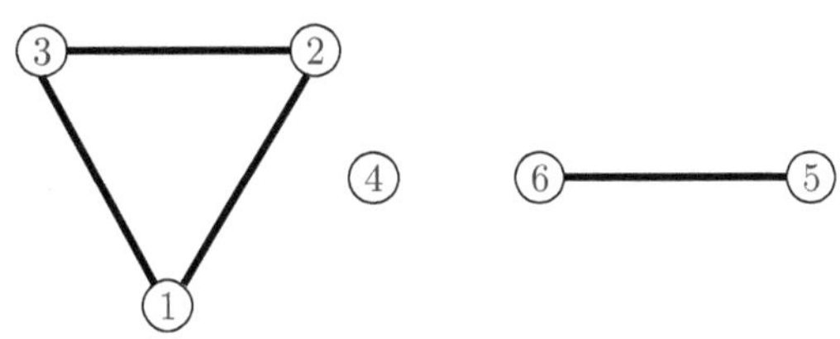

Fig. 4 Adjacency graph for the matrix U_0 given in Eq. (13)

For the matrix H given in Eq. (8), the adjacency matrix is

$$U = \begin{pmatrix} 1 & 1 & 1 & 1 & 0 & 1 \\ 1 & 1 & 1 & 0 & 0 & 0 \\ 1 & 1 & 1 & 0 & 0 & 0 \\ 1 & 0 & 0 & 1 & 0 & 0 \\ 0 & 0 & 0 & 0 & 1 & 1 \\ 1 & 0 & 0 & 0 & 1 & 1 \end{pmatrix}, \tag{15}$$

resulting in

$$A = \sum_{i=1}^{6} U^i = \begin{pmatrix} 836 & 604 & 604 & 354 & 178 & 426 \\ 604 & 453 & 453 & 250 & 106 & 284 \\ 604 & 453 & 453 & 250 & 106 & 284 \\ 354 & 250 & 250 & 158 & 72 & 178 \\ 178 & 106 & 106 & 72 & 90 & 142 \\ 426 & 284 & 284 & 178 & 142 & 268 \end{pmatrix}, \tag{16}$$

which is fully populated. The corresponding adjacency graph is given in Fig. 5. The accumulated adjacency matrix A is fully populated, demonstrating the irreducibility of H.

4 2*S*–2*S* Interaction in Hydrogen

The aim is to analyze the interaction of two excited hydrogen atoms in the metastable 2*S* state. We note that the 2*S*–2*S* van der Waals interaction has been analyzed before in Refs. [7, 8], but without any reference to the resolution of the hyperfine splitting. The Hamiltonian for the two-atom system is

$$H = H_{\mathrm{LS},A} + H_{\mathrm{LS},B} + H_{\mathrm{HFS},A} + H_{\mathrm{HFS},B} + H_{\mathrm{vdW}}\,. \tag{17}$$

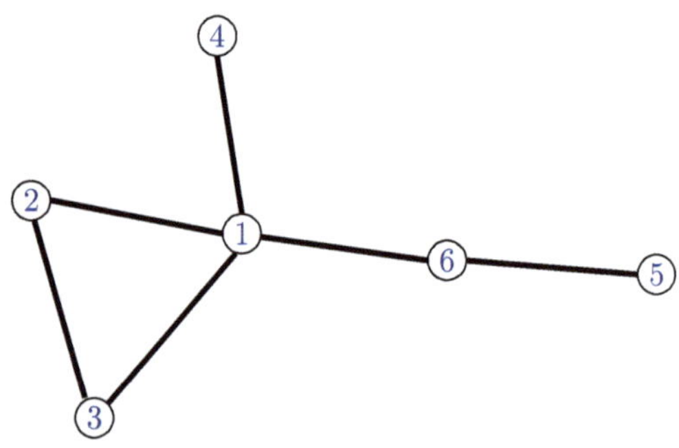

Fig. 5 Adjacency graph for the matrix U given in (15)

Here, $H_{\rm LS}$ is the Lamb shift Hamiltonian, while $H_{\rm HFS}$ describes hyperfine effects; these Hamiltonians have to be added for atoms A and B. In SI units, they are given as follows:

$$
\begin{aligned}
H_{\rm HFS} = {} & \frac{\mu_0}{4\pi}\mu_B\mu_N\, g_s g_p \sum_{i=A,B} \left[\frac{8\pi}{3} \mathbf{S}_i \cdot \mathbf{I}_i\, \delta^3(\mathbf{r}_i) \right. \\
& \left. + \frac{3(\mathbf{S}_i \cdot \mathbf{r}_i)(\mathbf{I}_i \cdot \mathbf{r}_i) - \mathbf{S}_i \cdot \mathbf{I}_i\, \mathbf{r}_i^2}{|\mathbf{r}_i|^5} + \frac{\mathbf{L}_i \cdot \mathbf{I}_i}{|\mathbf{r}_i|^3} \right],
\end{aligned} \tag{18a}
$$

$$
H_{\rm LS} = \frac{4}{3}\alpha^2\, mc^2 \left(\frac{\hbar}{mc}\right)^3 \ln\left(\alpha^{-2}\right) \sum_{i=A,B} \delta^3(\mathbf{r}_i), \tag{18b}
$$

$$
H_{\rm vdW} = \alpha\, \hbar c\, \frac{x_A\, x_B + y_A\, y_B - 2\, z_A\, z_B}{R^3}. \tag{18c}
$$

The symbols are explained as follows: α is the fine-structure constant, and m denotes the electron mass. The operators $\mathbf{r}_i$, $\mathbf{p}_i$, and $\mathbf{L}_i$ are the position (relative to the respective nuclei), linear momentum, and orbital angular momentum operators for electron i, while $\mathbf{S}_i$ is the spin operator for electron i and $\mathbf{I}_i$ is the spin operator for proton i [both are dimensionless]. Electronic and protonic g factors are $g_s \simeq 2.002\,319$ and $g_p \simeq 5.585\,695$, while $\mu_B \simeq 9.274\,010 \times 10^{-24}\,\mathrm{Am}^2$ is the Bohr magneton and $\mu_N \simeq 5.050\,784 \times 10^{-27}\,\mathrm{Am}^2$ is the nuclear magneton. Of course, the subscripts A and B refer to the relative coordinates within the two atoms. R is the interatomic distance. $H_{\rm LS}$ shifts S states relative to P states by the Lamb shift, which is given in Eq. (18b) in the Welton approximation [9], which is convenient within the formalism used for the evaluation of matrix elements. The important property of $H_{\rm LS}$ is that it shifts S states upward in relation to P states. The prefactor multiplying the Dirac-δ can be adjusted to the observed Lamb shift splitting. Indeed, for the final calculation of energy shifts, one conveniently replaces

$$
\langle 2S_{1/2}|H_{LS}|2S_{1/2}\rangle - \langle 2P_{1/2}|H_{LS}|2P_{1/2}\rangle = \frac{4\alpha}{3\pi}\frac{\alpha^4}{8} m\, c^2 \ln(\alpha^{-2}) \to \mathcal{L}, \tag{19}
$$

where $\mathcal{L} = h \times 1057.845(9)$ MHz is the "classic" $2S$–$2P_{1/2}$ Lamb shift [10] (m is the electron mass, c is the speed of light, and h is Planck's constant). In the Hamiltonian (17), the origin of energies is taken at the hyperfine center of the 2 $P_{1/2}$ levels.

The coupling scheme for the atomic levels entails that the orbital angular momentum $\mathbf{L}_i$ should be added to the electron spin to give the total angular momentum $\mathbf{J}_i$; then, $\mathbf{J}_i$ is added to the nuclear spin $\mathbf{I}_i$ to give $\mathbf{F}_i$. This vector coupling has to be done for both atoms $i = A, B$, and then, $\mathbf{F} = \mathbf{F}_A + \mathbf{F}_B$ (orbital+ spin+nuclear angular momentum, summed over both atoms A and B). One can show relatively easily that the z component F_z of the total angular momentum is conserved, i.e., F_z commutes with the Hamiltonian.

We restrict the discussion to states with total angular momentum $J = 1/2$, i.e., to the $2S$ and $2P_{1/2}$ states which are displaced from each other only by the Lamb shift. States displaced by the fine structure are subdominant because $\mathcal{F} \gg \mathcal{L}$ where $\mathcal{F} = \alpha^4 mc^2/32$ is the $2P$ fine-structure interval.

Let each atom be in a state $|\ell_i, F_i, F_{z,i}\rangle$, with $i = A, B$ (here, ℓ_i is the orbital angular momentum quantum number). The two-atom system occupies the states $|(\ell_A, F_A, F_{z,A})_A \, (\ell_B, F_B, F_{z,B})_B\rangle$. We have four S states ($F = 0$ and $F = 1$), and four P states ($F = 0$ and $F = 1$), for each atom, making for a total of eight states. For two atoms, one thus has 64 states in the $(n = 2)$–$(n = 2)$ manifold with $J = 1/2$.

Now, since $F_z = F_{z,A} + F_{z,B}$ is a conserved quantity, we should classify states according to $F_z = \pm 2$, $F_z = \pm 1$, and $F_z = 0$. There are 4 states in the $F_z = \pm 2$ manifolds, 16 states in the $F_z = \pm 1$ manifolds, and a total of 24 states in $F_z = 0$, adding up to a total of $64 = 24 + 2 \times 16 + 2 \times 4$. For $F_z = 0$, the matrix with 24 $\times 24 = 576$ entries is hard to analyze. The question is whether or not one can find an additional symmetry that simplifies the analysis. Such an additional symmetry would naturally lead to a separation of the Hamiltonian into further irreducible submatrices, thus reducing the complexity of the computational task drastically. It is precisely at this point that the methods discussed in Sect. 2 become useful.

To this end, we first order the states in the $F_z = 0$ manifold according to increasing quantum numbers. The states where atom A is in an S state with $F_A = 0$ are given by

$$
\begin{aligned}
|\Psi_1\rangle &= |(0,0,0)_A \, (0,0,0)_B\rangle \,, \\
|\Psi_2\rangle &= |(0,0,0)_A \, (0,1,0)_B\rangle \,, \\
|\Psi_3\rangle &= |(0,0,0)_A \, (1,0,0)_B\rangle \,, \\
|\Psi_4\rangle &= |(0,0,0)_A \, (1,1,0)_B\rangle \,.
\end{aligned}
\tag{20}
$$

With atom A in an S state with $F_A = 1$, we have

$$
\begin{aligned}
|\Psi_5\rangle &= |(0,1,-1)_A \, (0,1,1)_B\rangle \,, \\
|\Psi_6\rangle &= |(0,1,-1)_A \, (1,1,1)_B\rangle \,, \\
|\Psi_7\rangle &= |(0,1,0)_A \, (0,0,0)_B\rangle \,, \\
|\Psi_8\rangle &= |(0,1,0)_A \, (0,1,0)_B\rangle \,, \\
|\Psi_9\rangle &= |(0,1,0)_A \, (1,0,0)_B\rangle \,, \\
|\Psi_{10}\rangle &= |(0,1,0)_A \, (1,1,0)_B\rangle \,, \\
|\Psi_{11}\rangle &= |(0,1,1)_A \, (0,1,-1)_B\rangle \,, \\
|\Psi_{12}\rangle &= |(0,1,1)_A \, (1,1,-1)_B\rangle \,.
\end{aligned}
\tag{21}
$$

The states with atom A in a $P_{1/2}$ state (hyperfine singlet) are given as follows:

$$
\begin{aligned}
|\Psi_{13}\rangle &= |(1,0,0)_A \, (0,0,0)_B\rangle \,, \\
|\Psi_{14}\rangle &= |(1,0,0)_A \, (0,1,0)_B\rangle \,, \\
|\Psi_{15}\rangle &= |(1,0,0)_A \, (1,0,0)_B\rangle \,, \\
|\Psi_{16}\rangle &= |(1,0,0)_A \, (1,1,0)_B\rangle \,.
\end{aligned}
\tag{22}
$$

The states with atom A in a $2P_{1/2}$ hyperfine triplet are given by

$$
\begin{aligned}
|\Psi_{17}\rangle &= |(1,1,-1)_A\,(0,1,1)_B\rangle\,, \\
|\Psi_{18}\rangle &= |(1,1,-1)_A\,(1,1,1)_B\rangle\,, \\
|\Psi_{19}\rangle &= |(1,1,0)_A\,(0,0,0)_B\rangle\,, \\
|\Psi_{20}\rangle &= |(1,1,0)_A\,(0,1,0)_B\rangle\,, \\
|\Psi_{21}\rangle &= |(1,1,0)_A\,(1,0,0)_B\rangle\,, \\
|\Psi_{22}\rangle &= |(1,1,0)_A\,(1,1,0)_B\rangle\,, \\
|\Psi_{23}\rangle &= |(1,1,1)_A\,(0,1,-1)_B\rangle\,, \\
|\Psi_{24}\rangle &= |(1,1,1)_A\,(1,1,-1)_B\rangle\,.
\end{aligned} \tag{23}
$$

The adjacency matrix $U_{F_z=0}$ of the Hamiltonian (17) in the $F_Z = 0$ manifold is equal to

$$
\begin{pmatrix}
1&0&0&0&0&0&0&0&0&0&0&0&0&0&0&0&0&1&0&0&0&1&0&1\\
0&1&0&0&0&0&0&0&0&0&0&0&0&0&0&0&0&1&0&0&1&0&0&1\\
0&0&1&0&0&0&0&0&0&0&0&0&0&0&0&0&1&0&0&1&0&0&1&0\\
0&0&0&1&0&0&0&0&0&0&0&0&0&0&0&0&1&0&1&0&0&0&1&0\\
0&0&0&0&1&0&0&0&0&0&0&0&0&0&1&1&0&1&0&0&1&1&0&0\\
0&0&0&0&0&1&0&0&0&0&0&0&1&1&0&0&1&0&1&1&0&0&0&0\\
0&0&0&0&0&0&1&0&0&0&0&0&0&0&0&1&0&1&0&0&0&0&0&1\\
0&0&0&0&0&0&0&1&0&0&0&0&0&0&1&0&0&1&0&0&0&0&0&1\\
0&0&0&0&0&0&0&0&1&0&0&0&0&1&0&0&1&0&0&0&0&0&1&0\\
0&0&0&0&0&0&0&0&0&1&0&0&1&0&0&0&1&0&0&0&0&0&1&0\\
0&0&0&0&0&0&0&0&0&0&1&0&0&0&1&1&0&0&0&0&1&1&0&1\\
0&0&0&0&0&0&0&0&0&0&0&1&1&1&0&0&0&0&1&1&0&0&1&0\\
0&0&0&0&0&1&0&0&0&1&0&1&1&0&0&0&0&0&0&0&0&0&0&0\\
0&0&0&0&0&1&0&0&1&0&0&1&0&1&0&0&0&0&0&0&0&0&0&0\\
0&0&0&0&1&0&0&1&0&0&1&0&0&0&1&0&0&0&0&0&0&0&0&0\\
0&0&0&0&1&0&1&0&0&0&1&0&0&0&0&1&0&0&0&0&0&0&0&0\\
0&0&1&1&0&1&0&0&1&1&0&0&0&0&0&0&1&0&0&0&0&0&0&0\\
1&1&0&0&1&0&1&1&0&0&0&0&0&0&0&0&0&1&0&0&0&0&0&0\\
0&0&0&1&0&1&0&0&0&0&0&1&0&0&0&0&0&0&1&0&0&0&0&0\\
0&0&1&0&0&1&0&0&0&0&0&1&0&0&0&0&0&0&0&1&0&0&0&0\\
0&1&0&0&1&0&0&0&0&0&1&0&0&0&0&0&0&0&0&0&1&0&0&0\\
1&0&0&0&1&0&0&0&0&0&1&0&0&0&0&0&0&0&0&0&0&1&0&0\\
0&0&1&1&0&0&0&0&1&1&0&1&0&0&0&0&0&0&0&0&0&0&1&0\\
1&1&0&0&0&0&1&1&0&0&1&0&0&0&0&0&0&0&0&0&0&0&0&1
\end{pmatrix}\,. \tag{24}
$$

The accumulated adjacency matrix has the structure

$$A_{F_z=0} = \sum_{i=1}^{24} U^i \simeq \left(\begin{array}{cccccccccccccccccccccccc}
\chi & \chi & 0 & 0 & \chi & 0 & \chi & \chi & 0 & 0 & \chi & 0 & 0 & 0 & \chi & \chi & 0 & \chi & 0 & 0 & \chi & \chi & 0 & \chi \\
\chi & \chi & 0 & 0 & \chi & 0 & \chi & \chi & 0 & 0 & \chi & 0 & 0 & 0 & \chi & \chi & 0 & \chi & 0 & 0 & \chi & \chi & 0 & \chi \\
0 & 0 & \chi & \chi & 0 & \chi & 0 & 0 & \chi & \chi & 0 & \chi & \chi & \chi & 0 & 0 & \chi & 0 & \chi & \chi & 0 & 0 & \chi & 0 \\
0 & 0 & \chi & \chi & 0 & \chi & 0 & 0 & \chi & \chi & 0 & \chi & \chi & \chi & 0 & 0 & \chi & 0 & \chi & \chi & 0 & 0 & \chi & 0 \\
\chi & \chi & 0 & 0 & \chi & 0 & \chi & \chi & 0 & 0 & \chi & 0 & 0 & 0 & \chi & \chi & 0 & \chi & 0 & 0 & \chi & \chi & 0 & \chi \\
0 & 0 & \chi & \chi & 0 & \chi & 0 & 0 & \chi & \chi & 0 & \chi & \chi & \chi & 0 & 0 & \chi & 0 & \chi & \chi & 0 & 0 & \chi & 0 \\
\chi & \chi & 0 & 0 & \chi & 0 & \chi & \chi & 0 & 0 & \chi & 0 & 0 & 0 & \chi & \chi & 0 & \chi & 0 & 0 & \chi & \chi & 0 & \chi \\
\chi & \chi & 0 & 0 & \chi & 0 & \chi & \chi & 0 & 0 & \chi & 0 & 0 & 0 & \chi & \chi & 0 & \chi & 0 & 0 & \chi & \chi & 0 & \chi \\
0 & 0 & \chi & \chi & 0 & \chi & 0 & 0 & \chi & \chi & 0 & \chi & \chi & \chi & 0 & 0 & \chi & 0 & \chi & \chi & 0 & 0 & \chi & 0 \\
0 & 0 & \chi & \chi & 0 & \chi & 0 & 0 & \chi & \chi & 0 & \chi & \chi & \chi & 0 & 0 & \chi & 0 & \chi & \chi & 0 & 0 & \chi & 0 \\
\chi & \chi & 0 & 0 & \chi & 0 & \chi & \chi & 0 & 0 & \chi & 0 & 0 & 0 & \chi & \chi & 0 & \chi & 0 & 0 & \chi & \chi & 0 & \chi \\
0 & 0 & \chi & \chi & 0 & \chi & 0 & 0 & \chi & \chi & 0 & \chi & \chi & \chi & 0 & 0 & \chi & 0 & \chi & \chi & 0 & 0 & \chi & 0 \\
0 & 0 & \chi & \chi & 0 & \chi & 0 & 0 & \chi & \chi & 0 & \chi & \chi & \chi & 0 & 0 & \chi & 0 & \chi & \chi & 0 & 0 & \chi & 0 \\
0 & 0 & \chi & \chi & 0 & \chi & 0 & 0 & \chi & \chi & 0 & \chi & \chi & \chi & 0 & 0 & \chi & 0 & \chi & \chi & 0 & 0 & \chi & 0 \\
\chi & \chi & 0 & 0 & \chi & 0 & \chi & \chi & 0 & 0 & \chi & 0 & 0 & 0 & \chi & \chi & 0 & \chi & 0 & 0 & \chi & \chi & 0 & \chi \\
\chi & \chi & 0 & 0 & \chi & 0 & \chi & \chi & 0 & 0 & \chi & 0 & 0 & 0 & \chi & \chi & 0 & \chi & 0 & 0 & \chi & \chi & 0 & \chi \\
0 & 0 & \chi & \chi & 0 & \chi & 0 & 0 & \chi & \chi & 0 & \chi & \chi & \chi & 0 & 0 & \chi & 0 & \chi & \chi & 0 & 0 & \chi & 0 \\
\chi & \chi & 0 & 0 & \chi & 0 & \chi & \chi & 0 & 0 & \chi & 0 & 0 & 0 & \chi & \chi & 0 & \chi & 0 & 0 & \chi & \chi & 0 & \chi \\
0 & 0 & \chi & \chi & 0 & \chi & 0 & 0 & \chi & \chi & 0 & \chi & \chi & \chi & 0 & 0 & \chi & 0 & \chi & \chi & 0 & 0 & \chi & 0 \\
0 & 0 & \chi & \chi & 0 & \chi & 0 & 0 & \chi & \chi & 0 & \chi & \chi & \chi & 0 & 0 & \chi & 0 & \chi & \chi & 0 & 0 & \chi & 0 \\
\chi & \chi & 0 & 0 & \chi & 0 & \chi & \chi & 0 & 0 & \chi & 0 & 0 & 0 & \chi & \chi & 0 & \chi & 0 & 0 & \chi & \chi & 0 & \chi \\
\chi & \chi & 0 & 0 & \chi & 0 & \chi & \chi & 0 & 0 & \chi & 0 & 0 & 0 & \chi & \chi & 0 & \chi & 0 & 0 & \chi & \chi & 0 & \chi \\
0 & 0 & \chi & \chi & 0 & \chi & 0 & 0 & \chi & \chi & 0 & \chi & \chi & \chi & 0 & 0 & \chi & 0 & \chi & \chi & 0 & 0 & \chi & 0 \\
\chi & \chi & 0 & 0 & \chi & 0 & \chi & \chi & 0 & 0 & \chi & 0 & 0 & 0 & \chi & \chi & 0 & \chi & 0 & 0 & \chi & \chi & 0 & \chi
\end{array}\right), \tag{25}$$

where χ stands for any entry different from zero (the χs are not all equal). The adjacency graph given in Fig. 6 confirms the presence of two irreducible submatrices of $H_{F_z=0}$. Indeed, the two uncoupled subspaces are spanned by the states $|\Psi_i\rangle$ with $i = 1, 2, 5, 7, 8, 11, 15, 16, 18, 21, 22, 24$ (subspace I) and $|\Psi_j\rangle$ with $j = 3, 4, 6, 9, 10, 12, 13, 14, 17, 19, 20, 23$ (subspace II). An ordering of the eigenvalues reveals that one can have coupling among the *S*–*S* and *P*–*P* states (distributed among atoms *A* and *B*), forming submanifold I, and among all *S*–*P* and *P*–*S*

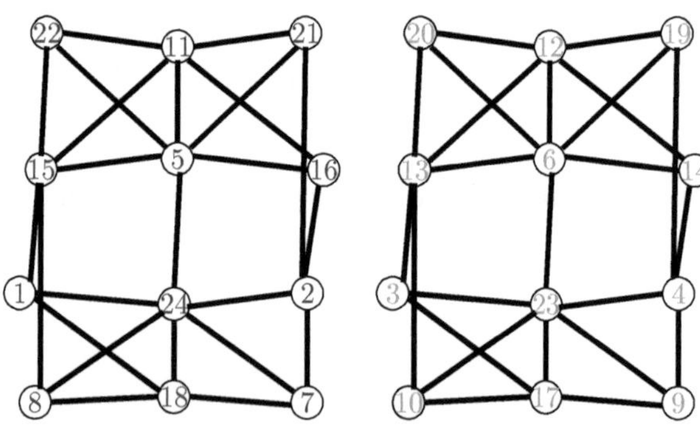

Fig. 6 Adjacency graph for the matrix $U_{F_z=0}$ given in Eq. (24)

states (distributed among atoms A and B), forming submanifold II. In retrospect, the separation is perhaps clear, but it is less obvious at first glance.

It is then possible to redefine the levels from which the 12×12 Hamiltonian matrix is constructed, in the first submanifold of $F_z = 0$ (the S–S coupled states). Specifically, one defines $|\Psi_1^{(\mathrm{I})}\rangle = |\Psi_1\rangle$, $|\Psi_2^{(\mathrm{I})}\rangle = |\Psi_2\rangle$, $|\Psi_3^{(\mathrm{I})}\rangle = |\Psi_5\rangle$, $|\Psi_4^{(\mathrm{I})}\rangle = |\Psi_7\rangle$, $|\Psi_5^{(\mathrm{I})}\rangle = |\Psi_8\rangle$, $|\Psi_6^{(\mathrm{I})}\rangle = |\Psi_{11}\rangle$, $|\Psi_7^{(\mathrm{I})}\rangle = |\Psi_{15}\rangle$, $|\Psi_8^{(\mathrm{I})}\rangle = |\Psi_{16}\rangle$, $|\Psi_9^{(\mathrm{I})}\rangle = |\Psi_{18}\rangle$, $|\Psi_{10}^{(\mathrm{I})}\rangle = |\Psi_{21}\rangle$, $|\Psi_{11}^{(\mathrm{I})}\rangle = |\Psi_{22}\rangle$, and $|\Psi_{12}^{(\mathrm{I})}\rangle = |\Psi_{24}\rangle$. Within the space spanned by the $|\Psi_i^{(\mathrm{I})}\rangle$ with $i = 1, 2, \cdots, 12$, the Hamiltonian matrix is

$$H^{(\mathrm{I})}_{F_z=0} = \begin{pmatrix}
2\mathcal{L}-\frac{9}{2}\mathcal{H} & 0 & 0 & 0 & 0 & 0 & 0 & 0 & -\mathcal{V} & 0 & -2\mathcal{V} & -\mathcal{V} \\
0 & 2\mathcal{L}-\frac{3}{2}\mathcal{H} & 0 & 0 & 0 & 0 & 0 & 0 & \mathcal{V} & -2\mathcal{V} & 0 & -\mathcal{V} \\
0 & 0 & 2\mathcal{L}+\frac{3}{2}\mathcal{H} & 0 & 0 & 0 & -\mathcal{V} & \mathcal{V} & 2\mathcal{V} & -\mathcal{V} & \mathcal{V} & 0 \\
0 & 0 & 0 & 2\mathcal{L}-\frac{3}{2}\mathcal{H} & 0 & 0 & 0 & -2\mathcal{V} & -\mathcal{V} & 0 & 0 & \mathcal{V} \\
0 & 0 & 0 & 0 & 2\mathcal{L}+\frac{3}{2}\mathcal{H} & 0 & -2\mathcal{V} & 0 & \mathcal{V} & 0 & 0 & \mathcal{V} \\
0 & 0 & 0 & 0 & 0 & 2\mathcal{L}+\frac{3}{2}\mathcal{H} & -\mathcal{V} & -\mathcal{V} & 0 & \mathcal{V} & \mathcal{V} & 2\mathcal{V} \\
0 & 0 & -\mathcal{V} & 0 & -2\mathcal{V} & -\mathcal{V} & -\frac{3}{2}\mathcal{H} & 0 & 0 & 0 & 0 & 0 \\
0 & 0 & \mathcal{V} & -2\mathcal{V} & 0 & -\mathcal{V} & 0 & -\frac{1}{2}\mathcal{H} & 0 & 0 & 0 & 0 \\
-\mathcal{V} & \mathcal{V} & 2\mathcal{V} & -\mathcal{V} & \mathcal{V} & 0 & 0 & 0 & \frac{1}{2}\mathcal{H} & 0 & 0 & 0 \\
0 & -2\mathcal{V} & -\mathcal{V} & 0 & 0 & \mathcal{V} & 0 & 0 & 0 & -\frac{1}{2}\mathcal{H} & 0 & 0 \\
-2\mathcal{V} & 0 & \mathcal{V} & 0 & 0 & \mathcal{V} & 0 & 0 & 0 & 0 & \frac{1}{2}\mathcal{H} & 0 \\
-\mathcal{V} & -\mathcal{V} & 0 & \mathcal{V} & \mathcal{V} & 2\mathcal{V} & 0 & 0 & 0 & 0 & 0 & \frac{1}{2}\mathcal{H}
\end{pmatrix}, \tag{26}$$

where

$$\mathcal{V} = 3\,\alpha\,\hbar c\,\frac{a_0^2}{R^3} \tag{27}$$

is a parameter that describes the strength of the van der Waals interaction. Furthermore,

$$\mathcal{H} = \frac{\alpha^4}{18}\,g_N\,\frac{m}{m_p}\,m\,c^2 \tag{28}$$

with $\mathcal{H} \approx h \times 59.1856114(22)$ MHz parameterizes the hyperfine splitting (m_p is the proton mass). A close-up of the six energetically highest, distance-dependent S–S state energy levels, coupled through virtual P–P states, is given in Fig. 7 (Born–Oppenheimer potential energy curves). We have verified that the crossing between the second and third level (counted in ascending order of the unperturbed energy for $R \to \infty$) persists under a drastic increase in the numerical accuracy, much like for

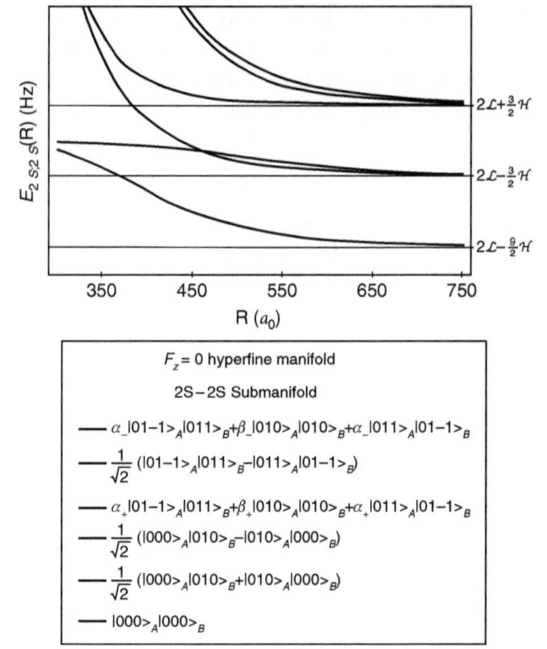

Fig. 7 Energy levels of the $2S$–$2S$ states within the $F_z = 0$ hyperfine manifold as a function of atomic separation R (given in units of the Bohr radius a_0). The eigenstates in the legend are those relevant to the $R \to \infty$ asymptotic limit; for finite separation these states mix. There is one remaining level crossing even if the Hamiltonian matrix is irreducible. The coefficients $\alpha_\pm$ and $\beta_\pm$ are determined from second-order perturbation theory and given by Eq. (29). The states are labeled from *top* to *bottom* in the legend, in the same order as they are relevant to the long-range asymptotics

our model problem (Fig. 3). The coefficients used in the legend for this figure are given by

$$\alpha_\pm = 2\sqrt{\frac{2}{33 \pm \sqrt{33}}}, \tag{29a}$$

$$\beta_\pm = \mp \frac{\sqrt{33} \pm 1}{\sqrt{2\left(33 \pm \sqrt{33}\right)}}. \tag{29b}$$

Despite the fact that subspace I is irreducible, one observes one level crossing, much in line with the discussion presented in Sect. 2. Finer details of the calculation will be presented in an upcoming work [3]. Specifically, for large R, we can point out that all of the level shifts of the states in Fig. 7 are found to be of order $\mathcal{V}^2/\mathcal{L}$ and are thus of second order in $\mathcal{V}$, proportional to $1/R^6$ but drastically enhanced in their numerical magnitude as compared to "normal" van der Waals shifts due to the $1/\mathcal{L}$ denominator. For the $1S$–$1S$ interaction, the well-known result involves a shift of order $\mathcal{V}^2/E_h$, where E_h is the Hartree energy. In the limit of large R, the $2S$–$2S$ interaction is seen to be larger by a factor $1/\alpha^3 \sim 10^6$, in view of the smaller energy denominator which only involves the Lamb shift.

5 Conclusions

Often, in physics, we need to resort to mathematical sophistications in order to uncover properties of a physical system hidden from us at first glance. In our case, we find that adjacency matrices and adjacency graphs help determine the reducibility of a matrix and, in the analysis of the hyperfine-resolved $2S$–$2S$ interaction, help determine the irreducible subspaces into which we may break the total Hamiltonian. We were able to identify an additional selection rule, which is relatively obvious *a posteriori*, namely, that couplings occur between *S*–*S* and *P*–*P* levels, and between *S*–*P* and *P*–*S* levels, but there are no coupling joining the two submanifolds (see Sect. 4). The size of the matrix is reduced from 24×24 to 12×12. It is somewhat surprising that the seemingly easy problem of identifying the irreducible submatrices of a Hamiltonian involves a rather sophisticated concept like an adjacency matrix.

Our model problem, studied in Sects. 2 and 3, reveals that level crossings can occur even in well-behaved quantum mechanical systems, described by interlevel couplings varying with some parameter. For the long-range interaction between atoms, the inverse interatomic distance $1\,/\,R$ is such a coupling parameter. In Fig. 2 (model problem) and in Fig. 7 ($2S$–$2S$ states within the $F_z = 0$ submanifold of the hydrogen long-range interaction), level crossings are clearly visible even if the Hamiltonian matrix is irreducible. Our extended precision numerical calculations (Fig. 3) and the analytic structure of the "crossing" eigenvectors in Eq. 11 together with the adjacency matrices in Figs. 4 and 5 indicate that the no-crossing theorem breaks down in higher-dimensional systems. Furthermore, we observe that our crossings, both for the model problem and for the $2S$–$2S$ system, involve situations where the couplings are indirect and the admixtures at the crossing point are between levels which are displaced from each other in the adjacency graph by at least two elementary steps. These observations could be of interest beyond the concrete problem studied here, in the context of a breakdown of the no-crossing theorem in higher-dimensional quantum mechanical systems. An improved understanding of the 2S–2S interaction is important for progress in the 2S hyperfine measurement by optical methods, using an atomic beam [11–13].

Attempts to study the hyperfine-resolved interaction have been made, but no reference has been made to the resolution of the hyperfine structure [7, 8]. The current approach leads to a solution, with partial results being presented in Eq. (26) and Fig. 7 and finer details being relegated to Ref. [3].

Acknowledgements The authors acknowledge support from the National Science Foundation (Grant PHY–1403973).

References

1. C. Cohen-Tannoudji, B. Diu, F. Laloë, *Quantum Mechanics*, vol. 1, 1st edn. (Wiley, New York, 1978)
2. C. Cohen-Tannoudji, B. Diu, F. Laloë, *Quantum Mechanics*, vol. 2, 1st edn. (Wiley, New York, 1978)
3. U. D. Jentschura, V. Debierre, C. M. Adhikari, A. Matveev, and N. Kolachevsky, *Long-range interactions of excited hydrogen atoms. II*. Hyperfine-resolved $(2S; 2S)$-system, submitted to Physical Review A (2016)
4. S. Wolfram, *The Mathematica Book*, 4th edn. (Cambridge University Press, Cambridge, 1999)
5. See the URL http://en.m.wikipedia.org/wiki/ Avoided_crossing
6. L.D. Landau, E.M. Lifshitz, *Quantum Mechanics, Volume 3 of the Course on Theoretical Physics* (Pergamon Press, Oxford, 1958)
7. S. Jonsell, A. Saenz, P. Froelich, R.C. Forrey, R. Côté, A. Dalgarno, Long-range interactions between two 2s excited hydrogen atoms. Phys. Rev. A **65**, 042501 (2002)
8. S.I. Simonsen, L. Kocbach, J.P. Hansen, Longrange interactions and state characteristics of interacting Rydberg atoms. J. Phys. B **44**, 165001 (2011)
9. C. Itzykson, J.B. Zuber, *Quantum Field Theory* (McGraw-Hill, New York, 1980)
10. S.R. Lundeen, F.M. Pipkin, Measurement of the Lamb shift in hydrogen, n=2. Rev. Lett. **46**, 232–235 (1981)
11. M. Fischer, N. Kolachevsky, S.G. Karshenboim, T. W. Hänsch, Optical measurement of the $2S$ hyperne interval in atomic hydrogen. Can. J. Phys. **80**, 1225–1231 (2002)
12. N. Kolachevsky, M. Fischer, S.G. Karshenboim, T.W. Hänsch, High-precision optical measurement of the $2S$ hyperfine interval in atomic hydrogen. Phys. Rev. Lett. **92**, 033003 (2004)
13. N. Kolachevsky, A. Matveev, J. Alnis, C.G. Parthey, S.G. Karshenboim, T.W. Hänsch, Measurement of the $2S$ hyperfine interval in atomic hydrogen. Phys. Rev. Lett. **102**, 213002 (2009)

Frequency-Comb Referenced Collinear Laser Spectroscopy of Be$^+$ for Nuclear Structure Investigations and Many-Body QED Tests

A. Krieger, W. Nörtershäuser, Ch. Geppert, K. Blaum, M.L. Bissell, N. Frömmgen, M. Hammen, K. Kreim, M. Kowalska, J. Krämer, R. Neugart, G. Neyens, R. Sánchez, D. Tiedemann, D.T. Yordanov, and M. Zakova

Abstract Transition frequencies of the $2s\,^2S_{1/2} \rightarrow 2p\,^2P_{1/2,\,3/2}$ transitions in Be$^+$ were measured in stable and short-lived isotopes at ISOLDE (CERN) using collinear laser spectroscopy and frequency-comb-referenced dye lasers. Quasi-simultaneous measurements in copropagating and counterpropagating geometry were performed to become independent from acceleration voltage determinations for Doppler-shift corrections of the fast ion beam. Isotope shifts and fine-structure splittings were obtained from the transition frequencies measured with a frequency

---

This article is part of the topical collection "Enlightening the World with the Laser" - Honoring T. W. Hänsch guest edited by Tilman Esslinger, Nathalie Picqué, and Thomas Udem.

A. Krieger
Institut für Kernphysik, Technische Universität Darmstadt, 64289 Darmstadt, Germany

Institut für Kernchemie, Johannes Gutenberg-Universität Mainz, 55128 Mainz, Germany

Helmholtz-Institut Mainz, Johannes Gutenberg-Universität Mainz, 55128 Mainz, Germany

W. Nörtershäuser (✉) • C. Geppert • J. Krämer
Institut für Kernphysik, Technische Universität Darmstadt, 64289 Darmstadt, Germany

Institut für Kernchemie, Johannes Gutenberg-Universität Mainz, 55128 Mainz, Germany
e-mail: wnoertershaeuser@ikp.tu-darmstadt.de

K. Blaum • K. Kreim
Max-Planck-Institut für Kernphysik, 69117 Heidelberg, Germany

M.L. Bissell • G. Neyens
Instituut Voor Kern- En Stralingsfysica, KU Leuven, 3001 Leuven, Belgium

N. Frömmgen • D. Tiedemann • M. Zakova
Institut für Kernchemie, Johannes Gutenberg-Universität Mainz, 55128 Mainz, Germany

M. Hammen
Institut für Kernchemie, Johannes Gutenberg-Universität Mainz, 55128 Mainz, Germany

Helmholtz-Institut Mainz, Johannes Gutenberg-Universität Mainz, 55128 Mainz, Germany

M. Kowalska
Physics Department, CERN, 1211, Geneva 23, Switzerland

D. Meschede et al. (eds.), *Exploring the World with the Laser*,
https://doi.org/10.1007/978-3-319-64346-5_12

comb with accuracies better than 1 MHz and led to a precise determination of the nuclear charge radii of $^{7,10-12}$Be relative to the stable isotope ^{9}Be. Moreover, an accurate determination of the *2p* fine-structure splitting allowed a test of high-precision bound-state QED calculations in the three-electron system. Here, we describe the laser spectroscopic method in detail, including several tests that were carried out to determine or estimate systematic uncertainties. Final values from two experimental runs at ISOLDE are presented, and the results are discussed.

1 Introduction

Laser spectroscopy provides a detailed insight into atomic structure including all subtle effects that contribute to the exact energy and the splittings of individual energy levels. Many of these effects are of great relevance in fundamental physics problems, as, for example, quantum electrodynamics, nuclear structure and weak interaction. Nowadays, laser spectroscopy combined with theoretical calculations is an indispensable tool to explore many-body QED in weak and strong fields and the search for a time or spatial dependence of fundamental constants like the fine-structure constant. It provides important information for the analysis of spectra from stars and quasars, for studies of the nuclear structure and for determining the weak charge of a nucleus.

The technique we present here has provided new data in two of the mentioned fields, namely the determination of nuclear charge radii and moments of beryllium isotopes [1, 2] and the test of many-body bound-state QED calculations in three-electron systems [3]. It is based on collinear laser spectroscopy, a technique that has been contributing to these fields considerably and is one of the workhorses for investigations of nuclear spins and moments, which is witnessed by a long series of review papers [4–11] over the last decades. In parallel, it has also been used to investigate the fine-structure splittings in helium-like ions as a test of bound-state QED. Such tests were carried out using boron B^{3+} [12], nitrogen N^{5+} [13] and fluorine F^{7+} [14]. In these experiments counter- and copropagating beams have been used to determine transition frequencies, while for the spectroscopy of short-lived neon isotopes a similar approach was used to calibrate the acceleration voltage of the ions [15, 16].

R. Neugart
Institut für Kernchemie, Johannes Gutenberg-Universität Mainz, 55128 Mainz, Germany

Max-Planck-Institut für Kernphysik, 69117 Heidelberg, Germany

R. Sánchez
Institut für Kernchemie, Johannes Gutenberg-Universität Mainz, 55128 Mainz, Germany

GSI Helmholtzzentrum für Schwerionenforschung, 64291 Darmstadt, Germany

D.T. Yordanov
Max-Planck-Institut für Kernphysik, 69117 Heidelberg, Germany

Physics Department, CERN, 1211, Geneva 23, Switzerland

For the measurements on beryllium isotopes we have further developed this technique and combined it with a frequency comb to provide high-precision measurements of the transition frequencies. A photon–ion coincidence detection provided the sensitivity required for the detection of the 20-ms isotope ^{12}Be. These investigations were motivated twofold, by the nuclear structure aspect and the possibility to provide an important benchmark for bound-state QED calculations in three-electron systems.

Observables of the nuclear ground state that can be measured with high accuracies like masses, charge radii, magnetic dipole and electric quadrupole moments are extremely valuable for nuclear structure physics. The change of the nuclear charge radius along a chain of isotopes is extracted with high precision from optical isotope shifts. This provides insight into differences of the radial distribution of protons and the underlying collective effects of soft or rigid deformation or cluster structures, which are often observed for the few-nucleon systems of light nuclei. Only during the last decade new experimental techniques and precise atomic structure calculations for few-electron systems gave access to the determination of charge radii of low-Z nuclei ($Z < 10$) with unprecedented precision. In 2000, first calculations of the mass shift in three-electron systems [17] provided sufficient accuracy to extract the very small finite nuclear size effect from high-precision isotope shift measurements. Since then, calculational precision for three-electron systems has been improved by two orders of magnitude [18–22]. Pachucki et al. [23, 24] published first results for four-electron systems and recently even showed results that pave the way toward boron-like five-electron systems [25].

Laser spectroscopy experiments on helium and lithium isotopes were strongly motivated by the existence of the so-called halo nuclei. These are nuclear systems with the last neutron(s) being bound by only a few 100 keV, compared to typical nuclear binding energies of the order of 5–7 MeV/nucleon. Due to this weak binding, the neutrons are allowed to tunnel far away from the central core, having a large part of their wavefunctions beyond the classical interaction length of the strong force. These nuclei have been a hot topic in nuclear structure research since their discovery in 1983 [26]. Isotope shifts for such systems were measured previously in helium and lithium isotopes. Single atoms of the short-lived two-neutron and four-neutron halo nuclei ^{6,8}He were confined in a magneto-optical trap and probed by laser light [27, 28]. The lithium isotopes including the two-neutron halo nucleus ^{11}Li were investigated by applying two-photon resonance ionization spectroscopy [22, 29, 30]. The beryllium isotope chain contains the one-neutron halo nucleus ^{11}Be and the isotope ^{12}Be which in the traditional shell model should have a closed neutron shell. First on-line measurements of transition frequencies and isotope shifts based on laser-cooled ^{7,10}Be^{+} ions in a Paul trap were reported in 2006 from the SLOWRI facility at RIKEN, but were limited to accuracies on the order of a few 10 MHz [31]. Further enhancements of this technique were discussed in [32], but improved values of transition frequencies or isotope shifts are not reported so far.

With regard to atomic structure, the vast progress in nonrelativistic few-electron bound-state QED has opened the possibility of additional tests of many-body QED in these light systems. The helium fine structure was recently calculated up to the order $m\alpha^7$ and is now one of the most precise QED tests in two-electron systems [33]. The application of such calculations to three-electron systems proved to be much harder since the extension of the respective computational methods with explicitly correlated functions turned out to be considerably more difficult. Only recently it became possible to perform a complete calculation of $m\alpha^6$ and $m\alpha^7 \ln \alpha$ contributions to the fine structure [34] of a three-electron atom. On the experimental side, measurements of the $2p$ fine-structure splitting in light three-electron systems are limited in accuracy for isotopes with nonzero nuclear spin due to the unresolved hyperfine structure (hfs) in the $2p\,^2\mathrm{P}_{3/2}$ level. This has been the reason for the fluctuating results on the fine-structure splittings in lithium [22, 35] being reported for a long time. These turned out to be caused by quantum interference effects in the observation of the unresolved resonance lines [36]. Once this issue had been resolved experimentally, good agreement with ab initio calculations was obtained [37]. Since relativistic and QED contributions grow in size with increasing Z, it became important to study the fine-structure splitting also in Be^+ to further test bound-state QED.

Both aspects have been addressed with the technique described in this paper. Besides giving a detailed description of the experiment implemented at ISOLDE (CERN), we will present an overview of the spectroscopic results obtained in two beam times (called Run I and Run II). Compared with the techniques used to study helium and lithium isotopes, the collinear approach has the advantage of being more generally applicable and providing high-precision isotope shift data for short-lived isotopes of elements in the so far inaccessible region $4 < Z < 10$.

2 Theory

It is well known that the isotope shift $\delta\nu^{A,A'}$ between two isotopes A and A' can be separated into the mass shift $\delta\nu_{\mathrm{MS}}^{A,A'}$ and the field shift $\delta\nu_{\mathrm{FS}}^{A,A'}$ according to

$$\delta\nu_{\mathrm{IS}}^{A,A'} = \nu^{A'} - \nu^{A} \tag{1}$$

$$= \underbrace{K_{\mathrm{MS}} \frac{M_{A'} - M_A}{M_A M_{A'}}}_{\delta\nu_{\mathrm{MS}}^{A,A'}} + \underbrace{F^{A,A'} \delta\langle r_c^2 \rangle^{A,A'}}_{\delta\nu_{\mathrm{FS}}^{A,A'}} . \tag{2}$$

The mass shift contribution (MS) is related to the center-of-mass motion of the atomic nucleus. For light elements this is the major part of the isotope shift, while the small nuclear volume shift $\delta\nu_{\mathrm{FS}}^{A,A'}$, being typically at the 10^{-5} level of the mass shift, contains the information about the change $\delta\langle r_c^2\rangle$ in the mean-square nuclear charge radius. Extraction of nuclear charge radii from experimental isotope shifts in the lightest elements requires accurate mass shift calculations. Semi-empirical techniques that have often been applied for heavier elements to evaluate the atomic parameters K_{MS} and F are not sufficiently accurate. Only state-of-the-art ab initio calculations can provide the accurate mass shift and field shift coefficients. Detailed descriptions of these calculations can be found, e.g.. in [17, 19, 22, 39]. Briefly, the starting point is the nonrelativistic Schrödinger equation which is solved with high numerical accuracy in the basis of Hylleraas coordinates that explicitly take electron–electron correlations into account. The wavefunctions obtained are then used to calculate relativistic and QED corrections perturbatively as a power series in terms of the fine-structure constant α. The results for the Be^+ isotopes as taken from [20, 21, 40, 41] are listed in Table 1. It is worthwhile to note that the calculations performed by two independent groups agree within uncertainties for all isotopes. The only significant difference concerns the case of $^{11}\mathrm{Be}$, where the nuclear polarizability correction of 211 kHz has been calculated and included in [41] but not in [40]. The field shift factor $F^{9,A}$ has been calculated for each isotope individually and is almost constant along the isotopic chain, besides a small difference in the relativistic correction. Using the mass shift values from the table and the measured isotope shifts, the change in the mean-square nuclear charge radius can be determined using

$$\delta\langle r_c^2\rangle^{9,A} = \frac{\delta\nu_{\mathrm{IS}}^{9,A} - \delta\nu_{\mathrm{MS}}^{9,A}}{F^{9,A}}. \tag{3}$$

The total charge radius $R_c(A) = \sqrt{\langle r_c^2\rangle^A}$ of at least one (or more) stable isotope(s) determined by other methods is required to obtain total charge radii of the radioactive isotopes. In the case of beryllium the nuclear charge radius of the stable $^9\mathrm{Be}$ nucleus was determined from elastic electron scattering [43] and thus

Table 1 Theoretical mass shifts $\delta\nu_{\mathrm{MS}}^{9,A}$ and field shift factors $F^{9,A}$ for the D1 and D2 transitions $2s\,^2\mathrm{S}_{1/2} \rightarrow 2p\,^2\mathrm{P}_{1/2,\,3/2}$ in Be^+ with respect to $^9\mathrm{Be}^+$ obtained in two independent calculations [20, 21, 38, 41] with updated values presented in [2] based on [40, 42]

Isotope	$\delta\nu_{\mathrm{MS}}^{9,A}$ D1 (MHz)	$\delta\nu_{\mathrm{MS}}^{9,A}$ D2 (MHz)	$F^{9,A}$ (MHz/fm^2)	References
$^7\mathrm{Be}^+$	−49 225.744 (35) (9)	−49 231.779 (35) (9)	−17.021 (31)	[38, 41, 42]
	−49 225.779 (38)	−49 231.828 (38)	−16.912	[20, 21, 40]
$^{10}\mathrm{Be}^+$	17 310.459 (13) (11)	17 312.553 (13) (11)	−17.027 (31)	[38, 41, 42]
	17 310.442 (12)	17 312.569 (12)	−16.912	[20, 21, 40]
$^{11}\mathrm{Be}^+$	31 560.245 (31) (12)	31 564.207 (31) (12)	−17.020 (31)	[38, 41, 42]
	31 559.990 (24)	31 563.868 (24)	−16.912	[20, 21, 40]
$^{12}\mathrm{Be}^+$	43 390.180 (30) (180)	43 395.480 (30) (180)	−17.022 (31)	[38, 41, 42]
	43 390.168 (39)	43 395.499 (39)	−16.912	[20, 21, 40]

The listed uncertainties are an estimation of unknown higher order terms. The calculations from [41] include another uncertainty that originates from the atomic mass. The deciphered contributions to the mass shift can be found, for example, in [38]

$$R_c(^A\mathrm{Be}) = \sqrt{R_c^2(^9\mathrm{Be}) + \delta\langle r_c^2\rangle^{9,A}}. \tag{4}$$

The many-electron Dirac equation poses some difficulties for the inclusion of relativistic effects and correlations between electrons in atomic systems. According to QED the equation has to include multiple electron–positron pairs, which leads to numerical instabilities. This problem limited the relativistic calculation of the lithium $2p\,^2\mathrm{P}_{1/2} - 2p\,^2\mathrm{P}_{3/2}$ splitting to one significant digit [44]. Forty years after first numerical calculations using explicitly correlated basis sets with Hylleraas and Gaussian functions for two electrons [45], Puchalski and Pachucki extended such calculations to three-electron systems [34]. Nonrelativistic QED can perturbatively account for relativistic, retardation, electron self-interaction, and vacuum polarization contributions by an expansion of the level energy in powers of the fine-structure constant α

$$E = m\alpha^2 E^{(2)} + m\alpha^4 E^{(4)} + m\alpha^5 E^{(5)} + m\alpha^6 E^{(6)} + \cdots, \tag{5}$$

where the expansion coefficients $E^{(i)}$ may include powers of $\ln\alpha$. In this expansion, the fine structure arises at the order of $m\alpha^4$, together with the nuclear recoil term, which in this order is comparable in size to $m\alpha^6$ contributions, but of opposite sign. For all details of the calculations and the individual contributions we refer to [46]. In the splitting isotope shift (SIS), i.e., the difference in fine-structure splitting between isotopes, all mass-independent terms cancel and only the mass-dependent terms remain, which can be calculated with very high accuracy.

The SIS therefore provides a valuable consistency check of the experimental results [47]. For isotopes with nuclear spin, hyperfine-induced fine-structure mixing can lead to an additional level shift that also contributes to the SIS. This in combination with the unresolved hyperfine splittings in the $2p\,^2\mathrm{P}_{3/2}$ level in light three-electron systems makes even–even isotopes with nuclear spin $I = 0$ an exceptionally suitable case to perform tests of the calculations. While there is no such isotope for lithium, the beryllium chain with $^{10}\mathrm{Be}$ and $^{12}\mathrm{Be}$ includes two spinless isotopes that are accessible to the measurement.

3 Experimental Setup

A schematic overview of the experimental setup applied for collinear laser spectroscopy on Be^+ ions in the $2s\,^2\mathrm{S}_{1/2} \rightarrow 2p\,^2\mathrm{P}_{1/2}$ (D1) and $2s\,^2\mathrm{S}_{1/2} \rightarrow 2p\,^2\mathrm{P}_{3/2}$ (D2) transitions is shown in Fig. 1. A mass-separated ion beam of a stable or a radioactive beryllium isotope at an energy up to 60 keV was transported to the laser beam line. Two frequency-stabilized dye laser systems delivered UV beams that superposed the beryllium ion beam in opposite directions. The resonance fluorescence photons were detected via photomultipliers. The resonance condition was

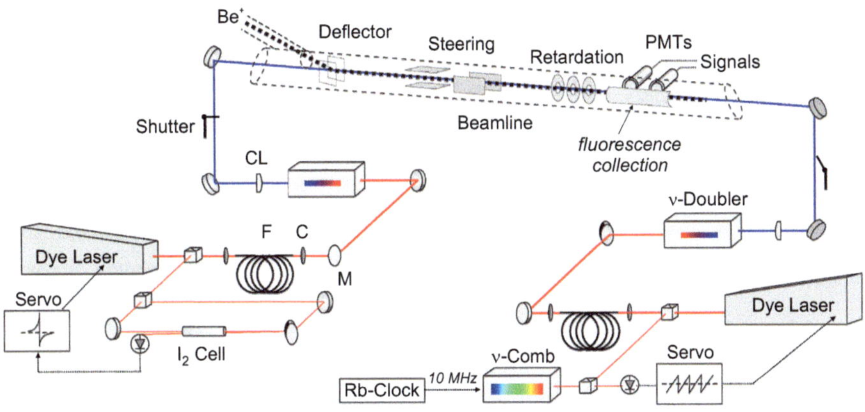

Fig. 1 Experimental setup for the beryllium measurements at ISOLDE. Two dye laser systems were used to excite the $2s\,^2S_{1/2} \rightarrow 2p\,^2P_{1/2,\,3/2}$ transitions in Be^+. The dye laser for collinear excitation (*left*) was operated at a fundamental wavelength of 624 nm and stabilized to a hyperfine transition of molecular iodine. The output beam is frequency doubled to 312 nm and guided into the beam line. The other laser (*right*) is locked to a frequency comb. After frequency-doubling to 314 nm the UV laser beam is anticollinearly superposed with the ion beam. The resonance fluorescence is detected by a pair of photomultipliers. A photon–ion coincidence detection unit increases the detection efficiency if the ion beam rate is low (not shown, see Fig. 3)

established by tuning the Doppler-shifted frequency with an electrical potential applied to the fluorescence detection chamber. The individual parts of the experimental setup as well as the scanning procedure are described in detail in the following subsections.

3.1 *Production of Radioactive Beryllium Isotopes*

The stable and radioactive beryllium isotopes were produced at the on-line isotope separator facility ISOLDE at CERN. High-energy (1.4 GeV) protons from the PS Booster synchrotron impinge on a uranium carbide target. The atoms are photo-ionized using the resonance ionization laser-ion source RILIS [49]. Resonant excitation at 234.9 nm from the atomic ground state in the $2s^2\ ^1S_0 \rightarrow 2s2p\,^1P_1$ transition, followed by excitation at 297.3 nm to the auto-ionizing $2p^2\,^1S_0$ level was employed to ionize the Be atoms which have the rather large ionization potential of 9.4 eV.

Table 2 lists the ion beam intensities decreasing from ^{7}Be to ^{12}Be by seven orders of magnitude [48]. In the final stage of our experiment an upgraded solid-state pump laser system [49] was used. This gave a ^{11}Be yield of up to $2.7 \cdot 10^7$ ions/s, about 4 times larger than reported previously.

The yields are sufficient to perform collinear laser spectroscopy on $^{7-11}$Be solely based on a standard fluorescence detection system. However, for a beam intensity

Table 2 Nuclear properties and production rates of the beryllium isotopes at the ISOLDE facility at CERN

	$T_{1/2}$	I	$\mu_I(\mu_N)$	Ions/μC
^{7}Be	53 days	3/2	−1.39928 (2) [69]	$1.4 \cdot 10^{10}$
^{9}Be	Stable	3/2	−1.177432 (3) [70]	
^{10}Be	$1.6 \cdot 10^{6}$ a	0	–	$6.0 \cdot 10^{9}$
^{11}Be	13.8 s	1/2	−1.68166 (11) [71]	$7.0 \cdot 10^{6}$
^{12}Be	23.8 ms	0	–	$1.5 \cdot 10^{3}$
^{14}Be	4.35 ms	0	–	$4.0 \cdot 10^{0}$

The table includes the half-life ($T_{1/2}$), nuclear spin I, magnetic dipole moment (μ_I) in nuclear magnetons (μ_N), and the yields using a 1.4-GeV proton beam from the PS booster and RILIS for ionization [68]

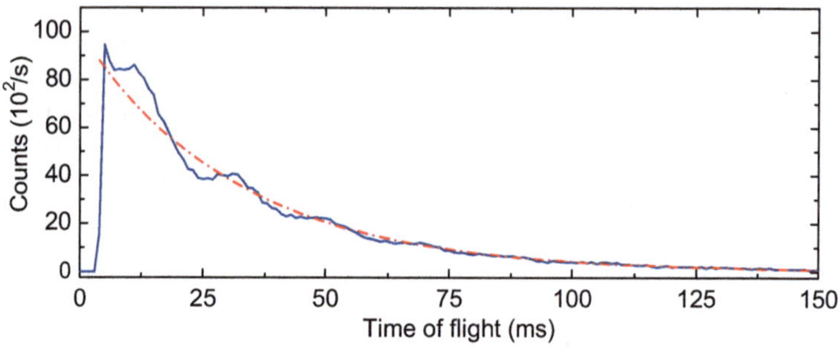

Fig. 2 Release of beryllium ions (*solid blue line*) from ISOLDE as a function of time after the proton pulse hit the target container, measured with a secondary electron multiplier at the end of the COLLAPS beam line. The release curve, integrated over 100 proton pulses with a resolution of 0.2 ms/channel, is modeled with an exponential decay curve (*dash-dotted red line*)

of $< 10^4$ ions/s, i.e., for measurements on ^{12}Be, the sensitivity had to be enhanced. This was achieved by detecting ion–photon coincidences and thus rejecting the stray light background which usually determines the sensitivity limit. Coincidence detection requires an isobarically clean ion beam. For that reason the pulse structure and possible contamination of the beam was investigated and optimized for ^{12}Be.

3.2 *Beryllium Ion Beam Structure*

During our experiment at ISOLDE, pulses of $3 \cdot 10^{13}$ protons impinged on a UC$_x$ target typically every 4 s. The release of resonantly ionized ^{12}Be was tracked using a secondary electron multiplier installed at the end of the laser spectroscopy beam line. The proton pulses triggered a multichannel analyzer that recorded the ion events as a function of time. Figure 2 shows such a release curve summed over 100 proton pulses with a resolution of 0.2 ms/channel. The integral corresponds to a release of 12 000 ions per proton pulse. This is almost a factor of 10 more than listed in the yield table (Table 2). During the measurements on ^{12}Be the typical ion yield was about 8000 ions/pulse.

The release curve of Fig. 2 demonstrates a characteristic feature of the ISOLDE HV supply: First ions are detected about 2–3 ms after the proton pulse hit the target. This delay is determined by the recovery time of the high voltage, which is pulsed down right before a proton pulse hits the target, in order to reduce the current load from ionized air [51]. After an initial steep rise the release curve follows essentially the exponential decay of ^{12}Be. The extracted half-life of $T_{1/2} \approx 21.9(8)$ ms agrees well with the literature value of 21.50 (4) ms [50]. The single exponential does not exhibit any significant offset. This demonstrates that practically no beam contamination from the isobar $^{12}C^+$ is present and $^{12}B^+$ (having a similar lifetime as ^{12}Be) is also not expected due to the relatively high ionization potential. This situation is prerequisite for the application of a photon–ion coincidence technique, which otherwise would suffer from random coincidences between scattered laser light and isobaric ions. With the rapid decay of ^{12}Be the fluorescence detection can be limited to about 100 ms after the proton pulses.

3.3 Experimental Beam Line

The COLLAPS collinear spectroscopy beam line at the ISOLDE facility was commissioned in the early eighties [52–54] and has been improved continuously [8, 15, 55–58] with the objective of widening the range of accessible elements and isotopes. An important aspect was the development of highly sensitive alternatives to the traditional fluorescence photon detection technique. For conventional collinear spectroscopy the ions are accelerated to a beam energy of typically 50 keV, with the corresponding positive potential applied to the ion source, while the mass separator and the experimental beam line are on ground potential. The ion beam is merged with a laser beam by a pair of deflector plates as shown in Figs. 1 and 3.

A quadrupole triplet collimates the ion beam, matching it to the laser beam profile, and a second set of deflector plates aligns it with the laser beam axis which is defined by two apertures at a distance of about 2 m. Two UV-sensitive photomultiplier tubes with 45 mm active aperture and 15% quantum efficiency at 313 nm are used for fluorescence detection. The light collection system consists of two fused-silica lenses of 75 mm diameter and a cylindrical mirror opposite to them. Stray laser light is suppressed by sets of apertures with diameters decreasing with distance from the optical detection region and Brewster-angle quartz windows at both ends.

Collinear laser spectroscopy is usually performed with the laser running at a fixed frequency, while the absorption frequency of the ions is tuned by changing their velocity (Doppler-tuning). This means that a variable electrical potential has to be applied to the interaction region. For applying post-acceleration/deceleration voltages up to 10 kV a set of four electrodes provides a smoothly variable potential

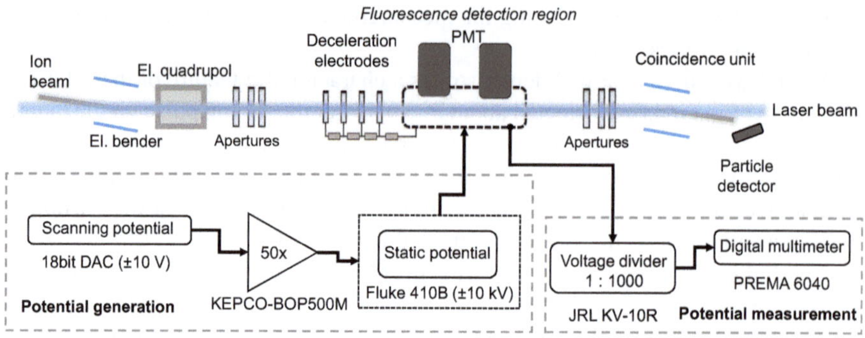

Fig. 3 Schematic view of the COLLAPS beam line at ISOLDE/CERN and corresponding high-voltage circuits: A mass-separated ion beam is directed along the axis of the vacuum beam line with the help of an electrostatic deflector. An electric dipole and quadrupole collimate and steer the beam through the apparatus. The fluorescence detection region is a cage floated on a variable potential against ground to enable Doppler-tuning. At the end of the beam line a photon–ion coincidence detection chamber is installed, whereby a secondary electron multiplier is used to count the ions. The generation and measurement of the high-voltage potential is explained in the text

along the beam axis. In order to avoid optical pumping into dark states, the final ion velocity is reached just in front of the detection region by applying a small fixed offset voltage between the last electrode and the detection chamber. The lower part of Fig. 3 illustrates the generation of the voltage between the detection region and ground as a combination of a static high voltage in the range of ± 10 kV and a scanning voltage of ± 500 V. The latter is created by amplification ($\times 50$) of the ± 10 V dc output of an 18-bit DAC controlled by the measuring computer. This voltage defines the floating offset potential of a stabilized ± 10 kV power supply. The combination of power supplies makes it possible to perform measurements on a series of isotopes with different Doppler-shifts and for each of them scan small frequency ranges covering the hyperfine structure with high resolution.

Both the static high voltage and the scanning voltage are measured with a high-precision 1:1000 voltage divider and a digital voltmeter. A comparison of the measured voltages with those obtained using a precision voltage divider calibrated at PTB (Braunschweig, Germany) [59] has demonstrated an uncertainty of $\Delta U/U < 3 \cdot 10^{-5}$ which corresponds to about 0.3 V at a maximum voltage of 10 kV applied to the excitation region [60]. Still, the knowledge of the ion beam velocity is limited by the uncertainty of the ion source potential which is determined by the main acceleration voltage power supply. Also, the specified accuracy $\Delta U/U < 1 \cdot 10^{-4}$ of the voltage measurement on the operational voltage of 60 kV was verified by calibration with the precision voltage divider. It translates to <6 eV uncertainty in the ion beam energy.

As in the laboratory frame the transition frequency of the ions in collinear geometry scales as

$$\nu_c = \nu_0 \gamma (1 + \beta) \tag{6}$$

where the dimensionless ion velocity is $\beta = v/c$, the relativistic factor $\gamma = \sqrt{1/(1-\beta^2)}$ and the transition frequency ν_0. Any uncertainty in β arising from the ion source potential results in an uncertainty of measured transition frequencies or isotope shifts, especially for light ions. In the particular case of beryllium a deviation of 6 V from the measured voltage result in an artificial isotope shift of $\delta\nu^{9,11}(^9\text{Be}, ^{11}\text{Be}) = 18$ MHz. To overcome these limitations, we have introduced the (quasi-)simultaneous excitation by a collinear and an anticollinear laser beam. The method is based on the fact that in this geometry the measured resonance frequencies, $\nu_c = \nu_0\gamma(1+\beta)$ for collinear and $\nu_a = \nu_0\gamma(1-\beta)$ for anticollinear excitation are simply related to the rest-frame frequency ν_0 by

$$\nu_c \cdot \nu_a = \nu_0^2\gamma^2 \cdot (1+\beta)(1-\beta) = \nu_0^2. \tag{7}$$

This provides a method to determine the transition frequency independently of the knowledge of the ion beam energy which depends on assumptions about the ion source potential and on measured voltages. However, in contrast to conventional collinear laser spectroscopy, this approach requires two laser systems instead of one and, additionally, the capability to determine the laser frequencies with an accuracy better than 10^{-9}. Similar approaches were proposed and demonstrated for the measurement of transition frequencies [61] and used for, e.g., precision spectroscopy in the fine structure of helium-like Li^+, yielding an accurate value of the Lamb shift [62]. Here we have developed a procedure which is widely applicable in cases where high precision is required for the spectroscopy of unstable isotopes.

3.4 Setup and Specification of the Frequency-Comb-Referenced Laser System

The transition wavelength of the $2s\,^2\text{S}_{1/2} \rightarrow 2p\,^2\text{P}_{1/2,\,3/2}$ transitions in Be^+ is about 313 nm corresponding to an energy splitting of ≈ 4 eV. The laser system installed at COLLAPS is schematically shown in Fig. 1. For anticollinear excitation a frequency doubled Nd : YVO_4 laser (Verdi V18) was operated at 9 W to pump a Coherent 699-21 dye laser. Using a dye solution of Sulforhodamine B in ethylene glycol, a typical output power of 700 mW was achieved at the wavelength of 628 nm. Another dye laser, Sirah MATISSE DS, was installed for collinear excitation and operated with a dye solution of DCM in 2-phenoxy-ethanol. With the 8-W pump beam from a Verdi V8, about 1.2 W were achieved at the fundamental wavelength of 624 nm. Each laser beam was then coupled into a 25-m long photonic crystal fiber (LMA-20) to transport the laser light to one of the second harmonic generators installed in the ISOLDE hall. A two-mirror delta cavity (Spectra Physics Wavetrain) and a four-mirror bow-tie cavity (Tekhnoscan

FD-SF-07) were located nearby the COLLAPS beam line. A Brewster-cut and an antireflection coated BBO crystal, respectively, converted the laser beams of 628 and 624 nm into their second harmonics at 314 and 312 nm, in both cases with an output of more than 10 mW. The elliptical UV beams were reshaped to circular beams with diameters of 3–4 mm to match the transversal profile of the ion beam and finally attenuated to powers below 5 mW. Two remote-controlled fast beam shutters blocked alternatively the collinear or the anticollinear laser beam. This enabled us to perform scans of 3–30 s duration in collinear or anticollinear configuration in a fast sequence.

The backbone of the laser system was the precise frequency stabilization and frequency measurement required for the application of Eq. (7). In practice, the transition rest-frame frequency ν_0 depends on the frequencies of both dye lasers in the laboratory frame which have to be known with a relative accuracy better than $\Delta\nu/\nu \leq 10^{-9}$ to yield the isotope shifts with an accuracy better than 10^{-5}. Therefore, a Menlo Systems frequency comb (FC 1500) with a repetition frequency of 100 MHz was employed. A Stanford Research rubidium clock (PRS10) provided the 10-MHz reference for the stabilization of the carrier-envelope-offset (CEO) and the repetition frequency. This clock was long-term stabilized using a GPS receiver tracking the 1-pps signal.

The MATISSE dye laser for collinear excitation was stabilized to its internal reference cavity for short-term stability. In this case frequency drifts were further reduced by locking the laser to a hyperfine transition in molecular iodine using frequency-modulated saturation spectroscopy. In total 12 hyperfine transitions of $^{127}I_2$ match the desired Doppler-shifted frequencies for a wide range of acceleration voltages between 30 and 60 kV. The demodulated dispersion signal from the phase-sensitive detection was fed into a 16-bit National Instruments DAQ card (NI-DAQ 6221) and further processed with the MATISSE control software to provide a counter-drift for the MATISSE reference cavity. In regular time intervals the laser frequency was measured with the frequency comb and recorded for a few 100 s to ensure the stability of the locking point and to provide the laboratory-frame frequency for the application of Eq. (7). A histogram of 1-s averaged beat signals measured over 2 h is depicted in Fig. 4a. It exhibits a Gaussian distribution with standard deviation of about 75 kHz. The frequency of the Matisse laser stabilized to the various iodine lines was repeatedly measured during the beam times. The averaged results are listed in Table 3 and compared with the calculated frequencies from [63]. Reasonable agreement is obtained in all cases.

The Coherent 699 dye laser for anticollinear excitation was internally stabilized to its own reference cavity of Fabry-Perot type; long-term frequency drifts were corrected by an additional stabilization to the frequency comb. Therefore, the beat signal between the dye laser and the nearest frequency comb mode was detected on a fast photo diode and fed into the Menlo Systems phase comparator DXD 100. A low-noise PI regulator (PIC 210) processed the signal from the phase comparator and provided a servo-voltage to counteract all frequency excursions of the dye laser

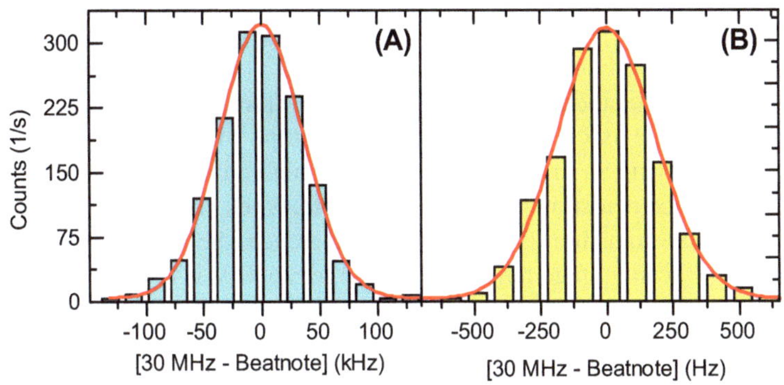

Fig. 4 Beat frequency histograms of the frequency-stabilized dye lasers. The beat was averaged for 1 s, and the distribution of the beat frequencies over a period of 2 h is shown in the histogram. Graph (**a**) shows the histogram of the MATISSE DS laser stabilized to a hyperfine transition of molecular iodine. The corresponding uncertainty was estimated as the FWHM of a Gaussian fit (*red*) of about 75 kHz. The results of the frequency-comb stabilized Coherent 699-21 dye laser are depicted in Graph (**b**). In this case the FWHM is approximately 400 Hz

Table 3 Frequencies of the a_1 hyperfine component in various transitions of iodine $^{127}I_2$ utilized and determined during the experiment

HFS a_1 transition	Frequency (theory)	Frequency (experiment)	Deviation
R(62)(8-3)	479 804 354.67	479 804 355.09	−0.42
R(70)(10-4)	479 823 072.75	479 823 072.58	0.17
P(64)(10-4)	479 835 709.4	479 835 708.96	0.44
R(60)(8-3)	479 870 011.92	479 870 012.20	−0.28
R(58)(8-3)	479 933 416.07	479 933 416.36	−0.29
R(56)(8-3)	479 994 568.08	479 994 568.11	−0.03
R(54)(8-3)	480 053 468.95	480 053 469.06	−0.11
R(52)(8-3)	480 110 119.57	480 110 119.59	−0.02
R(50)(8-3)	480 266 578.9	480 266 578.59	0.31
R(48)(8-3)	480 314 237.19	480 314 236.83	0.36
R(42)(8-3)	480 359 649.42	480 359 649.13	0.29
R(40)(8-3)	480 402 816.3	480 402 815.78	0.52

All values are in MHz. The total uncertainty of the experimental values is about 190 kHz. The calculated frequencies (theory) are afflicted with an uncertainty of approximately 3 MHz [63].

by correcting the length of the reference cavity. As a measure of the long-term stability a beat signal with the frequency comb was detected. The result is shown in Fig. 4b. The standard deviation over 2 h measuring time and 1-s averaging time is about 400 Hz.

4 Measurement Procedure

The ion beam acceleration voltage at the ISOLDE front end was fixed to 40 kV. A suitable iodine line is chosen such that the isotope under investigation can be recorded by applying an offset voltage in the available range of $U_{\text{Offset}} = \pm 10$ kV at the fluorescence detection region. For example, when choosing the a_1 hyperfine component in the transition R(56)(8-3) as a reference, the isotopes $^{9-12}$Be can be addressed. The scan voltage range U_{scan} of up to ±500 V is then adjusted to cover the full hyperfine structure in the collinear direction, and the expected position of the center of gravity is calculated. The required offset voltage as well as the scan voltage range was estimated based on previous measurements of the ^{9}Be transition frequency [66] and nuclear moments [15] in combination with the precisely calculated mass shift [40, 41]. Once the resonance position of ^{9}Be was found, the laser frequencies ν_c and ν_a could be predicted for all radioactive beryllium isotopes with an accuracy of a few MHz, which is the size of the expected field shift contribution. This knowledge allowed us to calculate the required frequency of the second dye laser to simultaneously cover the full hyperfine structure in anticollinear geometry within the same Doppler-tuning voltage range and even to ensure that the centers of gravity of both hyperfine spectra practically coincide within a few 100 mV, corresponding to the size of the field shift contribution. This is only possible because this dye laser is locked to the frequency comb and thus can be stabilized at any arbitrarily chosen frequency.

Fast laser beam shutters placed in front of the Brewster windows of the apparatus were controlled by the data acquisition software in order to allow only one of the two laser beams to enter. For the isotopes with half-lives longer than the typical 4-s repetition time of proton pulses, fast scans of the Doppler-tuning voltage U_{scan} were performed with alternating laser beams. The scanning range was chosen depending on the hyperfine splitting of the respective isotope, and spectra were taken in 200 channels for ^{10}Be and up to 800 channels for the odd-A isotopes ^{7,9,11}Be. The common dwell time was 22 ms per voltage step. Depending on the ion beam intensity, a single spectrum is the sum of 50–800 individual scans for each direction. This procedure was applied using about 3–4 different iodine lines for each isotope.

Because of the short 21.5-ms half-life of ^{12}Be, photon counts had to be accumulated for typically 60 ms after each proton pulse. The laser shutters for collinear and anticollinear beams were switched between consecutive pulses, and the voltage steps were triggered by every second pulse. Given the extremely low ion beam intensity, the single-line spectrum of ^{12}Be was taken in only 20 channels with a total measuring time of about 8 hours, corresponding to 200 scans.

Detection of the weak ^{12}Be signals required the additional rejection of background from scattered laser light reaching the photomultiplier tubes. This was achieved by implementing a photon–ion coincidence: Photomultiplier signals were accepted only if an ion was simultaneously traversing the detection unit. Downstream of the photon detection chamber the ions were deflected onto the

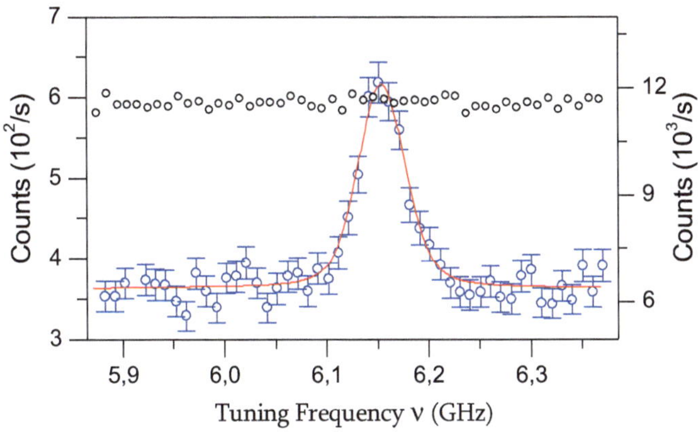

Fig. 5 Comparison between conventional optical fluorescence detection (*upper trace*, *right y* axis) and photon–ion coincidence detection (*lower trace*, *left y* axis) at an ion beam rate of 30,000 $^{10}Be^+$ ions/s. The optical spectrum (*black circles*) in the conventional detection is covered by stray light of the laser beam. In the photon–ion coincidence spectrum a clear resonance (*blue circles*) is observed, fitted with a Voigt profile

cathode of a secondary electron multiplier (SEM) installed off-axis. Discriminated pulses from the photomultipliers were delayed by the appropriate time of flight (TOF) (3–4 μs) of the ion to the SEM. To avoid electronic dead times, the delay was realized logically in a first-in first-out (FIFO) queue structure on a field-programmable gate array (FPGA) with a resolution of 10 ns, based on the FPGA's internal clock. Signals leaving the queue were transformed back into a TTL pulse and fed together with the SEM pulses into a standard coincidence unit. The photon–ion coincidence detection was optimized using a $^{10}Be^+$ ion beam, attenuated to about 30,000 ions/s by detuning the RILIS laser. The time of flight for ^{9}Be was determined with a multichannel analyzer, and the respective TOF for ^{10}Be was calculated. Figure 5 shows a comparison between the conventional ungated spectrum (gray circles) and the optical spectrum detected in delayed coincidence (blue circles). The resonance peak is only visible in the gated spectrum. The background induced by laser stray light was reduced by a factor of 35. However, it must be noted that this reduction factor strongly increases with a reduction of the ion beam rate.

5 Analysis and Results

Two beam times were performed to investigate first the isotopes $^{7-11}$Be (Run I) [1] and then concentrate on ^{12}Be (Run II) [2] after installing the ion–photon coincidence setup. The stable isotope ^{9}Be and the even–even isotope ^{10}Be were used as reference isotopes, respectively. We concentrate here on results and procedures from Run II and provide differences to Run I only when it is of importance.

5.1 Line Shape Studies on ^{9}Be and ^{10}Be

Resonance spectra of ^{9}Be in the $2s\,^{2}S_{1/2} \rightarrow 2p\,^{2}P_{1/2}$ transition are shown in the upper trace of Fig. 6 taken in collinear (left) and anticollinear geometry (right) as a function of the Doppler-tuned laser frequency. Each spectrum is the sum of 20 individual scans. To avoid saturation broadening, the laser beam was attenuated to 3 mW and collimated to a beam diameter of about 3–4 mm to match approximately the size of the ion beam. Similar spectra for ^{10}Be are shown in the lower traces. These are the integral of 50 single scans at an ion beam current of 10 pA. A best fit of the resonance was obtained for a double Voigt profile with a full width half maximum (FWHM) of 40 MHz. It becomes apparent that each peak in the hyperfine structure is actually a composition of two components: A satellite peak, with a typical intensity of <5% of the corresponding main peak, appears on the

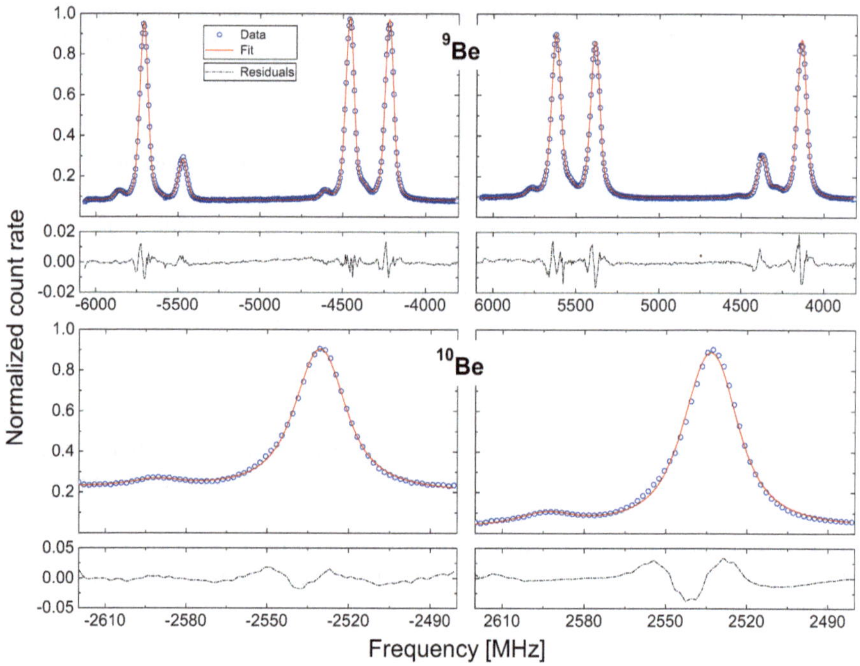

Fig. 6 *Top* optical hyperfine spectrum of the $2s\,^{2}S_{1/2} \rightarrow 2p\,^{2}P_{1/2}$ transition in $^{9}Be^{+}$ for collinear (*left*) and anticollinear excitation (*right*). Spectra were taken at 35-kV ISOLDE voltage and are the sum of 20 individual scans with a resolution of 400 channels. Hyperfine spectra are fitted using a multiple Voigt profile for each component (*red line*, for further details see text). Striking is the appearance of a small satellite peak on the *left* of each component which is ascribed to energy loss of the ions in inelastic collisions in flight. The differential Doppler-tuning parameter is about 39 MHz/V. *Bottom* resonance spectra of the $2s\,^{2}S_{1/2} \rightarrow 2p\,^{2}P_{1/2}$ transition in $^{10}Be^{+}$ again in collinear and anticollinear geometry also fitted with a Voigt doublet. A small structure in the residua (shown on the *bottom* of each graph) remains in all cases, which is discussed in the text

low-energy tail in each spectrum independent of the direction of excitation. It is induced by a class of ions which have lost some of their kinetic energy. The loss is almost exactly 4 eV and can be explained by inelastic collisions with residual gas atoms that lead to excitations into the $2p$ states. The energy required for this excitation is taken from the kinetic energy of the ion and is lost when the excited ion decays to the ground state by emitting a photon. The overall line shape is reasonably well fitted using a Voigt doublet and only small structures remain in the residua, depicted below each spectrum. The remaining small asymmetry seen in this structure is similar for the different peaks. It is an asset of the technique that asymmetries in the collinear and the anticollinear spectra shift the peak center to slightly lower and slightly larger frequencies, respectively. Hence, these shifts largely cancel when calculating the rest-frame frequency.

5.2 Hyperfine Fitting Procedure

Fitting was performed as follows: Each voltage information was converted into the corresponding Doppler-shifted laser frequency to account for the small nonlinearities in the voltage–frequency relation. Hyperfine peak positions relative to the center of gravity ν_{cg} were calculated based on the Casimir formula.

The position of each hfs sublevel with total angular momentum $\mathbf{F} = \mathbf{I} + \mathbf{J}$, composed of electronic angular momentum J and nuclear spin I, and $C = F(F+1) - I(I+1) - J(J+1)$ is given to first order by the hyperfine energy

$$\Delta E_{\text{hfs}} = \frac{A}{2}C + B \cdot \frac{\frac{3}{4}C(C+1) - I(I+1)J(J+1)}{2I(2I-1)J(2J-1)} . \tag{8}$$

In the fit function, these shifts determine the spectral line positions relative to the center of gravity. The factors A and B (only for $2p\,^2\mathrm{P}_{3/2}$) of the upper and the lower fine-structure state of the transition and the center of gravity ν_{cg} are the free-fitting parameters for the peak positions. The line shape of each component was modeled by two Voigt resonance terms representing the main peak and the satellite peak as discussed above. The distance between the two peaks was fixed to 4 V on the voltage axis. The Gaussian (Doppler) line width parameter and the intensity ratio between the main peak and the satellite were free parameters but constrained to be identical for all hyperfine components, while the total intensity of each component was also a free parameter. The Lorentzian line width was kept fixed at the natural line width of 19.64 MHz since significant saturation broadening was not observed. Nonlinear least-square minimization of χ^2 was performed using a Levenberg–Marquardt algorithm.

Fitting the collinear and the anticollinear spectra independently, we obtain in both cases the centroid frequency ν_{cg} of the hyperfine structure. However, for calculating the transition rest-frame frequency ν_0 we must take into account that

Eq. (7) requires ν_c and ν_a to be measured at the same ion velocity. This is only the case if the center of gravity appears in both spectra at the same voltage. This was accomplished approximately by changing the frequency of the laser used for anticollinear excitation until the deviation between the corresponding centers of gravity was typically smaller than 3 V, facilitated by the fact that the comb-stabilized laser can be locked at any arbitrary frequency. The remaining small shift δU was considered in the analysis by correcting the collinear frequency using the linear approximation $\delta\nu = \frac{\partial\nu}{\partial U}\cdot\delta U$, where U is the total acceleration voltage of the ions which have entered the optical detection region. Hence, the transition rest-frame frequency was calculated according to

$$\nu_0 = \sqrt{\left(\nu_c - \frac{\partial\nu_\mathrm{D}}{\partial U}\cdot\delta U\right)\cdot\nu_a} - \delta\nu_\mathrm{rec}. \tag{9}$$

with the differential Doppler-shift

$$\frac{\partial\nu_\mathrm{D}}{\partial U} = \frac{\nu_0}{mc^2}\left(e + \frac{e(mc^2+eU)}{\sqrt{eU(2mc^2+eU)}}\right) \tag{10}$$

and the recoil correction term

$$\delta\nu_\mathrm{rec} = \frac{h\nu^2_\mathrm{photon}}{m_0c^2}. \tag{11}$$

The latter takes energy and momentum conservation during the absorption/emission process into account. It contributes with about 200 kHz to the transition frequency and is slightly isotope dependent. Each measurement of ν_0 was repeated at least five times for each isotope. Statistical fitting uncertainty of the center of gravity was usually $<$100 kHz. For each pair of collinear/anticollinear spectra the rest-frame transition frequency was calculated, and the final statistical uncertainty was then derived as the standard error of the mean of all measurements being usually of the order of 100–500 kHz.

5.3 *Investigations of Systematic Uncertainties*

Sources of systematic errors were investigated on-line in Run I and Run II as well as in an additional test run, when only a previously irradiated target was used to extract the long-lived isotope ^{10}Be. For each isotope about 3–4 different iodine hyperfine transitions were used as reference points for the collinear laser frequency, which implies different locking frequencies of the comb-locked anticollinear laser as well as different offset voltages at the fluorescence detection region. It should be noted that the actual locking frequency of the iodine-locked laser was regularly checked with the frequency comb during each block of measurements.

In Run I, the Rb reference clock for the frequency comb was not long-term stabilized on the 1-pps signal and contributed with about 350 kHz to the systematic uncertainty of the transition frequency [1].

Additional uncertainties related to the applied acceleration voltages could only arise from the center-of-gravity correction according to Eq. (9), which was typically <3 V. The HV-amplification factor, calibrated regularly to better than $3 \cdot 10^{-4}$, leads to uncertainties clearly below the 3-mV level corresponding to approximately 100 kHz in transition frequency. This contribution can be safely neglected compared to other systematic uncertainties discussed below.

Additionally, ion and laser beam properties were modified on purpose for investigating a possible influence on the measured transition frequencies and isotope shifts. In deviation from a parallel collimation the ion beam was focused close to the fluorescence detection region with the available electrostatic quadrupole lenses. Similarly, additional convex lenses were added into the light path to focus the laser beams inside the beam line. It was found that these modifications merely changed the signal-to-noise ratio but had no significant influence on the determined resonance frequencies.

5.3.1 Laser-Ion-Beam Alignment

The parallel and antiparallel alignment of the respective laser beams with the ion beam was ensured using two apertures inside the beam line. Hence, the range of a possible angle misalignment between laser and ion beam was estimated taking the full aperture of 5 mm and their distance to each other of 2 m into account. A conservative estimate with beam diameters of about 4 mm results in an angle of $\alpha = \arctan(\Delta z/\Delta x) \approx 1$ mrad. However, during the preparation of the experiment, both laser beams were superimposed 2 m after the collinear and anticollinear exit windows, respectively. Two extreme cases are to be discussed: The Doppler-shifted frequencies $\nu_{c,a}$ get angle dependent if both laser beams are well superposed, but are misaligned relative to the ion beam

$$\nu_{c,a} = \nu_0 \gamma (1 \pm \beta \cdot \cos \alpha). \tag{12}$$

Then the transition rest-frame frequency ν_0 becomes angle dependent as well

$$\nu_0 = \frac{1}{\gamma} \sqrt{\frac{\nu_c \cdot \nu_{ac}}{1 - \beta^2 \cdot \cos^2 \alpha}} \tag{13}$$

$$= \sqrt{\frac{1 - \beta^2}{1 - \beta^2 \cos^2 \alpha}} \sqrt{\nu_c \cdot \nu_{ac}}. \tag{14}$$

Even though the Doppler-shift is reduced for both beams, the collinear–anticollinear geometry almost leads to a cancellation of the effect; the anticollinear resonance is less blue shifted while the collinear resonance is less red shifted. For an angle misalignment of 1 mrad, ν_a and ν_c will each be shifted by as much as 1.4 MHz, whereas the effect on ν_0 is only of the order of about 1 kHz.

The other extreme is a misalignment of one laser relative to a well superposed laser-ion–beam pair. Then the transition frequency ν_0 becomes

$$\nu_0 = \frac{1}{\gamma}\sqrt{\frac{\nu_c \cdot \nu_{ac}}{(1-\beta \cdot \cos\alpha)(1+\beta)}} \tag{15}$$

$$= \sqrt{\frac{1-\beta}{(1-\beta\cos\alpha}})\sqrt{\nu_c \cdot \nu_{ac}} \tag{16}$$

$$\approx \left(1-\beta\alpha^2\right)\sqrt{\nu_c \cdot \nu_{ac}}. \tag{17}$$

In contrast to Eq. (13) this angle-dependence can lead under unfavorable conditions to an appreciable shift. The influence of the laser-ion-beam alignment was extensively studied using a stable ^{9}Be ion beam by misaligning one of the laser beams so that a deviation was clearly visible in the horizontal or vertical direction. With the typical beam diameter a deviation of about 2 mm across a distance of ≈ 8 m ($\alpha = 0.25$ mrad) was detectable, corresponding to a total effect of about 600 kHz. Misalignment and realignments were repeated several times but the results of the measurements with misalignment scattered similarly as the measurements with optimized alignment, and in both cases the scatter was in accordance with the standard deviation of all regular ^{9}Be measurements. During the experiment, the counterpropagating alignment of the laser beams was inspected visually several times per day. A systematic uncertainty of 300 kHz, corresponding to half the full scattering amplitude was conservatively estimated.

5.3.2 Photon Recoil Shift

Repeated interaction with a laser beam can influence the external degrees of motion of an ion or atom as it is well known from laser cooling and laser deceleration in a Zeeman slower. In collinear laser spectroscopy the repeated directed absorption and isotropic reemission of photons will have the consequence that the ions are either accelerated (collinear excitation) or decelerated (anticollinear excitation). With every absorbed photon, the Doppler-shifted resonance frequency is shifted toward higher frequencies (ν_c and ν_a) for both directions and this systematic shift results in a transition frequency ν_0 that is too large. The combination of light ions and ultraviolet photons leads to an exceptionally large photon recoil, and the possible influence of this effect must be studied. Due to the absence of hyperfine splitting the $2s \rightarrow 2p$ transitions in the even isotopes $^{10,12}\mathrm{Be}^+$ are closed two-level systems. Hence, the possibility of repeated photon scattering is enhanced compared to the odd-mass isotopes which are pumped into a dark hyperfine state after a few absorption–emission cycles. To investigate whether the photon recoil has a

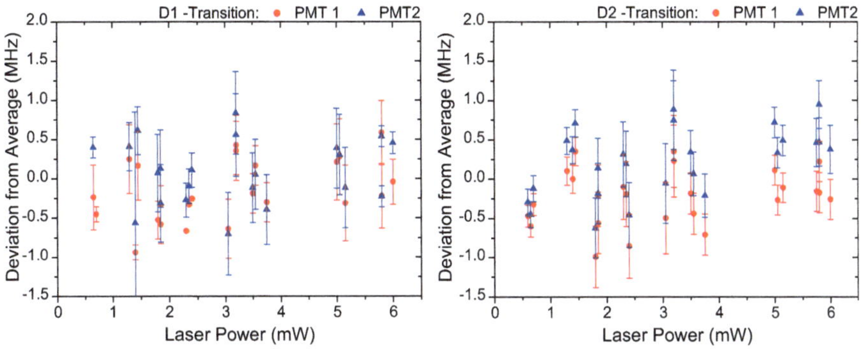

Fig. 7 Power dependence of the extracted frequency of $^{10}Be^{+}$ in the D1 (*left*) and the D2 transition (*right*). Plotted are the differences to the total mean frequency as a function of laser power. Copropagating and counterpropagating laser beams were adjusted to approximately equal power. Measurements were performed in three series varying the power from the highest to the lowest values. Uncertainties of the individual data points were estimated as the standard deviation for each individual set of three measurements at approximately the same power from this series

measurable effect, the power dependence of the transition frequency of ^{10}Be was determined as a function of the laser power. The laser power in both beams was increased stepwise and simultaneously from below 1 mW up to 6 mW. The deviation of the extracted transition frequencies from the mean frequency determined for the D1- and D2-transition is plotted in Fig. 7 as a function of laser power. Each data point is associated with an uncertainty estimated as the standard deviation of a block of three measurements at similar power. In both transitions the peak positions scatter but do not show a common trend upwards or downwards. It appears, however, that we observe at photomultiplier tube 2 (PMT2), located about 15 cm downstream from PMT1, resonances that are systematically higher in frequency than at PMT1.

As a consequence of this observation we have included only data from PMT1 in the analysis and have estimated an additional uncertainty for the remaining effect. The average difference between PMT1 and PMT2 is about 300 and 450 kHz in the D1 and D2 transition, respectively. Since the distance between the two PMTs is slightly larger than the path of the ions before reaching PMT1, we estimate conservatively a maximal shift of about 400 kHz for the systematic uncertainty $\Delta\nu_{\rm Ph}$ caused by photon recoil.

5.4 Spectra of the Short-Lived Isotopes ^{7,11,12}Be

A typical spectrum of ^{7}Be is depicted in Fig. 8. It is the sum of 50 individual scans, taken in the first beam time in 2008. Here, line shapes are slightly broader than observed in the second beam time [65]. Since the measurements on ^{7}Be were not repeated in 2010, the uncertainties of the fitted line positions are larger than for the

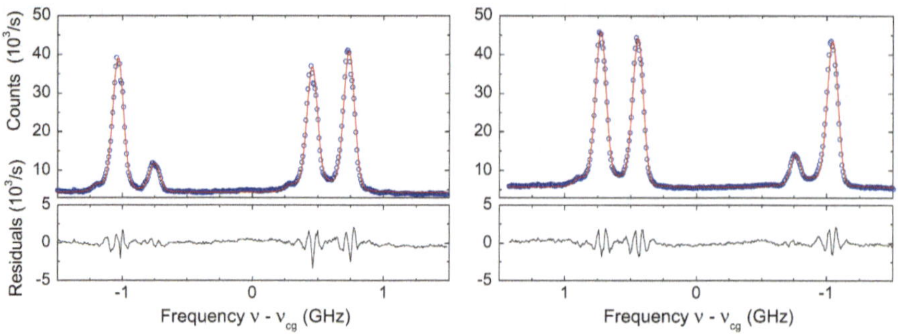

Fig. 8 Spectra obtained in the $2s\,^2S_{1/2} \rightarrow 2p\,^2P_{1/2}$ transition in $^7Be^+$ for copropagating (*left*) and counterpropagating laser excitation (*right*) as a sum of 50 individual scans. The data points are fitted with a multiple Voigt profile as discussed in the text. In the lower trace the residua of the fit are displayed. For more details see [65]

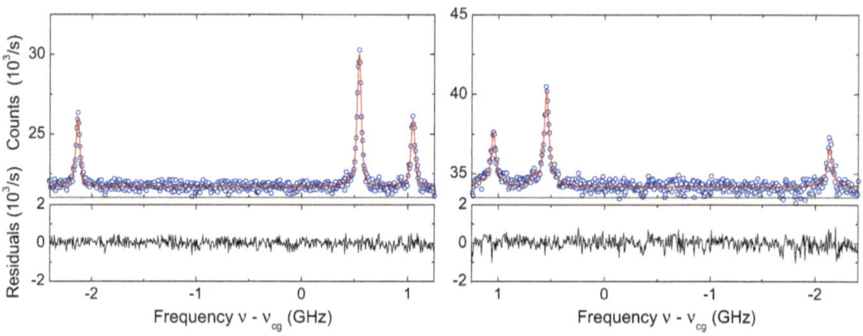

Fig. 9 Resonance spectra of the $2s\,^2S_{1/2} \rightarrow 2p\,^2P_{1/2}$ transition in $^{11}Be^+$ in collinear (*left*) and anticollinear direction (*right*). The production rate was about 10^6 ions / pulse, and thus 400 individual scans were accumulated. The data points are fitted with a multiple Voigt profile as discussed in the text. Fitting residua are displayed in the lower trace

other isotopes. The reduced line width in the second beam time is clearly visible in the spectrum of ^{11}Be depicted in Fig. 9. This is the sum of 400 individual scans. Spectra of the unresolved hyperfine structure in the $2s\,^2S_{1/2} \rightarrow 2p\,^2P_{3/2}$ transitions of both these isotopes can be found in [65].

For the investigation of $^{12}Be^+$, proton pulses impinged on the target every 3–5 s. In this case, spectra in co- and counterpropagating geometry were taken by switching the laser beams after each proton pulse and photon detection was limited to 60 ms ($\approx 3\,T_{1/2}(^{12}\mathrm{Be})$) after the pulse to reduce the number of random coincidence events. Figure 10 shows a typical spectrum which is accumulated over 180 individual scans. Here a detection efficiency of about 1 photon per 800 ions was obtained. The Doppler-tuning voltage range was restricted to 6 V corresponding to about 200 MHz frequency span around the main peak, to limit the time required to record a single resonance to a few hours. Reference measurements

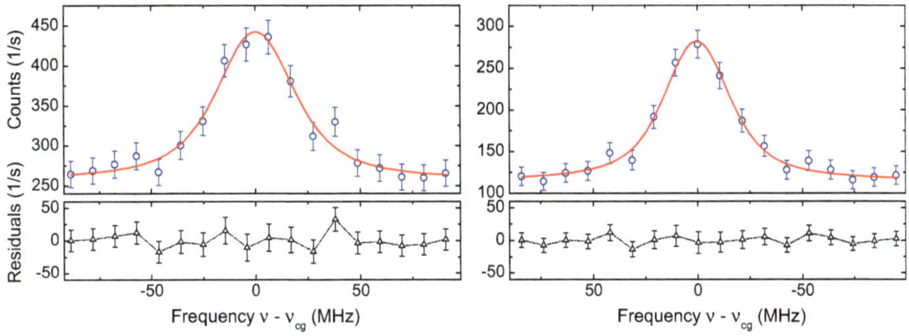

Fig. 10 Resonance spectra of the $2s\,^2S_{1/2} \rightarrow 2p\,^2P_{1/2}$ transition in $^{12}Be^+$ for collinear (*left*) and anticollinear excitation (*right*) plotted as a function of the Doppler-tuning frequency. In total 180 single scans were summed up for 8 h. The symmetric spectra were fitted with a single Voigt profile (*red line*). Residua are displayed in the lower trace

of $^{10}Be^+$ were interspersed after every 40 single scans to ensure stability of all conditions. Due to the limited statistics not allowing the observation of the small satellite peak, only a single Voigt profile was used for fitting the resonances. Statistical uncertainty obtained from the fit was usually <100 kHz.

5.5 *Transition Frequencies*

Each pair of spectra was fitted as discussed in Sec. 5.2 to determine the centers of gravity and to extract the respective rest-frame transition frequency ν_0 according to Eq. (9). For each isotope and beam time the weighted mean of all measurements was calculated, and results are listed in Table 4. All transition frequencies from both beam times agree within their $1-\sigma$ error bars confirming the reproducibility of the measurement. There is a small change compared to Table I of [3], namely a slightly larger uncertainty for ^{10}Be from Run II in the D2 line due to a transfer error in the statistical uncertainty. Statistical uncertainties in Run II are based on the standard error of the mean for typically 4–5 measurements for ^{11,12}Be and 20–30 measurements for the less exotic isotopes ^{9,10}Be. Where available, values from both runs are combined weighted with the respective uncertainty. Systematic uncertainties from the second run cannot be further reduced. While our ^{7}Be D2 transition frequency is in very good agreement with the value reported in [31], the frequencies of ^{9}Be and ^{10}Be differ by approximately 7σ and 3σ, respectively, from our results, which are 1–2 orders of magnitude more precise. In particular, the large discrepancy of the ^{9}Be value is surprising since this has been obtained in [31] from laser-cooled ions in the Paul trap, whereas the other two isotopes were measured with buffer-gas cooled ions exhibiting line widths of approximately 2 GHz. It seems that the systematic uncertainties of the trap measurements were underestimated in both cases. The

Table 4 Transition frequencies ν_0 for the $2s\,^2S_{1/2} \rightarrow 2p\,^2P_{1/2}$ (D1) and the $2s\,^2S_{1/2} \rightarrow 2p\,^2P_{3/2}$ (D2) transition in beryllium isotopes obtained in Run I (Beam time 2008) and Run II (Beam time 2010)

Isotope	Run Ref	ν_0 D1	ν_0 D2
^{7}Be	I	**957 150 316.2 (0.8) (0.9)**	**957 347 374.5 (0.9) (1.1)**
	[31]	–	957 347 369 (124)
^{9}Be	I	957 199 552.9 (0.8) (1.0)	957 396 616.6 (1.4)(1.5)
	II	957 199 553.40 (0.12)(0.52)	–
	comb.	**957 199 553.28 (0.12)(0.52)**	**957 396 616.6 (1.4)(1.5)**
	[31]	–	957 396 515 (14)
	[66]	957 199 652 (120)	957 396 802 (135)
^{10}Be	I	957 216 876.9 (1.4)(1.5)	957 413 943.9 (0.8) (1.0)
	II	957 216 876.84 (0.42)(0.66)	957 413 942.17 (0.44)(0.70)
	comb.	**957 216 876.85 (0.42)(0.66)**	**957 413 942.74 (0.44)(0.67)**
	[31]	–	957 413 839 (35)
^{11}Be	I	957 231 118.1 (1.1)(1.2)	957 428 185.2 (1.5)(1.6)
	II	957 231 118.11 (0.10)(0.52)	–
	comb.	**957 231 118.11 (0.10)(0.52)**	**957 428 185.2 (1.5)(1.6)**
^{12}Be	II	**957 242 944.86 (0.33)(0.61)**	**957 440 013.60 (0.28)(0.58)**

The first uncertainty represents the statistical, the second one the total uncertainty including systematic uncertainties as discussed in Sect. 5.3. Systematic uncertainties of the results of Run I may be reduced from those published in [1, 65] due to the information gained in Run II. Final results for all isotopes are printed bold. All values are in MHz

transition frequencies can now be used to evaluate differential observables, like the isotope shift, the fine-structure splitting and the splitting isotope shift.

5.6 *Isotope Shifts and Nuclear Charge Radii*

Isotope shifts are easily obtained as difference of the transition frequency of the isotope of interest and the reference isotope, which in our case is the stable isotope ^{9}Be. The field shift $\delta\nu_{\mathrm{FS}}$, also known as the finite nuclear size or nuclear volume effect, can then be extracted according to Eq. (1). The corresponding mass shifts $\delta\nu_{\mathrm{MS}}^{9,A}$ as well as the field shift constant $F^{9,A}$ were theoretically evaluated to an accuracy that exceeds the experimental uncertainty by about an order of magnitude and are compiled in Table 1. We have been using the results from [41, 67] to calculate $\delta\nu_{\mathrm{FS}}$, which is then combined with the reference radius $R_{\mathrm{c}} = 2.519(12)$ fm [43] of ^{9}Be to provide total charge radii along the chain using Eq. (4). It should be noted that the uncertainty of $R_{\mathrm{c}}(^9\mathrm{Be})$ is probably underestimated since C2 scattering from the quadrupole distribution has been omitted, which might change the radius by about 3% [72].

Table 5 Compilation of experimental isotope shifts $\delta\nu_{IS}^{9,A}$ obtained from the transition frequencies ν_0 in Table 4, field shift $\delta\nu_{FS}$ extracted as the difference to the theoretical mass shifts listed in Table 1 [41] and the corresponding change in the mean-square nuclear charge radius $\delta\langle r^2\rangle$ according to Eq. (3)

Isotope and transition	$\delta\nu_{IS}^{9,A}$ (MHz)	$\delta\nu_{FS}$ (MHz)	$\delta\langle r^2\rangle$ (fm^2)	R_c (fm)
$^7Be^+$ D1	−49 237.1 (1.1)	−11.4 (1.1)	0.67 (6)	
$^7Be^+$ D2	−49 242.1 (1.8)	−10.3 (1.8)	0.61 (11)	
$^7Be^+$ Mean			0.65 (5)	2.646 (15)
$^9Be^+$ D1/D2	0	0	0	2.519 (12)
$^{10}Be^+$ D1	17 323.57 (84)	13.11 (84)	−0.77 (5)	
$^{10}Be^+$ D2	17 326.1 (1.6)	13.6 (1.6)	−0.80 (10)	
$^{10}Be^+$ Mean			−0.78 (4)	2.360 (14)
$^{11}Be^+$ D1	31 564.82 (74)	4.58 (74)	−0.27 (4)	
$^{11}Be^+$ D2	31 568.6 (2.2)	4.4 (2.2)	−0.26 (13)	
$^{11}Be^+$ Mean			−0.27 (4)	2.465 (15)
$^{12}Be^+$ D1	43 391.58 (80)	1.40 (82)	−0.08 (5)	
$^{12}Be^+$ D2	43 397.0 (1.6)	1.5 (1.6)	−0.09 (10)	
$^{12}Be^+$ Mean			−0.08 (4)	2.502 (15)

Results from the $2s\,^2S_{1/2} \rightarrow 2p\,^2P_{1/2}$ (D1) and the $2s\,^2S_{1/2} \rightarrow 2p\,^2P_{3/2}$ (D2) transitions are compatible and were combined before the total charge radius R_c is calculated according to Eq. (4)

The results are listed in Table 5. Changes in the mean-square charge radii deduced from the isotope shifts in the D1- and D2-transition are of similar accuracy in case of even isotopes, while for odd-mass isotopes the unresolved hyperfine splitting in the D2 lines leads to larger uncertainties. Figure 11 depicts the development of the rms charge radius along the isotopic chain as extracted from the experiment (•) by combining all available data. We have also included results from Fermionic Molecular Dynamics (FMD) calculations [2] that follow the observed trend quite closely, but the charge radii are generally somewhat too small. The two triangles shown for ^{12}Be are results of two additional calculations performed under the assumption that the two outermost neutrons occupy either a pure p^2 state (▽), as expected in the traditional shell model or a pure $(sd)^2$ state (△). The latter is expected to contribute to the ground state only if the $N = 8$ shell gap between the p shell and the sd shell is significantly reduced. This prediction shows that the charge radius of ^{12}Be is extremely sensitive to the admixture of sd shell states to the ground state, which was a strong motivation for the measurement of the ^{12}Be isotope shift.

The trend of the charge radii along the isotopic chain can be understood in a simplified picture based on the cluster structure of light nuclei [64]. This is visualized in the small panels below the graph in Fig. 11. ^{7}Be can be thought of as a two-body cluster consisting of an α particle and a helion (pnp = ^{3}He) nucleus that are bound together and exhibit a considerable center-of-mass motion. This motion blurs the proton distribution and leads to an increased charge radius. ^{8}Be is missing since the two α particles constituting this nucleus are not bound and the

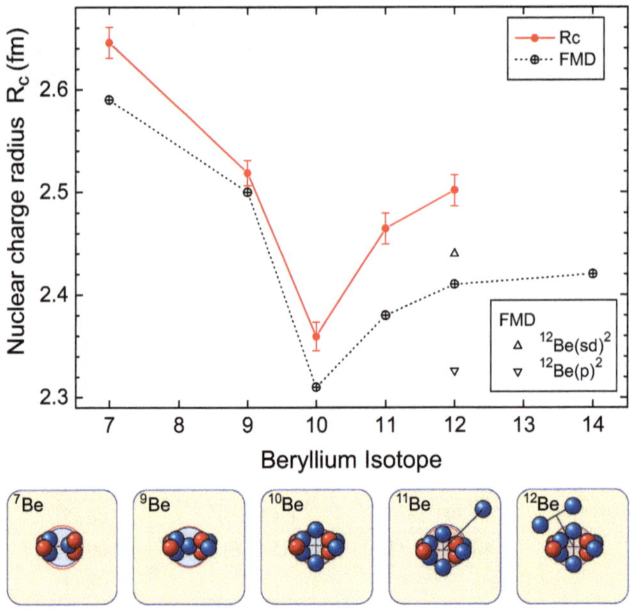

Fig. 11 Nuclear charge radii along the beryllium isotopic chain. The reference radius of ^{9}Be was determined from electron scattering experiments [43]. The *red bullets* (•) represent the experimental results with uncertainties dominated by the uncertainty of the reference charge radius of ^{9}Be. Therefore, all uncertainties are similar in size. Additionally shown are results of Fermionic Molecular Dynamic (FMD) calculations [2]. For ^{10}Be, calculations were also performed forcing the neutrons into a p^2 and an $(sd)^2$ orbit. The *bottom row* shows the structure of the isotopes interpreted in a cluster picture. For details see text

nucleus only exists as a resonance. The stable isotope ^{9}Be, which has a $\alpha + \alpha + n$ structure, is more compact than ^{7}Be because the α particles themselves are very compact and well bound by the additional neutron. This effect is even enhanced with the second neutron added in ^{10}Be. The sudden upward trend to ^{11}Be is attributed to the one-neutron halo character of ^{11}Be which can be disentangled into a ^{10}Be core and a loosely bound neutron. This halo character not only increases the matter radius, but also affects the charge radius due to the center-of-mass motion of the core caused by the halo neutron. The fact that the charge radius of ^{12}Be is even larger has been related to the fact that the two outermost neutrons exhibit a strongly mixed *sd* character rather than belonging to the *p* shell as expected in the simplified shell-model picture. This mixture leads to an increased probability density outside the ^{10}Be core, pulling the α particle apart due to the attractive $n - \alpha$ interaction. Theory predicts an $(sd)^2$ admixture of about 70% for this nucleus, being a clear indication for the disappearance of the classical $N = 8$ shell closure. For a more detailed discussion of the nuclear charge radii, the comparison with ab initio microscopic nuclear structure calculations and the conclusions about the shell closure see Refs. [1, 2, 65].

5.7 *Fine-Structure Splitting and Splitting Isotope Shifts*

From the information provided in Table 4 we can furthermore extract the fine-structure splitting as a function of the atomic number. The splitting has previously been measured for ^{9}Be with a relative uncertainty of about $3 \cdot 10^{-4}$ [66]. The present value obtained simply from the difference in ν_{D1} and ν_{D2} of this isotope is already 40 times more accurate although its accuracy is limited by the unresolved hyperfine structure in the D2 transition. Only recently the fine-structure splitting in three-electron atoms became calculable with high precision as demonstrated for the case of lithium [34]. Our result confirmed first calculations for the $Z = 4$ three-electron system of Be^+ [3]. Moreover, the change in the fine-structure splitting along the chain of isotopes provides a useful check of the so-called splitting isotope shift which can be calculated theoretically to very high accuracy, because the value is nearly independent of both QED and nuclear volume effects. This mass dependence is shown in Fig. 12: The blue data points represent the experimental fine-structure splittings and the red line with red crosses the theoretically expected mass dependence based on the calculated splitting isotope shift and the measured splitting of ^{9}Be. The excellent agreement between the theoretical curve and the experimental data can be interpreted as a reliable check of the consistency of the mass shift

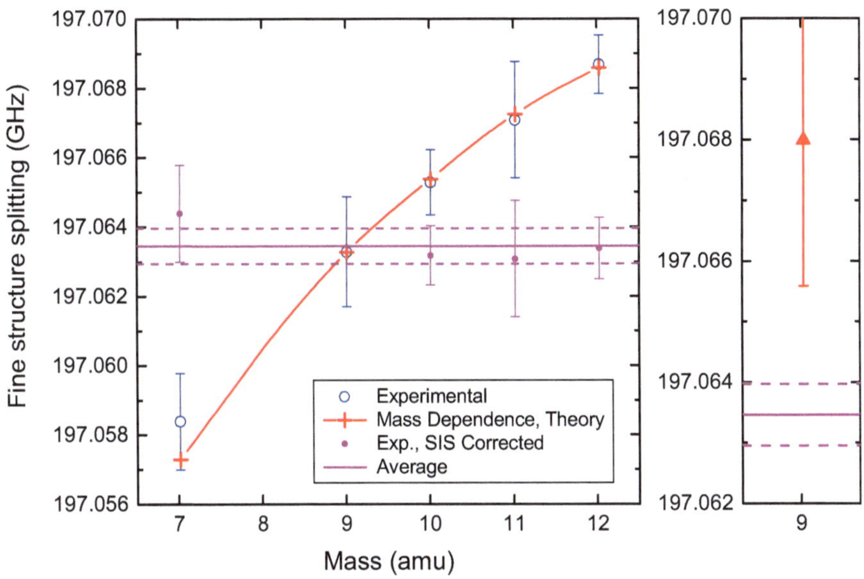

Fig. 12 *Left*: Mass dependence of the fine-structure splitting along the beryllium isotopic chain. The *red curve* with crosses shows the theoretically expected mass dependence—the splitting isotope shift (sis)—with respect to ^{9}Be. The experimental values (*blue circles*) and the sis-corrected values according to Eq. (18) (*magenta dots*) are included. The *solid* and *dotted lines* represent the mean and standard deviation of the corrected values, respectively. *Right*: Fine-structure splitting in ^{9}Be from experiment (*line*) and theory (data point)

calculations. On the other hand, it proves the consistency of the experimental data, because the fine-structure splittings are based on the combination of transition frequencies obtained independently in two beam times, even with independent optimization of the experimental conditions. The observation that the data points scatter much less than expected from their error bars can be ascribed to the fact that our systematic uncertainties largely cancel out in considering differential effects.

The fine-structure splitting can most reliably be determined for the even–even isotopes ^{10}Be and ^{12}Be since there is no hyperfine structure which in the D2 transition obscures the determination of the center of gravity. This is clearly visible in Fig. 12 where the error bars are smallest for these two cases. For these isotopes, the fine-structure splitting was determined sequentially in a short-time interval and therefore the best cancellation of all systematic uncertainties should occur. Indeed, the splitting isotope shift between the two isotopes $\delta\nu_{\text{sis}}^{10,12} = 3.43(78)$ MHz agrees very well with the theoretical value of 3.203 MHz. Here the uncertainty of the experimental value is purely statistical assuming that all dominant systematic contributions are cancelled out.

With the confidence gained about the reliability of the theoretical estimates, we can combine the calculated splitting isotope shifts of all isotopes to extract an improved value for the fine-structure splitting of ^{9}Be. To this aim, we correct all measured fine-structure splittings by subtracting the theoretical splitting isotope shift

$$\Delta\nu_{\text{fs},{}^A\text{Be}\rightarrow{}^9\text{Be}} = \Delta\nu_{\text{fs}}\left({}^A\text{Be}\right) - \delta\nu_{\text{sis, Theory}}^{A,9}. \tag{18}$$

The results are included in Table 6 and plotted in Fig. 12 as magenta bullets. The weighted mean for the "isotope-projected" fine-structure splitting of ^{9}Be represented by the horizontal line is 197 063.47 (53) MHz. The uncertainty is

Table 6 Fine-structure splittings $\Delta\nu_{\text{fs}}$, the experimental and theoretical [67] splitting isotope shifts $\delta\nu_{\text{sis}}$, and the transferred fine-structure splittings $\Delta\nu_{\text{fs},{}^A\text{Be}\rightarrow{}^9\text{Be}}$ for ^{9}Be based on the measured splittings in the radioactive isotopes according to Eq. (18) are listed

	$\Delta\nu_{\text{fs}}$	$\delta\nu_{\text{sis}}^{A,9}$		
Isotope	D2–D1	Exp	Theory	$\Delta\nu_{\text{fs},{}^A\text{Be}\rightarrow{}^9\text{Be}}$
^{7}Be	197 058.4 (1.4)	5.0 (2.1)	6.036(1)	197 064.4 (1.4)
^{9}Be	197 063.2 (1.6)	0.0	0.0	197 063.3 (1.6)
^{9}Be [a]	197 150 (64)			
^{10}Be	197 065.3 (0.9)	−2.0 (1.8)	−2.096(1)	197 063.2 (0.9)
^{11}Be	197 067.1 (1.7)	−3.8 (2.3)	−3.965(1)	197 063.1 (1.7)
^{12}Be	197 068.7 (0.9)	−5.4 (1.8)	−5.300(1)	197 063.4 (0.8)
$^{12-10}$Be	$\delta\nu_{\text{sis}}^{12,10} = -3.4\,(6)$		−3.203(1)	

The bottom row shows the splitting isotope shift between the two even isotopes ^{10}Be and ^{12}Be. For $\delta\nu_{\text{sis}}^{A,10}$ and $\Delta\nu_{\text{fs},{}^A\text{Be}\rightarrow{}^9\text{Be}}$ for ^{7,9,11}Be after Run II required information from Run I since D2 lines of these isotopes were not measured in Run II. All values are in MHz

[a] Bollinger et al. [66]

shown by the dashed lines. We believe that a small remaining systematic uncertainty of the fine-structure splitting is still covered by the size of this uncertainty. In the right hand part of the figure, the theoretical prediction represented by the filled red triangle is compared with experiment. The calculated splitting in ^{9}Be amounts to 197 068.0 (25) MHz, which is about 4.5 MHz larger than the experimental value. This difference corresponds to about $1.5\,\sigma$ of the combined uncertainties. The theoretical uncertainty is based on an estimation of the size of the uncalculated nonlogarithmic terms in $m\alpha^7$. These terms are expected to be $<50\%$ of the calculated leading logarithmic terms. To further test the QED calculations in lithium-like light systems, similar measurements on ions with higher Z, e.g., in boron B^{2+} or carbon C^{3+} are of great interest. At least for B^{2+} the wavelength of about 206 nm is still achievable, and measurements with the collinear laser spectroscopy technique presented here are planned.

6 Summary

We have demonstrated that quasi-simultaneous collinear–anticollinear laser spectroscopy on stable and radioactive isotopes can be used to perform precision measurements from which nuclear charge radii even of light and very short-lived species can be extracted. At the same time the data can be used to perform high-precision tests of fundamental atomic structure calculations and bound-state QED. The experimental technique relies on the accurate determination of the laser frequency, which has become possible with the invention and availability of frequency combs. We will apply the technique for further studies: An important case is the determination of the charge radius of the proton-halo candidate ^{8}B for which an experiment is currently under preparation at the ATLAS facility at the Argonne National Laboratory.

Acknowledgements We acknowledge enlightening discussions with I. Sick about the charge radius of ^{9}Be from elastic electron scattering. This work was supported by the Helmholtz Association (VH-NG 148), the German Ministry for Science and Education (BMBF) under contracts 05P12RDCIC and 05P15RDCIA, the Helmholtz International Center for FAIR (HIC for FAIR) within the LOEWE program by the State of Hesse, the Max-Planck Society, the European Union 7th Framework through ENSAR, and the BriX IAP Research Program No. P6/23 (Belgium). A. Krieger acknowledges support from the Carl-Zeiss-Stiftung (AZ:21-0563-2.8/197/1).

References

1. W. Nörtershäuser et al., Phys. Rev. Lett. **102**, 062503 (2009)
2. A. Krieger et al., Phys. Rev. Lett. **108**, 142501 (2012)
3. W. Nörtershäuser et al., Phys. Rev. Lett. **115**, 033002 (2015)

4. R. Neugart, Hyp. Int. **24**, 159 (1985)
5. E.W. Otten, Nuclear radii and moments of unstable isotopes, in *Treatise on Heavy Ion Science*, vol. 8, ed. by D.A. Bromley (Plenum Publishing Corp. (Springer), New York, 1989), p. 517
6. J. Billowes, P. Campbell, J. Phys. G **21**, 707 (1995)
7. R. Neugart, Eur. Phys. J. A **15**, 35 (2002)
8. R. Neugart, G. Neyens, Lect. Notes Phys. **700**, 135 (2006)
9. B. Cheal, K. Flanagan, J. Phys. G **37**, 113101 (2010)
10. K. Blaum, J. Dilling, W. Nörtershäuser, Phys. Scr. **T152**, 014017 (2013)
11. P. Campbell, I.D. Moore, M.R. Pearson, Prog. Part. Nucl. Phys. **86**, 127 (2016)
12. T.P. Dinneen, N. Berrah-Mansour, H.G. Berry, L. Young, R.C. Pardo, Phys. Rev. Lett. **66**, 2859 (1991)
13. J.K. Thompson, D.J.H. Howie, E.G. Myers, Phys. Rev. A **57**, 180 (1998)
14. E.G. Myers, H.S. Margolis, J.K. Thompson, M.A. Farmer, J.D. Silver, M.R. Tarbutt, Phys. Rev. Lett. **82**, 4200 (1999)
15. W. Geithner et al., Phys. Rev. Lett. **83**, 3792 (1999)
16. K. Marinova et al., Phys. Rev. C **84**, 034313 (2011)
17. Z.-C. Yan, G.W.F. Drake, Phys. Rev. A **61**, 022504 (2000)
18. Z.-C. Yan, G.W.F. Drake, Phys. Rev. Lett. **91**, 113004 (2003)
19. M. Puchalski, A.M. Moro, K. Pachucki, Phys. Rev. Lett. **97**, 133001 (2006)
20. Z.-C. Yan, W. Nörtershäuser, G.W.F. Drake, Phys. Rev. Lett. **100**, 243002 (2008)
21. Z.-C. Yan, W. Nörtershäuser, G.W.F. Drake, Phys. Rev. Lett. **102**, 249903(E) (2009)
22. W. Nörtershäuser et al., Phys. Rev. A **83**, 012516 (2011)
23. K. Pachucki, J. Komasa, Phys. Rev. Lett. **92**, 213001 (2004)
24. M. Puchalski, K. Pachucki, J. Komasa, Phys. Rev. A **89**, 012506 (2014)
25. M. Puchalski, J. Komasa, K. Pachucki, Phys. Rev. A **92**, 062501 (2015)
26. I. Tanihata et al., Phys. Rev. Lett. **55**, 2676–2679 (1985)
27. L.-B. Wang et al., Phys. Rev. Lett. **93**, 142501 (2004)
28. P. Müller et al., Phys. Rev. Lett. **99**, 252501 (2007)
29. G. Ewald et al., Phys. Rev. Lett. **93**, 113002 (2004)
30. R. Sánchez et al., Phys. Rev. Lett. **96**, 033002 (2006)
31. T. Nakamura et al., Phys. Rev. A **74**, 052503 (2006)
32. A. Takamine et al., Eur. Phys. J. A **42**, 369 (2009)
33. K. Pachucki, V.A. Yerokhin, Phys. Rev. Lett. **104**, 070403 (2010)
34. M. Puchalski, K. Pachucki, Phys. Rev. Lett. **113**, 073004 (2014)
35. G.A. Noble, B.E. Schultz, H. Ming, W.A. van Wijngaarden, Phys. Rev. A **74**, 012502 (2006)
36. C.J. Sansonetti, C.E. Simien, J.D. Gillaspy, J.N. Tan, S.M. Brewer, R.C. Brown, S. Wu, J.V. Porto, Phys. Rev. Lett. **107**, 023001 (2011)
37. R.C. Brown, S.J. Wu, J.V. Porto, C.J. Sansonetti, C.E. Simien, S.M. Brewer, J.N. Tan, J.D. Gillaspy, Phys. Rev. A **87**, 032504 (2013)
38. M. Puchalski, K. Pachucki, Phys. Rev. A **78**, 052511 (2008)
39. Z.T. Lu, P. Mueller, G.W.F. Drake, W. Nörtershäuser, S.C. Pieper, Z.C. Yan, Rev. Mod. Phys. **85**, 1383 (2013)
40. G.W.F. Drake, Windsor University, Priv. Commun. (2010)
41. M. Puchalski, K. Pachucki, Hyp. Int. **196**, 35 (2010)
42. K. Pachucki, Institute of Theoretical Physics, University of Warsaw, Priv. Commun. (2011)
43. J.A. Jansen, R.T. Peerdeman, C. de Vries, Nucl. Phys. A **188**, 337 (1972)
44. A. Derevianko, S.G. Porsev, K. Beloy, Phys. Rev. A **78**, 010503(R) (2008)
45. M. Douglas, N.M. Kroll, Ann. Phys. (N. Y.) **82**, 89 (1974)
46. M. Puchalski, K. Pachucki, Phys. Rev. A **92**, 012513 (2015)
47. Z.C. Yan et al., Phys. Rev. A **66**(042504), 1–8 (2002)
48. U. Koester et al., Enam 98 Exot. Nucl. Atomic Masses **455**, 989 (1998)
49. V.N. Fedosseev et al., Nucl. Instrum. Methods B **266**, 4378 (2008)
50. G. Audi et al., Nucl. Phys. A **624**, 1–124 (1997)

51. D.C. Fiander et al., in *CERN/PS92-38, Proceedings of the of 20th Power Modulator Symposium* (1992)
52. R. Neugart, Nucl. Instrum. Methods **186**, 165–175 (1981)
53. F. Buchinger et al., Nucl. Instrum. Methods B **202**, 159 (1982)
54. A.C. Müller et al., Nucl. Phys. A **403**, 234 (1983)
55. R. Neugart et al., Nucl. Instrum. Methods B **17**, 354 (1986)
56. W. Geithner et al., Phys. Rev. C **71**, 064319 (2005)
57. G. Neyens et al., Phys. Rev. Lett. **94**, 022501 (2005)
58. M. Kowalska et al., Eur. Phys. J. A **25**(s01), 193 (2005)
59. T. Thümmler et al., New J. Phys. **11**, 103007 (2009)
60. A. Krieger et al., Nucl. Instrum. Methods A **632**, 23 (2011)
61. O. Poulsen, E. Riis, Metrologia **25**, 147 (1988)
62. E. Riis, A.G. Sinclair, O. Poulsen, G.W.F. Drake, W.R.C. Rowley, A.P. Levick, Phys. Rev. A **49**, 207 (1994)
63. Program *IodineSpec* from Toptica Photonics (2004)
64. N.I. Ashwood, Phys. Lett. B **580**, 129 (2004)
65. M. Zakova et al., J. Phys. G **37**, 055107 (2010)
66. J.J. Bollinger, J.S. Wells, D.J. Wineland, W.M. Itano, Phys. Rev. A **31**, 2711 (1985)
67. M. Puchalski, K. Pachucki, Phys. Rev. A **79**, 032510 (2009)
68. https://oraweb.cern.ch/pls/isolde/query_tgt, ISOLDE database (2016)
69. K. Okada et al., Phys. Rev. Lett. **101**, 212502 (2008)
70. D.J. Wineland, J.J. Bollinger, W.M. Itano, Phys. Rev. Lett. **50**, 628 (1983)
71. A. Takamine et al., Phys. Rev. Lett. **112**, 162502 (2014)
72. I. Sick, Universität Basel, Priv. Commun. (2013)

High-Accuracy Deep-UV Ramsey-comb Spectroscopy in Krypton

Sandrine Galtier, Robert K. Altmann, Laura S. Dreissen, and Kjeld S.E. Eikema

Abstract In this paper, we present a detailed account of the first precision Ramsey-comb spectroscopy in the deep UV. We excite krypton in an atomic beam using pairs of frequency-comb laser pulses that have been amplified to the millijoule level and upconverted through frequency doubling in BBO crystals. The resulting phase-coherent deep-UV pulses at 212.55 nm are used in the Ramsey-comb method to excite the two-photon $4p^6 \rightarrow 4p^5 5p[1/2]_0$ transition. For the ^{84}Kr isotope, we find a transition frequency of 2829833101679(103) kHz. The fractional accuracy of 3.7×10^{-11} is 34 times better than previous measurements, and also the isotope shifts are measured with improved accuracy. This demonstration shows the potential of Ramsey-comb excitation for precision spectroscopy at short wavelengths.

1 Introduction

Quantum electrodynamics (QED) theory is a cornerstone of the standard model and one of the best tested fundamental theories in physics. Its predictions have been verified with extreme precision, e.g., by measuring the fine structure constant α derived from measurements of the electron g-factor [1, 2], interferometric recoil experiments [3, 4], and bound-state QED tests based on precision spectroscopy in atoms, molecules and highly charged ions (see, e.g., [5–10]). Moreover, in the pursuit of testing QED ever better, substantial efforts have been made to extract fundamental quantities such as the Rydberg constant R_∞ and the proton charge radius. Both can be obtained from spectroscopy of atomic hydrogen, assuming that QED is sufficiently precise. However, when the CREMA collaboration determined the proton charge radius from spectroscopy in muonic hydrogen (consisting of a proton and a muon), it leads to a considerable (7σ) mismatch with the value extracted from normal (electronic) hydrogen [11–13]. This mismatch, known as

This article is part of the topical collection "Enlightening the World with the Laser" - Honoring T. W. Hänsch guest edited by Tilman Esslinger, Nathalie Picqué, and Thomas Udem.

S. Galtier • R.K. Altmann (✉) • L.S. Dreissen • K.S.E. Eikema
LaserLaB, Department of Physics and Astronomy, VU University Amsterdam, De Boelelaan 1081, 1081 HV Amsterdam, The Netherlands
e-mail: rkaltmann@gmail.com

D. Meschede et al. (eds.), *Exploring the World with the Laser*,
https://doi.org/10.1007/978-3-319-64346-5_13

the *proton radius puzzle*, remains to be explained and requires more spectroscopic measurements, e.g., in systems other than (muonic) hydrogen. Recent results on muonic deuterium [14] reveal that also the deuteron radius is significantly smaller (7.5 σ) than the radius based on normal deuterium spectroscopy.

Interesting candidates for precision spectroscopy to solve this puzzle need to be sufficiently simple for precise theoretical treatment. One example is molecular hydrogen, made possible by recent improvements in theory [15]. Another is He^+ [16], which can be compared to muonic-He^+ spectroscopy [13]. The experimental challenge is the short wavelengths required for excitation, which ranges from the deep UV (≈ 200 nm) for H_2 to extreme ultraviolet (XUV, $\lambda < 60$ nm) for He^+.

Such short wavelengths are typically obtained by frequency upconversion of near-infrared lasers in nonlinear crystals or noble gases. One can use a frequency-comb (FC) laser as the fundamental laser and take advantage of its excellent spectral resolution and pulse peak power, to perform direct frequency-comb spectroscopy (DFCS) [17–19]. To achieve sufficient upconversion to the UV or XUV range, several approaches have been investigated. This includes the use of enhancement resonators with intra-cavity high harmonic generation [19, 20], and amplification at full repetition rate [21, 22], typically leading to infrared pulse energies in the μJ range.

We developed a different approach, based on amplification of only two FC pulses, which combines precision spectroscopy with high-power laser pulses. Amplification of only two pulses leads to orders of magnitude higher pulse energy in the mJ range. The accuracy and resolution are recovered in this case by recording a series of Ramsey-type signals, leading to *Ramsey-comb* spectroscopy (RCS). It features a strong suppression of systematic shifts such as the AC-Stark shift and chirp-induced effects, which are common problems in metrology experiments. The Ramsey-comb method has previously been demonstrated in the infrared on two-photon transitions in rubidium and cesium [23]. Recently we reported on the extension to the deep-UV region [24], and here we present a detailed account of this experiment. We demonstrate its potential by exciting the $4p^6 \rightarrow 4p^5 5p[1/2]_0$ two-photon transition in ^{84}Kr at 212.55 nm, and we also report improved isotope shift measurements for the other isotopes.

2 Ramsey-comb Spectroscopy Principle

We recall here the key features of Ramsey-comb spectroscopy; a more detailed description can be found in [23, 25]. As the name indicates, the method combines Ramsey spectroscopy [26] with frequency-comb (FC) lasers [27, 28]. Excitation of an atom or molecule is performed with two (short) coherent laser pulses that are selected from a frequency-comb laser. The pulses can be amplified and therefore easily frequency upconverted to cover a wide spectral range (see Sect. 4). By taking the original pulses from a FC, their coherence can be maintained over long intervals (even seconds), and it becomes possible to record a series of Ramsey signals using different pulse pairs with a delay that is a multiple of the FC pulse repetition time. This we call Ramsey-comb spectroscopy (RCS), leading to interesting new properties compared to standard Ramsey spectroscopy.

It starts with excitation using two coherent pulses, leading to interference between the coherent superpositions induced by each pulse. If the time delay ΔT is changed between the two resonant pulses, while keeping the relative optical phase $\Delta\phi$ constant, we observe an oscillation of the population $|\rho_e|^2$ of the excited state according to:

$$|\rho_e|^2 \propto \cos\left(2\pi f_{tr} \times \Delta T + \Delta\phi\right) \tag{1}$$

This Ramsey signal from two separate excitations is widely used in, e.g., atomic clocks [29]. To extract the atomic frequency f_{tr}, one must exactly know the time delay ΔT and the relative phase $\Delta\phi$ between the two pulses. These conditions are conveniently met by taking pulses from a frequency-comb laser. A frequency comb is a mode-locked laser whose spectrum comprises equally spaced optical frequencies (modes) separated by the repetition frequency f_{rep}. The modes start at an offset f_{ceo} due to the dispersion and nonlinear effects in the cavity, leading to a spectrum of $f_n = f_{ceo} + nf_{rep}$, where n is an integer. In the time domain, the modes interfere to pulses with a time interval of $T_{rep} = 1/f_{rep}$, and with a carrier-envelope phase shift from pulse to pulse equal to $\Delta\phi_{ceo} = 2\pi f_{ceo}/f_{rep}$. Both f_{ceo} and f_{rep} of the laser can be actively stabilized to an ultra-stable frequency reference, e.g., a cesium clock, leading to highly reproducible phase-coherent pulses, as required for Ramsey spectroscopy.

For the Ramsey-comb method, we select two pulses from the FC at multiples N of the repetition time T, initially equal to $T_{rep} = 1/f_{rep}$. At each macro-delay, NT, the time delay can be adjusted on a (much) smaller scale in steps of tens of attoseconds by tuning the repetition time T_{rep} of the laser. Therefore, the time delay ΔT between the two pulses can be written as:

$$\Delta T = NT + \delta t \tag{2}$$

The micro-delay δt scan provides the right time scale to probe the Ramsey oscillations of the atomic transition frequency. The ability to measure and combine Ramsey fringes from multiples of NT permits a much more precise frequency determination because the Ramsey fringes can be probed on much longer timescales. In addition, the frequency is determined from the phases of the time domain Ramsey signals at each NT. Any phase shift of the Ramsey signals that is constant for each NT will simply shift all the Ramsey fringes together, but will not affect the extracted the frequency. Therefore, any common phase shift does not influence the frequency determination which also enhances the accuracy (see [25]). This includes an optical phase shift between the excitation pulses. This is initially equal to $N \times \Delta\Phi_{ceo}$ from the FC, but is perturbed during the subsequent amplification process. The laser system is constructed such that this phase perturbation is constant (typically on the mrad level) as function of the macro-delay *NT*. Moreover, this also includes any constant phase shift of the atomic phase. For example, the AC-Stark shift manifests itself as a phase shift imprinted on the state superposition during the excitation. As long as the excitation pulse energy is kept constant for different macro-delays, the AC-Stark effect results in a constant phase shift of the Ramsey signals, which is therefore eliminated in the analysis procedure. The light-shift insensitive nature of the Ramsey-comb spectroscopy combined with high pulse energy for easy frequency conversion are two key advantages compared to conventional direct frequency-comb spectroscopy.

3 Laser System

The experimental setup is shown in Fig. 1. The pulse pairs are extracted from a home-built mode-locked Ti:sapphire frequency comb, delivering 4 nJ femtosecond pulses at a repetition rate of 126.6 MHz. The frequency comb is fully referenced to an atomic Cs clock. A pulse pair is selectively amplified in a noncollinear optical parametric chirped-pulse amplifier (NOPCPA). The NOPCPA is designed to provide high gain over a broad spectrum (hundreds of nanometers) and consists of three passes in two BBO crystals [30, 31] (see Fig. 2). The system bandwidth is convenient for easy tunability, as the actual amplified spectrum of the comb laser for this experiment is only a few nm wide.

The pump laser is detailed in Fig. 2. The starting point is a SESAM mode-locked $Nd:YVO_4$ laser producing 10 ps pulses. The pulse length is increased to 75 picoseconds by spectral clipping; two of them are then selected using fast modulators. Both pulses are then amplified to the 1-mJ level in a grazing-incidence side-pumped $Nd:YVO_4$ bounce amplifier (detailed in [32, 33]).

In the post-amplifier, the pulse pair is further amplified to 27 mJ per pulse, using a quasi-CW diode pumped Nd:YAG module (6.35 mm rod diameter, type number

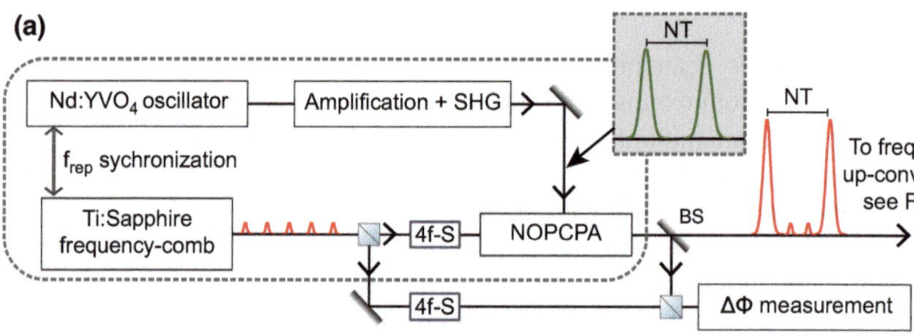

Fig. 1 Ramsey-comb laser system. Details of part (*a*) about on the generation of the pump pulses and on the noncollinear optical parametric chirped-pulse amplification (NOPCPA) are given in Fig. 2. *4f-S* 4f grating stretcher, *BS* beamsplitter

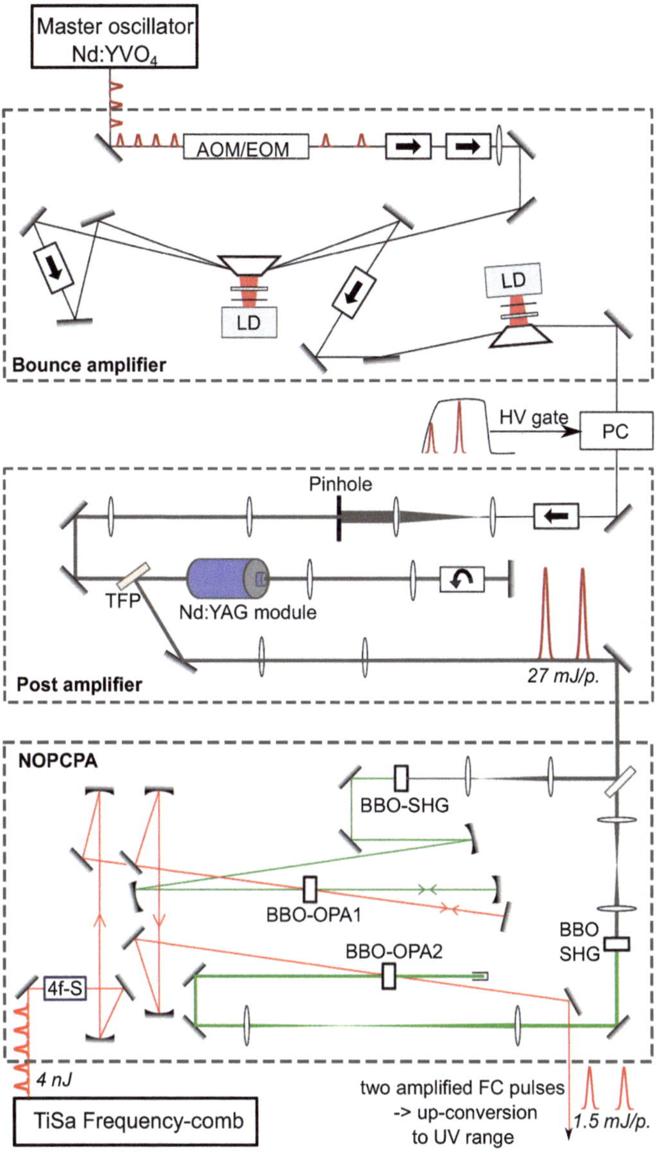

Fig. 2 Pump laser and parametric amplification stages. Two pulses from a master oscillator (Nd : YVO_4) are selected and amplified in a double-pass bounce amplifier followed by a single-pass bounce amplifier. The post-amplifier stage consists of two passes in a Nd:YAG module gain medium producing pulses of 27 mJ each, which are used as the pump laser of the NOPCPA. The amplitude of the first pump pulse is adjusted via by the rising-slope timing of a PC gate, such as the two frequency-comb pulses have equal intensities. *PC* Pockels cell, *LD* laser diode, *TFP* thin-film polarizer, *4f-S* 4f grating stretcher

REA6308-3P200H from Northrop Grumman). To obtain a top-hat seed intensity profile for this amplification stage, the beam from the bounce amplifier is enlarged to 6 mm, and a center part is selected with a 3.5 mm diameter iris. To prevent diffraction hot spots, the uniform intensity profile at the iris position is relay imaged to the Nd:YAG rod and throughout the whole amplifier. The amplifier is set up in a two-pass configuration as illustrated in Fig. 2. A Faraday rotator is placed in between the two passes to compensate for birefringence and to couple out the back-reflected beam with a thin-film polarizer. The amplified beam is further relay imaged and doubled to 532 nm to be used as the pump beam for the parametric amplifier.

At each amplification stage of the pump laser, the second pulse is less efficiently amplified compared to the first one due to energy depletion. There is also gain loss from unwanted pulses that are not fully suppressed by the modulators that select the pulses at the start of the amplifier chain. To ensure a pump pulse pair with equal intensities in the NOPCPA, the amplitude of the first pulse is injected with less energy compared to the second one in the bounce amplifier using the modulators. In between the bounce amplifier and post-amplifier, the energy ratio is adjusted (cut) again by changing the timing of the rising slope of a Pockels cell. With active feedback on the Pockels cell timing, the intensity ratio can be controlled and kept constant. Before seeding the NOPCPA, the Ti:sapphire laser pulses are stretched to 12 ps by adding a chirp of 1.2×10^6 fs^2 combined with wavelength selection of the spectrum via a movable slit in the Fourier plane of a 4f grating stretcher. The sharp spectral clipping of the seeding beam results in a sinc-like pulse in the time domain, offering spectral power at its wings. These wings are strongly amplified in the NOPCPA operating in a saturated regime. Because we use chirped pulses, the amplification in the wings of the pulses corresponds to an enhanced amplification of the 'red' front edge and the 'blue' trailing edge of the pulse, resulting in a spectrum with two peaks as shown in Fig. 3. It also leads to a separation in time of the two spectral regions and increases the total amplified pulse duration to approximately 20 ps.

Only the FC pulses that overlap in time with the pair of pump pulses are amplified in the NOPCPA. Therefore, the choice of the pump pulse pair determines the time interval between the pair of amplified FC pulses. The FC pulses reach an energy of 1.5 mJ each, with a spectral bandwidth of approximately 3.5 nm centered around 850.2 nm.

The parametric amplification process and self- and cross-phase modulation between all interacting beams can imprint a phase shift on the amplified pulses, depending mostly on the pump intensity [31, 34, 35]. Any differential phase shift $\Delta\Phi$ between different pulse pairs can lead to a frequency shift of the excited two-photon transition according to:

$$\Delta f = \frac{8 \times \Delta\Phi}{2\pi \times NT} \tag{3}$$

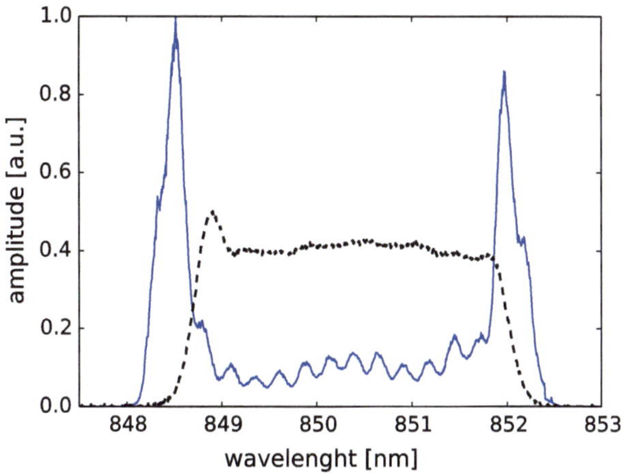

Fig. 3 Spectrum of the selected FC spectrum before amplification (*in dash black line*) and after amplification (*solid blue line*) in the NOPCPA

The factor 8 accounts for the two doubling stages to the UV ($\times$ 4) and the use of two photons to drive the atomic transition ($\times$ 2). To measure the influence of the amplification process on the phase of the FC pulses, we use a setup consisting of a Mach–Zehnder interferometer in combination with spectral interferometry [31, 35]. The two amplified pulses are combined with their respective non-amplified FC pulses which are extracted before the amplification process using a combination of a wave plate and polarizing cube (see Fig. 1). The spectral interference between the amplified and original FC pulses is recorded with a camera. From these interference patterns, the relative phase induced by the amplification can be determined.

4 Excitation Scheme

We excite the $4p^6 \rightarrow 4p^5 5p[1/2]_0$ two-photon transition in Kr using 50 μJ deep-UV pulses at 212.5 nm. Two-photon spectroscopy allows the use of multiple pairs of comb modes leading to the same total energy over many nanometers of bandwidth leading to efficient use of power and increased signal to noise compared to one-photon transitions. The deep-UV pulses are generated from the amplified FC laser after successive doubling stages using thin BBO crystals (beta barium borate—β-BaB_2O_4). The first doubling stage is performed in a 1-mm-thick BBO crystal. At this stage, the entire spectrum of the fundamental light is doubled with an efficiency of approximately 35%. We perform the second doubling with two different BBO crystals of 0.5 mm thickness, and a phase-matching bandwidth lower than 0.5 nm. The phase-matching angle of the two last crystals can be adjusted such that they double independently the 'red' or 'blue' edge of the spectrum (see Fig. 4).

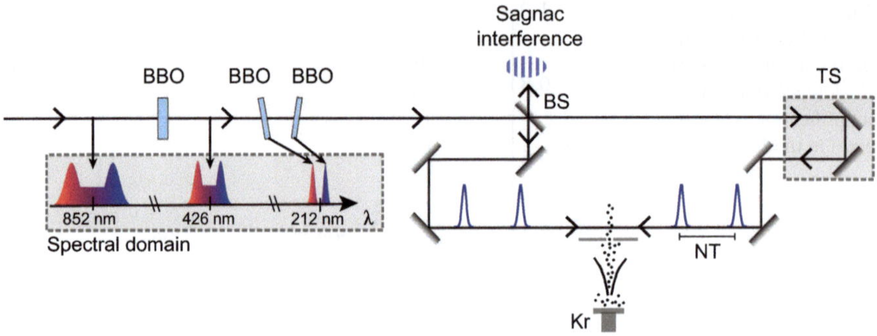

Fig. 4 Fourth harmonic of the amplified FC pulses is obtained using three different BBO crystals. The first crystal doubles the entire spectrum, the second and the third crystal double, respectively, the '*blue*' part and the '*red*' part of the spectrum (see *inset spectral domain*). The krypton atoms are excited in a counter-propagating configuration to reduce the first-order Doppler effect. After ionization with a 532 nm pulse (not shown), the ions are detected after a time-of-flight mass selector with a channel electron multiplier. *BS* metallic beamsplitter, *TS* translation stage

Because of the cathedral-like shape of the input spectrum, this configuration still results in sufficient intensity in the ultraviolet. The doubling in the last stage is adjusted such that at the center wavelength of the atomic transition no light is produced. Therefore, excitation is only possible if the 'red' front edge and the 'blue' trailing edge of the pulses overlap in time to match the transition frequency. This is fulfilled by a combination of 'red' and 'blue' parts of the spectrum from pulses that travel from opposite sides. It leads to two spatial positions (where 'red' meets 'blue,' and 'blue' meets 'red') where excitation takes place with a suppressed first-order Doppler shift. At the same time, it eliminates the background signal one would otherwise get due to (Doppler broadened) two-photon excitation from one side.

The krypton atoms perpendicularly cross the two counter-propagating UV beams to minimize residual first-order Doppler effects. The atoms are produced in a highly collimated beam with a the combination of a pulsed valve and a double skimmer arrangement (with 0.5 and 3 mm wide aperture at 3, and 25 cm distance to the nozzle, respectively). Because of the chirp in the pulses, the 'red' and 'blue' part of the spectrum are also spatially separated by about 6 mm. This is bigger than the 3 mm diameter atomic beam, and therefore we are able to observe the two collision points where 'red' meets 'blue,' and 'blue' meets 'red' separately (by using a delay line in one of the UV paths to position one of them into the atomic beam). Measurements for both collision points (which were in agreement with each other) have been combined for the final result. After the two-photon excitation, the atoms are state selectively ionized with a 532 nm pulse of a few mJ. It always arrives 4 ns after the second excitation pulse because it is derived from the second pump pulse of the NOPCPA. The generated ions are extracted by applying a 300 V electric potential. This extraction field operates in a pulsed mode to avoid any DC-Stark effect. Because of the bandwidth of the excitation pulses, all isotopes of krypton (^{80}Kr, ^{82}Kr, ^{83}Kr, ^{84}Kr, and ^{86}Kr) are excited at the same time. Their respective ions are separated in time by a time-of-flight (ToF) detection scheme, after which they are individually detected with an extended dynamical range, channel electron multiplier (EDR-CEM) detector (SJUTS ®). We used the most abundant ^{84}Kr isotope for the determination of the absolute transition frequency and measured the isotopes relative to this.

5 Experimental Results and Analysis

One RCS measurement consists of the observation of two scans of the Ramsey oscillation at two different macro-delay between the deep-UV pulses. We start at $N = 2$ to avoid any transient effects from the first pulse influencing the signal of the second pulse, e.g., in monitoring the pulse energy using the signal from a photodiode. We choose $N = 7$ for the second scan as a compromise between a large time interval ΔT which is beneficial for the accuracy (as for any Ramsey-type experiment), and a good signal-to-noise ratio (that deteriorates for longer delays due to the finite lifetime of the excited state of 27 ns).

An individual scan is performed by varying the repetition time of the frequency comb over a range of δt such that two periods of the Ramsey oscillations are recorded with a sampling of 14 points. Each point represents an average of the signal amplitude over 350 laser shots. We obtain a complete RCS measurement (two Ramsey oscillations at two different N) in approximately 6 minutes, as shown in Fig. 5. Such a recording is performed in a back-and-forth sequence, where even numbered data points at $N = 2$ and $N = 7$ are measured first followed by the odd-numbered data points at $N = 7$ then $N = 2$, to reduce the influence of drift effects (e.g., beam pointing).

At each N, a cosine is fitted to the Ramsey-fringe based on Eq. 1. The atomic frequency is then found based solely on the phases extracted from these fits [25]. In the simple case when only a single transition is excited, the phase evolves linearly with time, so that only two values of N are needed to extract the atomic frequency. In the RCS method, the frequency is extracted modulo the frequency sampling interval (the biggest interval corresponds to a frequency spacing of f_{rep}). However, the previous determination [17] is precise enough to resolve the ambiguity. Otherwise, repeating the measurement at different repetition rate would straightforwardly solve it.

We will now discuss several potential sources of systematic errors.

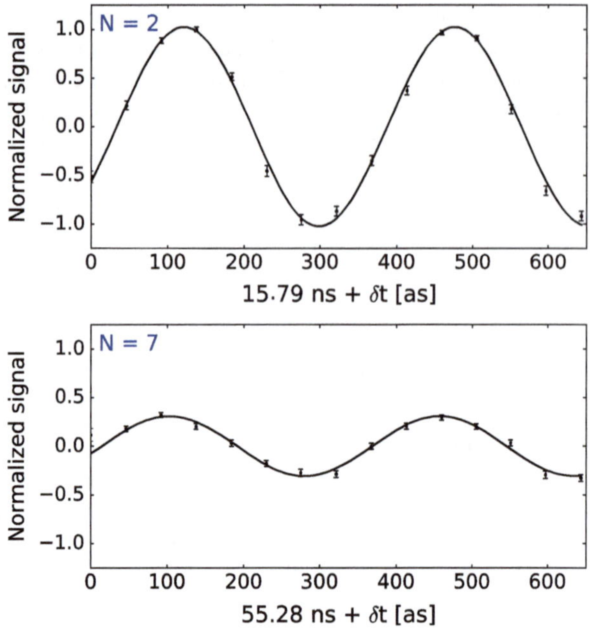

Fig. 5 Ramsey oscillations observed after the excitation of the atomic transition in ^{84}Kr by two different pairs of pulses. The two UV pulses are delayed by $2 \times T = 15.79$ ns in the upper graph and by $7 \times T = 55.28$ ns in the lower graph. The micro-delay δt is scanned by adjusting the repetition rate of the frequency comb. The amplitude of the Ramsey signal decreases for longer delays because of the finite lifetime ($\tau = 27$ ns) of the excited state

5.1 Amplification-Induced Phase Shifts

In order to determine that the amplification process did not alter the relative phase between the excitation-pulse pairs, the relative phase was measured with spectral interferometry (see Sect. 3). Even though absolute phase shifts between the amplified pulses can be on the order of 100 mrad (depending on, e.g., laser alignment), we measured no relative phase shift within a statistical uncertainty of 1 mrad between $N = 2$ and $N = 7$. This translates into an uncertainty of 35 kHz in the final determination of the atomic frequency.

5.2 The AC- and DC-Stark Shift

The AC-Stark shift (or light shift) is a common major source of uncertainty for high-resolution spectroscopy using high laser intensities. The atomic frequency extracted by the RCS method is not affected by the AC-Stark shift as long as the induced phase-shift is the same for every Ramsey scan. This is ensured by keeping the amplitude of the infrared light-pulses stable within 1% using active feedback via the timing of the Pockels cell gate between the bounce and post-amplifier, as explained previously. To detect any possible residual AC-Stark effect in the UV, we performed a conventional test by varying the UV light pulse intensity with a factor

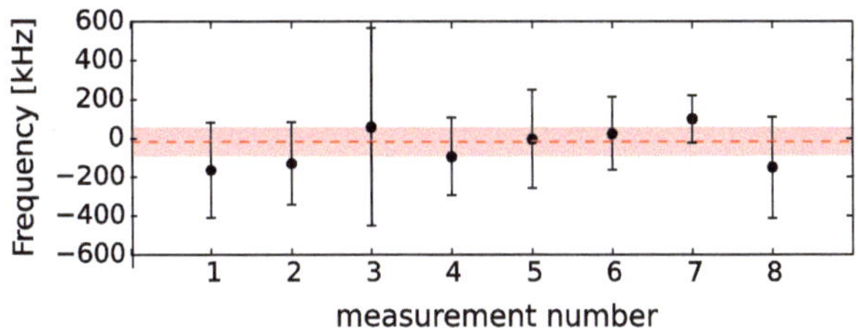

Fig. 6 Light-shift determination. Every point in this graph represents an AC-Stark shift determination (each based on typically 20 scans) by extrapolating the transition frequency to zero pulse energy from pairs of Ramsey-comb measurements with 25 and 50 μJ UV pulse energy. No frequency shift is observed within the statistical uncertainty of this measurement of 72 kHz

two (using a polarizer and half-wave plate just before the doubling stages). This procedure also tests potential phase shifts induced in the doubling stages itself [17]. The result, illustrated in Fig. 6, indicates no AC-Stark shift or amplitude-related shift, within a uncertainty of 72 kHz. This uncertainty is comparatively large in view of the other sources of uncertainty due to a relatively low number of measurements that were taken to determine this effect.

To extract the ionized krypton atoms for detection, electric fields are applied that can potentially lead to a DC-Stark shift. To enable (nearly) field-free excitation, the extraction field was operated in a pulsed mode. For this mode, we calculated a residual electric field of at most 0.17 V/cm. No significant shift was seen when the transition frequency was also measured in a DC electric field of 29.4 V/cm. Therefore, a negligible effect on the final transition frequency of <1 kHz is expected under the conditions of the pulsed extraction field.

5.3 *Doppler Shift*

The two-photon excitation is performed in a counter-propagating beam configuration in order to reduce first-order Doppler effects. To create the counter-propagating beams, the UV pulses are split using a metallic 50% beamsplitter (see Fig. 4). When the UV pulses combine again on the beamsplitter after one round trip through the vacuum apparatus, the exit port shows an interference pattern depending on the alignment of the UV beams. Perfectly counter-propagating beams cancel each other at the beamsplitter, leading to a single dark fringe. This procedure guarantees the collinearity of the two excitation beams with a precision better than 0.01 mrad [36]. Such an angular mismatch can give rise to a frequency shift of 18 kHz according to the following equation [36]:

$$\Delta f = \theta_1 \frac{||v||}{\lambda} \tag{4}$$

where $\theta_1 \ll 1$ is the angle between the two UV beams, v is the atomic velocity and λ is the excitation wavelength. Another source of first-order Doppler shift stems from our particular resonance condition which requires two excitation pulses with different frequencies (to cancel excitation from a single side). Therefore, even if the light beams perfectly overlap, they must also be exactly perpendicular to the atomic beam. An angular mismatch of θ_2 between the light and the atomic beams leads to a frequency error:

$$\Delta f = \Delta\nu_{\mathrm{UV}} \times \frac{||v||| \cos(\pi/2 + \theta_2)}{c} \tag{5}$$

where c is the speed of light and $\Delta\nu_{\mathrm{UV}}$ is the frequency difference between 'blue' and the 'red' component of the spectrum. Moreover, the use of chirped pulses can lead to a systematic effect as explained in [6], named chirp-induced first-order Doppler shift (CIFODS). All these effects can be minimized by finding the angle θ_2 for which the average first-order Doppler shift vanishes (when the UV beams are effectively perpendicular to the atomic beam).

To find this situation, we measure the atomic transition frequency for different velocities of the atomic beam. The velocity can be increased by using a mixture of krypton with neon (in a 1:5 ratio) or a mixture of krypton and helium (in a 1:9 ratio). The velocities of pure Kr and the mixtures have been measured using the approximately 40 cm long time-of-flight detection setup. After the deep-UV light excitation, the created ions are directed upward using an extraction field in the vertical direction. Two deflection plates, placed on both sides of the ions trajectory, steer the atomic beam on the detector. The velocity of the krypton atoms can then be calculated based on the geometry of the detection setup and the applied electric fields to steer the ions, see Fig. 7. We determined a velocity of the pure krypton beam of 380(80), 686(60) m/s for the krypton–neon mixture, and 931(134) m/s for the krypton–helium mixture.

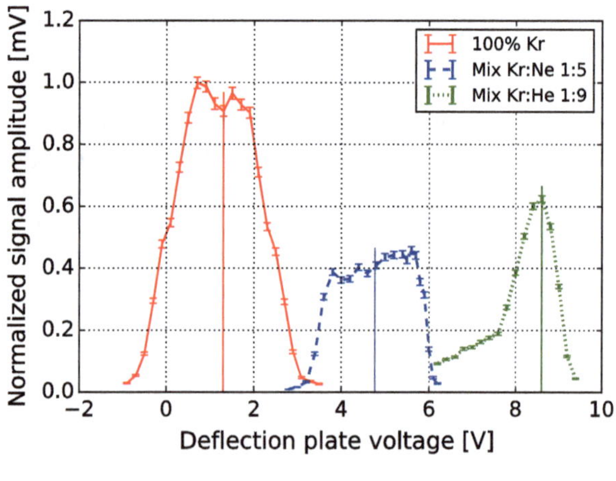

Fig. 7 Determination of the forward velocity of krypton atoms in the different gas mixtures, based on the voltage to deflect the ions in the ToF setup. The vertical lines show the voltage used to estimate the velocity of the two different gas mixtures. The signal amplitudes are normalized on the highest signal of pure Kr

The angle between the UV beams and the atomic beam is tuned until the difference in transition frequency between the different gas mixtures is undetectable within the statistical uncertainty. The atomic frequency is then extracted by extrapolating to zero velocity based on pairs of measurements with different atomic beam velocities. Also the second-order Doppler shift was taken into account for each velocity class, corresponding to 2.3 kHz for pure krypton, 7.4 kHz for the Kr:Ne mixture, and 13.6 kHz for the Kr:He mixture. With this procedure, we determined the Doppler-free transition frequency to be 2820833101688 kHz with a statistical uncertainty of 58 kHz.

6 Isotope Shift Measurements

Our time-of-flight measurement enables to resolve the signals of the different isotopes (see Fig. 8). Therefore, we can measure the signal for each isotope simultaneously with the ^{84}Kr signal using multiple boxcar integrators, each set on one isotope (see Fig. 9). The isotope shift $f_{84\mathrm{Kr}} - f_{\mathrm{XKr}}$ is straightforwardly extracted from the observed relative phase at one single N. The ambiguity of the isotope shift frequency due to the comb-mode spacing is solved by a comparison with previous measurements that have sufficient accuracy [17]. Some of the potential phase shifts that could influence the measurement are common mode in the isotope shift measurement, such as the AC-Stark shift and (to a large extent) the Doppler shift.

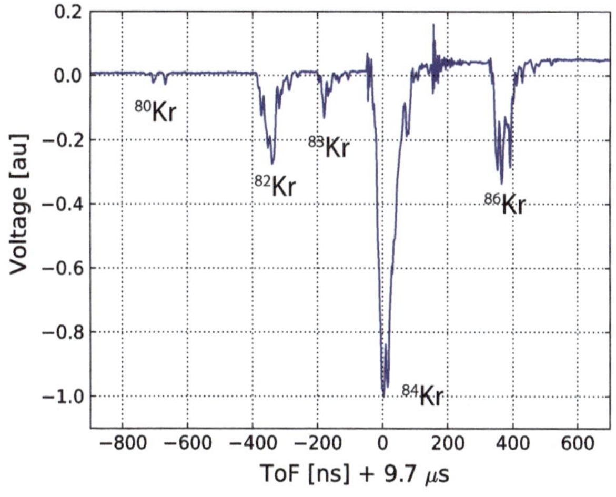

Fig. 8 Typical time-of-flight detection of the isotopes of krypton. The relative signal depends on the state of the phase of their respective Ramsey oscillation. We have chosen a time-of-flight snapshot which pattern is a close representation of their natural abundance. The large signal of the most abundant isotope ^{84}Kr strongly affects the operation of the detector, and therefore the signal of isotopes arriving later. This effect can also be seen as a shift of the detector signal baseline after ^{84}Kr

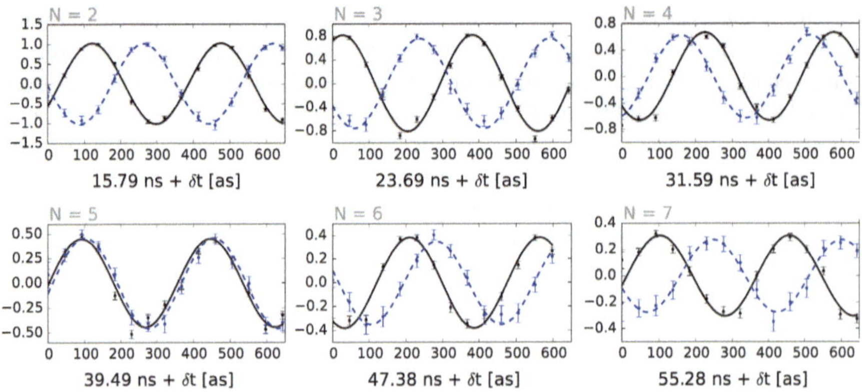

Fig. 9 Example of the isotope shift measurements. Multiple Ramsey scans for ^{84}Kr (*solid black line*) and ^{82}Kr (*dashed blue line*) for pulse pairs ranging from $N = 2$ to $N = 7$ are shown. For each N, the signals of the two isotopes have been acquired simultaneously. Because of the difference in transition frequency, the phases of the Ramsey scans evolve differently in time. The isotope shift can be determined from the relative phase difference at each value of N

There is, however, an effect that does influence the measured isotope shift due to saturation effects of the ion detector. The different bunches of ions for each of the five isotopes are resolved in time and give rise to five peaks spanning over roughly 800 ns, as shown in Fig. 8. Each peak is equivalent to the detection of approximately 1–10 ions within roughly 20 ns. The EDR-CEM detector is not specified for such high count rates. As a result, the signal from isotopes arriving early in the ToF might influence the signal from isotopes arriving later. This effect potentially distorts the Ramsey signal of the heavier isotopes. The situation is even more subtle, because the amplitudes of the five isotopes signals evolve differently as function of time. To estimate the influence of the detector we performed the Ramsey-comb analysis procedure with different pulse delays (i.e., different N). Figure 9 shows the simultaneous acquisition of the Ramsey fringes for the isotopes ^{82}Kr and ^{84}Kr for $N = 2$ to 7. The isotope shift $f_{84\text{Kr}} - f_{82\text{Kr}}$ can be extracted from every single N scan. No systematic dependence on N has been observed for the isotopes ^{80}Kr, ^{82}Kr, ^{83}Kr and ^{84}Kr within the statistical uncertainty of 25 kHz. This uncertainty is quadratically added with the individual statistical errors giving rise to the following isotopes shifts: 301 847(43) kHz ($f_{84\text{Kr}} - f_{80\text{Kr}}$), 152 403(35) kHz ($f_{84\text{Kr}} - f_{82\text{Kr}}$) and 98 527(45) kHz ($f_{84\text{Kr}} - f_{83\text{Kr}}$).

However, the ^{86}Kr isotope is strongly affected by the detector saturation effect due to the high abundance and therefore strong signal of the preceding ^{84}Kr. Significant time was spend on characterizing this effect, and an attempt was made to model it. However, the modeling did not have enough predicting power to improve the accuracy. Therefore, we conservatively base our error on the variation seen between the different measurements at different pulse delays and obtain an ^{86}Kr isotope shift of $f_{84\text{Kr}} - f_{86\text{Kr}} = -136.40(0.45)$ MHz.

The isotope shift uncertainties have been improved by a factor of 6.5, 4.2 and 3.7 for the isotopes ^{80}Kr, ^{82}Kr and ^{83}Kr, respectively, compared to [17]. The present ^{86}Kr isotope shift determination is less precise than the previous measurement, due to the saturation effect discussed above.

7 Conclusion

Table 1 lists the different contributions to the atomic-frequency determination of ^{84} Kr. The final result of the $4p^6 \rightarrow 4p^5 5p[1/2]_0$ transition in ^{84}Kr, taking all corrections into account, is ^{84}Kr is 2,820,833,101,679(103) kHz. This corresponds to an improvement of the uncertainty by a factor of 34. It shows that high-accuracy spectroscopy with powerful short laser pulses down to deep-UV wavelengths is possible, and with a suppressed AC-Stark effect.

The accuracy of the present deep-UV spectroscopy demonstration is mainly limited by the short lifetime (27 ns) of the excited state. If Ramsey-comb pulses with a longer time delay could be used (e.g., on longer lived states), then the accuracy improves accordingly. This feature gives good prospects for much more accurate spectroscopic measurements in, e.g., molecular hydrogen on the EF←X transition, where the longest lived states (with a low vibrational and rotational quantum number) have a lifetime of ≈200 ns. Likewise, Ramsey-comb measurements could improve the determination of the fundamental vibrational splitting of H_2 by at least an order of magnitude compared to previous measurements [37].

The demonstrated amplified pulse energy of 1.5 mJ per pulse is also enough for efficient HHG in gas jets, which makes RCS an interesting candidate to measure the 1S–2S transition in He^+. The long lifetime of the 2S excited state (1.9 ms) is very useful in this respect, as phase effects due to ionization in the HHG process can be avoided by choosing the pulses at least 200 ns apart. Each laser pulse then sees a fresh gas sample from the jet, and therefore a constant phase shift in the HHG process. Such a measurement could provide important new data to the proton radius puzzle, as the transition can then be compared to spectroscopy of muonic He^+ ions [13].

Table 1 Summary of the measurement result of the $4p^6 \rightarrow 4p^5 5p[1/2]_0$ two-photon transition in ^{84}Kr. The values listed here are in kHz, and the value in brackets denotes the 1σ statistical uncertainty. The listed ^{84}Kr transition frequency is Doppler-free as explained in Sect. 5.3

^{84}Kr transition freq.	2820833101688	(58)
AC-Stark shift	−13	(72)
Laser phase shift	1	(35)
Zeeman shift	3	(13)
Detector saturation	0	(25)
Total	2820833101679	(103)

Acknowledgements The work in this paper was funded by the Foundation for Fundamental Research on Matter (FOM) through its Program 125: 'Broken Mirrors and Drifting Constants' and Projectruimte 12PR3098: 'Exploring the Boundaries of QED with Helium$^+$'.

References

1. D. Hanneke, S. Fogwell, G. Gabrielse, New measurement of the electron magnetic moment and the fine structure constant. Phys. Rev. Lett. **100**, 120801 (2008)
2. P.J. Mohr, B.N. Taylor, D.B. Newell, CODATA recommended values of the fundamental physical constants: 2010. Rev. Mod. Phys. **84**(4), 1527–1605 (2012)
3. A. Wicht, J.M. Hensley, E. Sarajlic, S. Chu, A preliminary measurement of the fine structure constant based on atom interferometry. Phys. Scr. **T102**(1), 82–88 (2002)
4. R. Bouchendira, P. Cladé, S. Guellati-Khélifa, F. Nez, F. Biraben, New determination of the fine structure constant and test of the quantum electrodynamics. Phys. Rev. Lett. **106**, 080801 (2011)
5. A. Matveev, C.G. Parthey, K. Predehl, J. Alnis, A. Beyer, R. Holzwarth, Th Udem, T. Wilken, N. Kolachevsky, Mi Abgrall, D. Rovera, C. Salomon, P. Laurent, G. Grosche, O. Terra, T. Legero, H. Schnatz, S. Weyers, B. Altschul, T.W. Hänsch, Precision measurement of the hydrogen 1S–2S frequency via a 920-km fiber link. Phys. Rev. Lett. **110**, 230801 (2013)
6. D.C. Yost, A. Matveev, A. Grinin, E. Peters, L. Maisenbacher, A. Beyer, R. Pohl, N. Kolachevsky, K. Khabarova, T.W. Hänsch, Th Udem, Spectroscopy of the hydrogen 1S–3S transition with chirped laser pulses. Phys. Rev. A **93**, 042509 (2016)
7. S. Galtier, H. Fleurbaey, S. Thomas, L. Julien, F. Biraben, F. Nez, Progress in spectroscopy of the 1S–3S transition in hydrogen. J. Phys. Chem. Ref. Data **44**(3), 031201 (2015)
8. D.Z. Kandula, C. Gohle, T.J. Pinkert, W. Ubachs, K.S.E. Eikema, Extreme Ultraviolet Frequency Comb Metrology. Phys. Rev. Lett. **105**, 063001 (2010)
9. J. Biesheuvel, J-Ph Karr, L. Hilico, K.S.E. Eikema, W. Ubachs, J.C.J. Koelemeij, Probing QED and fundamental constants through laser spectroscopy of vibrational transitions in HD+. Nat. Commun. **7**, 10385 (2016)
10. P. Beiersdorfer, Testing QED and atomic–nuclear interactions with high-Z ions. J. Phys. B At. Mol. Opt. **43**(7), 074032 (2010)
11. R. Pohl, A. Antognini, F. Nez, F.D. Amaro, F. Biraben, J.M.R. Cardoso, D.S. Covita, A. Dax, S. Dhawan, L.M.P. Fernandes, A. Giesen, T. Graf, T.W. Hänsch, P. Indelicato, L. Julien, C. Kao, P. Knowles, E. Le Bigot, Y. Liu, J.A.M. Lopes, L. Ludhova, C.M.B. Monteiro, F. Mulhauser, T. Nebel, P. Rabinowitz, J.M.F. Dos Santos, L.A. Schaller, K. Schuhmann, C. Schwob, D. Taqqu, J.F.C.A. Veloso, F. Kottmann, The size of the proton. Nature **466**, 213–216 (2010)
12. A. Antognini, F. Nez, K. Schuhmann, F.D. Amaro, F. Biraben, J.M.R. Cardoso, D.S. Covita, A. Dax, S. Dhawan, M. Diepold, L.M.P. Fernandes, A. Giesen, A.L. Gouvea, T. Graf, T.W. Hänsch, P. Indelicato, L. Julien, C.-Y. Kao, P. Knowles, F. Kottmann, E.-O. Le Bigot, Y.-W. Liu, J.A.M. Lopes, L. Ludhova, C.M.B. Monteiro, F. Mulhauser, T. Nebel, P. Rabinowitz, J.M.F. dos Santos, L.A. Schaller, Ca Schwob, D. Taqqu, J.F.C.a Veloso, J. Vogelsang, R. Pohl, Proton structure from the measurement of 2S–2P transition frequencies of muonic hydrogen. Science **339**(6118), 417–420 (2013)
13. A. Antognini, K. Schuhmann, F.D. Amaro, P. Amaro, M. Abdou-Ahmed, F. Biraben, T.-L. Chen, D.S. Covita, A.J. Dax, M. Diepold, L.M.P. Fernandes, B. Franke, S. Galtier, A.L. Gouvea, J. Götzfried, T. Graf, T.W. Hänsch, M. Hildebrandt, P. Indelicato, L. Julien, K. Kirch, A. Knecht, F. Kottman, J.J. Krauth, Y.-W. Liu, J. Machado, C.M.B. Monteiro, F. Mulhauser, F. Nez, J.P. Santos, J.M.F. dos Santos, C.I. Szabo, D. Taqqu, J.F.C.A. Veloso,

A. Voss, B. Weichelt, R. Pohl, Experiments towards resolving the proton charge radius puzzle. EPJ Web Conf. **113**, 01006 (2016)
14. R. Pohl, F. Nez, L.M.P. Fernandes, D.A. Fernando, F. Biraben, J.M.R. Cardoso, D.S. Covita, A. Dax, S. Dhawan, M. Diepold, A. Giesen, A.L. Gouvea, T. Graf, T.W. Hänsch, P. Indelicato, L. Julien, P. Knowles, F. Kottmann, E.-O. Le Bigot, Y.-W. Liu, J.A.M. Lopes, L. Ludhova, C.M.B. Monteiro, F. Mulhauser, T. Nebel, P. Rabinowitz, J.M.F. dos Santos, L.A. Schaller, K. Schuhmann, C. Schwob, D. Taqqu, J.F.C.A. Veloso, A. Antognini, Laser spectroscopy of muonic deuterium. Science **353**(6300), 669–672 (2016)
15. K. Pachucki, J. Komasa, Schrodinger equation solved for the hydrogen molecule with unprecedented accuracy. J. Chem. Phys. **144**, 164306 (2016)
16. M. Herrmann, M. Haas, U.D. Jentschura, F. Kottmann, D. Leibfried, G. Saathoff, C. Gohle, A. Ozawa, V. Batteiger, S. Knünz, N. Kolachevsky, H.A. Schüssler, T.W. Hänsch, Th Udem, Feasibility of coherent XUV spectroscopy on the 1S–2S transition in singly ionized helium. Phys. Rev. A **79**, 052505 (2009)
17. S. Witte, R.T. Zinkstok, W. Ubachs, W. Hogervorst, K.S.E. Eikema, Deep-ultraviolet quantum interference metrology with ultrashort laser pulses. Science **307**, 400–403 (2005)
18. E. Peters, D.C. Yost, A. Matveev, T.W. Hänsch, Th Udem, Frequency-comb spectroscopy of the hydrogen 1S–3S and 1S–3D transitions. Ann. Phys. **525**(7), L29–L34 (2013)
19. A. Cingoz, D.C. Yost, T.K. Allison, A. Ruehl, M.E. Fermann, I. Hartl, J. Ye, Direct frequency comb spectroscopy in the extreme ultraviolet. Nature **482**, 68–71 (2012)
20. J. Lee, D.R. Carlson, R.J. Jones, Optimizing intracavity high harmonic generation for XUV fs frequency combs. Opt. Express **19**(23), 23315 (2011)
21. A. Ozawa, W. Schneider, T.W. Hänsch, Th Udem, P. Hommelhoff, Phase-stable single-pass cryogenic amplifier for high repetition rate few-cycle laser pulses. N. J. Phys. **11**, 083029 (2009)
22. A. Ruehl, A. Marcinkevicius, M.E. Fermann, I. Hartl, 80 W, 120 f. Yb-fiber frequency comb. Opt. Lett. **35**(18), 3015–3017 (2010)
23. J. Morgenweg, I. Barmes, K.S.E. Eikema, Ramsey-comb spectroscopy with intense ultrashort laser pulses. Nat. Phys. **10**, 30–33 (2014)
24. R.K. Altmann, S. Galtier, L.S. Dreissen, K.S.E. Eikema, High-precision Ramsey-comb spectroscopy at deep ultraviolet wavelengths. Phys. Rev. Lett. **117**(17), 173201 (2016). doi:10.1103/PhysRevLett.117.173201
25. J. Morgenweg, K.S.E. Eikema, Ramsey-comb spectroscopy: theory and signal analysis. Phys. Rev. A **89**, 052510 (2014)
26. N.F. Ramsey, A molecular beam resonance method with separated oscillating fields. Phys. Rev. **78**(6), 695–699 (1950)
27. S.A. Diddams, D.J. Jones, J. Ye, S.T. Cundiff, J.L. Hall, J.K. Ranka, R.S. Windeler, R. Holzwarth, Th Udem, T.W. Hänsch, Direct link between microwave and optical frequencies with a 300 THz femtosecond laser comb. Phys. Rev. Lett. **84**, 5102–5105 (2000)
28. R. Holzwarth, Th Udem, T.W. Hänsch, J.C. Knight, W.J. Wadsworth, P.S.J. Russell, Optical frequency synthesizer for precision spectroscopy. Phys. Rev. Lett. **85**, 2264–2267 (2000)
29. R. Wynands, S. Weyers, Atomic fountain clocks. Metrologia **42**, S64–S79 (2005)
30. S. Witte, RTh Zinkstok, A.L. Wolf, W. Hogervorst, W. Ubachs, K.S.E. Eikema, A source of 2 terawatt, 2.7 cycle laser pulses based on noncollinear optical parametric chirped pulse amplification. Opt. Express **14**(18), 8168–8177 (2006)
31. J. Morgenweg, K.S.E. Eikema, Multi-delay, phase coherent pulse pair generation for precision Ramsey-frequency comb spectroscopy. Opt. Express **21**(5), 5275–5286 (2013)
32. J. Morgenweg, K.S.E. Eikema, Tailored pulse sequences from an 880 nm pumped Nd:YVO_4 bounce amplifier. Opt. Lett. **37**(2), 208–210 (2012)
33. J. Morgenweg, K.S.E. Eikema, A 1.8 mJ, picosecond Nd:YVO4 bounce amplifier pump front-end system for high-accuracy XUV-frequency comb spectroscopy. Laser Phys. Lett. **5**(11), 781–785 (2012)

34. I.N. Ross, P. Matousek, Geoffrey, H.C. New, K. Osvay, Analysis and optimization of optical parametric chirped pulse amplification. J. Opt. Soc. Am. B **19**(12), 2945–2956 (2002)
35. A. Renault, D.Z. Kandula, S. Witte, A.L. Wolf, RTh Zinkstok, W. Hogervorst, K.S.E. Eikema, Phase stability of terawatt-class ultrabroadband parametric amplification. Opt. Lett. **32**(16), 2363–2365 (2007)
36. S. Hannemann, E.J. Salumbides, W. Ubachs, Reducing the first-order Doppler shift in a Sagnac interferometer. Opt. Lett. **32**(11), 1381–1383 (2007)
37. G.D. Dickenson, M.L. Niu, E.J. Salumbides, J. Komasa, K.S.E. Eikema, K. Pachucki, W. Ubachs, Fundamental vibration of molecular hydrogen. Phys. Rev. Lett. **110**, 193601 (2013)

Phase-Stable, Multi-μJ Femtosecond Pulses from a Repetition-Rate Tunable Ti: Sa-Oscillator-Seeded Yb-Fiber Amplifier

T. Saule, S. Holzberger, O. De Vries, M. Plötner, J. Limpert, A. Tünnermann, and I. Pupeza

Abstract We present a high-power, MHz-repetition-rate, phase-stable femtosecond laser system based on a phase-stabilized Ti:Sa oscillator and a multi-stage Yb-fiber chirped-pulse power amplifier. A 10-nm band around 1030 nm is split from the 7-fs oscillator output and serves as the seed for subsequent amplification by 54 dB to 80 W of average power. The μJ-level output is spectrally broadened in a solid-core fiber and compressed to ~30 fs with chirped mirrors. A pulse picker prior to power amplification allows for decreasing the repetition rate from 74 MHz by a factor of up to 4 without affecting the pulse parameters. To compensate for phase jitter added by the amplifier to the feed-forward phase-stabilized seeding pulses, a self-referencing feed-back loop is implemented at the system output. An integrated

This article is part of the topical collection "Enlightening the World with the Laser" - Honoring T. W. Hänsch guest edited by Tilman Esslinger, Nathalie Picqué, and Thomas Udem.

T. Saule
Max-Planck-Institut für Quantenoptik, Hans-Kopfermann-Str. 1, 85748 Garching, Germany

Ludwig-Maximilians-Universität München, Am Coulombwall 1, 85748 Garching, Germany
e-mail: Tobias.saule@physik.uni-muenchen.de

S. Holzberger
Max-Planck-Institut für Quantenoptik, Hans-Kopfermann-Str. 1, 85748 Garching, Germany

Ludwig-Maximilians-Universität München, Am Coulombwall 1, 85748 Garching, Germany

Present Address: Menlo Systems GmbH, Am Klopferspitz 19a, 82152 Martinsried, Germany
e-mail: S.holzberger@menlosystems.com

O. De Vries • M. Plötner
Fraunhofer Institute for Applied Optics and Precision Engineering, Albert-Einstein-Str. 7, 07745 Jena, Germany
e-mail: Oliver.deVries@iof.fraunhofer.de; Marco.Ploetner@iof.fraunhofer.de

J. Limpert
Institut für Angewandte Physik, Friedrich-Schiller-Universität Jena, Alber-Einstein-Str. 15, 07745 Jena, Germany

Helmholtz-Institut Jena, Fröbelstieg 3, 07743 Jena, Germany

Active Fiber Systems GmbH, Wildenbruchstr. 15, 07745 Jena, Germany
e-mail: Jens.limpert@uni-jena.de

D. Meschede et al. (eds.), *Exploring the World with the Laser*,
https://doi.org/10.1007/978-3-319-64346-5_14

out-of-loop phase noise of less than 100 mrad was measured in the band from 0.4 Hz to 400 kHz, which to the best of our knowledge corresponds to the highest phase stability ever demonstrated for high-power, multi-MHz-repetition-rate ultrafast lasers. This system will enable experiments in attosecond physics at unprecedented repetition rates, it offers ideal prerequisites for the generation and field-resolved electro-optical sampling of high-power, broadband infrared pulses, and it is suitable for phase-stable white light generation.

1 Introduction

Around the turn of the last century, the generation of visible/near-infrared few-cycle pulses became a matter of course in laser laboratories. The ability to control the electric field of such pulses enabled ground-breaking applications like high-precision frequency comb spectroscopy [1] and led to the establishment of new research fields like attosecond science [2, 3]. Titanium-sapphire (Ti:Sa) oscillators can be considered the workhorse in this field, since they are the most widespread technology employed to seed laser systems generating phase-stable few-cycle pulses. Important contributions to the success of these oscillators are their large optical bandwidth, leading to the direct generation of few-cycle pulses, and the excellent phase stability achievable with feed-back [4] and feed-forward stabilization schemes [5]. However, a major drawback of Ti:Sa amplifiers is the high thermal absorption and thermal lensing in the gain medium, which restricts the average powers to a few tens of Watts even with cryogenic cooling [6], thus limiting high-pulse-energy operation to repetition rates significantly lower than 1 MHz. Recently, Yb-based laser technology has rapidly progressed as a powerful competitor to the well-established Ti:Sa technology. The superior thermal properties of Yb-doped active media enable an improvement of several orders of magnitude in terms of average power [7–11]. To overcome the disadvantage of a significantly narrower gain bandwidth, several post-amplification nonlinear pulse compression techniques have been developed [9–14]. Recently, from an Yb:YAG thin-disk oscillator followed by two nonlinear compression stages, pulses of 2.2 cycles with 6 W of average power at a repetition rate of 38 MHz and with a phase jitter of 270 mrad (out-of-loop phase noise integrated in the band between 1 Hz and 500 kHz) were demonstrated [10].

A. Tünnermann
Fraunhofer Institute for Applied Optics and Precision Engineering, Albert-Einstein-Str. 7, 07745 Jena, Germany

Institut für Angewandte Physik, Friedrich-Schiller-Universität Jena, Alber-Einstein-Str. 15, 07745 Jena, Germany
e-mail: Andreas.tuennermann@uni-jena.de

I. Pupeza (✉)
Max-Planck-Institut für Quantenoptik, Hans-Kopfermann-Str. 1, 85748 Garching, Germany
e-mail: Ioachim.pupeza@mpq.mpg.de

In this study, we combine a feed-forward stabilized Ti:Sa master oscillator with an Yb-fiber power amplifier, resulting in a high-power, ultrashort-pulse laser system with several unique advantages for electric-field-resolved metrology. First, the high phase stability achieved with the Ti:Sa frontend is largely preserved upon amplification of a 10-nm band around 1030 nm by about 54 dB to 80 W. Additional phase fluctuations introduced by the multi-stage, chirped-pulse amplifier (CPA) and by subsequent nonlinear compression to about 30 fs were compensated for by a feed-back loop, resulting in an unprecedentedly small overall phase jitter of the high-power pulse train of less than 100 mrad (out-of-loop phase noise, integrated in the band between 0.4 Hz and 400 kHz). Due to its phase stability, this source is particularly well suited to drive cavity-enhanced high-order harmonic generation (HHG) for the generation of extreme ultraviolet (XUV) frequency combs [15, 16] and of XUV attosecond pulses [16–18]. Second, the master oscillator power amplifier (MOPA) approach readily enables the use of a high-frequency pulse picker after the low-power oscillator [19], allowing for a tunable repetition frequency (18.5, 24.7, 37 and 74 MHz). And third, the Ti:Sa oscillator produces a 74-MHz train of 7-fs pulses, inherently synchronized with the pulses amplified in the Yb-fiber CPA. These could be used for instance as a pump in time-resolved photoelectron emission microscopy [20], where attosecond XUV probe pulses are generated with the CPA output. Another application could be electro-optical sampling of broadband infrared radiation [21] generated with the CPA output, where a repetition-rate ratio of a factor of 2 between the sampling pulses and the long-wavelength field enables ultralow-noise lock-in detection [22].

2 Experimental Setup

The experimental setup is shown in Fig. 1a. The Ti:Sa frontend provides 7-fs pulses with 200 mW of average power at 74-MHz repetition rate. The 10-nm band around 1030 nm carries 300 μW of average power and acts as the seed for the Yb-fiber amplifier. In a first step, the repetition rate is (optionally) reduced by a fast acousto-optic modulator (AOM) pulse picker, and the resulting pulses are amplified to 150 mW and stretched to around 200 ps by a chirped fiber Bragg grating [19]. To minimize additional intensity and phase noise, the pulse picker is operated in synchronicity with the repetition rate as described in [19]. After phase stabilization by an acousto-optic frequency shifter (AOFS) operated in the feed-forward stabilization scheme [5], the subsequent amplification stages and the compression comprise two Yb-large-pitch-fiber (LPF) amplifiers [23], two grating compressors and a rod-type large-mode-area (LMA) fiber for spectral broadening, followed by a chirped-mirror compressor.

To fine-tune the pulse duration at the CPA output, the pre-compressor is motorized to control the distance of the gratings. It only influences the pulse duration of the stretched pulses by up to 10 ps and allows for stable and precise control of the output pulse duration. The estimated B-integral of the system amounts to around

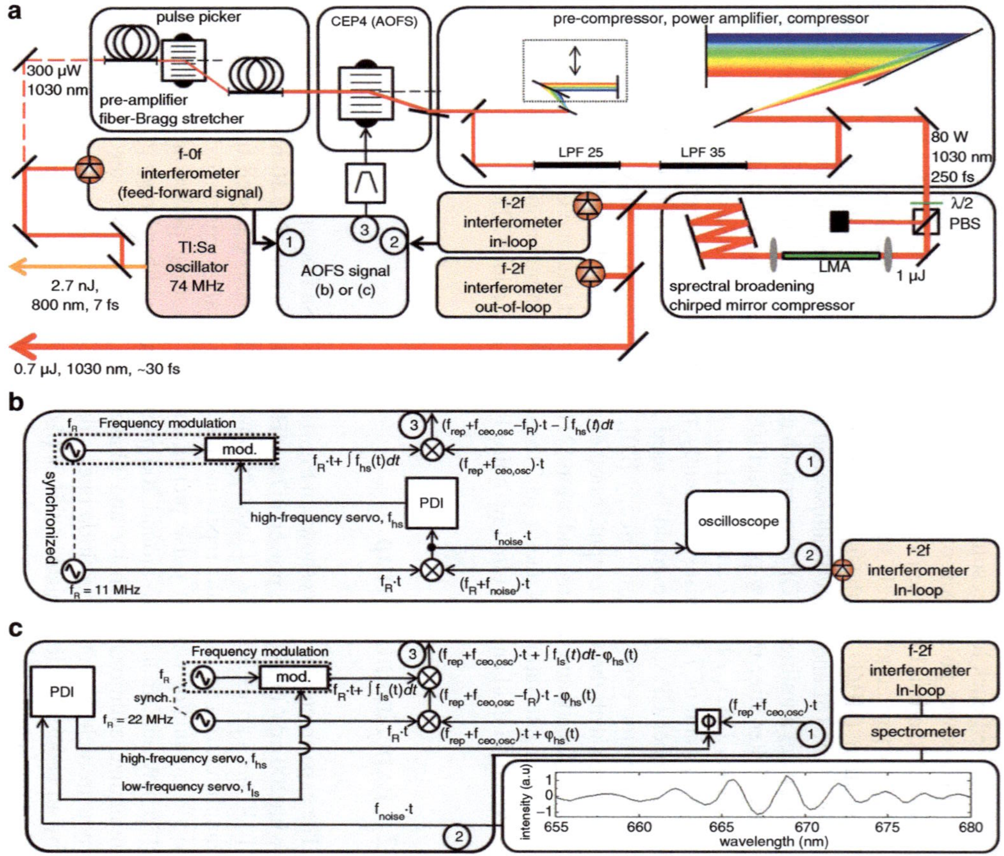

Fig. 1 **a** Schematic of the experimental setup. A 300-µW, 10-nm band around 1030 nm of the Ti:Sa output is split from the main beam and serves as the seed

0.1 rad irrespective of the setting of the pre-compressor. The dispersion of the gratings is matched up to the sixth order.

The overall phase noise at the exit of the system is measured with two independent $f-2f$ interferometers [24]. The first interferometer (in-loop) is used to generate a feed-back error signal which is combined with the original feed-forward stabilization by adding a frequency modulation to the RF signal used to drive the AOFS (signal 3 in Fig. 1) and thus also a phase modulation (see phase along the black lines in Fig. 1b, c). The second (out-of-loop) interferometer is used to determine the actual phase noise outside of the locking loop.

The amplification process contains several potential phase noise sources, including amplification noise [25], grating jitter (mechanical and air fluctuations) [26] and residual noise during the pulse picking in the AOM [19]. Furthermore, the mechanical jitter in the interferometers themselves and the supercontinuum generation necessary for $f-2f$ detection add to the measured phase noise [27].

The MOPA system is capable of delivering a 250-fs (all pulse durations are given in full width at half their intensity maximum), 80-W pulse train at repetition rates ranging from 74 MHz down to 18.5 MHz, which corresponds to a maximum of 4.3 μJ. The solid-core-fiber-based spectral broadening employed here is limited to around 1-μJ input pulse energy by self-focusing-induced damage in the fiber. This is, however, not a fundamental limitation in view of the recently demonstrated power scalable compression schemes in bulk [9, 28] as well as in kagome fibers [14]. The beam profile and autocorrelation traces for the four repetition rates are shown in Fig. 2a. The pulse durations, assuming a Gaussian-shaped pulse, are measured to be 32 fs for all considered repetition rates, which is close to the Fourier limit of ~28 fs. The spectra for the seed, amplified and compressed pulses are shown in Fig. 2b.

The value of the carrier-envelope offset frequency f_{ceo} of the amplified pulse train can be adjusted by manipulating the signal used for the stabilization in the

Fig. 1 (continued) pre-chosen carrier-envelope-offset frequency f_{ceo} by a feed-forward scheme using an AOFS (see text for details) using a CEP4 stabilization unit (Spectra-Physics, Femtolasers). The main amplification to 80 W is accomplished by two amplification steps in Yb-doped large-pitch fibers (LPF) [23] and two grating compressors. Subsequently, a rod-type LMA fiber and chirped mirrors are used to generate 0.7-μJ, ~30-fs pulses. The f_{ceo} is then measured with a $f-2f$ interferometer and stabilized by a feed-back scheme using frequency modulation of the feed-forward AOFS signal by the servo output of a proportional-derivative integral (PDI) lock box. **b** AOFS signal generation for stabilization to $f_R = f_{ceo} = 11$ MHz. The frequency of the common feed-forward signal $f_{rep} + f_{ceo,osc}$ (1) is reduced by mixing with a 11-MHz reference signal and then stabilized by frequency modulation. The error signal is derived from a heterodyne detection scheme with the beatnote (2). For characterization, the phase noise is recorded by sampling this signal with an oscilloscope. The phase of the RF signals is indicated along the *black lines*. **c** Stabilization scheme for $f_{ceo} = 0$ MHz. The $f-2f$ interferometer is modified to a spectrally resolved configuration. The stabilization is achieved by first decreasing the AOFS carrier frequency by $f_R = 22$ MHz and afterwards increasing it by the same synchronized frequency with a small modulation derived from the interferogram (2), containing the feedback signal. For stability reasons, a phase shifter (Φ) was also applied in a faster servo loop in order to fine-tune the lock. The *inset* shows an interferogram with subtracted background

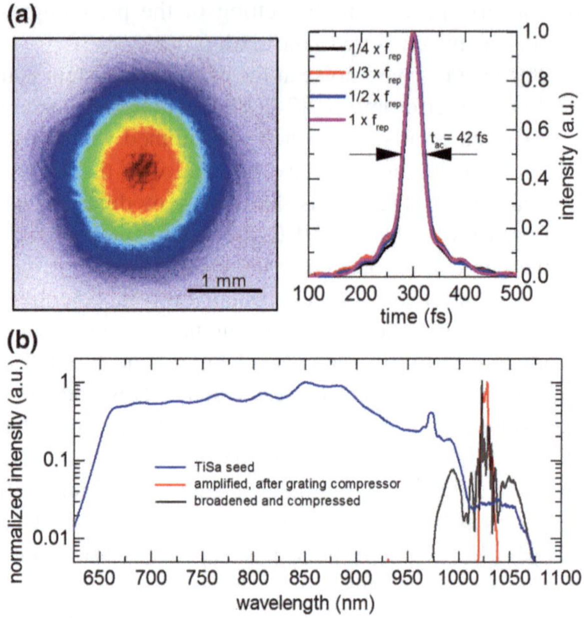

Fig. 2 **a** Beam profile and autocorrelation traces at the output of the MOPA system. Different repetition rates show the same autocorrelation trace duration of 42 fs, corresponding to 32 fs for a Gaussian pulse. **b** Spectra of the seed (*blue*), amplified (*red*) and compressed pulses (*black*)

AOFS, which consists of three contributions, as shown in Fig. 1b, c. Signal 1 is the feed-forward signal consisting of the repetition frequency and the carrier-envelope offset frequency of the oscillator ($f_{\mathrm{rep}} + f_{\mathrm{ceo,osc}}$) which, in normal operation, shifts f_{ceo} to 0 MHz in the -first diffraction order of the AOFS. The value of f_{ceo} can optionally be tuned by mixing an additional frequency with an RF mixer and a signal generator (signal 2). The combined signal is then band-pass filtered, amplified and fed to the AOFS (signal 3).

3 Results

3.1 Carrier-Envelope-Offset Frequency of 11 MHz

In a first configuration, we set the driving frequency of the AOFS to $f_{\mathrm{rep}} + f_{\mathrm{ceo,osc}}$- 11 MHz (a value of 71 MHz, the free running oscillator offset frequency $f_{\mathrm{ceo,osc}}$ is kept around 8 MHz by means of intracavity wedges and crystal temperature). This is depicted in Fig. 1b. Using this scheme, any value of the carrier-envelope-offset frequency of the output pulse train can be chosen, provided that the AOFS is suited for that frequency.

The feed-back loop is achieved by a heterodyne detection scheme using the detected beatnote and a synchronized RF generator operating at the same frequency as the generator used to shift the feed-forward signal. This provides the error signal for the feed-back servo loop. The output of the proportional-derivative-integral

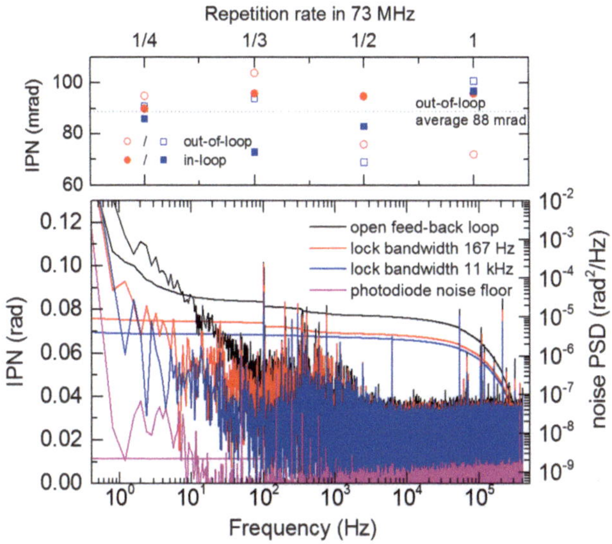

Fig. 3 Phase noise measurements (integrated phase noise, IPN, and noise power spectral density, PSD) for measurement times up to 2.5 s and for different pulse repetition rates. *Black line* noise at open feed-back locking loop using only the feed-forward stabilization. *Red and blue* closed-loop operating with a bandwidth of 167 Hz and 11 kHz, respectively. The upper panel summarizes the performance at different repetition rates: 18.5, 24.7, 37 and 74 MHz

(PDI) controller is used to modulate the AOFS driving frequency. The beatnote used in this work was at least 30 dB above the noise floor at 100 kHz resolution and video bandwidth in order to achieve a tight lock and a low-noise measurement. The results for different repetition rates are shown in Fig. 3. The lower part of the figure depicts a typical out-of-loop measurement including the photodiode noise floor (magenta), the phase noise with disabled feed-back but still enabled feed-forward locking servo (black) and two different closed-loop operation modes with locking bandwidths of 167 Hz (red) and 11 kHz (blue). The MOPA system shows an integrated phase noise (IPN) of 76 mrad and 69 mrad, respectively, at closed-loop operation and 145 mrad at open-loop operation in the frequency range of 0.4 Hz to 400 kHz, corresponding to an improvement by roughly a factor of 2 compared to state-of-the-art phase-stable multi-MHz Yb-based systems [10, 29]. The upper part is a summary of the in-loop and the out-of-loop measurements for different repetition rates (18.5, 24.7, 37 and 74 MHz), showing that within the measurement accuracy of ∼10% the IPN is independent on the pulse picking factor. Furthermore, the values for the out-of-loop results are very close to those for the in-loop measurements, showing that the lock does indeed correct for real phase noise and that these are not measurement artifacts, as the lock would add noise in the out-of-loop configuration in such a case. To the best of our knowledge, the average value of (88 ± 7) mrad RMS phase noise integrated between 0.4 Hz and 400 kHz for an out-of-loop measurement is the best value achieved so far with ultrashort high-power multi-MHz-repetition-rate systems. The 7-mrad error is the standard deviation at the repetition rate of 18.5 MHz, taken for several measurements over the course of half an hour, and it mainly depends on the alignment of the interferometers and on changes in the environment. Average open-loop measurements (black trace) yield a phase noise around 2 times higher than the closed-loop operation, demonstrating the benefit of the additional feed-back loop. The 0.4- to 10-Hz fluctuation in the open-loop trace originates from within the laser system, e.g., from small mechanical jitter in the grating compressors and/or air fluctuations. One noise source in the measurement is in the >100 kHz frequency range. Most probably, this originates from intensity fluctuations which are transformed into phase noise during the octave generation in the $f-2f$ interferometer as reported in [27]. Another noise source, located at 100 Hz, was nearly completely suppressed by a housing around the interferometers, meaning that the noise originates from air fluctuations and mechanical vibrations in the interferometer and not from the laser system. The long-term stability is shown in Fig. 4, in-loop (magenta) and out-of-loop (blue). The performance over 7.5 min was investigated for different configurations. The values of 60-mrad in-loop and 225-mrad out-of-loop for the long-term IPN are both unparalleled for high-power multi-MHz systems. The real out-of-loop phase noise is even lower as the statistical fluctuations (<1 Hz, stemming from,

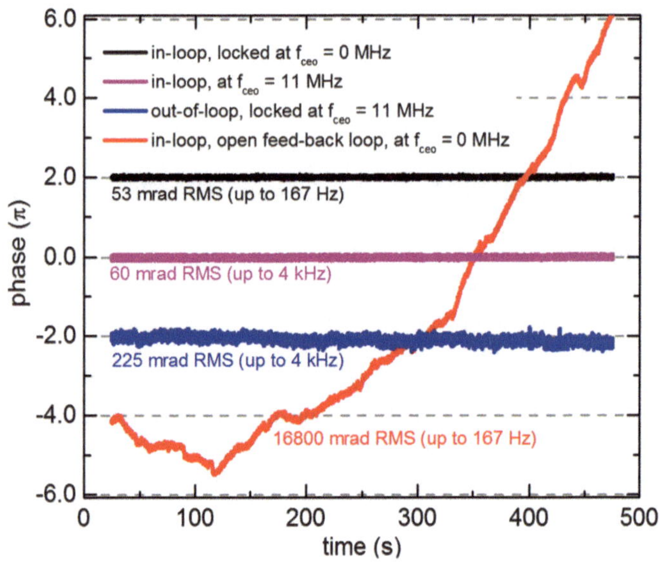

Fig. 4 Long-term phase noise measurement at 18.5 MHz repetition rate for $f_{ceo} = 0$ MHz and $f_{ceo} = 11$ MHz. *Black and red* in-loop measurements for the 0-MHz case, the *red* one was taken with an open feed-back loop (i.e., only feed-forward stabilization). *Magenta and blue* in-loop and out-of-loop phase fluctuations when $f_{ceo} = 11$ MHz. The traces are offset for clarity

e.g., beam pointing, delay jitter in the interferometer, air fluctuations, etc.) in the two interferometers add up and increase the value drastically. For comparison, one of the most stable state-of-the-art 10-kHz Ti:Sa systems reaches a 140-mrad out-of-loop long-term stability at an average power of 8 W [30]. For high-power ultra-short-pulse multi-MHz systems, previous out-of-loop measurements were taken on the order of a few seconds only and yielded a value of 270 mrad (out-of-loop jitter integrated in the bandwidth between 1 Hz and 500 kHz) [10]. For uncompressed pulses, 140 mrad (out-of-loop jitter integrated in the bandwidth between 0.4 Hz and 100 kHz) was achieved [29], albeit this measurement was taken before a grating compressor.

3.2 *Carrier-Envelope-Offset Frequency of 0 MHz*

By adapting the AOFS signal generation and phase detection in the $f - 2f$ interferometer, the value of the comb offset can also be set to 0 MHz, as shown in Fig. 1c. The main difference in the $f - 2f$ interferometer setup is the spectrally resolved detection and the signal processing. The interferometer delivers an interferogram which is digitally evaluated and fed to a PDI controller. The feed-back loop consists of two stages, a faster loop using a 360° phase shifter [30] and a slower one to compensate for drifts over several radians (see red trace in Fig. 4). The slow drift compensation is accomplished by decreasing the AOFS frequency by 22 MHz and afterwards increasing it by the same synchronized frequency with a small additional modulation, corresponding to the feed-back signal. Due to the digital evaluation of the interferogram, the lock was limited to a bandwidth of 167 Hz. The previous measurements at 11-MHz offset frequency, however, showed that the bandwidth of 167 Hz is sufficient as there was no major difference when using a locking bandwidth of 11 kHz (Fig. 3). The results for the long-term measurements are shown in Fig. 4. As no second spectrally resolved 0-MHz $f - 2f$ interferometer was available, only in-loop measurements can be shown. However, the 11-MHz measurements corroborate that the out-of-loop performance should be on the same order. The necessity of a slow drift compensation for more than 2 rad presents itself in the open-loop trace (red) as the phase drift is larger than 30 rad during 7.5 min. This drift is not possible to lock with only a single 360° phase shifter as demonstrated in [30]. The in-loop closed-loop measurement is shown in black and manifests an extraordinary value of 53 mrad within a bandwidth of 167 Hz which is consistent with the in-loop value of 60 mrad measured at $f_{ceo} = 11$ MHz (magenta trace, bandwidth 4 kHz). This operation is stable over tens of minutes.

4 Conclusion and Outlook

In conclusion, we have demonstrated an Yb-fiber MOPA system seeded by a Ti:Sa oscillator, achieving excellent phase stability. The repetition rate of the 0.7-μJ, ~32-fs output pulses can be reduced from 74 MHz (fundamental repetition rate of the oscillator) by an integer factor of up to 4 without affecting the pulse parameters. In principle, the repetition rate can be further reduced at the same average power by stronger pulse stretching in the CPA [31] and the pulse duration can be further compressed with alternative, pulse-energy-scalable schemes [9–14]. For all repetition rates, an integrated phase noise of 60-mrad in-loop and 225-mrad out-of-loop was measured in a bandwidth from 2 mHz to 4 kHz. These results have a threefold significance for ultrafast laser technology. First, they demonstrate that the residual phase noise introduced by Yb-fiber amplifiers (including CPA), even for an amplification factor of > 50 dB, is low enough to be readily compensated by a simple feed-back scheme. On the other hand, they indicate that for a high phase stability, a feed-back loop is necessary for the compensation of these phase fluctuations. Second, these results prove that feed-forward-stabilized Ti:Sa oscillators are well suited as low-phase-noise seeders for Yb:fiber systems. Third, they show that a fast AOM pulse picker after the low-power frontend mostly preserves the phase stability irrespective of the picking factor.

The laser system demonstrated here opens up new opportunities in several fields. For instance, the zero-offset-frequency pulse train can be used to drive HHG in a suitable femtosecond enhancement cavity [16, 17] for the generation of attosecond pulse trains and, ultimately, for the generation of isolated attosecond pulses at multi-MHz repetition rates [18]. This will dramatically decrease the measurement times in photoelectron emission microscopy and spectroscopy, in particular allowing for the study of plasmonic fields with a unique combination of nm-scale spatial resolution with sub-femtosecond temporal resolution [20]. Furthermore, the adjustable repetition rate allows for direct studies of cumulative effects such as those observed in HHG in gases at high repetition rates [16]. Another application which will tremendously profit from this MOPA is field-resolved detection of broadband infrared pulses [21], employing the 7-fs pulses generated by the Ti:Sa oscillator for electro-optical sampling with a lock-in detection at half the fundamental oscillator repetition rate [22]. Furthermore, the phase-stable output could be employed as a frequency comb or to drive the generation of ultrabroadband supercontinua, which can be used to obtain high-energy field-synthesized femtosecond transients [32].

Acknowledgements Open access funding provided by Max Planck Society. This work was funded by the Fraunhofer and Max-Planck cooperation within the project "MEGAS" and the Deutsche Forschungsgemeinschaft (DFG) (MAP). I.P. acknowledges funding by European Research Council (Grant Agreement No. 617173, "ACOPS"). We thank Ferenc Krausz and Fabian Lücking for fruitful discussions.

References

1. T. Udem, R. Holzwarth, T.W. Hänsch, Nature **416**, 6877 (2002)
2. F. Krausz, M. Ivanov, Rev. Mod. Phys. **81**, 163 (2009)
3. M. Hentschel, R. Kienberger, Ch. Spielmann, G.A. Reider, N. Milosevic, T. Brabec, P. Corkum, U. Heinzmann, M. Drescher, F. Krausz, Nature **414**, 509 (2001)
4. T. Fuji, J. Rauschenberger, A. Apolonski, V.S. Yakovlev, G. Tempea, T. Udem, C. Gohle, T.W. Hänsch, W. Lehnert, M. Scherer, F. Krausz, Opt. Lett. **30**, 332 (2005)
5. F. Lücking, A. Assion, A. Apolonski, F. Krausz, G. Steinmeyer, Opt. Lett. **37**, 2076 (2012)
6. S. Backus, R. Bartels, S. Thompson, R. Dollinger, H.C. Kapteyn, M.M. Murnane, Opt. Lett. **26**, 465 (2001)
7. H. Fattahi, A. Alismail, H. Wang, J. Brons, O. Pronin, T. Buberl, L. Vámos, G. Arisholm, A.M. Azzeer, F. Krausz, Opt. Lett. **41**, 1226 (2016)
8. C. Jauregui, J. Limpert, A. Tünnermann, Nat. Photonics **7**, 861 (2013)
9. P. Russbueldt, D. Hoffmann, M. Hofer, J. Löhring, J. Luttmann, A. Meissner, J. Weitenberg, M. Traub, T. Sartorius, D. Esser, R. Wester, P. Loosen, R. Poprawe, IEEE J. Sel. Top. Quantum Electron. **21**, 447–463 (2015)
10. O. Pronin, M. Seidel, F. Lücking, J. Brons, E. Fedulova, M. Trubestskov, V. Pervak, A. Apolonski, Th Udem, F. Krausz, Nat. Commun. **6**, 6988 (2015)
11. S. Hädrich, M. Kienel, M. Müller, A. Klenke, J. Rothhardt, R. Klas, T. Gottschall, T. Eidam, A. Drozdy, P. Jojart, Z. Várallyay, E. Cormier, K. Osvay, A. Tünnermann, and J. Limpert: Optics Letters (submitted)
12. T. Eidam, F. Röser, O. Schmidt, J. Limpert, A. Tünnermann, Appl. Phys. B **92**, 9 (2008)
13. K.F. Mak, M. Seidel, O. Pronin, M.H. Frosz, A. Abdolvand, V. Pervak, A. Apolonski, F. Krausz, J.C. Travers, P. St, J. Russell Opt. Lett. **40**, 1238 (2015)
14. S. Hädrich, M. Krebs, A. Hoffmann, A. Klenke, J. Rothhardt, J. Limpert, A. Tünnermann, Light Sci. Appl. **4**, e320 (2015)
15. A. Cingöz, D.C. Yost, T.K. Aliison, A. Ruehl, M.E. Fermann, I. Hartl, J. Ye, Nature **482**, 68 (2012)
16. H. Carstens, M. Högner, T. Saule, S. Holzberger, N. Lilienfein, A. Guggenmos, C. Jocher, T. Eidam, D. Esser, V. Tosa, V. Pervak, J. Limpert, A. Tünnermann, U. Kleineberg, F. Krausz, I. Pupeza, Optica **3**, 366 (2016)
17. S. Holzberger, N. Lilienfein, M. Trubetskov, H. Carstens, F. Lücking, V. Pervak, F. Krausz, I. Pupeza, Opt. Lett. **40**, 2165 (2015)
18. M. Hoegner et al., Generation of isolated attosecond pulses with enhancement cavities—a theoretical study, in preparation (2016)
19. O. de Vries, T. Saule, M. Plötner, F. Lücking, T. Eidam, A. Hoffmann, A. Klenke, S. Hädrich, J. Limpert, S. Holzberger, T. Schreiber, R. Eberhardt, I. Pupeza, A. Tünnermann, Opt. Express **23**, 19586 (2015)
20. M.I. Stockman, M.F. Kling, U. Kleineberg, F. Krausz, Nat. Photonics **1**, 539 (2007)
21. I. Pupeza, D. Sánchez, J. Zhang, N. Lilienfein, M. Seidel, N. Karpowicz, T. Paasch-Colberg, M. Znakovskaya, W. Pescher, V. Schweinberger, M. Pervak, E. Fill, O. Pronin, Z. Wei, F. Krausz, A. Apolonski, J. Biegert, Nat. Photonics **9**, 721 (2015)
22. C. Riek, D.V. Seletskiy, A.S. Moskalenko, J.F. Schmidt, P. Krauspe, S. Eckart, S. Eggert, G. Burkard, A. Leitenstorfer, Science **350**(6259), 420 (2015)
23. J. Limpert, F. Stutzki, F. Jansen, H.-J. Otto, T. Eidam, C. Jauregui, A. Tuennermann, Light Sci. Appl. **1**, e8 (2012)
24. J. Reichert, R. Holzwarth, Th Udem, T.W. Hänsch, Opt. Commun. **172**, 59 (1999)
25. I. Ricciardi, S. Mosca, P. Maddaloni, L. Santamaria, M. De Rosa, P. De Natale, Opt. Express **21**, 14618 (2013)
26. Z. Chang, Appl. Opt. **45**, 8350 (2006)
27. C. Li, E. Moon, H. Mashiko, H. Wang, C.M. Nakamura, J. Tackett, Z. Chang, Appl. Opt. **48**, 1303 (2009)

28. M. Seidel, G. Arisholm, J. Brons, V. Pervak, O. Pronin, Opt. Express **24**, 9412 (2016)
29. K. Yang, W. Li, X. Schen, J. Zhao, D. Bai, Q. Ru, H. Zeng, Laser Phys. **24**, 125101 (2014)
30. F. Lücking, V. Crozatier, N. Forget, A. Assion, F. Krausz, Opt. Lett. **39**, 3884 (2014)
31. M. Müller, M. Kienel, A. Klenke, T. Gottschall, E. Shestaev, M. Plötner, J. Limpert, A. Tünnermann, Opt. Lett. **41**, 3439 (2016)
32. H. Fattahi, H.G. Barros, M. Gorjan, T. Nubbemeyer, B. Alsaif, C.Y. Teisset, M. Schultze, S. Prinz, M. Haefner, M. Ueffing, A. Alismail, L. Vámos, A. Schwarz, O. Pronin, J. Brons, X.T. Geng, G. Arisholm, M. Ciappina, V.S. Yakovlev, D. Kim, A.M. Azzeer, N. Karpowicz, D. Sutter, Z. Major, T. Metzger, F. Krausz, Optica **1**, 45 (2014)

Quantum Electrodynamics, High-Resolution Spectroscopy and Fundamental Constants

Savely G. Karshenboim and Vladimir G. Ivanov

Abstract Recent progress in high-resolution spectroscopy has delivered us a variety of accurate optical results, which can be used for the determination of the atomic fundamental constants and for constraining their possible time variation. We present a brief overview of the results discussing in particular, the determination of the Rydberg constant, the relative atomic weight of the electron and proton, their mass ratio and the fine structure constant. Many individual results on those constants are obtained with use of quantum electrodynamics, and we discuss which sectors of QED are involved. We derive constraints on a possible time variation of the fine structure constants and m_e/m_p.

1 Introduction

The frequency measurements are the most accurate ones. For a while, such a statement was related to only microwave measurements, but with the recent discovery [1–3] and practical realization of the frequency comb technology the precision laser spectroscopy also became a base of the most accurate results. What do we need the most accurate results for? There are few areas where such an accuracy is crucial. One of them is the determination of the values of the fundamental constants, which helps to build new standards and to check certain fundamental theories. The most accurate fundamental theory is quantum electrodynamics. It is the most accurate in the sense that we are capable to produce very accurate results for the values, which can be measured accurately as well. The most accurate of those results have been achieved for free particles (an electron or a

This article is part of the topical collection "Enlightening the World with the Laser" - Honoring T. W. Hänsch guest edited by Tilman Esslinger, Nathalie Picqué, and Thomas Udem.

S.G. Karshenboim (✉)
Max-Planck-Institut für Quantenoptik, 85748 Garching, Germany

Pulkovo Observatory, Saint Petersburg, Russia, 196140
e-mail: savely.karshenboim@mpq.mpg.de

V.G. Ivanov
Pulkovo Observatory, Saint Petersburg, Russia, 196140

D. Meschede et al. (eds.), *Exploring the World with the Laser*,
https://doi.org/10.1007/978-3-319-64346-5_15

muon), for two-body (hydrogen, deuterium, muonic hydrogen, etc.) and three-body (ordinary and antiprotonic helium) atomic system and for three-body molecular systems (the HD^+ ion). Sometimes the accuracy of a calculation is limited by non-QED contributions, but in many situations the computation of well-defined theoretical expressions has rather practical than fundamental limitations.

Another application of precision measurements may allow us to look for a violation of certain symmetries or 'natural expectations.' In principle, those tests do require no accurate theory of the involved atoms. The symmetry principles claim by themselves that the results of various experiments on the same quantity should be exactly the same under certain conditions. One of the questions of this type is whether the values of the fundamental constants vary with time.

In this paper, we briefly give the overview of determination of the fundamental atomic constants, such as the Rydberg constant, the fine structure constants and the electron-to-proton mass ratio, their application to test quantum electrodynamics (QED) and the laboratory constraints on their variations. In all these fields, the high-resolution spectroscopy plays an important role.

2 Different Sectors of QED

Many individual values of the fundamental constants are determined with use of quantum electrodynamics (see Table 1). For some of them, the QED-related values (i.e., the values, in the interpretation of which QED is heavily involved) dominate.

What is QED in practice? Quantum electrodynamics is a theory described by a Lagrangian. No actual non-trivial value can be calculated exactly. One has to apply a perturbation theory. Fortunately, the expansion parameter $\alpha \simeq 1/137$ is small and the perturbative expansion is efficient in many situations. That is the QED Lagrangian that generates the rigorous rules how to calculate. It determines both the 'unperturbed approximation' and all the perturbation corrections. There is no separate theory of the vacuum polarization, electron's self-energy, electron–positron annihilation, etc. The details of all those phenomena are determined by the same Lagrangian.

Graphically, the QED perturbation theory is customarily described by the Feynman diagrams (see Fig. 1). The diagrams are not just pictures that describe the QED phenomena qualitatively. They are a complete quantitative description of the perturbation theory. So-called Feynman rules give rigorous and straightforward instructions to generate the diagrams for a required order in α and relate them to certain analytic expressions.

The QED Lagrangian for electrons and photons is of the form

$$\mathcal{L} = -\frac{1}{4}F_{\mu\nu}F^{\mu\nu} + \overline{\psi}\left(i\frac{\partial}{\partial x_\mu}\gamma_\mu - eA^\mu\gamma_\mu - m_e\right)\psi\,, \qquad (1)$$

Table 1 Individual values of some fundamental constants obtained by means of precision spectroscopy and/or based on QED calculations

Constant	Values	Method	References
R_∞	$10\,973\,731.568\,517(82)\ \mathrm{m}^{-1}$	Hydrogen and deuterium spectroscopy, two-body QED	[4–13]
m_e	$5.485\,799\,090\,70(17) \times 10^{-4}$ u	Bound g factor (C^{5+}), two-body QED	[14]
m_e	$5.485\,799\,0919(46) \times 10^{-4}$ u	Bound g factor (Si^{13+}), two-body QED	[15]
m_e	$5.485\,799\,0932(29) \times 10^{-4}$ u	Bound g factor (C^{5+}), two-body QED	[16, 17]
m_e	$5.485\,799\,0957(42) \times 10^{-4}$ u	Bound g factor (O^{7+}), two-body QED	[14, 18]
m_p/m_e	1836.152 674(5)	pHe spectroscopy, three-body QED	[19, 20]
m_p/m_e	1836.152 6736(23)	pHe spectroscopy, three-body QED	[20, 21]
m_p/m_e	1836.152 6781(98)	HD^+ spectroscopy, three-body QED	[22–24]
m_p/m_e	1836.152 6815(41)	HD^+ spectroscopy, three-body QED	[24–26]
m_p/m_e	1836.152 6695(53)	HD^+ spectroscopy, three-body QED	[24, 27–29]
m_p/m_e	1836.152 6695(24)	h / m(Rb), two-body QED	[30]
m_p/m_e	1836.152 6657(76)	h / m(Cs, UCB-13), two-body QED	[31]
α^{-1}	137.035 999 160(33)	$g_e - 2$, free-particle QED	[32, 33]
α^{-1}	137.035 998 27(50)	$g_e - 2$, free-particle QED	[33, 34]
α^{-1}	137.035 998 67(27)	h / m(Cs, UCB-13), two-body QED	[31]
α^{-1}	137.035 998 996(85)	h / m(Rb), two-body QED	[30]
α^{-1}	137.036 0000(11)	h / m(Cs, St-02), two-body QED	[35]
α^{-1}	137.035 999 139(31)	Helium fine structure, two-electron QED	[36, 37]
α^{-1}	137.036 0013(79)	Muonium HFS, two-body QED	[38]
r_p	0.8759(77) fm	Hydrogen and deuterium spectroscopy, two-body QED	[4–13]
r_p	0.840 29(55) fm	Lamb shift in muonic hydrogen, two-body QED	[39, 40]

where we use the standard notation for the operators of the electron–positron and electromagnetic fields. (Similar contributions with muons, τ-leptons and quarks are to be added.) The rules which relate an element of the Feynman graphs to an appropriate analytic expression are presented in Fig. 2. (The complete rules also include the instructions to build the diagrams with their combinatoric coefficients and the all involved integrations.)

The Lagrangian (1) is not complete. It completely describes neither all the interactions of the electrons and photons nor their actual electromagnetic

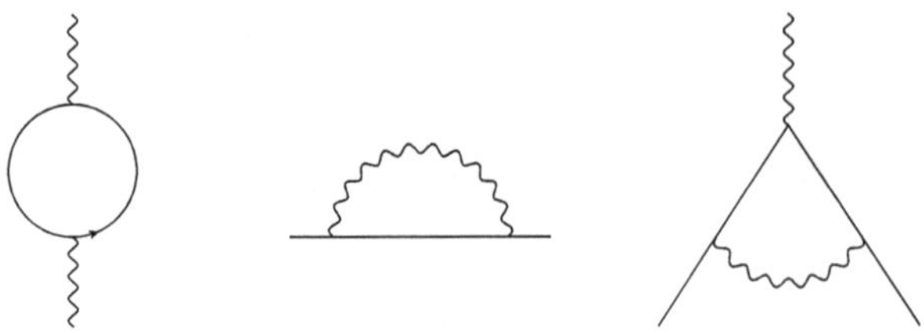

Fig. 1 Leading contributions to the electron's self-energy, vacuum polarization and correction to the vertex, including the anomalous magnetic moment, are derived from the Feynman rules following the well-defined instructions

photon propagator	$\frac{-ig_{\mu\nu}}{k^2+i0}$	
electron propagator	$\frac{i}{p_\mu\gamma^\mu-m+i0}$	
vertex	$ie\gamma_\mu$	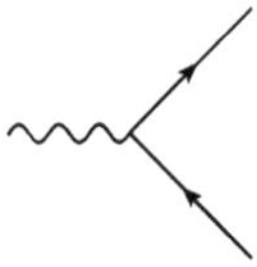

Fig. 2 Feynman rules completely describe the free propagation of a photon and an electron and the coupling of the electromagnetic field to an electron. Anything else, being more complicated, is to be expressed in the terms of these three building blocks. The photon propagator is given here in the Feynman gauge

interactions. But it produces the most of the contributions with a pretty high accuracy. Introducing the muons and τ-leptons into the consideration is not a problem. There is also a missing part, which comes from the weak interactions; however, it is small and well under control.

A more important problem for certain quantities is with the involvement of the strong interaction. The photons interact with all charged particles, including the hadrons, such as protons, neutrons and pions. In principle, we also have an *ab initio* theory to deal with that. That is quantum chromodynamics (QCD). It deals with quarks and gluons, not with actual particles, such as protons. Similarly to QED, it does not have any exact solution, and one needs to build an efficient perturbation theory to calculate the actual values. Unfortunately, in contrast to QED, the would-be expansion parameter is not always small and in the particular case of the proton structure the perturbative approach fails.

The characteristics of the proton and nuclei related to the strong interaction, knowledge of which is necessary for atomic physics, cannot be calculated *ab initio*. Instead, one has to measure the required parameters. Since the photons are coupled to the hadrons, the latter appear as the intermediate states in non-hadronic problems, such as the anomalous magnetic moment of the electron and muon, and muonium hyperfine structure. That is another kind of the non-QED contributions to the properties of photons and leptons. To deal with this problem, one has to use various experimental data as the input. However, the contribution is relatively small, because it involves the QED perturbation parameter, α, since we have to produce intermediate states, and the contribution is also suppressed at the low momentum transfer because the hadronic intermediate states are 'heavy.'

One has to remember that it is not the magnitude of the contribution which matters, but its uncertainty. The small hadronic contribution is known by themselves not very accurately, and therefore, they often dominate in the uncertainty budget of various QED-related values. The hadronic vacuum polarization and hadronic light-by-light scattering contributions are responsible for the theoretical uncertainty in the calculation of the anomalous magnetic moment of the muon, while the accuracy of the calculation of the energy levels of simple atoms suffers from the finite nuclear size effects and the nuclear polarizability.

The problem of the hadronic contributions is presented in QED calculations for free and bound particles. The QED of free particles basically deals with the precision theory and experiment on the anomalous magnetic moments and various phenomena, such as scattering, which are studied not very accurate. The bound-state physics offers an access to many other QED-related values, but it involves additional problems. Deriving the analytic expressions for bound systems is less straightforward. One has to solve the bound-state problem which effectively means a certain summation of an infinite number of the diagrams. (The very existence of the bound states is not a perturbative effect!)

There are a number of efficient approaches on how to solve the bound-state problem and systematically build the perturbation series. For instance, efficient two-body approaches include the effective Dirac equation (EDE) [41–43] (see also [44–46]) and the nonrelativistic quantum electrodynamics (NRQED) [47]. The former was successfully applied to hyperfine splitting in muonium, the Lamb shift in hydrogen (see, e.g., [45]) and muonic hydrogen (see, e.g., [40]). The latter has been efficiently used for the bound electron g factor in the two-body systems [48–50] and also for three-body systems, such as helium [37], antiprotonic helium [20] and molecular ion HD^+ [23, 26, 28, 29]. All these approaches may have a different efficiency or limited applicability (e.g., only to two-body systems), but all of them are well defined and allows a straightforward prescription for calculations of the pure QED part, with the accuracy limited only by the efficiency of our practical ability to deal with complicated expressions.

There are many QED-related values of the fundamental constants (see Table 1). It is important that the related experiments 'probe' different sectors of QED. We have already mentioned that there are free-QED tests, such as determination of the anomalous magnetic moments of the charged leptons, and there are bound-state QED experiments with atomic spectra or other characteristics of simple atomic systems.

The reference to the 'different sectors' means a much more broad variety. Already, anomalous magnetic moments of the electron and muon present two different QED sectors. They have different regimes of calculations, and therefore, the same QED contributions have a different relative importance for the electron and muon. The muon's theory has a number of enhanced QED contributions. The electron's and muon's theories of the anomalous magnetic moment also have the different sources of the uncertainty and the different importance of the non-QED contributions. The latter cover both 'standard physics' and 'new physics.' The standard physics contributions, originated from the hadronic vacuum polarization and hadronic light-by-light scattering, are of marginal importance for the electron and are crucial for the uncertainty for the muon. Speaking about the new physics issues, one has to remember that despite the essentially higher accuracy in the electron's case, that is the muon's $g-2$ that delivers the constraints on the new physics, to which it is more sensitive.

The bound-state QED values also belong to few distinguished sectors. Most of bound-state QED with precision opportunities is with the two-body hydrogen-like atomic systems. First, we have to mention that the theory of muonium and positronium is marginally sensitive to the hadronic effects and can be treated as a pure QED one, since the contributions beyond QED are small and well under control. In contrast to that, theory of any atomic state with ordinary nucleus is sensitive to the nuclear structure. An exception is the helium fine structure, where the specific transition under study is insensitive to the nuclear effects.

The second crucial feature which is important to distinguish the different sectors is whether we study nuclear-spin-independent or nuclear-spin-dependent intervals, or roughly speaking, whether we study the Lamb shift or the hyperfine-structure (HFS) interval. As it is known from the basic physics, the magnetic interaction is a relativistic correction to the Coulomb interaction in order $(v/c)^2$. It increases with the momentum ($(v/c)^2 = (p/mc)^2$), and therefore, the hyperfine interval by itself relates in general to higher momenta contribution than the [nuclear-spin-independent] Lamb shift. That means that in the HFS theory the relativistic corrections are enhanced.

The nucleus in two-body systems is used to be slow and we can neglect its motion which corresponds to the external-field approximation. The recoil contributions are the correction to such a picture. The hyperfine interactions enhance contributions of the relative atomic velocity and therefore of the higher momenta. The higher characteristic momentum is, the faster is the nuclear motion. As a result, the recoil correction is enhanced in the HFS theory in comparison with the Lamb shift.

A QED theory is usually an expansion in three important QED parameters: α (the number of QED loops), $Z\alpha$ (the strength of the Coulomb interaction) and m/M (the mass ratio for the orbiting particle and the nucleus). The series in all three parameters is not an analytic one (see, e.g., [51]). The most of the metrological applications of the two-body systems are currently with the nuclear charge $Z = 1$ (muonium, hydrogen, deuterium, muonic hydrogen). The only exception is the theory of the g factor of bound electron. The actual experimental data are for a hydrogen-like carbon ($Z = 6$), oxygen ($Z = 8$) and silicon ($Z = 14$). The higher Z values require more accurate calculations of the expansion in $Z\alpha$ (preferably an exact one for main contributions), but it also makes the overall $Z\alpha$ dependence more smooth than around $Z = 1$.

One more sector of QED which recently becomes of the 'metrological' importance is the three-body QED. While in the case of the two-body QED one may apply different approaches, such as the effective Dirac equation, the choice for the three-body QED is rather limited. In the most important metrological cases, we have to deal with weakly charged particles (the value of their charge is equal to single or double electron charge). Such systems are ordinary helium, antiprotonic helium (one of the electrons is substituted for an antiproton) and the HD^+ molecular ion. In all these situations, the natural starting point is the three-body Schrödinger equation. The main problem is not QED, but a numerical solution of the unperturbed three-body Schrödinger equation and the subsequent application of the three-body Green's functions to the QED perturbation series.

To conclude the consideration of QED, let's discuss a certain conceptual difference between quantum mechanics in general and quantum electrodynamics. Quantum mechanics, similarly to the Newtonian laws, is a powerful framework which allows us to successfully describe the motion of free particles and their motion under interactions. However, the interactions themselves are introduced as an input. They are not much constraint by quantum mechanics and can be very different. Quantum electrodynamics is a more specific theory, which describes all the effective interactions in a certain generic way. There is not many parameters of those interactions. In the case of the point-like particles, the only interaction parameter is the charge. For the not point-like particles, we may consider higher momenta, such as the [anomalous] magnetic moment and various structure functions. It is important that those parameters are the same for any electromagnetic interaction. That is why we can measure the structure effects in the scattering and apply to the bound states, or to restore some functions (such as the vacuum polarization contribution into the photon propagator) through their imaginary part, determined through very different phenomena.

The only remaining possible variety for the details of the interactions lies in the unknown details of the interaction parameters and structure functions of the hadronic objects. That means that we can discuss a modification of the structure parameters, but we cannot modify a particular electromagnetic contribution once the parameters are fixed. 'Testing QED' basically means to check the efficiency and reliability of our practical approaches to calculate the QED-related values and of our knowledge of the phenomenological parameters involved through the hadronic sector. We do not really test the QED Lagrangian. (It is incomplete and we do not doubt whether it is correct by itself, but we may wonder about the effects beyond it.) With such a general picture, we understand that while comparing two QED-related values, the values used to have the uncertainties from completely different sources. There is no way that a certain shortage of QED could consistently shift all the value of say, α or m_e/m_p, in the same direction by the same value. Comparing two QED-related values, we check the practical side of QED.

3 Determination of Fundamental Constants

To perform a QED test, one can compare theory versus experiment for a certain quantity. Alternatively, one can compare a specific individual value of a certain quantity extracted with help of a particular QED theory with other values extracted with or without QED. The atomic fundamental constants are the Rydberg constant R_∞, the fine structure constant α, the electron-to-proton mass ratio m_e/m_p. As one can see from Table 1, there are a number of QED-involved values of those quantities. Below we gave the overview of determination of those fundamental constants. (There are also some other atomic constants, such as the g factor of the proton, which we do not discuss here.)

Already for over than forty years, the determination of the practically important fundamental constants has been performed under the auspice of the CODATA[1] task group on fundamental constants [4, 52–57]. The progress in their determination is summarized in Figure 3.

We use in the plot a special color legend to distinguish different methods of the determination of the fundamental constants to stress to what extend the experiment are based on the high-resolution spectroscopy and theory on QED. Painting the colors we relate them to the most recent results. In the past, different methods were sometimes important. The color legend is as following: the blue points are for all the high-resolution [optical] spectroscopic results, while the red ones are for the QED-related values obtained without any application of the precision optical measurements. The rest is given in black. Indeed, some values such as the atomic weights of the electron and proton $A_r(e), A_r(p)$, and the electron-to-proton mass ratio m_e/m_p are strongly correlated. In this case, the color reflects only the 'direct' determinations.

One can note that the accuracy of the determination of the various constants in Fig. 3 is very different. The least-square-adjustment (LSA) of the values of the fundamental constants is realized as a 'single-shot' code which minimizes the related χ^2 value and determines all the values of the constants simultaneously. However, the large spread in the accuracy of the constants indicates that the evaluation of the constants can be realized step by step, working first on more accurate values. In practice, even applying a code which evaluates the data simultaneously, the data analysis in the CODATA LSA evaluation is done set by set (see discussion, e.g., in [58, 59, 60, 61]).

The accuracy of the CODATA-recommended values of the most important atomic constants and their correlation coefficients are presented in Table 2. In practice, the correlation coefficient smaller than 0.01 can be ignored. Besides, they are not really stable and by including marginally important data one could shift the values of the small coefficients significantly, but would not change the central values and the uncertainty of the constants. There is no controversy in that:

[1]The Committee on Data for Science and Technology of the International Council for Science.

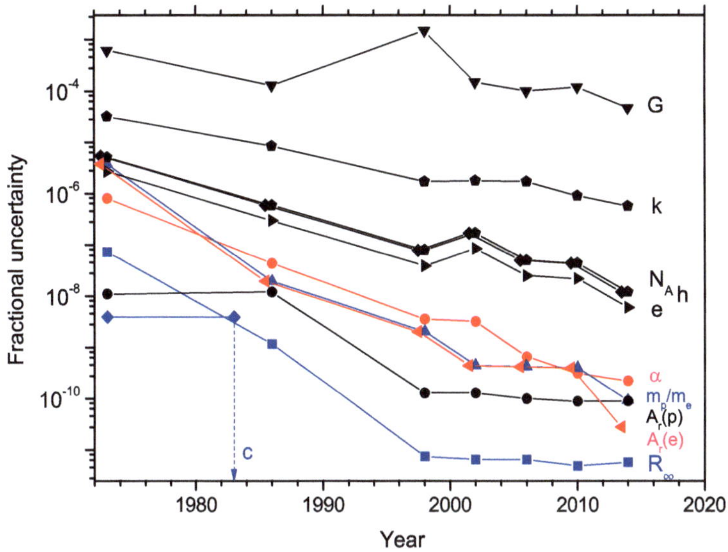

Fig. 3 Progress in CODATA's LSA determination of values of fundamental physical constants [4, 52–57]. (The 1983s point for the speed of light is the same value as of 1973; in 1983, the *meter* was redefined and the numerical value of c has got the zero uncertainty)

Table 2 Fractional uncertainty ($u(r)$) and the correlation coefficients (ρ_{ij}) for a part of the hierarchy basis set

Value	u_r	R_∞	$A_r(e)$	$A_r(p)$	α
R_∞	$[5.9 \times 10^{-12}]$	1	0.0024	−0.0058	0.0005
$A_r(e)$	$[2.9 \times 10^{-11}]$	0.0024	1	−0.0011	−0.0090
$A_r(p)$	$[9.0 \times 10^{-11}]$	−0.0058	−0.0011	1	0.0014
α	$[2.3 \times 10^{-10}]$	0.0005	−0.0090	0.0014	1

The numerical values of some coefficients are presented in [4]; the other can be found at the http://physics.nist.gov/cuu/Constants/index.html of the NIST's Physical Reference Data Web site

the uncertainty used to be a two-digit one; the rounding of the central values is related to the unity of the last digit of the uncertainty. In other words, the effects which are essentially below of one percent of the uncertainty are washed out by the rounding of the uncertainty and the central value. The correlation at the level of one percent roughly speaking affects the combined uncertainty of the sum or a product of two values with the same uncertainty roughly at the level of one percent. If the uncertainty is not comparable, the influence is even smaller. Nevertheless, for some specific reasons the correlation coefficients are given [4] with four digits (see, e.g., Table 2). Giving the small coefficients with superficial accuracy makes their numerical values 'fragile.' While affecting the values of the small coefficients this way, one could not change the order of their magnitude. They would remain very small and negligible.

Those constants form a part of the basis set for the determination of the related constants and their combinations such as m_e/m_p are not included into the table. (A complete atomic basis set should also include the deuteron mass, proton magnetic moment, proton and deuteron charge radii).

The appropriate choice of the basis set for LSA is very important for the presentation of the results. Dimensional constants, such as h, e, and m_e, have never really entered the atomic equations for the measurable quantities on their own, only in combinations. That is why their determination cannot be done purely by means of atomic physics and laser spectroscopy and does not have a comparable accuracy. (One may note that their definition involves the kilogram and indeed a laser could not help to related the international prototype [of the kilogram] to an atom, without involving macroscopic experiments which are to be much less accurate.) The choice of the basis which explicitly takes the hierarchy of the accuracy into account allows to present the atomic constants separately from the less accurate ones, the determination of which involve macroscopic measurements. The most important non-atomic fundamental constants of the hierarchy basis set are the Planck constant h, the Boltzmann constant k and the Newtonian gravitation constant G (see, e.g., [58, 60] for detail).

It has happened that the accuracy of determination of the electron and proton mass in the unified atomic mass units evolved very much both absolutely and relatively in respect of each to the other. The situation of 2002–2014 is summarized in Fig. 4. While in LSA-2002–2010 [55–57] the relative atomic weight of the electron $A_r(e)$ was known worse in fractional units than the atomic weight of the proton $A_r(p)$, the current situation (as in LSA-2014 [4]) is opposite.

The changes also took place for the comparison of the accuracy of $A_r(e)$ and α.

One can see that while the relation between the individual uncertainties changed very much, the accuracy of the 'direct' determinations (where the values of $A_r(p), A_r(e)$, and α were determined without any correlations (or with marginal correlations) of each with the others) essentially superseded the accuracy of the 'indirect' determinations (where the measured quantity was a certain combination of these three constants, such as $\alpha^2 A_r(e)$ and $A_r(e)/A_r(p) = m_e/m_p$). The 'indirect' determinations play a marginal role in the central values and accuracies of all three constants. In the meanwhile, the direct determinations are correlated only marginally. That allows us to determine $A_r(p), A_r(e)$, and α near independently even if their accuracy is comparable.

4 Determination of the Atomic Fundamental Constants

The determination of the fundamental constants can be done set by set (cf. [4, 62]) and here we focus on the atomic constants, which form a few sets of the most precision data. A precision description of atomic properties is based on QED and allows to express all the QED contributions in the terms of e, m_e and h and the

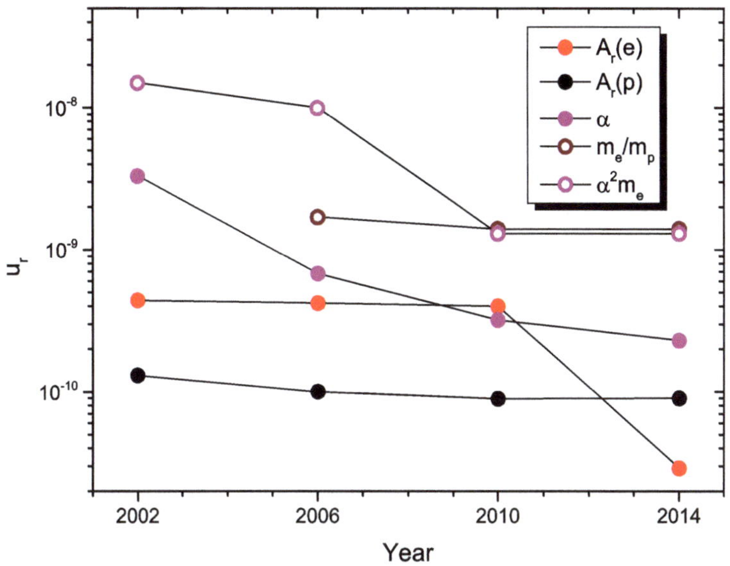

Fig. 4 Evolution of the accuracy of the determination of $A_r(e), A_r(p)$, and α [4, 55–57]: the uncertainty of 'direct' (*closed circles*) and 'indirect' (*open circles*) determinations. The 'direct' determination are the experiments targeting a measurement of the value of one of those three constants with a near zero correlation to the others. The 'indirect' determination stands here for the experiments targeting a measurement of a combination of those constants. The accuracy is given for the most accurate experiments of each kind. All the indirect determinations involve high-resolution spectroscopy and QED

involved nuclear masses. These constants are dimensional, their dimensions involve the kilogram, which is defined as the mass of the international prototype kept at BIPM. Therefore, their numerical values in the SI system need macroscopic measurements to bridge the distance between the atomic phenomena and the prototype. Such cannot be measured with a high accuracy. In the meantime, the most accurately measured properties of many atomic objects and free particles are the frequencies. The combination of e, m_e and h which has such a dimension is cR_∞, where

$$R_\infty = \frac{e^4 m_e}{8\epsilon_0^2 h^3 c} \,. \tag{2}$$

The energy of the nonrelativistic Bohr levels is expressed in the terms of chR_∞ and that value does depend on e (the levels are for a Coulomb bound system) and h (the discrete spectrum is a quantum effect) and does not depend on c (that indicates the nonrelativistic origin of the energy levels). Already the nonrelativistic Schrödinger levels involve rather the reduced mass

$$m_r = \frac{m_e}{1 + \frac{m_e}{M}},$$

than the electron mass, and therefore, we need to know the appropriate nuclear masses M to describe to the energy levels. The relativistic corrections are proportional to an effective $(v/c)^2$ value of the electron, which is expressed in hydrogen-like atoms in the terms of the fine structure constant

$$\alpha = \frac{e^2}{4\pi\epsilon_0\hbar c}. \tag{3}$$

In the meanwhile, both parameters, m_e/M (for the nuclei of interest) and α, contribute only to small corrections and may be determined separately (see Table 1). That practically disentangles their determination from the determination of the Rydberg constant (see, e.g., [58, 60]). However, still there are corrections due to the nuclear structure which relate the determination of the proton and deuteron charge radius to the determination of R_∞.

The state of the art in the determination of the proton charge radius is summarized in Fig. 5. We present there the results utilized in the LSA-2014 evaluation [4]. The closed circles are for the data utilized in the LSA-2014 [4] and the open circles are for those which are not taken into consideration. (Note, some results have been analyzed in [4], but on the base of the analysis they have been dismissed. Therefore, they were taken into consideration, but not included in the eventual set of data, the χ^2 value of which was minimized. The muonic hydrogen value is one of the values which was considered, but not included.) The data point in the bottom of the plot used to be not mention in the CODATA's evaluations (see [4, 56, 57]). It

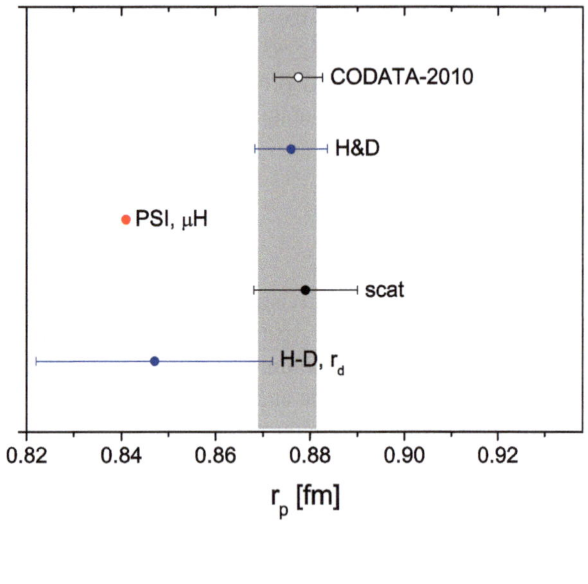

Fig. 5 Determination of the proton rms charge radius r_p. We follow the notation of [4]. The gray vertical belt is for the CODATA-recommended value [4]

presents a link between the determination of the proton and deuteron radii. The link has been established through the determination of the hydrogen–deuterium isotopic shift in the $1s-2s$ transition [7–10]. Because of the different accuracy in the direct determinations of the proton and deuteron radii (from the scattering), it is rather to produce an accurate value of r_d. Nevertheless, the link works in both directions, and therefore, we can use a scattering value of the deuteron radius [63] to obtain a value of the proton radius (see, e.g., [60]), which is shown in the plot.

The atomic results with their references have been already presented in Table 1. The main spectroscopic result is obtained by evaluation of the data of 24 different transitions in [4]). The references in Table 1 are given to the most accurate optical measurements [5–13], which dominate in the evaluations. The individual contributions are discussed in detail in [64].

The scattering value in Fig. 5 is obtained from [65]. It comes from a certain overall evaluation of the world data on absolute and relative measurements of the elastic electron–proton scattering.

We see that there is a serious problem in the consistency of the data on the determination of the proton charge radius. The decision taken for LSA-2014 was to exclude the value from the muonic hydrogen from the final evaluation [4]. Our personal opinion is that the accuracy of the determination of r_p by spectroscopic means (see [64]) and by the scattering is overestimated and the correct result is close to the muonic hydrogen one.

The electron-to-nucleus mass ratios, and in particular the electron-to-proton one, are crucial parameters of atomic calculations. However, the direct experimental access to the properties of free particles makes the parameter less important than the direct determination of the relative atomic weights of electron ($A_r(e)$), proton ($A_r(p)$), and some other nuclei.

The determination of the atomic weight of the proton is performed within a kind of atomic mass evaluation, which is denoted in our plots as 'AME.' The AME procedure is somewhat similar to the LSA evaluation. It deals with various mass radios determined as ratios of the related ion cyclotron frequency. One may find the ratio of the proton mass to the carbon atom mass through many relations, and the simplest of them is the ratio of the proton cyclotron frequency and the related frequency of a carbon ion (usually, C^{6+} or C^{5+}). There are many other possibilities. One can compare the proton mass and the deuteron mass and the deuteron cyclotron frequency with that for a carbon ion. To compare the proton and deuteron masses, one can deal with cyclotron frequencies of the related atomic ions, or to use molecular ions. The overall least-square evaluation of those data delivers us the AME value (see [66, 67] for details). However, the plotted results are not from the standard AME, performed by AMDC[2] [66, 67]. The problem is that the standard AMDC's AME are not well synchronized with the CODATA's LSA evaluations, and when the time of the LSA-2014 came there had been some recent data

[2]The Atomic Mass Data Center of the Nuclear Data Services of the International Atomic Energy Agency.

available, such as [68, 69]. They could not be taken into account in AME-2012 [66, 67], but they have been accommodated with the LSA-2014 procedure [4].

The determination of the atomic weight of the electron used to stay apart from the AME procedure. The current situation with the determination of the electron-to-proton mass ratio and the atomic weight of the electron is summarized in Figs. 6 and 7. The plots look like twins. The most accurate results for the both quantities are presented (with a magnification) in Fig. 8. The only difference is the accuracy of the MPIK-15 value and the width of the CODATA's belt. (We follow the notation of [4, 56, 57]; the results 'GSI-XX' and 'MPIK-XX' are in fact the results of the same

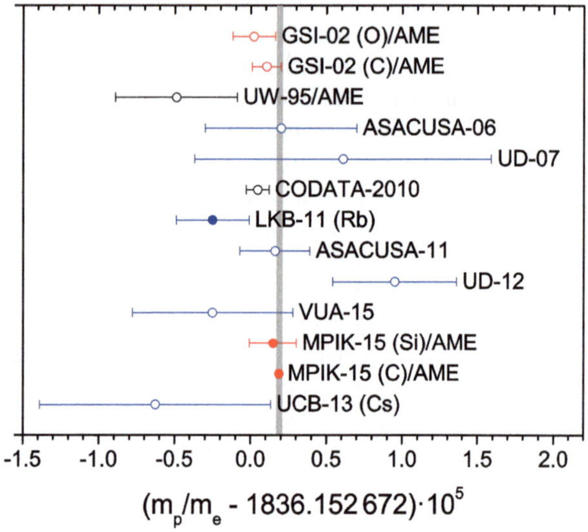

Fig. 6 Determination of the electron-to-proton mass ratio m_p/m_e. We follow the notation of [4]. The gray vertical belt is for the CODATA-recommended value [4]

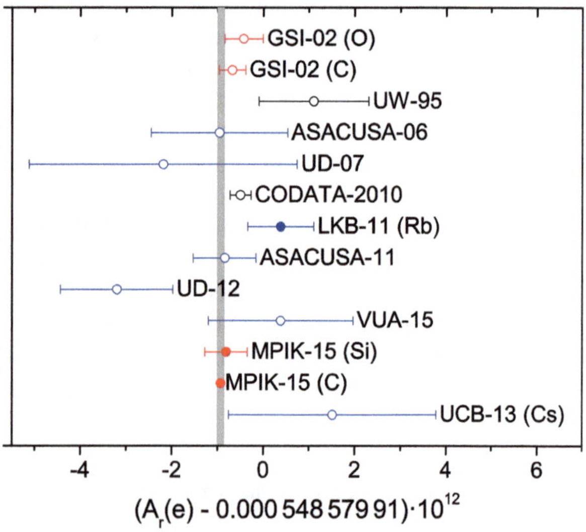

Fig. 7 Determination of the relative atomic weight of the electron $A_r(e)$. We follow the notation of [4]. The gray vertical belt is for the CODATA-recommended value [4]

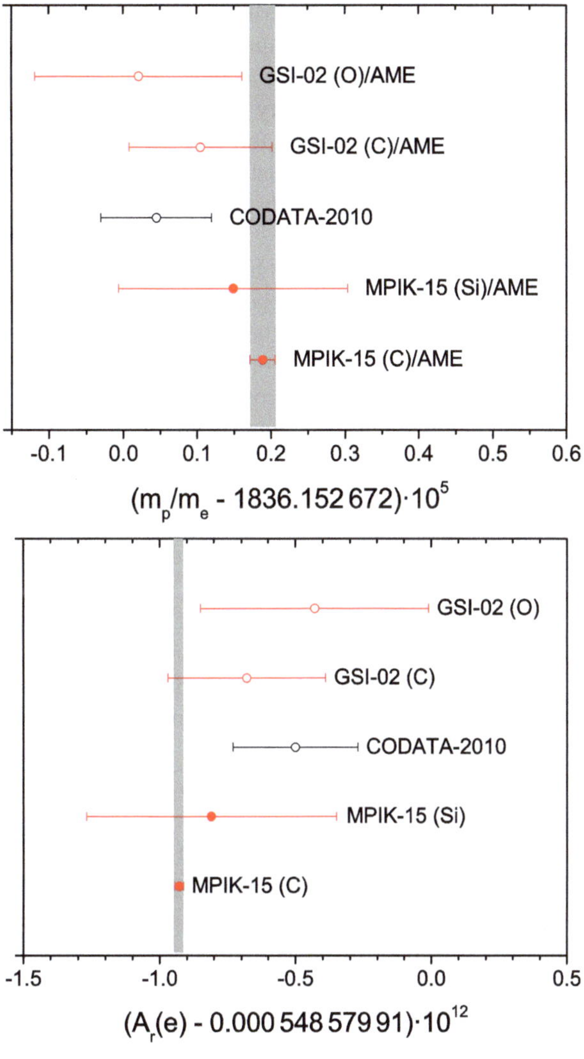

Fig. 8 Determination of m_e/m_p and $A_r(e)$: the most accurate results. We follow the notation of [4]. The gray vertical belts are for the CODATA-recommended value [4]

experiment performed at the University of Mainz (Johannes Gutenberg-Universität Mainz). The collaboration includes the University scientists and researchers from GSI and MPIK.) The mentioned difference with the fractional accuracy reflects an important property of the data. All the data on the atomic weight of the electron, but the MPIK-15 (C) result, are in fractional units less accurate than the determination of the atomic weight of the proton, as well as the results of the determination of m_e/m_p. As a result, an interpretation of the results on m_e/m_p in the terms of $A_r(e)$ is produced without loosing of accuracy, as well as the interpretation of $A_r(e)$ results in the terms of m_e/m_p. The accuracy of $A_r(p)$ does not enter the interpretation.

The only exclusion is the MPIK-15 (C) data points. The determination of $A_r(e)$ reached the accuracy higher than that of $A_r(p)$ and the interpretation of the $A_r(e)$ result as a result for $m_e/m_p = A_r(e)/A_r(p)$ has a larger fractional uncertainty, determined by the uncertainty of the determination of the atomic weight of the proton. The CODATA's belt follows the MPIK-15 (C) values which are far more accurate than the rest of the data.

In the plots in Fig. 8 with the most accurate results, the only values derived from the experiments on the g factor of bound electron in various hydrogen-like ions are presented. They are obtained in the same laboratory, by the same method and the theory is very correlated. Nevertheless, their comparison delivers very important information about the internal consistency in both theoretical and experimental approaches. It is important that many corrections (both experimental and theoretical) strongly depend on the nuclear charge Z or the ion charge $Z-1$. With the corrections behaving as Z^5 (which is the actual behavior of some uncalculated higher-order QED contributions), even a comparison of the carbon ($Z = 6, Z^5 = 7776$) and oxygen ($Z = 8, Z^5 = 32\,768$) is a sensitive test. The silicon experiment ($Z = 14$) is even much more sensitive, than the oxygen one, that is why the recent LSA evaluation [4] included the most accurate carbon result and the silicon one, but not the oxygen's.

One may like to 'absolutely' test the theory of the bound electron g factor [70, 71]. That requires a value of $A_r(e)$ or m_e/m_p obtained beyond the bound g factor's experiments. Those experiments are not that accurate as the g factor's ones (see Figs. 6, 7). Their weighted average value is

$$\left.\frac{m_p}{m_e}\right|_{\text{ave. excl. } g} = 1836.152\,6721(13) \tag{4}$$

with the fractional uncertainty $u_r = 7.1 \times 10^{-10}$. The related values of χ^2 is 9.8 which is consistent with the expected value of 7 ± 3.7 for seven degrees of freedom. Such a value is not caused by any particular outliner, but rather by the overall spread of the data. To consider it seriously, one has to discuss the possible correlations (since the eight results were obtained by only different four methods) and which data are to be used and with what uncertainty. As we mention, we use here the early ASACUSA's result in its original version [19], while the reconsideration of the averaging procedure over individual transitions in [56] allows to reduce the uncertainty. We use here the UCB value [31], which was dismissed in [4], being considered there as a preliminary result with an incomplete evaluation of the systematic effects. That values has a marginal effect on the calculation of the average. One also may suggest extending the fractional uncertainty of the average by a factor of $\sqrt{9.8/7} \simeq 1.2$. While the accuracy of the separate measurements does not exceed 1 ppb, the overall uncertainty is below it. The farther progress in the area is expected.

One can also find a related value of the relative atomic weight of the proton

$$A_r(p)\big|_{\text{ave. excl. } g} = 1.007\,276\,467\,36(95), \tag{5}$$

which is characterized by $u_r = 9.4 \times 10^{-10}, \chi^2 = 4.1$ for four degrees of freedom.

The experiments, others than that with the bound g factor, used to calculate these two averages are of four types. Some of them involve QED, but different sector of it, some are non-QED experiments. The most of the values in Figs. 6 and 7 have been numerously described in the literature and were used in part in the previous LSA evaluations, but some need more comments.

1. The UW value [72] is obtained by a comparison of the electron's and carbon ion's cyclotron frequencies and does not require any QED for its evaluation (except for small corrections due to the carbon ionization energy with a marginal uncertainty).
2. The one more method is due to the recoil spectroscopy of Rb and Cs. The former value from LKB [30] was used in [4], while the Cs ones (from UCB) [31] was dismissed (see discussion there). However, neither is mentioned in [4] as a partial contribution to m_e/m_p or $A_r(e)$. The reason is that the experiments rather determine the value of $\alpha^2 A_r(e)$, which presents a link between the determination of $A_r(e)$ and α. It used to be discussed in the LSA papers [4, 57] rather as a way to determine α (see Fig. 11). Nevertheless, once the link is 'activated' it also produces a value of $A_r(e)$ and m_e/m_p (cf. [58, 60]).
3. The ASACUSA data points [19, 21] were used in [56, 57]. They are obtained by a comparison of the transitions between certain Rydberg antiprotonic states in the antiprotonic helium (see [73, 74]). In the most recent evaluation, they are neglected because they are not really compatible with the determination of $A_r(e)$ and $A_r(p)$. However, they are important for our purpose to make an 'outside' test of the bound electron g factor values. Here, we use the original average over the experimental lines following [19], rather than that made in [56].
4. The current situation of the HD^+ results is somewhat similar to the antiprotonic helium ones. They are ignored for a different reason. Two UD papers present data sufficient to extract the related m_e/m_p values, but still they have not given explicitly those values. The most recent value, the VUA one, was published too late to be included into the LSA-2014. The results are tabulated here in Table 3.

We have to mention the use of so-called sensitivity coefficients. They appear here because the problem in the leading approximation for the three-body molecular system was solved numerically by minimization of the energy. The solution was made for a certain value of m_e/m_p. To extract the real value of the ratio, we have to equalize the theoretical and experimental value assuming a different value of m_e/m_p. To adjust the calculated numerical results for the transition frequencies to new numerical values of m_e/m_p one has to know how the leading term is sensitive to the input parameters. The related derivatives are referred to as the sensitivity coefficients. Similar coefficients are also necessary for the antiprotonic helium. The latter are calculated in [75] and were quoted in details in [56, 57, 74].

Table 3 Proton-to-electron mass ratio from the precision measurement of the rovibrational transitions of HD^+

Data	Transition $(\nu,L) \to (\nu',L')$	f_{exp} [MHz]	f_{th} [MHz]	$(m_p/m_e)^{(0)}$	$\partial \ln f_{th}/\partial \ln (m_p/m_e)$	m_p/m_e	References
UD-07	$(0,2) \to (4,3)$	214 978 560.6(5)	214 978 560.88(7)	1836.152 6726(9)	−2.28	1836.152 6781(98)	[22]
UD-12	$(0,0) \to (1,1)$	58 605 052.00(6)	58 605 052.14(2)	1836.152 6726(9)	−2.05	1836.152 6815(41)	[25]
VUA-15	$(0,2) \to (8,3)$	383 407 177.4(4)	383 407 177.15(2)	1836.152 6725(8)	−2.66	1836.152 6695(53)	[27]

In the case of UD-07 the theory follows [23], for UD-12 it follows [26]. The use of the more recent theoretical corrections [28, 29] could only marginally shift the related theoretical values. The values of the ratio m_p/m_e are not given there. They are derived here with the sensitivity coefficients from [24]. The value of m_d/m_p did not change much during the CODATA evaluations. It was known for LSA 2002, 2006 and 2010 more accurate than the value of m_e/m_p and the transitions are less sensitive to it. In [27], the derived value of the proton-to-electron mass ratio is presented as well as the theoretical value and the sensitivity coefficient. The theoretical value was taken from [28, 29], while the sensitivity coefficient was calculated in [27]. The value for the sensitivity coefficient for all three transitions deviates from −2 mostly because of anharmonic contributions to the effective potential within the Born–Oppenheimer approximation

Two kinds of these values (from the electron's cyclotron frequency and from the recoil spectroscopy) produce results for $A_r(e)$ and to reach an m_e/m_p value one has to combine them with $A_r(p)$ from AME. Two other kinds of the experiments (from antiprotonic helium and HD$^+$ spectroscopy), on contrary, deliver us m_e/m_p and to transform it into $A_r(e)$ one has to use a value of $A_r(p)$. Since $A_r(p)$ is known more accurately than any individual values of $A_r(e)$ and m_e/m_p mentioned in this paragraph (with the g factor's experiment are excluded), it does not much matter which of these two values, $A_r(e)$ and m_e/m_p, are produced 'directly' and which are 'derived' afterward, when we study their value. To find the average value of m_p/m_e in (4), we use all of these results, while finding $A_r(p)$ in (5) only values, which directly deliver m_e/m_p have been used.

The issue of correlations between the determination of the atomic weights of the electron and proton and their mass ratio is illustrated in Fig. 9. As we already mention, the determinations of the electron mass (in the atomic mass units) and of the fine structure constants is also correlated, which is shown in Fig. 10. We consider the both plots as an illustration and for the sake of simplicity we use

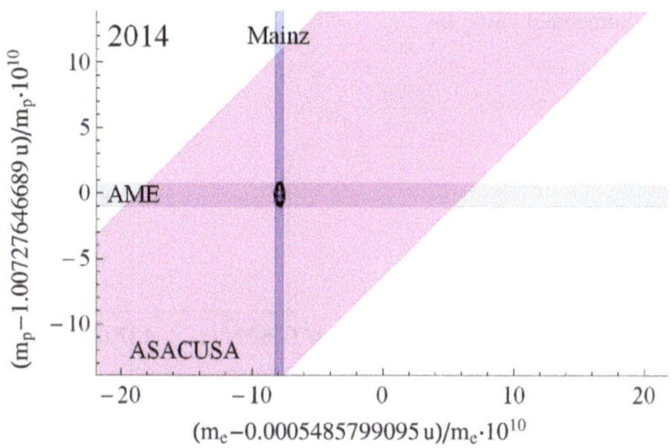

Fig. 9 Determination of atomic weights of the electron and proton: the correlations. To make the plot simple we include the average values of $A_r(e)$ and $A_r(p)$, rather than all of them, while the indirect measurement is presented by the more recent ASACUSA's result [21]

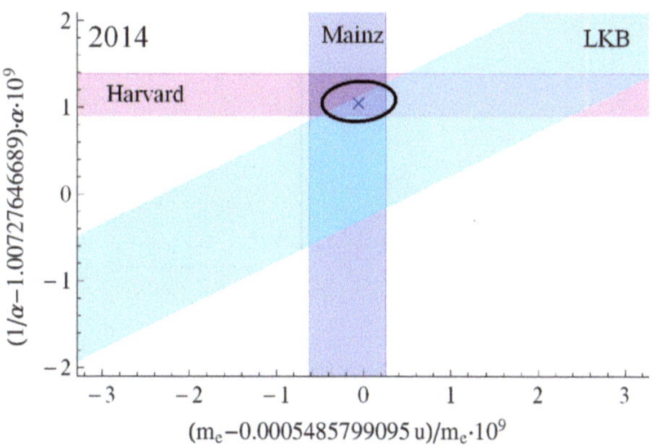

Fig. 10 Determination of m_e and α: the correlations. The results for the fine structure constant, for the electron's mass and for $\alpha^2 A_r(e)$ are presented by their dominant contributions

average values for the direct measurements of $A_r(e)$ (which coincides with the Mainz value combined of GSI's and MPIK's ones) and $A_r(p)$, and of α (which coincides with the Harvard value), while the indirect measurements are presented by a single measurement. In the case of the determination of m_e/m_p that is the more recent ASACUSA experiment [21], while for $\alpha^2 A_r(e)$ that is the LKB recoil experiment [30]. In the both cases the 'direct' uncorrelated determinations dominate (cf. Fig. 4), which means that the 'indirect' values have a marginal effect on the final results, but still are of interest as an independent test of the direct values.

To complete the overview of the determination of the atomic fundamental constants, we consider below the current situation with the fine structure constant. The state of the art is summarized in Fig. 11. We see in the plot a number of QED-related values; some of them also involve high-resolution spectroscopy.

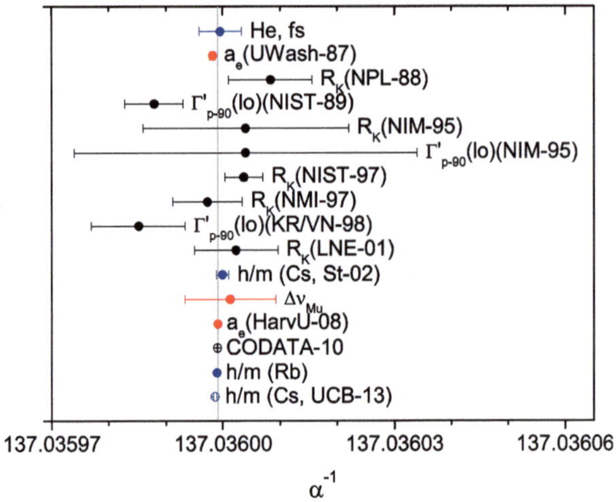

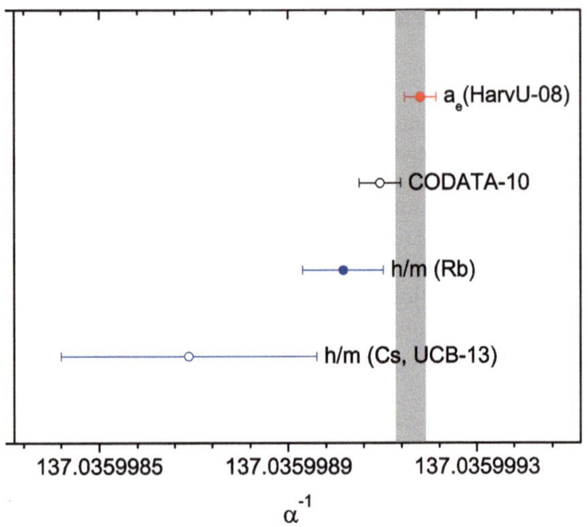

Fig. 11 Determination of the fine structure constant α. We follow the notation of [4]. The gray vertical belts are for the CODATA-recommended value [4]

Table 4 QED theory for the anomalous magnetic moment of the electron

Order in α/π	Number of Feynman graphs	Coefficient c_i
1	1	1 / 2
2	7	$-0.328\,478\,444\cdots$
3	72	$1.181\,234\cdots$
4	891	$-1.912\,06(84)$
5	12 672	7.79(34)

The most accurate value of α is a QED-based one. It is derived from study of the anomalous magnetic moment of the electron. The QED expression for the later can be written down as

$$\frac{g_e - 2}{2} = c_1\frac{\alpha}{\pi} + c_2\left(\frac{\alpha}{\pi}\right)^2 + c_3\left(\frac{\alpha}{\pi}\right)^3 + c_4\left(\frac{\alpha}{\pi}\right)^4 + c_5\left(\frac{\alpha}{\pi}\right)^5 + \cdots \qquad (6)$$

The value of the coefficients and the number of the Feynman diagrams to calculate, which is very impressive by itself, are presented in Table 4. The most advanced calculation was performed in [33]. There are also small corrections beyond QED, but they are well under control. To produce a theoretical prediction from so advanced QED theory one has not only to calculate the related QED coefficients with a high accuracy, but also to have in hands an accurate value of the fine structure constant α. As well as in the case of the determination of m_e/m_p we see the importance of 'less accurate' values. Those values may play not a great role in the determination of the values of a constant, because the weighted average is close to the most accurate value. However, those 'less accurate' values are crucial for the reliability of the determination and for a check of the most accurate values and therefore of the underlying theories.

The 'second in accuracy' value of α is the $h\,/\,m$ (Rb) value [30]. It is obtained by measuring a recoil due to the absorbtion of a photon and its subsequent induced re-emission by an atom. The experiment allows to determine the atomic mass in the frequency units, i.e., to measure the frequency equivalent mc^2/h of the mass of interest. The most accurate experiment of this type has been performed on rubidium-87 atoms at LKB. Combining the frequency equivalent of the mass with the relative atomic weight of rubidium [66, 67] and with a value of the Rydberg constant, one can find

$$\frac{h}{m(^{87}\mathrm{Rb})\,c^2}A_r(^{87}\mathrm{Rb})\,\frac{\alpha^2 m_e c^2}{2hc} = \frac{1}{2c}\alpha^2 A_r(e), \qquad (7)$$

which establishes a link between $A_r(e)$ and α. We have already used this link while discussing m_e/m_p and $A_r(e)$ (see Figs. 6, 7). Since the second in accuracy value is rather a link between two constants, than the determination of α alone, let's return to the determination of $A_r(e)$ and consider which QED theory is included into the various values.

Table 5 CODATA 2014-recommended values of some atomic fundamental constants [4]

Constant	CODATA 2014 values	u_r
R_∞	10 973 731.568 508(65) m $^{-1}$	$[5.9 \times 10^{-12}]$
m_e	$5.485\,799\,090\,70(16) \times 10^{-4}$ u	$[2.9 \times 10^{-11}]$
m_p	1.007 276 466 879(91) u	$[9.0 \times 10^{-11}]$
m_p/m_e	1836.152 673 89(17)	$[9.5 \times 10^{-11}]$
α^{-1}	137.035 999 139(31)	$[2.3 \times 10^{-10}]$

The most accurate atomic constant considered above (see Table 5) is indeed the Rydberg constant. As we see from the plot in Fig. 5 its determination (which mirrors the determination of the proton radius) is not free of the problems. The problems with the determination of the Rydberg constant are clearly seen at the level of 6 ppt, which is the fractional uncertainty of the Rydberg constant. This level is not the level at which we check 'essential QED'[3]. The leading term of the energy levels, which are used for the determination of the Rydberg constant, is the result of a quantum mechanics calculation. The essential QED contributions, which are the Lamb shift ones, are small. Combining the experimental and theoretical results, we check them at the level of a few ppm at best.

The rivalry experiment on muonic hydrogen is capable to check the Lamb shift theory at the level of a part in 10^3, being limited by our knowledge of the proton size, which value should be taken from outside as an input.

The accuracy of the determination of the other constants, where a value of the Rydberg constant is involved, ranges from one part to few parts in 10^9. At this level of accuracy, the determination of the Rydberg constant is perfectly consistent. The inconsistency in the determination of R_∞ 'does not propagate' into the determination of the other values. In particular, the value of the Rydberg constant is involved into the interpretation of the data of the recoil spectroscopy (to determine $\alpha^2 A_r(e)$) and spectroscopy of antiprotonic helium and HD^+ molecular ion (to determine m_e/m_p).

The second in accuracy atomic constant is the relative atomic weight of the electron. Its uncertainty is about 30 ppt. The QED theory is deeply involved. And again, that is not the level of accuracy to check the essential QED. The leading term $g_0 = 2$ follows from the Dirac equation for a free electron and the largest contributions still come from quantum mechanics (from the Dirac theory of a bound electron) and from theory of free electron g factor. We will discuss the second issue in the next paragraph and now we look for essential bound-state QED contributions, which are an analog of the Lamb shift. Such a correction is only about 0.1 ppm for the carbon ion. Therefore, that is equivalent to a test of the Lamb shift theory at the level of 300 ppm.

[3]The QED is the 'overall' theory of the electromagnetic interactions of the electron. That means that the Schrödinger and Dirac energy levels in hydrogen could be (and have numerously been) rederived from QED. They are indeed a part of QED, but they can be found in more simple approaches than QED.

The most accurate test of the essential QED theory is the one with the anomalous magnetic moment of the electron. The uncertainty of this test is 0.6 ppb and that is neither uncertainty of the $g-2$ experiment [32] nor of the theory [33]. That is the uncertainty of the 'second in accuracy' value of α from recoil spectroscopy [30], which is the most important non-$g - 2$ one. As we mention, the one of the leading contributions to the g factor of the bound electron is due to the $g-2$ of the free electron. The contribution is 10^{-3} of the g factor and that means that the bound g factor experiments can check it at the level of 3×10^{-8}, which is indeed not competitive with the free g factor test.

The value of α from $g-2$ is used to determine $A_r(e)$ in combination with the recoil experiments [30, 31].

The remaining values of the $A_r(e)$ and/or m_e/m_p includes a QED-free value of $A_r(e)$ [70] and values from two kinds of experiments with three-body systems. Both the experiments with antiprotonic helium [19, 21] and with HD^+ ion [22, 25, 27] need for their interpretation the knowledge of the Rydberg constant, which is based on the theory of the Lamb shift. The theory of the three-body systems is similar to that in hydrogen in the terms of which effects to take into account, but it differs in one very important respect. The leading term in the three-body case is to be solved numerically. The precision tests of such a theory are equivalent a very rough test of the essential QED (similar to the Lamb shift theory) at the level of few parts in 10^4. But they are a severe test of our ability to calculate those QED effects on the top of the numerically solved three-body problem.

Therefore, we see that

1. the determination of the fundamental constants touches very different sectors of QED;
2. the results are rather consistent, except for a certain problem in the determination of the proton charge radius and the Rydberg constant;
3. the accuracy of the QED tests is not the same as the accuracy of the related theory and experiment because the leading terms often come from quantum mechanics, and not from essential QED;
4. the most accurate test of the essential QED contributions (beyond the standard quantum mechanics) are with the anomalous magnetic moment; in this sense it is save to use QED at the level of accuracy up to 1 ppb.

Overall we see that the QED values from the different sectors are rather in a fair agreement as well as the QED and non-QED values.

5 Laboratory Constraints on Variation of Atomic Constants

The QED theory suggests that the basic parameters of the theoretical Lagrangian (1) are the time-independent constants. The standard model suggests that the vacuum average v of the Higgs field determines the electron mass. Cosmological evolution could in principle allow a certain variation of v. In other words, the standard model does not exclude any variation of the constants, and in particular, of m_e/m_p. Below we consider to what extend the atomic fundamental constants can vary at the present epoch from the pure experimental point of view.

As was pointed out by Savedoff [76] and Thompson [77], one may set an experimental constraint on the time variation of fundamental constants by

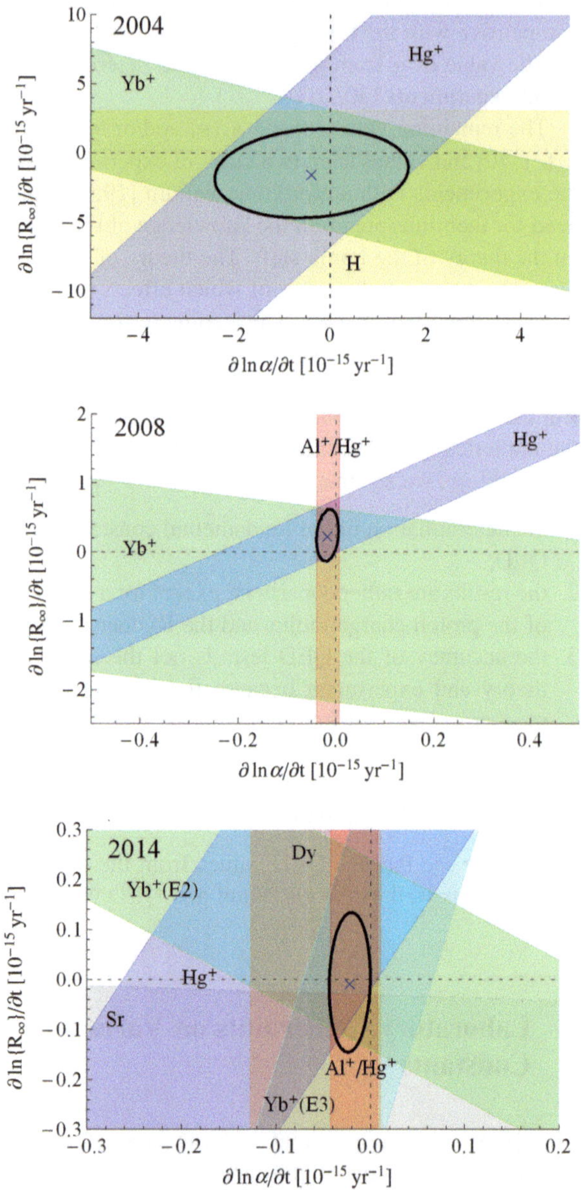

Fig. 12 Constraint on time variation of atomic fundamental constants of 2004 (following [82, 83]), 2008 (following [84]), and of 2014 using data summarized in [85], but following [82, 83]

comparing atomic and molecular transitions of different kind, e.g., by a comparison of rovibrational molecular transitions and optical gross-structure atomic transitions. That does not leave too much room for the precision spectroscopy, because it allows mostly to study optical gross-structure atomic transitions. As was later realized by Prestage et al. [78], the transitions of the same kind can still be useful if the relativistic corrections are taken into account. First it was suggested for the hyperfine intervals, but later on in the pioneering work of Flambaum and coauthors the idea has been extended to optical transitions, first for astrophysical observations [79] and later [80, 81] for the optical clocks.

Different values of the relativistic contributions set a different sensitivity to the α variation and comparing two pure optical transitions one may set a constraint on $\partial \ln \alpha / \partial t$. In the meantime, measuring the optical transitions in the SI units, i.e., against the cesium clocks one can set a model-independent constraint on either $\partial \ln \{R_\infty\} / \partial t$ [82, 83, 86, 87] (Fig. 12) or on $\partial \ln (g_{\rm Cs}(m_e/m_p))/\partial t$ [6]. (The curly brackets stand for the numerical value [in the SI units].) The model-independent constraints are summarized in Table 6.

Further on, applying the data on the comparison of the HFS intervals in ^{87}Rb and ^{133}Cs [88], one can derive $\partial \ln (m_e/m_p)/\partial t$. Unfortunately, there is no complete theory of the strong interaction, which could relate the nuclear g factor to the fundamental constants. Therefore, such constraints should exploit model assumptions. The model-dependent constraints on $\partial \ln (m_e/m_p)/\partial t$ are presented in Table 7. General scaling of the molecular transitions with m_e/m_p has been known and can be applied to produce some model-independent constraints on the m_e/m_p variation (cf. [77]). The most accurate of one from the molecular spectroscopy [89] is still much weaker than the model-dependent constraints (see Table 7).

Table 6 Progress in constraining a possible time variation of the fundamental constants: the results of 2004 (following [82, 83]), 2008 (following [84]), and of 2014 using data summarized in [85], but following [82, 83]

Year	$\partial \ln \alpha/\partial t(\text{year}^{-1})$	$\partial \ln \{R_\infty\}/\partial t(\text{year}^{-1})$	ρ
2004	$-0.4(2.0) \times 10^{-15}$	$-1.5(3.2) \times 10^{-15}$	0.13
2008	$-0.018(23) \times 10^{-15}$	$0.20(37) \times 10^{-15}$	0.17
2014	$-0.022(22) \times 10^{-15}$	$-0.01(14) \times 10^{-15}$	0.08

Here ρ is the correlation coefficient $\rho(\partial \ln \alpha/\partial t, \partial \ln \{R_\infty\}/\partial t)$

Table 7 Constraints on a possible time variation of m_p/m_e

Constraint $\partial \ln (m_p/m_e)/\partial t(\text{yr}^{-1})$	References	References to model
$(-3.8 \pm 5.6) \times 10^{-14}$	[89]	Model-independent
$(-0.5 \pm 1.6) \times 10^{-16}$	[85]	[90, 91]
$(0.5 \pm 1.6) \times 10^{-16}$	This work	[92]

6 Conclusions

The progress in high-resolution spectroscopy has made possible to accurately determine the atomic fundamental constants, such as $R_\infty, \alpha, m_e/m_p$ and to search their possible time variation. The former has allowed to make accurate tests of QED theory for free particles, two-body and three-body systems, while the latter sets the strong constraints.

Acknowledgements The authors are grateful to Victor Flambaum, Masaki Hori, Vladimir Korobov, Krzysztof Pachucki and Wolfgang Quint for useful discussions.

References

1. J. Reichert, M. Niering, R. Holzwarth, M. Weitz, Th. Udem, T.W. Hänsch, Phys. Rev. Lett. **84**, 3232 (2000)
2. R. Holzwarth, Th. Udem, T.W. Hänsch, J.C. Knight, W.J. Wadsworth, P.St.J. Russell, Phys. Rev. Lett. **85**, 2264 (2000)
3. S.A. Diddams, D.J. Jones, J. Ye, S.T. Cundiff, J.L. Hall, J.K. Ranka, R.S. Windeler, R. Holzwarth, Th. Udem, T.W. Hänsch, Phys. Rev. Lett. **84**, 5102 (2000)
4. P.J. Mohr, D.B. Newell, B.N. Taylor, CODATA recommended values of the fundamental physical constants: 2014. Rev. Mod. Phys. **88**, 035009 (2016)
5. M. Niering, R. Holzwarth, J. Reichert, P. Pokasov, M. Th Udem, T.W. Weitz, P. Hänsch, G. Lemonde, M. Santarelli, P. Abgrall, C. Laurent, C. Salomon, A.Clairon, Phys. Rev. Lett. **84**, 5496 (2000)
6. M. Fischer, N. Kolachevsky, M. Zimmermann, R. Holzwarth, T. Udem, T.W. Hänsch, M. Abgrall, J. Grünert, I. Maksimovic, S. Bize, H. Marion, F.P. Dos Santos, P. Lemonde, G. Santarelli, P. Laurent, A. Clairon, C. Salomon, M. Haas, U.D. Jentschura, C.H. Keitel, Phys. Rev. Lett. **92**, 230802 (2004)
7. A. Huber, T. Udem, B. Gross, J. Reichert, M. Kourogi, K. Pachucki, M. Weitz, T.W. Hänsch, Phys. Rev. Lett. **80**, 468 (1998)
8. C.G. Parthey, A. Matveev, J. Alnis, R. Pohl, T. Udem, U.D. Jentschura, N. Kolachevsky, T.W. Hänsch, Phys. Rev. Lett. **104**, 233001 (2010)
9. C.G. Parthey, A. Matveev, J. Alnis, B. Bernhardt, A. Beyer, R. Holzwarth, A. Maistrou, R. Pohl, K. Predehl, T. Udem, T. Wilken, N. Kolachevsky, M. Abgrall, D. Rovera, C. Salomon, P. Laurent, T.W. Hänsch, Phys. Rev. Lett. **107**, 203001 (2011)
10. A. Matveev, C.G. Parthey, K. Predehl, J. Alnis, A. Beyer, R. Holzwarth, T. Udem, T. Wilken, N. Kolachevsky, M. Abgrall, D. Rovera, C. Salomon, P. Laurent, G. Grosche, O. Terra, T. Legero, H. Schnatz, S. Weyers, B. Altschul, T.W. Hänsch, Phys. Rev. Lett. **110**, 230801 (2013)
11. R. de Beauvoir, F. Nez, B. Cagnac, F. Biraben, D. Touahri, L. Hilico, O. Acef, A. Clairon, J.J. Zondy, Phys. Rev. Lett. **78**, 440 (1997)
12. C. Schwob, L. Jozefowski, B. de Beauvoir, L. Hilico, F. Nez, L. Julien, F. Biraben, O. Acef, A. Clairon, Phys. Rev. Lett. **82**, 4960 (1999)
13. C. Schwob, L. Jozefowski, B. de Beauvoir, L. Hilico, F. Nez, L. Julien, F. Biraben, O. Acef, A. Clairon, Phys. Rev. Lett. **86**, 4193 (2001). Errata
14. S. Sturm, F. Köhler, J. Zatorski, A. Wagner, Z. Harman, G. Werth, W. Quint, C.H. Keitel, K. Blaum, Nature **506**, 467 (2014)

15. S. Sturm, A. Wagner, M. Kretzschmar, W. Quint, G. Werth, K. Blaum, Phys. Rev. A **87**, 030501 (2013)
16. T. Beier, H. Häffner, N. Hermanspahn, S.G. Karshenboim, H.-J. Kluge, W. Quint, S. Stahl, J. Verdú, G. Werth, Phys. Rev. Lett. **88**, 011603 (2002)
17. H. Häffner, T. Beier, S. Djekić, N. Hermanspahn, H.-J. Kluge, W. Quint, S. Stahl, J. Verdú, T. Valenzuela, G. Werth, Eur. Phys. J. D **22**, 163 (2003)
18. J. Verdú, S. Djekić, S. Stahl, T. Valenzuela, M. Vogel, G. Werth, Phys. Rev. Lett. **92**, 093002 (2004)
19. M. Hori, A. Dax, J. Eades, K. Gomikawa, R. Hayano, N. Ono, W. Pirkl, E. Widmann, H.A. Torii, B. Juhász, D. Barna, D. Horváth, Phys. Rev. Lett. **96**, 243401 (2006)
20. V.I. Korobov, Phys. Rev. A **77**, 042506 (2008)
21. M. Hori, A. Sótér, D. Barna, A. Dax, R. Hayano, S. Friedreich, B. Juhász, Th. Pask, E. Widmann, D. Horváth, L. Venturelli, N. Zurlo, Nature **475**, 484 (2011)
22. J.C.J. Koelemeij, B. Roth, A. Wicht, I. Ernsting, S. Schiller, Phys. Rev. Lett. **98**, 173002 (2007)
23. V.I. Korobov, Phys. Rev. A **74**, 052506 (2006)
24. S. Schiller, V. Korobov, Phys. Rev. A **71**, 032505 (2005)
25. U. Bressel, A. Borodin, J. Shen, M. Hansen, I. Ernsting, S. Schiller, Phys. Rev. Lett. **108**, 183003 (2012)
26. V.I. Korobov, Phys. Rev. A **77**, 022509 (2008)
27. J. Biesheuvel, J-Ph Karr, L. Hilico, K.S.E. Eikema, W. Ubachs, J.C.J. Koelemeij, Nat. Commun. **7**, 10385 (2015)
28. V.I. Korobov, L. Hilico, J-Ph Karr, $m\alpha^7$-order corrections in the hydrogen molecular ions and antiprotonic helium. Phys. Rev. Lett. **112**, 103003 (2014)
29. V.I. Korobov, L. Hilico, J-Ph Karr, Phys. Rev. A **89**, 032511 (2014)
30. R. Bouchendira, P. Cladé, S. Guellati-Khélifa, F. Nez, F. Biraben, Phys. Rev. Lett. **106**, 080801 (2011)
31. S.-Y. Lan, P.-C. Kuan, B. Estey, D. English, J.M. Brown, M.A. Hohensee, H. Müller, Science **339**, 554 (2013)
32. D. Hanneke, S. Fogwell Hoogerheide, G. Gabrielse, Phys. Rev. A **83**, 052122 (2011)
33. T. Aoyama, M. Hayakawa, T. Kinoshita, M. Nio, Phys. Rev. D **91**, 033006 (2015)
34. R.S. Van Dyck, Jr, P.B. Schwinberg, H.G. Dehmelt, Phys. Rev. Lett. **59**, 26 (1987)
35. A. Wicht, J.M. Hensley, E. Sarajlic, S. Chu, Phys. Scr. **T102**, 82 (2002)
36. J.S. Borbely, M.C. George, L.D. Lombardi, M. Weel, D.W. Fitzakerley, E.A. Hessels, Phys. Rev. A **79**, 060503 (2009)
37. K. Pachucki and V.A. Yerokhin, Phys. Rev. A **79**, 062516 (2009), A**81**, 039903(E) (2010)
38. W. Liu, M.G. Boshier, S. Dhawan, O. van Dyck, P. Egan, X. Fei, M. Grosse Perdekamp, V.W. Hughes, M. Janousch, K. Jungmann, D. Kawall, F.G. Mariam, C. Pillai, R. Prigl, G. zu Putlitz, I. Reinhard, W. Schwarz, P.A. Thompson, K.A. Woodle, Phys. Rev. Lett **82**, 711 (1999)
39. A. Antognini, F. Nez, K. Schuhmann, F.D. Amaro, F. Biraben, J.M.R. Cardoso, D.S. Covita, A. Dax, S. Dhawan, M. Diepold, L.M.P. Fernandes, A. Giesen, A.L. Gouvea, T. Graf, T.W. Hänsch, P. Indelicato, L. Julien, Cheng-Yang Kao, P. Knowles, F. Kottmann, E.-O. Le Bigot, Yi-Wei Liu, J.A.M. Lopes, L. Ludhova, C.M.B. Monteiro, F. Mulhauser, T. Nebel, P. Rabinowitz, J.M.F. dos Santos, L.A. Schaller, C. Schwob, D. Taqqu, J.F.C.A. Veloso, J. Vogelsang, R. Pohl, Science **339**, 417 (2013)
40. S.G. Karshenboim, E.Y. Korzinin, V.A. Shelyuto, V.G. Ivanov, J. Phys. Chem. Ref. Data **44**, 031202 (2015)
41. G.P. Lepage, Phys. Rev. A **16**, 863 (1977)
42. G.T. Bodwin, D.R. Yennie, M.A. Gregorio, Rev. Mod. Phys. **57**, 723 (1985)
43. M.I. Eides, S.G. Karshenboim, V.A. Shelyuto, Ann. Phys. **205**, 231 (1991)
44. J. Sapirstein, D.R. Yennie, in *Quantum Electrodynamics*, ed. by T. Kinoshita (World Science, Singapore, 1990), pp. 560–672

45. M.I. Eides, H. Grotch, V.A. Shelyuto, *Theory of Light Hydrogenic Bound States* (Springer, Berlin, 2007)
46. H. Grotch, D.R. Yennie, Z. Phys. **202**, 425 (1967)
47. W.E. Caswell, G.P. Lepage, Phys. Lett. B **167**, 437 (1986)
48. K. Pachucki, U.D. Jentschura, V.A. Yerokhin, Phys. Rev. Lett. **93**, 150401 (2004)
49. K. Pachucki, U.D. Jentschura, V.A. Yerokhin, Phys. Rev. Lett. **94**, 229902 (2005)
50. K. Pachucki, A. Czarnecki, U.D. Jentschura, V.A. Yerokhin, Phys. Rev. A **72**, 022108 (2005)
51. S.G. Karshenboim, Phys. Rep. **422**, 1 (2005)
52. E.R. Cohen, B.N. Taylor, J. Phys. Chem. Rev. Data **2**, 663 (1973)
53. E.R. Cohen, B.N. Taylor, Rev. Mod. Phys. **59**, 1121 (1987)
54. P.J. Mohr, B.N. Taylor, Rev. Mod. Phys. **72**, 351 (2000)
55. P.J. Mohr, B.N. Taylor, Rev. Mod. Phys. **77**, 1 (2005)
56. P.J. Mohr, B.N. Taylor, D.B. Newell, Rev. Mod. Phys. **80**, 633 (2008)
57. P.J. Mohr, B.N. Taylor, D.B. Newell, Rev. Mod. Phys. **84**, 1527 (2012)
58. S.G. Karshenboim, Phys. Usp. **48**, 255 (2005)
59. S.G. Karshenboim, Eur. Phys. J. Special Top. **172**, 385 (2009)
60. S.G. Karshenboim, Phys. Usp. **56**, 883 (2013)
61. S.G. Karshenboim, Ann. Phys. **525**, 472 (2013)
62. S.G. Karshenboim, P.J. Mohr, D.B. Newell, J. Phys. Chem. Ref. Data **44**, 031101 (2015)
63. I. Sick, in *Precision Physics of Simple Atoms and Molecules*, ed. by S.G. Karshenboim (Springer, Berlin, 2007), pp. 57–77
64. S.G. Karshenboim, Phys. Rev. A **91**, 012515 (2015)
65. J. Arrington, I. Sick, J. Phys. Chem. Ref. Data **44**, 031204 (2015)
66. G. Audi, M. Wang, A.H. Wapstra, F.G. Kondev, M. MacCormick, X. Xu, B. Pfeiffer, The AME2012 atomic mass evaluation (I). Evaluation of input data, adjustment procedures. Chin. Phys. C **36**, 1287 (2012)
67. M. Wang, G. Audi, A.H. Wapstra, F.G. Kondev, M. MacCormick, X. Xu, B. Pfeiffer, The AME2012 atomic mass evaluation. (II). Tables, graphs and references. Chin. Phys. C **36**, 1603 (2012)
68. S.L. Zafonte, R.S. Van Dyck, Metrologia **52**, 280 (2015)
69. E.G. Myers, A. Wagner, H. Kracke, B.A. Wesson, Phys. Rev. Lett. **114**, 013003 (2015)
70. V.A. Yerokhin, U.D. Jentschura, Phys. Rev. Lett. **100**, 163001 (2008)
71. V.A. Yerokhin, U.D. Jentschura, Phys. Rev. A **81**, 012502 (2010)
72. D.L. Farnham, R.S. Van Dyck Jr., P.B. Schwinberg, Phys. Rev. Lett. **75**, 3598 (1995)
73. T. Yamazaki, N. Morita, R.S. Hayano, E. Widmann, J. Eades, Phys. Rep. **366**, 183 (2002)
74. R.S. Hayano, M. Hori, D. Horváth, E. Widmann, Rep. Prog. Phys. **70**, 1995 (2007)
75. V.I. Korobov, in *Precise Spectroscopy of Antiprotonic Helium and Sensitivity of Transitions to the Antiproton Mass*, ed. A. Hirtl et al. Proceedings of International Conference on Exotic Atoms and Related Topics (EXA05) (Vienna, 2005), (Vienna: Austrian Academy of Sciences Press, 2005), pp. 391–400
76. M.P. Savedoff, Nature (London) **178**, 688 (1956)
77. R.I. Thompson, Astrophys. Lett. **16**, 3 (1975)
78. J.D. Prestage, R.L. Tjoelker, L. Maleki, Phys. Rev. Lett. **74**, 3511 (1995)
79. V.A. Dzuba, V.V. Flambaum, J.K. Webb, Phys. Rev. Lett. **82**, 888 (1999)
80. V.A. Dzuba, V.V. Flambaum, J.K. Webb, Phys. Rev. A **59**, 230 (1999)
81. V.A. Dzuba, V.V. Flambaum, Phys. Rev. A **61**, 034502 (2000)
82. E. Peik, B. Lipphardt, H. Schnatz, T. Schneider, Chr Tamm, S.G. Karshenboim, Phys. Rev. Lett. **93**, 170801 (2004)
83. S.G. Karshenboim, V.V. Flambaum, E. Peik, in *Springer Handbook of Atomic, Molecular and Optical Physics, ed. by G.W.F.Drake* (Springer, NY, 2005), pp. 455-463
84. S.G. Karshenboim, E. Peik, Eur. Phys. J. Special Top. **163**, 1 (2008)
85. N. Huntemann, B. Lipphardt, C. Tamm, V. Gerginov, S. Weyers, E. Peik, Phys. Rev. Lett. **113**, 210802 (2014)

86. S.G. Karshenboim, E. Peik (eds.), *Astrophysics, Clocks and Fundamental Constants* (Springer, Berlin, 2004)
87. S.G. Karshenboim, Gen. Rel. Grav. **38**, 159 (2006)
88. J. Guéna, M. Abgrall, D. Rovera, P. Rosenbusch, M.E. Tobar, Ph. Laurent, S.Bize, A. Clairon, Phys. Rev. Lett. **109**, 080801 (2012)
89. A. Shelkovnikov, R.J. Butcher, C. Chardonnet, A. Amy-Klein, Phys. Rev. Lett. **100**, 150801 (2008)
90. V.V. Flambaum, A.F. Tedesco, Phys. Rev. C **73**, 055501 (2006)
91. T.H. Dinh, A. Dunning, V.A. Dzuba, V.V. Flambaum, Phys. Rev. A **79**, 054102 (2009)
92. S.G. Karshenboim, Can. J. Phys. **78**, 639 (2000)

High-Precision Spectroscopy of the HD^+ Molecule at the 1-p.p.b. Level

J. Biesheuvel, J.-Ph. Karr, L. Hilico, K.S.E. Eikema, W. Ubachs, and J.C.J. Koelemeij

Abstract Recently we reported a high-precision optical frequency measurement of the (v, L): $(0, 2) \rightarrow (8, 3)$ vibrational overtone transition in trapped deuterated molecular hydrogen (HD^+) ions at 10 mK temperature. Achieving a resolution of 0.85 parts-per-billion (p.p.b.), we found the experimental value [ν_0 = 383, 407, 177.38 (41) MHz] to be in agreement with the value from molecular theory [ν_{th} 383, 407, 177.150 (15) MHz] within 0.6 (1.1) p.p.b. (Biesheuvel et al. in Nat Commun 7:10385, 2016). This enabled an improved test of molecular theory (including QED), new constraints on the size of possible effects due to 'new physics,' and the first determination of the proton–electron mass ratio from a molecule. Here, we provide the details of the experimental procedure, spectral analysis, and the assessment of systematic frequency shifts. Our analysis focuses in particular on deviations of the HD^+ velocity distribution from thermal (Gaussian) distributions under the influence of collisions with fast ions produced during (laser-induced) chemical reactions, as such deviations turn out to significantly shift the hyperfine-less vibrational frequency as inferred from the saturated and Doppler-broadened spectrum, which contains partly unresolved hyperfine structure.

This article is part of the topical collection "Enlightening the World with the Laser" - Honoring T. W. Hänsch guest edited by Tilman Esslinger, Nathalie Picqué, and Thomas Udem.

J. Biesheuvel
LaserLaB, Department of Physics and Astronomy, Vrije Universiteit, De Boelelaan 1081, 1081 HV Amsterdam, The Netherlands

Present Address: Statistics Netherlands (CBS), P.O. Box 24500, 2490 HA The Hague, The Netherlands

J.-P. Karr • L. Hilico
Laboratoire Kastler Brossel, CNRS, Collège de France, ENS-PSL Research University, UPMC-Sorbonne Universités, 4 Place Jussieu, 75005 Paris, France

Departement de physique, Université d'Evry Val d'Essonne, Boulevard François Mitterrand, 91025 Evry Cedex, France

K.S.E. Eikema • W. Ubachs • J.C.J. Koelemeij (✉)
LaserLaB, Department of Physics and Astronomy, Vrije Universiteit, De Boelelaan 1081, 1081 HV Amsterdam, The Netherlands
e-mail: j.c.j.koelemeij@vu.nl

D. Meschede et al. (eds.), *Exploring the World with the Laser*,
https://doi.org/10.1007/978-3-319-64346-5_16

1 Introduction

Because of their relative simplicity, molecular hydrogen ions such as H_2^+ and HD^+ are benchmark systems for molecular theory [1, 2], and suitable probes of fundamental physical models [3, 4]. Already four decades ago, Wing et al. [5] performed measurements of ro-vibrational transitions in HD^+ and conjectured that such measurements could be used to test quantum electrodynamics (QED) theory in molecules and lead to an improved value of the proton-to-electron mass ratio, μ. Today, QED corrections to ro-vibrational transition frequencies in H_2^+ and HD^+ have been calculated up to the order $m_e\alpha^8$, with m_e the electron mass and α the fine-structure constant, leading to relative uncertainties of the order of 3–4 $\times$ 10^{-11} [1, 2]. The frequencies of these transitions are typically in the (near-) infrared with linewidths below 10 Hz, which makes them amenable to high-resolution laser spectroscopy. Optical clocks based on trapped atomic ions have shown that laser spectroscopy at very high accuracy (below one part in 10^{17}) is possible [6]. Recent theoretical studies point out that also for molecular hydrogen ions experimental uncertainties in the 10^{-16} range should be possible [7, 8].

Ro-vibrational energy intervals have been determined experimentally using several spectroscopic techniques. One method involves high-resolution THz spectroscopy of transitions between selected Rydberg states of neutral H_2, which converge to the lowest rotational levels of the H_2^+ molecular ion. Extrapolating the Rydberg series using multichannel quantum-defect theory, the first rotational interval in para-H_2^+ was determined with a relative uncertainty of 0.4 parts-per-million [9]. Another technique, laser spectroscopy of trapped HD^+ ions, has led to frequency measurements of ro-vibrational transitions with a reported uncertainty of 1–2 p.p.b. [10, 11]. However, the most precise measurement of a ro-vibrational transition in HD^+ so far resulted in a 2.7 p.p.b. (2.4σ) discrepancy with more accurate theoretical data [11]. The question whether this offset is a statistical outlier, or caused by an experimental systematic effect or possible 'new physics' has remained unanswered. Therefore, additional experimental data on HD^+ are needed. In this article, we present the result of a high-precision frequency measurement of the ro-vibrational transition (v, L):(0, 2)→(8, 3) in the HD^+ molecule and compare it with state-of-the-art molecular theory to find excellent agreement. From this comparison, we subsequently derive new constraints on the size of effects due to possible new physics (in the form of hypothetical fifth forces or compactified higher dimensions [3, 4, 12]) in HD^+, and we determine a value of μ for the first time from a molecular system [12].

This article is organized as follows. In Sect. 2, we briefly review the theory of HD^+ relevant to this experiment. In Sect. 3, we describe a setup for spectroscopy of trapped HD^+ ions, sympathetically cooled by laser-cooled beryllium ions. The analysis of the spectroscopic data, systematic frequency shifts, and the results are discussed in Sect. 4, followed by the conclusions in Sect. 5.

2 Theory

2.1 *Calculation of Ro-Vibrational Frequency Transitions in HD$^+$*

The calculation of ro-vibrational energies in quantum mechanical three-body systems is usually divided into a nonrelativistic part, complemented by the relativistic and radiative parts, and finite nuclear size contributions. The nonrelativistic part is calculated through solving the three-body Schrödinger equation, which can be done with practically infinite precision [13] (up to a relative precision of $\sim 10^{-32}$ [14, 15]). The resulting wavefunctions allow an analytical evaluation of the Breit-Pauli Hamiltonian and the leading-order radiative corrections.

Recently, the precision of relativistic and radiative energy corrections to the nonrelativistic energies was strongly improved. With the inclusion of the full set of contributions of order $m\alpha^7$ and leading-order terms of order $m\alpha^8$, the relative uncertainty is now below 4×10^{-11}. For example, the theoretically determined value of the HD$^+$ (v, L):(0, 2)→(8, 3) transition frequency, ν_{th}, is 383,407,177.150 (15) MHz and has a relative uncertainty of 4×10^{-11} [1, 2]. Note that the specified error (within parentheses) does not include the uncertainty of the fundamental constants used. By far the largest contribution is due to the 4.1×10^{-10} uncertainty of the 2010 Committee on Data for Science and Technology (CODATA) value of μ [16], which translates to a relative frequency uncertainty of 1.5×10^{-10} (and an absolute frequency uncertainty of 59 kHz) [12].

2.2 *Hyperfine Structure and Rotational States*

Since the HD$^+$ constituents possess nonzero spin, the ro-vibrational transition spectra contain hyperfine structure due to spin–spin and spin–orbit couplings. In [17] the hyperfine energy levels and eigenstates in HD$^+$ are calculated by diagonalization of the effective spin Hamiltonian

$$\begin{aligned} H_{\text{eff}} &= E_1(\mathbf{L}\cdot\mathbf{s_e}) + E_2(\mathbf{L}\cdot\mathbf{I_p}) + E_3(\mathbf{L}\cdot\mathbf{I_d}) + E_4(\mathbf{I_p}\cdot\mathbf{s_e}) \\ &\quad + E_5(\mathbf{I_d}\cdot\mathbf{s_e}) + E_6K_d(\mathbf{L},\mathbf{I_p},\mathbf{s_e}) + E_7K_d(\mathbf{L},\mathbf{I_d},\mathbf{s_e}) \\ &\quad + E_8K_d(\mathbf{L},\mathbf{I_p},\mathbf{I_d}) + E_9K_Q(\mathbf{L},\mathbf{I_d}), \end{aligned} \tag{1}$$

where the spin coefficients, E_i, are obtained by averaging the Breit–Pauli Hamiltonian over the nonrelativistic wavefunctions, $\mathbf{L}$ is the rotational angular momentum operator, and $\mathbf{s_e}$, $\mathbf{I_p}$ and $\mathbf{I_d}$ are the electron, proton and deuteron spin operators. K_{d} and K_{Q} are spherical tensors composed of angular momenta, whose explicit form is given in [17]. The strongest coupling is the proton–electron spin–spin interaction [the term in E_4 in Eq. (1)], and the preferred angular momentum coupling scheme is

$$\mathbf{F} = \mathbf{I_p} + \mathbf{s_e} \qquad \mathbf{S} = \mathbf{F} + \mathbf{I_d} \qquad \mathbf{J} = \mathbf{L} + \mathbf{S}. \tag{2}$$

For states with $L = 0$, $L = 1$ and $L \geq 2$ this leads to the presence of 4, 10, and 12 hyperfine levels, respectively, which are distributed over an energy range of ~1 GHz (Fig. 2b). Diagonalization produces eigenstates $\phi_{\widetilde{F}\,\widetilde{S}\,JM_J}$, with the magnetic quantum number M_J corresponding to the projection of J onto the laboratory-fixed z axis. Note that after diagonalization, the quantum numbers $\widetilde{F}$, $\widetilde{S}$ are only approximately good, and a hyperfine eigenstate can be expressed in the 'pure' basis states $|FSJM_J\rangle$ in the form

$$\phi_{\widetilde{F}\,\widetilde{S}\,JM_J} = \sum_{F,S} \beta_{FSJ} |FSJM_J\rangle, \tag{3}$$

with real-valued coefficients β_{FSJ}. In [17] the hyperfine levels in vibrationally excited states in HD^+ are calculated with an error margin of ~50 kHz. This uncertainty might propagate through a spectral analysis employing a lineshape model which includes the theoretical hyperfine structure (as we do below). However, the uncertainty contribution to the determined hyperfine-less (v, L):(0, 2)→(8, 3) transition frequency is expected to be at least five times smaller due to strong correlation and partial cancelation of the error (see also Sect. 4.6.3).

2.3 Determination of Transition Rates

Because of the hyperfine structure, the spectrum of the (0, 2)→(8, 3) electric dipole transition consists of a large number of hyperfine components. Together with Doppler broadening, this leads to an irregular lineshape. In addition, the excitation laser may address multiple hyperfine states simultaneously, which are furthermore coupled to other rotational states by the ambient 300 K blackbody radiation (BBR) field. Therefore, for the analysis of the (0, 2)→(8, 3) signal we develop a model based on Einstein rate equations which takes all resonant molecule–electric field interactions into account. We note that at 300 K, states with $v = 0$ and $L = 1$–6 are significantly populated, with 27% in $L = 2$, and 2.4% in states with $L \geq 6$. Below, we calculate the Einstein rate coefficients at the level of individual hyperfine states for transitions driven by the laser and BBR fields.

Following the approach of Koelemeij [18], we first ignore hyperfine structure and obtain transition dipole moments given by

$$\mu_{vv'LL'} = \int_0^\infty \chi_{v',L'}(R) D_1(R) \chi_{v,L}(R) \mathrm{d}R, \tag{4}$$

where $\chi_{v,L}$ denotes the radial wavefunction of nuclear motion obtained by numerical solution of the radial Schrödinger equation, R stands for the internuclear separation, and $D_1(R)$ denotes the permanent electric dipole moment function of the $1s\sigma$ electronic ground state.

To take hyperfine structure into account, we evaluate the dipole transition matrix element between two states $\phi_{\tilde{F}\,\tilde{S}JM_J}\chi_{v,L}$ and $\phi_{\tilde{F'}\,\tilde{S'}J'M'_J}\chi_{v',L'}$ [18–20]

$$\begin{aligned}\langle\phi_{\tilde{F}\,\tilde{S}JM_J}\chi_{v,L}|\mathbf{E}\cdot\mu|\phi_{\tilde{F'}\,\tilde{S'}J'M'_J}\chi_{v',L'}\rangle = & \sum_{F'S'}\beta_{FSJ}\beta^*_{F'S'J'}E_p(-1)^{J+J'+S-M_J+1+L'+L} \\ & \times[(2J+1)(2J'+1)(2L+1)(2L'+1)]^{1/2} \\ & \times\begin{pmatrix} L & 1 & L' \\ M_J & p & M_J-p \end{pmatrix}\begin{pmatrix} L & 1 & L' \\ 0 & 0 & 0 \end{pmatrix}\begin{Bmatrix} L' & J' & S \\ J & L & 1 \end{Bmatrix} \\ & \times\mu_{vv'LL'}.\end{aligned} \tag{5}$$

Equation (5) includes a transformation from the molecule-fixed frame to the laboratory-fixed frame [19], with p denoting the polarization state of the electric field with respect to the laboratory-fixed frame ($p \in \{-1, 0, 1\}$).

The linestrength is subsequently calculated by squaring and summing over the M_J states [21]:

$$S_{\alpha\alpha'} = \sum_{M_J}\left|\langle\phi_{\tilde{F}\,\tilde{S}JM_J}\chi_{v,L}|\mathbf{E}\cdot\boldsymbol{\mu}|\phi_{\tilde{F'}\,\tilde{S'}J'M'_J}\chi_{v',L'}\rangle\right|^2, \tag{6}$$

with $\alpha \equiv vL\tilde{F}\,\tilde{S}J$. The Einstein rate coefficients for spontaneous emission, absorption and stimulated emission are subsequently obtained as described in [22] and used in Sect. 4.1 to calculate the expected (0, 2)→(8, 3) spectrum.

3 Experiment

3.1 Trapping and Cooling Be⁺ and HD⁺

To achieve narrow linewidths and small systematic shifts, we choose to perform spectroscopy on small samples of HD^+ molecules in a radiofrequency (rf) ion trap. We reduce the motional temperature of the HD^+ ions by storing them together with Be^+ ions which are Doppler-cooled by a continuous wave (cw) 313 nm laser beam (see [18, 23, 24] for details). The rf trap is placed inside an ultrahigh vacuum chamber with a background pressure of 1×10^{-10} mbar. The rf trap operates at a frequency of 13.2 MHz, leading to Be^+ radial trap frequencies of $\omega_r = 2\pi \times 290$ kHz. The trap geometry and rf circuitry are designed so as to minimize unwanted rf fields and phase differences between the rf electrodes. The two dc electrodes are segmented into two endcaps and a center electrode (Fig. 1). The dc voltages of the center electrodes, rf electrodes and endcap pairs can be individually adjusted to

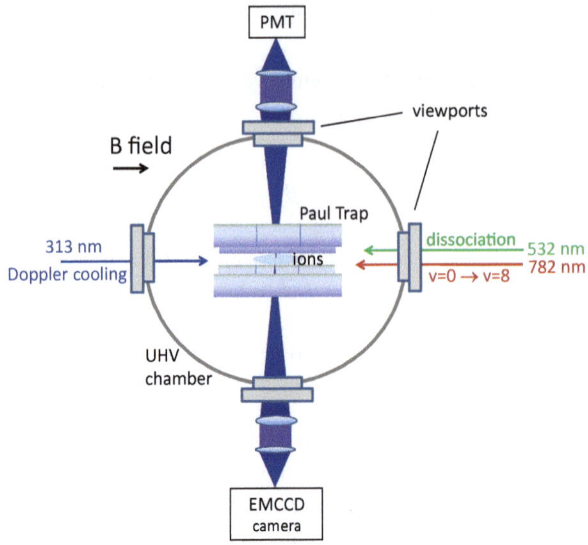

Fig. 1 Schematic view of the trap setup. An ultrahigh vacuum (UHV) chamber houses a linear rf trap in which Be^+ and HD^+ ions are loaded by electron-impact ionization. Be^+ ions are Doppler-cooled by the 313 nm laser, which is directed parallel to the trap axis and B-field direction. The 313 nm Be^+ fluorescence is imaged onto and detected with a PMT and an EMCCD camera. The 782 and 532 nm cw REMPD lasers are overlapped with the 313 nm laser and the ions, but propagate in the opposite direction

compensate stray electric fields. Be^+ and HD^+ are loaded by electron-impact ionization as done by Blythe et al. [25] and monitored with a photomultiplier tube (PMT) and an electron-multiplied charge-coupled-device (EMCCD) camera. EMCCD images show ellipsoidal mixed-species Coulomb crystals, with a dark core of molecular hydrogen ions surrounded by several shells of fluorescing Be^+ ions (Fig. 6). The apparatus and procedures for loading and compensation of stray electric fields are described in more detail in [18].

3.2 Spectroscopy of HD^+

The $(0, 2)\rightarrow(8, 3)$ transition in HD^+ is detected destructively through resonance enhanced multi-photon dissociation (REMPD), see Fig. 2a.

A 782 nm cw titanium:sapphire laser is used to excite HD^+ from its vibrational ground state to the $v = 8$ state, which is efficiently dissociated by the field of a co-propagating 532 nm cw laser beam. Both lasers are directed along the trap axis and counter propagate the 313 nm laser (see Fig. 1). Since all HD^+ ions are initially in the vibrational ground state (which is too deeply bound to be dissociated by the 532 nm laser), dissociation only takes place if the 782 nm laser is resonant with the

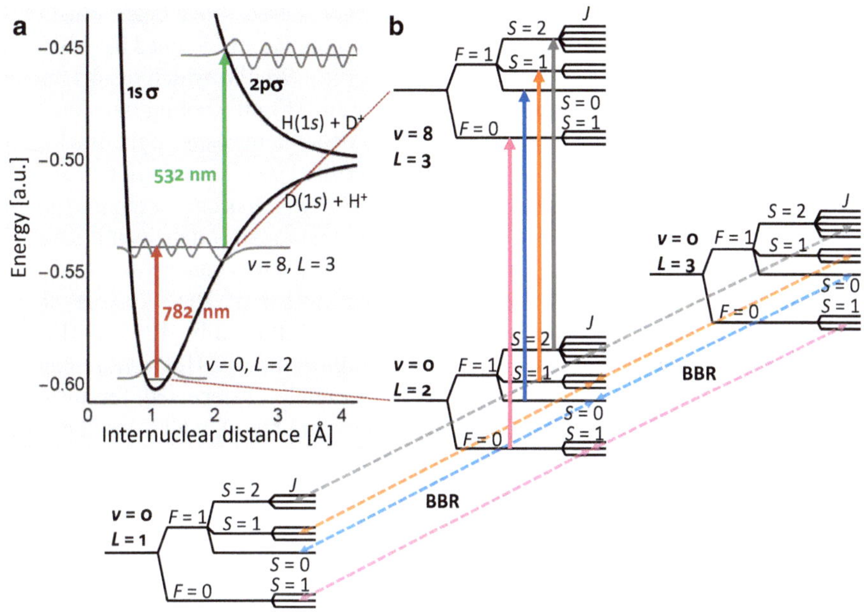

Fig. 2 **a** Potential energy curves of the 1sσ and 2pσ electronic states are plotted together with the radial nuclear wavefunctions for $v = 0$, $v = 8$, and for the dissociative wavefunction in the 2pσ state. The REMPD scheme is also shown, with the *red arrow* indicating the (v, L):(0, 2)→(8, 3) transition by the 782 nm laser and the *green arrow* corresponding to the 532 nm photon which dissociates the molecule. **b** Detailed partial level scheme (including hyperfine levels) showing the (0, 2)→(8, 3) transition together with the BBR interaction which couples the rotational levels $L = 1, 2, 3$ through electric dipole transitions. Coupling to $L = 0$ and $L = 4, 5, 6$ also occurs, but is not shown here

(0, 2)→(8, 3) transition. We used 90 mW of 532 nm radiation, focused to a beam waist of 140 μm, which is sufficient to dissociate an HD^+ molecule from this level within a few ms, much faster than the spontaneous decay of the $v = 8$ state (lifetime 12 ms). During each measurement cycle, the HD^+ ions are exposed to the REMPD lasers for 10 s. During the first seconds, the majority of the HD^+ in $L = 2$ are dissociated, leading to depletion of the $L = 2$ state (Fig. 2). During the remainder of the REMPD period, BBR repopulates the $L = 2$ state from other rotational levels, which enhances the number of dissociated HD^+ ions by approximately a factor of two.

The 782 nm laser has a linewidth of ∼0.5 MHz and is frequency-stabilized by locking its frequency to a nearby mode of a self-referenced optical frequency comb laser. To this end, a 63.5 MHz beat note is created by mixing the light of the 782 nm laser and the frequency comb. The frequency comb itself is locked to a rubidium atomic clock for short-term stability, which is disciplined to the 1-pps signal of a GPS receiver for long-term accuracy and traceability to the SI second with 2×10^{-12} relative uncertainty.

During REMPD, about 300 mW of 782 nm light is used with a beam waist of $\sim$ 120 μm at the location of the ions. The 313 nm cooling laser is detuned by −80 MHz from the cooling resonance and reduced in power to $\sim$70 μW, which results into an intensity of two times the saturation intensity of the 313 nm cooling transition and leads to an ion temperature of about 10 mK. To obtain a measure of the fraction of HD^+ ions lost due to REMPD, we employ the so-called secular excitation [26]. This procedure is based on the indirect heating of the Be^+ which occurs when the motion of the HD^+ ions is resonantly excited by an additional radial rf field as it is scanned over the resonance frequency. The heating of Be^+ leads to a change of the 313 nm fluorescence level, which is connected to the number of HD^+ ions. A single measurement cycle consists of secular scan (10 s), followed by 10 s of REMPD and another 10 s secular scan (see Fig. 3) to infer the number of HD^+ ions remaining after REMPD. To obtain a spectrum, this procedure is repeated for different frequencies of the 782 nm laser (indicated by the variable ν), which are chosen as follows. The $\sim$25 strongest hyperfine components of the $(0,2) \rightarrow (8,3)$ transition are located in the range (−110, 140 MHz) around the hyperfine-less frequency. As we expect Doppler broadening to be $\sim$16 MHz, we divide this range into a set of 140 evenly spaced frequencies (spacing $\sim$1.8 MHz) at which REMPD spectroscopy is performed. To convert possible time-varying systematic effects into random noise, we randomize the ordering of the frequency list. For each frequency point, 6–7 REMPD measurements are made, which results into a spectrum consisting of 886 REMPD measurements in total.

Scanning the frequency of the radial rf field over the secular motional resonance of HD^+ ($\sim$830 kHz) temporarily heats up the Coulomb crystal to a few Kelvin. At a detuning of the 313 nm cooling laser of −300 MHz, and using a few mW of laser

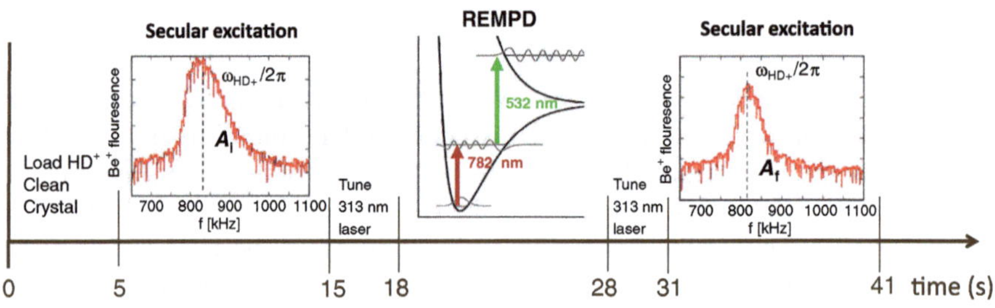

Fig. 3 The secular scan—REMPD—secular scan detection scheme. At the start of each cycle, a new sample of HD^+ ions is loaded into the ion trap. During loading the Be^+ ions are exposed to neutral HD gas, leading to the formation of a small number of BeH^+ and BeD^+ ions (see also Table 1). These are expunged from the trap by applying a dc quadrupole potential of $\sim$0.9 V, which reduces the trap depth such that only ions with mass $\leq$9 amu remain trapped. Five seconds after loading, the HD^+ is motionally excited by scanning an rf electric field over the secular motional resonance frequency at 830 kHz. A 313 nm laser detuning of −300 MHz is used, in which case the secular excitation results in a rise in the Be^+ fluorescence. Ten seconds later, the 313 nm laser is tuned to −80 MHz from the Be^+ cooling transition, and its intensity reduced to $I \sim 2I_{sat}$. After 10 s of REMPD, the 313 nm laser settings are restored to their values as used for the first secular scan, and a second secular scan is executed. A smaller 313 nm fluorescence peak indicates loss of HD^+

power (corresponding to ~80 times the saturation intensity), such a 'secular scan' shows up as a peak in the 313 nm beryllium fluorescence versus rf frequency (Fig. 3). In Appendix 2, we show that the area under this peak, A, scales with the number of HD^+ ions, although not linearly. We define a spectroscopic signal as the relative difference between the areas of the initial and final secular scan peaks:

$$S = \frac{A_i - A_f}{A_i} \tag{7}$$

Repeating this procedure for all 140 preselected frequencies of the 782 nm laser while recording fluorescence traces with both the PMT and EMCCD camera, we obtain a spectrum consisting of 1772 data points (ν_j, S_j) (with $j = 1, \ldots, 1772$).

4 Results and Discussion

4.1 Spectral Lineshape Model

In order to determine the hyperfine-less ro-vibrational (0, 2)→(8, 3) transition frequency, ν_0, we need a realistic lineshape model which includes the relevant physics present during REMPD. Here, the aim is to obtain a spectral lineshape function in which all effects are parameterized. Parameter values are estimated by independent means where possible, or included as a fit parameter otherwise. Importantly, the fit function will contain a variable $\nu - \nu_{0,\mathrm{fit}}$, where $\nu_{0,\mathrm{fit}}$ is a fit parameter from which we later deduce the value of ν_0. Before fitting, the (ν, S) data are corrected for reactions with background gas (primarily H_2). This procedure is described in Sect. 4.5.

We start with building a state vector $\boldsymbol{\rho}(t)$ which contains the population in all 62 hyperfine levels in the rotational states ranging from $L = 0$ to $L = 5$ with $v = 0$. This includes 97.6% of the total internal states of $v = 0$ given a BBR temperature of 300 K. We neglect the Zeeman splitting due to 0.19 mT B field at the location of the trapped ions, as this splitting is negligibly small compared to the Doppler linewidth and the width of the BBR spectrum. The lineshift to ν_0 due to the Zeeman effect is considered separately in Sect. 4.6.2. Also the Stark effect, the electric-quadrupole shift and second-order Doppler shift are not included in this model, but addressed in Sect. 4.6.2.

During the 10 s of REMPD, the hyperfine levels in the $L = 2$ initial state interact with the 782 nm laser and BBR. We here make the simplifying assumption that any population in the $v = 8$ target state will be dissociated by the 532 nm laser. The interaction with the 782 nm laser is therefore modeled as a simple loss process. The evolution of the state vector $\boldsymbol{\rho}(t)$ is obtained by solving the set of coupled rate equations:

$$\frac{\mathrm{d}\rho(t)}{\mathrm{d}t} = M_{\mathrm{R}}\,\rho(t) + M_{\mathrm{BBR}}\,\rho(t), \tag{8}$$

where M_{R} and M_{BBR} are the matrices describing the interaction with the REMPD lasers and the BBR field. Generally, the diagonal elements of these matrices contain depopulating (negative) terms, whereas the off-diagonal elements describe the population a certain state gains from another. Since M_{R} describes losses, it only contains diagonal entries of the form

$$M_{\mathrm{R},\alpha\alpha} = -\sum_{\alpha'} B_{\alpha\alpha'} D_z((\omega - \omega_{\alpha\alpha'}), T_{\mathrm{HD}^+}) I_{\mathrm{laser}}/c, \tag{9}$$

where $B_{\alpha\alpha'}$ denotes the Einstein coefficient for absorption from a lower hyperfine-ro-vibrational level α to an upper level α' (with for M_{R} the states α and α' being restricted to those having $v = 0, L = 2$ and $v' = 8, L' = 3$, respectively). The corresponding transition frequency is $\omega_{\alpha\alpha'}$, I_{laser} is the intensity of the 782 nm laser, and D_z represents a normalized response function averaged over the distribution of z-velocities of the HD^+ ions. This involves an integration over all Doppler shifts $-k_z v_z$ observed by the HD^+ ensemble, where k_z is the wavevector of the 782 nm laser and v_z the velocity in the z-direction. If the HD^+ velocity distribution is thermal (Gaussian), D_z depends only on the temperature in the z-direction, T_{HD^+}. However, the effects of micromotion and chemistry in the Coulomb crystal during REMPD lead to deviations from a thermal distribution. This implies that D_z cannot be described by a single Gaussian lineshape. In Sects. 4.3 and 4.4, these processes are explained in detail, and the precise shapes of D_z are determined.

The matrix M_{BBR} contains both diagonal and off-diagonal elements, taking into account the rate of exchange of population between all involved levels α and α' mediated through all possible electric dipole transitions,

$$\begin{aligned} &A_{\alpha'\alpha},\\ &B_{\alpha'\alpha} W(\omega_{\alpha\alpha'}, T_{\mathrm{BBR}}),\\ &B_{\alpha\alpha'} W(\omega_{\alpha\alpha'}, T_{\mathrm{BBR}}), \end{aligned}$$

which correspond to spontaneous emission, stimulated emission and absorption by BBR, respectively. The relationship with Eq. (6) is given in [22]. $W(\omega_{\mathrm{BBR}}, T_{\mathrm{BBR}})$ denotes the BBR energy distribution function at frequency ω_{BBR}, which is given by:

$$W(\omega_{\mathrm{BBR}}, T_{\mathrm{BBR}}) = \frac{\hbar\omega_{\mathrm{BBR}}^3}{\pi^2 c^3}\left(\mathrm{e}^{\frac{\hbar\omega_{\mathrm{BBR}}}{k_B T_{\mathrm{BBR}}}} - 1\right)^{-1}. \tag{10}$$

Since the typical frequency of the internal degrees of freedom (>1 THz) differs from that of the external degrees of freedom (<1 MHz) by many orders of magnitude, any energy transfer mechanism between them must be of extremely high order and consequently negligibly small. Laser cooling of the external degrees of freedom may therefore be expected to have no significant effect on the temperature of the internal degrees of freedom, which are coupled strongly to (and in equilibrium with) the BBR field [27].

We use MATHEMATICA to solve Eq. (8) in order to obtain $\boldsymbol{\rho}(t)$. We subsequently find the relative HD$^+$ loss, ϵ, by summing over the hyperfine state populations (62 hyperfine states in $(v, L) = (0, 2)$) before (i.e., $t = 0$ s) and after ($t = 10$ s) REMPD and computing:

$$\epsilon = \frac{N_i - N_f}{N_i} = \frac{\sum_{j=1}^{62} \rho_j(0\ \mathrm{s}) - \rho_j(10\ \mathrm{s})}{\sum_{j=1}^{62} \rho_j(0\ \mathrm{s})}. \tag{11}$$

Here N_i and N_f are the numbers of HD$^+$ ions present in the trap directly before and after REMPD, respectively. For a thermal ensemble of HD$^+$ ions, ϵ is a function of the variables $\nu - \nu_{0,\mathrm{fit}}, T_{\mathrm{HD}}^+$, and I_{laser}. In what follows we furthermore assume that $T_{\mathrm{BBR}} = 300$ K, which is the average temperature in our experimental setup.

The question arises what the relation is between the signal S defined in Eq. (7) and the fractional loss ϵ defined above. In previous work, it was assumed that S and ϵ are interchangeable [10, 18, 26, 28]. In Appendix 2, we study the dependence of the signal S on the initial number of ions N_i and the dissociated fraction ϵ in detail using realistic molecular dynamics (MD) simulations. We find that the fraction ϵ (which is a theoretical construct) can be mapped to the signal S by use of a slightly nonlinear function,

$$S_{\mathrm{fit}} = f_{\mathrm{NL}}(\bar{T}_0, \epsilon) \tag{12}$$

where $\bar{T}_0$ is defined as the effective Be$^+$ temperature along the z axis during the initial secular scan of the REMPD cycle (see Appendix 2). This means we have to use a five-dimensional fit function

$$\begin{aligned} &S_{\mathrm{fit}}(\bar{T}_0, \epsilon(\nu - \nu_{0,\mathrm{fit}}, T_{\mathrm{HD}^+}, I_{\mathrm{laser}})) \\ &= S_{\mathrm{fit}}(\nu - \nu_{0,\mathrm{fit}}, T_{\mathrm{HD}^+}, I_{\mathrm{laser}}, \bar{T}_0). \end{aligned} \tag{13}$$

An analytical solution of the fit function proves difficult to find, whereas a numerical implementation of the fit function takes excessively long to compute. Therefore, we compute values of S_{fit} on a sufficiently dense grid of values $\nu - \nu_{0,\mathrm{fit}}$ (encompassing 271 values between -120 and 150 MHz), T_{HD^+} (20 values between 1 and 20 mK), and I_{laser} (9 values between 0.65×10^7 and 2.1×10^7 W m^{-2}), which we interpolate (again using MATHEMATICA) to find a fast, continuous and smooth approximation to the function ϵ, ϵ', which we insert into the function S_{fit}. The result is a five-dimensional (5D) fit function which is suitable for nonlinear least-squares fitting. A 3D projection of S_{fit} assuming fixed values of I_{laser}, $\bar{T}_0$ and $\nu_{0,\mathrm{fit}}$ is plotted in Fig. 4.

The reason for treating I_{laser} as a fit parameter instead of inserting a single fixed value is the following. The entire spectroscopic measurement is divided into 15 sessions each taken at a different day. To ensure reproducible laser beam intensities from session to session, we used diaphragms to overlap all laser beams with the 313 nm cooling laser, which itself is aligned with the Be$^+$ Coulomb crystal visually using the EMCCD camera. Using a mock version of this setup, we verified

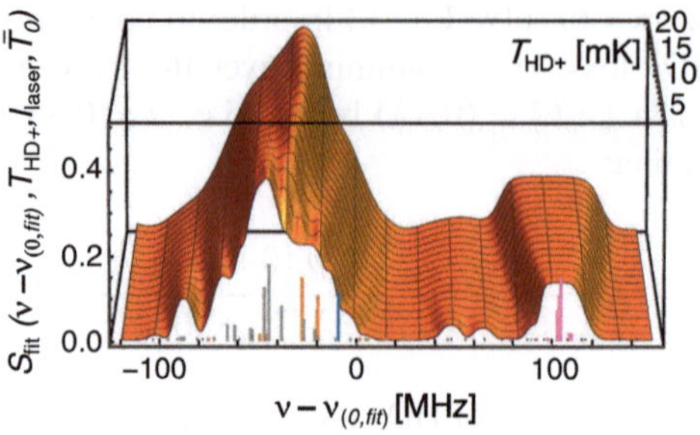

Fig. 4 A plot of the fit function S_{fit} with $I_{laser}= 1.0\times10^7$ W m^{-2}, $\bar{T}_0$=4 K and $\nu_{0,\mathrm{fit}} = \nu_{\mathrm{th}}$. On the frequency axis, the theoretical values of transitions between the particular hyperfine levels are depicted as sticks. The heights of the sticks correspond to their linestrengths. The stick colors are the same as those used in Fig. 2 and distinguish different groups of transitions with similar quantum numbers F and S. For decreasing HD^+ temperature, Doppler broadening is reduced and the hyperfine structure becomes more resolved. Effects of saturation are also visible in the spectrum

that this procedure leads to beam pointing errors up to 40 μm at the location of the HD^+ ions. Regarding the location of the HD^+ ions, we note that these always accumulate along the trap axis, as radial static electric fields are effectively shielded by the surrounding and more mobile Be^+ ions. MD simulations point out that HD^+ ions either line up in a string (i.e., zero radial displacement) or in a shell structure with at most 7 μm radial displacement. Assuming a Gaussian distribution of beam pointing errors, we find an intensity which varies from the intensity in the center of the beam by a factor of 0.6 to 1. Since the spectral line shape is strongly saturated, these variations of the intensity only lead to small signal changes. It is therefore allowed to treat I_{laser} as a free-fit parameter which represents the average 782 nm laser intensity for all data points. Similarly, the variables T_{HD^+} and $\bar{T}_0$ cannot be determined accurately a priori and are treated as free-fit parameters as well.

4.2 Estimation of Absolute Trapped Ion Numbers

As explained in Sect. 4.4, effects of chemistry in the Coulomb crystal significantly influence the measured lineshape of the $(0, 2)\rightarrow(8, 3)$ transition. The impact of such effects depends on (and can be estimated from) the absolute numbers of beryllium ions and molecular hydrogen ions. In order to estimate absolute numbers, we combine results from MD simulations and spectroscopy. Similar as observed by Blythe et al. [25], our mixed-species ion ensembles contain not only Be^+ and HD^+, but also BeH^+, BeD^+, H_2D^+, and HD_2^+. In this paragraph, we focus on the latter two species which are created during the HD^+ loading procedure through the exothermic reactions

$$HD + HD^+ \rightarrow H_2D^+ + D \tag{14}$$

and

$$HD + HD^+ \rightarrow HD_2^+ + H. \tag{15}$$

After loading, the excess HD gas is pumped out of the vacuum chamber within a few seconds, after which the background vapor resumes its steady-state composition (predominantly H_2). Reactions with HD therefore only play a significant role during and just after loading.

Triatomic hydrogen ions can be detected by secular excitation. An example is shown in Fig. 5, where the peak in the PMT signal at the left is attributed to the overlapping peaks belonging to species with charge-to-mass ratios 1:4 and 1:5, corresponding to H_2D^+ and HD_2^+, respectively.

In rf traps, lighter species experience stronger confinement. This is evident from EMCCD camera images which exhibit fluorescing Be^+ ions surrounding a dark core of lighter species (Fig. 6). The size of the core reflects the total number of light ions.

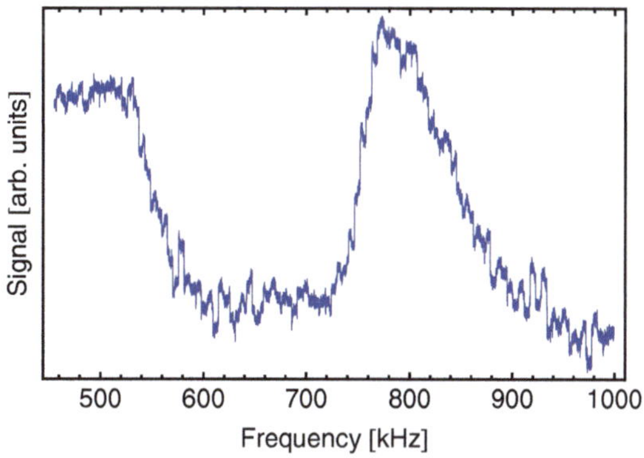

Fig. 5 Fluorescence at 313 nm recorded by the PMT during a secular scan from 450 to 1000 kHz. The peak at 800 kHz indicates the presence of trapped particles with charge-to-mass ratio 1:3 (mostly HD^+). The fluorescence signal at 500 kHz indicates the presence of particles with charge-to-mass ratios of 1:4 and 1:5 which are attributed to H_2D^+ and HD_2^+. Note that peaks belonging to the two species overlap and are not resolved

Fig. 6 **a** An EMCCD image of the Be^+ fluorescence just before a secular scan. **b** An image of ions obtained from a MD simulation based on 750 Be^+ ions and 50 singly charged ions with masses 3, 4, and 5 amu

We analyze this by means of MD simulations (Appendix 1), from which a relation is obtained between the size of the dark core and the number of trapped ions.

By comparing simulated and real EMCCD images, an average number of ~750 Be^+ ions is obtained. From the analysis of the image intensity profiles, we cannot distinguish the HD^+ from the triatomic molecular species. To solve this, we use a collection of 140 EMCCD images taken before and after REMPD while the 782 nm laser was tuned at the same fixed frequency near the maximum of the $(0, 2)\rightarrow(8, 3)$ spectrum. From the lineshape model, we estimate that the average relative HD^+ loss is 57% at this frequency. By comparing the above-mentioned set of 140 EMCCD images taken before and after REMPD with the simulated images, we also determine the total initial and final numbers of light ions. Combining this with the expected loss of 57% of the HD^+ ions, we infer the ratio of HD^+ numbers to heavier molecular species (H_2D^+, HD_2^+). An average number of 43 HD^+ ions is obtained together with 60 ions of heavier species. We cannot determine the relative abundance of H_2D^+ and HD_2^+, but previous observations indicate a branching ratio between Eqs. (14) and (15) of 1:1 and, thus, equal abundances [29, 30]. The set of EMCCD images shows an appreciable spread in the size of the dark core and, in particular, the ratio of HD^+ to heavier triatomic hydrogen ions. Variations in both are due to uncontrolled shot-to-shot fluctuations of the HD background pressure during HD^+ loading. We find somewhat asymmetrical distributions of HD^+ ions and ions of heavier species, with a wider tail toward higher numbers. The resulting standard deviation of the number of HD^+ is 41 ions. This also indicates that analysis of EMCDD images (under the present conditions) is not suitable to replace the signal obtained by mass-selective secular excitation (Eq. 7).

For the treatment of effects of chemistry on the lineshape in Sect. 4.4 below, we consider two scenarios:

- Scenario a: $N_{HD^+} = 43, N_{H_2D^+} = N_{HD_2^+} = 30$
- Scenario b: $N_{HD^+} = 84, N_{H_2D^+} = N_{HD_2^+} = 60,$

where scenario b reflects the one sigma upper variation (which well represents the width of the upper tail of the ion number distribution).

4.3 Effect of Micromotion

The rf quadrupole field of the ion trap inevitably leads to micromotion of ions with nonzero displacement from the trap *z* axis. In addition, excess micromotion may be caused by unwanted rf fields arising from geometrical imperfections of the trap electrode structure or phase differences between the rf electrodes [31]. In an ideal linear rf trap, micromotion is strictly radially oriented, but small imperfections in the trap geometry can cause excess micromotion with a component along the trap axis and laser direction, thus adding phase-modulation sidebands to each hyperfine component in the $(0, 2)\rightarrow(8, 3)$ spectrum. Due to the combination of an asymmetric and saturated lineshape of the (0, 2)→(8, 3) spectrum, these sidebands can lead to a shift of ν_0. Therefore, the micromotion amplitude along the 782 nm laser needs to be determined. As the laser propagates virtually parallel to the trap axis, and since the HD$^+$ ions are always located near the trap axis, we are primarily concerned with the possible axial micromotion component.

The HD$^+$ axial micromotion amplitude x_{HD^+} can be determined through fluorescence measurements of a trapped string of beryllium ions using a modified version of the photon-rf field correlation technique [31]. The idea here is to radially displace a string of about 10 Be$^+$ ions by ~100 μm by applying a static offset field. This will induce significant radial micromotion, in addition to the axial micromotion. The 313 nm cooling laser propagates at an unknown but small angle θ (θ <10 mrad) with respect to the trap axis and may therefore have a nonnegligible projection along the radial direction. In Appendix 3, we show that if the rf voltage amplitude, V_0, is varied, the projection of the axial micromotion amplitude scales linearly with V_0, while the radial one varies as θ/V_0. The latter behavior stems from the V_0-dependent confinement and the concomitant variation of the Be$^+$ radial displacement with V_0. Thus, measuring the micromotion amplitude for various values of V_0 allows separating the radial and axial contributions.

To determine the micromotion amplitude, we use a similar setup as described in [31]. Photons detected with the PMT are converted to electrical pulses and amplified by an amplifier–discriminator, which generates a START pulse at time t_i. Subsequently, a STOP pulse is generated at time t_f at the first downward zero crossing of the rf signal. A time-to-amplitude converter (TAC) converts the duration between the START and STOP pulses to a voltage. We record the TAC output voltage with a digital phosphor oscilloscope for 400 ms. We subsequently process the stored voltage trace with a computer algorithm to obtain a histogram of START–STOP time delays, employing 1-ns bins in a range 0–76 ns (i.e., one rf cycle). The bin heights thus reflect the scattering rate as a function of the rf phase, and micromotion will lead to a modulation of the scattering rate about its mean value (Fig. 7). The Be$^+$ scattering rate (indicated as R^{MB}, where MB stands for Maxwell–Boltzmann) is given by:

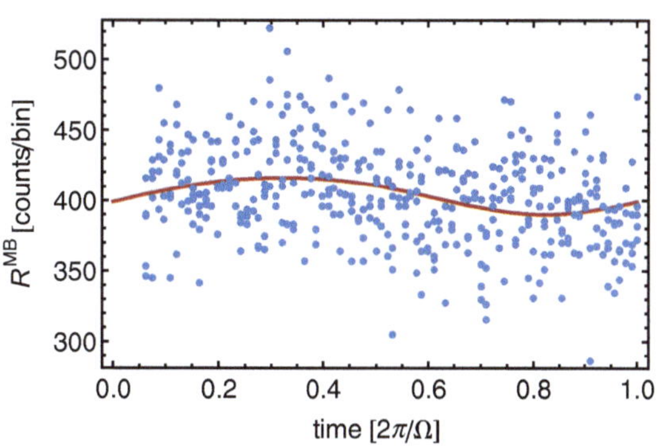

Fig. 7 Micromotion-induced Be$^+$ fluorescence modulation detected with the TAC setup. The data are fitted using Eqs. (16) and (17)

$$R^{\mathrm{MB}} = \frac{\Gamma}{2}\sqrt{\frac{m_{\mathrm{Be}}}{2\pi k_B T}} \int \frac{I/I_{\mathrm{sat}}}{I/I_{\mathrm{sat}} + 1 + (2(\Delta - \mathbf{k} \cdot (\mathbf{v}_\mu - \mathbf{v}))/\Gamma)^2} \mathrm{e}^{-\frac{m_{\mathrm{Be}} v^2}{2k_B T}} \mathrm{d}v \tag{16}$$

where m_{Be} is the Be^+ mass, T the Be^+ temperature, I the 313 nm laser intensity, I_{sat} the saturation intensity of the 313 nm cooling transition in Be^+, $\Delta = 2\pi \times -25\,\mathrm{MHz}$ the detuning of the 313 nm laser light, $\mathbf{k}$ the wavevector of the 313 nm laser, and $\mathbf{v}$ and $\mathbf{v}_\mu$ the secular and micromotion velocities of the Be^+ ions. While the rf voltage is being varied, the 313 nm laser is displaced so that the ions always are at the maximum of the Gaussian laser intensity profile. The $\mathbf{k} \cdot \mathbf{v}_\mu$ term can be written as

$$\mathbf{k} \cdot \mathbf{v}_\mu = k x_{0,k} \Omega \sin(\Omega(t - t_0)), \tag{17}$$

where $x_{0,k}$ is the amplitude of Be^+ micromotion along the direction of the laser wavevector and t_0 is a time offset.

We extract $x_{0,k}$ by fitting Eq. (16) to the acquired fluorescence histogram. Repeating this procedure for various values of V_0, a list of data points of the form $(V_0, x_{0,k})$ is obtained. We subsequently extract the radial and axial micromotion components by fitting a model function to these data. The model function is derived in Appendix 3.

The procedure of displacing a string of Be^+ ions and varying V_0 is carried out for both the horizontal and vertical directions. The data and fit functions are shown in Fig. 8, and an average axial micromotion amplitude x_{HD^+} (the projection along the 782 nm wavevector) of 11 (4) nm is found. As explained in Appendix 3, the radial micromotion contribution (due to a possible small angle of the 782 nm laser with the trap axis) averages to zero.

We incorporate the micromotion effect by extending the lineshape function D_z in Eq. (9) with sidebands at frequencies $m\Omega$ with amplitudes $J_m^2(k_{782} x_{\mathrm{HD}^+})$. Here, k_{782} denotes the wavevector of the 782 nm laser, and the J_m are Bessel functions of the first kind, with m an integer in the range $[-3, 3]$.

4.4 Effects of Chemistry in the Coulomb Crystal

During REMPD, H_2 molecules in the background gas can react with the ions in the Coulomb crystal. Such reactions can be divided into two classes: (1) elastic collisions and (2) inelastic collisions, during which a chemical reaction or charge exchange occurs, and chemical energy is converted into kinetic energy. In general, the kinetic-energy transfer to the ion from elastic collisions with room-temperature particles is much lower than the chemical energy released from inelastic collisions.

At close range r, the electric field of the ion polarizes the neutral molecule which results in an attractive interaction potential $U(r) = -\alpha Q^2/(8\pi\epsilon_0 r)^4$. Here α denotes

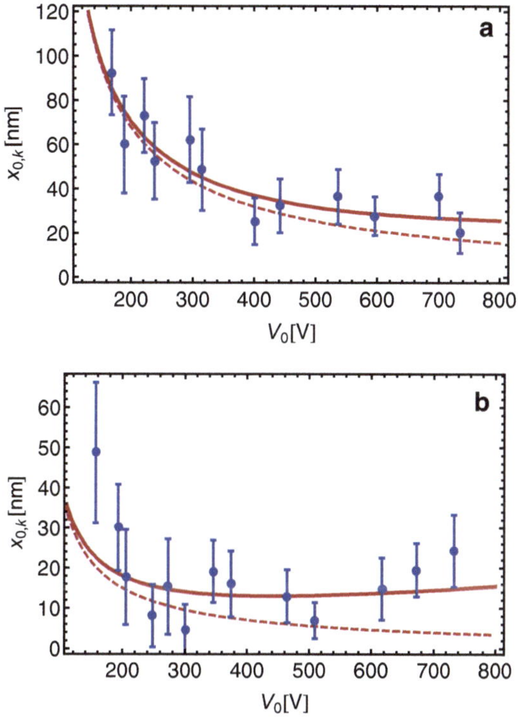

Fig. 8 Be$^+$ micromotion amplitude along the 313 nm wavevector versus rf voltage amplitude for an offset in the *horizontal* (**a**) and *vertical* (**b**) direction. The *red solid curve* is the model fit function (Eq. 36) which includes x_{HD^+} as a fit parameter. For comparison, the fitted curve with $x_{\mathrm{HD}^+} = 0$ is also shown (*dashed curve*)

the polarizability volume (in m^3)[1] of the molecule, ϵ_0 is the electric constant, and Q is the elementary charge. If we integrate the interaction force over the trajectory of the neutral near the ion (assuming a relatively large impact parameter $b > b_{\mathrm{crit}}$; see below), we obtain a change of momentum which corresponds to a velocity kick of tens of meters per second for most species.

If a neutral atom or molecule and an ion approach each other within a critical impact parameter $b_{\mathrm{crit}} = (\alpha Q^2/\pi\epsilon_0\mu_{\mathrm{red}}v)^{1/4}$, where μ_{red} and v are the reduced mass and relative velocity of the pair, a so-called Langevin collision occurs, during which the collisional partners spiral toward each other and a chemical reaction can occur at very short range [32]. The chemical reaction products contain hundreds of meV of kinetic energy, which is dissipated into the ion crystal which itself only contains about 2 meV of kinetic energy (at 10 mK). A possible adverse side effect is that such collisions may lead to time-averaged velocity distributions which deviate from thermal distributions. Table 1 shows the relevant reactions during REMPD along with the released chemical energy and their reaction rates.

The reaction numbered 1 corresponds to the REMPD process itself. From observations reported in [33], we infer that the ratio of HD$^+ \rightarrow$ D$^+$ + H and HD$^+$ $\rightarrow$ H$^+$ + D is approximately 1:1. The charge-to-mass ratio of H$^+$ is too large for this

[1]The polarizability volume and the polarizability in SI units are related through $\alpha = \alpha_{\mathrm{SI}}/(4\pi\epsilon_0)$.

Table 1 Chemical reactions occurring in the Coulomb crystal during REMPD

Number	Reaction	Energy of ionic product (eV)	Rate (s^{-1})
1	$HD^+ + h\nu + h\nu' \rightarrow D^+ + H$	0.41	0–10[a]
2	$HD^+ + H_2 \rightarrow H_2D^+ + H$	0.36	0.0042
3	$HD^+ + H_2 \rightarrow H_3^+ + D$	0.66	0.0014
4	$H_2D^+ + H_2 \rightarrow HD + H_3^+$	0.016	0.0019
5	$HD_2^+ + H_2 \rightarrow D_2 + H_3^+$	0.017	0.0004
6	$HD_2^+ + H_2 \rightarrow HD + H_2D^+$	0.022	0.0015
7	$Be^+(^2P_{3/2}) + H_2 \rightarrow BeH^+ + H$	0.25	0.0019/0.005[b]

Rates are computed for the average H_2 background pressure observed during the measurements
[a]The rate (in s^{-1} per molecule) of D^+ production is dependent on REMPD time and frequency of the 782 nm laser
[b]The rate is dependent on the fraction of time a Be^+ ion spends in the excited $^2P_{3/2}$ state, which is dependent on the 313 nm laser intensity and detuning (−80 MHz or −300 MHz, respectively)

product to be stably trapped, but the D^+ ions can stay trapped and can orbit the Coulomb crystal for many seconds. The reaction rate of 1 is calculated from the REMPD model described in Sect. 4.1 and is dependent on the frequency of the 782 nm laser.

Reaction 7 occurs most frequently due to the large number of Be^+ ions present in the trap. The reaction rate is obtained from the exponential decay of the measured 313 nm fluorescence emitted by a Coulomb crystal of Be^+ ions and is in good agreement with the rate estimated from the Langevin cross section given the background pressure of 1×10^{-8} Pa in our apparatus [34]. The different rate constants of reactions 2 and 3 illustrate the fact that HD^+ can react with H_2 in two ways: either the H_2 breaks apart, donating an H atom to the HD^+ molecule, or the HD^+ breaks apart, after which an H^+ or D^+ is added to the neutral molecule. According to [29] and [30], the probability of each scenario is approximately 50%. In case the ion breaks apart, the probability that either a H^+ or a D^+ is donated to the H_2 molecule is also 50%. This leads to a ratio of reaction 2 to 3 of 3:1. The rate of reaction 2 can be measured (keeping in mind that HD^+ and H_3^+ in reaction 3 have the same charge-to-mass ratio and therefore cannot be distinguished) by applying the measurement scheme depicted in Fig. 3 without 782 nm laser, which is further described in Sect. 4.5. The rates of reactions 4, 5, and 6 are obtained from [35]. The kinetic energies of the chemical products are calculated by using the binding energies and energy and momentum conservation laws.

MD simulations show that the fast ionic chemical products may heat up the Coulomb crystal by 1–2 mK (depending on the REMPD rate) and that the HD^+ velocity distribution becomes slightly nonthermal. A MD simulation of a Coulomb crystal containing 750 laser-cooled Be^+ ions, 40 sympathetically cooled HD^+ ions and 14 fast D^+ ions produces the HD^+ z-velocity distribution shown in Fig. 9.

Details of the MD code are given in Appendix 1. It turns out that the z-velocity distribution deviates from a Gaussian curve and is better described by a q-Gaussian

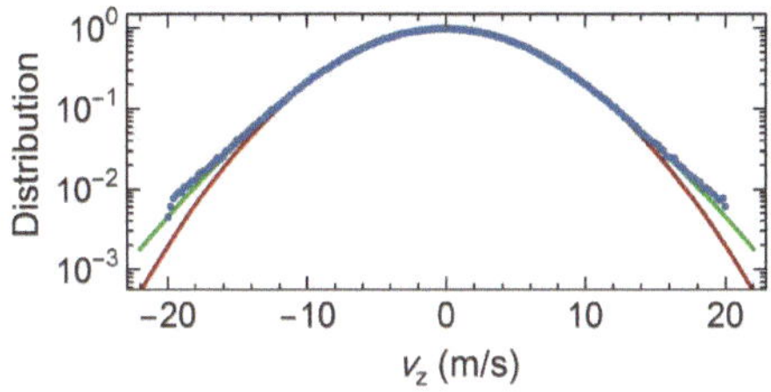

Fig. 9 The z-velocity distribution of 43 HD$^+$ molecules obtained from 100 ms simulation. 14 D$^+$ ions with an initial velocity of 6300 ms^{-1} heat up the ion crystal and give rise to a velocity distribution which differs slightly from a Gaussian (*red curve*, fitted temperature 11.60 (3) mK)) and is better described by a q-Gaussian (*green curve*, fitted temperature 11.00 (3) mK, and fitted q parameter 1.070 (3))

[36], which is a Gaussian curve with higher wings parameterized by a parameter q. For $1 < q < 3$, the q-Gaussian is written as

$$Q_G(\omega, \beta, q) = \frac{\sqrt{q-1}\left(1+\frac{(q-1)(\omega_0-\omega)^2}{2\beta^2}\right)^{\frac{1}{1-q}} \Gamma_E\left(\frac{1}{q-1}\right)}{\sqrt{2\pi}\ \beta\ \Gamma_E\left(-\frac{q-3}{2(q-1)}\right)}. \tag{18}$$

Here, ω and ω_0 are the frequency and center frequency, Γ_E is the gamma function, and β is analogous to the standard deviation of a Gaussian distribution, which is related to the Doppler width. For $q = 1$, the function reduces to a regular Gaussian distribution. While we use q-Gaussians to describe the simulated data, the observation that q-Gaussians describe the simulated velocity distributions well is merely empirical, and we did not derive this velocity distribution from a particular physical model. Since the value $\nu_{0,\text{fit}}$ turns out to be sensitive to the shape of the velocity distribution, it is important that we insert the lineshape based on the correct velocity distribution in Eq. (9) and specify the bounds to within this distribution is valid. An initial analysis reveals that $\nu_{0,\text{fit}}$ may shift by several hundreds of kHz by implementing a q-Gaussian with q varying between 1.0 and 1.1. Note that another recent study of MD simulations independently confirmed the nonthermal character of velocity distributions of laser-cooled ion crystals due to collisions with background gas molecules [37].

In the remainder of this section, we determine the q-values applicable to our REMPD measurement with the help of MD simulations. The value of q is dependent on the number of trapped fast ions, N_{fast}. The more fast ions, the higher the q-value. We note that N_{fast} is frequency dependent (more D$^+$ are produced near the peak of the REMPD spectrum) as well as time dependent [the production rate of D$^+$ is governed by the rate equations, Eq. (8)].

When a fast ion collides with a cold ion, each particle may undergo a nonadiabatic transition to a different solution of the Mathieu equation which governs its motion in the trap [38]. This implies that a fast ion can either lose or gain energy during such a collision. Since the energy change per collision is relatively small, fast ions can retain their high speeds in the trap for many seconds. Values for N_{fast} can be obtained by solving the rate equation:

$$\frac{\partial N_{\text{fast}}}{\partial t} = \alpha_{\text{prod}} N_{\text{source}} - \alpha_{\text{relax}} N_{\text{fast}} - \alpha_{\text{esc}} N_{\text{fast}} \tag{19}$$

The term α_{prod} is the production rate from a number of N_{source} particles, such as Be^+ or HD^+. In Sect. 4.2, we determined $N^+_{\text{source, Be}} = 750$ and $N^+_{\text{source, HD}} = 43$ or $N^+_{\text{source, HD}} = 84$. α_{relax} is the rate at which fast ions are cooled and become embedded within the Coulomb crystal, while α_{esc} is the rate at which (fast) ions escape from the trap. The values of α_{prod} correspond to the reaction rates in Table 1. However, obtaining α_{relax} and α_{esc} requires a multitude of individual MD simulations, with simulation periods of several seconds each. Even with current academic supercomputers, the total time to perform such simulations is prohibitively long. Therefore, we consider two extreme scenarios for relaxation and escape rates of trapped fast ions:

- Scenario 1: A minimum number of fast ions are present in the trap. α_{relax} and α_{esc} are set to their maximum value of one per second for all species, which is based on the observation that no ion loss and no relaxation are observed over simulated times up to 800 ms. This scenario will produce the smallest value of q.
- Scenario 2: A maximum number of fast ions are present in the trap, which is realized by setting α_{relax} and α_{esc} to their minimum value of zero. All fast ions remain in the trap at high speed for the entire 10 seconds of REMPD. This scenario leads to the largest value of q.

We make the assumption that fast ions do not mutually interact when present in numbers of ten or less, so that the observations based on MD simulations with ten fast ions are also valid for smaller numbers of fast ions. Note that BeH^+ ions are already created during the first secular scan before the REMPD phase starts. Fast specimen of H_2D^+ and H_3^+ occur less frequently in the trap, and in line with the extreme scenarios introduced above, we assume zero H_2D^+ and H_3^+ for scenario 1, and 3 H_2D^+ and 1 H_3^+ for scenario 2. Together with scenarios a and b described in Sect. 4.2, this gives us four scenarios in total, and therefore four different spectral fit functions and four different $\nu_{0,\text{fit}}$ results.

For all possible combinations of fast ions present during REMPD (e.g., 1 D^+, 2 BeH^+ and 1 H_2D^+ or 3 D^+, 3 BeH^+ and 3 H_2D^+) an MD simulation is carried out. From these simulations, the HD^+ z-velocity distribution is determined and a q-Gaussian is fitted which results in one q value for each simulated case. As mentioned before, the production rate of fast D^+ depends on the REMPD rate (and thus on the 782 nm laser frequency ν), which itself depends on the time-dependent number of available HD^+ ions in the target state. To take this properly into account, we introduce a time- and frequency-dependent parameter $q(t, \nu)$ as follows. For each of the four scenarios, the number of fast D^+ is simulated on a grid of different REMPD durations, t_j, and of different frequencies, ν_j of the 782 nm laser. On each

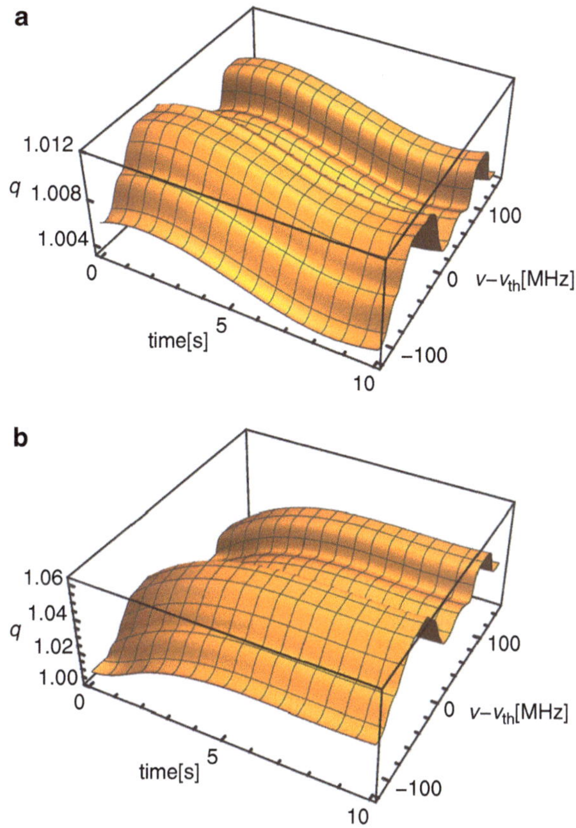

Fig. 10 3D $q(t,\nu)$ plots for scenario 1a (**a**) and 2b (**b**). Note that scenario 1a nearly corresponds to the case $q = 1$, i.e., the case in which all chemistry-induced effects can be ignored. See text for further details

point of this two-dimensional grid, the number of fast D$^+$ is combined with the number of other fast ions, and the corresponding value of $q(t_j, \nu_j)$ is looked up in a library of q values, obtained from many MD simulations performed with various combinations and abundances of fast ion species. Third-order Hermite interpolation of the grid $q(t_j, \nu_j)$ leads to a smooth continuous function $q(t, \nu)$, which is subsequently inserted into the lineshape function D_z used in Eqs. (8) and (9). Figure 10 shows the 3D plots of $q(t, \nu)$ for the different scenarios.

Besides the q value, also the ion temperature T^+_{HD} is frequency and time dependent. Due to a larger number of D$^+$ ions at the top of the spectrum than at the wings, the temperature differences between top and wings can reach a few mK. We note that the increase of q and T^+_{HD} share the same origin (namely collisions with fast ions), and in Fig. 11 we show the relation between q and T^+_{HD}, obtained from fitting q-Gaussians to simulated velocity distributions. We find a linear dependence of the form $T^+_{\mathrm{HD}}[1 + h(q(t,\nu) - 1)]$, with the slope h obtained from the fit. In scenario 1, the temperature difference between top and wings is found to be 0.5 mK. For scenario 2, the estimated temperature difference is 2.5 (5) mK,

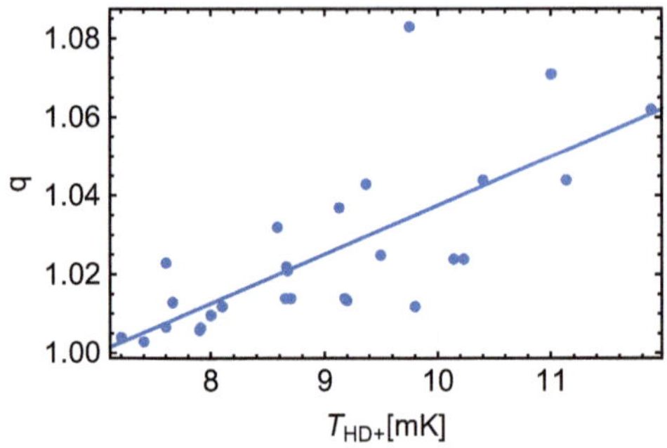

Fig. 11 The relation between the temperature of a simulated Coulomb crystal and the q value. Each point represents a simulation of approximately 100 ms. The data are best fitted with a linear function (*blue line*)

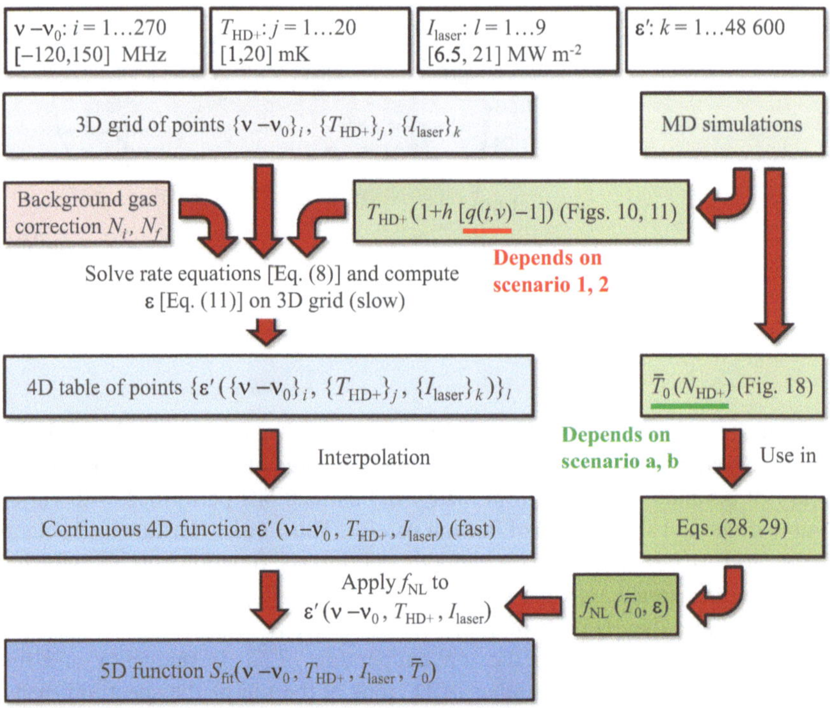

Fig. 12 Flow diagram showing the various steps involved in the construction of S_{fit}, and the role of the four scenarios considered here. As indicated at the top, the starting point is a 3D grid containing 48,600 entries, on which the rate equations are solved. While this computation takes about 1–2 days, the resulting 4D grid is readily interpolated. With the interpolation stored in computer memory, it is available for fast evaluation. MD simulations provide input on the effect of chemistry in the form of the function $q(t, \nu)$, which is different for scenarios 1 and 2. MD simulations are also used to find the relationship between $\bar{T}_0$ and the number of trapped HD^+ ions during secular excitation, which determines the shape of f_{NL} (see Appendix 2). This relationship (and therefore f_{NL}) depends on the different ion numbers used in scenarios a and b

where the uncertainty of 0.5 mK is treated as one standard deviation. The t and ν dependent temperature is also included in D_z.

A schematic overview of the various steps involved in the construction of the fit function S_{fit}, as well as the role of the four scenarios, is presented in Fig. 12.

4.5 Background Gas Reactions

During the REMPD phase, the number of HD$^+$ ions is not only reduced through photodissociation by the lasers. As described in Sect. 4.4, trapped ions can react with residual H_2 molecules in the vacuum setup. In order to correct the spectroscopic signal for these so-called background losses, the rates of reactions 2 and 3 in Table 1 are measured and included into the rate equations, Eq. (8), as an additional loss channel.

The spectral data were acquired over the course of several months during 15 independent measurement sessions lasting several hours each (Fig. 13). During this period, the background pressure varied from session to session. For each session, the HD$^+$ background signal is obtained by using the measurement scheme depicted in Fig. 3, but with a shutter blocking the 782 nm laser, thus preventing REMPD. Typically, a few HD$^+$ ions react with H_2 which is detected as a small difference between the secular scan peak areas $A_{\text{bg,i}}$ and $A_{\text{bg,f}}$. During each measurement session, a series of $\sim$7 background loss measurements is carried out three times, providing a data set of about 20 background measurements per session. From the average background loss signal per session, a reaction rate γ_{bg} is extracted from the relation

$$1 - e^{-\gamma_{\text{bg}} t} = \frac{N_{\text{bg,i}} - N_{\text{bg,f}}}{N_{\text{bg,i}}} = f_{\text{NL}}^{-1}\left(\bar{T}_0, \frac{A_{\text{bg,i}} - A_{\text{bg,f}}}{A_{\text{bg,i}}}\right) \qquad (20)$$

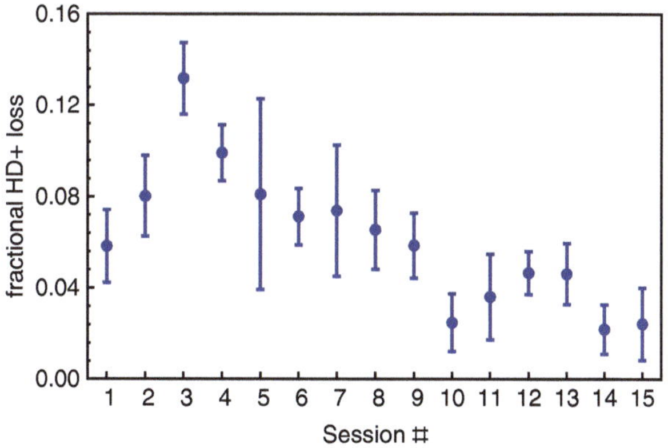

Fig. 13 Signals corresponding to background gas losses per measurement session. During some sessions, the background pressure of the vacuum was higher, which results in a higher signal. The dots are the averages of a set of measurements, the error bars represent a $\pm 1\sigma$ standard deviation

with $t = 10$ s and f_{NL}^{-1} mapping $(A_{\mathrm{bg,i}} - A_{\mathrm{bg,f}})/A_{\mathrm{bg,i}}$ onto $(N_{\mathrm{bg,i}} - N_{\mathrm{bg,f}})/N_{\mathrm{bg,i}}$ (see Appendix 2). The values of γ_{bg} are inserted into a modified set of rate equations, which include the process of background loss reactions:

$$\frac{\mathrm{d}\boldsymbol{\rho}_{bg}(t)}{\mathrm{d}t} = \left(M_{\mathrm{rempd}} + M_{\mathrm{BBR}} + M_{\mathrm{bg}}\right)\boldsymbol{\rho}_{bg}(t). \tag{21}$$

Here, the vector $\boldsymbol{\rho}_{bg}$ equals $\boldsymbol{\rho}$ of Eq. (8) extended by two additional rows which describe the occurrence of ions in the form of H_2D^+ or H_3^+. M_{bg} is a diagonal matrix containing γ_{bg} which describes the HD^+ losses. We take into account the fact that conversion of HD^+ to H_3^+ in reaction 3 in Table 1 is not detected by the method of detection through secular excitation because HD^+ and H_3^+ have the same mass-to-charge ratio. This also means that the measured values of γ_{bg} only represent the rates for reaction 2. As explained in Sect. 4.4, the reaction rate of 3 is a factor of three lower, which is taken into account as well.

We correct the data set of each session individually for the background signal following an iterative procedure. During the first step, we insert an initial (coarse) estimate of the value of $\bar{T}_0$ in Eq. (20), and we simply subtract the signal predicted by the model without background losses (based on Eq. 8) from the signal prediction including background losses [based on Eq. (21)]. In this way, an estimate of the background signal is obtained which is subsequently subtracted from the raw measurement data. Then the spectral fit function S_{fit} (see Sect. 4.1) is fitted to the corrected data points with free-fit parameters $\bar{T}_0$, T_{HD}^+, I_{laser} and $\nu_{0,\mathrm{fit}}$. To find an improved estimate of the background signal, the thus found values of $\bar{T}_0$, T_{HD}^+, I_{laser} are reinserted into Eqs. (8) and (21), and the above procedure is repeated. After a few iterations, convergence is achieved.

The apparent loss of HD^+ ions is also influenced by reactions 4 and 5 in Table 1, which result in the production of additional H_3^+ during REMPD, leading to a reduction of the observed REMPD and background signals mentioned above. This effect cannot be easily assessed experimentally, and instead we estimate it for both scenarios a and b using the reaction rates of Table 1, leading to slightly modified values of N_i and N_f. This translates to a small correction to ϵ in Eqs. (11) and (20); see also Fig. 12). In addition, reactions with background gas also take place during the secular scans, thus influencing the determination of the initial and final numbers of particles with charge-to-mass ratio 1:3 itself. For example, the conversion of HD^+ to H_2D^+ (reaction 2) and to H_3^+ (reaction 3) has a small effect on the number of estimated HD^+ ions from the secular scan. Again, we can readily correct for these effects, knowing the values of all the relevant reaction rates, and noting that the reactions take place on a much longer timescale than the secular scan itself so that constant production rates can be assumed. Altogether, the corrections to N_i and N_f are at the level of a few percent and are applied through Eqs. (11) and (20) (see also Fig. 12).

4.6 Spectrum, Systematic Effects and Final Result

The function S_{fit} (Eq. 13) is fitted to the REMPD data set after correction for the background signal. Figure 14 shows the REMPD data set and fit function for scenario 1a. The noise in the spectrum has several origins. Firstly, the number of trapped ions is relatively small and varies from shot-to-shot. Secondly, the population in the various hyperfine states of the $L = 2$ state varies from shot-to-shot, as expected for hyperfine states with a mean occupancy of order unity. The (stochastic) BBR interaction, which couples the states with $L = 2$ with other rotational states, introduces additional random signal variations. Furthermore, the variation of the number of reactions of HD$^+$ with the background gas is in the order of a few per shot, which dominates the noise for low REMPD signals. Finally, part of the noise originates from random intensity variations due to spatial alignment variations of the 313, 782, and 532 nm lasers.

For each of the four scenarios (1a, 1b, 2a, and 2b), we obtain a particular set of fit parameters, which are listed in Table 2. The correlation coefficients of the fit parameters are presented in Table 3, and the values for $\nu_{0,\text{fit}} - \nu_{\text{th}}$ are graphically shown in Fig. 15. The error bars represent the $\pm 1\sigma$ fit uncertainty which can be considered as the purely statistical precision of the spectroscopy measurement. The 0.33 MHz statistical uncertainty includes a 5 kHz error due to the finite grid spacing used to obtain S_{fit}, as well as a 9 kHz uncertainty associated with the piecewise polynomial interpolation method (found by comparing Hermite interpolation with

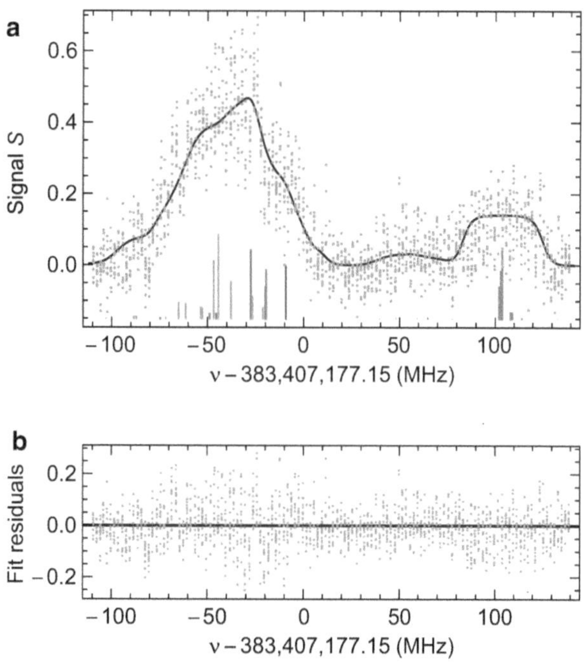

Fig. 14 **a** Measured spectrum (*dots*) of the (v, L): $(0, 2) \rightarrow (8, 3)$ transition and a least-squares fit (*solid curve*). All 1772 data points are plotted together with the fit function corresponding to scenario 1a. The fitted functions corresponding to the different scenarios are visually indistinguishable from the result shown here. The *colored* sticks represent the positions and linestrengths of the theoretical individual hyperfine components and are plotted using the same color coding as used in Fig. 2b. **b** Fit residuals

Table 2 The fit results of the free-fit parameters per scenario

Fit parameter	Scenario 1a	1b	2a	2b
$\nu_{0,\mathrm{fit}} - \nu_{\mathrm{th}}$(MHz)	0.46 (33)	0.37 (34)	−0.03 (32)	−0.12 (33)
T_{HD^+}(mK)	10.9 (8)	11.0 (8)	10.6 (8)	10.35 (80)
I_{laser} ($\times 10^7$ Wm^{-2})	0.99 (14)	1.04 (15)	0.95 (14)	0.98 (15)
$\bar{T}_0$(K)	2.79 (3)	3.85 (5)	2.82 (5)	3.84 (5)

Scenarios 1 and 2 (corresponding to small and large values of q, respectively) result in significantly different values of $\nu_{0,\mathrm{fit}}$. Scenarios a and b (small and large numbers of molecular hydrogen ions, respectively) give different results for $\bar{T}_0$. This can be explained by the different amounts of trapped molecular species, which result in different Be^+ temperatures during a secular scan

Table 3 The correlation coefficients for the fit of S_{fit} (scenario 1a) to the data

	$\nu_{0,\mathrm{fit}} - \nu_{\mathrm{th}}$	T_{HD^+}	I_{laser}	$\bar{T}_0$
$\nu_{0,\mathrm{fit}} - \nu_{\mathrm{th}}$	1	−0.609	0.483	0.049
T_{HD^+}	−0.609	1	−0.690	0.038
I_{laser}	0.483	−0.690	1	0.579
$\bar{T}_0$	0.049	0.038	0.579	1

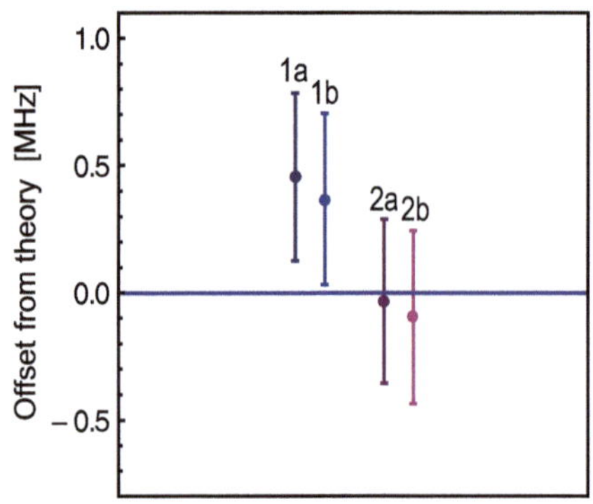

Fig. 15 Values of $\nu_{0,\mathrm{fit}} - \nu_{\mathrm{th}}$ found for the four different scenarios 1a, 1b, 2a, and 2b. *Error bars* indicate the 1σ fit uncertainty. All values are plotted with respect to the theoretical frequency (*blue horizontal line*) [2]

B-spline interpolation results). We remark that the sensitivity to the chemistry processes (scenarios 1 and 2) is much stronger than the sensitivity to the numbers of HD^+ ions in the trap (scenarios a and b) and that the values $\nu_{0,\mathrm{fit}}$ for 1a and 2a represent extreme upper and lower limits (with respect to the line shift due to chemistry) for the average number of HD^+ ions. We therefore chose to obtain our final result for $\nu_{0,\mathrm{fit}}$ by taking the mean of these two values, while interpreting the mean single-fit error of (which is virtually the same for all four results) as the statistical uncertainty of the final result. We subsequently quantify the 'which-scenario' uncertainty as follows. For scenarios 1a and 1b (and similarly for 2a and 2b), the difference is due to a 1σ variation in the number of HD^+ ions. Therefore, we treat the frequency interval between the two values corresponding to scenario a and b as the corresponding 1σ interval, which amounts to 80 kHz when averaged over the two scenario's 1 and 2. To find the error corresponding to scenarios 1 and 2, we

Table 4 Systematic shifts and uncertainty budget

Origin	Shift (MHz)	Uncertainty (MHz)	(ppb)
Resolution (statistical fit error)	0	0.33	0.85
Uncertainty q value	−0.25[a]	0.23	0.61
Uncertainty N_{HD^+}	0	0.080	0.21
Ignoring populations $L = 6$	0	0.032	0.083
T_{HD^+} variation in spectrum	0	0.028	0.072
Doppler effect due to micromotion	−0.055[a]	0.020	0.052
Frequency measurement	0	0.010	0.026
BBR temperature	0	0.005	0.013
Zeeman effect	−0.0169	0.003	0.008
Stark effect	−0.0013	0.0001	0.0004
Electric-quadrupole shift	0[b]	0.0001	0.0003
2nd order Doppler effect	0[b]	0.000005	0.00001
Uncertainty E_4	0[b]	0.000001	0.000003
Total	−0.0182	0.41	1.1

[a]This is a shift with respect to a scenario with a zero effect of q or micromotion and serves to illustrate the size of the effect. The shift itself, however, is absorbed in the value of $\nu_{0,\text{fit}}$
[b]The value of these shifts is actually nonzero but negligibly small, and therefore ignored here

take the frequency interval between the values found for scenarios 1a and 2a (which are essentially extreme limits) and conservatively equate the interval to a 68% confidence interval. The interval thus corresponds to 2σ, with $\sigma = 0.23$ MHz. The uncertainty of 0.5 mK in the temperature difference between top and wings of the spectrum (see Sect. 4.4) results in 28 kHz difference in $\nu_{0,\text{fit}}$, which is treated as a 1σ variation. The frequency shifts due to these systematic effects are listed together with their uncertainties in Table 4.

4.6.1 Frequency Uncertainty of the 782 nm Laser

The beat note of the frequency-locked 782 nm laser with the optical frequency comb is counted during REMPD. We use the beat-note frequencies to compute the Allan deviation, which is of the order of 0.1 MHz after 10 s averaging. The uncertainty of the 782 nm laser frequency can be transferred to an uncertainty in the (0, 2)→(8, 3) fit result $\nu_{0,\text{fit}}$ by taking the Allan deviation as a measure of the standard deviation of a Gaussian noise distribution, describing the laser frequency offset from the set frequency during each REMPD cycle. We perform a Monte Carlo simulation in which each of the 140 measurement frequencies is assigned a frequency offset, selected at random from the Gaussian distribution. Repeating this 100 times generates 100 different spectral data sets. Fitting S_{fit} to each of the data sets gives 100 different values of $\nu_{0,\text{fit}}$. A histogram of the resulting distribution of $\nu_{0,\text{fit}}$ values is shown in Fig. 16. From the histogram, we find a mean offset 0.5 kHz

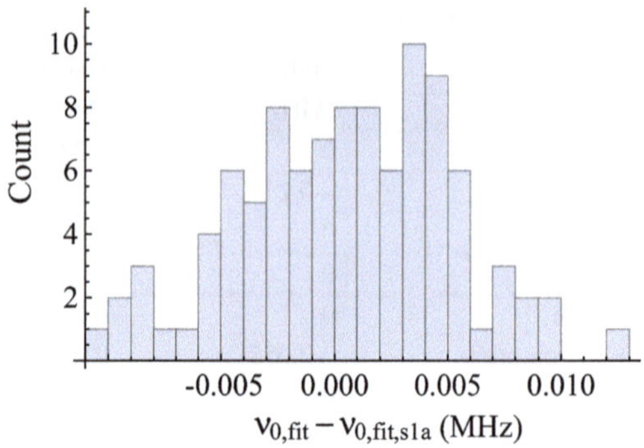

Fig. 16 Histogram of 100 fitted frequencies of the (0, 2)→(8, 3) transition, obtained from a Monte Carlo simulation involving 100 data sets with added random frequency noise (with the noise level corresponding to the measured noise level; see text). On the horizontal axis, zero corresponds to $\nu_{0,\mathrm{fit}}$ obtained for scenario 1a. The mean offset and standard deviation are 0.5 and 8 kHz, respectively

from the frequency value $\nu_{0,\mathrm{fit}}$ found for scenario 1a, and a standard deviation of 8 kHz. We conclude that frequency noise introduces no significant bias, and we conservatively assume the uncertainty of the frequency measurement to be 0.01 MHz.

The 782 nm laser has a Gaussian lineshape with a width of ∼0.5 MHz. The convolution of this lineshape with the (Gaussian) Doppler-broadened line (16 MHz) will give rise to another Gaussian lineshape. Since the linewidths add up quadratically, the increase in linewidth is smaller then the uncertainty of the linewidth due to the fit uncertainty of the temperatures, which is ∼0.8 mK. Therefore, we consider the laser linewidth to be completely absorbed into the fitted temperature T_{HD^+} with no significant effect on its value.

4.6.2 Zeeman, Stark and Other Shifts

So far we have neglected the Zeeman splitting of the lines in the spectrum. Incorporating the Zeeman effect makes the hyperfine transition matrices very large, and MATHEMATICA is only able to solve the rate equations (Eq. 8) effectively if lineshapes are not too complicated. We circumvent these issues as follows. First, we calculate the lineshifts and linestrengths of the magnetic subcomponents of individual hyperfine lines. This is done by diagonalizing the sum of the hyperfine and Zeeman Hamiltonians (using the 0.19 mT field value used throughout the experiments), after which the eigenvectors and energy values are used to compute the stick spectrum of the (v, L):(0, 2)→(8, 3) transition. This procedure is similar to that followed in Refs. [10, 39] and can readily be done for the case of σ^+, σ^- and π transitions and superpositions thereof. As the Zeeman splitting is small compared to the Doppler width, the magnetic subcomponents belonging to the same hyperfine

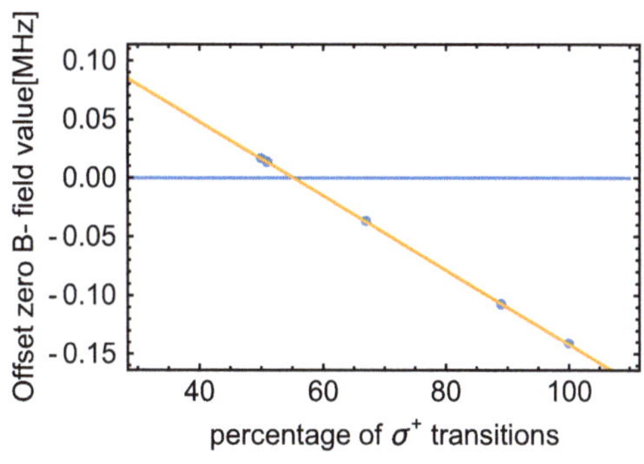

Fig. 17 Zeeman shift of $\nu_{0,\mathrm{fit}}$ found with a spectral fit function including the Zeeman effect due to a 0.19 mT magnetic field, for different ratios of σ^+/σ^- transitions. The *horizontal line* at 0 MHz corresponds to the value obtained using a fit function assuming zero magnetic field

component overlap well within the profile of the lineshape function D_z, forming a new composite (and Zeeman-shifted) lineshape function D'_z. This new lineshape function is subsequently used in Eq. (9). For simplicity, we do not implement effects of micromotion and chemistry in this analysis, and we compare $\nu_{0,\mathrm{fit}}$ fit results based on versions with $B = 0.19$ mT with a version with zero B field. It is important to note that the linear polarization of the 782 nm laser is practically perpendicular to the B field, so that during the (0, 2)→(8, 3) excitation only σ^+ and σ^- transitions are driven. Due to polarization imperfections, caused by the polarization optics and the slightly birefringent viewports of the vacuum chamber, the two circularly polarized components are estimated to have a maximum possible intensity imbalance of 2%. Figure 17 shows the offset between the $\nu_{0,\mathrm{fit}}$ fit results, obtained with the above model for several σ^+/σ^- intensity ratios, and the fit result we found previously assuming zero magnetic field. For the 0.19 mT field used in our experiment, a small shift of -0.017 (3) MHz is obtained, with the uncertainty due to the possible maximum polarization imbalance.

The ac Stark shifts due to the 782, 532, and 313 nm lasers are -869, -452, and 8 Hz, respectively. These values represent the shift of the center of gravity of the spectral line and can therefore be considered as weighted means of all the shifts of the single hyperfine components. The calculation of the Stark shifts is shortly explained in Appendix 4. Stark shifts due to the BBR and trap rf field are calculated in [20] and are smaller than 1 Hz. Together, this gives us a total Stark shift of 1.3 (1) kHz. The uncertainty in this value stems almost exclusively from the accuracy to which the laser beam intensities are known.

A conservative upper limit of 100 Hz to the electric-quadrupole shift for the (v, L):(0, 2)→(8, 3) transition is obtained from [40]. The second-order Doppler effect is dominated by average micromotion velocity of the HD⁺ ions and is found to be less than 5 Hz. The values of the aforementioned shifts and their uncertainties are listed in Table 4.

If we compare $\nu_{0,\mathrm{fit}}$ from scenario 1a based on the model containing rotational states $L \leq 5$ with an extended version containing $L \leq 6$ states, we find a shift of $\nu_{0,\mathrm{fit}}$ of 28 kHz, which we treat as the uncertainty due to the neglect of population in $L = 6$. The rate-equation model furthermore includes the BBR temperature, which we estimate to be 300 K with an uncertainty of about 5 K, caused by day-to-day variations of the temperature in the laboratory, and by a possibly elevated temperature of the trap electrodes due to rf current dissipation. If we compare the $\nu_{0,\text{fit}}$ values after inserting BBR temperatures of 300 and 305 K, we obtain a difference of 5 kHz, which we include in the uncertainty budget. Micromotion also causes a shift: by comparing the $\nu_{0,\text{fit}}$ value from scenario 1a that includes micromotion (amplitude 11 (4) nm) with a the result of a version with zero micromotion, a shift of 0.055 (20) MHz is obtained. Finally, we investigated the effect due to the uncertainty of the spin coefficient E_4 (see Eq. 1), which is estimated to be 50 kHz [17]. Comparing fits with E_4 values differing by 50 kHz, we find that this has a negligibly small effect of ~1 Hz on the result for ν_0.

4.6.3 Frequency of the (v, L):(0, 2)→(8, 3) Transition

We take the average of the fit values obtained from scenarios 1a and 2a, corrected by the systematic shifts described above, to find the value of the (v, L):(0, 2)→(8, 3) transition frequency, $\nu_0 = 383,407,177.38$ MHz. Table 4 shows the error budget, with a total uncertainty of 0.41 MHz that corresponds to 1.1 ppb. This result differs 0.21 MHz (0.6 ppb) from the more accurate theoretical value, ν_{th}= 383,407,177.150 (15) MHz [2]. The two main uncertainty contributions are the statistical fit error of 0.33 MHz and the uncertainty in the q-factor scenario of 0.23 MHz.

5 Conclusion, Implications and Outlook

We have measured the (v, L):(0, 2)→(8, 3) transition in the HD^+ molecule with 0.85 p.p.b (0.33 MHz) precision, which is the first sub-p.p.b. resolution achieved in molecular spectroscopy. A thorough analysis of systematic effects points out that the total uncertainty is 1.1 p.p.b., and the result ($\nu_0 = 383,407,177.38(41)$ MHz) differs by only 0.6 p.p.b. from the theoretically predicted value ($\nu_{\text{th}} = 383,407,177.150(15)$ MHz). A large contribution to the systematic uncertainty is the effect of chemical reactions in the Coulomb crystal, of which the 1σ uncertainty is 0.61 p.p.b (0.23 MHz). This effect, which had not been recognized before, causes a nonthermal velocity distribution that can be approximated by a q-Gaussian function, and which significantly influences the accuracy of laser spectroscopy of composite lineshapes in the presence of strong saturation and depletion of the HD^+ sample. This is a situation regularly encountered in laser spectroscopy of Doppler-broadened transitions in finite samples of trapped molecular ions.

The agreement between experimental and theoretical data has several implications. First, it enables a test of molecular theory at the 1-p.p.b. level, and a test of molecular QED at the level of 2.7×10^{-4}, which are the most stringent tests performed so far. Second, it allows us to put new bounds on the existence of hypothetical fifth forces, and put new limits on the compactification radius of higher dimensions, as described in detail in [3, 4, 12]. Third, the result presented here can be used to obtain a new value the proton-to-electron mass ratio with a precision of 2.9 p.p.b. [12], for the first time from molecular spectroscopy as proposed already four decades ago [5].

Our analysis clearly demonstrates that the first-order Doppler effect is responsible for the largest contribution to the uncertainty. In fact, removing the first-order Doppler effect would render the frequency measurement as the largest source of error (0.026 p.p.b), thereby immediately improving the uncertainty by about a factor of 40. Such—and even larger—improvements are possible using two-photon spectroscopy. For example, in Refs. [22, 41] an experiment was proposed in which the $(v, L) = (0, 3) \rightarrow (9, 3)$ line in HD$^+$ is addressed through a two-photon transition with nearly degenerate photons. Using counter-propagating laser beams with a narrow linewidth, the Lamb Dicke regime may be reached in the present apparatus, such that first-order Doppler broadening is entirely eliminated, while all other systematic effects could be controlled below the 1×10^{-13} uncertainty level. Such spectroscopy would allow more stringent tests of molecular theory and QED, tighter bounds on new physics at the Ångström scale, a competitive determination of the proton–electron mass ratio and even contribute to the determination of several other fundamental constants including the Rydberg constant, the deuteron-electron mass ratio, and the proton and deuteron charge radii [41].

Acknowledgements We are indebted to J. Bouma, T. Pinkert and R. Kortekaas for technical assistance, and to V. Korobov, E. Hudson and R. Gerritsma for fruitful discussions. This research was funded through the Netherlands Foundation for Fundamental Research on Matter (FOM), the COST action MP1001 IOTA, and the Dutch-French bilateral Van Gogh Programme. J.C.J.K. thanks the Netherlands Organisation for Scientific Research (NWO) and the Netherlands Technology Foundation (STW) for support. SURFsara (www.surfsara.nl) is acknowledged for the support in using the Lisa Compute Cluster for MD simulations.

Appendix

Appendix 1: Molecular Dynamics Simulations

MD simulations are implemented in FORTRAN code in order to realistically describe the dynamics of trapped and laser-cooled ions in the presence of the time-dependent trapping field, 313 nm photon scattering by the Be$^+$ ions, and fast ionic products from chemical reactions. A cycle of one time step starts by computing the sum of the forces acting on each ion, which consists of the Coulomb force, $\mathbf{F}_{\mathrm{C}}$, the time-dependent force from the trapping field, $\mathbf{F}_{\mathrm{trap}}$, and an optional rf electric field, $\mathbf{F}_{\mathrm{SS}}$, which drives the secular motion:

$$\mathbf{F}_{\text{tot}} = \sum \mathbf{F} = \mathbf{F}_{\text{C}} + \mathbf{F}_{\text{trap}} + \mathbf{F}_{\text{SS}}. \tag{22}$$

The radial and axial part of $\mathbf{F}_{\text{trap}}$ are given by

$$F_{\text{trap},x,y} = -\frac{QV_0}{R^2}(x\hat{x} - y\hat{y})\cos(\Omega t) + \frac{1}{2}Q\omega_z^2(x\hat{x} + y\hat{y}) \tag{23}$$

and

$$F_{\text{trap},z} = az + bz^3 + cz^5, \tag{24}$$

where ω_z is the secular angular frequency in the z-direction. The constants a, b, and c depend on the trap geometry, which are determined through a finite-element analysis performed with the software package SIMION. The forces exerted on each ion are calculated, and trajectories are obtained using the leapfrog method [42]. Doppler-cooling is included at the level of single-photon scattering. Photon momentum kicks are simulated as velocity changes where absorption only takes place in the laser direction. In order to include ion motional heating which occurs in the trap, we implemented additional stochastic velocity kicks with a size of the recoil momentum of a single 313 nm photon with random directions. If an average kick rate of 75 MHz is used, ion temperatures of around 10 mK are obtained.

The processes of elastic and inelastic neutral-ion collisions are simulated as velocity kicks in random directions. For example, simulating reaction 7 in Table 1, a Be^+ ion is substituted with a BeH^+ ion at 10 mK, after which its speed is modified so as to give it 0.25 eV of kinetic energy.

Simulation of particles which in some cases have high velocities requires the use of a variable time step size, $\Delta\tau$. If the proper step size is not observed, two particles with a high velocity difference at close distance could 'skip' each other within one time step instead of colliding. The default step size is $\Delta\tau = 0.2$ ns. However, if for any of the trapped particles the condition $v_j\Delta\tau > 10 \times \min\{\Delta x_{jk}\}$ is met, where min $\{\Delta x_{jk}\}$ is the distance between particle j with velocity v_j and the nearest particle k, the time step $\Delta\tau$ is reduced by a factor of 10. Likewise, the step size is increased if the colliding particles separate again and $v_j\Delta\tau < 100 \times \min\{\Delta x_{jk}\}$.

To simulate EMCCD images, we made use of a simpler MD implementation, which treats the motion of the ions in the pseudopotential approximation and which does not include high-energy ions. This allows for a larger integration time step (10 ns) and, thus, faster MD simulations.

Appendix 2: Derivation of the Nonlinear Fluorescence Function

The relative HD^+ loss during REMPD, ϵ, is related to the spectroscopic signal S through the nonlinear function f_{NL}, which we derive here. The fluorescence yield during a secular scan depends on the Be^+ temperature, T, and is described by the scattering rate formula integrated over a Maxwell–Boltzmann velocity distribution. Neglecting micromotion effects, we take the scattering rate $R^{MB} = R^{MB}(T, \Delta, I/I_{sat})$ defined in Eq. (16). During a secular scan in the experiment, we use the values $\Delta = 2\pi \times -300$ MHz and $I/I_{sat} = 67$, and in what follows we drop these variables from the function argument of R^{MB}. While performing the secular scan, the temperature T varies, which leads to a fluorescence peak as described by Eq. (16). The spectroscopic signal S is the relative difference between the areas under the fluorescence peaks (see Eq. 7). We may rewrite the area, A, as

$$A = C\left[R^{MB}(\bar{T})\Delta t - R^{MB}(\bar{T}_{bl})\Delta t\right], \tag{25}$$

where C is a constant taking into account the collection and quantum efficiencies of the PMT or EMCCD imaging system, Δt denotes the duration of the secular scan (10 s), and $\bar{T}$ stands for the 'effective' value of T during a secular scan. The effective temperature $\bar{T}$ is defined through Eq. (25) and can be interpreted as follows. If we would fix the secular excitation field frequency and amplitude at a certain value during a scan, the Be^+ temperature and the fluorescence level would remain constant. $\bar{T}$ is the constant temperature that leads to the same area under the fluorescence trace as during a true secular scan. Likewise, $\bar{T}_{bl}$ stands for the effective baseline temperature during Δt (corresponding to the fluorescence level that results if no HD^+ ions are present). In the experiment, the baseline may have a small slope due to the wing of the secular resonance of particles with mass 4 and 5 amu (see, for example, Fig. 5). This slope is detected and removed by the MATHEMATICA code we use to analyze the PMT signal traces.

Both in the experiment and in the MD simulations described below, we observe the area A under experimental or simulated fluorescence traces, to which we subsequently can assign an effective temperature $\bar{T}$ through Eq. (25). We emphasize that no attempt is made to derive $\bar{T}$ directly from, for example, the simulated velocities of Be^+ ions. Also note that in practice, we only use Eq. (25) to assign effective temperatures to simulated fluorescence traces, as in this case the constant C is known ($C = 1$).

Inserting Eq. (25) into Eq. (7), we obtain

$$S = \frac{R^{MB}(\bar{T}_i) - R^{MB}(\bar{T}_f)}{R^{MB}(\bar{T}_i) - R^{MB}(\bar{T}_{bl})}, \tag{26}$$

which is a relationship between the spectroscopic signal S and the effective ion temperatures during the initial and final secular scan, $\bar{T}_i$ and $\bar{T}_f$, respectively.

The relationship between the number of trapped HD^+ molecules and $\bar{T}$ can be obtained from MD simulations. A Coulomb crystal is simulated containing 750 trapped Doppler-cooled Be^+ ions and with HD^+ numbers varying from 0 to 100. This is done once using a number of additional H_2D^+ and HD_2^+ ions equal to that of scenario a, and once using the numbers of H_2D^+ and HD_2^+ of scenario b (see Sect. 4.2). The simulated secular scans over the HD^+ secular resonance frequency

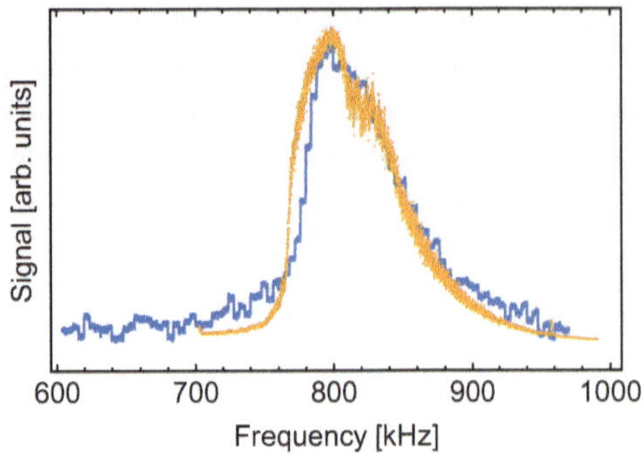

Fig. 18 A simulated (*yellow*) and a real (*blue*) secular scan peak plotted on the same frequency axis. The signals are scaled vertically to achieve matching peak heights

produce fluorescence peaks which agree qualitatively with those obtained in the laboratory as shown in Fig. 18.

In the experiment, secular scans are acquired over a time span of 10 s. The dynamics due to the time-varying frequency of the ac electric field take place at a timescale much longer than the timescale of fluorescence dynamics during laser cooling, which takes place at timescales of the order of 10 μs [43], and also longer than typical ion oscillation periods and the frequency of the ac electric field itself (1–20 μs). However, due to limited computational resources, the simulated duration of a secular scan is approximately 100 ms, which implies that the simulated dynamics take place at a considerably faster rate than in the experiment, typically on the scale of milliseconds. However, this still is much longer than the timescale of fluorescence dynamics and the motional dynamics. Therefore, we assume that the simulated secular scan peaks provide a reliable model of the experimentally observed secular scans.

The MD simulations reveal a linear relationship between the number of trapped HD^+ ions and $\bar{T}$. This agrees with the intuitive picture of Be^+ ions with frictionally damped motion (because of the laser cooling), whose temperature rise during secular excitation is directly proportional to the number of HD^+ ions. Figure 19 shows the $(N_{HD^+}, \bar{T})$ relationship for the two scenarios a and b. Having established that $\bar{T}$ is a linear measure of the number of trapped HD^+ molecules, we now combine the relations $\epsilon = (N_i - N_f)/N_i$, $\bar{T}_i = c_1 N_i + c_2$ and $\bar{T}_f = c_1 N_f + c_2$, where c_1 and c_2 are constants derived from MD simulations (Fig. 19), to obtain

$$\bar{T}_f(\epsilon) = \bar{T}_i(1 - \epsilon) + c_2\epsilon. \tag{27}$$

$\bar{T}_i$ can also be defined as the effective temperature with zero HD^+ loss ($\bar{T}_i = \bar{T}_f(\epsilon = 0) \equiv \bar{T}_0$), while the term c_2 can be considered as the effective baseline temperature ($c_2 = \bar{T}_f(\epsilon = 1) = \bar{T}_{bl}$). Inserting Eq. (27) into Eq. (26) results in the nonlinear function

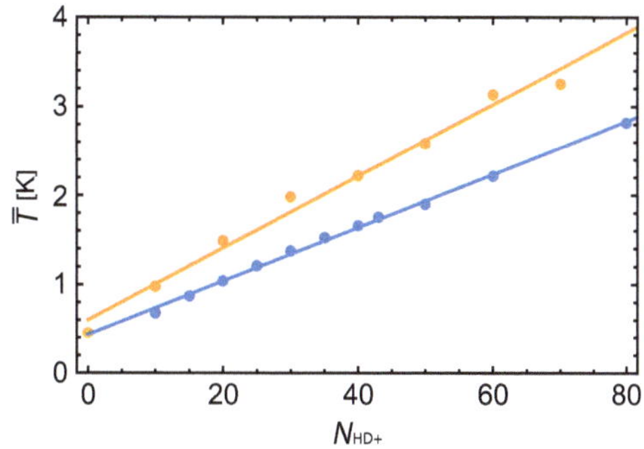

Fig. 19 Results from MD simulations showing the effective Be$^+$ ion temperature during a secular scan $\bar{T}$ versus the number of HD$^+$ ions N_{HD^+}, assuming numbers of H_2D^+ and HD_2^+ ions as in scenario a (*blue dots*) and scenario b (*yellow dots*). The *blue* and *yellow lines* represent least-squares fits, revealing a linear relationship between $\bar{T}$ and N_{HD^+}

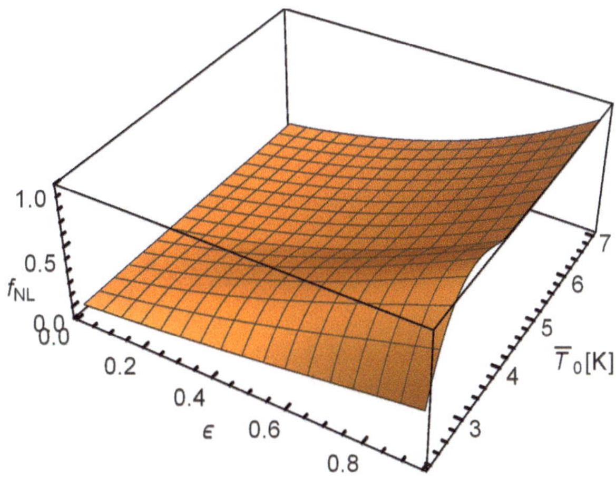

Fig. 20 3D plot of the function $f_{\mathrm{NL}}(\bar{T}_0, \epsilon)$ (here plotted for scenario a) which connects the raw measurement signal S to the actual fractional loss of HD$^+$, ϵ

$$f_{\mathrm{NL}}(\bar{T}_0, \epsilon) \equiv \frac{R^{\mathrm{MB}}(\bar{T}_0) - R^{\mathrm{MB}}((\bar{T}_{\mathrm{bl}} - \bar{T}_0)\epsilon + \bar{T}_0)}{R^{\mathrm{MB}}(\bar{T}_0) - R^{\mathrm{MB}}(\bar{T}_{\mathrm{bl}})}, \tag{28}$$

which is plotted for scenario a in Fig. 20. Note that the nonlinear dependence on ϵ originates from the nonlinear dependence of R^{MB} on T in Eq. (16).

In the analysis $\bar{T}_0$ is treated as a free fit parameter. $\bar{T}_{\mathrm{bl}}$ is kept at a fixed value which is obtained from MD simulations. From Fig. 19, it can be seen that $\bar{T}_{\mathrm{bl}} \simeq 0.5$ K for scenarios a and b. This indicates that the rf field used for secular excitation already induces heating of Be$^+$ while the field is still far away from the Be$^+$ resonance (at ~ 300 kHz). This effect is also seen in the experiment.

The nonlinear function f_{NL} is used to map the relative HD$^+$ loss ϵ onto the spectroscopic signal S. However, for the correction of background signals (Sect.

4.5) we need to map S to ϵ, which requires the inverse nonlinear function f_{NL}^{-1}. This inverse function is obtained numerically by use of MATHEMATICA.

Appendix 3: Micromotion Fit Function

The time-dependent electric field of the trap, $\mathbf{E}_{\mathrm{t}}$, can be expressed as [31]

$$\begin{aligned}\mathbf{E}_{\mathrm{t}}(x,y,z,t) \cong \; & -\frac{V_0}{R^2}(x\hat{x} - y\hat{y})\cos(\Omega t) \\ & -\frac{\kappa U_0}{Z_0^2} \times (2z\hat{z} - x\hat{x} - y\hat{y}),\end{aligned} \tag{29}$$

where R is half the distance between two diagonally opposing electrodes, U_0 is the endcap voltage, Z_0 stands for half the distance between the end caps, and κ is a shielding factor. The Be^{+} micromotion amplitude can be written as

$$\mathbf{x}_0 = \frac{Q}{m_{\mathrm{Be}}\Omega^2}\mathbf{E}_{\mathrm{t}}(x,y,z,0). \tag{30}$$

The measured micromotion amplitude can be written as

$$x_{0,k} = \frac{\mathbf{k}\cdot\mathbf{x}_0}{||\mathbf{k}||}, \tag{31}$$

which is the projection of the 313 nm laser direction onto $\mathbf{x}_0$. From simulations of the rf trap circuitry with the simulation software SPICE, we find a small possible phase difference ϕ_{ac} of 4 mrad in between the rf electrodes, which has a negligible effect on the ion micromotion and is ignored here. Using the program SIMION, we calculate the shielding factor κ and the static electric field $\mathbf{E}_{\mathrm{dc}}$ as a function of the dc voltages applied to the trap electrodes. From the static electric field, the radial ion displacement $\mathbf{r}_d$ is obtained by balancing the ponderomotive force and static E-field in the radial direction,

$$m_{\mathrm{Be}}\omega_r^2\mathbf{r}_d = -Q\mathbf{E}_{\mathrm{dc}}, \tag{32}$$

where ω_r is the radial secular trap frequency. By inserting the x- and y-components of $\mathbf{r}_d$ into Eq. (29), the field vector $\mathbf{E}_{\mathrm{t}}$ at the location of the ions is obtained.

A geometric imperfection of the trap could lead to an axial rf field, which can be written as (here we ignore the small modification of the radial rf field of the trap due to the same imperfection):

$$E_{\mathrm{ax,HD^+}}(V_0,t) = \frac{1}{Q}\frac{V_0}{V_{0,\mathrm{e}}} m_{\mathrm{HD}} x_{\mathrm{HD}} \Omega^2 \cos(\Omega t), \tag{33}$$

where x_{HD} is the HD$^+$ micromotion amplitude along the trap z axis, m_{HD} is the mass of HD$^+$, and $V_{0,\mathrm{e}}$ is the rf voltage used during the spectroscopic measurements, which is 270 V.

Now, we turn to the case of a linear string of Be$^+$ ions, which is the configuration used to determine the axial rf field amplitude. Adding $E_{\mathrm{ax,HD^+}}$ to the z-component of $\mathbf{E}_\mathrm{t}$ gives a new expression for $\mathbf{E}_\mathrm{t}$ which is inserted into Eq. (30). We then obtain the following expression for $\mathbf{x}_0$:

$$\mathbf{x}_0 = \left(\frac{2E_{\mathrm{dc},x} Q R^2 V_0 Z_0^2}{Q V_0^2 Z_0^2 - 2m_{\mathrm{Be}} R^4 U_0 \kappa \Omega^2}, \frac{2E_{\mathrm{dc},y} Q R^2 V_0 Z_0^2}{Q V_0^2 Z_0^2 - 2m_{\mathrm{Be}} R^4 U_0 \kappa \Omega^2}, \frac{m_{\mathrm{HD}} V_0 x_{\mathrm{HD}}}{m_{\mathrm{Be}} V_{0,\mathrm{e}}} \right), \tag{34}$$

which is subsequently inserted into Eq. (31), together with the wavevector, which is written as

$$\mathbf{k} = \frac{2\pi}{\lambda}(\sin(\theta)\cos(\phi), \sin(\theta)\sin(\phi), \cos(\theta)). \tag{35}$$

Here θ is the angle between $\mathbf{k}$ and the trap z axis, and ϕ is the angle between $\mathbf{k}$ and the trap y axis, which is very close to $\pi/4$ in our setup. The value of θ lies between ± 10 mrad and is treated as a free fit parameter. We insert Eqs. (35) and (34) into Eq. (31) and then expand the expression in powers of θ. This gives us the following fit function:

$$\begin{aligned} x_{0,k}(V_0) &= \frac{m_{\mathrm{HD}}}{m_{\mathrm{Be}}} \frac{V_0 x_{\mathrm{HD}}}{V_{0,\mathrm{e}}} \\ &\quad - \frac{8(E_{\mathrm{h,dc}} - E_{\mathrm{v,dc}} + \delta E_\mathrm{h} - \delta E_\mathrm{v}) Q^2 V_0}{m_{\mathrm{Be}}^2 R^2 \Omega^4 (2a_\mathrm{M} + q_\mathrm{M}^2)} \theta \\ &\quad + \mathcal{O}(\theta^2). \end{aligned} \tag{36}$$

Here $E_{\mathrm{h,dc}}$, $E_{\mathrm{v,dc}}$ are the applied static electric fields (corresponding to $\mathbf{E}_{\mathrm{dc}}$) in the horizontal and vertical directions, respectively, and δE_h, δE_v are the unknown offset electric fields (due to *e.g.* charging of electrodes). The Mathieu parameters a_M and q_M are given by

$$a_\mathrm{M} = \frac{-4Q\kappa U_0}{m_{\mathrm{Be}} Z_0^2 \Omega^2}, \quad q_\mathrm{M} = \frac{2QV_0}{m_{\mathrm{Be}} R^2 \Omega^2}. \tag{37}$$

The displacement of the Be$^+$ string in the vertical direction can be accurately determined with images of the EMCCD camera, and therefore δE_h can be zeroed (for example, by minimizing the displacement of the Be$^+$ string while the radial confinement of the trap is modulated by varying the rf amplitude). However, the displacement in the horizontal direction (i.e., perpendicular to the EMCCD image plane) is not accurately known, and therefore, we treat δE_h as another free fit parameter. In summary, we use Eq. (36) as a fit function with x_{HD^+}, θ and δE_{h} as free fit parameters while neglecting higher orders of θ. The fitted curves and the result for $x_{0,k}$ are shown in Sect. 4.3.

The question arises what happens if the 782 nm laser propagates at a small angle with respect to the trap axis, while the HD^+ ions form a shell structure around the trap axis. In this case, a small fraction of the radial micromotion is projected onto the wavevector. However, from Eqs. (30–29) it follows that the sign of this additional micromotion alternates for each quadrant in the (x, y) plane. As long as the radial micromotion component does not exceed the axial micromotion amplitude (which is the case here), the former averages out to zero given the radial symmetry of the HD^+ crystal.

Appendix 4: Stark Shift Calculations

Here, we summarize the formulas that are used to calculate the ac Stark shift of a ro-vibrational transition $(v, L) \rightarrow (v', L')$ in the HD^+ molecule induced by a laser with intensity I and polarization state p. A general expression for the second-order energy shift depending on the angle θ between the polarization direction and the quantization axis is:

$$\begin{aligned} \Delta E &= -\frac{1}{2}\frac{I}{c}[\alpha_{vL}^{(0)}(\omega) \\ &\quad + P_2(\cos\theta)\frac{3M^2 - L(L+1)}{L(2L-1)}\alpha_{vL}^{(2)}(\omega)], \end{aligned} \tag{38}$$

where $P_2(x) = \frac{1}{2}(3x^2 - 1)$ is a Legendre polynomial. This expression contains the scalar and tensor polarizabilities

$$\begin{aligned} \alpha_{vL}^{(0)}(\omega) &= 4\pi a_0^3 Q_s, \\ \alpha_{vL}^{(2)}(\omega) &= 4\pi a_0^3 \sqrt{\frac{L(2L-1)}{(L+1)(2L+3)}} Q_t, \end{aligned} \tag{39}$$

where a_0 is the Bohr radius and Q_s and Q_t stand for the two-photon scalar and tensor matrix elements:

$$\begin{aligned} Q_s &= \frac{\langle vL \| Q^{(0)} \| v'L \rangle}{\sqrt{2L+1}} \\ Q_t &= \frac{\langle vL \| Q^{(2)} \| vL \rangle}{\sqrt{2L+1}}. \end{aligned} \tag{40}$$

Here $Q^{(0)}$ and $Q^{(2)}$ are the irreducible scalar and tensor components that belong to the two-photon operator (in atomic units):

$$Q_{pp}(E) = \mathbf{d} \cdot \boldsymbol{\epsilon}_p \frac{1}{H - E} \mathbf{d} \cdot \boldsymbol{\epsilon}_p, \tag{41}$$

with Hamiltonian H, dipole moment operator, $\mathbf{d}$ and polarization vector $\boldsymbol{\epsilon}_p$. The matrix elements Q_s and Q_t were calculated numerically using the three-body variational wave functions described in [8].

Since the hyperfine structure is partially resolved in this spectrum, we also have to consider the contribution of the Stark shifts to off-resonant coupling to hyperfine levels in $v = 0$ and $v = 8$ by the 782 nm laser during spectroscopy. Here, the situation is more complicated as the 782 nm laser also nonresonantly couples $v = 8$ states to continuum states above the dissociation limit of the $1s\sigma$ electronic ground state. The 782-nm contribution to the Stark shift was calculated at the hyperfine level, and will be published elsewhere. The corresponding shifts turn out to be negligible for our experiment, contributing only at the level of a few Hertz.

References

1. V.I. Korobov, L. Hilico, J-Ph Karr, $m\alpha^7$-order corrections in the hydrogen molecular ions and antiprotonic helium. Phys. Rev. Lett. **112**, 103003 (2014)
2. V.I. Korobov, L. Hilico, J-Ph Karr, Theoretical transition frequencies beyond 0.1 ppb accuracy in H_2^+, HD$^+$, and antiprotonic helium. Phys. Rev. A **89**, 032511 (2014)
3. E.J. Salumbides, J.C.J. Koelemeij, J. Komasa, K. Pachucki, K.S.E. Eikema, W. Ubachs, Bounds on fifth forces from precision measurements on molecules. Phys. Rev. D **87**, 112008 (2013)
4. E.J. Salumbides, A.N. Schellekens, B. Gato-Rivera, W. Ubachs, Constraints on extra dimensions from precision molecular spectroscopy. New J. Phys. **17**(3), 033015 (2015)
5. W.H. Wing, G.A. Ruff, W.E. Lamb, J.J. Spezeski, Observation of the infrared spectrum of the hydrogen molecular ion HD$^+$. Phys. Rev. Lett. **36**, 1488–1491 (1976)
6. C.W. Chou, D.B. Hume, J.C.J. Koelemeij, D.J. Wineland, T. Rosenband, Frequency comparison of two high-accuracy Al$^+$ optical clocks. Phys. Rev. Lett. **104**, 070802 (2010)
7. S. Schiller, D. Bakalov, V.I. Korobov, Simplest molecules as candidates for precise optical clocks. Phys. Rev. Lett. **113**, 023004 (2014)
8. J-Ph Karr, H_2^+ and HD$^+$: Candidates for a molecular clock. J. Mol. Spectr. **300**, 37–43 (2014)
9. C. Haase, M. Beyer, Ch. Jungen, F. Merkt, The fundamental rotational interval of para-H_2^+ by MQDT-assisted Rydberg spectroscopy of H_2. J. Chem. Phys. **142**, 064310 (2015)
10. J.C.J. Koelemeij, B. Roth, A. Wicht, I. Ernsting, S. Schiller, Vibrational spectroscopy of HD$^+$ with 2-ppb accuracy. Phys. Rev. Lett. **98**, 173002 (2007)
11. U. Bressel, A. Borodin, J. Shen, M. Hansen, I. Ernsting, S. Schiller, Manipulation of individual hyperfine states in cold trapped molecular ions and application to HD$^+$ frequency metrology. Phys. Rev. Lett. **108**, 183003 (2012)
12. J. Biesheuvel, J-Ph Karr, L. Hilico, K.S.E. Eikema, W. Ubachs, J.C.J. Koelemeij, Probing QED and fundamental constants through laser spectroscopy of vibrational transitions in HD$^+$. Nat Commun **7**, (2016)
13. V.I. Korobov, Coulomb three-body bound-state problem: Variational calculations of nonrelativistic energies. Phys. Rev. A **61**, 064503 (2000)
14. H. Li, J. Wu, B.-L. Zhou, J.-M. Zhu, Z.-C. Yan, Calculations of energies of the hydrogen molecular ion. Phys. Rev. A **75**, 012504 (2007)

15. Y. Hijikata, H. Nakashima, H. Nakatsuji, Solving non-Born-Oppenheimer Schrödinger equation for hydrogen molecular ion and its isotopomers using the free complement method. J. Chem. Phys. **130**, 024102 (2009)
16. P.J. Mohr, B.N. Taylor, D.B. Newell, CODATA recommended values of the fundamental physical constants: 2010. Rev. Mod. Phys. **84**, 1527–1605 (2012)
17. D. Bakalov, V.I. Korobov, S. Schiller, High-precision calculation of the hyperfine structure of the HD^+ ion. Phys. Rev. Lett. **97**, 243001 (2006)
18. J.C.J. Koelemeij, D.W.E. Noom, D. de Jong, M.A. Haddad, W. Ubachs, Observation of the $v' = 8 \leftarrow v = 0$ vibrational overtone in cold trapped HD^+. Appl. Phys. B **107**(4), 1075–1085 (2012)
19. J.M. Brown, A. Carrington, *Rotational spectroscopy of diatomic molecules* (Cambridge University Press, Cambridge, 2003)
20. J.C.J. Koelemeij, Infrared dynamic polarizability of HD^+ rovibrational states. Phys. Chem. Chem. Phys. **13**(42), 18844–18851 (2011)
21. R.C. Hilborn, Einstein coefficients, cross sections, f values, dipole moments, and all that. Am. J. Phys. **50**, 982–986 (1982)
22. V.Q. Tran, J-Ph Karr, A. Douillet, J.C.J. Koelemeij, L. Hilico, Two-photon spectroscopy of trapped HD^+ ions in the Lamb-Dicke regime. Phys. Rev. A **88**, 033421 (2013)
23. J. Biesheuvel, D.W.E. Noom, E.J. Salumbides, K.T. Sheridan, W. Ubachs, J.C.J. Koelemeij, Widely tunable laser frequency offset lock with 30 GHz range and 5 THz offset. Opt. Express **21**(12), 14008–14016 (2013)
24. F.M.J. Cozijn, J. Biesheuvel, A.S. Flores, W. Ubachs, G. Blume, A. Wicht, K. Paschke, G. Erbert, J.C.J. Koelemeij, Laser cooling of beryllium ions using a frequency-doubled 626 nm diode laser. Opt. Lett. **38**, 2370–2372 (2013)
25. P. Blythe, B. Roth, U. Fröhlich, H. Wenz, S. Schiller, Production of ultracold trapped molecular hydrogen ions. Phys. Rev. Lett. **95**, 183002 (2005)
26. B. Roth, J.C.J. Koelemeij, H. Daerr, S. Schiller, Rovibrational spectroscopy of trapped molecular hydrogen ions at millikelvin temperatures. Phys. Rev. A **74**, 040501 (2006)
27. J.C.J. Koelemeij, B. Roth, S. Schiller, Blackbody thermometry with cold molecular ions and application to ion-based frequency standards. Phys. Rev. A **76**, 023413 (2007)
28. T. Schneider, B. Roth, H. Duncker, I. Ernsting, S. Schiller, All-optical preparation of molecular ions in the rovibrational ground state. Nat. Phys. **6**, 275–278 (2010)
29. T. Oka, The infrared spectrum of H_3^+ in laboratory and space plasmas. Rev. Mod. Phys. **64**, 1141–1149 (1992)
30. J.E. Pollard, L.K. Johnson, D.A. Lichtin, R.B. Cohen, State selected reactive scattering. H_2^++H $_2$ $\rightarrow H_3^+$+H. J. Chem. Phys. **95**(7), 4877–4893 (1991)
31. D.J. Berkeland, J.D. Miller, J.C. Bergquist, W.M. Itano, D.J. Wineland, Minimization of ion micromotion in a Paul trap. J. Appl. Phys. **83**(10), 5025–5033 (1998)
32. J.B. Hasted, *Physics of Atom Collisions* (Butterworths, London, 1964)
33. B.D. Esry, H.R. Sadeghpour, Adiabatic formulation of heteronuclear hydrogen molecular ion. Phys. Rev. A **60**, 3604–3617 (1999)
34. D.J. Wineland, C. Monroe, W.M. Itano, D. Leibfried, B.E. King, D.M. Meekhof, Experimental issues in coherent quantum state manipulation of trapped atomic ions. J. Res. Natl. Inst. Stand. Tech. **103**, 259 (1998)
35. K. Giles, N.G. Adams, D. Smith, A study of the reactions of trihydrogen(1+), H_2D^+, HD_2^+, and D_3^+ with hydrogen molecule, HD, and D_2 using a variable-temperature selected ion flow tube. J. Phys. Chem. **96**(19), 7645–7650 (1992)
36. S. Umarov, C. Tsallis, S. Steinberg, On a q-central limit theorem consistent with nonextensive statistical mechanics. Milan Journal of mathematics **76**(1), 307–328 (2008)
37. I. Rouse, S. Willitsch, Superstatistical velocity distributions of cold trapped ions in molecular-dynamics simulations. Phys. Rev. A **92**, 053420 (2015)
38. K. Chen, S.T. Sullivan, W.G. Rellergert, E.R. Hudson, Measurement of the coulomb logarithm in a radio-frequency Paul trap. Phys. Rev. Lett. **110**, 173003 (2013)

39. D. Bakalov, V.I. Korobov, S. Schiller, Magnetic field effects in the transitions of the HD^+ molecular ion and precision spectroscopy. J. Phys. B: At. Mol. Opt. Phys. **44**, 025003 (2011)
40. D. Bakalov, S. Schiller, The electric quadrupole moment of molecular hydrogen ions and their potential for a molecular ion clock. Appl. Phys. B **114**(1–2), 213–230 (2014)
41. J.-Ph. Karr, L. Hilico, J.C.J. Koelemeij, V.I. Korobov, Hydrogen molecular ions for improved determination of fundamental constants. Phys. Rev. A **94**, 050501(R) (2016)
42. L. Verlet, Computer "experiments" on classical fluids. I. Thermodynamical properties of Lennard-Jones molecules. Phys. Rev. **159**(1), 98 (1967)
43. J.H. Wesenberg, R.J. Epstein, D. Leibfried, R.B. Blakestad, J. Britton, J.P. Home, W.M. Itano, J.D. Jost, E. Knill, C. Langer, R. Ozeri, S. Seidelin, D.J. Wineland, Fluorescence during Doppler cooling of a single trapped atom. Phys. Rev. A **76**, 053416 (2007)

Orientational Dependence of Optically Detected Magnetic Resonance Signals in Laser-Driven Atomic Magnetometers

Simone Colombo, Vladimir Dolgovskiy, Theo Scholtes, Zoran D. Grujić, Victor Lebedev, and Antoine Weis

Abstract We have investigated the dependence of lock-in-demodulated M_x-magnetometer signals on the orientation of the static magnetic field $\mathbf{B}_0$ of interest. Magnetic resonance spectra for 2400 discrete orientations of $\mathbf{B}_0$ covering a 4π solid angle have been recorded by a PC-controlled steering and data acquisition system. Off-line fits by previously derived lineshape functions allow us to extract the relevant resonance parameters (shape, amplitude, width, and phase) and to represent their dependence on the orientation of $\mathbf{B}_0$ with respect to the laser beam propagation direction. We have performed this study for two distinct M_x-magnetometer configurations, in which the rf-field is either parallel or perpendicular to the light propagation direction. The results confirm well the algebraic theoretical model functions. We suggest that small discrepancies are related to hitherto uninvestigated atomic alignment contributions.

1 Introduction

Optically pumped atomic magnetometers, also known as optical magnetometers (OM), are based on resonant magneto-optical effects in atomic (usually alkali-metal) vapors [1]. We refer the reader to the comprehensive overview of various OM methods and their applications in Ref. [2]. Magnetometers based on optically detected magnetic resonance (ODMR) have the longest history in the field of atomic magnetometry, and the so-called M_x-magnetometer using a single light beam has proven to be a highly sensitive and robust device. The theoretical modeling of the signals generated by ODMR-based magnetometers is addressed in great detail in a recently published textbook (Chapter 13 in Ref. [3]).

This article is part of the topical collection "Enlightening the World with the Laser" - Honoring T. W. Hänsch guest edited by Tilman Esslinger, Nathalie Picqué, and Thomas Udem.

S. Colombo (✉) • V. Dolgovskiy • T. Scholtes • Z.D. Grujić • V. Lebedev • A. Weis
Physics Department, University of Fribourg, Chemin du Musée 3, CH-1700 Fribourg, Switzerland
e-mail: simone.colombo@unifr.ch

D. Meschede et al. (eds.), *Exploring the World with the Laser*,
https://doi.org/10.1007/978-3-319-64346-5_17

ODMR magnetometers infer the modulus $B_0 = \omega_0/\gamma_F$ of the magnetic field vector $\mathbf{B}_0$ from the (driven) Larmor precession frequency ω_0 of an atomic vapor's magnetization, where γ_F is the gyromagnetic ratio of the used atom ($\gamma_F/2\pi \approx 3.5$ Hz/nT for ^{133}Cs). The precession is driven by a much weaker additionally applied oscillating field $\mathbf{B}_1(t)$, called the 'rf-field'. In the standard M_x-magnetometer, a single circularly polarized light beam whose frequency is resonant with an atomic transition is used both to create the medium's spin polarization by optical pumping [4] and to read out the spin precession signal.

The magnetometric sensitivity of an OM, i.e., the smallest magnetic field change that the device can detect (in a given bandwidth), depends on many parameters, such as the light intensity, the atomic number density, the size of the atomic sample, the spin coherence time, and the amplitude of the rf-field. Moreover, the sensitivity critically depends on the relative orientations of the light propagation direction $\hat{k}$, the rf-field $\hat{\mathbf{B}}_1$, and the field of interest $\hat{\mathbf{B}}_0$. The latter dependencies imply that there are, on the one hand, orientation(s) that optimize the device's sensitivity, and, on the other hand, orientations (so-called dead-zones) for which the sensitivity vanishes. The quantitative understanding of these dependencies is crucial when designing a magnetometer, be it for a laboratory application in which the orientation $\hat{\mathbf{B}}_0$ of $\mathbf{B}_0$ is mostly known *a priori*, or for field applications where the knowledge of the dead-zones is of great importance.

The problem of the OM sensitivity's orientation dependence is closely related to the so-called 'heading error' that has already been addressed in the very early accounts on optically pumped atomic magnetometers [5]. Several attempts have been made to overcome those fundamental effects and realize dead-zone-free OM [6–10].

The object of the present paper is an experimental verification of the theoretically predicted [3] orientation dependencies of lock-in-detected signals in different M_x-magnetometers of two distinct geometrical configurations, viz., rf-field either parallel or perpendicular to the light's $\mathbf{k}$-vector. For this, we have developed a computer-controlled experimental setup allowing the rotation of a static magnetic field vector of constant modulus over the full 4π solid angle. We record magnetic resonance spectra at 2400 discrete (θ_B, ϕ_B) orientations of the field, and off-line analysis permits then three-dimensional representations of the results.

2 Experimental Setup

The experimental setup (Fig. 1) is mounted inside of a cubic five-layer μ-metal shield (produced by Sekels GmbH) with inner dimensions of $\sim 50^3$ cm^3.

The central part of the magnetometer is a spherical (30 mm diameter) Pyrex cell with paraffin-coated inner surface which is connected by a capillary to a reservoir stem containing a droplet of solid cesium producing a room-temperature saturated atomic vapor [11]. Laser light is guided to the setup by a multimode fiber which

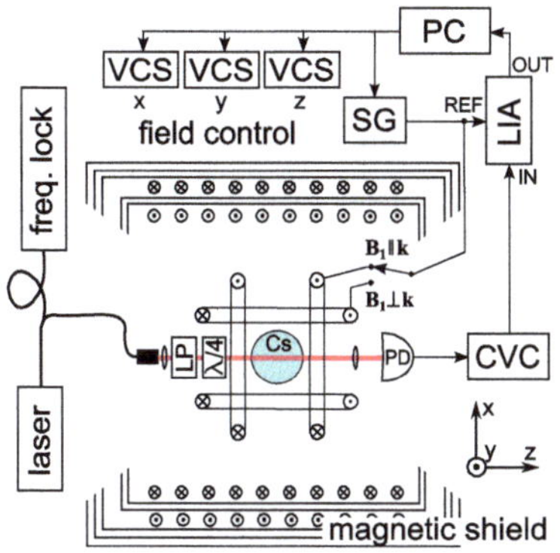

Fig. 1 Schematic of the experimental setup. The coils producing the $\mathbf{B}_0$ field's x, y, and z components are wound around the two innermost μ-metal layers. Two of the four coils producing B_z are shown as illustration. *VCS* voltage-controlled current source, *SG* signal generator, it LIA lock-in amplifier, *LP* linear polarizer, *PD* photodiode, *CVC* current-to-voltage converter, *PC* computer running control and data acquisition code

effectively scrambles the light polarization. The out-coupled (≈ 2 mm diameter) light beam's polarization is made circular using a linear polarizer and a quarter-wave plate. We use a polarimeter (Thorlabs, model PAX5710IR1-T) for the precision control of the light's polarization (cf. Sect. 7.1).

The frequency of the extended cavity diode laser (Toptica, model DL100 pro) is actively stabilized to the center of the $F = 4 \rightarrow F' = 3$ hyperfine transition of the Cs D_1 line ($\lambda = 894.6$ nm) by means of a separate saturation-absorption spectroscopy unit. The power of the laser beam is kept constant by an active stabilization circuit using an intensity modulator (Jenoptik, model AM894) driven by a slow PI controller with a 10 Hz cutoff frequency. The power of the transmitted light beam is detected by a photodiode whose photocurrent is amplified by a current-to-voltage converter (Femto, model DLPCA-200, 10^6 V/A gain, 200 kHz bandwidth) and fed to a lock-in detector (Zurich Instruments, model HF2LI).

Magnetic resonance spectra are recorded by automated frequency sweeps of the rf-coil current produced by the built-in oscillator of the lock-in amplifier. The frequency dependence of both the in-phase and quadrature signals obtained by phase-sensitive demodulation of the photodiode signal is stored for off-line processing.

The amplifier's bandwidth and the finite inductivity of the rf-coil imply that the photocurrent's Fourier component of interest (oscillating at ~ 35 kHz) is phase-shifted by a (frequency dependent) value φ of $\approx 20°$ with respect to the coil-driving voltage.

Data taking is fully computer-controlled by a dedicated LabView code allowing control of the rf-field amplitude and the three components of the offset field $\mathbf{B_0}$.

2.1 Magnetic Field Control and Calibration

The static magnetic field inside the shield is produced by a triaxial coil system wound onto the two innermost μ-metal layers. The coils (resistance 4.5 Ω) are driven by voltages from three programmable arbitrary waveform generators (Agilent, model 33500B) via 50 Ω series resistors. After demagnetization, we measure the remnant field components in the shield using level-crossing (Hanle) resonances as described in Refs. [12, 13]. We nullify these components—that are typically below 70 nT—and calibrate the field producing coils in the following way: For calibrating the B_z coil, we use an M_z geometry, in which we scan B_z from negative to positive values, while irradiating the atoms with a 1.5 kHz rf-field. From the positions of the two magnetic resonances observed in this configuration, we infer both the residual field δB_z and the coil calibration constant k_z. The B_x and B_y coils are calibrated using a 10 μT offset field $(0, B_{0y}, B_{0z})$ direction. We measure changes of the corresponding Larmor frequency

$$f_0 = \gamma_F \sqrt{(k_x I_x)^2 + (B_{0y} + k_y I_y)^2 + B_{0z}^2} \tag{1}$$

when powering the x- and y- coils individually. Fits allow then to infer k_x and k_y as well as the residual fields δB_x and δB_y.

We have observed current scan direction-related (sub-%) differences of the k_x and k_y constants. These differences may be attributed to the nonlinear ferromagnetic response of the μ-metal onto which the coils are wound. The calibration constant of a given coil may thus be affected by the orientation and the magnitude of the offset field. Based on this, we believe that we control the field orientation at the $\lesssim 1\%$ level.

Using this calibration, the generators are programmed such as to vary the θ_B- and ϕ_B- orientations of the magnetic field vector $\mathbf{B_0}$ in a step-wise manner, while keeping the field modulus B_0 (nominally) constant. The field is varied on a full sphere evolving from the north pole $(0, 0, B_0)$ to the south pole $(0, 0, -B_0)$ in the coordinates of Fig. 1.

We record magnetic resonance spectra for typically 2400 pairs of discrete values of the field orientation angles (Fig. 2). The duration of each spectrum scan is $\approx$ 23 s, so that a complete full-sphere orientation scan ('$\theta - \phi$-scan') takes $\approx$ 15 h.

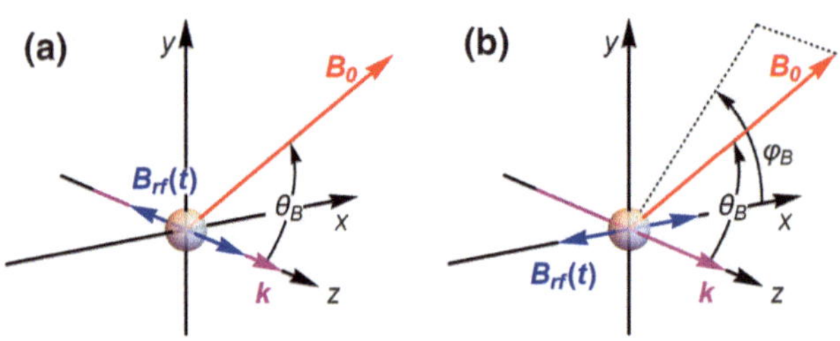

Fig. 2 Definition of the orientation angles θ_B and ϕ_B of the magnetic field in the geometries $\mathbf{B}_1 \| \mathbf{k}$ (**a**) and $\mathbf{B}_1 \perp \mathbf{k}$ (**b**)

3 Theory of the M_x-Magnetometer

The so-called M_x-magnetometer relies on reading out (by optical means) the frequency at which an atomic medium's spin polarization precesses around a static magnetic field $\mathbf{B_0}$. The precession is coherently driven by a magnetic field (rf-field) of small amplitude $B_1 \ll B_0$ that oscillates at a frequency $\omega_{\rm rf}$ close to the Larmor frequency $\omega_0 = \gamma_F B_0$. The ensuing phase-synchronized precession of all individual spins leads to a modulation of the transmitted light power at the frequency $\omega_{\rm rf}$. The amplitude (and phase) of the induced power modulation depends in a resonant manner on the detuning $\delta\omega \equiv \omega_{\rm rf} - \omega_0$ of the rf frequency from the Larmor frequency. Weis, Grujić ,and Bison have presented an exhaustive discussion of the mathematical expressions for the observed lineshapes in magnetic resonance-based atomic magnetometers in a recently published textbook [3].

The simplest M_x-magnetometer implementation uses a single circularly polarized laser beam (in resonance with an atomic transition) that serves for both the creation of the spin polarization and for the readout of the coherently driven spin precession. The laser power transmitted through the atomic vapor, which is assumed to be optically thin ($\kappa L \ll 1$), is given by

$$P = P_0 \exp[-\kappa L] \approx P_0(1 - \kappa L)\,, \tag{2}$$

where P_0 is the incident light power and L the length of the traversed vapor column. The absorption coefficient κ for circularly polarized light depends on the medium's spin polarization and can be expressed as

$$\kappa = \kappa_0^{\rm unpol}\left[1 - \alpha^{(1)}_{F,F'} S_z - \alpha^{(2)}_{F,F'} A_{zz}\right], \tag{3}$$

where the coefficients $\alpha^{(K)}_{F,F'}$ are transition-specific constants, and where $\kappa_0^{\rm unpol}$ is the absorption coefficient of the unpolarized medium, S_z and A_{zz} being the medium's longitudinal spin orientation (vector polarization) and alignment (tensor polarization), respectively. Since alignment-related contributions on the $4 \rightarrow 3$ transition studied here are small [14], we will neglect in the following the last term in Eq. (3). The subscript z refers to the quantization axis, chosen along the light propagation direction $\mathbf{k}$. In that frame, the longitudinal orientation can be expressed in terms of the sublevel populations p_m as

$$S_z \propto \sum_{m=-F}^{F} p_m\, m\,, \tag{4}$$

where m is the magnetic quantum number (eigenvalue of F_z). In what follows S_0 is the value of S_z achieved by optical pumping in a (polarization-stabilizing) magnetic field $\mathbf{B_0}$ oriented along $\mathbf{k}$.

The joint action of the torques exerted by the offset magnetic field $\mathbf{B}_0$ and the oscillating rf-field $\mathbf{B}_{\mathrm{rf}}(t) = B_1\hat{\mathbf{B}}_1 \sin\omega_{\mathrm{rf}}t$ on the spin polarization causes a modulation of $S_z(t)$, and hence a time-dependent modulation

$$\delta P(t) = P_0\kappa_0^{\mathrm{unpol}} L\alpha_{F,F'}^{(1)}\, S_z(t) \tag{5}$$

$$\equiv P_{\mathrm{R}}(\delta\omega)\,\sin\left[\omega_{\mathrm{rf}}t + \varphi(\delta\omega)\right] \tag{6}$$

$$\equiv P_{\mathrm{IP}}(\delta\omega)\sin\omega_{\mathrm{rf}}t + P_{\mathrm{QU}}(\delta\omega)\cos\omega_{\mathrm{rf}} \tag{7}$$

of the transmitted laser power, where $\delta\omega = \omega_{\mathrm{rf}} - \omega_0$ is the rf frequency detuning, and $P_{\mathrm{IP}} = P_{\mathrm{R}}\cos\varphi$ the in-phase and $P_{\mathrm{QU}} = P_{\mathrm{R}}\sin\varphi$ the quadrature amplitude, respectively. As shown in [3], the power modulation is phase-shifted with respect to the rf-field oscillation by

$$\varphi(\delta\omega) = \varphi\left(\delta\omega;\hat{\mathbf{B}}_1,\hat{\mathbf{B}}_0\right) = \varphi^{(0)}\left(\hat{\mathbf{B}}_1,\hat{\mathbf{B}}_0\right) - \arctan\frac{\delta\omega}{\gamma}\,, \tag{8}$$

where the on-resonance phase shift $\varphi^{(0)} = \varphi(\delta\omega = 0)$ depends both on the orientation, $\hat{\mathbf{B}}_1$, of the rf-field and on the orientation, $\hat{\mathbf{B}}_0$, of the magnetic field of interest.

We choose $\hat{z} = \hat{\mathbf{k}}$ as polar axis of a spherical coordinates system, so that the magnetic field orientation is determined by the polar and the azimuthal angles, θ_B and ϕ_B, respectively. In this coordinate frame, the amplitude of the detected light power oscillation ('R'-signal) is given by

$$P_R(\delta\omega) = \widetilde{P}\,\frac{\sqrt{\gamma^2 + \delta\omega^2}}{\delta\omega^2 + \gamma^2 + \widetilde{\Omega}_i^2}\,\widetilde{\Omega}_i|\sin\theta_B\cos\theta_B|\,, \tag{9}$$

where γ is the spin relaxation rate, which is assumed to be isotropic (setting $\gamma = \gamma_1 = \gamma_2$), and where the signal calibration factor is given by $\widetilde{P} = S_0P_0\kappa_0^{\mathrm{unpol}}L\alpha_{F,F'}^{(1)}$.

As discussed in Ref. [3], all M_x-magnetometer signals can be represented in the form of Eqs. (8) and (9). We therefore refer to these expressions as a 'universal representation' of the M_x-magnetometer signal, whose spectral dependence (but not its absolute value) is independent of the orientation of the rf-field. The interest of writing the phase signal in the universal form lies in the fact that—after electronic subtraction of the offset phase $\varphi^{(0)}$ depending on $\hat{\mathbf{B}}_0$ and $\hat{\mathbf{B}}_1$—the frequency dependence of the phase has a pure arctan dependence on the detuning with $\varphi = 0$ on resonance and a negative slope $\mathrm{d}\varphi/\mathrm{d}\delta\omega$.

We define the Rabi frequency associated with the rf-field as $\Omega = \frac{1}{2}\gamma_{\mathrm{F}}\mathrm{B}_1$. Since the component of $\mathbf{B}_1$ along $\mathbf{B}_0$ does not induce magnetic resonance transitions, we have introduced in (9) an 'effective' Rabi frequency $\widetilde{\Omega}$ that is associated with the component of $\mathbf{B}_1$ that is orthogonal to $\mathbf{B}_0$, viz.,

$$\widetilde{\Omega}_{\hat{\mathbf{B}}_1}(\hat{\mathbf{B}}_0) = \Omega\left|\hat{\mathbf{B}}_1 - \left(\hat{\mathbf{B}}_1\cdot\hat{\mathbf{B}}_0\right)\hat{\mathbf{B}}_0\right|\,. \tag{10}$$

Below we will describe experiments carried out with M_x-magnetometers having two distinct orientations of the rf-field with respect to the light propagation direction, as shown in Fig. 2:

- The $\mathbf{B}_1 \| \mathbf{k}$ geometry shown in Fig. 2a has a cylindrical symmetry around the $\mathbf{k}$-vector, so that one expects the magnetometer signals only to depend on the polar angle θ_B. One easily sees that in this case the effective Rabi frequency is given by

$$\widetilde{\Omega}_{\|}(\theta_B, \phi_B) = \Omega\,|\sin\theta_B|\,, \tag{11}$$

while the on-resonance phase is given by

$$\varphi_{\|}^{(0)} \equiv \varphi^{(0)}\left(\hat{z}, \{\theta_B, \phi_B\}\right) = -\frac{\pi}{2}\,\mathrm{sign}(\cos\theta_B)\,. \tag{12}$$

We note that the offset phase is erroneously given in Ref. [3] (the referred work appeared while preparing the present text) as $\varphi_{\|}^{(0)} = 0$, while the above result is the correct expression using the 4-quadrant definition of the arctan function for inferring the phase.

- The $\mathbf{B}_1 \perp \mathbf{k}$ geometry shown in Fig. 2b is no longer rotationally invariant around $\mathbf{k}$, and it is shown in [3] that in that case one has

$$\widetilde{\Omega}_{\perp}(\theta_B, \phi_B) = \Omega\sqrt{\cos^2\theta_B + \sin^2\theta_B\,\sin^2\phi_B} \tag{13}$$

$$\begin{aligned}\varphi_{\perp}^{(0)} &= \varphi^{(0)}\left(\hat{x}, \{\theta_B, \phi_B\}\right)\\ &= \arctan(\cos\theta_B\,\cot\phi_B).\end{aligned} \tag{14}$$

Figure 3 represents the orientation dependencies given by Eqs. (11) and (13).

The on-resonance amplitudes of the R-signals for the two cases (distinguished by the direction $\hat{\mathbf{B}}_1$) are then given by

$$P_R^{\hat{\mathbf{B}}_1}(0) = \widetilde{P}\,\frac{\gamma\,\widetilde{\Omega}_{\hat{\mathbf{B}}_1}(\theta_B, \phi_B)}{\gamma^2 + \widetilde{\Omega}^2_{\hat{\mathbf{B}}_1}(\theta_B, \phi_B)}\,|\sin\theta_B\cos\theta_B|. \tag{15}$$

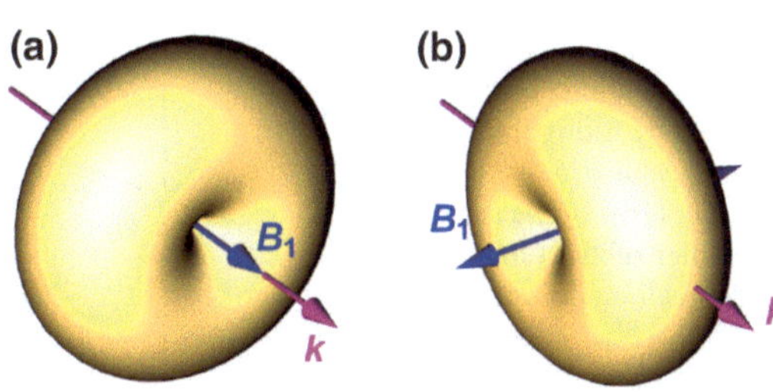

Fig. 3 Anticipated angular dependencies of the effective Rabi frequencies $\widetilde{\Omega}_{\hat{\mathbf{B}}_1}$ for the $\mathbf{B}_1 \| \mathbf{k}$ (**a**) and $\mathbf{B}_1 \perp \mathbf{k}$ (**b**) geometries

4 Data Recording

After the careful calibration of the coils generating the x-, y- and z- components of the field vector $\mathbf{B}_0$, we orient the latter in a systematic manner along 2400 discrete directions, while keeping the field modulus B_0 constant. For this, we feed suitable voltages to the three coils via three series resistors. The voltages are generated by computer-controlled programmable signal generators. A full 4π scan is achieved by varying θ_B from 0° to 180° in 40 equidistant steps. For each value of θ_B, we vary the azimuthal angle ϕ_B from 0° to 360° in 60 equidistant steps.

The field modulus B_0 is chosen to be $\sim 10\,\mu$T, which corresponds to a Larmor frequency $f_0 = \frac{\omega_0}{2\pi}$ of $\sim$ 35 kHz. For each field orientation (θ_B, ϕ_B), we scan the rf frequency $\omega_{\rm rf}$ across a range of $\pm$400 Hz (in 8 Hz steps) around the Larmor frequency ω_0. To ensure that the resonance line is always well centered in the scan range, we do a coarse (automated) determination of the line center after each recording in order to define the start- and stop-frequencies for the scan at the subsequent $\theta_B - \phi_B$ orientation.

Figure 4a shows the spatial distribution of the Larmor frequencies extracted from the off-line analysis of the data discussed in the next section. The average frequency of the displayed points is 36.0(3) kHz. However, we find that during the angular scans the Larmor frequency varies in a systematic manner over a range of $\pm$1 kHz with respect to that average. This variation has an rms value of 0.9% (compatible with the 1% estimated uncertainty of the calibration constants k_i) and is represented in Fig. 4b, where we show the data of Fig. 4a after subtraction of 35.4 kHz.

5 Data Analysis

The digital lock-in amplifier records and stores the in-phase and quadrature signals (Cartesian coordinates) as well as the R-signal and phase (polar coordinates). We have discovered a synchronous (but phase-shifted) photodiode signal (oscillating at the rf frequency) that is present even in absence of light. We assign this signal to inductive pick-up of the rf-field by the photodiode's electrical circuit. We have

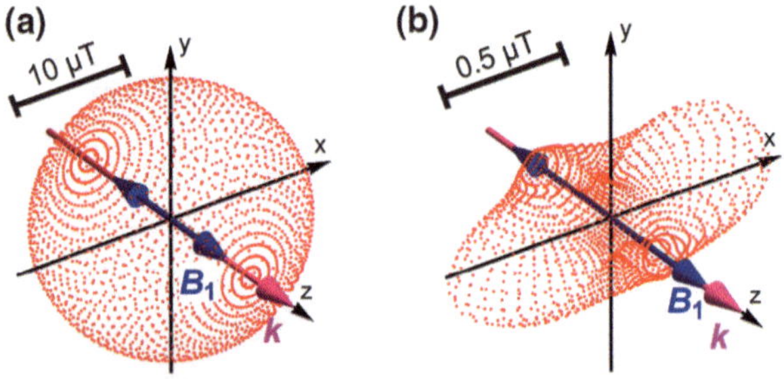

Fig. 4 **a** Angular distribution of the magnetic field modulus B_0 as inferred from the parameters $\omega_{\rm fit}$ obtained by fitting the experimental $\varphi(\delta\omega)$ curves with Eq. (16). **b** The same data, after subtracting $\sim$ 10 μT from all moduli, reveal a slight (0.9 $\%_{\rm rms}$) anisotropy of the distribution

recorded this signal with the laser beam blocked and subtract its in-phase and quadrature components from the corresponding magnetometer signals in the off-line analysis. After this correction we calculate the amplitude (P_R) and phase (φ) signals from the in-phase and quadrature signals. From the phase signal we subtract the frequency-dependent phase shift of $\approx 20°$ mentioned in Sect. 2. Typical results are shown in Fig. 5.

We then fit the theoretical expressions for $\varphi(\delta\omega)$ and $P_R(\delta\omega)$ given by Eqs. (8) and (9), respectively, to the data. We note that the experimental phase signal is determined by the ratio of the quadrature and the in-phase signals, so that $\varphi_{\text{exp}}(\delta\omega)$ does not depend on the common prefactor of those signals. We therefore first fit the function

$$\varphi_{\text{fit}}\left(\omega_{\text{rf}};\omega_{\text{fit}},\gamma_{\text{fit}},\varphi_{\text{fit}}^{(0)}\right)=\varphi_{\text{fit}}^{(0)}-\arctan\frac{\omega_{\text{rf}}-\omega_{\text{fit}}}{\gamma_{\text{fit}}} \tag{16}$$

to the experimental phase signals $\varphi_{\text{exp}}(\delta\omega)$. Besides 'horizontal' ($\omega_{\text{fit}}$) and 'vertical' ($\varphi_{\text{fit}}^{(0)}$) offsets, the shape of the phase signals is fully determined by the relaxation rates γ_{fit}. The fits yield the resonance (Larmor) frequency, ω_{fit}, the (light power broadened) linewidth, γ_{fit}, as well as the offset phase, $\varphi_{\text{fit}}^{(0)}$ for each field orientation. We note that the resonance frequencies ω_{fit} determined from such fits are the values used in Figs. 4.a,b.

Figure 6a shows the angular dependence of the fitted relaxation rates on the field orientation. In Fig. 6b we represent a superposition of the 30 cuts through the sphere of Fig. 6a, one for each of the 30 ϕ_B angles. Each given cut thus contains 80 $\gamma(\theta_B)$ points. The average value of all data is $\gamma/(2\pi)$=7.3(4) Hz.

The slight deviation from spherical symmetry observed near the dead-zones occurring at $\theta_B = n\pi/2$ may be due to spin alignment effects addressed in Sect. 7.2. The dead-zones of the alignment contributions occur at different field orientations than those of the orientation contributions and may thus become more prominent

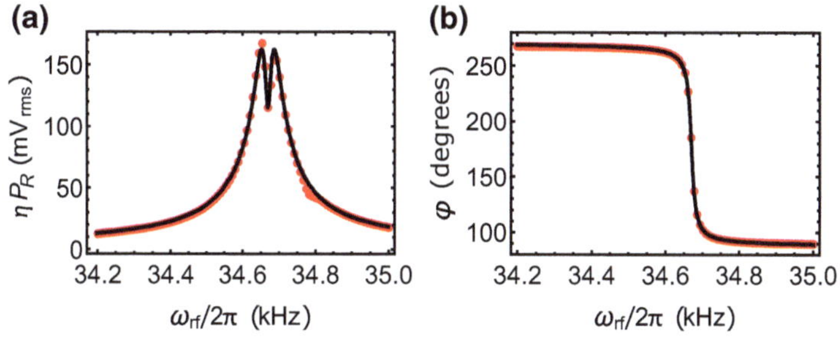

Fig. 5 Typical rf frequency dependence of the amplitude (**a**) and phase (**b**) signals of the demodulated detected laser power. Data shown as *red points* together with fitted functions [Eqs. (17), (16)] shown as solid lines. Data taken with the magnetometer operated in the transverse ($\mathbf{B}_1 \perp \mathbf{k}$) geometry with $\hat{\mathbf{B}}_0$ oriented along $\theta_B = 45.67°$, $\phi_B = 90°$ and an effective Rabi frequency of $\widetilde{\Omega} = 2.0\gamma$

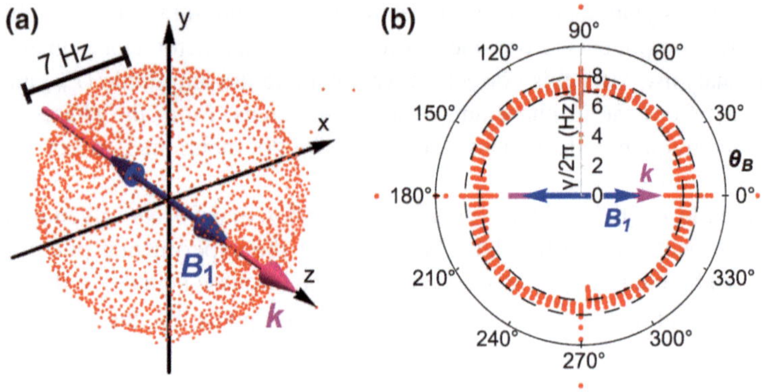

Fig. 6 **a** Angular distribution of the relaxation rates (magnetic resonance linewidths) $\gamma/2\pi$ as inferred from the parameters γ_{fit} obtained by fitting the experimental $\varphi(\delta\omega)$ curves with Eq. (16); **b**: The same data as a function of θ_B, with data for all ϕ_B values superposed

when the (otherwise dominant) orientation contribution vanishes. The increased scatter of the data points near these zones is obviously a consequence of the lower signal/noise ratio near these zones.

In the next step, we fit the R-signals by the function

$$R_{\mathrm{fit}}(\omega_{\mathrm{rf}};\omega_{\mathrm{fit}},\widetilde{\Omega}_{\mathrm{fit}},A_{\mathrm{fit}}) = \frac{A_{\mathrm{fit}}\widetilde{\Omega}_{\mathrm{fit}}\sqrt{\gamma_{\mathrm{exp}}^2+(\omega_{\mathrm{rf}}-\omega_{\mathrm{fit}})^2}}{(\omega_{\mathrm{rf}}-\omega_{\mathrm{fit}})^2+\gamma_{\mathrm{exp}}^2+\widetilde{\Omega}_{\mathrm{fit}}^2}, \tag{17}$$

where we have fixed the linewidths γ_{exp} to the values γ_{fit} inferred from the phase fits, but leave the resonance frequencies ω_{fit} again as free parameters. We note that these fits yield resonance frequency values ω_{fit} that agree with those of the phase fits. Those fits also yield the scale factors A_{fit} and the effective Rabi frequencies $\widetilde{\Omega}_{\mathrm{fit}}$. A_{fit} is related to the theoretical amplitude $\widetilde{P}$ in Eq. (9) by $A_{\mathrm{fit}} = \eta\widetilde{P}$, where η is a calibration factor (measured in V/W) that converts laser power levels to recorded voltages.

We have performed the above analysis for all field orientations (θ_B, ϕ_B), obtaining a set of on-resonance R-signal values that can be expressed by the fitted parameters via

$$R_{\mathrm{fit}}^{(0)}(\theta_B,\phi_B) \equiv R_{\mathrm{fit}}(\delta\omega=0) = A_{\mathrm{fit}}\frac{\gamma_{\mathrm{fit}}\widetilde{\Omega}_{\mathrm{fit}}}{\gamma_{\mathrm{fit}}^2+\widetilde{\Omega}_{\mathrm{fit}}^2}, \tag{18}$$

where γ_{fit} results from the fit by Eq. (16) and $\widetilde{\Omega}_{\mathrm{fit}}$ and A_{fit} from the fit by Eq. (17). We recall that those equations are universal fit functions that can be applied to M_x-magnetometer geometries with any orientation of the rf-field.

6 Results

The fit of the magnetic resonance spectrum at each field orientation point (θ_B, ϕ_B) yields the fit parameters $\omega_{\rm fit}, \gamma_{\rm fit}, \varphi_{\rm fit}^{(0)}, \widetilde{\Omega}_{\rm fit}, A_{\rm fit}$ as well as the inferred on-resonance signal amplitude $R_{\rm fit}^{(0)}$ [defined by Eq. (18)] for each grid point. This allows graphical representations of the $\hat{\mathbf{B}}_0$ orientation dependencies of all parameters, the orientation dependence of the applied field modulus having already been shown in Fig. (4).

We complement the latter dependence by showing in Fig7a, b the dependencies of the fitted effective Rabi frequencies on the orientation of the field $\mathbf{B}_0$ in the $\mathbf{B}_1 \| \mathbf{k}$ and $\mathbf{B}_1 \perp \mathbf{k}$ geometries. Note that in the plot we have excluded some points near the dead-zones occurring for $\theta_B = 0°, 90°$, and $180°$ for which the signals vanish in the noise. The results reflect very well the anticipated dependencies shown in Fig. 3.

In view of magnetometric applications, the resonance amplitudes $R_{\rm fit}^{(0)}$ and the observed linewidths $\gamma_{\rm fit}$ are the quantities of main interest, since they determine the magnetometric sensitivity as discussed in Sect. 7.4. While the angular dependence of the relaxation rates has already been shown in Fig. 6, we will focus in the following subsections on the angular distributions of the $R_{\rm fit}^{(0)}$ signals, treating separately the $\mathbf{B}_1 \| \mathbf{k}$ and $\mathbf{B}_1 \perp \mathbf{k}$ geometries.

6.1 $\mathbf{B}_1 \| \mathbf{k}$ *Geometry*

We first address the orientation dependence of the on-resonance phase $\varphi_{\|}^{(0)}$. Equation (12) predicts that the $\varphi_{\|}^{(0)}$ can only assume the values $-\pi/2$ (for $\theta_B < \pi/2$) and $+\pi/2$ (for $\theta_B > \pi/2$) in the $\mathbf{B}_1 \| \mathbf{k}$ geometry. The experimental findings shown in Fig. 8 reflect very well this anticipated behavior.

In Fig. 9a, b we compare the theoretical prediction and the experimental findings for the R-signal in the $\mathbf{B}_1 \| \mathbf{k}$ geometry. The plots reflect well the rotational symmetry

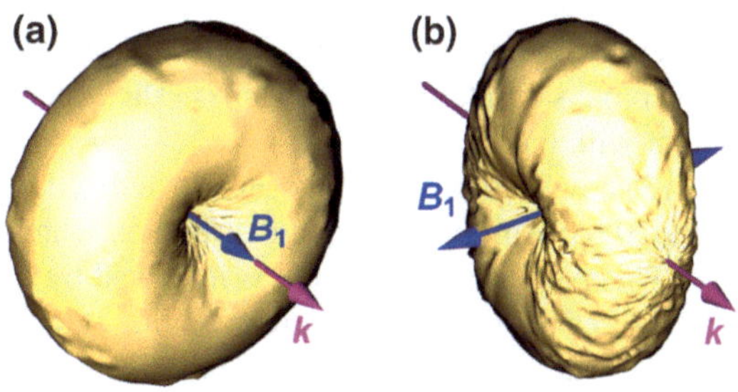

Fig. 7 Measured angular dependencies of the effective Rabi frequencies $\widetilde{\Omega}_{\rm fit}$ for the $\mathbf{B}_1 \| \mathbf{k}$ (**a**) and $\mathbf{B}_1 \perp \mathbf{k}$ (**b**) geometries. The surface plots have been obtained by numerical interpolation of ~2400 discrete data points (some outliers near the dead-zones have been removed). The results are to be compared with the anticipated dependencies shown in Fig. 3

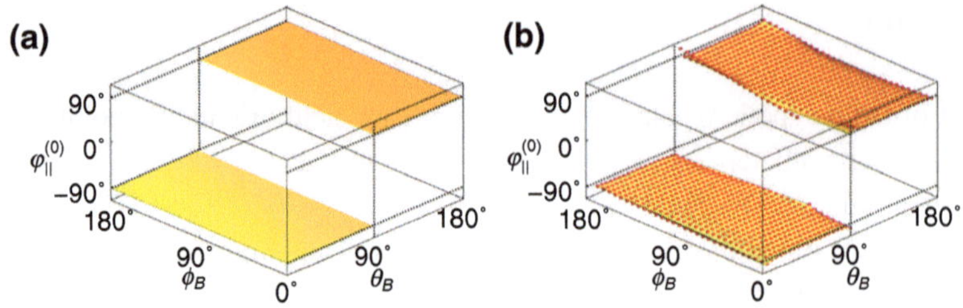

Fig. 8 $\mathbf{B}_1 \| \mathbf{k}$ **geometry**: Anticipated (**a**) and measured (**b**) θ_B-ϕ_B dependencies of the on-resonance phases $\varphi_{\|}^{(0)}$. The surfaces in the right graph have been obtained by an interpolation algorithm using the discrete data points (shown as *red dots*)

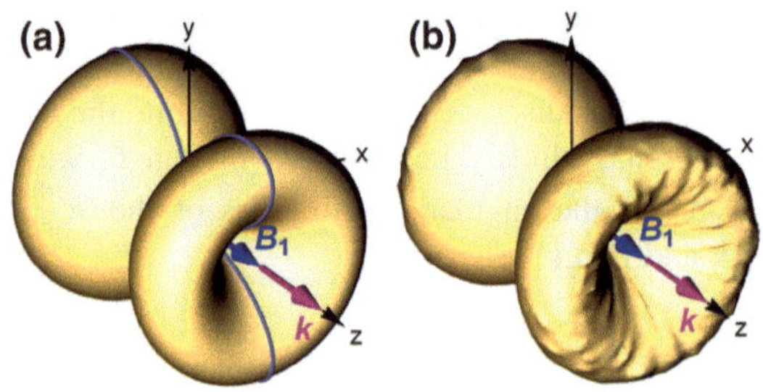

Fig. 9 $\mathbf{B}_1 \| \mathbf{k}$ **geometry**: Theoretical (**a**) and experimental (**b**) angular dependencies of the fitted on-resonance R-signals $R_{\text{fit}}^{(0)}$ for $\Omega \approx 2\gamma$. The *solid line* on the theoretical graph represents the cut leading to the polar plots shown in Fig. 10

around the **k**-direction, and show—at least on the displayed scale—a good qualitative agreement.

In order to explore the agreement in a more quantitative manner, we have performed scans of the θ_B angle along the trajectory shown in Fig. 9a at a fixed value of ϕ_B. The results are shown in Fig. 10. Since the anticipated patterns of $R_{\text{fit}}^{(0)}$ (θ_B) depend on the rf saturation parameter $G_{\text{rf}} = \Omega^2/\gamma^2$, we have measured these dependencies for different settings of the Rabi frequency Ω, ranging from below-saturation ($\Omega < \gamma$) to above-saturation ($\Omega > \gamma$) values. The experimental results are in excellent agreement with theory.

Equations 11 and 15 show that for $\Omega \ll \gamma$ the anticipated θ_B dependence of the R-signal is given by $P_{R\hat{z}}(0) \propto |\sin^2\theta_B \cos\theta_B|$, a distribution which has a maximum value at $\theta_B \sim 54.7°$. This particular orientation is indicated by dashed lines in the figure. The data points reflect well this behavior. When the Rabi frequency is increased to saturating values $\Omega \geq \gamma$, the angular distribution becomes more stretched along the $\hat{\mathbf{k}}$ direction.

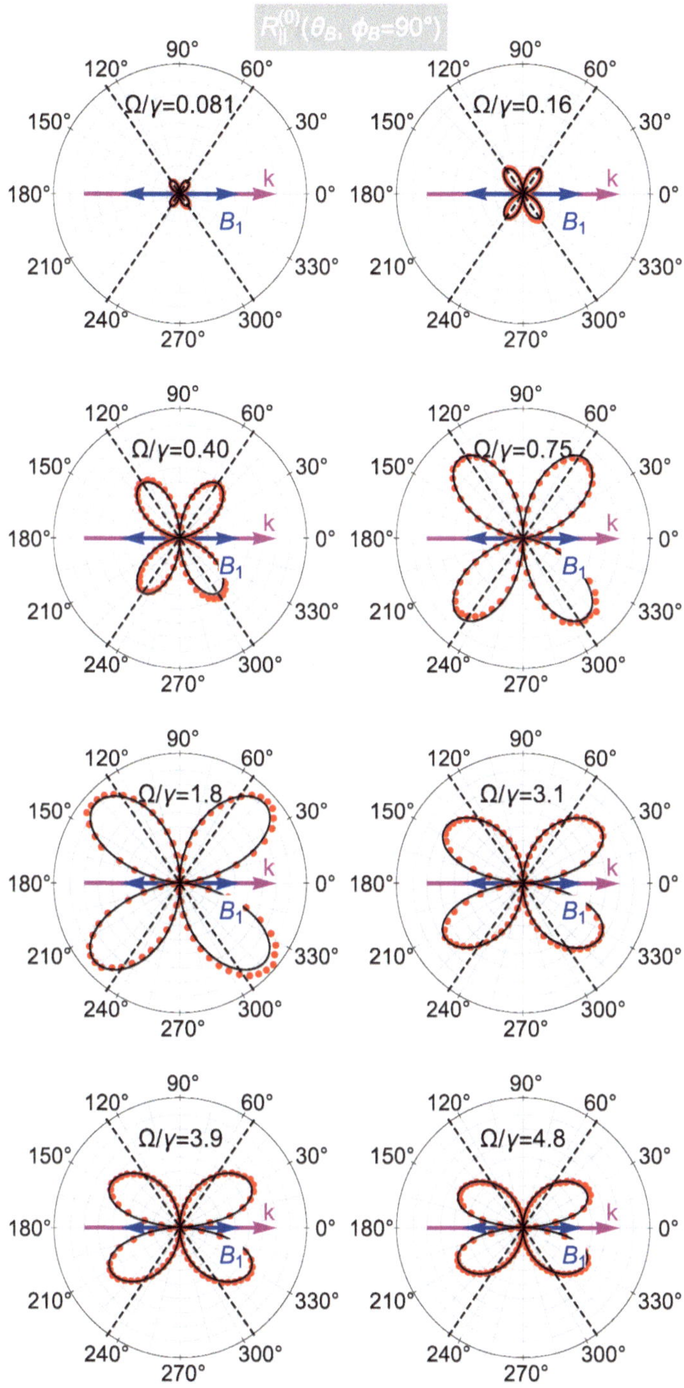

Fig. 10 $\mathbf{B}_1 \| \mathbf{k}$ **geometry**: Cuts through the angular distributions of Fig. 9b for $\phi_B = 90°$ (y–z plane). The polar angle θ_B is measured with respect to the **k** axis. The Rabi frequency (expressed in

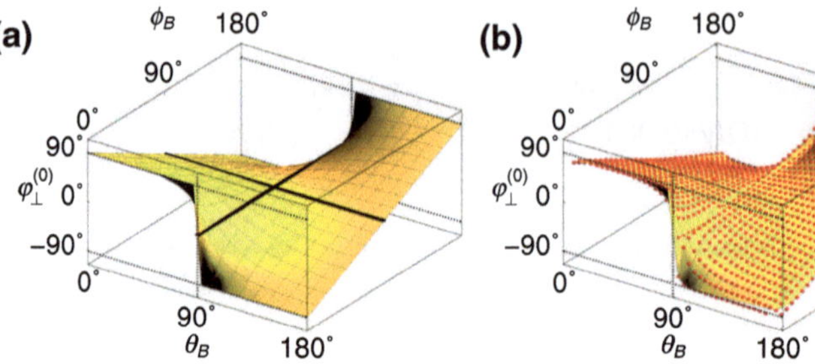

Fig. 11 $\mathbf{B}_1 \perp \mathbf{k}$ **geometry**: Anticipated (**a**) and measured (**b**) θ_B - ϕ_B dependencies of the on-resonance phases $\varphi_\perp^{(0)}$. The surfaces in the *right graph* have been obtained by an interpolation algorithm using the discrete data points, shown as *red dots*. The *black lines* on the left graph indicate $\mathbf{B}_0$ orientations for which $\varphi_\perp^{(0)} = 0$. The *red dots* on the right graph are data points

6.2 $\mathbf{B}_1 \perp \mathbf{k}$ *Geometry*

We have performed the same type of measurements in the $\mathbf{B}_1 \perp \mathbf{k}$ geometry. The magnetometer geometry, determined by the orientations $\hat{\mathbf{k}}$ and $\hat{\mathbf{B}}_1$, has no longer a rotational symmetry axis, and one expects this broken symmetry to be reflected in the orientational dependencies of the magnetometer signals. The latter obey a reflection symmetry with respect to the plane determined by $\hat{\mathbf{k}}$ and $\hat{\mathbf{B}}_1$.

We start again by addressing first the on-resonance phase, which—because of the broken symmetry—has a non-trivial angular distribution (see Fig. 11). Disregarding some outliers near the dead-zones, which result from large fit errors, the experimental data reflect well the anticipated orientational dependence given by Eq. (14).

The angular dependence of the on-resonance R-signal shown in Fig. 12 exhibits well the broken symmetry leading to a 'squeezing/stretching' along the direction $\hat{\mathbf{k}} \times \hat{\mathbf{B}}_1$. In order to illustrate the broken ϕ_B symmetry, we have performed angular scans in which we have varied ϕ_B at a fixed angle $\theta_B \sim 45°$. The results displayed in Fig. 13a show how the squeezing along $\hat{\mathbf{B}}_1$ evolves into a stretching along that direction as Ω is increased from below- to above-saturation values.

We also show in Fig. 13b, θ_B-scans in the plane $\phi_B = 0$, defined by $\hat{\mathbf{k}}$ and $\hat{\mathbf{B}}_1$ and in Fig. 13c θ_B-scans in the plane $\phi_B = \pi/2$, defined by $\hat{\mathbf{k}}$ and $\hat{\mathbf{k}} \times \hat{\mathbf{B}}_1$. Data points are shown together with theoretical fits (solid lines).

Fig. 10 (continued) units of γ) is varied from below- to above-saturation values. The *red dots* are data points and the solid (*black*) lines represent fits with the model function of Eq. (15). The *dashed lines* represent the orientation θ_B at which the low-saturation data are expected to show a maximum signal

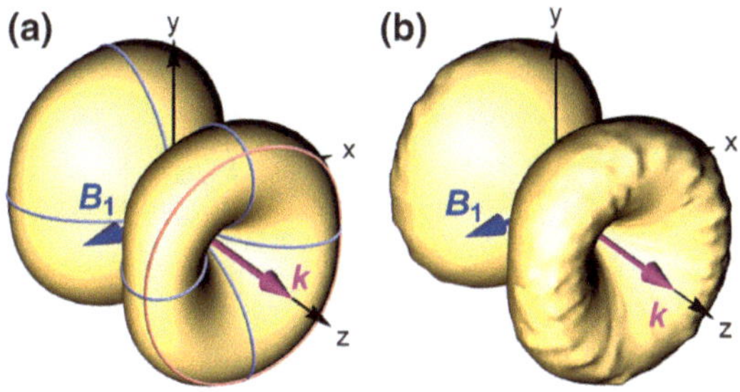

Fig. 12 $\mathbf{B}_1 \perp \mathbf{k}$ **geometry**: Theoretical (**a**) and experimental (**b**) angular dependence of the on-resonance R-signals $R_{\text{fit}}^{(0)}$ for $\Omega \approx \gamma$. The *reddish* and *bluish* solid lines on the *left graph* represent the trajectories of the polar plots shown in Fig. 13

7 Discussion

We have investigated in detail the dependence of lock-in detected signals produced by spin-orientation (vector magnetization)-based M_x-magnetometers on the orientation of the applied magnetic field of interest $\mathbf{B}_0$ and the orientation of the rf-field $\mathbf{B}_1$ used to drive the magnetic resonance in the magnetometer. The lock-in signals are fully described by the amplitude (R-signal) and the phase φ of the photocurrent oscillation. In all investigated geometries, we find a very good agreement between theoretical predictions and our observations. We can thus claim a successful verification of the general theoretical expressions for M_x-magnetometer signals that have been presented recently [3].

7.1 Experimental Asymmetries

We wish to state nonetheless that achieving the excellent quality of the presented experimental results has been a non-trivial task. One of our goals was to clearly demonstrate the anisotropic angular dependence in the two investigated geometries, viz., the $\mathbf{B}_1 \| \mathbf{k}$ case (rotational symmetry with respect to $\hat{\mathbf{k}}$, and reflection symmetry with respect to the plane perpendicular to $\hat{\mathbf{k}}$), compared to the $\mathbf{B}_1 \perp \mathbf{k}$ case which lacks the rotational symmetry while keeping the reflection symmetry. In the former case with rotational symmetry, satisfactory results have only been obtained after ensuring that the experimental setup contained no symmetry-breaking elements. On our way to the final results, we have identified and successively eliminated the following perturbing effects:

- Even a tiny elliptical polarization of the light beam leads to the creation of transverse alignment (tensor polarization) in the atomic medium that manifests itself in an anisotropic angular distribution. Special care has therefore been taken

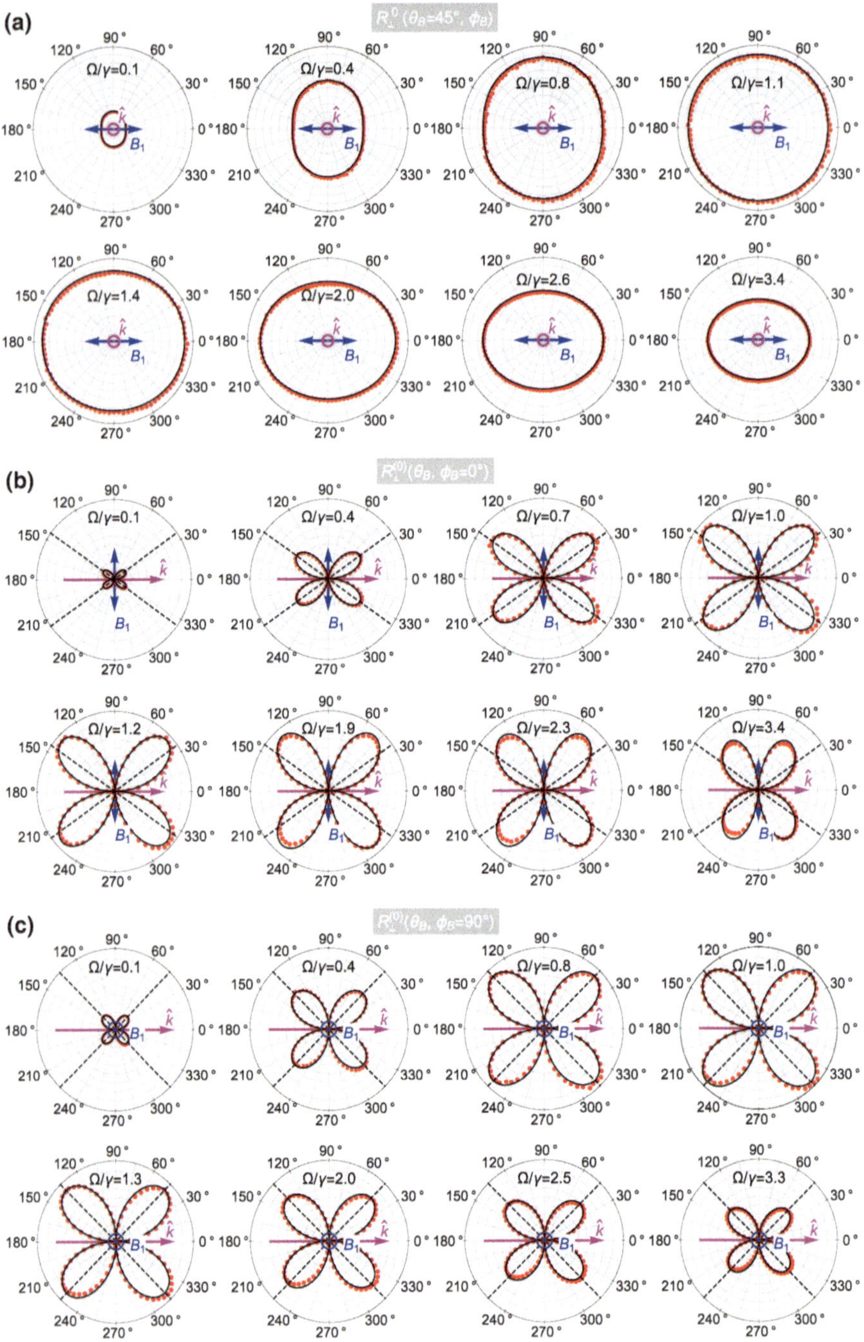

Fig. 13 $\mathbf{B}_1 \perp \mathbf{k}$ **geometry**: Cuts through the angular distributions of Fig. 12b along the trajectories shown in Fig. 12a. *Top graphs* (**a**): $R_{\rm fit}^{(0)}(\phi_B)$ for $\theta_B = \pi/4$; *Middle graphs* (**b**): $R_{\rm fit}^{(0)}(\theta_B)$ for $\phi_B = \pi/2$; *Bottom graphs* (**c**): $R_{\rm fit}^{(0)}(\theta_B)$ for $\phi_B = 0$. In each group of graphs, the Rabi frequency is varied

to ensure the highest possible degree of circular polarization. For the experiments reported in the paper, the light beam (prior to entering the cell) has a degree of linear polarization (DOLP) of 0.70%, which corresponds to a degree of circular polarization (DOCP) of >99.99%. In an earlier experiment, we had observed a pronounced anisotropy and discovered later that the light entering the cell in that experiment has had a DOLP of 6%, which corresponded to a DOCP of 99.8%. We note that the DOCP value is given by the corresponding Stokes parameter and that $\text{DOCP}^2 + \text{DOLP}^2 = 1$ assuming 100% polarized light. Stress-induced birefringence of the spherical glass cell may slightly alter the light polarization inside of the cell. Any spurious contamination by linear polarization will lead to the creation of transverse spin alignment (tensor polarization), see Sect. 7.2.

- Any magnetic component placed near the Cs vapor cell will perturb the angular distributions in a similar way. We have been able to eliminate such effects by a careful choice and screening of all deployed components for magnetic contaminants.
- In an early stage we used rf-coils of rectangular shape ($38 \times 29\,\text{mm}^2$, compared to the vapor cell diameter of $\sim 30\,\text{mm}$), which led to a pronounced breaking of the rotational ϕ_B-symmetry in the $\mathbf{B}_1 \| \mathbf{k}$ geometry. Replacing those coils by 100 mm diameter circular coils has led us to the presented results.

7.2 *Alignment Effects*

Despite the taken precautions, some minor imperfections remain in the data:

(a) A slight ($< 1\%$) anisotropy of the measured Larmor frequencies (Fig. 4b). The maximum quadratic Zeeman shift in the used field of $10\,\mu\text{T}$ is $\ll$ nT and can thus be ruled out as cause of the observed deviations at the level of $\lesssim 100\,\text{nT}$.
(b) A slight asymmetry of all recorded R-signals under saturating conditions $\Omega \geq \gamma$ (Fig. 5a).
(c) A slight anisotropy of the relaxation rates γ (Fig. 6b).
(d) A slight forward–backward (with respect to $\mathbf{k}$) asymmetry of the θ_B scans in the $\mathbf{B}_1 \| \mathbf{k}$ geometry (Fig. 10).

The effect (a) may mainly be assigned to field-dependent calibration constants as discussed in Sect. 4. We believe that the origin of the other effects relies in the following imperfection of the magnetometer model [3]: The model assumes that the atomic medium carries only vector spin polarization, commonly referred to

Fig. 13 (continued) from below- to above-saturation values. The *red dots* are data points, and the solid *black lines* represent fits by the model functions. In graphs **b** and **c**, the *dashed lines* represent the field orientations yielding maximum signals in the low-saturation limit as 'orientation,' which, in the language of atomic multipole moments represents a $K = 1$ multipole. It is well known [1] that optical pumping with circularly polarized light produces multipole moments $m_{K,Q}$ of ranks $K = 1, 2, \ldots, 2F$ ($2F = 8$ in the case of the $F = 4$ ground state investigated here), while optical pumping with linearly polarized light produces only even multipole moments $K = 2, 4, \ldots, 2F$. It is also known that light interacting with the atoms via an electric dipole transition is only sensitive to the rank $K = 1$ and 2 multipole moments [4].

In view of this, pumping with light of perfect circular polarization will not only produce a longitudinal (with respect to $\hat{\mathbf{k}}$) vector polarization ($m_{1,0}$), but also a *longitudinal alignment* (second rank tensor moment $m_{2,0}$), whose contribution has magnetic resonance line shapes and angular distributions that differ from the orientation contributions discussed in the paper. By dropping the last term in Eq. (3), we have explicitly ignored alignment contributions. To our knowledge, the effect of the ignored alignment contribution on the M_x signals has not been discussed in the literature, despite that the M_x-magnetometer has been known for more than half a century. One of the reasons for this may be the fact that for many decades M_x-magnetometers have been operated with discharge lamp light sources, whose broad spectrum does not allow resolving the atomic hyperfine structure, making lamp pumped magnetometers insensitive to alignment contributions in $J = 1/2$ atoms, such as the alkalis. We recall that alignment contributions have already been observed in [14] and that they are particularly small on the 4–3 hyperfine components of the D_1 line used here. We are in process of developing a theoretical model allowing us to study this effect in more detail in the near future.

Moreover, the *transverse alignment* produced by imperfect circular light polarization is also not included in our model.

7.3 *Self-Oscillating Magnetometer*

We note that the $\mathbf{B}_1 \perp \mathbf{k}$ geometry is interesting for realizing a so-called self-oscillating magnetometer [2], in which the AC part of the detected photocurrent is used to drive (after suitable amplification and phase shifting) the rf coil. While broadband amplifiers are easy to implement, a broadband phase shifter with fixed amplification is not. We recall that in the $\mathbf{B}_1 \| \mathbf{k}$ geometry the phase shift φ between the photocurrent and the rf-field is $\pm 90°$ (Fig. 8), so that a self-oscillating magnetometer in that configuration will always need a phase shifter.

In the $\mathbf{B}_1 \perp \mathbf{k}$ geometry, on the other hand, the data of Fig. 11 show that arbitrary phase shifts φ—among them $\varphi = 0$—can be achieved. In that geometry, a suitable orientation of $\mathbf{B}_0$ can therefore be used to ensure self-oscillation without the need of a phase shifter. From Fig. 11, one sees that the phase shift φ vanishes for fields $\mathbf{B}_0$ lying on the circle $\phi_B = 90°$. The θ_B orientation on that circle yielding the maximum

R-signal depends on the Rabi frequency Ω as illustrated by the graphs in Fig. 13b and shall be chosen appropriately in order to maximize the signal/noise ratio.

One should note, however, that the $\mathbf{B}_1 \| \mathbf{k}$ geometry, in which the phase is basically independent of the field orientation is more reliable in terms systematic readout errors: Changing the orientation of the field will not change the phase and will hence not affect the oscillation frequency. The $\mathbf{B}_1 \perp \mathbf{k}$ geometry, on the other hand, offers the possibility of zero-phase shift operation. However, because of the strongly curved phase vs. orientation surface (Fig. 11), any tilt of the field will change the phase and hence the oscillation frequency. The zero-phase advantage can therefore be brought to full profit only in experiments, in which the direction of $\mathbf{B}_0$ does not change.

7.4 *Magnetometric Sensitivity*

We end the discussion by addressing the magnetometric sensitivity (or noise-equivalent magnetic field) δB_{NEM} of the M_x-magnetometer, which can be expressed as

$$\delta B_{\mathrm{NEM}} \equiv \frac{1}{\gamma_{\mathrm{F}}} \frac{1}{\mathrm{SNR}} \left| \frac{\mathrm{d}\varphi}{\mathrm{d}\delta\omega} \right|^{-1} \Bigg|_{\delta\omega=0} = \frac{\gamma}{\gamma_{\mathrm{F}}} \frac{\delta R}{R^{(0)}} \propto \frac{\gamma}{R^{(0)}}, \tag{19}$$

where $\mathrm{SNR} = R^{(0)}/\delta R$ denotes the signal/noise density ratio, δR being the spectral noise density of the total detected light power. Since the on-resonance phase slope $\mathrm{d}\varphi/\mathrm{d}\delta\omega = -\gamma^{-1}$ does not depend on the magnetic field orientation (as supported by our data in Fig. 6), the sensitivity's angular dependence varies only with $R^{(0)}$. Obviously, the larger the amplitude, the higher will be the magnetometric sensitivity, i.e., the smaller will be the detectable field changes.

From Eq. (15), we conclude that the best magnetometric sensitivity is achieved for $\theta_B = 45^\circ$ when the value of $|\sin\theta_B \cos\theta_B|$ is maximal for both $\mathbf{B}_1 \| \mathbf{k}$ and $\mathbf{B}_1 \perp \mathbf{k}$ case. This argument uses the fact that by adjusting Ω one can always ensure that the expression

$$\frac{\gamma \widetilde{\Omega}_{\hat{\mathbf{B}}_1}(\theta_B = 45^\circ, \phi_B)}{\gamma^2 + \widetilde{\Omega}^2_{\hat{\mathbf{B}}_1}(\theta_B = 45^\circ, \phi_B)} \tag{20}$$

can be made to reach its maximum value of 1/2 for any ϕ_B. Under optimized conditions $(\theta_B, \phi_B, \Omega)$ our system has an SNR of $1.5 \times 10^5\ \sqrt{\mathrm{Hz}}$ assuming shotnoise-limited detection of the 4 μA DC photocurrent. This yields $\delta B_{\mathrm{NEM}} = 13\,\mathrm{fT}/\sqrt{\mathrm{Hz}}$, assuming $\gamma/2\pi = 7\,\mathrm{Hz}$.

8 Outlook

We have demonstrated that our algebraic theory of optically detected magnetic resonances is supported with very good accuracy by experimental observations—at least for the two most commonly deployed configurations of the magnetometer. An essential result is the now well-understood interdependent effect of the rf-field strength and static field orientation on the resonance amplitudes and phases. This success encourages us to go one step further and to derive a universal algebraic expression that accounts both for an arbitrary orientation (θ_B, ϕ_B) of the static field $\mathbf{B}_0$ and—this will be new—for an arbitrary orientation $(\theta_{\mathrm{rf}}, \phi_{\mathrm{rf}})$ of the rf-field $\mathbf{B}_1$. Such an expression may be easily adapted to all-optical designs with Bell-Bloom type of pumping [14–16] as well as to two-beam (pump-probe) experiments. Work in this direction is in progress in our laboratory.

The motivation of the study reported here has been the optimization of an M_x magnetometer in view of reaching the highest sensitivity for measuring $\mathbf{B}_0$ (and/or changes thereof). On the other hand, the M_x principle is also applicable to so-called rf-magnetometry. In that case, a static field $\mathbf{B}_0$ of known strength and variable direction is applied to the sensor, and the task is to measure the amplitude and orientation of an unknown rf-field oscillating at a known (or eventually unknown) frequency. $\hat{\mathbf{B}}_0$ orientation scans—as the ones deployed here—can then be used to infer the properties of the rf-field. We have performed preliminary investigations along this direction in the frame of our ongoing application of atomic magnetometry to the detection of the magnetic response of small ($\lesssim \mu$g) samples of magnetic nanoparticles (MNPs). A problem encountered in such experiments is the detection (in view of minimization) of the stray field from a coil system used to harmonically excite the MNP sample [17, 18]. Details of this research shall be published elsewhere.

Acknowledgements We acknowledge financial support by the Physics Department and the *Pool de Recherche* of the University of Fribourg as well as by Grant No. 200020_162988 of the Swiss National Science Foundation.

References

1. D. Budker, W. Gawlik, D.F. Kimball, S.M. Rochester, V.V. Yashchuk, A. Weis, Rev. Mod. Phys. **74**, 1153 (2002)
2. D. Budker, D.F. Jackson Kimball, *Optical magnetometry* (Cambridge University Press, Cambridge, 2013)
3. A. Weis, G. Bison, Z.D. Grujić, in *High Sensitivity Magnetometers*, ed. by A. Grosz, M.J. Haji-Sheikh, S.C. Mukhopadhyay Springer, Berlin (2017)
4. W. Happer, Rev. Mod. Phys. **44**, 169 (1972)
5. A.L. Bloom, Appl. Opt. **1**, 61 (1962)
6. A. Ben-Kish, M.V. Romalis, Phys. Rev. Lett. **105**, 193601 (2010)
7. T. Wu, X. Peng, Z. Lin, H. Guo, Rev. Sci. Instr. **86**, 103105 (2015)

8. C. Hovde, B. Patton, O. Versolato, E. Corsini, S. Rochester, D. Budker, Heading error in an alignment-based magnetometer, Proc. SPIE **8046**, 80460Q (2011)
9. J. Kitching, S. Knappe, V. Shah, P. Schwindt, C. Griffith, R. Jimenez, J. Preusser, L.-A. Liew, J. Moreland, in *IEEE International Frequency Control Symposium* (2008), pp. 789–794
10. E. Breschi, Z.D. Grujić, P. Knowles, A. Weis, Appl. Phys. Lett. **104**(2), 023501 (2014)
11. N. Castagna, G. Bison, G. Di Domenico, A. Hofer, P. Knowles, C. Macchione, H. Saudan, A. Weis, Appl. Phys. B **96**(4), 763 (2009)
12. N. Castagna, A. Weis, Phys. Rev. A **84**, 053421 (2011)
13. E. Breschi, Z.D. Gruijć, A. Weis, Appl. Phys. B **115**, 85 (2014)
14. Z.D. Grujić, A. Weis, Phys. Rev. A **88**, 012508 (2013)
15. I. Fescenko, P. Knowles, A. Weis, E. Breschi, Opt. Express **21**(13), 15121 (2013)
16. E. Breschi, Z.D. Gruijć, P. Knowles, A. Weis, Phys. Rev. A **88**, 022506 (2013)
17. S. Colombo, V. Lebedev, Z.D. Grujić, V. Dolgovskiy, A. Weis, Int. J. Magn. Part. Imaging **2**, 1606002 (2016)
18. S. Colombo, V. Lebedev, Z.D. Grujić, V. Dolgovskiy, A. Weis, Int. J. Magn. Part. Imaging **2**, 1604001 (2016)

All Polarization-Maintaining Fiber Laser Architecture for Robust Femtosecond Pulse Generation

Wolfgang Hänsel, Heinar Hoogland, Michele Giunta, Sebastian Schmid, Tilo Steinmetz, Ralf Doubek, Peter Mayer, Sven Dobner, Carsten Cleff, Marc Fischer, and Ronald Holzwarth

Abstract We report on a novel architecture for robust mode-locked femtosecond fiber lasers using a nonlinear optical loop mirror with all polarization-maintaining fibers. Due to a nonreciprocal phase shift, the loop mirror can be operated in a compact and efficient reflection mode, offering the possibility to reach high repetition rates and easy implementation of tuning elements. In particular, longitudinal mode spacing and carrier-envelope offset frequency may be controlled in order to operate the laser as an optical frequency comb. We demonstrate femtosecond pulse generation at three different wavelengths (1030, 1565, and 2050 nm) using Ytterbium, Erbium, and co-doped Thulium–Holmium as gain media, respectively. Robust operation is achieved for a wide range of parameters, including repetition rates from 10 to 250 MHz.

This article is part of the topical collection "Enlightening the World with the Laser" - Honoring T. W. Hänsch guest edited by Tilman Esslinger, Nathalie Picqué, and Thomas Udem.

Electronic supplementary material The online version of this chapter (doi:10.1007/978-3-319-64346-5_18) contains supplementary material, which is available to authorized users.

W. Hänsel (✉) • T. Steinmetz • R. Doubek • P. Mayer • S. Dobner • C. Cleff • M. Fischer
Menlo Systems GmbH, Am Klopferspitz 19a, 82152 Martinsried, Germany
e-mail: w.haensel@menlosystems.com

H. Hoogland
Menlo Systems GmbH, Am Klopferspitz 19a, 82152 Martinsried, Germany

Department of Physics, University of Erlangen-Nuremberg, Staudtstrasse 1, 91058 Erlangen, Germany

M. Giunta • R. Holzwarth
Menlo Systems GmbH, Am Klopferspitz 19a, 82152 Martinsried, Germany

Max-Planck-Institute of Quantum Optics, Hans-Kopfermann-Strasse 1, 85748 Garching, Germany

S. Schmid
Menlo Systems GmbH, Am Klopferspitz 19a, 82152 Martinsried, Germany

Present Address: TSED2, Airbus Defence and Space GmbH, Robert Koch Str. 1, 82024 Taufkirchen, Germany

D. Meschede et al. (eds.), *Exploring the World with the Laser*,
https://doi.org/10.1007/978-3-319-64346-5_18

1 Introduction

Over the last decades, mode-locked laser sources emitting ultrashort pulses have enabled an ever growing number of applications, covering diverse fields such as precision spectroscopy, synchronization and timing, two-photon spectroscopy, generation of THz radiation, study of attosecond dynamics, pure microwave generation, and material processing. In particular, as a source for a frequency comb [1], mode-locked lasers have initiated a new era of technology with diverse applications. While early applications of femtosecond pulses were limited to well-controlled laboratory environments, a huge number of applications require pulse sources which work under less stable, even harsh conditions as might be present in industrial manufacturing halls, vehicles, air planes, and even satellites. Fiber lasers naturally shield the beam path from the environment to a large extent and hence present a promising technology for such applications. However, polarization-maintaining (PM) fiber has to be used to sufficiently decouple the laser from perturbations such as vibration or variation of temperature, humidity, and air pressure.

Two main strategies are typically used to achieve mode-locking in fiber lasers: intrinsic saturable absorbers such as semiconductor saturable absorber mirrors (SESAM), carbon nanotubes or graphene on the one hand, and additive pulse mode-locking (APM) [2] on the other hand. While intrinsic absorbers are easily combined with PM fiber, a low damage threshold and potential degradation over time need to be carefully considered in the laser design [3]. Even more, recent studies of an optical frequency comb based on a SESAM mode-locked fiber laser suggest that intrinsic effects of the SESAM such as the slow relaxation cause a noise floor dominating all other noise contributions like pump or technical noise [4, 5]. APM schemes, on the other hand, have been used in fiber lasers since the early 1990s and show the advantage of fast response and therefore potentially lower intrinsic noise. The most prominent examples of APM in fiber lasers are nonlinear polarization rotation (NPR) [6, 7], and a Sagnac loop with intensity-dependent transmission [8], often termed nonlinear optical loop mirror (NOLM). To create a mode-locked laser from a NOLM, the loop is naturally operated in transmission. In such a geometry, the connecting fibers form the shape of an "8" (see Fig. 1a), and such lasers are also known as figure-eight lasers. The intensity dependence of the NOLM is caused by a differential phase shift between the two counter-propagating pulses (nonlinear nonreciprocal phase shift) which itself is induced by an asymmetry in the loop. While in the original demonstration the asymmetry is induced by the splitting ratio at the NOLM entrance, a popular enhancement of this scheme consists of a loop with an asymmetrically placed amplifying fiber, which is termed nonlinear amplifying loop mirror (NALM) [9].

Due to its nature, NPR cannot be realized in PM fibers. Instead, efforts have recently been taken to implement NALM lasers using PM fiber. While NALM loops with standard (non-PM) fiber offer the possibility to introduce an intrinsic nonreciprocity into the loop using polarization controllers [10], NALM loops with

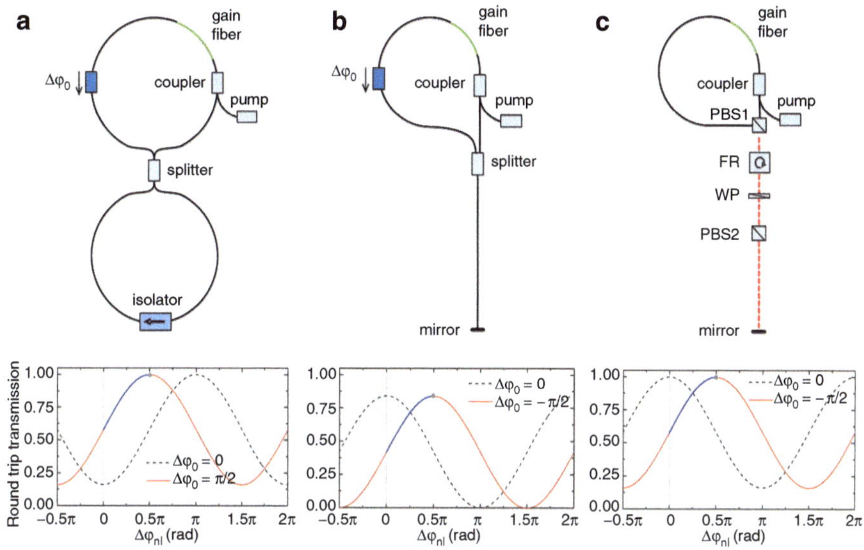

Fig. 1 Schematic of three different laser configurations for additive pulse mode-locking with a nonreciprocally biased NALM. **a** NALM configuration in transmission, **b** reflective NALM configuration with reciprocal beam splitter, **c** NALM configuration with nonreciprocal beam splitter. For each configuration (*top row*), the theoretical round-trip transmission is plotted (*bottom row*) without (*black, dashed*) and with (*red/blue, straight*) nonreciprocal phase bias, assuming an exemplary splitting ratio of 70:30. For the transmission graphs, an exemplary phase bias $\Delta\varphi_0 = (\pi/2, -\pi/2, -\pi/2)$ in graphs **a–c** has been used, and the corresponding location of mode-locked operation has been marked by a *filled circle*. The direction of the nonreciprocal phase bias has been defined to be in the direction along which the nonlinear phase shift occurs. The *blue* part of the round-trip transmission marks the evolution of the laser from continuous-wave (low-intensity) to pulsed (maximum peak power) operation

PM fibers are strictly reciprocal for low intensity unless an additional nonreciprocal phase shift element is introduced. Without such nonreciprocal phase bias, the transmission for low intensity is close to zero and does not offer any intensity dependence. This strongly impedes self-starting and has led to efforts to actively start the mode-locking [11, 12]. Recently, it has been shown by several groups that self-starting ability can be achieved by introducing a nonreciprocal phase bias in the NOLM loop [4, 13, 14] using a proper arrangement of Faraday rotators and waveplates. Even more, the phase bias facilitates operation of the PM-NOLM loop in transmission as well as in reflection. Lasers mode-locked with a reflective NOLM or NALM loop are offered by Menlo Systems GmbH under the trademark figure 9™.

Unlike a Sagnac loop operated in transmission, a reflective Sagnac loop exhibits a reduced maximum transmission if an asymmetric splitter is used at the junction. Unfortunately, for increasing repetition rates, the asymmetry introduced into a NALM by the amplifying fiber becomes ever smaller while the loss from an asymmetric splitting, which might be raised for compensation, becomes increasingly severe. Here we present a novel configuration of a mode-locked fiber laser with a reflective all-PM Sagnac loop that combines the low losses of a Sagnac loop in transmission with the advantageous form factor of a reflective configuration. This has led to the demonstration of repetition rates as high as 250 MHz, which is about three times higher than results previously reported for all-PM fiber lasers using NALM mode-locking.

2 NOLM Lasers with Nonreciprocal Phase Bias

A NOLM loop can be integrated into a mode-locked laser in a variety of configurations. Figure 1 compares properties of three different layouts. In the context of this paper, these configurations are considered for single polarization propagation as it is achieved using PM fiber. The upper row shows the laser configuration while the lower row shows the interferometric contribution of the NOLM to the round-trip transmission of a light pulse as a function of the nonlinear differential phase shift $\Delta\varphi_{NL}$. For the plots, we have assumed an exemplary splitting ratio of 70:30 at the entrance of the NOLM loop. Each of the plots shows the round-trip transmission without a nonreciprocal phase bias (dashed) and with an exemplary nonreciprocal phase bias appropriate for mode-locking operation (straight line). The round-trip transmission of the laser is calculated using the fundamental properties of standard reciprocal beam splitters and, in the third case, using Jones matrices of polarization optics.

The first column (Fig. 1a) shows the popular figure-eight configuration, where the NOLM is operated in transmission, while the other two columns (Fig. 1b, c) represent different reflective configurations. The main difference between the first configuration and the two reflective configurations is that, without any nonreciprocal phase bias (dashed line in the transmission plot), the figure-eight variant exhibits an increasing transmission as a function of the nonlinear phase difference, hence intensity, while the reflective variants both show decreasing transmission. This is why figure-eight lasers are naturally used for passive mode-locking. However, in the presence of an appropriate nonreciprocal phase bias $\Delta\varphi_0$, herein chosen as $\Delta\varphi_0 = \pm\,\pi/2$, all configurations can be tuned to increasing intensity (solid line), thus supporting passive mode-locking of the laser. Even more, the phase bias permits to choose a low-intensity point with a nonvanishing slope, which enhances the self-starting of the mode-locking.

It is interesting to note that for all three configurations the contrast of the intensity modulation computes to $c = 4(R - 0.5)^2$, where R represents the reflectance of the NOLM splitter. When using a standard beam splitter for the NOLM (first two configurations), the modulation ranges of the figure-eight and the reflective configurations are offset by $1 - c$ with respect to one another, reaching a maximum round-trip transmission of unity for the figure-eight case (Fig. 1a) and of c in the reflective configuration with a standard beam splitter (Fig. 1b). In the third column (Fig. 1c), we show our novel reflective configuration which replicates the maximum transmission of 1 inherent to the figure-eight configuration. This feature is achieved by a nonreciprocal beam splitter consisting of a 45°-Faraday rotator and a waveplate placed between two polarizing beam splitters (PBS). The fiber connecting the ports of PBS1 at the NOLM side is twisted by 90°, so that the pulse contribution transmitted at the entrance of the NOLM will be fully reflected at the exit, and vice versa. Therefore, the pulse contribution with linear polarization at ±45° will return onto itself when passing the NOLM loop through the Faraday rotator. The phase shift $\Delta\varphi$ of the waveplate WP, which is itself oriented at 45°, then leads to a nonreciprocal phase bias of $\Delta\varphi_0 = 2\Delta\varphi$ per round trip. The splitting ratio of the NOLM may be tuned by changing the angular orientation of PBS2 or by an additional half wave plate in front of PBS2. This configuration presents the advantage that the splitting ratio can be tuned away from 50:50 without compromising the round-trip transmission of the laser. Furthermore, it can be used to fine-tune the splitting ratio according to the performance of the fiber loop which may vary due to manufacturing tolerances of fiber components or variable splice quality.

3 Experimental Results

In the following, we report on three exemplary mode-locked fiber lasers that have all been built in the novel reflective configuration as displayed in Fig. 1c. The lasers are operated at center wavelengths of 1030, 1565, and 2050 nm, using Ytterbium (Yb), Erbium (Er), and co-doped Thulium–Holmium (Tm/Ho) as gain medium, respectively.

For all of the lasers, we have monitored the circulating laser light using a reflective beam splitter in the free-space optical path. Figure 2a shows the optical spectrum of an Erbium fiber laser with a repetition rate of 250 MHz. The laser is operated in the vicinity of zero dispersion with a small positive net value. Therefore, the optical spectrum is partially influenced by solitonic pulse shaping and partially by wavelength-dependent transmission of the fiber gain and optical components. The net dispersion is tailored using fibers of dispersion with opposite sign.

The spectrum has been taken with a resolution of 0.02 nm (2.46 GHz); therefore, the absence of spectral modulation excludes double pulses in the oscillator with a separation smaller than ~400 ps. Furthermore, a fast photodiode has been used to sample the optical power in the time domain. Figure 2b shows the corresponding RF spectrum from 0 to 5 GHz. Here, the absence of sinusoidal modulation excludes double pulses with a separation of more than 200 ps. Together, these measurements confirm single-pulse operation of the laser. Independently, the use of this laser as a frequency comb generator [15] attests single-pulse operation.

The observed relative intensity noise (RIN) shows a mostly constant floor between −130 and −120 dB/Hz for low frequencies, which we attribute to power and/or frequency fluctuations of the pump light. Above 10 kHz, these fluctuations are suppressed by the slow time constant of excitation relaxations in Erbium (Fig. 3).

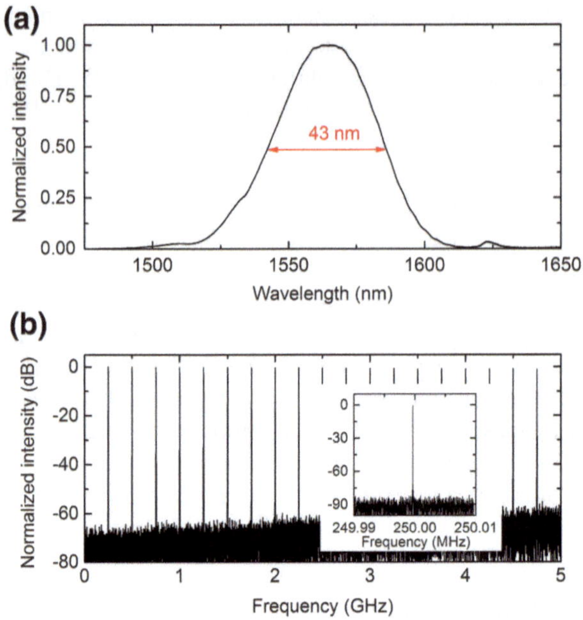

Fig. 2 Output characterization of a mode-locked Er fiber laser with a repetition rate of 250 MHz. **a** Optical spectrum (normalized). **b** RF spectrum of the laser power, measured with a resolution bandwidth of 1 MHz. The *inset* shows a close-up of the fundamental peak with a span of 20 kHz

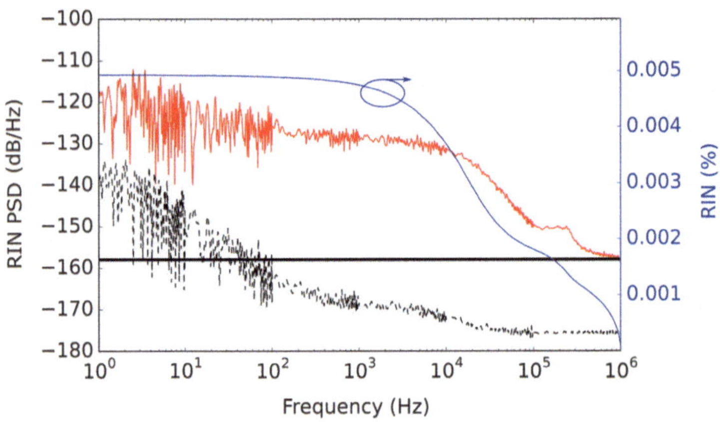

Fig. 3 Relative intensity noise (RIN) of the mode-locked Er fiber laser. *Left axis* power spectral density of the RIN (*red*, *straight*) and measurement noise floor (*black*, *dashed*), *right axis* integrated RIN (*blue*) as computed from the corresponding x-axis value to 1 MHz. The *black horizontal line* indicates the shot-noise limit corresponding to the average optical power impinging on the photodiode

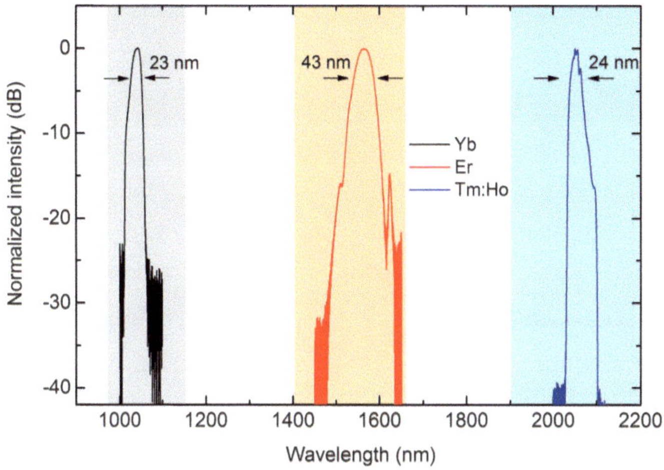

Fig. 4 Normalized optical spectra of three fiber lasers mode-locked with the novel reflective Sagnac loop, using Yb, Er, and Tm/Ho as gain media. The *shadowed area* behind the spectra indicates the typical gain region of the corresponding gain medium. A close-up of all spectra in logarithmic scale can be found in the online material

As mentioned above, the novel reflective configuration has been successfully applied with other gain media and wavelengths. In the graph of Fig. 4, we show the optical spectrum of the Er-based laser together with exemplary spectra of an Yb-based laser at 1030 nm and Tm/Ho-based laser at 2050 nm, respectively. The laser at 1030 nm is operated at a repetition rate of 100 MHz, and the laser at 2050 nm features a repetition rate of 10 MHz. While tuning of net dispersion at telecom wavelength or in the 2-μm regime is easily achieved by combination of fibers with normal and anomalous dispersion, standard fibers in the vicinity of 1 μm do not offer this possibility. Therefore, the free-space section of the Yb laser has been complemented by a grating pair and tuned to a dispersion regime similar to the one described for the Erbium laser. The individual spectra exhibit a 3-dB width of 23 nm (Yb), 43 nm (Er) and 24 nm (Tm/Ho) and support a Fourier-limited pulse width of (84, 72, 240) fs, respectively. More details about the Tm/Ho system can be found in [16].

All of these oscillators are scaled to produce an output in the milliwatt regime [Yb: 10 mW, Er: 3 mW, Tm/Ho: 4.6 mW with respective pump powers of approximately 215 mW (974 nm), 280 mW (974 nm), and 1.7 W (790 nm, double-clad pumped)]. Single-pulse operation is typically obtained in a pump range of $\pm 15\%$ around a center value, and for pump power above the single-pulse regime, either cw-breakthrough or double-pulse operation is observed. For the Yb laser with repetition rate of 100 MHz, the peak intensity is high enough that an autocorrelation signal can be obtained without prior amplification (see Fig. 5). The signal is coupled out by a free-space component and compressed in 16 cm of SF6 glass. The measured autocorrelation width of 137 fs points to a pulse width of 98 fs (FWHM) assuming a Gaussian pulse shape. The deviation from the Fourier limit is explained by a slight third-order dispersion which also causes the small side peaks in the autocorrelation.

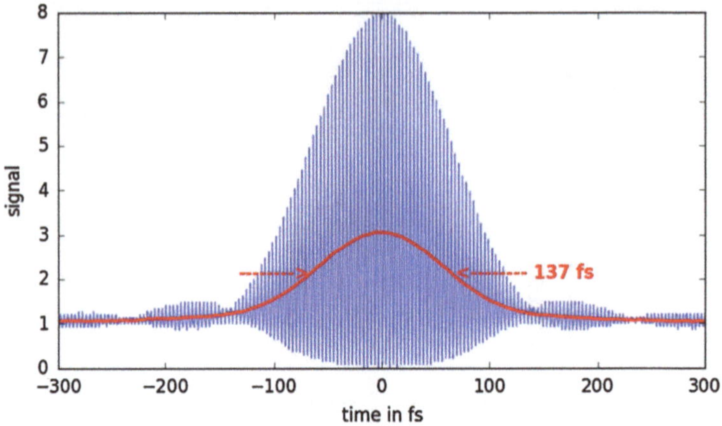

Fig. 5 Normalized autocorrelation trace of a femtosecond pulse from a 100 MHz Yb laser with reflective Sagnac loop. *Thin*, *blue* interferometric autocorrelation trace; *thick*, *red* intensity autocorrelation trace. The output from the laser has been compressed by 16 cm of SF6 glass before analysis. Assuming a Gaussian pulse shape, the width of the intensity autocorrelation trace of 137 fs corresponds to a pulse duration of 98 fs

4 Conclusion

All lasers have proven to work reliably over months, serving as stable sources for the generation of terahertz radiation, as seed oscillators for material processing, for dielectric laser acceleration of electrons [17] and for the generation of highly pure microwave radiation [18]. At Menlo, we are particularly proud that these oscillators have also become the back bone of our new frequency comb systems. The reflective loop configuration with polarization-maintaining fiber provides operation with very low intrinsic phase noise [15], surpassing the performance of fiber lasers mode-locked through NPR or semiconductor saturable absorbers [19]. Together with a novel electro-optic actuator for the carrier-envelope-offset frequency, the new frequency comb systems rival the performance of Ti: Sa-based systems [20], while the robustness of the PM fiber technology permits operation in rough environments and has even facilitated the first optical frequency comb in space [21].

We know it has always been a dream of Ted Hänsch that frequency comb technology becomes available in a simple and compact device. At Menlo, we have been inspired by his visions and his passion, and we hope that our efforts are helping to make his dream come true. Therefore, we dedicate this paper to Ted Hänsch in honor of his 75th birthday.

Acknowledgements We gratefully acknowledge funding from the BMWi (Project ZIM-TERA), the BMBF (Projects ExtruTera, DiaTumor, and Mirandus), the DLR (Projects FOKAL and IRASSI), the EU FP7 initial training network FACT, and the DARPA PULSE program (Project PμreComb).

References

1. T. Udem, J. Reichert, R. Holzwarth, T.W. Hänsch, Absolute optical frequency measurement of the cesium D1 Line with a mode locked laser. Phys. Rev. Lett. **82**, 3568–3571 (1999)
2. E.P. Ippen, H.A. Haus, L.Y. Liu, Additive pulse mode locking. J. Opt. Soc. Am. B **6**, 1736–1745 (1989)
3. K. Kieu, F.W. Wise, All-fiber normal-dispersion femtosecond laser. Opt. Express **16**, 11453–11458 (2008)
4. N. Kuse, J. Jiang, C.-C. Lee, T.R. Schibli, M.E. Fermann, All polarization-maintaining Er fiber-based optical frequency combs with nonlinear amplifying loop mirror. Opt. Express **24**, 3095–3102 (2016)
5. N. Raabe, M. Mero, Y. Song, W. Hänsel, R. Holzwarth, A. Sell, A. Zach, and G. Steinmeyer, Detecting determinism in laser noise: a novel diagnostic approach for ultrafast lasers. In *Conference on Lasers and Electro-Optics* (*p. SM3I.5*). (OSA, Washington, DC, 2016)
6. M. Hofer, M.E. Fermann, F. Haberl, M.H. Ober, A.J. Schmidt, Mode locking with cross-phase and self-phase modulation. Opt. Lett. **16**, 502–504 (1991)
7. K. Tamura, H.A. Haus, E.P. Ippen, Self-starting additive pulse mode-locked erbium fibre ring laser. Electron. Lett. **28**, 2226–2228 (1992)
8. I.N. Duling III, All-fiber ring soliton laser mode locked with a nonlinear mirror. Opt. Lett. **16**, 539–541 (1991)
9. M.E. Fermann, F. Haberl, M. Hofer, H. Hochreiter, Nonlinear amplifying loop mirror. Opt. Lett. **15**, 752–754 (1990)
10. H.A. Haus, E.P. Ippen, K. Tamura, Additive-pulse modelocking in fiber lasers. IEEE J. Quantum Electron. **30**, 200–208 (1994)
11. J.W. Nicholson, M. Andrejco, A polarization maintaining, dispersion managed, femtosecond figure-eight fiber laser. Opt. Express **14**, 8160–8166 (2006)
12. E. Baumann, F.R. Giorgetta, J.W. Nicholson, W.C. Swann, I. Coddington, N.R. Newbury, High-performance, vibration-immune, fiber-laser frequency comb. Opt. Lett. **34**, 638–640 (2009)
13. W. Hänsel, R. Holzwarth, R. Doubek, M. Mei, Laser with non-linear optical loop mirror. Eur. Pat. Appl. EP **2637265**, A1 (2013)
14. T. Jiang, Y. Cui, P. Lu, C. Li, A. Wang, Z. Zhang, All PM fiber laser mode locked with a compact phase biased amplifier loop mirror. IEEE Photonics Technol. Lett. **28**, 1786–1789 (2016)
15. W. Hänsel, M. Giunta, K. Beha, M. Lezius, M. Fischer, R. Holzwarth, Ultra-low phase noise all-PM Er: fiber optical frequency comb. In *Advanced Solid State Lasers* (*p. ATh4A.2*). (OSA, Washington, DC, 2015)
16. H. Hoogland, S. Wittek, W. Hänsel, S. Stark, R. Holzwarth, Fiber chirped pulse amplifier at 2.08 μm emitting 383-fs pulses at 10 nJ and 7 MHz. Opt. Lett. **39**, 6735–6738 (2014)
17. H. Hoogland, J. McNeur, M. Kozák, P. Hommelhoff, R. Holzwarth, Compact ultrashort pulsed 2.05 μm all-PM fiber laser for dielectric laser acceleration of non-relativistic electrons. In *Conference on Lasers and Electro-Optics, OSA Technical Digest* (*p. SF1I.7*). (OSA, Washington, DC, 2016)
18. X. Xie, R. Bouchand, D. Nicolodi, M. Giunta, W. Hänsel, M. Lezius, A. Joshi, S. Datta, C. Alexandre, M. Lours, P.-A. Tremblin, G. Santarelli, R. Holzwarth, Y. Le Coq, Photonic microwave signals with zeptosecond-level absolute timing noise. Nat Photonics. doi:10.1038/nphoton.2016.215
19. N. Kuse, C.-C. Lee, J. Jiang, C. Mohr, T.R. Schibli, M.E. Fermann, Ultra-low noise all polarization-maintaining Er fiber-based optical frequency combs facilitated with a graphene modulator. Opt. Express **23**, 24342–24350 (2015)
20. M. Giunta, W. Hänsel, K. Beha, M. Fischer, M. Lezius, R Holzwarth, Ultra low noise Er: fiber frequency comb comparison. In *Conference on Lasers and Electro-Optics* (*p. STh4H.1*). (OSA, Washington, DC, 2016)

21. M. Lezius, T. Wilken, C. Deutsch, M. Giunta, O. Mandel, A. Thaller, V. Schkolnik, M. Schiemangk, A. Dinkelaker, A. Kohfeldt, A. Wicht, M. Krutzik, A. Peters, O. Hellmig, H. Duncker, K. Sengstock, P. Windpassinger, K. Lampmann, T. Hülsing, T. W. Hänsch, and R. Holzwarth, Space-borne Frequency Comb Metrology. Optica, **3**, 1381–1387 (2016)

A >2-MJ, 10^{14}-W Laser System for DT Fusion—NIF: A Note in Celebration of the 75th Birthday of Prof. Theodore Haensch

John F. Holzrichter and Kenneth R. Manes

Abstract In 1970, Dr. Theodore Haensch joined A.L. Schawlow's group in the physics department at Stanford, as a NATO postdoctoral researcher. Within a short time, he and his colleagues had invented a new, high-resolution, tunable laser system using expanded reflection gratings and an N_2 laser for pumping the fluorescing dyes. This work resulted in a high-brightness, high-repetition-rate, narrow-band laser probe for conducting optical spectroscopy at extreme levels of precision. Dr. Haensch, and his many colleagues, particularly Prof. Arthur Schawlow and their students at Stanford, then proceeded to revolutionize optical spectroscopy and to train several generations of exceptional young scientists. At the same time, the Siegman, Harris, and Byer laboratories also at Stanford were making major contributions to the laser and quantum electronics fields. Several students from both groups joined the Livermore Laboratory. That early work, and that of others, encouraged teams at the Lawrence Livermore National Laboratory to design and build a series of increasing complicated, high-power multi-beam laser systems to investigate the potential of laser fusion. The National Ignition Facility, recently completed, is enabling investigations of matter at very high temperatures, $T > 1$ million K and densities 100-1000× normal. In addition, researchers are creating 10^{15} DT fusion neutrons per fusion experiment and generating new knowledge about unusual and important conditions of matter.

1 Introduction

In this paper, we describe many of the dominant and also unusual physical scientific and technical concepts enabling the series of very high-power lasers to have been built at the Livermore Laboratory *(LLNL1964 to*

This article is part of the topical collection "Enlightening the World with the Laser" - Honoring T. W. Hänsch guest edited by Tilman Esslinger, Nathalie Picqué, and Thomas Udem.

J.F. Holzrichter (✉) • K.R. Manes
Lawrence Livermore National Laboratory, Livermore, CA, USA
e-mail: jfholz@gmail.com

D. Meschede et al. (eds.), *Exploring the World with the Laser*,
https://doi.org/10.1007/978-3-319-64346-5_19

Fig. 1 Ted Haensch and Arthur Schawlow 1970

present).[1] Up to 2 MJ of UV output from the multi-beam NIF laser is being used for irradiating mm- to cm-size materials with up to 4 × 10^{14} W of UV light. Because of the special interest in fusion research [1, 2, 3] and the cost of the system, the NIF laser system was designed like other similarly complex scientific-user systems such as large accelerators, telescopes, gravity detectors.

Those of us who had the good fortune to be at Stanford when A.L. Schawlow and Theodor Haensch (Fig. 1) were developing modern laser spectroscopic techniques and when the Siegman, Harris, Byer groups were creating new quantum electronics devices, were very fortunate indeed. Those influences continue to this day.

Shortly after the laser's description [4] and the first lasing demonstration in 1960 [5] at the Hughes Laboratory, researchers around the world began experimental and theoretical studies to understand the breath of this new optical creation. At Stanford, A.L. Schawlow and his group began investigating the laser's uses for spectroscopy, first using accidental overlaps of laser emission lines with atomic absorption lines. Later, they developed a flashlamp-pumped dye laser for spectroscopic investigations [6]. See Fig. 2.

In the 1970,[2] Ted Haensch joined the Schawlow group and developed host of new tunable lasers and instrumentation techniques [7–9]. These techniques were

[1]1964 to present: LLNL and several other laboratories begin using lasers to heat plasmas to study potential fusion applications. LLNL developed a sequence of ever more capable lasers starting with Long Path 1968, Janus 1974, Cyclops 1975, Argus 1978, Shiva 1980, Novette 1984, Nova 1986, and NIF 2010.

[2]1965–2010: Much early laser understanding for fusion related research stems from the 1960s at Stanford and the Hughes research laboratories, the UC Livermore Laboratory, the CEA laboratories in France, and the Lebedev in Moscow and also at the MaxPlanck Institute of Quantum Optics, University of Rochester, the US Naval Research Laboratory, Ecole Polytechnique in France, the Rutherford and AWRE laboratories in England, Kurchatov Laboratory, University of Osaka, and by many other contributors worldwide.

Fig. 2 Holzrichter with tunable dye laser and MnF_2 sample

important to early fusion laser designers by helping them to understand the spectroscopy of many materials, especially Nd^{3+} ions in optical crystals and glasses [10, 11].

Also in the mid 1960s, Emmett and Schawlow began conducting high-intensity experiments using the then most powerful ruby laser, to understand longitudinal and transverse Brillouin and Raman scattering from liquids and gases [12]. Their laser caused stimulated transverse optical beams to become intense enough to occasionally break their glass-holding vessels, especially if feedback of some sort from windows or edges were present. This nonlinear conversion work highlighted some of the important parasitic processes always present in very high-power laser systems, which remain a major constraint to increasing single-aperture laser power. At the same time, Siegman, Harris, Byer, and Manes [13] at Stanford, as well as many colleagues at Bell Labs and other laboratories, began studying frequency stability and communications qualities for laser communications, and also used today in our laser master oscillators. Also, other groups began investigating both pulsed and high CW power lasers for materials heating, cutting, and (surprisingly for that time) DT fusion applications (*Fusion Investigators 1965*).[3] Today, many prior limitations have been overcome and several large laser systems have been constructed. The largest of them, the National Ignition Facility (NIF) at the Livermore Laboratory, is being used for high-temperature materials and fusion research. It was finished in 2009 [14].

Optical coherent light generators (lasers) and corresponding optical systems have proven to be perfect instruments for communications and for delivering pin-point optical power to targeted atoms or molecules, at precise wavelengths, bandwidths, and power vs. time patterns. They have become commonplace tools for

[3]1975 J.B. Trenholme et al. Optimizing Laser Performance: Private Communication. Dr. John Trenholme joined the Emmett group at the Naval Research laboratory in 1970, upon graduating from Cal. Tech. He has played a major role in the design of all of the LLNL high-power lasers.

processing and assembling industrial materials, for studying materials at temperatures exceeding several million Kelvin, as well as under conditions seen only in astronomical events and nuclear explosions. Lasers have revolutionized materials cutting, drilling, welding, stress conditioning, just to mention a few applications. Exotic ideas employing lasers and coherent delivery systems for remote heating of rocket engines, mineral extraction, for transmitting space power to earth, and for nuclear fusion have been discussed. These "exotic" ideas have on the whole not yet been developed, except for laser-driven nuclear fusion [2, 3]. This topic has been under development for almost 50 years, with steady progress on this very difficult problem [15].

2 Issues Regarding the Generation of Very High-Power Laser Light

The physics of light propagation in vacuum permits coherent light to be generated, amplified, propagated, and focused to amazing extremes, thereby enabling enormous brightness levels on targets. However in the laboratory world, the propagation of light takes place over many meter distances through physical media such as air, glasses, reflecting and transmitting films, and crystals. Through combinations of these materials, a beam of coherent light can be created, amplified, transmitted, wavelength-converted, and focused to reach a target object.

However, these conversion processes are limited by surface and bulk materials damage along the beam path, by transverse and longitudinal reflections, and by many types of nonlinear parasitic modes. At the same time it is important to achieve efficient amplification of the laser beams, to reduce power costs, as well as to maintain low optical losses, and to use as high a fluence as possible per laser beam. This minimizes the number of beams (and costs) required to deliver a required total power to a target. To design a "best system," experienced designers make use of past data from prior LLNL lasers. Figure 3 shows the Nova laser, prior to NIF (e.g., [16, 17]). The designers conduct detailed computer simulations, obtain specific laboratory data, and employ new technologies to minimize the costs for each laser design (they usually consider 1000 s of variations) to reach a power and energy requested by those planning future experiments.

One limitation of high-power pulsed lasers is that the production of high-power laser light is an inefficient process—with the total beam energy comprising only a few percent of the amplifier's input electrical energy. This process is presently expensive, but is slowly being reduced by using arrays of semiconductor lasers instead of xenon flashlamps. Also, the propagation of laser light to a distant target (often 100 s of meters from the oscillator) is governed by optical physics such as aberrations, linear and nonlinear losses, and the use of special inventions such as relay imaging and high-power spatial filtering [18]. These also require clean space, special optics, controls, etc., which add to the total costs.

Fig. 3 The 100 kJ, 10-beam, IR, *Green*, or UV Nova Laser 1985

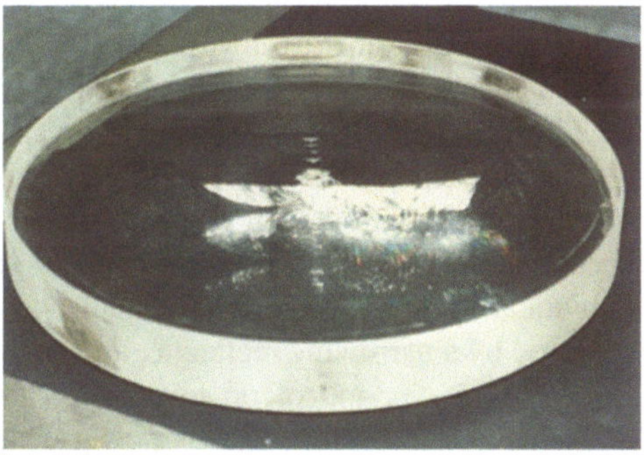

Fig. 4 Nova optical window damaged by stimulated, transverse Brillouin scattering from perpendicular edges

For example, in present lasers the laser beam's transverse dimensions are restricted to about 40 cm (but 80 cm long) due to amplified spontaneous fluoresce inside the Brewster's angled Nd:glass laser amplifier plates. (This is often called ASE.) There are typically 40 such plates and many other types of glass and crystal optics. Damage of transparent optics due to inclusions, pits, and stimulated transverse scattering via Raman or Brillouin processes occurs (see Fig. 4). Today, these limits restrict single-beam NIF laser output to about 20,000 joules of 1.05-micron laser light or $\sim 10^{10}$ W/beam, which become 10,000 joules of UV light per beam upon harmonic conversion.

3 Large Laser Design

Consider a very high-energy laser pulse which is contained inside a long, narrow virtual rectangular container (i.e., a photon package that contains the electromagnetic energy) The container is about 6 m in length and 1/3 m wide on its two

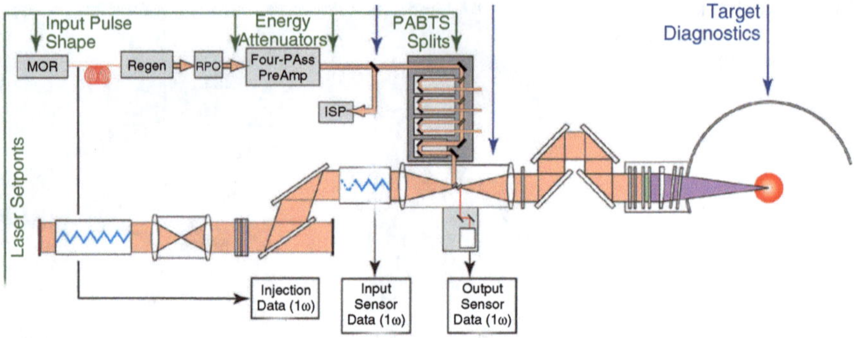

Fig. 5 One NIF Beam line (tan *color*) showing a 2-pass regenerative amplifier on the *left*, a power amplifier, and focusing system on *right*

rectangular sides. It contains ~20,000 J of infrared electromagnetic energy, and it travels at the local speed of light $\sim 3 \times 10^8$ m/s or slower through the laser components described in this paper (see Fig. 5). For fusion, the electromagnetic photon density is low at the front of the "box" called the "foot" and rises to a very high level in the last 1 m of this moving spatial frame.

The total ensemble of photons from 192 beams is divided into 48 beam quads of 4 beams each, with 2 such clusters of 96 beams from each side. The 192 beams contain up to 1.9 MJ of UV energy and strike the target from two sides. Each individual beam (i.e., a virtual container of photons) remains about 6 M in length enclosed by a narrowing rectangle, as it impinges on a ~1-mm-diameter surface of matter, over a period of up to 30 ns. The two large cones, each composed of 96 beam clusters, deliver a peak power density of $\sim 10^{16}$ W/cm^2 to two sides of a target in the final several nanoseconds of the irradiation.

In summary, a pulse from the master oscillator (top left of Fig. 5) is expanded, split into 192 beams, pre-amplified, and then apodized (i.e., meaning transversely shaped) to develop a square-like, flat-topped shape, while retaining its single transverse spatial mode character. It is also pre-shaped in time to correct for amplifier saturation and other effects, and its bandwidth is "spread" to 0.1 nm to avoid Brillouin scattering in glasses and in the final crystals. It is then amplified further by a sequence of Nd:glass disk amplifiers. Then, after a sequence of single-beam amplifiers, which increase the beam energy by a factor of 10^9, a 20,000-J beam is transferred to the optics package to the entrance to the target chamber window. There the 1.053 micron beam, 35 cm × 35 cm in transverse dimension, exits from the transfer telescope, and it traverses a vacuum window and encounters two KDP harmonic plates, which convert the output to 351 nm (3510 angstroms). The pulse for a fusion experiment has a pre-planned time shape of a initial "foot" and final "peak," containing about 10,000 J of UV energy. Each beam then passes through a lens, then a debris shield, and is focused to a preset location at the target. Figure 5 shows a single-beam layout, and Fig. 6 shows a sketch of the entire NIF laser system and building.

Fig. 6 NIF schematic showing blocks of amplifiers (*white*), whose beams meander through a labyrinth of turning mirrors surrounding the *blue* target chamber. The enlarged target is shown to the *right*

4 Present Status

The output area of NIF's 192 beams is 24 m^2, somewhat smaller than the 30 m^2 defining aperture of the optical elements. The total exit area is comparable to that of other large-aperture optical systems such as a Keck I aperture at about 75 m^2. However, NIF contains approximately 7648-m-scale optical elements in its beam lines whose combined area is about 2300 m^2. There are also $\sim$40 such elements in each NIF laser beam line, which exceed the number of large optics used in most telescope optical paths by $\sim 10\times$. Hence, the NIF final cost which includes both the extra optics and a target system was $3B. This is about 7–10 times the cost of a Keck class telescope due to many more large optics, a building, target system, etc. Compared to other large instrumental systems, NIF is in the same cost and complexity class category as the Hubble space telescope and the LHC accelerator at CERN.

The system is complicated, mounting and aligning 7648-m-scale optics and needing about 60,000 control points to operate. Yet, the system is highly automated and easily managed for target experiments, using modern control and optical sensor technologies.

In addition, inertial fusion experiments continue to be conducted, using remarkable imaging systems, particle probes, non-imaging detectors, a 30-picosecond backlighting system (ARC) based on petawatt laser technology [19], and other diagnostics. The NIF laser is a very successful laser design and has become a very successful experimental tool. It enables high-temperature and high-density plasmas and materials experiments to be conducted on a regular basis, by users from across the USA and abroad [20, 21].

Acknowledgements The staff of the Laser program at Livermore and our colleagues around the world have been working to develop these systems since the early 1960s. The lasers described herein could not have been contemplated without their creativity and enormous dedication. At Livermore, John Emmett, Carl Haussman (deceased), and John Nuckolls deserve a great deal of credit for creating and building the Livermore inertial fusion program in 1971, to today's level of performance. Thanks also to the Livermore teams who explored early target concepts and started experiments in the 1960s [1]. Dr. Edward Moses and his team, led by Ralph Patterson, managed the NIF construction project to its completion in 2009.

This work was performed under the auspices of the U.S. Department of Energy by Lawrence Livermore National Laboratory under Contract DE-AC52-07NA27344. The document submitted for publication is LLNL-JRNL-701399.

References

1. S. Colgate, R.E. Kidder, J.H. Nuckolls, R.F. Zabawski, E. Teller, Lawrence Livermore Laboratory, Livermore, Calif., unpublished inertial fusion calculations (1961)
2. J. Nuckolls, L. Wood, A. Thiessen, G. Zimmerman, Laser compression of matter to super-high densities: thermonuclear (CTR) applications. Nature **239**, 139 (1972)
3. J.L. Emmett, J. Nuckolls, L. Wood, Fusion power by laser implosion. Sci. Am. **230**(6), 24–37 (1974)
4. A.L. Schawlow, C.H. Townes, Infrared and optical masers. Phys. Rev. **112**, 1940 (1958)
5. T.H. Maiman, Stimulated optical radiation in ruby. Nature **187**, 493 (1960)
6. J.F. Holzrichter, R.M. Macfarlane, A.L. Schawlow, Magnetization induced by optical pumping in antiferromagnetic, MnF_2. Phys. Rev. Lett. **26**, 652 (1971)
7. T.W. Haensch, M.D. Levenson, A.L. Schawlow, Complete hyperfine structure of a molecular iodine line. Phys. Rev. Lett. **26**(16), 946 (1971)
8. T.W. Haensch, I.S. Shahin, A.L. Schawlow, Optical resolution of the lamb shift in atomic hydrogen by laser saturation spectroscopy. Nature **235**(56), 63–65 (1972)
9. T.W. Haensch, Applications of Dye Lasers. Springer Series of topics in Applied Physics, p. 194 (1973)
10. D.W. Hall, R.A. Haas, W.F. Krupke, M.J. Weber, Spectral and polarization hole burning in neodymium glass lasers. IEEE J. Quantum Electron. **19**, 1704 (1983)
11. R.M. Macfarlane, R.M. Shelby, Homogeneous line broadening of optical transitions of ions and molecules in glasses. J. Lumin. **36**, 179 (1987)
12. J.L. Emmett, A.L. Schawlow, Transverse stimulated emission in liquids. Phys. Rev. **170**, 358 (1968)
13. K.R. Manes, A.E. Siegman, Observation of quantum phase fluctuations in infrared gas lasers. Phys. Rev. A **4**, 373 (1971)
14. B.M. Van Wonterghem, S.C. Burkhart, C.A. Haynam, K.R. Manes, C.D.Marshall, J.E. Murry, M.L. Spaeth, D.R. Speck, S.B. Sutton, P.J. Wegner, National Ignition Facility Commissioning and Performance, in *Proceedings of the SPIE*, vol. 5341. Optical Engineering at the Lawrence Livermore National Laboratory II: The National Ignition Facility, 55. 28 May 2004. doi:10.1117/12.53846
15. J.D. Lindl et al., Review of the national ignition campaign 2009–2012. Phys. Plasmas **21**, 020501 (2014)
16. J.F. Holzrichter, High-power solid state lasers. Nature **316**, 309–313 (1985)

17. K.R. Manes, R.G. Ozarski, W.F. Hagen, J.F. Holzrichter, Multimegajoule Laser Design Lawrence Livermore National Laboratory UCID-20682, August 1985
18. W. Simmons, J.T. Hunt, W.E. Warren, Light propagation through large laser systems. IEEE J. Quantum Electron. **17**, 1727 (1981)
19. D.M. Pennington, M.D. Perry, B.C. Stuart, R.D. Boyd, J.A. Britten, C.G. Brown, S.M. Herman, J.L. Miller, H.T. Nguyen, B.W. Shore, G.L. Tietbohl, V.P. Yanovsky, Petawatt laser system, in *SPIE Proceedings*, vol. 3047, 8 Dec 1997
20. B.A. Remington, R.E. Rudd, J.W. Wark, Observation of magnetic field generation via the Weibel instability in interpenetrating plasma flows. Nat. Phys. 090501 (2015)
21. M.L. Spaeth, K.R. Manes, M. Bowers, P. Celliers, J.-M. Di Nicola, P. Di Nicola, S. Dixit, G. Erbert, J. Heebner, D. Kalantar, O. Landen, B. MacGowan, B. Van Wonterghem, P. Wegner, C. Widmayer, S. Yang, National ignition facility laser system performance. Fusion Sci. Technol. **69**, 366–394 (2016)

Topological Spin Models in Rydberg Lattices

Martin Kiffner, Edward O'Brien, and Dieter Jaksch

Abstract We show that resonant dipole–dipole interactions between Rydberg atoms in a triangular lattice can give rise to artificial magnetic fields for spin excitations. We consider the coherent dipole–dipole coupling between *np* and *ns* Rydberg states and derive an effective spin-1/2 Hamiltonian for the *np* excitations. By breaking time-reversal symmetry via external fields, we engineer complex hopping amplitudes for transitions between two rectangular sub-lattices. The phase of these hopping amplitudes depends on the direction of the hop. This gives rise to a staggered, artificial magnetic field which induces non-trivial topological effects. We calculate the single-particle band structure and investigate its Chern numbers as a function of the lattice parameters and the detuning between the two sub-lattices. We identify extended parameter regimes where the Chern number of the lowest band is $C = 1$ or $C = 2$.

1 Introduction

Regular arrays of ultracold neutral atoms [1, 2] are a versatile tool for the quantum simulation [3–5] of many-body physics [6]. Recent experimental progress allows one to control and observe atoms with single-site resolution [7–12] which makes dynamical phenomena experimentally accessible in these systems. One promising perspective is to use this set-up for investigating the rich physics of quantum magnetism [13–15] and strongly correlated spin systems that are extremely challenging to simulate on a classical computer. However, the simulation of magnetic

---

This article is part of the topical collection "Enlightening the World with the Laser" - Honoring T. W. Hänsch guest edited by Tilman Esslinger, Nathalie Picqué, and Thomas Udem.

M. Kiffner (✉) • D. Jaksch
Centre for Quantum Technologies, National University of Singapore, 3 Science Drive 2, Singapore 117543, Singapore

Clarendon Laboratory, University of Oxford, Parks Road, Oxford OX1 3PU, UK
e-mail: martin.kiffner@physics.ox.ac.uk

E. O'Brien
Clarendon Laboratory, University of Oxford, Parks Road, Oxford OX1 3PU, UK

D. Meschede et al. (eds.), *Exploring the World with the Laser*,
https://doi.org/10.1007/978-3-319-64346-5_20

phenomena with cold atoms faces two key challenges. First, neutral atoms do not experience a Lorentz force in an external magnetic field. In order to circumvent this problem, tremendous effort has been made to create artificial gauge fields for neutral atoms [16–32]. For example, artificial magnetic fields allow one to investigate the integer [33] and fractional quantum Hall effects [27–29] with cold atoms, and the experimental realization of the topological Haldane model was achieved in Ref. [30]. Second, cold atoms typically interact via weak contact interactions. Spin systems with strong and long-range interactions can be achieved by admixing van der Waals interactions between Rydberg states [34, 35] or by replacing atoms with dipole–dipole interacting polar molecules [36–38]. In particular, it has been shown that the dipole–dipole interaction can give rise to topological flat bands [39, 40] and fractional Chern insulators [41]. The creation of bands with Chern number $C = 2$ via resonant exchange interactions between polar molecules has been explored in Ref. [40].

Recently, an alternative and very promising platform for the simulation of strongly correlated spin systems has emerged [42]. Here resonant dipole–dipole interactions between Rydberg atoms [43] enable quantum simulations of spin systems at completely different length scales compared with polar molecules. For example, the experiment in Ref. [42] demonstrated the realization of the *XY* Hamiltonian for a chain of atoms and with a lattice spacing of the order of 20 μm. At these length scales, light modulators allow one to trap atoms in arbitrary, two-dimensional geometries and to apply custom-tailored light shifts at individual sites [44–46]. The resonant dipole–dipole interaction is also ideally suited for the investigation of transport phenomena [47–49] and can give rise to artificial magnetic fields acting on the relative motion of two Rydberg atoms [50–52].

Here we show how to engineer artificial magnetic fields for spin excitations in two-dimensional arrays of dipole–dipole interacting Rydberg atoms. More specifically, we consider a triangular lattice of Rydberg atoms as shown in Fig. 1 where the resonant dipole–dipole interaction enables the coherent exchange of excitations between atoms in *np* and *ns* states. We derive an effective spin-1/2 Hamiltonian for the *np* excitations with complex hopping amplitudes giving rise to artificial, staggered magnetic fields. This results in nonzero Chern numbers of the single-particle band structure, and the value of the Chern number in the lowest band can be adjusted to $C = 1$ or $C = 2$ by changing the lattice parameters.

Note that in our system all atoms comprising the lattice are excited to a Rydberg state. This is in contrast to the work in Refs. [34, 35], where the atoms mostly reside in their ground states and the population in the Rydberg manifold is small. Consequently, our approach is in general more vulnerable towards losses through spontaneous emission. On the other hand, the magnitude of the resonant dipole–dipole interaction is much stronger compared with a small admixing of van der Waals interactions, and hence, the coherent dynamics takes place on much shorter timescales. In addition, the distance between the atoms can be much larger in our approach which facilitates the preparation and observation of the excitations.

This paper is organized as follows. We give a detailed description of our system in Sect. 2 where we engineer an effective Hamiltonian for the *np* excitations. We

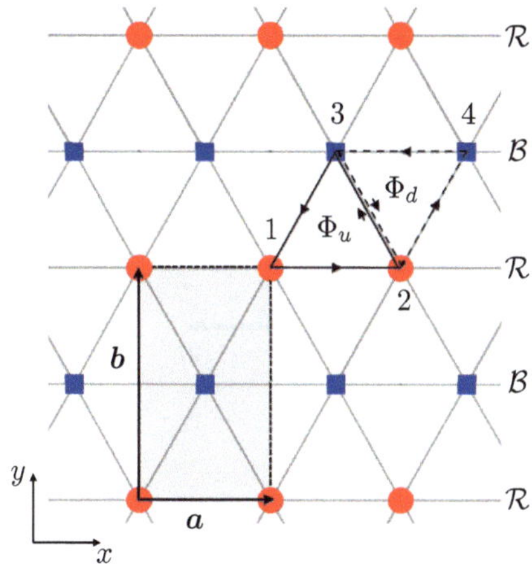

Fig. 1 (Color online) Triangular lattice of Rydberg atoms in the $x-y$ plane. The lattice is comprised of two rectangular sub-lattices $\mathcal{R}$ and $\mathcal{B}$ that are shifted by $\mathbf{a}/2+\mathbf{b}/2$ with respect to each other, where $\mathbf{a}=a\mathbf{e}_x$ and $\mathbf{b}=b\mathbf{e}_y$ are the primitive basis vectors of each sub-lattice. The sites of the $\mathcal{B}$ ($\mathcal{R}$) lattice are indicated by *blue squares* (*red dots*). The unit cell of the whole lattice is shown by the shaded area and contains two lattice sites. Φ_u is the flux through the upward pointing triangle $1\to2\to3\to1$, and Φ_d is the flux through the downward pointing triangle $2\to4\to3\to2$

then investigate the single-particle band structure and provide a systematic investigation of the topological features of these bands as a function of the system parameters in Sect. 3. A brief summary of our work is presented in Sect. 4.

2 Model

We consider a two-dimensional triangular lattice of Rydberg atoms in the $x-y$ plane as shown in Fig. 1. Each lattice site contains a single Rydberg atom which we assume to be pinned to the site. The triangular lattice is comprised of two rectangular sub-lattices $\mathcal{B}$ and $\mathcal{R}$ that are labelled by blue squares and red dots in Fig. 1, respectively. Each sub-lattice is described by two orthogonal primitive basis vectors $\mathbf{a}=a\mathbf{e}_x$ and $\mathbf{b}=b\mathbf{e}_y$, and the two sub-lattices are shifted by $\mathbf{a}/2+\mathbf{b}/2$ with respect to each other. In the following, we derive an effective spin-1/2 model for Rydberg excitations in the *np* manifold over a background of *ns* states with principal quantum number $n\gg1$. After introducing the general Hamiltonian of the system, we first engineer an effective Hamiltonian for *np* excitations on the $\mathcal{B}$ sub-lattice. We then apply the same procedure to the $\mathcal{R}$ sub-lattice but choose a different *np* state compared to the $\mathcal{B}$ atoms. Finally, we show that the dipole–dipole interaction couples the two sub-lattices and the corresponding Hamiltonian contains complex hopping amplitudes giving rise to artificial magnetic fields.

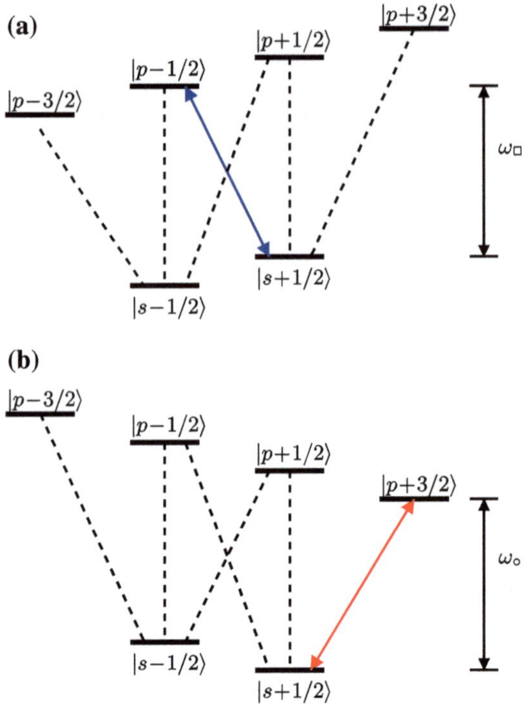

Fig. 2 (Color online) The level scheme of each atom consists of the $ns_{1/2}$ and $np_{3/2}$ manifolds. *Dashed lines* denote allowed dipole transitions. **a** The effective spin-1/2 system at sites $\mathcal{B}$ is formed by states $|p_{3/2}-1/2\rangle$ and $|s_{1/2}1/2\rangle$. The corresponding dipole transition with transition frequency $\omega_{\square}$ is indicated in *blue*. **b** The effective spin-1/2 system at sites $\mathcal{R}$ is formed by states $|p_{3/2}3/2\rangle$ and $|s_{1/2}1/2\rangle$. The associated dipole transition with transition frequency $\omega_{\circ}$ is indicated in *red*

The atomic level scheme of each atom is comprised of two angular momentum manifolds $ns_{1/2}$ and $np_{3/2}$ with principal quantum number $n \gg 1$ as shown in Fig. 2. The Zeeman sub-levels of each multiplet are denoted by $|l_j m\rangle$, where l labels the orbital angular momentum, j is the total angular momentum and the projection of the electron's angular momentum onto the z-axis is denoted by m. The Hamiltonian of a single atom at site α is given by

$$\begin{aligned} H_{\alpha}^{(0)} = \hbar\omega_{\rm p} \sum_{m=-3/2}^{3/2} |p_{3/2}m\rangle_{\alpha}\langle p_{3/2}m|_{\alpha} \\ +\hat{L}_{\alpha}[p_{3/2}] + \hat{L}_{\alpha}[s_{1/2}], \end{aligned} \quad (1)$$

where the first line is the Hamiltonian for the degenerate $np_{3/2}$ manifold in the absence of external fields, $\hbar\omega_{\rm p}$ is the energy of the $np_{3/2}$ multiplet and we set the frequency of the $ns_{1/2}$ multiplet $\omega_{\rm s} = 0$. In the second line of Eq. (1), $\hat{L}_{\alpha}[l_j]$ are level shift operators removing the Zeeman degeneracy of the multiplet l_j at site α. An example for the operators $\hat{L}_{\alpha}[l_j]$ is given in Eq. (17) at the end of Sect. 2. In the following, we assume that all atoms in rows labelled by $\mathcal{B}$ and indicated by a blue square in Fig. 1 experience the same level shifts. Similarly, all atoms in rows labelled by $\mathcal{R}$ and indicated by a red dot in Fig. 1 have equivalent level schemes. However, atoms in sites $\alpha \in \mathcal{R}$ have a different internal level structure compared with atoms in sites $\alpha \in \mathcal{B}$. The full Hamiltonian for the system shown in Fig. 1 is then given by

$$H = \sum_{\alpha} H_{\alpha}^{(0)} + \frac{1}{2} \sum_{\substack{\alpha,\beta \\ \alpha \neq \beta}} V_{\alpha\beta}, \tag{2}$$

where $V_{\alpha\beta}$ is the dipole–dipole interaction [53] between atoms at sites α and β,

$$V_{\alpha\beta} = \frac{1}{4\pi\varepsilon_0 R^3} [\hat{\mathbf{d}}^{(\alpha)} \cdot \hat{\mathbf{d}}^{(\beta)} - 3(\hat{\mathbf{d}}^{(\alpha)} \cdot \widetilde{\mathbf{R}})(\hat{\mathbf{d}}^{(\beta)} \cdot \widetilde{\mathbf{R}})]. \tag{3}$$

Here ε_0 is the dielectric constant, $\hat{\mathbf{d}}^{(\alpha)}$ is the electric-dipole-moment operator of atom α, $\mathbf{R} = \mathbf{R}_{\alpha} - \mathbf{R}_{\beta}$ is the relative position of the two atoms located at $\mathbf{R}_{\alpha}$ and $\mathbf{R}_{\beta}$, respectively, and $\widetilde{\mathbf{R}} = \mathbf{R}/R$ is the corresponding unit vector. In the following, we consider only near-resonantly coupled states and neglect all matrix elements between two-atom states differing in energy by $\Delta E_{\mathrm{FS}} = \hbar\omega_{\mathrm{p}}$ or more. This is justified if the dipole–dipole coupling strength V_0 is much smaller than the fine structure interval ΔE_{FS}, which is the case for the typical parameters based on rubidium atoms (see Sect. 3).

Next we focus on the $\mathcal{B}$ lattice and reduce the level scheme at each site to a two-level system by a suitable choice of the shift operators in Eq. (1). To this end, we assume that the level shifts break the degeneracy of the Zeeman sub-levels as shown in Fig. 2a such that all dipole transitions can be addressed individually. In particular, we require that the strength of the dipole–dipole coupling between nearest neighbours is much smaller than the splitting between Zeeman sub-levels. For all $\mathcal{B}$ atoms, we choose the states $|p_{3/2} - 1/2\rangle$ and $|s_{1/2} 1/2\rangle$ as the effective spin-1/2 system. The dipole matrix element of the $|p_{3/2} - 1/2\rangle \leftrightarrow |s_{1/2} 1/2\rangle$ transition with transition frequency $\omega_{\square}$ is (see "Appendix 1")

$$\mathbf{d}_{\mathcal{B}} = \langle p_{3/2} - 1/2 | \hat{\mathbf{d}} \, | s_{1/2} 1/2 \rangle = \mathcal{D} \frac{1}{\sqrt{6}} (\mathbf{e}_x + \mathrm{i}\mathbf{e}_y), \tag{4}$$

where $\mathcal{D}$ is the reduced dipole matrix element of the $s_{1/2} \leftrightarrow p_{3/2}$ transition. In an interaction picture with respect to the bare atomic energies, the Hamiltonian H in Eq. (2) restricted to all $\mathcal{B}$ atoms can thus be written as

$$H_{\mathcal{B}} = -\frac{1}{6}\sum_{\substack{\alpha\in\mathcal{B}\\ \beta\in\mathcal{B}}} \mathcal{C}_{\alpha\beta}\left(S_\alpha^+ S_\beta^- + S_\beta^+ S_\alpha^-\right), \tag{5}$$

where

$$\mathcal{C}_{\alpha\beta} = \frac{|\mathcal{D}|^2}{4\pi\varepsilon_0 |\mathbf{R}_\alpha - \mathbf{R}_\beta|^3} \tag{6}$$

describes the coupling strength between two $\mathcal{B}$ atoms located at $\mathbf{R}_\alpha$ and $\mathbf{R}_\beta$, respectively. In the following, it will be useful to characterize the strength of the dipole–dipole interaction between two atoms separated by a, and hence, we introduce the parameter

$$V_0 = \frac{|\mathcal{D}|^2}{4\pi\varepsilon_0 a^3}. \tag{7}$$

The raising operator for a spin excitation in Eq. (5) is defined as

$$S_\alpha^+ = |p_{3/2} - 1/2\rangle\langle s_{1/2} 1/2|, \qquad \alpha \in \mathcal{B}, \tag{8}$$

and its adjoint is the corresponding lowering operator, $S_\alpha^- = [S_\alpha^+]^\dagger$.

Next we follow a similar procedure within the $\mathcal{R}$ lattice. In contrast to $\mathcal{B}$ atoms, we choose the states $|p_{3/2} 3/2\rangle$ and $|s_{1/2} 1/2\rangle$ as an effective spin-1/2 system as shown in Fig. 2b. We assume that all other transitions within $\mathcal{R}$ atoms are so far-detuned that the dipole–dipole interaction remains restricted to this sub-system. We find that the dipole matrix element of the corresponding transition $|p_{3/2} 3/2\rangle \leftrightarrow |s_{1/2} 1/2\rangle$ is (see "Appendix 1")

$$\mathbf{d}_{\mathcal{R}} = \langle p_{3/2} 3/2|\hat{\mathbf{d}}\,|s_{1/2} 1/2\rangle = -\mathcal{D}\frac{1}{\sqrt{2}}(\mathbf{e}_x - \mathrm{i}\mathbf{e}_y). \tag{9}$$

The raising operator of this transition with resonance frequency $\omega_\circ$ is defined as

$$S_\alpha^+ = |p_{3/2} 3/2\rangle\langle s_{1/2} 1/2|, \qquad \alpha \in \mathcal{R}, \tag{10}$$

and $S_\alpha^- = [S_\alpha^+]^\dagger$ is the lowering operator. In a rotating frame where S_α^+ oscillates with the frequency $\omega_\square$ of excitations in the $\mathcal{B}$ lattice, the Hamiltonian for excitations in the $\mathcal{R}$ lattice can be written as

$$H_{\mathcal{R}} = \hbar\Delta \sum_{\alpha\in\mathcal{R}} S_\alpha^+ S_\alpha^- - \frac{1}{2}\sum_{\substack{\alpha\in\mathcal{R}\\ \beta\in\mathcal{R}}} \mathcal{C}_{\alpha\beta}\left(S_\alpha^+ S_\beta^- + S_\beta^+ S_\alpha^-\right), \tag{11}$$

where $\Delta = \omega_\circ - \omega_\square$ is the detuning between excitations in the $\mathcal{R}$ and $\mathcal{B}$ lattices and $\mathcal{C}_{\alpha\beta}$ is defined in Eq. (6).

For our given geometry and chosen transitions, we find that the dipole–dipole coupling between the two sub-lattices is different from zero. If the detuning Δ between $\mathcal{B}$ and $\mathcal{R}$ excitations is smaller than the strength of the dipole–dipole coupling between the two sub-lattices, the *np* excitations can hop between the $\mathcal{B}$ and $\mathcal{R}$ sites. With the expressions for the dipole matrix elements in Eqs. (4) and (9), the Hamiltonian governing the coupling between the two sub-lattices is given by

$$H_{\mathcal{BR}} = \frac{\sqrt{3}}{2} \sum_{\substack{\alpha\in\mathcal{R}\\ \beta\in\mathcal{B}}} C_{\alpha\beta} \left(e^{-2\mathrm{i}\phi_{\alpha\beta}} S_\alpha^+ S_\beta^- + e^{2\mathrm{i}\phi_{\alpha\beta}} S_\beta^+ S_\alpha^- \right), \tag{12}$$

where

$$e^{\mathrm{i}\phi_{\alpha\beta}} = \left(\widetilde{\mathbf{R}}_\alpha - \widetilde{\mathbf{R}}_\beta \right) \cdot \left(\mathbf{e}_x + \mathrm{i}\mathbf{e}_y \right). \tag{13}$$

Note that the phase $\phi_{\alpha\beta}$ of excitation hopping between sites $\alpha \in \mathcal{R}$ and $\beta \in \mathcal{B}$ is determined by the azimuthal angle of the relative position vector $\widetilde{\mathbf{R}}_\alpha - \widetilde{\mathbf{R}}_\beta$ between the two sites.

In summary, by restricting the effective level scheme on each site to a two-level system we obtain

$$H_{\mathrm{eff}} = H_{\mathcal{B}} + H_{\mathcal{R}} + H_{\mathcal{BR}} = \hbar\Delta \sum_{\alpha\in\mathcal{R}} S_\alpha^+ S_\alpha^- - \frac{1}{6} \sum_{\substack{\alpha\in\mathcal{B}\\ \beta\in\mathcal{B}}} C_{\alpha\beta} \left(S_\alpha^+ S_\beta^- + S_\beta^+ S_\alpha^- \right)$$
$$- \frac{1}{2} \sum_{\alpha\in\mathcal{R}} \beta \in \mathcal{R} C_{\alpha\beta} \left(S_\alpha^+ S_\beta^- + S_\beta^+ S_\alpha^- \right) + \frac{\sqrt{3}}{2} \sum_{\substack{\alpha\in\mathcal{R}\\ \beta\in\mathcal{B}}} C_{\alpha\beta} \left(e^{-2\mathrm{i}\phi_{\alpha\beta}} S_\alpha^+ S_\beta^- + e^{2\mathrm{i}\phi_{\alpha\beta}} S_\beta^+ S_\alpha^- \right), \tag{14}$$

where the definition of the spin operators $S_\alpha^\pm$ depends on the lattice site as described by Eqs. (8) and (10). The operators $S_\alpha^\pm$ obey Fermi anticommutation relations on the same site,

$$S_\alpha^+ S_\alpha^- + S_\alpha^- S_\alpha^+ = \mathbb{1}, \quad S_\alpha^+ S_\alpha^+ = S_\alpha^- S_\alpha^- = 0, \tag{15}$$

and Bose commutation relations between different sites,

$$\left[S_\alpha^-, S_\beta^+ \right] = \left[S_\alpha^+, S_\beta^+ \right] = \left[S_\alpha^-, S_\beta^- \right] = 0, \quad \alpha \neq \beta. \tag{16}$$

It follows that the raising and lowering operators S_α^+ and S_α^- are equivalent to hard-core bosonic creation and annihilation operators $a_\alpha^\dagger$ and a_α, respectively. The Hamiltonian in Eq. (14) describes the hopping dynamics of these hard-core bosons on the two coupled sub-lattices $\mathcal{A}$ and $\mathcal{B}$.

An example for the dipole–dipole coupling strengths in rubidium atoms and the magnitude of the level shifts required for realizing the effective Hamiltonian in

Eq. (14) is provided in "Appendix 2". Here we outline two physical implementations of the level shifts $\hat{L}_\alpha[l_j]$ in Eq. (1). First, we consider linear Zeeman shifts induced by an external magnetic field B_α in z direction,

$$\hat{L}_\alpha[l_j] = \frac{g[l_j]}{\hbar}\mu_B B_\alpha \hat{J}_z[l_j] \,, \tag{17}$$

where μ_B is the Bohr magneton, $\hat{J}_z[l_j]$ is the z component of the angular momentum operator restricted to the multiplet l_j, and $g[l_j]$ is the Landé g-factor,

$$g[l_j] = \frac{3}{2} + \frac{3/4 - l(l+1)}{2j(j+1)} \,. \tag{18}$$

Since $g[s_{1/2}] = 2$ and $g[p_{3/2}] = 4/3$, the magnitude of the Zeeman shifts is different for the $s_{1/2}$ and $p_{3/2}$ manifolds, respectively. We assume that atoms in lattices $\mathcal{B}$ and $\mathcal{R}$ experience different magnetic field strengths,

$$B_\alpha = \begin{cases} B_\mathcal{B}, & \alpha \in \mathcal{B}\,, \\ B_\mathcal{R}, & \alpha \in \mathcal{R}\,, \end{cases} \tag{19}$$

where $B_\mathcal{B} \neq B_\mathcal{R}$. Exact resonance $\Delta = 0$ between the two sub-lattices can be achieved for $B_\mathcal{B} = -5B_\mathcal{R}/3$, and periodic magnetic fields could be engineered by a regular array of micromagnets [54, 55].

Second, the effective Hamiltonian in Eq. (14) can be realized with a uniform magnetic field across all lattice sites and static or AC Stark shifts that are different for the $\mathcal{B}$ and $\mathcal{R}$ lattices. For example, one could employ AC Stark shifts using a standing wave with periodicity b such that all $\mathcal{B}$ and $\mathcal{R}$ atoms are located at the nodes and antinodes, respectively. Since the magnitude of the AC Stark shifts depends on $|m_j|$, a relative shift between the $|p_{3/2} 3/2\rangle \leftrightarrow |s_{1/2} 1/2\rangle$ and $|p_{3/2} - 1/2\rangle \leftrightarrow |s_{1/2} 1/2\rangle$ transitions can be induced such that the resonance condition $\Delta \approx 0$ holds.

3 Results

The effective Hamiltonian in Eq. (14) exhibits complex hopping amplitudes for exciton transitions between the $\mathcal{B}$ and $\mathcal{R}$ lattices which correspond to an artificial vector potential $\mathbf{A}$ according to the Peierls substitution [56]. This result can be understood as follows. Excitations in the $\mathcal{B}$ and $\mathcal{R}$ lattices couple to different dipole transitions with complex dipole moments $\mathbf{d}_\mathcal{B}$ and $\mathbf{d}_\mathcal{R}$, respectively. The two different transitions on sites $\mathcal{B}$ and $\mathcal{R}$ are tuned into resonance through external fields that break time-reversal symmetry. Since $\mathbf{d}_\mathcal{B}$ and $\mathbf{d}_\mathcal{R}$ have a well-defined relative phase, hopping between the two sub-lattices gives rise to a complex hopping amplitude that depends on the azimuthal angle of the relative position vector $\mathbf{R}_\alpha - \mathbf{R}_\beta$ between sites α and β (see Eq. (13)). The total magnetic flux Φ_u through the upward pointing triangle $1 \to 2 \to 3 \to 1$ is shown in Fig. 1. For nearest-neighbour interactions only, the total flux is determined by the sum of the phases along the edges of the triangle,

$$\Phi_\mathrm{u} = 2(\phi_{32} - \phi_{13}) + \pi \,. \tag{20}$$

We find that the total flux is in general different from zero and can be adjusted by varying the lattice parameters. This is shown by the red solid line in Fig. (3), where Φ_u is depicted as a function of ratio $b\,/\,a$. Φ_u is different from zero except for $b/a = 1$ and attains all possible values between $-\pi$ and π, which is the maximal range for the flux defined mod 2π. Similarly, the total magnetic flux Φ_d through the downward pointing triangle $2 \to 4 \to 3 \to 2$ in Fig. 1 is given by

$$\Phi_\mathrm{d} = 2(\phi_{34} - \phi_{23}) + \pi \,, \tag{21}$$

and Φ_d is shown by the blue dot-dashed line in Fig. 3. Since $\Phi_\mathrm{d} + \Phi_\mathrm{u} = 0$ for all values $b\,/\,a$, the flux in neighbouring triangles has the same magnitude but the opposite sign, and hence, the complex transition amplitudes in our system correspond to a staggered artificial magnetic field. This result is consistent with the assumed translational symmetry of the lattice, which requires that all magnetic fluxes within the unit cell must add up to zero.

Next we investigate the single-particle band structure of H_eff using a rectangular unit cell containing two lattice sites as shown by the shaded area in Fig. 1. It follows that the k-space Hamiltonian $\mathcal{H}(\mathbf{k})$ is represented by a 2×2 matrix, where $\mathbf{k}$ describes a point in the first Brillouin zone of the reciprocal lattice (see "Appendix 3"). We include all hopping terms between sites separated by $R \leq r_\mathrm{D}$. Through a numerical study we find that $\mathcal{H}(\mathbf{k})$ describes the bulk properties of our system well for $r_\mathrm{D} \geq 6a$

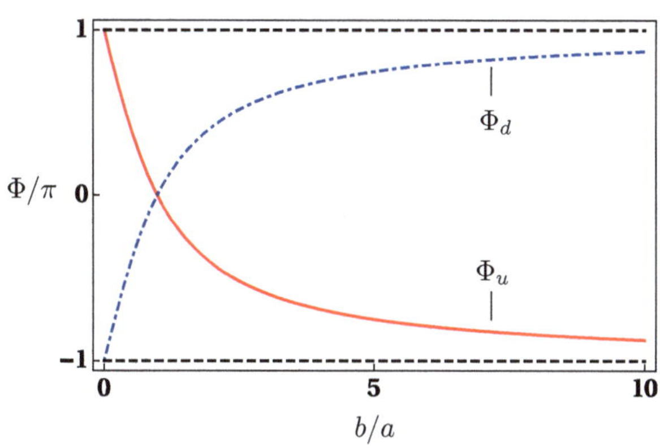

Fig. 3 (Color online) Magnetic flux Φ enclosed in an elementary triangle of the lattice in Fig. 1 as a function of $b\,/\,a$, where b and a are the lattice constants of the rectangular sub-lattices. Φ_u (Φ_d) is the flux through the upward (downward) pointing triangle $1 \to 2 \to 3 \to 1$ ($2 \to 4 \to 3 \to 2$) in Fig. 1 and for the effective Hamiltonian in Eq. (14) with only nearest-neighbour interactions taken into account

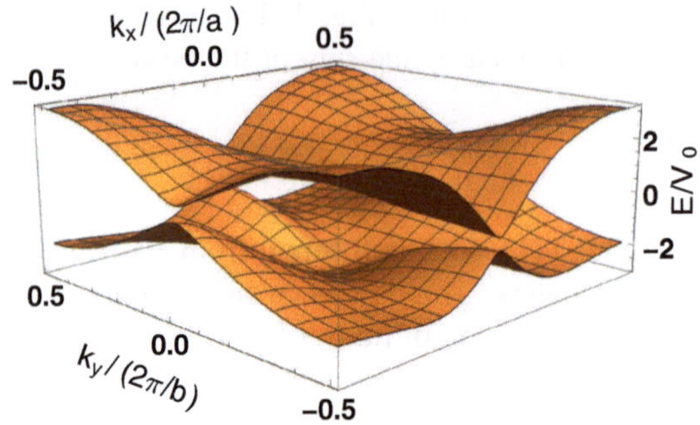

Fig. 4 (Color online) Single-excitation band structure for $\Delta = 0$ and $b/a = \sqrt{3}$. All hopping terms between sites within a radius of $r_{\mathrm{D}} = 6a$ are taken into account. The two bands are separated by a gap and the *lower* (*upper*) band has Chern number $C = 1$ ($C = -1$)

and $a \geq b/2$. The band structure for the special case of equilateral triangles as in Fig. 1 (i.e. $b/a = \sqrt{3}$) is shown in Fig. 4. There are two separate bands, and the band gap varies in size across the Brillouin zone. The gap is the smallest near the following points at the zone boundary,

$$\mathbf{k}_1 = (0, \pi/b) \quad \mathbf{k}_2 = (0, -\pi/b) \tag{22a}$$
$$\mathbf{k}_3 = (\pi/a, 0) \quad \mathbf{k}_4 = (-\pi/a, 0)\,. \tag{22b}$$

The magnitude of the band gap near these points is of the order of $V_0/2$, where V_0 is defined in Eq. (7). The broken time-reversal symmetry in our system endows the band structure with non-trivial topological properties. We numerically calculate the Chern number as described in Ref. [57] and find that the lower and upper bands have Chern numbers $C = 1$ and $C = -1$, respectively.

Various topological regimes can be realized in our system by adjusting the lattice parameters and the detuning between the excitations on the sub-lattices $\mathcal{B}$ and $\mathcal{A}$. This is illustrated in Fig. 5, where we show the Chern number of the lower band as a function of the ratio $b\,/\,a$ and Δ. First, we note that the phase diagram in Fig. 5 exhibits extended regions with nonzero Chern numbers that are robust with respect to small variations in Δ and $b\,/\,a$. The solid lines in Fig. 5 indicate topological phase transitions where the lower and upper bands touch in at least two of the $\mathbf{k}$ points in Eq. (22) which then represent a Dirac point.

The qualitative features of the $C = 1$ region marked in orange in Fig. 5 can be understood by noting that nonzero Chern numbers require an efficient coupling between the sub-lattices $\mathcal{B}$ and $\mathcal{R}$. In particular, the dipole–dipole coupling needs to be larger or comparable to the detuning Δ. For fixed lattice constant a, reducing $b\,/\,a$ corresponds to an increased dipole–dipole coupling between the sub-lattices and hence the region with $C = 1$ broadens along the Δ axis for $b/a < 1$. The narrowing of the $C = 1$ region near $b/a = 1$ can be understood from Fig. 3. For nearest-neighbour interactions only, the magnetic flux vanishes for $b/a = 1$ and hence the

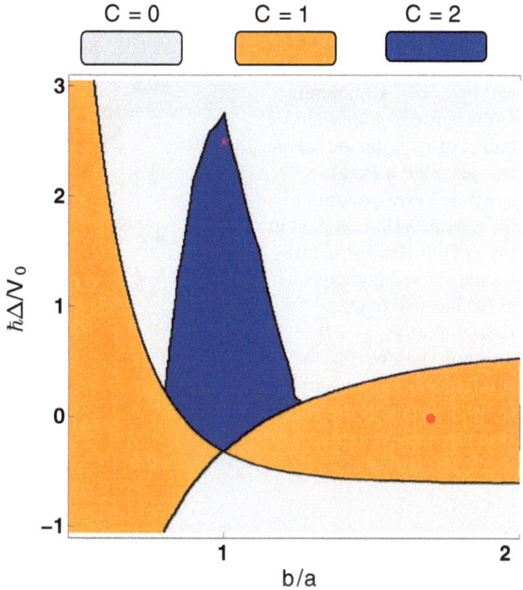

Fig. 5 (Color online) Topological regimes for the lower band for $r_D = 8$. The *black lines* indicate topological phase transitions where the Chern number of the lower band changes. The *red dot* and magenta star correspond to the parameters of the band structures in Figs. 4 and 6, respectively

corresponding bands would have Chern number $C = 0$. Taking into account interactions beyond nearest neighbours gives rise to modifications as shown in Fig. 5. In particular, these interactions are responsible for the blue wedged area with Chern number $C = 2$. The single-particle band structure for the parameters corresponding to the magenta star inside the blue wedged area in Fig. 5 is shown in Fig. 6a. The lower and upper bands have Chern numbers $C = 2$ and $C = -2$, respectively. The two bands are gapped, but in contrast to the parameters in Fig. 4 the gap is the smallest near the Brillouin zone centre $\mathbf{k} = (0, 0)$ where it is approximately given by $0.1V_0$.

The asymmetry of the phase diagram in Fig. 5 with respect to the $\Delta = 0$ axis can be traced back to the fact that the dipole–dipole interaction differs in strength for the $\mathcal{B}$ and $\mathcal{R}$ lattices. In order to illustrate this, we focus on the blue wedge with $C = 2$ in Fig. 5 and show the band structures of the uncoupled, individual sub-lattices in Fig. 6b for $\hbar\Delta = 2.5V_0$ and $b/a = 1$. Both band structures are convex surfaces with their minimum at $\mathbf{k} = 0$, but the depth of the potential well is significantly larger for the upper band. The reason is that the strength of the dipole–dipole interaction is three times stronger for the $\mathcal{R}$ lattice compared to the $\mathcal{B}$ lattice for $b/a = 1$ (see Eqs. (5) and (11)). A necessary condition for non-trivial topological bands is that the two sub-lattices are efficiently coupled by the Hamiltonian $H_{\mathcal{BR}}$ in Eq. (12), which depends on the magnitude of the dipole–dipole interaction connecting the $\mathcal{B}$ and $\mathcal{R}$ lattices and the energy spacing between $\mathcal{B}$ and $\mathcal{R}$ excitations at each $\mathbf{k}$ point. As shown in Fig. 6b, the two surfaces touch near $\mathbf{k} \approx \mathbf{0}$, and hence, the relatively weak next-nearest-neighbour coupling in $H_{\mathcal{BR}}$ can give rise to nonzero Chern numbers for $\hbar\Delta = 2.5V_0$ and $b/a = 1$. On the other hand, the distance between the two uncoupled bands increases quickly if Δ is decreased from zero to negative values. This explains why $H_{\mathcal{BR}}$ cannot induce a $C = 2$ band for $\hbar\Delta \lesssim -0.3V_0$.

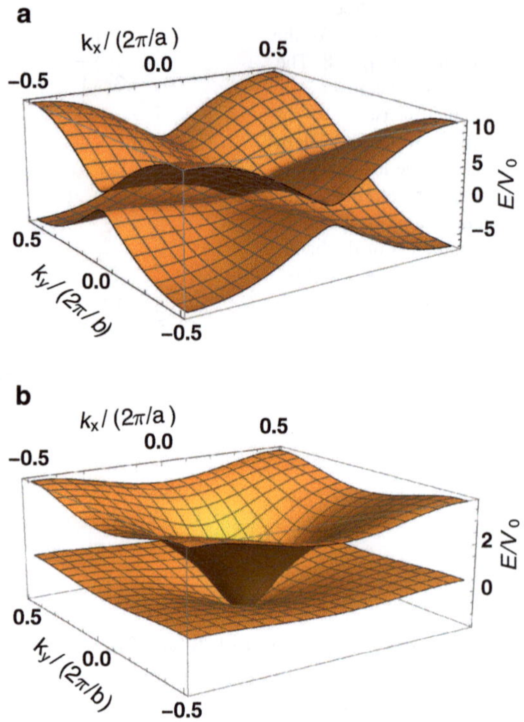

Fig. 6 (Color online) Single-excitation band structure for $\hbar\Delta = 2.5V_0$ and $b/a = 1$. All hopping terms between sites within a radius of $r_D = 6a$ are taken into account. **a** Band structure corresponding to the effective Hamiltonian in Eq. (14). **b** Band structure for the same parameters as in (**a**) but without the Hamiltonian $H_{\mathcal{BR}} = 0$ coupling the two sub-lattices. The *upper* (*lower*) surface is the band structure for excitations on the $\mathcal{R}$ ($\mathcal{B}$) lattice

Finally, we discuss the physical realization of our system and the observation of its topological features. The experimental realization of a one-dimensional chain of resonantly coupled Rydberg atoms has been reported in Ref. [42]. Here the excitation of all atoms to a Rydberg state is achieved within $\tau \approx 0.5\,\mu s$ [42]. Note that this process is not hampered by the dipole blockade since the van der Waals shifts are small for the considered lattice constants a. For example, for Rubidium *ns* states with $n = 70$ and $a = 20\,\mu m$, the van der Waals shift is $\Delta_{vdW} \approx 13\,kHz$ [58], which is small compared with the Rabi frequency of the lasers exciting the Rydberg state [42]. The time interval ΔT where excitation hopping can take place is limited by the lifetime of the Rydberg states and the residual atomic motion. For atomic temperatures of the order of $10\,\mu K$, motional effects are negligible for $\Delta T \approx 10\,\mu s$ [42]. This is typically much smaller than the Rydberg state lifetime and large compared with the inverse hopping amplitude such that many coherent hops can take place (see "Appendix 2"). Note that these considerations also show that autoionization processes due to Rydberg atom collisions can be neglected [59, 60] since the initial positions of the atoms in the lattice change only very slightly during ΔT. Recently, tremendous experimental progress towards the extension of the experiment in Ref. [42] to two dimensions and arbitrary lattice geometries has been made [44, 45]. In particular, it is now possible to create arbitrary lattice structures where each site is filled with exactly one atom [46]. It follows that our system can be realized with a combination of state-of-the-art experimental techniques.

A direct signature of the artificial magnetic fields associated with the complex hopping amplitudes in Eq. (14) can be obtained by investigating the quantum dynamics of a single excitation as shown in Fig. 7. We consider a lattice with 53 sites where only the site in the middle of the lower edge is excited at time $t = 0$. The excitation probability of the lattice sites at a later time is shown in Fig. 7a, b, where Fig. 7a the dynamics according to the effective Hamiltonian in Eq. (14). We find that the largest excitation probabilities can be found along the lower edge and to the right of the initially excited site. Figure 7b is generated by setting all phases $\phi_{\alpha\beta}$ in Eq. (14) to zero. In this case, the distribution of excitation probabilities is symmetric with respect to the dashed line. The latter result is expected since the magnitude of the hopping amplitudes does only depend on the distance between two sites. It follows that the marked asymmetry in Fig. 7a is a direct consequence of the complex hopping amplitudes and the associated artificial magnetic field. More specifically, the magnetic flux through the upward pointing triangles shown in Fig. 7 and for the considered parameters is negative (see Fig. 3). The force associated with the artificial magnetic field thus favours an anticlockwise motion

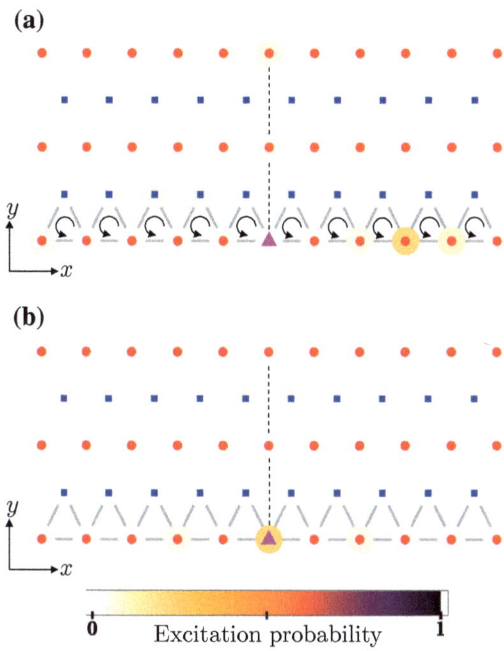

Fig. 7 (Color online) Quantum dynamics of a single excitation on a lattice with 53 atoms, $b/a = 2$, $\Delta = 0$ and $r_D = 6a$. At time $t = 0$, only the atom in the *bottom row* indicated by a *magenta triangle* is excited. The population of each lattice site at time $t = 4\hbar/V_0$ is indicated by the *colour* of the halo around each site. The *dashed line* is used as a guide to the eye (see text). **a** Quantum dynamics according to the Hamiltonian H_{eff} in Eq. (14). The magnetic flux through the indicated *triangular* plaquettes is negative and thus favours counterclockwise motion of the excitation. **b** Same as in (**a**), but with all phases $\phi_{\alpha\beta}$ in Eq. (14) set to zero

around each triangular plaquette. This explains why the propagation moves along the edge in an anticlockwise direction. Note that this asymmetry develops within a few hopping events such that the residual motion of the atoms hosting these excitations can be neglected. Our results are also consistent with the fact that a semi-infinite version of our lattice exhibits chiral edge states for nonzero Chern numbers according to the bulk-edge correspondence [61, 62].

The Chern number of the individual bands can be determined by observing the motional drifts due to the nonzero Berry curvature in each band. To this end, the excitations need to be selectively prepared in either the upper or lower band. This can be achieved in different ways. First, one could prepare an excitation in one of the sub-lattices with a large detuning Δ such that the $\mathcal{R}$ and $\mathcal{B}$ lattices are uncoupled. This is followed by an adiabatic reduction of $|\Delta|$ in order to adjust the required parameter regime. Second, one could prepare all atoms in the $|ns_{1/2}1/2\rangle$ state and apply a weak microwave field such that only a single $\mathbf{k}$ mode is resonantly excited. Efficient methods to extract the local Berry curvature from motional drifts are described in Ref. [63] and require an external force acting on the particle. In our set-up, this could be realized by making the detuning Δ position-dependent through magnetic field gradients along a certain direction.

4 Summary

We have shown that the resonant dipole–dipole interaction between Rydberg atoms allows one to engineer effective spin-1/2 models where the spin excitations experience a staggered magnetic field in a triangular lattice. A necessary condition for engineering artificial magnetic fields is that time-reversal symmetry of the system is broken. In our system, this is achieved by external fields shifting the Zeeman sub-levels of the considered $ns_{1/2}$ and $np_{3/2}$ manifolds. In this way, we ensure that the spin excitation couples to different dipole transitions on the $\mathcal{B}$ and $\mathcal{R}$ lattices with dipole moments $\mathbf{d}_{\mathcal{B}}$ and $\mathbf{d}_{\mathcal{R}}$, respectively. These dipole moments have a well-defined relative phase which is different from zero. Since $\mathbf{d}_{\mathcal{B}}$ and $\mathbf{d}_{\mathcal{R}}$ are orthogonal, we find that the phase of the hopping amplitude is determined by the azimuthal angle associated with the relative position of the two sites connected by the hop.

We find that the magnitude of the magnetic flux through an elementary triangular plaquette can be controlled by changing the ratio b / a of the rectangular sub-lattices. In addition, the staggered magnetic field endows the single-particle band structure with non-trivial Chern numbers. The Chern number of the lower band can be adjusted between $C = 0$, 1 and 2 and its value depends on the lattice parameters and the detuning Δ between the $\mathcal{B}$ and $\mathcal{R}$ lattices.

The quantum simulation of the dynamics of a single excitation shows that an excitation placed at an edge of the lattice will propagate along the edge in a specific direction. This effect is a direct consequence of the artificial magnetic field. The topological features of the bands can be explored by monitoring the deflection of the exciton motion due to the nonzero Berry curvature in either the lower or upper band. An intriguing prospect for future studies is the investigation of quantum many-body states. Here the hard-core interaction between the particles is expected to modify the single-particle picture considerably, and the interplay of strong interactions and complex hopping amplitudes may give rise to exotic quantum phases like fractional Chern insulators.

Acknowledgements M.K. thanks the National Research Foundation and the Ministry of Education of Singapore for support and Tilman Esslinger for helpful discussions. The authors would like to acknowledge the use of the University of Oxford Advanced Research Computing (ARC) facility in carrying out this work (http://dx.doi.org/10.5281/zenodo.22558).

Appendix 1: Dipole Matrix Elements

We evaluate the matrix elements of the electric-dipole-moment operator $\hat{\boldsymbol{d}}$ of an individual atom via the Wigner–Eckert theorem [64, 65] and find

$$\langle nl'_{j'}m'|\hat{\boldsymbol{d}}|nl_jm\rangle = \mathcal{D}\sum_{q=-1}^{1} C_{jm1q}^{j'm'}\boldsymbol{\epsilon}_q\,, \tag{23}$$

where $C_{jm1q}^{j'm'}$ are Clebsch–Gordan coefficients and the spherical unit vectors $\boldsymbol{\epsilon}_q$ in Eq. (23) are defined as

$$\epsilon_1 = -\frac{\mathbf{e}_x - i\mathbf{e}_y}{\sqrt{2}}, \quad \epsilon_0 = \mathbf{e}_z, \quad \epsilon_{-1} = \frac{\mathbf{e}_x + i\mathbf{e}_y}{\sqrt{2}}. \tag{24}$$

The reduced dipole matrix element is [64, 65]

$$\begin{aligned}\mathcal{D} = \; & (-1)^{\mathrm{j}+\mathrm{l}'-1/2}\sqrt{2j+1}\sqrt{2l+1} \\ & \begin{Bmatrix} l' & l & 1 \\ j & j' & 1/2 \end{Bmatrix} C_{l0 1 0}^{l'0} e\langle n'l'|r|nl\rangle\,,\end{aligned} \tag{25}$$

where the 3×2 matrix in curly braces is the Wigner $6-j$ symbol, e is the elementary charge, and $\langle n'l'|r|nl\rangle$ is a radial matrix element.

Appendix B: Rubidium Parameters

Here we calculate the strength of the dipole–dipole interaction for rubidium atoms and estimate the magnitude of the level shifts required for realizing our model. For $ns_{1/2} \leftrightarrow np_{3/2}$ transitions in rubidium with principal quantum number $n = 70$, the reduced dipole moment $\mathcal{D}$ in Eq. (25) is given by

$$\mathcal{D} \approx 2909 e a_0\,, \tag{26}$$

where e is the elementary charge and a_0 is the Bohr radius. It follows that the strength of the dipole–dipole coupling V_0 in Eq. (7) for $a = 20\,\mu\text{m}$ is

$$V_0/\hbar \approx 2\pi \times 1.03\,\text{MHz}\,. \tag{27}$$

The lifetime of the $ns_{1/2}$ and $np_{3/2}$ states at temperature $T = 300\text{K}$ and for $n = 70$ is $T_\text{s} \approx 151.6\,\mu\text{s}$ and $T_\text{p} \approx 191.3\,\mu\text{s}$, respectively [66]. Note that these values take into account the lifetime reduction due to blackbody radiation. The hopping rates vary with the lattice parameters but are typically of the order of V_0. It follows that in principle many coherent hopping events can be observed before losses due to spontaneous emission set in. This finding is consistent with the experimental observations in Ref. [42]. Note that the magnitude of V_0 can be increased by reducing the size of the lattice constant a or by increasing n.

Next we discuss the requirements for reducing the general Hamiltonian in Eq. (2) to our model in Eq. (14). First, we note that the level shifts induced between Zeeman sub-states must be large compared to V_0 and hence of the order of 10 MHz. Shifts of this magnitude can be realized with weak magnetic fields [67] or AC stark shifts [42]. Furthermore, the fine structure splitting between the $ns_{1/2}$ and $np_{3/2}$ manifolds is $\Delta E_\text{FS} \approx 2\pi \times 10.8\,\text{GHz}$ [68], which is much larger than V_0, and hence, it is justified to neglect off-resonant terms in Eq. (3). Finally, we note that the energy difference between the $np_{3/2}$ manifold and the nearby $np_{1/2}$ manifold is approximately 285 MHz [68], which is also much larger than V_0. It follows that the $np_{1/2}$ states can be safely neglected.

Appendix C: *k*-space Hamiltonian

The k-space Hamiltonian can be obtained by considering the single-excitation sub-space $\mathcal{E}_1$ spanned by the basis states

$$|\alpha\rangle = S_\alpha^+|0\rangle\,, \tag{28}$$

where $|\alpha\rangle$ denotes one p excitation at site α and $|0\rangle$ is the "vacuum" state with zero excitations, i.e. the atoms at all lattice sites are in state $|s_{1/2}1/2\rangle$. In order to solve the eigenvalue equation

$$H_\text{eff}|\psi\rangle = E|\psi\rangle \tag{29}$$

with $|\psi\rangle \in \mathcal{E}_1$, we describe the lattice in Fig. 1 by a rectangular Bravais lattice with a two-atomic basis. More specifically, the direct lattice points are given by the $\mathcal{R}$ atoms such that the basis is comprised of one $\mathcal{R}$ atom at $\mathbf{0}$ and one $\mathcal{B}$ atom at

$(\mathbf{a}+\mathbf{b})/2$. According to Bloch's theorem [69], we can solve Eq. (29) with the Ansatz

$$|\psi\rangle = \sum_\alpha u_\alpha |\alpha\rangle\,, \tag{30}$$

where the coefficients u_α can be written as

$$u_\alpha = \begin{cases} \psi_{\mathcal{R}} e^{i\mathbf{k}\cdot\mathbf{U}(\alpha)}, & \alpha \in \mathcal{R}\,, \\ \psi_{\mathcal{B}} e^{i\mathbf{k}\cdot\mathbf{U}(\alpha)}, & \alpha \in \mathcal{B}\,, \end{cases} \tag{31}$$

and $\mathbf{k}$ is a point in the first Brillouin zone of the direct lattice. The vector $\mathbf{U}(\alpha)$ in Eq. (31) is the Bravais lattice point associated with site α,

$$\mathbf{U}(\alpha) = \begin{cases} \mathbf{R}_\alpha, & \alpha \in \mathcal{R}\,, \\ \mathbf{R}_\alpha - (\mathbf{a}+\mathbf{b})/2, & \alpha \in \mathcal{B}\,. \end{cases} \tag{32}$$

With Eqs. (30) and (31), Eq. (29) can be reduced to the following matrix equation for the amplitudes $\psi_{\mathcal{R}}$ and $\psi_{\mathcal{B}}$,

$$\mathcal{H}(\mathbf{k}) \begin{pmatrix} \psi_{\mathcal{R}} \\ \psi_{\mathcal{B}} \end{pmatrix} = E \begin{pmatrix} \psi_{\mathcal{R}} \\ \psi_{\mathcal{B}} \end{pmatrix}, \tag{33}$$

where the 2×2 matrix $\mathcal{H}(\mathbf{k})$ is the k-space Hamiltonian. We find $\mathcal{H}(\mathbf{k})$ using the software package MATHEMATICA [70] for each set of lattice parameters a and b. In general, the resulting expressions are too complicated to display here. In the special case of nearest-neighbour interactions only, we find

$$[\mathcal{H}(\mathbf{k})]_{11} = -V_0 \cos(\mathbf{k}\cdot\mathbf{a}) + \hbar\Delta\,, \tag{34a}$$

$$[\mathcal{H}(\mathbf{k})]_{12} = \frac{V_0\sqrt{48}}{[1+(b/a)^2]^{3/2}} \Big(e^{-i(\mathbf{k}\cdot\mathbf{a}+\mathbf{k}\cdot\mathbf{b})} e^{-2i\alpha} + e^{-2i\alpha} + e^{-i\mathbf{k}\cdot\mathbf{a}} e^{2i\alpha} + e^{-i\mathbf{k}\cdot\mathbf{b}} e^{2i\alpha} \Big)\,, \tag{34b}$$

$$[\mathcal{H}(\mathbf{k})]_{22} = -\frac{1}{3} V_0 \cos(\mathbf{k}\cdot\mathbf{a})\,, \tag{34c}$$

where $\cos(\alpha) = 1/[1+(b/a)^2]^{1/2}$ and $[\mathcal{H}(\mathbf{k})]_{21} = [\mathcal{H}(\mathbf{k})]_{12}^*$.

References

1. D. Jaksch, C. Bruder, J.I. Cirac, C.W. Gardiner, P. Zoller, Phys. Rev. Lett. **81**, 3108 (1998)
2. I. Bloch, Nature **1**, 23 (2005)
3. T. Calarco, H.J. Briegel, D. Jaksch, J.I. Cirac, P. Zoller, J. Mod. Opt. **47**, 2137 (2000)
4. J.I. Cirac, P. Zoller, Nat. Phys. **8**, 264 (2012)
5. T.H. Johnson, S.R. Clark, D. Jaksch, EPJ Quantum Technol. **1**, 10 (2014)
6. I. Bloch, J. Dalibard, W. Zwerger, Rev. Mod. Phys. **80**, 885 (2008)

7. W.S. Bakr, J.I. Gillen, A. Peng, S. Fölling, M. Greiner, Nature **462**, 74 (2009)
8. W.S. Bakr, A. Peng, M.E. Tai, R. Ma, J. Simon, J.I. Gillen, S. Fölling, L. Pollet, M. Greiner, Science **329**, 547 (2010)
9. J.F. Sherson, C. Weitenberg, M. Endres, M. Cheneau, I. Bloch, S. Kuhr, Nature **467**, 68 (2010)
10. C. Weitenberg, M. Endres, J.F. Sherson, M. Cheneau, P. Schauß, T. Fukuhara, I. Bloch, S. Kuhr, Nature **471**, 319 (2011)
11. T. Gericke, P. Würtz, D. Reitz, T. Langen, H. Ott, Nat. Phys. **4**, 949 (2008)
12. P. Würtz, T. Langen, T. Gericke, A. Koglbauer, H. Ott, Phys. Rev. Lett. **103**, 080404 (2009)
13. A. Auerbach, *Interacting Electrons and Quantum Magnetism* (Springer, New York, 1994)
14. J. Simon, W.S. Bakr, R. Ma, M.E. Tai, P.M. Preiss, M. Greiner, Nature **472**, 307 (2011)
15. C. Sanner, E.J. Su, W. Huang, A. Keshet, J. Gillen, W. Ketterle, Phys. Rev. Lett. **108**, 240404 (2012)
16. J. Ruseckas, G. Juzeliūnas, P. Öhberg, M. Fleischhauer, Phys. Rev. Lett. **95**, 010404 (2005)
17. J. Dalibard, F. Gerbier, G. Juzeliūnas, P. Öhberg, Rev. Mod. Phys. **83**, 1523 (2011)
18. R. Dum, M. Olshanii, Phys. Rev. Lett. **76**, 1788 (1996)
19. Y.-J. Lin, R.L. Compton, A.R. Perry, W.D. Phillips, J.V. Porto, I.B. Spielman, Phys. Rev. Lett. **102**, 130401 (2009a)
20. Y.-J. Lin, R.L. Compton, K. Jiménez-García, J.V. Porto, I.B. Spielman, Nature **462**, 628 (2009b)
21. Y.-J. Lin, R.L. Comption, K. Jiménez-García, W.D. Phillips, J.V. Porto, I.B. Spielman, Nat. Phys. **7**, 531 (2011)
22. M. Aidelsburger, M. Atala, S. Nascimbène, S. Trotzky, Y.-A. Chen, I. Bloch, Phys. Rev. Lett. **107**, 255301 (2011)
23. D. Jaksch, P. Zoller, New J. Phys. **5**, 56 (2003)
24. J. Struck, C. Ölschläger, R.L. Targat, P. Soltan-Panahi, A. Eckardt, M. Lewenstein, P. Windpassinger, K. Sengstock, Science **333**, 996 (2011)
25. J. Struck, C. Ölschläger, M. Weinberg, P. Hauke, J. Simonet, A. Eckardt, M. Lewenstein, K. Sengstock, P. Windpassinger, Phys. Rev. Lett. **108**, 225304 (2012)
26. K. Jiménez-García, L.J. LeBlanc, R.A. Williams, M.C. Beeler, A.R. Perry, I.B. Spielman, Phys. Rev. Lett. **108**, 225303 (2012)
27. R.N. Palmer, D. Jaksch, Phys. Rev. Lett. **96**, 180407 (2006)
28. R.N. Palmer, A. Klein, D. Jaksch, Phys. Rev. A **78**, 013609 (2008)
29. N.R. Cooper, J. Dalibard, Phys. Rev. Lett. **110**, 185301 (2013)
30. G. Jotzu, M. Messer, R. Desbuquois, M. Lebrat, T. Uehlinger, D. Greif, T. Esslinger, Nature **515**, 237 (2014)
31. P. Hauke, O. Tieleman, A. Celi, C. Ölschläger, J. Simonet, J. Struck, M. Weinberg, P. Windpassinger, K. Sengstock, M. Lewenstein et al., Phys. Rev. Lett. **109**, 145301 (2012)
32. H. Zhang, Q. Guo, Z. Ma, X. Chen, Phys. Rev. A **87**, 043625 (2013)
33. A. Sterdyniak, N.R. Cooper, N. Regnault, Phys. Rev. Lett. **115**, 116802 (2015)
34. A.W. Glaetzle, M. Dalmonte, R. Nath, I. Rousochatzakis, R. Moessner, P. Zoller, Phys. Rev. X **4**, 041037 (2014)
35. A.W. Glaetzle, M. Dalmonte, R. Nath, C. Gross, I. Bloch, P. Zoller, Phys. Rev. Lett. **114**, 173002 (2015)
36. D. Peter, S. Müller, S. Wessel, H.P. Büchler, Phys. Rev. Lett. **109**, 025303 (2012)
37. A.V. Gorshkov, S.R. Manmana, G. Chen, J. Ye, E. Demler, M.D. Lukin, A.M. Rey, Phys. Rev. Lett. **107**, 115301 (2011a)
38. A.V. Gorshkov, S.R. Manmana, G. Chen, E. Demler, M.D. Lukin, A.M. Rey, Phys. Rev. A **84**, 033619 (2011b)
39. N.Y. Yao, C.R. Laumann, A.V. Gorshkov, S.D. Bennett, E. Demler, P. Zoller, M.D. Lukin, Phys. Rev. Lett. **109**, 266804 (2012)
40. D. Peter, N.Y. Yao, N. Lang, S.D. Huber, M.D. Lukin, H.P. Büchler, Phys. Rev. A **91**, 053617 (2015)
41. N.Y. Yao, A.V. Gorshkov, C.R. Laumann, A.M. Läuchli, J. Ye, M.D. Lukin, Phys. Rev. Lett. **110**, 185302 (2013)

42. D. Barredo, H. Labuhn, S. Ravets, T. Lahaye, A. Browaeys, C. Adams, Phys. Rev. Lett. **114**, 113002 (2015)
43. T.F. Gallagher, *Rydberg Atoms* (Cambridge University Press, Cambridge, 1994)
44. F. Nogrette, H. Labuhn, S. Ravets, D. Barredo, L. Béguin, A. Vernier, T. Lahaye, A. Browaeys, Phys. Rev. X **4**, 021034 (2014)
45. H. Labuhn, D. Barredo, S. Ravets, S. de Léséleuc, T. Macrì, T. Lahaye, A. Browaeys, Nature **534**, 667 (2016)
46. D. Barredo, S. de Léséleuc, V. Lienhard, T. Lahaye, and A. Browaeys, arXiv:1607.03042
47. F. Robicheaux, N.M. Gill, Phys. Rev. A **89**, 053429 (2014)
48. S. Bettelli, D. Maxwell, T. Fernholz, C.S. Adams, I. Lesanovsky, C. Ates, Phys. Rev. A **88**, 043436 (2013)
49. H. Schempp, G. Güner, S. Wüster, M. Weidemüller, S. Whitlock, Phys. Rev. Lett. **115**, 093002 (2015)
50. B. Zygelman, Phys. Rev. A **86**, 042704 (2012)
51. M. Kiffner, W. Li, D. Jaksch, Phys. Rev. Lett. **110**, 170402 (2013a)
52. M. Kiffner, W. Li, D. Jaksch, J. Phys. B **46**, 134008 (2013b)
53. C. Cohen-Tannoudji, J. Dupont-Roc, G. Grynberg, *Atom-Photon Interactions* (J. Wiley & Sons, New York, 1998)
54. P.D. Ye, D. Weiss, R.R. Gerhardts, M. Seeger, K. von Klitzing, K. Eberl, H. Nickel, Phys. Rev. Lett. **95**, 3013 (1995)
55. A. Nogaret, J. Phys. Condens. Matter **22**, 253201 (2010)
56. D.R. Hofstadter, Phys. Rev. B **14**, 2239 (1976)
57. T. Fukui, Y. Hatsugai, H. Suzuki, J. Phys. Soc. Jpn. **74**, 1674 (2005)
58. K. Singer, J. Stanojevic, M. Weidemüller, R. Cote, J. Phys. B At. Mol. Opt. Phys. **38**, S295 (2005). ISSN issn0953-4075. http://iopscience.iop.org/0953-4075/38/2/021
59. T. Amthor, M. Reetz-Lamour, S. Westermann, J. Denskat, M. Weidemüller, Phys. Rev. Lett. **98**, 023004 (2007)
60. M. Kiffner, D. Ceresoli, D. Jaksch, J. Phys. B **49**, 204004 (2016)
61. Y. Hatsugai, Phys. Rev. B **48**, 11851 (1993)
62. X. Qi, Y. Wu, S. Zhang, Phys. Rev. B **74**, 45125 (2006)
63. H.M. Price, N.R. Cooper, Phys. Rev. A **85**, 033620 (2012)
64. T.G. Walker, M. Saffman, Phys. Rev. A **77**, 032723 (2008)
65. A.R. Edmonds, *Angular Momentum in Quantum Mechanics* (Princeton University Press, Princeton, 1960)
66. I.I. Beterov, I.I. Ryabtsev, D.B. Tretyakov, V.M. Entin, Phys. Rev. A **79**, 052504 (2009)
67. D.A. Steck, Rubidium 87 D Line Data. http://steck.us/alkalidata (revision 2.1.4, 23 December 2010)
68. W. Li, I. Mourachko, M.W. Noel, T.F. Gallagher, Phys. Rev. A **67**, 052502 (2003)
69. N.W. Ashcroft, N.D. Mermin, *Solid State Physics* (Saunders, Philadelphia, 1976)
70. Wolfram Research, Inc., *Mathematica Version 10.1* (Wolfram Research, Inc., Irvine, 2015)

Ultrafast Optomechanical Pulse Picking

Nikolai Lilienfein, Simon Holzberger, and Ioachim Pupeza

Abstract State-of-the-art optical switches for coupling pulses into and/or out of resonators are based on either the electro-optic or the acousto-optic effect in transmissive elements. In high-power applications, the damage threshold and other nonlinear and thermal effects in these elements impede further improvements in pulse energy, duration, and average power. We propose a new optomechanical switching concept which is based solely on reflective elements and is suitable for switching times down to the ten-nanosecond range. To this end, an isolated section of a beam path is moved in a system comprising mirrors rotating at a high angular velocity and stationary imaging mirrors, without affecting the propagation of the beam thereafter. We discuss three variants of the concept and exemplify practical parameters for its application in regenerative amplifiers and stack-and-dump enhancement cavities. We find that optomechanical pulse picking has the potential to achieve switching rates of up to a few tens of kilohertz while supporting pulse energies of up to several joules.

1 Introduction

Picking individual pulses from the MHz-repetition-rate pulse trains produced by (amplified) laser oscillators is particularly important in the context of high-pulse-energy lasers. Here, pulse pickers are necessary to reduce the repetition rate of the seed oscillator pulse train before amplification in average-power-limited laser systems. Other prominent applications are the direct extraction of pulses from laser resonators (cavity dumping) [1], and the coupling of pulses into and out of

This article is part of the topical collection "Enlightening the World with the Laser" - Honoring T. W. Hänsch guest edited by Tilman Esslinger, Nathalie Picqué, and Thomas Udem.

N. Lilienfein (✉) • I. Pupeza
Max-Planck-Institut fuer Quantenoptik, Hans-Kopfermann-Strasse 1, 85748 Garching, Germany
e-mail: nikolai.lilienfein@mpq.mpg.de

S. Holzberger
Max-Planck-Institut fuer Quantenoptik, Hans-Kopfermann-Strasse 1, 85748 Garching, Germany

Present Address: Menlo Systems GmbH, Am Klopferspitz 19a, 82152 Martinsried, Germany

D. Meschede et al. (eds.), *Exploring the World with the Laser*,
https://doi.org/10.1007/978-3-319-64346-5_21

the cavities of regenerative amplifiers [2–4]. Cavity dumping has also been demonstrated in passive external resonators known as enhancement cavities (ECs) [5, 6]. While this concept has not yet found widespread application, it has recently encountered renewed interest in the context of high-pulse-energy Yb-based amplifiers [7, 8]. The relevant properties of pulse pickers are the switching time and switching rate, efficiency, contrast, and optical bandwidth. Equally important for intracavity applications are properties which affect the transmitted pulses, i.e., losses, chromatic dispersion, and, for high-power applications, self-induced nonlinear and thermal effects.

Cavity dumping from oscillators and regenerative amplifier cavities typically relies on Pockels cells. These devices use the electro-optic Pockels effect in nonlinear crystals to rapidly rotate the polarization of the intracavity pulse upon application of a high voltage. Depending on its polarization, the pulse can be subsequently coupled out by means of a polarizer. The length of the crystals necessary to reach a polarization rotation sufficient for efficient switching is typically tens of millimeters. Pockels cells include several antireflection-coated surfaces which create losses. Since the electro-optic effect scales linearly with the electric field, the half-wave voltage increases with the aperture width. This, together with the availability of large crystals of sufficient quality, makes increasing the aperture of the cell technologically challenging [3]. The damage threshold of crystals and antireflective coatings, as well as nonlinear and thermal effects, are the limiting factors for the extractable peak and average power from regenerative amplifiers [3, 9]. State-of-the-art systems achieve pulse energies of 30 mJ at 10 kHz repetition rate and 200 mJ at 1 kHz [10]. For amplification, the pulses are typically stretched to durations in the nanosecond range [4, 11].

In contrast to the cavities of laser oscillators and of regenerative amplifiers, which include gain media, the increase in pulse energy in ECs depends on the correct temporal and spatial overlap of circulating and seeding pulses, and the growth per round-trip is typically small. Thus, losses, dispersion, and other distortions of the intracavity pulse are critical, and rule out the use of Pockels cells for efficient dumping. For proof-of-principle demonstrations of the stack-and-dump concept, acousto-optic modulators (AOMs) have been employed [5, 12, 13]. Here, the beam is coupled out by a transient grating, induced by an acoustic wave propagating in a solid, e.g., a fused silica plate. The switching time of these devices is determined by the beam size and by the speed of sound in the material. A sufficiently short switching time to pick pulses from a megahertz pulse train requires the beam to be tightly focused in the AOM. Recently, the extraction of 0.16 mJ pulses from an EC has been demonstrated [14]. The pulses were chirped to a duration of 2.5 ns, and the output repetition rate was 30 kHz. In particular, the high nonlinearity caused by the small beam size in the AOM represents a bottleneck for the further scaling of such systems [14].

Recently, a mechanical pulse picker based on a concept similar to a chopper wheel has been proposed as a possible way to circumvent these limitations [7]. Here, a moving mirror that periodically intercepts the circulating pulse would act as the switching element, avoiding the losses, dispersion, nonlinearities, and thermal lensing associated with intracavity transmissive elements. However, the centrifugal force that would occur in a chopper wheel spinning fast enough to couple out single pulses even from a tightly focused MHz-repetition-rate pulse train would be close to the limits given by the tensile strength of potential rotor materials. This poses a major technological challenge and would severely limit the capabilities of the such devices. Even if feasible, this output coupler would represent a high-precision element of considerable size, complexity, and cost.

In this article, we present a novel concept for an optomechanical pulse picker consisting of exclusively reflective optics. It uses rotating mirrors and stationary imaging mirrors in order to move an entire section of a beam path without affecting the beam path outside of this section. While sharing the advantages of the chopper wheel, the concept drastically reduces the mechanical demands on the rotor by using the beam path as a lever. We present three particular geometries and discuss their advantages and drawbacks. We find that devices based on this concept could be suitable to pick joule-level pulses from optical cavities with repetition rates of several tens of megahertz, at rates of several kilohertz.

2 Circular Single-Mirror Geometry

Figure 1a shows the most basic variant of the concept. It is based on a mirror that rotates at a high angular velocity. The normal of its flat surface is tilted by a small angle from its rotation axis, such that the rotating mirror reflects a stationary input beam into a time-dependent deflection beam path. The rotating mirror surface is positioned in the center of curvature of a large spherical mirror. The deflection beam path impinges orthogonally on the stationary mirror surface for all rotation angles, and is reflected back along its incident path. Irrespective of the deflection angle, the second pass on the rotating mirror deflects the output beam along a path coinciding with the input path. Since the stationary spherical mirror images the surface of the rotating mirror onto itself, we refer to it as the imaging mirror. To achieve collimated input and output beams, a curved folding mirror can be used to focus the beam on the imaging mirror. Figure 1b illustrates the points where individual pulses from a pulse train impinge on the imaging mirror. If the deflection paths of subsequent pulses are separated by a distance Δs that is larger than the beam diameter, mirrors in the deflection beam path or openings in the imaging mirror can be used to pick pulses from the pulse train.

The switching time τ of the pulse picker is given by the velocity with which the beam moves along the circle of radius r, and the $1/e^2$ beam radius w in that plane. If we require that $\Delta s = 2w$, we find that

$$\tau = \frac{w}{\pi r f_{\mathrm{rot}}}, \tag{1}$$

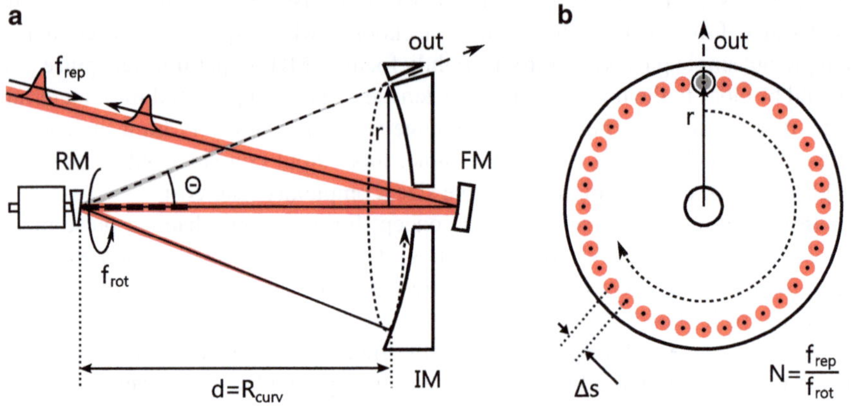

Fig. 1 **a** Side view of the circular single-mirror geometry comprising one rotating mirror (RM), a spherical imaging mirror (IM, radius of curvature R_{curv}) with an opening, and a curved folding mirror (FM). The beam path is shown for two rotation angles (*dashed line* beam path for output coupling). For a full rotation, the deflected beam path describes a cone. **b** Front view of the imaging mirror with N points of incidence of individual pulses from the incoming pulse train being distributed along a circle of radius r

with f_{rot} being the revolution rate of the rotating mirror. This equation also holds for a chopper-wheel-based pulse picker. In this case, the value of r, is given by the size of the chopper wheel. For our optomechanical pulse picker, on the other hand, it depends on the deflection angle from the rotating mirror Θ and on the distance to the imaging mirror d, according to $r = d \sin(\Theta)$. The beam path is used as a lever, and the diameter of the rotor needs to be just large enough to provide a sufficient aperture for the transverse beam size. The switching rate is equal to the revolution rate. A magnetic bearing of the rotor is desirable to achieve a high rotation speed as well as a high stability of the rotor and of its revolution rate, a long lifetime, and to allow for operation in vacuum. In [15], a suitable self-bearing motor reaching a speed of 8.4 kHz intended for laser scanning applications has been demonstrated. With a length of 55 mm and a diameter of about 30 mm, the motor is quite compact. With this motor, and, e.g., an imaging mirror of 15 cm diameter and a beam radius of 50 μm in the focal plane, the switching time is 25.3 ns, corresponding to a maximum pulse repetition rate of close to 40 MHz. Precise holes with diameters in the range of several tens of μm can be manufactured, e.g., via laser drilling [16]. To allow for reliable pulse picking, the points of incidence of the pulses on the imaging mirror have to be fixed. To this end, the repetition rate of the pulse train has to be locked to an integer multiple of the rotation frequency. In addition, the pulse train has to be stabilized to a specific fixed phase with respect to the rotation phase. Both of these requirements are achievable with standard technologies.

A pulse picker of this kind could be used in a linear resonator cavity, with the imaging mirror being one of the end mirrors. The number of round-trips during one rotation period is $N = f_{\text{rep}}/f_{\text{rot}}$, with f_{rep} being the repetition rate of the resonator. The transverse separation of the paths of successive pulses is given by

$$\Delta s_1 = 2\pi d \sin(\Theta)\frac{f_{\text{rot}}}{f_{\text{rep}}}. \tag{2}$$

When the input beam path coincides with the rotation axis of the rotating mirror, the angle of incidence on its surface is constant, and equals half the deflection angle Θ. The polarization of the beam with respect to the imaging mirror surface turns with the rotating mirror. The spectral phase and reflectivity of dielectric mirror coatings typically show increasing polarization dependence with a larger angle of incidence and with an increasing spectral bandwidth. Even though the imaging setup does not affect the polarization geometrically, this effect will cause a birefringent spectral phase modulation increasing with the angle of incidence and the required bandwidth. Any rotation-angle-dependent modulation is periodic with f_{rot}, leading to a constant output of the system at f_{rot}. For some applications , however, small angles of incidence and, thus, a long distance d may be necessary.

For large values of d, the propagation time $\Delta t = 2d/c$ between the first and the second pass of the pulse on the rotating mirrors becomes relevant. During this time, the mirror rotates by $\Delta\phi = 2\pi f_{\text{rot}}\Delta t$. This rotation results in an effective tilt β between the vectors normal to the mirror surface for the first and second passes given by

$$\sin\left(\frac{\beta}{2}\right) = \sin\left(\frac{\Theta}{2}\right)\sin\left(\frac{\Delta\phi}{2}\right). \tag{3}$$

For the example mentioned above, a distance $d = 0.5\text{m}$ corresponding to $\Theta = 8.6^\circ$, results in an effective tilt of 13.5 μrad.

A limitation of this type of geometry for high-power applications is presented by the high intensity on the imaging mirror. The beam radius w on the imaging mirror depends on the focusing power of the curved folding mirror and on the caustic of the surrounding resonator. A small beam waist reduces the necessary separation of adjacent spots, but increases the peak and average intensity on the imaging mirror. While no transmissive elements prone to thermal and nonlinear effects are used, mirror damage will occur for high peak intensities or fluences. Additionally, thermal lensing in mirrors can affect the operation of cavities at high average powers [17, 18]. While the overall size of the pulse picker can be scaled up with the required peak and average power, the necessity of focussing the beam on a cavity mirror is generally disadvantageous for high-power applications.

3 Circular Double-Mirror Geometry

Figure 2 shows a second variant of the concept which mitigates the above problem. Here, two rotating mirrors and two imaging mirrors are used in a setup which is symmetric with respect to a central plane. The rotating mirrors are fixed on a single shaft which is driven by the motor. The curved imaging mirrors are arranged such that their focal points coincide with the center of the rotating mirror surfaces, analogously to a 4-f imaging configuration. The first rotating mirror reflects the input beam into a deflection beam path that depends on its rotation angle. The two imaging mirrors guide the deflected beam to the second rotating mirror, which produces a stationary output beam. Between the rotating mirrors, some space is needed for the bearing and power unit of the rotor, separating their surfaces by a distance a. When the distance between the imaging mirrors L is large with respect to a and the deflection angle Θ is small, the imaging mirror configuration is close to a

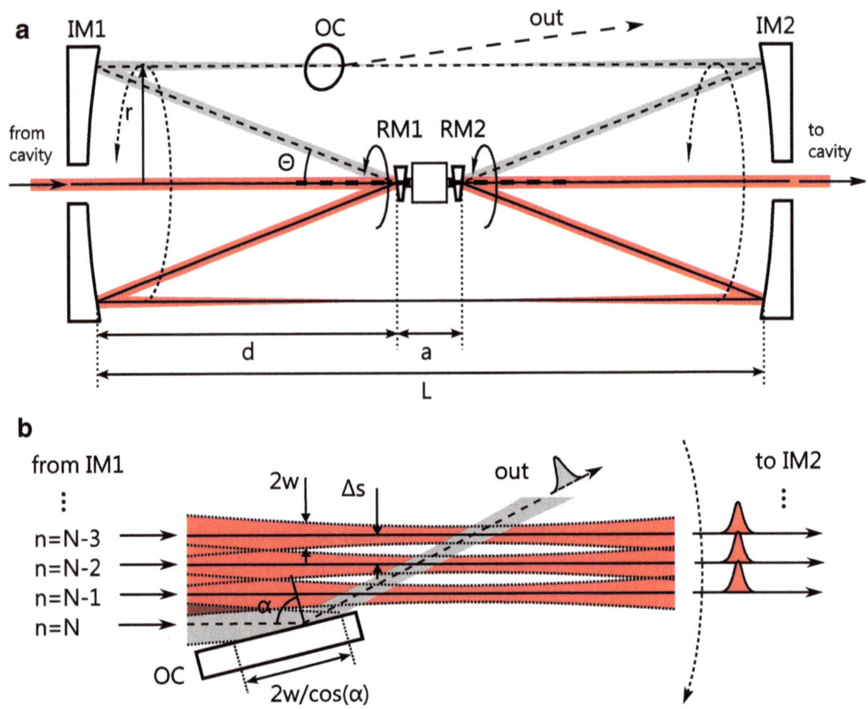

Fig. 2 **a** Side view of the circular two-mirror geometry with two rotating mirrors (RM1, RM2), two imaging mirrors (IM1, IM2) and an output coupling mirror (OC). The beam path is shown for two rotation angles (*dashed line* beam path just before output coupling). For a full rotation, the deflected beam path describes a cylinder between the imaging mirrors. **b** Schematic view of the grazing-incidence output coupling mirror. The beam path is shown for the three final round-trips of the pulse before the output coupling event (*dashed line*)

true 4-f imaging. In this case, the deflection beam is focused in the symmetry plane of the setup when the input and output beams are nearly collimated. For larger values of a, the beam has to be slightly divergent at the entrance of the pulse-picker setup and slightly convergent at the exit. A mirror placed at some point along the deflected beam path between the imaging mirrors can be used for input or output coupling of the pulse (Fig. 2b). For some applications, the resonator cavity may remain blocked for a small fraction of the rotation period after output coupling and/or before input coupling. Then, the picking mirrors can be larger than the beam size at the position of input/output coupling. The space just before or behind the coupling mirrors can also be used for the mechanical support of the rotor unit. The surface of the stationary mirrors can be either spherical or parabolic. Spherical mirrors are simpler to align and manufacture, but exhibit astigmatism when hit at nonzero angles of incidence. Note that the sagittal and tangential planes with respect to the rotating and imaging mirrors, and thus the astigmatism, are rotating together with the beam path. Particularly in resonators operated close to an edge of the stability range [19], this astigmatism will result in a rotating ellipticity of the cavity mode. Similar to the birefringent effects in dielectric mirror coatings mentioned before, this can be mitigated by decreasing the angle of incidence at the cost of increased size.

An optomechanical pulse picker of the second variant could be used in both linear and ring resonator cavities. For small angles of incidence on the mirrors, the following equation gives the pulse path separation for the second variant:

$$\Delta s_2 = \pi(L - a)\tan(\Theta)\frac{f_{\mathrm{rot}}}{f_{\mathrm{rep}}}. \tag{4}$$

It is similar to the one for the single-mirror geometry, but the total length, given here by the distance of the imaging mirrors L, doubles for otherwise identical parameters. In this geometry, the rotation of the mirrors during the propagation time along the deflection beam path can be compensated by adjusting the orientation of the rotating mirrors with respect to each other. In contrast to the single-mirror geometry, the spot size of the beam is large on both the rotating and the imaging mirrors. On the pulse-picking mirrors (output or input couplers), the beam size can be chosen corresponding to the required switching time by changing the position along the deflection beam path (Fig. 2b).

While the average power impinging on the picking mirrors is far lower than on the other optics, the small spot size necessary to achieve switching times below 100 ns causes a considerably higher peak intensity. The intensity on the output coupler can be reduced by placing it under grazing incidence. For instance, at an extreme angle of incidence α of 89°, the irradiated area would be increased by a factor of about 57. Generally, the peak fluence F of a beam with a Gaussian profile on the output coupler is

$$F = \frac{2\cos(\alpha)}{\pi w^2} E_p, \tag{5}$$

where E_p denotes the pulse energy. At $\alpha = 89°$, and a wavelength of 1 μm, the reflectivity of an uncoated quartz surface for s-polarized light is 94%. By avoiding a mirror with a coating, which contain high-refractive-index material and typically come with small deposition errors and contaminants, the highest possible damage threshold can be achieved. The output coupler, being a quartz (or sapphire, diamond, etc.) plate, can be readily and cost-effectively replaced. For a given switching time and rate, a larger spot size can be accommodated by increasing the overall size of the system. Thus, the maximum peak power scales with the square of the system size (Eqs. 4, 5).

4 Planar Double-Mirror Geometry

In principle, the pulse-picker geometries discussed so far allow for an unobstructed deflection beam path for all rotation angles (apart from input or output couplers). Thus, they can provide a resonator with an open beam path for most of the rotation period, which corresponds to the input and/or output switching rate. However, in regenerative amplifiers for instance, the time between input coupling to and output coupling from the cavity is often only a fraction of the repetition period. Thus, the cavity needs to be closed for just a small range of rotation angles of the rotating mirrors. In the case of the two pulse-picker variants discussed so far, smaller imaging mirrors covering only the necessary rotation angles could be used. In addition, applications of this kind would allow for the use of a particularly advantageous third geometry, as shown in Fig. 3a. In contrast to the previous implementation, the axis of rotation of the rotating mirrors is perpendicular to the incoming beam path. The deflection beam paths for different rotation angles form a plane. The mirror surfaces of the rotor are parallel to the rotation axis and to each other. The imaging mirror setup is similar to the second variant, with the focal planes coinciding with the rotating mirror surfaces. However, the central axis along which the imaging mirrors are aligned is offset from the incoming beam and the rotating mirrors by a distance Δx. Figure 3b illustrates the beam path in the system. The 4-f configuration of the imaging mirrors creates an inverted image in the central plane, after the beam has passed each of the imaging mirrors once. In this plane, the beam is offset by $-\Delta x$ from the central axis. If Δx is larger than the maximum radius of the rotor, the beam passes the rotor without being clipped and enters the imaging setup for a second time. Two passes through the 4-f imaging setup create an upright image of the first rotating mirror surface on the parallel second surface. The reflection from this surface produces a stationary output beam from the rotation-angle-dependent deflection beams for a continuous range of rotation angles. As in the second variant, the (close to) collimated input beam is focused in the Fourier planes in the symmetry plane, and input or output coupling elements can be placed

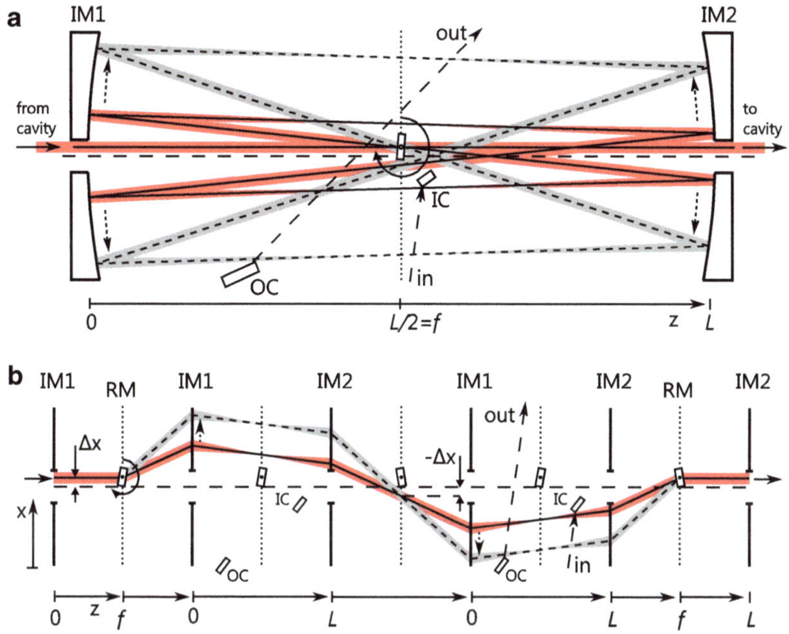

Fig. 3 **a** Top view of the planar two-mirror geometry with a two-faceted rotor (RM), two imaging mirrors (IM1, IM2), an input coupling mirror (IC), and an output coupling mirror (OC). The beam path is shown for two rotation angles (*solid line* just after input coupling, *dashed line* just before output coupling). **b** Schematic view of the beam path. The beam passes each of the imaging mirrors twice

at some position of the deflection beam having a suitable beam size. A major advantage of this geometry is its inherent insensitivity to vibrations and other deviations from a perfect rotation of the rotor: Any kind of angular or positional error of the first surface is reproduced by the second surface and, due to the imaging system, its effects on the output beam cancel out.

The rotor can be a two-faced substrate, or have a geometry with a larger number of parallel surface pairs, i.e., square, hexagonal, and so forth. A simple two-faced rotor would allow for a switching rate corresponding to twice the rotation frequency. Rotors with a higher number of facets would further increase the maximum switching rate. For small deflection angles, the transverse separation of the paths of subsequent pulses is

$$\Delta s_3 = 2\pi L \frac{f_{\mathrm{rot}}}{f_{\mathrm{rep}}}. \tag{6}$$

In contrast to the first two variants, this separation does not depend on the deflection angle. Consequently, a much larger separation or, alternatively, a much shorter switching time can be achieved in a system with the same footprint. The deflection angle Θ equals twice the rotation angle, i.e., the angle of incidence on the rotating mirrors. The maximum range of deflection angles $\Delta\Theta$ for which the cavity is open is given by the minimum angle $\Theta_{\min}$ for which the beam is not clipped by the rotating mirror in the first Fourier plane (Fig. 3b), and a maximum angle $\Theta_{\max}$ given by the dimensions of the imaging mirrors. The maximum number of round-trips N which a pulse can undergo in a cavity containing this pulse picker is

$$N_{\text{plan}} = \frac{\Theta_{\max} - \Theta_{\min}}{4\pi} \frac{f_{\text{rep}}}{f_{\text{rot}}}. \tag{7}$$

Since the angle of incidence on the rotating mirrors is not constant, rotation-angle-dependent astigmatism will occur when spherical imaging mirrors are used. These changes in the effective focal length of the imaging mirrors with the rotation angle will cause a varying ellipticity of the cavity mode. If large deflection angles are required, the use of parabolic imaging mirrors might be necessary (also depending on the sensitivity of the surrounding cavity). In this geometry, the rotation of the mirrors during the propagation time in the device causes a small tilt of the output beam that is constant for a given rotation frequency.

5 Nonplanar Double-Mirror Geometry

Figure 4 shows a simple modification of the planar double-mirror geometry with which the maximum deflection angle to achieve a specific number of round-trips can be reduced by more than a factor of two. Here, the rotating mirror and its rotation axis are inclined by an angle α_y in the $z-y$ plane with respect to the symmetry plane of the imaging setup, offsetting the beam by Δy in the first Fourier plane, and $-\Delta y$ in the second Fourier plane (Fig. 4b). With this modification, the deflection beam circumvents the rotor and its bearing and the holes in the imaging mirrors, even for zero deflection angle in the $x-z$ plane. Thus, there is an unobstructed path for the deflection beam for all deflection angles between $-\Theta_{\max}$ and $\Theta_{\max}$, and $\Delta\Theta$ is equal to $2\Theta_{\max}$. The maximum number of round-trips is now

$$N_{\text{nonplan}} = \frac{\Theta_{\max}}{2\pi} \frac{f_{\text{rep}}}{f_{\text{rot}}}. \tag{8}$$

The minimum width X of the imaging mirrors is

$$X_{\text{nonplan}} = N \Delta s_3. \tag{9}$$

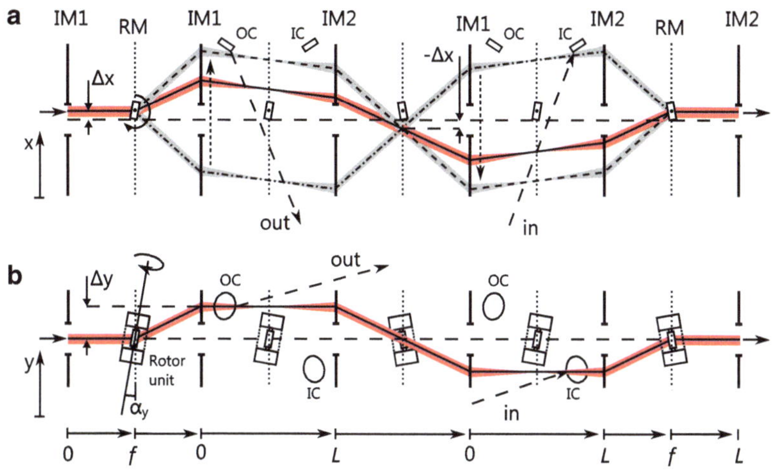

Fig. 4 **a** Schematic top view of the beam path in the nonplanar two-mirror geometry for three rotation angles (*dash-dotted line* path of a pulse just after input coupling, *dashed line* just before output coupling, *solid line* in between). **b** Schematic side view of the beam path. The inclined rotation axis allows the beam path to circumvent the rotor unit and the opening in the imaging mirrors

6 Application Example 1: Regenerative Amplifier

The nonplanar double-mirror geometry would be particularly suited for regenerative amplifiers. With rotation frequencies of several kHz and multifaceted rotors, switching rates of up to a few 10 kHz should be achievable. We exemplify practical pulse-picker parameters based on an amplifier build by Nubbemeyer et al. [20], who have recently achieved an average power of about 1 kW at an output repetition rate of either 5 or 10 kHz, with 40 round-trips and a cavity repetition rate of 20 MHz. A nonplanar double-mirror optomechanical switch with a two-surface rotor at a rotation frequency of 5 kHz would allow an output repetition rate of both 5 or 10 kHz. A magnetically beared motor which could be suitable for this application is commercially available [21]. Its length along the rotation axis is 85 mm. To achieve 40 round-trips, $\Delta\Theta$ has to be 7.2° (Eq. 8), resulting in a maximum angle of incidence of 1.8° in the $z - x$ plane. With a distance of 1 m between the imaging mirrors, a pulse path separation of 1.57 mm would be achieved (Eq. 6). The dimensions of the motor would imply a Δy of about 45 mm, and thus an angle of incidence of 3° in the $z - y$ plane. The input/output coupling plate could be placed at a position where the beam radius is 0.7 mm. With an angle of incidence of 88°, the spot on the plate would be elongated to a w of 20 mm. For a Gaussian profile with this spot size and a pulse energy of 200 mJ, the peak fluence is about 0.9 J cm^{-2} (Eq. 5). At a central wavelength of 1053 nm and a pulse duration of 1 ns, the damage fluence of fused silica is about 40 J cm^{-2}; for a pulse duration of 1 ps, it is 2 J cm^{-2} [22]. Shorter amplifier cavities would allow for a larger number of round-trips, but increase the fluence on the output coupler in a system with identical dimensions. Higher output rates can be achieved by using multifaceted rotors. In conclusion, the outlined optomechanical pulse picker could tolerate higher pulse energies, and/or far shorter pulses than state-of-the-art Pockels cells. Apart from the higher damage threshold, the losses, nonlinearity, thermal lensing, and dispersion associated with the transmission through the Pockels cell would be avoided. An optomechanical pulse picker would likely require vacuum for stable operation.

7 Application Example 2: Stack-and-Dump Cavity

In Ref. [7], a stack-and-dump cavity with a repetition rate of 10 MHz is proposed as a part of a future fiber-laser-based particle accelerator. From this cavity, an output pulse energy of 1.2 J with pulses stretched to 4 ns is envisaged. The pulse train seeding the cavity is continuous, making a pulse picker that provides an unobstructed beam path for most of the output repetition period necessary. We outline practical parameters for an optomechanical pulse picker of a circular double-mirror geometry as discussed in Sect. 4, and a rotation frequency of 8.3 kHz [15]. With this output switching rate, the maximum number of stacked pulses is 1200. With a length of 1.5 m, a distance between the rotating mirrors of 70 mm and an angle of incidence of 2°, the diameter of the deflection beam circle on the imaging mirrors is about 100 mm. At an angle of incidence of 2°, birefringent effects in typical broadband highly reflective mirrors are negligible even for the largest bandwidths yet demonstrated in high-finesse ECs [23]. The separation of the optical axes of successive pulses is 0.26 mm (Eq. 4). For a pulse energy of 1.2 J and a beam radius of 130 μm at an output coupler with 89° angle of incidence, the peak fluence would be 79 J cm^{-2}. Using the $\tau^{0.5}$-law for the pulse width dependence of the damage threshold, a damage threshold of fused silica of about 80 J cm^{-2} for a pulse duration of 4 ns can be extrapolated from the measurements in [22], putting the highly ambitious pulse energy targeted in [7] within reach.

8 Misalignment Sensitivity

The feasibility of the concept critically depends on the perturbations of the beam caused by the rotation of the mirrors, and on the sensitivity of the application to these. While the imaging configuration ideally results in a stationary output beam, misalignments, manufacturing tolerances, and astigmatism in spherical imaging mirrors can cause a rotation-angle-dependent translation and ellipticity of the mode. An important question for intra-resonator applications is whether the resulting periodic change in the beam path and shape causes any significant effects that would not be present for a static misalignment or astigmatism. If so, such effects

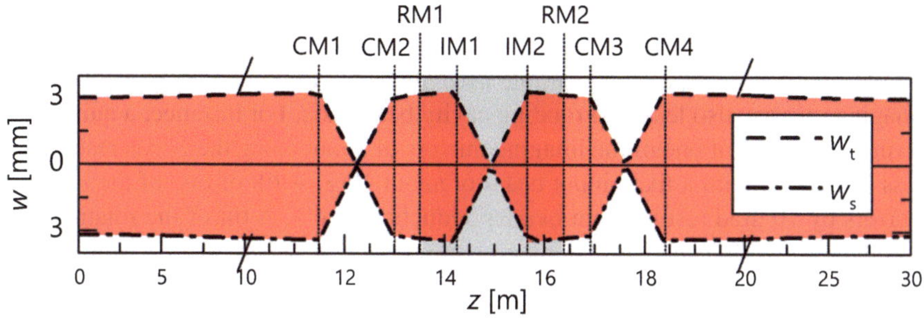

Fig. 5 Calculated sagittal and tangential $1/e^2$-intensity radius (w_s, w_t) for the 10-MHz EC outlined in Sect. 8 containing a pulse picker of the circular two-mirror geometry (RM1, IM1, IM2, RM2), and four additional curved mirrors (CM1-4)

would be particularly severe for enhancement cavities, where the circulating pulse train needs to overlap spatially, temporally, and in terms of its polarization with an incoming pulse train. To address this question, we simulate the enhancement in a 10-MHz EC including an optomechanical pulse picker as outlined in the previous section, giving a maximum number of 1200 round-trips between two switching events. To be able to tune the position in the cavity's stability range, the cavity should include additional concave mirrors. In a cavity with a round-trip Gouy phase close to an odd multiple of π, large spots on the mirrors can be achieved while avoiding a high misalignment sensitivity [19]. In our cavity example, two pairs of concave mirrors with a radius of curvature of 1500 mm are used. With a distance of 1470 mm between each of the two curved mirror pairs and a wavelength of 1040 nm, the beam radius is about 3 mm on all cavity mirrors (Fig. 5). Assuming the same footprint as in [14], 20 flat folding mirrors producing overall losses of about 400 ppm are needed. To achieve a good stacking efficiency with 1150 round-trips, an input coupler of 99.75% reflectivity is chosen.

Because of the rotating deflection beam path, the cavity cannot be described using "global" sagittal and tangential planes. Here, we simulate the buildup in the cavity numerically in 3D using a modified Fox-Li algorithm [24]. The beam size and divergence of a round input beam is optimized for the cavity with an inactive pulse picker (rotating mirrors are at rest). In this case, the simulation yields an enhancement of about 780 after 1150 round-trips, corresponding to a stacking efficiency of 0.65. The angle of incidence of 2° on the imaging mirrors causes an eccentricity of the beam of 0.27 at the curved cavity mirrors. The simulation for an active pulse picker shows a periodic rotation of the major axis of the elliptic intracavity mode profile with the deflection beam path, but no change in the enhancement level. When a tilt of one of the rotating mirrors is introduced, the central axis of the cavity beam is misaligned. For the active pulse picker, the misaligned beam axis periodically rotates around the axis of the unperturbed beam. The enhancement level drops by the same amount for both the active and inactive pulse picker. The loss of enhancement stems exclusively from the decrease in the spatial overlap of the intracavity beam with the seeding beam. Thus, we can follow the approach taken in [19], which uses far less computation time, to investigate the misalignment sensitivity of the system in detail.

We find that in the cavity example used above, positioning errors or tilts of the imaging mirrors also lead to a rotating cavity beam axis. For instance, a quite large positioning error of one of the imaging mirrors of 5 mm along the z-axis results in a loss of overlap with a fixed input beam of about 10%. A tilt of one of the imaging mirrors by 10 μrad results in an overlap reduction of 4%. A tilt of the rotation axis and the rotor with respect to the imaging mirrors causes a stationary misalignment, but no rotation of the cavity beam axis. For pulse pickers of the circular two-mirror geometry, the rotating mirror surfaces should have identical inclination angles and orientations with respect to the axis of rotation. Errors in the inclination angle can be compensated by adjusting the distances of the rotating to the stationary mirrors to change the magnification of the imaging system. For example, for an error of the inclination angle of 100 μrad, a nearly static cavity beam can be achieved by repositioning both imaging mirrors by about 20 mm along the z-axis. However, the imaging setup cannot compensate for different orientations of the mirrors, i.e., when the planes of the inclination angles of the mirrors are not identical. The effective mirror tilt resulting from this effect is described by Eq. 3, with $\Delta\phi$ now representing the orientation mismatch. In the cavity simulation, an orientation mismatch of 500 μrad results in a rotating offset of the intracavity mode axis of about 0.2 mm at the curved cavity mirrors and the overlap decreases by 10%. Since the rotor has to meet high mechanical demands, and its mirrors will most probably not be adjustable, achieving the necessary accuracy in the orientation of the surfaces will be critical. Importantly, the sensitivity of the EC with respect to misalignments of the mirrors comprising the pulse picker is not higher than for the other cavity mirrors. Furthermore, all perturbations are periodic and synchronized to the output repetition rate, resulting in a stable output. While a detailed analysis of all possible perturbations exceeds the scope of the paper, we conclude that we see no fundamental limitations related to the alignment sensitivity of the optomechanical pulse picker.

9 Sensitivity to Rotor Motion Irregularities

Another important aspect of the technical implementation of this concept will be deviations of the actual rotor motion from a perfect rotation caused by, e.g., unbalance. These can be classified in three categories: parallel deviations (shifts) and angular deviations (tilts) of the rotation axis, and axial displacements of the rotor along the rotation axis. The planar and nonplanar double-mirror geometries are inherently insensitive to these effects. Here, the upright imaging of the front surface of the compact rotor onto its back surface results in a self-compensation of positional and angular errors of the rotor position. Axial displacements are parallel to the rotor surfaces. For the circular one-mirror geometry, in contrast, axial movements would directly translate to changes in the cavity length and, therefore, render its use for applications requiring interferometric stability challenging. Parallel rotation errors would, to a much lesser extent, also affect the cavity length. The imaging configuration results in self-compensation of angular rotation errors. The circular two-mirror geometry is insensitive to length changes due to axial rotor positioning errors, since the displacement of the first rotating mirror surface is compensated by an identical displacement of the second surface. However, due to the inverted imaging configuration in this geometry, angular errors of the rotor position result in a displacement of the cavity beam path. For the stack-and-dump-cavity example described in the previous section, an angular error of 10 μrad corresponding to a rotor tip displacement of 0.35 μm would result in a 14% decrease in overlap. In [15], a maximum displacement of the rotor tip of 17 μm is measured at a rotation frequency of 8.7 kHz. These measurements were taken with an unbalanced rotor, and it is not stated how much of the displacement is due to angular deviations. Still, the results suggest that the balancing of the rotor and the optimization of its active stabilization may be the main technological challenges for an application of the concept in enhancement cavities. For the regenerative amplifier application discussed in Sect. 6, the effects of the rotor displacements reported in [15] would be negligible.

10 Conclusion

In this article, we have outlined a concept for a family of ultrafast all-reflective pulse pickers. Discussing three specific geometries, we have shown that such optomechanical pulse pickers could particularly benefit high-power and high-pulse energy applications. Designs suitable for regenerative amplifiers could support pulse energies of several joules and repetition rates of up to a few tens of kHz, while circumventing the losses, dispersion, and the nonlinear and thermal effects in state-of-the-art Pockels cells. Optomechanical pulse picking could additionally render the temporal stretching of pulses in such systems unnecessary. The concept also holds promise for cavity dumping in passive enhancement cavities. Here, output repetition rates of up to several kHz could potentially be reached for joule-level output pulse energies. Other applications may emerge in wavelength ranges where conventional pulse pickers are not available, in particular in the terahertz [25] and X-ray [26] ranges.

Acknowledgements Open access funding provided by Max Planck Society. We thank Henning Carstens for help with the simulations. We thank Sven Breitkopf, Jens Limpert, Thomas Nubbemeyer, and Moritz Ueffing for fruitful discussions. This work was funded by the European Research Council under Grant Agreement No. [617173] ACOPS.

References

1. A. Killi, J. Dörring, U. Morgner, M.J. Lederer, J. Frei, D. Kopf, Opt. Expr. **13**(6), 1916 (2005). doi:10.1364/OPEX.13.001916
2. C. Hönninger, I. Johannsen, M. Moser, G. Zhang, A. Giesen, U. Keller, Appl. Phys. B **65**(3), 423 (1997). doi:10.1007/s003400050291
3. A. Beyertt, D. Nickel, A. Giesen, Appl. Phys. B **80**(6), 655 (2005). doi:10.1007/s00340-005-1796-3
4. T. Metzger, A. Schwarz, C.Y. Teisset, D. Sutter, A. Killi, R. Kienberger, F. Krausz, Opt. Lett. **34**(14), 2123 (2009). doi:10.1364/OL.34.002123
5. T. Heupel, M. Weitz, T.W. Hänsch, Opt. Lett. **22**(22), 1719 (1997). doi:10.1364/OL.22.001719
6. R.J. Jones, J. Ye, Opt. Lett. **27**(20), 1848 (2002). doi:10.1364/OL.27.001848
7. S. Breitkopf, T. Eidam, A. Klenke, Lv Grafenstein, H. Carstens, S. Holzberger, E. Fill, T. Schreiber, F. Kraus, A. Tünnermann, I. Pupeza, J. Limpert, Light Sci. Appl. **3**(10), e211 (2014). doi:10.1038/lsa.2014.92
8. G. Mourou, B. Brocklesby, T. Tajima, J. Limpert, Nat. Photon. **7**(4), 258 (2013). doi:10.1038/nphoton.2013.75
9. J.P. Negel, A. Voss, M.A. Ahmed, D. Bauer, D. Sutter, A. Killi, T. Graf, Opt. Lett. **38**(24), 5442 (2013). doi:10.1364/OL.38.005442
10. K. Michel, S. Klingebiel, M. Schultze, C.Y. Tesseit, R. Bessing, M. Häfner, S. Prinz, D. Sutter, T. Metzger: *SPIE LASE*, ed. by W.A. Clarkson, R.K. Shori (SPIE 2016), SPIE Proc. doi:10.1117/12.2208127
11. H. Fattahi, A. Alismail, H. Wang, J. Brons, O. Pronin, T. Buberl, L. Vamos, G. Arisholm, A.M. Azzeer, F. Krausz, Opt. Lett. **41**(6), 1126 (2016). doi:10.1364/OL.41.001126
12. Y. Vidne, M. Rosenbluh, T.W. Hansch, Opt. Lett. **28**(23), 2396 (2003). doi:10.1364/OL.28.002396
13. R.J. Jones, J. Ye, Opt. Lett. **29**(23), 2812 (2004). doi:10.1364/OL.29.002812
14. S. Breitkopf, S. Wunderlich, T. Eidam, E. Shestaev, S. Holzberger, T. Gottschall, H. Carstens, A. Tünnermann, I. Pupeza, J. Limpert, Appl. Phys. B. **122**(12), 297 (2016). doi: 10.1007/s00340-016-6574-x
15. T. Baumgartner, R.M. Burkart, J.W. Kolar, IEEE Trans. Ind. Electron. **61**(8), 4326 (2014). doi:10.1109/TIE.2013.2284159
16. D. Esser, J. Weitenberg, W. Broring, I. Pupeza, S. Holzberger, H.D. Hoffmann, Opt. Expr. **21**(22), 26797 (2013). doi:10.1364/OE.21.026797
17. H. Carstens, N. Lilienfein, S. Holzberger, C. Jocher, T. Eidam, J. Limpert, A. Tünnermann, J. Weitenberg, D.C. Yost, A. Alghamdi, Z. Alahmed, A. Azzeer, A. Apolonski, E. Fill, F. Krausz, I. Pupeza, Opt. Lett. **39**(9), 2595 (2014)
18. N. Lilienfein, H. Carstens, S. Holzberger, C. Jocher, T. Eidam, J. Limpert, A. Tünnermann, A. Apolonski, F. Krausz, I. Pupeza, Opt. Lett. **40**(5), 843 (2015). doi:10.1364/OL.40.000843
19. H. Carstens, S. Holzberger, J. Kaster, J. Weitenberg, V. Pervak, A. Apolonski, E. Fill, F. Krausz, I. Pupeza, Opt. Expr. **21**(9), 11606 (2013)
20. T. Nubbemeyer, M. Kaumanns, M. Ueffing, M. Gorjan, A. Alismail, H. Fattahi, J. Brons, O. Pronin, H.G. Barros, Zs. Major, T. Metzger, D. Sutter, F. Krausz, Opt. Lett. (2016) (**submitted**)
21. http://www.celeroton.com/fileadmin/user_upload/produkte/motoren/datasheets/Datasheet CM-AMB-400
22. B.C. Stuart, M.D. Feit, A.M. Rubenchik, B.W. Shore, M.D. Perry, Phys. Rev. Lett. **74**(12), 2248 (1995)
23. S. Holzberger, N. Lilienfein, M. Trubetskov, H. Carstens, F. Lucking, V. Pervak, F. Krausz, I. Pupeza, Opt. Lett. **40**(10), 2165 (2015). doi:10.1364/OL.40.002165
24. A.G. Fox, T. Li, Bell Syst. Techn. J. **40**(2), 453 (1961). doi:10.1002/j.1538-7305.1961.tb01625.x

25. S. Takahashi, G. Ramian, M. Sherwin, Appl. Phys. Lett. **95**(23), 234102 (2009). doi:10.1063/1.3270041
26. M. Cammarata, L. Eybert, F. Ewald, W. Reichenbach, M. Wulff, P. Anfinrud, F. Schotte, Rev. Sci. Instr. **80**(1), 015101 (2009). doi:10.1063/1.3036983

Shifting the Phase of a Coherent Beam with a $^{174}Yb^+$ Ion: Influence of the Scattering Cross Section

Martin Fischer, Bharath Srivathsan, Lucas Alber, Markus Weber, Markus Sondermann, and Gerd Leuchs

This article is dedicated to Theodor W. Hänsch at the occasion of his 75th birthday. We are grateful to him for all he has done for our field and we wish him all the best for the next 25 years.

Abstract We discuss and measure the phase shift imposed onto a radially polarized light beam when focusing it onto an $^{174}Yb^+$ ion. In the derivation of the expected phase shifts, we include the properties of the involved atomic levels. Furthermore, we emphasize the importance of the scattering cross section and its relation to the efficiency for coupling the focused light to an atom. The phase shifts found in the experiment are compatible with the expected ones when accounting for known deficiencies of the focusing optics and the motion of the trapped ion at the Doppler limit of laser cooling (Hänsch and Schawlow in Opt Commun 13:68–69, 1975).

Martin Fischer, Bharath Srivathsan and Lucas Alber have contributed equally.

This article is part of the topical collection "Enlightening the World with the Laser" - Honoring T. W. Hänsch guest edited by Tilman Esslinger, Nathalie Picqué, and Thomas Udem.

M. Fischer • L. Alber • M. Weber • M. Sondermann
Max-Planck-Institute for the Science of Light, Staudtstr. 2, 91058 Erlangen, Germany

Department of Physics, Friedrich-Alexander University Erlangen-Nürnberg (FAU), Staudtstr. 7/B2, 91058 Erlangen, Germany

B. Srivathsan
Max-Planck-Institute for the Science of Light, Staudtstr. 2, 91058 Erlangen, Germany

G. Leuchs (✉)
Max-Planck-Institute for the Science of Light, Staudtstr. 2, 91058 Erlangen, Germany

Department of Physics, Friedrich-Alexander University Erlangen-Nürnberg (FAU), Staudtstr. 7/B2, 91058 Erlangen, Germany

Department of Physics, University of Ottawa, 25 Templeton, Ottawa, ON K1N 6N5, Canada
e-mail: gerd.leuchs@mpl.mpg.de

D. Meschede et al. (eds.), *Exploring the World with the Laser*,
https://doi.org/10.1007/978-3-319-64346-5_22

1 Recollection by One of Us (GL)

Birthdays often evoke memories of the one who is celebrating. Sometimes it is a single question they have asked you that has stuck in your mind for a long time. Of the many times I met Ted Hänsch one comes to my mind in particular. It was when I saw him in a corridor at the Max Planck Institute of Quantum Optics, about thirty years ago—the building was quite new at the time. I vividly remember the question he asked me: 'Do you have a good explanation why the cross section of an atom for scattering light is as large as it is?' He was referring to the classical on-resonance cross section of an atom, $\sigma_{sc} = 3\lambda^2/2\pi$, being so much larger—i.e. many orders of magnitude—than the cross section of the atomic charge distribution. Naturally, I knew the phenomenon and answered that in scattering processes the larger of the two following values dominates: the cross section of the atom as a massive object or the cross section of the particle you send in to probe the atom, namely a photon in the case under consideration. Obviously, the smallest cross section of an optical beam is limited by diffraction and this, I had thought, should define the cross section of the photon. I was surprised to see that Ted Hänsch did not seem satisfied as he slowly turned away. At the time, this made me think, and throughout the years since then I have returned to this thought every now and then.

Ten years later, after I moved to Erlangen, this 'thinking' became more intense when within my group we started to first discuss spontaneous emission and the possibility of observing its time-reversed counterpart. In spontaneous emission, the energy is initially concentrated in a tiny volume, which is orders of magnitude smaller than the wavelength cubed—partially still stored in the atom—and begins to travel outwards. At first, the energy is both in the evanescent and propagating components of the field. Then, as the outgoing dipole wave travels further, the evanescent components decay away leaving only the propagating part of the dipole wave. The idea arose that the evanescent field is more part of the atom than of the outgoing dipole wave. Consulting any book on electromagnetism, one can calculate the outward going energy flux of the near field and that of the propagating field. The near-field part quickly decreases to zero as the distance to the source increases, whereas the far-field component is constant. It is interesting to note that the radial position r, at which the energy flux of the near field has reduced to half the far-field portion, is given by $(2\pi/\lambda)^2 r^2 \sim 6$. This r value corresponds exactly to the above-mentioned cross section, indicating that in terms of cross section the near field can be considered part of the atom. Was this good enough to tell Ted Hänsch? Without the atom, light would produce a diffraction-limited spot, but when an atom is at the origin of the dipole wave, one instead expects time-reversed spontaneous emission to occur such that the energy density of the field increases far beyond the diffraction-limited value in free space. One might speculate that the evanescent field is excited via the atom's reaction to the incident field. If one considers the evanescent field as part of the atom, its extent defines the cross section of the atom, resulting in a cross section almost matching the classical textbook value quoted above. Nevertheless, at that point I still felt it was too early to go back to Ted Hänsch. There was still something that puzzled me.

The incoming dipole wave with its evanescent and propagating components is an exact solution of Maxwell's equations, but it has a singularity. Accordingly, when one excites an inward propagating dipole wave in the far field, one would expect the singularity to develop—this is part of the rigorous solution after all—up to the point when the wave reaches the atomic charge distribution. We know however that this is not what happens. Thus, it was a great relief to me when Simon Heugel, a doctoral student in our group at the time, came to me about seven years ago suggesting that I look at problem C_I.6 in the text book by Cohen-Tannoudji, Dupont-Roc and Grynberg [2]. There it is stated that in free space the inward propagating dipole will continue as an outward propagating dipole once it has passed the origin and will thus interfere with itself. The task given to the students is to calculate the energy density of the resulting standing wave and—alas—the result is the diffraction-limited field distribution, provided one takes into account a phase shift at the origin which is in a way the Gouy phase shift under this extreme full solid angle focusing scenario. Looking at the problem in this way everything seems to fall into place: (1) when focusing in free space, the singular terms in the dipole wave solution interfere destructively and (2) suppressing the outward going wave via full absorption at the origin by a sub-wavelength antenna such as an atom gives rise to the well-known field enhancement. We asked ourselves whether there are other ways to restore the singular behavior. One way we found theoretically was by studying the time evolution of the energy distribution for focusing in free space near the origin when the inward going dipole wave has a sharp rising leading edge, rising over a distance significantly smaller than the wavelength. This indeed also gave a transient enhancement [3]. Other experiments are under way.

Encouraged by these considerations and findings we hope this anniversary is the right moment to give Ted Hänsch an update on our, by now decades long, attempt to answer the question he posed such a long time ago.

2 Introduction

The *scattering cross section* is a quantity used in many areas in physics, relating the rate of particles scattered by a target to the flux of particles incident onto it. In quantum optics, the conceptually simplest target is a single atom and the incident particles are photons. For this scenario, the resonant scattering cross section for a two-level atom is determined to be [4, 5]

$$\sigma_{\mathrm{sc}} = \frac{3}{2\pi}\lambda^2 \tag{1}$$

for an atomic transition with resonance wavelength λ provided the oscillator strength [6] is equal to one.

As mentioned above, the area given by σ_{sc} is large: It is by far larger than the spatial extent of an atom given by the Bohr radius and also larger than the smallest spot sizes achievable via diffraction-limited focusing of light with lenses of sufficient numerical aperture [7, 8]. The term cross section was created to describe scattering of particles, but in wave mechanics there is also the interference of fields. As pointed out in Ref. [9], absorption can be described as the interference of the (non-attenuated) incident field and the scattered field. In this model, attenuation in forward direction is caused by the destructive interference between these two fields, which requires a power of the scattered field which may seem counter-intuitive at first sight: full attenuation, and only back-scattered light requires the power of the scattered field to be twice that of the incident field because of the destructive interference with the incident light in the forward direction, in order to fulfill energy conservation. Along those lines, the rate of scattered photons, which is not to be confused with the detected photons, expressed in terms of cross sections is given by [10]

$$\gamma_{sc} = \frac{\sigma_{sc}}{A} \cdot \gamma_{inc}, \tag{2}$$

with A denoting the effective mode area [10, 11] of the incident stream of photons γ_{inc}. The remarkable scenario of more photons being scattered than photons arriving, both per unit time [12], arises when σ_{sc} becomes larger than A. Due to the interference of the different outward going partial waves, energy conservation is, of course, maintained. Within this reasoning, several intriguing phenomena occurring in the interaction of light and single quantum emitters have been investigated in recent years; see Ref. [13] for a review. However, as reported in Ref. [14] it was found already in the early 1980s by Bohren [15] and Paul and Fischer [16] that an atom can scatter more light than incident onto its massive cross section, which is on the order of the Bohr radius squared. As also discussed in more recent publications, the key step in these papers was indeed the examination of the superposition of incident and scattered fields. Refs. [15, 16] revealed that within a certain area larger than the size of the scatterer the resulting lines of energy flux end up at the scatterer's position. Within a similar reasoning and as outlined in the first section of this paper, one could attribute the spatial extent of the non-propagating near-field components of the field re-radiated by the atom to the size of the atom, leading to the expression for σ_{sc} given by Eq. 1.

Here, we relate to such concepts by investigating the phase shift imprinted onto a tightly focused light beam by a single $^{174}Yb^{+}$ ion. In the next section, the importance of the magnitude of the effective mode area of the incident beam to the obtained phase shift is revisited. With simple arguments, we modify the equation obtained in Refs. [13, 17] describing the achievable phase shift to account for the level structure of the used ion species. Explicitly, we make use of the dependence of the scattering cross section on the angular momenta of the involved atomic levels. In Sect. 4, we describe our experimental apparatus, present the phase shift observed in our experiments and compare the obtained results to the predictions of Sect. 3. At the end of the paper, we give concluding remarks.

3 Relation of Scattering Cross Section and Phase Shift

In order to emphasize the role of the scattering cross section σ_{sc} in phase shifting a weak coherent beam, we briefly recall some essential aspects. Typically, the induced phase shift is considered as the phase difference of the superposition of the incident electric field E_{inc} and the scattered field E_{sc} relative to the phase of the incident one, i.e. the phase of the incident field leaving the interaction region when no atom is present [18–20]. The phase shift $\Delta\varphi$ can then be written as [18]

$$\Delta\varphi = \arg\left(\frac{E_{inc} + E_{sc}}{E_{inc}}\right) \tag{3}$$

with arg( ) denoting the argument of its complex variable.

Since one is considering a coherent process in this situation, it is detrimental to saturate the atomic transition, i.e. to produce incoherent components in the scattered radiation. We therefore assume negligible saturation. For this case, the phase shift imprinted by a *two-level* atom is found to be [17]

$$\Delta\varphi = \arg\left(1 - 2G \cdot \frac{1 + i \cdot 2\Delta/\Gamma}{1 + 4\Delta^2/\Gamma^2}\right). \tag{4}$$

Here, Γ denotes the spontaneous emission rate and Δ is the detuning between the laser and the atomic resonance frequency. At fixed detuning, the crucial parameter determining the magnitude of the imprinted phase shift is G, describing the extent to which the atom experiences the highest possible electric field at constant input power which is allowed for by diffraction: $G = E_{inc}^2/E_{max}^2$, where $0 \leq G \leq 1$. add E_{max} is the field amplitude obtained by focusing a dipole wave in free space [21], i.e. G determines how efficiently the incident field couples to the atomic dipole transition. Assuming an atom at rest, G is solely determined by the properties of the focusing optics and the spatial mode of the incident field which has an overlap of η with the field of the driven transition [13, 22], $G \propto \eta^2$. It also accounts for phase front aberrations that are induced by imperfect focusing optics [23, 24]. Therefore, G is a measure for the quality of the mode matching of the incident mode to the atomic dipole-radiation pattern.

The role of G becomes obvious when relating it to the so-called *scattering ratio* on resonance, which is defined as $R = \gamma_{sc}/\gamma_{inc}$ [10, 18]. One can show that in general $G = R/4$ [13], resulting in

$$G = \frac{\sigma_{sc}}{4A}. \tag{5}$$

Hence, in order to reach unit coupling efficiency and thus the maximum phase shift at a fixed, nonzero detuning, the effective mode area of the focused beam must not be larger than a quarter of the scattering cross section. One can actually show that $\sigma_{sc}/4$ is the minimum possible mode area in free space. We interpret the effective mode area A as the power P of the incident light divided by the intensity I at the position of the atom. With the help of Eqs. 11 and 12 from Ref. [13] one can directly obtain $A = 3\lambda^2/(8\pi G)$ and thus $A = \sigma_{sc}/(4G)$, which is minimized in free space by $G = 1$.

Inserting Eq. 5 into 4 results in

$$\Delta\varphi = \arg\left(1 - \frac{\sigma_{sc}}{2A} \cdot \frac{1 + i \cdot 2\Delta/\Gamma}{1 + 4\Delta^2/\Gamma^2}\right), \tag{6}$$

similar to the findings of Ref. [18]. On resonance, the phase of the outgoing field can only take two values: zero if $A \geq \sigma_{sc}/2$ and π as soon as the electric field is focused to a spot smaller than $\sigma_{sc}/2$. This representation reveals that the obtainable phase shift is not only limited by imperfect focusing, as expressed by a too large A. But also choosing the 'ideal' atom is of importance, i.e. an atom for which Eq. 1 is valid. Deviations could originate from a degenerate ground state as is the case for $^{174}Yb^+$ or from an atom not being at rest [5]. Both obstacles occur in the experiment presented in the next section.

In the remainder of this section, we explicitly treat the level structure. In general, when accounting for the sub-structure of the atomic levels involved, the resonant scattering cross section of an atomic transition can be written as [5]

$$\sigma_{sc} = \frac{\lambda^2}{2\pi} \cdot \frac{2J' + 1}{2J + 1} \tag{7}$$

with J' and J being the total angular momentum of upper and lower level, respectively. For our experiment involving the $P_{1/2} \leftrightarrow S_{1/2}$ transition of $^{174}Yb^+$ (cf. Fig. 1), we have $J' = J$ and hence $\sigma_{sc} = \lambda^2/(2\pi)$, which is only 1 / 3 of the value used so far. We explicitly account for this reduction of the scattering cross section in writing

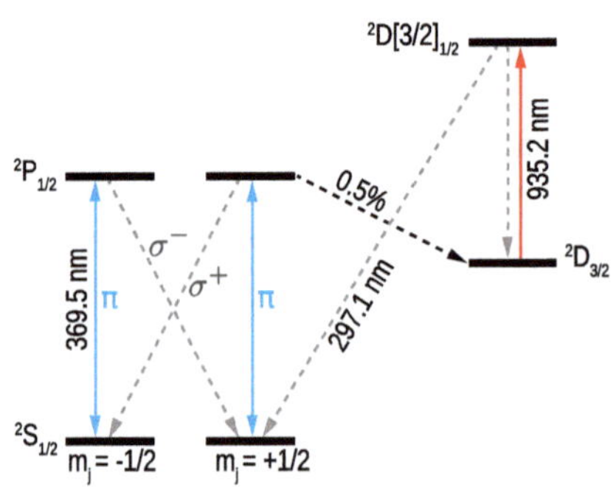

Fig. 1 Level scheme of $^{174}Yb^+$. In the phase-shift experiments, we drive the π-transition between the $S_{1/2}$ and the $P_{1/2}$ state. Furthermore, we use optical pumping to prepare the ion in the metastable $D_{3/2}$ (dark) state for obtaining a reference phase. The branching ratio from the $P_{1/2}$ state into the $D_{3/2}$ state is 0.5 % [25]

$$\Delta\varphi_{J=J'} = \arg\left(1 - \frac{2G}{3} \cdot \frac{1 + i \cdot 2\Delta/\Gamma}{1 + 4\Delta^2/\Gamma^2}\right). \tag{8}$$

Consequently, G from now on only accounts for imperfect focusing and atomic motion.

The result of Eq. 8 can also be obtained from a solution of the Bloch equations for a $J' = 1/2 \leftrightarrow J = 1/2$ system driven only by a π-polarized light field. The modification $G \rightarrow G/3$ can be understood as follows. First, the dipole moment in excitation is reduced by a factor $1/\sqrt{3}$ in comparison with a two-level atom. Second, the amplitude of the coherently scattered field that can interfere with the incident radiation is reduced by the same factor, because the $\sigma^{\pm}$-components of the scattered field cannot interfere with the incident light. A detailed calculation will be presented somewhere else.

4 Setup and Experiment

In our setup, we utilize a parabolic mirror as the focusing device [23, 24, 26]. The parabolic mirror tightly focuses a radially polarized donut mode to a field that is linearly polarized along the optical axis [7, 27]. This field drives a linear dipole oriented in the same direction.

We position the $^{174}Yb^{+}$ ion in the focus of the mirror by means of a movable open-access ion trap [26]. The focused donut mode continuously drives the linear dipole of the $S_{1/2}$ to $P_{1/2}$ transition of the ion, with a linewidth of $\Gamma/2\pi = 19.6$ MHz [25], at a wavelength of 369.5 nm. The power of this beam is chosen such that saturation effects are negligible. In Ref. [24], it was found that the aberrations of the parabolic mirror used are so strong in the outer parts that it is favorable to focus only from the 'backward' half space when not correcting for these aberrations. We therefore decided to use this focusing configuration in the experiment reported here, by inserting a suitable iris in the excitation beam path, cf. Fig. 2. The iris has a radius of two times the focal length of the paraboloid. We refer to this configuration as focusing from half solid angle, since the bore in the vertex of the parabolic mirror, housing the trap, reduces the solid angle, relevant for the linear dipole, by less than 0.5%. The focused donut mode also provides Doppler cooling for the ion. Auxiliary beams needed, e.g. for the initial ionization and trapping as well as the repumping beam (cf. Figs. 1, 2), are entering the focal region of the mirror through a small auxiliary hole close to the vertex of the parabola.

Each phase-shift measurement is preceded by the following sequence: First the ion is Doppler-cooled by an auxiliary beam red detuned by half a linewidth from the $S_{1/2} \leftrightarrow P_{1/2}$-transition. Then, this auxiliary beam is switched off and the donut mode drives the ion at half linewidth detuning. The ion is scanned through the focal region while monitoring the count rate of photons at 297 nm; see Fig. 1. The ion is positioned such that this count rate is maximized. Afterward, the auxiliary beam at

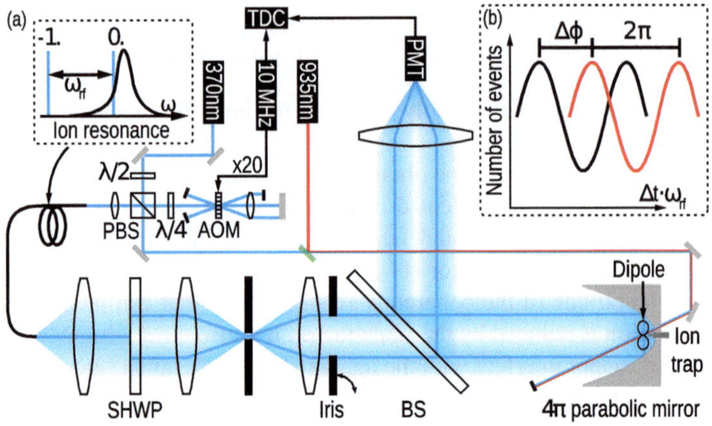

Fig. 2 **a** Setup for measuring the phase shift imprinted by a single $^{174}Yb^{+}$-ion. With an additional laser beam at a wavelength of 935.2 nm, we can pump the ion back from the metastable $D_{3/2}$ state into the $S_{1/2}$ ground state. This laser is sent onto the ion from a hole at the backside of the parabolic mirror. The same is done for a second laser beam at a wavelength of 369.5 nm that is used for ionization and for cooling the ion in certain steps of the experimental procedure (see text). *SHWP* segmented half-wave-plate, *(P)BS* (polarizing) beam splitter (other abbreviations in the text). **b** Intensity signal $I_{\mathrm{TDC}}(t)$ obtained from the statistics of photon detection times on the TDC for the ion being in the bright state (*red*) and in the $D_{3/2}$ dark state (*black*)

369.5 nm is switched on again for Doppler cooling. Switching this beam off again and setting the donut beam to the desired detuning, the phase-shift measurement is initiated.

In this measurement interval, the temperature of the ion is governed by the interaction with the donut beam. Hence, the temperature is explicitly depending on the detuning of the donut beam, as also discussed later. For a detuning of $\Delta = -\Gamma/2$, Doppler cooling theory [28] predicts a minimal temperature of the ion of about $T = \hbar\Gamma/2k_{\mathrm{B}} = 470\,\mu\mathrm{K}$, where $\hbar$ is Planck's constant and k_{B} the Boltzmann constant. From experimentally measured point spread functions (see Ref. [24]) and the characteristics of our ion trap (trap frequencies of 480 and 1025 kHz in radial and axial direction, respectively), we determine an upper bound of the ion's temperature to be 50% above the Doppler limit at half linewidth detuning.

The phase shift imprinted by the ion is measured in a common path interferometer by heterodyne detection. We illuminate the ion with the near-resonant carrier donut mode and an off-resonant sideband donut, similar to the technique applied in Ref. [19]. The sideband donut is red-detuned from the $S_{1/2} \leftrightarrow P_{1/2}$-transition by $\omega_{\mathrm{rf}}/2\pi = 400$ MHz (amounting to about 20 linewidths) by using the diffraction order '-1' of an acousto-optical modulator (AOM) in double-pass configuration ($\omega_{\mathrm{rf}} = 2\,\omega_{\mathrm{AOM}}$, see Fig. 2). Except for the frequency difference, both beams have exactly the same properties and are in the same spatial mode that is focused onto the ion.

After interaction with the trapped ion, the beams are retro-reflected and recollimated by the parabolic mirror. We measure the beating signal of the two beams with a correlation measurement (Fig. 2) involving a photomultiplier tube (PMT), a time to digital converter (TDC), and a 10 MHz trigger signal that is synchronized to the AOM, respectively. The intensity signal $I_{\mathrm{TDC}}(\Delta t)$ obtained from the statistics of photon detection times on the TDC is fitted with a function proportional to $\cos(\omega_{\mathrm{rf}}\,\Delta t + \phi_1)$ with phase offset ϕ_1. To infer the relative phase shift $\Delta\varphi$ of the near-resonant beam, we repeat the measurement and fitting procedure after preparing the ion in the metastable $D_{3/2}$ (dark) state by optical pumping (see Fig. 1). This second measurement delivers the reference phase offset ϕ_2. The phase shift $\Delta\varphi$ of the near-resonant beam is finally calculated via $\Delta\varphi = \phi_1 - \phi_2$. The acquisition of sufficient statistics for each data point takes approximately ten seconds.

The results for measuring the phase shift as a function of detuning are shown in Fig. 3. We achieve a phase shift of $2.2^\circ \pm 0.5^\circ$ at approximately half linewidth detuning. These values are compared to the theoretically predicted values of Eq. 8 expected for a coupling efficiency of $G = 13.7 \pm 1.4\,\%$, found in an independent experiment based on a saturation measurement [24]. For detunings $-\Gamma/2 \leq \Delta \leq 0$ the measured phase shift shows a systematic deviation from the theoretical model, which assumes a detuning independent coupling parameter G. There are three possible reasons for these deviations. Firstly, saturation effects are neglected for the theoretical curve based on Eq. 8. However, since the saturation of the transition was kept low during the measurements ($S < 0.1$), the reduction of the measured phase due to saturation effects is expected to be less than 2%. Secondly, the observed phase shift drops faster than expected when going closer to resonance. A possible reason for this is that the temperature of the atom diverges when the detuning approaches zero [29] and consequently the size of the ion's wave function increases [30]. This leads to a stronger averaging of the experienced electric field by the extent of the ions wave function [31] entailing a reduction of the coupling efficiency and therefore also of the measured phase shift. Lastly, measuring the phase shift via heterodyne detection leads not only to a phase shift of the close to resonant part of the two light fields focused on the ion but also to a nonzero phase

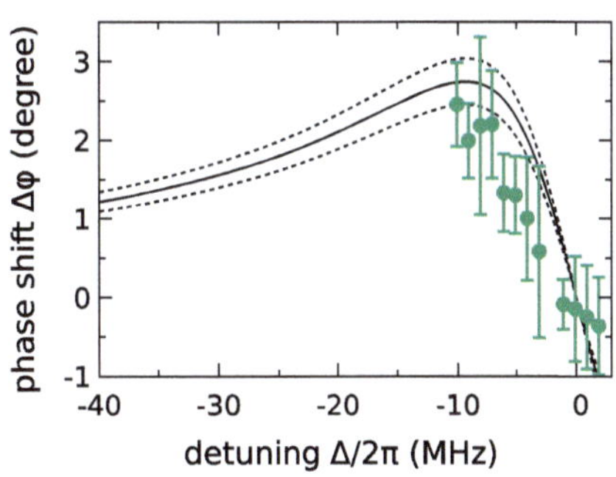

Fig. 3 Measured phase shift $\Delta\varphi$ for different detunings (*symbols*) and phase shift according to Eq. 8 for a coupling efficiency of $G = 13.7 \pm 1.4\,\%$ (*solid and dashed lines*). The value used for G is the one found in a saturation measurement in Ref. [24]

shift of the 400 MHz detuned sideband, acting as a phase reference. At about 400 MHz detuning, this phase shift of the reference beam can be assumed constant over the measured data points and takes a value of approximately $0.13°$ at a coupling efficiency of $G = 13.7\%$, leading to an effective offset of the zero phase value, which is neglected in Fig. 3.

5 Concluding Remarks

The phase shift obtained in our experiments is among the largest phase shifts measured for a coherent beam interacting with a single emitter in free space so far [18–20]. Nevertheless, it still is far below the maximum possible value $\Delta\varphi = \pi$ which can be obtained on resonance for $G > 0.5$ [13, 17, 32]. The lower phase shift demonstrated in our experiments is in parts due to the motion of the ion in the trap and the aberrations imprinted by the parabolic mirror, which made it necessary to focus only from half solid angle. The latter restriction limits the coupling to $G \leq 0.5$ [13, 22]. But the more severe limitation is the choice of our atomic species with its reduced scattering cross section. Even for optimum focusing and cooling the ion to its motional ground state the imprinted phase shift will never be larger than $30°$ — what still appears to be a fairly large value. Therefore, besides compensating mirror aberrations we aim at repeating our experiments with $^{174}Yb^{2+}$ [33], which offers the desired $J' = 1 \leftrightarrow J = 0$ transition that enables the maximum scattering cross section.

Acknowledgements Open access funding provided by Max Planck Society. G.L. gratefully acknowledges financial support from the European Research Council via the Advanced Grant 'PACART.' We acknowledge the contributions of R. Maiwald, A. Golla, S. Heugel and M. Bader in the earlier stages of our experimental endeavors and useful comments on the manuscript by M. R. Foreman.

References

1. T. Hänsch, A. Schawlow, Cooling of gases by laser radiation. Opt. Commun. **13**, 68–69 (1975)
2. C. Cohen-Tannoudji, J. Dupont-Roc, G. Grynberg, *Photons and Atoms* (Wiley, Hoboken, 1989)
3. I. Gonoskov, A. Aiello, S. Heugel, G. Leuchs, Dipole pulse theory: maximizing the field amplitude from 4π focused laser pulses. Phys. Rev. A **86**, 053836 (2012)
4. L. Mandel, E. Wolf, *Optical Coherence and Quantum Optics* (Cambridge University Press, Cambridge, 1995)
5. D. Budker, D.F. Kimball, D.P. DeMille, *Atomic Physics* (Oxford University Press, Oxford, 2004)

6. I.I. Sobel'man, *Introduction to the Theory of Atomic Spectra* (Pergamn Press Inc, Oxford, 1972)
7. S. Quabis, R. Dorn, M. Eberler, O. Glöckl, G. Leuchs, Focusing light to a tighter spot. Opt. Commun. **179**, 1–7 (2000)
8. R. Dorn, S. Quabis, G. Leuchs, Sharper focus for a radially polarized light beam. Phys. Rev. Lett. **91**, 233901 (2003)
9. M. Cray, M.-L. Shih, P.W. Milonni, Stimulated emission, absorption, and interference. Am. J. Phys. **50**, 1016–1021 (1982)
10. G. Zumofen, N.M. Mojarad, V. Sandoghdar, M. Agio, Perfect reflection of light by an oscillating dipole. Phys. Rev. Lett. **101**, 180404 (2008)
11. P. Domokos, P. Horak, H. Ritsch, Quantum description of light-pulse scattering on a single atom in waveguides. Phys. Rev. A **65**, 033832 (2002)
12. P. Kochan, H.J. Carmichael, Photon-statistics dependence of single-atom absorption. Phys. Rev. A **50**, 1700–1709 (1994)
13. G. Leuchs, M. Sondermann, Light matter interaction in free space. J. Mod. Opt. **60**, 36–42 (2013)
14. S. Tretyakov, Maximizing absorption and scattering by dipole particles. Plasmonics **9**, 935–944 (2014)
15. C.F. Bohren, How can a particle absorb more than the light incident on it? Am. J. Phys. **51**, 323–327 (1983)
16. H. Paul, R. Fischer, Light absorption by a dipole. Sov. Phys. Uspekhi **26**, 923 (1983)
17. M. Sondermann, G. Leuchs, The phase shift induced by a single atom in free space. J. Eur. Opt. Soc. Rapp. Public. **8**, 13502 (2013)
18. S.A. Aljunid, M.K. Tey, B. Chng, T. Liew, G. Maslennikov, V. Scarani, C. Kurtsiefer, Phase shift of a weak coherent beam induced by a single atom. Phys. Rev. Lett. **103**, 153601 (2009)
19. M. Pototschnig, Y. Chassagneux, J. Hwang, G. Zumofen, A. Renn, V. Sandoghdar, Controlling the phase of a light beam with a single molecule. Phys. Rev. Lett. **107**, 063001 (2011)
20. G. Hétet, L. Slodička, N. Röck, R. Blatt, Free-space read-out and control of single-ion dispersion using quantum interference. Phys. Rev. A **88**, 041804 (2013)
21. I.M. Bassett, Limit to Concentration by Focusing. Opt. Acta Int. J. Opt. **33**, 279–286 (1986)
22. A. Golla, B. Chalopin, M. Bader, I. Harder, K. Mantel, R. Maiwald, N. Lindlein, M. Sondermann, G. Leuchs, Generation of a wave packet tailored to efficient free space excitation of a single atom. Eur. Phys. J. D **66**, 190 (2012)
23. M. Fischer, M. Bader, R. Maiwald, A. Golla, M. Sondermann, G. Leuchs, Efficient saturation of an ion in free space. Applied Physics B **117**, 797–801 (2014)
24. L. Alber, M. Fischer, M. Bader, K. Mantel, M. Sondermann, and G. Leuchs, Focusing characteristics of a 4π parabolic mirror light-matter interface, arXiv:1609.06884 (2016)
25. S. Olmschenk, K.C. Younge, D.L. Moehring, D.N. Matsukevich, P. Maunz, C. Monroe, Manipulation and detection of a trapped Yb^{+} hyperfine qubit. Phys. Rev. A **76**, 052314 (2007)
26. R. Maiwald, A. Golla, M. Fischer, M. Bader, S. Heugel, B. Chalopin, M. Sondermann, G. Leuchs, Collecting more than half the fluorescence photons from a single ion. Phys. Rev. A **86**, 043431 (2012)
27. M. Sondermann, R. Maiwald, H. Konermann, N. Lindlein, U. Peschel, G. Leuchs, Design of a mode converter for efficient light-atom coupling in free space. Appl. Phys. B **89**, 489–492 (2007)
28. S. Stenholm, The semiclassical theory of laser cooling. Rev. Mod. Phys. **58**, 699–739 (1986)
29. R. Chang, A.L. Hoendervanger, Q. Bouton, Y. Fang, T. Klafka, K. Audo, A. Aspect, C.I. Westbrook, D. Clément, Three-dimensional laser cooling at the Doppler limit. Phys. Rev. A **90**, 063407 (2014)
30. J. Eschner, Sub-wavelength resolution of optical fields probed by single trapped ions: Interference, phase modulation, and which-way information. Eur. Phys. J D At. Mol. Opt. Plasma Phys. **22**, 341–345 (2003)

31. C. Teo, V. Scarani, Lenses as an atom photon interface: a semiclassical model. Opt. Commun. **284**, 4485–4490 (2011)
32. G. Zumofen, N.M. Mojarad, M. Agio, Light scattering by an oscillating dipole in a focused beam. Nuovo Cimento C **31**, 475–485 (2009)
33. S. Heugel, M. Fischer, V. Elman, R. Maiwald, M. Sondermann, G. Leuchs, Resonant photoionization of Yb^+ to Yb^{2+}. J. Phys. B **49**, 015002 (2016)

Thirty Years of Hydrogen Spectroscopy in Paris

François Biraben, Lucile Julien, and François Nez

Abstract This paper gives a review of the experiments performed since the 1980s at the Laboratoire Kastler Brossel in Paris on two-photon spectroscopy of atomic hydrogen. Firstly devoted to the 2S–*n*S and 2S–*n*D transitions, they are currently running on the 1S–3S transition at 205 nm. During all that time, they were inspired by the plentiful ideas proposed by Ted Hänsch and were complementary with the measurements developed in parallel in his groups.

1 Introduction

Hydrogen spectroscopy in Paris is a long story which began in 1983. One of us (FB) decided to use the tunable cw monomode laser he had himself developed [1] to excite two-photon transitions in atomic hydrogen. At this time, Ted Hänsch's work on hydrogen—the simplest and the most fascinating of the atoms—were worldwide known: in particular, the first observation of the 2S Lamb shift in an optical spectrum recorded by saturated absorption [2] and the one of the 1S–2S two-photon transitions using a pulsed laser [3].

The Paris idea was then to choose other transitions than the 1S–2S one, actually transitions starting from the 2S metastable state, to excite Rydberg S or D states. After several years, we also moved to the 1S–3S transition, a transition from the ground state, which is nowadays still under study in our group.

We show in this paper how Paris works in atomic hydrogen have been stimulated by the ones of Ted Hänsch's groups, first in Stanford and then in Garching, and how they developed in a context where an healthy competition turned often into collaboration and joint efforts for a common goal.

This article is part of the topical collection "Enlightening the World with the Laser" - Honoring T. W. Hänsch guest edited by Tilman Esslinger, Nathalie Picqué, and Thomas Udem.

F. Biraben • L. Julien • F. Nez (✉)
Laboratoire Kastler Brossel, UPMC-Sorbonne Universités, CNRS, Collège de France, ENS-PSL Research University, 4 Place Jussieu, 75005 Paris, France
e-mail: francois.nez@lkb.upmc.fr

D. Meschede et al. (eds.), *Exploring the World with the Laser*,
https://doi.org/10.1007/978-3-319-64346-5_23

2 The Beginning of the Paris Story

At the beginning, we (FB and LJ) were then only two people on the hydrogen project in Paris, with the scientific support of Bernard Cagnac who founded a few years before in our laboratory a group devoted to Doppler-free two-photon spectroscopy. He himself had pointed out for the first time the advantage of this spectroscopy for the study of the 1S–2S transition in hydrogen [4] but without undertaking such an experiment since Ted Hänsch seized this subject very quickly. It is why we decided to study other hydrogen lines.

One of us (F.B.) had already a very good expertise in high-resolution spectroscopy, having studied two-photon transitions in sodium [5] and rare gas atoms [6, 7]. The other one (L.J.) had previously measured atomic structures and Lamb shifts in excited hydrogen states by an anticrossing method [8].

Our target was to excite the $n \geq 8$ states because the wavelength range needed 730–778 nm was easily obtained with our homemade dye laser.

In a first step, we had to build a 2S metastable atomic beam, which was obtained by the following method: molecular hydrogen is dissociated by a RF discharge in a pyrex tube; an effusive 1S beam flows into a first vacuum chamber through a Teflon nozzle; 2S state is excited by electronic bombardment which bends the atomic beam and then makes it collinear with the two laser beams used for the two-photon excitation. The grid of the electron gun creates an equipotential volume around the metastable beam to shield it from quenching electric fields. The interaction between atoms and laser beams takes place in a second vacuum chamber delimited by two holes 96 cm apart from each other. The atoms remaining in the 2S state are detected in a third chamber where two electrodes quench the 2S state and the resulting Lyman-α fluorescence is detected. The two-photon transition signal is recorded through the decrease of the number of metastable atoms at the end of the beam. The schematized geometry of the excitation is shown in Fig. 1.

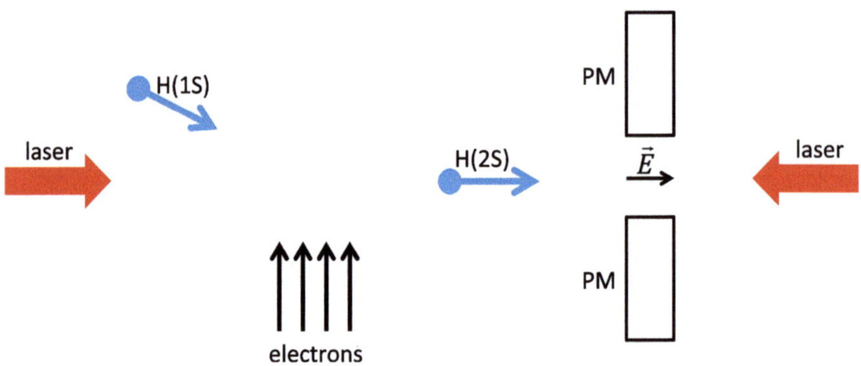

Fig. 1 Geometry of the two-photon excitation for the observation of 2S–nS/nD signals. After interaction with the two counter-propagating laser beams, the remaining 2S atoms are detected through electric quenching and Lyman alpha fluorescence

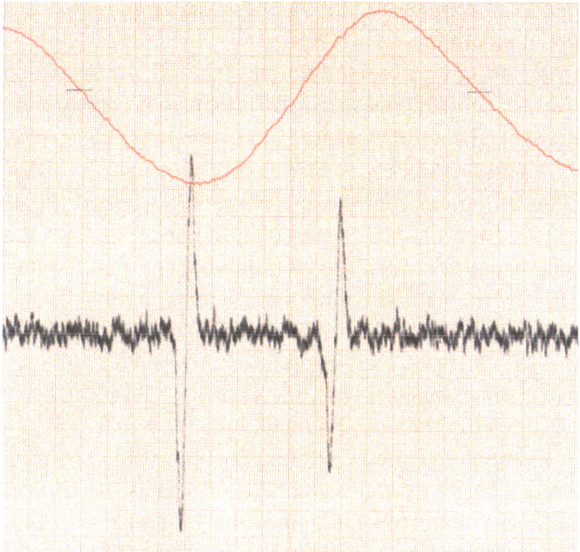

Fig. 2 2S–8D signal recorded on November 9, 1984. The two fine structure components of the 8D level are visible, separated by about 57 MHz. The laser frequency is modulated and the 2S signal demodulated by a lock-in amplifier. The *red trace* is the transmission of a FP cavity used for the calibration, having a low finesse at this wavelength

To enhance the two-photon excitation probability, the whole metastable beam was placed inside a Fabry–Perot cavity, the length of which was locked on the laser frequency by monitoring the reflected polarization, using the famous method proposed by Hänsch and Couillaud [9].

The search for $n = 8$ signal was not straightforward. Our laser frequency was scanned by locking it to a pressure swept Fabry–Perot cavity and known by comparison with a reference cavity whose length was not perfectly determined. Moreover, the Rydberg constant was actually known with a poorer exactness than believed, so that we did not look for our resonance at the correct frequency. After several months of efforts and changing the curvature of the enhancing cavity mirrors to optimize the excitation probability along the beam, we finally obtained signals (see Fig. 2) in both hydrogen and deuterium and were able to perform a preliminary measurement of the $n = 8$ isotopic shift [10]. The way to absolute measurements was now open for us.

3 Wavelength Measurements and First Determinations of the Rydberg Constant

The largest signal amplitude was the $2S_{1/2}(F = 1) \rightarrow 8D_{5/2}$ one, with a 10% decrease of the metastable beam intensity, and an experimental width of 1.4 MHz (relative linewidth 1.8×10^{-9}), which was at that time the narrowest one obtained in hydrogen. This width has to be compared to the 550 kHz natural width of the 8D states. The main broadening and shift effects were the light shift, the saturation of the 2S depopulation, the second-order Doppler effect and the finite transit time of the atoms in the laser beams.

Rapidly, we observed not only the 2S–8D transitions in H and D at 778 nm but also the 2S–10D ones in H at 760 nm, and we undertook to determine the Rydberg constant from the measurement of the three $2S - nD_{5/2}$ transition wavelengths.

The most recent determinations of the Rydberg constant were previously deduced either from the one-photon 2S–3P and 2P–3S Balmer-α transitions [11–13], or from the 1S–2S two-photon transition, both methods in which Ted Hänsch took a pioneer role. The first one has the disadvantage of being limited to about 10^{-9} by the natural width of the P level involved. The second one was much more promising, but needs the development of a tunable laser source at 243 nm. Although the 1S–2S signal had been already observed with a relative linewidth of 5×10^{-9} using cw excitation [14], the first wavelength measurement had been performed with a pulsed laser and a much broader signal [15].

Compared to the 1S–2S transition, the 2S–nD ones ($n \geq 8$ or 10) were limited by a much larger natural width but were competitive at that time for the determination of the Rydberg constant, since the 2S Lamb shift was precisely measured by RF spectroscopy [16, 17].

We measured our transition wavelengths [18] by interferometric comparison with an iodine-stabilized helium–neon laser at 633 nm. The wavelength of this laser was known at 2×10^{-10} since it had been itself compared to one of the "Institut National de Métrologie." We deduced our first value of the Rydberg constant with an uncertainty of 5.2×10^{-10}, but this value was in slight disagreement with the other ones published. As often in such a situation, such a discrepancy is uncomfortable for a new team in the field, but is a strong encouragement to pursue it.

During the following years, we studied also the 2S–8S transition and performed a detailed study of the line profiles taking into account for each transition all shifting and broadening effects for various values of the excitation power. Calculated profiles were then fitted to experimental signals in order to determine the line positions corrected from these effects. Examples of such fits are shown in Fig. 3 for the $2S_{1/2} - 8S_{1/2}$ and $2S_{1/2} - 8D_{5/2}$ transitions. Here the quenching field at the end of the atomic beam is modulated and the 2S signal demodulated by a lock-in amplifier. A good agreement is observed for both transitions, even if their line profiles have different asymmetries.

During this period, we also worked to reduce systematic effects and were able to extend our method to higher n levels. For that purpose, our interaction region was drastically shielded from stray electric fields by painting the walls of the vacuum chamber with Aquadag, a graphite liquid mixture: we were then able to record 2S–20D two-photon signals and to estimate an upper limit for the electric field experienced by the atoms ($\leq$2 mV/cm), and for the resulting Stark effect on our signals [19]. Another improvement was to measure the velocity distribution of our metastable atomic beam, by probing the 2S–3P Balmer-α one-photon transition with a collinear 656 nm laser beam [20].

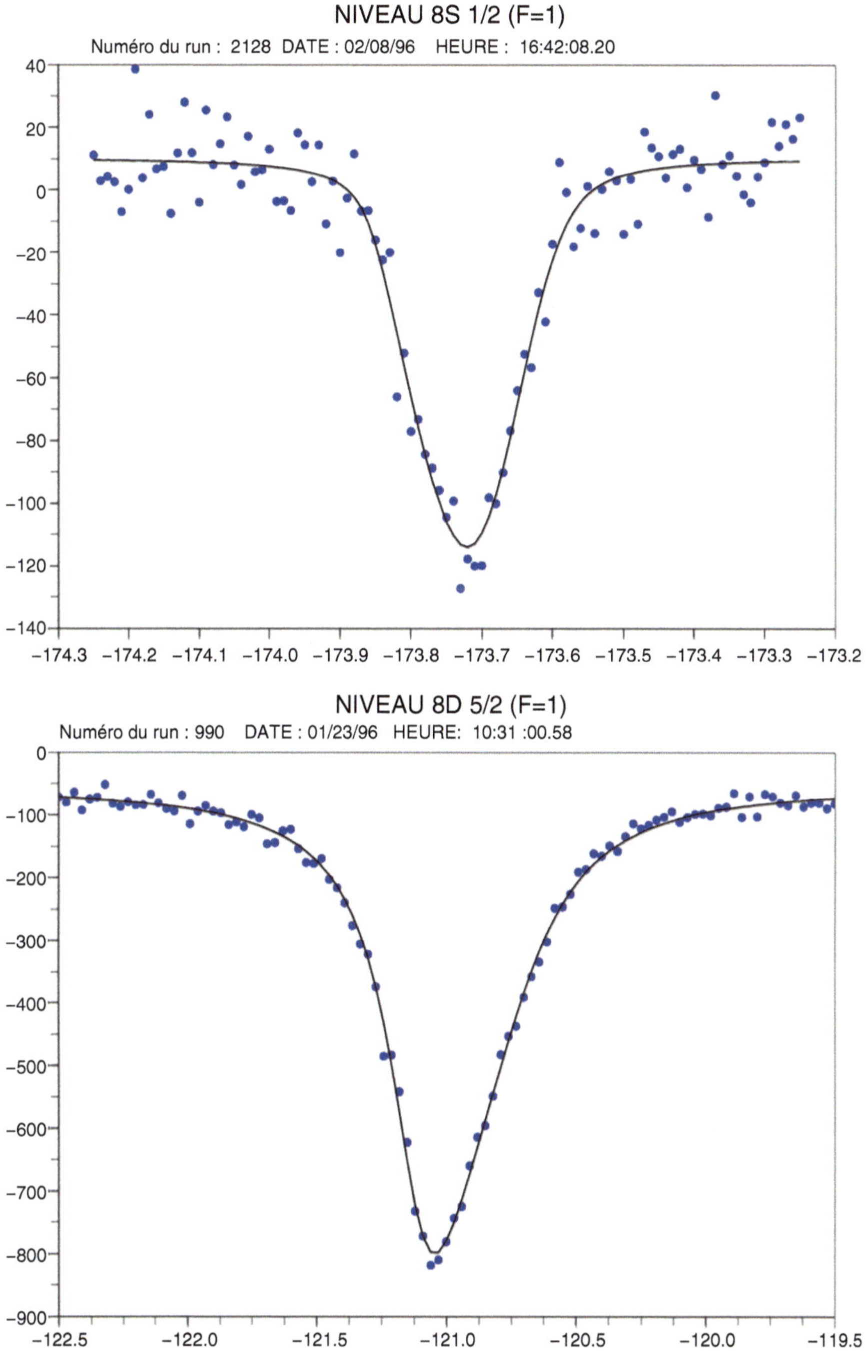

Fig. 3 $2S_{1/2} - 8S_{1/2}$ (*upper part*) and $2S_{1/2} - 8D_{5/2}$ (*lower part*) signals recorded in 1996, with fitted calculated profiles superimposed. *Vertical scale* number of 2S atoms recorded after demodulation (see *text*) by a lock-in and offset. The signal amplitudes are, respectively, 4.4 and 21.0 % of

In 1989, our result for the Rydberg constant with an uncertainty of 1.7×10^{-10} was mainly limited by the standard He–Ne laser [21]. During the same time, a good agreement of the results obtained on other transitions in hydrogen was found [22–24], especially after elimination of chirping effects in pulsed amplifiers, previously used for the 1S–2S excitation.

However, in these measurements, as in others, our laser wavelength was measured by interferometric comparison to the one of the standard lasers using the so-called virtual mirrors method, and the resulting uncertainty arose both from the knowledge of the reference laser frequency and from the comparison method itself. The further reduction of the uncertainty needed to abandon wavelength measurement for direct frequency comparisons.

Let us also point out here that in 1990 our homemade excitation dye laser was turned into a Ti:sapphire laser (TiSa), perfectly suitable for our 730–778 nm range and much more convenient in terms of stability, efficiency and easiness of running.

4 Direct Comparison of Two Optical Frequencies

As explained in the following, absolute optical frequency measurements were a quite complicated issue in the early 1990s. On the other hand, the determination of the Rydberg constant from the study of an optical transition was limited by the knowledge of the Lamb shifts of the levels involved. Since in atomic hydrogen the ratio between transition frequencies is very close to simple rational fractions, both Ted's group in Garching and our Paris group decided to complete their experimental setup in order to directly compare transition frequencies with a ratio close to 4:1: the 1S–2S and 2S–4S/4D transitions in Garching, the 1S–3S and 2S–6S/6D ones in Paris.

In Garching, a new 2S metastable beam was then developed with a Ti:sapphire laser at 972 nm for the atom excitation. In a similar way as in Paris, residual 2S atoms were detected at the end of the interaction region. The numerical calculation of the 2S–4S/4D signal profile was an opportunity for a fruitful collaboration between our two groups [25].

In Paris, the experiment was not easy to achieve, but during one year we benefited from the precious help of Derek Stacey, who had an expertise in the UV domain since he had studied the 1S–2S transition in the Oxford group [27]. A 1S beam was build and we had to develop a new laser source at 205 nm. For that purpose we chose to perform two successive frequency doubling stages of the Ti–sapphire radiation at 820 nm used for the 2S–6S/6D excitation. The first doubling in

---

Fig. 3 (continued) the 2S population. *Horizontal scale* frequency of the acousto-optic modulator used to scan the excitation laser. The different asymmetries of the lines arise from the respective weight of saturation and light shift. The saturation is almost negligible for the 8S line and important for the 8D one an enhancement external cavity was quite efficient, and we obtained more than 400 mW at 410 nm with our LBO crystal [28]. However, the second doubling step in a BBO crystal was far more challenging since 205 nm is the shortest wavelength which can be generated by SHG in such a crystal and because of photochemical reactions on the crystal faces and photorefractive effects in its bulk. We finally got only about 1 mW in a quasi-continuous mode where the enhancement cavity length is modulated [29].

Compared to the 1S–2S transition, the 1S–3S one is much broader and the signal much smaller, so that it was not straightforward for us to see the signal. It was finally observed in 1995. On the other hand the 2S–6S/6D transitions are much narrower than the 2S–4S/4D ones and, following the Garching and Yale groups [25, 26], we were able to determine the 1S Lamb shift with an uncertainty of 46 kHz [30], using the experimental value of the 2S one. The results obtained by the three groups were in good agreement with each other and, with comparable uncertainties, they notably improved the knowledge of the 1S Lamb shift. However, they were limited by the precision of the 2S Lamb shift, measured in the RF domain. To go further, it was necessary to perform the absolute frequency measurement of two different hydrogen optical frequencies.

5 Absolute Optical Frequency Measurements in Paris

The first absolute frequency measurement of an hydrogen transition was the one of the 2S–8D transitions [31], performed in our group during the PhD thesis of the third of us (F.N., also called F2). At that time, such a measurement needed to have a good frequency reference in the laboratory and a convenient frequency chain to compare the frequency to be measured to the one of the reference.

5.1 Frequency References

Our wavelength measurements had been previously performed by interferometric comparison with an I_2-stabilized He–Ne laser at 633 nm. This reference laser had been itself calibrated by comparison with the Bureau International des Poids et Mesures (BIPM) standard lasers through an intermediate laser, so that its frequency was known with an uncertainty of 1.6×10^{-10}.

In order to implement direct frequency measurements in hydrogen, we decided to look for new references around 778 nm, the wavelength of the 2S–8S/nD two-photon transitions. After a tentative work with the hyperfine components of the IBr molecule located in this frequency range [32], we finally chose the ones of the 5S–5D two-photon transitions in Rb [33]. Laser diodes could easily be locked to these transitions, giving a new optical frequency standard at 778 nm. Its frequency was measured, thanks to a collaboration with the Laboratoire Primaire du Temps et des Fréquences (LPTF, presently LNE-SYRTE) and the BIPM, with an uncertainty of 2 kHz [34]. This frequency standard with high metrological features [35] is still used in our hydrogen setup.

At the same time, after several years of effort and thanks to the tenacity of B. Cagnac, a 3-km-long optical fiber has been placed underground between our laboratory and the LPTF (LNE-SYRTE). This opened up the possibility to transfer optical frequencies between our two laboratories with an accuracy of a few Hz [36] and then to connect our experiment to the primary frequency standard, the Cs atomic clock.

5.2 *Frequency Chains*

The first frequency chain built in Paris to measure the 2S–8S/8D transitions of hydrogen used actually two frequency standards: the I_2-stabilized He–Ne laser at 473 THz (633 nm) and the CH_4-stabilized He–Ne laser at 88 THz (3.39 μm). It took advantage of the quasi-coincidence of the frequency difference between these two frequency standards with the one of our transitions (2 photons at 385 THz). Both standard frequencies were precisely known, thanks to recent calibrations done at the LPTF, and the 89 GHz frequency gap was measured in our laboratory firstly using a Fabry–Perot cavity [37], secondly using a Schottky diode which means that we realized for the first time a direct link to the Cs clock without any interferometry [31]. The Rydberg constant was deduced with an uncertainty of 2.2×10^{-11}.

In the following, this result could be improved to 9×10^{-12} thanks to the Rb frequency standard and the optical fiber discussed above, and the measurement was extended to deuterium, giving a new determination of the 2S Lamb shift in this atom [38].

As the direct frequency comparison with our Rb standard was only possible for the 2S–8S/8D transitions, we were obliged to build-up again a specific frequency chain to measure other transitions in hydrogen. As a new target, we chose the $n = 12$ levels which are very sensitive to stray electric fields (the quadratic Stark effect varies as n^7) and then give complementary information to possible systematic effects on the $n = 8$ transitions. This choice was motivated by the possibility to compare the transition frequency at 799 THz (2 TiSa photons at 750 nm to excite the $n = 12$ levels in H) to twice the one of the Rb standard at 385 THz, using a standard OsO_4-stabilized CO_2 laser at 29 THz (≈10 μm) to measure the gap. One has indeed:

$$799\,\text{THz} = 2 \times 385\,\text{THz} + 29\,\text{THz} \qquad (1)$$

In this new chain, we used an optical frequency divider to reduce by a factor 2 the frequency difference between our TiSa laser and the Rb standard, as suggested for the first time by Hänsch et al. [39].

The experiment was simultaneously carried out in our laboratory and in the LPTF, thanks to our link by a double optical fiber. An auxiliary 809 nm radiation (370.5 THz) delivered by a laser diode in our laboratory was sent to the LPTF to be compared to the frequency difference between a laser diode at 750 nm (385 THz) and the CO_2/OsO_4 laser. A radiation at 750 nm was also send to our laboratory to be compared to our TiSa laser, whose frequency was mixed with the one of the auxiliary laser diodes at 809 nm. The frequency sum obtained was compared to twice the one of the Rb standard. The two following equations were then realized, respectively, in LPTF and in LKB:

$$370.5\,\text{THz}\,(\text{auxiliaryDL}) = 385\,\text{THz} - 29\,\text{THz}\,/2 \tag{2}$$

$$399.5\,\text{THz}\,(\text{Ti}-\text{Sa}) + 370.5\,\text{THz}\,(\text{auxiliaryDL}) = 2 \times 385\,\text{THz} \tag{3}$$

The four 2S–12D transitions in hydrogen and deuterium were measured and a careful analysis of the line profiles was carried out, including various systematic effects, in particular the Stark effect. Using this result and the ones obtained by other groups, especially the Garching one on the 1S–2S transition, the Rydberg constant could be deduced with an uncertainty lower than 8×10^{-12} [40, 41].

5.3 *Frequency Combs*

As in Paris, the study of hydrogen transitions in Ted Hänsch's group, first in Stanford and then in Garching, needed during the same period successively the design of an auxiliary standard laser [42], the development of a frequency chain with a frequency divider [43], the use of a transportable calibrated standard [44] and finally a direct link to a metrology institute through an optical fiber [45]. All these means were used to measure the 1S–2S frequency with an increasing and unbeatable precision.

Moreover, during the same time occurred the so-called frequency comb revolution, the main father of which is Ted Hänsch, rewarded by the well-deserved Nobel Prize jointly awarded to John Hall and Ted Hänsch in 2005. The life in our laboratories was totally changed. No more need of complicated frequency chains depending on the transition to be measured: a fs laser coupled to a photonic fiber was the unique instrument needed to measure optical frequencies, allowing one to compare them directly to the Cs frequency standard. What we dreamed about was realized.

Of course, the 1S–2S transition was the first one in hydrogen to be measured in this new way [46], and other improvements followed [47, 48]. For us, it was time to leave 2S–nS/nD and focus on 1S–3S.

6 From 2S–nS/nD to 1S–3S Spectroscopy

At the turn of the century, it seemed to us that it was no more possible for us to push further our precision on the transitions starting from the 2S metastable state. Let us point out the advantages and the limitations of these transitions.

They are in a frequency range easy to reach with a TiSa laser and have small natural widths of a few hundreds of kHz. The stray electric field was reduced at best and estimated by recording the transitions toward $n = 20$ levels, and the velocity distribution measured (see above).

However, the number of metastable atoms in our atomic beam was small 2×10^6 at/s being a limiting factor for our signal-to-noise ratio. We then used a high laser power to excite the transitions: up to 150 W in each direction of propagation inside the enhancement cavity. Due to the conjugated effect of saturation of the 2S depopulation and light shifts, the line profiles are asymmetric (see Fig. 3) but well understood by numerical simulations. Extrapolation to null power of the center of the line is performed using a procedure detailed in [41].

The features of the 1S–3S transition are quite different. The number of 1S atoms is about eight orders of magnitude larger than the metastable one. The natural width of 1 MHz is similar, but the excitation wavelength which lies in the UV range (205 nm) is responsible for a better quality factor for the line. The laser power is only a few milliwatts and the light shift much smaller. Moreover, the residual Stark shift effect is negligible.

However, since we have no tunable laser source to probe the velocity distribution of the 1S beam through a one-photon transition, we have no simple way to determine the second-order Doppler effect which shifts the line by $-v^2/2c^2$, almost 150 kHz. This is why we have proposed and implemented an original method allowing to measure this shift [49] in our thermal atomic beam. The basic idea is to apply a transverse magnetic field to the atoms, inducing a quadratic motional Stark effect varying also as v^2 and able to partially compensate the second-order Doppler effect near a 3S–3P anticrossing [50].

With this method and using a frequency comb, we measured for the first time the absolute frequency of the 1S–3S transition [51], with an uncertainty of 13 kHz (4.5×10^{-12}). After the measurement of the 1S–2S transition in Garching, our result was in 2010 the second most precise one of the optical transitions in hydrogen. This experiment is still being improved. We have in particular developed a new laser source at 205 nm, by frequency mixing of two radiations at 266 and 894 nm, respectively generated by frequency doubling of a Verdi laser (Coherent Inc.) at 532 nm and by our TiSa laser. By this method, we obtain up to 15 mW of cw operation at 205 nm [52].

In 2013, the 1S–3S signal was observed with a signal-to noise ratio up to 170 after 3.5 h of integration, and with a width of about 1.5 MHz, to be compared to the 1 MHz natural width of the line. The absolute frequencies of the two radiations at 532 nm and 894 nm were measured with an optical frequency comb referenced to the Cs clock of the LNE-SYRTE laboratory. The statistical

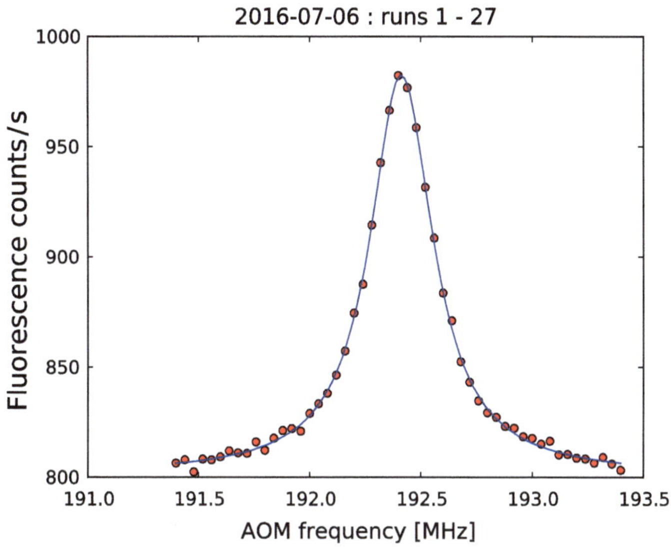

Fig. 4 1S–3S signal recorded on July 6, 2016. The excitation laser at 205 nm is swept through an acousto-optic modulator. The integration time is here 4.5 h and the signal linewidth 1.36 MHz in atomic frequency units. The signal background is mainly due to stray fluorescence induced by the UV excitation light

uncertainty on the deduced transition frequency has been significantly improved to reach a value of 2.2 kHz [53]. However, a complete analysis of systematic shifts of the line, in particular collisional shifts, was still needed to determine the 1S–3S transition frequency.

This problem is presently under study. Our experiment is in progress and a recent signal is recorded in Fig. 4.

7 Hydrogen Spectroscopy and the Rydberg Constant

When we began the study of 2S–nD transitions during the 1980s, our first goal was the determination of the Rydberg constant. It was simply deduced from our transition frequencies, using the experimental value of the 2S Lamb shift. Such a method was limited to an uncertainty of 1.2×10^{-11} due to the 2S Lamb shift measurement.

To go further in precision, two different optical transitions were needed, which were found in the 1S–2S and 2S–8D transitions. As the quantity $L(1\mathrm{S}) - 8L(2\mathrm{S})$ had been precisely calculated [54–56], the comparison of the two optical frequency intervals allows the determination of the Rydberg constant with an uncertainty of 8.6×10^{-12}. The advantage of this method is that it needs neither the 2S Lamb shift nor the proton radius and is applicable both to hydrogen and deuterium.

Finally, the most precise value of the Rydberg constant is given by the least squares adjustment of the fundamental constants performed by the Committee on Data for Science and Technology (CODATA). It takes into account data concerning all the fundamental constants, but is limited to spectroscopic data on hydrogen (e-p) only and e-p diffusion experiments. The uncertainty deduced for the Rydberg constant by the last adjustment is 5.5×10^{-12} [57] resulting mostly from the uncertainties obtained by the Paris and the Garching groups on the 2S–nS/nD and 1S–2S transitions.

In the same manner as for the Rydberg constant, hydrogen 1S and 2S Lamb shifts can be deduced from the same measurements, using the scaling law quoted above. The value of the 2S–2P interval obtained is more precise than the one given by direct microwave spectroscopy. The 1S Lamb shift with an uncertainty of 24 kHz only would give a test of the QED two-loop corrections in QED if the proton charge radius $r_{\rm p}$ was perfectly known. On the other hand, the QED calculations are a way to determine $r_{\rm p}$, which is complementary with the e-p scattering experiments.

8 The Hydrogen Atom, a Still Hot Topic

At the end of the 1990s, it was then clear that the proton radius became a limiting factor to test theoretical predictions of QED on the hydrogen atom. It is why we joined in 1998 the international collaboration, now called CREMA, constituted to determine $r_{\rm p}$ from 2S–2P spectroscopy of muonic hydrogen at the Paul Scherrer Institute (PSI) in Switzerland. The Paris and the Garching groups were active together in this new adventure, which is not closed at the present time.

The idea was to measure the 2S–2P Lamb shift interval in muonic hydrogen, an atom made of a proton and a negative muon (μ^-). As the muon is 207 times more massive than the electron, the correction due to the finite size of the nucleus is largely enhanced. Indeed, the sensitivity of the frequency of the 2S–2P transition to $r_{\rm p}$ in such a spectroscopy at the 10^{-6} level is enough to get a determination of $r_{\rm p}$ 10 times better than the best determination of $r_{\rm p}$ from the electronic hydrogen spectroscopy and quantum electrodynamics.

The principle of the experiment can be summarized in three steps: production of muonic hydrogen (μ-p) atoms in the metastable 2S state, excitation of the 2S–2P transition with a laser at 6 μm and then detection of the 2P–1S fluorescence photon at 2 keV. Even if it looks simple, each step was challenging. The source of metastable atoms has been realized thanks to the expertise of F. Kottmann, D. Taqqu and R. Pohl. Members of the Paris group took part in the development of the laser chain which produced the 6 μm laser pulse in collaboration with A. Antognini. After three unsuccessful beam times (2002, 2003, 2007), in 2009, we finally observed two transitions of muonic hydrogen and three of muonic deuterium [58–60], but they were not at the frequency predicted by the theory.

This was the beginning of the so-called proton size puzzle which has triggered and stimulated a lot of activities in our community but also outside of it, both theoretical and experimental ones.

On the side of experiments in hydrogen atom, a new RF measurement of the 2S–2P hydrogen interval is currently in progress in the group of Hessels [61] while the Paris and Garching groups pursue their efforts to improve their optical measurements: in the one-photon 2S–4P transition [62] (cw laser in Garching) and in the two-photon 1S–3S transition [53, 63] (pulsed laser in Garching and cw laser in Paris). At the level of precision needed to clarify the proton puzzle, all stray shifting effects have to be carefully studied, including quantum cross-damping interference effects [64–66].

New results are looked forward in the near future, not only on the above hydrogen experiments, but also on hydrogenic systems [67–70].

9 Conclusion

This review of more than thirty years of hydrogen spectroscopy in Paris and also in Ted Hänsch's groups in Stanford and Garching shows how the study of this simple atom has stimulated the development of laser technology and new frequency measurement methods in parallel with QED calculations.

Thanks to the development of frequency combs, ultra-stable cavities and laser references, the optical part of an hydrogen experiment is almost no more an issue, except for the lack of a powerful cw 121 nm laser to optically cool atomic hydrogen, even if 1S–2P spectroscopy has been already performed with a low power source in Ted Hänsch's group [71]. The last experimental frontier in hydrogen spectroscopy is certainly the atomic source: hydrogen spectroscopy is still performed on atomic beams.

The other frontier to overcome is in the perturbative QED calculations. Nowadays, the 1S–2S transition frequency is measured with an uncertainty of 10 Hz that is 250 times smaller than the theoretical uncertainty!

From the beginning of quantum mechanics, the hydrogen atom is a source of new discoveries in physics and can be considered as "the Rosetta stone of modern Physics" as pointed out in a paper of Schawlow et al. [72]. Coming back to the Paris adventure, two of us who were at the beginning of this story are now retired, although still active in research. We did not imagine that after more than thirty years, the hydrogen atom would be always so fascinating and rich of secrets still to discover.

Acknowledgements We are pleased to acknowledge here all people who worked with us on our "hydrogen experiments" in Paris: Jean-Claude Garreau, Sophie Bourzeix, Béatrice Chatel (De Beauvoir), Gaëtan Hagel, Olivier Arnoult, Sandrine Galtier and Hélène Fleurbaey as Ph.D. students, Mark D. Plimmer, F. de Tomasi and L. Jozefowski as post-docs and our colleagues M. Allegrini, D.N. Stacey and C. Schwob. The authors are indebted to the people of LNE-SYRTE laboratory for the time reference and constant help during all these years.

References

1. F. Biraben, P. Labastie, Opt. Commun. **41**, 49 (1982)
2. T.W. Hänsch, I.S. Shahin, A.L. Schawlow, Nature **235**, 63 (1972)
3. T.W. Hänsch, S.A. Lee, R. Wallenstein, C. Wieman, Phys. Rev. Lett. **34**, 307 (1975)
4. B. Cagnac, G. Grynberg, F. Biraben, J. Phys. **34**, 845 (1973)
5. F. Biraben, B. Cagnac, G. Grynberg, Phys. Rev. Lett. **32**, 643 (1974)
6. F. Biraben, E. Giacobino, G. Grynberg, Phys. Rev. A **12**, 2444 (1975)
7. E. Giacobino, F. Biraben, J. Phys. B: At. Mol. Phys. **15**, L385 (1982)
8. M. Glass-Maujean, L. Julien, T. Dohnalik, J. Phys. B. **11**, 421 (1978)
9. T.W. Hänsch, B. Couillaud, Opt. Commun. **35**, 441 (1980)
10. F. Biraben, L. Julien, Opt. Commun. **53**, 320 (1985)
11. J.E.M. Goldsmith, E.W. Weber, T.W. Hänsch, Phys. Rev. Lett. **41**, 1525 (1978)
12. B.W. Petley, K. Morris, R.E. Shawyer, J. Phys. B: At. Mol. Phys. **13**, 3099 (1980)
13. S.R. Amin, C.D. Caldwell, W. Lichten, Phys. Rev. Lett. **47**, 1234 (1981)
14. C. Foot, B. Couillaud, R.G. Beausoleil, T.W. Hänsch, Phys. Rev. Lett. **54**, 1913 (1985)
15. E.A. Hildum, U. Boesl, D.H. McIntyre, R.G. Beausoleil, T.W. Hänsch, Phys. Rev. Lett. **56**, 576 (1986)
16. YuL Sokolov, V.P. Yakovlev, Sov. Phys. JETP **56**, 7 (1982)
17. S.R. Lundeen, F.M. Pipkin, Phys. Rev. Lett. **46**, 232 (1981)
18. F. Biraben, J.C. Garreau, L. Julien, Europhys. Lett. **2**(12), 925 (1986)
19. J.C. Garreau, M. Allegrini, L. Julien, F. Biraben, J. Phys. **51**, 2275 (1990)
20. F. Biraben, J.C. Garreau, L. Julien, M. Allegrini, Rev. Sci. Instrum. **61**, 1468 (1990)
21. F. Biraben, J.C. Garreau, L. Julien, M. Allegrini, Phys. Rev. Lett. **62**, 621 (1989)
22. P. Zhao, W. Lichten, H.P. Layer, J.C. Bergquist, Phys. Rev. Lett. **58**, 1293 (1987)
23. R.G. Beausoleil, D.H. McIntyre, C.J. Foot, E.A. Hildum, B. Couillaud, T.W. Hänsch, Phys. Rev. A **35**, 4878 (1987)
24. M.G. Boshier, P.E.G. Baird, C.J. Foot, E.A. Hinds, M.D. Plimmer, D.N. Stacey, J.S. Swan, D.A. Tate, D.M. Warrington, G.K. Woodgate, Nature **330**, 463 (1987)
25. M. Weitz, A. Huber, F. Schmidt-Kaler, D. Leibfried, W. Vassen, C. Zimmermann, K. Pachucki, T.W. Hänsch, L. Julien, F. Biraben, Phys. Rev. A **52**, 2664 (1995)
26. D.J. Berkeland, E.A. Hinds, M.G. Boshier, Phys. Rev. Lett. **75**, 2470 (1995)
27. M.G. Boshier, P.E.G. Baird, C.J. Foot, E.A. Hinds, M.D. Plimmer, D.N. Stacey, J.B. Swan, D.A. Tate, D.M. Warrington, G.K. Woodgate, Phys. Rev. A **40**, 6169 (1989)
28. S. Bourzeix, M.D. Plimmer, F. Nez, L. Julien, F. Biraben, Opt. Commun. **99**, 89 (1993)
29. S. Bourzeix, B. de Beauvoir, F. Nez, F. de Tomasi, L. Julien, F. Biraben, Opt. Commun. **133**, 239 (1997)
30. S. Bourzeix, B. de Beauvoir, F. Nez, M.D. Plimmer, F. de Tomasi, L. Julien, F. Biraben, Phys. Rev. Lett. **76**, 384 (1996)
31. F. Nez, M.D. Plimmer, S. Bourzeix, L. Julien, F. Biraben, R. Felder, Y. Millerioux, P. de Natale, Europhys. Lett. **24**, 635 (1993)
32. F. Biraben, D. Jasmin, L. Julien, F. Nez, J. Vigué, Opt. Commun. **92**, 135 (1992)
33. F. Nez, F. Biraben, R. Felder, Y. Millerioux, Opt. Commun. **102**, 432 (1993)
34. D. Touahri, O. Acef, A. Clairon, J.J. Zondy, R. Felder, L. Hilico, B. de Beauvoir, F. Biraben, F. Nez, Opt. Commun. **133**, 471 (1997)
35. L. Hilico, R. Felder, D. Touahri, O. Acef, A. Clairon, F. Biraben, Europhys. J. Appl. Phys. **4**, 219 (1998)
36. B. de Beauvoir, F. Nez, L. Hilico, L. Julien, F. Biraben, B. Cagnac, J.J. Zondy, D. Touahri, O. Acef, A. Clairon, Eur. Phys. J. D **1**, 227 (1998)
37. F. Nez, M.D. Plimmer, S. Bourzeix, L. Julien, F. Biraben, Phys. Rev. Lett. **69**, 2326 (1992)
38. B. de Beauvoir, F. Nez, L. Julien, B. Cagnac, F. Biraben, D. Touahri, L. Hilico, O. Acef, A. Clairon, J.J. Zondy, Phys. Rev. Lett. **78**, 440 (1997)
39. H.R. Telle, D. Meschede, T.W. Hänsch, Opt. Lett. **15**, 532 (1990)

40. C. Schwob, L. Jozefowski, B. de Beauvoir, L. Hilico, F. Nez, L. Julien, F. Biraben, Phys. Rev. Lett. **82**, 4960 (1999)
41. B. de Beauvoir, C. Schwob, O. Acef, L. Jozefowski, L. Hilico, F. Nez, L. Julien, A. Clairon, F. Biraben, J.J. Zondy, Eur. Phys. J. D **12**, 61 (2000)
42. D.H. McIntyre, R.G. Beausoleil, C.J. Foot, E.A. Hildum, B. Couillaud, T.W. Hänsch, Phys. Rev. A **39**, 4591 (1989)
43. Th Udem, A. Huber, B. Gross, J. Reichert, M. Prevedelli, M. Weitz, T.W. Hänsch, Phys. Rev. Lett. **79**, 2646 (1997)
44. M. Niering, R. Holzwarth, J. Reichert, P. Pokasov, Th Udem, M. Weitz, T.W. Hänsch, P. Lemonde, G. Santarelli, M. Abgrall, P. Laurent, C. Salomon, A. Clairon, Phys. Rev. Lett. **84**, 5496 (2000)
45. A. Matveev, C.G. Parthey, K. Predehl, J. Alnis, A. Beyer, R. Holzwarth, T. Udem, T. Wilken, N. Kolachevsky, M. Abgrall, D. Rovera, C. Salomon, P. Laurent, G. Grosche, O. Terra, T. Legero, H. Schnatz, S. Weyers, B. Altschul, T.W. Hänsch, Phys. Rev. Lett. **110**, 230801 (2013)
46. J. Reichert, M. Niering, R. Holtzwarth, M. Weitz, Th Udem, T.W. Hänsch, Phys. Rev. Lett. **84**, 3232 (2000)
47. M. Fischer, N. Kolachevsky, M. Zimmermann, R. Holzwarth, Th Udem, T.W. Hänsch, M. Abgrall, J. Grnert, I. Maksimovic, S. Bize, H. Marion, F. Pereira Dos Santos, P. Lemonde, G. Santarelli, P. Laurent, A. Clairon, C. Salomon, M. Haas, U.D. Jentschura, C.H. Keitel, Phys. Rev. Lett. **92**, 230802 (2004)
48. C.G. Parthey, A. Matveev, J. Alnis, B. Bernhardt, A. Beyer, R. Holzwarth, A. Maistrou, R. Pohl, K. Predehl, Th Udem, T. Wilken, N. Kolachevsky, M. Abgrall, D. Rovera, Ch. Salomon, Ph Laurent, T.W. Hänsch, Phys. Rev. Lett. **107**, 203001 (2011)
49. F. Biraben, L. Julien, J. Plon, F. Nez, Opt. Commun. **15**, 831 (1991)
50. G. Hagel, R. Battesti, F. Nez, L. Julien, F. Biraben, Phys. Rev. Lett. **89**, 203001 (2002)
51. O. Arnoult, F. Nez, L. Julien, F. Biraben, Eur. Phys. J. D **60**, 243 (2010)
52. S. Galtier, F. Nez, L. Julien, F. Biraben, Opt. Commun. **324**, 34 (2014)
53. S. Galtier, H. Fleurbaey, S. Thomas, L. Julien, F. Biraben, F. Nez, J. Phys. Chem. Ref. Data **44**, 031201 (2015)
54. S.G. Karshenboim, J. Phys. B **29**, L29 (1996)
55. S.G. Karshenboim, Z. Phys. D **39**, 109 (1997)
56. A. Czarnecki, U.D. Jentschura, K. Pachucki, Phys. Rev. Lett. **95**, 180404 (2005)
57. P.J. Mohr, D.B. Newell, B.N. Taylor, arXiv:1507.07956v1 (2015)
58. R. Pohl, A. Antognini, F. Nez, F.D. Amaro, F. Biraben, J.M.R. Cardoso, D. Covita, A. Dax, S. Dhawan, L. Fernandes, A. Giesen, T. Graf, T.W. Hänsch, P. Indelicato, L. Julien, C.-Y. Kao, P.E. Knowles, E. Le Bigot, Y.-W. Liu, J.A.M. Lopes, L. Ludhova, C.M.B. Monteiro, F. Mulhauser, T. Nebel, P. Rabinowitz, J.M.F. dos Santos, L. Schaller, K. Schuhmann, C. Schwob, D. Taqqu, J.F.C.A. Veloso, F. Kottmann, Nature **466**, 213 (2010)
59. A. Antognini, F. Nez, K. Schuhmann, F.D. Amaro, F. Biraben, J.M.R. Cardoso, D. Covita, A. Dax, S. Dhawan, M. Diepold, L. Fernandes, A. Giesen, A. Gouvea, T. Graf, T.W. Hänsch, P. Indelicato, L. Julien, C.-Y. Kao, P.E. Knowles, F. Kottmann, E. Le Bigot, Y.-W. Liu, J.A.M. Lopes, L. Ludhova, C.M.B. Monteiro, F. Mulhauser, T. Nebel, P. Rabinowitz, J.M.F. dos Santos, L. Schaller, C. Schwob, D. Taqqu, J.F.C.A. Veloso, J. Vogelsang, R. Pohl, Science **339**, 417 (2013)
60. R. Pohl, F. Nez, L.M.P. Fernandes, F.D. Amaro, F. Biraben, J.M.R. Cardoso, D.S. Covita, A. Dax, S. Dhawan, M. Diepold, A. Giesen, A.L. Gouvea, T. Graf, T.W. Hänsch, P. Indelicato, L. Julien, P. Knowles, F. Kottmann, E.O. Le Bigot, Y.W. Liu, J.A.M. Lopes, L. Ludhova, C.M.B. Monteiro, F. Mulhauser, T. Nebel, P. Rabinowitz, J.M.F. dos Santos, L.A. Schaller, K. Schuhmann, C. Schwob, D. Taqqu, J.F.C.A. Veloso, A. Antognini, Science **353**, 669 (2016)
61. E. Hessels, in *Frontiers in Optics 2015*. OSA Technical Digest (online) (Optical Society of America, 2015), paper LTu2G.2

62. A. Beyer, J. Alnis, K. Khabarova, A. Matveev, C.G. Parthey, D.C. Yost, R. Pohl, T. Udem, T.W. Hänsch, N. Kolachevsky, Anna. Phys. **525**, 671 (2013)
63. E. Peters, D.C. Yost, A. Matveev, T.W. Hänsch, T. Udem, Ann. Phys. **525**, L29 (2013)
64. M. Horbatsch, E.A. Hessels, Phys. Rev. A **82**, 052519 (2010)
65. D.C. Yost, A. Matveev, E. Peters, A. Beyer, T.W. Hänsch, Th Udem, Phys. Rev. A **90**, 012512 (2014)
66. P. Amaro, B. Franke, J.J. Krauth, M. Diepold, F. Fratini, L. Safari, J. Machado, A. Antognini, F. Kottmann, P. Indelicato, R. Pohl, J.P. Santos, Phys. Rev. A **92**, 022514 (2015)
67. M. Herrmann, M. Haas, U.D. Jentschura, F. Kottmann, D. Leibfried, G. Saathoff, C. Gohle, A. Ozawa, V. Batteiger, S. Knnz, N. Kolachevsky, H.A. Schssler, T.W. Hänsch, Th Udem, Phys. Rev. A **79**, 052505 (2009)
68. U.D. Jentschura, P.J. Mohr, J.N. Tan, B.J. Wundt, Phys. Rev. Lett. **100**, 160404 (2008)
69. J.P Karr, L. Hilico, J.C.J. Koelemeij, V.I. Korobov, arXiv:1605.05456 [physics.atom-ph] (2016)
70. R. Pohl, F. Nez, L.M.P. Fernandes, M. Abdou Ahmed, F.D. Amaro, P. Amaro, F. Biraben, J.M.R. Cardoso, D.S. Covita, A. Dax, S. Dhawan, M. Diepold, B. Franke, S. Galtier, A. Giesen, A.L. Gouvea, J. Götzfried, T. Graf, T.W. Hänsch, M. Hildebrandt, P. Indelicato, L. Julien, K. Kirch, A. Knecht, P. Knowles, F. Kottmann, J.J. Krauth, E-O. Le Bigot, Y-W. Liu, J.A.M. Lopes, L. Ludhova, J. Machado, C.M.B. Monteiro, F. Mulhauser, T. Nebel, P. Rabinowitz, J.M.F. dos Santos, J. Paulo Santos, L.A. Schaller, K. Schuhmann, C. Schwob, C.I. Szabo, D. Taqqu, J.F.C.A. Veloso, A. Voss, B. Weichelt, A. Antognini, in *Proceedings of LEAP 2016* (**submitted to Japanese Physical Society**)
71. K.S.E. Eikema, J. Walz, T.W. Hänsch, Phys. Rev. Lett. **86**, 5679 (2001)
72. A.L. Schawlow, T.W. Hänsch, G.W. Series, Sci. Am. **240**, 94 (1979)

A Dense Gas of Laser-Cooled Atoms for Hybrid Atom–Ion Trapping

Bastian Höltkemeier, Julian Glässel, Henry López-Carrera, and Matthias Weidemüller

Abstract We describe the realization of a dark spontaneous-force trap of rubidium atoms. The atoms are loaded from a beam provided by a two-dimensional magneto-optical trap yielding a capture efficiency of 75%. The dense and cold atomic sample is characterized by saturated absorption imaging. Up to 10^9 atoms are captured with a loading rate of 3×10^9 atoms/s into a cloud at a temperature of 250 μK with the density exceeding 10^{11} atoms/cm^3. Under steady-state conditions, more than 90% of the atoms can be prepared into the absolute atomic ground state, which provides favorable conditions for the investigation of sympathetic cooling of ions in a hybrid atom–ion trap.

This article is part of the topical collection "Enlightening the World with the Laser" - Honoring T. W. Hänsch guest edited by Tilman Esslinger, Nathalie Picqué, and Thomas Udem.

B. Höltkemeier • H. López-Carrera
Physikalisches Institut, Ruprecht-Karls-Universität Heidelberg, Im Neuenheimer Feld 226, 69120 Heidelberg, Germany

J. Glässel
Physikalisches Institut, Ruprecht-Karls-Universität Heidelberg, Im Neuenheimer Feld 226, 69120 Heidelberg, Germany

Present Address: Department of Physics and Astronomy, Ny Munkegade 120, 8000 Aarhus C, Denmark

M. Weidemüller (✉)
Physikalisches Institut, Ruprecht-Karls-Universität Heidelberg, Im Neuenheimer Feld 226, 69120 Heidelberg, Germany

Hefei National Laboratory for Physical Sciences at the Microscale and Department of Modern Physics, CAS Center for Excellence and Synergetic Innovation Center in Quantum Information and Quantum Physics, University of Science and Technology of China, Hefei 230026, Anhui, China
e-mail: weidemueller@uni-heidelberg.de

D. Meschede et al. (eds.), *Exploring the World with the Laser*,
https://doi.org/10.1007/978-3-319-64346-5_24

1 Theodor Hänsch's Legacy

In their visionary paper of 1975, Hänsch and Schawlow [1] propose cooling of atoms using the radiation forces of near-resonant laser beams.[1] The cooling scheme employs sets of counter-propagating laser beams which are slightly red-detuned from an atomic transition. The resulting velocity-dependent force exhibits a strong damping term which can be used to cool particles to sub-Kelvin temperatures. Laser cooling of atomic ions was realized only three years after the original proposal [3, 4]. Using ions instead of neutral atoms had the advantage that charged particles could be efficiently trapped which provided enough time for the cooling process. For atoms, however, all proposed traps had trap depths in the microKelvin regime [5] which required extensive cooling before the atoms could be trapped. Therefore, the first laser-cooling experiments on neutral atoms were performed on atomic beams, the main challenge being to keep the atoms in resonance with the light field as they are being decelerated. This problem was addressed by using either a chirped laser beam [6, 7] or by Zeeman shifting the atomic energy levels using a superimposed magnetic field (Zeeman slower) [8]. Both of these techniques are still used in modern experiments as a source of cold atoms.

The next important milestone in the preparation of ultra-cold atomic samples was the development of the magneto-optical trap which allowed to efficiently trap and simultaneously cool neutral atoms to sub-mK temperatures [9, 10]. A magneto-optical trap is based on the light-pressure forces with the addition of using a combination of circularly polarized light and a magnetic quadrupole field to create spatially dependent force trapping the atoms at the center of the quadrupole field. Based on these techniques, atoms as well as ions are nowadays routinely cooled to ultra-cold temperatures. This development has revolutionized atomic physics and triggered completely new fields of research, in particular the field of ultra-cold quantum gases.

One of the authors (M.W.) had first heard about the essential concepts of cooling and trapping of atoms as a 4th-year student in an undergraduate course given by Theodor Hänsch to a small audience at the University of München in 1988. These ideas were presented in such an inspiring way that M.W. decided to abandon his original plan to become a theorist, but to put the focus of his further studies on experimental physics, in particular atomic, molecular and optical physics including quantum optics. After the diploma thesis in molecular physics and an 1-year internship in Serge Haroche's group at Ecole Normale Supérieure in Paris doing experiments on creating a laser-cooled beam of Rydberg atoms, he joined Theodor Hänsch's group in 1992. At that time, most physicists considered laser cooling and trapping a rather esoteric subfield of AMO physics with little scientific impact. Together with fellow PhD students Tilman Esslinger, Leonardo Ricci, and Vladan Vuletic, and senior postdocs Andreas Hemmerich and Claus Zimmermann, a small team was formed to explore the world of ultra-cold atomic gases with the

[1] Wineland and Dehmelt independently proposed the same idea [2].

laboratories situated in Munich's Schellingstraße, part of München's culturally vibrant quarter Schwabing, while the major part of the Hänsch group resided in the far better equipped laboratoriess at the Max-Planck Institute for Quantum Optics in Garching, 15 km outbound of München. As it happened, Theodor Hänsch had also decided to place his private laboratory in Schellingstraße, and once in a while, when he had setup a new gadget there, he would allow the members of the team to take a sneak view into some exciting new idea. For M.W., the years of intense scientific exchange in München turned out to be utmost important for his future career, as the unique environment created by Theodor Hänsch allowed him to actively participate in a rapidly developing, exciting research field in its early infancy. The PhD thesis was devoted to cooling and trapping ultra-cold atoms in optical lattices [11–13], which, at that time, were still rather sparsely populated (only one out of 1000 lattice sites). As a final result, Bragg scattering was demonstrated with atoms in optical lattices to explore the structure of these exotic atomic samples [14–16].

2 Sympathetic Cooling of Ions with Laser-Cooled Atoms

Meanwhile, a large variety of atomic species and positively charged atomic ions have been cooled to microKelvin temperatures based on the original schemes by Hänsch/Schawlow and Wineland/Dehmelt. Anions, however, carrying a negative charge, have so far resisted all attempts to get cooled into the ultra-cold regime. Except a few examples [17–20], direct laser cooling falls short for most anions due to the lack of excited electronic bound states. This limitation might be overcome by sympathetically cooling anions using neutral atoms as a cooling agent. Cooling of anions with noble-atom buffer gases has been demonstrated in various laboratories, thus allowing one to reach temperatures in the low-Kelvin regime [21, 22]. In order to reach even lower temperatures, laser-cooled atomic gases have to be employed as a cooling agent. In the recent years, an increasing number of experiments have explored the combined trapping of cations with ultra-cold atoms to study atom–ion interactions in great detail [23–26].

Based on early predictions of Dehmelt [27], it had been common wisdom that sympathetic cooling would only lead to cooling in a radiofrequency quadrupole trap, if the mass of the cooling agent is lower than the mass of the ion. For larger mass ratios, elastic ion–atom collisions lead to an energy transfer from the ion's micromotion into the secular motion, thus effectively increasing the ion's energy even if the atom is at rest. Actually, recent experiments on hybrid ion–atom traps using Bose–Einstein condensates as buffer gas demonstrated this energy transfer process and found good agreement with theoretical prediction [28–30]. The mass criterion would exclude many atomic species, which can conveniently be laser cooled such as rubidium or cesium, as cooling agents for ions, in particular anions. However, we could theoretically show how to overcome the limitation of a critical mass ratio. By developing a general theoretical framework to describe the effect of collisions of ultra-cold atoms on trapped ions, it was found that ions can be efficiently cooled even beyond the critical mass ratio, if the buffer gas is spatially confined and higher-order radiofrequency traps are being used [31, 32]. In addition, by dynamically changing the trap parameters and/or the spatial extension of the buffer gas, the ions can further be cooled to even lower temperatures ("Forced Sympathetic Cooling").

Following these considerations, we are setting up an experiment to demonstrate sympathetic cooling of anions with laser-cooled atoms using a hybrid ion–atom trap, in particular for cooling the hydroxyl anion OH^- with optically cooled and trapped rubidium atoms. Efficient sympathetic cooling requires a dense cloud of laser-cooled atoms containing a large number of atoms. Based upon experience gained with a previous version of a hybrid ion–atom trap based on an octupole ion trap joined with a MOT [33], we are currently working on an improved setup. Here, we report on the realization and characterization of the ultra-cold atomic buffer gas for the hybrid ion–atom trap. Instead of a standard MOT, a dark spontaneous-force optical trap (Dark SPOT, [34]) is employed. Firstly, this variant of a MOT leads to an increase in the density by an order of magnitude. Secondly, the trapped atoms populate the electronic ground state, which suppresses inelastic ion–atom collisions involving atoms in the first excited electronic state. In our realization of the Dark SPOT, the atoms are loaded from a two-dimensional MOT [35, 36] which provides a cold and slow atomic beam, which is favorably matched to the reduced capture velocity imposed by the Dark SPOT. This configuration allows us to create a cloud of 1.5×10^9 atoms at a density in 1.1×10^{11} atoms/cm^3 range extending over 1.4 mm (1/e diameter) at a temperature of 250 μs, thus being ideally suited for being merged with ions trapped in a wire-based octupole ion trap for sympathetic cooling [33].

3 Two-Dimensional Magneto-Optical Trap

In our experiments we are using ^{85}Rb, which provides the closed transition $5^2S_{1/2}$ $(F = 3) \rightarrow 5^2P_{3/2}(F' = 4)$ for the application of efficient laser cooling (see inset of Fig. 3). Atoms might eventually be transferred into the lower hyperfine state $5^2S_{1/2}$ $(F = 2)$ by a spontaneous Raman process. This state cannot be addressed by the laser pumping the cooling transition as the splitting of the lower hyperfine levels is too large ($\approx$3 GHz). In order to pump the atoms back to the cooling transition, a second laser is used on the transition $5^2S_{1/2}(F = 2) \rightarrow 5^2P_{3/2}(F' = 3)$. Typical capture velocities for rubidium atoms entering a magneto-optical trap are on the order of 30–40 m/s [37] which is about an order magnitude smaller than typical velocities of thermal atoms at room temperature. Dark SPOTs exhibit even lower capture velocities as the capture volume is limited to the parts of the trap illuminated by the repumping beams (see below). This issue can be addressed by loading the trap from a slow and cold atom beam. Different techniques to create such kind of slow atom beams have been developed, e.g., the Zeeman slower discussed in the introduction. In our setup, another widely used technique, the two-dimensional magneto-optical trap (2D MOT), has been implemented. A 2D MOT is based on the same principles as a MOT with the atoms only being cooled in two dimensions creating a cold atom beam in the third direction.

Figure 1a shows a schematic of the 2D MOT used in our experiment to load the Dark SPOT. The same setup was previously characterized using a recoil-ion-momentum spectrometer [38]. The 2D MOT is build as a compact module which is connected to the main chamber via a CF16 flange with a tube-shaped differential pumping stage in between. All laser beams are transferred to the setup via optical fibers from a separate optical table. The fiber couplers and all the other optics are placed on a metal cage which surrounds an AR-coated glass cell containing two Rb dispensers. The setup consists of three separately adjustable cooling regions with a total length of 6.5 cm. It has been shown previously [39] that using multiple cooling regions instead of a single large cooling volume results in nearly the same flux as the beam transfer between different cooling regions is very efficient. The

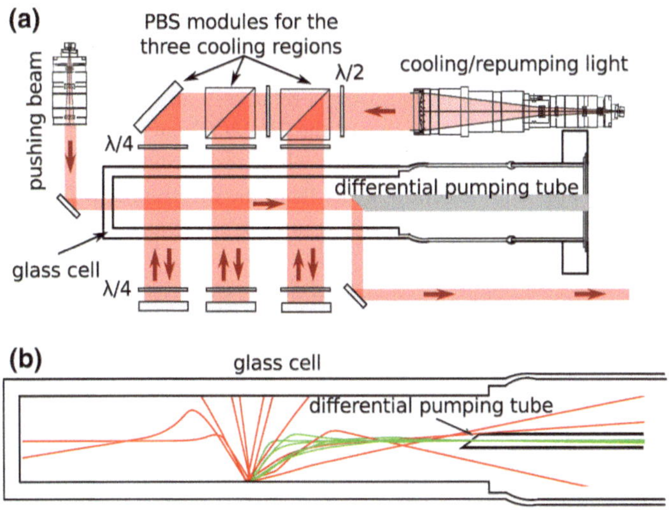

Fig. 1 2D MOT setup used in the experiment. **a** Cut through the center plane of the setup, illustrating the path of all used laser beams. The 2D MOT is a build as a single module which is connected to the main chamber via a CF16 flange with a tube-shaped differential pumping stage in between. All laser beams are transferred to the setup via optical fibers from a separate optical table. The fiber couplers and all the other optics are placed on a metal cage which surrounds an AR-coated glass cell which contains two Rb dispensers. The setup consists of three separately adjustable cooling regions with a total length of 6.5 cm. An additional pushing beam coincides with the center axis of the glass cell before it hits the polished surface of the differential pumping tube. **b** Exemplary trajectories in the glass cell of the 2D MOT. The shown trajectories were calculated by solving the atoms equation of motion in the light field of the 2D MOT setup. All atoms were chosen to start on the surface of glass cell, assuming that inter-atomic collisions can be neglected. The atom flux out of the 2D MOT can be estimated by comparing the number of trajectories passing the exit hole (*green*) to the number trajectories ending in another wall (*red*)

two-dimensional quadrupole field required for the 2D MOT is created by a total of 12 permanent magnets which are placed on two metal holders around the glass cell.

With this setup an atom beam of up to 4×10^9 atoms/s at a mean velocity of 14 m/s is created which compares favorably to most other 2D MOTs [35, 36, 40–43]. At first glance it might be surprising that the average velocity of the atom beam is far lower than the average thermal velocity of the rubidium vapor in the glass cell, even though the atoms are not actively decelerated in transversal direction. This can be explained by an indirect selection of slower atoms as indicated in Fig. 1b. Fast atoms (indicated in red) pass the light field of the 2D MOT very quickly and are therefore nearly unaffected by the light forces. Only slow atoms (indicated in green) interact long enough with the light field to be compressed into the beam and can therefore pass the small exit hole in the differential pumping tube toward the main science chamber.

Due to this velocity selection, the diameter of the exit hole has a direct influence on the mean velocity of the atom beam. The hole in our setup has an diameter of 800 μm resulting in an average velocity of 14 m/s. For larger exit holes, faster average velocities of up to 100 m/s [41] (for a 2 mm exit hole) have been observed.

Figure 2 shows the transversal velocity distribution of our atom beam measured by a time-of-flight technique [41]. The larger of the two distributions was created using an additional pusher beam which is aligned with the center axis of the glass cell (cf. Fig. 1). We find that by choosing a relatively large detuning for this beam (approximately 6Γ), the total atom flux of the 2D MOT can be increased by nearly a factor of two. We attribute this to the atoms which are confined to the center axis but initially travel in opposite direction, away from the differential pumping tube. The pusher beam can turn this second beam around as illustrated in the inset of Fig. 2

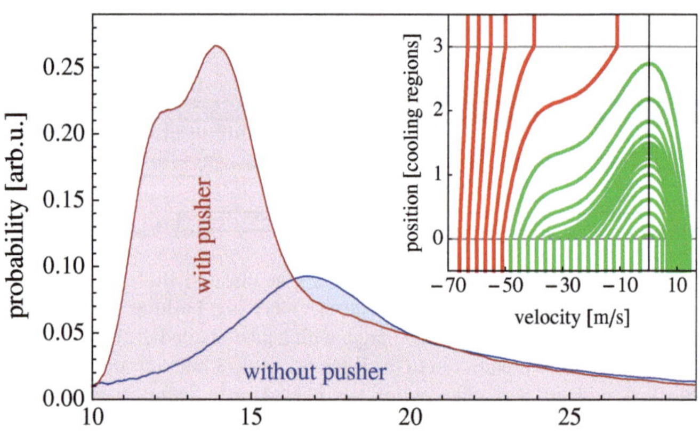

Fig. 2 Velocity distribution of the atom beam. The two distributions correspond to an operation of the 2D MOT with and without the pushing beam. The *inset* shows calculated trajectories in the light field of the pushing beam for atoms initially moving away from the differential pumping tube. The *colors* indicate if the atoms can be turned around by the pushing beam (*green*) or simply pass the cooling region (*red*)

showing the atoms trajectories in phase space. The green trajectories correspond to the velocity components which are turned around by the pusher beam over the distance of the three cooling regions, assuming that the atoms only interact with the pusher beam as long as illuminated with repumping light. These trajectories accumulate around a final velocity on the order of 10 m/s which agrees reasonably with the position of the additional peak observed in the measured time-of-flight distribution.

4 Dark Spontaneous-Force Optical Trap

The cold atom beam from the 2D MOT is used to efficiently load a Dark SPOT [34] which is a slightly modified version of a magneto-optical trap (MOT) enabling increased atom densities and a large population of atoms in the ground state. In order to understand the working principle of a Dark SPOT, one has to understand the three main loss processes limiting the density in a magneto-optical trap. First, multiple photon scattering disturbs the trapping forces, as photons are scattered in random directions with polarizations not consistent with the circular polarization required for a MOT. These secondary photons lead to a repulsive force if re-absorbed by other atoms [44]. Second, inelastic collisions of an atom with an excited atom lead to losses from the trap, as the trap depth of a MOT is typically smaller than kinetic energy gained from the inelastic collision [45]. Third, elastic collisions with room-temperature background gas also lead to a loss of atoms from the trap.

The last loss mechanism can simply be avoided by a sufficiently low background pressure. In our vacuum chamber, the background pressure is on the order of 10^{-9} mbar which results in an atom lifetime of approximately 3–4 s which is considerably longer than the time required to load the trap which is on the order of 300 ms. We are therefore not limited by the background pressure. The other two loss mechanisms, however, directly depend on the atom density in the trap. This limits the maximum achievable densities in a MOT to a few 10^{10} atoms/cm^3 [45–47].

The main improvement of the Dark SPOT is to reduce both of these loss mechanisms by increasing the population of atoms in the dark state and thus removing them from the cooling cycle. Less atoms in the cooling cycle result in fewer scattered photons and less excited atoms which are the main factors limiting the density. The problem of this scheme is that once the atoms fall into the dark state they are no longer actively cooled which results in a reduced capture velocity and thus a reduced loading rate of the trap. This is why many Dark SPOTs are operated in two phases [45]. First, in the initial trapping phase the trap is operated with full laser powers to achieve efficient loading of a large number of atoms. In the second phase, the trap is switched to Dark SPOT operation leading to a fast increase in the atom density on a typical time scale of a few tens of milliseconds. This is particularly useful if the Dark SPOT is used only as a first cooling step and is for example used to load a dipole trap. In this case, a momentary increase in the phase space density is all that is required.

For experiments such as ours, however, where the atoms serve as a target for other particles, a constant high-density operation is desired. In this case, a spatial separation of low and high dark state fraction can be realized [34]. In our experiment, this is achieved by using a repumping beam with a hollow core. With this optical setup, most atoms within the "dark region" fall into the dark state and they are only pumped back to the cooling transition once they reach the outer parts of the trap where repumping light is present.

Figure 3 shows a schematic of the Dark SPOT setup. The quadrupole field for the MOT is created by a pair of anti-Helmholtz coils placed inside the vacuum chamber creating a maximum magnetic field gradient of 60 G/cm at 100 A. The required laser beams are transferred to the experiment from a separate optical table via polarization-maintaining single-mode optical fibers. All relevant laser frequencies used in the setup are indicated in the level scheme in Fig. 3.

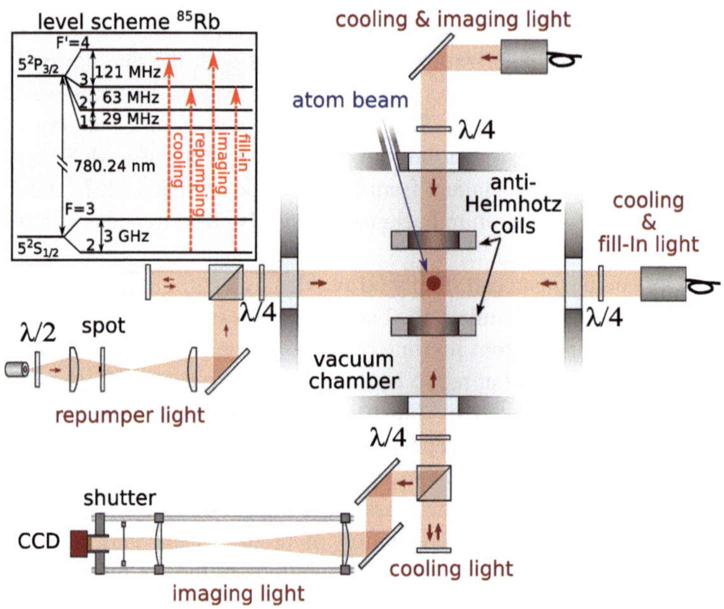

Fig. 3 Optical setup of the Dark SPOT setup and the imaging system. The light components contained in all beams are labeled in *red* with the corresponding transitions being indicated in the *inset*. A repumping beam with a hollow center is overlapped with the cooling light in the horizontal plane using a polarizing beam splitter (PBS). To suppress the light at the beam center, the repumping beam passes a glass plate with a circular spot which is imaged to the center of the MOT using two lenses. In the vertical plane, the cooling light is mixed with imaging light of orthogonal polarization which is separated below the vacuum chamber using another PBS. The imaging light falls onto a CCD camera which captures an absorption image of the atom cloud

Three large fiber couplers provide beams with a diameter of 24 mm ($1/e^2$) for the MOT which are retro-reflected after passing the vacuum chamber. Every beam contains light on the cooling transition and one additional wavelength which is coupled into the same fiber with crossed polarization. The vertical beam contains light on the imaging transition which is used to probe the atom cloud. Both horizontal beams include light on the "fill-in" transition (see inset in Fig. 3) which is used to pump all atoms into $5^2S_{1/2}(F = 3)$ prior to the imaging sequence (see Sect. 5).

The repumping light required for the trap is provided by two additional fibers. These beams pass a glass plate with a circular spot creating a beam with a hollow core. In order to reduce the amount of light being diffracted back into this dark region, the spot is imaged to the center of the trap using two lenses (cf. Fig. 3) resulting in an extinction ratio better than 1:200.[2] The hollow repumping beams are overlapped with the horizontal trapping beams using a PBS which is placed between the chamber and the mirror retro-reflecting the cooling light.

Using the atomic beam provided by the 2D MOT, the atoms can be efficiently transferred into the Dark Spot with a loading rate corresponding to 75% of the total measured flux from the 2D MOT. By removing the spot blocking, the repumping light at the trap center and thus operating the trap as a normal MOT the loading rate even reaches about 95% of the total flux. This illustrates that the capture velocity of the Dark SPOT is indeed lower than the one of the MOT but that the better part of the atoms from our 2D MOT exhibits even lower velocities.

The resulting peak atom densities are shown in Fig. 4 as a function of the bright state fraction. There is a total of 1.1×10^9 atoms in the trap at a temperature of 250 μ K, both of which only exhibit a small dependence on the bright state fraction. As expected, the density is lower for a large fraction of atoms in the bright state leading to the aforementioned density-limiting mechanisms (multiple-scattering regime). We find the optimal bright state fraction to be on the order of 5–10% which is in good agreement with previous results [45, 48]. The decrease in the density at small bright state fractions p occurs as the system enters the so-called temperature limited regime. In this regime, the atoms can be seen as independent particles and the peak density is proportional to $N(\kappa p/T)^{3/2}$ with N being the total number of atoms inside the trap, T being the atoms temperature and κ being the spring constant of the radiation force in the MOT [48]. As κ is proportional to the magnetic field gradient, larger peak densities can be reached by increasing the magnetic field [47]. This dependence was confirmed as the measured peak density could be increased twofold by doubling the magnetic field gradient from 15 to 30 G/cm.

[2]This extinction ratio was measured outside the vacuum chamber. The actual ratio might be reduced due to reflections from the wires of the ion trap.

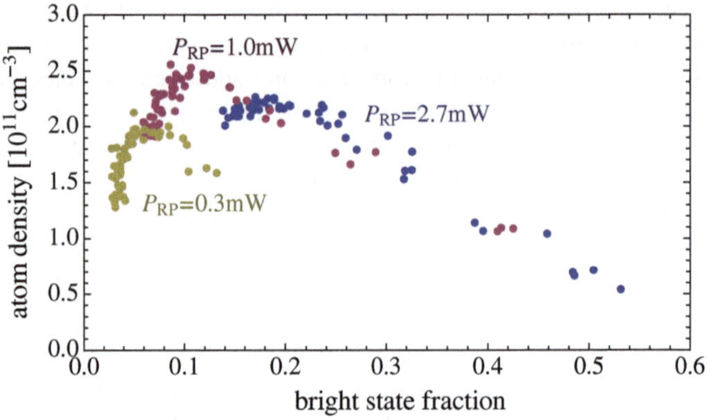

Fig. 4 Peak atom density as a function of the bright state fraction. The bright state fraction was varied by changing the detuning of the cooling beam and the power of the repumping beam (indicated by the *different colors*). The maximum peak density was achieved for a total repumping power (both beams) of 1 mW and a detuning of the cooling beam of 3Γ. Details on the determination of the atom density can be found in Sect. 5

5 Characterizing the Atomic Cloud at High Optical Density

In order to investigate the shape and density of laser-cooled atomic clouds, there are two main techniques, fluorescence and absorption imaging [49]. In the first approach, one detects the light scattered by an atom cloud illuminated by a resonant laser beam, the latter one detects the shadow in the probe beam caused by its absorption in the atomic cloud. With both techniques, a two-dimensional projection of the atom density profile can be calculated. The analysis is based on the precise knowledge of the photon scattering rate and angular scattering distribution of all atoms in the cloud.

In order to prepare the system in a well-known state, different steps have to be taken [49]. Ideally all atoms are pumped into the same m_F level and are probed with a well-known polarization. This requires a well-defined quantization axis given by the direction of the magnetic field. Otherwise, it is not possible to define meaningful m_F levels or the polarization states of the light fields ($\sigma^{\pm}$ or π). Therefore, at the beginning of every imaging sequence all trapping lasers and the magnetic quadrupole field are turned off. Afterward, a homogeneous magnetic field aligned with the propagation direction of the probe laser is turned on. Consequently, the probe beam has a well-defined σ^+ polarization. At this point, the atoms are still distributed over all m_F levels of $5^2S_{1/2}(F = 3)$ as well as $5^2S_{1/2}(F = 2)$ state. In order to optically pump all atoms into the same magnetic sub-level, the probe laser and the so-called "fill-in" beam (see Fig. 3) are turned on. The duration of this pumping pulse should be chosen short enough that the atom cloud is not strongly influenced by the light forces, but long enough to ensure all atoms to reach the outermost m_F level $|F = 3, m_F = 3\rangle$, also known as the stretched state.

After this state preparation, the system is in a well-defined state and can now be used to perform density measurements with either fluorescence or absorption imaging. However, this precise state preparation fails for clouds, which are optically thick, i.e., the optical absorption length is smaller than the size of the cloud. As a result, atoms located on the side of the cloud facing the laser scatter far more photons than the ones located on the far side. For the pumping pulse, this poses a serious problem as it will take orders of magnitude longer times to pump the far atoms into the stretched state than the ones in front. Consequently, the above-mentioned condition that the pumping pulse should be short enough for the cloud not to be influenced by the light forces of the probe beam can no longer be fulfilled. In the following, we will discuss how this problem can be overcome by saturated absorption imaging.

For an optically thick cloud, the atoms will always be distributed over multiple m_F levels with p_{m_F} denoting the population of the different levels. A resonant probe beam with an initial intensity profile $I_i(x,y)$ passing through such a cloud is attenuated as

$$\frac{\mathrm{d}I(x,y,z)}{\mathrm{d}z} = -n(x,y,z)\left(\frac{\Gamma}{2}\right)\sum_{m_F} p_{m_F}\frac{\hbar\omega_\gamma\, I(x,y,z)/I_{\mathrm{sat}}^{m_F}}{1+I(x,y,z)/I_{\mathrm{sat}}^{m_F}}, \tag{1}$$

with $n(x, y, z)$ being the density profile of the cloud, Γ and ω_γ being the natural line width and the optical frequency of the transition, $\hbar$ being the Planck constant and $I_{\mathrm{sat}}^{m_F}$ being saturation intensities of the different m_F levels. Calculating the final intensity profile of the probe beam therefore requires exact knowledge of the sub-level population. This can be circumvented by assuming that the sum over all magnetic sub-levels can be approximated by an average absorption term

$$\sum_{m_F} p_{m_F}\frac{I/I_{\mathrm{sat}}^{m_F}}{1+I/I_{\mathrm{sat}}^{m_F}} \approx \frac{I/(\alpha I_{\mathrm{sat},0})}{1+I/(\alpha I_{\mathrm{sat},0})}, \tag{2}$$

with $I_{\mathrm{sat},0}$ being the saturation intensities of the stretched state and α being a dimensionless parameter. We have verified numerically using arbitrary sub-level population that using the effective absorption intensity $\alpha I_{\mathrm{sat},0}$ typically agrees within a 10%-margin with the result obtained by using the full sum over all states. Plugging Eq. (2) into Eq. (1) and integrating over the propagation direction of the laser (z) results in [50, 51]

$$\left(\frac{\Gamma}{2}\right)\frac{\hbar\omega_\gamma}{I_{\mathrm{sat},0}}\int_{-\infty}^{\infty} n(x,y,z)\mathrm{d}z = \alpha \ln\left[\frac{I_f(x,y)}{I_i(x,y)}\right] + \frac{I_i(x,y)-I_f(x,y)}{I_{\mathrm{sat},0}}, \tag{3}$$

with $I_f(x,y)$ being the final intensity profile of the probe beam. The expression on the left side is also known as the optical density $\mathrm{od}(x,y)$ of the cloud with the integral expressing the column density of the atom distribution.

The two terms on the right fundamentally differ by the fact that only the second term depends on the absolute probe beam intensity. For standard absorption imaging, a probe intensity $I_i(x, y) \ll I_{sat}$ is used in which case the second term can be neglected and the optical density becomes independent of the absolute probe intensity. With increasing probe intensity, however, the second term becomes relevant until for $I \gg I_{sat}$ the first term can be neglected and the optical density becomes independent of the parameter α. Consequently, without any knowledge about the sub-level population of the atom cloud and thus α, the actual optical density and thus the atom density can be measured by using sufficiently large probe intensities.

The idea behind saturation absorption imaging [50, 51] is to measure the absorption of the probe beam as a function of the probe beam intensity. Figure 5a shows the intensity dependence of the measured optical density $\overline{od}$, which is

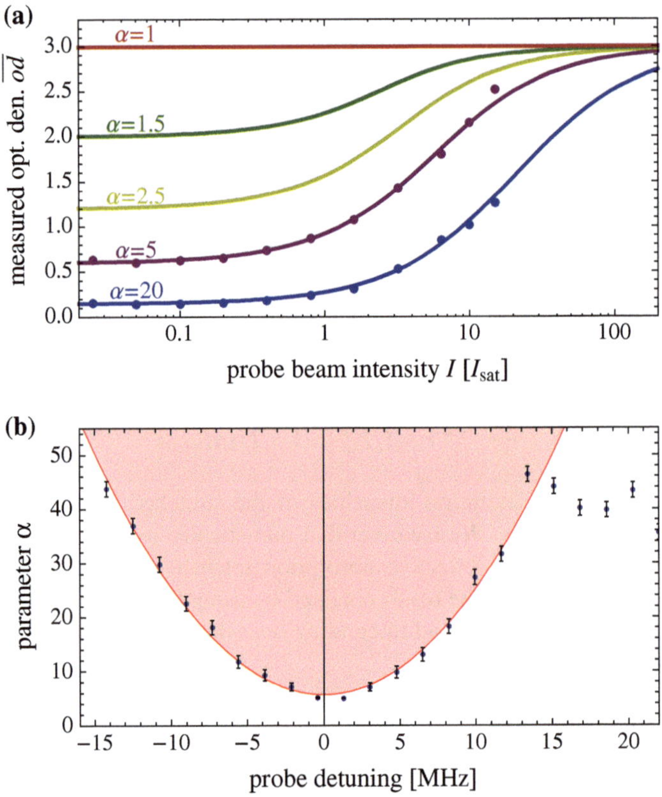

Fig. 5 Determination of the parameter α. **a** Dependence of the measured optical density $\overline{od}$ as a function of the probe beam intensity for different values of α. Also shown are two data sets which were performed for $\alpha \approx 5$ and 20 (probe detuning of 3 and 9 MHz). **b** Measured values of α for different probe beam detunings. *The solid curve* corresponds to a Lorentzian-shaped absorption cross section $\sigma(\delta)$ leading to a dependence of $\alpha \propto 1/\sigma(\delta)$

calculated assuming $\alpha = 1$. This would correspond to the true optical density *od* if all atoms were prepared in the stretched state. If the atoms are distributed over multiple m_F-levels ($\alpha > 1$), $\overline{od}$ underestimates the true optical density *od* at small probe intensities and only approaches the real value at large intensities (cf. Fig. 5a).

Figure 5a shows the intensity dependence of the measured optical density for different values of α including two measurements which were performed for different probe laser detunings. Measuring the intensity dependence of the optical density of an atom cloud therefore allows to estimate α without any prior knowledge of the specific sub-level populations or the probe laser detuning. Artificially increasing α, e.g., through detuned absorption imaging, the measured optical density $\overline{od}$ is reduced and atomic clouds with large optical densities can be quantitatively probed.

Figure 5b shows measured values of α as a function of the probe beam detuning. The solid curve corresponds to the expected distribution assuming a Lorentzian line-shape of the absorption cross section. At large detunings, there is a significant deviation of the measured points from the expected curve as our fitting algorithm appears to fail because the signal-to-noise ratio becomes too small. It also has to be taken into account that the detuning leads to a phase shift of the probe laser caused by the complex refractive index of the atom cloud. For a non-resonant probe beam, the atom cloud acts like a lens causing changes in the cloud shape on the camera. This effect can be minimized by choosing sufficiently large detunings, as the induced phase shift peaks at $\delta = \Gamma/2$ falling off for larger detunings. Additionally, the distortion due to the phase shift is minimized by using imaging systems with a large depth of focus. Especially for very small clouds such as atoms inside a dipole trap this might not be feasible in which case the only solution is to measure the phase of the probe beam, for example using phase-contrast imaging [52].

The atom density in Fig. 4 was determined using detuned saturated absorption imaging. The highest densities in this measurement would correspond to a peak optical density of nearly 120 if all atoms were in the stretched state and the probe beam was on resonance ($\alpha = 1$). Obviously, this condition is not fulfilled in an actual cloud of atoms (see discussion above).

6 Conclusion

We have presented the setup for a Dark SPOT which is efficiently loaded from a 2D MOT. Using a far-detuned pushing beam and a narrow differential pumping tube, a cold atom beam with a mean velocity well below the capture velocity of the Dark SPOT could be realized. The main challenge in characterizing the atom cloud is the very high optical density of the cloud which is on the order of 100. At this density, most standard imaging techniques fail as the cloud is optically thick and can no longer be probed by a resonant laser beam. We therefore implemented saturation absorption imaging scheme with the probe laser being detuned from the resonance frequency. Using this technique, we found the total number of atoms in the trap to be as large as 1.5×10^9 at a peak density exceeding 10^{11} atoms/cm^3.

For future experiments in the hybrid atom–ion trap, this precise determination of the atoms density distribution will be of great importance. First, it allows one to perform absolute measurements of the atoms collision cross section with the simultaneously trapped ions. Second, the efficiency of sympathetic ion cooling inside a radiofrequency trap strongly depends on the size, density, and relative position of the atom cloud [31, 32]. A large fraction of the atoms populates the absolute electronic ground state which is favorable to suppress inelastic collisions with the ions. With the setup presented here, our theoretical model predicts final ion temperatures comparable to the lowest temperatures achievable in cryogenic ion traps and even lower. This might open the door to a completely new regime in the experimental investigation of anions and their chemistry at ultra-low temperatures.

Acknowledgements This work is supported in part by the Heidelberg Center for Quantum Dynamics and the BMBF under contract number 05P12VHFA6. B.H. acknowledges support by HGSHire.

References

1. T.W. Hänsch, A.L. Schawlow, Cooling of gases by laser radiation. Opt. Commun. **13**, 68–69 (1975)
2. D. Wineland, H. Dehmelt, Proposed $10^{14}\delta\nu < \nu$ laser fluorescence spectroscopy on Tl^+ mono-ion oscillator III, Bull. Am. Phys. Soc. **20**, 637 (1975)
3. D.J. Wineland, R.E. Drullinger, F.L. Walls, Radiation-pressure cooling of bound resonant absorbers. Phys. Rev. Lett. **40**, 1639–1642 (1978)
4. W. Neuhauser, M. Hohenstatt, P. Toschek, H. Dehmelt, Optical-sideband cooling of visible atom cloud confined in parabolic well. Phys. Rev. Lett. **41**, 233–236 (1978)
5. A. Ashkin, Trapping of atoms by resonance radiation pressure. Phys. Rev. Lett. **40**, 729–732 (1978)
6. W.D. Phillips, J.V. Prodan, in *Cooling atoms with a frequency chirped laser*, ed By V.L. Mandel, E. Wolf, Coherence and Quantum Optics V (Plenum, New York, 1984), p. 15
7. W. Ertmer, R. Blatt, J.L. Hall, M. Zhu, Laser manipulation of atomic beam velocities: demonstration of stopped atoms and velocity reversal. Phys. Rev. Lett. **54**, 996–999 (1985)
8. W.D. Phillips, H. Metcalf, Laser deceleration of an atomic beam. Phys. Rev. Lett. **48**, 596–599 (1982)
9. E.L. Raab, M. Prentiss, A. Cable, S. Chu, D.E. Pritchard, Trapping of neutral sodium atoms with radiation pressure. Phys. Rev. Lett. **59**, 2631–2634 (1987)
10. C. Monroe, W. Swann, H. Robinson, C. Wieman, Very cold trapped atoms in a vapor cell. Phys. Rev. Lett. **65**, 1571–1574 (1990)
11. M. Weidemüller, T. Esslinger, M.A. Ol'shanii, A. Hemmerich, T.W. Hänsch, A novel scheme for efficient cooling below the photon recoil limit. Europhys. Lett. **27**(2), 109–114 (1994)
12. A. Hemmerich, M. Weidemüller, T. Hänsch, Four-wave mixing in a 3d optical lattice. Europhys. Lett. **27**, 427–432 (1994)
13. A. Hemmerich, M. Weidemüller, T. Esslinger, C. Zimmermann, T.W. Hänsch, Trapping atoms in a dark optical lattice. Phys. Rev. Lett. **75**, 37–40 (1995)

14. M. Weidemüller, A. Hemmerich, A. Görlitz, T. Esslinger, T.W. Hänsch, Bragg diffraction in an atomic lattice bound by light. Phys. Rev. Lett. **75**, 4583–4586 (1995)
15. A. Görlitz, M. Weidemüller, T.W. Hänsch, A. Hemmerich, Observing the position spread of atomic wave packets. Phys. Rev. Lett. **78**, 2096–2099 (1997)
16. M. Weidemüller, A. Görlitz, T. Hänsch, A. Hemmerich, Local and global properties of light-bound atomic lattices investigated by bragg diffraction. Phys. Rev. A **58**, 4647–4661 (1998)
17. R.C. Bilodeau, H.K. Haugen, Experimental studies of Os^-: observation of a bound-bound electric dipole transition in an atomic negative ion. Phys. Rev. Lett. **85**, 534 (2000)
18. A. Kellerbauer, A. Fischer, U. Warring, Measurement of the zeeman effect in an atomic anion: prospects for laser cooling of Os^-. Phys. Rev. A **89**, 43430 (2014)
19. C.W. Walter, N.D. Gibson, D.J. Matyas, C. Crocker, K.A. Dungan, B.R. Matola, J. Rohlén, Candidate for laser cooling of a negative ion: observations of bound-bound transitions in La^-. Phys. Rev. Lett. **113**, 63001 (2014)
20. A. Kellerbauer, G. Cerchiari, E. Jordan, C.W. Walter, High-resolution laser spectroscopy on bound transitions in La^-. Phys. Scr. **90**, 54014 (2015)
21. D. Gerlich, Ion-neutral collisions in a 22-pole trap at very low energies. Phys. Scr. **1995**, 256 (1995)
22. R. Wester, Radiofrequency multipole traps: tools for spectroscopy and dynamics of cold molecular ions. J. Phys. B At. Mol. Opt. Phys. **154001** (2009)
23. O.P. Makarov, R. Côté, H. Michels, W.W. Smith, Radiative charge-transfer lifetime of the excited state of $NaCa^+$. Phys. Rev. A **67**, 42705 (2003)
24. W.W. Smith, O.P. Makarov, J. Lin, Cold ion-neutral collisions in a hybrid trap. J. Mod. Opt. **52**, 2253–2260 (2005)
25. E.R. Hudson, Method for producing ultracold molecular ions. Phys. Rev. A **79**, 32716 (2009)
26. S. Willitsch, Ion–atom hybrid systems. ArXiv:1401.1699 (2014)
27. H.G. Dehmelt, Radiofrequency spectroscopy of stored ions. Adv. At. Mol. Phys. **3**, 53–72 (1967)
28. C. Zipkes, L. Ratschbacher, C. Sias, M. Köhl, Kinetics of a single trapped ion in an ultracold buffer gas. New J. Phys. **13**, 53020 (2011)
29. M. Cetina, A.T. Grier, V. Vuletic, Micromotion-induced limit to atom–ion sympathetic cooling in paul traps. Phys. Rev. Lett. **109**, 253201 (2012)
30. K. Chen, S.T. Sullivan, E.R. Hudson, Neutral gas sympathetic cooling of an ion in a paul trap. Phys. Rev. Lett. **112**, 143009 (2014)
31. B. Höltkemeier, P. Weckesser, H. López-Carrera, M. Weidemüller, Buffer-gas cooling of a single ion in a multipole radio frequency trap beyond the critical mass ratio. Phys. Rev. Lett. **116**, 233003 (2016)
32. B. Höltkemeier, P. Weckesser, H. López-Carrera, M. Weidemüller, Dynamics of a single trapped ion immersed in a buffer gas. Phys. Rev. A **94**, 062703 (2016)
33. J. Deiglmayr, A. Göritz, T. Best, M. Weidemüller, R. Wester, Reactive collisions of trapped anions with ultracold atoms. Phys. Rev. A **86**, 43438 (2012)
34. W. Ketterle, K.B. Davis, M.A. Joffe, A. Martin, D.E. Pritchard, High densities of cold atoms in a dark spontaneous-force optical trap. Phys. Rev. Lett. **70**, 2253–2256 (1993)
35. S. Weyers, E. Aucouturier, C. Valentin, N. Dimarcq, A continuous beam of cold cesium atoms extracted from a two-dimensional magneto-optical trap. Opt. Commun. **143**, 30–34 (1997)
36. K. Dieckmann, R.J.C. Spreeuw, M. Weidemüller, J.T.M. Walraven, Two-dimensional magneto-optical trap as a source of slow atoms. Phys. Rev. A **58**, 3891–3895 (1998)
37. K.M.F. Magalhaes, S.R. Muniz, G.D. Telles, P.W. Courteille, V.S. Bagnato, L.G. Marcassa, The escape velocity in a magneto-optical trap and its importance to trap loss investigation. Laser Phys. **12**, 145–151 (2002)
38. S. Götz, B. Höltkemeier, C.S. Hofmann, D. Litsch, B.D. DePaola, M. Weidemüller, Versatile cold atom target apparatus. Rev. Sci. Instrum. **83**, 043107 (2012)
39. J. Ramirez-Serrano, N. Yu, J.M. Kohel, J.R. Kellogg, L. Maleki, Multistage two-dimensional magneto-optical trap as a compact cold atom beam source. Opt. Lett. **31**, 682–684 (2006)

40. P. Berthoud, A. Joyet, G. Dudle, N. Sagna, P. Thomann, A continuous beam of slow, cold cesium atoms magnetically extracted from a 2d magneto-optical trap. Europhys. Lett. **41**, 141 (1998)
41. J. Schoser, A. Batär, R. Löw, V. Schweikhard, A. Grabowski, Y.B. Ovchinnikov, T. Pfau, Intense source of cold Rb atoms from a pure two-dimensional magneto-optical trap. Phys. Rev. A **66**, 23410 (2002)
42. J. Catani, P. Maioli, L. De Sarlo, F. Minardi, M. Inguscio, Intense slow beams of bosonic potassium isotopes. Phys. Rev. A **73**, 33415 (2006)
43. T.G. Tiecke, S.D. Gensemer, A. Ludewig, J.T.M. Walraven, High-flux two-dimensional magneto-optical-trap source for cold lithium atoms. Phys. Rev. A **80**, 13409 (2009)
44. T. Walker, D. Sesko, C. Wieman, Collective behavior of optically trapped neutral atoms. Phys. Rev. Lett. **64**, 408–411 (1990)
45. C.G. Townsend, N.H. Edwards, C.J. Cooper, K.P. Zetie, C.J. Foot, A.M. Steane, P. Szriftgiser, H. Perrin, J. Dalibard, Phase-space density in the magneto-optical trap. Phys. Rev. A **52**, 1423–1440 (1995)
46. D.W. Sesko, T.G. Walker, C.E. Wieman, Behavior of neutral atoms in a spontaneous force trap. J. Opt. Soc. Am. B **8**, 946 (1991)
47. W. Petrich, M.H. Anderson, J.R. Ensher, E.A. Cornell, Behavior of atoms in a compressed magneto-optical trap. J. Opt. Soc. Am. B **11**, 1332 (1994)
48. C. Townsend, N. Edwards, K. Zetie, C. Cooper, J. Rink, C. Foot, High-density trapping of cesium atoms in a dark magneto-optical trap. Phys. Rev. A **53**, 1702–1714 (1996)
49. W. Ketterle, D.S.S. Durfee, D.M.M. Stamper-Kurn, Making, probing and understanding Bose–Einstein condensates. ArXiv:cond-mat/9904034(1999)
50. G. Reinaudi, T. Lahaye, Z. Wang, D. Guéry-Odelin, Strong saturation absorption imaging of dense clouds of ultracold atoms. Opt. Lett. **32**, 3143–3145 (2007)
51. W. Kwon, J. Choi, Y. Shin, Calibration of saturation absorption imaging of ultracold atom clouds. J. Korean Phys. Soc. **61**, 1970–1974 (2012)
52. L.D. Turner, K.P. Weber, D. Paganin, R.E. Scholten, Off-resonant defocus-contrast imaging of cold atoms. Opt. Lett. **29**, 232–234 (2004)

A Compact Echelle Spectrograph for Characterization of Astro-Combs

Rafael A. Probst, Tilo Steinmetz, Yuanjie Wu, Frank Grupp, Thomas Udem, and Ronald Holzwarth

Abstract We present an echelle spectrograph that is optimized for characterization of frequency combs for astronomical applications (astro-combs). In spite of its very compact and cost-efficient design, it allows viewing the spectrum of a frequency comb in nearly the same way as a full-sized high-resolution echelle spectrograph as used at astronomical observatories. This is of great value for testing and characterizing astro-combs during their assembly phase. The spectrograph can further be utilized to effectfully demonstrate the remarkable capabilities of astro-combs.

1 Introduction

At the beginning of this millennium, the advent of laser frequency combs (LFCs) [1] has paved the way to pushing precision spectroscopy to a new extreme. Through this invention, the uncertainty of 1S–2S transition measurements in atomic hydrogen was brought down to the 10^{-15} level, for a precision test of quantum

This article is part of the topical collection "Enlightening the World with the Laser" - Honoring T. W. Hänsch guest edited by Tilman Esslinger, Nathalie Picqué, and Thomas Udem.

Electronic supplementary material The online version of this chapter (doi:10.1007/978-3-319-64346-5_25) contains supplementary material, which is available to authorized users.

R.A. Probst (✉) • T. Udem
Max-Planck-Institute for Quantenoptik (MPQ), Hans-Kopfermann-Str. 1, 85748 Garching, Germany
e-mail: rprobst@mpq.mpg.de

T. Steinmetz
Menlo Systems GmbH, Am Klopferspitz 19a, 82152 Martinsried, Germany

Y. Wu • R. Holzwarth
Max-Planck-Institute for Quantenoptik (MPQ), Hans-Kopfermann-Str. 1, 85748 Garching, Germany

Menlo Systems GmbH, Am Klopferspitz 19a, 82152 Martinsried, Germany

F. Grupp
Universitäts-Sternwarte München, Scheinerstr. 1, 81679 München, Germany

D. Meschede et al. (eds.), *Exploring the World with the Laser*,
https://doi.org/10.1007/978-3-319-64346-5_25

electrodynamics [2]. It was largely for this achievement that the pioneers of laser spectroscopy T. W. Hänsch and J. L. Hall were awarded the Nobel Prize of Physics in 2005.

As the LFC quickly found manifold applications, it was recognized in 2007 that also astronomy would greatly profit from a vast leap in precision enabled by the LFC [3]. An LFC provides a multitude of sharp, evenly spaced spectral lines (modes), whose optical frequencies can be phase-coherently linked to an atomic clock. The frequency of the nth mode is $f_n = n \times f_r + f_0$, with the mode spacing f_r and the offset frequency f_0. For astronomy, this provides a nearly ideal calibrator for high-resolution echelle spectrographs. Conventional calibration sources, such as thorium-argon arc lamps, presently limit the precision of astronomical spectrographs through their irregular line pattern and line drifts with aging of the lamp.

LFCs optimized for astronomical applications (astro-combs) have thus been developed [4–6], which promise discoveries of Earth-like exoplanets through radial-velocity measurements and a more precise constraint on a potential cosmic variability of fundamental constants [7]. A direct measurement of the acceleration of the cosmic expansion also appears within reach [8]. Several important spectrographs at major observatories, such as HARPS [9, 10], HARPS-N [4], VTT [11] and FOCES [12, 13], have recently been equipped with astro-combs, and others, such as ESPRESSO [14], are about to follow.

For spectrograph calibration tests, several astro-combs have temporarily been deployed at astronomical observatories. Such campaigns represent a major effort, as astro-combs are complicated and sensitive systems to be transported to remote places, and measurement time at observatories is in high demand. Some aspects of astro-combs can easily and accurately be characterized without the need for an echelle spectrograph: the mode spacing f_r is exactly the repetition rate of the emitted laser pulses detectable with a photodiode. The spectral coverage can be measured with a small low-resolution spectrometer, which is already integrated in many astro-combs [15]. For the characterization of a few other aspects, however, there is hardly an alternative to the use of an echelle spectrograph: studying line-by-line intensity fluctuations or a possible continuum background between the lines requires resolving individual lines over the full spectral width. Yet, the mode spacing is adapted to a high-resolution echelle spectrograph, and the selection of lab instruments with a similar resolution and bandwidth is very limited. Investigations of these properties are however highly relevant for applications [10].

It would thus be convenient to have a spectrograph directly in place, where astro-combs are being built, to characterize them before shipping them to their destinations. The state-of-the-art astronomical spectrographs are, however, elaborate machines of respectable cost and size. Yet, as shown in this article, a spectrograph explicitly built for astro-comb characterization can be compact and inexpensive.

Theodor W. Hänsch, to whom this journal issue is dedicated on the occasion of his 75th birthday, has always sought to find simple yet effective solutions. This spirit has led him to inventing phase-stabilized LFCs. Sometimes, his experiments playfully involve everyday items, such as jelly as an active laser material [16] or a toy railroad for motion control of optical elements [17], while at the same time yielding important scientific results. Following this school of thought, we opt for elegantly simple solutions for our spectrograph. We show how off-the-shelf laboratory components and photographic equipment can be used to create a valuable measurement tool.

2 Spectrograph Design

2.1 A note on Spectrograph Size

What allows us to drastically shrink down our spectrograph is the possibility of using a single-mode optical fiber as a spectrograph input. Being a fiber-laser system, the output of the astro-comb is intrinsically single mode. This is different for stars, which are blurred into a non-diffraction-limited disk by atmospheric seeing. For applications other than solar astronomy, where optical power is abundant, this light is thus commonly fed into multimode fibers to preserve efficient coupling to the spectrograph. The fiber exit is the spectrograph entrance slit, whose multimode nature precludes the spectrograph from reaching a diffraction-limited resolution. A typical astronomical spectrograph hence needs to be much bigger than ours, to illuminate a larger number of grating grooves, to ultimately reach the same resolution.

Since bigger telescopes capture more seeing elements, and thus greater numbers of spatial modes, the size of the spectrograph scales with that of the telescope. The resulting trend towards bigger spectrographs may soon be stopped by continuing progress in adaptive optics [18]. Several single-mode spectrographs are currently under development for red to near-infrared wavelengths [19–21], where "extreme" adaptive optics systems are starting to allow single-mode fiber injection with reasonable efficiency [22].

As an input fiber for our spectrograph, we use a single-mode fiber (Thorlabs S405-XP) designed for a wavelength range of 400–680 nm. The fiber core has a mode-field diameter of 4 μm and a numerical aperture (NA) of 0.12. For comparison, HARPS uses a 70 μm fiber core, which is collimated into a beam of 21 cm in diameter. Assuming a similar fiber NA and echelle grating, our ∼20 times smaller fiber core means that we can reach a similar resolution as HARPS with a beam of just 1 cm. We can thus conveniently build our spectrograph from standard 1 inch optics.

2.2 Optical Setup

We construct our spectrograph following the optical layout shown in Fig. 1. The fiber output within the spectrograph is polished into a ceramic ferrule (without connector), which is held in place by a small clamp mount, and connected to the fiber jacket with a heat shrink tube. The fiber output is collimated into a Gaussian beam of 9.5 mm full-width at half-maximum (FWHM). The collimator is a 90°

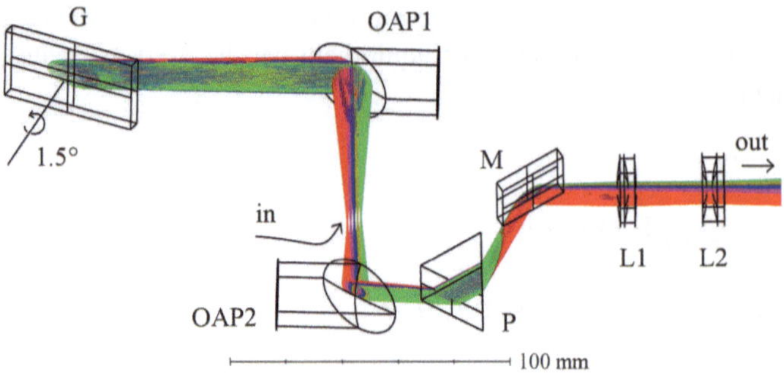

Fig. 1 Spectrograph optical layout. Rays in different *colors* represent different wavelengths within the same diffraction order. *In* Fiber input, *OAP1*/*OAP2* 90° off-axis parabolic mirror with 101.6/50.8 mm effective focal length, *G* echelle grating, turned by 1.5° around its surface normal, *P* N-SF11 equilateral prism, *M* plane mirror, *L1*/*L2* lens telescope consisting of two achromatic doublets with 125/−100 mm focal length

off-axis parabolic mirror (OAP1 in Fig. 1) with a 101.6 mm effective focal length (EFL). The 90° angle was chosen, because it is the most common in optics catalogs, so mirrors with suitable focal lengths were available off-the-shelf with good optical quality (surface accuracy $\lambda/8$ at 630 nm, roughness <50 Å root-mean-square, supplier: Edmund Optics).

Ten centimeters after OAP1, we place a 50×25 mm^2 echelle grating with 31.6 grooves/mm and a 63° blaze angle (Thorlabs GE2550-0363). The grating is positioned, such that at the center of each diffraction order, the beam is precisely reflected back to OAP1 (Littrow configuration). From this initial position, the grating is turned by 1.5° around its surface normal. This creates an upward displacement of the beam of 2.6 mm after OAP1, allowing it to pass above the input fiber without getting obstructed. A dispersed intermediate focus is formed above the fiber end, which can be utilized to intercept stray light from the grating by inserting a slit.

After the intermediate focus, the light is collimated by another 90° off-axis parabolic mirror (OAP2), with an EFL of 50.8 mm. The shorter EFL of OAP2 demagnifies the beam diameter to 4.7 mm FWHM, which facilitates the handling of the dispersed beam with 1 in wide optical components without clipping it. Furthermore, it steepens the dispersion angle, which magnifies the echellogram image in horizontal direction. This can be understood as being a direct consequence of the reduced beam diameter, as the magnification of the image is required to preserve the Lagrange invariant. In our optical setup, this effect allows us to obtain an image of equal size with a shorter focal length of the subsequent lens system.

As an echelle grating produces numerous diffraction orders, a cross disperser is an essential part of any echelle spectrograph. Its dispersion is perpendicular to that of the grating to separate the echelle orders from one another. For this, we employ an equilateral prism that we place after OAP2. As this is after the location where the angular dispersion is doubled, we need a relatively strong cross-dispersion. We thus choose N-SF11 as a prism material, for its high refractive index and low Abbe number. This choice comes with the minor disadvantage of a relatively unequal angular separation of the orders, with the blue ones being separated more widely than the red ones. The strongly dispersive prism also causes the echelle orders to have a pronounced curvature. Light on the side of an echelle order passes through the prism in a different angle, and is, therefore, deflected differently.

As the beam is deflected upwards by the prism, we insert a fold mirror to get it back into the horizontal plane. This is followed by a sequence of two achromatic doublet lenses with 125 and -100 mm focal length, respectively, mounted in an expandable tube. The total focal length of the lens system is controlled by the relative distance of the two lenses. With this, we can project the echellogram on a laboratory wall. At a distance of 3 m, this yields a roughly $30 \times 20\ \text{cm}^2$ large image, creating an impressive and enjoyable visual demonstration for lab visitors. For measurements, we replace the lens telescope by a commercial camera system (see Sect. 3).

2.3 *Aberrations*

The 4.7 mm FWHM Gaussian beam focused over a 3 m distance should ideally yield a spot of 154 μm FWHM at a wavelength of 550 nm. For an astro-comb with an 18 GHz mode spacing, the spot separation becomes 778 μm, which can be resolved with the naked eye. A focal length of at least 3 m is thus appropriate for the purpose of a visual demonstration. The visibility can be further enhanced by holding a piece of paper at an angle into the image to stretch it. The spectral resolution $R = \lambda/\delta\lambda$ should theoretically amount to 154,000. It depends on aberrations in our optical system, whether this resolution is in fact reached. Therefore, we analyze the optical layout of Fig. 1 with Zemax. The simulated rays cover the beam FWHM. We compare the resulting spots to the FWHM as calculated from the theory of Gaussian beams (see Fig. 2). Usually, illumination of the full aperture is simulated and compared to the airy disk, but in our case, the Gaussian FWHM provides a clearer picture of the expected spectral resolution.

The main source of aberrations in our system is the fact that the beam bundles propagate off the optical axis between OAP2 and OAP1. Aberrations thus arise on both OAPs, which would add up if the OAPs were oriented parallel to one another (resulting in a U-shaped beam path). In our configuration, where the OAPs are antiparallel, aberrations from OAP1 are largely canceled by those from OAP2. For this compensation to work, it is irrelevant whether or not OAP1 and OAP2 have the same EFL. The residual aberrations are least pronounced for beams at the center of the diffraction orders (Fig. 2a) as they are only displaced vertically from the optical axis by 2.6 mm. The simulated focal spot is a tilted ellipse, which is well within the diffraction limit. In addition, we see a structure extending upwards from the ellipse caused by spherical aberration of the lens system. Beams that additionally suffer from a lateral displacement of 1.2 mm (Fig. 2b) start being non-diffraction-limited,

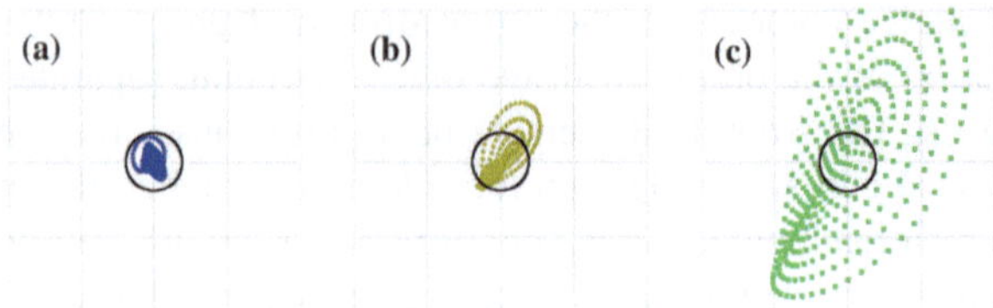

Fig. 2 Spot diagrams for diffraction order 103, going from the center at 547.5 nm (**a**) towards the right side of the order at 549.2 nm (**b**) and 550.8 nm (**c**). The rays sample the beam FWHM. The *circles* represent the diffraction-limited FWHM in the focus of a perfect Gaussian beam

and after another 1.2 mm (Fig. 2c, outer wing of echelle order) the comb lines start blending.

The main difference between the lateral and vertical beam displacement is that the vertical displacement is the same for all beams, so the lens system can be adjusted for the resulting defocusing. The lateral displacement changes across the echelle order, which causes the image plane to be curved. In many spectrographs, this effect is corrected by the camera optics [20]. The effect is also smaller in most spectrographs, because of the use of OAPs with smaller angles, which diminishes the aberrations. Our simulations show that with a 30° OAP angle, we could reach a 20–30% smaller spot size in most places where the spots are non-diffraction limited. It appears reasonable for us to trade this relatively minor improvement in for the easier availability of 90° OAPs with the desired focal lengths and optical quality. In an early attempt to find the simplest acceptable solution, spherical mirrors were also tested, whose spherical aberration and coma would, however, decrease the spectrograph's resolution by a factor of 20.

3 Camera Integration

With a science-grade camera, the most costly part of our setup by far would be the image sensor. Astronomical cameras use CCDs with very high quantum efficiency. The CCDs are cooled, often down to −90° C or even −120° C to suppress dark current. This has to be done in a dry atmosphere to suppress condensation of water. However, for the sole purpose of characterizing an astro-comb, high quantum efficiency is unnecessary, since the laser system delivers vast amounts of light. Detector cooling is useful for very faint sources requiring long exposures, as the number of dark counts scales with exposure time. With the optical power available from the astro-comb, we can work with even extremely short exposures in the μs range.

For these reasons, we want to explore the use of a consumer-grade photographic camera in our spectrograph. We use a Canon EOS M3, featuring a CMOS image sensor with 6000×4000 px^2 and an active area of 22.3×14.9 mm^2 (APS-C format, 3.72 μm pixel pitch). Each color channel has 14 bit of depth, which is preserved by recording the images in raw format (.CR2) and later converting them to FITS. The battery is replaced by a line adapter provided by Canon, and images are transferred via USB. Exposures are triggered remotely via WiFi or remote control. Many models also offer the possibility to both control and read out the camera via USB ("tethered shooting"), which would be simpler.

As a camera lens, we mount a prime lens with a 200 mm focal length (Canon EF 200mm) on the camera, which replaces the lens telescope of our setup. The complete setup, including the camera system, proves to be compact enough to be accommodated on a $30 \times 45\ \text{cm}^2$ base plate, as shown Fig. 3. The 200 mm focal length of the camera lens roughly matches the size of the echellogram image to that of the image sensor. The lens aperture is set to its maximum of 7 cm (f/2.8), which is more than enough to avoid clipping of the beam. With this configuration, we can image 33 echelle orders onto the detector, spanning a spectral range of 477–658 nm (diffraction order 87–120). At around 550 nm, a single echelle order spans 7 nm on the detector yielding a 32% spectral overlap from one order to the next.

For our measurements, we use the shortest available exposure time of 250 µs and set the detector sensitivity to ISO 100 to minimize noise. For the measurements presented here, we use an astro-comb with an 18 GHz mode spacing as a light source. This astro-comb is presently in preparation for its duty on ESPRESSO. An echellogram recorded with it can be seen in Fig. 4, zoomed-in towards the center of the image to make individual comb lines visible. The full image is included as a JPEG in the supplementary material to this article. Analysis of the full frame confirms the result of the simulation that the image plane is curved cylindrically: in all echelle orders, the spots are well focused and nearly diffraction limited at the center, while moving out of focus towards the outer wings of the orders.

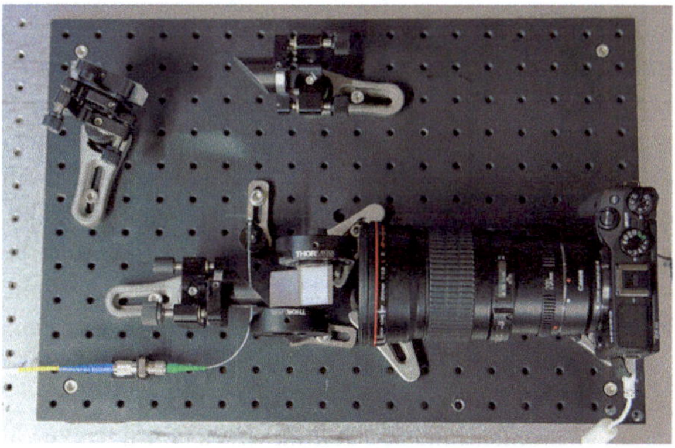

Fig. 3 Photo of the spectrograph, including camera. The physical dimensions of the *black* base plate are $30 \times 45\ \text{cm}^2$

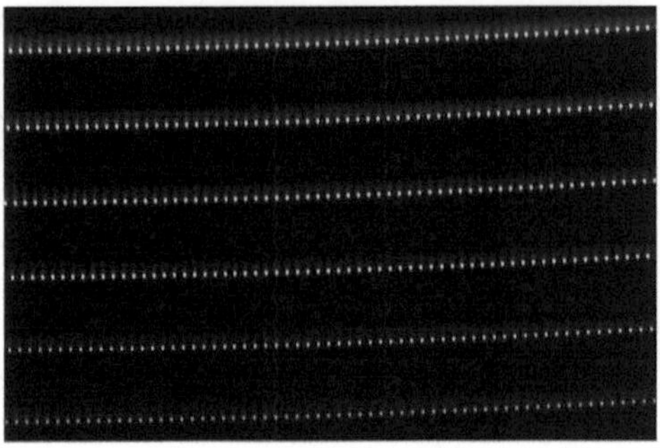

Fig. 4 Part of the echellogram near the center of the image sensor

4 Results

To extract data from the recorded images, we have implemented an automatic data pipeline in Python. The pipeline first subtracts a dark image, and then sums up the three color channels, by imposing a 2×2 binning on the pixels arranged in a Bayer pattern. This yields a 3000×2000 px^2 gray-scale image. As expected from the spot diagram (Fig. 2), the image is strongly defocused at the borders. Therefore, we crop the image to a 2000×2000 px^2 region of interest with reasonable resolution. An echelle order at 550 nm spans 4.7 nm over this range, with 0.6 nm wide gaps between the orders, where spectral information is discarded. Next, the positions of the echelle orders are recognized, and each echelle order is summed up vertically over 5 (binned) pixels into a 1D data array.

The resulting 1D spectrum can be seen in Fig. 5a for order 102, centered at 553 nm. The instrumental line shape is very well approximated by a Gaussian function, as seen from the Gaussian line fit in Fig. 5b. This is because the single-mode fiber provides a very clean Gaussian beam profile. In the center of the echelle order, the line FWHM is 1.81 px, while the line spacing is 7.44 px (binned). This translates into a spectral resolution of 124,000, which is about the same as that of HARPS and ESPRESSO in the standard mode. At the borders of the extracted region of the echelle order, the spectrograph resolution declines to about 65,000 due to the increasing aberrations, causing the spectral lines to slightly overlap. The maximum in resolution is somewhat lower than the 154,000 calculated above, which likely results from the fact that the Gaussian beam is truncated by the 1 in optics after 2.7 times its FWHM. Another reason might be the low-pass filter on the image sensor. This is usually a plate consisting of cemented, birefringent layers of quartz in different orientations. Its purpose to blur out the image over the 2×2 px^2 large Bayer elements to avoid Moiré effects.

As an application, we examine the spectral background of the astro-comb. It was seen in [10] that some spectral broadening schemes create an unexpectedly strong continuum background in addition to the comb lines. If an excessively strong background is not properly taken into account by the data processing pipeline of an astronomical spectrograph, it can have a severe impact on its calibration [10]. It

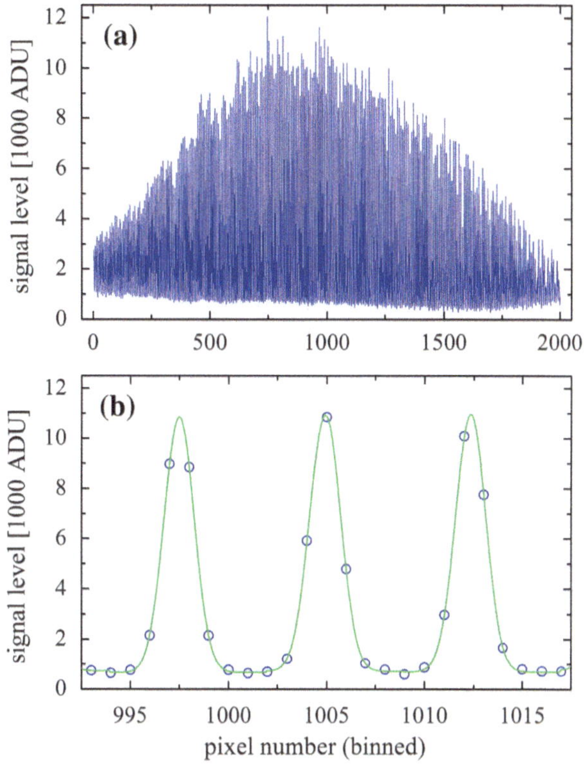

Fig. 5 Diffraction order 102 (centered at 553 nm) after data extraction. **a** Measured data (*blue line*) over the full region of interest of 2000 px, after binning over $2 \times 2\,\mathrm{px}^2$ in the Bayer matrix. **b** Smaller region at the center of the echelle order, with Gaussian functions (*green line*) fitted to the data points (*blue circles*). *ADU* analog-to-digital unit

would, therefore, be desirable to test an astro-comb for such properties before delivery to the observatory. This is, however, difficult without an echelle spectrograph because of its very particular resolution. The background can only clearly be seen when the comb lines are fully resolved without any residual overlap. An excessively high resolution, on the other hand, also makes it hard to see the background when the dynamic range is limited. This is because with higher resolution, the comb lines grow taller relative to the background, as their photon flux becomes more concentrated. This difficulty in measuring the background has led us to the conclusion that a laboratory echelle spectrograph was needed. The background characterization was hence our primary objective for this project. We demonstrate this as a first but nonetheless important application.

Our spectrograph allows viewing the spectrum with the same resolution as HARPS or ESPRESSO in the middle of the echelle orders, thus yielding the same relation between background level and comb line amplitudes. In the center of order 102 (Fig. 5b), the background is at 6% of the amplitudes of the comb lines, as determined by the fit. The background is expected to be most intense at the center of an echelle order and to decrease along with the line intensities towards the borders (see [10]). In our spectrograph, we see a different behavior (Fig. 5a), which is supposedly the result of the changing resolution across the echelle order. In the outer wings of the order, the comb lines start to overlap, which masks the background.

For an accurate assessment of the background, it must be questioned to what extent the measured background level might be affected by a possible stray-light contribution, arising e.g. from imperfections of the echelle grating or mirror surfaces. It can easily be seen that the acquired images globally contain very little stray light: between the echelle orders, the dark-corrected signal level is more than 25 dB below the peak values of the comb lines. Furthermore, the spectral shaping reported in the next section attenuates the signal within an echelle order with a contrast of at least 20 dB. However, each comb line might be surrounded by a small stray-light halo. We have investigated this question on a single isolated line by feeding the spectrograph with a frequency-doubled Nd:YAG laser at 532 nm. Fitted with an offset-free Gaussian function, the data points follow the fit down to below -20 dB from its peak. A possible stray-light contribution must, therefore, be limited to $<1\%$, confirming that the background level measured above must be dominated by an incoherent spectral component.

The spectral background can systematically distort the calibration results, depending on its structure and how it is modeled in the data analysis. A recent study on HARPS used an astro-comb with a background level very similar to ours [10]. By fitting the slope of the background, the related systematic effect could be canceled at the cm/s level. The background also comes with an increased statistical uncertainty, because it adds photon noise, but does not contribute to the signal of the comb lines, degrading their signal-to-noise ratio. In addition, less of the dynamic range is available for the comb lines. We calculate that a 6% background increases the photon noise by 21% as compared to the background-free case. For HARPS, this would imply an increase of the photon noise from typically 2.8 cm/s in a single exposure to 3.4 cm/s.

5 Happy Birthday!

Many astro-combs employ a spatial light modulator (SLM) as a programmable spectral filter [23, 15, 24]. This allows generating a comb spectrum with a flat-top shape, such that all lines can simultaneously reach the optimum signal level on the image sensor. The spectral sensitivity of a given spectrograph can be factored in by the flattening algorithm, which we might characterize for our spectrograph. We could then use our spectrograph to judge the attainable spectral flatness. Being able to resolve all spectral lines individually and simultaneously, the spectrograph can take into account line-by-line intensity variations. Such fine spectral structures lie below the resolution of the SLM. It is hence very important to assess how far they limit the attainable flatness. This has been done with HARPS [10], and can readily be done with our spectrograph.

However, since we dedicate this article to Ted Hänsch in honor of his 75th birthday, measuring a flat spectrum appears somewhat too uninteresting to us. With

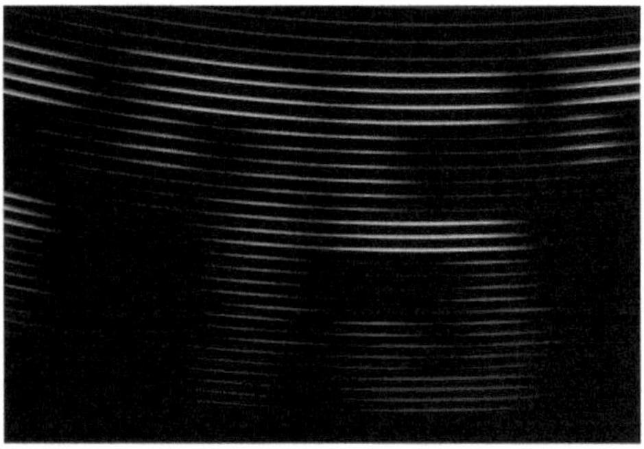

Fig. 6 75, written into the echellogram of an astro-comb by reshaping its spectrum with a spatial light modulator. Happy birthday Ted Hänsch!

the SLM being able to create very general spectral shapes with a 0.4 nm resolution and a 27 dB dynamic range, we would rather use it to modulate the spectrum so as to make a 75 appear in the echellogram. This idea has been inspired by a work published for Ted Hänsch's 60th birthday, where the motion of a Bose–Einstein condensate was steered into writing a 60 [25]. What better way is there to congratulate the inventor of the self-referenced LFC than using an LFC?

For this experiment to work, we must calibrate the wavelength scale of our spectrograph with the spectral flattening unit. The wavelength scale of this unit might be slightly inaccurate, which is canceled if the spectrograph data use the same inaccurate scale. We do this using the SLM to create dips in the spectrum and see where these dips show up in the echellogram, to obtain the center wavelengths of the echelle orders. The wavelength scale within the echelle orders is derived from the Gaussian fit to the comb lines. With this calibration, we compute a look-up table that specifies a fake spectrograph sensitivity to the flattening software. For places that we want to light up, we specify a very low sensitivity at the corresponding wavelengths, and in places that we want to remain dark, we specify a very high sensitivity.

Figure 6 shows the result. The straightness of the vertical lines is in most places nearly limited by the pixelation of the SLM that allows placing transmission windows in discrete steps only. Horizontal lines are brighter than vertical lines, because narrow spectral transmission windows inflict diffraction losses on the beam. This could be adjusted for with a further refinement of the SLM pattern. Left and right of the 75, we see how the pattern repeats into the next order, shifted up or down, respectively, by one echelle order. The image has been overexposed to make the 75 more clearly visible. A full-resolution JPEG version of this image is included in the supplementary material.

6 Possible Upgrades and Future Applications

Accuracy and long-term stability are essential features of any astro-comb, and represent the main advantage over alternative calibration methods. A stringent test of these crucial properties generally requires a comparison of two independent astro-combs on a two-channel spectrograph, as it was done during a recent campaign on HARPS [10]. Such expensive and time-consuming campaigns might become unnecessary with a high-resolution echelle spectrograph for laboratory use. Directly after assembly, each astro-comb could be compared to a reference astro-comb, to ensure that it fulfills its specifications. A laboratory-based spectrograph may also facilitate stability tests on very long time scales up to years, as needed for searching long-period exoplanets and other applications.

Making our spectrograph suitable for these important and ambitious goals requires a number of upgrades, such as extending it to a two-channel spectrograph. The large separation of the echelle orders leaves more than enough space in between them to accommodate additional orders from a second input fiber. This could be implemented using a ferrule at the spectrograph input with a hole large enough to comprise two fibers. The distance of the echelle orders from the two fibers could be controlled by turning the ferrule around its axis. For characterizing the stability down to 1 cm/s as commonly desired in the search for Earth-like exoplanets, pressure and temperature stabilization of the echelle spectrograph are likely to be a requirement: A change in ambient air pressure of 1 hPa alters the wavelengths equivalent to a Doppler shift of 82.5 m/s. This change might be too large to track the relative drift of two astro-combs down to 1 cm/s. Fortunately, both pressure and temperature stabilization are greatly facilitated by the small footprint of our spectrograph.

Similar to the stability, a set of two independent astro-combs can be tested for their consistency. For this purpose, two astro-combs are used for a simultaneous calibration of a two-channel spectrograph showing how well they agree on an absolute basis [10]. For an absolute calibration of a spectrograph, the comb modes need to be identified. The coarse calibration with the flattening setup used in this paper was not accurate enough for this. To identify the modes in one order, a single-frequency laser can be used, whose wavelength is measured with a wavemeter. The obtained calibration can be scaled to other orders using $\lambda_2 = \lambda_1 m_1 / m_2$, which relates the wavelengths λ_1 and λ_2 at the same position for two different orders m_1 and m_2. To account for any possible lateral displacement of the orders relative to one another, the prism dispersion direction can be precisely measured on the image sensor by replacing the echelle grating with a plane mirror.

7 Conclusion

In this article, we have shown how to build a high-resolution echelle spectrograph in a simple, compact and cost-efficient way. In the center of each echelle order, our spectrograph reaches the same resolution as state-of-the-art astronomical echelle spectrographs used for exoplanet hunting. As an application, we have shown how the spectrograph can be used to characterize temporally incoherent components within the astro-comb spectrum. This aspect is highly relevant for calibrating astronomical spectrographs. Its characterization was the major motivation for the development of the spectrograph we have described, next to the characterization of line-by-line intensity fluctuations. We have also discussed several upgrades that may open up the door to more advanced applications, such as the relative characterization of two astro-combs in terms of stability and consistency. The requirements for this include adding a second input fiber, as well as temperature and pressure stabilization of the spectrograph.

Acknowledgements Open access funding provided by Max Planck Society. We gratefully acknowledge Florian Kerber, Gaspare Lo Curto, Gerardo Avila, Luca Pasquini, and Antonio Manescau from the European Southern Observatory, and Hanna Kellermann from the University Observatory Munich, for collaborating with us on astro-combs, and for helpful discussions on spectrograph design.

References

1. Th. Udem, R. Holzwarth, T.W. Hänsch, Nature **416**, 233–237 (2002)
2. C.G. Parthey et al., Phys. Rev. Lett. **107**, 203001 (2011)
3. M.T. Murphy, Th. Udem, R. Holzwarth, A. Sizmann, L. Pasquini, C. Araujo-Hauck, H. Dekker, S. D'Odorico, M. Fischer, T.W. Hänsch, A. Manescau, Mon. Not. R. Astron. Soc. **380**, 839–847 (2007)
4. A.G. Glenday, C.-H. Li, N. Langellier, G. Chang, L.-J. Chen, G. Furesz, A.A. Zibrov, F. Kärtner, D.F. Phillips, D. Sasselov, A. Szentgyorgyi, R.L. Walsworth, Optica **2**, 250–254 (2015)
5. G.G. Ycas, F. Quinlan, S.A. Diddams, S. Osterman, S. Mahadevan, S. Redman, R. Terrien, L. Ramsey, C.F. Bender, B. Botzer, S. Sigurdsson, Opt. Express **20**, 6631–6643 (2012)
6. T. Wilken, G. Lo Curto, R.A. Probst, T. Steinmetz, A. Manescau, L. Pasquini, J.I. González Hernández, R. Rebolo, T.W. Hänsch, Th. Udem, R. Holzwarth, Nature **485**, 611–614 (2012)
7. J.K. Webb, J.A. King, M.T. Murphy, V.V. Flambaum, R.F. Carswell, M.B. Bainbridge, Phys. Rev. Lett. **107**, 191101 (2011)
8. J. Liske et al., Mon. Not. R. Astron. Soc. **386**, 1192–1218 (2008)
9. M. Mayor et al., The Messenger **114**, 20–24 (2003)
10. R.A. Probst et al., Proc. SPIE **9908**, 990864 (2016)
11. R.A. Probst, L. Wang, H.-P. Doerr, T. Steinmetz, T.J. Kentischer, G. Zhao, T.W. Hänsch, Th. Udem, R. Holzwarth, W. Schmidt, New J. Phys. **17**, 023048 (2015)
12. M.J. Pfeiffer, C. Frank, D. Baumüller, K. Fuhrmann, T. Gehren, Astron. Astrophys. Suppl. Ser. **130**, 381–393 (1998)
13. A. Brucalassi, F. Grupp, H. Kellermann, L. Wang, F. Lang-Bardl, N. Baisert, S.M. Hu, U. Hopp, R. Bender, Proc. SPIE **9908**, 99085W (2016)

14. D. Mégevand et al., Proc. SPIE **8446**, 84461R (2012)
15. R.A. Probst et al., Proc. SPIE **9147**, 91471C (2014)
16. T.W. Hänsch, Optics Photon. News **16**, 14–16 (2005)
17. H.-R. Xia, S.V. Benson, T.W. Hänsch, Laser Focus **17**, 54–58 (1981)
18. J. Ge, J.R.P. Angel, D.G. Sandler, J.C. Shelton, D.W. McCarthy, J.H. Burge, Proc. SPIE **3126**, 343–354 (1997)
19. J.R. Crepp et al., Proc. SPIE **9908**, 990819 (2016)
20. C. Schwab et al., Proc. SPIE **9912**, 991274 (2016)
21. A.D. Rains, M.J. Ireland, N. Jovanovic, T. Feger, J. Bento, C. Schwab, D.W. Coutts, O. Guyon, A. Arriola, S. Gross, Proc. SPIE **9908**, 990876 (2016)
22. A. Bechter et al., Proc. SPIE **9909**, 99092X (2016)
23. R.A. Probst, T. Steinmetz, T. Wilken, G.K.L. Wong, H. Hundertmark, S.P. Stark, P.St.J. Russell, T.W. Hänsch, R. Holzwarth, Th. Udem, Proc. SPIE **8864**, 88641Z (2013)
24. R.A. Probst, Y. Wu, T. Steinmetz, S.P. Stark, T.W. Hänsch, Th. Udem, R. Holzwarth, CLEO **2015**, SW4G.7 (2015)
25. J. Reichel, W. Hänsel, *Laser physics at the limits* (Springer, Berlin, 2002)

Mode Locking Based on the Temporal Talbot Effect

Thomas Udem and Akira Ozawa

Abstract We propose a new laser mode locking state in which the pulse disperses quickly and then revives after a certain time. This mechanism is based on the temporal Talbot effect and requires a large amount of intra-cavity group velocity dispersion. Similar to the usual mode locking it should be possible to employ the Kerr effect to force the laser into this mode, even when the cold cavity dispersion is not exactly matched. We show that the mode spectrum of such a laser is not equidistant but increases linearly with very high precision. This Talbot frequency comb can be self referenced. The beating with the adjacent modes uniquely defines the optical mode frequency, which means that the optical spectrum is directly mapped into the radio frequency domain. This is similar to the dual frequency comb technique, albeit without the limiting relative jitter between two combs.

1 Introduction

The Talbot effect has been described for the first time in 1836 as a peculiar phenomenon observed in the near field of an optical grating. Summing over contributions of the individual rulings to the total field in the Fresnel approximation, a term of the form $\exp(-ikl^2a^2/2z)$ appears with the wave number k, the rulings numbered by l and spaced by a and the distance from the grating z. Summing over l generally yields a rather chaotic intensity distribution. Talbot noted though that all these terms reduce to $\exp(-i2\pi l^2) = 1$ at a distance of $z = ka^2/4\pi$. The remaining terms add up to the intensity at $z = 0$ provided that this intensity is periodic with a [1]. The same phenomenon can be observed in the time domain with a periodic pulse train signal that is subject to group velocity

This article is part of the topical collection "Enlightening the World with the Laser" - Honoring T. W. Hänsch guest edited by Tilman Esslinger, Nathalie Picqué, and Thomas Udem.

T. Udem (✉) • A. Ozawa
Max-Planck-Institute for Quantenoptik (MPQ), Hans-Kopfermann-Str. 1, 85748 Garching, Germany
e-mail: thu@mpq.mpg.de

D. Meschede et al. (eds.), *Exploring the World with the Laser*,
https://doi.org/10.1007/978-3-319-64346-5_26

dispersion ϕ'' that provides the quadratic phase evolution. The pulses first spread out in time, an then reassemble after propagation the distance $t_r^2/2\pi|\phi''|$, where t_r is the pulse repetition time [2, 3]. The same behaviour should be observable with a single pulse that is on a repetitive path in an optical cavity. In contrast to a free space pulse train, higher order dispersion is required as we will show below.

2 Talbot Comb

Consider the modes of a laser cavity with a mode spectrum

$$\omega_n = \omega_0 + \left(n + \frac{n^2}{m}\right)\omega_r. \tag{1}$$

Here the modes are numbered around the optical carrier frequency ω_0 with integers $n = 0, \pm 1, \pm 2 \ldots$. Compared to the usual regularly spaced frequency comb [4] there is a quadratic term in n which leads to an equally spaced comb of radio frequency (RF) components:

$$\Delta_n = \omega_{n+1} - \omega_n = \omega_r\left(1 + \frac{2n+1}{m}\right) \tag{2}$$

These RF components are the result of beating between adjacent optical modes and can be seen in the power spectrum of the laser output. Higher order mode beatings like $\omega_{n+2} - \omega_n$ etc. are separated in the power spectrum by about ω_r. This is illustrated in Fig. 1 and similar to the harmonics of the repetition rate in a regular frequency comb. The optical mode spacing at the carrier frequency ω_0 is given approximately by ω_r for large m. It becomes the spacing for all modes for $m \to \infty$ for which (1) turns into a regular frequency comb [4]. Note however that ω_r is not the usual repetition frequency. Nevertheless ω_r and ω_0 can be measured in very much the same way as for a regular frequency comb (see Sect. 3). Besides representing an all new mode locking regime, the interesting aspect of (1) is that each mode beating uniquely belongs to on particular pair of modes. For example the RF signal at $\Delta_0 = \omega_r(1 + 1/m)$ belongs to the beating between ω_0 and ω_1 and so on. Hence a RF spectrum recorded with a photo detector and a radio frequency spectrum analyzer directly displays a scaled down version of the optical spectrum of the laser. By placing a sample between the laser and the photo detector and recording the change of the RF spectrum one gets the absorption spectrum of the sample. This is similar to a dual frequency comb setting with a linearly increasing spacing between the modes of the two frequency combs [5], albeit with a single laser avoiding problems due to the relative jitter of the combs.

To see how the laser spectrum might be forced to the modes defined by (1) it is instructive to compute the electric field at a fixed point inside the cavity. Assuming that the modes oscillate with some complex amplitudes a_n we get:

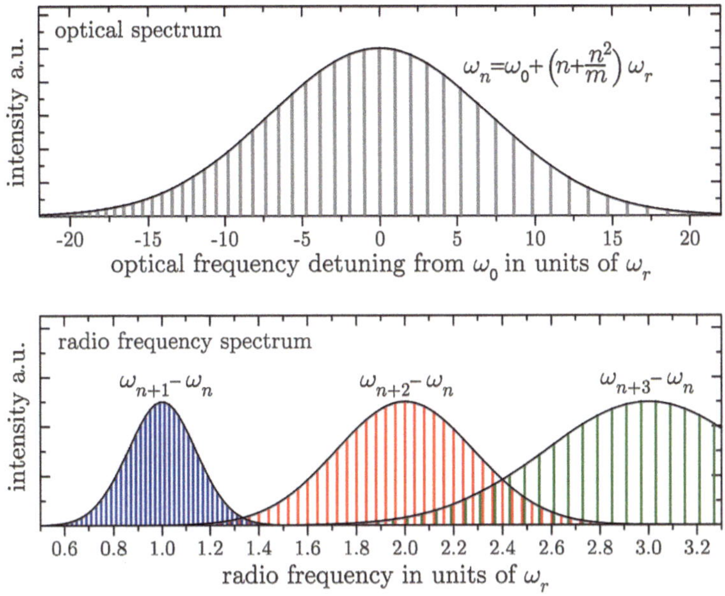

Fig. 1 *Upper* The optical spectrum of the Talbot-comb according to (1) for $m = 100$ assuming a Gaussian spectral envelope. With this rather small value it is possible to illustrate the linearly increasing mode spacing. *Lower* Equidistant radio frequency combs obtained with a photo detector. The *left* (*blue*) part is due to the first order ($k = 1$) beating of adjacent modes $\omega_{n+k} - \omega_n$ (Eq. (2)) which would be used for spectroscopy. The *center* (*red*) part are the second order ($k = 2$) beatings and the *right* (*green*) part is the $k = 3$ radio frequency comb and so on. All of them have a uniform, equidistant spacing of $2k\omega_r/m$ (hence the name radio frequency comb). The width of the spectral envelopes are the same when measured by the number of components. Increasing m prevents the overlapping regions, at least for the lower orders. Even if the radio frequency combs would overlap, each component is still uniquely linked to a particular pair of optical modes

$$E(t) = E_o e^{-i\omega_0 t} \sum_{n=-\infty}^{+\infty} a_n e^{-i\left(n+\frac{n^2}{m}\right)\omega_r t}. \tag{3}$$

This can not represent a stable pulse in the time domain. Assuming that m is an exact integer however, the pulse will revive up to a carrier–envelope phase of $\varphi_{ce} = 2\pi m\omega_0/\omega_r$ after the time $T = 2\pi m/\omega_r$:

$$\begin{aligned} E(t+T) &= E_o e^{-i\omega_0(t+T)} \sum_n a_n e^{-i\left(n+\frac{n^2}{m}\right)\omega_r t - 2\pi i\left(mn+n^2\right)} \\ &= E_o e^{-i\varphi_{ce}} e^{-i\omega_0 t} \sum_n a_n e^{-i\left(n+\frac{n^2}{m}\right)\omega_r t} \\ &= e^{-i\varphi_{ce}} E(t) \end{aligned} \tag{4}$$

This is similar to the regular mode locking scheme, except that the pulse repetition rate is replaced by the much lower revival rate $\omega_r/(2\pi m)$. The revival time is the

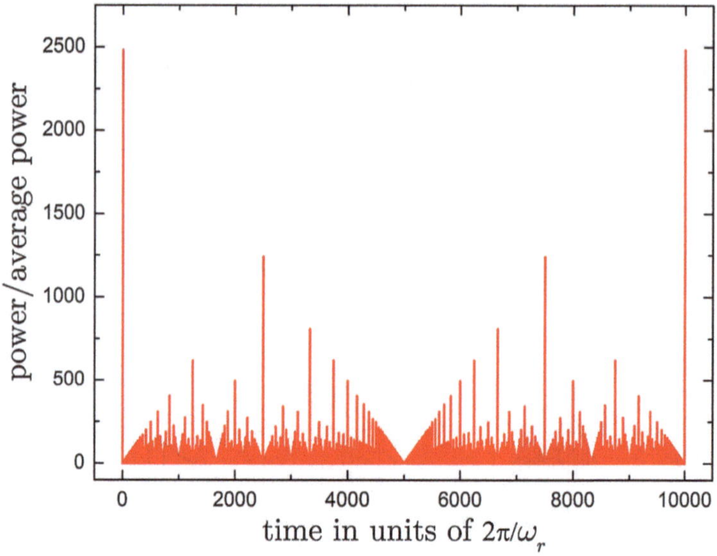

Fig. 2 The power $\propto |E(t)|^2$ at a fixed point inside the laser cavity according to (3) normalized to the time averaged power. The recurrence coefficient is $m = 10,000$ and the amplitudes follow a Gaussian distribution $a_n = e^{-(n/1000)^2}$ (which is not exactly a Gaussian spectral envelope). After m cavity round trips the original pulse revives. The *plot* looks very similar if one scales m and the number of modes up. The brute force computation using (3) with sufficient resolution for larger m is computational challenging

m multiple of the cavity round trip phase delay of the $n = 0$ mode at ω_0. Figure 2 shows an example of the resulting laser power for a special case as a function of time. Like in a regular mode locked laser, the peak power enhancement over the time averaged power is roughly given by the number of active modes. In contrast to the latter, the large peak power occurs not every cavity round trip but only every m-th cavity round trip. The usual Kerr lens mode locking mechanism might be used to enforce an integer m by reduced loss of the high peak intensity pulse. In the case of the Talbot comb, that large peak intensity occurs only every m-th round trip. We therefore expect that the self amplitude modulation of the Kerr effect has to be stronger for successful mode locking. The mode spacing nominally becomes negative for $n < -m/2$ (see Fig. 3). Physically this means that the corresponding spectral region possesses a negative group velocity. While this is possible in principle we exclude this for a reasonable laser design and assume that the active modes of the laser are limited to $n > -m/2$.

To find the required dispersion that results in the mode spectrum of (1), we first resolve it for n:

$$n = \frac{m}{2}\left(\sqrt{1 + 4\frac{\omega_n - \omega_0}{m\omega_r}} - 1\right) \tag{5}$$

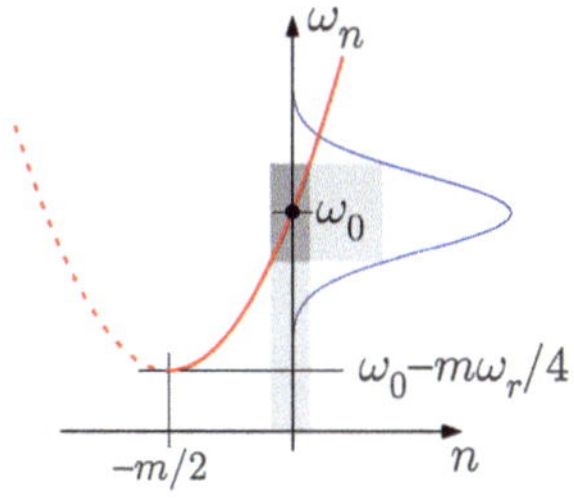

Fig. 3 Frequencies of the modes of the Talbot comb according to (1) shown as the *red solid curve*. The *dashed part* belongs to the negative mode spacing section (i.e. negative group velocity). We do not consider this possibility and therefore also discarded the other sign solution in (5). This *curve* defines the dispersion properties of the cavity whose expansion is shown in (11). The vertex of the *curve* at $[-m/2; \omega_0 - m\omega_r/4]$ can be chosen without restriction by selecting m and ω_r. The active modes, i.e. the laser spectrum, is assumed to be centered at ω_0. It covers a certain range in ω_n and n-space (grey area). The curvature of the parabola reflects the required group velocity dispersion which can be minimized by large values of m and ω_r in accordance with (11)

Like in any other laser with cavity length L, the round trip phase $\phi(\omega)$ at frequency ω has to fulfill the boundary condition:

$$\phi(\omega) = 2\pi n + \frac{L}{c}\omega_0 \tag{6}$$

$$= \pi m\left(\sqrt{1 + 4\frac{\omega - \omega_0}{m\omega_r}} - 1\right) + \frac{L}{c}\omega_0 \tag{7}$$

Since ω_0 is the resonant mode with $n = 0$, the last term has to be added to obtain the total round trip phase. Without loss of generality we assumed in (1) that the parameter ω_0 is the center of the emitted spectrum (see Fig. 3). Using (7) and computing the derivatives at ω_0 we obtain the dispersion required to generate the mode spacing of (1):

$$\phi''_{\omega_0} = -\frac{4\pi}{m\omega_r^2} \tag{8}$$

$$\phi'''_{\omega_0} = +\frac{24\pi}{m^2\omega_r^3} \tag{9}$$

$$\phi''''_{\omega_0} = -\frac{240\pi}{m^3\omega_r^4}\cdots \tag{10}$$

$$\phi^{(k)}_{\omega_0} = (-1)^{k+1}\frac{2^k(2k-3)!!\,\pi}{m^{k-1}\omega_r^k} \tag{11}$$

Here !! represents the double factorial. For a real laser the requirements on the dispersion are quite extreme but not impossible (see Sect. 5 and Fig. 4). These requirements are mitigated by large values of m, i.e. a long pulse revival time (for a given ω_r). However we expect that this would lead to a reduced Kerr effect and hence weakens the mode locking mechanism. We expect that once the laser is set up

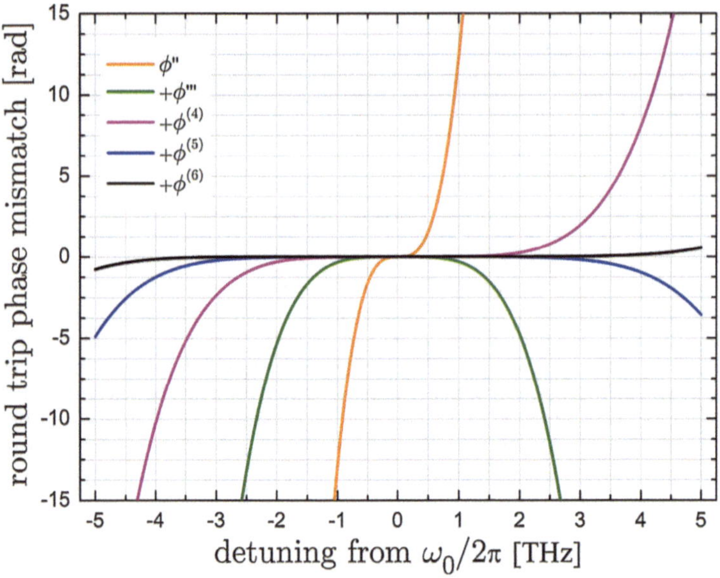

Fig. 4 Comparing the exact round trip phase given by (5) and (6) with the expansion given by (11) for different order with $m = 10^6$ and $\omega_r = 2\pi \times 100$ MHz. Successively compensating dispersion up to the order $\phi^{(6)}(\omega)$, the phase mismatch reduces to 0.77 rad at the edges of the spectrum (width 10 THz), i.e. about 0.12 free spectral ranges. This mismatch has to be compensated by the Kerr effect like in a regular mode locked laser

for a particular value of m, it will be reproduced every time it is put in the mode locked state.

3 Self-Referencing

For self-referencing the two parameters of the Talbot comb, ω_r and ω_0 need to be measured and ideally stabilized. We assume that the recurrence index m is known. One might get an estimate of it and then fix it to be an integer, by measuring the recurrence time and compare it to the cavity length. A more reliable method would be to measure a known optical frequency with a self-referenced Talbot comb and then identify the proper m compatible with that measurement. This is the same method often applied with regular combs to determine the correct mode number.

The parameter ω_r can be determined by observing the mode beating dependence on n as expressed by (2), i.e. by the second order mode differences:

$$\frac{2}{m}\omega_r = \Delta_{n+1} - \Delta_n \tag{12}$$

In practical terms this frequency is readily generated by mixing two adjacent RF modes Δ_n, say by driving something non-linear with the photo detector signal. The mixing result can then be locked to a precise reference frequency such as an atomic clock by feeding back on the cavity length and presumably also on the pump laser power. In that sense ω_r is determined almost as simple as the repetition rate of a regular frequency comb.

The second parameters of the Talbot comb, ω_0 can be measured in very much the same way as done with the carrier envelope offset frequency of a regular comb [4], i.e. with an $f - 2f$ interferometer. A part from the red side of the Talbot comb with mode number n_1 is frequency doubled and superimposed the blue part of mode number n_2. According to (1) the generated beat notes have the frequencies:

$$2\omega_{n_1} - \omega_{n_2} = \omega_0 - \left(2n_1 - n_2 + \frac{2n_1^2 - n_2^2}{m}\right)\omega_r \tag{13}$$

The condition for a signal in the radio frequency domain is that out of the combinations of integers in the bracketed term there is one, that is large enough to multiply ω_r all the way up to the optical frequency ω_0. For $m \to \infty$ this condition is identical of the comb spanning an optical octave. If the combs bandwidth is sufficient there are many combinations of integers that fulfill this requirement, i.e. several radio frequency beat notes may be taken as ω_0. Again this is very similar to regular frequency combs where the offset frequency is only determined modulo the repetition rate. Which of the beat notes is taken for ω_0 does not matter as long as the mode numbering is adapted to that choice. Frequency doubling the Talbot comb however will generate even more frequencies as assumed in (13), as we will see in the next section. The challenge then might be to identify the correct beat note that determines ω_0.

4 Non-linear Interactions

Both, frequency doubling for self-referencing and spectral broadening to obtain an octave spanning bandwidth require non-linear processes. Driving a $\chi^{(2)}$ non-linearity the resulting field can be written as

$$E^2(t) = E_0^2 e^{-i2\omega_0 t} \sum_{n,n'} a_n a_{n'} e^{-i\left(n+n'+\frac{n^2+n'^2}{m}\right)\omega_r t}. \tag{14}$$

Just like (4), this field reproduces itself after the recurrence time $T = 2\pi m/\omega_r$ by virtue of the same argument made in (3). This should not surprise because whatever the $\chi^{(2)}$ does in the time domain, a periodic input should result in an output of the same periodicity. The mode distribution of terms with $n = n'$ belongs to a frequency doubled Talbot comb as used in (13). Since there is only one such combination in the sum for each n, the $\chi^{(2)}$ process is expected to be less efficient than for a regular frequency comb. Of course this can also be understood in the time domain where a short pulse is formed only after m cavity round trips. The mode combinations in (14) with $n' \neq n$ do not belong to the doubled Talbot comb. This probably means that the frequency doubled Talbot comb are not good for the type of spectroscopy described above. However, the usual $f - 2f$ self referencing is possible using these processes if one finds ways for a more efficient non-linear interaction. Since the doubling process is expected to be weak, it might be advisable to work with an auxiliary frequency doubled cw laser beating the fundamental and second harmonic with the Talbot comb and the doubled Talbot comb respectively. It should be mentioned that even without self-referencing the Talbot comb could be a useful tool. One might instead reference one of its modes to a wavemeter or to a atomic or molecular line. In fact for the intended broad band spectroscopy referencing to an atomic clock is usually not necessary.

With a similar argument as above, driving a $\chi^{(3)}$ non-linearity for spectral broadening the combination of modes

$$\omega_0 + \left(n + n' - n'' + \frac{n^2 + n'^2 - n''^2}{m}\right)\omega_r \tag{15}$$

are generated. In a regular frequency comb this process simply extends spectrum by adding new modes on the extended grid of original modes. In the case of the Talbot comb this does not take place. One can see this by requiring the modes in (15) to be a Talbot comb of the same m: $(n + n' - n'')^2 = n^2 + n'^2 - n''^2$. Solving this equation yields two solution $n'' = n$ and $n'' = n'$. In neither case additional modes are added to the initial Talbot comb.

5 Example Design

One of the possibilities to demonstrate the Talbot comb is to employ a fiber-laser-based design. Because of the small mode-volume and long interaction length, the fiber-laser could introduce strong nonlinear effects such as nonlinear polarization rotation which is often used for conventional mode-locked fiber lasers [6]. In the case of Talbot comb, the mode-locking mechanism is effectively m times smaller than the conventional lasers. Therefore either strong nonlinear effects or matching the overall dispersion to (6) with very good accuracy is required to enforce the Talbot mode-locking. Intracavity dispersion can be introduced through a fiber Bragg grating (FBG) that can be designed with very large values for the group velocity dispersion and precise values for the higher order dispersions. To manufacture a FBG a grating is written into a photo sensitive fiber with a UV laser. Up to the 6th order dispersion is commercially available.

One or even several FBGs can be conveniently implemented into the fiber-laser cavity using an optical circulator. The cavity should include a gain fiber, a pump beam combiner, an optical isolator and an output coupler as shown in Fig. 5. An

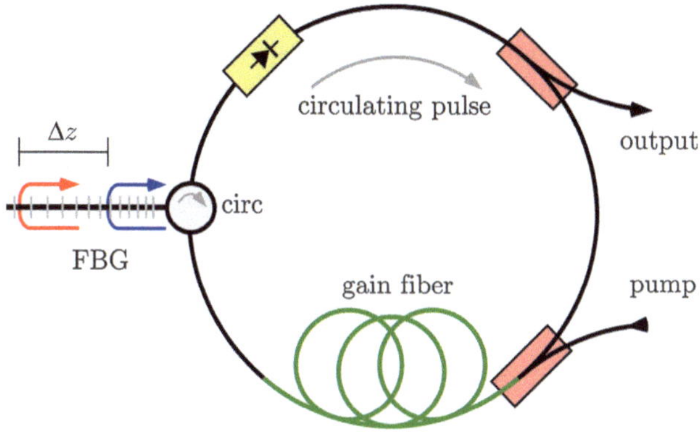

Fig. 5 Possible set-up of a Talbot laser build with fiber components. A fiber Bragg grating (FBG) included into the ring resonator with a circulator (circ) might be used to set the required dispersion. For smaller values of m and hence larger dispersion several FBGs might be part of the resonator

example for design parameters might be $m = 10^6$, $\omega_r = 2\pi \times 100$ MHz assuming a spectral width of $\delta\omega = 2\pi \times 10$ THz and an optical carrier frequency of $\omega_0 = 2\pi \times 300$ THz ($\lambda = 1$ μm) which is close to the Ytterbium gain maximum. With this we obtain from (11) $\phi''_{\omega_0} = -3.2 \times 10^7$ fs^2, $\phi'''_{\omega_0} = 3.0 \times 10^8$ fs^3 and $\phi^{(4)}_{\omega_0} = -4.8 \times 10^9$ fs^4 etc. Rather than this expansion one might use the first term in (7) to directly compute the required dispersion function.

To estimate the order of magnitude of the required length of the FBG Δz we calculate the difference of the round trip phase delay for the two ends of the spectrum (full width $\Delta\omega$) using (7): $\Delta\phi = \phi(\omega_0 + \Delta\omega/2) - \phi(\omega_0 - \Delta\omega/2)$. This phase difference has to be divided by the wavenumber $2\pi/\lambda$ to obtain the required path length difference between the two extreme colors. With the parameters above, dividing by the refractive index of 1.5 and taking into account that the light travels twice through the FBG we obtain $\Delta z = 3.3$ cm. The real length might then also depend on the requirements for the reflectivity. Fiber lasers generally come with a large optical gain so that it may be possible to compromise on that parameter. The first design may not be the rather optimistic one of this proposal but could be a trade off between large m (=low dispersion) and small m (=stronger mode locking). To start operation at a very large value of m it may also conceivable to include an intracavity modulator that mimics all or parts of the temporal envelope shown in Fig. 2.

6 Conclusions

This article is dedicated to Theodor Hänsch on the occasion of his 75th birthday. Among the many other laser tricks that he has invented is the optical frequency comb. Originally intended as a tool to measure laser frequencies, it has found several other applications for example in attosecond science and astronomy. Further it is used for large bandwidth direct comb spectroscopy. The current work tries to extends these possibilities, even though we have to admit that more ideas are required to turn it into a useful tool. Nevertheless we hope that this idea provides a new playground and entertainment for those who like curiosity driven research. In this spirit Theodor Hänsch has been our guide for many decades and we hope that there will be many more to follow. Happy Birthday!

Acknowledgements Open access funding provided by Max Planck Society.

References

1. S. Teng, J. Wang, F. Li, W. Zhang, Opt. Commun. **315**, 103 (2014)
2. J. Azaña, M.A. Muriel, Appl. Opt. **38**, 6700 (1999)
3. T. Suzuki, M. Katsuragawa, Opt. Expr. **18**, 23088 (2010)
4. Th Udem, R. Holzwarth, T.W. Hänsch, Nature **416**, 233 (2002)
5. T. Ideguchi et al., Nat. Commun. **5**, 3375 (2014)
6. M. Hofer et al., Opt. Lett. **16**, 502 (1991)

T^3-Interferometer for Atoms

M. Zimmermann, M.A. Efremov, A. Roura, W.P. Schleich, S.A. DeSavage, J.P. Davis, A. Srinivasan, F.A. Narducci, S.A. Werner, and E.M. Rasel

Abstract The quantum mechanical propagator of a massive particle in a linear gravitational potential derived already in 1927 by Kennard [2, 3] contains a phase that scales with the third power of the time T during which the particle experiences the corresponding force. Since in conventional atom interferometers the internal atomic states are all exposed to the same acceleration a, this T^3-phase cancels out

Our proposal of a T^3-interferometer for atoms was inspired by the seminal experiment of Hänsch and collaborators [1] to test the equivalence principle of general relativity based on a matter-wave interferometer for two different isotopes of rubidium. Whereas Ref. [1] profits from quadratic phases reminiscent of the Talbot effect we employ the cubic phase of the quantum mechanical propagator associated with a particle moving in a linear potential combined with a Ramsey interferometer. It is with great pleasure that we dedicate this article to Theodor W. Hänsch on the occasion of his 75th birthday.

This article is part of the topical collection "Enlightening the World with the Laser" - Honoring T. W. Hänsch guest edited by Tilman Esslinger, Nathalie Picqué, and Thomas Udem.

M. Zimmermann • M.A. Efremov (✉) • A. Roura
Institut für Quantenphysik and Center for Integrated Quantum Science and Technology (IQST), Universität Ulm, D-89081 Ulm, Germany
e-mail: max.efremov@gmail.com

W.P. Schleich
Institut für Quantenphysik and Center for Integrated Quantum Science and Technology (IQST), Universität Ulm, D-89081 Ulm, Germany

Texas A & M University Institute for Advanced Study (TIAS), Institute for Quantum Science and Engineering (IQSE) and Department of Physics and Astronomy, Texas A & M University, College Station, TX 77843-4242, USA

S.A. DeSavage • F.A. Narducci
Naval Air Systems Command, EO Sensors Division, Patuxent River, MD 20670, USA

J.P. Davis
AMPAC, North Wales, PA 19154, USA

A. Srinivasan
St. Mary's College of Maryland, St. Mary's City, MD 20686-3001, USA

S.A. Werner
Physics Laboratory, NIST, Gaithersburg, MD 20899, USA

E.M. Rasel
Institut für Quantenoptik, Leibniz Universität Hannover, D-30167 Hannover, Germany

D. Meschede et al. (eds.), *Exploring the World with the Laser*,
https://doi.org/10.1007/978-3-319-64346-5_27

and the interferometer phase scales as T^2. In contrast, by applying an external magnetic field we prepare two different accelerations a_1 and a_2 for two internal states of the atom, which translate themselves into two different cubic phases and the resulting interferometer phase scales as T^3. We present the theoretical background for, and summarize our progress towards experimentally realizing such a novel atom interferometer.

1 Introduction

Phases play an extraordinary role in quantum theory. On one hand, they represent the central ingredient of wave mechanics à la Schrödinger, and on the other, they build a bridge to classical mechanics à la Hamilton–Jacobi [4]. For these reasons they constitute a crucial ingredient of matter-wave interferometers [5–7] which nowadays represent standard tools for precision measurements. In this article, we propose a new type of atom interferometer whose phase scales with the cube of the time T the atom spends in the interferometer.

1.1 In a Nutshell

In 1927, Kennard [2, 3] showed that a particle exposed for a time T to a linear potential accumulates a phase proportional to T^3. Since then the Kennard phase has been rediscovered by many authors [8–10], especially in the context [11] of atom interferometry. However, no suggestion for its measurement in this field has been given.

It is interesting that recently an experiment [12] on an atom interferometer displaying a cubic phase [13] was reported. Our proposal is different in three aspects: (*i*) We obtain a closed interferometer with four rather than three pulses, and no momentum transfer from the laser beams to the atoms is required. (*ii*) In contrast to Ref. [12] our interferometer relies solely on the application of time-independent forces, and (*iii*) the interferometer of Ref. [12] leads to a phase shift that is quadratic *and* cubic in T, whereas the one in our device is solely cubic in T, and is proportional to the difference of the squares of these two constant accelerations.

Our scheme is reminiscent of the Ramsey–Bordé interferometer [14, 15] and other four-pulse configurations frequently employed as gradiometers and gyrometers [16–28]. However, we emphasize that in our setup there is no exchange of photon momenta.

In atom interferometry cubic phases are not rare. They appear for example as corrections due to gravity gradients [29–31], but also in the butterfly interferometer which is sensitive to rotation [27, 28]. However, again these devices rely on the exchange of the photon momentum.

There exist many different techniques to calculate the phase shift in an atom interferometer ranging from the semi-classical action [11] associated with the Feynman path integral [8], via the intriguing formalism of Christian Bordé based on a five-dimensional theory [7, 32] to the use of analogies to neutron optics [33, 34]. Throughout this article, we pursue a representation-free approach based on unitary time evolution operators [28, 35–38] and outline our interferometer, evaluate the resulting phase shift and review the present status of the experiment.

1.2 At the Interface of Quantum and Gravity

During an impromptu seminar at the NATO Advanced Summer Institute at Bad Windsheim in 1981, Eugene Paul Wigner expressed his discomfort with general relativity [39] in view of quantum mechanics [40]. He emphasized that the notion of a space-time trajectory which is a crucial element of gravity is incompatible with the uncertainty relation of quantum mechanics. Guided by the work of Niels Bohr and Leon Rosenfeld on the limitations of the electromagnetic field [41, 42] due to quantum fluctuations he argued that quantum theory puts severe restrictions on the measurement of the metric tensor representing the gravitational field.

Wigner's thoughts expressed in this seminar were a consequence of his work several decades earlier. Indeed, already in 1958 together with Helmut Salecker he had constructed a clock [43] based on the reflection of light signals from two mirrors and had analyzed the restrictions of the uncertainty relation on the weight of the mirrors.[1]

In this context, it is also worth mentioning that H. Salecker at the *Conference on the Role of Gravitation in Physics* in Chapel Hill in 1957 triggered [45] a discussion on the equivalence principle by a gedanken experiment involving a stream of particles being scattered off a diffraction grating. Greenberger [46] a decade later considered a similar arrangement and even argued that mass in quantum mechanics should be an operator.

The celebrated Colella–Overhauser–Werner (COW) experiment [47] performed in 1975 propelled these and many other thoughts about the interface of quantum and gravity to the real world. Indeed, based on the de Broglie wave nature of neutrons [33] the COW experiment could measure for the first time the phase shift between two arms of a neutron interferometer induced by the gravitational potential of the earth [47–49].

The development of new sources of cold atoms [6] as well as molecules [50], and in particular, the realization of Bose–Einstein condensates [51, 52] has ushered in a new era of experiments to probe quantum mechanics and gravity. Indeed, novel

[1]John Archibald Wheeler frequently emphasized in conversations about this topic and in print [44] that these estimates were too conservative. However, to the best of our knowledge they have never been improved.

tests of the equivalence principle based on matter-wave interferometry of different isotopes of the same atom [1] such as ^{85}Rb and ^{87}Rb or even different species [53] such as ^{39}K and ^{87}Rb could be performed. Even the detection of gravitational waves based on atom interferometry is pursued today [54]. Recently, Ref. [55] suggested that gravitational decoherence gives rise to a universal decoherence. Moreover, the atom laser [56] utilizes gravity to form an Airy mode. Furthermore, it is mind-boggling that nowadays measurements of atomic transitions [57, 58] are sensitive to the redshift of the gravitational field.

An extremely interesting development in this realm is again taking place in the field of neutron optics due to the experimental realization of the quantum bouncer [59, 60]. Here neutrons exposed to the gravitational field of the earth are reflected from a surface and oscillate up and down. In particular, they experience a potential consisting of the linear ramp and an infinitely steep wall. It is amazing that a measurement of the transitions between the resulting discrete energy levels can put upper bounds [61] on dark energy and dark matter scenarios.

1.3 Drive for Enhanced Sensitivity

Hopefully these examples convey the excitement in this field of quantum optics and gravity. Notwithstanding the fact that we still do not have a complete understanding of quantum gravity [62] we have come a long way since the early Salecker–Wigner discussions but many questions remain. Indeed, goals such as gravitational wave detection or a compact gravimeter [63] based on atom optics drive the strive for higher sensitivity of these devices.

The Kasevich–Chu atom interferometer [64, 65] is analogous to the neutron interferometer employed by the COW experiment. The Bragg diffraction of the neutron from three crystal planes of a silicon slab are replaced by Raman diffraction of the atom from three standing light crystals. As a result, the phase shift introduced by the gravitational potential is *quadratic* in the time $2T$ the particle spends in the interferometer.

Needless to say a different scaling of the phase shift, for example with T^3, would be desirable. In the present article, we propose such an interferometer and describe our progress towards realizing it.

Our device rests on three principles: (*i*) We take advantage of the cubic dependence of the phase in the propagator in a linear potential. (*ii*) We employ two different internal states of the atom which experience different time-independent accelerations, and (*iii*) we close the interferometer in position *and* velocity by a sequence of four laser pulses.

1.4 Outline

Our article is organized as follows. Section 2 serves as a motivation. Here we recall that the propagator of a particle in a linear potential contains a phase which is cubic in time.

In Sect. 3 we introduce our interferometer capable of measuring the cubic phase accumulated by a particle during its motion in a linear potential provided by a constant gravitational field and a magnetic field gradient. The three-level atom probing these fields has two ground states associated with two different magnetic moments and experiences a sequence of four Raman pulses. To close the interferometer in position *and* velocity, we choose the separation of $T - 2T - T$ between the pulses. The resulting probability for the atom to exit the interferometer in one of the two ground states is the familiar oscillatory function. However, in contrast to standard interferometers its argument now depends solely on the phase cubic in T and the discrete third derivative of the laser phase.

We dedicate Sect. 4 to a comparison of our scheme to the Kasevich–Chu interferometer [36, 64–67] and distinguish the cubic phase shift from the ones caused by a gravity gradient or the Continuous-Acceleration Bloch (CAB) technique [12]. In Sect. 5 we discuss a possible experimental implementation of our proposal. We conclude in Sect. 6 by briefly summarizing our results and providing an outlook.

To keep our article self-contained but focused on the central ideas, we include lengthy calculations in five appendices. In Appendix A we show that the Kennard phase depends on the initial wave function. However, we emphasize that in our arrangement the resulting interferometer phase which is cubic in time is independent of the initial state of the center-of-mass motion. In Appendix B we recall the technique of creating coherent superpositions of the two atomic ground states and interchanging their populations using Raman pulses. We then turn in Appendix C to a description of our interferometer as a sequence of unitary operators and derive an expression for the phase of our T^3-interferometer. In Appendix D we derive the conditions to close our interferometer and obtain in Appendix E an explicit formula for the resulting phase. It is interesting that this result also follows from the formalism of Refs. [37, 38].

2 From Global to Interferometer Phase

The propagator of a quantum particle experiencing a linear potential is determined by a phase factor governed by the corresponding classical action. Since the relevant classical motion involves time in the coordinate and velocity in a quadratic and a linear way, both the kinetic as well as potential energies bring in time quadratically. As a result, the action being the integral over time must contain a term proportional to the cube of time. This cubic phase which is independent of the coordinate is at the center of our interest in the present section.

We first recall the essential features of the propagator for the wave function in a linear potential. Here we focus especially on this cubic phase. Moreover, in Appendix A we show that due to the Huygens principle for matter waves the integration over the initial coordinate leads to a dependence of this phase on the initial wave function.

Although we find this property interesting we emphasize that it is of no importance to the present discussion. Indeed, due to the Born rule we cannot measure the global phase factor of a *single* quantum system. However, an interferometric measurement of the difference of *two* different global phase factors of *two* systems is possible. Especially, for a closed interferometer in which the phase shift does not depend on the initial state also the cubic phase is independent [37, 38] of the initial wave function. We dedicate the second part of this section to this topic.

2.1 Emergence of T^3-Phase in the Propagator

We start our analysis by discussing the propagator of a particle in a linear potential. Here we emphasize especially the emergence of the phase factor cubic in time.

We consider a particle of mass m moving in a linear potential $V(z) \equiv -Fz$ corresponding to a constant force $\mathbf{F} \equiv F\mathbf{e}_z$ directed along the z-axis with the unit vector $\mathbf{e}_z$. This problem occurs for (1) a particle, which experiences a constant gravitational acceleration g with $F \equiv -mg$ as indicated in Fig. 1a, and (2) a charge $-e$ in an ideal capacitor with the constant electric field E, for which $F \equiv -eE$ as shown in Fig. 1b. Throughout this article we focus on the example of a linear gravitational potential.

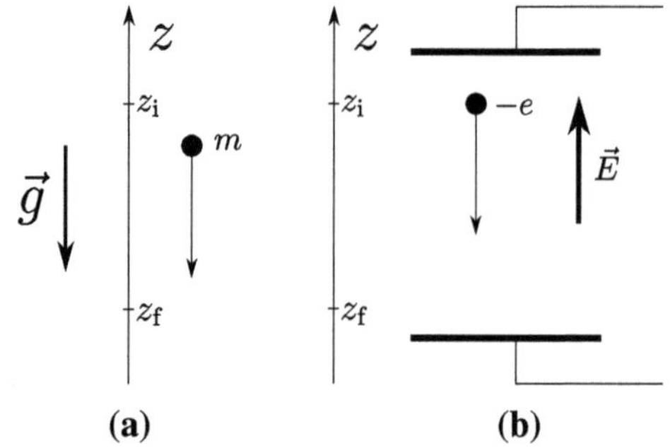

Fig. 1 Two physical systems with a linear potential $V(z) \equiv -Fz$ corresponding to a constant force $\mathbf{F} = F\mathbf{e}_z$ directed along the z-axis: **a** a particle of mass m which experiences the constant gravitational acceleration g with $F \equiv -mg$, and **b** a charge $-e$ in an ideal capacitor with the constant and homogeneous electric field E with $F \equiv -eE$

The wave function

$$\psi(z_\mathrm{f}, t_\mathrm{f}) = \int_{-\infty}^{+\infty} G(z_\mathrm{f}, t_\mathrm{f}|z_\mathrm{i}, t_\mathrm{i})\psi(z_\mathrm{i}, t_\mathrm{i})\mathrm{d}z_\mathrm{i} \tag{1}$$

representing the probability amplitude to find the particle at the final position z_f at time t_f is determined by the propagator [2, 3, 8]

$$G(z_\mathrm{f}, t_\mathrm{f}|z_\mathrm{i}, t_\mathrm{i}) \equiv \left\langle z_\mathrm{f}\left|\exp\left[-\frac{i}{\hbar}\left(\frac{\hat{p}_z^2}{2m} - F\hat{z}\right)(t_\mathrm{f} - t_\mathrm{i})\right]\right|z_\mathrm{i}\right\rangle, \tag{2}$$

where $\psi(z_\mathrm{i}, t_\mathrm{i})$ is the value of the wave function at the initial position z_i and time t_i, and $\hat{z}$ and $\hat{p}_z$ denote the position and momentum operators, respectively.

The propagator G defined by Eq. (2) can be cast [8] in terms of the classical action

$$\begin{aligned} S_\mathrm{cl}(z_\mathrm{f}, t_\mathrm{f}|z_\mathrm{i}, t_\mathrm{i}) &\equiv \int_{t_\mathrm{i}}^{t_\mathrm{f}} L\big(z_\mathrm{cl}(t), \dot{z}_\mathrm{cl}(t)\big)\mathrm{d}t \\ &= \frac{m}{2}\frac{(z_\mathrm{f} - z_\mathrm{i})^2}{t_\mathrm{f} - t_\mathrm{i}} + \frac{F}{2}(z_\mathrm{f} + z_\mathrm{i})(t_\mathrm{f} - t_\mathrm{i}) - \frac{F^2}{24m}(t_\mathrm{f} - t_\mathrm{i})^3 \end{aligned} \tag{3}$$

along the classical trajectory given by

$$z_\mathrm{cl}(t) \equiv z_\mathrm{i} + \frac{z_\mathrm{f} - z_\mathrm{i}}{t_\mathrm{f} - t_\mathrm{i}}(t - t_\mathrm{i}) + \frac{F}{2m}(t - t_\mathrm{i})(t - t_\mathrm{f}) \tag{4}$$

and

$$\dot{z}_\mathrm{cl}(t) \equiv \frac{d}{dt}z_\mathrm{cl}(t) = \frac{z_\mathrm{f} - z_\mathrm{i}}{t_\mathrm{f} - t_\mathrm{i}} + \frac{F}{m}\left(t - \frac{t_\mathrm{i} + t_\mathrm{f}}{2}\right). \tag{5}$$

Here we have used the Lagrangian

$$L(z, \dot{z}) \equiv \frac{m}{2}\dot{z}^2 + Fz \tag{6}$$

of a particle in a linear potential.

Indeed, the representation

$$G(z_\mathrm{f}, t_\mathrm{f}|z_\mathrm{i}, t_\mathrm{i}) = N(t_\mathrm{f} - t_\mathrm{i})\exp\left[\frac{i}{\hbar}S_\mathrm{cl}(z_\mathrm{f}, t_\mathrm{f}|z_\mathrm{i}, t_\mathrm{i})\right] \tag{7}$$

with the normalization

$$N(\tau) \equiv \sqrt{\frac{m}{2i\pi\hbar\tau}} \tag{8}$$

brings out most clearly that G contains the phase

$$\phi(\tau) \equiv -\frac{1}{24}\frac{F^2}{\hbar m}\tau^3, \tag{9}$$

which is independent of the initial and final positions z_{i} and z_{f}, and scales with the third power of the time difference

$$\tau \equiv t_{\mathrm{f}} - t_{\mathrm{i}}, \tag{10}$$

that is the time during which the particle experiences the constant force F.

2.2 How to Observe the T^3-Phase?

The propagator G defined in Eq. (7) contains a global phase ϕ given by Eq. (9) which is cubic in time, proportional to the square of the constant force F, and inversely proportional to the mass m of the particle. However, due to the integration over the initial position in the Huygens integral, Eq. (1), this phase depends, as exemplified in Appendix A, on the initial wave function. We now briefly outline our strategy for measuring this phase and emphasize that for our closed interferometer the resulting phase is independent of the initial state.

Obviously a setup providing us only with the probability density $|\psi(z_{\mathrm{f}}, t_{\mathrm{f}})|^2$ is insensitive to any global phase like the T^3-phase. Therefore, we need to involve an interferometric measurement either with a path-dependent strength of the constant force, or a path-dependent mass of the particle.

Throughout this article we focus on the first alternative although we can imagine possibilities[2] to utilize the dependence of the mass on the internal state. Key elements of our technique are: (*i*) an atom with two internal states associated with two different magnetic moments, and (*ii*) an external time-independent magnetic field with a constant gradient along one direction. Due to the Zeeman effect, the atom experiences a constant force determined by its internal state. It is the same force that acts on a classical magnetic dipole in a non-uniform magnetic field.

[2]According to Ref. [68] mass and proper time are conjugate variables and two internal states of the atom correspond [7, 32] to two different masses giving rise to an additional phase shift [54] for the atom prepared in a superposition state. Although this effect is minute the improved scaling of the Kennard phase might help to identify this effect.

3 Atom interferometer with Four Light Pulses

In this section, we introduce the two crucial elements of our T^3-interferometer: (*i*) the population dynamics of the two resonant states of the atom driven by the Raman laser pulses, and (*ii*) the Zeeman shift of the atomic levels induced by the external time-independent magnetic field with a constant gradient along one direction. The description of our interferometer is based on the representation-free approach described in Refs. [28, 35, 36].

3.1 Population Dynamics

We are now in the position to present our atom interferometer capable of measuring the cubic phase. The general scheme depicted in Fig. 2 consists of two distinct "building blocks": (*i*) four Raman pulses, that is two $\frac{\pi}{2}$- and two π-pulses, which form a $\frac{\pi}{2} - \pi - \pi - \frac{\pi}{2}$ sequence, and (*ii*) three regions of the atomic center-of-mass motion with constant accelerations a_1 and a_2.

We consider a three-level atom consisting of the ground state $|g_1\rangle$, the state $|g_2\rangle$, and the excited state $|e\rangle$. Here $|g_1\rangle$ and $|g_2\rangle$ are chosen such that the mean values of their magnetic moment are different, which leads to a state-dependent acceleration and magnetically induced phase shifts [69, 70]. The center-of-mass motion of the atom is assumed to be along the z-axis, which is the direction of the constant gravitational acceleration.

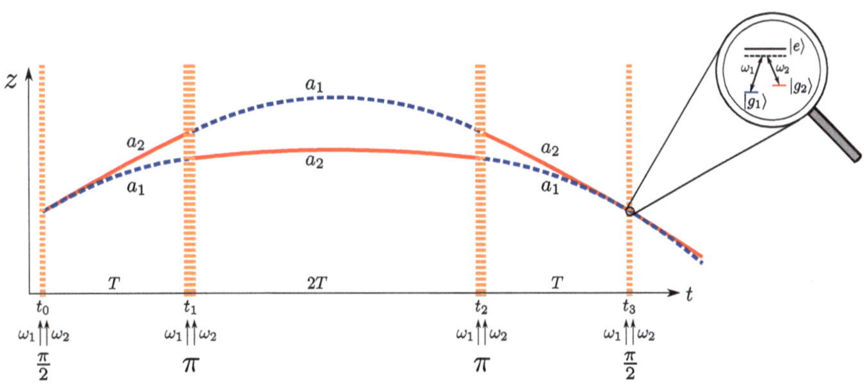

Fig. 2 Space-time diagram of the T^3-interferometer for a three-level atom consisting of the states $|g_1\rangle$, $|g_2\rangle$, and $|e\rangle$, and interacting with four short Raman laser pulses at $t = t_0$, $t = t_1 \equiv t_0 + T$, $t = t_2 \equiv t_0 + 3T$ and $t = t_3 \equiv t_0 + 4T$. The laser frequencies ω_1 and ω_2 are assumed to only drive the transitions $|g_1\rangle \longleftrightarrow |e\rangle$ and $|g_2\rangle \longleftrightarrow |e\rangle$, respectively

Thus, we arrive at the Hamiltonian

$$\begin{aligned}\hat{H}_{\text{at}} \equiv 1_3 \otimes \frac{\hat{p}_z^2}{2m} &+ |e\rangle\langle e| \otimes (E_e 1_z - m a_e \hat{z}) \\ &+ |g_1\rangle\langle g_1| \otimes (E_{g_1} 1_z - m a_1 \hat{z}) \\ &+ |g_2\rangle\langle g_2| \otimes (E_{g_2} 1_z - m a_2 \hat{z}),\end{aligned} \tag{11}$$

of the three-level atom, where

$$1_3 \equiv |e\rangle\langle e| + |g_1\rangle\langle g_1| + |g_2\rangle\langle g_2|$$

and

$$1_z \equiv \int_{-\infty}^{+\infty} \mathrm{d}z \, |z\rangle\langle z|$$

are the identity operators corresponding to the Hilbert space of the internal atomic states, and the center-of-mass motion along the z-axis, respectively. Here E_e, E_{g_1}, and E_{g_2} are the energies of the atom in the internal states $|e\rangle$, $|g_1\rangle$, and $|g_2\rangle$, respectively, with a_e being the constant acceleration corresponding to the excited state.

Each block of our T^3-interferometer is described in terms of an appropriate unitary operator. As discussed in more detail in Appendix B, the atom–light interaction is accounted for by the evolution operator

$$\begin{aligned}\hat{U}_{\text{p}}(\theta) \equiv & \left(|g_1\rangle\langle g_1| + |g_2\rangle\langle g_2| \right) \cos\left(\frac{\theta}{2}\right) \\ & - i\left(e^{i\phi_{\text{L}}}|g_1\rangle\langle g_2| + e^{-i\phi_{\text{L}}}|g_2\rangle\langle g_1|\right) \sin\left(\frac{\theta}{2}\right),\end{aligned} \tag{12}$$

where $\phi_{\text{L}}(t) \equiv \phi_2 - \phi_1$ is the difference of the phases ϕ_1 and ϕ_2 of the two lasers used for the Raman transition and θ denotes the total pulse area. Moreover, we have assumed that each Raman pulse consists of two co-propagating laser beams with almost identical wavelengths which makes $\hat{U}_{\text{p}}$ given by Eq. (12) independent of z. Hence, there is no momentum transfer from the light to the atom.

On the other hand, the operator

$$\hat{U}_a(t_{\text{f}}, t_{\text{i}}) \equiv \exp\left[-\frac{i}{\hbar}\left(\frac{\hat{p}_z^2}{2m} - m a \hat{z}\right)(t_{\text{f}} - t_{\text{i}})\right] \tag{13}$$

with $a = a_1$ or $a = a_2$ provides us with the center-of-mass motion.

In Appendices C and D we analyze the interferometer of Fig. 2 as a sequence of these unitary operators and find the expression

$$P_{g_2} = \frac{1}{2}[1 + \cos(\varphi_\mathrm{i} + \varphi_\mathrm{L})] \tag{14}$$

for the probability of observing the atoms in the state $|g_2\rangle$ after the action of the four Raman pulses. According to Appendix E, the interferometer phase φ_i reads

$$\varphi_\mathrm{i} \equiv \frac{m}{\hbar}(a_1^2 - a_2^2)T^3, \tag{15}$$

and in Appendix C we derive the formula

$$\varphi_\mathrm{L} \equiv \phi_\mathrm{L}(t_0) - 2\phi_\mathrm{L}(t_0 + T) + 2\phi_\mathrm{L}(t_0 + 3T) - \phi_\mathrm{L}(t_0 + 4T) \tag{16}$$

for the contribution due to the phases ϕ_L of the four laser pulses.

We emphasize that this result is independent of the initial state of the center-of-mass motion. This property of the *interferometer phase* is in sharp contrast to the dependence of *global phase* of Appendix A corresponding to a *single* degree of freedom, and is a consequence [28, 37, 38] of the fact that our interferometer is closed in both position *and* velocity.

3.2 Zeeman Effect: Control of External Degrees of Freedom

Next, we turn to a possible realization of our interferometer scheme, and in particular, of the accelerations a_1 and a_2. To make contact with the experiment discussed in Sect. 5, we consider a special case.

For this purpose we focus on the interaction of the atom with a time-independent magnetic field having locally the form[3]

$$\mathbf{B}(\mathbf{r}) \cong (B_0 + z\,\nabla_z B_z)\mathbf{e}_z \tag{17}$$

which results in the linear Zeeman shifts

$$\Delta E_{g_1}^Z = \mu_B g_{F_1} m_{F_1}(B_0 + z\,\nabla_z B_z) \tag{18}$$

and

[3]Throughout the article, we use the notation $\nabla_z B_z \equiv \frac{\partial B_z}{\partial z}(\mathbf{r}=0)$ for the derivative of the z-component of the magnetic field $\mathbf{B} = \mathbf{B}(\mathbf{r})$ along the z-direction at the origin $\mathbf{r} = 0$. This derivative is assumed to be small compared to B_0, such that $L|\nabla_z B_z| \ll |B_0|$, where L is the total length of the interferometer. Moreover, we note that the form of the magnetic field given by Eq. (17) is an approximate one. Indeed, according to the Maxwell equation $\nabla \cdot \mathbf{B} = 0$, which is valid everywhere, a non-zero value of $\nabla_z B_z$ induces non-zero values of $\nabla_x B_x$ and $\nabla_y B_y$, such that $\nabla_x B_x + \nabla_y B_y = -\nabla_z B_z$, where B_x and B_y are the components of $\mathbf{B}$ along the x- and y-axis. However, in the limit of $L|\nabla_z B_z| \ll |B_0|$ the magnetic field $\mathbf{B}$ given by Eq. (17) is approximately directed along the z-axis.

$$\Delta E^Z_{g_2} = \mu_B g_{F_2} m_{F_2} (B_0 + z \nabla_z B_z) \tag{19}$$

of the energies of $|g_1\rangle$ and $|g_2\rangle$. Here μ_B, g_{F_1}, g_{F_2}, m_{F_1}, and m_{F_2} denote the Bohr magneton, the Landé g-factors of $|g_1\rangle$ and $|g_2\rangle$, and the magnetic quantum numbers associated with the z-component of the angular momentum corresponding to $|g_1\rangle$ and $|g_2\rangle$.

The homogeneous magnetic field B_0 leads to an energy shift of the magnetive sensitive states and together with the constant gravitational field we arrive at the expressions

$$a_1 \equiv -g - \frac{\mu_B}{m} g_{F_1} m_{F_1} \nabla_z B_z \tag{20}$$

and

$$a_2 \equiv -g - \frac{\mu_B}{m} g_{F_2} m_{F_2} \nabla_z B_z \tag{21}$$

for the accelerations of the atomic center-of-mass corresponding to $|g_1\rangle$ and $|g_2\rangle$, respectively.

In our experiment, we use the D2 transition of ^{85}Rb and choose $|g_1\rangle$ and $|g_2\rangle$ from the $F = 2$ and $F = 3$ hyperfine state manifolds. For the atomic transition $|F_1 = 2, m_{F_1} = 1\rangle \to |F_2 = 3, m_{F_2} = 1\rangle$ the interferometer phase given by Eq. (15) reduces to

$$\varphi_i = \frac{4}{3} \frac{\mu_B}{\hbar} \nabla_z B_z g T^3 , \tag{22}$$

where we have used the fact [71] that $g_{F_2} = -g_{F_1} = 1/3$.

4 Discussion

In the preceding section, we have derived an expression for the probability of finding the atom in the state $|g_2\rangle$ at one exit of our interferometer. In the present section, we compare and contrast our device with the Kasevich–Chu interferometer, and make contact with other cubic phases such as the ones caused by a gravity gradient, or arising in the CAB technique.

4.1 Comparison with Kasevich–Chu Interferometer

We first recall the expressions corresponding to Eqs. (14), (15) and (16) and then discuss the similarities and differences between these two devices. A brief analysis of the respective scale factors concludes this comparison.

4.1.1 General Considerations

In the Kasevich–Chu interferometer, the probability corresponding to Eq. (14) reads [11, 29, 36, 64, 65, 67]

$$P_{g_2}^{(KC)} = \frac{1}{2}\left[1 - \cos\left(\varphi_{\rm i}^{(KC)} + \varphi_{\rm L}^{(KC)}\right)\right] \tag{23}$$

with the interferometer phase

$$\varphi_{\rm i}^{(KC)} \equiv (k_1 + k_2)gT^2 \tag{24}$$

and the total laser phase

$$\varphi_{\rm L}^{(KC)} \equiv \phi_{\rm L}(t_0) - 2\phi_{\rm L}(t_0 + T) + \phi_{\rm L}(t_0 + 2T). \tag{25}$$

We note four major differences between our scheme and that of Kasevich–Chu: (*i*) The *four* rather than the *three* Raman pulses create the *sum* of the two terms appearing in the square brackets of P_{g_2} given by Eq. (14) rather than the *difference* in $P_{g_2}^{(KC)}$ defined by Eq. (23). In the absence of any potentials the different pulse sequences correspond to a 2π-rotation on the Bloch sphere in the case of the Kasevich–Chu interferometer, and a 3π-rotation for our T^3-interferometer giving rise to the opposite signs. (*ii*) The phase $\varphi_{\rm i}$ induced by the linear potentials and given by Eq. (15) depends on the separation T of the pulses in a *cubic* rather than *quadratic* way as in $\varphi_{\rm i}^{(KC)}$ expressed by Eq. (24). (*iii*) *Co-propagating* laser beams together with a constant magnetic field gradient lead to a proportionality of $\varphi_{\rm i}$ to $\nabla_z B_z$ as reflected by Eq. (22), while in the case of Kasevich–Chu the use of *counter-propagating* laser beams results in a momentum transfer of $\pm\hbar(k_1 + k_2)$ which reflects itself in $\varphi_{\rm i}^{(KC)}$. (*iv*) The total laser phase $\varphi_{\rm L}$ defined by Eq. (16) is a discrete *third* derivative rather than the *second* one for $\varphi_{\rm L}^{(KC)}$ given by Eq. (25). Indeed, this feature becomes obvious when we consider the limit of $T \to 0$, for which $\varphi_{\rm L} \cong -2\dddot{\varphi}_{\rm L}(t_0)T^3$ while [36, 67] $\varphi_{\rm L}^{(KC)} \cong \ddot{\varphi}_{\rm L}(t_0)T^2$, with $\dddot{\varphi}_{\rm L}(t_0)$ and $\ddot{\varphi}_{\rm L}(t_0)$ being the third and second continuous derivatives of the phase $\varphi_{\rm L} = \varphi_{\rm L}(t)$ of the Raman pulse, respectively.

4.1.2 Scale Factors and Limitations

Next, we turn to a comparison between the scale factors of the two interferometers. We first note that the phases φ_i, Eq. (22), and $\varphi_i^{(KC)}$, Eq. (24), are both (*i*) linear proportional to the gravitational acceleration g, (*ii*) independent of the atomic mass m, and (*iii*) determined only by classical quantities, that is Eqs. (22) and (24) do not include the Planck constant $\hbar$. For the latter statement we keep in mind that the Bohr magneton μ_B is linearly dependent on $\hbar$.

However, we emphasize that the difference in the scale factors due to the appearance of the gradient of the magnetic field in the phase φ_i, Eq. (22), rather than the sum of the wave vectors in $\varphi_i^{(KC)}$, Eq. (24), is crucial for the scale factor stability. Indeed, in our experiment discussed in more detail in Sect. 5 we have found that the stability of the magnetic field as measured by the frequency difference of the two transitions $|F_1 = 2, m_{F_1} = 0\rangle \rightarrow |F_2 = 3, m_{F_2} = 0\rangle$ and $|F_1 = 2, m_{F_1} = 0\rangle \rightarrow |F_2 = 3, m_{F_2} = 1\rangle$ did not vary more than 0.1 % without sophisticated electronic feedback or magnetic shielding. In contrast, the scale factor stability of the Kasevich–Chu interferometer can be estimated to be 1 part in 10^{10} due to the use of the wave vector of the light. Hence, in this respect the Kasevich–Chu interferometer is superior compared to our device.

Although our interferometer scales with T^3 rather than with T^2 as the Kasevich–Chu interferometer, φ_i is larger than $\varphi_i^{(KC)}$ only for $T > 0.23$ s with $\nabla_z B_z = 600$ μT/m and $k_1 + k_2 \approx 1.6 \times 10^7 \text{m}^{-1}$ [72]. Moreover, it is important to note that nowadays a beamsplitter with large momentum transfer [73, 74] allows to increase $\varphi_i^{(KC)}$ by a few orders of magnitudes.

In the presence of a magnetic field gradient $\nabla_z B_z = 600\,\mu\text{T/m}$ as measured in our experiment, we have one complete oscillation of the excited state population P_{g_2}, Eq. (14), for $T \gtrsim 2.1$ ms. However, if the signal-to-noise ratio allows it one could see the different dependence on T even for short times as demonstrated in Ref. [12].

4.2 Other Origins of Cubic Phases

We now compare and contrast the T^3-phase in our interferometer induced by the propagator of a particle in a linear potential to other phases cubic in time. Here we focus on two different situations: (*i*) the presence of a gravity gradient, or (*ii*) the application of the CAB technique.

4.2.1 Gravity Gradient

In the presence of a gravity gradient Γ a phase cubic in time appears [28–31, 37] in the Kasevich–Chu interferometer as a consequence of a *single quadratic* potential

$$V_\Gamma(z) \equiv mgz + \frac{1}{2}m\Gamma z^2 \tag{26}$$

rather than *two linear* potentials. In particular, it results from an expansion of the atomic center-of-mass motion in the limit of a weak gravity gradient, that is $\Gamma T^2 \ll 1$. Indeed, a gravity gradient leads to a *position-dependent* acceleration, while the two linear potentials in our interferometer lead to a *state-dependent* acceleration.

Furthermore, we emphasize that the Kasevich–Chu interferometer is no longer closed in position *and* velocity in the presence of a gravity gradient. This deficiency leads to a loss of contrast [37] and a dependence of the phase shift on the initial state.

The interferometer can be closed (at least approximately) by additional laser pulses [17] or by suitably adjusting the laser wavelength of the intermediate pulse [75]. However, this procedure also eliminates the cubic contributions to the phase shift, in contrast to the situation considered here.

4.2.2 Continuous-Acceleration-Bloch Technique

Our T^3-interferometer shares the underlying idea of CAB [12, 13], that is applying different accelerations along each interferometer arm. However, instead of achieving these accelerations via state-dependent linear potentials, a beam splitter based on Bragg diffraction is used to load one of the two exit ports into an optical lattice, which is accelerated subsequently by the use of Bloch oscillations.

While in our scheme we can close the interferometer easily in position *and* velocity by simply choosing the correct timing between the pulses, the CAB scheme requires a sophisticated control of the time-dependent acceleration of the optical lattice. Moreover, we emphasize that in contrast to the T^3-interferometer, in the CAB scheme not only a phase proportional to T^3, but also one proportional to T^2 emerges.

5 Towards an Experimental Realization

In this section, we discuss a possible experimental implementation of our proposal for a T^3-interferometer based on our current laboratory apparatus depicted in Fig. 3. We first summarize the key features of our setup, describe our method to deduce the magnetic fields from the observed Raman spectra, and conclude by briefly analyzing the present limitation of our device due to decoherence.

5.1 Experimental Setup

We use the D2 transition of ^{85}Rb and choose $|g_1\rangle$ and $|g_2\rangle$ from the $F = 2$ and $F = 3$ hyperfine state manifolds with a frequency separation of approximately 3 GHz [76]. The atoms are loaded into a three-dimensional magneto-optical trap (3D MOT) emerging from a two-dimensional trap (2D MOT) as shown in Fig. 3. The 3D MOT consists of standard cooling and repump beams as well as magnetic coils. We rely on an all-glass chamber as our vacuum system.

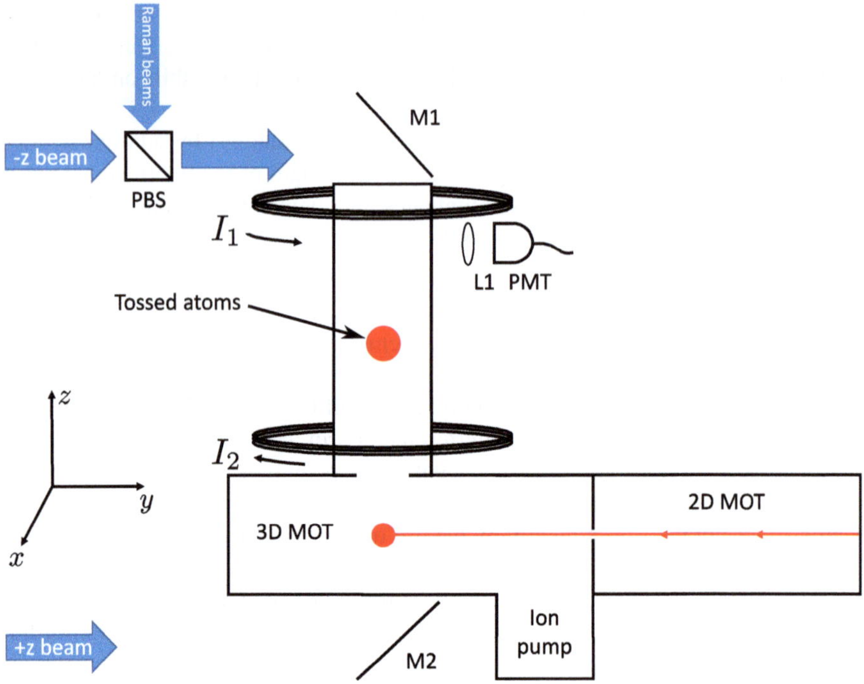

Fig. 3 Experimental setup for the T^3 -interferometer. Our ^{85}Rb atoms emerge from a two-dimensional magneto-optical trap (2D MOT), pass through an aperture into the 3D MOT as indicated by the *red line*, and are then launched into a glass tower in a moving molasses configuration. The mirrors M_1 and M_2 reflect the $+z$ and $-z$ beams of the 3D MOT, respectively, as well as the two co-propagating Raman beams with the parallel circular polarizations, used for the control of the internal atomic states. We employ two coils in an anti-Helmholtz configuration with currents I_1 and I_2 to create a magnetic field with constant gradient in the z-direction along the glass tower. A solenoid not depicted here surrounds the tower to provide an additional nominally uniform magnetic field. The lens L_1 collects the fluorescence from the tossed atoms at the top of the tower and focuses this light onto a photomultiplier (PMT)

After 1 s of loading to obtain a sufficient signal-to-noise ratio, the atoms are launched upwards along the z-axis in a moving optical molasses configuration with a velocity of approximately 3 m/s such that they strike the top of a 10 cm tall glass tower. It takes the atoms between 20 and 40 ms to reach this point depending on the launch velocity which can be adjusted by the voltage of the launch signal. We emphasize that the top of the tower does not coincide with the apex of the trajectory.

After launch, the atoms are freely moving in the dark and we are able to apply a single or several Raman pulses involving two co-propagating laser beams along the z-axis with the same circular polarization. During their motion the atoms interact with a magnetic field which varies linearly along the z-axis due to coils in an anti-Helmholtz configuration which are not the ones used to trap the atoms. Moreover, they feel the field of a solenoid of finite length not depicted in Fig. 3 surrounding the glass tower. The ability to change the current independently in each of the gradient coils and the solenoid provides us with the control of the location of the zero crossing of the magnetic field, or an effective way to adjust the bias field.

On the top of the tower a photomultiplier tube (PMT) detector performs a projective measurement of the population in the state $|g_2\rangle$ by collecting the fluorescence emitted by the atoms caused by the vertical trapping beams, that is the $+z$ and $-z$ beams, which are switched back on for the measurement. Observing the atomic population always at the top of the tower provides us with a convenient way of varying the position where the Raman pulse is applied to the atoms along their flight path.

5.2 Raman Spectrum

With this setup we can map out the magnetic field along the atom trajectory by measuring Raman spectra of the type shown in Fig. 4. For this purpose we first use an optical pumping stage to transfer all atoms to the ground state $|g_1\rangle$, that is the state with $F = 2$. Then we apply a Raman light pulse whose intensity and duration are chosen to be close to a zero-detuning π-pulse. We have found that this condition is satisfied for a pulse duration of 25–100 μs, using our typical total Raman power of approximately 80 mW in a beam of diameter 2.5 cm and a single photon detuning of

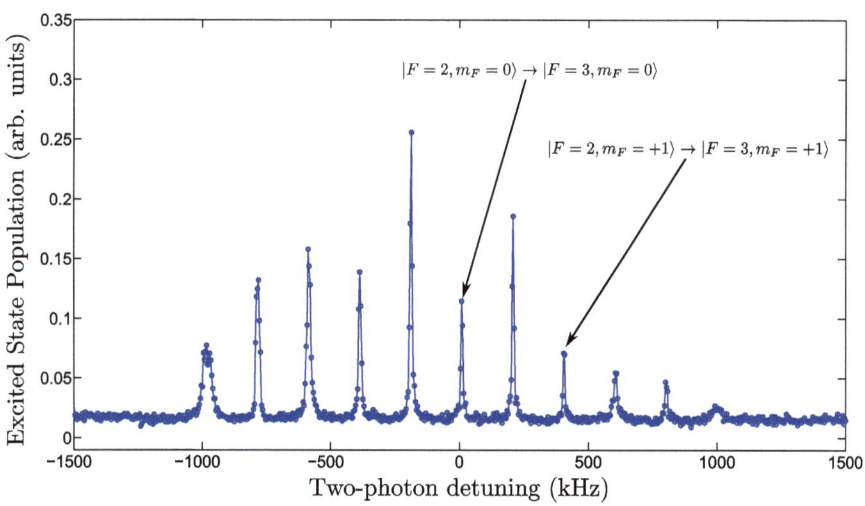

Fig. 4 Typical Raman spectrum containing the transitions between the states of the $F = 2$ and $F = 3$ manifolds for an arbitrarily aligned magnetic field in their dependence on the two-photon detuning $\widetilde{\Delta}$ defined by Eq. (27) with respect to the clock transition, for which resonance occurs at $\widetilde{\Delta} = 0$. We observe 11 resonances corresponding to 5 or 6 transitions with $\Delta m_F = 0$ or $\Delta m_F = \pm 1$, respectively. Transitions with $\Delta m_F = \pm 2$ are heavily suppressed [76, 77]

1–2 GHz. Finally, we observe the number of atoms transferred from $|g_1\rangle$ to $|g_2\rangle$, that is the state with $F = 3$.

In this manner, we obtain Raman spectra such as the one presented in Fig. 4, that is the population of $|g_2\rangle$ versus the two-photon detuning

$$\tilde{\Delta} \equiv \frac{E_{3,0} - E_{2,0}}{\hbar} + \omega_2 - \omega_1 \tag{27}$$

of the Raman pulse of the frequencies ω_1 and ω_2 with respect to the "clock" transition $|F_1 = 2, m_{F_1} = 0\rangle \to |F_2 = 3, m_{F_2} = 0\rangle$, for which the resonance occurs at $\tilde{\Delta} = 0$. Here E_{F,m_F} denotes the energy of the hyperfine state.

For a magnetic field pointing in an arbitrary direction, the observed Raman spectrum displays up to 11 peaks, that is 5 peaks for $\Delta m_F = 0$ transitions and 6 peaks for $\Delta m_F = \pm 1$ transitions, where $\Delta m_F \equiv m_{F,2} - m_{F,1}$ is the change of the magnetic quantum number. For a magnetic field directed along the z-axis as required for the T^3-experiment, transitions with $\Delta m_F = \pm 1$ are suppressed in our experimental setup [76, 77].

5.3 *Magnetic Field Measurement*

The position of the relative two-photon resonance in the Raman spectrum corresponding to the magnetically sensitive transition $|F_1, m_{F_1}\rangle \to |F_2, m_{F_2}\rangle$ is determined by the first-order Zeeman shift

$$\tilde{\Delta}^{F_2, m_{F_2}}_{F_1, m_{F_1}} \equiv \frac{\mu_B}{2\pi\hbar}(g_{F_2} m_{F_2} - g_{F_1} m_{F_1}) B(z) \tag{28}$$

in frequency, where $B(z)$ is the value of the magnetic field at the center-of-mass coordinate z of the atom cloud during the interaction with the Raman pulse.

Using the $+2$ transition, that is $|F_1 = 2, m_{F_1} = 1\rangle \to |F_2 = 3, m_{F_2} = 1\rangle$, as indicated in Fig. 4, we determine from the Zeeman shift, Eq. (28), the magnetic field[4]

$$B(z) = 0.11\, \tilde{\Delta}^{3,1}_{2,1} \frac{\mu\text{T}}{\text{kHz}}, \tag{29}$$

where we have used the fact [71] that $m_{F_2} = m_{F_1} = 1$ and $g_{F_2} = -g_{F_1} = 1/3$. It is the opposite signs of g_{F_2} and g_{F_1} that result in the non-degenerate spectrum.

By measuring the resonance frequencies corresponding to the clock and first (or second) peak in the Raman spectrum, we automatically correct for any possible

[4]Only relative positions and magnitudes of the different peaks in the Raman spectrum allow us to determine the different components of the magnetic field [76].

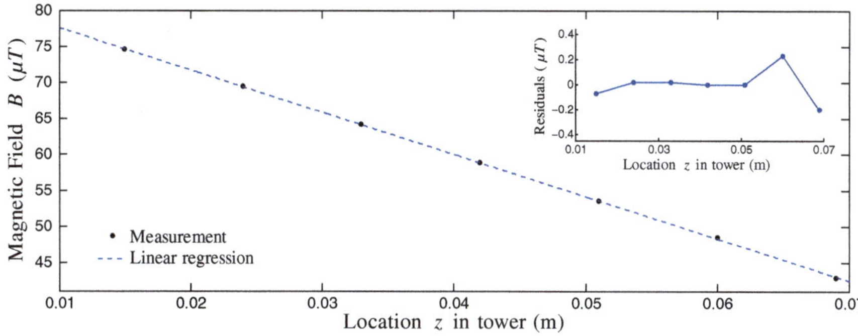

Fig. 5 Measurement of the magnetic field gradient along the vertical direction in the glass tower. A linear regression of our data points deduced from the Raman spectra with the help of Eq. (29) yields the magnetic field $B(z) = (83.5\text{–}587\ \text{m}^{-1}\,z)\,\mu\text{T}$ along the tower and corresponds to a magnetic field gradient of approximately 600 $\mu\text{T/m}$. The inset shows the residuals of the linear fit. For this measurement we have used the frequency difference between the clock and the +2 transition induced by the Zeeman effect. The magnetic field is generated by currents of 90 mA for the gradient coils and 40 mA for the solenoid

drift in the AC Stark shift caused by a drift in the intensity or frequency of the Raman fields.

We map out the magnetic field experienced by the atoms as a function of their location within the tower, by repeated launching them and applying a Raman pulse at different times after their launch. The corresponding location z of the atom cloud is determined beforehand by time-of-flight photography.

Figure 5 presents the measured magnetic field with a gradient of approximately 600 $\mu\text{T/m}$ generated by a current of 90 mA in the gradient coils and 40 mA in the solenoid.

5.4 *Imperfect Pulses and Decoherence*

Our proposal for a T^3-interferometer presented in the preceding sections assumes that the Raman pulses applied to the atoms are perfect $\frac{\pi}{2}$- and π-pulses. By definition, such pulses can occur only when the Raman fields are in a two-photon resonance. Since this resonance shifts as the atoms travel up the tower, a rapid detuning of the relative frequency of the Raman fields is necessary. Fortunately, we have been able to achieve this task using a combination of a high-frequency acousto-optic modulator and digital frequency synthesizer, thereby keeping the atoms in resonance during their flight [78].

Unfortunately, a more severe restriction is decoherence and at the moment we identify three culprits: (*i*) magnetic field noise, (*ii*) a non-linear dependence of the magnetic field on z, and (*iii*) non-zero components of the magnetic field along the x- and y-axis.

The upper bound of the magnetic field noise in our setup has been observed to be 50 nT leading to a noise in the magnetic field gradient of about $1\,\mu\mathrm{T/m}$. The non-linear terms in the dependence of the magnetic field on z lead to an open geometry of our T^3-interferometer, and therefore to a loss of contrast [37].

Enforced by the Maxwell equations, the non-zero components of the magnetic field in the x- and y-direction result in (*i*) a three-dimensional rather than a one-dimensional center-of-mass motion of the atom, and (*ii*) incoherent dynamics of the atomic states involved.

We are currently working on finding an optimal arrangement for the magnetic field and perform numerical simulations of our T^3-interferometer.

6 Summary and Outlook

In the present article, we have proposed an atom interferometer that is sensitive to the quantum mechanical T^3-phase emerging in the propagator of a particle in a linear potential. For this purpose we have considered an atom with two magnetic sensitive internal states being exposed to a constant gravitational field as well as a magnetic field with a constant gradient. By applying a sequence of four co-propagating Raman pulses, the atom interferometer can be closed in position *and* velocity. The resulting interferometer phase φ_i displays the cubic scaling in T but also depends on the gravitational acceleration and the magnetic field gradient.

We have compared and contrasted this cubic term to the one appearing in the phase of the Kasevich–Chu interferometer in the presence of a gravity gradient, and to the one obtained using the CAB technique. Furthermore, we have outlined a possible experimental realization of our interferometer.

Cubic phases appear frequently in quantum physics and give rise to mind-boggling effects. For example, the energy wave function of a linear potential is given by the Airy function [79] whose standard integral representation involves a cubic phase. This term emerges from the eigenvalue equation in momentum space due to the integration of the kinetic energy which is quadratic in the momentum.

When we suddenly turn-off the potential the so-created Airy wave packet accelerates and its probability density keeps its shape [80] during the free propagation. Deeper insight [66] into this surprising phenomenon springs from Wigner phase space [81] and the fact that the Wigner function of the Airy wave packet is again an Airy function.

Closely related to the cubic phase in the Airy integral and the dispersionless free propagation of the Airy wave packet is the oscillatory probability density created by a point source [9] located in a linear potential and continuously emitting particles into all three space directions with an identical speed. These oscillations appearing in the plane orthogonal to the gravitational force are a consequence of the interference between two classical trajectories of different inclinations. The knowledge of the two distinct paths encoded in the different arrival times is erased by the continuous stream of particles. Again the origin of this particular interference pattern can be traced back to the cubic phase in the Green's function.

Due to the analogy between the constant gravitational field and the constant electric field between two plates of a capacitor discussed in the beginning of this article one might wonder if it is possible to construct a similar charged particle fountain. Indeed, in the case of electrons in a uniform electric field such type of fountain has already been realized in photoionization and photodetachment microscopes [82–84].

It would be fascinating to illuminate the similarities and differences between these three examples of cubic phases and our T^3-interferometer. Unfortunately, this task goes beyond the scope of the present article and has to be postponed to future publications.

Acknowledgements We are grateful to E. Giese, M. A. Kasevich, S. Kleinert, H. Müller, G. Welch, and W. Zeller for many fruitful discussions on this topic. Moreover, we thank N. Ashby for pointing out Ref. [2] to us. This work is supported by DIP, the German-Israeli Project Cooperation, as well as the German Space Agency (DLR) with funds provided by the Federal Ministry for Economic Affairs and Energy (BMWi) due to an enactment of the German Bundestag under Grants No. DLR 50WM1152-1157 (QUANTUS-IV) and the Centre for Quantum Engineering and Space-Time Research (QUEST). We appreciate the funding by the German Research Foundation (DFG) in the framework of the SFB/TRR-21. W.P.S. is grateful to Texas A&M University for a Texas A&M University Institute for Advanced Study (TIAS) Faculty Fellowship. S.A.D., J.P.D., A.S., and F.A.N. gratefully acknowledge funding from the Office of Naval Research and a grant from the Naval Air Systems Command Chief Technology Office.

Appendix A: Dependence of T^3-Phase on Initial Wave Function

According to Sect. 2.1 the cubic phase ϕ in the propagator of a quantum particle moving in a linear potential manifests itself in every wave function exposed to this situation. Indeed, since ϕ is independent of the initial coordinate z_i it can be factored out of the Huygens integral for matter waves, Eq. (1).

Nevertheless, the integration of the remaining parts of the propagator in combination with the initial wave function can change the dependence of the phase of the final wave function on the propagation time $\tau = t_f - t_i$. To bring out this fact most clearly we consider the normalized initial wave function

$$\psi(z_\mathrm{i}, t_\mathrm{i}) \equiv \frac{1}{\left(\sqrt{\pi}\Delta z_0\right)^{1/2}} \exp\left(-\frac{z_\mathrm{i}^2}{2\Delta z_0^2}\right) \tag{30}$$

in the form of a real-valued Gaussian of width Δz_0 and note that wave functions of this kind can be easily realized nowadays in an experiment with cold atoms in a harmonic potential trap.

When we substitute this expression into the Huygens integral of matter waves, Eq. (1), and use the expressions, Eqs. (3) and (7), for the propagator we arrive at the final wave function

$$\psi(z_{\mathrm{f}}, t_{\mathrm{f}}) = \frac{1}{\left[\sqrt{\pi}\Delta z(\tau)\right]^{1/2}} \exp\left\{-\frac{\left(z_{\mathrm{f}} - \frac{F}{2m}\tau^2\right)^2}{2[\Delta z(\tau)]^2} + i\beta\right\} \tag{31}$$

with the time-dependent width

$$\Delta z(\tau) \equiv \Delta z_0 \sqrt{1 + \frac{\tau^2}{\tau_{\mathrm{s}}^2}} \tag{32}$$

and phase

$$\beta(\tau) \equiv \frac{F z_{\mathrm{f}} \tau}{\hbar} + \frac{\tau}{\tau_{\mathrm{s}}} \frac{\left(z_{\mathrm{f}} - \frac{F}{2m}\tau^2\right)^2}{2[\Delta z(\tau)]^2} - \frac{F^2 \tau^3}{6\hbar m} - \frac{1}{2}\arctan\left(\frac{\tau}{\tau_{\mathrm{s}}}\right). \tag{33}$$

Here

$$\tau_{\mathrm{s}} \equiv \frac{m\Delta z_0^2}{\hbar} \tag{34}$$

denotes the spreading time of the wave packet.

We combine the terms in Eq. (33), which are determined by the strength F of the force and independent of the final position z_{f}, and find the total phase

$$\widetilde{\phi}\,(\tau) \equiv -\alpha\left(\frac{\tau}{\tau_s}\right)\frac{F^2}{\hbar m}\tau^3. \tag{35}$$

Here we have introduced the time-dependent numerical factor

$$\alpha(\eta) \equiv \frac{1}{24}\frac{\eta^2 + 4}{\eta^2 + 1} \tag{36}$$

depending on the dimensionless ratio η of time difference τ, Eq. (10), and spreading time τ_s, which according to Eq. (34) is proportional to square of the initial width Δz_0.

For a plane wave we find $\Delta z_0 \to \infty$, and thus $\tau_s \to \infty$ leading us to

$$\alpha(\eta \to 0) = \frac{1}{6}. \tag{37}$$

However, for an infinitely narrow Gaussian with $\Delta z_0 \to 0$, and thus $\tau_s \to 0$, we obtain

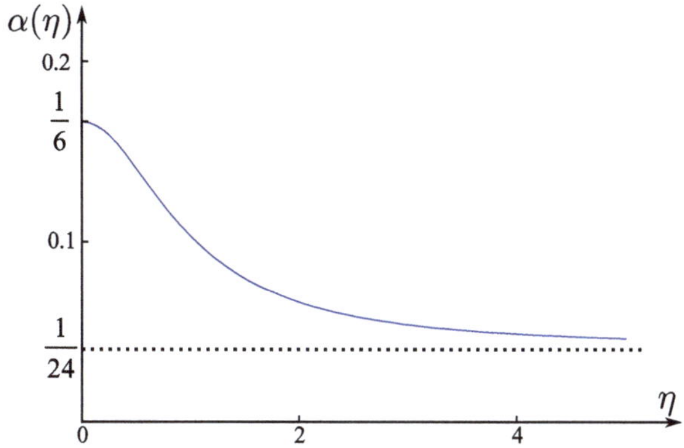

Fig. 6 Dependence of the numerical factor $\alpha = \alpha(\eta)$ defined by Eq. (36) on the dimensionless ratio η of the time difference τ and spreading time τ_s given by Eqs. (10) and (34), respectively. For $\eta \to 0$ and $\eta \to \infty$ the factor α is almost constant and given by 1 / 6 and 1 / 24 (*dashed line*), respectively. However, for values of η between these extremes α changes rapidly and thus in this transition domain the phase $\tilde{\phi} = \tilde{\phi}(\tau)$, Eq. (35), is not strictly cubic.

$$\alpha(\eta \to \infty) = \frac{1}{24}. \tag{38}$$

So far we have restricted ourselves to the extreme limits of η. Only in the domains where α is approximately constant do we find a pure cubic phase dependence on τ. Indeed, between the extremes the time dependence is more complicated as expressed in Fig. 6.

Appendix B: Raman Pulses: Superpositions and Exchanges

In this Appendix, we describe the population dynamics [36, 64, 65] of the two resonant atomic states driven by the Raman laser pulses. For this purpose we consider the interaction between a three-level atom and two laser pulses of the form

$$\mathbf{E}_1(z,t) \equiv \boldsymbol{\varepsilon}_1(t)\cos(k_1 z - \omega_1 t + \phi_1)$$

and

$$\mathbf{E}_2(z,t) \equiv \boldsymbol{\varepsilon}_2(t)\cos(k_2 z - \omega_2 t + \phi_2), \tag{39}$$

where $\boldsymbol{\varepsilon}_j, k_j, \omega_j$, and ϕ_j with $j = 1, 2$ denote the time-dependent envelope, frequency, wave vector, and phase of the jth field, respectively.

The laser frequencies ω_1 and ω_2 are assumed to only drive the transitions $|g_1\rangle \longleftrightarrow |e\rangle$ and $|g_2\rangle \longleftrightarrow |e\rangle$, respectively. Moreover, we assume that the laser pulses are so short that the atom does not move significantly during the interaction. Therefore, the position of the center-of-mass of the atom is considered to be fixed during the laser pulses.

Within the rotating-wave approximation [81] and in the limit of far-detuned laser pulses with identical detunings, that is when the Rabi frequencies $\Omega_j(t) \equiv \mathbf{d}_{g_je} \cdot \boldsymbol{\varepsilon}_j(t)/\hbar$ of the transitions $|g_j\rangle \longleftrightarrow |e\rangle$ are much smaller than the detuning $\Delta_j \equiv \omega_j - \omega_{eg_j}$ of the two laser pulses, $|\Delta_j| \gg |\Omega_j|$, and $\Delta \equiv \Delta_1 = \Delta_2$, we can eliminate the excited state $|e\rangle$ and neglect the Stark shifts $|\Omega_j(t)|^2/(4\Delta)$. The resulting effective Hamiltonian [36]

$$\begin{aligned} \hat{H}_\mathrm{p} = \hbar \frac{\Omega_1(t)\Omega_2(t)}{4\Delta} \Big(& e^{i[\Delta k z + \phi_\mathrm{L}]} |g_1\rangle\langle g_2| \\ & + e^{-i[\Delta k z + \phi_\mathrm{L}]} |g_2\rangle\langle g_1| \Big) \end{aligned} \tag{40}$$

describes the transitions between the states $|g_1\rangle$ and $|g_2\rangle$ due to the Raman pulses. Here $\mathbf{d}_{g_je} \equiv \langle g_j|\mathbf{d}|e\rangle$ and $\omega_{eg_j} \equiv (E_e - E_{g_j})/\hbar$ are the dipole moment matrix element and the frequency of the transition $|g_j\rangle \longleftrightarrow |e\rangle$, respectively, with $\Delta k \equiv k_2 - k_1$ and the slowly varying laser phase $\phi_\mathrm{L}(t) \equiv \phi_2 - \phi_1$.

To avoid a momentum transfer during the Raman transitions, we assume that the laser pulses propagate in the same directions along the z-axis, Eq. (39), and that the difference Δk of the two wave vectors is small compared to the size δz of the atomic wave packet, that is $|\Delta k|\delta z \ll 1$. In this case the dependence on z in Eq. (40) can be neglected and we arrive at

$$\hat{H}_\mathrm{p} \cong \hbar \frac{\Omega_1(t)\Omega_2(t)}{4\Delta} \left(e^{i\phi_\mathrm{L}} |g_1\rangle\langle g_2| + e^{-i\phi_\mathrm{L}} |g_2\rangle\langle g_1| \right). \tag{41}$$

The interaction of the atom with the two far-detuned Raman pulses, corresponding to Eq. (39), during the time interval $t_\mathrm{i} < t < t_\mathrm{f}$, and with $\Omega_j(t_\mathrm{i}) = \Omega_j(t_\mathrm{f}) = 0$, is given by the evolution operator [36]

$$\begin{aligned} \hat{U}_\mathrm{p} & \equiv 1 + \left(-\frac{i}{\hbar}\right) \int_{t_\mathrm{i}}^{t_\mathrm{f}} \mathrm{d}t \hat{H}_\mathrm{p}(t) \\ & + \left(-\frac{i}{\hbar}\right)^2 \int_{t_\mathrm{i}}^{t_\mathrm{f}} \mathrm{d}t \int_{t_\mathrm{i}}^{t} \mathrm{d}t' \hat{H}_\mathrm{p}(t) \hat{H}_\mathrm{p}(t') + \dots , \end{aligned} \tag{42}$$

which can be expressed as

$$\begin{aligned} \hat{U}_\mathrm{p}(\theta) & = \left(|g_1\rangle\langle g_1| + |g_2\rangle\langle g_2| \right) \cos\left(\frac{\theta}{2}\right) \\ & - i\left(e^{i\phi_\mathrm{L}} |g_1\rangle\langle g_2| + e^{-i\phi_\mathrm{L}} |g_2\rangle\langle g_1|\right) \sin\left(\frac{\theta}{2}\right), \end{aligned} \tag{43}$$

where

$$\theta \equiv \frac{1}{2\Delta}\int_{t_i}^{t_f} dt\, \Omega_1(t)\Omega_2(t) \tag{44}$$

denotes the total pulse area.

The case $\theta = \frac{\pi}{2}$, which is a $\frac{\pi}{2}$-pulse, gives rise [67] to the coherent superpositions

$$\hat{U}_{\mathrm{P}}\left(\frac{\pi}{4}\right)|g_1\rangle = \frac{1}{\sqrt{2}}\left(|g_1\rangle - ie^{-i\phi_{\mathrm{L}}}|g_2\rangle\right)$$

and

$$\hat{U}_{\mathrm{P}}\left(\frac{\pi}{4}\right)|g_2\rangle = \frac{1}{\sqrt{2}}\left(|g_2\rangle - ie^{i\phi_{\mathrm{L}}}|g_1\rangle\right). \tag{45}$$

In contrast, the case $\theta = \pi$, known as a π-pulse describes an exchange

$$\hat{U}_{\mathrm{P}}\left(\frac{\pi}{2}\right)|g_1\rangle = -ie^{-i\phi_{\mathrm{L}}}|g_2\rangle$$

and

$$\hat{U}_{\mathrm{P}}\left(\frac{\pi}{2}\right)|g_2\rangle = -ie^{i\phi_{\mathrm{L}}}|g_1\rangle \tag{46}$$

of the level populations.

Appendix C: Interferometer: Sequence of Unitary Operators

Unitary operators describe both the interaction of the atom with the four Raman pulses and the time evolution associated with the center-of-mass motion. In the present Appendix, we derive the complete quantum state of the atom consisting of the internal states as well as the center-of-mass in the two exit ports of our interferometer following the procedure outlined in Refs. [28, 36, 67].

The dynamics in our interferometer consists of the following steps:

1. Before the first $\frac{\pi}{2}$-pulse, at $t = t_0 - \varepsilon$, the initial state

$$|\Psi(t_0 - \varepsilon)\rangle \equiv |g_1\rangle|\psi_0\rangle \tag{47}$$

consists of the center-of-mass motion $|\psi_0\rangle$ and the internal state $|g_1\rangle$. Here and throughout this Appendix ε is an infinitesimally small and positive number.

2. After the first $\frac{\pi}{2}$-pulse at $t = t_0 + \varepsilon$, the state reads

$$\begin{aligned}|\Psi(t_0+\varepsilon)\rangle &= \hat{U}_{\mathrm{p}}\Big(\frac{\pi}{4}\Big)|\Psi(t_0-\varepsilon)\rangle\\ &= \Big(\frac{1}{\sqrt{2}}|g_1\rangle - \frac{i}{\sqrt{2}}e^{-i\phi_{\mathrm{L}}(t_0)}|g_2\rangle\Big)|\psi_0\rangle,\end{aligned} \tag{48}$$

where we have used Eq. (45).

3. Before the first π-pulse at $t = t_1 - \varepsilon$, we find

$$\begin{aligned}|\Psi(t_1-\varepsilon)\rangle = \hat{U}_a(t_1,t_0)|\Psi(t_0+\varepsilon)\rangle &= \frac{1}{\sqrt{2}}|g_1\rangle\hat{U}_{a_1}(t_1,)t_0|\psi_0\rangle\\ &-\frac{i}{\sqrt{2}}e^{-i\phi_{\mathrm{L}}(t_0)}|g_2\rangle\hat{U}_{a_2}(t_1,)t_0|\psi_0\rangle.\end{aligned} \tag{49}$$

4. After the first π-pulse at $t = t_1 + \varepsilon$, we obtain

$$\begin{aligned}|\Psi(t_1+\varepsilon)\rangle = \hat{U}_{\mathrm{p}}\Big(\frac{\pi}{2}\Big)|\Psi(t_1-)\varepsilon) &= -\frac{i}{\sqrt{2}}e^{-i\phi_{\mathrm{L}}(t_1)}|g_2\rangle\hat{U}_{a_1}(t_1,t_0)|\psi_0\rangle\\ &-\frac{1}{\sqrt{2}}e^{-i[\phi_{\mathrm{L}}(t_0)-\phi_{\mathrm{L}}(t_1)]}|g_1\rangle\hat{U}_{a_2}(t_1,t_0)|\psi_0\rangle.\end{aligned} \tag{50}$$

5. Before the second π-pulse at $t = t_2 - \varepsilon$, the state takes the form

$$\begin{aligned}|\Psi(t_2-\varepsilon)\rangle = \hat{U}_a(t_2,t_1)|\Psi(t_1+\varepsilon)\rangle &= -\frac{i}{\sqrt{2}}e^{-i\phi_{\mathrm{L}}(t_1)}|g_2\rangle\hat{U}_{a_2}(t_2,t_1)\hat{U}_{a_1}(t_1,t_0)|\psi_0\rangle\\ &-\frac{1}{\sqrt{2}}e^{-i[\phi_{\mathrm{L}}(t_0)-\phi_{\mathrm{L}}(t_1)]}|g_1\rangle\hat{U}_{a_1}(t_2,t_1)\hat{U}_{a_2}(t_1,t_0)|\psi_0\rangle.\end{aligned} \tag{51}$$

6. After the second π-pulse at $t = t_2 + \varepsilon$, we arrive at the state

$$\begin{aligned}|\Psi(t_2+\varepsilon)\rangle = \hat{U}_{\mathrm{p}}\Big(\frac{\pi}{2}\Big)|\Psi(t_2-\varepsilon)\rangle &= -\frac{1}{\sqrt{2}}e^{-i[\phi_{\mathrm{L}}(t_1)-\phi_{\mathrm{L}}(t_2)]}|g_1\rangle\hat{U}_{a_2}(t_2,t_1)\hat{U}_{a_1}(t_1,t_0)|\psi_0\rangle\\ &+\frac{i}{\sqrt{2}}e^{-i[\phi_{\mathrm{L}}(t_0)-\phi_{\mathrm{L}}(t_1)+\phi_{\mathrm{L}}(t_2)]}\times|g_2\rangle\hat{U}_{a_1}(t_2,t_1)\hat{U}_{a_2}(t_1,t_0)|\psi_0\rangle.\end{aligned} \tag{52}$$

7. Before the second $\frac{\pi}{2}$-pulse at $t = t_3 - \varepsilon$, the state reads

$$\begin{aligned}|\Psi(t_3-\varepsilon)\rangle = \hat{U}_a(t_3,t_2)|\Psi(t_2+\varepsilon)\rangle &= -\frac{1}{\sqrt{2}}e^{-i[\phi_{\mathrm{L}}(t_1)-\phi_{\mathrm{L}}(t_2)]}\\ &\times|g_1\rangle\hat{U}_{a_1}(t_3,t_2)\hat{U}_{a_2}(t_2,t_1)\hat{U}_{a_1}(t_1,t_0)|\psi_0\rangle + \frac{i}{\sqrt{2}}e^{-i[\phi_{\mathrm{L}}(t_0)-\phi_{\mathrm{L}}(t_1)+\phi_{\mathrm{L}}(t_2)]}\\ &\times|g_2\rangle\hat{U}_{a_2}(t_3,t_2)\hat{U}_{a_1}(t_2,t_1)\hat{U}_{a_2}(t_1,t_0)|\psi_0\rangle.\end{aligned} \tag{53}$$

8. Finally, after the second $\frac{\pi}{2}$-pulse at $t = t_3 + \varepsilon$, we conclude with the state

$$|\Psi(t_3+\varepsilon)\rangle = \hat{U}_{\mathrm{p}}\left(\frac{\pi}{4}\right)|\Psi(t_3-\rangle\varepsilon) = \frac{1}{2}e^{-i[\phi_{\mathrm{L}}(t_1)-\phi_{\mathrm{L}}(t_2)]}|g_1\rangle\left(e^{-i\varphi_{\mathrm{L}}}\hat{U}_{\mathrm{u}} - \hat{U}_{\mathrm{l}}\right)|\psi_0\rangle$$
$$+\frac{i}{2}e^{-i[\phi_{\mathrm{L}}(t_1)-\phi_{\mathrm{L}}(t_2)+\phi_{\mathrm{L}}(t_3)]}|g_2\rangle\left(e^{-i\varphi_{\mathrm{L}}}\hat{U}_{\mathrm{u}} + \hat{U}_{\mathrm{l}}\right)|\psi_0\rangle, \tag{54}$$

where

$$\hat{U}_{\mathrm{l}} \equiv \hat{U}_{a_1}(t_3,t_2)\hat{U}_{a_2}(t_2,t_1)\hat{U}_{a_1}(t_1,t_0)$$

and

$$\hat{U}_{\mathrm{u}} \equiv \hat{U}_{a_2}(t_3,t_2)\hat{U}_{a_1}(t_2,t_1)\hat{U}_{a_2}(t_1,t_0) \tag{55}$$

are the unitary evolution operators associated with the center-of-mass motion for the lower and the upper paths of the interferometer shown in Fig. 2, and

$$\varphi_{\mathrm{L}} \equiv \phi_{\mathrm{L}}(t_0) - 2\phi_{\mathrm{L}}(t_1) + 2\phi_{\mathrm{L}}(t_2) - \phi_{\mathrm{L}}(t_3) \tag{56}$$

is the total phase resulting from the action of the four laser pulses.

Appendix D: Conditions for a Closed T^3-Interferometer

In the preceding Appendix, we have derived an expression for the complete quantum state of the atom in the exit ports of the interferometer. Here we have allowed arbitrary times for the interactions with the laser pulses. In the present Appendix, we choose these times in such a way as to maximize the contrast.

The probability P_{g_1} to observe atoms in the ground state $|g_1\rangle$ after the action of the four Raman pulses at $t = t_3 + \varepsilon$, follows from the quantum state $|\Psi(t_3+\varepsilon)\rangle$ given by Eq. (54) and contains the state

$$|\psi_{g_1}\rangle \equiv \langle g_1|\Psi(t_3+\varepsilon)\rangle \tag{57}$$

of the center-of-mass motion of atom in $|g_1\rangle$. It takes the form

$$P_{g_1} \equiv \langle\psi_{g_1}|\psi_{g_1}\rangle = \frac{1}{2}[1 - C\cos(\varphi_{\mathrm{i}} + \varphi_{\mathrm{L}})], \tag{58}$$

where the contrast C and the phase φ_{i} of the interferometer are the modulus and the argument of the matrix element

$$\langle\psi_0|\hat{U}_{\mathrm{u}}^{\dagger}\hat{U}_{\mathrm{l}}|\psi_0\rangle \equiv Ce^{i\varphi_{\mathrm{i}}}. \tag{59}$$

We maximize C, that is we have $C = 1$, when we close our interferometer. In this case P_{g_1} given by Eq. (58) is independent of the initial velocity and position of the atom.

To close the interferometer we have to find the time intervals $t_{j+1,j} \equiv t_{j+1} - t_j$ with $j = 0, 1, 2$ between the Raman pulses shown in Fig. 2, such that the final velocities $v_u(t_3)$ and $v_l(t_3)$, as well as the final positions $z_u(t_3)$ and $z_l(t_3)$ on the upper and lower paths of the interferometer are identical.

Indeed, for the velocity we derive the following formulae:

(i) for the upper path

$$\begin{aligned} v_0 &\to v_u(t_1) = v_0 + a_2 t_{10} \\ &\to v_u(t_2) = v_u(t_1) + a_1 t_{21} \\ &\to v_u(t_3) = v_u(t_2) + a_2 t_{32} \end{aligned}$$

(ii) for the lower path

$$\begin{aligned} v_0 &\to v_l(t_1) = v_0 + a_1 t_{10} \\ &\to v_l(t_2) = v_l(t_1) + a_2 t_{21} \\ &\to v_l(t_3) = v_l(t_2) + a_1 t_{32}. \end{aligned}$$

As a result, the interferometer is closed in *velocity* space, if $v_u(t_3) = v_l(t_3)$, that is,

$$\begin{aligned} &v_0 + a_2 t_{10} + a_1 t_{21} + a_2 t_{32} \\ &= v_0 + a_1 t_{10} + a_2 t_{21} + a_1 t_{32}, \end{aligned}$$

or, equivalently,

$$t_{10} - t_{21} + t_{32} = 0. \tag{60}$$

As for the position, we obtain the following rather lengthy expressions:

(i) for the upper path

$$\begin{aligned} z_0 &\to z_u(t_1) = z_0 + v_0 t_{10} + \frac{1}{2} a_2 t_{10}^2 \\ &\to z_u(t_2) = z_u(t_1) + v_u(t_1) t_{21} + \frac{1}{2} a_1 t_{21}^2 \\ &\to z_u(t_3) = z_u(t_2) + v_u(t_2) t_{32} + \frac{1}{2} a_2 t_{32}^2 \\ &= z_0 + v_0 (t_{10} + t_{21} + t_{32}) + \frac{1}{2}(a_2 t_{10}^2 + a_1 t_{21}^2 + a_2 t_{32}^2) \\ &\quad + a_2 t_{10}(t_{21} + t_{32}) + a_1 t_{21} t_{32}, \end{aligned}$$

(ii) for the lower path

$$
\begin{aligned}
z_0 \quad &\to z_1(t_1) = z_0 + v_0 t_{10} + \frac{1}{2} a_1 t_{10}^2 \\
&\to z_1(t_2) = z_1(t_1) + v_1(t_1) t_{21} + \frac{1}{2} a_2 t_{21}^2 \\
&\to z_1(t_3) = z_1(t_2) + v_1(t_2) t_{32} + \frac{1}{2} a_1 t_{32}^2 \\
&= z_0 + v_0(t_{10} + t_{21} + t_{32}) + \frac{1}{2}(a_1 t_{10}^2 + a_2 t_{21}^2 + a_1 t_{32}^2) \\
&+ a_1 t_{10}(t_{21} + t_{32}) + a_2 t_{21} t_{32}.
\end{aligned}
$$

As a result, the interferometer is closed in *position* space if $z_{\mathrm{u}}(t_3) = z_{\mathrm{l}}(t_3)$, that is,

$$
t_{10}^2 - t_{21}^2 + t_{32}^2 + 2t_{10}(t_{21} + t_{32}) - 2t_{21}t_{32} = 0. \tag{61}
$$

When we solve the system of the two algebraic Eqs. (60) and (61), for t_{21} and t_{32} in terms of t_{10}, we obtain

$$
t_3 - t_2 = t_1 - t_0 = T \quad \text{and} \quad t_2 - t_1 = 2T. \tag{62}
$$

Hence, to close the interferometer, the four Raman pulses must be separated in time by T, $2T$, and T as indicated in Fig. 2.

Appendix E: Interferometer Phase

In the preceding Appendix, we have used classical trajectories to find the separation $T - 2T - T$ between the four Raman pulses leading to a closed interferometer. We now show that in this case the product $\hat{U}_{\mathrm{u}}^{\dagger}\hat{U}_{\mathrm{l}}$ of the evolution operators $\hat{U}_{\mathrm{l}}$ and $\hat{U}_{\mathrm{u}}$ defined by Eq. (55) is proportional [28, 37, 67] to the identity operator, that is

$$
\hat{U}_{\mathrm{u}}^{\dagger}\hat{U}_{\mathrm{l}} = e^{i\varphi_{\mathrm{i}}}\, \mathbb{1}, \tag{63}
$$

where φ_{i} is the interferometer phase.

Therefore, a normalized state $|\psi_0\rangle$ leads by virtue of Eq. (59) to a perfect contrast, that is $C = 1$, indicating that the interferometer is independent of $|\psi_0\rangle$. Moreover, this calculation provides us with an explicit expression for φ_{i}.

To evaluate the evolution operator

$$
\hat{U}_{\mathrm{l}} \equiv \hat{U}_{a_1}(T)\hat{U}_{a_2}(2T)\hat{U}_{a_1}(T) \tag{64}
$$

for the lower path of our interferometer, shown in Fig. 2, we use the Baker–Campbell–Hausdorff and Zassenhaus formulas [85] to represent the operator $\hat{U}_a(T)$ given by Eq. (13) in the form of a product

$$\hat{U}_a(T) = \exp\left(i\frac{ma^2T^3}{12\hbar}\right)\hat{\mathcal{D}}\left(\frac{1}{2}aT^2, maT\right)\hat{U}_0(T) \tag{65}$$

consisting of a phase factor, the displacement operator

$$\hat{\mathcal{D}}(Z,P) \equiv \exp\left[-\frac{i}{\hbar}(Z\hat{p}_z - P\hat{z})\right], \tag{66}$$

and the unitary operator

$$\hat{U}_0(T) \equiv \exp\left(-\frac{i}{\hbar}\frac{\hat{p}_z^2}{2m}T\right) \tag{67}$$

of a free particle.

The decomposition, Eq. (65), allows us to rewrite Eq. (64) as

$$\begin{aligned}\hat{U}_1 = &\exp\left[i\frac{m(a_1^2+4a_2^2)}{6\hbar}T^3\right]\hat{\mathcal{D}}\left(\frac{1}{2}a_1T^2, ma_1T\right)\hat{U}_0(T)\\ &\times\hat{\mathcal{D}}\left(2a_2T^2, 2ma_2T\right)\hat{U}_0(2T)\hat{\mathcal{D}}\left(\frac{1}{2}a_1T^2, ma_1T\right)\hat{U}_0(T).\end{aligned} \tag{68}$$

With the help of the commutation relation

$$\hat{U}_0(T)\hat{\mathcal{D}}(Z,P) = \hat{\mathcal{D}}\left(Z + \frac{P}{m}T, P\right)\hat{U}_0(T)$$

and the addition identity

$$\hat{U}_0(T_1)\hat{U}_0(T_2) = \hat{U}_0(T_1+T_2)$$

for the operators $\hat{\mathcal{D}}$ and $\hat{U}$ given by Eqs. (66) and (67), we can shift all free-evolution operators $\hat{U}_0$ in Eq. (68) to the right and we arrive at

$$\begin{aligned}\hat{U}_1 = &\exp\left[i\frac{m(a_1^2+4a_2^2)}{6\hbar}T^3\right]\hat{\mathcal{D}}\left(\frac{1}{2}a_1T^2, ma_1T\right)\\ &\times\hat{\mathcal{D}}\left(4a_2T^2, 2ma_2T\right)\hat{\mathcal{D}}\left(\frac{7}{2}a_1T^2, ma_1T\right)\hat{U}_0(4T),\end{aligned}$$

or

$$\begin{aligned}\hat{U}_1 &= \exp\left[i\frac{mT^3}{3\hbar}\left(5a_1^2 + 9a_1a_2 + 2a_2^2\right)\right]\\ &\times\hat{\mathcal{D}}\left(4(a_1+a_2)T^2, 2m(a_1+a_2)T\right)\hat{U}_0(4T).\end{aligned} \tag{69}$$

In the last step we have made use of the addition identity

$$\hat{\mathcal{D}}(Z_1, P_1)\hat{\mathcal{D}}(Z_2, P_2) = e^{i\widetilde{\varphi}}\hat{\mathcal{D}}(Z_1 + Z_2, P_1 + P_2)$$

with

$$\widetilde{\varphi} \equiv \frac{1}{2\hbar}(P_1 Z_2 - P_2 Z_1),$$

to combine all three displacement operators into a single one.

Since the evolution operator $\hat{U}_u$ defined by Eq. (55) for the upper path of our interferometer follows directly from the operator $\hat{U}_l$ given by Eq. (69) for the lower path by an exchange of the accelerations a_1 and a_2, we arrive at

$$\hat{U}_u = \exp\left[i\frac{mT^3}{3\hbar}\left(2a_1^2 + 9a_1 a_2 + 5a_2^2\right)\right] \times \hat{\mathcal{D}}\left(4(a_1 + a_2)T^2, 2m(a_1 + a_2)T\right)\hat{U}_0(4T). \tag{70}$$

When we substitute Eqs. (69) and (70) into the left-hand side of Eq. (63) and use the property that the operators $\hat{\mathcal{D}}$ and $\hat{U}_0$ are unitary, the interferometer phase reads

$$\varphi_i = \frac{m}{\hbar}(a_1^2 - a_2^2)T^3. \tag{71}$$

Hence, φ_i is independent of the initial position z_0 and velocity v_0 as well as of the initial state. Moreover, it scales with the third power of the time interval $T \equiv t_1 - t_0$ between the first and the second Raman pulses.

References

1. S. Fray, C. Alvarez Diez, Th.W. Hänsch, M. Weitz, Phys. Rev. Lett. **93**, 240404 (2004)
2. E.H. Kennard, Zeitschrift für Physik **44**, 326 (1927)
3. E.H. Kennard, J. Frank. Inst. **207**, 47 (1929)
4. W.P. Schleich, D.M. Greenberger, D.H. Kobe, M.O. Scully, Proc. Nat. Acad. Sci. **110**, 5374 (2013)
5. P.R. Berman (ed.) *Atom Interferometry* (Academic Press, San Diego, 1997)
6. A.D. Cronin, J. Schmiedmayer, D.E. Pritchard, Rev. Mod. Phys. **81**, 1051 (2009)
7. G.M. Tino, M.A. Kasevich (eds.) Proceedings of the International School of Physics "Enrico Fermi" Course 188 "Atom Interferometry" (IOS, Amsterdam, 2014)
8. R.P. Feynman, A.R. Hibbs, *Quantum Mechanics and Path Integrals* (McGraw-Hill, New York, 1965)
9. M.V. Berry, J. Phys. A **15**, L385 (1982)
10. F. Fratini, L. Safari, Physica Scripta **89**, 085004 (2014)
11. P. Storey, C. Cohen-Tannoudji, J. Phys. II **4**, 1999 (1994)
12. G.D. McDonald, C.C.N. Kuhn, S. Bennetts, J.E. Debs, K.S. Hardman, J.D. Close, N.P. Robins, EPL **105**, 63001 (2014)
13. G.D. McDonald, C.C.N. Kuhn, arXiv:1312.2713 (2013)
14. Ch.J. Bordé, Ch. Salomon, S. Avrillier, A. van Lerberghe, Ch. Bréant, D. Bassi, G. Scoles, Phys. Rev. A **30**, 1836 (1984)

15. Ch.J. Bordé, Phys. Lett. A **140**, 10 (1989)
16. J.F. Clauser, Physica B **151**, 262 (1988)
17. K.-P. Marzlin, J. Audretsch, Phys. Rev. A **53**, 312 (1996)
18. T.L. Gustavson, *Precision rotation sensing using atom interferometry*, Ph.D. thesis, (Stanford University, 2000)
19. J.M. McGuirk, G.T. Foster, J.B. Fixler, M.J. Snadden, M.A. Kasevich, Phys. Rev. A **65**, 033608 (2002)
20. F.Y. Leduc, *Caractérisation d'un capteur inertiel à atomes froids*, Ph.D. thesis, (Université Paris XI Orsay, 2004)
21. B. Canuel, F. Leduc, D. Holleville, A. Gauguet, J. Fils, A. Virdis, A. Clairon, N. Dimarcq, Ch. J. Bordé, A. Landragin, P. Bouyer, Phys. Rev. Lett. **97**, 010402 (2006)
22. B. Dubetsky, M.A. Kasevich, Phys. Rev. A **74**, 023615 (2006)
23. A. Tonyushkin, M. Prentiss, Phys. Rev. A **78**, 053625 (2008)
24. S. Wu, E. Su, M. Prentiss, Phys. Rev. Lett. **99**, 173201 (2007)
25. K. Takase, *Precision rotation rate measurements with a mobile atom interferometer*, Ph.D. thesis, (Stanford University, 2008)
26. T. Lévèque, *Développement d'un gyromètre à atomes froids de haute sensibilité fondé sur une géométrie repliée*, Ph.D. thesis, (Observatoire de Paris, 2010)
27. J.K. Stockton, K. Takase, M.A. Kasevich, Phys. Rev. Lett. **107**, 133001 (2011)
28. S. Kleinert, E. Kajari, A. Roura, W.P. Schleich, Phys. Rep. **605**, 1 (2015)
29. J. Audretsch, K.-P. Marzlin, J. Phys. II **4**, 2073 (1994)
30. P. Wolf, P. Tourrenc, Phys. Lett. A **251**, 241 (1999)
31. A. Peters, K.Y. Chung, S. Chu, Metrologia **38**, 25 (2001)
32. Ch.J. Bordé, Eur. Phys. J. Spec. Top. **163**, 315 (2008)
33. H. Rauch, S.A. Werner, *Neutron Interferometry*, 2nd edn. (Oxford University Press, New York, 2015)
34. D.M. Greenberger, W.P. Schleich, E.M. Rasel, Phys. Rev. A **86**, 063622 (2012)
35. W.P. Schleich, D.M. Greenberger, E.M. Rasel, Phys. Rev. Lett. **110**, 010401 (2013)
36. W.P. Schleich, D.M. Greenberger, E.M. Rasel, New J. Phys. **15**, 013007 (2013)
37. A. Roura, W. Zeller, W.P. Schleich, New J. Phys. **16**, 123012 (2014)
38. W. Zeller, *The impact of wave-packet dynamics in long-time atom interferometry*, PhD thesis, (Ulm University, 2016)
39. C. Misner, K. Thorne, J.A. Wheeler, *Gravitation* (W. H. Freeman, San Francisco, 1973)
40. D. Bohm, *Quantum Theory* (Prentice Hall, Englewood Cliffs, 1951)
41. N. Bohr, L. Rosenfeld, Kgl. Danske Videnskab. Selskab, Mat.-fys. Medd. **12,** No. 8 (1933)
42. N. Bohr, L. Rosenfeld, Phys. Rev. **78**, 794 (1950)
43. H. Salecker, E.P. Wigner, Phys. Rev. **109**, 571 (1958)
44. J.A. Wheeler, in Proceedings of the International School of Physics "Enrico Fermi" Course 72 "Problems in the foundations of physics" (North-Holland, Amsterdam, 1979)
45. C.M. DeWitt, D. Rickle (eds.) *Conference on the Role of Gravitation in Physics*, (Wright Air Development Center, 1957). These proceedings have been reprinted under the title *The Role of Gravitation in Physics* by the Max-Planck Research Library for the History and Development of Knowledge
46. D.M. Greenberger, J. Math. Phys. **11,** 2329 (1970); *ibid.***11,** 2341 (1970)
47. R. Colella, A.W. Overhauser, S.A. Werner, Phys. Rev. Lett. **34**, 1472 (1975)
48. J.-L. Staudenmann, S.A. Werner, R. Colella, A.W. Overhauser, Phys. Rev. A **21**, 1419 (1980)
49. K.C. Littrell, B.E. Allman, S.A. Werner, Phys. Rev A **56**, 1767 (1997)
50. K. Hornberger, S. Gerlich, P. Haslinger, S. Nimmrichter, M. Arndt, Rev. Mod. Phys. **84**, 157 (2012)
51. E.A. Cornell, C.E. Wieman, Rev. Mod. Phys. **74**, 875 (2002)
52. W. Ketterle, Rev. Mod. Phys. **74**, 1131 (2002)
53. D. Schlippert, J. Hartwig, H. Albers, L.L. Richardson, C. Schubert, A. Roura, W.P. Schleich, W. Ertmer, E.M. Rasel, Phys. Rev. Lett. **112**, 203002 (2014)

54. S. Dimopoulos, P.W. Graham, J.M. Hogan, M.A. Kasevich, Phys. Rev. D **78**, 042003 (2008)
55. I. Pikovski, M. Zych, F. Costa, Č. Brukner, Nat. Phys. **11**, 668 (2015)
56. M. Köhl, Th.W. Hänsch, T. Esslinger, Phys. Rev. Lett. **87**, 160404 (2001)
57. C.W. Chou, D.B. Hume, T. Rosenband, D.J. Wineland, Science **329**, 1630 (2010)
58. A. Matveev, Ch.G. Parthey, K. Predehl, J. Alnis, A. Beyer, R. Holzwarth, Th. Udem, T. Wilken, N. Kolachevsky, M. Abgrall, D. Rovera, Ch. Salomon, Ph. Laurent, G. Grosche, O. Terra, Th. Legero, H. Schnatz, S. Weyers, B. Altschul , Th.W. Hänsch, Phys. Rev. Lett. **110**, 230801 (2013)
59. T. Jenke, P. Geltenbort, H. Lemmel, H. Abele, Nat. Phys. **7**, 468 (2011)
60. H. Abele, H. Leeb, New J. Phys. **14**, 055010 (2012)
61. T. Jenke, G. Cronenberg, J. Burgdörfer, L.A. Chizhova, P. Geltenbort, A.N. Ivanov, T. Lauer, T. Lins, S. Rotter, H. Saul, U. Schmidt, H. Abele, Phys. Rev. Lett. **112**, 151105 (2014)
62. C. Kiefer, *Quantum Gravity* (Oxford University Press, Oxford, 2007)
63. S. Abend, M. Gebbe, M. Gersemann, H. Ahlers, H. Müntinga, E. Giese, N. Gaaloul, C. Schubert, C. Lämmerzahl, W. Ertmer, W.P. Schleich, E.M. Rasel, Phys. Rev. Lett. **117**, 203003 (2016)
64. M.A. Kasevich, S. Chu, Phys. Rev. Lett. **67**, 181 (1991)
65. M.A. Kasevich, S. Chu, Appl. Phys. B **54**, 321 (1992)
66. E. Kajari, N.L. Harshman, E.M. Rasel, S. Stenholm, G. Süssmann, W.P. Schleich, Appl. Phys. B **100**, 43 (2010)
67. E. Giese, W. Zeller, S. Kleinert, M. Meister, V. Tamma, A. Roura, W.P. Schleich, Proceedings of the International School of Physics "Enrico Fermi" Course 188 "Atom Interferometry", ed. by G.M. Tino, M.A. Kasevich (IOS, Amsterdam, 2014)
68. D.M. Greenberger, Phys. Rev. Lett **87**, 100405 (2001)
69. J. Schmiedmayer, Ch. Ekstrom, M. Chapman, T. Hammond, D. Pritchard, J. Phys. II **4**, 2029 (1994)
70. M. Jacquey, A. Miffre, M. Büchner, G. Trénec, J. Vigué, EPL **77**, 20007 (2007)
71. D.A. Steck, Rubidium 85 D Line Data. http://steck.us/alkalidata (revision 2.1.6, 20 September 2013)
72. J.M. Hogan, D.M.S. Johnson, M.A. Kasevich, in Proceedings of the International School of Physics "Enrico Fermi" on Atom Optics and Space Physics, ed. by E. Arimondo, W. Ertmer, and W.P. Schleich (Amsterdam, 2009)
73. G.D. McDonald, C.C.N. Kuhn, S. Bennetts, J.E. Debs, K.S. Hardman, M. Johnsson, J.D. Close, N.P. Robins, Phys. Rev. A **88**, 053620 (2013)
74. S. Chiow, T. Kovachy, H. Chien, M.A. Kasevich, Phys. Rev. Lett. **107**, 130403 (2011)
75. A. Roura. arXiv:1509.08098 (2015)
76. S.A. DeSavage, J.P. Davis, F.A. Narducci, J. Mod. Opt. **60**, 95 (2013)
77. S.A. DeSavage, K.H. Gordon, E.M. Clifton, J.P. Davis, F.A. Narducci, J. Mod. Opt. **58**, 2028 (2011)
78. C.L. Adler, R. Johnson, A. Srinivasan, F.A. Narducci, to be submitted (2016)
79. G. Breit, Phys. Rev. **32**, 273 (1928)
80. M.V. Berry, N.L. Balazs, Am. J. Phys. **47**, 264 (1979)
81. W.P. Schleich, *Quantum Optics in Phase Space* (Wiley-VCH, Weinheim, 2001)
82. Y.N. Demkov, V.D. Kondratovich, V.N. Ostrovskii, Pis'ma Zh. Eksp. Teor. Fiz. **34**, 425 (1981)
83. C. Bracher, T. Kramer, M. Kleber, Phys. Rev. A **67**, 043601 (2003)
84. C. Blondel, C. Delsart, F. Dulieu, Phys. Rev. Lett. **77**, 3755 (1996)
85. R.M. Wilcox, J. Math. Phys. **8**, 962 (1967)

Double-Electron Ionization Driven by Inhomogeneous Fields

A. Chacón, L. Ortmann, F. Cucchietti, N. Suárez, J.A. Pérez-Hernández, M.F. Ciappina, A.S. Landsman, and M. Lewenstein

Abstract Electron–electron correlation effects play a crucial role in our understanding of sequential (SDI) and non-sequential double ionization (NSDI) mechanisms. Here, we present a theoretical study of NSDI driven by plasmonic-enhanced spatial inhomogeneous fields. By numerically solving the time-dependent Schrödinger equation for a linear reduced model of He and a double-electron time-evolution probability analysis, we provide evidence for enhancement effects in NSDI showing that the double ionization yield at lower laser peak intensities is increased due to the spatial inhomogeneous character of plasmonic-enhanced field.

This article is part of the topical collection "Enlightening the World with the Laser" - Honoring T. W. Hänsch guest edited by Tilman Esslinger, Nathalie Picqué, and Thomas Udem.

A. Chacón • N. Suárez
ICFO-Institut de Ciencies Fotoniques, The Barcelona Institute of Science and Technology, Av. Carl Friedrich Gauss 3, 08860 Castelldefels, Barcelona, Spain
e-mail: alexis.chacon@icfo.eu

L. Ortmann
Max Planck Institute for the Physics of Complex Systems, Nöthnitzer Straße 38, 01187 Dresden, Germany

F. Cucchietti
Barcelona Supercomputing Center (BSC), C/Jordi Girona 29, 08034 Barcelona, Spain

J.A. Pérez-Hernández
Centro de Láseres Pulsados (CLPU), Parque Científico, 37185 Villamayor, Salamanca, Spain

M.F. Ciappina (✉)
Institute of Physics of the ASCR, ELI-Beamlines, Na Slovance 2, 182 21 Prague, Czech Republic
e-mail: marcelo.ciappina@eli-beams.eu

A.S. Landsman
Max Planck Institute for the Physics of Complex Systems, Nöthnitzer Straße 38, 01187 Dresden, Germany

Max Planck Postech/Department of Physics, Pohang, Gyeongbuk 37673, Republic of Korea

M. Lewenstein
ICFO-Institut de Ciencies Fotoniques, The Barcelona Institute of Science and Technology, Av. Carl Friedrich Gauss 3, 08860 Castelldefels, Barcelona, Spain

ICREA, Pg. Lluís Companys 23, 08010 Barcelona, Spain

D. Meschede et al. (eds.), *Exploring the World with the Laser*,
https://doi.org/10.1007/978-3-319-64346-5_28

The change in the emission direction of the double-ion as a function of the field inhomogeneity degree demonstrates that plasmonic-enhanced fields could configure a reliable instrument to control the ion emission. Furthermore, our quantum mechanical model, as well as classical trajectory Monte Carlo simulations, show that inhomogeneous fields are as well as a useful tool for splitting the binary and recoil processes in the rescattering scenario.

1 Introduction

Since 1982, when L'Huillier presented the first experimental observation of a large enhancement in the double-charge ionization yield of Xe driven by an intense infrared (IR) laser-field, a number of questions about electron–electron (e-e) correlation effects and their mechanisms have arisen [1, 2]. The fact that those results could not be explained in the framework of SDI, where e-e correlation effects are assumed negligible, opened the path of considering the importance of such correlation effects in the ionization processes [1–5]. It was then that the concept of Non-Sequential Double Ionization (NSDI) arose as an explanation of the 1982 experiment [6–9]. However, in the NSDI mechanism there are several processes such as the shake-off, laser-field-assisted rescattering ionization, Rescattering Impact Double Ionization (RIDI) [10–13] and Rescattering Excitation with Subsequent Ionization (RESI), which might take place during the double ionization (DI) of atoms. The question of how to disentangle RIDI and RESI (and within RIDI the binary and recoil processes) are therefore still under investigation in the attosecond science community [3].

The most important mechanisms behind the NSDI process driven by a spatially homogeneous strong laser field in the mid-infrared regime are the RIDI and the RESI [9]. The importance of each of them basically depend on: (1) the gas atomic (or molecular) target and (2) the field features. The RIDI mechanism occurs within the so-called rescattering scenario. According to Corkum [10], once the first electron is launched into the continuum, this process happens about the maximum of the driven laser field via tunneling, this electron accumulates a kinetic energy E_{1k} and has certain probability to come back to the vicinity of ion core. At this rescattering time, and if the electron kinetic energy is larger than the second ionization potential (I_{2p}) of the remaining electron ($E_{1k} \geq I_{2p}$), the second electron is kicked out from the target "instantaneously" (see Fig. 1a).

However, in case that the collided first electron does not have enough energy to knock out the second electron, (i.e. $E_{1k} < I_{2p}$), this remaining electron should have a certain probability to be excited from its ground state to another excited state and will not be instantaneously ionized. Nevertheless, at a subsequent maximum of the oscillating laser field, this second excited electron can be indeed ionized via tunneling (see Fig. 1b). The latter process is known as RESI.

Prior studies addressing e–e correlation effects in laser-driven multiple ionization processes were done considering only spatially homogeneous fields, i.e. fields that do not present spatial variations in the region where the electron dynamics

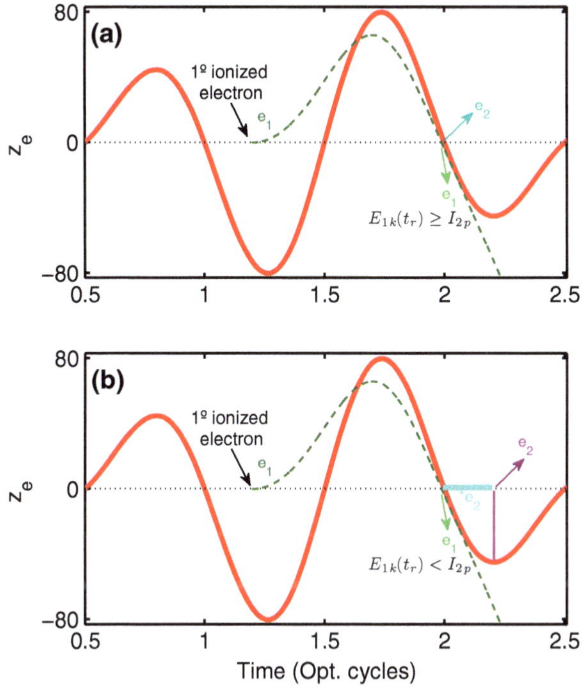

Fig. 1 *Panels* **a** and **b** show the classical pictures of the rescattering impact double ionization (RIDI) and rescattering excitation with subsequent double-ionization (RESI) scenarios, respectively. The *red continuum* and *green dashed lines* denote the IR laser field oscillations and the trajectory of the first ionized electron about $t \approx 1.25T_0$ (T_0 is the period of the laser optical field), respectively. In **a** the *light green* and *light blue arrows* indicate the first and second ionized electrons. Note that the second electron is launched into the continuum about $t \approx 2T_0$ when $E_{1k} \geq I_{2p}$. In **b**, at the recollision time, denoted by a *light green arrow* ($t \approx 2T_0$), the second electron is excited and remains in this state, noted by a *light blue line*, until it is laser-ionized by tunneling at a subsequent maximum of the laser field (denoted by a *magenta arrow*) about $t \approx 2.25T_0$

takes place. This is a legitimate assumption considering that in conventional laser-matter experiments the laser electric field changes in a region on the orders of micrometers, whereas the electron dynamics develops on a nanometric scale. However, since recent studies of post-ionization dynamics in spatially inhomogeneous fields [14] provides new physical effects and insights, a question arises as to the influence of spatial variation on the DI process. The aim of this work is to present a complete study of DI driven by plasmonic-enhanced spatially inhomogeneous fields with an investigation of NSDI in general, and the RESI and RIDI mechanisms in particular. Plasmonic-enhanced spatially inhomogeneous fields appear when a short laser pulse, of low intensity, illuminates a metallic or dielectric nanostructure. As a consequence of the formation of surface plasmon polaritons (SPPs), an amplification (enhancement) of the incoming laser electric field takes place. This enhancement is high enough to allow the development of strong laser-matter phenomena, as high-order harmonic generation (HHG), above-threshold ionization (ATI) and DI, amongst others. In addition, due the nanometric dimensions of the so-called hot spots, both the laser electric field and the associated vector potential present spatial variations in a scale comparable to the one where the electron develops its motion. As a consequence, the theoretical approaches should have to incorporate this new feature in the laser-matter coupling (for a comprehensive review of recent theoretical and experimental developments see e.g. [15]).

We will employ both quantum mechanical approaches, based on the numerical solution of the time-dependent of the Schrödinger equation (TDSE) for two electrons in reduced dimensions, and classical schemes employing classical trajectories Monte Carlo (CTMC) simulations to deal with DI driven by plasmonic-enhanced spatially inhomogeneous fields. Within the quantum framework, we employ a linear model for the helium atom, where the motion of both electrons is restricted to the direction of the laser polarization. Experience has shown that 1D models qualitatively reproduce strong-field phenomena such as the double-ionization knee structure [16, 17] or above-threshold ionization [18] and intense-field double ionization mechanisms [19].

The rest of this paper[1] is organized as follows. In the next section we present our theoretical tools, namely the TDSE and CTMC for two electrons in reduced dimensions. Then, in Sect. 3, we show a comparative study between DI driven by conventional laser pulses and DI governed by plasmonic-enhanced spatially inhomogeneous fields. We put particular emphasis on the two-electron momentum distribution, considering it represents one of the most detailed observables and is fully experimentally accessible. We end up with concluding remarks and a brief outlook.

2 Numerical Model

We study the two-electron dynamics driven by plasmonic-enhanced fields via a fully quantum mechanical linear model of the helium atom and the integration of Newton's equations in the framework of the CTMC method. The *ab*-initio quantum mechanical calculations allow us to address the whole electron–electron (e-e) correlated dynamics by means of the numerical solution of the TDSE similarly to

[1]This contribution is dedicated to Ted Hänsch on the occasion of his 75th birthday. Although Prof. Hänsch is mostly regarded for "contributions to the development of laser-based precision spectroscopy, including the optical frequency comb technique", and his contributions to laser cooling and physics of ultracold atoms, his influence on attosecond physics is hard to underestimate. For instance, he predicted at very early stages the possibility of generating attosecond pulse trains from phase locked harmonics [20, 21] and pioneered and contributed to the initial studies of the high-order harmonics coherence [21, 22]. His group has also developed decisive steps extending the frequency combs toward high frequencies regimes.

those used by Lein [19] and Watson [18, 23]. The Hamiltonian of our 2e system reads (atomic units are used throughout the paper unless specified otherwise)

$$H = \sum_{j=1}^{2}\left[\frac{p_j^2}{2} + V(z_j) + V_{\text{int}}(z_j, t)\right] + V(z_1, z_2), \quad (1)$$

where $p_j = -i\frac{\partial}{\partial z_j}$ is the momentum operator corresponding to the j-th electron (jth-e), $j = 1, 2$. $V(z_j) = -\frac{Z}{\sqrt{z_j^2 + a}}$ and $V(z_1, z_2) = \frac{1}{\sqrt{(z_1 - z_2)^2 + b}}$ are the attractive potential of the interaction of the jth electron with the nucleus of charge Z and the repulsive e–e potential, respectively. The potential describing the interaction of the jth electron with the spatially dependent laser electric field in length gauge is [24]

$$V_{\text{int}}(z_j, t) = \left(z_j + \frac{\beta}{2}z_j^2\right)E_{\text{h}}(t), \quad (2)$$

where β denotes the inhomogeneity strength (see e.g. [24, 25] for more details) of the plasmonic field and $E_h(t) = E_0 \sin^2(\omega_0 t/2N)\sin(\omega_0 t + \varphi_0)$ is the spatially homogenous or conventional laser electric field. Here, E_0, ω_0, N and φ_0 are the laser electric peak amplitude, laser frequency, total number of cycles and carrier envelope phase (CEP), respectively.

The numerical algorithm used to solve the TDSE for our linear 1Dx1D He model is the Split Operator method described in Refs. [26, 27]. This algorithm takes advantage of the Fast Fourier Transform (FFT) paradigm to evaluate the kinetic energy operators of Eq. (1) in the Fourier space. To speed up our calculations and redistribute the whole 2e wavefunction in position space—with a total number of points $N_{\text{T}} = N_1 \times N_2 \approx 4 \times 10^4 \times 4 \times 10^4 = 1.6 \times 10^9$—on different computational nodes, N_p, we employ the message passing interface MPI parallelized version of the FFTW [28]. This implementation allows us to reach large electron excursions $z_j \gg \frac{E_0}{\omega_0^2}$, which is typical for electrons driven by spatially inhomogeneous fields [24, 25]. Each 1Dx1D TDSE calculation took about 11735 CPU-hours on $N_p = 1024$ cores in the Barcelona Supercomputer Center.

For the He linear model, we have fixed the soft-core parameters and the nucleus-charge to $a = b = 1$ a.u., and $Z = 2$, respectively. With these values, we obtain a 2e ground state energy of $E_{1,2} = -2.238$ a.u. Although the matching with the experimental data is not perfect, it is sufficient to qualitatively reproduce the 2e-dynamics driven by a linearly polarized laser-field [18, 19, 23, 29]. The 2e ground state wavefunction $\Psi_0(z_1, z_2)$ is obtained via imaginary-time propagation—with an imaginary time step $\delta t = -i0.025$ a.u.—switching off the interaction potential in Eq. (2), i.e. $V_{\text{int}}(z_j, t) = 0$. The first electron ionization and the second electron ionization potentials are then $I_{1p} = 0.751$ a.u. and $I_{2p} = 1.487$ a.u., respectively.

To follow the 2e dynamics driven by the plasmonic-enhanced spatially inhomogeneous fields, encoded in the the time-dependent wavefunction $\Psi(z_1, z_2, t)$, the wavefunction $\Psi_0(z_1, z_2)$ of the ground state is propagated in real time via TDSE with the Hamiltonian defined in Eq. (1). In addition, we compute the single-electron ionization, $P_{1e}(t)$, and 2e-ionization, $P_{2e}(t)$, as a function of time t. The 2e position distribution $|\Psi(z_1, z_2, t)|^2$ is split into three parts: (i) $\{|z_1|, |z_2| < z_a\}$, (ii) $\{|z_1| < z_a, |z_2| \geq z_a\}$ or $\{|z_1| \geq z_a, |z_2| < z_a\}$, and (iii) $\{|z_1|, |z_2| \geq z_a\}$ with $z_a = 90$ a.u.

As it is illustrated in Fig. 2, the first region (*i*) describes the 2e bound wavefunction, $\Psi_b(z_1, z_2, t)$, part of $\Psi(z_1, z_2, t)$. The second one (*ii*) defines the single-electron ionization (SI) $\Psi_{1e}(z_1, z_2, t)$, which is the time dependent He^+ yield. And the third region (*iii*) includes the double-electron ionization (DI) $\Psi_{2e}(z_1, z_2, t)$ part, which represents the He^{2+} time-dependent yield production. Then, by integrating over regions (*ii*) and (*iii*) the single- and double-electron ionization $P_{1e}(t)$ and $P_{2e}(t)$ rates (He^+ and He^{2+} production yields) are computed, respectively.

The final two-electron momentum distribution $S_{2e}(p_1, p_2) = |\Psi_{2eM}(p_1, p_2, t_F)|^2$ is evaluated half a laser cycle after the end of the IR laser field as the absolute square of the projection of the final 2e wavepacket $\Psi_{2e}(z_1, z_2, t_F)$ on the double-electron plane waves $\Phi_{p_1,p_2}(z_1, z_2) = \frac{1}{2\pi}\exp[i(z_1 p_1 + z_2 p_2)]$ [11, 19]. Convergence tests on time propagation at $\frac{1}{3}T_0, \frac{1}{2}T_0$ and T_0 after the end of the oscillating laser field show that the 2e probability distribution does not change considerably between $\frac{1}{2}T_0$ and T_0 (T_0 is the cycle period of the IR laser beam). To reduce the computational calculation time and position number of grid points, we chose as detection time half a cycle after the end of our oscillating field. Furthermore, the correlated ion $S_{He^{2+}}(p_{ion})$ momentum distribution is calculated by projecting $S_{2e}(p_1, p_2)$ on the diagonal $p_1 = p_2$, which corresponds to the total 2e momentum $p = p_1 + p_2$.

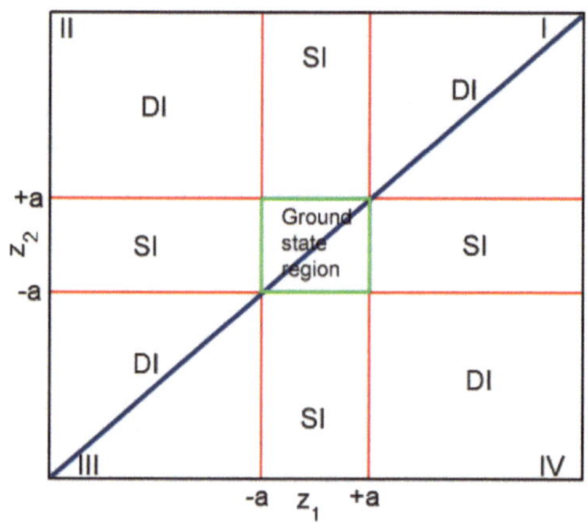

Fig. 2 Diagram of the different regions for the whole two-electron wavefunction in the position space (z_1, z_2). The entire region can be split in (*i*) the spatial ground state area, (*ii*) the single-electron ionization (SI) area and (*iii*) the double-electron ionization (DI) area (see the text and consider $z_a = a$). Note that the (*i*) spatial region is graphically described by the *green central square*, the (*ii*) is the inner region within the *red horizontal* and *vertical lines* without considering (*i*), and the (*iii*) region is the remaining area, including parts in the *I*, *II*, *III* and *IV* quadrants. The *diagonal dark blue line* points out the antisymmetric condition for the two-electron systems

Thereby, via momentum conservation of the system, the ion momentum reads $p_{\text{ion}} = -(p_1 + p_2)$.

To supplement the quantum mechanical calculations and better understand the physical origin of the effects of the plasmonic-enhanced field, we implement CTMC simulations to investigate electron trajectories after ionization of helium under the so-called RIDI mechanisms. The simulations are restricted to one dimension, namely the direction of field polarization, in which also the field inhomogeneity develops. The trajectories are launched at a starting time t_0, which is distributed probabilistically following the Ammosov–Delone–Krainov (ADK) formula [30, 31], typically used to model strong field ionization [32–35]

$$P(t_0, v_\perp) = \exp\left(-\frac{2(2I_p(t_0))^{3/2}}{3E_\text{h}(t_0)} \right) \tag{3}$$

corresponding to an atom centered at the origin. I_p denotes the Stark shifted ionization potential [36]

$$I_p(t_0) = I_{1p} + \frac{1}{2}(\alpha_N - \alpha_I)E_\text{h}(t_0)^2, \tag{4}$$

with α_N and α_I representing the polarizability of the atom and ion, respectively. The tunnel exit radius is assumed to be zero following the simple man's model [10]. The dynamics of each electronic trajectory after ionization is solved numerically by integrating the Newton's equations of motion, which take into account the laser field, but not the Coulomb potential following the model in [11].

If the electron returns to the ion's position ($z = 0$) at time t_r with kinetic energy $E_{k1}(t_r)$ larger than the ionization potential I_{2p} of the second electron [18], this second electron can be ionized as well. In this ionization process the kinetic energy of the first electron is reduced by I_{2p} and the second electron is born in the continuum with zero velocity

$$\begin{aligned} p_1(t_r) &= \pm\sqrt{2(E_{k1}(t_r) - I_{2p})} \\ p_2(t_r) &= 0. \end{aligned} \tag{5}$$

Here, the two different signs in p_1 describe the possibility of scattering the first electron into forward, binary, or backward, recoil, direction with respect to its momentum directly before the ionization of the second remaining electron.

For each double ionization event, we calculated both options. The dynamics after the second ionization is again determined by the propagation in the laser field, where the Coulomb force is completely neglected [11]. The reason for doing so is the fact that close to the ion quantum effects play a crucial role, which cannot be captured in our classical model. Therefore, we restrict ourselves to the classical dynamics in the laser field. For the electron dynamics far away from the ion, this is not a problem anyway since here the Coulomb force is negligible, which is why it is common practice to treat the Coulomb force in this regime perturbatively [37].

3 Double-Electron Ionization

To study the e-e correlation effects we first compute the final single- and double-electron ionization yields as a function of the peak laser field intensity for a few-cycle IR pulse. This allows us to identify the intensity regions where the spatially inhomogeneous field substantially modifies the double-electron ionization process. Because of possible damage and ablation of the nanostuctures, we restrict ourselves to relatively low peak plasmonic-enhanced intensities: $I_0 < 10^{15}$ W/cm^2 (for more details about the parameters range and experimental constraints see e.g. [15]). Second, we compute the two-electron momentum distribution as a function of the inhomogeneity strength at a fixed laser intensity. This scan on the β parameter provides enough evidence about the role of the spatial inhomogeneous field in the 2e ionization process. Furthermore, we scrutinise if the e–e correlation features are affected by the plasmonic-enhanced driven field.

3.1 He^+ and He^{2+} Ion Yields

We numerically compute the final 2e-ionization yield by the procedure described in Sect. 2. The grid parameters used in those calculations are $N_1 = N_2 = 40960$ points and $\delta z_1 = \delta z_2 = 0.25$ a.u. The integration time step was chosen $\delta t = 0.025$ a.u. The results of the single- and double-electron ionization yield as a function of the peak laser field intensity for the homogeneous ($\beta = 0$) and inhomogeneous fields with $\beta = 0.005$ a.u., are depicted in Fig. 3.

An enhancement of the final 2e ionization $P_{2e}(t_F, I_0)$ is observed for the inhomogeneous field case when compared to the conventional one. Similar effects are also obtained in the comparison with the single-electron ionization yields. This

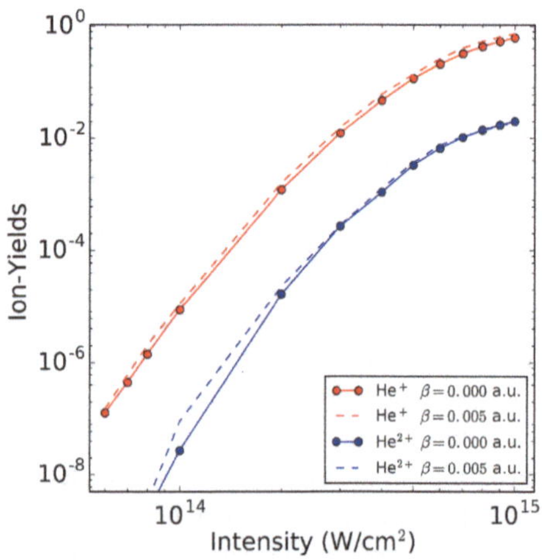

Fig. 3 Numerical TDSE calculations of the single- (*red*) and double-electron (*blue*) ionization yields of our linear 1D × 1D He 2e-model driven by homogeneous (*line with circles*) and inhomogeneous (*dashed line*) fields as a function of the laser peak intensity. The mean frequency of the IR laser field is $\omega_0 = 0.057$ a.u. (1.55 eV), the CEP is $\varphi_{CEP} = 0°$ and the total number of cycles is N = 4 under a $\sin^2$ envelope

feature is hardly visible in the Fig. 3, due to the logarithmic scale, although we have found an increment up to a factor of 4 in the double ionization yield. This enhancement clearly shows that the spatially inhomogeneous fields play an instrumental role in the NSDI of helium.

Naturally, the question about the origin of this enhancement arises. To answer it, we compute the single- $P_{1e}(t)$ and double-electron $P_{e2}(t)$ ionization yield as a function of time at a fixed peak intensity of $I_0 = 2 \times 10^{14}$ W/cm^2. Here we focus our attention on the intensity region where the double ion yield, He^{2+}, is enhanced by the inhomogeneous field. According to Fig. 3, one such region is $I_0 = 1 - 5 \times 10^{14}$ W/cm^2. The results of the time-evolved probabilities are depicted in Fig. 4.

For the single-electron ionization $P_{1e}(t)$ shown in Fig. 4a, the "inhomogeneous" ionization yield is larger than the conventional one, in particular, at about 2.5 cycles of the IR laser. We could trace out the origin of this observation in a much stronger distortion of the laser-atomic potential barrier, which raises the probability of the first bound electron to 'escape' from the atom.

Figure 4b shows a comparison of $P_{2e}(t)$ for conventional and inhomogeneous fields. About 3.4 cycles of the IR laser oscillations, the 2e ionization yield largely increases for the inhomogeneous field case with respect to the conventional one by more than 5-times. At this very low inhomogeneity degree of $\beta = 0.005$ a.u., and low IR peak intensity, this enhancement of the 2e-ionization rate is a very surprising result. Similar behaviour was previously observed in [14], where the double-electron ionization reaches higher yields leading to an enhancement in the intensity of the HHG signal. However, in that latter case a larger inhomogeneity degree of $\beta = 0.02$ a.u was used.

An hypothesis that might explain that result is based on the three step Corkum's model [6, 9, 10] where: first, the first electron ionizes via tunnelling, second, this electron propagates in the continuum gaining energy from the laser field—in our case a spatially inhomogeneous field—and then when the field changes its sign the electron has a probability to re-collide with the ion core He^+. As a third step, this colliding electron can kick out the second electron if and only if the first electron

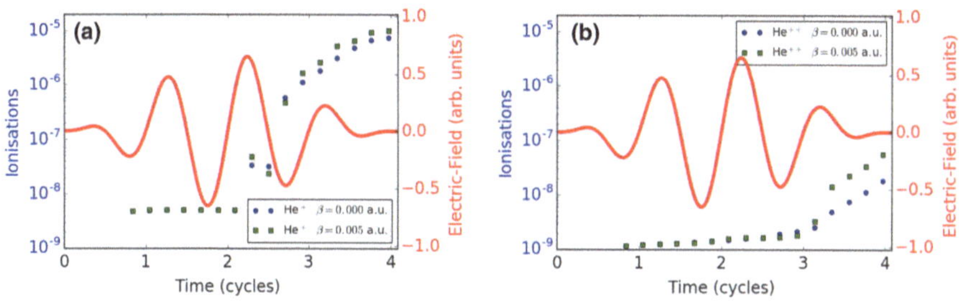

Fig. 4 Single- (**a**) and double-electron (**b**) ionization yields of our linear 1D × 1D He 2e-model driven by conventional and spatially inhomogeneous fields as a function time (see *left axis*). The IR laser field oscillations are depicted in *red solid line*. The laser peak intensity used to follow the two-electron dynamics is $I_0 = 2 \times 10^{14}$ W/cm^2. The other laser parameters are the same than those used in Fig. 3

kinetic energy is larger than I_{2p}, the ionization potential of the second remaining electron in the ion core: RIDI or (e, 2e) mechanism. For the conventional field cases at *low* laser peak intensities (about 0.8–3 $\times 10^{14}$ W/cm^2), the probability that the first electron reaches a larger enough energy as to overcome the ionization potential of the second electron is negligible. Thus, it is rather unlikely that the double ionization process be mediated by the so-called RIDI mechanism (see Sect. 1 for more details). However, from the behaviour of electrons driven by spatially inhomogeneous fields (see e.g. [15]), it is very likely that the first-ionized electron gains a much larger energy compared to the conventional case. Thus, at the instant of re-collision, the second electron would have a higher chance to be ionized in a spatially inhomogeneous field, which corresponds to an enhancement of the double electron ionization probability.

Note that according to Refs. [10, 11] the RIDI process is limited by the energy that the first electron can accumulate from the laser field. The calculations depicted in Figs. 3 and 4 provide enough evidence to confirm that the spatial inhomogeneous field could open this channel at a much lower laser peak intensity. This happens because the first electron gains much more energy in the field, increasing then the probability to ionize the second electron.

3.2 Correlated Two-Electron Momentum Maps

Another interesting observable, which contains information about the e–e correlation, is the 2e-momentum distribution. This observable has allowed to disentangle the common sequential and non-sequential double RESI, rescattering impact ionization and laser-field assisted rescattering ionization mechanisms [12, 13, 38]. Fig. 5 depicts $S_{2e}(p_1, p_2)$ for different β parameters at the same fixed laser peak intensity of $I_0 = 2 \times 10^{14}$ W/cm^2. The double-electron map in Fig. 5a exhibits two large probability lobe peaks on the first quadrant of the correlation region—in almost perfect concordance with the results published in Ref. [19]. This probability distribution indicates that both electrons prefer to leave on the same (positive) direction. It is understood that the repulsive e–e Coulomb potential plays an important role at those relative low peak intensity for the He model [19].

Note that a classical rescattering electron scenario (e, 2e) is not good enough for describing this NSDI mechanism of our He model at this peak intensity. From a classical viewpoint, the rescattering energy $E_{\mathrm{k,max}} = 3.17U_p = 1.4$ a.u., is lower than the second ionization potential $I_{2p} \sim 1.5$ a.u. This is the main reason to not compute the double electron ionization maps by means of CTMC simulations. This is so because the classical rescattering energy of the first electron is not enough to exceed the ionization potential of the second electron. Instead, these double-electron ionization maps, Fig. 5a, b, could be understood as a laser field-assisted rescattering process for which such a constraint does not apply [11–13, 19, 38]. As

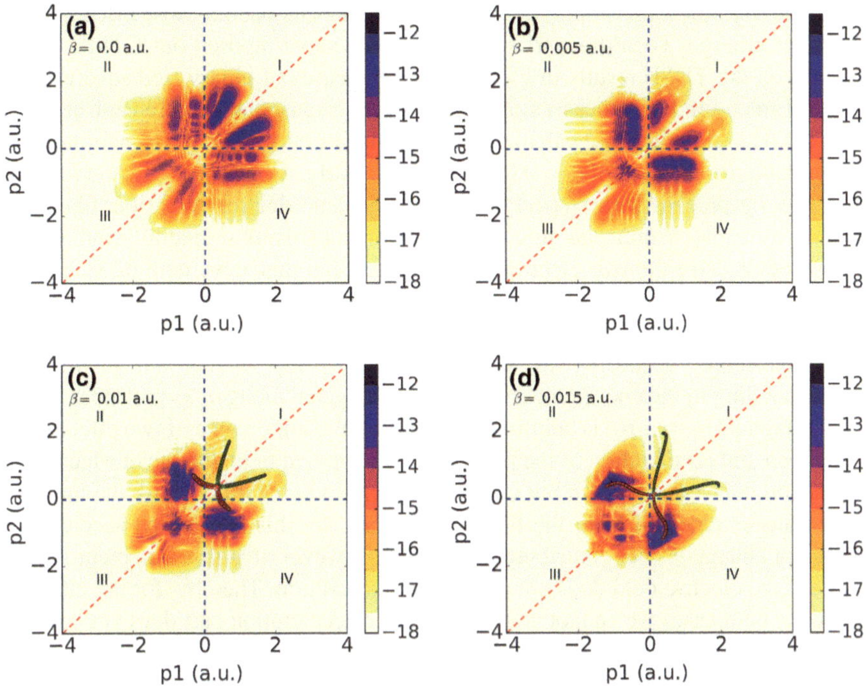

Fig. 5 Numerical two-electron momentum distribution for various inhomogeneity degrees: $\beta = 0.000$ (**a**), 0.005 (**b**), 0.010 (**c**) and 0.015 a.u. (**d**). The *color scale* is $\log_{10}[S_{2e}(p_1, p_2)]$.*Vertical* and *horizontal blue dashed lines* denote the 2e momentum axes which help us to distinguish between correlated, (*i*) and (*iii*) quadrants, and anticorrelated, (*ii*) and (*iv*) quadrants, regions. The *diagonal red dashed line* $p_1 = p_2$ represents the max e–e correlation momentum points or the total 2e momentum $p = p_1 + p_2$. CTMC for the (e, 2e) mechanisms are superimposed in *green-circles* (recoil process) and in red-squares (binary process) in **c** and **d** *panels*. The laser peak intensity used for these numerical calculations is $I_0 = 2 \times 10^{14}$ W/cm^2. The other laser parameters are the same as those used in Fig. 3

pointed out in [9, 19], the driving laser field provides the rest of the required energy to remove the second electron at the instant of recollision.

For further interpretations of Fig. 5 we recall that finding double ionization in quadrants I and III corresponds to both electron momenta pointing in the same direction. In contrast, quadrants II and IV contain the cases of the electrons' momenta pointing into opposite directions. When both electrons leave the atom in the same direction, we say they are correlated. Comparing the 2e-momentum distributions in Fig. 5a, b, we find that the two electrons prefer to detach in opposite directions when driven by plasmonic-enhanced spatially inhomogeneous fields. This effect is even larger for an inhomogeneity degree of $\beta = 0.01$ and 0.015 a.u., as can be seen in Fig. 5c, d. We note, however, the appearance of a small 2e-probability also in the correlated regions.

Naturally, questions about the physical mechanisms behind those effects in the 2e maps emerge. To address those questions, we superimposed our CTMC calculations on the TDSE results in Fig. 5c, d for the cases of binary (red-squares) and recoil (green-circles) processes. As is observed, a reasonable agreement between the TDSE and the CTMC calculations is found. In particular, the concordance is remarkable for the case of $\beta = 0.015$ a.u. This clearly corroborates that the forward rescattering process with respect to the first incident electron direction, binary, is highly probable within that so-called (e, 2e) mechanisms if spatially inhomogeneous fields drive the two-electron system. Note that this agreement of TDSE and CTMC supports our previous observation that the 2e-particles are likely to prefer leaving the atom in opposed directions.

Furthermore, according to Weber et al. [8] the momentum distribution corresponding to the coordinates $p = p_1 + p_2$ (diagonal along $p_1 = p_2$) and $p^- = p_1 - p_2$ (diagonal $p_1 = -p_2$) is helpful for describing the importance of two effects: e–e repulsion and acceleration of the particles by the optical field. On the one hand, e–e repulsion does not change p but contributes to p^-. On the other hand, the momentum transfer received from the field is identical. So, this part of the acceleration does not change p^- but contributes to p. Note, however, that this statement is only valid if the electric field does not depend on the position. Thereby, for the inhomogeneous field cases we cannot conclude that the acceleration part does not contribute to p^- as it is the case in conventional fields. This is in absolute concordance with what we observe in the 2e momentum maps for $\beta = 0.010$ and 0.015 a.u.

Additionally, in Fig. 6 we show the correlated ion He^{2+} momentum distributions corresponding to the $S_{2e}(p_1, p_2)$ panels of Fig. 5. A first observation is that a large momentum-shift is found for the recoiling ion as the inhomogeneity degree β increases. For the conventional field case depicted in Fig. 6a, the full momentum width of the distribution is about $\pm 2A_0 = \pm 2.6$ a.u., where $A_0 = E_0/\omega_0$ ($A_0 = 1.3$ a.u.), is the maximum peak vector potential strength [7, 19, 39]. An asymmetry in the amplitude of the ion distribution $S_{\mathrm{He}^{2+}}(p_{\mathrm{ion}})$ is observed at $\beta = 0$ a.u. This is due to the employed laser field being within the few-cycle regime, $N = 4$, see e.g. Ref. [3] about the CEP effects. However, three peaks at about $p_{\mathrm{ion}}^{(\max)} = \{-A_0, 0, +A_0\}$ are found. These might suggest that the laser field-assisted rescattering double ionization mechanism and the RESI mechanism take place simultaneously in such a complex correlated momentum map.

In the case of inhomogeneous fields, the ion distribution shape strongly depends on the parameter β. While the inhomogeneity increases, the expectation value of the ion momentum $\langle p_{\mathrm{He}^{2+}} \rangle$ is shifted from negative to positive momentum values, i.e. the momentum expectation value changes from $\langle p_{\mathrm{He}^{2+}} \rangle = -0.92$ to +0.49 a.u. (see the top left of each panel). This indicates that the ion recoils in a completely opposite direction compared to the conventional field case. This strong modification in the ion direction emission, in principle experimentally detectable, is a signature of the spatial inhomogeneous character of the driven field in the DI process. In addition, the several peaks that appear in the ion distributions, suggest

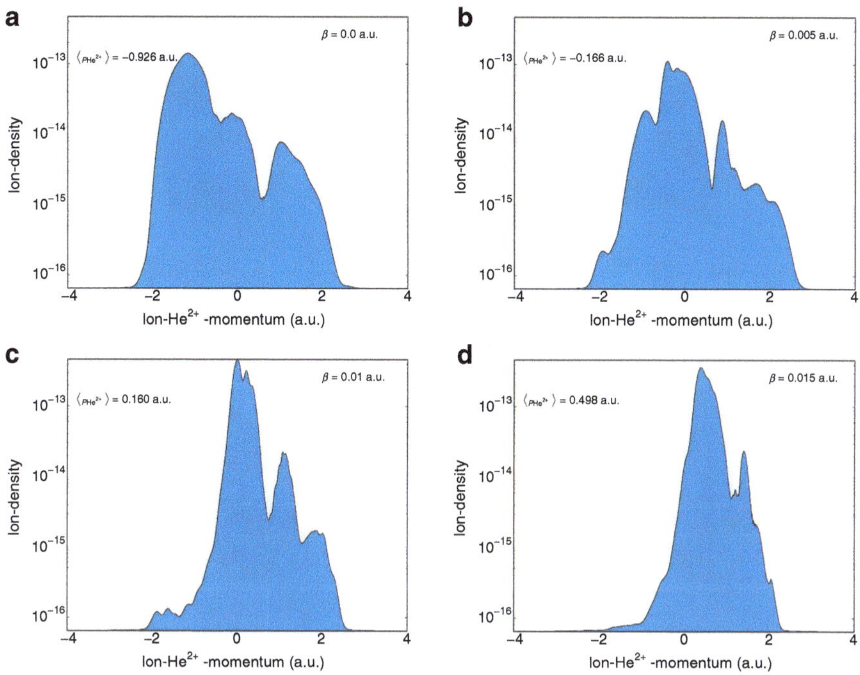

Fig. 6 Correlated ion He^{2+} momentum distributions, $S^2_{\mathrm{He}} + (p_{\mathrm{ion}})$, corresponding to the *panels* **a**–**d** of Fig. 5. We have also computed the ion expectation values $\langle p_{\mathrm{He}^+2}\rangle$ as a function of β and show it in each plot (see the *top left of each panel*). The difference is remarkable: from negative values to positive ones. This clearly points out that the spatial inhomogeneous field configures an instrumental tool to control the ion direction emission

the possibility of different interference paths in the DI process driven by the spatially inhomogeneous field.

So far, we have studied double ionization in He via scanning the 2e-momentum distribution over the inhomogeneity parameter β at a fixed laser intensity. To obtain an insight about the 2e ionization when the laser-peak intensity increases, we compute and compare the momentum–momentum distributions for the conventional $\beta = 0$ and inhomogeneous $\beta = 0.005$ a.u. fields. Additionally, our *ab*-initio TDSE calculations are compared with the CTMC simulations. The results are depicted in Fig. 7. While the peak intensity increases from 3 to 7×10^{14} W/cm^2 for conventional fields, some pronounced lobes in the correlation regions are observed. Furthermore, large probability lobes in the anticorrelated region are also visible. This is a signal that the e–e Coulomb repulsion force is losing its importance while the laser field peak intensity increases. In particular, that effect is larger for the highest intensity. In addition, note that a better agreement between TDSE and CTMC is found in the cases of Fig. 7c, e as it is expected [7, 11, 19]. This indicates that the (e, 2e) processes are the main mechanisms behind those calculations. However, in Fig. 7a, a laser field-assisted rescattering DI process still

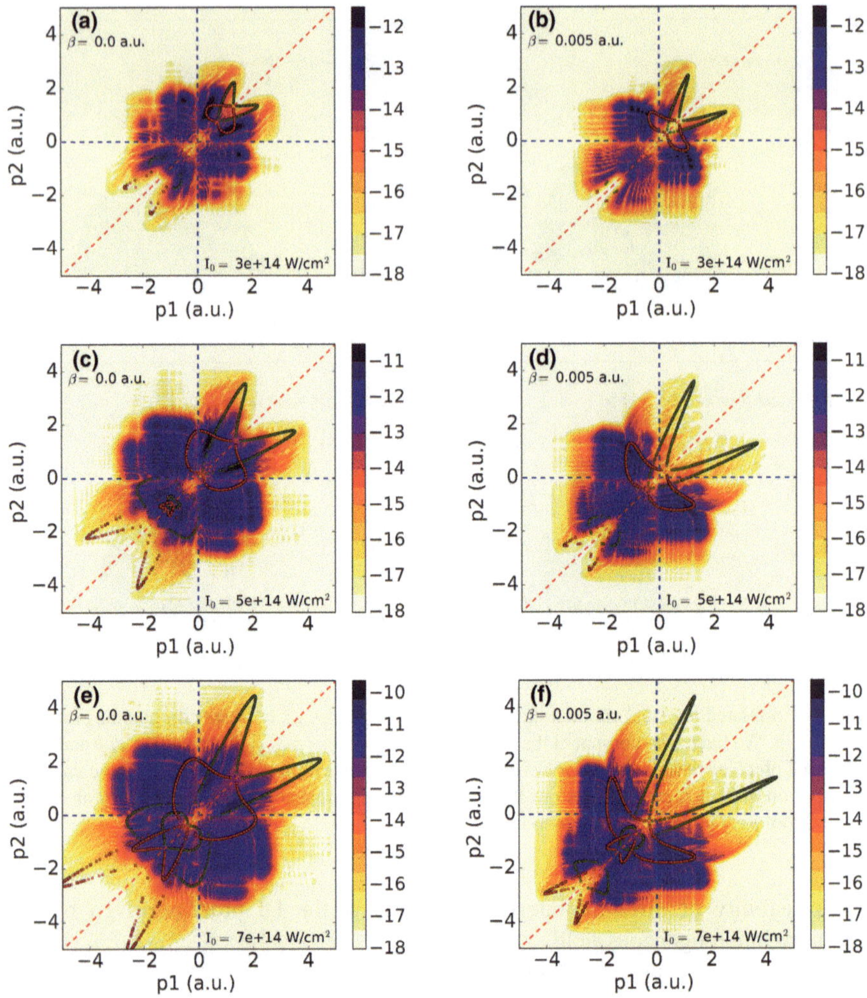

Fig. 7 Numerical two-electron momentum distributions driven by homogeneous $\beta = 0$ (**a**, **c**, **e**), and inhomogeneous fields (**b**, **d**, **f**) with $\beta = 0.005$ a.u., for three different laser-peak intensities: $I_0 = 3, 5, 7 \times 10^{14}$ W/cm^2. The CTMC calculations for binary (*red-squares*) and recoil (*green-circles*) processes are superimposed on the 2e momentum maps. Other laser parameters are the same as in Fig. 3

dominates over the RIDI mechanisms. This is concluded from the poor agreement between the quantum mechanical and the classical calculations for the binary and recoil processes.

On the other hand, for inhomogeneous fields, Fig. 7b, d, f, the probability of 2e ionization with opposite momenta increases. This clearly indicates that the propagation of electrons under the influence of plasmonic field changes completely the 2e-dynamics. Note, that a signal in quadrant III of the correlation region is also observed, which is an indication that both electrons, independently of the incident direction of the first colliding particle, prefer to leave with negative momenta directions. Furthermore, while the laser peak intensity increases, the V-like shape in the quadrant III tends to be much closed, and also a strong signal along the diagonal $p_1 = p_2$ for $p_1 < 0$ is clearly observed. These facts are the signature that e–e correlation effects rapidly losing importance while the particles are propagating in the plasmonic field. Note, however, that e–e repulsion somehow is still present because of the large momentum density width along the diagonal $p_1 = -p_2$.

Finally, it is interesting to point out that up to 7×10^{14} W/cm^2 the NSDI by inhomogeneous fields is still within the rescattering (e, 2e) scenario. This statement is supported by the CTMC simulations that agree very well with the TDSE calculations for all studied cases. This demonstrates that the isolation of binary and recoil processes is very sensitive to the laser peak intensity. We should note, however, that between 2 and 5×10^{14} W/cm^2, we ensure that those backward and forward rescattering processes could be separated, just by observing the anticorrelating and correlating regions of the momentum–momentum distribution.

4 Conclusions

Non-sequential double ionization of helium atoms driven by a plasmonic-enhanced spatially inhomogeneous fields has been theoretically investigated. By means of the fully numerical solution of the time-dependent Schrödinger equation, we observed that ion yield of He^{2+} substantially increases while the inhomogeneous field drives the system. An analysis of the single- and double-electron time-evolution probabilities and the two-electron momentum distribution simulations of the binary an recoil mechanisms support that the main reason for this enhancement corresponds to a high accumulated energy of the first re-colliding electron when it is moving in the spatially inhomogeneous field.

An unexpected (e, 2e) mechanism at very low intensity, i.e. $I_0 = 2 \times 10^{14}$ W/cm^2 is observed with increasing the inhomogeneity strength. Note that the double electron ionization effects induced by the plasmonic-enhanced fields will depend on (1) the peak intensity; (2) the spatial properties of the field and (3) the applied target. The latter is so because of the different ionization potentials for different atomic and molecular species. This means that both by engineering the inhomogeneous field and controlling the laser intensity the two different mechanisms, namely the laser field-assisted re-scattering and the RIDI process can be isolated. Furthermore, our interpretation of the fully ab-initio TDSE for the two-electron momentum distributions by comparing to CTMC simulations allowed us to distinguish between binary and recoil processes if and only if the spatially inhomogeneous field drives the system.

Furthermore, the spatial characteristics of the plasmonic-enhanced field break the symmetry of the 2e acceleration in the anticorrelated region. This is noted in the pronounced peaked probability of the two-electron momentum distribution located in the II and IV quadrants. Physically this is translated by the fact that the two-electron propagation is much more affected by the plasmonic field than by the e–e correlation at the double-ionization time. Thereby, plasmonic-enhanced fields configure an interesting alternative to control correlation effects in the double ionization process.

Still there are open questions, e.g. concerning the role of the e–e Coulomb potential while both identical particles are propagating within the spatially inhomogeneous field and how this effect is related to the 2e momentum distribution maps for the largest laser peak intensities here used. We plan to address these questions in a subsequent work.

Acknowledgements This work was supported by the project ELI-Extreme Light Infrastructure-phase 2 (Project No. CZ.02.1.01/0.0/0.0/15_008/0000162) from European Regional Development Fund, Spanish MINECO (National Plan grants FIS2011-30465-C02-01, FOQUS No. FIS2013-46768-P, FISICATEAMO FIS2016-79508-P and Severo Ochoa Excellence Grant No. SEV-2015-0522), the Generalitat de Catalunya (SGR 874 and CERCA/Program) and Fundació Privada Cellex Barcelona. N.S. was supported by the Erasmus Mundus Doctorate Program Europhotonics (Grant No. 159224-1-2009-1-FR-ERA MUNDUS-EMJD). N.S., A.C., and M.L. acknowledge ERC AdG OSYRIS, EU FETPRO QUIC and National Science Centre, Poland—Symfonia Grant 2016/20/W/ST4/00314. A. S. L. acknowledges Max Planck Center for Attosecond Science (MPC-AS). J. A. P.-H. acknowledges to the Spanish Ministerio de Economía y Competitividad (FURIAM Project No. FIS2013-47741-R and PALMA project FIS2016- 81056-R) and Laserlab-Europe (EU-H2020 654148). L.O. acknowledges valuable input from Andre Staudte. The authors thankfully acknowledge the computer resources at MareNostrum, technical expertise and assistance provided by the Barcelona Supercomputing Center and the Red Española de Supercomputación (RES).

References

1. A. L'Huillier, L.A. Lompre, G. Mainfray, C. Manus, Multiply charged ious formed by multiphoton absorption processes in the continuum. Phys. Rev. Lett. **48**(26), 1814–1817 (1982)
2. P. Lambropoulos, X. Tang, P. Agostini, G. Petite, A. L'Huillier, Multiphoton spectroscopy of doubly excited, bound, and autoionizing states of strontium. Phys. Rev. A **38**(12), 6165–6179 (1988)
3. B. Bergues, M. Kübel, N.G. Kling, C. Burger, M.F. Kling, Single-cycle non-sequential double ionization. IEEE J. Sel. Top. Quantum Electron. **21**(5), 8701009–8701009 (2015)
4. B. Bergues, M. Kübel, N.G. Johnson, B. Fischer, N. Camus, K.J. Betsch, O. Herrwerth, A. Senftleben, A.M. Sayler, T. Rathje, T. Pfeifer, I. Ben-Itzhak, R.R. Jones, G.G. Paulus, F. Krausz, R. Moshammer, J. Ullrich, M.F. Kling, Attosecond tracing of correlated electron-emission in non-sequential double ionization. Nat. Comm. **3**(813), 1–6 (2012)
5. M. G. Pullen, B. Wolter, X. Wang, X.-M. Tong, M. Sclafani, M. Baudisch, H. Pires, C.D. Schröter, J. Ullrich, T. Pfeifer, R. Moshammer, J.H. Eberly, J. Biegert, *Transition from non-sequential to sequential double ionisation in many-electron systems*. ArXiv:1602.07840 (2016)
6. B. Walker, B. Sheehy, L.F. DiMauro, P. Agostini, K.J. Schafer, K.C. Kulander, Precision measurement of strong field double ionization of helium. Phys. Rev. Lett. **73**(9), 1227–1230 (1994)

7. Th Weber, M. Weckenbrock, A. Staudte, L. Spielberger, O. Jagutzki, V. Mergel, F. Afaneh, G. Urbasch, M. Vollmer, H. Giessen, R. Dörner, Recoil-ion momentum distributions for single and double ionization of helium in strong laser fields. Phys. Rev. Lett. **84**(3), 443–446 (2000)
8. Th Weber, H. Giessen, M. Weckenbrock, G. Urbasch, A. Staudte, L. Spielberger, O. Jagutzki, V. Mergel, M. Vollmer, R. Dörner, Correlated electron emission in multiphoton double ionization. Nature **405**(6787), 658–661 (2000)
9. W. Becker, X.J. Liu, P.J. Ho, J.H. Eberly, Theories of photoelectron correlation in laser-driven multiple atomic ionization. Rev. Mod. Phys. **84**(3), 1011–1043 (2012)
10. P.B. Corkum, Plasma perspective on strong-field multiphoton ionization. Phys. Rev. Lett. **71** (13), 1994–1997 (1993)
11. A. Staudte, C. Ruiz, M. Schöffler, S. Schössler, D. Zeidler, Th Weber, M. Meckel, D.M. Villeneuve, P.B. Corkum, A. Becker, R. Dörner, Binary and recoil collisions in strong field double ionization of helium. Phys. Rev. Lett. **99**(26), 263002 (2007)
12. R. Kopold, W. Becker, H. Rottke, W. Sandner, Routes to nonsequential double ionization. Phys. Rev. Lett. **85**(18), 3781–3784 (2000)
13. E. Eremina, X. Liu, H. Rottke, W. Sandner, A. Dreischuh, F. Lindner, F. Grasbon, G.G. Paulus, H. Walther, R. Moshammer, B. Feuerstein, J. Ullrich, Laser-induced non-sequential double ionization investigated at and below the threshold for electron impact ionization. J. Phys. B: At. Mol. Phys **36**(15), 3269–3280 (2003)
14. A. Chacón, M.F. Ciappina, M. Lewenstein, Double-electron recombination in high-order-harmonic generation driven by spatially inhomogeneous fields. Phys. Rev. A **94**(4), 043407 (2016)
15. M.F. Ciappina, J.A. Pérez-Hernández, A.S. Landsman, W. Okell, S. Zherebtsov, B. Förg, J. Schötz, L. Seiffert, T. Fennel, T. Shaaran, T. Zimmermann, R. Guichard A. Chacón, A. Zaïr, J.W.G. Tisch, J.P. Marangos, T. Witting, A. Braun, S.A. Maier, L. Roso, M. Krüger, P. Hommelhoff, M.F. Kling, F. Krausz, M. Lewenstein. Attosecond physics at the nanoscale. Rep. Prog. Phys. **80**(5), 054401 (2017)
16. D. Bauer, Two-dimensional, two-electron model atom in a laser pulse: Exact treatment, single-active-electron analysis, time-dependent density-functional theory, classical calculations, and nonsequential ionization. Phys. Rev. A **56**(4), 3028–3039 (1997)
17. D.G. Lappas, R. van Leeuwen, Electron correlation effects in the double ionization of he. J. Phys. B **31**(6), L249–L256 (1998)
18. D.G. Lappas, A. Sanpera, J.B. Watson, K. Burnett, P.L. Knight, R. Grobe, J.H. Eberly, Two-electron effects in harmonic generation and ionization from a model he atom. J. Phys. B: At. Mol. Phys **29**(16), L619 (1996)
19. M. Lein, E.K.U. Gross, V. Engel, Intense-field double ionization of helium: Identifying the mechanism. Phys. Rev. Lett. **85**(22), 4707–4710 (2000)
20. S.E. Harris, J.J. Macklin, T.W. Hänsch, Atomic scale temporal structure inherent to high-order harmonic generation. Opt. Comm. **100**(5–6), 487–490 (1993)
21. C.G. Wahlström, M.B. Gaarde, A. L'Huillier, C. Lyngå, I. Mercer, E. Mevel, R. Zerne, Ph Antoine, M. Bellini, T.W. Hänsch, Manipulations of high-order harmonics, in *Multiphoton Processes 1996*, ed. by P. Lambropoulos, H. Walther (Institute of Physics Publishing, Bristol, 1997), p. 160
22. R. Zerne, C. Altucci, M. Bellini, M.B. Gaarde, T.W. Hänsch, A. L'Huillier, C. Lyngå, C.-G. Wahlström, Phase-locked high-order harmonic sources. Phys. Rev. Lett. **79**(6), 1006–1009 (1997)
23. J.B. Watson, A. Sanpera, D.G. Lappas, P.L. Knight, K. Burnett, Nonsequential double ionization of helium. Phys. Rev. Lett. **78**(10), 1884–1887 (1997)
24. A. Chacón, M.F. Ciappina, M. Lewenstein, Numerical studies of light-matter interaction driven by plasmonic fields: The velocity gauge. Phys. Rev. A **92**(6), 063834 (2015)
25. M.F. Ciappina, J. Biegert, R. Quidant, M. Lewenstein, High-order-harmonic generation from inhomogeneous fields. Phys. Rev. A **85**(3), 033828 (2012)

26. M.D. Feit, J.A. Fleck, A. Steiger, Solution of the schrdinger equation by a spectral method. J. Comput. Phys. **47**(3), 412–433 (1982)
27. C. Ruiz, A. Chacón, *QFISHBOWL library* (2008). http://code.google.com/p/qfishbowl
28. M. Frigo, S.G. Johnson. *FFTW library* (1998). http://www.fftw.org
29. R. Grobe, J.H. Eberly, One-dimensional model of a negative ion and its interaction with laser fields. Phys. Rev. A **48**(6), 4664–4681 (1993)
30. M.V. Ammosov, N.B. Delone, V.P. Krainov, Tunnel ionization of complex atoms and of atomic ions in an alternating electromagnetic field. *Sov. Phys.-JETP*, **64**(6), 1191–1194 (1986)
31. N.B. Delone, V.P. Krainov, Energy and angular electron spectra for the tunnel ionization of atoms by strong low-frequency radiation. J. Opt. Soc. Am. B **8**(6), 1207–1211 (1991)
32. L. Arissian, C. Smeenk, F. Turner, C. Trallero, A.V. Sokolov, D.M. Villeneuve, A. Staudte, P.B. Corkum, Direct test of laser tunneling with electron momentum imaging. Phys. Rev. Lett. **105**(13), 133002 (2010)
33. A.S. Landsman, A.N. Pfeiffer, C. Hofmann, M. Smolarski, C. Cirelli, U. Keller, Rydberg state creation by tunnel ionization. New J. Phys. **15**(1), 013001 (2013)
34. A.S. Landsman, U. Keller, Tunnelling time in strong field ionisation. J. Phys. B: At. Mol. Phys **47**(20), 204024 (2014)
35. T. Nubbemeyer, K. Gorling, A. Saenz, U. Eichmann, W. Sandner, Strong-field tunneling without ionization. Phys. Rev. Lett. **101**(23), 233001 (2008)
36. C. Hofmann, A.S. Landsman, C. Cirelli, A.N. Pfeiffer, U. Keller, Comparison of different approaches to the longitudinal momentum spread after tunnel ionization. J. Phys. B: At. Mol. Phys **46**(12), 125601 (2013)
37. S.P. Goreslavski, G.G. Paulus, S.V. Popruzhenko, N.I. Shvetsov-Shilovski, Coulomb asymmetry in above-threshold ionization. Phys. Rev. Lett. **93**(2), 233002 (2004)
38. Y. Liu, S. Tschuch, A. Rudenko, M. Dürr, M. Siegel, U. Morgner, R. Moshammer, J. Ullrich, Strong-field double ionization of ar below the recollision threshold. Phys. Rev. Lett. **101**(5), 053001 (2008)
39. A. Rudenko, K. Zrost, B. Feuerstein, V.L.B. de Jesus, C.D. Schröter, R. Moshammer, J. Ullrich, Correlated multielectron dynamics in ultrafast laser pulse interactions with atoms. Phys. Rev. Lett. **93**(25), 253001 (2004)

Parity-Nonconserving Interaction-Induced Light Shifts in the $7S_{1/2}$–$6D_{3/2}$ Transition of the Ultracold ^{210}Fr Atoms to Probe New Physics Beyond the Standard Model

T. Aoki, Y. Torii, B.K. Sahoo, B.P. Das, K. Harada, T. Hayamizu, K. Sakamoto, H. Kawamura, T. Inoue, A. Uchiyama, S. Ito, R. Yoshioka, K.S. Tanaka, M. Itoh, A. Hatakeyama, and Y. Sakemi

Abstract We present an experimental technique to measure light shifts due to the nuclear spin independent (NSI) parity-nonconserving (PNC) interaction in the $7S_{1/2}$–$6D_{3/2}$ transition in ultracold ^{210}Fr atoms. The approach we propose is similar to the one by Fortson (Phys Rev Lett 70:2383, 10) to measure the PNC-induced light shift which arises from the interference of parity nonconserving electric dipole

---

This article is part of the topical collection "Enlightening the World with the Laser" - Honoring T. W. Hänsch guest edited by Tilman Esslinger, Nathalie Picqué, and Thomas Udem.

T. Aoki (✉) • Y. Torii
Institute of Physics, Graduate School of Arts and Sciences, The University of Tokyo, Tokyo 153-8902, Japan
e-mail: aoki@phys.c.u-tokyo.ac.jp

B.K. Sahoo
Atomic, Molecular and Optical Physics Division, Physical Research Laboratory, Ahmedabad 380009, India

B.P. Das
International Education and Research Center of Science and Department of Physics, Tokyo Institute of Technology, Tokyo 152-8550, Japan

K. Harada • T. Hayamizu • K. Sakamoto • A. Uchiyama • S. Ito • R. Yoshioka • K.S. Tanaka • M. Itoh
Cyclotron and Radioisotope Center, Tohoku University, Sendai, Miyagi 980-8578, Japan

H. Kawamura • T. Inoue
Cyclotron and Radioisotope Center, Tohoku University, Sendai, Miyagi 980-8578, Japan

Frontier Research Institute for Interdisciplinary Sciences, Tohoku University, Sendai, Miyagi 980-8578, Japan

A. Hatakeyama
Department of Applied Physics, Tokyo University of Agriculture and Technology, Tokyo 184-8588, Japan

Y. Sakemi
Center for Nuclear Study, The University of Tokyo, Wako, Saitama 351-0198, Japan

D. Meschede et al. (eds.), *Exploring the World with the Laser*,
https://doi.org/10.1007/978-3-319-64346-5_29

transition and electric quadrupole transition amplitudes. Its major advantage is that it can treat more than 10^4 ultracold ^{210}Fr atoms to enhance the shot noise limit. A relativistic coupled-cluster method has been employed to calculate the electric dipole transition amplitudes arising from the PNC interaction. Based on these calculations, we have evaluated the PNC-induced light shifts for transitions between the hyperfine levels of the $7S_{1/2}$ and $6D_{3/2}$ states and suitable transitions are identified for carrying out PNC measurements. It is possible in principle to probe new physics beyond the standard model with our proposed experimental scheme.

1 Introduction

Studies of atomic parity nonconservation effects have the potential to probe new physics beyond the standard model of particle physics and to measure the nuclear anapole moment (NAM) which is an unusual property of the nucleus [1, 2]. There are mainly two sources of parity nonconservation (PNC) interactions in atomic systems. One is due to the neutral current weak (NCW) interactions by the exchange of the Z_0 boson between the nucleus and electrons, while the other is due to the interaction of the NAM with the electrons [1–5]. The NCW interaction can be classified into nuclear spin independent (NSI) and nuclear spin-dependent (NSD) PNC interactions depending on whether the vector and axial-vector currents come from the nuclear or electronic parts of the atomic system. In addition, the interaction of electrons with the NAM can also contribute to the NSD PNC interaction [1, 2]. Since the electron–nucleon NSI interaction propagates coherently in an atomic system, its contribution to the atomic PNC effect is significantly larger than that of the NSD interactions, which is primarily contributed by the unpaired nucleons of the atomic nucleus.

It is possible in principle to gain insights into new physics beyond the standard model (SM) of particle interactions by investigating the NSI PNC interaction, and the NSD PNC interaction can serve to measure the NAM in atomic systems. It is possible to extract the nuclear weak charge (Q_W) and compare it to the corresponding value in the SM by combining the results of high precision measurements and calculations of NSI PNC observables in these systems. A discrepancy between these two values could reveal possible new physics beyond the SM [1, 2, 6, 7]. The most accurate data on atomic PNC currently comes from the $6S_{1/2} \rightarrow 7S_{1/2}$ transition in cesium [8]. To obtain a more accurate value of Q_W, it would be desirable to consider other candidates. In this context, many other atomic systems such as Yb [9], Ba^+ [10], Ra^+ [11], and Yb^+ [12] have been proposed for the PNC studies. The last successful NSI PNC measurement was carried out on atomic Yb by detecting the fluorescence in transitions using the technique of Stark mixing [9], but its accuracy was about 10%. On the basis of theoretical [13–15] and experimental analyses [16–19], it has been argued that Fr atom is a favorable candidate for performing PNC studies. A detection method by the measurement of the light shift induced by a PNC effect has been proposed in a single trapped and laser-cooled Ba^+ ion by Fortson [10]. Motivated by this proposal, similar approaches have been proposed to investigate PNC interactions in Yb^+ [12] and Ra^+ [11]. Application of the light-induced technique to these ions seems to be favorable owing to the fact that they have ground S and metastable $D_{3/2}$ states, unlike the alkali atoms. The PNC effects in these systems are enhanced in these transitions compared to the $S - S$ transitions. The reason for this is that in the S state, the valence s orbitals have large overlap with the atomic nucleus, and in the $S - D_{3/2}$ transition the large contribution of the PNC amplitude arising from the ground state is not canceled unlike in the $S - S$ transitions, where there are substantial cancellations between the contributions from the ground and excited states (see the discussions in [11, 20]). It was also later realized that by applying the same method to the $S - D_{5/2}$ transitions of Ba^+ and Ra^+ ions, it is possible to measure the NAM unambiguously [21, 22]. We had proposed in a recent work that it is possible to use a similar technique for the $S - D_{5/2}$ transition in ultracold ^{210}Fr atoms to observe a possible signature due to the NAM [15, 23]. Since the $D_{5/2}$ state in Fr is not metastable in contrast to the aforementioned singly charged ions, we had suggested that a similar level of shot-noise-limit like the ions can be attained in the ^{210}Fr isotope by considering a large number ($\sim 10^4$) of atoms in an optical lattice for the measurement [23].

In this paper, we investigate an analogous method for the detection scheme of the NSI PNC interaction in the $7S_{1/2} - 6D_{3/2}$ transition of the ^{210}Fr atoms. We plan to carry out this measurement using the Cyclotron and Radioisotope Center (CYRIC) facility at Tohoku University, Sendai, Japan along with the NSD PNC effect measurement in the $S - D_{5/2}$ transition. As both the NSI and NSD PNC interactions will contribute to the $S_{1/2} - D_{3/2}$ transitions, it is not possible to discriminate between the two light shifts induced by the NSI and NSD effects in a PNC experiment. However, our approach would be to get information about the contribution of the NAM from the NSD interactions from the PNC measurement in the $7S_{1/2} - 6D_{5/2}$ transition in ^{210}Fr atoms, and then from this knowledge we would like to infer the NSI observable very precisely by carrying out PNC measurement in the $7S_{1/2} - 6D_{3/2}$ transition of these atoms trapping using the same facility. Our method would enable us to precisely measure the NSI PNC effect to probe new physics beyond the SM of particle physics.

The paper is organized as follows: In the next section, we introduce the sources of NSI and NSD PNC interactions in atomic systems, and their implications for elementary particle physics. In the subsequent section, we discuss the measurement procedure for PNC-induced light shifts in ^{210}Fr . We then present our theoretical results and discuss them before making our concluding remarks.

2 Parity Nonconservation Interaction and New Physics Beyond the SM

Here we discuss briefly the PNC interactions in an atomic system. The atomic Hamiltonian due to the NSI PNC interaction is given by [3, 4, 24]

$$H_{\rm PNC}^{\rm NSI} = \frac{G_{\rm F}}{2\sqrt{2}} Q_{\rm W} \gamma_5 \rho_{\rm nuc}(r), \tag{1}$$

where $G_{\rm F}$ is the Fermi constant, $Q_{\rm W}$ is the weak charge, $\rho_{\rm nuc}$ is the nuclear density, and γ_5 is the Dirac matrix. The weak charge in the SM is described as [1–5]

$$\begin{aligned} Q_{\rm W}^{\rm SM} &\equiv ZQ_{\rm W}^{(p)} + NQ_{\rm W}^{(n)} \\ &= (2Z+N)Q_{\rm W}^{(u)} + (Z+2N)Q_{\rm W}^{(d)}, \end{aligned} \tag{2}$$

where $Q_{\rm W}^{\rm SM}$ is the weak charge due to the SM, Z is the number of protons, and N is the number of neutrons in atomic nuclei. $Q_{\rm W}^{(u)}$ and $Q_{\rm W}^{(d)}$ are the weak charges characterizing the interactions of the electron-axial vector-current with the vector currents of up and down quarks, respectively. $Q_{\rm W}^{(p)}$ and $Q_{\rm W}^{(n)}$ are the proton and neutron weak charges, respectively. They can be expressed as $Q_{\rm W}^{(p)} = 2Q_{\rm W}^{(u)} + Q_{\rm W}^{(d)}$ and $Q_{\rm W}^{(n)} = Q_{\rm W}^{(u)} + 2Q_{\rm W}^{(d)}$. At the lowest order of electroweak interactions, these parameters can be expressed as [1, 2]

$$Q_{\rm W}^{(p)} = 1 - 4\sin^2\theta_{\rm W} \text{ and } Q_{\rm W}^{(n)} \approx -1, \tag{3}$$

where $\theta_{\rm W}$ is the Weinberg angle. Thus, we can have $Q_{\rm W}^{\rm SM} \approx -[(4\sin^2\theta_{\rm W} - 1)Z + N] \approx -N$. In fact, this is the reason why effects due to $H_{\rm PNC}^{\rm NSI}$ are often expressed in unit of $(-Q_{\rm W}/N)$. It is to be noted that in the framework of the SM, the expression for the NSI interaction Hamiltonian given by Eq. (1) is derived by taking into account only the exchange of the Z_0 boson in the neutral current weak interaction between the nucleus and the electrons. However, several new theories of particle physics argue that the SM is an intermediate manifestation of a more complete theory. These theories predict many extra-new particles. Some of these particles are similar to the Z_0 boson in nature but are predicted to possess larger masses than Z_0. In such a scenario there will be shift in the value of $Q_{\rm W}$ and it is possible to probe the signature of the existence of these new interactions by inferring the deviation in the value of $Q_{\rm W}$ from its predicted SM value. Measurements of the PNC effects in the atomic systems in combination with relativistic many-body calculations are capable of providing accurate values of these quantities.

Assuming that there exist new interactions that can contribute to the NSI PNC interactions in an atomic system, we can rewrite the weak charge $Q_{\rm W}$ of an atom as

$$Q_{\rm W} \equiv \quad Q_{\rm W}^{\rm SM} + Q_{\rm W}^{\rm New\ Physics}, \tag{4}$$

where $Q_{\rm W}^{\rm SM}$ and $Q_{\rm W}^{\rm New\ Physics}$ are now the weak charge contributions from the SM and new physics beyond the SM, respectively. If the weak charge that is obtained from the atomic studies ($Q_{\rm W}^{\rm atomic}$) is different from $Q_{\rm W}^{\rm SM}$, then one can yield

$$\begin{aligned}\Delta Q_{\mathrm{W}} &\equiv Q_{\mathrm{W}}^{\mathrm{atomic}} - Q_{\mathrm{W}}^{\mathrm{SM}} = Q_{\mathrm{W}}^{\mathrm{New\ Physics}} \\ &= Z\Delta Q_{\mathrm{W}}^{(p)} + N\Delta Q_{\mathrm{W}}^{(n)} \\ &= (2Z+N)\Delta Q_{\mathrm{W}}^{(u)} + (Z+2N)\Delta Q_{\mathrm{W}}^{(d)},\end{aligned} \tag{5}$$

and this difference can be used to find the signature for the existence of new physics. $\Delta Q_{\mathrm{W}}^{(p)}$, $\Delta Q_{\mathrm{W}}^{(n)}$, $\Delta Q_{\mathrm{W}}^{(u)}$, and $\Delta Q_{\mathrm{W}}^{(d)}$ are the new physics contributions for $Q_{\mathrm{W}}^{(p)}$, $Q_{\mathrm{W}}^{(n)}$, $Q_{\mathrm{W}}^{(u)}$, and $Q_{\mathrm{W}}^{(d)}$, respectively.

We mentioned below some scenarios for new physics that are relevant for NSI PNC in atomic systems [6, 7].

(i) Extra Z_χ boson in SO(10): As a grand unified theory (GUT), the superstring-inspired E6 symmetry is spontaneously broken in SO(10). This can give rise an extra Z_χ boson [25] and would modify the weak charge of an atom by [6]

$$\Delta Q_{\mathrm{W}} = 0.4(2N+Z)\left(\frac{M_{\mathrm{W}}}{M_{Z_\chi}}\right)^2, \tag{6}$$

where the $M_{\mathrm{W}} \simeq 80\ \mathrm{GeV}/\mathrm{c}^2$ and M_{Z_χ} are the masses of $W^{\pm}$ boson, and the extra Z_χ boson, respectively. With ΔQ_{W} of 0.65 for ^{133}Cs with $Z = 55$ and $N = 78$ [26], we can obtain from Eq. (6) as $M_{Z_\chi} \simeq 912\ \mathrm{GeV}/c^2$. This shows that atomic PNC studies can put the stringent limit on the mass of the extra Z_χ boson.

(ii) Leptoquark and supersymmetry (SUSY): The leptoquark resonance state in the lepton to quark conversion process as BSM is also suggested. It arises naturally in the R-parity violating SUSY theory [27], which involves the electron–quark–squark interaction. This can also lead to PNC effect in the atomic systems. In this case, the coupling λ_s between the down-quark and squark with charge $+2/3$ can be related with the mass of squark M_ϕ and ΔQ_{W} by [7]

$$\lambda_s \leq 0.4\sqrt{\frac{\Delta Q_{\mathrm{W}}}{2N+Z}}\left(\frac{M_\phi}{100\mathrm{GeV}/c^2}\right). \tag{7}$$

Here $2N+Z$ in the denominator corresponds to the number of down quarks involved in the process. Similarly, for the interactions involving the up quarks can be obtained by replacing $2N+Z$ with $2Z+N$ in the above expression. The finite value of λ_s will indicate physics beyond the SM, which can be inferred from the knowledge of ΔQ_{W}. With the assumption of $\lambda_s = 0.6$ as in Ref. [7], substituting ΔQ_{W} of 0.65 for Cs into Eq. (7) results in giving $M_\phi \geq 2.70\ \mathrm{TeV}/c^2$.

(iii) Compositeness: The SM assumes leptons and quarks to be pointlike. However, these particles might have a composite structure. Manifestation of the composite structure includes the presence of fermion form factor, and the existence of contact interaction [28–30]. The contact interaction between a lepton and a quark gives rise to the contribution to the weak charge as discussed in Refs. [7, 31].

(iv) Others: The radiative isospin conserving S and violating T parameters, that are related to masses of existing Higgs bosons, can also contribute to ΔQ_{W}

[6, 32]. After the recent discovery of SM Higgs particle, values of the above parameters have been revised in Ref. [33]. Possibility of existence of other light U boson has been discussed in [34]. Such light boson can also give rise dark parity violation at the lower energy scale and is a possible candidate to explain the muon g-2 factor anomaly and astronomical observation of positron excesses [35, 36].

Nevertheless, the large hadron collider (LHC) facility, which operates at the TeV energy scale, can in principle rule out some of these discussed models and may also validate the existence of some of the predicted extra particles. Alternatively, an experiment involving the scattering of longitudinally polarized electrons from the unpolarized protons has also been under consideration to infer the $Q_{\rm W}^{(p)}$ value [37, 38]. Our proposed atomic PNC studies can be the complementary approaches to these accelerator based methods to probe the aforementioned new physics.

We can find from Eqs. (6) and (7) that an atom with a large N is more sensitive to the mass of the new bosonic particles. In this work, we have considered the ^{210}Fr atom with $Z = 87$ and $N = 123$ to probe PNC due to the weak interactions. Using Eq. (2), we can estimate nuclear weak charge of this atom in the SM framework as

$$\begin{aligned} Q_{\rm W}^{\rm SM}(^{210}{\rm Fr}) \equiv & \;\; 87 \times Q_{\rm W}^{(p)} + 123 \times Q_{\rm W}^{(n)} \\ \approx & \;\; -115.47(11), \end{aligned} \tag{8}$$

where we have used $Q_{\rm W}^{(p)} = 0.0710(7)$ and $Q_{\rm W}^{(n)} = -0.9890(7)$ [37, 39]. Let us assume that we observe a value of $|\Delta Q_{\rm W}(^{210}{\rm Fr})| \approx 0.23$ from our study, then using Eq. (8) this will correspond to $|\Delta Q_{\rm W}(^{210}{\rm Fr})/Q_{\rm W}^{\rm SM}(^{210}{\rm Fr})| \approx 0.2\%$. Therefore, it is imperative to achieve the total error less than 0.2% from the combined theoretical and experimental results for the NSI interaction in ^{210}Fr to be able to probe existence of squark and Z_χ boson with the lower limits on their masses as 1.8 TeV/c^2 for M_{Z_χ} and 5.5 TeV/c^2 for M_ϕ as per the relations given in Eqs. (6) and (7).

In our previous works [15, 23], we had carried out rigorous analysis on the light shifts one would observe in the $7S_{1/2} - 6D_{5/2}$ transition of ^{210}Fr that will arise only due to the NSD PNC interaction for which the interaction Hamiltonian is given by

$$H_{\rm PNC}^{\rm NSD} = \frac{G_{\rm F}}{\sqrt{2}} \mathcal{K}_{\rm W} \boldsymbol{\alpha} \cdot \mathbf{I} \, \rho_{\rm nuc}(r) = \sum_q (-1)^q I_q K_{\rm PNC}^q, \tag{9}$$

where $\boldsymbol{\alpha}$ is the Dirac matrix, $\mathbf{I} = I\hat{\mathbf{I}}$ is the nuclear spin with qth component I_q, the dimensionless quantity $\mathcal{K}_{\rm W}$ is related to the NAM and $K_{\rm PNC}$ is an effective tensor operator representing the NSD PNC interaction Hamiltonian in the electronic space. In heavy atomic systems, the dominant contribution to the NSD PNC interactions comes from the NAM. In the present work, we, however, focus on the PNC effect in the $7S_{1/2} - 6D_{3/2}$ transition of ^{210}Fr mainly due to the NSI interaction.

3 PNC-Induced Light Shift

We consider the transitions between the ground state $7S_{1/2}$ and excited state $6D_{3/2}$ of ^{210}Fr. The usual electromagnetic transitions between these states are through the magnetic dipole (M1) and electric quadrupole (E2) channels. The transition probability via the E2 channel dominates over the M1 channel, and the lifetime of this excited state is about 540 (10) ns [40]. When weak interactions are taken into account, we can also get a finite electric dipole (E1) transition amplitude between these states as shown in Fig. 1. Here, we refer to E1 amplitudes due to NSI PNC and NSD PNC interactions by $E1^{\mathrm{NSI}}_{\mathrm{PNC}}$ and $E1^{\mathrm{NSD}}_{\mathrm{PNC}}$, respectively. The transition wavelengths from the $7S_{1/2}$ to the $6D_{3/2}$ and to the $6D_{5/2}$ states of ^{210}Fr are 616 and 609 nm, respectively, as shown in Fig. 1. The selection of suitable hyperfine levels for the PNC observation will be decided by analyzing the magnitudes of PNC-induced effects among all possible transitions.

Now the $E1^{\mathrm{NSI}}_{\mathrm{PNC}}$ and $E1^{\mathrm{NSD}}_{\mathrm{PNC}}$ amplitudes between an initial (i) state and final (f) state are given by

$$E1^{\mathrm{NSI}}_{\mathrm{PNC}} = (-1)^{J_{\mathrm{f}}-M_{J,\mathrm{f}}} \begin{pmatrix} J_{\mathrm{f}} & 1 & J_{\mathrm{i}} \\ -M_{J,\mathrm{f}} & q_{M_J} & M_{J,\mathrm{i}} \end{pmatrix} \mathcal{X} \tag{10}$$

and

$$E1^{\mathrm{NSD}}_{\mathrm{PNC}} = (-1)^{F_{\mathrm{f}}-M'} \begin{pmatrix} F_{\mathrm{f}} & 1 & F_{\mathrm{i}} \\ -M' & q_M & M \end{pmatrix} \mathcal{Y}, \tag{11}$$

respectively, where $J_{\mathrm{i}}(J_{\mathrm{f}})$ are the total angular momenta with magnetic quantum number $M_{J,\mathrm{i}}(M_{J,\mathrm{f}})$ in the initial (final) state, $F_{\mathrm{i}}(F_{\mathrm{f}})$ are the hyperfine angular momenta with magnetic quantum number $M(M')$ in the initial (final) state, $q_{M_J}(q_M)$

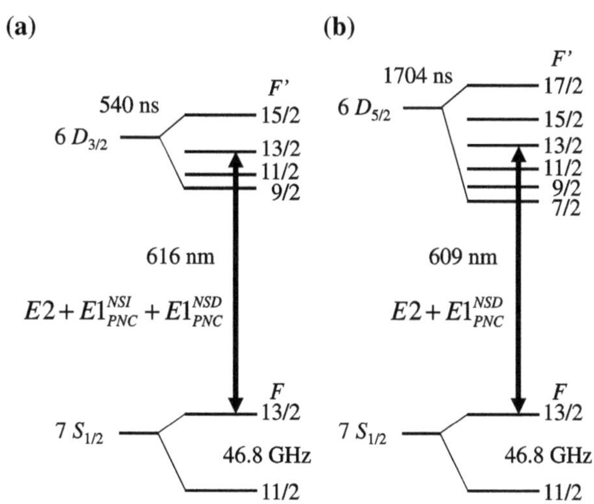

Fig. 1 Schematic hyperfine energy level diagrams of ^{210}Fr. *Arrows* indicate the laser-induced transitions to observe the E2 light shifts and PNC-induced light shifts from the $7S_{1/2}$ state to **a** the $6D_{3/2}$ state and **b** the 6 $D_{5/2}$ state

values are $q_{J,M}(q_M) = -1, 0, 1$ depends on the selection rule of $q_{M_J} \equiv M_{J,\mathrm{f}} - M_{J,\mathrm{i}}$, and $q_M \equiv M' - M$, and the reduced matrix element $\mathcal{X}$ is given by

$$\begin{aligned}\mathcal{X} = &\sum_{k \neq i} \frac{\langle J_\mathrm{f}||D||J_k\rangle\langle J_k||H_{\mathrm{PNC}}^{\mathrm{NSI}}||J_\mathrm{i}\rangle}{\sqrt{2J_\mathrm{i}+1}(E_\mathrm{i}-E_k)} \\ &+ \sum_{k \neq f} \frac{\langle J_\mathrm{f}||H_{\mathrm{PNC}}^{\mathrm{NSI}}||J_k\rangle\langle J_k||D||J_\mathrm{i}\rangle}{\sqrt{2J_\mathrm{f}+1}(E_\mathrm{f}-E_k)}\end{aligned} \tag{12}$$

and by expressing in a form $H_{\mathrm{PNC}}^{\mathrm{NSD}} = \sum_q (-1)^q I_q K_{\mathrm{PNC}}^q$ to separate out the nuclear and electronic space, we get an expression for $\mathcal{Y}$, which depends on the angular momenta (J, F) of the initial and final states, the nuclear spin, the reduced matrix elements of the electric dipole operators, and K_{PNC} [15]. Evaluation of $\mathcal{X}$ and $\mathcal{Y}$ values demands consideration of E1 matrix elements and matrix elements of the K_{PNC} operator among all possible intermediate k states. It is not possible to account for all the states in a sum-over-states approach to evaluate these quantities. However, it is advantageous to use a sum-over-states approach here as in this case we can use experimental energies that appear in the denominators. Usually, uncertainties associated with the energies could introduce large errors in the evaluation of the above quantities in an ab initio method. In this work, we consider the sum-over-states approach. We calculate the E1 matrix elements and the matrix elements of the K_{PNC} operator using a relativistic coupled-cluster (RCC) method only for the 7 $P_{1/2;3/2} - 11P_{1/2;3/2}$ intermediate P states, since they are the leading contributors to the values of $\mathcal{X}$ and $\mathcal{Y}$. This is due to the fact that the energy differences between these states and the $7S_{1/2}$, $6D_{3/2}$ and $6D_{5/2}$ states are comparatively small. Though it may seem justified to neglect contributions from the higher intermediate $P_{1/2;3/2}$ states on the basis of the above argument, the continuum $P_{1/2;3/2}$ states can have large overlap with the atomic nucleus and may contribute substantially to the values of $\mathcal{X}$ and $\mathcal{Y}$. It is also important to evaluate the contributions from the core orbitals to improve the accuracy of the results. We have employed a RCC method considering all possible single, double and important triple excitations (CCSD$_{t3}$ method), as discussed in [15], to calculate the $\mathcal{X}$ and $\mathcal{Y}$. We have also reported earlier E2 matrix elements for the $7S_{1/2} \to 6D_{3/2}$ and $7S_{1/2} \to 6D_{5/2}$ transitions of Fr in [40] using the CCSD$_{t3}$ method.

We assume that ^{210}Fr atoms, which are trapped in an optical lattice, interact with an oscillating electric field given by

$$\mathcal{E}(r,t) = \frac{1}{2}\hat{\epsilon}\left[\mathcal{E}(r)e^{-i\omega t} + c.c.\right], \tag{13}$$

where ω is the laser frequency, $\hat{\epsilon}$ is the unit vector of polarization, and $\mathcal{E}(r)$ is the amplitude of electric field. Following the principle discussed in Ref. [10], the Rabi frequencies due to the PNC interaction and E2 amplitude couplings are given by

$$\Omega^{\text{PNC}}_{MM'} = -\frac{1}{2\hbar}\sum_{i}\left(E1^{\text{PNC}}_{MM'}\right)_i \mathcal{E}_i(0), \tag{14}$$

and

$$\Omega^{E2}_{MM'} = -\frac{1}{2\hbar}\sum_{i,j}(E2_{MM'})_{ij}\left[\frac{\partial \mathcal{E}_i(r)}{\partial x_j}\right]_{r=0}, \tag{15}$$

where $E1^{\text{PNC}}_{MM'}$ are either $E1^{\text{NSI}}_{\text{PNC}}$ or $E1^{\text{NSD}}_{\text{PNC}}$ or the total PNC amplitude depending on the transition considered, and $E2_{MM'}$ is the E2 transition amplitude in that particular transition. Considering interference between both the forbidden transition amplitudes, the total Rabi frequency $\Omega_{MM'}$ would yield $|\Omega_{MM'}|^2 = |\Omega^{E2}_{MM'} + \Omega^{\text{PNC}}_{MM'}|^2 \sim |\Omega^{E2}_{MM'}|^2 + 2Re\left(\Omega^{PNC*}_{MM'}\Omega^{E2}_{MM'}\right)$. Normally Rabi frequency due to the E2 transition is much larger than the detuning frequency. Using the above quantities, the light shifts due to PNC and E2 transitions are given by

$$\Delta\omega^{\text{PNC}}_{M} \approx -\frac{Re\sum_{M'}\left(\Omega^{PNC*}_{MM'}\Omega^{E2}_{MM'}\right)}{\sqrt{\sum_{M'}|\Omega^{E2}_{MM'}|^2}}, \tag{16}$$

and

$$\Delta\omega^{E2}_{M} \approx \frac{(\omega_0 - \omega)}{2} - \sqrt{\sum_{M'}|\Omega^{E2}_{MM'}|^2}, \tag{17}$$

where ω_0 is the resonant frequency of the considered transition before applying the laser field. This approximation is valid if the laser frequency ω is close to the resonance.

Consideration of hyperfine levels and magnetic sublevels is restricted by the angular momentum selection rules of the E2 transition and by the selection rules for Eqs. (10) and (11). Our further discussions on the choice of transition among the hyperfine levels for performing PNC amplitude measurements in the $7S_{1/2} \to 6D_{3/2}$ transition of ^{210}Fr will be based on these restrictions and on the magnitudes of the $\mathcal{X}$ and $\mathcal{Y}$ values.

4 Results and Discussions

In our previous work reported in [15], we have given important E1 matrix elements and the matrix elements of the K_{PNC} operator between the ground state and the $7P_{1/2;3/2} - 11P_{1/2;3/2}$ excited states and also for between the $6D_{3/2;5/2}$ state and the $7P_{1/2;3/2} - 11P_{1/2;3/2}$ excited states using the CCSD$_{t3}$ method. We had found enhancement in the matrix elements of the K_{PNC} operator using our method compared to the previously reported values. The main reason for this enhancement

Table 1 Reduced E1 matrix elements in atomic unit (a.u.), $H_{\rm PNC}^{\rm NSI}$ matrix elements in $-ie\, a_0(Q_{\rm W}/N) \times 10^{-11}$ and $K_{\rm PNC}$ matrix elements in $-iea_0K_{\rm W} \times 10^{-11}$ in the top part of the table

$f \to i$	E1	$\langle J_f \|\| H_{\rm PNC}^{\rm NSI} \|\| J_i \rangle$	$\langle J_f \|\| K_{\rm PNC} \|\| J_i \rangle$
$7P_{1/2} \to 7S_{1/2}$	4.26	−10.53	25.03
$8P_{1/2} \to 7S_{1/2}$	0.34	5.98	14.64
$9P_{1/2} \to 7S_{1/2}$	−0.11	4.03	−9.93
$10P_{1/2} \to 7S_{1/2}$	0.06	−2.98	7.39
$11P_{1/2} \to 7S_{1/2}$	−0.04	−2.38	−5.92
$8S_{1/2} \to 7P_{1/2}$	−4.27	4.83	−12.96
$8S_{1/2} \to 8P_{1/2}$	−10.08	−2.74	−6.60
$8S_{1/2} \to 9P_{1/2}$	−1.00	1.85	4.53
$8S_{1/2} \to 10P_{1/2}$	0.41	−1.37	−3.38
$8S_{1/2} \to 11P_{1/2}$	0.24	−1.09	2.71
$7P_{3/2} \to 7S_{1/2}$	5.98		2.54
$8P_{3/2} \to 7S_{1/2}$	0.95		1.02
$9P_{3/2} \to 7S_{1/2}$	0.44		0.61
$10P_{3/2} \to 7S_{1/2}$	0.28		0.45
$11P_{3/2} \to 7S_{1/2}$	0.18		0.33
$6D_{3/2} \to 7P_{1/2}$	−7.45		2.60
$6D_{3/2} \to 8P_{1/2}$	2.75		0.49
$6D_{3/2} \to 9P_{1/2}$	−0.83		−0.22
$6D_{3/2} \to 10P_{1/2}$	−0.45		0.14
$6D_{3/2} \to 11P_{1/2}$	−0.29		−0.10
$6D_{3/2} \to 7P_{3/2}$	−3.44	−0.23	−0.17
$6D_{3/2} \to 8P_{3/2}$	0.88	−0.16	−0.45
$6D_{3/2} \to 9P_{3/2}$	0.28	−0.12	−0.35
$6D_{3/2} \to 10P_{3/2}$	0.15	−0.09	−0.28
$6D_{3/2} \to 11P_{3/2}$	0.09	−0.07	−0.21

$f \to i$	λ_{fi}	E2	M1
$6D_{3/2} \to 7S_{1/2}$	613.5	34.06	0.001
$6D_{5/2} \to 7S_{1/2}$	608.7	42.27	
$6D_{5/2} \to 6D_{3/2}$	77,059	32.03	1.55

Wavelengths (in nm) and reduced E2 (in a.u.) and M1 (in a.u.) matrix elements in the bottom part of the table

was due to the inclusion of pair-correlation and core-polarization effects to all orders and the addition of contributions from the triples and higher level excitations through the non-linear terms of the CCSD$_{t3}$ method in the determination of atomic wave functions. Furthermore, we had also observed that core-polarization effects were making very large contributions to the matrix elements of the $K_{\rm PNC}$ operator between the $P_{1/2;3/2}$ and $D_{3/2;5/2}$ states, which were not expected due to the small overlap of the valence orbitals states with the atomic nucleus. The other factor that is responsible for such an enhancement is due to the small energy difference between the $P_{1/2;3/2}$ and $D_{3/2;5/2}$ states in contrast to their isoelectronic ions [15]. In Table 1, we give all the quantities required for analyzing PNC-induced light shifts in the hyperfine levels of the $7S_{1/2} \rightarrow 6D_{3/2}$ transition in ^{210}Fr using the CCSD$_{t3}$ method. Some of these quantities were already reported in Ref. [15] for analyzing PNC-induced light shifts of the $7S_{1/2} \rightarrow 6D_{5/2}$ transition of the same atom; however, we present them again here for completeness.

Using the matrix elements from Table 1, the lifetimes of the $6D_{3/2}$ state and 6 $D_{5/2}$ state in Fr are estimated about to be 540 (10) ns and 1704 (32) ns, respectively. These lifetimes are long enough to carry out PNC measurements for the $7S_{1/2} - 6D_{3/2}$ and $7S_{1/2} - 6D_{5/2}$ transitions of the ^{210}Fr atom. In Ref. [15], we have also given $E1^{\mathrm{NSD}}_{\mathrm{PNC}}$ amplitudes between different hyperfine sublevels of the ^{210}Fr, ^{211}Fr, and ^{223}Fr isotopes using the above matrix elements and had found that PNC-induced light shifts are sufficiently large in the $7S_{1/2}, F = 11/2, M = 1/2$ to $6D_{5/2}, F' = 11/2, M' = 1/2$ and $7S_{1/2}, F = 13/2, M = 1/2$ to $6D_{5/2}, F' = 13/2, M' = 1/2$ transitions in ^{210}Fr to be detected. Now using the matrix elements of $H^{\mathrm{NSI}}_{\mathrm{PNC}}$ and reduced E1 matrix elements from Table 1 and experimental energy values listed in the National Institute of Standards and Technology (NIST) database [41], we determine $E1^{\mathrm{NSI}}_{\mathrm{PNC}}$ amplitudes of the $7S_{1/2} - 8S_{1/2}$ and $7S_{1/2} - 6D_{3/2}$ atomic transitions in the ^{210}Fr, ^{211}Fr, and ^{223}Fr isotopes and give them in Table 2. We have also given results for the $7S_{1/2} - 8S_{1/2}$ transition to highlight enhancement of $E1^{\mathrm{NSI}}_{\mathrm{PNC}}$ in the $7S_{1/2} - 6D_{3/2}$ transition for the measurement. Moreover, previously several calculations of $E1^{\mathrm{NSI}}_{\mathrm{PNC}}$ for the $7S_{1/2} - 8S_{1/2}$ transition in ^{223}Fr have been reported earlier in Refs. [42–44] employing a variety of many-body methods. Thus our work provides an idea about the differences in the results due to the CCSD method which accounts for contributions from non-linear terms representing higher level excitations that are very important in a heavy system like Fr atom [45]. These quantities can be estimated between the hyperfine levels with the help of the Wigner–Eckart theorem. For these evaluations we have used the matrix elements involving only the $7P_{1/2;3/2} - 11P_{1/2;3/2}$ states from the CCSD$_{t3}$ method and have referred them as "Main" contributions. Other smaller contributions from the core orbitals ("Core") and higher level excited states ("Tail") are estimated using the Dirac–Fock (DF) method. It is also evident from Table 2 that these contributions are small and a sophisticated method is not necessary to evaluate them. It is also observed from this table that the $E1^{\mathrm{NSI}}_{\mathrm{PNC}}$ amplitude is larger in heavier isotopes. Since we are planning to measure the NSD PNC-induced light shift for the $7S_{1/2} - 6D_{5/2}$ transition in the ^{210}Fr isotope, we would also like to measure the NSI

Table 2 Comparison of the $E1^{\mathrm{NSI}}_{\mathrm{PNC}}$ amplitudes in $-iea_0(Q_W/N) \times 10^{-11}$ in the $7S_{1/2} - 8S_{1/2}$ and $7S_{1/2} - 6D_{3/2}$ atomic transitions of the ^{210}Fr, ^{211}Fr, and ^{223}Fr isotopes

	$8s\ ^2S_{1/2} \rightarrow 7s\ ^2S_{1/2}$			$6d\ ^2D_{3/2} \rightarrow 7s\ ^2S_{1/2}$		
Isotope	Main	Core	Final	Main	Core	Final
^{210}Fr	13.49	−0.03	13.53	47.15	1.84	49.13
^{211}Fr	13.60	−0.03	14.64	45.68	1.86	49.55
^{223}Fr	14.91	−0.03	14.96	50.10	2.04	54.45

Table 3 Estimated light shifts in the hyperfine levels of the $7S_{1/2}(F) \rightarrow 6D_{3/2}(F')$ transition of ^{210}Fr due to the E2 (in MHz), NSI PNC (in Hz) and NSD PNC (in mHz) interactions with the amplitude of the radiation field 2×10^6 V/m

F_f	F_i	M	$\Delta\omega^{E2}_{\lvert M\rvert}/2\pi$	$\Delta\omega^{\mathrm{NSI}}_M/2\pi$	$\Delta\omega^{\mathrm{NSD}}_M/2\pi$
9/2	11/2	1/2	−2.37	−5.191	−51.0
11/2	11/2	1/2	−16.2	0.546	6.5
13/2	11/2	1/2	−4.14	−3.504	−50.9
11/2	13/2	1/2	−4.51	2.778	38.6
13/2	13/2	1/2	−20.5	−0.432	−4.9
15/2	13/2	1/2	−2.57	5.622	47.4
13/2	13/2	1/2	−20.5	−0.432	−4.9
13/2	13/2	3/2	−18.0	−1.297	−14.7
13/2	13/2	5/2	−12.8	−2.162	−24.6
13/2	13/2	7/2	−5.15	−3.027	−34.4
13/2	13/2	9/2	−5.15	3.892	44.2
13/2	13/2	11/2	−18.0	4.757	54.0
13/2	13/2	13/2	−33.3	5.622	63.8

Here we have used E2 amplitude as 34.06 ea_0^2

PNC-induced light shift in the $7S_{1/2} - 6D_{3/2}$ transition of ^{210}Fr. Below, we discuss the estimated magnitude due to both the NSI and NSD PNC-induced light shifts between all possible hyperfine levels of the $7S_{1/2} - 6D_{3/2}$ transition in ^{210}Fr using the quantities calculated above.

The calculated light shifts induced by E2 ($\Delta\omega^{E2}_M$), NSI PNC ($\Delta\omega^{\mathrm{NSI}}_M$) and NSD PNC ($\Delta\omega^{\mathrm{NSD}}_M$) for all possible hyperfine levels of the $7S_{1/2}, F, M = 1/2$ to $6D_{3/2}$, $F', M' = 1/2$ transition, considering a fixed value of magnetic sublevel $M = M' = 1/2$, using Eqs. (16) and (17) are given in Table 3. The amplitude of the irradiated electric field $\mathcal{E}(0)$ at the center of the trapped atomic cloud is assumed to be 2×10^6 V/m, which can be attained with a 100 mW beam focused to 10 μm diameter spot size at atoms through a single pass (or focused to 100 μm diameter with 100 times enhancement in the ring cavity configuration). In these evaluations, we assume that the E2 laser beam propagates in a direction which corresponds to tilting the z axis by an angle of 45° in the direction of the x axis, and the polarization of E2 laser beam with a direction that corresponds to tilting the z axis by an angle of −45° in the direction of the x axis, as shown in Fig. 2. The selection rule $\Delta M = 0$ is maximized and a smaller $\Delta M = \pm 2$ amplitude also exists. The selection rule of $\Delta M = \pm 1$ is not allowed in this configuration. The maximum E2 Rabi frequency for $|\Delta M| = 2$ is obtained as 10.37 MHz. To avoid the $\Delta M = \pm 2$ transition, it is required to apply a magnetic field larger than 85.98 G, which corresponds to a first order Zeeman shift of 20.74 MHz with a g-factor for the $6D_{3/2}$ state of 28/325 to separate the $|\Delta M| = 2$ states. This Zeeman shift between the $|\Delta M| = 2$ states is double for the largest E2 ($|\Delta M| = 2$) Rabi frequency, which is equivalent to an interaction time width. This width is half of the first period in the Rabi oscillation of the E2 ($|\Delta M| = 2$) transition. Therefore, $\Delta M = \pm 2$ transitions can be ignored if the magnetic field is larger than 85.98 G.

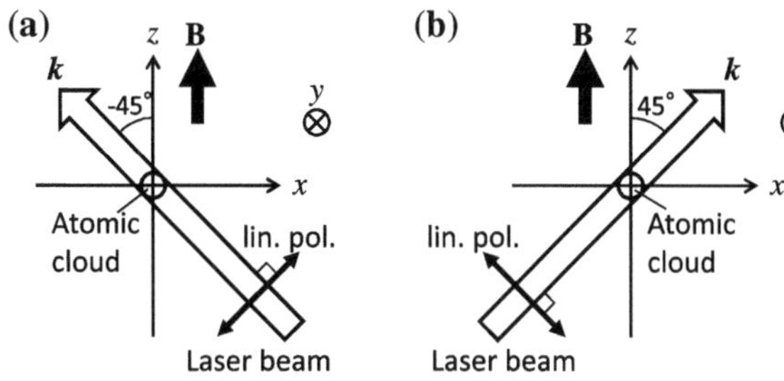

Fig. 2 A schematic diagram of the proposed trap geometry with the linearly polarized (lin. pol.) E2 laser beam and the applied magnetic field **B**. The wave vector **k** of the laser beam and polarization vector are assumed to be on the *x-z* plane (without any *y* component) and the magnetic field **B** is applied along the *z* axis. The center of the atomic cloud is located at the origin of the *x*, *y*, and *z* axes

The $7S_{1/2} - 6D_{3/2}$ transition also involves the M1 amplitude, while the $7S_{1/2} - 6D_{5/2}$ transition does not allow this channel. The selection rule of the M1 transition is restricted to $\Delta M = \pm 1$ because the direction of $\mathbf{k} \times \mathbf{E}$ is parallel to the *y* axis (perpendicular to the *z* axis). From Table 1, we find that the M1 transition amplitude for this transition is much smaller than that of the E2 channel by a factor of 1/3406. We neglect contributions due to the M1 polarization configuration in this analysis because of the fact that the E2 laser frequency cannot excite electrons to the $\Delta M = \pm 1$ transitions.

From Table 3, the E2 light shifts for both the $M = 1/2$ and $M = -1/2$ states are -20.5 MHz, and the NSI PNC-induced light shift of the $M = 1/2$ ($M = -1/2$) state is -0.432 ($+0.432$) Hz. Therefore, measuring the RF frequency of the transition between the $M = 1/2$ and $M = -1/2$ states gives twice the PNC-induced light shift and cancels the E2 light shift, i.e. the fluctuation of the E2 light shift due to its laser intensity fluctuation can also be canceled out. Similar to the $7S_{1/2} - 6D_{5/2}$ transition reported in Ref. [23], the E2 light shifts for the $F = 13/2, M = 7/2$ and 9 / 2 states with $\Delta M = 0$ for $7S_{1/2} - 6D_{3/2}$ transition also have the same values as -5.15 MHz, as can be seen in Table 3. This coincidence is due to the same value of the Wigner 3J factors for the E2 transition between the hyperfine levels $F = 13/2, M = 7/2$ and 9/2 with $q = 0$ for both the states. Moreover, the signs of the NSI PNC-induced light shifts in the $F = 13/2, M = 7/2$ and 9/2 states are opposite. Therefore, a pair consisting of $M = 7/2$ and 9/2 values is favorable for measuring the NSI PNC-induced effect because in this case the E2 light shift can get canceled out.

We define now the shift of frequency as $\delta\omega_M \equiv \Delta\omega_M^{E2} + \Delta\omega_M^{\rm PNC} + E_{\rm Zeeman}$ for the transition between the $7S_{1/2}, F = 13/2, M$ and $6D_{3/2}, F' = 13/2, M'$ states with $\Delta M = 0$, where $E_{\rm Zeeman}$ is the Zeeman shift. Figure 3 shows that these energy shift for the $M = -1/2, 1/2, 7/2$, and 9 / 2 values with two RF transitions. The frequency difference between two different magnetic sublevels in the hyperfine ground state with $F = 13/2$ is defined as $\delta\omega_{M1,M2} \equiv \omega_{M1} - \omega_{M2}$. The frequency difference between the $M = 1/2$ and $-1/2$ levels is written as $\delta\omega_{1/2,-1/2}$. Likewise, that of the $M = 9/2$ and 7 / 2 values is written as $\delta\omega_{9/2,7/2}$. The difference of the NSI

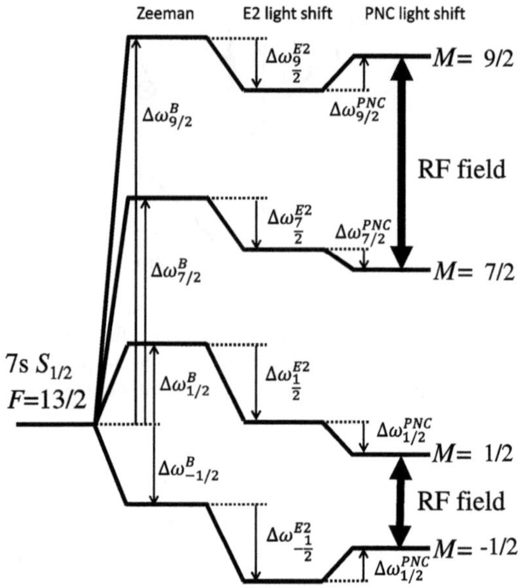

Fig. 3 Magnetic sublevels (shown only for $M = 9/2$, $7/2$ and $M = \pm 1/2$) of the $F = 13/2$ level of the $7S_{1/2}$ state with the corresponding RF transitions. The E2 light shifts are $\Delta\omega^{E2}_{9/2} = \Delta\omega^{E2}_{7/2}$ and $\Delta\omega^{E2}_{1/2} = \Delta\omega^{E2}_{-1/2}$. Energy differences between $M = 9/2$ and $7/2$, and between $M = 1/2$ and $-1/2$ without NSI PNC light shifts have the same first order Zeeman shift $g_F \mu_B B/\hbar$

PNC-induced shift $\omega^{\mathrm{PNC}}/2\pi$ for $\delta\omega_{1/2,-1/2}$ is estimated to be about -0.864 Hz, and that for $\delta\omega_{9/2,7/2}$ is about 6.919 Hz, as given in Table 3. Subtracting the frequency difference, i.e. $\delta\omega_{9/2,7/2}$ from $\delta\omega_{1/2,-1/2}$, the first order Zeeman shift can be canceled, and the second order Zeeman shift can be reduced. This is because both the signs of the curves representing the Zeeman shifts as a function of magnetic field for these transitions are positive.

The difference of $\delta\omega_{1/2,-1/2}$ and $\delta\omega_{9/2,7/2}$ is written as

$$\begin{aligned} \delta\omega &\equiv \delta\omega_{1/2,-1/2} - \delta\omega_{9/2,7/2} \\ \delta\omega/2\pi &= -7\,783\ (\mathrm{NSI}) - 88.4\ (\mathrm{NSD}) \\ &\quad +58\,509 \times 10^3\ (\mathrm{Zeeman@85.98\ G}) \\ &\quad +11.7 \left(\frac{\delta B}{8.598 \times 10^{-6} G}\right)^2 \mathrm{mHz}, \end{aligned} \tag{18}$$

where the difference of Zeeman shift at 85.98 G is $\sim$58.509 kHz, estimated using the Breit–Rabi formula [46], considering the hyperfine coupling constant A of the $7S_{1/2}$ state as 7 195.1 MHz [47], the nuclear spin $I = 6$, the g-factor of nuclear spin g_I of 0.0003913 for ^{210}Fr [48] and here, δB is the fluctuation of the magnetic field. Therefore, if the magnitude of the magnetic field is stabilized to better than 10^{-7}, the NSI PNC-induced light shifts can be detected with uncertainty of less than 11.7 mHz, which corresponds to $|\Delta Q_{\mathrm{W}}(^{210}\mathrm{Fr})/Q_{\mathrm{W}}(^{210}\mathrm{Fr})| \sim 0.15\%$. It requires that the currents of coils generating the bias field of 85.98 G should be stabilized to better than 10^{-7}, and the earth magnetic field of 0.3 G should be suppressed by 1/35,000 using magnetic shields. The experiment in these conditions is feasible using currently available technology at the CYRIC facility, Tohoku University, Japan [49–52].

The present experimental procedure will be almost similar to the technique discussed in our previous works [15, 23] to measure PNC-induced light shifts in the $7S_{1/2} \rightarrow 6D_{5/2}$ transition. We briefly describe this scheme for the $7S_{1/2} \rightarrow 6D_{3/2}$ transition below. First, we would like to prepare the magnato-optical trapping (MOT) of ^{210}Fr atoms. These trapped atoms will be loaded into a 3D optical lattice. Then, the atoms will be spin polarized in the $7S_{1/2}, F = 13/2, M = 13/2$ state by applying a bias magnetic field and laser beam resonant for the $7S_{1/2}, F = 13/2$ and $7P_{3/2}, F^{''} = 13/2$ transition with σ^+ polarization, and a repumping beam with $\sigma^{\pm}$ polarization. Next, we will create the superposition state of $F = 13/2, M = 9/2$ and $F = 13/2, M = 1/2$ using the stimulated Raman adiabatic passage (STIRAP) in three-levels [53–55] and in five levels [56, 57] as described in our previous work [23]. The atoms will be irradiated with the E2 laser beam with a wavelength of 616 nm. The RF fields will be applied that are resonant with the $F = 13/2, M = 9/2$ and 7 / 2, and with $M = 1/2$ and $-1/2$ transitions as shown in Fig. 3. The resonance frequencies of these RF transitions will include the PNC-induced light shifts. After the interaction time of 540 ns, the estimated lifetime of the $6D_{3/2}$ state, we will apply the state selective detection technique for the $M = 7/2$ and $M = -1/2$ transition using a microwave (MW) field with a frequency of 46.8 GHz that will transfer the population in the $F = 13/2, M = 7/2$ state to the $F = 11/2$, $M = 7/2$ state. Then the trapping beam near resonant with the transition between the $7S_{1/2}, F = 13/2$ and $7P_{3/2}, F^{''} = 15/2$ transition (corresponding to the wavelength of 718 nm) will be applied and the fluorescence of atoms during this interaction will be detected by the photo multiplier tube (PMT). From the selective detection of the $M = 7/2$ and $M = 1/2$ transition in combination with the MW field, blaster beams, and trapping beams, we will obtain the $\delta\omega_{1/2,-1/2}$ and $\delta\omega_{9/2,7/2}$ values [23]. Substituting the $\delta\omega_{1/2,-1/2}$ and $\delta\omega_{9/2,7/2}$ values into Eq. (18), it will yield frequency difference $\delta\omega$. Repeating the above measurement for different magnetic fields and fitting the difference of the Zeeman shifts with the Breit–Rabi formula to the data, we shall obtain the theoretical curve as a function of B as shown in Fig. 4. From our theoretical prediction, we are expecting to obtain the

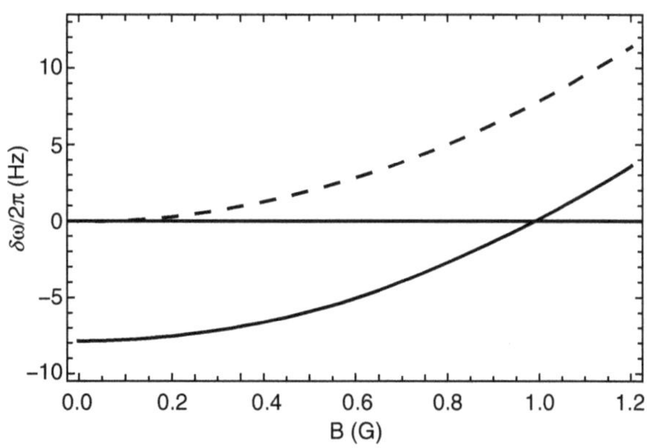

Fig. 4 Energy curve as a function of the magnetic field B. The *dashed* and *solid lines* represent the frequency differences of $\delta\omega = \delta\omega_{1/2,-1/2} - \delta\omega_{9/2,7/2}$ without and with NSI PNC-induced light shifts, respectively

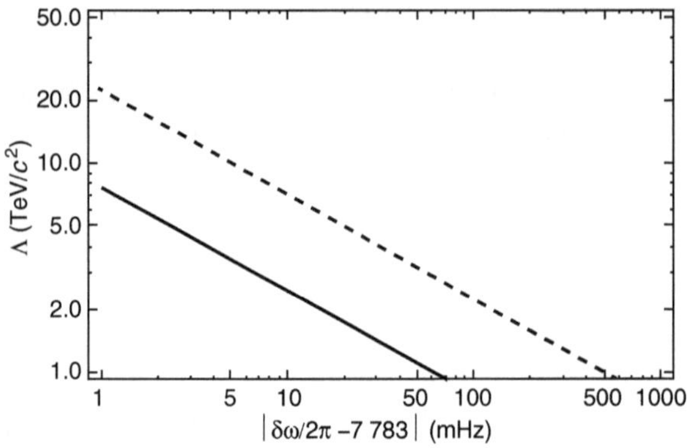

Fig. 5 Estimated range of mass of an extra particle Λ as a function of changes in the PNC-induced light shift ($\delta\omega$) in the $7S_{1/2} - 6D_{3/2}$ transition of ^{210}Fr from the standard model predicted value of 7783 mHz. The *solid* and *dashed lines* represent masses M_{Z_χ} and M_ϕ of an extra Z_χ boson in SO (10) and of the squark in the R-parity violating SUSY theory with $\lambda_s \sim 0.6$, respectively, that can be attained from the above PNC measurement

signature of the NSI PNC-induced light shift with a negative value of −7783 mHz for $B = 0$.

Figure 5 shows the sensitivity of mass depending on $\Delta Q_{\rm W}$ as a function of frequency shift. Λ is the mass of an extra particle. If there is a shift of 11.7 mHz from the value due to the SM, frequency measurement with a magnetic fluctuation of 10^{-7} will result in detection of the Λ as 2.1 TeV/c^2 for M_{Z_χ} and 6.4 TeV//c^2 for M_ϕ, as shown in Fig. 5. To detect the frequency shift of 11.7 mHz in Fig. 5, the NSD PNC-induced light shift of 88.4 mHz in Eq. (18) needs to be eliminated. To remove this, we can rewrite the frequency difference $\delta\omega$ of the $7S_{1/2} - 6D_{3/2}$ transition accounting the NSD contribution after compensation of the magnetic field using Eq. (18) as

$$\begin{aligned}\delta\omega/2\pi &\simeq -7\,783\ (\text{NSI}) - 88.4\ (\text{NSD})\ \text{mHz} \\ &= 67.40 Q_{\rm W}(^{210}\text{Fr}) - 156 K_{\rm W}\ \text{mHz},\end{aligned} \tag{19}$$

where the value of $Q_{\rm W}(^{210}\text{Fr})$ is expected to depend on new physics. Knowledge of precise value $K_{\rm W}$ is required to extract this new physics. Since we cannot distinguish these two kinds of shifts experimentally from Eq. (19), we intend to extract out $K_{\rm W}$ value from another experiment. Thus, to cancel out the NSD component from the above expression, we would also like to carry out measurement of the frequency difference $\delta\omega'$ in the $7S_{1/2} - 6D_{5/2}$ transition using the procedure suggested in our previous work [23]. Now the frequency shift of $\delta\omega'$ can be expressed as

$$\begin{aligned}\delta\omega'/2\pi \equiv \; & \frac{1}{2\pi}\left[\delta\omega_{1/2,-1/2} - \delta\omega_{9/2,7/2}\right]_{7S1/2-6D5/2} \\ = \; & -78.3K_{\rm W}\ \text{mHz}.\end{aligned} \quad (20)$$

By combining the measured $\delta\omega'$ with Eq. (20), we can infer the value of $K_{\rm W}$. Then, substituting this experimentally determined $K_{\rm W}$ value into Eq. (19), we can estimate the light shifts due to the NSD component in the $7S_{1/2} - 6D_{3/2}$ transition. This will ultimately yield the frequency difference given in Eq. (19) corresponding to only the NSI component. By this procedure, we shall be able to extract a precise value of $Q_{\rm W}(^{210}{\rm Fr})$ at the end to fulfill our objective. Therefore, it seems to us that a measurement up to an estimated frequency shift of 11.7 mHz, as shown in Fig. 5, in the $7S_{1/2} - 6D_{3/2}$ transition of ^{210}Fr can be made. This in combination with high precision relativistic coupled cluster calculations has the potential to test the possible existence of new particles.

The uncertainty in the measurement would be restricted by the shot noise limit. The mathematical expression to estimate this quantity is given by [10]

$$\delta\omega/2\pi = 1/2\pi\sqrt{\tau N_{\rm atom} T}, \quad (21)$$

where τ is the interaction time with the RF pulse, $N_{\rm atom}$ is the number of trapped atoms, and T is the total measurement time. If τ is 540 ns, the lifetime of the $6D_{3/2}$ state [40], and $N_{\rm atom} = 10^4$, the number of atoms that is expected to be trapped at CYRIC, T should be more than 34,300 s to measure the NSI PNC-induced light shift with an uncertainty of 11.7 mHz. This seems to be reasonable from a practical point of view.

5 Conclusion

We have investigated the suitability of the $7S_{1/2} \rightarrow 6D_{3/2}$ transition of ^{210}Fr to measure the NSI PNC interaction-induced light shift. We have explored the feasibility of conducting such an experiment using ultracold ^{210}Fr isotopes, and have investigated typical orders of magnitudes of the light shifts induced by the electric quadrupole transitions and NSI and NSD PNC interactions. Procedures to measure the frequency differences of $\delta\omega_{1/2,-1/2}$ and $\delta\omega_{9/2,7/2}$, (subscripts representing hyperfine levels of the respective ground and excited states, of the above transition) due to both the NSI and NSD PNC-induced light shifts without the first order Zeeman shift have been outlined. We have also suggested the measurement of the NSD PNC-induced light shift in the $7S_{1/2}, F = 13/2$ to $6D_{5/2}, F = 13/2$ transition to eliminate contributions due to the NSD component of the PNC-induced light shift in the $7S_{1/2} \rightarrow 6D_{3/2}$ transition to determine the nuclear weak charge $Q_{\rm W}$ (^{210}Fr) with an uncertainty of less than 11.7 mHz. Such measurements could in principle probe new physics beyond the standard model. A plan to carry out experiments at CYRIC is underway.

Acknowledgements We thank Dr. M. Mukherjee of National University of Singapore, Prof. S. Tojo of Chuo University, and Dr. U. Tanaka of Osaka University for many useful discussions on light shifts and E2 transitions. We also thank Dr. N. Yamanaka in Riken for discussions. This work was supported by a Grant-in-Aid for Scientific Research (B) (No. 25287100), a Grant-in-Aid for Scientific Research (S) (No. 26220705), and a Grant-in-Aid for Scientific Research on Innovative Areas (No. 21104005) from the Japan Society for the Promotion of Science (JSPS). BKS acknowledges access to the Vikram-100 HPC cluster of the Physical Research Laboratory (PRL), India for carrying out computations.

References

1. M.A. Bouchiat, C. Bouchiat, Rep. Prog. Phys. **60**, 1351 (1997)
2. J.S.M. Ginges, V. Flambaum, Phys. Rep. **397**, 63 (2004)
3. M.A. Bouchiat, C. Bouchiat, J. Phys. (Paris) **35**, 899 (1974)
4. M.A. Bouchiat, C.C. Bouchiat, Phys. Lett. B **48**, 111 (1974)
5. V.V. Flambaum, I.B. Khriplovich, J. Exp. Theor. Phys. **52**, 835 (1980)
6. W.J. Marciano et al., Phys. Rev. Lett. **65**, 2963 (1990)
7. M.J. Ramsey-Musolf, Phys. Rev. C **60**, 015501 (1999)
8. C.S. Wood, S.C. Bennett, D. Cho, B.P. Masterson, J.L. Roberts, C.E. Tanner, C.E. Wieman, Science **275**, 1759 (1997)
9. K. Tsigutkin, D. Dounas-Frazer, A. Family, J.E. Stalnaker, V.V. Yashchuk, D. Budker, Phys. Rev. Lett. **103**, 071601 (2009)
10. N. Fortson, Phys. Rev. Lett. **70**, 2383 (1993)
11. L.W. Wansbeek, B.K. Sahoo, R.G.E. Timmermans, K. Jungmann, B.P. Das, D. Mukherjee, Phys. Rev. A **78**, 050501(R) (2008)
12. B.K. Sahoo, B.P. Das, Phys. Rev. A **84**, 010502(R) (2011)
13. B.M. Roberts, V.A. Dzuba, V.V. Flambaum, Phys. Rev. A **89**, 012502 (2014)
14. W.R. Johnson, M.S. Safronova, U.I. Safronova, Phys. Rev. A **67**, 062106 (2003)
15. B.K. Sahoo, T. Aoki, B.P. Das, Y. Sakemi, Phys. Rev. A **93**, 032520 (2016)
16. M.-A. Bouchiat, Phys. Rev. Lett. **100**, 123003 (2008)
17. E. Gomez, S. Aubin, G.D. Sprouse, L.A. Orozco, D.P. DeMille, Phys. Rev. A **75**, 033418 (2007)
18. D. Sheng, L.A. Orozco, E. Gomez, J. Phys. B At. Mol. Opt. Phys. **43**, 074004 (2010)
19. G. Stancari, S.N. Atutov, R. Calabrese, L. Corradi, A. Dainelli, C. de Mauro, A. Khanbekyan, E. Mariotti, P. Minguzzi, L. Moi, S. Sanguinetti, L. Tomassetti, S. Veronesi, Eur. Phys. J. Special Topics **150**, 389 (2007)
20. B.K. Sahoo, B.P. Das, R.K. Chaudhuri, D. Mukherjee, Phys. Rev. A **75**, 032507 (2007)
21. K.P. Geetha, A.D. Singh, B.P. Das, C.S. Unnikrishnan, Phys. Rev. A **58**, R16(R) (1998)
22. B.K. Sahoo, P. Mandal, M. Mukherjee, Phys. Rev. A **83**, 030502(R) (2011)
23. T. Aoki, Y. Torii, B.K. Sahoo, B.P. Das, K. Harada, T. Hayamizu, K. Sakamoto, H. Kawamura, T. Inoue, A. Uchiyama et al., Asian J. Phys. **25**, 1247 (2016)
24. E.D. Commins, P.H. Bucksbaum, *Weak Interactions of Leptons and Quarks* (Cambridge University Press, Cambridge, 1983)
25. D. London, J.L. Rosner, Phys. Rev. D **34**, 1530 (1986)
26. V.A. Dzuba, J.C. Berengut, V.V. Flambaum, B. Roberts, Phys. Rev. Lett. **109**, 203003 (2012)
27. V. Barger, G.F. Guidice, T. Han, Phys. Rev. D **40**, 2987 (1989)
28. E.J. Eichten et al., Phys. Rev. Lett. **50**, 811 (1983)
29. A.E. Nelson, Phys. Rev. Lett. **78**, 4159 (1997)
30. A. Deandrea, Phys. Lett. B **409**, 277 (1997)
31. D. Wang et al., Nature **506**, 67 (2014)

32. M.E. Peskin, T. Takeuchi, Phys. Rev. Lett. **65**, 964 (1990)
33. K.S. Kumar, S. Mantry, W.J. Marciano, P.A. Souder, Annu. Rev. Nucl. Part. Sci. **63**, 237 (2013)
34. C. Bouchiat, P. Fayet, Phys. Lett. **608**, 87 (2005)
35. H. Davoudiasl, H.-S. Lee, W.J. Marciano, Phys. Rev. D **85**, 115019 (2012)
36. H. Davoudiasl, H.-S. Lee, W.J. Marciano, Phys. Rev. Lett. **109**, 031802 (2012)
37. D. Androic et al. (Q_{weak} Collaboration), Phys. Rev. Lett. **111**, 141803 (2013)
38. J. Erler, A. Kurylov, M.J. Ramsey-Musolf, Phys. Rev. D **68**, 016006 (2003)
39. J. Beringer et al. (Particle Data Group), Phys. Rev. D **86**, 010001 (2012)
40. B.K. Sahoo, B.P. Das, Phys. Rev. A **92**, 052511 (2015)
41. http://physics.nist.gov/cgi-bin/ASD/energy1.pl
42. M.S. Safronova, W.R. Johnson, Phys. Rev. A **62**, 022112 (2000)
43. V.A. Dzuba, V.V. Flambaum, M.S. Safronova, Phys. Rev. A **73**, 022112 (2006)
44. V.M. Shabaev, I.I. Tupitsyn, K. Pachucki, G. Plunien, V.A. Yerokhin, Phys. Rev. A **72**, 062105 (2005)
45. B.K. Sahoo, D.K. Nandy, B.P. Das, Y. Sakemi, Phys. Rev. A **91**, 042507 (2015)
46. N.F. Ramsey, *Molecular Beams* (Oxford University Press, Oxford, 1956)
47. J.E. Sansonetti, J. Phys. Chem. Ref. Data **36**, 497 (2007)
48. C. Ekström, L. Robertsson, A. Rosán, Physica Scripta. **34**, 624 (1986)
49. Y. Sakemi et al., J. Phys. Conf. Ser. **302**, 012051 (2011)
50. H. Kawamura et al., JPS Conf. Proc. **6**, 030068 (2015)
51. T. Inoue et al., JPS Conf. Proc. **6**, 030070 (2015)
52. K. Harada et al., Appl. Opt. **55**, 1164 (2016)
53. U. Gaubatz, P. Rudecker, M. Becker, S. Schiemann, M. Külz, K. Bergmann, Chem. Phys. Lett. **149**, 463 (1988)
54. J.R. Kuklinski, U. Gaubatz, F.T. Hioe, K. Bergmann, Phys. Rev. A **40**, R6741 (1989)
55. M. Weitz, B.C. Young, S. Chu, Phys. Rev. A **50**, 2438 (1994)
56. P. Marte, P. Zoller, J.L. Hall, Phys. Rev. A **44**, R4118 (1991)
57. T. Nakajima, Phys. Rev. A **59**, 559 (1999)

52. M.L. Peskin, T. Takeuchi, Phys. Rev. Lett. 65, 964 (1990)
53. [illegible] Annu. Rev. Nucl. Part. Sci. 54
    237 (2005)
54. [illegible] Phys. Lett. 608, 87 (2005)
[illegible]

Single-Slit Focusing and Its Representations

Manuel R. Gonçalves, William B. Case, Ady Arie, and Wolfgang P. Schleich

Abstract We illustrate the phenomenon of the focusing of a freely propagating rectangular wave packet using three different tools: (1) the time-dependent wave function in position space, (2) the Wigner phase-space approach, and (3) an experiment using laser light.

This article is part of the topical collection "Enlightening the World with the Laser" - Honoring T. W. Hänsch guest edited by Tilman Esslinger, Nathalie Picqué, and Thomas Udem.

The TWH productions on classical wave optics illustrating, for example, a pinhole caleidoscope, a Fresnel iris or different diffraction gratings are legendary. They can be found on Dropbox and YouTube at https://dl.dropboxusercontent.com/u/87280051/pinhole%20diffraction%204-15-2014%20720p.mov, https://www.youtube.com/watch?v=llevPEEd4L4 and https://www.youtube.com/watch?v=jzmqeRp_tmk. Over the last decades we have had great fun discussing wave phenomena such as Talbot carpets, Fresnel lenses and the diffraction from a single slit with Theodor W. Hänsch. For this reason we find it appropriate to dedicate to him this article on an elementary example of diffractive focusing on the occasion of his 75th birthday.

M.R. Gonçalves (✉)
Institute of Experimental Physics, Ulm University, 89081 Ulm, Germany
e-mail: manuel.goncalves@uni-ulm.de

W.B. Case
Department of Physics, Grinnell College, P.O. Box 805, Grinnell, IA 50112, USA

A. Arie
Department of Physical Electronics, Faculty of Engineering, Tel Aviv University, 6997801 Tel Aviv, Israel

W.P. Schleich
Institut für Quantenphysik and Center for Integrated Quantum Science and Technology (IQST), Ulm University, 89081 Ulm, Germany

Department of Physics and Astronomy, Hagler Institute for Advanced Study at Texas A&M University, Texas A&M AgriLife Research, Institute for Quantum Science and Engineering (IQSE), Texas A&M University, College Station, TX 77843-4242, USA

D. Meschede et al. (eds.), *Exploring the World with the Laser*,
https://doi.org/10.1007/978-3-319-64346-5_30

1 Introduction

In July 1816, the civil engineer Augustin-Jean Fresnel published his preliminary results [1] confirming the wave theory of light. Three years later he participated with his *Mémoire sur la Diffraction de la Lumière* in the Grand Prix of the French Academy of Sciences [2]. It was on this occasion that Siméon Poisson predicted that an opaque disc illuminated by parallel light would create a bright spot in the center of a shadow. This phenomenon was experimentally confirmed by Francois Arago and led to the victory of the wave over the particle theory. In the present article we discuss an effect related to the Poisson spot which is the one-dimensional analogue of the *camera obscura* [3, 4].

Indeed, we have recently found [5] that a rectangular matter wave packet which undergoes free time evolution according to the Schrödinger equation focuses before it spreads. This phenomenon has been confirmed for light [6], water and surface plasmon waves [7]. In the present article we illustrate this effect in Wigner phase space and verify it using classical light in real space.

Our article is organized as follows: in Sect. 2 we first give a brief history of the diffraction of waves, and then review several focusing effects especially those associated with the phenomenon of diffraction in time introduced in Moshinsky [8].

We dedicate Sect. 3 to the discussion of the focusing of a rectangular wave packet from the point of view of the time-dependent wave function. In particular, we show this effect manifests itself in the time-dependent probability density as well as the Gaussian width [5] of the wave packet. For this purpose we derive exact as well as approximate analytical expressions for the time-dependent probability amplitude and density.

In Sect. 4 we verify these predictions reporting on an experiment using laser light diffracted from a single slit. Here we take advantage of the analogy between the paraxial approximation of the Helmholtz equation of classical optics and the time-dependent Schrödinger equation of a free particle. We measure the intensity distributions of the light in the near-field of the slits and obtain the Gaussian width of the intensity field. Moreover, we make contact with the predictions of non-paraxial optics.

Section 5 illuminates this focusing effect from quantum phase space using the Wigner function. In particular, we show that the phenomenon of focusing which reflects itself in a dominant maximum of the probability density on the optical axis follows from radial cuts through the initial Wigner function at different angles with respect to the momentum axis. Moreover, we analyze the rays and envelopes of the Wigner function in more detail.

We conclude in Sect. 6 by summarizing our results and by providing an outlook. Here we allude to the influence of the number of dimensions on the focusing and emphasize the importance of corrections to paraxial optics.

To keep our article self-contained while focusing on the central ideas we have included three appendices. Indeed, "Appendix A" contains the calculations associated with the Gaussian width of our wave packet and "Appendix B" presents a detailed discussion of the Wigner function approach towards diffractive focusing.

As an outlook we compare in "Appendix C" the paraxial and non-paraxial results obtained for diffraction by slits and circular apertures.

2 Diffraction Theory

In this section we first provide a historical overview of diffraction and then address the phenomenon of diffractive focusing. Due to their different nature we distinguish in this discussion between light and matter waves. Moreover, we briefly review the concept of diffraction in time.

2.1 A Brief History

Following the experimental demonstration of the wave nature of light by Thomas Young [9] and the first theory on diffraction by Fresnel [1] the nineteenth century was extremely successful in the investigation of wave phenomena, specially in optics. The unifying electromagnetic theory of James Clerk Maxwell [10] was the culmination of all previous developments on electromagnetism. Gustav Kirchhoff readdressed the diffraction of scalar waves and put it on a rigorous mathematical foundation [11]. The Fresnel diffraction arises now as a special case of the Kirchhoff diffraction. Arnold Sommerfeld and Lord Rayleigh [4, 12, 13] improved the Kirchhoff theory correcting the boundary conditions at the aperture and with that eliminating the discrepancy arising between the solutions and the boundary conditions chosen by Kirchhoff. Friedrich Kottler proposed another reason for this discrepancy by showing that the Kirchhoff integral can be interpreted not as a solution of the boundary value problem but as a solution for the "saltus" at the boundary [14, 15]. Moreover, he extended the scalar theory to electromagnetic waves [16, 17].

During the twentieth century numerous theoretical and experimental contributions to diffraction theory emerged. Julius Stratton and Lan Jen Chu extended the scalar Kirchhoff diffraction theory to vector waves [18] accounting for polarization. Hans Bethe found analytical solutions for the diffraction of electromagnetic waves by an aperture much smaller than the wavelength [19]. His theory and the corrections later introduced by Christoffel Bouwkamp [20] became important because of the invention of the near-field scanning microscope (SNOM or NSOM) [21] and the developments related to near-field optics [22–24]. In 1998, Thomas Ebbesen and collaborators observed that light transmission through an array of subwavelength apertures drilled in noble metal thin films can largely surpass the value predicted by Bethe [25]. This extraordinary optical transmission is dependent on the geometry of the array, on the illumination conditions and on the size and shape of the apertures [26, 27]. It results from the excitation of surface plasmon modes near the aperture. In plasmonic gratings with narrow slits it may also lead to an attenuation of the transmitted light stronger than that predicted by the Bethe–Bouwkamp theory [28].

2.2 Focusing of Waves

Focusing of waves by diffraction due to slits or apertures falls into two categories: (1) near-field focusing effects arising mainly in the diffraction of electromagnetic waves, and (2) focusing resulting from diffraction of slits or apertures larger than the wave length, where the focus is located in the Fresnel zone.

In the first category we include the focusing of light resulting from the confinement of surface plasmons in nanostructured apertures in plasmonic materials [25, 26, 29]. Frequently scalar and electromagnetic diffraction theories assume the apertures to be located in infinite and perfectly absorbing screens, and thus surface plasmons are ignored. Hence, these theories cannot account for plasmonic modes and their optical effects. To accurately describe the effects produced by the excitation of surface plasmons a full electromagnetic theory using the optical properties of real materials is required.

The focusing of light by apertures smaller than the wave length has been investigated theoretically several times in the last decades [30, 31]. However, this near-field focusing is dependent on the polarization of light and restricted to small apertures.

The properties of the focus in laser beams and atomic beams is of interest in microscopy and atom optics. Standard laser beams such as Laguerre–Gaussian, or Hermite–Gaussian beams can be strongly focused. The smallest order Hermite–Gaussian beam called TEM_{00} has the highest confinement and is, therefore, preferred in confocal microscopy. Other beams such as Airy and Bessel beams [32, 33] have non-diffracting properties.

To increase the field confinement, and thus the resolution of a microscope, novel laser beams and illumination mechanisms have been proposed [22, 29, 34–36].

In parallel, a similar interest exists in the confinement and squeezing of matter wave packets [37–39]. Focusing effects in atomic beams resulting from the interaction with laser fields diffracted by apertures in metallic screens were investigated recently [40]. Indeed, the interaction of matter waves with light fields has been the subject of intensive research in atom and quantum optics [41]. However, this spatial confinement, or focusing is of different nature than that of diffraction. In the latter, the field confinement created is solely determined by the properties of the incoming wave and the aperture. No other optical element is involved.

Self-focusing of light may also arise in nonlinear media [42]. However, we will not discuss this phenomenon in this article, but rather concentrate in our analysis in the focusing effects arising from diffraction in free space due to slits larger than the wavelength.

In 2012, the focusing of light waves by a slit larger than the wavelength was experimentally observed [6]. The diffraction pattern is similar to that of a circular aperture of several wavelengths in diameter [43, 44]. The main difference between a slit and a circular aperture is the value of the dominant maximum, relative to the intensity of the incident wave. For a circular aperture it reaches 4.0, whereas for a slit is only 1.8 stronger than the incoming wave [6, 43].

The diffraction patterns of slits and circular apertures for scalar waves and non-polarized electromagnetic waves can be accurately calculated using the Rayleigh–Sommerfeld diffraction integrals, even in the case of apertures of the size of the wavelength, without using any mathematical approximation. Moreover, analytical solutions for the on-axis field intensity were found for the circular aperture [43, 45], and the oscillations of the intensity on-axis were confirmed for electromagnetic waves [44, 46].

2.3 Diffraction in Time

Moshinsky [8] introduced the concept of diffraction in time using matter waves. Remarkably the time evolution of the probability density of a wave packet suddenly released by a shutter is mathematically identical to the intensity pattern behind a semi-infinite plane. This analogy stands out most clearly when we substitute the time coordinate of the wave packet by the corresponding space coordinate in diffraction. Then the solution of the Schrödinger equation for the problem of the Moshinsky shutter is identical to that of the Fresnel diffraction by a semi-infinite plane, and the probability density reaches a maximum of 1.3. Moreover, Moshinsky analyzed later the time–energy uncertainty associated with the shutter arrangement [47] and Godoy investigated the Fresnel and Fraunhofer diffraction in time of initially stationary states [48].

Recently, the diffraction in time of the double-shutter problem was analyzed [5]. An initially confined rectangular wave packet in one dimension is suddenly released and evolves in time. Again, the solution of the corresponding Schrödinger equation has the same form as the Fresnel diffraction of scalar waves by a single slit of infinite length.

However, we emphasize that Fresnel diffraction only holds true in the paraxial approximation of optics. The general solution of the diffraction by a slit is found by solving the Kirchhoff, or the Rayleigh–Sommerfeld diffraction integrals.

The mathematical analysis of the classical diffraction problems makes use of wave functions expressed in real, or reciprocal space. It is also very common in the investigation of the diffraction of matter waves [8, 49–51].

However, since Wigner introduced his famous distribution function [52] an increasing number of publications has used the Wigner phase space representation to study the dynamics of light beams [53–57] and matter waves [58–65]. Other phase space distribution functions related to the Wigner function have been also used in matter waves phenomena. They are interrelated and belong to the Cohen class [66]. In this article we employ both the wave function and the Wigner representations of matter wave packets.

The evolution in time of matter waves with zero angular momentum, so-called s-waves, strongly depends on the number of space dimensions [67]. For instance, in two dimensions, an initial ring-shaped wave packet first contracts reaching a minimum, reducing the radius of the ring, and then monotonically expands. In three dimensions, the radius only increases. This effect is attributed to a quantum anti-centrifugal force [67–69]. This example shows that the focusing effect of a free wave packet is a more general phenomenon than that arising from the free time evolution of a one-dimensional rectangular wave packet.

We conclude with a brief reference to the type of boundaries of the slit, or shutter. Most of the classical treatments of diffraction problems define the edges of the slit, or of other aperture shape as sharp transitions between a perfectly absorbing surface and a homogeneous fully transmitting medium. In quantum matter waves, the Moshinsky shutter or the sudden release of rectangular wave packet is also an example of sharp boundaries. The effects arising in the diffraction patterns due to non-sharp boundaries have been investigated recently [5, 50, 70–73].

3 Wave Function Approach

In this section we use the solution of the time-dependent Schrödinger equation, that is the wave function, to show that a freely propagating rectangular wave packet exhibits the phenomenon of focusing. For this purpose we first express the time-dependent wave function in terms of a Fresnel integral, and then derive analytic approximations for the wave function as well as the probability density. To bring out most clearly the focusing effect we finally calculate the Gaussian width [5] of the wave packet and demonstrate that it exhibits a clear minimum at the time of the focusing.

3.1 *Time Evolution*

Central to our discussion is the free propagation of a wave packet corresponding to a non-relativistic particle of mass M. The initial wave function

$$\psi_0(x) \equiv \psi(x, t=0) \equiv \frac{1}{\sqrt{L}} \Theta\left(\frac{L}{2} - |x|\right) \tag{1}$$

is of rectangular form with a length L. Here Θ denotes the Heaviside step function.

With the help of the propagator [74]

$$G(x, t|y, 0) \equiv \sqrt{\frac{\alpha(t)}{i\pi}} e^{i\alpha(t)(x-y)^2} \tag{2}$$

of a free particle connecting the initial coordinate y with x at time t, and the abbreviation

$$\alpha(t) \equiv \frac{M}{2\hbar t} \tag{3}$$

containing the reduced Planck constant $\hbar$, we find from the Huygens principle of matter waves

$$\psi(x,t) = \int_{-\infty}^{\infty} \mathrm{d}y\, G(x,t|y,0)\psi_0(y) \tag{4}$$

the expression

$$\psi(x,t) = \sqrt{\frac{1}{i\pi L}} \int_{\sqrt{\alpha(t)}(x-L/2)}^{\sqrt{\alpha(t)}(x+L/2)} \mathrm{d}\xi\, \mathrm{e}^{i\xi^2}, \tag{5}$$

for the time-dependent probability amplitude. Here we have introduced the integration variable $\xi \equiv \alpha^{1/2}(x-y)$.

When we decompose the integral in Eq. 5 into two parts each starting from $x=0$, the wave function

$$\begin{aligned}\psi(x,t) &= \sqrt{\frac{1}{2iL}}\{F[\sqrt{\alpha(t)}(x+L/2)] \\ &\quad -F\left[\sqrt{\alpha(t)}(x-L/2)\right]\}\end{aligned} \tag{6}$$

consisting of the difference of two Fresnel integrals

$$F(w) = \sqrt{\frac{2}{\pi}} \int_0^w \mathrm{d}\xi\, \mathrm{e}^{i\xi^2}, \tag{7}$$

is thus determined by the interference of the diffraction patterns originating from two semi-infinite walls located at $x = L/2$ and $x = -L/2$. The amplitude and phase of each contribution are given by the Fresnel integral F, whose real and imaginary parts

$$C(w) \equiv \sqrt{\frac{2}{\pi}} \int_0^w \mathrm{d}\xi\, \cos\xi^2 \tag{8}$$

and

$$S(w) \equiv \sqrt{\frac{2}{\pi}} \int_0^w \mathrm{d}\xi\, \sin\xi^2 \tag{9}$$

follow from the Cornu spiral [75] represented in the complex plane.[1]

[1]The Cornu spiral was studied for the first time by Jacques Bernoulli in the context of elastic deformations and Leonhard Euler defined it in more rigorous terms. Alfred Cornu associated this curve with the Fresnel integrals C and S and achieved excellent numerical accuracy. Due to the work of the Italian mathematician Ernesto Cesaro it is also called clothoid.

In Fig. 1 we present the probability density $|\psi(x,t)|^2$ as a function of space and time. Here and in the remainder of our article we represent the coordinate x in units of L and the time t in units of the characteristic time

$$T \equiv \frac{ML^2}{2\pi\hbar} \equiv \frac{ML^2}{h} \tag{10}$$

The curious inclusion of the factor 2π is motivated by the asymptotic expressions of $|\psi|^2$ discussed in the appendices. Moreover, the probability density $|\psi|^2$ is always in units of $1\,/\,L$.

Whereas on the top of Fig. 1 we show $|\psi(x,t)|^2$ in continuous space-time in the bottom panel we select specific time slices corresponding to (1) short times where the probability distribution oscillates strongly, (2) intermediate times leading to focusing, and (3) longer times representing the ballistic regime.

At $t = 0$ the probability density starts from its initial rectangular shape and immediately develops two peaks at the edges decorated with fringes. However, after this transitional phase the two peaks disappear and a dominant maximum at the origin $x = 0$ forms. It is most pronounced at $t \approx 0.342$ corresponding to the focus when the width of the wave packet assumes a minimum. Indeed, here the probability density assumes a maximum, which is about a factor 1.8 larger then at $t = 0$ where it is unity. After the focus, that is for larger times, the wave packet displays the familiar spreading effect.

3.2 Analytic Approximations

Next we give approximate but analytical expressions for the time-dependent probability amplitude and probability density. Since our interest is to obtain the behavior at early times, that is before the ballistic expansion occurs, we shall consider small values of t, corresponding to the regime where $\alpha(t)$ is large.

With the help of the asymptotic expansion [76]

$$\int_0^a d\xi\, e^{i\xi^2} \cong \frac{\sqrt{i\pi}}{2} + \frac{e^{ia^2}}{2ia} \tag{11}$$

valid for $1 \ll a$ the expression Eq. 5 for the probability amplitude reduces to

$$\begin{aligned}\psi(x,t) \;&\cong \frac{1}{\sqrt{i\pi L}}\left[\sqrt{i\pi} + \frac{e^{i\alpha(t)(L/2-x)^2}}{2i\sqrt{\alpha(t)}(L/2-x)}\right.\\ &\left.\quad + \frac{e^{i\alpha(t)(L/2+x)^2}}{2i\sqrt{\alpha(t)}(L/2+x)}\right],\end{aligned} \tag{12}$$

and the probability density reads

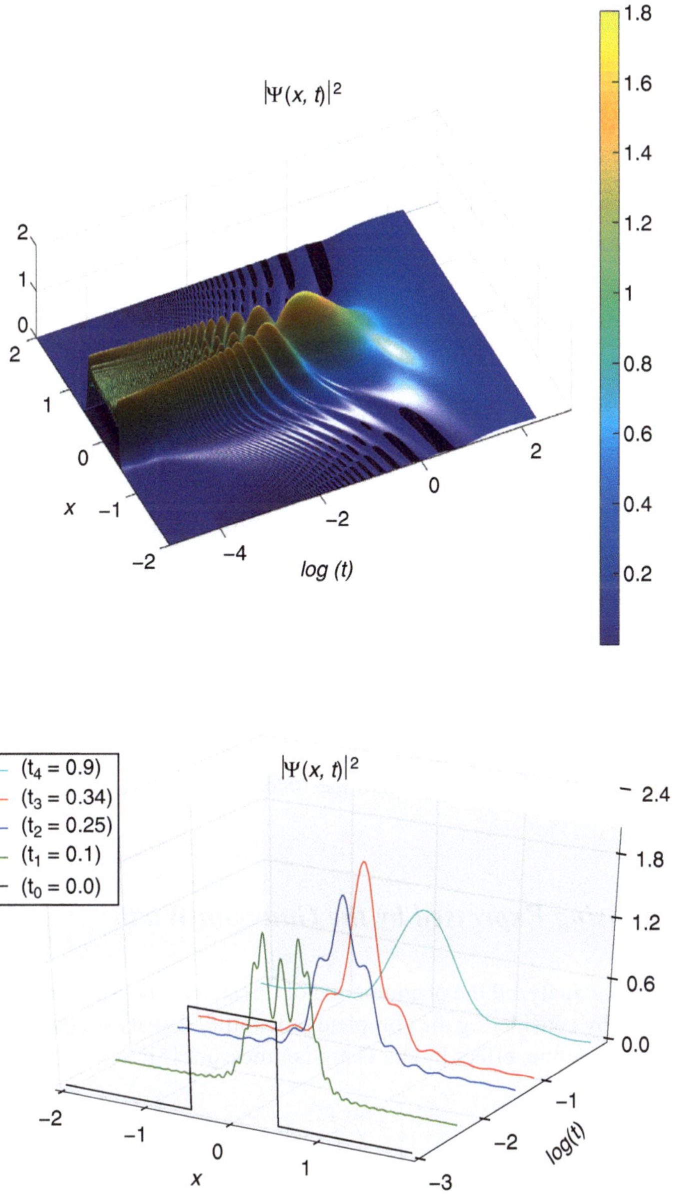

Fig. 1 Time evolution of a rectangular wave packet represented by its probability density $|\psi(x,t)|^2$, given by Eq. 5, depicted in continuous space-time (*top*) and at specific times (*bottom*) corresponding to a strongly oscillatory behavior, the focus, and the ballistic expansion. To bring out the characteristic features in the early-time evolution we have represented the time axis by a logarithmic scale. For a better comparison the initial wave packet at $t = 0$, that is for $\log(t = 0) = -\infty$ is moved to $\log(t) = -3.0$ since our time axis extends only to this value

$$|\psi(x,t)|^2 \cong \frac{1}{L\pi}\left|\left[\sqrt{i\pi} + \frac{e^{i\alpha(t)(L/2-x)^2}}{2i\sqrt{\alpha(t)}(L/2-x)} + \frac{e^{i\alpha(t)(L/2+x)^2}}{2i\sqrt{\alpha(t)}(L/2+x)}\right]\right|^2. \tag{13}$$

This expression simplifies further when we neglect terms $\mathcal{O}\left[t^2/(1 \pm 2x/L)^2\right]$ and takes the form

$$|\psi(x,t)|^2 \cong \frac{1}{L}\left[1 + \frac{\sin\left[\alpha(t)(L/2-x)^2 - \pi/4\right]}{\sqrt{\pi\alpha(t)}(L/2-x)} + \frac{\sin\left[\alpha(t)(L/2+x)^2 - \pi/4\right]}{\sqrt{\pi\alpha(t)}(L/2+x)}\right]. \tag{14}$$

In Fig. 2 we compare and contrast the resulting probability densities at $x = 0$ as a function of time and find excellent agreement between the numerical result following from the evaluation of the integral of Eq. 5, and the approximations based on Eqs. 13 and 14. We emphasize that our approximations break down for very large values of t, but they succeed in giving the maximum of the distribution corresponding to the focusing effect.

We also test in Fig. 3 our approximate but analytic expressions, Eqs. 13 and 14, for the probability density against the exact numerical result given by Eq. 5 at characteristic times confirming again the focusing effect.

3.3 *Focusing Expressed by the Gaussian Width*

So far we have analyzed the phenomenon of diffractive focusing of our rectangular wave packet by considering the complete probability density in space and time. We now characterize this effect by the Gaussian measure [5]

$$\delta x^2(t) \equiv \frac{1}{\kappa^2}\left[1 - \int_{-\infty}^{\infty} \mathrm{d}x\, \mathrm{e}^{-(\kappa x)^2} |\psi(x,t)|^2\right] \tag{15}$$

discussed in more detail in "Appendix A". Here κ is a constant with units of an inverse length.

For our rectangular wave packet we obtain the Gaussian width by numerically evaluating Eq. 15 using the integral representation Eq. 5 of the probability amplitude. In Fig. 4 we depict the corresponding curve normalized to its initial value $\delta x^2(0)$ for $\kappa = 6.0$ which displays a clear minimum at $t \approx 0.39$, thus confirming the focusing effect.

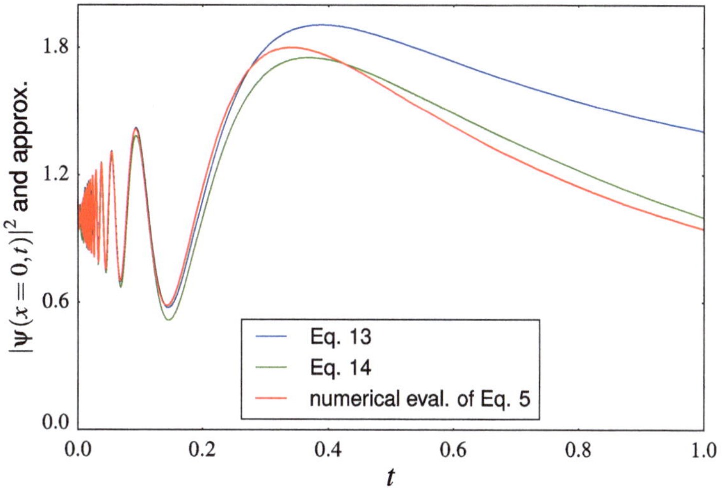

Fig. 2 Comparison between the exact numerical (*red*) and two approximate analytical expressions (*blue* and *green*) for the probability density $|\psi(x=0,t)|^2$ as a function of time. Our *curves* are based on Eqs. 5, 13 and 14, respectively. The oscillations near $t=0$ are well approximated both in amplitude and phase by the *blue* and *green curves*. A maximum occurs for $t \approx 0.342$. Both approximations fail for large times

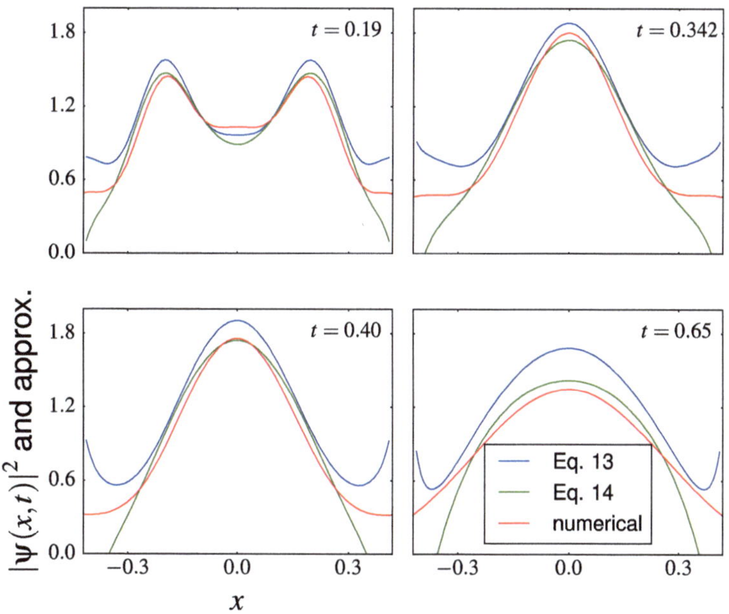

Fig. 3 Comparison between the exact numerical position-dependent probability density (*red*) given by Eq. 5 with the approximate formulae (*blue* and *green*) represented by Eqs. 13 and 14, respectively. In the *four panels* the axes cover identical domains

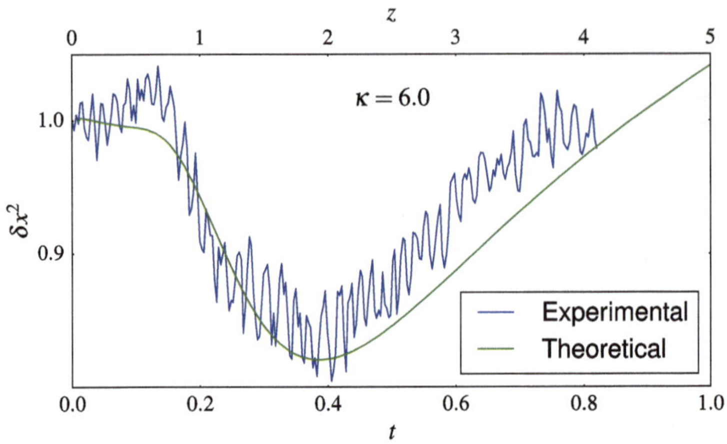

Fig. 4 Comparison between the numerical and experimental Gaussian width $\delta x^2(t)/\delta x^2(0)$ of a focusing rectangular wave packet. To have the same domain in the abscissa the z-coordinate of the experimental data (*blue curve*) was scaled using $L = 2.44$ μ m. The time coordinate t of the theoretical evaluation (*green curve*) based on Eq. 5 was scaled according to Eq. 18 with $\lambda = 0.2$. For *both curves* we have chosen $\kappa = 6.0$

However, we note that the location of the minimum of $\delta x^2 = \delta x^2(t)$ deviates slightly from the location of the maximum of $|\psi(x = 0, t)|^2$ which occurs at $t = 0.342$ as indicated in Fig. 2. This deviation is the result of the integration in Eq. 15.

4 Experimental Approach

In the preceding section we have shown that a rectangular matter wave packet first focuses before it spreads. We now describe an experiment to observe the diffraction pattern and, in particular, the focusing arising close to the slit. Here we take advantage of the familiar analogy between the Schrödinger equation

$$i\hbar \frac{\partial \psi}{\partial t} = -\frac{\hbar^2}{2M} \frac{\partial^2 \psi}{\partial x^2} \tag{16}$$

of a free particle, and the paraxial wave equation

$$2ik \frac{\partial \psi}{\partial z} = -\frac{\partial^2 \psi}{\partial x^2} \tag{17}$$

of classical optics. In this situation z denotes the coordinate of propagation and k the wave vector of the electromagnetic wave.

Indeed, the two wave equations are identical when we make the substitution

$$z \equiv \frac{\hbar k}{M} t = \frac{2\pi\hbar}{ML^2} \frac{L}{\lambda} Lt = \frac{L}{\lambda} L \frac{t}{T} \tag{18}$$

where λ is the wave length. In the last step we have recalled the definition, Eq. 10 of the characteristic time T. Hence, time in the Schrödinger equation translates into propagation distance z.

As a consequence of the analogy between Eqs. 16 and 17 with the scaling given by Eq. 18, it suffices to perform our experiment with light rather than matter waves.

4.1 Setup

We use a confocal microscope for measuring the light intensity diffracted by a one-dimensional slit milled in an aluminum film.[2] The slit of length 50 μm and width 2440 nm is illuminated by laser light of wavelength 488 nm as shown in Fig. 5a. A single mode optical fiber with collimator was used for the illumination. The collimated laser beam has a diameter of approximately 1 mm and is, therefore, much larger than the slit. For this reason we assume the illumination of the slit as a plane wave.

To generate images in different planes above the slit we employ a confocal laser scanning microscope (WITec GmbH). The objective used for light collection was an infinitely corrected Olympus MPlan with 100× magnification and numerical aperture NA = 0.9. The collected light was focused into a multimode optical fiber connected to an avalanche photodiode. The light diffracted by the slit is measured in a rasterized way. Each point corresponds to a pixel of an image generated by scanning in the horizontal, or in the vertical direction, as outlined in Fig. 5a.

4.2 Results

In Fig. 5b, c, we present images of the light intensity in the plane of the slit, and perpendicular to the sample, respectively. In the latter case, we scan the confocal microscope in the vertical direction with a minimum scan step of $\Delta z \approx 50$ nm. The pixel size in the horizontal direction corresponds to a dislocation of $\Delta x = \Delta y = 10$ μm/512−−20 nm. Thus, each pixel is much smaller than the wavelength and the optical resolution, which according to the Rayleigh criterion is $\Delta r \equiv 0.61 \times \lambda/\mathrm{NA}$. The horizontal fringes appearing in the light intensity of Fig. 5c are due to the mechanical motion of the microscope when scanning vertically the diffracted light.

[2]The slit was fabricated using focused ion beam milling (FIB) of a 75 nm Al thin film, evaporated at a pressure of approximately 10^{-6} mbar on top of a glass substrate of 1 mm thickness.

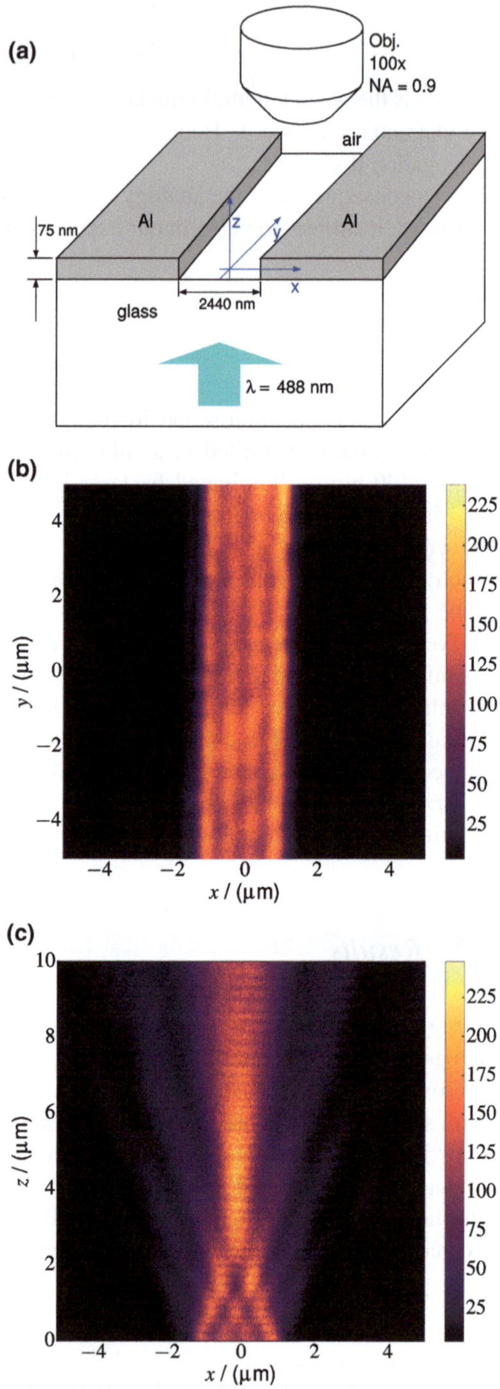

Fig. 5 Experimental verification of diffractive focusing from a single slit: setup (**a**) based on a confocal microscope viewing a section of the slit, and light intensity measured in the $x - y$ plane of the slit (**b**), and in the $x - z$ plane perpendicular to the substrate and slit (**c**)

We note the dominant maximum of the intensity at $z \approx 4.1$ µm which confirms our prediction of the focusing effect.

To analyze the phenomenon in a quantitative way we use the experimental intensity distribution of Fig 5c to obtain the Gaussian width δx^2 defined by Eq. 15. The so-calculated curve, now displayed in Fig. 4 as a function of the propagation distance z and scaled according to Eq. 18, follows nicely the theoretical prediction. In particular, it displays the characteristic minimum indicating focusing at the same location as the theoretical curve.

The rapid modulation of the experimental curve is a consequence of the horizontal fringes emerging due to the mechanical motion of the microscope as mentioned above, and thus of the measurement technique. An average over these oscillations leads to a smooth curve following the theoretical curve.

4.3 *Paraxial Versus Non-paraxial Optics*

To compare our experimental results to the theoretical predictions of classical optics we have calculated the intensity and phase for a slit of width $L = 2.44$ µm illuminated by light of wave length $\lambda = 0.488$ µm corresponding to the same ratio $L/\lambda = 5$ as in the experiment. For this purpose we use the Fresnel and the Rayleigh–Sommerfeld diffraction integrals familiar from paraxial and non-paraxial optics, respectively. In particular, we have chosen the Rayleigh–Sommerfeld integral of the first kind (RS-I) [4, 12, 13][3] and made use of numerical integration and algorithms reported in [77][4].

In Fig. 6a, b we depict the intensities following from the Rayleigh–Sommerfeld and Fresnel integrals, respectively. Moreover, in Fig. 7a, b we show the corresponding phases. For a direct comparison with Fig. 5 we have refrained from using normalized coordinates for both axis.

We note three characteristic features: (1) the number of intensity lobes is finite for RS-I, whereas it is infinite for Fresnel. (2) The phase pattern predicted by RS-I evolves almost as a plane wave in the propagation direction. In contrast, the Fresnel phase shows small oscillations around zero radians in the propagation direction, but

[3]The integral RS-I is often preferred to the integral of the second kind, RS-II, because it describes more accurately the value of the lobes close to the aperture. However, RS-II predicts locations of the lobes that are identical to that of RS-I.

[4]Mielenz developed numerical algorithms for the calculation of the diffraction by slits and circular apertures in paraxial and non-paraxial optics [77–83]. We note, however, that the Mielenz definition of the Lommel functions employed in the calculation of the Fresnel diffraction of a circular aperture is not correct. Correct definitions were provided by Lommel [84], Born and Wolf [85] and Daly et al. [86]. Moreover, we note that the abbreviations of the Rayleigh–Sommerfeld integrals RS-I and RS-II by Mielenz as $\mathrm{RS}^{(s)}$ and $\mathrm{RS}^{(p)}$ are misleading, since the indices usually refer to s- and p-polarization of vector waves, and the Rayleigh–Sommerfeld theory applies only to scalar waves.

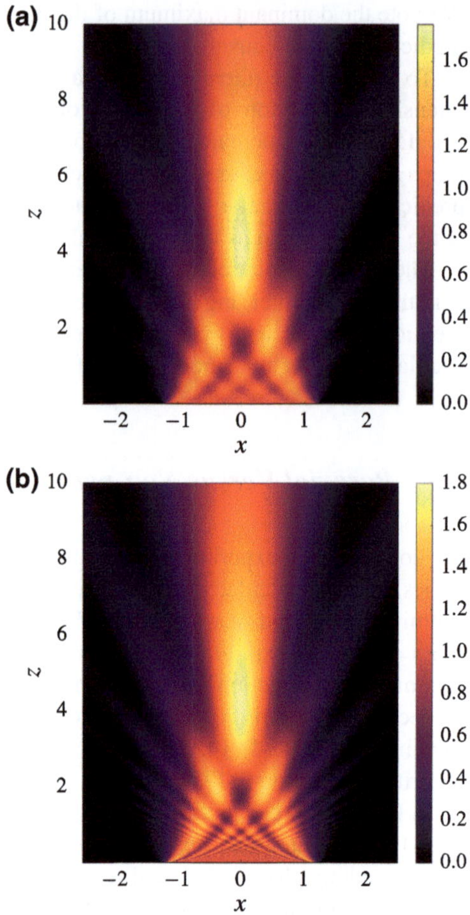

Fig. 6 Comparison of the intensities of a plane wave of wave length $\lambda = 0.488\,\mu m$ diffracted by a slit of width $L = 2.44$ μm calculated either from the Rayleigh–Sommerfeld integral RS-I (**a**), or the Fresnel integral (**b**). In both cases focusing occurs at $z \approx 4$, where the intensity reaches the value of 1.8. However, in contrast to the Fresnel diffraction the number of lobes for the RS-I diffraction is finite

rapid oscillations in the transverse direction, and (3) the Fresnel and RS-I intensity tend to agree as we move away from the slit. In particular, the focus occurs approximately in the same position for both calculations, as we show later.

We conclude this section by noting that although RS-I is only valid for scalar waves its intensity map still fits well with the optical intensity obtained experimentally.

5 Wigner Function Approach

In the preceding sections we have analyzed the focusing of a rectangular wave packet with the help of the time-dependent Schrödinger equation and have confirmed the effect using optical waves. We now consider this phenomenon from a different point of view, that is from phase space taking advantage of the Wigner

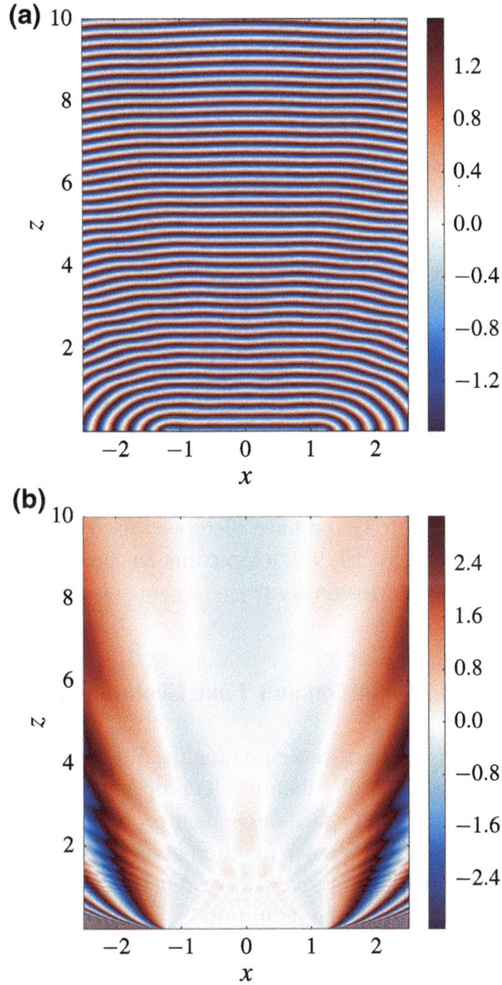

Fig. 7 Comparison of the phase patterns of a plane wave of wavelength $\lambda = 0.488\,\mu\text{m}$ diffracted by a slit of width $L = 2.44$ μm calculated either from the Rayleigh–Sommerfeld integral RS-I (**a**), or the Fresnel integral (**b**). The wavefronts for the non-paraxial case evolve in the forward direction almost parallel with small perturbations. However, the paraxial phase shows a much more complicated behavior with small fluctuations around zero radians on the optical axis and rapid oscillations in the lateral direction close to the slit

function. This formalism has the remarkable feature that the time evolution is identical to the classical one, consisting of a shearing of the initial Wigner function. The latter contains the interference nature of quantum mechanics.

5.1 Probability Density from Tomographic Cuts

The Wigner function has several unique properties. The ones most relevant for the present discussion are: (1) the time evolution of a free particle follows by a replacement of the phase space variables according to the classical motion, and (2) the corresponding probability densities originate from the integration of the Wigner function over the conjugate variable, that is by an appropriate tomographic cut. We now discuss these features in more detail and derive an expression for the probability density which is different from, but equivalent to the one discussed in Sect. 2.

5.1.1 A Brief Introduction

Wigner functions are quasi-probability distributions first introduced by Wigner [52] in the context of quantum correlations in statistical mechanics. They belong to the Cohen class of distributions [66] and are, therefore, related to other quadratic kernel distributions.

There exist extensive applications of the Wigner function, not only in quantum physics, but also in optics and signal processing [57, 87]. Indeed, Wigner functions are frequently used to represent quantum systems in phase space [60, 62, 88, 89]. Moreover, they describe the time evolution of wave packets in a similar way as classical phase space distributions in terms of the trajectories of classical particles [63] and have also been used in the study of diffraction in time of wave packets [51, 61, 90], a topic that has been addressed several times since the seminal article of Moshinsky [8].

5.1.2 Definition and Time Evolution

In this section we first define the Wigner function and discuss its time evolution in phase space. We then use its property to provide us with the marginals to compute the time- and space-dependent probability density. Here we take advantage of the fact that the Wigner function at arbitrary times is easy to obtain [63].

Indeed, for a given wave function $\psi = \psi(x)$ we find the corresponding Wigner function from the definition

$$W(x,p) \equiv \mathcal{N} \int_{-\infty}^{\infty} \mathrm{d}\xi\, \mathrm{e}^{ip\xi/\hbar} \psi\left(x - \frac{1}{2}\xi\right) \psi^*\left(x + \frac{1}{2}\xi\right) \tag{19}$$

with the normalization factor $\mathcal{N} \equiv 1/(2\pi\hbar)$, and a free particle undergoing time evolution is described in phase space by the time-dependent Wigner function

$$W(x,p;t) = W_0\left(x - \frac{pt}{M}, p\right). \tag{20}$$

Here $W_0 \equiv W(x,p;t=0)$ denotes the Wigner function of the initial wave function $\psi_0 \equiv \psi(x,t=0)$.

The probability density

$$|\psi(x,t)|^2 = \int_{-\infty}^{\infty} \mathrm{d}p\, W_0\left(x - \frac{pt}{M}, p\right) \tag{21}$$

at a given point in space and time is obtained by integration over p which after the change of variables $y \equiv x - (p/M)t$ reads

$$|\psi(x,t)|^2 = \frac{M}{t} \int_{-\infty}^{\infty} \mathrm{d}y\, W_0(y, M(x-y)/t). \tag{22}$$

This representation is particularly useful in the discussion of the asymptotic behavior of the probability density addressed in more detail in “Appendix B”.

5.2 *Time Evolution: Rectangular Wave Function*

We are now in the position to consider the time evolution of our rectangular wave function. For this purpose we first calculate the Wigner function corresponding to this wave, and then analyze the shearing in phase space. Although we mainly study the motion of the complete Wigner function we also concentrate on those contributions in phase space which lead to the focusing.

5.2.1 General Case

According to “Appendix B” the Wigner function of the rectangular wave packet, Eq. 1 reads

$$W_0(x,p) = \frac{1}{\pi L} \Theta\left(\frac{L}{2} - |x|\right) \frac{1}{p} \sin\left[\frac{p}{\hbar}(L - 2|x|)\right] \tag{23}$$

and is shown in Fig. 8a. Here we identify three characteristic features: (1) since the rectangular wave function is restricted to the interval $|x| \leq L/2$ also the Wigner function is limited in phase space to this domain. (2) The Wigner function assumes positive as well as negative values. (3) We recognize a dominant positive contribution along the x-axis with a maximum at the origin.

The time evolution given by Eq. 20 manifests itself in a shearing of the initial Wigner function W_0 shown in Figs. 8b, c. Indeed, every point (x, p) of the Wigner phase space moves according to the Newton law, that is $x \to x + pt/M$ while p is a constant of the motion.

Especially the Wigner function of Fig. 8c is interesting since it represents the moment of focusing. Indeed, by this time all negative contributions of W have moved away[5] from the positive peak centered at the x-axis but now subtract from its wings. This fact stands out more clearly in the position distribution shown in Fig. 8d

[5]We emphasize that our interpretation is different from [57], in page 320, which states: “... a peak in the axial intensity is achieved, associated with the vertical alignment of some of the main positive regions of the Wigner function”.

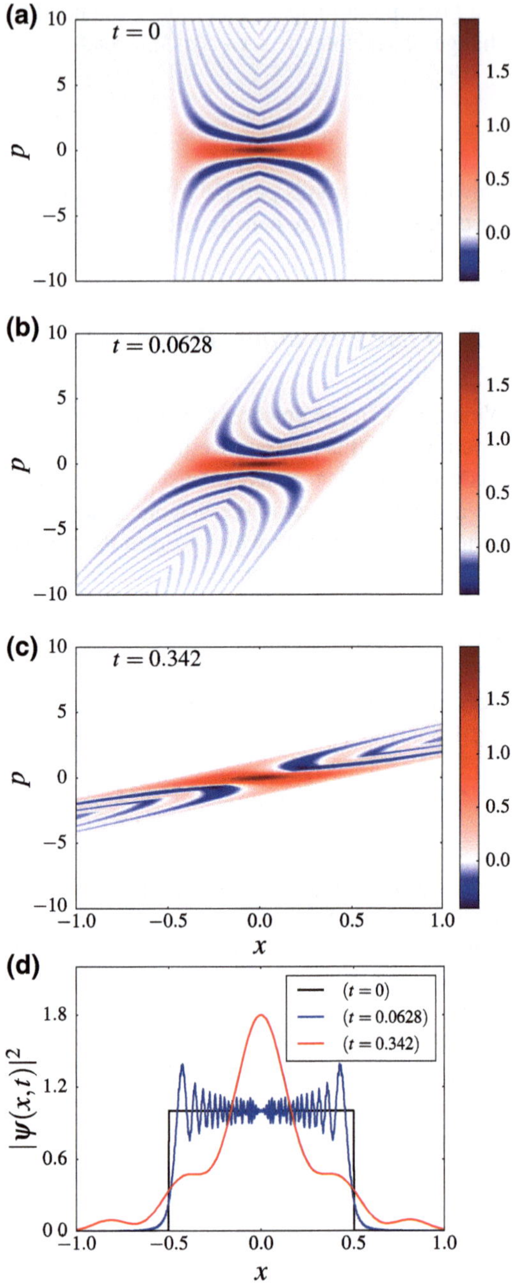

Fig. 8 Time evolution of the Wigner function corresponding to a rectangular wave packet represented by three density plots illustrating the shearing in phase space together with the associated probability densities in position space. The Wigner function of the initially rectangular wave packet is depicted in **a** at $t = 0$. The distribution shears with time, as shown in **b** at time $t = 0.0628$, and the moment of focussing **c** at $t = 0.342$. At the *bottom* **d** we display the corresponding probability densities in the coordinate x obtained by integration over the conjugate variable p. The momentum p is expressed in units of $\hbar/L$ as suggested by the argument of the sine function in the expression, Eq. 23, for the Wigner function

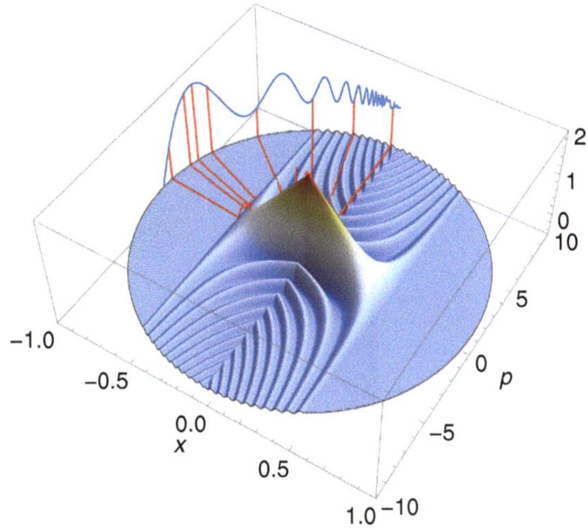

Fig. 9 Diffractive focusing of a rectangular wave packet explained from Wigner phase space. At the center we depict the Wigner function of the initial rectangular wave packet as a surface on top of the $x - p$ plane. The *blue curve* projected onto the surface of a cylinder represents the probability density $|\psi(0,t)|^2$, and is the result of an integration along the *red lines* $x + pt/M = 0$ at different times corresponding to different angles with respect to the x–axis. Clearly, the focusing occurs for a line sweeping only positive values of the Wigner function at the approximate time $t \approx 0.342$

which can be understood as the result of an integration of W over the momentum variable.

In this active approach the Wigner function evolves in time but the axes of phase space which define the tomographic cuts, that is integration over x or p remain fixed. An alternative formulation follows from Eq. 22. Here the Wigner function remains static, while the line of integration evolves in time according to $x + (p/M)t = 0$. In Fig. 9, we show the probability density at $x = 0$ as obtained from the Wigner function through integration from this *passive* point of view.

The cut at $t = 0$ runs along the momentum axis. Here the oscillations in the Wigner function W_0 along p average out and the main contribution to the probability density arises from the positive domain along the x-axis. For small times the lines of integration enclose small angles with respect to the p-axis and the oscillations in W_0 translate in an oscillatory probability density. A dominant maximum occurs when the cut feels the central positive ridge along the x-axis. The density decays for larger times, that is as the cut approaches the x-axis, since there is a decreasing overlap.

5.2.2 Rays and Envelopes

According to Eq. 20 each point in Wigner phase space follows its classical trajectory. Thus, the value of W_0 at (x_0, p_0) will move from x_0 to

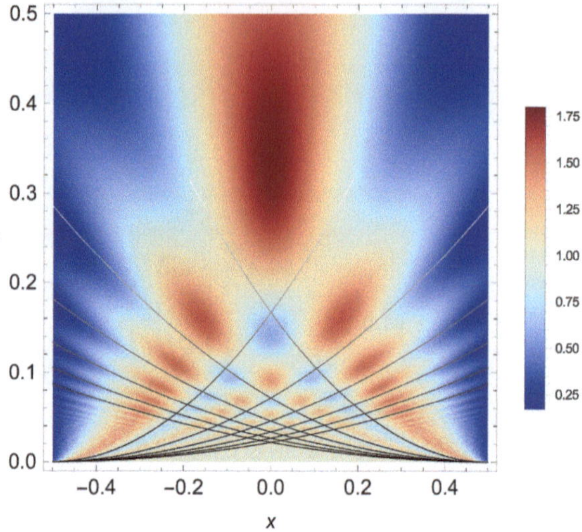

Fig. 10 Probability density $|\psi(x,t)|^2$ represented in space-time and overlaid with the envelopes (*gray lines*) of the diffraction in time created by a rectangular wave packet of initial width $L = 1$. The time t has been scaled using the characteristic time T. Here the envelopes given by Eqs. 32 and 36 approximate well the motion of the minima of $|\psi(x,t)|^2$

$$x = x_0 + \frac{p_0}{M}t \tag{24}$$

at time t. We now use these rays to show that the minima of the Wigner function generate the regions in space-time where the probability density $|\psi(x,t)|^2$ assumes small values as exemplified by Fig 10. Thus, they indicate the x values of the minima of $|\psi(x,t)|^2$ as the system evolves in time.

The explicit expression Eq. 23 for the initial Wigner function W_0 tells us that the minima of W_0 are given by the condition

$$\frac{p_0^{(n)}}{\hbar}(L - 2|x_0|) = \pi\left(2n + \frac{3}{2}\right) \tag{25}$$

for p positive, and

$$\frac{p_0^{(n)}}{\hbar}(L - 2|x_0|) = -\pi\left(2n + \frac{3}{2}\right) \tag{26}$$

for p negative, with $n = 0, 1, 2, \cdots$

We substitute these expressions for $p_0^{(n)}$ into the motion, Eq. 24, and rearrange the terms, which yield

$$(x - x_0)(L - 2x_0) \mp \left(2n + \frac{3}{2}\right)\frac{\pi\hbar}{M}t = 0 \tag{27}$$

for x_0 positive, and

$$(x - x_0)(L + 2x_0) \mp \left(2n + \frac{3}{2}\right)\frac{\pi\hbar}{M}t = 0 \tag{28}$$

for x_0 negative. The upper signs in Eqs. 27 and 28 are for positive $p_0^{(n)}$, that is for a positive slope in the (x, t) plane, and the lower signs for negative $p_0^{(n)}$.

The envelope of a set of curves $F(x, t; x_0) = 0$, each determined by a parameter x_0, follows [91] from the requirements

$$\frac{\partial F}{\partial x_0} = 0 \quad \text{and} \quad F = 0. \tag{29}$$

We consider the two cases, $x_0 > 0$ and $x_0 < 0$ in turn.

For positive x_0, the first condition gives

$$x_0 = \frac{L + 2x}{4}, \tag{30}$$

which, when substituted into Eq. 27, yields the relation

$$-\frac{1}{2}\left(\frac{L}{2} - x\right)^2 = \pm\left(2n + \frac{3}{2}\right)\frac{\pi\hbar}{M}t. \tag{31}$$

Thus, only negative values of $p_0^{(n)}$ contribute to this envelope and we obtain the space-time trajectories

$$t = \frac{(L/2 - x)^2}{2n + 3/2}\frac{M}{2\pi\hbar} \tag{32}$$

of the minima.

According to this expression they are parabolas[6] emerging from $x = L/2$ with a steepness inversely proportional to $(2n + 3/2)$. Hence, the largest steepness corresponds to the case of $n = 0$ with the parabola crossing the t-axis, that is $x = 0$, at the time

$$t = \frac{1}{6}\frac{ML^2}{2\pi\hbar} = \frac{1}{6}T, \tag{33}$$

where in the last step we have recalled the definition, Eq. 10 of the characteristic time T.

For negative x_0, the first condition in Eq. 29 provides us with the relation

[6]For a circular aperture these lines are rectilinear [92].

$$x_0 = -\frac{L - 2x}{4}, \tag{34}$$

which, when substituted into Eq. 28, gives

$$\frac{1}{2}\left(\frac{L}{2} + x\right)^2 = \pm\left(2n + \frac{3}{2}\right)\frac{\pi\hbar}{M}t. \tag{35}$$

Since in this case only positive values of $p_0^{(n)}$ contribute to this envelope we arrive at the space-time trajectory

$$t = \frac{(L/2 + x)^2}{2n + 3/2} \frac{M}{2\pi\hbar} \tag{36}$$

corresponding to parabolas emerging from $x = -L/2$. They are the mirror image of the ones given by Eq. 32.

In Fig. 10 we present the envelopes, Eqs. 32 and 36, along with the corresponding probability density $|\psi(x,t)|^2$. Hence, these minima coincide with the minima of the intensity pattern and are harbingers of the maxima and, in particular, of the focus.

6 Summary and Outlook: Beyond Slits

The spreading of a wave packet in the absence of a potential is a well-known phenomenon in quantum physics. However, also the opposite effect, that is a focusing of the wave packet can be achieved. In this case we have to imprint an appropriate position-dependent phase onto the initial wave function. Is it possible to achieve focusing even in the absence of any phase factors, that is for a real-valued initial wave function?

In the present article we have provided such an example in the form of a rectangular wave packet and have illuminated the resulting focusing in position as well as in phase space. Here we have used the time-dependent wave function and the Wigner distribution. Moreover, we have confirmed this phenomenon using laser light being diffracted from a slit building on the analogy between classical Fresnel optics and Schrödinger wave mechanics.

Throughout our article we have concentrated on rectangular wave packets created by a slit, that is by a one-dimensional aperture. However, similar effects appear in the diffraction from a two-dimensional aperture. In "Appendix C" we briefly summarize the literature on this problem, and discuss the similarities and differences between the one- and two-dimensional case using the examples of a slit and a circular aperture.

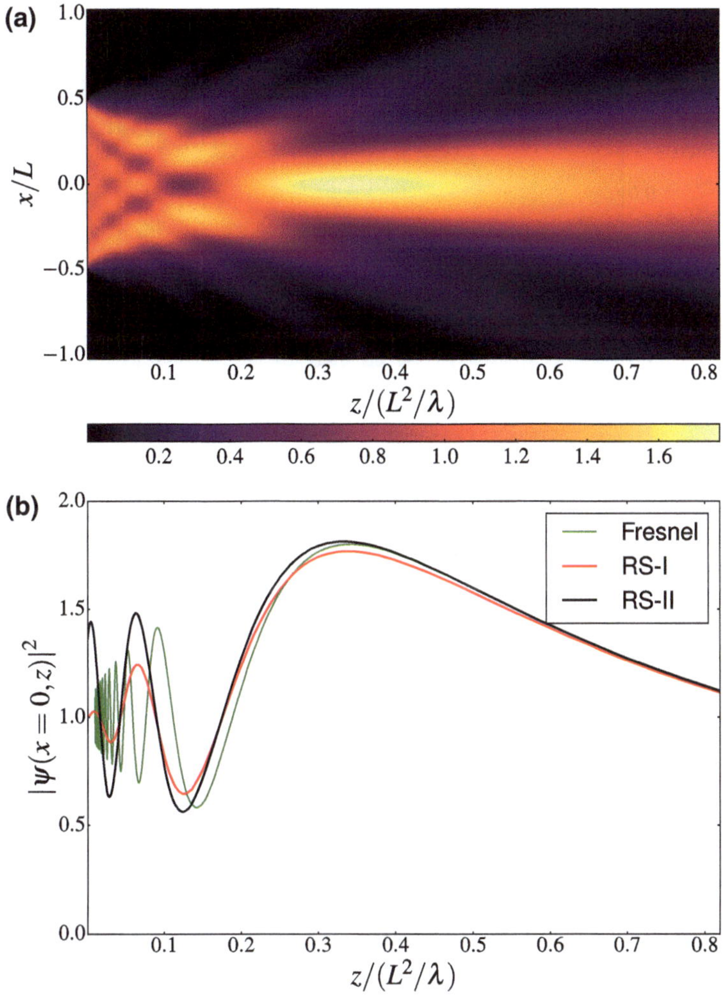

Fig. 11 Rayleigh–Sommerfeld intensity pattern for a slit of width $L = 5\lambda$ and illumination wave length $\lambda = 0.488$ μm (**a**), together with a comparison between the on-axis intensities predicted by the Rayleigh–Sommerfeld and Fresnel integrals (**b**). In contrast to Figs. 5, 6 and 7 the horizontal axes are normalized

Figures 11 and 12 bring out most clearly the characteristic features of this dependence on the number of dimensions: (1) for the slit the relative intensity at the focus reaches the value of 1.8, whereas for the circular aperture the maximum is 4.0. (2) The position of the focus in the case of the circular aperture is closer to the opaque screen than in the slit, and (3) the focus originating from a circular aperture is more confined, since its decay towards the far-field region is faster.

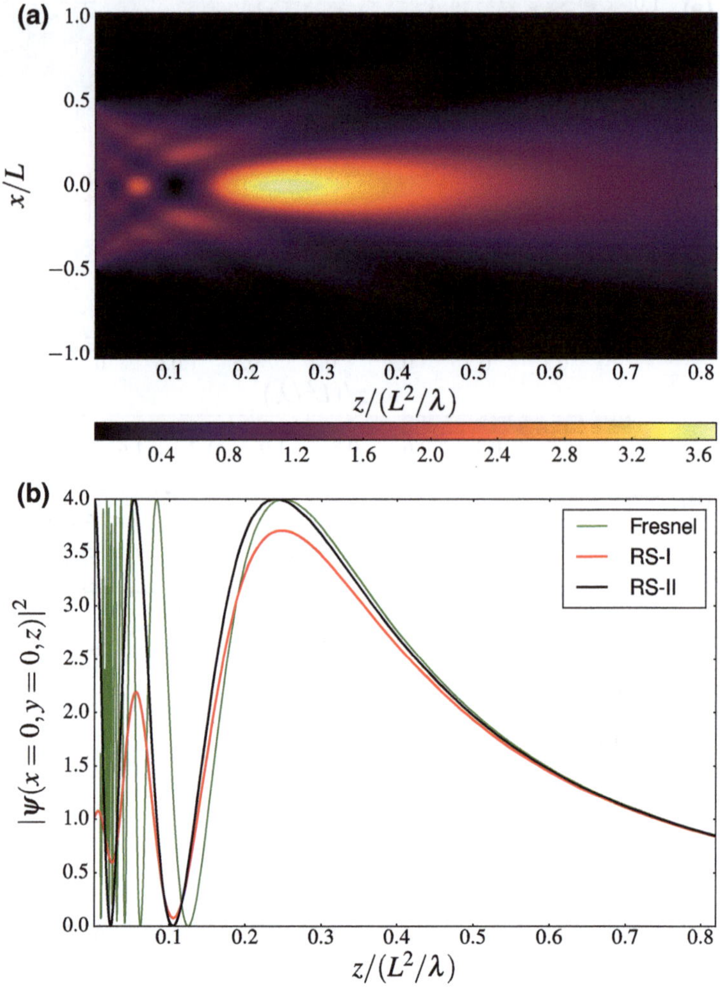

Fig. 12 Rayleigh–Sommerfeld intensity pattern for a circular aperture of diameter $L = 2a = 5\lambda$ and illumination wavelength $\lambda = 0.488$ μm (**a**), together with a comparison between the on-axis intensities predicted by the Rayleigh–Sommerfeld and Fresnel integrals (**b**). Again the axes are normalized

This analysis shows that in two dimensions the focusing effect is more pronounced. Moreover, it is possible to optimize this effect by choosing different apertures, either by appropriate apodization, or by creating an optimal wave packet. Unfortunately, this topic goes beyond the scope of the present article and has to be postponed to a future publication.

Acknowledgements Our interest in diffractive focusing dates back many years when it was at the center of a very productive collaboration with Emerson Sadurni. Unfortunately, in the mean time his interests have shifted and he has decided not to join us in this article walking down memory lane. Although we regret his decision we respect it and are most grateful to him for working with us on this topic during his period at Ulm University and for allowing us to use some of the material prepared in those days. We have immensely profited from and enjoyed many discussions with M. Arndt, S. Fu, N. Gaaloul, G. Leuchs, E. M. Rasel, L. Shemer, D. Weisman, E. Wolf, J. Zhou, and M. S. Zubairy. In particular, we thank T. W. Hänsch for a fruitful exchange of emails concerning this topic. We thank Ch. Kranz and collaborators of the Institute of Analytical and Bioanalytical Chemistry, Ulm University, for the preparation of the slits on the thin aluminum film using FIB. This work is supported by DIP, the German–Israeli Project Cooperation. WPS is most grateful to Texas A&M University for a Faculty Fellowship at the Hagler Institute for Advanced Study at the Texas A&M University as well as to the Texas A&M AgriLife Research for its support.

Appendix A: Gaussian Width

In this appendix we summarize the evaluation of the time-dependent Gaussian width defined by Eq. 15 of our rectangular wave packet. To gain some insight into this unusual definition of width we first discuss several of its general properties and then rederive the familiar spreading of a Gaussian wave packet undergoing free expansion employing this measure. We conclude by considering the case of the rectangular wave packet.

A.1: General Properties

We start our discussion of the Gaussian width

$$\delta x^2 \equiv \frac{1}{\kappa^2}\left[1 - \langle e^{-(\kappa x)^2}\rangle\right] \tag{37}$$

of the probability density $P = P(x)$ with the average

$$\langle e^{-(\kappa x)^2}\rangle \equiv \int_{-\infty}^{\infty} dx\, e^{-(\kappa x)^2} P(x) \tag{38}$$

by noting that for $\kappa \to 0$ the familiar expansion

$$e^{-(\kappa x)^2} \cong 1 - (\kappa x)^2 \tag{39}$$

leads us to

$$\delta x^2 \cong \langle x^2 \rangle. \tag{40}$$

Hence, in this limit the Gaussian width δx^2 reduces to the second moment $\langle x^2 \rangle$ of $P = P(x)$.

Whenever P displays oscillations in the position variable x around an average value $\bar{P}$ that are rapid on the decay length $1/\kappa$ of the Gaussian in the definition, Eq. 37, of δx^2 the oscillatory part averages out, and the familiar integral relation

$$\int_{-\infty}^{\infty} \mathrm{d}\xi\, \mathrm{e}^{-\gamma\xi^2} = \sqrt{\frac{\pi}{\gamma}} \tag{41}$$

yields

$$\delta x^2 \cong \frac{1}{\kappa^2}\left[1 - \frac{\sqrt{\pi}}{\kappa}\bar{P}\right]. \tag{42}$$

The other extreme occurs when $P = P(x)$ is slowly varying compared to $\exp[-(\kappa x)^2]$. In this case we can evaluate P at $x = 0$, factor it out of the integral and perform the integral with the help of Eq. 41. Thus, we arrive at the approximate expression

$$\delta x^2 \cong \frac{1}{\kappa^2}\left[1 - \frac{\sqrt{\pi}}{\kappa}P(x=0)\right], \tag{43}$$

which implies that the dependence of δx^2 on an additional parameter, such as time is governed by the dependence of the probability density P at the origin on that parameter.

Obviously, the most interesting case emerges when P and $\exp\left[-(\kappa x)^2\right]$ vary on approximately the same length scale. This situation is of special interest when P exhibits a dominant maximum at $x = 0$. Indeed, with the help of the Taylor expansion

$$P(x) \cong P(0) - \frac{1}{2}\left|P''(0)\right|x^2 \tag{44}$$

around this point we find from the identity

$$P(x) = \exp\{\ln\left[P(x)\right]\} \tag{45}$$

the Gaussian approximation

$$P(x) \cong P(0)\exp\left[-\frac{1}{2}\frac{\left|P''(0)\right|}{P(0)}x^2\right] \tag{46}$$

of P. Here prime denotes differentiation with respect to x.

As a result, we obtain the expression

$$\delta x^2 \cong \frac{1}{\kappa^2}\left[1 - \frac{\sqrt{\pi}P(0)}{\sqrt{\kappa^2 + \dfrac{|P''(0)|}{2P(0)}}}\right], \tag{47}$$

where we have used again the integral relation, Eq. 41.

A.2: Gaussian Wave Packet

Next we evaluate the Gaussian width for a freely spreading Gaussian wave packet of initial form

$$\psi_0(x) \equiv \mathcal{N}\mathrm{e}^{-\beta x^2} \tag{48}$$

with $\mathcal{N} \equiv (2\beta/\pi)^{1/4}$ and a real-valued constant β.

From the Huygens integral, Eq. 4, together with the propagator, Eq. 2, we obtain with the help of the integral relation

$$\int_{-\infty}^{\infty} \mathrm{d}\xi\, \mathrm{e}^{-\gamma\xi^2 + 2\sigma\xi} = \sqrt{\frac{\pi}{\gamma}}\mathrm{e}^{\sigma^2/\gamma} \tag{49}$$

the wave function

$$\psi = \mathcal{N}\sqrt{\frac{\alpha}{\alpha + i\beta}}\exp\left(i\frac{\alpha\beta}{\beta - i\alpha}x^2\right), \tag{50}$$

and thus the probability density

$$|\psi|^2 = \mathcal{N}^2\frac{\alpha}{\sqrt{\alpha^2 + \beta^2}}\exp\left(-2\frac{\alpha^2\beta}{\alpha^2 + \beta^2}x^2\right). \tag{51}$$

When we now substitute this expression into the definition, Eq. 37 of the Gaussian measure δx^2 of width and recall the identity, Eq. 41, we arrive at the result

$$\delta x^2 = \frac{1}{\kappa^2}\left[1 - \sqrt{\frac{2\alpha^2\beta}{\kappa^2(\alpha^2 + \beta^2) + 2\alpha^2\beta}}\right], \tag{52}$$

or

$$\delta x^2 = \frac{1}{\kappa^2}\left(1 - \left\{1 + \frac{\kappa^2}{2\beta}\left[1 + \left(\frac{\beta}{\alpha}\right)^2\right]\right\}^{-1/2}\right). \tag{53}$$

We note that for $\kappa^2/(2\beta) \ll 1$ and $(\beta/\alpha)^2 < 1$ corresponding to early times this expression reduces to

$$\delta x^2 \cong \frac{1}{4\beta}\left[1 + \left(\frac{\beta}{\alpha}\right)^2\right], \tag{54}$$

that is to the second moment $\langle x^2 \rangle$ of the Gaussian, Eq. 51, as predicted by Eq. 40.

Even more interesting is the limit $1 \ll (\beta/\alpha)^2$ corresponding to large times where Eq. 53 takes the form

$$\delta x^2 \cong \frac{1}{\kappa^2}\left[1 - \sqrt{\frac{2}{\beta}}\frac{\alpha(t)}{\kappa}\right]. \tag{55}$$

According to the definition, Eq. 3, of α we find $\alpha \propto 1/t$ and thus δx^2 increases in a monotonic way tending towards the constant $1/\kappa^2$. This time dependence of the Gaussian width is in sharp contrast to the one of the second moment x^2 given by Eq. 54 which increases quadratically with time without a bound.

A.3: Rectangular Wave Packet

Finally we turn to the Gaussian width δx^2 of the rectangular wave packet. Here we use the approximations developed in Sect. A.1 together with the properties of the approximate expressions, Eqs. 14 and 89 for the probability density $P(x,t) = |\psi(x,t)|^2$.

Indeed, for early times Eq. 14 predicts rapid oscillations around $\bar{P} = 1/L$ due to the fact that α, which according to Eq. 3 is inversely proportional to t, is large. Hence, we find from Eq. 42 the approximation

$$\delta x^2 \cong \frac{1}{\kappa^2}\left(1 - \frac{\sqrt{\pi}}{\kappa L}\right). \tag{56}$$

In the other extreme, that is for large times Fig. 1 shows that P is slowly varying. As a result, we obtain from Eq. 43 the formula

$$\delta x^2(t) \cong \frac{1}{\kappa^2}\left[1 - \frac{\sqrt{\pi}}{\kappa}|\psi(x=0,t)|^2\right], \tag{57}$$

and the time dependence of δx^2 is determined by the one of the probability density at the origin. Hence, a maximum in $|\psi(x=0,t)|^2$ translates into a minimum of δx^2 and vice versa.

However, in the neighborhood of the focus it is necessary to use the approximation Eq. 47 which requires $P(0)$ and $P''(0)$. According to Fig. 3 the expression for P, Eq. 14, approximates well the exact distribution. Therefore, it suffices to expand Eq. 14 up to the second order in x and we find

$$P(0,t) = \frac{1}{L}\left[1 + 4\sqrt{\frac{2\hbar t}{\pi M L^2}}\sin\left(\frac{ML^2}{8\hbar t} - \frac{\pi}{4}\right)\right] \tag{58}$$

and

$$\begin{aligned} P''(0,t) &= \frac{32}{L^3}\sqrt{\frac{2\hbar t}{\pi M L^2}}\left\{\left[2\left(\frac{ML^2}{8\hbar t}\right)^2 - 1\right]\right. \\ &\quad \times \sin\left(\frac{ML^2}{8\hbar t} - \frac{\pi}{4}\right) \\ &\quad \left. + \frac{ML^2}{8\hbar t}\cos\left(\frac{ML^2}{8\hbar t} - \frac{\pi}{4}\right)\right\}. \end{aligned} \tag{59}$$

In Fig. 13 we compare and contrast the three approximations Eqs. 56, 57, and 47 together with 58 and 59 to the numerically obtained curve based on the Fresnel integral, Eq. 5. We note that Eq. 56 provides us with the correct starting point of δx^2.

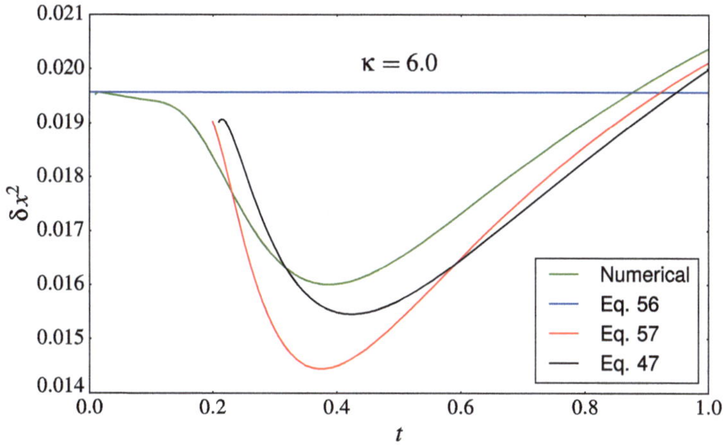

Fig. 13 Comparison of the numerically obtained Gaussian width δx^2 as a function of time for the freely spreading rectangular wave packet (*green curve*), and its approximations for early and large times given by Eqs. 56 and 57 and depicted by blue and red curves, respectively, together with the Gaussian approximation, Eq. 47, around the focus using Eqs. 58 and 59. Here we have chosen the value $\kappa = 6.0$ as in Fig. 4. Equation 57 was evaluated using the approximation for large times, Eq. 89. In the graphic representation of Eq. 47 we have used Eqs. 58 and 59

Moreover, when we use Eq. 89 for $|\psi(x=0,t)|^2$ the approximation Eq. 57 describes well the long-time behavior but breaks down for short times. It is slightly off close to the focus. Here, only Eq. 47 together with Eqs. 58 and 59 yields a good fit to the exact curve.

Appendix B: Wigner Dynamics of Rectangular Wave Packet

In this appendix we analyze the time evolution of the rectangular wave packet from the point of view of Wigner phase space. For this purpose we first rederive the Wigner function of the initial rectangular wave packet and then obtain by integration over the momentum variable the time-dependent probability density in position space. The resulting integral representation for the probability density at $x=0$ allows us to derive analytical approximations in three different time domains: (1) early and intermediate times, (2) around the focus, and (3) very long times. We conclude by presenting an exact expression as well as several approximations for the location of the focus.

B.1: Initial Wigner Function

We start by computing the Wigner function

$$W_0(x,p) \equiv \mathcal{N} \int_{-\infty}^{\infty} \mathrm{d}\xi \, \mathrm{e}^{ip\xi/\hbar} \psi_0\left(x-\frac{1}{2}\xi\right) \psi_0^*\left(x+\frac{1}{2}\xi\right) \tag{60}$$

with the normalization constant $\mathcal{N} \equiv 1/(2\pi\hbar)$ for the square packet given by the wave function

$$\psi_0(x) \equiv \frac{1}{\sqrt{L}} \Theta\left(\frac{1}{2}L - |x|\right). \tag{61}$$

When we substitute this expression into the definition of W_0, Eq. 60, we find

$$\begin{aligned} W_0(x,p) &= \frac{\mathcal{N}}{L} \int_{-\infty}^{\infty} \mathrm{d}\xi \, \mathrm{e}^{ip\xi/\hbar} \\ &\times \Theta\left(\frac{1}{2}L - \left|x - \frac{1}{2}\xi\right|\right) \Theta\left(\frac{1}{2}L - \left|x + \frac{1}{2}\xi\right|\right). \end{aligned} \tag{62}$$

The step functions imply the inequalities $-\frac{L}{2} < x - \frac{\xi}{2} < \frac{L}{2}$ and $-\frac{L}{2} < x + \frac{\xi}{2} < \frac{L}{2}$, which can be expressed as $-L+2x < \xi < L+2x$ and $-L-2x < \xi < L-2x$, leading us to the relation $-L+2|x| < \xi < L-2|x|$. These boundaries establish the limits of integration. Indeed, the integral vanishes for $2|x| - L \geq L - 2|x|$ giving rise to a factor $\Theta(L - 2|x|)$.

Hence, the Wigner function corresponding to the rectangular wave packet reads

$$W_0 = \frac{\mathcal{N}}{L} \int_{-L+2|x|}^{L-2|x|} \mathrm{d}\xi \, \mathrm{e}^{ip\xi/\hbar} \Theta(L - 2|x|), \tag{63}$$

which after integration takes the form

$$W_0(x,p) = \frac{1}{\pi L} \Theta\left(\frac{L}{2} - |x|\right) \frac{1}{p} \sin\left[(L - 2|x|)\frac{p}{\hbar}\right], \tag{64}$$

and is shown in Fig. 9.

B.2: Time-Dependent Probability Density

Next we derive the time-dependent probability density $|\psi(x,t)|^2$ by integrating the time-dependent Wigner function

$$W(x,p;t) \equiv W_0\left(x - \frac{p}{M}t, p\right) \tag{65}$$

over p, using the form

$$|\psi(x,t)|^2 = \frac{M}{t} \int_{-\infty}^{\infty} \mathrm{d}y \, W_0\left[y, \frac{M(x-y)}{t}\right]. \tag{66}$$

When we substitute the expression Eq. 64 for W_0 with the appropriate arguments into Eq. 66 we arrive at the integral

$$\begin{aligned} |\psi(x,t)|^2 &= \frac{1}{\pi L} \int_{-\infty}^{\infty} \mathrm{d}y \, \Theta\left(\frac{L}{2} - |y|\right) \\ &\quad \times \frac{\sin\left[(L - 2|y|)M(x-y)/(\hbar t)\right]}{x - y}, \end{aligned} \tag{67}$$

which reduces with the help of the Heaviside step function to

$$\begin{aligned} |\psi(x,t)|^2 &= \frac{1}{\pi L} \int_{-L/2}^{L/2} \mathrm{d}y \frac{1}{x-y} \\ &\quad \times \sin\left[\frac{ML^2}{\hbar t}\left(\frac{x-y}{L}\right)\left(1 - 2\left|\frac{y}{L}\right|\right)\right] \end{aligned} \tag{68}$$

We conclude by introducing the abbreviation

$$\tau(t) \equiv \frac{\hbar t}{ML^2} \tag{69}$$

and the change of variable $\eta \equiv 2y/L$ and arrive at the exact expression

$$\begin{aligned} |\psi(x,t)|^2 &= \frac{1}{\pi L}\int_{-1}^{1} \mathrm{d}\eta \frac{1}{2x/L-\eta} \\ &\times \sin\left[\frac{1}{2\tau(t)}\left(\frac{2x}{L}-\eta\right)(1-|\eta|)\right] \end{aligned} \tag{70}$$

for the time-dependent probability density.

B.3: Two Exact Expressions for $x = 0$

Next we evaluate the integral in Eq. 70 at $x = 0$ and note that the resulting integrand is symmetric with respect to $\eta = 0$. As a result, Eq. 70 takes the form

$$|\psi(0,t)|^2 = \frac{2}{\pi L}\int_0^1 \mathrm{d}\eta \frac{1}{\eta}\sin\left[\frac{\eta}{2\tau(t)}(1-\eta)\right]. \tag{71}$$

We can obtain an alternate form by the change of variables $\chi \equiv 4\eta(1-\eta)$ and χ covers the interval [0, 1] twice as η moves from 0 to 1. Indeed, for $0 < \eta < 1/2$ we have $\eta \equiv \frac{1}{2}(1-\sqrt{1-\chi})$, while for $1/2 < \eta < 1$ we have $\eta \equiv \frac{1}{2}(1+\sqrt{1-\chi})$ and the corresponding differentials are $d\chi = 4\sqrt{1-\chi}\mathrm{d}\eta$ and $d\chi = -4\sqrt{1-\chi}\,\mathrm{d}\eta$, respectively.

When we decompose the integral

$$I \equiv \int_0^1 \mathrm{d}\eta \frac{1}{\eta}\sin\left[\frac{\eta}{2\tau}(1-\eta)\right] \tag{72}$$

into these two domains we find

$$\begin{aligned} I &= \int_0^1 \mathrm{d}\chi \, \frac{\sin\left[\chi/(8\tau)\right]}{2\sqrt{1-\chi}(1-\sqrt{1-\chi})} \\ &- \int_1^0 \mathrm{d}\chi \, \frac{\sin\left[\chi/(8\tau)\right]}{2\sqrt{1-\chi}(1+\sqrt{1-\chi})}, \end{aligned} \tag{73}$$

or

$$I = \int_0^1 \mathrm{d}\chi \, \frac{\sin\left[\chi/(8\tau)\right]}{\chi\sqrt{1-\chi}}. \tag{74}$$

Hence, we obtain the alternate exact expression

$$|\psi(0,t)|^2 = \frac{2}{\pi L} \int_0^1 \mathrm{d}\chi \, \frac{\sin\left[\chi/(8\tau(t))\right]}{\chi\sqrt{1-\chi}}, \tag{75}$$

for the time-dependent probability density at $x = 0$.

B.4: Approximations

Now we analyze the integrals of Eqs. 71 and 75 in three different time regimes. Our main interest is to estimate the probability density around the time of focusing.

B.4.1: Early Times

In this domain we use the integral representation, Eq. 71, of $|\psi(0,t)|^2$ and note that for $t \to 0$, that is for $\tau \to 0$ two dominant terms emerge. Indeed, when we complete the square in the argument of the sine function we find

$$\eta(1-\eta) = \frac{1}{4} - \left(\eta - \frac{1}{2}\right)^2, \tag{76}$$

indicating a point of stationary phase at $\eta = 1/2$.

Moreover, due to the factor $1/\eta$ in the integrand another contribution arises. These observations suggest to decompose the integral into two parts: one around the origin and one containing the point of stationary phase, that is

$$\int_0^1 \mathrm{d}\eta f(\eta) = \int_0^a \mathrm{d}\eta f(\eta) + \int_a^1 \mathrm{d}\eta f(\eta) \tag{77}$$

with

$$f(\eta) \equiv \frac{1}{\eta} \sin\left[\frac{1}{2\tau}\eta(1-\eta)\right]. \tag{78}$$

Here a is a constant parameter such that $0 < a < 1/2$.

In the first integral, we approximate $\eta(1-\eta) \cong \eta$ and take the limit $\tau \to 0$. This procedure leads us to

$$\int_0^a \mathrm{d}\eta f(\eta) \cong \int_0^a \mathrm{d}\eta \, \frac{\sin(\eta/2\tau)}{\eta} \\ \cong \pi \int_0^a \mathrm{d}\eta \, \delta(\eta) \cong \frac{\pi}{2}, \tag{79}$$

where we have used the representation

$$\lim_{\epsilon \to 0} \frac{\sin(x/\epsilon)}{x} = \pi\delta(x) \tag{80}$$

of the Dirac delta function.

We evaluate the second integral in Eq. 77 with the help of the method of stationary phase [93] which yields

$$\int_a^1 \mathrm{d}\eta f(\eta) \cong 2\sqrt{2\pi\tau} \sin\left(\frac{1}{8\tau} - \frac{\pi}{4}\right). \tag{81}$$

Finally, we combine the two results and arrive at the approximate expression

$$|\psi(0,t)|^2 \cong \frac{1}{L}\left[1 + 4\sqrt{\frac{2\hbar t}{\pi M L^2}} \sin\left(\frac{ML^2}{8\hbar t} - \frac{\pi}{4}\right)\right], \tag{82}$$

which reproduces the initial value of the wave packet and yields a strongly oscillatory correction with a square root envelope. Indeed, these features stand out most clearly when we express this formula in terms of the characteristic time

$$T = \frac{ML^2}{2\pi\hbar} \tag{83}$$

which yields

$$|\psi(0,t)|^2 = \frac{1}{L}\left\{1 + \frac{4}{\pi}\sqrt{\frac{t}{T}} \sin\left[\frac{\pi}{4}\left(\frac{T}{t} - 1\right)\right]\right\}. \tag{84}$$

The simplicity of this asymptotic expression for $|\psi|^2$ clearly motivates the inclusion of the sometimes mysterious factor 2π in the definition of T.

Equation 82 is in perfect agreement with Eq. 58, that is with our expression of Eq. 14 for $x = 0$. Needless to say, for $t \to \infty$ this approximate formula, originally constructed for small times, breaks down. Indeed, we find

$$|\psi(0,t)|^2 \cong \frac{1}{L}\left[1 - \frac{4}{\pi}\sqrt{\frac{t}{T}} \sin\left(\frac{\pi}{4}\right)\right] \tag{85}$$

leading to unphysical negative probabilities.

B.4.2: Time of Maximum and Long-time Approximation

Next we use the representation of $|\psi(x=0,t)|^2$ given by Eq. 75 to obtain an estimate for long times, that is for $1 < \tau$, which captures the focusing effect as it manifests itself in a dominant maximum of the probability density. Since in this regime, the function $\sin(\chi/8\tau)$ does not display oscillations in the interval $0 < \chi < 1$, we may approximate it by a line secant to the curve $\sin(\chi/8\tau)/\chi$ at the end points, that is

$$\frac{1}{\chi}\sin\left(\frac{\chi}{8\tau}\right) \cong \frac{1}{8\tau} + \left[\sin\left(\frac{1}{8\tau}\right) - \frac{1}{8\tau}\right]\chi. \tag{86}$$

With this approximation and the integral relations

$$\int_0^1 \mathrm{d}\chi \frac{1}{\sqrt{1-\chi}} = 2 \tag{87}$$

and

$$\int_0^1 \mathrm{d}\chi \frac{\chi}{\sqrt{1-\chi}} = \frac{4}{3} \tag{88}$$

we can carry out the integration in Eq. 75 to obtain the formula

$$|\psi(0,t)|^2 \cong \frac{ML}{2\pi\hbar t} + \frac{8}{3\pi L}\left[\sin\left(\frac{ML^2}{8\hbar t}\right) - \frac{ML^2}{8\hbar t}\right], \tag{89}$$

or when expressed in terms of T

$$|\psi(0,t)|^2 \cong \frac{1}{L}\left\{\frac{T}{t} + \frac{8}{3\pi}\left[\sin\left(\frac{\pi}{4}\frac{T}{t}\right) - \frac{\pi}{4}\frac{T}{t}\right]\right\}, \tag{90}$$

This expression clearly shows the asymptotic behavior $T\,/\,t$, as well as the presence of a maximum due to the sign change of the second term.

B.4.3: Very Long Times

A rather straight-forward approximation valid for extremely long times results from a Taylor expansion of the integral representation, Eq. 66, of the probability density in $1\,/\,t$. Indeed, for $M(x-y)/t \cong 0$, we obtain

$$|\psi(x,t)|^2 \cong \frac{M}{t}\int_{-\infty}^{\infty} \mathrm{d}y\, W_0(y,0), \tag{91}$$

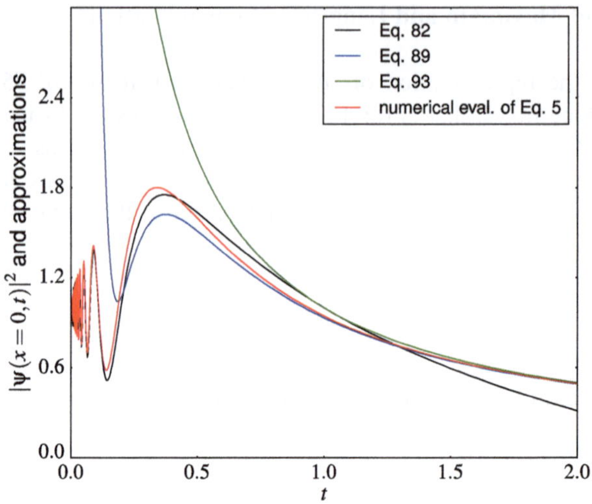

Fig. 14 Comparison between different approximations of the probability density $|\psi(x=0,t)|^2$. The *red line* represents the exact result obtained numerically from the Fresnel diffraction integral, Eq. 5, and the black curve shows the early-time estimate given by Eq. 82. In *blue* we depict the intermediate and long-time estimate defined by Eq. 89, and the *green curve* shows the $1\ /\ t$-behavior at very long times following from Eq. 93. The *different curves* show progressively the behavior of $|\psi(x=0,t)|^2$ as time increases, or equivalently, as one moves from the near-field to the far-field. Near-field: the square root envelope is visible and agrees perfectly with the exact result both in amplitude and in phase. Focusing: all approximations—except for the very long time curve in *green*—show a peak located close to the actual maximum. Far-field: all the approximations reproduce the asymptotics, except for the *black curve* at very long times

which together with the expression Eq. 64 of the Wigner function for $p=0$, that is with

$$W_0(x,0) = \frac{2}{\pi\hbar L}\Theta\left(\frac{1}{2}L - |x|\right)\left(\frac{1}{2}L - |x|\right) \tag{92}$$

yields after integration the approximate but analytical formula

$$|\psi(x,t)|^2 \cong \frac{ML}{2\pi\hbar t} = \frac{1}{L}\frac{T}{t}, \tag{93}$$

which is independent of the coordinate x. In particular, Eq. 93 is valid for $x=0$ and it coincides with the first term of our previous estimate, Eq. 89, as shown in Fig. 14 by the blue and green curves.

B.5: Location of the Focus

We are now in the position to determine the exact location of the dominant maximum of $|\psi(x=0,t)|^2$, that is of the focus by solving a transcendental equation. For this purpose we differentiate the exact expression for $|\psi(x=0,t)|^2$ in terms of the integral, Eq. 71, with respect to t which leads us to the equation

$$\int_0^1 \mathrm{d}\eta\,(1-\eta)\cos\left[\frac{\eta(1-\eta)}{2\tau}\right]=0, \tag{94}$$

or

$$\cos\left(\frac{1}{8\tau}\right)C\left(\frac{1}{2\sqrt{\pi\tau}}\right)+\sin\left(\frac{1}{8\tau}\right)S\left(\frac{1}{2\sqrt{\pi\tau}}\right)=0. \tag{95}$$

Here we have recalled the definitions Eqs. 8 and 9. When we solve Eq. 95 numerically we find $\tau_f \approx 0.054$.

A comparison between the definition, Eq. 83, of the characteristic time T with the definition, Eq. 69 of τ, yields the relation

$$\tau=\frac{1}{2\pi}\frac{t}{T}, \tag{96}$$

which translates into $t_f/T \equiv 2\pi\tau_f \cong 0.34$, in complete agreement with the analysis of Sect. 2, and in particular Figs. 1 and 2.

It is interesting that our estimates based on the various approximations for $|\psi(x=0,t)|^2$ discussed before are very close to this value. Indeed, our early-time approximation, Eq. 82, provides us with the condition

$$\cot\left(\frac{1}{8\tau}-\frac{\pi}{4}\right)=4\tau \tag{97}$$

leading to the value $\tau_e \approx 0.058$.

Remarkably, our long-time approximation, Eq. 89, which yields the transcendental equation

$$\cos\left(\frac{1}{8\tau}\right)+\frac{1}{2}=0 \tag{98}$$

with the solution $\tau_l \equiv 3/(16\pi) \approx 0.059$ is also accurate up to two digits.

Appendix C: Circular Apertures

Slits are frequently employed in diffraction problems because of the mathematical simplicity of the resulting diffraction integrals and the physical interpretation being limited to one space dimension only. However, circular apertures are often preferred due to their direct application in microscopy, photography and transmission between different regions of space. Indeed, early optics problems such as the *camera obscura* [3], the transmission by an aperture smaller than the wave length [19, 20, 22, 31], the Fourier optics of pupils and their optical transfer functions [94] are only but a few examples illustrating this point. In this Appendix, we first briefly review the history of diffraction from a circular aperture, and then discuss the similarities and differences between the diffractive focusing of scalar waves arising from slits and circular apertures.

C.1: Cusps and Multiple Diffraction

We start by recalling that the Poisson–Arago spot is the intensity map resulting from the illumination of a circular disk. Therefore, the field distribution is the complement of that generated by a circular aperture and can be calculated using the Babinet principle [95–98].

In 1922, Coulson and Becknell[7] have investigated experimentally a variant of the Poisson–Arago spot [101, 102] by measuring the intensity patterns generated by a disk rotated around its diagonal and illuminated by a point light source. The geometric shadow of this arrangement is an ellipse and the intensity pattern in the far-field is a diamond-like figure with four cusps being the evolute of the edge of the shadow. It is interesting that a few years earlier, Raman has also observed cusps in the diffraction patterns generated by an elliptical aperture [103].

Cusp are examples of catastrophe optics [99]. The interaction of light with a sharp-edge aperture was mathematically reformulated recently [104].

Finally, we turn to Letfullin and collaborators who have analyzed [105–109] theoretically and experimentally the phenomenon of multiple diffractive focusing, which occurs when we apply a second circular aperture of radius a_2 to the diffraction pattern generated by a first circular aperture of larger diameter a_1, placed in the symmetry axis. Provided the distance L between the second and the first apertures is such that the Fresnel number $N_1 \equiv a_1^2/(\lambda L)$ is an odd integer, the second diffraction pattern has a maximum that can reach 10.0 times the illuminating intensity of the first aperture. Based on this observation they proposed a lens for matter waves [110].

[7]Surprisingly, this pioneering work is neither referred in [99], nor in [100].

Slit and Circular Aperture: Paraxial Versus Non-paraxial Optics

In Sect. 4 we have studied the diffraction from a slit of width L illuminated by light of wave length λ using the Fresnel and the Rayleigh–Sommerfeld integrals. The evaluation of the RS-I integral requires the numerical integration of Hankel functions [77]. We now briefly summarize an analogous discussion for the case of a circular aperture of radius $a = L/2$.

In the Fresnel diffraction from a circular aperture we make use of Lommel functions in the calculation algorithm [77, 84–86]. For the non-paraxial calculation, the RS-I and RS-II integrals are evaluated numerically using the GNU Scientific Library (GSL) [111]. The results for the slit and the circular aperture are depicted in Figs. 11 and 12 in Sect. 6.

The fields and the corresponding normalized intensities on-axis based on the Rayleigh–Sommerfeld integrals for a circular aperture of radius a illuminated by a plane wave [43, 45, 77] read

$$\begin{cases} U_{\mathrm{RS-I}}(z) &= U_0\left(\mathrm{e}^{ikz} - \dfrac{z}{\sqrt{a^2+z^2}}\mathrm{e}^{ik\sqrt{a^2+z^2}}\right) \\ U_{\mathrm{RS-II}}(z) &= U_0\left(\mathrm{e}^{ikz} - \mathrm{e}^{ik\sqrt{a^2+z^2}}\right) \end{cases} \tag{99}$$

and

$$\begin{cases} \dfrac{I_{\mathrm{RS-I}}}{I_0} = 1 + \dfrac{z^2}{z^2+a^2} \\ \qquad -2\dfrac{z}{\sqrt{z^2+a^2}}\cos\left[k\left(\sqrt{a^2+z^2}-z\right)\right] \\ \dfrac{I_{\mathrm{RS-II}}}{I_0} = 2 - 2\cos\left[k\left(\sqrt{a^2+z^2}-z\right)\right]. \end{cases} \tag{100}$$

From these expressions,[8] especially the one for I_{RS-II}/I_0, several properties of the non-paraxial diffraction stand out: the number of maxima, or minima of each intensity function is finite. Indeed, the maxima follow from the condition

$$\cos\left[k(\sqrt{a^2+z^2}-z)\right] = -1, \tag{101}$$

or equivalently, $k(\sqrt{a^2+z^2}-z) = (2m+1)\pi$, with $m = 0, 1, 2, \cdots$ The minima occur when

$$k(\sqrt{a^2+z^2}-z) = 2n\pi, \tag{102}$$

with $n = 1, 2, \cdots$

The argument of the cosine function never vanishes, except for the trivial case of $a = 0$, or in the asymptotic limit $z \to \infty$. On the other hand, the intensity at the

[8]We emphasize that the z-values of the maxima and the minima coincide for I_{RS-I} and I_{RS-II}.

origin ($z = 0$) oscillates between 0 and 4.0, depending only on a. In contrast, the intensity $I_{\mathrm{RS-I}}/I_0$ always converges to 1.0.

The representation of the on-axis intensities $I_{\mathrm{RS-I}}$ and $I_{\mathrm{RS-II}}$ using a logarithmic scale for z permits us to find another interesting property of the non-paraxial diffraction: the separation between maxima, or minima along the $\log(z)$-axis is constant. This property is due to the argument of cosine function.

We note that Forbes has investigated the scaling properties of the Fresnel diffraction patterns for circular apertures [92], but he did not mention this feature as it arises only in non-paraxial diffraction.

The Fresnel diffraction intensity on-axis of a circular aperture reads

$$I_{\mathrm{Fc}}(0,0,z) = 2 - 2\cos\left(\frac{ka^2}{2z}\right), \tag{103}$$

with the argument of the cosine-function being inversely proportional to z. Thus, the number of maxima and minima is infinite.

We can find the location $z_{\mathrm{m}} = a/\lambda$ of the last maximum from Eq. 103. If we normalize this length by L^2/λ we obtain

$$z_f^{(2)}/(L^2/\lambda) = 0.25, \tag{104}$$

that is the position of the focus.

In the case of the slit no analytic expressions are known for the non-paraxial intensity, but the graphical representations of the RS-I intensities in Fig. 11 show that the number of lobes in the intensity map of a slit of width $L = 2a$ is equal to that for a circular aperture of the same diameter. It depends only on the ratio L/λ.

However, the maxima are differently distributed along the z-axis, and the dominant lobe containing the focusing maximum is much more elongated than that of the circular aperture.

The Fresnel intensity on-axis is approximated by

$$I_{\mathrm{Fs}}(0,0,z) \approx 1 - \frac{2\sqrt{z}}{\pi L}\cos\left(\frac{\pi}{4} + \frac{kL^2}{8z}\right), \tag{105}$$

which shows why the position of the focus is shifted towards larger z values, compared to the circular aperture. After the normalization by L^2/λ we find

$$z_f^{(1)}/(L^2/\lambda) = 0.33 \tag{106}$$

Moreover, the maxima of the Fresnel intensity decay in amplitude converging to 1.0 for $z \to 0$, unlike in the circular aperture.

References

1. A.-J. Fresnel, Mémoire sur la Diffraction de la Lumière, où l'on examine particulièrement le phénomène des franges colorées que présentent les ombres des corps éclairés par un point lumineux. Annal. Chim. Phys. 2nd Ser. **1**, 239–281 (1816)
2. A.-J. Fresnel, *Mémoire Sur la Diffraction de la Lumière*. Crochard (1819)
3. J. Petzval, Bericht über dioptrische Untersuchungen. Sitzungsberichte der Kaiserlichen Akademie der Wissenschaften. Mathematisch-Naturwissenschaftliche, Cl. XXVI. Bd. I. Hft. 3 (1857)
4. R. Lord, On the passage of waves through apertures in plane screens, and allied problems. Philos. Mag. Ser. 5 **43**(263), 259–272 (1897)
5. W.B. Case, E. Sadurni, W.P. Schleich, A diffractive mechanism of focusing. Opt. Express **20** (25), 27253–27262 (2012)
6. G. Vitrant, S. Zaiba, B.Y. Vineeth, T. Kouriba, O. Ziane, O. Stéphan, J. Bosson, P.L. Baldeck, Obstructive micro diffracting structures as an alternative to plasmonics nano slits for making efficient microlenses. Opt. Express **20**(24), 26542–26547 (2012)
7. D. Weisman, S. Fu, M. Gonçalves, L. Shemer, J. Zhou, W. P. Schleich, A. Arie, Diffractive focusing of waves in time and in space. Phys. Rev. Lett. (2017)
8. M. Moshinsky, Diffraction in time. Phys. Rev. **88**, 625–631 (1952)
9. T. Young, The Bakerian lecture: experiments and calculations relative to physical optics. Phil. Trans. R. Soc. Lond. **94**, 1–16 (1804)
10. J.C. Maxwell, A dynamical theory of the electromagnetic field. Phil. Trans. R. Soc. Lond. **155**, 459–512 (1865)
11. G. Kirchhoff, Zur Theorie der Lichtstrahlen. Ann. Phys. **254**(4), 663–695 (1883)
12. Lord Rayleigh, On the passage of waves through fine slits in thin opaque screens. Proc. R. Soc. A **89**(609), 194 (1913)
13. A. Sommerfeld, F. Bopp, J. Meixner, in *Vorlesungen über Theoretische Physik, Bd.4, Optik*, 3rd edn. Harri Deutsch Verlag (1989)
14. F. Kottler, Zur Theorie der Beugung an schwarzen Schirmen. Ann. Phys. **375**(6), 405–456 (1923)
15. F. Kottler, Diffraction at a black screen: part I: Kirchhoff's theory. Prog. Opt. **4**, 281–314 (1965)
16. F. Kottler, Elektromagnetische Theorie der Beugung an schwarzen Schirmen. Ann. Phys. **376** (15), 457–508 (1923)
17. F. Kottler, Diffraction at a black screen: part II: electromagnetic theory. Prog. Opt. **6**, 331–377 (1967)
18. J.A. Stratton, L.J. Chu, Diffraction theory of electromagnetic waves. Phys. Rev. **56**, 99–107 (1939)
19. H.A. Bethe, Theory of diffraction by small apertures. Phys. Rev. **66**, 163 (1944)
20. C.J. Bouwkamp, Diffraction theory. Rep. Prog. Phys. **17**(1), 35 (1954)
21. D.W. Pohl, W. Denk, M. Lanz, Optical stethoscopy: image recording with resolution $\lambda/20$. Appl. Phys. Lett. **44**, 651 (1984)
22. Y. Leviatan, Study of near-zone fields of a small aperture. J. Appl. Phys. **60**, 1577 (1986)
23. D. Courjon, *Near-Field Microscopy and Near-Field Optics* (World Scientific, Singapore, 2003)
24. L. Novotny, *Progress in Optics, Chapter 5—The History of Near-Field Optics*, vol. 50 (Elsevier, Amsterdam, 2007)
25. T.W. Ebbesen, H.J. Lezec, H.F. Ghaemi, T. Thio, P.A. Wolff, Extraordinary optical transmission through sub-wavelength hole arrays. Nature **391**, 667–669 (1998)
26. H.J. Lezec, A. Degiron, E. Devaux, R.A. Linke, L. Martin-Moreno, F.J. Garcia-Vidal, T.W. Ebbesen, Beaming light from a subwavelength aperture. Science **297**, 820–822 (2002)
27. F.J. Garcia-Vidal, L. Martin-Moreno, T.W. Ebbesen, L. Kuipers, Light passing through subwavelength apertures. Rev. Mod. Phys. **82**, 729–787 (2010)

28. Q. Cao, P. Lalanne, Negative role of surface plasmons in the transmission of metallic gratings with very narrow slits. Phys. Rev. Lett. **88**, 057403 (2002)
29. F.J. García-Vidal, L. Martlín-Moreno, H.J. Lezec, T.W. Ebbesen, Focusing light with a single subwavelength aperture flanked by surface corrugations. Appl. Phys. Lett. **83**, 4500–4502 (2003)
30. M.W. Kowarz, Homogeneous and evanescent contributions in scalar near-field diffraction. Appl. Opt. **34**(17), 3055–3063 (1995)
31. V.V. Klimov, V.S. Letokhov, A simple theory of the near field in diffraction by a round aperture. Opt. Commun. **106**(4–6), 151–154 (1994)
32. M.V. Berry, N.L. Balazs, Nonspreading wave packets. Am. J. Phys **47**(3), 264–267 (1979)
33. J. Durnin, J.J. Miceli, J.H. Eberly, Diffraction-free beams. Phys. Rev. Lett. **58**, 1499–1501 (1987)
34. T.R.M. Sales, Smallest focal spot. Phys. Rev. Lett. **81**, 3844–3847 (1998)
35. S. Quabis, R. Dorn, M. Eberler, O. Glöckl, G. Leuchs, Focusing light to a tighter spot. Opt. Commun. **179**(1–6), 1–7 (2000)
36. T. Grosjean, D. Courjon, Smallest focal spots. Opt. Commun. **272**(2), 314–319 (2006)
37. B.L. Schumaker, Quantum mechanical pure states with Gaussian wave functions. Phys. Rep. **135**(6), 317–408 (1985)
38. K. Vogel, F. Gleisberg, N.L. Harshman, P. Kazemi, R. Mack, L. Plimak, W.P. Schleich, Optimally focusing wave packets. Chem. Phys. **375**(2–3), 133–143 (2010)
39. V.V. Dodonov, Rotating quantum Gaussian packets. J. Phys. A: Math. Theor. **48**(43), 435303 (2015)
40. V.I. Balykin, V.G. Minogin, Focusing of atomic beams by near-field atom microlenses: the Bethe-type and the Fresnel-type microlenses. Phys. Rev. A **77**, 013601 (2008)
41. A.D. Cronin, J. Schmiedmayer, D.E. Pritchard, Optics and interferometry with atoms and molecules. Rev. Mod. Phys. **81**, 1051–1129 (2009)
42. S.A. Akhmanov, A.P. Sukhorukov, R.V. Khokhlov, Self-focusing and diffraction of light in a nonlinear medium. Sov. Phys. Uspekhi **10**(5), 609 (1968)
43. A. Dubra, J.A. Ferrari, Diffracted field by an arbitrary aperture. Am. J. Phys **67**(1), 87–92 (1999)
44. G.D. Gillen, S. Guha, Modeling and propagation of near-field diffraction patterns: a more complete approach. Am. J. Phys **72**, 1195 (2004)
45. H. Osterberg, L.W. Smith, Closed solutions of Rayleigh's diffraction integral for axial points. J. Opt. Soc. Am. **51**(10), 1050–1054 (1961)
46. S. Guha, G. Gillen, Description of light propagation through a circular aperture using nonparaxial vector diffraction theory. Opt. Express **13**(5), 1424–1447 (2005)
47. M. Moshinsky, Diffraction in time and the time-energy uncertainty relation. Am. J. Phys **44** (11), 1037–1042 (1976)
48. S. Godoy, Diffraction in time: Fraunhofer and Fresnel dispersion by a slit. Phys. Rev. A **65**, 042111 (2002)
49. C. Brukner, A. Zeilinger, Diffraction of matter waves in space and in time. Phys. Rev. A **56**, 3804–3824 (1997)
50. A. del Campo, G. García-Calderón, J.G. Muga, Quantum transients. Phys. Rep. **476**(1–3), 1–50 (2009)
51. A. Goussev, Diffraction in time: an exactly solvable model. Phys. Rev. A **87**, 053621 (2013)
52. E. Wigner, On the quantum correction for thermodynamic equilibrium. Phys. Rev. **40**, 749–759 (1932)
53. M.J. Bastiaans, Wigner distribution function and its application to first-order optics. J. Opt. Soc. Am. **69**(12), 1710–1716 (1979)
54. C.J. Román-Moreno, R. Ortega-Martínez, C. Flores-Arvizo, The Wigner function in paraxial optics II. Optical difffraction pattern representation. Revista Mexicana de Física **49**(3), 290–295 (2003)
55. A. Torre, *Linear Ray and Wave Optics in Phase Space* (Elsevier, Amsterdam, 2005)

56. M. Testorf, B. Hennelly, J. Ojeda-Castaneda, *Phase-Space Optics: Fundamentals and Applications* (McGraw-Hill Education, New York, 2009)
57. M.A. Alonso, Wigner functions in optics: describing beams as ray bundles and pulses as particle ensembles. Adv. Opt. Photon. **3**(4), 272–365 (2011)
58. G.A. Baker, Formulation of quantum mechanics based on the quasi-probability distribution induced on phase space. Phys. Rev. **109**, 2198–2206 (1958)
59. G.A. Baker, I.E. McCarthy, C.E. Porter, Application of the phase space quasi-probability distribution to the nuclear shell model. Phys. Rev. **120**, 254 (1960)
60. R.G. Littlejohn, The semiclassical evolution of wave packets. Phys. Rep. **138**(4–5), 193–291 (1986)
61. V. Man'ko, M. Moshinsky, A. Sharma, Diffraction in time in terms of Wigner distributions and tomographic probabilities. Phys. Rev. A **59**, 1809–1815 (1999)
62. W.P. Schleich, *Quantum Optics in Phase Space* (Wiley-VCH, Weinheim, 2001)
63. W.B. Case, Wigner functions and Weyl transforms for pedestrians. Am. J. Phys. **76**(10), 937 (2008)
64. W.P. Schleich, J.P. Dahl, S. Varró, Wigner function for a free particle in two dimensions: a tale of interference. Opt. Commun. **283**(5), 786–789 (2010)
65. ThL Curtright, D.B. Fairlie, C.K. Zachos, *A Concise Treatise on Quantum Mechanics in Phase Space* (World Scientific, Singapore, 2014)
66. L. Cohen, Time-frequency distributions—a review. Proc. IEEE **77**(7), 941–981 (1989)
67. I. Białynicki-Birula, M.A. Cirone, J.P. Dahl, M. Fedorov, W.P. Schleich, In- and outbound spreading of a free-particle s-wave. Phys. Rev. Lett. **89**, 060404 (2002)
68. M.A. Cirone, J.P. Dahl, M. Fedorov, D. Greenberger, W.P. Schleich, Huygens' principle, the free Schrödinger particle and the quantum anti-centrifugal force. J. Phys. B: Atomic Mol. Opt. Phys. **35**(1), 191 (2002)
69. M.A. Andreata, V.V. Dodonov, On shrinking and expansion of radial wave packets. J. Phys. A: Math. Gen. **36**(25), 7113 (2003)
70. E. Granot, A. Marchewka, Generic short-time propagation of sharp-boundaries wave packets. Europhys. Lett. **72**(3), 341–347 (2005)
71. A. del Campo, J.G. Muga, M. Moshinsky, Time modulation of atom sources. J. Phys. B: Atomic Mol. Opt. Phys. **40**(5), 975 (2007)
72. A. del Campo, J.G. Muga, M. Kleber, Quantum matter-wave dynamics with moving mirrors. Phys. Rev. A **77**, 013608 (2008)
73. A. Goussev, Manipulating quantum wave packets via time-dependent absorption. Phys. Rev. A **91**, 043638 (2015)
74. R.P. Feynman, A.R. Hibbs, *Quantum Mechanics and Path Integrals: Emended Edition* (Dover Publications, New York 2010)
75. R.W. Wood, *Physical Optics* (Macmillan Company, New York, 1911)
76. M. Abramowitz, I.A. Stegun (eds.), *Handbook of Mathematical Functions with Formulas, Graphs, and Mathematical Tables* (Dover Publications, New York, 1965)
77. K. Mielenz, Optical diffraction in close proximity to plane apertues: I. Boundary-value solutions for circular apertures and slits. J. Res. Natl. Inst. Stand. Technol. **107**(4), 355–362 (2002)
78. K. Mielenz, Computation of Fresnel integrals. J. Res. Nat. Inst. Stand. Technol. **102**(3), 363 (1997)
79. K. Mielenz, Algorithms for Fresnel diffraction at rectangular and circular apertures. J. Res. Natl. Inst. Stand. Technol. **103**(5), 497–509 (1998)
80. K. Mielenz, Computation of Fresnel integrals. II. J. Res. Nat. Inst. Stand. Technol. **105**(4), 589 (2000)
81. K. Mielenz, Optical diffraction in close proximity to plane apertures. II. Comparison of half-plane diffraction theories. J. Res. Natl. Inst. Stand. Technol. **108**(1), 57–68 (2003)
82. K. Mielenz, Optical diffraction in close proximity to plane apertures. III. Modified, self-consistent theory. J. Res. Natl. Inst. Stand. Technol. **109**(5), 457–464 (2004)

83. K. Mielenz, Optical diffraction in close proximity to plane apertures. IV. Test of a pseudo-vectorial theory. J. Res. Nat. Inst. Stand. Technol. **111**(1), 1–8 (2006)
84. E. Lommel, Die Beugungserscheinungen einer kreisrunden Oeffnung und eines kreisrunden Schirmchens. Abh. Bayer. Akad. Math. Naturwiss. XV. Bd. II. Abth. **31**, 233–328 (1885)
85. M. Born, E. Wolf, *Principles of Optics*, 7th edn. (Cambridge University Press, Cambridge, 1999)
86. C.J. Daly, T.W. Nuteson, N.A.H.K. Rao, The spatially averaged electric field in the near field and far field of a circular aperture. IEEE Trans. Antennas Propag. **51**(4), 700 (2003)
87. E. Sejdić, I. Djurović, J. Jiang, Time-frequency feature representation using energy concentration: an overview of recent advances. Digital Signal Process. **19**(1), 153–183 (2009)
88. N.L. Balazs, B.K. Jennings, Wigner's function and other distribution functions in mock phase spaces. Phys. Rep. **104**(6), 347–391 (1984)
89. H.-W. Lee, Theory and application of the quantum phase-space distribution functions. Phys. Rep. **259**(3), 147–211 (1995)
90. R. Mack, V.P. Yakovlev, W.P. Schleich, Correlations in phase space and the creation of focusing wave packets. J. Mod. Opt. **57**(14–15), 1437–1444 (2010)
91. J.W. Bruce, P.J. Giblin, *Curves and Singularities*, 2nd edn. (Cambridge University Press, Cambridge, 1992)
92. G.W. Forbes, Scaling properties in the diffraction of focused waves and an application to scanning beams. Am. J. Phys **62**(5), 434–443 (1994)
93. R. Courant, D. Hilbert, *Methode der mathematischen Physik*, 4th edn. (Springer, Berlin, 1993)
94. J.W. Goodman, *Introduction to Fourier Optics*, 2nd edn. (McGraw-Hill, Singapore, 1996)
95. J.E. Harvey, J.L. Forgham, The spot of Arago: new relevance for an old phenomenon. Am. J. Phys **52**, 243 (1984)
96. R.L. Lucke, Rayleigh–Sommerfeld diffraction and Poisson's spot. Eur. J. Phys. **27**(2), 193 (2006)
97. W.R. Kelly, E.L. Shirley, A.L. Migdall, S.V. Polyakov, K. Hendrix, First- and second-order Poisson spots. Am. J. Phys **77**(8), 713–720 (2009)
98. M. Gondran, A. Gondran, Energy flow lines and the spot of Poisson–Arago. Am. J. Phys. **78** (6), 598 (2010)
99. M.V. Berry, C. Upstill, *Progress in Optics, Chapter IV Catastrophe Optics: Morphologies of Caustics and their Diffraction Patterns*, vol. 18 (Elsevier, Amsterdam, 1980)
100. J.F. Nye, *Natural Focusing and Fine Structure of Light*, 1st edn. (IOP Publishing, Bristol, 1999)
101. J. Coulson, G.G. Becknell, Reciprocal diffraction relations between circular and elliptical plates. Phys. Rev. **20**, 594 (1922)
102. G.G. Becknell, J. Coulson, An extension of the principle of the diffraction evolute, and some of its structural detail. Phys. Rev. **20**, 607 (1922)
103. C.V. Raman, On the diffraction-figures due to an elliptic aperture. Phys. Rev. **13**, 259–260 (1919)
104. R. Borghi, Catastrophe optics of sharp-edge difffraction. Opt. Lett. **41**(13), 3114–3117 (2016)
105. R.R. Letfullin, T.F. George, Optical effect of diffractive multifocal focusing of radiation on a bicomponent diffraction system. Appl. Opt. **39**(16), 2545–2550 (2000)
106. R.R. Letfullin, O.A. Zayakin, Observation of diffraction multifocal radiation focusing. Quantum Electron. **31**(4), 339–342 (2001)
107. R.R. Letfullin, O.A. Zayakin, Diffractive focusing of a Gaussian beam. J. Russ. Laser Res. **23** (2), 148 (2002)
108. J.T. Foley, R.R. Letfullin, H.F. Arnoldus, T. George, The diffractive multifocal focusing effect and the phase of the optical field. Int. J. Theor. Phys. Group Theory Nonlinear Opt. **11** (3), 149–163 (2004)

109. J.R. Foley, R.R. Letfullin, H.F. Arnoldus, *Tribute to Emil Wolf: Science and Engineering Legacy of Physical Optics, Chapter 14—The Diffractive Multifocal Focusing Effect* (SPIE Press, Washington, DC, 2004)
110. R.R. Letfullin, T.F. George, A. Siahmakoun, M.F. McInerney, De Broglie-wave lens. Opt. Eng. **47**(2), 028001 (2008)
111. M. Galassi, J. Davies, J. Theiler, B. Gough, G. Jungman, P. Alken, M. Booth, F. Rossi, R. Ulerich. *GNU Scientific Library Reference Manual*, 2.3 edn. for GSL Version 2.3 (2016). https://www.gnu.org/software/gsl/

Nonadiabatic Molecular Alignment of Linear Molecules Probed by Strong-Field Ionization Yields of Photoelectrons

G. Kaya, N. Kaya, J. Strohaber, N.A. Hart, A.A. Kolomenskii, and H.A. Schuessler

Abstract The dynamics of rotational wave packets of laser-aligned linear molecules were studied with femtosecond laser-driven strong-field ionization (SFI). The dynamics were observed as a function of the delay between a femtosecond probe pulse and a linearly polarized aligning pump pulse. The induced nonadiabatic molecular alignment was directly monitored by the total SFI yield. The measured revival signatures were compared to the calculated degree of molecular alignment taking into account the effects of electronic structure and symmetry of the molecules. By fitting the calculated alignment parameter to the measured experimental data, we also determined the molecular rotational constants of N_2, CO, O_2, and C_2H_2 gas molecules.

1 Introduction

When a molecule is exposed to a strong laser pulse, the interaction creates an induced dipole moment due to the anisotropic polarizability of the molecule, which tends to align with its axis of highest polarizability along the laser polarization

This article is part of the topical collection "Enlightening the World with the Laser" - Honoring T. W. Hänsch guest edited by Tilman Esslinger, Nathalie Picqué, and Thomas Udem.

G. Kaya
Department of Physics, Texas A&M University, College Station, TX 77843, USA

Science Program, Texas A&M University at Qatar, Doha 23874, Qatar

N. Kaya
Science Program, Texas A&M University at Qatar, Doha 23874, Qatar

Department of Physics, Giresun University, Giresun 28200, Turkey

J. Strohaber
Department of Physics, Florida A&M University, Tallahassee, FL 32307, USA

N.A. Hart • A.A. Kolomenskii • H.A. Schuessler (✉)
Department of Physics, Texas A&M University, College Station, TX 77843, USA
e-mail: schuessler@physics.tamu.edu

D. Meschede et al. (eds.), *Exploring the World with the Laser*,
https://doi.org/10.1007/978-3-319-64346-5_31

direction [1]. If the laser pulse is short, i.e., its duration is less than the characteristic molecular rotational period, the laser pulse cannot align the molecule adiabatically, and the resulting alignment process is called nonadiabatic or field-free alignment [2, 3]. In nonadiabatic alignment, the short laser pulse creates in each molecule a rotational wave packet, i.e., a coherent superposition of the rotational states, which will dephase and rephase in time after the laser pulse has ended. After the impulsive interaction of molecules with the laser pulse, the absence of the aligning field prevents a distortion of the molecular rotational dynamics caused by the field, thus giving this advantage for the dynamics observation compared to adiabatic alignment [4–6]. Since the pioneering works on the theory [7, 8] and the first experimental demonstration of nonadiabatic molecular alignment [9], this phenomenon continues attracting the interest of researchers (see [10, 11]). A variety of applications of the molecular alignment was proposed, such as controlling chemical reactions [12], selectively controlled alignment of isotopes [13], imaging of molecular structures [14], nanolithography with molecular beams [15], pulse compression [16], and quantum information processing [17]. Femtosecond laser technology allows aligning molecules and observing the molecular dynamics by using a pump–probe technique [18]. Alignment initiated by a pump pulse creates a dynamically anisotropic medium evolving in time, and hence, concurrent variations of the interaction with such a medium of a probe pulse can be observed by ionization [19–21], fragmentation [22–26], high harmonic generation [14, 26–29] or by spatial and spectral modulation of the probe pulse [30–35].

It should be noted that rotational coherence spectroscopy (RCS) [36], which is based on quantum beats that arise from the coherent excitation of rotational levels, offered for the spectroscopy of the rotational states a rich variety of detection schemes, including a pump–probe photoionization approach. Even earlier [37], transient birefringence resulting from the time evolution of a coherent superposition of laser-induced rotational states was used for their observation. To measure the degree of alignment, a weak field polarization technique that employs Raman scattering was proposed [38]. The depolarization of the probe pulse induced by the transient birefringence due to field-free alignment was also employed for monitoring rotational dynamics [39]. In another approach, the coherences in the molecular rotational dynamics initiated by the nonresonant dynamic Stark effect and by stimulated Raman scattering were detected via refractive index modulation, resulting in the Stokes and anti-Stokes sidebands of the scattered probe pulse [40].

By using the combined effect of the laser field and a static electric extraction field, the orientation effect of quantum-state-selected polar molecules was achieved and was assessed by the probe-induced angular distribution of ions in Coulomb explosion [41] or of photoelectrons due to tunneling ionization [42]. A nonadiabatic field-free orientation of a heteronuclear diatomic molecule by femtosecond two-color laser fields was reported in Ref. [43]. The procedure to quantitatively extract the degree of alignment from measured strong-field ionization (SFI) yields was developed in [44]. It uses above threshold ionization (ATI) photoelectron spectroscopy and photoelectron–photofragment coincidences [45, 46]. The isotopic rotational constants and fragmentation channels were determined using correlated measurement of rotational and mass or photoelectron spectra by combining Fourier transform rotational coherence spectroscopy with resonance-enhanced multiphoton ionization [47].

The degree of alignment can be improved by reducing the temperature of the gas [7, 48, 49] and also by selecting single states with state separators. A combination of an electrostatic field, nonresonant femtosecond laser excitation, and the preparation of state-selected molecules using a hexapole was implemented [50] to improve the degree of alignment up to $\langle\cos^2\theta\rangle = 0.82$. The degree of the alignment can be further improved by using supersonic beams of polar molecules that are deflected using inhomogeneous electric fields. The quantum-state selectivity of the deflection is used to spatially separate molecules according to their quantum state [51], achieving a very high degree of alignment $\langle\cos^2\theta\rangle = 0.972$. With a Stark-selected beam of ground-state OCS molecules in an approach that uses a quantum-state-dependent deflection in an inhomogeneous electrostatic dipole field, also high alignment degrees were reached, $\langle\cos^2\theta\rangle = 0.84$ [52]. A combination of strong inhomogeneous static electric field and the ac field from the laser pulse are used for separation of polar molecules in a supersonic beam according to their quantum states to demonstrate an unprecedented degree of laser-induced one-dimensional alignment $\langle\cos^2\theta\rangle = 0.97$ and strong orientation of state-selected iodobenzene molecules with an estimated temperature of 1.0 K. [41].

The molecular ionization rate is often maximal, when the maximum of the initial electron density distribution corresponds to the laser polarization direction of the applied field. However, not all diatomic molecules have their HOMO, which donates electrons most readily, aligned along the molecular axis. For example, for an π electron configuration, it is preferentially lined up in the direction perpendicular to the molecular axis with no electron density along the internuclear axis. This is unlike the σ electron configuration, for which the maxima of the electron density are preferentially aligned in the direction parallel to the molecular (internuclear) axis due to the shared electron density distribution directly between the bonding atoms. Therefore, depending on the configuration of the highest occupied molecular orbital (HOMO), qualitatively different alignment signatures can be expected. The contributions to total ionization yield not only from HOMO but also from energetically lower orbitals such as HOMO-1 and HOMO-2 were investigated in high harmonic generation studies [53, 54]. The techniques to resolve SFI channels for polyatomic molecules were developed in Refs. [45, 46].

Our goal is to investigate under the same experimental conditions the nonadiabatic alignment process initiated with femtosecond pump pulses of linear molecules N_2, CO, O_2, and C_2H_2 by detecting SFI yields of photoelectrons produced by the probe pulses. Previously, the nonadiabatic alignment in CO [55] and C_2H_2 [56] was studied with a pump–probe scheme by detecting produced ions. The choice of the molecules in our study was determined by their variety of configurations of HOMOs. The measured molecular alignment is compared to the calculated time dependence of the molecular alignment parameter $\langle\langle\cos^2\theta\rangle\rangle$, and the effects of the electronic structure and symmetry of molecules on the signatures of alignment revivals are investigated.

2 Experimental Setup

The pump–probe setup is depicted in Fig. 1. A mode-locked Ti/sapphire laser oscillator provides ~20 fs pulses at a repetition rate of 80 MHz. These pulses are seeded into a regenerative laser amplifier, which outputs ~50 fs laser pulses at a repetition rate of 1 kHz with a central wavelength of 800 nm. Since shorter pulses have a higher peak intensity for a given pulse energy, temporal compression of the laser pulses in the focus was achieved by maximizing the integrated SFI yield, while adjusting the grating compressor in the laser amplifier. The maximum pulse energy from the amplified laser system was ~1 mJ. The linearly polarized incident light was split by a beam splitter into two beams with about equal intensities to serve as pump and probe pulses. A half-wave plate and a polarizer were placed into the pump arm assuring proper linear polarization and for adjusting the intensity of the pump pulse.

In the experiments, different gases (N_2, CO, O_2, or C_2H_2 from Matheson TRIGAS, with purities of 99.9995, 99.9, 99.997, and 99.6%, respectively) were admitted to the chamber through an adjustable leak valve. The pressure in the vacuum chamber for the various gases studied was adjusted to 6×10^{-6} mbar, which was about three orders of magnitude higher than the background pressure. The photoelectron signal from aligned molecules was measured as a function of the time delay between the linearly polarized pump and probe laser pulses. This time delay was precisely adjusted using an optical delay line with a translational stage controlled by a stepping motor (GTS150, ESP300, Newport), which provided a high temporal resolution of ~0.67 fs. Another beam splitter was used to recombine the two beams, and the laser beams were focused in a vacuum chamber by a 20-cm achromatic lens. The pump pulse created a rotational wave packet, while the delayed probe pulse ionized molecules in the focus producing electrons. During the experiment, the pump and probe beams were horizontally polarized along the axis connecting the centers of the left and right microchannel plate (MCP) detectors. The ionized electrons were ejected by the probing laser field toward the MCP detectors. These electrons traveled within a μ-metal TOF tube in a magnetic field-free region. The signals from the MCP detectors were amplified by a preamplifier (ZKL-2 Mini-Circuits), and a FAST ComTech MCS6 multiple-event time digitizer with 100-ps time resolution was used for data acquisition. Laser pulses were

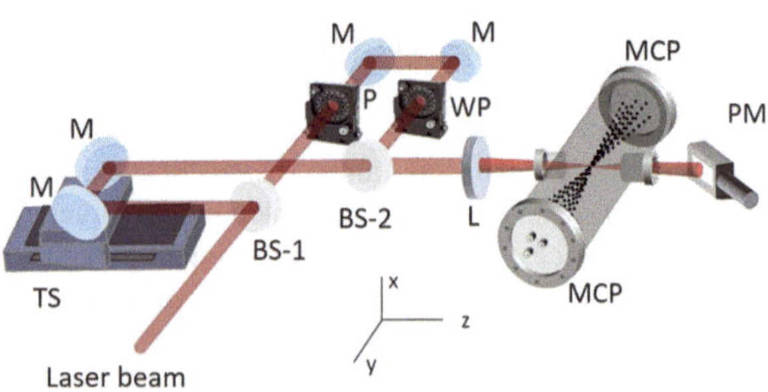

Fig. 1 Experimental setup. *BS* beam splitters, *TS* translational stage, *WP* wave plate, *M* flat mirrors, *L* achromatic focusing lens, *PM* power meter, *MCP* microchannel plates. The *dots* depict electrons flying toward the MCP detectors

detected behind one of the mirrors on the probe arm by a photodiode, and this signal was used to trigger the data acquisition software. The total number of counts, depending on the pump–probe delay, was registered as the output signal. We note that this setup was also used for ATI experiments; in particular, the intensity measurements were referenced to the well-defined kink in the ATI spectra of Xe at two times the ponderomotive energy, which was measured with the same setup [57]. The pulse duration was measured using a frequency-resolved optical gating autocorrelator (Grenouille 8–20, Swamp Optics).

When using linearly polarized pump pulses, the photoelectron momentum distribution showed a sharp maximum in the direction of the field polarization [58], and direct electrons have a narrower distribution than re-scattered electrons [59]. In ATI spectra, the total number of the direct electrons (typically with energies up to $2Up$, where Up is the ponderomotive energy of an electron in the oscillating field) is significantly larger than that of the re-scattered electrons in the plateau region, which have larger energies, namely from $2Up$ to$10Up$ [60, 61]. The solid angle of our MCP detector is relatively small, and therefore mostly direct electrons emitted close to the probe field polarization direction are detected.

3 Theory

The induced dipole potential of a linear molecule in a linearly polarized field along the Z-axis is $V_L(t) = -1/2[\Delta\alpha \cos^2\theta + \alpha_\perp]E^2(t)$ [62–64]. The interaction of a linearly polarized pump pulse with linear molecules can be described by the effective Hamiltonian $H = H_0 - 1/2(\Delta\alpha \cos^2\theta + \alpha_\perp)E^2(t)$, where $H_0 = hBc\hat{J}^2$ is the field-free Hamiltonian with the rotational molecular constant B and the angular momentum operator $\hat{J}$, θ is the angle between the molecular axis and the field polarization of the aligning pulse, $E(t)$ is the time-dependent electric field of the aligning pulse, and $\Delta\alpha = \alpha_\parallel - \alpha_\perp$ is the polarizability difference between the components parallel $\alpha_\parallel$ and perpendicular $\alpha_\perp$ to the molecular axis. Centrifugal distortion effects are not taken into account in our calculations.

The degree of molecular alignment is characterized by the quantity $\langle\langle \cos^2\theta\rangle\rangle$ which is calculated by a double averaging procedure: First, the Schrödinger equation, $i\hbar\frac{\partial}{\partial t}\Phi_{JM}(t) = H\Phi_{JM}(t)$, is solved with a Runge–Kutta procedure (typically taking into account terms up to $J_{\max} = 25$–30) for each initial molecular rotational state $|\Phi_{J_0M_0}(t)\rangle = \sum_{JM} d_{JM}^{J_0M_0}(t)\exp(-iE_Jt/\hbar)|JM\rangle$ in terms of a series in free-field rotor functions $|JM\rangle$ with eigenenergies $E_J = hBcJ(J+1)$ [10, 65, 66]. Then, the degree of molecular alignment at time t is found knowing population amplitudes $d_{JM}^{J_0M_0}(t)$ of the rotational states, as $\langle \cos^2\theta\rangle_{J_0M_0}(t) = \langle\Phi_{J_0M_0}(t)|\cos^2\theta|\Phi_{J_0M_0}(t)\rangle$ [3]. The evolution of the rotational wave packet in time can be described as $|\Phi_{J_0M_0}(\tau)\rangle = \sum\limits_{J,M} d_{JM}^{J_0M_0}|JM\rangle\exp[-i\pi\tau J(J+1)]$ [67], where we have written the time-dependent phase term explicitly, and the time τ is given in units of

$T_{rev} = 1/(2Bc)$ with a rotational constant B. Thus, for the time dependence of the observable of interest, it follows

$$\begin{aligned}\left\langle \cos^2\theta \right\rangle_{J_0 M_0}(\tau) &= \langle \Phi_{J_0 M_0}(\tau) | \cos^2\theta | \Phi_{J_0 M_0}(\tau) \rangle \\ &= \sum_{J',M'} \sum_{J,M} d^{*J_0 M_0}_{J'M'} d^{J_0 M_0}_{JM} \langle J', M' | \cos^2\theta | J, M \rangle \varphi_{J,J'}, \end{aligned} \quad (1)$$

where $\langle J'M' | \cos^2\theta | JM \rangle \propto \delta_{\left\{ \begin{array}{l} J,J' \\ J,J' \pm 2 \end{array} \right\}} \delta_{M,M'}$ [3] and $\varphi_{J,J'} = \exp(-i\pi\tau[J'(J'+1) - J(J+1)])$. For the terms with the same angular momentum quantum number, $J = J'$, the time-dependent phase disappears, but for the terms with $J = J' \pm 2$, the time dependence is given by the phase factors $\varphi_{J,J\pm 2} = \exp[-i\pi\tau(4J+6)]$. At integer values $\tau =$ 0, 1, 2, … which correspond to full revivals, the phase term $\varphi_{J,J\pm 2} = 1$ for all the J quantum numbers. At half revivals $\tau = 1/2, 3/2, \ldots$, the phase terms are π shifted for all of the J's, resulting in the phase term $\varphi_{J,J\pm 2} = -1$, and thus overall distribution has the same magnitude, but the opposite sign compared to full revivals. At quarter revivals, the phase term $\varphi_{J,J\pm 2}$ depends on the even or odd parity of J-state: At $\tau = 1/4, 5/4, \ldots$, the phase term $\varphi_{J,J\pm 2} = i$ for even J-states and $\varphi_{J,J\pm 2} = -i$ for odd J-states. At $\tau = 3/4, 7/4, \ldots$, the phase term $\varphi_{J,J\pm 2} = -i$ for even J-states and $\varphi_{J,J\pm 2} = i$ for odd J-states. The even and odd states contribute to the signal; however, their contributions are shifted by π, and therefore they interfere destructively. Indeed, the expression for the time dependence of Eq. (1) at quarter revivals has a phase shift of $\pi/2$ with regard to the signal observed at half and full revivals.

To understand the observed revival signatures, we considered the molecular ionization rate. Depending on the configuration of the HOMO, qualitatively different alignment signatures are expected based on the angular dependence of the SFI yield in the interaction of the aligned molecules with the probe pulse. To take into account the possible deviation of the direction of the maximal SFI yield relative to the molecular axis, we add an additional term Δ to the phase: $\varphi'_{J,J\pm 2} = \exp[-i(\pi\tau(4J+6) + \Delta)]$, where Δ is determined by the configuration of the HOMO.

Before the laser interaction, the gas ensemble is assumed to be in thermal equilibrium having a Boltzmann distribution with temperature T. In the quantum mechanical approach, an ensemble is described by a statistical mixture of states $|J_0 M_0\rangle$, having angular momenta $J_0 =$ 0, 1, 2,… and their projections on the laser polarization axis $M_0 = -J_0, -(J_0-1), \ldots 0, \ldots (J_0-1), J_0$ with the probabilities of these states following the Boltzmann distribution $P_{J_0} \sim g_{J_0}(2J_0+1)\exp(-E_{J_0}kT)$, where E_{J_0} is the rotational energy of $|J_0 M_0\rangle$ state and k is the Boltzmann constant. The $(2J_0+1)$ term accounts for the degeneracy within a given J_0 state. In the case of a molecule containing two identical nuclei, g_{J_0} is the relative weight between odd and even J-states, which comes from the nuclear spin statistics as an additional factor. For gerade symmetry states, bosonic nuclei have $g_{J_0} = (2I+1)(I+1)$, if

J is even and $g_{J_0} = (2I + 1)I$, if J is odd, and fermionic nuclei have $g_{J_0} = (2I + 1)I$, if J is even and $g_{J_0} = (2I + 1)(I + 1)$, if J is odd; I is the nuclear spin of identical particles exchanged during rotation [68, 69]. For ungerade electronic states, the ratio of odd to even states is reversed.

Finally, the degree of alignment of the ensemble at temperature T can be found by averaging the alignment degree of a single initial state over the Boltzmann distribution, including nuclear spin statistical factors, as [70]

$$\langle\langle \cos^2\theta\rangle\rangle(t) = \frac{\sum_{J_0}\sum_{M_0=-J_0}^{J_0} g_{J_0}\langle \cos^2\theta\rangle_{J_0M_0}(t)\exp(-hBcJ_0(J_0+1)kT)}{\sum_{J_0}\sum_{M_0=-J_0}^{J_0} g_{J_0}\exp(-hBcJ_0(J_0+1)kT)}, \quad (2)$$

where $\langle \cos^2\theta\rangle_{J_0M_0}(t)$ is the degree of alignment for a single initial state.

To illustrate the dependence of the degree of alignment on temperature, we calculated the N_2 temporal alignment signatures for a set of temperatures $T = 50$, 100, 200, and 300 K (Fig. 2a). The summation of the contributions of different rotational states up to $J_{max} = 30$ is shown in Fig. 2b.

We note that the duration of the used laser pulse of 50 fs is much shorter than revival periods for all investigated molecules. Consequently, the details of the pulse shape only slightly influence the alignment, and therefore the initial action of the pump pulse can be considered as a kick of transferred angular momentum proportional to $\Delta\alpha$ and the laser fluence [71, 72]. The correspondence between quantum and classical quantities in the molecular alignment is discussed in Ref. [73, 74].

The presented treatment is valid only for linear molecules and a linearly polarized laser field that aligns only the molecular axis leaving the rotation about this axis unrestricted. For molecules similar to an asymmetric top, the rotation about the molecular axis is not isotropic, and it was shown [75] that an intense, elliptically polarized, nonresonant laser field can simultaneously force all three axes of a

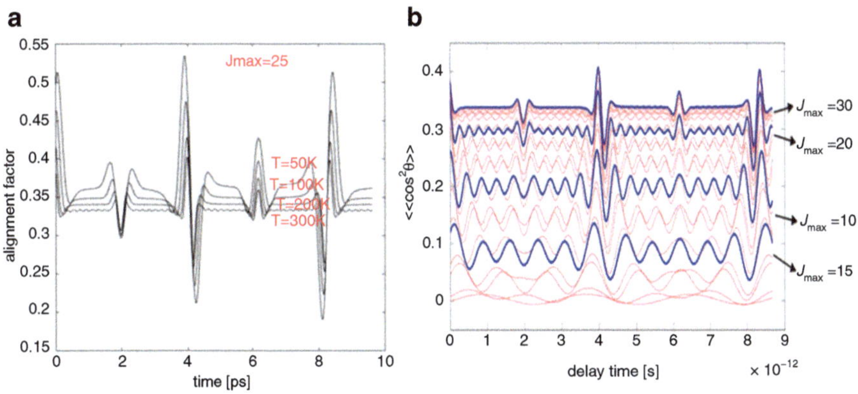

Fig. 2 **a** Dependence of the degree of alignment on the gas temperature calculated for N_2. J_{max} is the maximum number of rotational states taken into account in the calculation. **b** Summation of the contributions of different rotational states for different J_{max} for N_2 at $T = 300$ K

molecule to align along given axes fixed in space. An enhanced degree of field-free three-dimensional alignment was also obtained using two orthogonally polarized, time-separated laser pulses [76]. Field-free molecular revivals of 3D alignment of asymmetric top molecules using elliptically polarized laser pulses was also demonstrated [39] and is most efficient at low temperatures.

4 Results and Discussion

In this section, we present experimental results for N_2, CO, O_2, and C_2H_2 gases followed by the interpretation of the observed revival signal for each gas. For this interpretation, we consider the total wave function of a linear molecule, which according to the Born–Oppenheimer approximation can be factorized, $\Psi_{tot} = \psi_{el} \times \psi_{vib} \times \psi_{rot} \times \psi_{ns}$, as a product of the electronic wave function ψ_{el}, the vibrational wave function ψ_{vib}, the rotational wave function ψ_{rot}, and the nuclear spin wave function ψ_{ns}. We present observed temporal signatures of the alignment and compare them to the calculated time evolution of the alignment parameter $\langle\langle \cos^2\theta \rangle\rangle(t)$, which depends on the laser parameters and the molecular properties.

Figure 3a shows the calculated alignment parameter $\langle\langle \cos^2\theta \rangle\rangle$ at temperature $T = 300$ K with our laser parameters and the properties of N_2 molecules with the rotational constant $B = 1.9896\ \text{cm}^{-1}$ [77, 78]; Fig. 3b depicts the measured signal of the SFI yield for N_2 molecules aligned by a linearly polarized pump pulse. The SFI yields are smaller at $T_{rev}/4$ and $3T_{rev}/4$ and larger at $T_{rev}/2$ and T_{rev}. This can be explained in the following way. $^{14}N_2$ is a boson, and therefore, the total wave function Ψ_{tot} is symmetric [79]. At room temperatures, the electronic ψ_{el} $(^3\Sigma_g^+)$, and the vibrational ψ_{vib} wave functions of ^{14}N correspond to the ground state, and both are symmetric [79]. Therefore, to understand the ψ_{rot} behavior, we should consider the symmetric and antisymmetric forms of the nuclear spin wave function ψ_{ns} of $^{14}N_2$. For Ψ_{tot} to be symmetric, both ψ_{rot} and ψ_{ns} must be symmetric or antisymmetric. For ^{14}N, the nuclear spin is $I = 1$ and the nuclear spin of a N_2 molecule takes only values $I_{tot} = 0$, 1, or 2. Consequently, N_2 exists in even ($I_{tot} = 0$, 2) and odd ($I_{tot} = 1$) forms. A state degeneracy with I_{tot} is $2\,I_{tot} + 1$, so the even-N_2 and odd-N_2 statistical weights are 6 and 3, respectively. Due to the relative abundance of the even-N_2 versus odd-N_2, 2:1 for even/odd states, the two opposed $T_{rev}/4$ revival signals do not cancel completely; however, the signal has a reduced amplitude compared to the T_{rev} and $T_{rev}/2$ revivals, as is observed in the experiment (Fig. 3b). Consequently, our measured SFI yield in Fig. 3b varies in phase with the time dependence of the molecular alignment parameter in Fig. 3a. In Fig. 3b, we also presented the corresponding fit curve based on Eq. (2) for the alignment parameter $\langle\langle \cos^2\theta \rangle\rangle$. Thus, we can conclude that $\Delta = 0$ for the N_2 measurements.

As described, the molecular ionization rate is therefore maximal when molecules are aligned along the laser polarization direction and the configuration of the

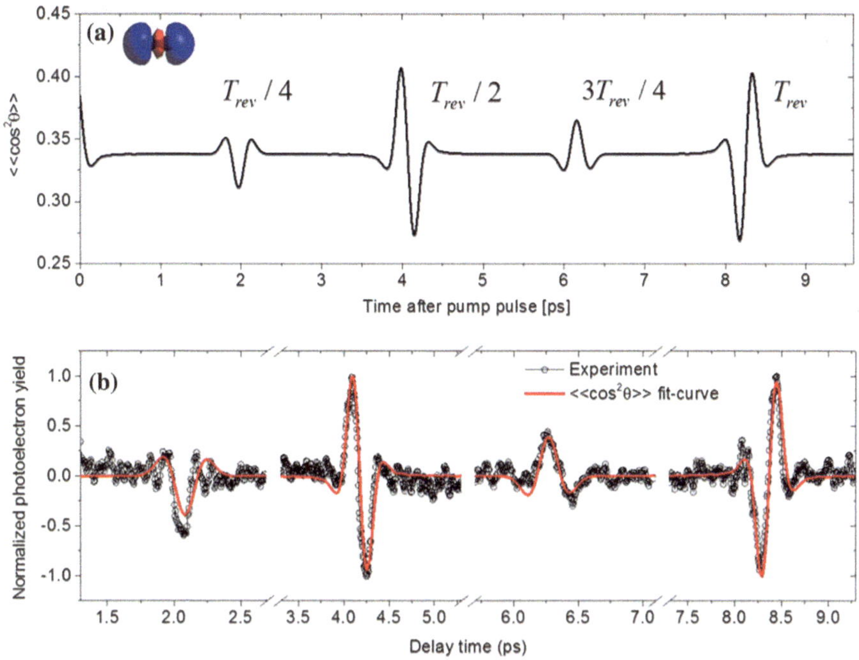

Fig. 3 **a** Calculated time dependence of the molecular alignment parameter induced by the pump pulse in N_2; the configuration of the HOMO is also shown; **b** the time evolution of the measured SFI signal for N_2 molecules and the fit curve. The pump and probe pulses were polarized collinearly and had peak intensities of 2×10^{13} and 7.8×10^{13} W/cm^2, respectively

HOMO is parallel to the molecular axis, because such configuration makes ejection of an electron easier. This agrees with the σ_g HOMO symmetry of the neutral nitrogen molecule [80], where the molecular ionization rate is maximal, when the molecules are aligned along the probe polarization direction. As seen in the inset of Fig. 3a, N_2 has its maximum electron density along the internuclear (molecular) axis in its σ_g HOMO configuration. By fitting the curve within one full revival period and taking the same laser parameters as experimental ones, we also found the characteristic rotational constant and the corresponding revival time for nitrogen gas. We note that no offset for the time axis was introduced. The theoretical values of the rotational constant 1.9896 cm^{-1} [77, 78] and the revival time 8.383 ps agree well with our experimental values of the rotational constant $B = 1.999(+0.01)$cm^{-1} and the corresponding revival time 8.342 ps; the number in parentheses here and below shows the deviation of the experimental value of B from the theoretical one.

In Fig. 4, we show the corresponding frequency spectrum amplitudes of the measured time-dependent SFI yield for N_2 obtained by a Fourier transform (FT) analysis. The spectrum reveals two sequences of allowed Raman transitions obtained from the matrix elements of $\langle\langle \cos^2\theta \rangle\rangle$, which are calculated as $(E_{J+2} - E_J)/h = (4J + 6)Bc$ with the selection rule $\Delta J = \pm 2$. Namely, these sequences

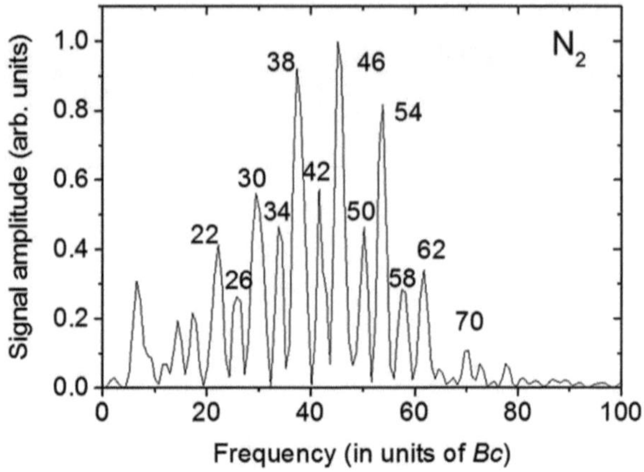

Fig. 4 Frequency spectrum amplitudes of the measured time-dependent SFI yield for N_2 shown in Fig. 3b. The numbers on spectral peaks, which correspond to the allowed transitions, show the frequencies in terms of $4J + 6$. The series (6, 14, 22, 30, 38, 46, 54, . . .) *Bc* and (10, 18, 26, 34, 42, 50, 58, . . .) *Bc* come from even and odd values of *J*'s, respectively

are as follows: (6, 14, 22, 30, 38, . . .)*Bc* for even *J*'s and (10, 18, 26, 34, 42,. . .) *Bc* for odd *J*'s. Recall that the nuclear spin of the nitrogen nucleus is 1, and both even and odd rotational states are permitted for the nitrogen molecule [13]. The relative ratio of the even and odd *J*-states is 2:1, which follows from the nuclear spin statistics and gives an additional factor of ~2 for even *J*'s in the spectrum [25, 81, 82].

Figure 5a shows the calculated alignment parameter $\langle\langle \cos^2\theta \rangle\rangle$ at temperature $T = 300$ K with our laser parameters and the properties of CO molecules with the rotational constant $B = 1.9226\ \text{cm}^{-1}$ [77]. The CO molecule has two nonidentical nuclei. Because of this lack of symmetry, even and odd states in an ensemble of CO molecules are equally populated, and the contributions that come from the even and odd states completely cancel each other at quarter revivals [69, 83]. As a result, no quarter revivals appear in the calculated time dependence of the molecular alignment parameter $\langle\langle \cos^2\theta \rangle\rangle$ and in the measured SFI signal, and only half and full revivals are observed, as shown in Fig. 5b. The calculated time dependence of molecular alignment in Fig. 5a changes in phase with the measured SFI yield, so there is no need for any additional term arising from the phase term in the calculation of the alignment parameter, thus $\Delta = 0$. This is similar to N_2, since CO also has σ-type HOMO [80] and no nodal plane along the internuclear axis. While the theoretical rotational constant for CO is 1.9226 cm^{-1} [77] and revival time 8.675 ps, we evaluated the characteristic rotational constant from the fit to our experimental data as $B = 1.939(+0.016)\text{cm}^{-1}$, and the corresponding periodic revival time is 8.601 ps.

In Fig. 6, we presented the frequency spectrum amplitudes of the time-dependent molecular alignment signal for CO, which is expected to have peaks at frequencies $(4J + 6)Bc$ for all *J*'s, giving the sequence (6, 10, 14, 18, 22, 26, 30, 34, 38, 42, . . .) *Bc*.

As an example of an π_g type HOMO configuration [80], we investigated the O_2 molecule, and Fig. 7a, b shows the time dependences of the molecular alignment

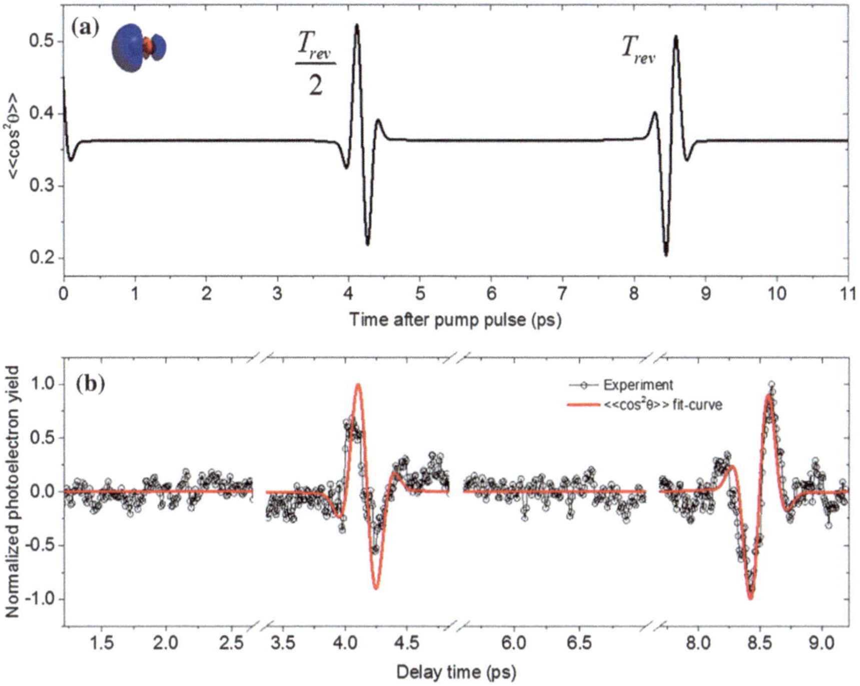

Fig. 5 **a** Calculated time dependence of the molecular alignment parameter for CO; the configuration of the HOMO is also shown; **b** the time evolution of the measured SFI signal for CO and the fit curve are depicted. Collinearly polarized pump and probe pulses were used with peak intensities of 2×10^{13} and 7.9×10^{13} W/cm^2, respectively

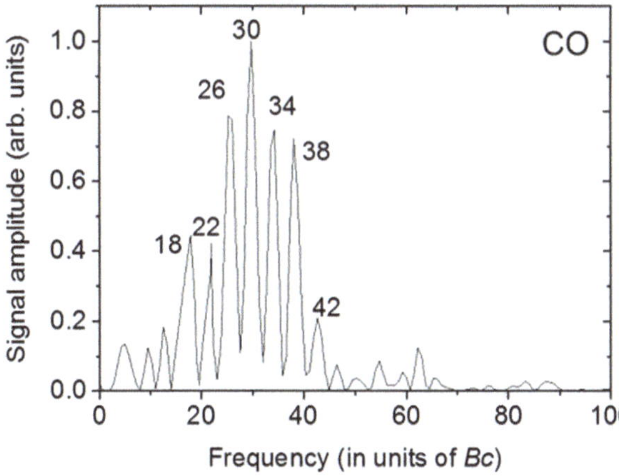

Fig. 6 Frequency spectrum amplitudes of the measured time-dependent SFI yield for CO shown in Fig. 5b. The number on each spectral peak shows the frequency from the series (6, 10, 14, 18, 22, 26, 30, 34, 38, 42, ...) *Bc*

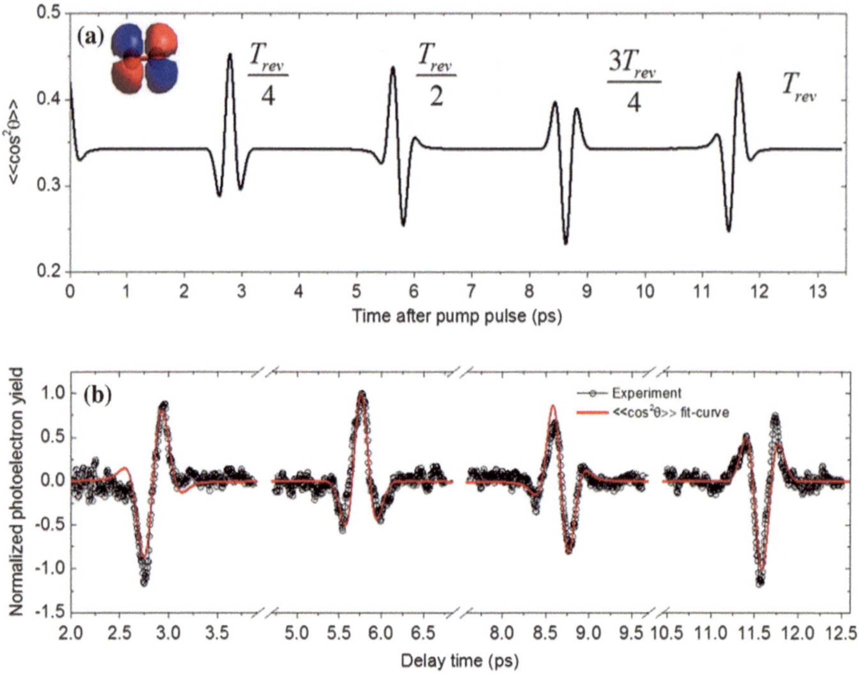

Fig. 7 **a** Calculated time dependence of molecular alignment parameter for O_2, following the action of the pump pulse; the configuration of the HOMO is also shown. **b** The time evolution of the measured SFI signal for O_2 and the fit curve. Linearly polarized pump and probe pulses were used with peak intensities of 2×10^{13} and 8.1×10^{13} W/cm^2, respectively

parameter $\langle\langle \cos^2\theta \rangle\rangle$ calculated at a temperature of $T = 300$ K for our laser parameters and O_2 molecule with a rotational constant $B = 1.4376$ cm^{-1} [77, 84] as well as the measured temporal alignment profile for O_2 gas. Oxygen ^{16}O is bosonic, so the total wave function Ψ_{tot} is symmetric. The ground state of the electronic wave function, ψ_{el}, of ^{16}O is $(^3\sum_g^-)$ and antisymmetric, while the vibrational wave function, ψ_{vib}, is symmetric. Then, for the ground state Ψ_{tot} to be symmetric, ψ_{rot} and ψ_{ns} must be symmetric and antisymmetric, respectively, or vice versa. For an O_2 molecule, the nuclear spin $I_{\text{tot}} = 0$, and hence only odd rotational J-states are populated [85]; consequently, the relative abundance of molecules with even and odd J's is 0:1. Therefore, only a wave packet with odd J exists, and strong effects of alignment and antialignment are observed at the quarter and the three-quarter revivals of O_2 with the amplitudes comparable to the full and half revivals, as shown in Fig. 7a. However, when the graph of Fig. 7a and experimentally measured dependence in Fig. 7b are compared, one can see that the shapes of the calculated and experimental dependences are different, which has the following explanation. The HOMO of O_2 is of π_g type and has nodes of the electron densities [86] in the direction of and also perpendicular to the molecular axis, as

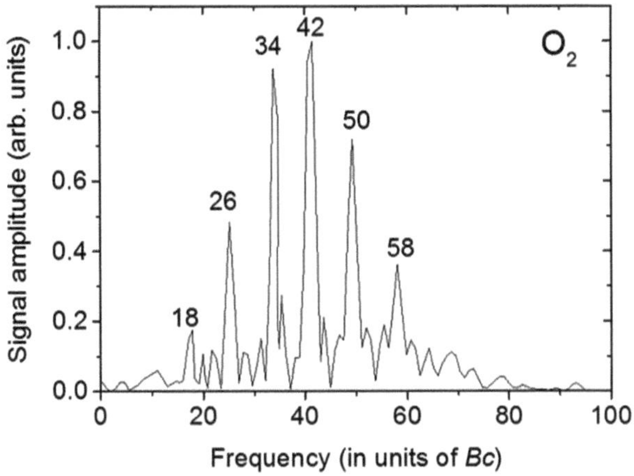

Fig. 8 Frequency spectrum amplitudes of the measured time-dependent SFI yield for O_2, shown in Fig. 7b. The number on each spectral peak shows the frequency in terms of $4J + 6$ from the series (10, 18, 26, 34, 42, 50, 58, ...) Bc for odd J

shown in the inset of Fig. 7a. Because of the exact twofold symmetry of this molecular orbital, the SFI signal from O_2 maximizes near 45°, and the SFI signal is minimal when the molecular axis is exactly parallel or perpendicular to the laser polarization. Indeed, the fitting function calculated with the value $\Delta = \pi/2$ gives a good agreement with the experiment (Fig. 7b). The theoretical value of the rotational constant $B = 1.4376\ \text{cm}^{-1}$ [77, 84] and the corresponding periodic revival time 11.601 ps compare favorably with our experimental values found by fitting $B = 1.461(+0.023)\ \text{cm}^{-1}$ and the revival period 11.415 ps.

In Fig. 8, we show the corresponding frequency spectrum amplitudes of the time dependence of the measured SFI yield for O_2 obtained by the Fourier transform (FT). This figure exhibits the experimental spectrum, showing series (10, 18, 26, 34, 42,...)Bc, since only odd J levels are permitted for O_2 due to the nuclear spin statistics.

Figure 9a shows the alignment parameter $\langle\langle \cos^2\theta \rangle\rangle$ calculated at temperature $T = 300$ K with our laser parameters for C_2H_2 molecules with the rotational constant $B = 1.1766\ \text{cm}^{-1}$ [87] and the measured time trace of the SFI yield with variations due to alignment of C_2H_2 molecules. Acetylene C_2H_2 has both even and odd J-states, which are populated with the ratio 1:3 [85]. As a result of this fact, even and odd wave packets partially cancel each other, and some alignment and antialignment are observed at quarter revivals as shown in Fig. 9a, b. The experimental results for C_2H_2 molecules at quarter revival, half revival, three-quarter revival, and full revival are in good agreement with the calculated time dependence of molecular alignment $\langle\langle \cos^2\theta \rangle\rangle$ parameter, but with the inverted polarity. This can be understood from the configuration of HOMO for acetylene, which is dominated by a π_u orbital [88]. Molecular orbitals of π_u type have higher electron density above/below the internuclear axis, with a node along the internuclear axis, as shown in the inset of Fig. 9a. Consequently, the temporal structure of C_2H_2 revivals in Fig. 9b has inverted polarity as compared to the calculated molecular alignment parameter $\langle\langle \cos^2\theta \rangle\rangle$ in Fig. 9a. With $\Delta = \pi$, the temporal signature of

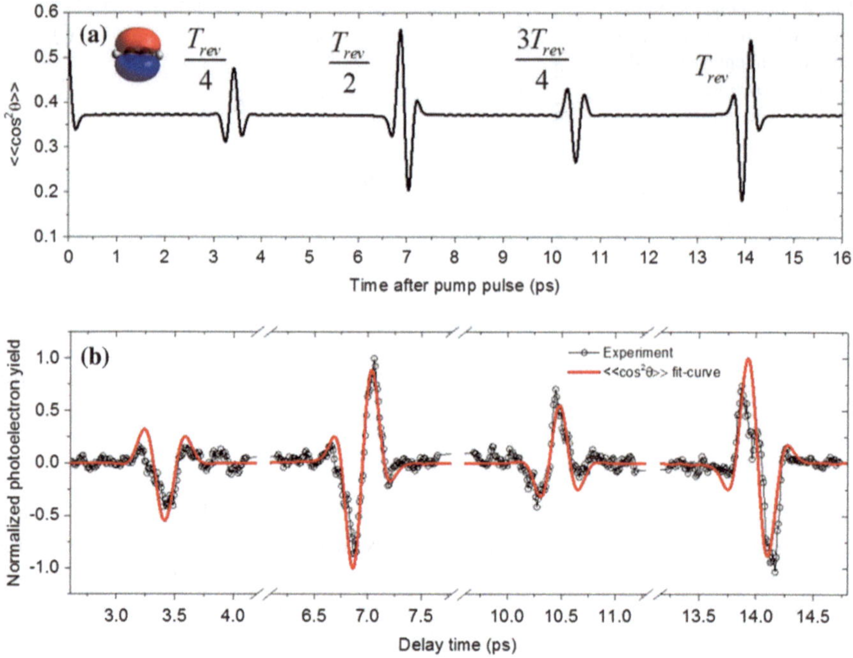

Fig. 9 **a** Calculated for C_2H_2 time dependence of the molecular alignment parameter; the configuration of the HOMO is also shown. **b** The time evolution of the measured SFI signal for C_2H_2 and the fit curve. We used linearly polarized pump and probe pulses with peak intensities of 3×10^{13} and 7.6×10^{13} W/cm^2, respectively

the molecular alignment parameter $\langle\langle \cos^2\theta\rangle\rangle$ is inverted, as shown in Fig. 9b, and the calculated function gives a shape similar to the experimental one. The theoretical value of the molecular constant for C_2H_2, $B = 1.1766$ cm^{-1} [87], and the respective revival time 14.175 ps match well the characteristic rotational constant $B = 1.180(+0.003)$ cm^{-1} and the revival period 14.133 ps following from the fitting procedure in our data.

The frequency spectrum of the time-dependent SFI yield for C_2H_2 is shown in Fig. 10. The number on each spectral peak shows the frequency in terms of $4J + 6$ which corresponds to allowed Raman transitions, given by the series (6, 14, 22, 30, 38, 46, 54, ...)Bc for even J and by the series (10, 18, 26, 34, 42, 50, 58, ...)Bc for odd J. The relative ratio of the amplitudes of the even and odd spectral components is close to 1:3, following from the nuclear spin statistics.

We note that although the qualitative interpretation of our results can be provided based only on considerations involving HOMO configurations, in more complex cases it can be necessary to take into account ionization occurring from several orbitals using a description involving Dyson orbitals, which represent the overlap between an initial wave function for the neutral system and the electron wave function of the corresponding ionized system [54, 89].

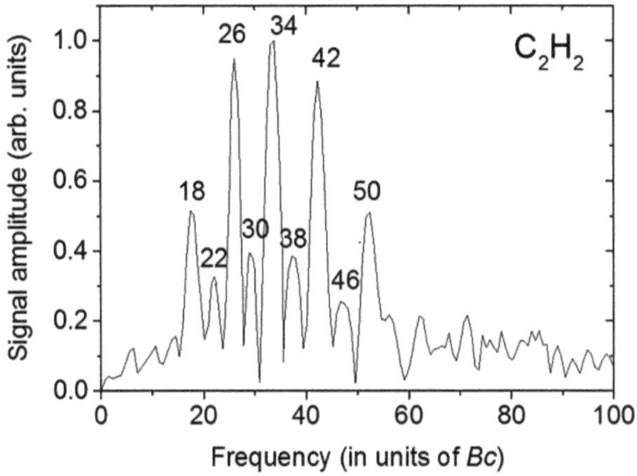

Fig. 10 Frequency spectrum amplitudes of the measured time-dependent SFI yield for C_2H_2, shown in Fig. 9**b**. The numbers on the spectral peaks show frequencies in terms of $4J + 6$, which correspond to allowed Raman transitions from the series $(6, 14, 22, 30, 38, 46, 54, \ldots)Bc$ for even J and the series $(10, 18, 26, 34, 42, 50, 58, \ldots)$ Bc for odd J. The relative ratio of the magnitudes of the spectral harmonics corresponding to even and odd J correlates with the nuclear spin statistics ratio 1:3

5 Conclusions

The revival signatures of the linear molecules N_2, CO, O_2, and C_2H_2 produced by linearly polarized femtosecond pump pulses have been studied under nonadiabatic alignment conditions and measured by the detection of the SFI photoelectron yield in a pump–probe setup with a variable delay. For N_2 we observed strong alignment effects at multiples of $T_{\text{rev}}/2$ and reduced amplitude signals (not completely suppressed) at odd multiples of $T_{\text{rev}}/4$ due to the ratio of even/odd states 2:1 for the nuclear spin statistics factors of N_2. For CO, containing nonidentical nuclei, there is no additional factor arising from the nuclear spin statistics. Thus, the revivals at odd multiples of $T_{\text{rev}}/4$ are completely canceled, whereas the revivals at multiples of $T_{\text{rev}}/2$ remain. Because the HOMO configuration donates electrons most readily, the HOMO symmetry of molecules is important for interpretation of the observed alignment signals. Therefore, the symmetry effect of the HOMO electron density distribution on the shape of the temporal dependence of the SFI yields of photoelectron measured with aligned molecules was discussed. N_2 and CO have their maximum electron densities along the internuclear (molecular) axis due to σ_g and σ HOMOs, respectively. The molecular ionization rate is maximal when molecules are aligned along the laser polarization direction, because such configuration makes ejection of an electron easier. Consequently, in case of N_2 and CO, our measured SFI yields were modulated in phase with the time dependency of the molecular alignment parameter. In the case of O_2, since only odd J-states are relevant, we observed strong alignment signals at $T_{\text{rev}}/4$ and $3T_{\text{rev}}/4$ revivals as well as at multiples of $T_{rev}/2$. For O_2 molecules, the ionization rate has minima when the molecular axis is parallel or perpendicular to the laser polarization due to the π_g HOMO configuration. The measured SFI yield is temporally modulated with a $(-\pi/2)$ phase shift with respect to the time dependence of the molecular alignment parameter, which agreed with the experimental data. For the C_2H_2, the nuclear spin statistics ratio of even and odd states is 1:3, and the strong alignment effects were observed at multiples of $T_{rev}/2$, and the reduced amplitude alignment signals (not completely suppressed) were at odd multiples of $T_{rev}/4$. Due to the π_u HOMO symmetry of C_2H_2, which has higher electron density above/below the internuclear axis with a node along the direction of this axis, our experimental SFI yields for C_2H_2 have a polarity-inverted dependence compared to the molecular alignment parameter [a $(-\pi)$ shift]. By taking into account these additional phase shifts, a good agreement between calculated and measured alignment signatures was obtained. In addition, the spectra of revivals reveal peaks corresponding to sequences of transitions between even and odd rotational states with amplitude ratios in agreement with the population ratio following from the nuclear spin statistics. Thus, we have demonstrated that the implemented alignment monitoring by measuring SFI photoelectron yields presents a viable and efficient way of studying rotational molecular dynamics. By fitting the molecular alignment parameter $\langle\langle \cos^2\theta\rangle\rangle$, modified by taking into account discussed additional phase shifts to the experimental data, we determined the characteristic rotational constants and corresponding revival periods, which are in good agreement with the theoretical values found in the literature.

5.1 *Personal Recollections of Hans Schuessler Dedicated to Ted Haensch*

As a friendly colleague, I was privileged to watch your many contributions to fundamental atomic spectroscopy. Early on, lasers were considered a passing fad, since they were not widely tuneable. In Heidelberg, your He–Ne laser tubes in Schmelzer's Applied Physics laboratory were long and fragile, while the small hollow cathode light sources in Kopfermann's Physikalischen Institut, when combined with radio frequency techniques, were well suited for hyperfine spectroscopy. Several years later, matters changed dramatically. You had gone to Stanford and I to Seattle. I guess we both were enjoying the stimulating atomic physics atmosphere at these West Coast universities, where a starting scientist could work independently and realize his dreams. On one of my visits to Stanford, you picked me up in a Karmann Ghia sports car, keeping with this famous Volkswagen brand. During this visit, I realized the playfulness in your research by watching your wavemeter, where the moving arm was on a rail track. This playful treat continued to be reflected in many of your subsequent works. I was also impressed by your widely applicable nitrogen pumped dye laser, so much that on my return to Seattle I wanted to build such a setup for spectroscopy with ion traps, but no such luck, Hans Dehmelt insisted for me to wait until I have my own laboratory at Texas A&M University. Decades full of wonderful experiments have passed, and still today thanks to your work and help, I use now frequency comb lasers for sensing and precision spectroscopy, and the wonderful journey continues.

Acknowledgements This work was supported by the Robert A. Welch Foundation Grant No. A1546, and the Qatar Foundation under the Grant NPRP 6-465-1-091.

References

1. N. Xu, J. Li, J. Li, Z. Zhang, Q. Fan, in *Lasers: Applications in Science and Industry*, ed. K. Jakubczak (InTech, 2011)
2. T. Seideman, J. Chem. Phys. **103**, 7887 (1995)
3. J. Ortigoso, M. Rodríguez, M. Gupta, B. Friedrich, J. Chem. Phys. **110**, 3870 (1999)
4. I.S. Averbukh, N.F. Perelman, Phys. Lett. A. **139**, 449 (1989)
5. R.W. Robinett, Phys. Rep. **392**, 1 (2004)
6. G.H. Lee, H.T. Kim, J.Y. Park, C.H. Nam, T.K. Kim, J.H. Lee, H. Ihee, J. Korean Phys. Soc. **49**, 337 (2006)
7. N.E. Henriksen, Chem. Phys. Lett. **312**, 196 (1999)
8. T. Seideman, Phys. Rev. Lett. **83**, 4971 (1999)
9. F. Rosca-Pruna, M.J.J. Vrakking, Phys. Rev. Lett. **87**, 153902 (2001)
10. H. Stapelfeldt, T. Seideman, Rev. Mod. Phys. **75**, 543 (2003)
11. T. Seideman, E. Hamilton, in *Advances in Atomic, Molecular, and Optical Physics*, ed. by P.R. Berman, C.C. Lin (Academic Press, New York, 2005), p. 289
12. S. Henrik, Phys. Scr. **2004**, 132 (2004)
13. S. Fleischer, I.S. Averbukh, Y. Prior, Phys. Rev. A. **74**, 041403 (2006)
14. J. Itatani, J. Levesque, D. Zeidler, H. Niikura, H. Pepin, J.C. Kieffer, P.B. Corkum, D.M. Villeneuve, Nature **432**, 867 (2004)
15. R.J. Gordon, L. Zhu, W.A. Schroeder, T. Seideman, J. Appl. Phys. **94**, 669 (2003)
16. R.A. Bartels, T.C. Weinacht, N. Wagner, M. Baertschy, C.H. Greene, M.M. Murnane, H.C. Kapteyn, Phys. Rev. Lett. **88**, 013903 (2001)
17. K.F. Lee, D.M. Villeneuve, P.B. Corkum, E.A. Shapiro, Phys. Rev. Lett. **93**, 233601 (2004)
18. A.H. Zewail, J. Phys. Chem. A **104**, 5660 (2000)
19. V. Loriot, E. Hertz, B. Lavorel, O. Faucher, J. Phys. B. **41**, 015604 (2008)
20. D. Pavičić, K.F. Lee, D.M. Rayner, P.B. Corkum, D.M. Villeneuve, Phys. Rev. Lett. **98**, 243001 (2007)
21. I.V. Litvinyuk, K.F. Lee, P.W. Dooley, D.M. Rayner, D.M. Villeneuve, P.B. Corkum, Phys. Rev. Lett. **90**, 233003 (2003)
22. E. Péronne, M.D. Poulsen, C.Z. Bisgaard, H. Stapelfeldt, T. Seideman, Phys. Rev. Lett. **91**, 043003 (2003)
23. M. Kenzo, S. Takayuki, N. Didier, J. Phys. B. **37**, 753 (2004)
24. I. Nevo, L. Holmegaard, J.H. Nielsen, J.L. Hansen, H. Stapelfeldt, F. Filsinger, G. Meijer, J. Kupper, PCCP **11**, 9912 (2009)
25. P.W. Dooley, I.V. Litvinyuk, K.F. Lee, D.M. Rayner, M. Spanner, D.M. Villeneuve, P.B. Corkum, Phys. Rev. A **68**, 023406 (2003)
26. R. Velotta, N. Hay, M.B. Mason, M. Castillejo, J.P. Marangos, Phys. Rev. Lett. **87**, 183901 (2001)
27. T. Kanai, S. Minemoto, H. Sakai, Nature **435**, 470 (2005)

28. C. Vozzi, F. Calegari, E. Benedetti, J.P. Caumes, G. Sansone, S. Stagira, M. Nisoli, R. Torres, E. Heesel, N. Kajumba, J.P. Marangos, C. Altucci, R. Velotta, Phys. Rev. Lett. **95**, 153902 (2005)
29. C.B. Madsen, A.S. Mouritzen, T.K. Kjeldsen, L.B. Madsen, Phys. Rev. A. **76**, 035401 (2007)
30. J.-F. Ripoche, G. Grillon, B. Prade, M. Franco, E. Nibbering, R. Lange, A. Mysyrowicz, Opt. Commun. **135**, 310 (1997)
31. S. Varma, Y.H. Chen, J.P. Palastro, A.B. Fallahkair, E.W. Rosenthal, T. Antonsen, H.M. Milchberg, Phys. Rev. A. **86**, 023850 (2012)
32. Y. Feng, H. Pan, J. Liu, C. Chen, J. Wu, H. Zeng, Opt. Express **19**, 2852 (2011)
33. F. Calegari, C. Vozzi, S. Gasilov, E. Benedetti, G. Sansone, M. Nisoli, S. De Silvestri, S. Stagira, Phys. Rev. Lett. **100**, 123006 (2008)
34. J. Wu, H. Cai, Y. Peng, Y. Tong, A. Couairon, H. Zeng, Laser Phys. **19**, 1759 (2009)
35. N. Kaya, G. Kaya, M. Sayrac, Y. Boran, S. Anumula, J. Strohaber, A.A. Kolomenskii, H.A. Schuessler, Opt. Express **24**, 2562 (2016)
36. P.M. Felker, J. Phys. Chem. **96**, 7844 (1992)
37. J.P. Heritage, T.K. Gustafson, C.H. Lin, Phys. Rev. Lett. **34**, 1299 (1975)
38. V. Renard, M. Renard, S. Guérin, Y.T. Pashayan, B. Lavorel, O. Faucher, H.R. Jauslin, Phys. Rev. Lett. **90**, 153601 (2003)
39. A. Rouzée, S. Guérin, O. Faucher, B. Lavorel, Phys. Rev. A **77**, 043412 (2008)
40. P.J. Bustard, R. Lausten, B.J. Sussman, Phys. Rev. A **86**, 053419 (2012)
41. L. Holmegaard, J.H. Nielsen, I. Nevo, H. Stapelfeldt, F. Filsinger, J. Küpper, G. Meijer, Phys. Rev. Lett. **102**, 023001 (2009)
42. L. Holmegaard, J.L. Hansen, L. Kalhoj, S. Louise Kragh, H. Stapelfeldt, F. Filsinger, J. Kupper, G. Meijer, D. Dimitrovski, M. Abu-samha, C.P.J. Martiny, L. Bojer Madsen, Nat. Phys. **6**, 428 (2010)
43. S. De, I. Znakovskaya, D. Ray, F. Anis, N.G. Johnson, I.A. Bocharova, M. Magrakvelidze, B.D. Esry, C.L. Cocke, I.V. Litvinyuk, M.F. Kling, Phys. Rev. Lett. **103**, 153002 (2009)
44. J. Mikosch, C.Z. Bisgaard, A.E. Boguslavskiy, I. Wilkinson, A. Stolow, J. Chem. Phys. **139**, 024304 (2013)
45. A.E. Boguslavskiy, J. Mikosch, A. Gijsbertsen, M. Spanner, S. Patchkovskii, N. Gador, M.J.J. Vrakking, A. Stolow, Science **335**, 1336 (2012)
46. J. Mikosch, A.E. Boguslavskiy, I. Wilkinson, M. Spanner, S. Patchkovskii, A. Stolow, Phys. Rev. Lett. **110**, 023004 (2013)
47. C. Schröter, K. Kosma, T. Schultz, Science **333**, 1011 (2011)
48. M. Machholm, N.E. Henriksen, Phys. Rev. Lett. **87**, 193001 (2001)
49. E. Péronne, M.D. Poulsen, H. Stapelfeldt, C.Z. Bisgaard, E. Hamilton, T. Seideman, Phys. Rev. A **70**, 063410 (2004)
50. O. Ghafur, A. Rouzee, A. Gijsbertsen, W.K. Siu, S. Stolte, M.J.J. Vrakking, Nat. Phys. **5**, 289 (2009)
51. F. Filsinger, J. Küpper, G. Meijer, L. Holmegaard, J.H. Nielsen, I. Nevo, J.L. Hansen, H. Stapelfeldt, J. Chem. Phys. **131**, 064309 (2009)
52. J.H. Nielsen, P. Simesen, C.Z. Bisgaard, H. Stapelfeldt, F. Filsinger, B. Friedrich, G. Meijer, J. Kupper, Phys. Chem. Chem. Phys. **13**, 18971 (2011)
53. B.K. McFarland, J.P. Farrell, P.H. Bucksbaum, M. Gühr, Science **322**, 1232 (2008)
54. O. Smirnova, Y. Mairesse, S. Patchkovskii, N. Dudovich, D. Villeneuve, P. Corkum, M.Y. Ivanov, Nature **460**, 972 (2009)
55. D. Pinkham, R.R. Jones, Phys. Rev. A **72**, 023418 (2005)
56. X. Xie, K. Doblhoff-Dier, H. Xu, S. Roither, M.S. Schöffler, D. Kartashov, S. Erattupuzha, T. Rathje, G.G. Paulus, K. Yamanouchi, A. Baltuška, S. Gräfe, M. Kitzler, Phys. Rev. Lett. **112**, 163003 (2014)
57. N.A. Hart, J. Strohaber, G. Kaya, N. Kaya, A.A. Kolomenskii, H.A. Schuessler, Phys. Rev. A **89**, 053414 (2014)

58. D. Comtois, D. Zeidler, H. Pépin, J.C. Kieffer, D.M. Villeneuve, P.B. Corkum, J. Phys. B At. Mol. Opt. Phys. **38**, 1923 (2005)
59. A. Rudenko, K. Zrost, E. Th, A.B. Voitkiv, B. Najjari, V.L.B.D. Jesus, B. Feuerstein, C.D. Schröter, R. Moshammer, J. Ullrich, J. Phys. B At. Mol. Opt. Phys. **38**, L191 (2005)
60. P. Colosimo, G. Doumy, C.I. Blaga, J. Wheeler, C. Hauri, F. Catoire, J. Tate, R. Chirla, A.M. March, G.G. Paulus, H.G. Muller, P. Agostini, L.F. DiMauro, Nat. Phys. **4**, 386 (2008)
61. G.G. Paulus, W. Nicklich, H. Xu, P. Lambropoulos, H. Walther, Phys. Rev. Lett. **72**, 2851 (1994)
62. T. Seideman, J. Chem. Phys. **107**, 10420 (1997)
63. J.D. Graybeal, *Molecular Spectroscopy* (McGraw-Hill, New York, 1988)
64. S.M. Purcell, P.F. Barker, Phys. Rev. Lett. **103**, 153001 (2009)
65. F.H.M. Faisal, A. Abdurrouf, Phys. Rev. Lett. **100**, 123005 (2008)
66. P.F. Bernath, P.F. Bernath, *Spectra of Atoms and Molecules* (Oxford University Press, New York, 1995)
67. S. Fleischer, Y. Khodorkovsky, E. Gershnabel, Y. Prior, I.S. Averbukh, Isr. J. Chem. **52**, 414 (2012)
68. B.H. Bransden, C.J. Joachain, *Physics of Atoms and Molecules* (Prentice-Hall, Englewood Cliffs, 2003)
69. D.A. McQuarrie, J.D. Simon, *Physical Chemistry: A Molecular Approach* (University Science Books, Sausalito, 1997)
70. H. Zeng, P. Lu, J. Liu, W. Li, in *Progress in Ultrafast Intense Laser Science VIII*, ed. by K. Yamanouchi, M. Nisoli, T.W. Hill (Springer, Berlin, 2012), p. 47
71. M. Leibscher, I.S. Averbukh, H. Rabitz, Phys. Rev. Lett. **90**, 213001 (2003)
72. J. Floß, I.S. Averbukh, J. Phys. Chem. A **120**, 3206 (2016)
73. Y. Khodorkovsky, K. Kitano, H. Hasegawa, Y. Ohshima, I.S. Averbukh, Phys. Rev. A **83**, 023423 (2011)
74. J.-M. Hartmann, C. Boulet, J. Chem. Phys. **136**, 184302 (2012)
75. J.J. Larsen, K. Hald, N. Bjerre, H. Stapelfeldt, T. Seideman, Phys. Rev. Lett. **85**, 2470 (2000)
76. J.G. Underwood, B.J. Sussman, A. Stolow, Phys. Rev. Lett. **94**, 143002 (2005)
77. W.M.L.D.R. Haynes, *CRC Handbook of Chemistry and Physics: A Ready-Reference Book of Chemical and Physical Data* (CRC Press, Boca Raton, 2011)
78. J. Bendtsen, J. Raman Spectrosc. **2**, 133 (1974)
79. P. Atkins, J. de Paula, R. Friedman, *Quanta, Matter, and Change: A Molecular Approach to Physical Chemistry* (Oxford University Press, Oxford, 2009)
80. X.M. Tong, Z.X. Zhao, C.D. Lin, Phys. Rev. A **66**, 033402 (2002)
81. F.H.M. Faisal, A. Abdurrouf, K. Miyazaki, G. Miyaji, Phys. Rev. Lett. **98**, 143001 (2007)
82. K. Miyazaki, M. Kaku, G. Miyaji, A. Abdurrouf, F.H.M. Faisal, Phys. Rev. Lett. **95**, 243903 (2005)
83. R.B. Philip, J. Per, *Molecular Symmetry and Spectroscopy*, 2nd edn. (NRC Research Press, Ottawa, 2006)
84. W.H. Fletcher, J.S. Rayside, J. Raman Spectrosc. **2**, 3 (1974)
85. H. Li, W. Li, Y. Feng, J. Liu, H. Pan, H. Zeng, Phys. Rev. A **85**, 052515 (2012)
86. J. Itatani, D. Zeidler, J. Levesque, M. Spanner, D.M. Villeneuve, P.B. Corkum, Phys. Rev. Lett. **94**, 123902 (2005)
87. M. Herman, A. Campargue, M.I. El Idrissi, Auwera J. Vander, J. Phys. Chem. Ref. Data **32**, 921 (2003)
88. R. Torres, N. Kajumba, J.G. Underwood, J.S. Robinson, S. Baker, J.W.G. Tisch, R. de Nalda, W.A. Bryan, R. Velotta, C. Altucci, I.C.E. Turcu, J.P. Marangos, Phys. Rev. Lett. **98**, 203007 (2007)
89. M. Spanner, S. Patchkovskii, Chem. Phys. **414**, 10 (2013)

Ultracold Metastable Helium: Ramsey Fringes and Atom Interferometry

W. Vassen, R.P.M.J.W. Notermans, R.J. Rengelink, and R.F.H.J. van der Beek

Abstract We report on interference studies in the internal and external degrees of freedom of metastable triplet helium atoms trapped near quantum degeneracy in a 1.5 μm optical dipole trap. Applying a single $\pi/2$ rf pulse we demonstrate that 50% of the atoms initially in the $m = +1$ state can be transferred to the magnetic field insensitive $m = 0$ state. Two $\pi/2$ pulses with varying time delay allow a Ramsey-type measurement of the Zeeman shift for a high precision measurement of the $2\,^3S_1$ $-2\,^1S_0$ transition frequency. We show that this method also allows strong suppression of mean-field effects on the measurement of the Zeeman shift, which is necessary to reach the accuracy goal of 0.1 kHz on the absolute transition frequencies. Theoretically the feasibility of using metastable triplet helium atoms in the $m = 0$ state for atom interferometry is studied demonstrating favorable conditions, compared to the alkali atoms that are used traditionally, for a non-QED determination of the fine structure constant.

1 Introduction

The helium atom has a long history as testing ground for fundamental atomic physics. With two electrons, helium is a three-body system and the nonrelativistic Schrödinger equation cannot be solved exactly. Level energies are therefore more difficult to calculate than for atomic hydrogen showing a more stringent test of atomic physics theory. Calculations of level energies and transition frequencies have pushed our understanding of atomic physics since the twenties of last century. A major breakthrough occurred in the nineties with the advent of variational calculations in a double basis set in correlated form for the electrons, adding relativistic and quantum electrodynamics (QED) terms in orders of the fine structure constant α and the reduced electron to helium mass ratio μ/M_{He} [1, 2]. As

This article is part of the topical collection "Enlightening the World with the Laser" - Honoring T. W. Hänsch guest edited by Tilman Esslinger, Nathalie Picqué, and Thomas Udem.

W. Vassen (✉) • R.P.M.J.W. Notermans • R.J. Rengelink • R.F.H.J. van der Beek
LaserLaB, Department of Physics and Astronomy, Vrije Universiteit Amsterdam, De Boelelaan 1081, 1081 HV Amsterdam, The Netherlands
e-mail: w.vassen@vu.nl

D. Meschede et al. (eds.), *Exploring the World with the Laser*,
https://doi.org/10.1007/978-3-319-64346-5_32

nonrelativistic calculations can now be performed to virtually arbitrary precision, measurements of level energies nowadays are sensitive to QED and nuclear size effects. As these effects are strongest for *S*-states and small principle quantum number n, the $n\,^{1,3}S$ states are theoretically the most promising to test QED. In particular the $n = 2$ states are important for high-resolution spectroscopy as these also show long lifetimes, 7800 s for the $2\,^3S_1$ state and 20 ms for the $2\,^1S_0$ state (natural linewidth 8 Hz), while the $2\,^3P$ state has, for an allowed electric dipole transition, a relatively long lifetime of 98 ns (natural linewidth 1.6 MHz). A helium level scheme is shown in Fig. 1.

Transition frequencies in helium can nowadays be measured more accurately than calculated, where the theoretical limitation is in the calculation of high-order QED terms. This hampers extraction of the charge radius of the helium nucleus (the alpha-particle for ^{4}He and the helion for ^{3}He) from transition frequencies with an accuracy that can compete with other experiments. However, in calculating transition isotope shifts between ^{4}He and ^{3}He, QED terms cancel to a large extent, allowing very accurate extraction of the difference in the (squared) nuclear charge radii of the alpha-particle and the helion. This is particularly interesting in relation to the proton size puzzle [3–5]. To help solving the proton size puzzle, Lamb shift measurements have recently been performed in muonic ^{4}He$^+$ and ^{3}He$^+$ ions, from which results are expected soon. The projected accuracy of the muonic helium experiments is around 0.5 am (0.03% relative accuracy in the nuclear charge radius) [4]. For electronic helium and assuming point nuclei, Pachucki and Yerokhin [6]

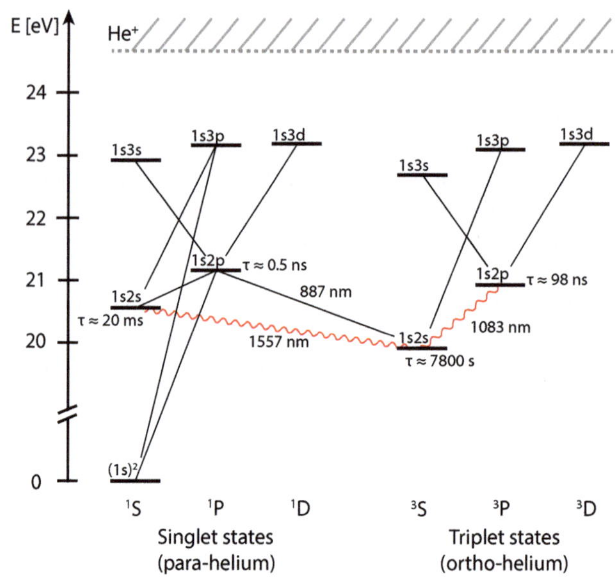

Fig. 1 Level scheme, lifetimes and transition wavelengths for low-lying $n\,^{1,3}L_J$ states of helium ($n < 4$). The $2\,^3S_1$ and $2\,^1S_0$ states are metastable and can be populated in a dc discharge. The $2\,^3S_1$ state is the ground state of orthohelium and is the starting point of experimental work in this paper

have performed QED calculations of the isotope shift with an accuracy of 0.7 and 3.9 kHz for, respectively, the $2\,^3S_1$–$2\,^1S_0$ transition at 1557 nm and the $2\,^3S_1$–$2\,^3P$ transition at 1083 nm. These QED limited accuracies allow extraction of the squared nuclear charge radius difference with an accuracy that will be similar to values deduced from the muonic helium experiments if the experimental accuracy of the isotope shift is of similar or higher accuracy. Presently, for the $2\,^3S_1$–$2\,^1S_0$ transition, the accuracy is 2.4 kHz [7] while for the $2\,^3S_1$–$2\,^3P$ transition the isotope shift accuracy is 3.2 kHz [8, 9]. Surprisingly, a four standard deviation difference exists between the nuclear charge radius difference extracted from both measurements.

The possibility to accurately calculate level energies and wavefunctions has allowed confrontation with several other experimental results. Radiative lifetimes of the $2\,^3S_1$ and $2\,^1P_1$ states have been measured in cold clouds of helium atoms initially prepared in the metastable $2\,^3S_1$ state of ^{4}He [10, 11], showing good agreement with theory. Also molecular potentials for two metastable helium atoms can be calculated very accurately in some cases. This has allowed a stringent test of quantum chemistry calculations from a measurement of the s-wave scattering length a between two $m = +1$ atoms in the metastable $2\,^3S_1$ state, $a_{\text{theory}} = 7.567\,(24)$ nm [12, 13], while $a_{\text{exp}} = 7.512\ (5)$ nm [14]. Examples of other confrontations between experiment and theory for cold collisions between metastable helium atoms can be found in Ref. [15].

Helium atoms in the metastable $2\,^3S_1$ state (He*) are also very interesting from the perspective of atomic matter wave physics. Being light, superposition states with different momenta spatially separate fast and detection of He* atoms can be performed on a microchannel plate (MCP) detector with high efficiency [15]. Actually one of the first experiments on atom interferometry (Young's double slit experiment) was performed with a beam of He* atoms [16]. Transversal Bragg scattering experiments in a well-collimated beam of He* atoms were performed around the turn of the century in Eindhoven [17]. Recent experiments include measurement of the Hanbury Brown Twiss effect for bosons [18] and fermions [19], the Hong-Ou-Mandel effect for matter waves [20] and realization of Wheelers delayed choice experiment for single massive particles [21].

In this paper we present recent results on interferometry in the internal and external degrees of freedom with He* atoms. First we discuss interference experiments to prepare ^{4}He* atoms in a superposition of the $m = +1, 0$ and -1 states. This is realized by a $\pi/2$ rf pulse transferring 50% of the $m = +1$ atoms, trapped in a dipole trap, to the $m = 0$ state and 25% to the $m = -1$ state. By performing a $\pi/2$ $-\pi/2$ pulse sequence we deduce the value of the magnetic field in our setup by measuring Ramsey-type oscillations between the states. This is crucial to correct for the Zeeman shifts in the $2\,^3S_1$–$2\,^1S_0$ transitions in ^{3}He, which has hyperfine structure and no $m = 0$ states. Furthermore, we show that this method allows a mean-field shift free measurement of the Zeeman shift, which is necessary to reach the absolute transition frequency accuracy goal of 0.1 kHz.

The second part of the paper focuses on the use of He* atoms in atom interferometry. We discuss the fundamental advantages of using helium for atom interferometry in general and for determining an accurate value of the fine structure constant. These advantages relate to the metastability of the atoms allowing efficient detection, the very small second-order Zeeman shift for $m = 0$ atoms, the 1083 nm laser wavelength, and the low mass, allowing macroscopic ($\sim 1\,\mathrm{m}$) wavepacket separation.

2 Spectroscopy

In an earlier publication we have measured the transition frequency of the $2\,{}^3S_1$–$2\,{}^1S_0$ transition in ^{3}He and ^{4}He with 1.5 kHz resp. 1.8 kHz accuracy [7]. The 2.3 kHz accuracy in the isotope shift, together with the 0.7 kHz theoretical accuracy in the point-nucleus isotope shift [6], has allowed a determination of the difference in the squared nuclear charge radius of 1.028 (11) fm^2 [6]. With the present knowledge of the charge radius of the alpha-particle, i.e., 1.681 (4) fm [22], this provided an evaluation of the helion charge radius with the same accuracy. However, the 0.5 am projected accuracy of the nuclear charge radii for both isotopes determined from muonic helium [4] constitutes a factor 4 smaller (0.0026 fm^2) error in the difference in squared nuclear charge radii. Our goal for a new measurement is at least to match this accuracy in order to compare the values of the squared nuclear charge radii for muonic and 'normal' matter with similar accuracy. For this to happen we need to measure the absolute frequencies of the ^{4}He and ^{3}He $2\,{}^3S_1-2\,{}^1S_0$ transition at 1557 nm with sub-kHz accuracy. A 0.2 kHz accuracy in the experimental isotope shift together with the 0.7 kHz accuracy in the theoretical isotope shift for point nuclei already allows an accuracy of 0.0033 fm^2 in the squared nuclear charge radius difference. If theory would be at the same 0.2 kHz accuracy (which seems feasible [23]), the squared nuclear charge radius difference would be at the 0.001 fm^2 level, better than anticipated with the muonic experiments and so providing crucial input to solve the proton size puzzle.

Aiming for a 0.1 kHz accuracy in the experimental transition frequency for both ^{4}He and ^{3}He several challenges have to be met. Looking at the error budget of our previous experiment [7] one major error source is the ac Stark shift as a result of the optical dipole trapping potential. This we will solve by working in a magic wavelength crossed dipole trap. At 319.8 nm the polarizabilities of the $2\,{}^3S_1$ and $2\,{}^1S_0$ states cancel [24]. We have generated 2 W narrowband radiation at this wavelength, more than enough to trap our atoms and have demonstrated trapping and Bose–Einstein condensation in a 319.8-nm dipole trap [25]. Our lock of the 1557-nm spectroscopy laser to the femtosecond frequency comb was a second error source. This we resolved by phase-locking a narrowband 1557-nm laser to an ultrastable laser (linewidth $<$2 Hz) using a frequency comb to bridge the 15 nm wavelength difference between both lasers. A third challenge is Zeeman shifts due to the presence of a small (earth-)magnetic field. $m = +1$ and $m = -1$ Zeeman states of the $2\,^3S_1$ state shift by ±2.8 MHz/Gauss. While for ^{4}He one could work with atoms in the $m = 0$ state, that shows a negligible second-order Zeeman shift (2.3 and 3.2 mHz/G^2 for, respectively, the $2\,^3S_1$ and $2\,^1S_0$ states [26]), ^{3}He does not allow this and one has to measure the magnetic field with sufficient accuracy and rely on sequential σ^- excitation (from $F = 3/2, m_F = +3/2$) and σ^+ (from $F = 3/2$, $m_F = -3/2$) excitations with opposite Zeeman shift to find the zero B-field transition frequency.

In Sect. 2.1 we give a short overview of our experimental setup, in Sect. 2.2 we discuss how we populate the $m = 0$ state for ^{4}He, and in Sect. 2.3 we discuss the present status of our Zeeman shift measurement, based on two $\pi/2$ pulses, including a full theoretical discussion of the frequency-degenerate three-level cascade system interacting with resonant rf radiation.

2.1 Experimental Setup

The main parts of our experimental setup have been described in Ref. [7]. In short, we use a liquid-nitrogen cooled dc discharge to excite helium atoms into the metastable triplet state. The He* atomic beam, with a most probable velocity of 1100 m/s, is collimated and slowed to allow trapping in a magneto-optical trap (MOT), where the atoms are transferred to a cloverleaf magnetic trap and further cooled toward BEC by evaporative cooling. About 10^6 atoms in the $m = +1$ state are transferred to a 1557-nm (sufficiently detuned from the transition) crossed dipole trap. A small magnetic field is used to keep the atoms in the spin-polarized $m = +1$ state as atoms in the $m = 0$ state show fast Penning ionization [15]. $m = -1$ states are populated by rapid adiabatic passage using a magnetic field sweep [7, 27]. In this way we can transfer nearly 100% of the atoms to $m = -1$. We detect the number of atoms as well as the temperature by either absorption imaging (on an InGaAs photodiodes camera) or by time-of-flight to a microchannel plate (MCP) detector mounted 17 cm below the trap. Ultracold ^{3}He atoms are laser cooled and trapped using a separate laser system and further cooled toward quantum degeneracy by sympathetic cooling with ^{4}He atoms [28, 29].

We detect the relative population of the magnetic substates $m = +1, m = 0$ and $m = -1$ by a Stern–Gerlach technique, where we release the atoms from the dipole trap and apply a magnetic field gradient. The substates separate in space and are observed using absorption imaging. We have determined the magnetic field strength during our experiment by inducing transitions from $m = +1$ to $m = 0$ and $m = -1$ as a function of rf frequency for a 40 μs pulse and measuring the spin-flip ratio (see Fig. S1 of Ref. [7]). Using this method and observing the resonance frequency over extended periods of time (to measure a linear trend) we were able to deduce the Zeeman shift with 0.5 kHz accuracy, which should be improved to the 0.1 kHz level.

2.2 *rf Excitation of m=0 States*

For spectroscopy of ^{4}He it is advantageous to work with $m = 0$ atoms in-stead of $m = +1$ atoms due to the absence of magnetic field shifts. It is not trivial to populate only the $m = 0$ state. Two Raman laser pulses slightly detuned from the $2\,^3S_1$–$2\,^3P_0$ transition [20] can be used and close to 100% transfer efficiency can be obtained at the cost of a separate laser system. A simpler and robust method is to apply an rf pulse as discussed in the next subsection.

2.2.1 Theory of Resonant Three-Level Rabi Oscillations

^{4}He has no nuclear spin, which causes the quadratic Zeeman shift to be extremely small. At the small fields that we typically work at (of the order of a few Gauss) it is completely negligible so to excellent approximation ^{4}He in the $2\,^3S_1$ state is a symmetric three-level system with a splitting given by the linear Zeeman shift. The special case of a *cascaded* multi-level system is a system of multiple symmetric Zeeman levels. This system has a very interesting symmetry: coupling together two levels with an rf field automatically couples the other levels as well in a stepwise (cascaded) fashion.

For the state vector

$$|\psi\rangle = C_+|m = +1\rangle + C_0|m = 0\rangle + C_-|m = -1\rangle \tag{1}$$

the Hamiltonian, in the rotating wave approximation and neglecting the z-component of $\boldsymbol{B}_{\rm rf}$, is

$$H = g_J\mu_B \boldsymbol{S}\cdot\boldsymbol{B}(t) = \hbar\begin{pmatrix} \omega_0 & \frac{\Omega}{\sqrt{2}}e^{-i\omega t} & 0 \\ \frac{\Omega}{\sqrt{2}}e^{i\omega t} & 0 & \frac{\Omega}{\sqrt{2}}e^{-i\omega t} \\ 0 & \frac{\Omega}{\sqrt{2}}e^{i\omega t} & -\omega_0 \end{pmatrix}. \tag{2}$$

Here $\boldsymbol{B}(t) = \boldsymbol{B}_0 + \boldsymbol{B}_{\rm rf}(t)$, ω is the rf frequency, $\omega_0 = g_J\mu_B B_0/\hbar$ the Larmor frequency and Ω the effective Rabi frequency, defined as $\Omega^2 = \Omega_R^2 + \delta^2$, with $\Omega_R = g_j\mu_B|B_{\rm rf}|/2\hbar$ the on-resonance Rabi frequency and δ the detuning from resonance.

The populations in the substates then follow from the coupled equations, derived from the time-dependent Schrödinger equation:

$$i\frac{\mathrm{d}C_+}{\mathrm{d}t} = \omega_0 C_+(t) + \frac{\Omega}{\sqrt{2}}\exp(-i\omega t)C_0(t) \tag{3}$$

$$i\frac{\mathrm{d}C_0}{\mathrm{d}t} = \frac{\Omega}{\sqrt{2}}\exp(i\omega t)C_+(t) + \frac{\Omega}{\sqrt{2}}\exp(-i\omega t)C_-(t) \tag{4}$$

$$i\frac{\mathrm{d}C_-}{\mathrm{d}t} = \frac{\Omega}{\sqrt{2}}\exp(i\omega t)C_0(t) - \omega_0 C_-(t). \tag{5}$$

These equations can be solved analytically. For the case of zero detuning ($\omega = \omega_0$), where $\Omega = \Omega_R$, and when the atoms start out spin-polarized ($m = +1$):

$$|C_+(t)|^2 = \cos^4\left(\frac{\Omega_R t}{2}\right) \tag{6}$$

$$|C_0(t)|^2 = \frac{1}{2}\sin^2(\Omega_R t) \tag{7}$$

$$|C_-(t)|^2 = \sin^4\left(\frac{\Omega_R t}{2}\right). \tag{8}$$

This result is shown in Fig. 2. It is thus possible to transfer 50% of the atoms to $m = 0$, while 100% can be transferred to $m = -1$.

2.2.2 Results on Three-Level Rabi Oscillations

Both for the experiments on spectroscopy and atom interferometry it is important to use metastable ^{4}He atoms in the $m = 0$ substate. Figure 2 shows that a $\pi/2$ pulse ($t = \pi/(2\Omega_R)$) is expected to transfer 50% of the atoms in the $m = +1$ state to $m = 0$. In the experiment we prepared 10^6 atoms, either slightly above the

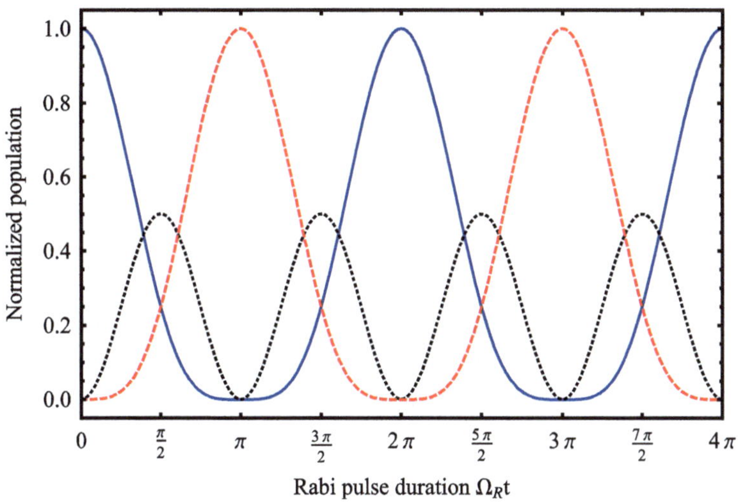

Fig. 2 Normalized population of the $m = +1$ (*blue line, full*), $m = 0$ (*black, dotted*) and $m = -1$ (*red, dashed*) substates according to Eqs. 6, 7 and 8. A $\pi/2$ pulse transfers 50% of the atoms to $m = 0$, a π pulse transfers all atoms to $m = -1$

transition to BEC or primarily in a BEC, in the $m = +1$ state at an earth-magnetic field strength of about 0.5 Gauss. An rf pulse was then applied for a varying time and the population in the three magnetic substates was measured by absorption imaging after expansion in a magnetic field gradient. In Fig. 3 we show results for the case of a BEC as this provides the best contrast in absorption imaging 8 ms after switching off the trap. As it is difficult to normalize the different pictures we have normalized to the total number of atoms at all measurement times. Poisson noise in the atom number then introduces an up to 10% error in the normalized population of the different m states.

Figure 3 clearly shows that Eqs. 6, 7 and 8, with a Rabi frequency $\Omega_R = 2\pi\times$ 23.348(3) kHz, represent the measurements very well in general. The signal-to-noise stays the same, but a change in the Rabi frequency is apparent for long Rabi pulse lengths. This is possibly caused by transient behavior of the rf amplifier system, but currently not limiting for creating $\pi/2$ and π pulses for which we use the shortest Rabi pulse length available ($\sim$ 10 and $\sim 20\,\mu$s, respectively). A π pulse transfers all atoms to the $m = -1$ state while for a $\pi/2$ pulse we find equal numbers of atoms in each of the spin-stretched states as expected. However, there is a clear deficit of atoms in the $m = 0$ state. Where 50% is expected only about 25% is observed. This is illustrated in the left part of Fig. 4, which shows an image of the three magnetic substates after a $\pi/2$ pulse on a BEC and 8 ms expansion time. We attribute this deficit of $m = 0$ atoms to Penning ionization within the expanding $m = 0$ cloud. Being strongly dependent on density [15] we tested this by preparing a thermal cloud with about a factor 10 lower density. Indeed we see (right part of Fig. 4b) that after the same expansion time now the number of $m = 0$ atoms

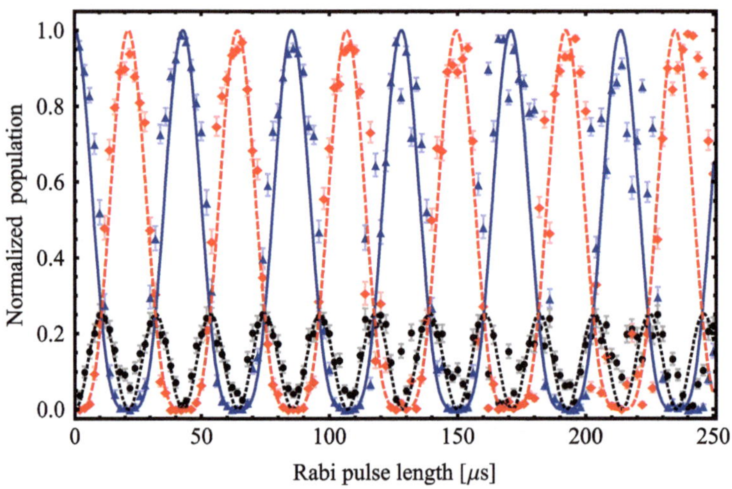

Fig. 3 Relative populations of the $m = +1$ (*blue triangles*), $m = 0$ (*black dots*) and $m = -1$ (*red diamonds*) state of He* as a function of pulse length, for atoms starting in the $m = +1$ state at $t = 0$. The *error bars* are largest near a population of 50% and zero at 0 and 100% as a consequence of normalization

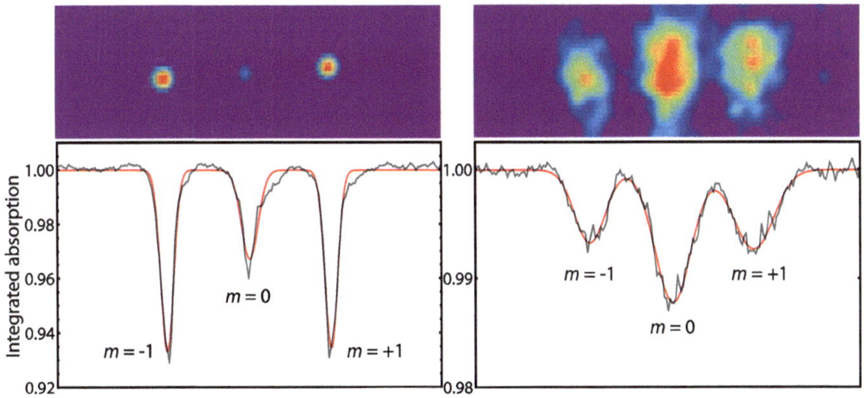

Fig. 4 Absorption images taken after a $\pi/2$ pulse in case of Bose–Einstein condensate (*left*) and a thermal cloud (*right*). The pictures and the integrated column density below it are taken 8 ms after switching off the dipole trap. This causes the thermal cloud to expand much faster diminishing Penning ionization losses. The expected relative populations of the three magnetic substates are thus recovered

remaining for imaging is much larger relative to the number of atoms in $m = 1$ and $m = -1$ and of the order of 50%.

Applying a magnetic field gradient after the $\pi/2$ pulse pushes the $m = \pm 1$ state atoms out of the dipole trap, leaving a pure $m = 0$ cloud of quantum degenerate ^{4}He*. This can be used for atom interferometry (see Sect. 3), but also provides a new possibility for spectroscopy in ^{4}He. Although Penning ionization losses do not allow sufficient time to perform spectroscopy in a Bose - Einstein condensate, we have observed the $2\,^3S_1\; m = 0 - 2\,^1S_0\; m = 0$ transition in a thermal cloud near quantum degeneracy. This allows for a Zeeman shift free measurement of the transition frequency in ^{4}He, especially if future measurements are performed in an optical lattice where Penning ionization losses are suppressed.

2.3 $m = 1$ *Zeeman Shift Measurements*

Although it is in principle best to work with $m = 0$ atoms, Penning ionization limits the lifetime of such a cloud. Moreover, for ^{3}He there is no $m = 0$ state. It is therefore mandatory to investigate ways to measure the magnetic field strength with sufficient accuracy. In the following subsection we discuss Ramsey spectroscopy as a means to measure the magnetic strength by observing the Larmor precession after creation of a three-level superposition state with a $\pi/2$ pulse and waiting as long as possible before introducing the second $\pi/2$ pulse. As we work with a magnetic field B_0 of the order of 0.5 G, this causes oscillations in the order of microseconds (Larmor frequency $\sim 1.5\,\text{MHz}$) and allows an accurate measurement of the magnetic field strength via $\omega_0 = g_J \mu_B B_0/\hbar$.

In between the two rf pulses mean-field interactions between the different Zeeman substates may play a role as these cause shifts of the chemical potential and therefore also of the transition frequency. These effects and their mitigation therefore need to be addressed, which is done in Sect. 2.3.2.

2.3.1 Theory of Ramsey Spectroscopy

To generate a Ramsey signal the rf field is switched on for a time τ, after which the spins precess freely for a time ΔT, and then the rf field is switched on again for a time τ. To describe this theoretically we need the time evolution operator for free precession $U_{\text{free}}(\Delta T)$ and for an rf pulse $U_{\text{Rabi}}(\tau)$. The wavefunction after a Ramsey measurement is then

$$|\psi(t + t_{\text{seq}})\rangle = U_{\text{Rabi}}(\tau)U_{\text{free}}(\Delta T)U_{\text{Rabi}}(\tau)|\psi(t)\rangle, \tag{9}$$

where $t_{\text{seq}} = 2\tau + \Delta T$. The initial phases for both rf pulses should be the same for all free evolution times ΔT.

To measure the population of the different magnetic substates with good signal-to-noise, $\tau = \pi/(2\Omega_R)$ (a $\pi/2$ pulse) is used. Scanning ΔT, oscillations should be observed with frequency given by the Larmor frequency.

The analytic form of the time evolution operator for the $\pi/2$ pulses with zero detuning have simple forms:

$$U_{\text{Rabi}}\left(\frac{\pi}{2\Omega_R}\right) = \begin{pmatrix} \frac{1}{2} & -\frac{i}{\sqrt{2}} & -\frac{1}{2} \\ -\frac{i}{\sqrt{2}} & 0 & -\frac{i}{\sqrt{2}} \\ -\frac{1}{2} & -\frac{i}{\sqrt{2}} & \frac{1}{2} \end{pmatrix}. \tag{10}$$

From Eq. 9 an analytical expression for the populations of the different spin components after a Ramsey measurement with $\pi/2$ pulses can be deduced which is very similar to the case of a single $\pi/2$ pulse (with C_+ and C_- interchanged, and ω_0 instead of Ω_R):

$$|C_+(\Delta T)|^2 = \sin^4\left(\frac{\omega_0 \Delta T}{2}\right), \tag{11}$$

$$|C_0(\Delta T)|^2 = \frac{1}{2}\sin^2(\omega_0 \Delta T), \tag{12}$$

$$|C_-(\Delta T)|^2 = \cos^4\left(\frac{\omega_0 \Delta T}{2}\right). \tag{13}$$

2.3.2 Mean-Field Effects on the Zeeman Shift

In the Ramsey spectroscopy experiment, set up to measure the Zeeman shift, we have not yet considered the fact that mean-field interactions between the different magnetic substates have an effect on the phase evolution of the individual states. We incorporate these effects to the total energy for the individual m states (expressed as the chemical potential μ for trapped Bose–Einstein condensates), and they can be directly used as a modification of the free evolution operator U_{free} introduced in the previous section.

In order to fully describe the problem, we use the Gross-Pitaevskii (GP) description of the three BECs that are present once we apply a $\pi/2$ Rabi pulse. For a state i out of the three states i, j, k (constituting the $m = +1, 0, -1$ states), the wavefunction ψ_i is described by the GP equation

$$\left(-\frac{\hbar^2\nabla^2}{2m} + V_{\text{int}} + N_i g_{i,i}|\psi_i|^2 + N_j g_{i,j}|\psi_j|^2 + N_k g_{i,k}|\psi_k|^2\right)\psi_i = \mu_i\psi_i, \tag{14}$$

where V_{int} is the internal energy of the state and N_i the atom number in state i. The interaction parameter $g_{i,j}$ is defined as

$$g_{i,j} = \frac{2\pi\hbar^2 a_{i,j}}{M}, \tag{15}$$

where $a_{i,j}$ is the s-wave scattering length between states i and j. From Fig. 1 of Ref. [30] we extract scattering lengths if we ignore the imaginary part related to Penning ionization or degenerate collisions $(0,0 \rightarrow -1, +1$ or vice versa): $a_{0,0} \approx 120$ $a_0, a_{1,1} = a_{-1,-1} = a_{0,1} = a_{0,-1} \approx 140\, a_0, a_{1,-1} \approx 60\, a_0$ (with a_0 the Bohr radius). Of these three scattering lengths, only $a_{1,1}$ is known from experiments to be 142.0 (1) a_0 [14, 15]. As the interaction strengths $g_{i,j}$ scale linearly with the scattering lengths, we find $g_{0,0} \approx \frac{6}{7} g_{1,1} \approx 2g_{1,-1}$.

In the Thomas-Fermi approximation the kinetic term in the GP equation is neglected with respect to other energies, and the problem simplifies to the scalar expression

$$V_{\text{int}} + N_i g_{i,i}|\psi_i|^2 + N_j g_{i,j}|\psi_j|^2 + N_k g_{i,k}|\psi_k|^2 = \mu_i. \tag{16}$$

From $\psi_i = \sqrt{n_i(\mathbf{r})/N_i}$ we deduce

$$\mu_1 = V_{\text{int}}^1 + g_{1,1}(n_1 + n_0) + n_{-1}g_{1,-1}, \tag{17}$$

$$\mu_0 = V_{\text{int}}^0 + g_{1,1}(n_1 + n_{-1}) + g_{0,0}n_0, \tag{18}$$

$$\mu_{-1} = V_{\text{int}}^{-1} + g_{1,1}(n_{-1} + n_0) + n_1 g_{1,-1}. \tag{19}$$

The internal energies all contain the same Stark shift from the optical dipole trap (as this is a scalar shift). It is therefore omitted from this calculation. The Zeeman shift is included as a linear shift. Defining the $m = 0$ state as reference energy for the

Zeeman shift, $V_{\rm int}^{0} = 0$ and $V_{\rm int}^{\pm 1} = \pm\hbar\omega_0$, where ω_0 is the Larmor frequency. The energies V_{int} would be the only relevant energies in the free evolution operator $U_{\rm free}$ as introduced in the previous section, and now they can be replaced by the full chemical potential μ of each state. As long as the timescale of interest is shorter than the Penning ionization timescale (which leads to atom number loss), the chemical potentials fully describe the phase evolution of the three components of the Bose–Einstein condensate.

When applying a perfect $\pi/2$ pulse, $n_1 = n_{-1} = n/4$ and $n_0 = n/2$, where n is the initial density of the BEC. In that case

$$\mu_1 = \hbar\omega_0 + \frac{n}{4}(3g_{1,1} + g_{1,-1}), \tag{20}$$

$$\mu_0 = \frac{n}{2}(g_{1,1} + g_{0,0}), \tag{21}$$

$$\mu_{-1} = -\hbar\omega_0 + \frac{n}{4}(3g_{1,1} + g_{1,-1}) \tag{22}$$

and the differential phases become

$$\mu_1 - \mu_{-1} = 2\hbar\omega_0, \tag{23}$$

$$\mu_1 - \mu_0 = \hbar\omega_0 + \frac{n}{4}(g_{1,1} + g_{1,-1} - 2g_{0,0}) \approx \hbar\omega_0 - \frac{n}{14}g_{1,1}. \tag{24}$$

To get an order of magnitude estimate of these effects on the chemical potential of the individual states, we calculate the mean-field energies $E_{\rm mf}^{m}$ of the m states for a typical BEC density of 10^{13} cm^{-3}. We get

$$E_{\rm mf}^{\pm 1} = \frac{n}{4}(3g_{1,1} + g_{1,-1}) \approx h \times 2.0 \text{ kHz}, \tag{25}$$

$$E_{\rm mf}^{0} = \frac{n}{2}(g_{1,1} + g_{0,0}) \approx h \times 2.2 \text{ kHz}, \tag{26}$$

which are typically $\sim 0.1\%$ of the Zeeman shift experienced by the states. At first sight this is problematic when trying to measure the Zeeman shift beyond the kHz level. However, the mean-field energy drops out as common mode when considering the phase evolution between the $m = +1$ and $m = -1$ states if the densities are equal. In terms of sensitivity, we can calculate the resulting frequency shift $\Delta E_{\rm mf}$ for a fractional population difference $\Delta n/n$ between the two states at the typical experimental density of 10^{13} cm^{-3} :

$$\Delta E_{\rm mf} = \frac{\Delta n}{n} \times h \times 1.3 \text{ kHz}. \tag{27}$$

If we can get the populations to be stable within a few percent the shift is < 0.1 kHz. Also, from a practical point of view this method only involves the $m = \pm 1$ states, which are much better imaged than the $m = 0$ state (see Sect. 3).

To summarize, at our typical BEC densities of 10^{13} cm^{-3}, the mean-field energy of the individual states is around $h \times 2$ kHz and an order of magnitude larger than the accuracy goal of the optical transition frequency measurement. However, when applying the Ramsey measurement method the mean-field energies of the $m = +1$ and $m = -1$ states result in a common phase evolution and they can cancel to a shift of ~ 0.1 kHz if both states are equally populated within 10% of each other using a $\pi/2$ pulse. From the results shown in Fig. 4 this seems doable.

2.3.3 Ramsey Spectroscopy Experimental Results

Extending the experiments to the Ramsey scheme, we applied two $\pi/2$ pulses according to the scheme of Eq. 9. The results are shown in Fig. 5, which shows the normalized (to the sum of the population of the $m = +1$ and $m = -1$ states) number of $m = +1$ atoms after the second $\pi/2$ pulse (with statistical error bars as in Fig. 3, while now ~50 ns timing noise in the $\pi/2$ pulse delay as well as magnetic field noise somewhat increase the error at all delay times). We were able to observe up to 80 oscillations after a $\pi/2$ pulse preparation of the superposition state. The origin of the observed decoherence is presently unkown. It cannot be due to Penning ionization as this does not cause decoherence and also does not occur yet at the 50 μs timescale. A possible source may be experimental rf amplifier noise. On the other hand, as every shot takes about 15 s we suspect the observed decoherence may also be due to slow drifts and shot-to-shot variations in the magnetic field strength.

3 Atom Interferometry

Macroscopic separation of atomic wave packets allows stringent tests of basic quantum mechanics and sensitive sensing of forces when wave packets are split and combined to form an atom interferometer. For the most sensitive experiments cold atoms are used, either from a MOT or BEC. We aim to build an atom interferometer for ultracold helium atoms in the metastable triplet state, optically trapped close to Bose–Einstein condensation at a temperature of ~ 0.2 μK.

3.1 Metastable Helium for Atom Interferometry

Cold atom interferometers generally use alkali atoms. The workhorse is Rb but Cs atoms are also used. Recently first results on Sr have been published [31]. In Table 1 a comparison of some atomic properties of these atoms is presented that shows why He* is a good candidate for several studies.

Despite the experimental complexity of working with atoms in a metastable state, He* has several advantages. From the experimental point of view, He* allows efficient detection of atoms, as a function of time and position, on an MCP detector [15]. Secondly, the required wavelength for manipulating the atoms and splitting

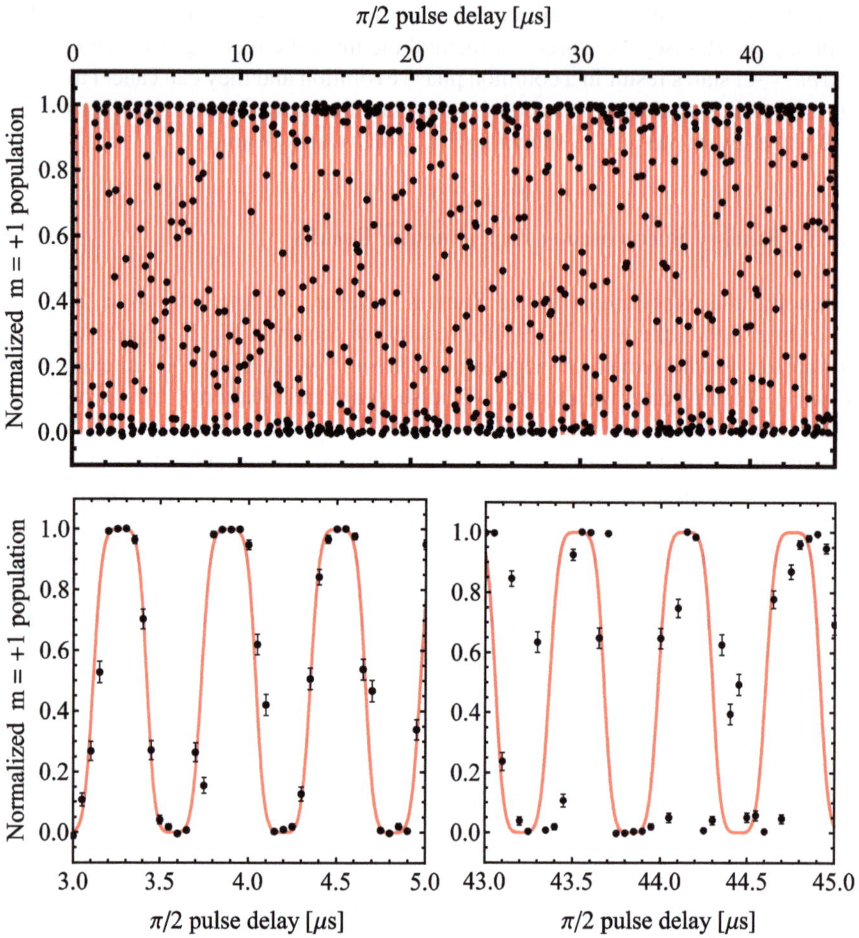

Fig. 5 Relative populations of the $m = +1$ state of He* as a function of pulse delay between two $\pi/2$ pulses, for atoms starting in the $m = +1$ state at $t = 0$. Below an expanded part for short and long time delay showing that the oscillations remain visible up to 45 μs. The *red line* is a fit to the signal applying Eqs. 11 and 13

the wavepacket, often requiring high laser power, is 1083 nm, a wavelength where high power fiber amplifiers and narrow band fiber lasers are commercially available. The required power for generating flat wavefronts over large dimension for wavepacket splitting is therefore readily available. A very important advantage of using He* atoms (in the $m = 0$ state) is the five orders of magnitude lower sensitivity to stray magnetic fields compared to Rb and Cs (see Table 1). This small second-order Zeeman shift originates from the absence of hyperfine structure and, as a consequence, very large energetic distance to nearby other $m = 0$ states that mix in at second order. Because of this low sensitivity no elaborate magnetic

Table 1 Atomic data for atoms used in cold atom interferometry

	Rb	Cs	Sr	He*
Atomic mass	87	133	88	4
Atomic mass precision (in ppb)[a]	0.08	0.07	14	0.016
Laser cooling wavelength (nm)	780	852	461	1083
Recoil velocity (m/s)	0.006	0.0035	0.010	0.092
Electronic state	$5s\,^2S_{1/2}$	$6s\,^2S_{1/2}$	$5s^{2\,1}S_0$	$1s2s\,^3S_1$
Nuclear spin	3/2	7/2	0	0
Total angular momentum	2	4	0	1
2nd order Zeeman shift [Hz/G^2]	288[b]	214[b]	?[c]	0.0023[d]

[a] Ref. [32]; [b] Ref. [33]; [c] For two-electron atoms no information is, to our knowledge, available. However, due to the absence of hyperfine structure, we expect the coefficient to be orders of magnitude smaller than for the alkalis. Note that data on quadratic Zeeman shifts are available for the optical lattice clock $^1S_0$–$^3P_0$ transitions, however, these relate to the shift of the upper state; [d] Ref. [26]

shielding is required, in contrast to the alkalis. A disadvangae of using $m = 0$ atoms is Penning ionization, which implies that low densities have to be used.

Although the low mass will not help in reducing systematic shifts there is still a large advantage in using low mass atoms that relates to the large recoil velocity. This allows easy wavepacket separation over macroscopic distances with only a few photon recoils. To illustrate this we compare the recent large 0.54 m wavepacket splitting obtained with Rb in the 10 m fountain in Stanford [34]. This was realized after 1.04 s ballistic flight following 45 consecutive Bragg pulses providing $90\hbar k$ wavepacket splitting. We estimate that we can realize a 1 m wavepacket splitting after 0.5 s in a fountain of no more than 1.6 m height applying one 12th order Bragg pulse generated using a 2-W fiber amplifier. An experimental setup dedicated to atom interferometry was designed in our laboratory to demonstrate such a large superposition state (see Fig. 6).

We have designed this setup analogous to the setup described in Sect. 2, with a vertical geometry for the atom interferometer. We will work with evaporatively cooled atoms, close to or below the temperature for BEC, initially trapped in a horizontal 1557-nm crossed dipole trap. We then intend to transfer the atoms to a vertical one-dimensional optical lattice at 1083 nm. Bragg diffraction will both be used to launch the atoms and for wavepacket splitting to realize an atom interferometer. Figure 6 shows we intend to detect the different vertical momentum states on an MCP detector after horizontal displacement by a push beam.

Apart from demonstrating the 1-m wavepacket splitting, with a possible test of atom neutrality as application [35], we aim for a measurement of the fine structure constant by measuring the one-photon recoil velocity analogous to the Rb experiment in Paris [36]. He* has additional advantages here as well [37]. The fine structure constant can be deduced from the relation

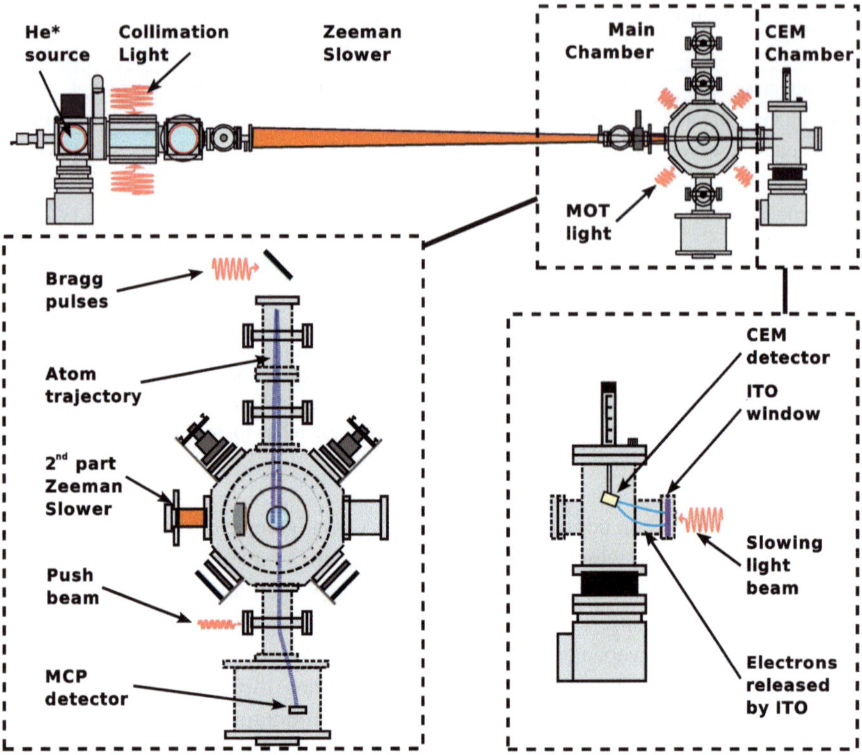

Fig. 6 Experimental setup for atom interferometry and measurement of the fine structure constant. Enlarged parts show the designed geometry for the fountain (*left*) and detection of the atomic beam (*right*). For details see text

$$\alpha^2 = \frac{2R_\infty}{c}\frac{h}{m_e} = \frac{2R_\infty}{c}\frac{M_X m_u}{m_u m_e}\frac{h}{M_X}. \tag{28}$$

The Rydberg constant R_∞ is presently known with 0.006 ppb accuracy, the electron mass in atomic units m_e/m_u with 0.03 ppb accuracy [38] and the atomic masses M_X/m_u for X = Rb, Cs and He with resp. 0.08, 0.07 and 0.016 ppb accuracy [32]. As the most precise measurement of $h\,/\,M$ today is for Rb, i.e., 1.3 ppb [36], a factor of ten improvement in this measurement will already hit the limit set by the knowledge set by Rb mass (measured in a Penning ion trap). Helium provides more room at the bottom in challenging the presently most accurate determination of α, i.e., by applying QED to the measurement of the g-factor of the electron [39]. Comparing an improved measurement of the fine structure constant from photon recoil measurements, almost independent of QED theory, with g-factor results (that require the most advanced high-order QED calculations [40]), may provide the most stringent test of QED theory possible in the near future.

3.2 *Characterization of the Zeeman Slower Using an ITO Window*

The experimental setup shown in Fig. 6 is presently being built. In an initial experiment demonstrating slowing in our Zeeman slower, we used a vacuum window with an indium tin oxide (ITO) coating on the vacuum side. ITO is a conductive layer, transmitting in our case > 80% of the circularly polarized light entering the Zeeman slower. ITO is known to release an electron after impact of a metastable helium atom [41] albeit with an unknown efficiency hampering quantitative measurements. As indicated in Fig. 6 we detect the detached electrons on a channeltron electron multiplier (CEM) detector (with exposed front side at +100 V), mounted 20 cm upstream from the vacuum window just above the 1083-nm Zeeman slower laser beam. Chopping the atomic beam with a rotating slit in a chopper wheel, mounted at the entrance of the first part of the Zeeman slower, allows measurement and optimization of the velocity distribution of the atomic beam. Results are shown in Fig. 7 demonstrating with high signal-to-noise that one can detect the He* atomic beam and measure its velocity.

From the time delay between the light peak, caused by XUV photons from the source, and the time the atoms arrive, the most probable velocity of the atomic beam is measured to be 1100 m/s, which agrees with earlier measurement in another setup [41]. The width of the peaks depends on the rotation velocity and the width of the slit in the chopper wheel. By ramping up the current in the second part of the

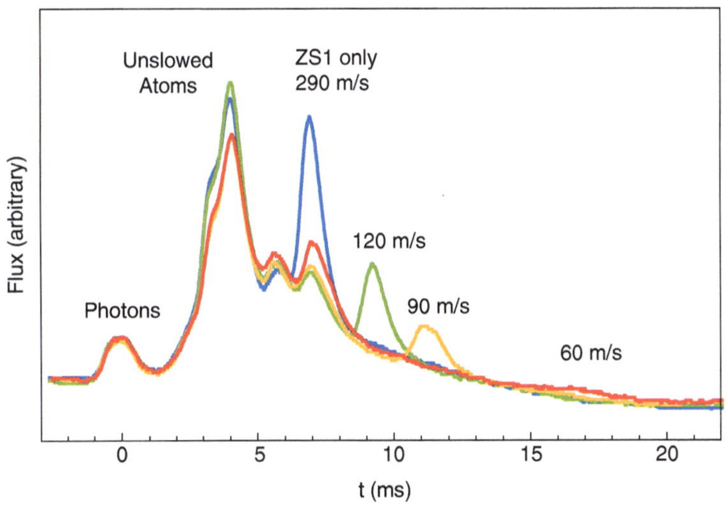

Fig. 7 Channeltron signal showing released electrons from the ITO window in case of a He* beam that is slowed to various final velocities varying the current in the second part of the Zeeman slower. The first peak, caused by high energy XUV photons from the He* plasma source, provides a t=0. Slowed atoms with a average velocity of 290 m/s appear after 7 ms when switching on only the first part of the Zeeman slower (ZS1 only), while lower velocities corresponding to later arrival times appear for increasing currents applied to the second part of the Zeeman slower

slower, the end velocity is tuned. The slowed atomic beam peaks in Fig. 7 correspond to a velocity of 290, 120, 90 and 60 m/s, in good agreement with calculations at the applied currents and the −250 MHz detuning of the Zeeman slower beam. The small peak in between the unslowed and slowed atom peaks in Fig. 6 is caused by atoms that drop out of the slowing process in the first part of the Zeeman slower. Due to transverse divergence of the atomic beam, the atomic flux at the ITO window after slowing to 50 m/s (the capture velocity of a He* MOT) is very small. To measure the flux and optimize the number of slowed atoms we tweaked the transversal collimation stage, which makes a quantitative comparison with the unslowed beam difficult. In the optimization procedure the measured most probable velocity actually reduces to 900 m/s demonstrating that the collimation stage also acts as a velocity selector. Because the slowed atoms cover a large area on the ITO window, the position of the CEM detector directly influences the amount of electrons that can be detected. Lowering the CEM to face more of the window drastically changes the amplitude ratios of the slowed and unslowed beams, where a practical limit was set at the point where it starts to clip the slowing light beam. We also observe a small reduction in the base line due to saturation of the CEM detector, which was used in current mode.

The ITO window thus allows for direct detection of the slowed atomic beam in a way that does not interfere with the rest of the experiment, an advantage made possible by the high internal energy of the metastable atoms.

4 Conclusions and Future Prospects

We have shown progress toward measuring the transition isotope shift of the $2\,^3S_1$–$2\,^1S_0$ transition at 1557 nm with an accuracy that may compete with the muonic helium Lamb shift experiment underway in Switzerland. We demonstrated that we can efficiently transfer our trapped atoms to the magnetic field insensitive $m = 0$ state and demonstrated the suitability of $m = 0$ atoms for both spectroscopy and atom interferometry. Densities below 10^{12} cm^{-3} are needed in order to neglect Penning ionization losses at time scales < 1 s. In the interferometry experiment this implies fast expansion of the cloud after transfer to the $m = 0$ state.

For the spectroscopy experiment we can use both $m = 0$ and $m = 1$ atoms. We can measure the transition in a thermal $m = 0$ gas close to the BEC transition where Doppler broadening increases the linewidth by only a factor of two and the lifetime is sufficiently large while Zeeman shifts are absent. However, to compare and look for systematic effects we will also use $m = 1$ atoms in a BEC (a BEC of $m = 0$ atoms has only a lifetime of $\sim$ 10ms). An $m = 1$ BEC shows a long ($\sim$ 10 s) lifetime also at high density while the earth magnetic field strength may be measured with sufficient accuracy. The mean-field shift of $\sim$ 2 kHz in the magnetic field measurements is a systematic shift which is an order of magnitude larger than the accuracy goal of 0.1 kHz (and is therefore a systematic effect on the measurement of a systematic effect). The Ramsey-type measurement as proposed and demonstrated in this paper provides a way to negate this effect to allow for a 0.1 kHz accurate measurement of the absolute optical transition frequency. As we will soon turn to full operation of the magic wavelength trap [25], the ac Stark shift will no longer be the dominant systematic effect. It should be noted that for ^{3}He, where a magnetic field insensitive measurement is impossible, the accuracy of the Zeeman shift correction will be the limiting factor in the final accuracy. Our proposed Ramsey-type measurement is currently the most promising way to suppress this systematic uncertainty. Ultimately we could work with a three-dimensional magic wavelength lattice with single atoms trapped in the Lamb-Dicke regime at the nodes. This would provide an ideal geometry to perform the experiment with $m = 0$ atoms, without ac Stark, mean-field, recoil and Doppler shifts, and approaching the 8 Hz natural linewidth.

The large recoil velocity, the extremely small second-order Zeeman shift, and unique spatial and temporal detection possibilities make metastable helium an excellent candidate for atom interferometry. For the fine structure constant measurement we showed the high potential to measure the recoil velocity accurately using Bragg diffraction and atom interference. The high accuracy of the helium mass promises a QED test at the 0.01 ppb level by comparison of a measured α with a value of α calculated from g-factor measurements, if systematic effects can be controlled well enough.

Acknowledgements This work was financially supported by the Dutch Foundation for Fundamental Research on Matter (FOM). We gratefully acknowledge Gordon Drake and Krzysztof Pachucki for ab-initio calculations of the coefficient for the second-order Zeeman effect in the 2 $^3S_1$ $m = 0$ and 2 $^1S_0$ $m = 0$ states of ^{4}He*. We acknowledge Rob Kortekaas for technical support, Steven Knoop for stimulating discussions and Kjeld Eikema for support in locking our spectroscopy laser to the femtosecond frequency comb and Hz-laser in his laboratory.

References

1. G.W.F. Drake, in *Handbook of Atomic, Molecular, and Optical Physics*, ed. by G.W.F. Drake (Springer, New York, 2006), p. 199
2. V.I. Yerokhin, K. Pachucki, Phys. Rev. A **81**, 022507 (2010)
3. R. Pohl et al., Nature **466**, 213 (2010)
4. A. Antognini et al., Can. J. Phys. **89**, 47 (2011)
5. C.E. Carlson, Prog. Part Nucl. Phys. **82**, 59 (2015)
6. K. Pachucki, V.A. Yerokhin, J. Phys. Chem. Ref. Data **44**, 031206 (2015)
7. R. van Rooij, J.S. Borbely, J. Simonet, M.D. Hoogerland, K.S.E. Eikema, R.A. Rozendaal, W. Vassen, Science **333**, 196 (2011)
8. P. Cancio Pastor, L. Consolino, G. Giusfredi, P. De Natale, M. Inguscio, V.A. Yerokhin, K. Pachucki, Phys. Rev. Lett. **108**, 143001 (2012)
9. D. Shiner, R. Dixson, V. Vedantham, Phys. Rev. Lett. **74**, 3553 (1995)
10. S.S. Hodgman, R.G. Dall, L.J. Byron, K.G.H. Baldwin, S.J. Buckman, A.G. Truscott, Phys. Rev. Lett. **103**, 053002 (2009)
11. R.P.M.J.W. Notermans, W. Vassen, Phys. Rev. Lett. **112**, 253002 (2014)
12. M. Przybytek, B. Jeziorski, J. Chem. Phys. **12**, 134315 (2005)
13. M. Przybytek, Ph.D. thesis, University of Warsaw (2008)
14. S. Moal, M. Portier, J. Kim, J. Dugué, U.D. Rapol, M. Leduc, C. Cohen-Tannoudji, Phys. Rev. Lett. **96**, 023203 (2006)

15. W. Vassen, C. Cohen-Tannoudji, M. Leduc, D. Boiron, C. Westbrook, A. Truscott, K. Baldwin, G. Birkl, P. Cancio, M. Trippenbach, Rev. Mod. Phys. **84**, 175 (2012)
16. O. Carnal, J. Mlynek, Phys. Rev. Lett. **66**, 2689 (1991)
17. A.E.A. Koolen, G.T. Jansen, K.F.E.M. Domen, H.C.W. Beijerinck, K.A.H. van Leeuwen, Phys. Rev. A **65**, 041601 (2002)
18. M. Schellekens, R. Hoppeler, A. Perrin, J. Viana Gomes, D. Boiron, A. Aspect, C.I. Westbrook, Science **310**, 648 (2005)
19. T. Jeltes, J.M. McNamara, W. Hogervorst, W. Vassen, V. Krachmalnicoff, M. Schellekens, A. Perrin, H. Chang, D. Boiron, A. Aspect, C.I. Westbrook, Nature **445**, 402 (2007)
20. R. Lopes, A. Imanaliev, M. Cheneau, D. Boiron, C.I. Westbrook, Nature **520**, 66 (2015)
21. A.G. Manning, R. Khakimov, R.G. Dall, A.G. Truscott, Nat. Phys. **11**, 539 (2015)
22. I. Sick, Phys. Rev. C **77**, 041302 (2008)
23. K. Pachucki, private communication (2014)
24. R.P.M.J.W. Notermans, R.J. Rengelink, K.A.H. van Leeuwen, W. Vassen, Phys. Rev. A **90**, 052508 (2014)
25. R.J. Rengelink, R.P.M.J.W. Notermans, W. Vassen, Appl. Phys. B **122**, 122 (2016)
26. G. Drake, K. Pachucki, private communication (2016)
27. J.S. Borbely, R. van Rooij, S. Knoop, W. Vassen, Phys. Rev. A **85**, 022706 (2012)
28. A.S. Tychkov, T. Jeltes, J.M. McNamara, P.J.J. Tol, N. Herschbach, W. Hogervorst, W. Vassen, Phys. Rev. A **73**, 031603(R) (2006)
29. J.M. McNamara, T. Jeltes, A.S. Tychkov, W. Hogervorst, W. Vassen, Phys. Rev. Lett. **97**, 080404 (2006)
30. P.J. Leo, V. Venturi, I.B. Whittingham, J.F. Babb, Phys. Rev. A **64**, 042710 (2001)
31. T. Mazzoni, X. Zhang, R. Del Aguila, L. Salvi, N. Poli, G.M. Tino, Phys. Rev. A **92**, 053619 (2015)
32. M. Wang, G. Audi, A.H. Wapstra, F.G. Kondev, M. MacCormick, X. Xu, B. Pfeiffer, Chin. Phys. C **36**, 1603 (2012)
33. D.A. Steck, http://steck.us/alkalidata (2015)
34. T. Kovachy, P. Asenbaum, C.A. Donnelly, S.M. Dickerson, A. Sugarbaker, J.M. Hogan, M. Kasevich, Nature **528**, 530 (2015)
35. A. Arvanitaki, S. Dimopoulos, A.A. Geraci, J. Hogan, M. Kasevich, Phys. Rev. Lett. **100**, 120407 (2008)
36. R. Bouchendira, P. Cladé, S. Guellati-Khélifa, F. Nez, F. Biraben, Phys. Rev. Lett. **106**, 080801 (2011)
37. F. Terranova, G.M. Tino, Phys. Rev. A **89**, 052118 (2014)
38. S. Sturm, F. Kohler, J. Zatorski, A. Wagner, Z. Harman, G. Werth, W. Quint, C.H. Keitel, K. Blaum, Nature **506**, 467 (2014)
39. D. Hanneke, S. Fogwell, G. Gabrielse, Phys. Rev. Lett. **100**, 120801 (2008)
40. T. Aoyama, M. Hayakawa, T. Kinoshita, M. Nio, Phys. Rev. Lett. **109**, 111807 (2012)
41. W. Rooijakkers, W. Hogervorst, W. Vassen, Opt. Commun. **135**, 149 (1997)

Monolayer Graphene as Dissipative Membrane in an Optical Resonator

Hendrik M. Meyer, Moritz Breyer, and Michael Köhl

Abstract We experimentally demonstrate coupling of an atomically thin, free-standing graphene membrane to an optical cavity. By changing the position of the membrane along the standing-wave field of the cavity, we tailor the dissipative coupling between the membrane and the cavity, and we show that the dissipative coupling can outweigh the dispersive coupling. Such a system, for which controlled dissipation prevails dispersion, will prove useful for novel laser-cooling schemes in optomechanics. In addition, we have determined the continuous-wave optical damage threshold of free-standing monolayer graphene of 1.8(4) MW/cm^2 at 780 nm.

The coupling between light and matter is a fundamental ingredient for many applications ranging from highly precise detection of mechanical motion [1, 2] to quantum information science [3]. Accurate control over the amplitude and phase of the coupling requires precise localization of the matter relative to the light field. Such control has been experimentally demonstrated using microscopic, point-like physical systems such as single atoms [4], trapped ions [5, 6], gold nanoparticles [7], semiconductor quantum dots [8], NV centers in diamond [9], and carbon nanotubes [10, 11]. Additionally, macroscopic objects, such as SiN membranes have been coupled locally to standing-wave light fields [12, 13]. They combine low absorption and high dispersion with very good handling, and they have been used, for example, in optomechanical experiments [14] in which the motion-dependent interaction between the membrane and a light field is studied.

Advances in the fabrication of two-dimensional materials, like graphene [15], have made it possible to fabricate atomically thin membranes, which combine the ease of use of macroscopic membranes with the positioning accuracy of a single

This article is part of the topical collection "Enlightening the World with the Laser" - Honoring T. W. Hänsch guest edited by Tilman Esslinger, Nathalie Picqué, and Thomas Udem.

Electronic supplementary material The online version of this chapter (doi:10.1007/978-3-319-64346-5_33) contains supplementary material, which is available to authorized users.

H.M. Meyer (✉) • M. Breyer • M. Köhl
Physikalisches Institut, University of Bonn, Wegelerstraße 8, 53115 Bonn, Germany
e-mail: hmeyer@physik.uni-bonn.de

D. Meschede et al. (eds.), *Exploring the World with the Laser*,
https://doi.org/10.1007/978-3-319-64346-5_33

atom. Owing to their low mass and high stiffness, they promise higher vibrational frequencies [16, 17] than SiN membranes, which is of great interest for future graphene–based optomechanical applications. Another key difference between membranes made of graphene and those made of SiN is the nature of the light-matter coupling. While for SiN membranes the coupling is dispersive, for graphene membranes it has been predicted to be mostly dissipative since a monolayer of graphene exhibits a single–pass absorption of $A \approx \pi\alpha = 2.3\%$ in the optical range [18]. Hence, a graphene membrane will cause a position-dependent dissipation of the intra-cavity field, which is wavelength-independent within the visible to near infrared spectral range. In contrast to standard dispersive optomechanical coupling [14], a predominantly dissipative coupling between membrane and cavity field could possibly allow for novel laser cooling of the membrane's motion along similar lines as recently discussed for dissipation external to the cavity [19, 20]. In the past, photothermal coupling has been used to achieve cooling of mechanical oscillators [21, 22].

Graphene has been embedded into rudimentary optical cavities [23], optomechanically coupled to superconducting microwave cavities [17, 24, 25], and to the evanescent field of an optical microsphere resonator [26]; however, in those cases the spatial structure of the electromagnetic field was not resolved in the coupling. Here we present an experimental study of a free-standing layer of graphene inside a Fabry–Perot resonator. We demonstrate that the position of the membrane with respect to the cavity standing-wave field leads to a strong and highly controlled modulation of the cavity losses due to dissipation. Furthermore, we find that the change of the cavity linewidth due to dissipation is three times larger than the dispersive frequency shift caused by the membrane, which realizes for the first time a regime dominated by dissipation rather than dispersion.

Our experimental setup is shown in Fig. 1. It is composed of a Fabry–Perot optical resonator with a free-standing layer of graphene in the "membrane-in-the-middle" configuration. The position of the graphene with respect to the standing wave can be experimentally controlled which allows us to influence the dissipation and hence the cavity loss rate. The latter is reflected by an increased linewidth and a reduced in-coupling into the cavity when the membrane is moved from a node to an anti-node of the standing-wave field. In order to achieve a good coupling between the cavity mode and the graphene layer, the mode waist of the cavity must be much smaller than the physical size of free-standing graphene flakes, which typically is on the order of 10 μm.

In order to address the challenge of working with small free-standing samples, we have designed an optical cavity featuring two integrated lenses to achieve an experimentally verified TEM_{00} mode waist of $w_0 = (1.2 \pm 0.2)$ μm. The cavity has a total length of $l_{cav} = 32$ cm, corresponding to a free spectral range of $\nu_{FSR} = c/(2l_{cav}) = 470$ MHz (see Fig. 1). We perform reflection spectroscopy on the cavity using light from a single-mode external-cavity diode laser at 780 nm. For reference, we first measure the finesse of the empty cavity to be $\mathcal{F}_e = 2\pi/\mathcal{L}_e = \nu_{FSR}/\delta\nu_e = 83 \pm 9$, where $\delta\nu_e = (5.6 \pm 0.6)$ MHz is the linewidth of the cavity and $\mathcal{L}_e = (7.5 \pm 0.8)\%$ are the total losses. The major contribution of

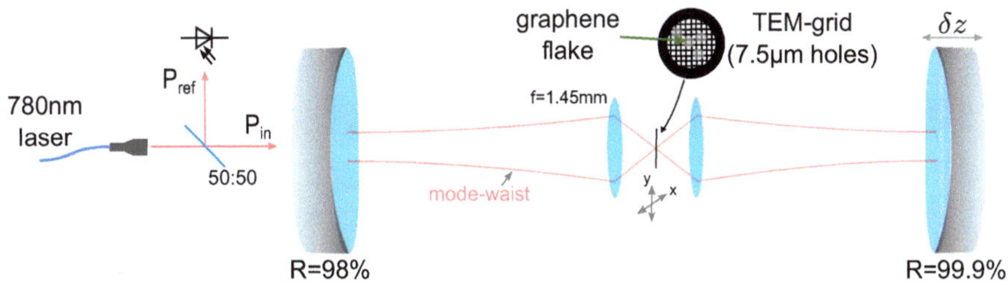

Fig. 1 (Color online) An optical Fabry-Perot resonator is loaded with a free-standing monolayer of graphene. The position of the graphene sample can be fine-tuned in the x-y-plane using a piezo stage. Light is coupled into the cavity from the low-reflectivity side and the reflection from the cavity is monitored using a 50:50 beam splitter and a photodiode. The high-reflectivity mirror is mounted on a piezo transducer in order to adjust the cavity length

the losses comes from the integrated lenses. The coating has a reflectance of $R = 0.5\%$ per surface leading to round trip losses of 4%. The mirror transmissions are $T_1 = 2\%$ and $T_2 = 0.01\%$. As calculations suggest that clipping losses can be neglected, we assume that the additional loss ($\sim 1.5\%$) is coming from lens misalignment. Using better coatings at 780 nm and improved alignment, the finesse can be improved by about a factor of two. The linewidth of the cavity determines the electric field decay rate by $\kappa_e = 2\pi \times \delta\nu_e/2 = 2\pi \times (2.8 \pm 0.3)$ MHz. In our measurements, the length of the cavity was not actively stabilized. Next, we introduce a free-standing graphene sample into the middle Fabry-Perot resonator. We use a commercially available sample grown by chemical vapor deposition (CVD) and transferred onto a transmission electron microscopy (TEM)-grid [27]. The graphene is mostly a single-layer sample but also a few two-layer flakes have been detected, see below.

Initially, we position the graphene sample at the node of the standing-wave field by minimizing the losses (linewidth) of the cavity. When subsequently sweeping the center frequency of the laser by one free spectral range of the cavity ($\Delta\nu_L = \nu_{FSR}$), the relative position between graphene and the standing-wave changes from a node to an anti-node as depicted in Fig. 2a. Resultingly, the losses of the resonator are different for the two laser frequencies and allow us to perform a differential measurement of the cavity losses caused by the graphene layer without the need for mechanical realignment. An example of such a measurement is shown in Fig. 2b. We can also access other locations of the graphene sample relative to the standing wave by changing the cavity length by δz using a piezo transducer at one of the cavity mirrors and perform a similar measurement. In Fig. 2c we show the complete results of how the cavity finesse changes when positioning the graphene sample at different locations within the standing wave. The results highlight that the periodic structure of the cavity losses can be clearly resolved. The observed periodicity of λ stems from moving the cavity mirror rather than the graphene sample.

We fit our experimental data with a theoretical model which computes the total cavity dispersion and losses from the dispersion and absorption of each element in

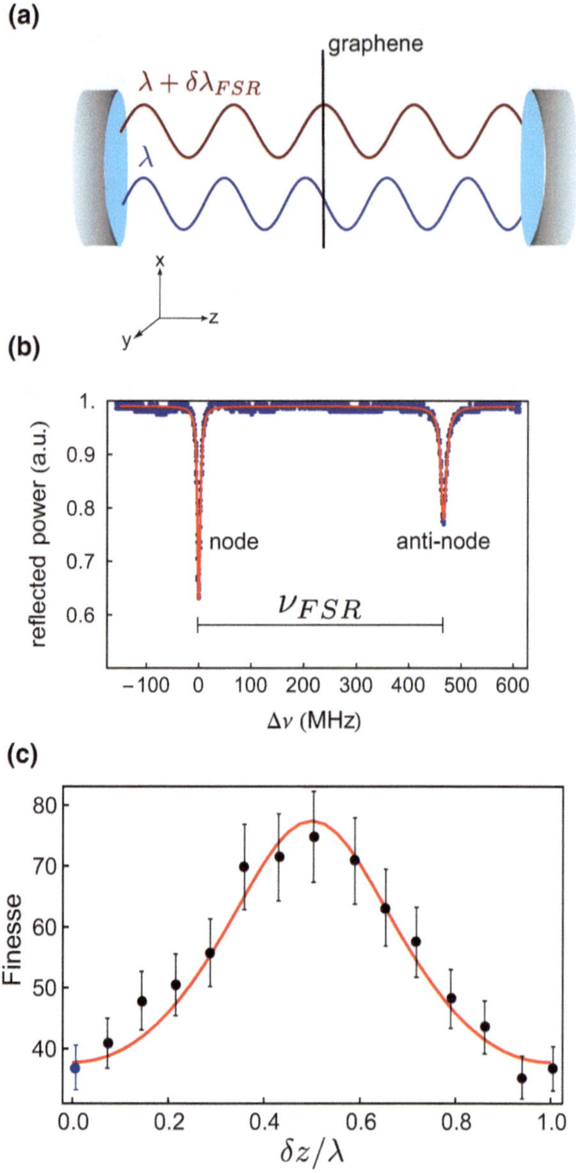

Fig. 2 (Color online) **a** Localization of a layer of graphene with respect to the cavity standing-wave field. When the frequency of the laser is changed by one free spectral range of the cavity and the graphene sample is placed in the middle of the cavity, the single layer shifts from a node into an anti-node of the standing-wave pattern. **b** Reflected power from the cavity while scanning the laser across one free spectral range. On the left resonance graphene is located at a node, while on the right resonance it is located at an anti-node. **c** Finesse of the cavity for different relative positions of the graphene layer and the anti-node of the standing-wave field. The point at $\delta z/\lambda = 0$ (*blue*) is extracted from the data shown in (**b**). Here δz is the mirror position as shown in Fig. 1. The *solid line* is derived from the numerical model discussed in the main text

the resonator and then solves for the condition of a stable resonator round trip [13]. In the model, we have used the reflection and transmission amplitudes for graphene calculated from the Fresnel formulas for a membrane of thickness $d_g = 0.345$ nm and with a complex index of refraction of $n_g = 2.71 + 1.41\,i$ [28]. Together with the empty cavity loss of $\mathcal{L}_e = 8\%$, which is in agreement with the data for the graphene at the node, this model fits the data well, as can be

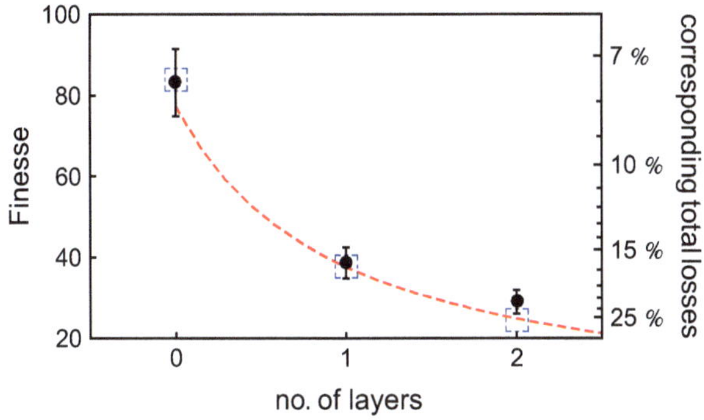

Fig. 3 (Color online) Cavity finesse versus number of graphene layers located at the cavity anti-node. The *dashed line* is obtained from the simulation with the parameters obtained in Fig. 2c. The *empty squares* (*blue*) assume a loss of $4N \cdot A$ for N layers

seen from the solid line in Fig. 2c. For a single layer of graphene, we find that the decay rate of the cavity electric field increases by $\Delta\kappa = 2\pi \times (3.5 \pm 0.4)$ MHz, which quantifies the dissipative coupling between the graphene flake and the resonator mode.

On different sections of our graphene sample, we have observed different values of the absorption. In particular, there are three different discrete values of the finesse, which we attribute to different number of layers. In order to analyze this effect, we extend the numerical model to analyze how many layers of graphene contribute to the absorption. In Fig. 3, the dashed line shows the behavior of the cavity finesse as a function of the (continuous) thickness of the absorber. The data points show the measurements and we find that they match the curve equidistantly, and hence count the numbers of layers. The behavior can also be understood in the following simplified way: an absorber with a single-pass loss A leads to an average loss of $2A$ in the standing-wave cavity. From the $\cos^2(2\pi x/\lambda)$ structure of the standing-wave field, it follows that the peak loss of a sub-wavelength thick absorber is $4A$. In lowest order, we assume that the single-pass loss is proportional to the number N of layers and hence the cavity finesse for the graphene sample at the anti-node behaves as $\mathcal{F} = 2\pi/(\mathcal{L}_e + 4N \cdot A)$. The squares in Fig. 3 display the result of this estimate using $A = 2.3\%$, which agrees well with the data.

So far, we have demonstrated that positioning of a graphene membrane at specific positions within the cavity field leads to controlled dissipation. We now set this into relation to the dispersive shift, which is the change of the resonance frequency of the cavity caused by the presence of the graphene layer. This is an important quantity since certain optomechanical laser-cooling schemes require optical dissipation–dominated coupling rather than dispersive coupling between membrane and cavity [19, 20]. However, membranes with such optical properties have not yet been demonstrated experimentally. Considering the change in the optical path length of the cavity by the graphene membrane, we theoretically estimate the dispersive frequency shift $\Delta\omega_D = -2\pi \times \nu_{\mathrm{FSR}} \frac{2d_g}{\lambda} \left[\mathrm{Re}(n_g) - 1\right] = -2\pi \times 0.71$ MHz. Hence, we expect the dispersive shift to be smaller than the total linewidth of our resonator and toward smaller absolute resonance frequency as the optical path length becomes longer when graphene is placed inside the resonator.

In order to experimentally measure the dispersive shift without the efforts of homodyne detection, we position the graphene at the anti-node of the cavity field and apply a laser power above the optical damage threshold of graphene when resonantly coupled into the resonator. When scanning the laser across the cavity resonance, the graphene layer gets destroyed when the light couples into the resonator. This leads to a quick change of the cavity resonance frequency by the dispersive shift $\Delta\omega_D = 2\pi \times \Delta\nu_D$ (see Fig. 4). It is important to note that the cavity response time is about $\kappa_e^{-1} \sim 50\,\text{ns}$, which is two orders of magnitude smaller than the separation between the data points and therefore the cavity response is instantaneous. In order to quantify the dispersive shift, the data are divided into two parts representing the loaded and empty cavity (red (diamonds) and blue (triangles) data in Fig. 4a). These parts can be identified by a strong change in the cavity in-coupling efficiency. We fit each of the wings with a Lorentzian profile using the linewidth obtained from the previous measurement and determine the line-centers of the Lorentzian. With this method, we find a dispersive shift of $\Delta\omega_D = -2\pi \times (1.0 \pm 0.4)$ MHz, in agreement with our expectation.

From the dispersive and the absorptive measurements, it is possible to calculate the corresponding optomechanical coupling constants, defined as $G = \mathrm{d}\omega/\mathrm{d}z$ and $\Gamma_{dp} = \mathrm{d}\kappa/\mathrm{d}z$, obtained when graphene is placed in the center between node and anti-node by $G \lesssim k \cdot |\Delta\omega_D| = 2\pi \times (8.3 \pm 2.5)$ kHz/nm and $\Gamma_{dp} = k \cdot \Delta\kappa = 2\pi \times (28.3 \pm 3.3)$ kHz/nm, where $k = 2\pi/\lambda$ is the wavenumber (see Appendix). From this, we find a ratio between dissipative and dispersive coupling of

$$\Gamma_{dp}/G \gtrsim \Delta\kappa/\Delta\omega_D = 3.4 \pm 1.1, \qquad (1)$$

which exceeds previous measurements in other systems [26, 29] by approximately one order of magnitude. Moreover, it shows the unusual features of the graphene membrane as compared to standard optomechanical membranes, such as SiN, for which this ratio typically is in the range of $\sim 10^{-4}$. Finally, we can estimate the single photon coupling rate g_0 by assuming a mechanical resonance frequency of $\Omega_M \sim 2\pi \times 20$ MHz, giving rise to a zero-point motion amplitude of $x_0 \sim 3 \times 10^{-13}$ m and hence $g_0 = G \cdot x_0 \simeq 2\pi \times 2$ Hz (and, correspondingly, a factor of three larger for the dissipative rate).

From our measurements, we also deduce the optical cw-damage threshold of graphene for $\lambda = 780$ nm, which is interesting for instance for applications in nonlinear optics [30] or to interface other nanoscopic system like nanodiamonds [31]. To this end, we calculate the intensity of the cavity field at the anti-node from a numerical model [13]. Our data yield a cw-damage threshold of $I_D = (1.8 \pm 0.4)$ MW/cm^2, which is the average of ten measurements. This result is comparable to the optical damage threshold of graphene mounted on silicon substrates where lattice modifications caused by bond breaking have been observed [32]. We note that below the damage threshold intensity, no saturation of the graphene absorption has been observed. This is in contrast to experiments with short (ps) laser pulses,

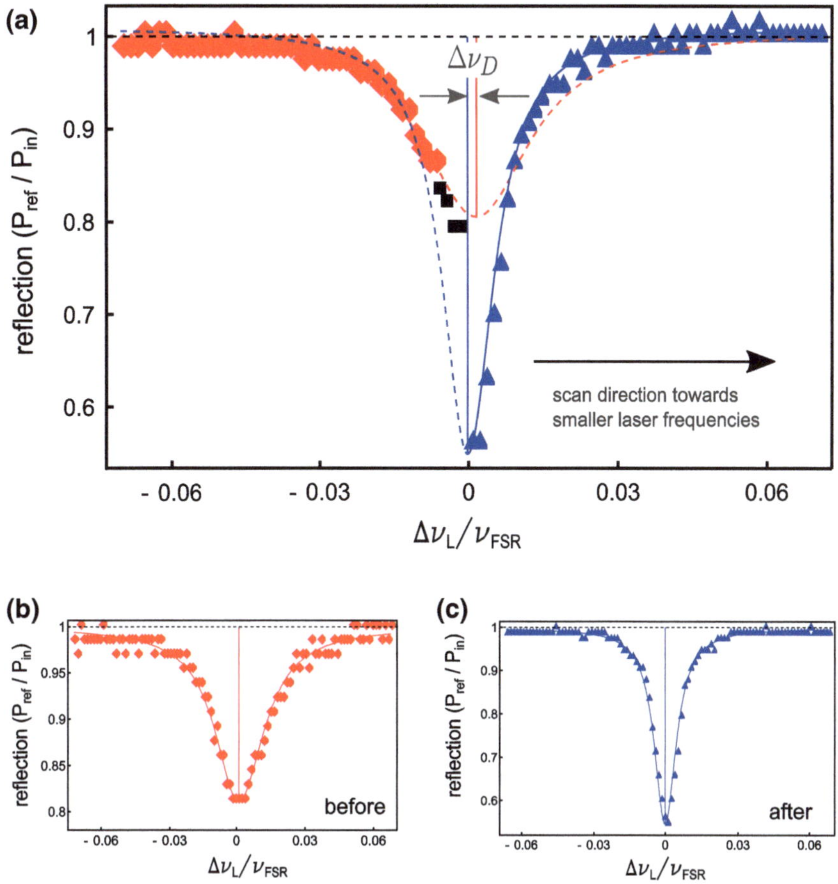

Fig. 4 (Color online) Measurement of the dispersive shift caused by a single layer of graphene. **a** Cavity transmission during the frequency scan of the laser at an intensity higher than the damage threshold of graphene. Near the peak coupling into the resonator, the graphene is optically damaged and becomes transparent which leads to a shift of the cavity resonance. *Red data* points (*diamonds*): undamaged graphene; *blue data* points (*triangles*): damaged graphene. The dispersive shift $\Delta\nu_D$ is determined from Lorentzian line fits to the wings of the data. The *black data* points near to the threshold have not been used for the fit. Also shown are reference scans at low intensity before (**b**) and after (**c**) the dispersive measurement to determine the width and amplitude of the Lorentzian fits

where the damage threshold [33] is several orders of magnitude larger than the saturation intensity [34].

We conclude that a single free-standing monolayer of graphene in a membrane-in-the-middle configuration is a promising experimental platform for cavity optomechanics in the dissipation-dominated regime. Along with the good mechanical properties of graphene, which have been verified by evanescent motional readout [26], theoretically predicted dissipative ground state cooling schemes

[35] might become possible in future experiments provided that the cooling rate exceeds the heating rate from absorption of laser light.

Acknowledgements This work has been supported by BCGS and the Alexander-von-Humboldt Stiftung.

Appendix

For optomechanical applications, the coupling constants are of fundamental interest. In general, they can be divided into dispersive coupling G, which is the change of the cavity resonance frequency $\Delta\omega$ with the membrane displacement along the cavity axis (z-axis) and into dissipative coupling Γ_{dp}, which is the equivalent change of the cavity electric field decay rate $\Delta\kappa$. In general these coupling constants can be written as [29]:

$$G(z) = \frac{\partial \omega_0(z)}{\partial z}, \quad \Gamma_{dp}(z) = \frac{\partial \kappa(z)}{\partial z}. \tag{2}$$

Because the electric field along the cavity axis forms a periodic standing-wave, both of these coupling constants also follow a periodic pattern along the cavity axis. In Fig. 5 the theoretical values of $\Delta\omega = \omega_0 - \omega_{0,e}$ and $\Delta\kappa = \kappa - \kappa_e$ are shown, where $\omega_{0,e}$ and κ_e denote the resonance frequency and the electric field decay rate of the empty cavity, respectively. Here the same simulation and parameters as given in the main text have been used. Because the length scale along the cavity axis is set by the wavelength, the slope and therefore the coupling constants are fully defined by the amplitudes $\Delta\omega_{\max}$ and $\Delta\kappa_{\max}$ when the membrane is moved from a node into an anti-node of the standing-wave field (see Fig. 5). In order to calculate the coupling constants, we fit the simulation using suitable fit-functions defined by

$$\Delta\omega = \frac{\Delta\omega_{\max}}{2} \cos(2kz) + y_{0,\omega} \tag{3}$$

$$\Delta\kappa = \frac{\Delta\kappa_{\max}}{2} \cos(2kz) + y_{0,\kappa}, \tag{4}$$

where $k = 2\pi/\lambda$ is the wavenumber of the intra-cavity field. From this follows:

$$G = \frac{\partial \omega_0(z)}{\partial z} = \frac{\partial \Delta\omega}{\partial z} = k \cdot \Delta\omega_{\max} \sin(2kz) = G_{\max} \sin(2kz), \tag{5}$$

and equivalently $\Gamma_{dp} = k \cdot \Delta\kappa_{\max} \sin(2kz) = \Gamma_{dp,\max} \sin(2kz)$. Therefore, the absolute values become largest when the membrane is placed in the center between node and anti-node ($z = \lambda/8$), as here the slope of the curves in Fig. 5 is largest. While each individual coupling constant even approaches zero when the membrane is

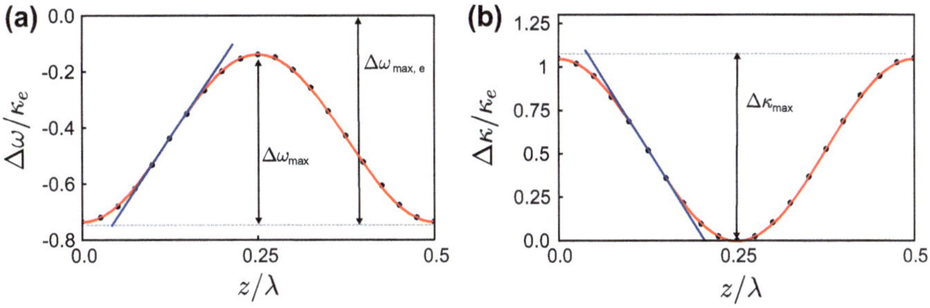

Fig. 5 (Color online) **a** Change of the resonance frequency $\Delta\omega$ and **b** change of the cavity electric field decay rate $\Delta\kappa$ with respect to the empty cavity when graphene is moved from an anti-node ($z/\lambda = \{0, 0.5\}$) to a node ($z/\lambda = 0.25$) of the cavity standing-wave field. Both values are given in units of the empty cavity electric field decay rate κ_e. The simulation data (*black dots*) are fitted with a suitable fit-function (*red curve*, see Equ. 3–4). The coupling constants reach their maximum in the center between node and anti-node ($z = \lambda/8$), where the linear approximation of the curve is shown (*blue line*)

placed close to a node or anti-node of the standing-wave field, the ratio of the coupling constants is independent of the exact location as

$$\frac{\Gamma_{dp}}{G} = \frac{\Delta\kappa_{\max}}{\Delta\omega_{\max}}. \tag{6}$$

In our experiment, we measured the shift of the resonance frequency of the cavity when the membrane is removed from an anti-node of the standing-wave field. In Fig. 5 this corresponds to the theoretical value indicated $\Delta\omega_{\max,e}$. If we use this value for calculating the coupling constant $G_{\max}$ using Eq. 5, our simulation suggests that we overestimate the dispersive coupling by about 20%. Therefore, generally $\Gamma_{dp}/G \gtrsim \Delta\kappa_{\max}/\Delta\omega_{\max,e}$, which is the value we quote in Eq. (1) of the main text.

References

1. G. Anetsberger, O. Arcizet, Q.P. Unterreithmeier, R. Rivière, A. Schliesser, E.M. Weig, J.P. Kotthaus, T.J. Kippenberg, Near-field cavity optomechanics with nanomechanical oscillators. Nat. Phys. **5**(12), 909–914 (2009)
2. I. Favero, K. Karrai, Optomechanics of deformable optical cavities. Nat. Photonics **3**(4), 201–205 (2009)
3. H.J. Kimble, The quantum internet. Nature **453**(7198), 1023–1030 (2008)
4. M. Lee, J. Kim, W. Seo, H.-G. Hong, Y. Song, R.R. Dasari, K. An, Three-dimensional imaging of cavity vacuum with single atoms localized by a nanohole array. Nat. Commun. **5**, 3441 (2014)
5. G. Guthöhrlein, M. Keller, K. Hayasaka, W. Lange, H. Walther, A single ion as a nanoscopic probe of an optical field. Nature **414**(6859), 49–51 (2001)

6. M. Steiner, H.M. Meyer, C. Deutsch, J. Reichel, M. Köhl, Single ion coupled to an optical fiber cavity. Phys. Rev. Lett. **110**(4), 043003 (2013)
7. M. Mader, J. Reichel, T.W. Hänsch, D. Hunger, A scanning cavity microscope. Nat. Commun. **6**, 7249 (2015)
8. J. Miguel-Sanchez, A. Reinhard, E. Togan, T. Volz, A. Imamoglu, B. Besga, J. Reichel, J. Estve, Cavity quantum electrodynamics with charge-controlled quantum dots coupled to a fiber fabryperot cavity. New J. Phys. **15**(4), 045002 (2013)
9. R. Albrecht, A. Bommer, C. Deutsch, J. Reichel, C. Becher, Coupling of a single nitrogen-vacancy center in diamond to a fiber-based microcavity. Phys. Rev. Lett. **110**, 243602 (2013)
10. I. Favero, S. Stapfner, D. Hunger, P. Paulitschke, J. Reichel, H. Lorenz, E.M. Weig, K. Karrai, Fluctuating nanomechanical system in a high finesse optical microcavity. Opt. Express **17**, 12813–12820 (2009)
11. S. Stapfner, L. Ost, D. Hunger, J. Reichel, I. Favero, E.M. Weig, Cavity-enhanced optical detection of carbon nanotube brownian motion. Appl. Phys. Lett. **102**, 151910 (2013)
12. B.M. Zwickl, W.E. Shanks, A.M. Jayich, C. Yang, A.C. Bleszynski Jayich, J.D. Thompson, J.G.E. Harris, High quality mechanical and optical properties of commercial silicon nitride membranes. Appl. Phys. Lett. **92**(10), 103125 (2008)
13. A. Jayich, J. Sankey, B. Zwickl, C. Yang, J. Thompson, S. Girvin, A. Clerk, F. Marquardt, J. Harris, Dispersive optomechanics: a membrane inside a cavity. New J. Phys. **10**(9), 095008 (2008)
14. M. Aspelmeyer, T.J. Kippenberg, F. Marquardt, Cavity optomechanics. Rev. Mod. Phys. **86**, 1391–1452 (2014)
15. J.C. Meyer, A.K. Geim, M. Katsnelson, K. Novoselov, T. Booth, S. Roth, The structure of suspended graphene sheets. Nature **446**(7131), 60–63 (2007)
16. C. Chen, S. Rosenblatt, K.I. Bolotin, W. Kalb, P. Kim, I. Kymissis, H.L. Stormer, T.F. Heinz, J. Hone, Performance of monolayer graphene nanomechanical resonators with electrical readout. Nat. Nanotechnol. **4**(12), 861–867 (2009)
17. V. Singh, S. Bosman, B. Schneider, Y.M. Blanter, A. Castellanos-Gomez, G. Steele, Optomechanical coupling between a multilayer graphene mechanical resonator and a superconducting microwave cavity. Nat. Nanotechnol. **9**(10), 820–824 (2014)
18. R. Nair, P. Blake, A. Grigorenko, K. Novoselov, T. Booth, T. Stauber, N. Peres, A. Geim, Fine structure constant defines visual transparency of graphene. Science **320**(5881), 1308–1308 (2008)
19. F. Elste, S.M. Girvin, A.A. Clerk, Quantum noise interference and backaction cooling in cavity nanomechanics. Phys. Rev. Lett. **102**, 207209 (2009)
20. A. Xuereb, R. Schnabel, K. Hammerer, Dissipative optomechanics in a michelson-sagnac interferometer. Phys. Rev. Lett. **107**, 213604 (2011)
21. J. Restrepo, J. Gabelli, C. Ciuti, I. Favero, Classical and quantum theory of photothermal cavity cooling of a mechanical oscillator. C. R. Phys. **12**, 860–870 (2011)
22. R.A. Barton, I.R. Storch, V.P. Adiga, R. Sakakibara, B.R. Cipriany, B. Ilic, S.P. Wang, P. Ong, P.L. McEuen, J.M. Parpia, H.G. Craighead, Photothermal self-oscillation and laser cooling of graphene optomechanical systems. Nano Lett. **12**(9), 4681–4686 (2012)
23. J.S. Bunch, A.M. van der Zande, S.S. Verbridge, I.W. Frank, D.M. Tanenbaum, J.M. Parpia, H.G. Craighead, P.L. McEuen, Electromechanical resonators from graphene sheets. Science **315**(5811), 490–493 (2007)
24. P. Weber, J. Gttinger, I. Tsioutsios, D.E. Chang, A. Bachtold, Coupling graphene mechanical resonators to superconducting microwave cavities. Nano Lett. **14**(5), 2854–2860 (2014)
25. X. Song, M. Oksanen, J. Li, P.J. Hakonen, M.A. Sillanpää, Graphene optomechanics realized at microwave frequencies. Phys. Rev. Lett. **113**, 027404 (2014)
26. R.M. Cole, G.A. Brawley, V.P. Adiga, R. De Alba, J.M. Parpia, B. Ilic, H.G. Craighead, W.P. Bowen, Evanescent-field optical readout of graphene mechanical motion at room temperature. Phys. Rev. Appl. **3**, 024004 (2015)
27. Graphene Platform Inc., CVD Single layer Graphene Transferred on TEM grid 2000 mesh

28. S. Cheon, K.D. Kihm, H. goo Kim, G. Lim, J.S. Park, J.S. Lee, How to reliably determine the complex refractive index (ri) of graphene by using two independent measurement constraints. Sci. Rep. 4, 6364 (2014)
29. M. Wu, A.C. Hryciw, C. Healey, D.P. Lake, H. Jayakumar, M.R. Freeman, J.P. Davis, P.E. Barclay, Dissipative and dispersive optomechanics in a nanocavity torque sensor. Phys. Rev. X **4**, 021052 (2014)
30. E. Hendry, P.J. Hale, J. Moger, A.K. Savchenko, S.A. Mikhailov, Coherent nonlinear optical response of graphene. Phys. Rev. Lett. **105**, 097401 (2010)
31. A. Brenneis, L. Gaudreau, M. Seifert, H. Karl, M.S. Brandt, H. Huebl, J.A. Garrido, F.H. Koppens, A.W. Holleitner, Ultrafast electronic readout of diamond nitrogen-vacancy centres coupled to graphene. Nat. Nanotechnol. **10**(2), 135–139 (2015)
32. B. Krauss, T. Lohmann, D.-H. Chae, M. Haluska, K. von Klitzing, J.H. Smet, Laser-induced disassembly of a graphene single crystal into a nanocrystalline network. Phys. Rev. B **79**, 165428 (2009)
33. A. Roberts, D. Cormode, C. Reynolds, T. Newhouse-Illige, B.J. LeRoy, A.S. Sandhu, Response of graphene to femtosecond high-intensity laser irradiation. Appl. Phys. Lett. **99** (5), 051912 (2011)
34. Q. Bao, H. Zhang, Y. Wang, Z. Ni, Y. Yan, Z.X. Shen, K.P. Loh, D.Y. Tang, Atomic-layer graphene as a saturable absorber for ultrafast pulsed lasers. Adv. Funct. Mater. **19**(19), 3077–3083 (2009)
35. L.-D. Xiao, Y.-F. Shen, Y.-C. Liu, M.-Y. Yan, Y.-F. Xiao, Dissipative optomechanics of a single-layer graphene in a microcavity. arXiv preprint arXiv:1411.2202 (2014)

Cavity Ring-Up Spectroscopy for Dissipative and Dispersive Sensing in a Whispering Gallery Mode Resonator

Yong Yang, Ramgopal Madugani, Sho Kasumie, Jonathan M. Ward, and Síle Nic Chormaic

Abstract In whispering gallery mode resonator sensing applications, the conventional way to detect a change in the parameter to be measured is by observing the steady-state transmission spectrum through the coupling waveguide. Alternatively, sensing based on cavity ring-up spectroscopy, i.e. CRUS, can be achieved transiently. In this work, we investigate CRUS using coupled mode equations and find analytical solutions with a large spectral broadening approximation of the input pulse. The relationships between the frequency detuning, coupling gap and ring-up peak height are determined and experimentally verified using an ultrahigh Q-factor silica microsphere. This work shows that distinctive dispersive and dissipative transient sensing can be realised by simply measuring the peak height of the CRUS signal, which may improve the data collection rate.

1 Introduction

Whispering gallery mode (WGM) resonators are widely used for a number of applications, such as frequency comb generation [1, 2], laser stabilization [3–5] and sensing [6, 7]. The high optical quality factor (Q-factor) and relatively small

This article is part of the topical collection "Enlightening the World with the Laser" - Honoring T. W. Hänsch guest edited by Tilman Esslinger, Nathalie Picqué, and Thomas Udem.

Y. Yang (✉)
Light-Matter Interactions Unit, Okinawa Institute of Science and Technology Graduate University, Onna, Okinawa 904-0495, Japan

National Engineering Laboratory for Fiber Optics Sensing Technology, Wuhan University of Technology, Wuhan 430070, China
e-mail: yong.yang@oist.jp

R. Madugani
Light-Matter Interactions Unit, Okinawa Institute of Science and Technology Graduate University, Onna, Okinawa 904-0495, Japan

Physics Department, University College Cork, Cork, Ireland

S. Kasumie • J.M. Ward • S. Nic Chormaic
Light-Matter Interactions Unit, Okinawa Institute of Science and Technology Graduate University, Onna, Okinawa 904-0495, Japan

D. Meschede et al. (eds.), *Exploring the World with the Laser*,
https://doi.org/10.1007/978-3-319-64346-5_34

mode volume of whispering gallery resonators (WGRs) render the modes very sensitive to subtle environmental changes. Until now, WGRs have been used to measure changes in a number of parameters such as refractive index [8, 9], temperature [10–12], pressure [13, 14] and stress [15, 16]. Aside from parameter change detection, ultrahigh Q resonators have also been used to detect nanoparticles [17, 18] and single viruses [19, 20]. The mechanism behind ultrahigh sensitivity sensing in WGRs is based on a reactive (i.e. dispersive) frequency shift of the whispering gallery modes [19] as a result of perturbations that may be present. Alternatively, a perturbation may increase the optical linewidth of the WGM by introducing more dissipation [21, 22], or may change the back-scattering strength [23] and subsequent mode splitting if modal coupling is present [17, 20]. The optomechanical properties of WGRs can also be used for force [24] or viscosity sensing [25].

Currently, in order to retrieve the dispersive, dissipative and mode splitting information, the transmission spectrum of a WGR through an externally coupled waveguide, such as a tapered optical fibre, is usually measured. Light from a tunable laser source is coupled into the tapered fibre and the transmission is monitored. Low powers are used in order to minimise thermal and nonlinear effects on the whispering gallery modes. By sweeping the laser frequency, the transmission spectrum through the fibre can be recorded. Any changes to the frequency, mode splitting or linewidth are used to monitor perturbations induced by the physical parameter that is being sensed. During measurements, the transmission spectrum represents a *steady state* of the coupled system; therefore, time response of the sensor [26–28] is related to the lifetime of the resonator which limits the scanning speed. For a WGR with an optical Q-factor $> 2 \times 10^7$, even when the pumping laser is shut off immediately, there is still light intensity remaining in the cavity due to the long lifetime of the high Q cavity. A transient ringing effect [29, 30] is observable even if the laser is scanned as quickly as 25 MHz/μs [31]. The ringing spectrum can be used to distinguish between the overcoupled and undercoupled cases [32]. When the scanning speed is faster than the *character speed*, as defined in [28], the steady-state treatment can no longer be used to describe the coupled mode system. Therefore, by recording lineshape changes in the ringing tail of an observed transmission spectrum, either by (1) a scanning probe laser or (2) a fixed laser in resonance with a high Q mode, transient sensing should be possible; this would allow us to not only transiently detect the existence of a nanoparticle but also provide information such as speed or size of the particle [28]. A proof-of-principle experiment based on the ringing phenomenon has recently been reported [33].

Another possible approach is to send light pulses, which are far detuned from a WGM resonance, through the optical coupler. The retrieved signal on the coupler's output shows an oscillatory lineshape similar to that in [28]; this effect is termed *cavity ring-up spectroscopy* (CRUS) [27] and the rising edge of the light pulse leads to transient broadening. Even though the light is far detuned from the WGM, a fraction can still be coupled into the cavity if the broadening is much larger than the detuning. The system is not affected by thermal or nonlinear processes which may arise due to the ultrahigh Q of the mode. The ringing effect occurs within the lifetime of the WGM and, therefore, can be used for ultrafast sensing. The transient capability of CRUS has already been demonstrated by measuring the time response for thermo-refractive effects, Kerr nonlinearity and optomechanical vibrations [27].

To date, there has been no thorough theoretical investigation of CRUS and details, such as the influence of the pulse's rise time on the observed spectra, are relatively unknown. In this manuscript, we use coupled mode theory to solve the related differential equations without relying on the steady-state assumption. An approximate analytical solution is obtained and compared to a precise numerical transient solution. The theoretical results fit well to experimental data that we obtained for a silica microsphere resonator. The influences of the pulse rise time, the coupling condition and the detuning on the ring-up spectrum are given. This provides a solid foundation for future applications in transient sensing.

2 Coupled Mode Theory

A typical CRUS setup is shown in Fig. 1a. The WGR is coupled evanescently to a tapered optical fibre through which light from a laser propagates. The light couples into the resonator and is monitored at the opposite end of the fibre. The coupling dynamics can be described using coupled mode theory. The amplitude of the intracavity electromagnetic field, $a(t)$, changes in time according to the following [28, 34, 35]

$$\frac{\mathrm{d}a(t)}{\mathrm{d}t} = -j\omega_0 a(t) - (\kappa_\mathrm{e} + \kappa_0)a(t) + \sqrt{2\kappa_\mathrm{e}}S_\mathrm{in}(t), \tag{1}$$

where $j = \sqrt{-1}$, the resonant frequency of the WGM is ω_0, and κ_e and κ_0 represent the external and intrinsic coupling rates, respectively. The total damping rate of the cavity is given by $\kappa = \kappa_\mathrm{e} + \kappa_0$ and $\tau = 1/\kappa$ is the intracavity lifetime. In order to arrive at the transient response of the WGR, the laser light is pulsed with a temporal profile, $S_\mathrm{in}(t)$. The laser frequency, ω_L, is far detuned, i.e. $\omega_\mathrm{L} - \omega_0 \gg \kappa$. The pulsed input field can be separated into a slowly varying and a fast varying term so that

$$S_\mathrm{in}(t) = a_\mathrm{in}(t)\mathrm{e}^{-j\omega_\mathrm{L}t}. \tag{2}$$

Here, the slowly varying part, $a_\mathrm{in}(t)$, represents the temporal profile of the pulse and, in the following discussions, it takes the form of a Guassian function where

$$a_\mathrm{in}(t) = \begin{cases} 0, & t < \tau_0 \\ \alpha_\mathrm{in}\exp\left(-\dfrac{4[t-\tau_0-t_\mathrm{r}]^2}{\ln 2 \cdot t_\mathrm{r}^2}\right), & \tau_0 < t < \tau_0 + t_\mathrm{r} \\ \alpha_\mathrm{in}, & t > \tau_0 + t_\mathrm{r} \end{cases} \tag{3}$$

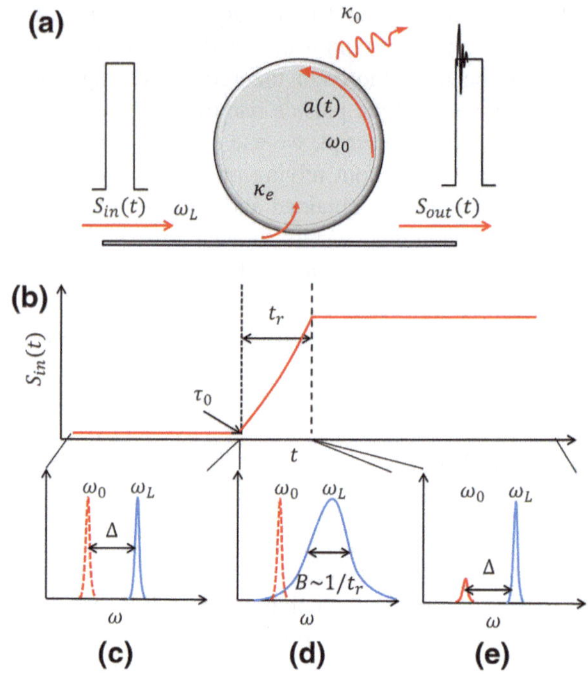

Fig. 1 **a** Schematic of a taper coupled WGR system for transient sensing using CRUS. A pump laser of frequency ω_L, far off resonance with the WGM, ω_0, is coupled through the tapered fibre with a temporal profile, $S_{in}(t)$. The pulse profile is depicted in (**b**); the laser pulse starts at a time τ_0 and rises up to its maximum within a time t_r. The mathematical description of the pulse is given in Eq. 3. **c–e** Transient frequency of the laser pulse for different time intervals. **c** $t = [0, \tau_0]$; **d** $t = [\tau_0, \tau_0 + t_r]$; **e** $t = [\tau_0 + t_r, +\infty)$. At the rising edge of the pulse, the laser source is transiently broadened (**d**), so a fraction of the pump signal couples to the WGM and contributes to a beat signal between it and the pump source (**e**)

The pulse is illustrated in Fig. 1b. The pulse starts at time τ_0 and follows a Gaussian profile with a rise time, $t_r \ll \tau$. At time $t = \tau_0 + t_r$, the total power of the pulse reaches its maximum, $|\alpha_{in}|^2$, and, for later times, the laser can be treated as a continuous light source over the lifetime of the cavity mode.

2.1 A Simple Model: Large Spectral Broadening Bandwidth

The temporal profile of $S_{in}(t)$ can be obtained from Eqs. 2 and 3 and Fourier expanded as follows:

$$S_{\text{in}}(t) = \begin{cases} 0, & t < \tau_0 \\ \sqrt{\frac{\ln 2}{\pi}} \frac{t_r \alpha_{\text{in}}}{4} \\ \times \int_{-\infty}^{+\infty} e^{-\frac{\ln 2 t_r^2 \omega^2}{16}}, & \tau_0 < t < \tau_0 + t_r \\ e^{-j[(\omega+\omega_L)(t-\tau_0)]} d\omega & \\ \alpha_{\text{in}} e^{-j\omega_L t}, & t > \tau_0 + t_r \end{cases} \tag{4}$$

It can be seen that, for a time interval $t \in [\tau_0, \tau_0 + t_r]$, the rise time of the pulse induces sideband frequencies even though the laser source is monochromatic. As the Fourier transform of a Gaussian function is a Gaussian, the laser pulse is expanded transiently with a bandwidth, $B = 1/(t_r \ln 2)$.

In our experiments, the pulse has a rise time ranging from ~1 ns to several tens of ps. The laser frequency broadening bandwidth, B, is of the order of GHz, and, for a WGM with $Q > 10^7, \kappa \sim$ MHz. As $B \gg \kappa$, we can assume that only the portion of the broadened laser source at the resonant frequency, ω_0, can be efficiently coupled to the WGM. As illustrated in Fig. 1c–e, we assume that the laser is far red-detuned from the resonant frequency, such that $\omega_L - \omega_0 = \Delta \gg \kappa$. The WGM acts as an infinitely narrow band pass filter so that we only need to consider the frequency at ω_0, see Fig. 2c. Thus, the input signal can be simplified as:

$$S_{\text{in}}(t) \approx \sqrt{\frac{\ln 2}{\pi}} \frac{t_r \alpha_{\text{in}}}{4} e^{-\frac{\ln 2 t_r^2 \Delta^2}{16}} e^{-j\omega_0 t}. \tag{5}$$

Note that, in the above expression, only the last factor is time dependent. The single frequency pulse as defined by Eq. 5 must satisfy Eq. 1. Thence, we get the dynamics of the intracavity amplitude for the transient interval at the rise time and for later times. Equation 1 has a general solution [28]:

$$a(t) = \sqrt{2\kappa_e} a_{\text{in}} e^{j\omega_0 t - \kappa t} \left[\frac{\tau}{1 + j(\omega_L - \omega_0)\tau} + \int_{\tau_0}^{t} e^{j\phi(t') - j\omega_0 t' + \kappa t'} dt' \right]. \tag{6}$$

The complete dynamics can be separated into two steps. First, at the rise time interval, the signal is broadened and light of frequency ω_0 is coupled into the cavity. As $t_r \ll \tau$, dissipation during this short time can be ignored and ω_L in Eq. 6 can be substituted by ω_0. If we consider the accumulated phase $\phi(t') = \int_{\tau_0}^{t'} \omega_0(t'') dt'' = \omega_0 t'$, then the accumulated amplitude at a time $t = \tau_0 + t_r$ is given by

$$a(t) = \frac{\sqrt{2\kappa_e}}{\kappa} \sqrt{\frac{\ln 2}{\pi}} \frac{t_r \alpha_{\text{in}}}{4} e^{-\frac{\ln 2 t_r^2 \Delta^2}{16}} e^{-j\omega_0 t}, \quad \tau_0 < t < t_r. \tag{7}$$

Next, as already explained, the frequency returns to ω_L and is far detuned. Therefore, for later times, there is no light coupled into the WGM and the system follows simple decay dynamics with a decay rate, κ, whose initial value is as in Eq. 7:

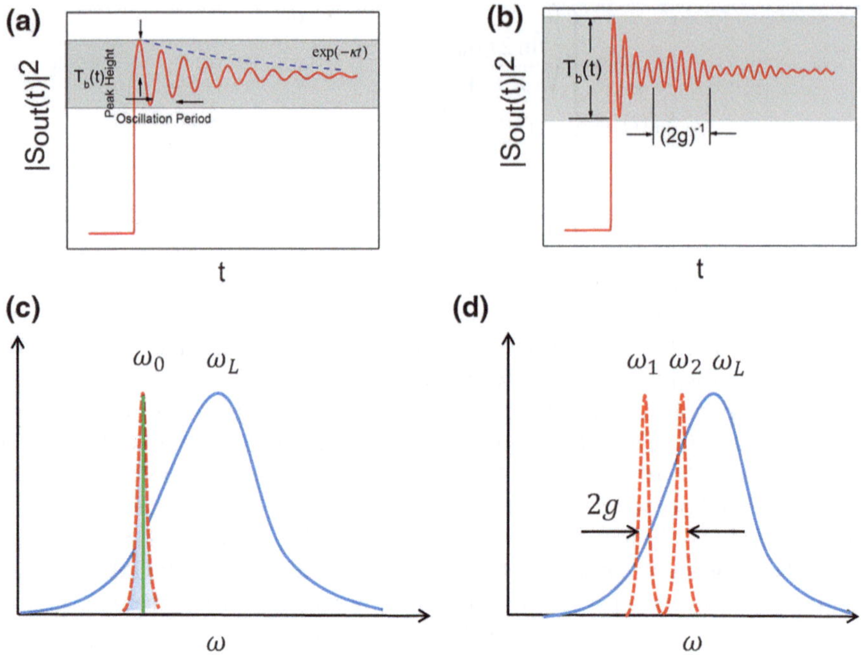

Fig. 2 **a** A typical CRUS signal. The curve in the *grey rectangle* is the beat signal deduced from Eq. 10. This is due to the frequency broadening at the rising edge of the input signal. We can treat the WGM as an ultra-narrow filter; therefore, only the on-resonance, $\omega = \omega_0$, component in the broadened pump laser signal can be stored in the WGM as shown in (**c**). **b** The CRUS signal when there is modal coupling with coupling strength, g. The beating is modulated by a cosine envelope of period $(2g)^{-1}$. This is due to the beating between the two normal standing wave modes, with resonant frequencies of ω_1 and ω_2, as shown in (**d**)

$$a(t) = \frac{\sqrt{2\kappa_e}}{\kappa}\sqrt{\frac{\ln 2}{\pi}}\frac{t_r\alpha_{in}}{4}e^{-\frac{\ln 2 t_r^2\Delta^2}{16}}e^{-j\omega_0 t-\kappa t}, \quad t > \tau_0 + t_r. \tag{8}$$

The output signal, $S_{out}(t)$, can be calculated using the input–output relationship:

$$S_{out}(t) = S_{in}(t) - \sqrt{2\kappa_e}a(t). \tag{9}$$

Here, we are interested in the output signal after the pulse reaches its maximum value. At this point the input signal changes to a continuous state and the broadening vanishes, yielding $S_{in}(t) = \alpha_{in}e^{-j\omega_L t}$. It can be seen from Eqs. 8 and 9 that the final signal is a superposition of the laser signal at ω_0, which is the residue of intracavity photons leaking out after the rise time, and ω_L of the pumping signal. Therefore, it gives transient transmission, $T(t)$, at a time after the rise up of the pulse as follows:

$$T(t) = \left|\frac{S_{out}}{S_{in}}\right|^2 = 1 + \frac{\kappa_e}{\kappa}\sqrt{\frac{\ln 2}{\pi}} e^{-\frac{\ln 2 t_r^2 \Delta^2}{16}} e^{-\kappa t} \sin\left(\Delta(t - \tau_0 - t_r)\right). \tag{10}$$

A typical calculated waveform is shown in Fig. 2a. There are three values that can be retrieved from the CRUS signal: peak height, oscillation period, and the decay rate. From Eq. 10, the oscillation frequency is the detuning, Δ and the decay rate is the lifetime, τ. Note that this is the loaded Q-factor of the system. The peak height is more complicated as it is related to both the dissipation rate and the detuning. This will be discussed in more detail later.

2.2 *Transient Response During the Rising Edge*

From the above discussion, we assume that the phase accumulation during the rise time can be ignored. This is not valid when considering a large value for the detuning. As shown in Eq. 10, the frequency of the obtained transient signal is determined by the detuning of the laser source from the resonant frequency. At large detuning, one period of oscillation is almost as long as the rise time; hence, the phase accumulated during the rise time can influence the output signal. This distorts the first few periods of the beat signal. To calculate this, we should use the full form of Eq. 7 at the rise time.

The output signal, $S_{out}(t)$, can be determined from Eq. 9, with the input amplitude defined in Eq. 3. Therefore, a beating frequency equalling $\Delta = \omega_L - \omega_0$ can be achieved during the rise time :

$$T(t) \approx \left[1 + \sqrt{\frac{\ln 2}{\pi}} \frac{\kappa_e}{2\kappa} e^{-\kappa(t-\tau_0)} \sin\left(\Delta(t - \tau_0)\right)\right] e^{-\frac{4(t-\tau_0-t_r)^2}{\ln 2 \cdot t_r^2}}. \tag{11}$$

The signal which arrives at the detector can be separated into two components. The first part has the same form as that in Eq. 3, i.e. the rising part of the retrieved signal maintains a Gaussian profile and the rise time is unchanged. The second part comes from the contribution of the intracavity amplitude and takes the form of a trigonometric function with a period $2\pi/\Delta$. Since this component only occurs for a time interval, t_r, if Δ is small it follows that $\sin\left(\Delta(t - \tau_0)\right) \approx 0$ and the rising edge waveform will not be disturbed.

2.3 *The Mode Mixing Case*

In practice, multiple WGM resonances may fall within the transient spectral broadening bandwidth. In the following, the case in which two modes mix will be investigated. There are two different types of mode mixing that we consider. The first situation is that where two modes exist quite close to each other in the spectrum. In the following discussion, we assume that both modes lie within the transiently broadened spectrum bandwidth, B, while satisfying the narrow band pass filter approximation made in Sect. 2.1. Defining the two modes as ω_1 and ω_2, with a separation between them of δ, the coupled mode equations for each mode are given by

$$\frac{\mathrm{d}a_i}{\mathrm{d}t} = -j\omega_i a_i(t) - (\kappa_{0,i} + \kappa_{\mathrm{e},i})a_i(t) + \sqrt{2\kappa_\mathrm{e}}a_{\mathrm{in},i}(t)\mathrm{e}^{-j\omega_i t}, \tag{12}$$

where $i = 1, 2$ and represents the two modes. Since there is no spectral overlap of the two modes, the mode equations are decoupled and can be solved separately. This yields the same expressions as in Eq. 8 and each mode beats with $S_{\mathrm{in}}(t)$ separately. The total beat signal, $T_\mathrm{b}(t)$, is

$$T_\mathrm{b}(t) = \sum_{i=1,2} T_{i,\mathrm{b}}(t) = \sum_{i=1,2} \frac{\kappa_{\mathrm{e},i}}{\kappa_i}\sqrt{\frac{\ln 2}{\pi}}\mathrm{e}^{-\frac{\ln 2 t_r^2 \Delta_i^2}{16}}\mathrm{e}^{-\kappa_i t}\sin(\Delta_i t). \tag{13}$$

$T_\mathrm{b}(t)$ is a linear combination of the two individual beat signals. The amplitudes of the signals with different frequencies are determined from the detunings, Δ_i, of the resonances relative to the laser source. If δ is large, then one of the beat signals will be much larger than the other since $T_{\mathrm{b},i}$ follows the Gaussian relationship with the detuning. In this case, the weaker signal can be ignored. In contrast, if the two modes are close to each other so that $\delta \ll \Delta_i$, then, for simplicity, we set $\Delta_1 \approx \Delta_2 = \Delta$ and the coupling rates are approximately equal. From Eq. 13, the total beat signal, $T_\mathrm{b}(t)$, is ($T(t) = 1 + T_\mathrm{b}(t)$):

$$T_\mathrm{b}(t) \approx \frac{\kappa_\mathrm{e}}{\kappa}\sqrt{\frac{\ln 2}{\pi}}\mathrm{e}^{-\frac{\ln 2 t_r^2 \Delta^2}{16}}\mathrm{e}^{-\kappa t}(1 + \cos(\delta t))\sin(\Delta t). \tag{14}$$

The other possible case to consider is that where the two modes are coupled. This is often the case in travelling wave resonators, such as WGRs. Degenerate modes in the resonator represent clockwise and counter clockwise propagation. Due to scattering along the propagation path, the two modes can indirectly couple to each other; this effect is called *modal coupling* and leads to normal mode splitting [36, 37]. Here, we define a modal coupling strength, g, and the coupled mode equations for the two modes are given by a_{cw} and a_{ccw} representing the two opposing propagation directions. The two coupled mode equations can be written as

$$\begin{aligned} \frac{\mathrm{d}a_{cw}}{\mathrm{d}t} &= -j\omega_0 a_{cw}(t) - (\kappa_0 + \kappa_\mathrm{e})a_{cw}(t) - jga_{ccw}(t) + \sqrt{2\kappa_\mathrm{e}}S_{\mathrm{in}}(t); \\ \frac{\mathrm{d}a_{ccw}}{\mathrm{d}t} &= -j\omega_0 a_{ccw}(t) - (\kappa_0 + \kappa_\mathrm{e})a_{ccw}(t) - jga_{cw}(t). \end{aligned} \tag{15}$$

The two modes should have the same values of $\kappa_0, \kappa_\mathrm{e}$ and detuning. By letting $A_1(t) = a_{cw}(t) + a_{ccw}(t)$ and $A_2(t) = a_{cw}(t) - a_{ccw}(t)$, these two equations can be transformed into two uncoupled equations [37]:

$$\begin{aligned} \frac{\mathrm{d}A_1(t)}{\mathrm{d}t} &= -j(\omega_0 + g)A_1(t) - (\kappa_0 + \kappa_\mathrm{e})A_1(t) + \sqrt{2\kappa_\mathrm{e}}S_\mathrm{in}(t) \\ \frac{\mathrm{d}B(t)}{\mathrm{d}t} &= -j(\omega_0 - g)A_2(t) - (\kappa_0 + \kappa_\mathrm{e})A_2(t) + \sqrt{2\kappa_\mathrm{e}}S_\mathrm{in}(t) \end{aligned} \tag{16}$$

Here, the original two travelling modes generate two equivalent standing wave modes with frequency shifts, $\pm g$, as illustrated in Fig. 2d. Usually, the modal coupling strength is in the MHz range for silica WGRs, i.e. the frequencies of the two standing waves are quite close to each other. Since $B \gg 2g$, we can assume that the two modes have the same transient components at the rising edge time, as in the previous case. Also, in the actual experiments, we measure the light transmitted through the tapered fibre, so that $T(t) = |(1 - \sqrt{2\kappa_\mathrm{e}}a_{cw}(t))/S_\mathrm{in}(t)|^2$, where $a_{cw} = 1/2(A_1(t) + A_2(t))$. $T_\mathrm{b}(t)$ is a beat signal between the initial signals with frequencies Δ and $\Delta \pm g$ and has the following form

$$T_\mathrm{b}(t) \approx \frac{\kappa_\mathrm{e}}{\kappa}\sqrt{\frac{\ln 2}{\pi}}\mathrm{e}^{-\frac{\ln 2 t_\mathrm{r}^2 \Delta^2}{16}}\mathrm{e}^{-\kappa t}\sin(\Delta t)\left\{1 + \frac{1}{2}\cos(2gt)\right\}. \tag{17}$$

From the above equation, when there is mode splitting due to intrinsic scattering, the transient signal has a cosine form of frequency Δ and is modulated by a slowly oscillating envelope with a frequency of $2g$. This yields a similar waveform to that obtained in the uncoupled modes' case. Comparing Eqs. 17 to 14, we see that there is a factor of 1/2 difference. When modal coupling is present, the two modes interfere with each other, whereas for the uncoupled case, the beating of the two modes has no coherent property.

2.4 Numerical Method

For a more precise simulation of the transient system, we should solve the coupled mode equations numerically. Equation 1 can be transformed into a rotating frame of reference, with an angular frequency ω_L such that

$$\frac{\mathrm{d}a(t)}{\mathrm{d}t} = -j\Delta a(t) - \kappa a(t) + \sqrt{2\kappa_\mathrm{e}}a_\mathrm{in}(t). \tag{18}$$

The input–output relationship for the rotating frame is $a_\mathrm{out}(t) = a_\mathrm{in}(t) - \sqrt{2\kappa_\mathrm{e}}a(t)$. For simplicity, we rescale the time by t_r in the following discussion and we assume that the WGR is critically coupled to the external coupler, unless we explicitly mention otherwise. First, let us assume that the lifetime of a WGM is about 250 ns. Three different responses under various detunings are plotted in Fig. 3a, ranging

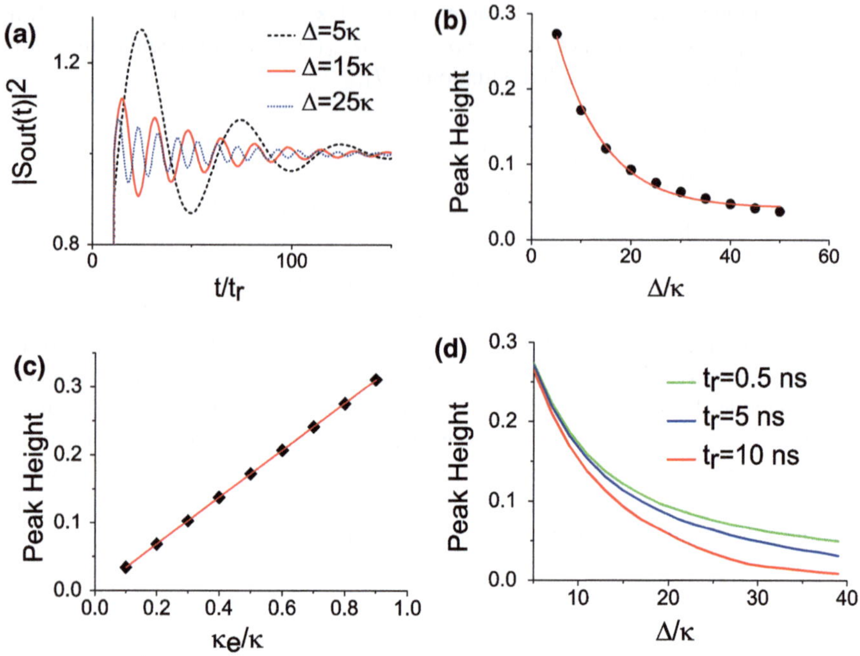

Fig. 3 **a** The normalised transient response of a pulse detuned to the high Q WGM. The legend shows the detuning (normalised to κ) for different curves. **b** The peak heights of the transient signals for different detunings in (**a**). **c** The peak heights at different coupling conditions κ_e/κ. **d** Peak heights as a function of detuning for certain coupling conditions with different pulse rise-up times, t_r

from $\Delta = 5\kappa$ to $\Delta = 35\kappa$. The oscillation period follows the detuning, so that a larger detuning yields a higher oscillation frequency, which represents the beating between the laser and WGM frequencies, as discussed in Sect. 2(A). However, the peak height *reduces* when the detuning increases. In Fig. 3b we plot the peak height as a function of detuning. The data points can be fitted quite well with a Gaussian function and this confirms the presence of the Gaussian term in Eq. 10. If the detuning, Δ, is fixed while the ratio κ_e/κ is changed, according to Eq. 10, the peak height should be proportional to this ratio. From the numerical simulations, this linear relationship is confirmed and depicted in Fig. 3c. Varying t_r and τ should not affect the peak height, as shown in both Eq. 10 and the numerical simulations. However, t_r determines the bandwidth of the transient broadening; therefore, it controls the peak height relationship to detuning, as illustrated in Fig. 3d. Significant bandwidth shrinkage is visible only if the rise time increases by more than a factor of 10. This verifies the behaviour we assumed in Sect. 2. In essence, it is the broadening from the rise time of a detuned pulse that allows light to couple into the WGM and leads to the subsequent beat signal.

In most cases, we find that the analytical method gives the same result as the numerical method, see Fig. 4a. Here, the lifetime of the WGM is chosen to be

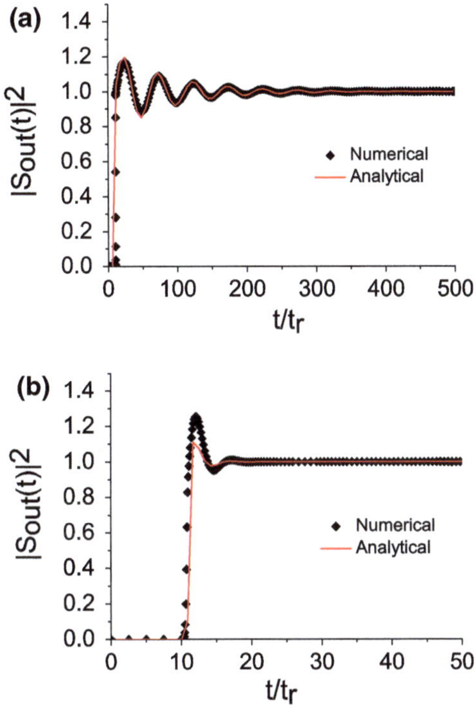

Fig. 4 Transient response of a WGR with a pulsed input signal. *Red curve* analytical result using Eqs. 10 and 11. *Black dots* numerical results by solving Eq. 18 directly. Two different cases were considered. **a** $\tau = 200t_r, \Delta = 10\kappa$; **b** $\tau = 10t_r, \Delta = 2\kappa$. Both cases satisfy the under coupling condition

$\tau = 100$ ns with a rise time of $t_r = 1$ ns. In this case, $B \gg \kappa$ is well satisfied and both methods yield the same results. However, when the lifetime decreases to $\tau = 10$ ns, the analytical result yields a peak height less than the more accurate numerical method (see Fig. 4b). In this case, $B \sim \kappa$ and the assumption of narrow filtering by the WGM for obtaining Eq. 10 is not justified. The broader linewidth of the WGM should permit more photons to enter into the cavity at the transient broadening time (shown as the shaded area in Fig. 2c), so that, at a later time, more light can beat with the transmitted pulse.

3 Experiment

In order to confirm the above theory, we performed an experiment using an ultrahigh Q silica microsphere. The experimental setup is depicted in Fig. 5. A 30 μW, 1550 nm laser was initially modulated using an intensity modulator with an EOM (Thorlabs model LN63S-FC, with rise-up time 50 ps). For this purpose, a pulse generator providing a pulse with a rise time of 5 ns, a width of 500 ns and a delay about 100 ns was used. The modulated light was coupled to the microsphere's WGM using a fibre taper. The transmission through the fibre was detected with a fast photodetector (Newport model 818-BB-35F) with a typical rise time of 500 ps.

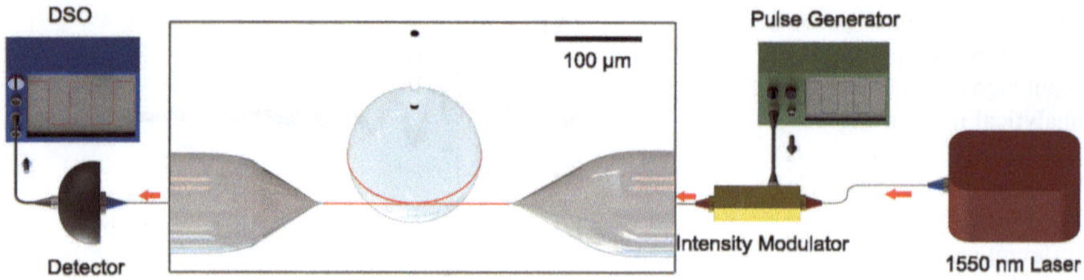

Fig. 5 Experimental setup used for cavity ring-up spectroscopy. A 1550-nm laser is intensity modulated and coupled to the microsphere cavity and the transmitted light pulses are detected using a fast detector, with the signals recorded on a fast digital storage oscilloscope (DSO)

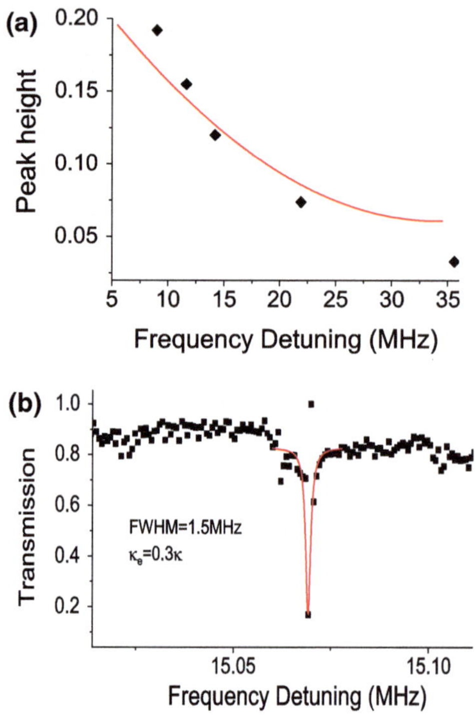

Fig. 6 **a** The peak height measured at different laser detunings simulating a dispersive shift of the microsphere (*black dots*). The *red curve* is the theoretical results calculated from the measured parameters. **b** The transmission spectrum of the microsphere which has an ultrahigh Q WGM

The signal was retrieved on a digital storage oscilloscope (DSO) and recorded at a sampling rate of >1 GS/s. The microsphere had a diameter of 80 μm and the fibre waist was ~1.2 μm. We chose a high Q silica microsphere with a lifetime of ~500 ns. To study the detuning effects on the CRUS, the microsphere and the fibre taper were aligned to be in contact coupling; therefore, the coupling condition κ_e/κ is fixed. The detuning of the laser with respect to the whispering gallery mode was changed so that its frequency approaches that of the WGM in finite steps. The results were normalised to get the peak heights and are plotted in Fig. 6a.

To match with the theoretical framework, we made a separate measurement of the transmission spectrum of the WGM being probed, as illustrated in Fig. 6b. From the transmission efficiency of the mode (80%), and assuming that the system is in the undercoupled regime, it can be deduced that $\kappa_e/\kappa = 0.3$. The FWHM of the mode is 1.5 MHz using Lorentz fitting to the dip in Fig. 6b. In fact, thermal broadening was present even at a low pump laser power of 30 μW; therefore, κ is over estimated. In practice, it is found that $\tau \approx 750$ ns yields a good fit to the experimental data. By the means mentioned in Sect. 2, the theoretical peak height relationship to the frequency detuning is plotted as the red curve in Fig. 6a. The trend of the peak height to the detuning follows a Gaussian profile.

We also evaluated the peak height with different coupling gaps by varying the gap using a closed-loop, piezo nanopositioner (Smaract SLC1730s-416). The relative position of the taper from the microsphere was determined using a nanopositioner controller (Smaract MCS-3D). The experimental results for different coupling gaps were fitted with the theory to determine κ_e/κ (see Fig. 7a–e). Since $\kappa_e \ll \kappa_0$, we assume that $\kappa_e/\kappa = \kappa_e/\kappa_0$. Here, κ_0 is a constant, while κ_e satisfies a near exponential curve to the coupling gap [34], as shown in the inset of Fig. 7f. The corresponding peak height for different coupling conditions plotted in Fig. 7f shows a near-linear relationship.

From Fig. 7a–e, the coupling gap is increased and the period of the CRUS becomes larger. The taper introduces a dispersive red shift to the microsphere's resonance [38]. In our experiments, the laser is blue-detuned relative to the resonance and fixed; the larger the distance between the WGR and the taper, the less the dispersion introduced; thus, the cavity mode shifts relative to the laser thereby decreasing the beat frequency. In Eq. 10, the peak height should be related to both the coupling condition and the detuning. However, since Δ appears in the Gaussian term (and assuming that it is large), slightly changing its value will not influence the peak height significantly (see Fig. 3b). In the experiment, we deliberately chose an initial large detuning; therefore, the peak height is still linear with κ_e/κ despite the dispersive disturbances. When the system is strongly overcoupled, the results deviate from the linear relationship, as seen when $\kappa_e/\kappa > 0.65$. In the supplementary material of Ref. [38], it was shown that the dispersive shift rate increases exponentially when moving to a strongly overcoupled regime. This means that the dispersive influence of the taper will induce a very large frequency shift for the cavity mode and the changes in the Gaussian term in Eq. 10 cannot be neglected. As a consequence, the peak height does not vary linearly with κ_e/κ.

4 Discussion

Similar to the work in [28], the deduced formula of CRUS in this manuscript shows that it also provides redundant information if the cavity's intrinsic Q factor is known; hence, it could be very useful for transient sensing. Instead of doing a time-consuming fast Fourier transform (FFT) of the transient response signal [27],

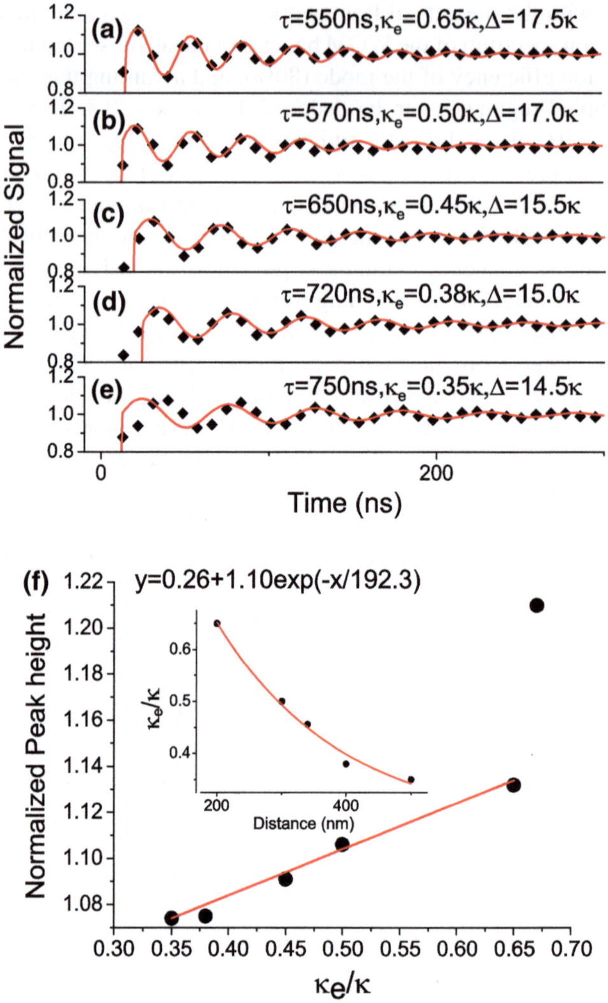

Fig. 7 **a–e** The experimental cavity ring-up signals from a silica microsphere resonator. *Black dots* Experimental data; *Red curves* numerical simulations with a given rise time, $t_r = 5$ ns. The fitting parameters are given in each case. **f** The peak heights defined as the maxima for plots **a–e** satisfy a linear relationship to κ_e/κ. The outlier in the plot is due to a large dispersive shift which breaks the linear relationship. The *inset* shows the external coupling coefficient as an exponential function of the coupling gap

one need only record the maximum of the transient signal to retrieve the information for sensing, assuming that one can measure all the other parameters, such as κ_e, κ_0 and t_r, from the steady-state transmission spectrum. This significantly reduces the complexity of the data processing and decreases the burden for data acquisition. In this sense, the acquisition speed can be further improved. For sensing based on reactive/dispersive interactions [19], sensitivity can be optimised by choosing the correct laser detuning. From Fig. 3b, the peak height has a Gaussian relationship to the detuning. In order to obtain the highest sensitivity, the pump laser frequency should be chosen so that it is closer to the WGM resonance. For example, in Fig. 3b, when $\Delta \sim 10\kappa$ the sensitivity is $\mathrm{d}H/\mathrm{d}\Delta \sim 0.01/\kappa$ (where H is the peak height). Also, from Fig. 3d, when t_{r} is longer, the Gaussian profile is steeper; this also improves the sensitivity.

For dissipative sensing [21], measuring the peak height for a fixed detuning will also yield valid results. As predicted by theory, the peak height changes linearly with $\kappa_{\mathrm{e}}/\kappa$. If the system experiences an intrinsic dissipation change due to environmental conditions, the peak height should maintain an inverse relationship to intrinsic dissipation under a certain coupling condition, which is the gap between the taper and the microsphere in our case (in more complicated situations, the coupling is not only determined by the gap but also the dissipation of the surrounding environment [22]). In a more complicated scenario, where both dispersion and dissipation exist, a measurement of the peak height may still be sufficient. In the experiment mentioned in the previous section, it was shown that, for large detuning, the peak height always satisfies a linear relationship to $\kappa_{\mathrm{e}}/\kappa$; therefore, it provides a dissipative sensing method immune to any changes to the laser detuning. By having different laser detuning configurations, contributions from dispersive and dissipative interactions can be well categorised. For instance, using a WDM technique, two lasers with independent configurations can be used for simultaneous dispersion and dissipation sensing.

5 Conclusion

In summary, the dynamical mechanisms behind CRUS were investigated by solving the coupled mode equations for a transient response to a Gaussian input pulse. The detailed relationship of the CRUS to laser detuning, coupling coefficient and rise time was determined using approximate analytical solutions. This is further verified by experimental measurements using an ultrahigh Q silica microsphere. Using this method, dispersive and dissipative sensing can be performed separately in the transient domain.

6 Personal Recollections: Síle Nic Chormaic

I had the pleasure of working in the Laser Spectroscopy Division of the MPQ during a very exciting time, from 1999 to 2000, when the mode-locked femtosecond laser frequency comb was being developed and used for precision spectroscopy. Though I was working on a different experiment, the energy in the laboratory at the time was contagious; the weekly group meetings and the Schloss Ringberg workshop ensured that we all felt part of this exciting project. I am not sure how many of us fully realised the impact that this early work would have on the field of precision spectroscopy over the following decades. Ted's insight and vision drove this work forward. I am extremely grateful for the exceptional opportunities he gave to me and for introducing me to *Rote Grütze*. Happy 75th birthday!

Acknowledgements This work was supported by the Okinawa Institute of Science and Technology Graduate University. Y.Y. and R.M. made equal contributions to this work.

References

1. C.Y. Wang, T. Herr, P. Del'Haye, A. Schliesser, J. Hofer, R. Holzwarth, T.W. Hänsch, N. Picqué, T.J. Kippenberg, Mid-infrared optical frequency combs at 2.5 μm based on crystalline microresonators. Nat Commun **4**, 1345 (2013)
2. Y. Yang, X. Jiang, S. Kasumie, G. Zhao, L. Xu, J. M. Ward, L. Yang, S. Nic Chormaic, Four-wave mixing parametric oscillation and frequency comb generation at visible wavelengths in a silica microbubble resonator. Opt. Lett. **41**, 5266 (2016)
3. J. Alnis, A. Schliesser, C.Y. Wang, J. Hofer, T.J. Kippenberg, T.W. Hänsch, Thermal-noise-limited crystalline whispering-gallery-mode resonator for laser stabilization. Phys Rev A **84**, 011804 (2011)
4. I. Fescenko, J. Alnis, A. Schliesser, C.Y. Wang, T.J. Kippenberg, T.W. Hänsch, Dual-mode temperature compensation technique for laser stabilization to a crystalline whispering gallery mode resonator. Opt. Express **20**, 19185–19193 (2012)
5. R. Madugani, Y. Yang, V.H. Le, J.M. Ward, S. Nic Chormaic, Linear laser tuning using a pressure-sensitive microbubble resonator. IEEE Photonics Technol. Lett. **28**, 1134–1137 (2016)
6. J.M. Ward, N. Dhasmana, S. Nic Chormaic, Hollow core, whispering gallery resonator sensors. Eur. Phys. J. Spec. Top. **223**, 1917–1935 (2014)
7. M.R. Foreman, J.D. Swaim, F. Vollmer, Whispering gallery mode sensors. Adv. Opt. Photonics **7**, 168 (2015)
8. N.M. Hanumegowda, C.J. Stica, B.C. Patel, I. White, X. Fan, Refractometric sensors based on microsphere resonators. Appl. Phys. Lett. **87**, 201107 (2005)
9. I.M. White, H. Oveys, X. Fan, Liquid-core optical ring-resonator sensors. Opt. Lett. **31**, 1319 (2006)
10. C.-H. Dong, L. He, Y.-F. Xiao, V.R. Gaddam, S.K. Ozdemir, Z.-F. Han, G.-C. Guo, L. Yang, Fabrication of high-Q polydimethylsiloxane optical microspheres for thermal sensing. Appl. Phys. Lett. **94**, 231119 (2009)
11. Y.-Z. Yan, C.-L. Zou, S.-B. Yan, F.-W. Sun, Z. Ji, J. Liu, Y.-G. Zhang, L. Wang, C.-Y. Xue, W.-D. Zhang, Z.-F. Han, J.-J. Xiong, Packaged silica microsphere–taper coupling system for robust thermal sensing application. Opt. Express **19**, 5753–5759 (2011)
12. J.M. Ward, Y. Yang, S. Nic Chormaic, Highly sensitive temperature measurements with liquid-core microbubble resonators. IEEE Photonics Technol. Lett. **25**, 2350–2353 (2013)
13. R. Henze, T. Seifert, J. Ward, O. Benson, Tuning whispering gallery modes using internal aerostatic pressure. Opt. Lett. **36**, 4536–4538 (2011)

14. Y. Yang, S. Saurabh, J.M. Ward, S. Nic Chormaic, High-Q, ultrathin-walled microbubble resonator for aerostatic pressure sensing. Opt. Express **24**, 294 (2016)
15. M. Sumetsky, Y. Dulashko, R.S. Windeler, Super free spectral range tunable optical microbubble resonator. Opt. Lett. **35**, 1866–1868 (2010)
16. R. Madugani, Y. Yang, J.M. Ward, J.D. Riordan, S. Coppola, V. Vespini, S. Grilli, A. Finizio, P. Ferraro, S. Nic Chormaic, Terahertz tuning of whispering gallery modes in a PDMS stand-alone, stretchable microsphere. Opt. Lett. **37**, 4762 (2012)
17. B.-B. Li, W.R. Clements, X.-C. Yu, K. Shi, Q. Gong, Y.-F. Xiao, Single nanoparticle detection using split-mode microcavity Raman lasers. Proc. Natl. Acad. Sci. **111**, 14657–14662 (2014)
18. S.K. Özdemir, J. Zhu, X. Yang, B. Peng, H. Yilmaz, L. He, F. Monifi, S.H. Huang, G.L. Long, L. Yang, Highly sensitive detection of nanoparticles with a self-referenced and self-heterodyned whispering-gallery Raman microlaser. Proc. Natl. Acad. Sci. **111**, E3836–E3844 (2014)
19. F. Vollmer, S. Arnold, D. Keng, Single virus detection from the reactive shift of a whispering-gallery mode. Proc. Natl. Acad. Sci. USA **105**, 20701–20704 (2008)
20. L. He, S.K. Ozdemir, J. Zhu, W. Kim, L. Yang, Detecting single viruses and nanoparticles using whispering gallery microlasers. Nat. Nanotechnol. **6**, 428–432 (2011)
21. Y. Hu, L. Shao, S. Arnold, Y.-C. Liu, C.-Y. Ma, Y.-F. Xiao, Mode broadening induced by nanoparticles in an optical whispering-gallery microcavity. Phys. Rev. A **90**, 043847 (2014)
22. B.-Q. Shen, X.-C. Yu, Y. Zhi, L. Wang, D. Kim, Q. Gong, Y.-F. Xiao, Detection of single nanoparticles using the dissipative interaction in a high-Q microcavity. Phys. Rev. Appl. **5**, 024011 (2016)
23. J. Knittel, J.D. Swaim, D.L. McAuslan, G.A. Brawley, W.P. Bowen, Back-scatter based whispering gallery mode sensing. Sci. Rep. **3**, 2974 (2013)
24. E. Gavartin, P. Verlot, T.J. Kippenberg, A hybrid on-chip optomechanical transducer for ultrasensitive force measurements. Nat. Nanotechnol. **7**, 509–514 (2012)
25. G. Bahl, K.H. Kim, W. Lee, J. Liu, X. Fan, T. Carmon, Brillouin cavity optomechanics with microfluidic devices. Nat. Commun. **4**, 1994 (2013)
26. L. Stern, I. Goykhman, B. Desiatov, U. Levy, Frequency locked micro disk resonator for real time and precise monitoring of refractive index. Opt. Lett. **37**, 1313–1315 (2012)
27. S. Rosenblum, Y. Lovsky, L. Arazi, F. Vollmer, B. Dayan, Cavity ring-up spectroscopy for ultrafast sensing with optical microresonators. Nat. Commun. **6**, 6788 (2015)
28. F.-J. Shu, C.-L. Zou, A.K. Özdemir, L. Yang, G.-C. Guo, Transient microcavity sensor. Opt. Express **23**, 30067–30078 (2015)
29. A.A. Savchenkov, A.B. Matsko, M. Mohageg, L. Maleki, Ringdown spectroscopy of stimulated Raman scattering in a whispering gallery mode resonator. Opt. Lett. **32**, 497–499 (2007)
30. A.A. Savchenkov, A.B. Matsko, V.S. Ilchenko, L. Maleki, Optical resonators with ten million finesse. Opt. Express **15**, 6768–6773 (2007)
31. C. Dong, C. Zou, J. Cui, Y. Yang, Z. Han, G. Guo, Ringing phenomenon in silica microspheres. Chin. Opt. Lett. **7**, 299–301 (2009)
32. A. Rasoloniaina, V. Huet, T.K.N. Nguyên, E. Le Cren, M. Mortier, L. Michely, Y. Dumeige, P. Féron, Controling the coupling properties of active ultrahigh-Q WGM microcavities from undercoupling to selective amplification. Sci. Rep. **4**, 4023 (2014)
33. M.-Y. Ye, M.-X. Shen, X.-M. Lin, Ringing phenomenon based whispering-gallery-mode sensing. Sci. Rep. **6**, 19597 (2016)
34. M.L. Gorodetsky, V.S. Ilchenko, Optical microsphere resonators: optimal coupling to high-Q whispering-gallery modes. J. Opt. Soc. Am. B **16**, 147 (1999)
35. C.-L. Zou, Y. Yang, C.-H. Dong, Y.-F. Xiao, X.-W. Wu, Z.-F. Han, G.-C. Guo, Taper–microsphere coupling with numerical calculation of coupled-mode theory. J. Opt. Soc. Am. B **25**, 1895 (2008)

36. T.J. Kippenberg, S.M. Spillane, K.J. Vahala, Modal coupling in traveling-wave resonators. Opt. Lett. **27**, 1669 (2002)
37. K. Srinivasan, O. Painter, Mode coupling and cavity-quantum-dot interactions in a fiber-coupled microdisk cavity. Phys. Rev. A **75**, 023814 (2007)
38. R. Madugani, Y. Yang, J.M. Ward, V.H. Le, S. Nic Chormaic, Optomechanical transduction and characterization of a silica microsphere pendulum via evanescent light. Appl. Phys. Lett. **106**, 241101 (2015)

Sensitivity and Resolution in Frequency Comb Spectroscopy of Buffer Gas Cooled Polyatomic Molecules

P. Bryan Changala, Ben Spaun, David Patterson, John M. Doyle, and Jun Ye

Abstract We discuss the use of cavity-enhanced direct frequency comb spectroscopy in the mid-infrared region with buffer gas cooling of polyatomic molecules for high-precision rovibrational absorption spectroscopy. A frequency comb coupled to an optical enhancement cavity allows us to collect high-resolution, broad-bandwidth infrared spectra of translationally and rotationally cold (10–20 K) gas-phase molecules with high absorption sensitivity and fast acquisition times. The design and performance of the combined apparatus are discussed in detail. Recorded rovibrational spectra in the CH stretching region of several organic molecules, including vinyl bromide (CH_2CHBr), adamantane ($C_{10}H_{16}$), and diamantane ($C_{14}H_{20}$) demonstrate the resolution and sensitivity of this technique, as well as the intrinsic challenges faced in extending the frontier of high-resolution spectroscopy to large complex molecules.

1 Introduction

The development of the optical frequency comb has dramatically changed the field of precision metrology and spectroscopy [1, 2]. While the use of combs as frequency references is now ubiquitous, they can also be directly employed for optical spectroscopy using techniques known as direct frequency comb spectroscopy [3]. By coupling frequency combs to high-finesse optical cavities, cavity-enhanced direct frequency comb spectroscopy (CE-DFCS) combines broad spectral bandwidth, high spectral resolution, and high detection sensitivity [4, 5].

This article is part of the topical collection "Enlightening the World with the Laser" - Honoring T. W. Hänsch guest edited by Tilman Esslinger, Nathalie Picqué, and Thomas Udem.

P.B. Changala (✉) • B. Spaun • J. Ye
JILA, National Institute of Standards and Technology and University of Colorado, Department of Physics, University of Colorado, Boulder, CO 80309, USA
e-mail: bryan.changala@colorado.edu

D. Patterson • J.M. Doyle
Department of Physics, Harvard University, Cambridge, MA 02138, USA

D. Meschede et al. (eds.), *Exploring the World with the Laser*,
https://doi.org/10.1007/978-3-319-64346-5_35

The extension of CE-DFCS to the mid-infrared spectral region has provided a powerful new tool for rovibrational molecular spectroscopy, with recent demonstrations of trace gas detection [6] and time-resolved spectroscopy of transient radicals [7]. However, up to now mid-infrared CE-DFCS studies have been limited to small, simple molecules in order to avoid spectral congestion. Consequently, such studies have not taken full advantage of the high spectral resolution afforded by the narrow linewidth of individual comb modes. One primary challenge to studying more complex molecules is that those with more than ~ten atoms have intractably congested absorption spectra at room temperature, precluding high-resolution studies. We have recently addressed this issue by combining mid-infrared CE-DFCS with buffer gas cooling of complex polyatomic molecules [8]. By cooling molecules' translational and rotational temperatures to 10–20 K, rovibrational spectra are drastically simplified, exhibiting narrow Doppler linewidths, reduced spectral congestion, and enhanced absorption cross sections. In contrast to supersonic expansion jets [9, 10], buffer gas cooling [11–14] provides a continuous source of cold molecules with slow laboratory frame velocities and long interaction times. The modest gas throughput obviates the need for significant pumping infrastructure. This method has permitted the measurement of the first rotationally resolved absorption spectra in the CH stretching region (~3 μm) of several large organic molecules, with all the benefits of sensitivity and fast acquisition times provided by CE-DFCS.

In this article, we provide a detailed description of the design and performance of our buffer gas cooling frequency comb spectrometer. Selected spectroscopic studies are discussed that illustrate both the capabilities of this technique, as well as more general prospects for high-resolution rovibrational spectroscopy of complex polyatomic molecules. We report spectra in the CH stretching region of vinyl bromide, CH_2CHBr, where we readily resolve small isotope splittings and nuclear hyperfine structure. Accurate relative transition intensities are shown to be measured over broad spectral bandwidths. We also discuss the spectra of two large molecules: adamantane, $C_{10}H_{16}$, and diamantane, $C_{14}H_{20}$, the simplest two diamonoid molecules, as an example of the onset of intramolecular vibrational redistribution (IVR) [15]. While the spectrum of adamantane displays well-resolved rotational structure at internal energies of $\sim$3000 cm^{-1}, intrinsic spectral congestion caused by IVR renders that of diamantane essentially continuous. This observation has important implications for high-resolution spectroscopy of molecules of this size or larger at the internal energies probed here.

2 Apparatus and Methods

2.1 Frequency Comb Light Source

The mid-infrared frequency comb is generated in a tunable optical parametric oscillator (OPO) synchronously pumped by a 1 μm ytterbium (Yb) fiber comb [16]. The 1 μm pump comb has a repetition rate $f_{\mathrm{rep}} \approx 136.6$ MHz and a maximum output power of 10 W. The OPO is based on a fan-out MgO-doped periodically poled lithium niobate (PPLN) crystal within a cavity that is singly resonant at the signal wavelength. By tuning the quasi-phase-matching period of the fan-out PPLN crystal, idler output is generated with a center wavelength from 2.8 to 4.8 μm, a maximum average power of 1.5 W, and a simultaneous bandwidth up to 0.3 μm. The experiments reported here use idler light from 3.0 to 3.3 μm with an average power ranging from 100 to 300 mW.

The optical frequency of each comb mode is determined by $\nu_n = f_{\text{ceo}} + nf_{\text{rep}}$, where n is the integer comb mode number of order 10^5–10^6. Stabilization of the optical frequency requires locking both the carrier-envelope offset frequency f_{ceo} and the repetition rate f_{rep} of the idler comb. We lock f_{rep} by picking off a small portion of the 1 μm ytterbium fiber comb and measuring its repetition rate directly on a fast photodiode. The seventh harmonic of the repetition rate $7f_{\text{rep}} \approx 956$ MHz is mixed with a stable 1 GHz Wenzel quartz oscillator, slowly slaved to a 10 MHz cesium (Cs) clock. The resulting beat note is used to generate a phase error signal with respect to the RF output of a direct digital synthesizer (DDS), which is phase locked to the same RF oscillator. This error signal is then used to feed back simultaneously on a fast piezo (PZT)-actuated cavity mirror and a slower PZT fiber stretcher in the fiber laser oscillator.

The f_{ceo} frequency of the mid-IR idler comb is measured by performing an optical beat note between the 1 μm pump light (p) and the parasitic sum frequency of the pump and idler ($p + i$) generated by the OPO crystal. A 150- to 200-mW portion of the pump light is picked off and coupled into a 15-cm-long piece of highly nonlinear supercontinuum fiber to broaden it to ~780 nm in order to be spectrally overlapped with the $p + i$ light. The output of the supercontinuum fiber and the $p + i$ light are then spatially and temporally overlapped onto a single photodiode to measure the $p - (p + i)$ optical beat note, which is equal to f_{ceo} of the idler comb. This RF is used to generate a phase error signal with another DDS set to the desired f_{ceo} value. The error signal is fed back onto PZT-actuated mirrors in the OPO cavity.

2.2 *Buffer Gas Cooled Molecule Source and Enhancement Cavity*

Cryogenic cooling is achieved with a liquid-helium cryostat refrigerator placed within a vacuum dewar chamber. Figure 1a shows a schematic of the apparatus. A $(6\ \text{cm})^3$ aluminum cell is anchored to the 4 K stage of the refrigerator and reaches wall temperatures of 5–10 K, depending on residual blackbody and molecule flow heat loads. The cold cell is surrounded by a 35 K copper shield to reduce radiative blackbody heating. A small continuous flow (~10 sccm) of helium buffer gas is pre-cooled to 4 K before entering the cold cell, where it builds up a steady-state density of $\sim 10^{14}\ \text{cm}^{-3}$. Helium that escapes the cell is cryopumped by charcoal

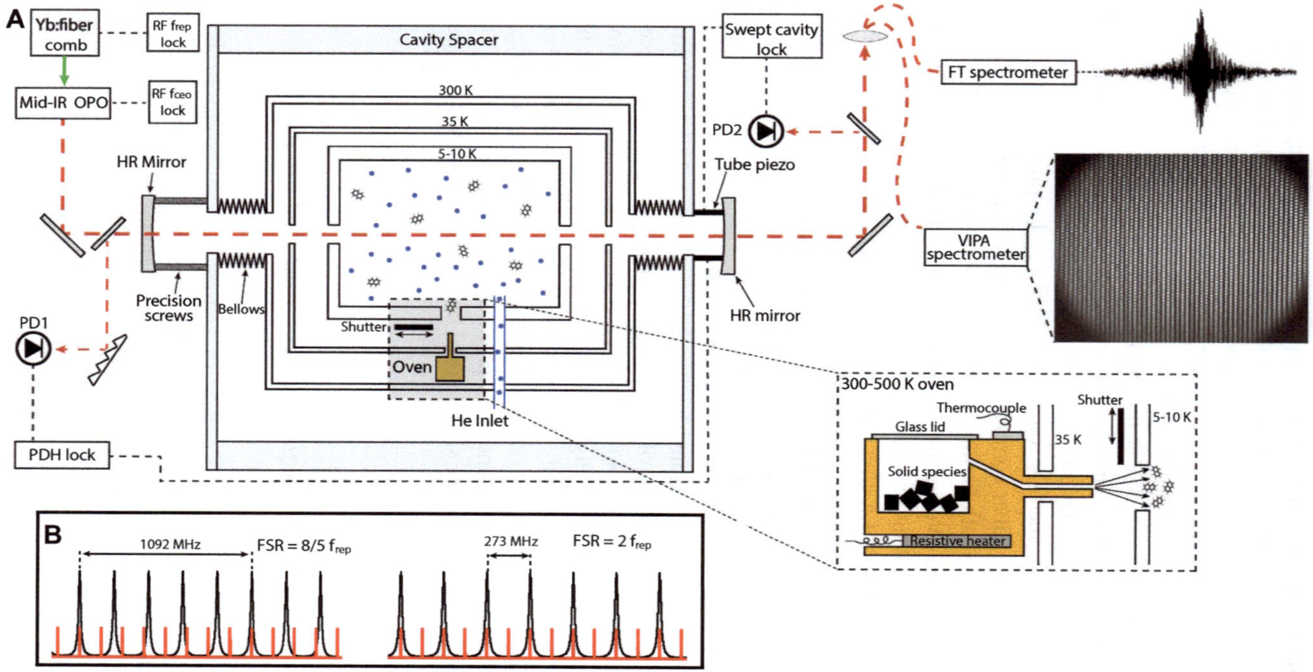

Fig. 1 Buffer gas cooling frequency comb spectrometer. **a** Schematic of the apparatus. Mid-infrared frequency comb light generated in the Yb fiber-pumped OPO is coupled to an enhancement cavity surrounding the cold cell dewar. RF locks stabilize the comb f_{ceo} and f_{rep} frequencies. The spectrum of the transmitted comb light is read out by an FT spectrometer, which measures a comb interferogram, or a VIPA spectrometer, which generates a two-dimensional image of dispersed comb modes. The cavity FSR is stabilized relative to the comb f_{rep} with a PDH or swept cavity lock. *Inset* illustrates the oven source used to introduce nonvolatile samples into the cold cell. **b** Cavity–comb filtering schemes. By adjusting the cavity FSR, an 8:5 (*left*) or 2:1 (*right*) f_{rep}-to-FSR ratio can be selected. Comb modes not resonant with the cavity are rejected, resulting in a sparser transmitted comb spectrum, which facilitates single comb mode resolution with the FTS or VIPA

sorbs attached to the 4 K stage. The sorbs have sufficient capacity to pump helium for several hours before requiring a warm-up to allow for helium desorption.

Warm gas-phase molecules enter the cold cell through a $\sim$1-cm aperture in the cell wall. After many helium-molecule and helium-cell collisions, the molecular translational and rotational degrees of freedom are brought into quasi-thermal equilibrium with the cold cell wall. Measured rotational and translational temperatures are typically 10–20 K. For volatile molecules with sufficient vapor pressure at room temperature, a flow of 1–10 sccm is introduced into the cell via a one-quarter-inch stainless steel tube that terminates 1–2 cm in front of the outer surface of the cell wall. For nonvolatile molecules, a small copper oven (2 cm $\times$ 2 cm $\times$ 4 cm) is used to heat samples to produce sufficiently high vapor pressures. The oven, with an inner cavity of $\sim$3 cm^3, is located *in vacuo* just outside the 35 K radiation shield, with a small tube outlet passing through the shield to guide molecules into the cold cell (see inset of Fig. 1a). The inner diameter (2 mm) and length (3 cm) of the outlet determine the output conductance of the oven and therefore what vapor pressure is necessary for sufficient molecule output flow. The output conductance is high enough to require only modest oven temperatures (100–200 $^\circ$C), but small enough to allow sufficient oven pressure buildup to provide stable vaporization and prevent boiling. A small glass window permits visual monitoring of the interior of the oven while in use. The oven is resistively heated by two embedded cartridge heaters and temperature monitored by a thermocouple.

A "molecule shutter" consisting of a thin aluminum paddle is placed between the molecule tube or oven outlet and the cold cell aperture in order to quickly turn on or off the molecule flow into the cold cell. The shutter is rotated into and out of the molecule flow with a small DC electric motor mounted to the 35 K shield. Short current pulses switch the shutter with an open/close time of 1 ms and a delay of 10 ms.

In order to increase the effective interaction length of the comb light and the molecular absorbers, an enhancement cavity consisting of two parallel mirrors with high reflectivity (HR) coatings surrounds the cold cell. Four one-inch-thick stainless steel rods, girdling the cryostat dewar, fix the gross cavity length. The cavity mirrors are mechanically isolated from the dewar by a system of edge-welded bellows that form the vacuum connection between the cavity mirrors and the dewar. The position and alignment of each mirror are adjusted macroscopically by a set of fine precision screws. One of the cavity mirrors is fitted with a tube piezo for fine length adjustment and feedback control. The HR mirrors have a broadband coating covering 3.1–3.5 μm and a maximum finesse of $\mathcal{F} \approx 6000$. We typically achieve on-resonance cavity transmission efficiency of a few percent over a simultaneous transmission bandwidth of $\sim$100 nm, limited by the HR mirror dispersion. Recent developments in mid-infrared mirror coatings [17] show that both the cavity finesse and transmission efficiency can be significantly improved, which provides direct benefits to the sensitivity of CE-DFCS.

2.3 Cavity–Comb Coupling and Spectral Readout Schemes

We have implemented two different methods for coupling the infrared frequency comb to the enhancement cavity and for reading out the transmitted comb light spectrum. The first of these uses a Pound–Drever–Hall (PDH) lock to match the cavity free spectral range (FSR) to the comb f_{rep}, while the spectrum of the continuous comb transmission is measured with a fast-scanning Fourier transform spectrometer (FTS) [6, 18, 19]. The second approach uses a swept cavity coupling scheme and a virtually imaged phased array (VIPA) spectrometer for readout [5, 20]. These approaches have different advantages in resolution and sensitivity, and depending on the conditions of the experiment one may be preferable over the other. The remainder of this section describes each method in detail and compares their performance.

2.3.1 PDH Cavity Lock and FTS Readout

A PDH lock permits the cavity and comb to be resonantly coupled continuously, resulting in the highest cavity transmission duty cycle. Phase modulation sidebands are generated by dithering the fiber laser oscillator cavity length at 760 kHz, which corresponds to a resonance in the fiber oscillator mirror piezo. Light reflected from the cold cell enhancement cavity is picked off with a magnesium fluoride flat, dispersed with a reflection grating, passed through a slit, and incident on a photodiode. The grating and slit allow only a $\sim$10-nm portion of the comb light to reach the photodiode. The photodiode signal is then demodulated at the 760 kHz dither frequency to generate a PDH error signal. This error signal is used to servo the enhancement cavity length via the tube piezo to ensure the cavity FSR matches the comb f_{rep}. The PDH error signal is used to feed back in parallel on the fiber laser oscillator piezo at fast timescales, which helps to extend the effective servo bandwidth beyond the response of the enhancement cavity tube piezo ($\sim$1 kHz). Due to dispersion in the cavity, its optical resonance frequencies have a DC offset $f_{0,\mathrm{cav}}$ ($0 \leq f_{0,\mathrm{cav}} < \mathrm{FSR}$), analogous to the comb f_{ceo} frequency. We obtain the widest PDH lock transmission bandwidth by setting $f_{\mathrm{ceo}} = f_{0,\mathrm{cav}}$. However, we typically detune f_{ceo} from $f_{0,\mathrm{cav}}$ by up to a few MHz, which we observe decreases the cavity transmission intensity noise, albeit at the cost of reduced transmission bandwidth.

The transmitted comb light is routed to a home-built fast-scanning FTS [19]. An example of the center burst of a comb interferogram is illustrated in Fig. 1a. Each arm of the interferometer is double passed on a translating corner-cube retroreflector. This geometry multiplies the physical scanning range of the FTS delay stage (0.7 m) by a factor of four, to yield interferograms with a net optical path length difference of $\ell = 2.8$ m. The corresponding spectrometer linewidth is $\Delta\nu_{\mathrm{FTS}} = c/\ell = 110\,\mathrm{MHz}$. For a traditional white light source, this would determine the fundamental resolution limit of the FTS. However, the situation can be very different with an optical frequency comb: If adjacent comb modes are individually resolved by the spectrometer, then the effective resolution is determined by the linewidth of the comb mode (typically ~50 kHz), instead of that of the spectrometer [21]. This condition is met if the spectrometer resolution is narrower than the comb mode spacing. As our repetition rate $f_{rep} = 136.6$ MHz is only slightly larger than $\Delta\nu_{FTS}$, we set the length of the enhancement cavity such that the cavity FSR $= 2f_{rep}$. This has the effect of filtering out every other comb mode, as illustrated in the right half of Fig. 1b, resulting in a transmitted comb mode spacing of $2f_{rep} = 273$ MHz, which is readily resolved by the FTS.

Because the cavity FSR is not an integer multiple of $\Delta\nu_{FTS}$, the transmitted comb mode frequencies and the center frequencies of the Fourier transform spectrum walk off with respect to one another. In order to measure the spectrum at the actual comb mode frequencies, we resample the complex-valued frequency spectrum via convolution with the known instrument lineshape function (a sinc function). In this manner, we can efficiently locate the center frequency and intensity of each comb mode. The absolute frequency calibration of the spectrum is performed by measuring known transition frequencies of calibrant molecules, typically CH_4, which provide a ~ 30 MHz absolute accuracy. The relative frequency accuracy is 5–10 MHz, limited by the f_{rep} scan step size (see below).

The filtered comb mode spacing of 273 MHz is much larger than the typical Doppler linewidths of the cold molecules ($\Delta\nu_{mol} \approx 20 - 60$ MHz). Therefore, for a given cavity FSR, a single FTS acquisition will only measure a fraction of the molecular absorption transitions. In order to fill in the gaps, multiple spectra are acquired at slightly different comb f_{rep} values (and corresponding cavity FSRs), spaced apart by Δf_{rep}. The shift in the optical frequency of each comb mode is $\Delta\nu_n = n\Delta f_{rep}$, where $n \approx 10^5 - 10^6$ is the comb mode number. We typically choose Δf_{rep} such that $n\Delta f_{rep} < \Delta\nu_{mol}/5$. FTS acquisitions at each value of Δf_{rep} are then interleaved together to form a single high-resolution spectrum. To obtain an absorption spectrum, we separately normalize out the slowly varying transmission spectrum for each FTS acquisition. This baseline is obtained by simply applying a low-pass filter to the measured FTS spectrum. In the absence of severe spectral congestion, molecular absorption features, with 20–60 MHz Doppler-limited linewidths, can be clearly distinguished from broader fluctuations across the comb spectrum. Etalons and shot-to-shot fluctuations in the measured comb spectrum, which are typically observed with characteristic widths of 2–300 GHz, are completely removed by this normalization process. A single interferogram can be collected and processed in 5–10 s. An entire f_{rep}-stepped spectrum, with four averages per f_{rep} value, requires 30 min of scanning time. Because of the massively multiplexed nature of the frequency comb, this represents an orders-of-magnitude improvement in "spectral velocity" relative to state-of-the-art single frequency cw laser spectroscopy [10].

The fundamental limit to the absorption sensitivity is set by the photon shot noise. For the comb powers used here, this limit is about 10^{-4} $Hz^{-1/2}$ fractional absorption for a single spectral element of the Fourier transform spectrum, which is well below our measured noise floor. In the PDH/FTS configuration, our dominant noise is intensity fluctuations of cavity-transmitted comb light, caused mostly by the unfavorable frequency-to-amplitude noise conversion by the cavity. An autobalanced detector circuit that measures both output arms of the FTS interferometer reduces the common mode intensity noise by up to 20 dB at the interferogram carrier frequency (100–200 kHz). However, residual intensity noise remains the largest noise source, resulting in an absorption sensitivity limit of 4.4×10^{-8} cm^{-1} Hz$^{-1/2}$ for a single comb mode. The typical bandwidth of the FTS spectra contains 3300 resolved comb lines, corresponding to 7.6×10^{-10} cm^{-1} Hz$^{-1/2}$ per spectral element (PSE). This is roughly an order of magnitude worse than previous trace detection experiments using mid-infrared CE-DFCS, which reported a sensitivity of 6.9×10^{-11} cm^{-1} Hz$^{-1/2}$ PSE [6]. This discrepancy is accounted for by our narrower cavity linewidth and transmission bandwidth, as well as increased cavity length noise. We note that while the enhancement cavity is ~60 cm long, we only make use of 6 cm (the length of the cold cell) where molecules are present. Consequently, we suffer from a narrow cavity linewidth and therefore increased frequency-to-amplitude noise conversion, without the compensation in sensitivity provided by a longer path length.

2.3.2 Swept Cavity Lock and VIPA Readout

Given that the PDH/FTS absorption sensitivity is limited by cavity transmission noise, an amplitude noise immune cavity locking scheme can potentially provide significant sensitivity improvements. Such a scheme is realized by the swept cavity lock method [5]. Here, the enhancement cavity length is scanned by applying a 10 kHz modulation signal to the cavity tube piezo. As the cavity length is swept, it passes through resonance with the incident comb light, permitting a transient buildup and transmission of power over the entire comb bandwidth. A regular sequence of transmission bursts, occurring at twice the sweep frequency, exits the cavity as the cavity length is scanned back and forth. The relative time spacing between sequential transmission bursts is determined by the offset between the resonant cavity length and the mean value of the cavity length scan range. A small fraction of the transmitted cavity light is measured by a photodiode, and this signal is demodulated at the sweep frequency via a lock-in amplifier, generating an error signal that is fed back onto the cavity length tube piezo. The feedback loop only requires that the resonant cavity length remains near the middle of the tube piezo scan range. Therefore, this locking scheme only requires a ~1 kHz servo loop bandwidth with no feedback on the fiber comb.

Unfortunately, the pulsed nature of the swept cavity transmission is incompatible with the FTS readout. The interferogram is digitized at a sampling rate of 8 MHz, which is significantly higher than the maximum cavity length sweep frequency. Consequently, there is no way to synchronize the FTS sampling with the intermittent cavity transmission. As an alternative to the FTS, we use a dispersive mid-IR VIPA spectrometer [20]. In brief, the spectrometer consists of an HR-coated parallel plate etalon, which disperses the comb light strongly in the vertical direction, resulting in spatially resolved comb modes, followed by a reflective grating, which separates out FSRs of the etalon in the horizontal direction. The two-dimensional pattern of comb light is then imaged onto a liquid N_2 cooled InSb array detector. We integrate the detector pixel counts for 2 ms per frame, corresponding to collecting light from 40 swept cavity transmission bursts.

The VIPA etalon has a resolution of 600 MHz. Therefore, the 2:1 f_{rep}-to-FSR cavity filtering used with the FTS is insufficient for comb mode-resolved spectra. Instead, we increase the enhancement cavity length to perform filtering with an 8:5 f_{rep}-to-FSR ratio, as illustrated in the left of Fig. 1b. The cavity-transmitted comb lines are spaced in this configuration by $8f_{rep} \approx 1093$ MHz, which can be well resolved by the VIPA spectrometer. A raw camera image displaying an array of individually resolved comb modes is shown in Fig. 1. Due to the Lorentzian lineshape of the VIPA etalon, the relatively wide tails of the imaged comb modes partially overlap with neighboring modes, resulting in spectral "crosstalk." After processing the image to extract the absorption spectrum, this crosstalk appears as artificial absorption lines spaced from the real lines by exactly the comb mode spacing ($8f_{rep}$). When optimally aligned, the amplitude of these features is approximately 3% relative to the real feature. We also apply an active correction filter to further reduce spectral crosstalk. This is performed by subtracting from the spectrum the expected crosstalk contributions from absorption features stronger than a given threshold. To do this, one must know the relative size of crosstalk peaks to the real peak, which is separately measured using a sparsely absorbing calibrant molecule such as methane. This active correction filter further reduces spectral crosstalk to 0.5%. In the absence of very strong absorption features, these artifacts are below our noise floor and do not obstruct the interpretation of our spectra.

The VIPA spectrometer camera frame rate (max. 435 Hz) permits rapid readout of successive spectra, enabling us to perform fast differential absorption measurements to further reduce the absorption noise baseline. Using the molecule shutter described above, we block molecules from entering the cold cell and acquire a reference light spectrum with the VIPA spectrometer. Then the molecule shutter is opened, permitting absorbers to enter the cold cell and another spectrum is acquired. The fractional absorption spectrum is computed from the relative change in the comb mode intensities. Given the diffusion time in the cell ($\sim$10 ms), we perform this differential measurement in 40 ms periods (50% molecules on, 50% molecules off). The reduction of the light intensity noise as a function of the switching time is illustrated in Fig 2a. At the switching times comparable to the camera frame rate, the noise curve begins to plateau, indicating that the camera background noise limit has been reached. At this limit, we achieve an absorption sensitivity $\sim$ten times smaller than the previous PDH/FTS limit. This represents a 20-fold improvement in the cavity transmission noise floor, but the swept cavity lock introduces a factor of 1/2 to the enhancement cavity's effective path length [5]. An example of the baseline noise reduction in actual spectra is shown in Fig 2b.

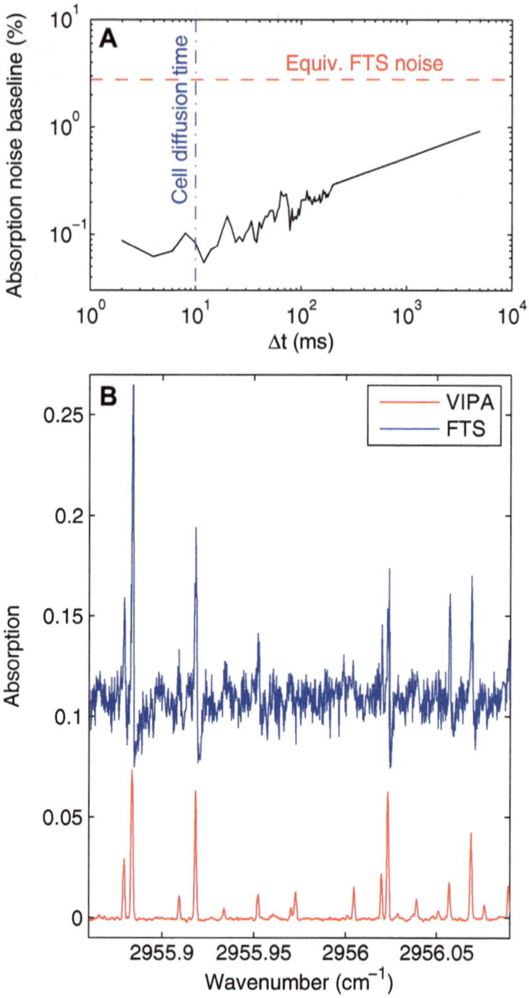

Fig. 2 Absorption noise baseline comparison of FTS versus VIPA. **a** The VIPA spectrometer absorption noise baseline as a function of the differential light intensity measurement period is plotted as the *black solid trace*. Each measurement used a total integration time of 5 ms. The equivalent time FTS noise baseline is also indicated (*red horizontal dashed line*), as well as the molecule diffusion time in the cold cell (*blue vertical dashed line*). **b** A small portion of the absorption spectrum of nitromethane (CH_3NO_2) near 2956 cm^{-1}, comparing the noise baseline of the FTS spectrum (*blue*, shifted up) and the VIPA spectrum (*red*, bottom trace). Both spectra were gathered with a 30 min total acquisition time

As with the PDH/FTS scheme, the comb f_{rep} and cavity FSR must be scanned to cover the spectral range between transmitted comb modes. One disadvantage of the poorer VIPA spectrometer resolution is that the 8:5 cavity filtering requires a 4× longer scanning range than the 2:1 filtering. However, because of the improved sensitivity of the swept cavity lock/VIPA spectrometer, and because the camera frame acquisition time is much shorter than the FTS scanning time, we average for less time per f_{rep} value, resulting in a similar total acquisition time of 30 min. Also, the rapid differential absorption measurements allowed by this spectrometer remove etalons from the computed absorption spectrum. Laser amplitude fluctuations produce small, broad variations across the baseline of the absorption spectrum. We remove these variations by separately fitting the baseline of each acquired spectrum.

3 Spectroscopy of Polyatomic Molecules

3.1 Vinyl Bromide

The 3 μm region of the absorption spectrum of vinyl bromide, CH_2CHBr, is dominated by the CH stretching fundamentals. All three CH stretching modes are IR active, and all three are readily observed in the spectrum spanning the region 3020–3120 cm^{-1}. As a demonstration of both the resolution and accurate intensities provided by the spectrometer, we discuss in detail only the spectrum of the ν_3 fundamental, which has the lowest frequency of the three CH stretches. An overview of the ν_3 band is shown in Fig. 3a. This mode is assigned as the in-phase stretching vibration of the two hydrogens on the CH_2 end of the molecule [22, 23].

Vinyl bromide is a relatively rigid, near-prolate top [24–26]. The otherwise straightforward appearance of its spectrum is, however, complicated by the bromine atom, which occurs as two main isotopes, ^{79}Br and ^{81}Br. Each has approximately 50% natural abundance, and thus, there are two equally strong vibrational bands superimposed in the spectrum, resulting in increased congestion and line density. Additionally, both bromine isotopes have a nuclear spin of $I = 3/2$ and a relatively large nuclear electric quadrupole moment. As a result, hyperfine structure in the spectrum is observable, especially in the low J transitions. An example of

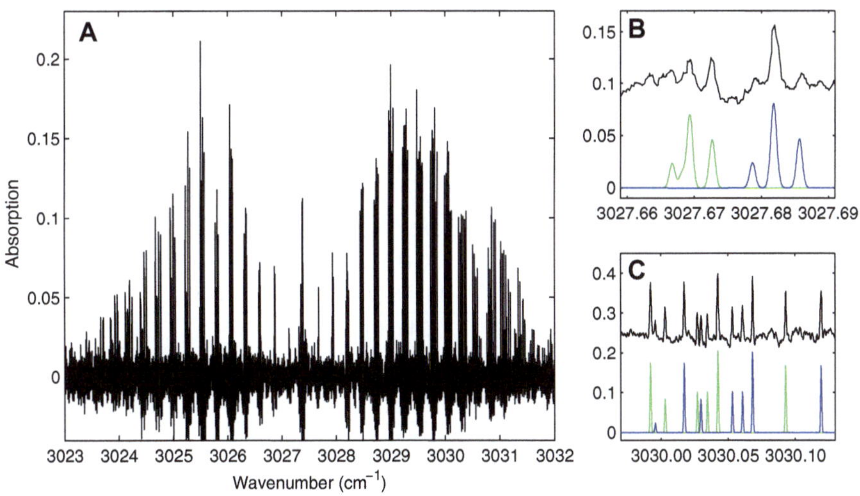

Fig. 3 Vinyl bromide ν_3 band. **a** Overview of the ν_3 spectrum acquired with the PDH/FTS technique. **b** Hyperfine splittings of the $R(0)$ transition. The measured spectrum is shown in *black*, with a vertical offset. The simulated spectra of $CH_2CH^{79}Br$ (*blue*) and $CH_2CH^{81}Br$ (*green*) are shown below. The three hyperfine components correspond to transitions from a common $F'' = 3/2$ lower state to upper states with $F' = 1/2$, $5/2$, and $3/2$, in order of increasing transition frequency. **c** Measured and simulated spectra of the $R(9)$ transitions. The multiple transitions belong to different K sub-bands and parity components with $K = 0$–4

these hyperfine splittings for the $R(0)$ transition of each isotopologue is shown in Fig. 3b.

With the aid of known ground-state combination differences measured by rotational microwave spectroscopy [26], we assigned several hundred rovibrational transitions involving rotational levels up to $J \leq 18$. Using the PGOPHER program [27], we performed a fit to an effective rotational Hamiltonian for each isotopologue. The results of these fits are summarized in Table 1. The average fit error of 6 MHz is consistent with our relative line center measurement uncertainty. Moreover, as shown in Fig. 3c, the simulated relative transition intensities agree very well with the experimental spectrum, and we observe no perturbations to the measured transitions.

The hyperfine splittings, observed most clearly in the $R(0)$ and $P(1)$ transitions of each isotopologue, indicate that the nuclear quadrupole coupling constants do not change significantly between the vibrational ground and excited states. This is consistent with the expectation that this CH stretching mode does not include relative motion between the bromine atom and the carbon atom it forms a bond with. The electric field gradient at the bromine nucleus is therefore relatively unaffected by excitation in ν_3.

All three CH stretching modes, being in-plane motions, can in principle have a transition dipole projection along both the a and b principal axes. However, we have not yet identified any b-type transitions, readily discerned by their ΔK_a = odd

Table 1 Vinyl bromide effective Hamiltonian fits for the ν_3 band of $CH_2CH^{79}Br$ and $CH_2CH^{81}Br$

		^{79}Br		^{81}Br	
Parameter		$v = 0$	ν_3	$v = 0$	ν_3
ν_0		0	3027.4152 (10)	0	3027.4041 (10)
A		1.810093	1.804461 (7)	1.809641	1.804053 (9)
B		0.1388471	0.1387183 (4)	0.1380359	0.1379096 (5)
C		0.1288373	0.1286733 (5)	0.1281362	0.1279755 (5)
Δ_J	$\times 10^7$	0.534	0.520 (12)	0.528	0.553 (13)
Δ_{JK}	$\times 10^5$	−0.10216	−0.1188 (29)	−0.10146	−0.1018 (47)
Δ_K	$\times 10^4$	0.4223	0.1624 (40)	0.4206	0.2835 (51)
δ_J	$\times 10^8$	0.57	[0.57]	0.57	[0.57]
δ_K	$\times 10^6$	0.3922	[0.3922]	0.3841	[0.3841]
χ_{aa}		470.98	469.55 (237)	393.58	394.13 (237)
χ_{bb-cc}		37.04	31.59 (733)	30.86	22.68 (716)
$\lvert\chi_{ab}\rvert$		246.14	[246.14]	204.17	[204.17]
RMS error $\times 10^4$			2.20		2.24

The standard Watson A-reduced quartic Hamiltonian (I^r representation) was used to fit the measured transition frequencies. All values are given in cm^{-1}, except for χ_{aa}, χ_{bb-cc}, and $|\chi_{ab}|$, which are given in MHz. All ground-state ($v = 0$) constants are taken from Ref. [24], again except for χ_{aa}, χ_{bb-cc}, and $|\chi_{ab}|$, which are taken from Ref. [26]. Values in [] brackets are held fixed during the fit. 1σ uncertainties are specified in parentheses; for ν_0 values, the uncertainty corresponds to the estimated absolute frequency accuracy of our calibration procedure selection rule, in the ν_3 spectrum. We estimate the ratio of the transition dipole components to be $|\mu_b/\mu_a|^2 < 0.1$, i.e., that the ν_3 transition dipole is aligned closely along the *a* axis. A simple model would be to assume that the vibrational transition dipole is approximately the vector sum of the oscillating dipoles of each local CH stretch. If both of the CH_2 stretches contribute equally to the ν_3 normal mode, then the transition dipole would be approximately parallel to the C=C double bond, which has a non-negligible projection along the *b* axis. Instead, the measured spectrum suggests that the ν_3 mode contains a larger contribution from the CH stretch *trans* to the C–Br bond and a lesser contribution from the CH stretch *cis* to the C–Br bond.

In summary, the frequency and intensity information contained in our CE-DFCS spectrum provides complementary insights into how the electronic properties of vinyl bromide are changed upon excitation in the ν_3 normal mode. The measured nuclear hyperfine structure indicates the C–Br bond is unaffected by this vibration, while the relative strengths of rovibrational transitions illustrate that the electric charge distribution oscillates nearly parallel to the *a* axis.

3.2 Large Hydrocarbons: Adamantane and Diamantane

Extending the domain of high-resolution infrared spectroscopy to large molecules is only possible if individual rovibrational transitions can be well resolved in a given vibrational band. Several factors determine the degree of spectral congestion. These can be classified as either extrinsic effects (those that depend on the experimental conditions) or intrinsic effects (those determined by inherent molecular properties, such as rotational line spacing and IVR).

The low translational and rotational temperatures afforded by buffer gas cooling significantly reduce spectral congestion caused by extrinsic effects, namely by reducing the Doppler broadened linewidth and the rotational partition function. At the conditions of our experiment, the linewidth decreases with temperature as $T^{1/2}$ and the rotational partition function as $T^{3/2}$. At 15 K, this yields improvements by factors of 4.5 and 90, respectively, relative to room temperature. Not only does this significantly reduce spectral congestion, but also offers an orders-of-magnitude increase in peak absorption cross sections.

We have exploited these benefits to measure the first rotationally resolved absorption spectra in the CH stretching region of adamantane, $C_{10}H_{16}$, a large hydrocarbon that serves as the basic building block of "diamonoids," carbon cage molecules with the structure of the diamond lattice. Spectra from our recent initial report [8] are shown in Fig. 4a. Instead of single *R*(J) or *P*(J) rotational transitions, as expected from a rigid spherical top, the rotational transitions are split into dense clusters of lines by non-spherical centrifugal distortion effects allowed by the lower tetrahedral symmetry of the molecule. The detailed patterns observed in the spectrum report on subtle aspects of the rovibrational structure. Due to the spectroscopic

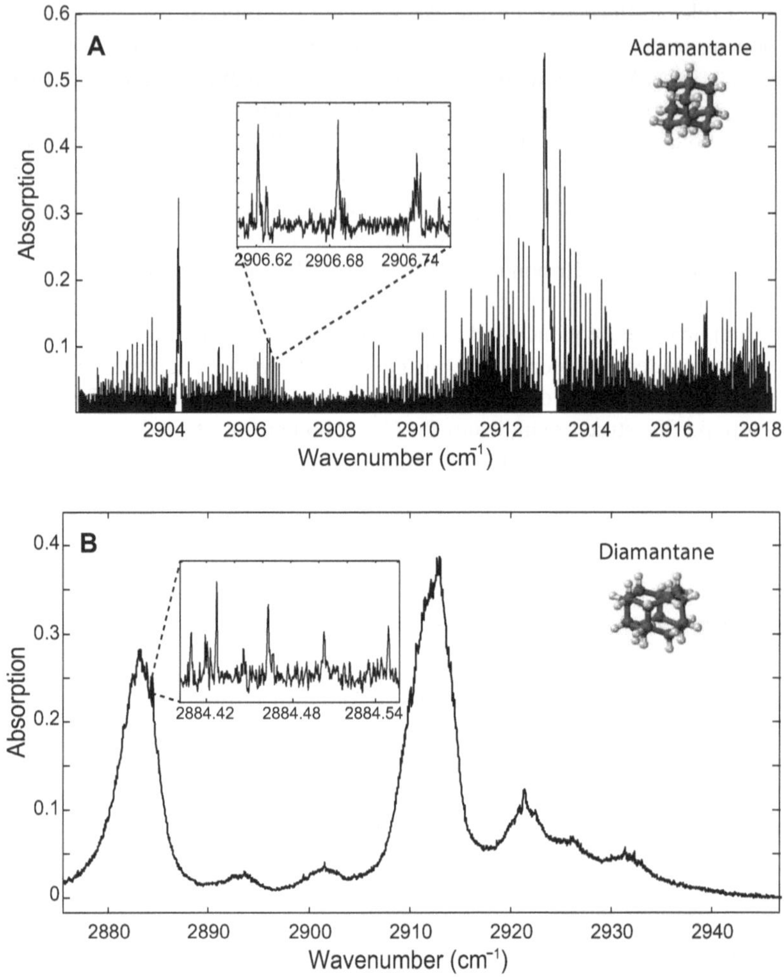

Fig. 4 Mid-infrared spectra of diamonoids. **a** A portion of the CH stretching manifold of adamantane. This spectrum was recorded with the PDH/FTS scheme. *Inset* shows resolved rotational fine structure. **b** The absorption spectrum of diamantane in the CH stretching region, acquired with the swept cavity/VIPA technique. At our experimental conditions, the spectrum is essentially continuous, with no well-resolved rotational structure. *Inset* shows the rare appearance of narrow features above the continuous background (*inset* spectrum has been high-pass filtered for visual clarity). The width of these features is consistent with the expected Doppler broadened linewidth, and the spacing between them is approximately equal to $2B$

selection rules of tetrahedral tops, transition combination differences cannot be obtained from the infrared active fundamentals [28], complicating the analysis. We thus forego further discussion of fitting transition frequencies of adamantane here.

The obvious question is: How large can one go? While the answer depends on detailed properties of an individual molecule in addition to just its size, a natural comparison to adamantane is diamantane, $C_{14}H_{20}$, the next largest member of the diamonoid family. Using an ab initio equilibrium geometry [29], we estimate diamantane's rotational constants to be $A = 0.0423$ cm^{-1} and $B = C = 0.0258$ cm^{-1}. In the absence of anharmonic or rovibrational perturbations, rigid symmetric top simulations using these rotational constants display dense, yet rotationally resolved, mid-infrared vibrational bands. The measured absorption spectrum of diamantane in the CH stretch region is shown in Fig 4b. While measured at the same experimental conditions as that of adamantane, the spectrum of diamantane shows no clearly resolved structure and is essentially continuous. A limited number of narrow features appearing above the background absorption indicate that the Doppler linewidth is as narrow as expected (20 MHz at 20 K). These features are spaced by approximately 0.05 cm^{-1} $\approx 2B$ and mostly likely correspond to sequential $R(J)$ transitions of a rotational branch. Furthermore, the widths of the rotational contours suggest a rotational temperature in the range 10−30 K. These observations indicate that the buffer gas cooling of this molecule is just as efficient as for adamantane.

We therefore attribute the failure to observe well-resolved rotational fine structure as an indication of intrinsic congestion caused by IVR. An important parameter in determining the severity of IVR is the vibrational density of states at the internal energy of the excited vibrational state. As illustrated in Fig. 5, the total density of vibrational states near 3000 cm^{-1} is roughly an order of magnitude larger for diamantane versus adamantane. This is apparently enough to move diamantane over the congestion threshold, with the spectroscopic bright state character of the CH stretching fundamental distributed over many densely spaced eigenstates, all of which contribute to the continuous absorption spectrum.

The caged structure of diamantane leads to a relatively rigid nuclear framework. A general molecule of this size or larger will typically be less rigid and have lower vibrational frequencies, resulting in significantly higher vibrational state densities at a given internal energy. While a few remarkable exceptions exist, like the highly symmetric dodecahedrane, $C_{20}H_{20}$ [31], our expectation is that in general

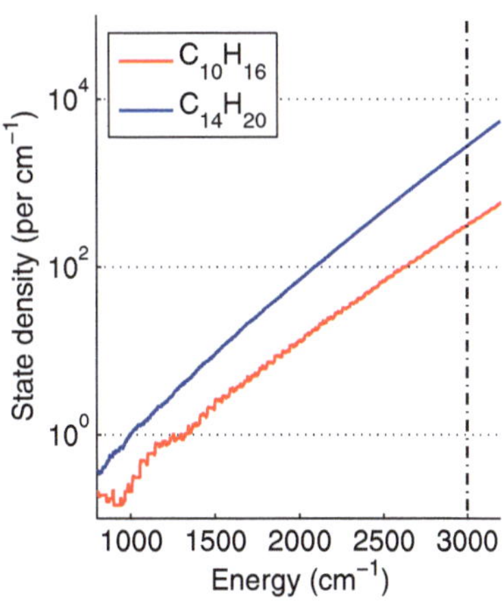

Fig. 5 Vibrational state density. The estimated density of vibrational states per cm^{-1} is plotted against vibrational energy for adamantane, $C_{10}H_{16}$, (*red*) and diamantane, $C_{14}H_{20}$, (*blue*). The *vertical dashed line* corresponds to the approximate internal energy probed in this study. The state densities are calculated using a direct state counting method [30] as done in Ref. [8]

high-resolution rovibrational spectroscopy of large molecules will have to be conducted at longer wavelengths, probing molecules at lower internal energies. Such approaches with cw laser spectroscopy have found success [10, 32]. Recent and ongoing development of frequency comb technology beyond 5 μm [33–37] will enable direct frequency comb spectroscopy in these wavelength regions.

4 Conclusions

We have used cavity-enhanced direct frequency comb spectroscopy of buffer gas cooled molecules to obtain the first high-resolution rovibrational spectra of several large complex molecules in the CH stretch region, providing new and detailed insights into their molecular structure. High-resolution spectroscopy in the 3 μm region is ultimately limited by intrinsic spectral congestion in the largest molecules studied here. We are currently developing new frequency comb systems in the longer IR wavelength regions, and we plan to use these to investigate even larger systems, such as C_{60}. Another promising future direction is the study of reactive species in the cryogenic buffer gas cell environment. Our high-sensitivity, broadband absorption technique presents the possibility of studying chemical kinetics at low thermal energies over long interrogation times via time-resolved frequency comb spectroscopy [7]. Such measurements would have direct and significant relevance to our understanding of atmospheric and interstellar chemistry.

Acknowledgements We dedicate this paper to Ted Hänsch, who has pioneered the field of laser spectroscopy in general and optical frequency combs in particular. This research was funded by DARPA SCOUT, AFOSR, NIST, and NSF-JILA PFC. P.B.C. is supported by the NSF GRFP (Award no. DGE1144083). B.S. is supported through an NRC Postdoctoral Fellowship. D.P. and J.M.D. acknowledge additional support from the NSF. We would like to thank Matthew Radzihovsky for experimental assistance at JILA.

References

1. T. Udem, R. Holzwarth, T.W. Hänsch, Optical frequency metrology. Nature **416**, 233–237 (2002)
2. S.T. Cundiff, J. Ye, Colloquium : femtosecond optical frequency combs. Rev. Mod. Phys. **75**, 325–342 (2003)
3. M.C. Stowe, M.J. Thorpe, A. Pe'er, J. Ye, J.E. Stalnaker, V. Gerginov, S.A. Diddams, Direct frequency comb spectroscopy. Adv. At. Mol. Opt. Phys. **55**, 1–60 (2008)
4. M. Thorpe, J. Ye, Cavity-enhanced direct frequency comb spectroscopy. Appl. Phys. B **91**, 397–414 (2008)
5. F. Adler, M.J. Thorpe, K.C. Cossel, J. Ye, Cavity-enhanced direct frequency comb spectroscopy: technology and applications. Annu. Rev. Anal. Chem. **3**, 175–205 (2010)
6. A. Foltynowicz, P. Masłowski, A.J. Fleisher, B.J. Bjork, J. Ye, Cavity-enhanced optical frequency comb spectroscopy in the mid-infrared application to trace detection of hydrogen peroxide. Appl. Phys. B **110**, 163–175 (2012)

7. A.J. Fleisher, B.J. Bjork, T.Q. Bui, K.C. Cossel, M. Okumura, J. Ye, Mid-infrared time-resolved frequency comb spectroscopy of transient free radicals. J. Phys. Chem. Lett. **5**, 2241–2246 (2014)
8. B. Spaun, P.B. Changala, D. Patterson, B.J. Bjork, O.H. Heckl, J.M. Doyle, J. Ye, Continuous probing of cold complex molecules with infrared frequency comb spectroscopy. Nature **533**, 517–520 (2016)
9. S. Davis, D.T. Anderson, G. Duxbury, D.J. Nesbitt, Jet-cooled molecular radicals in slit supersonic discharges: Sub-Doppler infrared studies of methyl radical. J. Chem. Phys. **107**, 5661 (1997)
10. B.E. Brumfield, J.T. Stewart, B.J. McCall, Extending the limits of rotationally resolved absorption spectroscopy: pyrene. J. Phys. Chem. Lett. **3**, 1985–1988 (2012)
11. J.K. Messer, F.C. De Lucia, Measurement of pressure-broadening parameters for the CO-He system at 4 K. Phys. Rev. Lett. **53**, 2555–2558 (1984)
12. D. Patterson, E. Tsikata, J.M. Doyle, Cooling and collisions of large gas phase molecules. Phys. Chem. Chem. Phys. **12**, 9736–9741 (2010)
13. D. Patterson, J.M. Doyle, Cooling molecules in a cell for FTMW spectroscopy. Mol. Phys. **110**, 1757–1766 (2012)
14. J. Piskorski, D. Patterson, S. Eibenberger, J.M. Doyle, Cooling, spectroscopy and non-sticking of trans-stilbene and Nile Red. Chemphyschem **15**, 3800–4 (2014)
15. D.J. Nesbitt, R.W. Field, Vibrational energy flow in highly excited molecules: role of intramolecular vibrational redistribution. J. Phys. Chem. **100**, 12735–12756 (1996)
16. F. Adler, K.C. Cossel, M.J. Thorpe, I. Hartl, M.E. Fermann, J. Ye, Phase-stabilized, 1.5 W frequency comb at 2.8–4.8 μm. Opt. Lett. **34**, 1330 (2009)
17. G.D. Cole, W. Zhang, B.J. Bjork, D. Follman, P. Heu, C. Deutsch, L. Sonderhouse, J. Robinson, C. Franz, A. Alexandrovski, M. Notcutt, O.H. Heckl, J. Ye, M. Aspelmeyer, High-performance near- and mid-infrared crystalline coatings. Optica **3**, 647 (2016)
18. J. Mandon, G. Guelachvili, N. Picqué, Fourier transform spectroscopy with a laser frequency comb. Nat. Photon. **3**, 99–102 (2009)
19. F. Adler, P. Masłowski, A. Foltynowicz, K.C. Cossel, T.C. Briles, I. Hartl, J. Ye, Mid-infrared Fourier transform spectroscopy with a broadband frequency comb. Opt. Express **18**, 21861–72 (2010)
20. L. Nugent-Glandorf, T. Neely, F. Adler, A.J. Fleisher, K.C. Cossel, B. Bjork, T. Dinneen, J. Ye, S.A. Diddams, Mid-infrared virtually imaged phased array spectrometer for rapid and broadband trace gas detection. Opt. Lett. **37**, 3285 (2012)
21. P. Maslowski, K.F. Lee, A.C. Johansson, A. Khodabakhsh, G. Kowzan, L. Rutkowski, A.A. Mills, C. Mohr, J. Jiang, M.E. Fermann, A. Foltynowicz, Surpassing the path-limited resolution of Fourier-transform spectrometry with frequency combs. Phys. Rev. A **93**, 021802 (2016)
22. D. McKean, CH stretching frequencies, bond lengths and strengths in halogenated ethylenes. Spectrochim. Acta Part A Mol. Spectrosc. **31**, 1167–1186 (1975)
23. N. Zvereva-Loëte, J. Demaison, H. Rudolph, Ab initio anharmonic force field and equilibrium structure of vinyl bromide. J. Mol. Spectrosc. **236**, 248–254 (2006)
24. D. de Kerckhove Varent, Contribution a l'etude de la molecule de bromure de vinyle en spectroscopie hertzienne. II. Bromure de vinyle monodeutere et substitue en 13C, second ordre du couplage quadripolaire, structure de la molecule. Ann. Soc. Sci. Brux. **T84**, 277–292 (1970)
25. A. Pietropolli Charmet, P. Stoppa, A. Baldacci, S. Giorgianni, S. Ghersetti, Diode laser spectrum and rovibrational study of the ν_6 fundamental of vinyl bromide. J. Mol. Struct. **612**, 213–221 (2002)
26. M. Hayashi, C. Ikeda, T. Inagusa, Microwave spectrum, structure, and nuclear quadrupole coupling constant tensor of vinyl bromide and iodide. J. Mol. Spectrosc. **139**, 299–312 (1990)
27. PGOPHER, A program for simulating rotational, vibrational and electronic spectra, C. M. Western, University of Bristol. http://pgopher.chm.bris.ac.uk

28. O. Pirali, V. Boudon, J. Oomens, M. Vervloet, Rotationally resolved infrared spectroscopy of adamantane. J. Chem. Phys. **136**, 024310 (2012)
29. J. Oomens, N. Polfer, O. Pirali, Y. Ueno, R. Maboudian, P.W. May, J. Filik, J.E. Dahl, S. Liu, R.M. Carlson, Infrared spectroscopic investigation of higher diamondoids. J. Mol. Spectrosc. **238**, 158–167 (2006)
30. T. Baer, W.L. Hase, *Unimolecular Reaction Dynamics* (Oxford University Press, New York, 1996)
31. B.S. Hudson, D.G. Allis, S.F. Parker, A.J. Ramirez-Cuesta, H. Herman, H. Prinzbach, Infrared, Raman, and inelastic neutron scattering spectra of dodecahedrane: an I(h) molecule in T(h) site symmetry. J. Phys. Chem. A **109**, 3418–24 (2005)
32. S.K. Tokunaga, R.J. Hendricks, M.R. Tarbutt, B. Darquié, High-resolution mid-infrared spectroscopy of buffer-gas-cooled methyltrioxorhenium molecules (2016), arXiv:1607.08741
33. S.M. Foreman, D.J. Jones, J. Ye, Flexible and rapidly configurable femtosecond pulse generation in the mid-IR. Opt. Lett. **28**, 370 (2003)
34. A. Ruehl, A. Gambetta, I. Hartl, M.E. Fermann, K.S.E. Eikema, M. Marangoni, Widely-tunable mid-infrared frequency comb source based on difference frequency generation. Opt. Lett. **37**, 2232 (2012)
35. A. Gambetta, N. Coluccelli, M. Cassinerio, D. Gatti, P. Laporta, G. Galzerano, M. Marangoni, Milliwatt-level frequency combs in the $8 - 14\ \mu m$ range via difference frequency generation from an Er:fiber oscillator. Opt. Lett. **38**, 1155 (2013)
36. Q.Y. Lu, M. Razeghi, S. Slivken, N. Bandyopadhyay, Y. Bai, W.J. Zhou, M. Chen, D. Heydari, A. Haddadi, R. McClintock, M. Amanti, C. Sirtori, High power frequency comb based on mid-infrared quantum cascade laser at $\lambda \sim 9 \mu m$. Appl. Phys. Lett. **106**, 051105 (2015)
37. K.F. Lee, C.J. Hensley, P.G. Schunemann, M.E. Fermann, Difference frequency generation in orientation-patterned gallium phosphide. In *Conf. Lasers Electro-Optics*, page STu1Q.3. OSA, Washington, D.C. (2016). ISBN 978-1-943580-11-8

Laser Phase Spectroscopy in Closed-Loop Multilevel Schemes

Ennio Arimondo

Abstract Atomic/molecular systems with closed-loop configurations have singular features associated with the creation of multiple interconnected coherences. The resonant condition is determined by the full-loop detuning and the full-loop phase of the laser excitation. The main theoretical features and experimental tests published so far are reviewed. Perspectives of laser spectroscopy based on closed loops are discussed.

1 Introduction

High-resolution laser spectroscopy makes use of the narrow bandwidth of lasers in order to measure with high-precision some atomic/molecular transitions. The basic idea is the selective modifications of the atomic/molecular population of the energy levels resonant with the laser. The resolution is associated with the well-defined frequency of the exciting laser radiation, as demonstrated in a pioneer experiment on sodium atoms by Hänsch et al. [1], and later extensively applied to all kind of systems, see for instance textbooks as [2]. Laser sources are characterized by their electric/magnetic field amplitude, frequency and phase. The spectroscopy tools presented above are based on the proper use of frequency and amplitude of the laser radiation to modify populations. Instead the laser phase, either for a single-frequency excitation or the phase difference in a dual-frequency excitation, does not play a role. In effect for a long time the laser phase was not a variable under easily accessible experimental control. This case was modified with the development of the frequency combs, where the phase of laser components with an arbitrary frequency difference can be precisely controlled, as presented in [3, 4].

This article is part of the topical collection "Enlightening the World with the Laser" - Honoring T. W. Hänsch guest edited by Tilman Esslinger, Nathalie Picqué, and Thomas Udem.

E. Arimondo (✉)
Dipartimento di Fisica "E. Fermi", Università di Pisa, Largo B. Pontecorvo 3, 56127 Pisa, Italy

INO-CNR, Via G. Moruzzi 1, 56124 Pisa, Italy
e-mail: ennio.arimondo@unipi.it

D. Meschede et al. (eds.), *Exploring the World with the Laser*,
https://doi.org/10.1007/978-3-319-64346-5_36

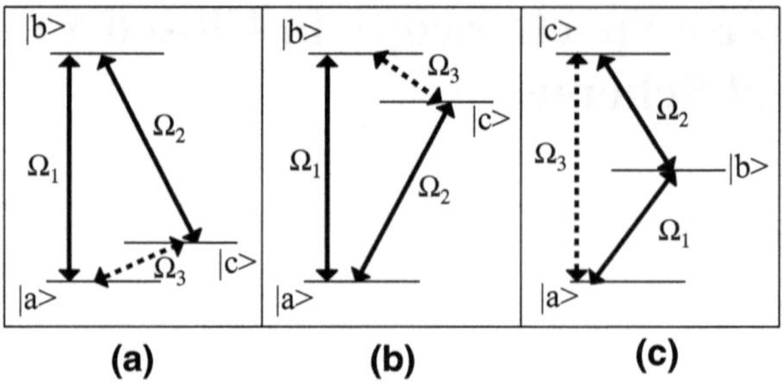

Fig. 1 Closed-loop three-level systems driven by three electromagnetic fields with $\Omega_1, \Omega_2, \Omega_3$ Rabi frequency. In **a**, **b** and **c**, triangle-Λ, -V and -cascade configurations, respectively. The laser-driven dipole allowed transitions denoted by *continuous lines* and by *dashed lines* the microwave-driven magnetic-dipole transition or laser-driven two-photon transition

Populations are not the only observables for a system interacting with light radiation. Atomic/molecular coherences are associated with coherent superpositions of the states. For a two-level system irradiated by laser radiation, the coherence oscillates at the laser frequency, usually in the optical range. Hertzian coherences are associated with superpositions of eigenstates whose frequency separation is lower than optical one, in the Hertz range where the name was coined. Pioneer experiments [5–7] in the creation of MHz Hertzian atomic coherences were based on classical light sources modulated at those frequencies and acquiring a phase relation between the light frequency components. Within three-level systems in the Λ or V configuration of Fig. 1, a Hertzian coherence is resonantly created when the modulation frequency matches the frequency separation between the lower or upper levels. For the Λ configuration such resonant excitation was rediscovered under a dual-frequency laser excitation with its frequency separation matching the hyperfine ground states of sodium atoms [8, 9]. This process was denoted as coherent population trapping (CPT), because that laser excitation combined with the spontaneous emission prepares the initially absorbing system in a not-absorbing (dark) coherent superposition. The quantum-interference effects in laser-driven three-level systems were already pointed out by Hänsch and Toschek [10] in 1970. The important connection between CPT and electromagnetic-induced transparency (EIT) was discussed by Fleischhauer et al. [11]. Within the above dual-frequency excitation schemes, the phase difference between the sources does not play a role determining only the phase of the atomic coherent superposition.

The present work reviews different spectroscopic schemes where the phase of three or more lasers plays a key role. All these schemes are based on multilevel closed-loop atomic/molecular configurations, i.e. configurations with three or more levels excited by several laser sources, whose number is at least equal to the level number. Several theoretical papers treated this topic, while only few experimental investigations were reported so far. Because frequency combs are becoming accessible to a large number of laboratories, this work targets to stimulate the experimentalist's interest into the processes relying on the multiple laser phase control.

Laser phase plays an important role for the optical nonlinear generation. In fact at a given time the laser electric field phase depends on the spatial position, with a spatial periodicity determined by the laser wavelength. When several laser fields interact with an atomic/molecular system, their relative phase changes along the positions of the atoms/molecules. This spatial phase change leads to the phase-matching condition [2]. The attention of the present work is not centred on the light propagation, but instead on the laser spectroscopy of a single atom/ion. Therefore, the theoretical investigations of laser propagation for closed-loop systems will not be considered. Nevertheless it should be pointed out that in a recent investigation [12] phase-resonant closed-loop optical transitions and phase-matching propagation were combined within a novel phase-by-phase control mechanism.

Phase spectroscopy (not on closed-loop systems!) is already an established technique where a phase shift produced by a laser-system interaction is measured either within a the sum-frequency generation process [13] or on the reemitted field [14]. That phase shift was determined in an interferometric set-up using a reference laser beam. No requirement on the relative phases of the exciting laser beams applies to these investigations. In addition Ref. [14] measured the wavefunction phase associated with a discrete autoionized state, i.e. a state interacting with a near-resonant continuum, originally pointed out by Fano [15]. This autoionized state phase issue is closely related to the discrete state phases determining the interference in the considered closed-loop spectroscopy.

Section 2 examines the basic properties of three-level systems within a triangle configuration and of four-level systems within double-Λ or diamond configurations driven by three lasers and four lasers, respectively. Only the main theoretical features are presented in order to introduce the key experimental results. Section 3 describes experimental investigations where the closed-loop characteristics were verified. The concluding Section presents the perspective of closed-loop phase-sensitive spectroscopy.

2 Three- and Four-Level Theoretical Features

2.1 *Three-Level Loop*

The three-level systems of Fig. 1, generally denoted as triangle, in either Λ, V or cascade configuration, are the simplest schemes where the closed-loop phase plays a role. In effect the closed-loop interference was discussed for the first time within the triangle-Λ system in connection with the Λ-scheme CPT [16]. The $|a\rangle \rightarrow |b\rangle$ and $|c\rangle \rightarrow |b\rangle$ transitions (continuous lines) are electric-dipole allowed optical transitions. The $|a\rangle$ and $|c\rangle$ states have the same parity, and the $|a\rangle \rightarrow |c\rangle$ dashed line transition of Fig. 1 may be driven by either a magnetic-dipole transition or a two-photon electric-dipole transition. This triangle configuration has received a large attention, see for instance [17–24] and references therein. This configuration, as all closed-loop configurations, is characterized by multiple interferences into the final states. In fact by switching off any one of the three exciting lasers, we recover a three-level system driven by two lasers only and experiencing an interference into the intermediate state excited by both lasers, leading to the CPT and EIT phenomena [11]. The additional laser closing the triangle introduces additional interferences and leads to a more complex response.

The triangle Hamiltonian is written as

$$H = H^0 + H_{\mathrm{int}}, \tag{1}$$

where the self-energy H_0 is given by

$$H^0 = \hbar\omega_b^0 |b\rangle\langle b| + \hbar\omega_c^0 |c\rangle\langle c|. \tag{2}$$

Here, $\hbar\omega_l^0, (l = a, b, c)$ corresponds to the energy of the $|l\rangle$ state referenced to the $\omega_a^0 = 0$ ground state value. For single-frequency monochromatic lasers at frequencies $\omega_i (i = 1, 3)$, the H_{int} interaction Hamiltonian is written as

$$\begin{aligned} H_{\mathrm{int}} &= -\hbar\frac{\Omega_1}{2}\mathrm{e}^{-i\omega_1 t}|b\rangle\langle a| - \hbar\frac{\Omega_2}{2}\mathrm{e}^{-i\omega_2 t}|c\rangle\langle b| \\ &\quad -\hbar\frac{\Omega_3 \mathrm{e}^{-i\Phi_T}}{2}\mathrm{e}^{-i\omega_3 t}|c\rangle\langle a| + \mathrm{H.c.} \end{aligned} \tag{3}$$

For the case of multiple-frequency lasers, other driving terms should be added. The atomic interaction with each electromagnetic field is described by the associated Rabi frequencies Ω_i, except for the two-photon driving case where an effective Rabi frequency should be introduced. That frequency is determined by either dipole or magnetic moment as

$$\Omega_i = -\frac{\mathbf{d} \cdot \mathbf{E}_i}{\hbar} = \pm\frac{d_{l,m}|\mathbf{E}_i|}{\hbar}, \tag{4}$$

where $\mathbf{E}_i, (i = 1, 3)$ is the electric (or magnetic) field amplitude, $\mathbf{d}$ the electric-dipole moment (or the magnetic one depending on the level scheme). $d_{l,m}$ represents the modulus of the dipole moment matrix element between states $|l\rangle$ and $|m\rangle$ driven by the ω_i laser. The electric field and dipole moment components are calculated within a Cartesian basis, or within a tensorial one in the case of circularly polarized light. Notice the $\pm$ sign imposed by the sign of the Clebsch–Gordan coefficients in calculating the matrix element [2]. We assume $\Omega_i \ (i = 1 - 3)$ to be real and associate a phase Φ_T to the ω_3 field, i.e. its Rabi frequency is $\Omega_3 \mathrm{e}^{-i\Phi_T}$. As shown in [16, 19] within the closed loop a single phase is the free parameter determining the process physics. That phase can be associated with any Rabi frequency, and in fact Φ_T represents the phase difference over the whole closed loop.

We introduce several frequency detunings with $\delta_1 = \omega_1 - \omega_b$, $\delta_2 = \omega_2 - (\omega_c - \omega_b)$, $\delta_3 = \omega_3 - \omega_c$, and finally a triangle-loop detuning Δ_T defined by

$$\Delta_T = \omega_3 - \omega_2 - \omega_1 = \delta_3 - \delta_2 - \delta_1. \quad (5)$$

The triangle closed-loop system is characterized by the Δ_T and Φ_T parameters.

1. As presented within the "Appendix", the Δ_T detuning leads to two different regimes imposed by $\Delta_T = 0$ and $\Delta_T \neq 0$, respectively. For the first case the rotating-wave approximation produces the standard result of time-independent populations and of coherences oscillating at the frequency of the driving nearly resonant laser. The atomic susceptibility oscillates at the laser driving frequency. Instead for $\Delta_T \neq 0$ all the density matrix elements (population and coherences) contain their $\Delta_T = 0$ temporal dependences and in addition sidebands at all $n\Delta_T$ frequencies, with n an integer. Therefore, we write the solution for each i, j element of the σ rotating frame density matrix (defined by Eq. 11 in the "Appendix") as

$$\sigma_{i,j} = \sum_{n=-\infty}^{\infty} \sigma_{i,j}^{n} \mathrm{e}^{in\Delta_T t} \quad (6)$$

   The atomic response becomes very complex, and the sideband generation has been investigated only in few cases [21, 23].
2. The important Φ_T parameter produces different behaviours depending on its value. Buckle et al. [16] demonstrated that for specific Φ_T values the Hilbert space for the three-level system Hamiltonian of Eq. 1 can be separated into a two-level closed system and into a dark state. Later Rangelov et al. [25] demonstrated how a properly chosen Hilbert-space transformation breaks the closed loop and reduces the system to a single chain, and how under certain conditions on the interaction parameters the chain is broken into the two-state system and the dark state. These results are not valid in presence of the density matrix damping mechanisms; nevertheless, they are very useful for analysing the system response for Rabi frequencies larger than the damping rates. As an example of the Φ_T dependence for the triangle-Λ system, Fig. 2 shows the excited state occupation as a periodic function of the Φ_T phase (continuous line). Instead for an open Λ loop, the excited state population denoted by the dashed line of Fig. 2 is constant vs Φ_T. The closed-loop interferences enhance or decrease the excited state occupation. For $\sin(\Phi_T) = 0$ the coupling between the dark state and the remaining states is null and the system is prepared into the triangle dark state with the open-loop occupation. An evaluation of the $\Delta\Phi_T$ sensitivity around the $\sin(\Phi_T) = 0$ points was defined in [17] as the distance of the phase values producing one-half dark state occupation. For the particular case of $|\Omega_1| = |\Omega_2| = |\Omega_3| = \Omega$ the sensitivity results

$$\Delta\Phi_T = 2\arcsin\sqrt{\frac{\gamma_b^2}{2\gamma_g^2 + 12\Omega^2}}, \quad (7)$$

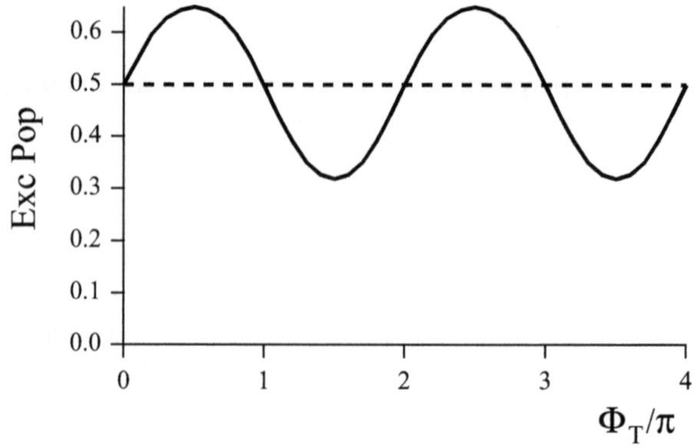

Fig. 2 The *continuous line* shows the $|c\rangle$ excited state occupation of the triangle-Λ system vs the Φ_T laser phase. Parameters $\Omega_1 = \Omega_2 = 0.5$, $\Omega_3 = 0.1$, $\gamma_g = 0.001$, assuming $\gamma_c = 1$. Relaxation rates are defined in "Appendix". The *dashed line* shows the excited state population for the $\Omega_3 = 0$ case, i.e. a not-closed system. The population oscillations have a 30% visibility

γ_g being the ground states damping rate within the triangle-Λ scheme. This sensitivity defines the required control in the laser phase or vice versa the accuracy in the determination of an atomic phase.

2.2 Four-Level System

The closed-loop four-level (FL) systems are based the double-Λ (DL) and diamond configurations of Fig. 3. Several theory treatments have examined the first configuration [26–29] including storing of a pair of light pulses in [30]. On the contrary the unique diamond study is in [31]. For both double-Λ and diamond configurations, the transitions are dipole allowed. The main difference between DL and diamond is on the spontaneous emission decays of $|a\rangle, |b\rangle, |c\rangle$. The laser Rabi frequencies Ω_i, phases ϕ_i and δ_i detunings, numbered $(i = 1, 4)$, are defined as for the triangle system. The overall $\Delta_{\rm FL}$ detuning and overall $\Phi_{\rm FL}$ phase of the four-level loop are defined as

$$\begin{aligned} \Delta_{\rm FL} &= \delta_1 + \delta_3 - \delta_2 - \delta_4, \\ \Phi_{\rm FL} &= \phi_1 + \phi_3 - \phi_2 - \phi_4. \end{aligned} \tag{8}$$

The $\Delta_{\rm FL}$ and $\Phi_{\rm FL}$ role is equivalent to that discussed for the triangle configuration [16]. Time-independent solutions are obtained for $\Delta_{\rm FL} = 0$. For $\Delta_{\rm FL} \neq 0$ Fourier sidebands for all density matrix elements are obtained following the Fourier expansion of Eq. 6. The system solution is periodic with the $\Phi_{\rm FL}$ phase, and results equivalent to those of Fig. 2 are obtained, except for differences listed below.

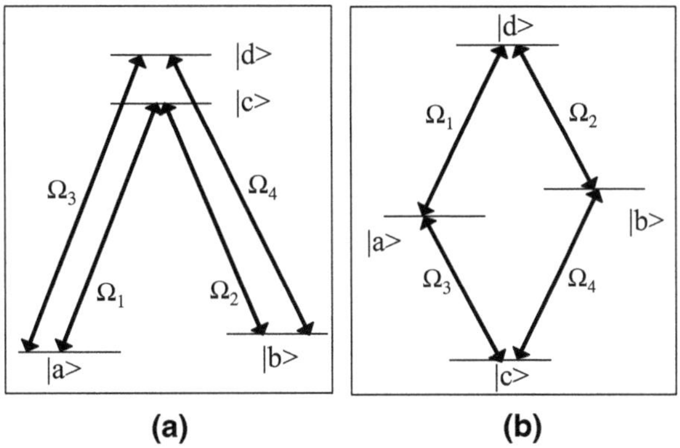

Fig. 3 Closed-loop FL systems driven by four lasers with Rabi frequencies $\Omega_i, (i = 1 - 4)$. In **a** and **b** double-Λ and diamond configurations, respectively

For the double-Λ an interference of the excited state occupation takes place at all values of the Rabi frequencies. This system has received a large attention because of the existence of a total dark state under the following conditions:

$$\begin{aligned} \frac{\Omega_1}{\Omega_2} &= \frac{\Omega_3}{\Omega_4}, \\ \Phi_{\rm FL} &= n2\pi, \end{aligned} \tag{9}$$

the second one is valid when all Rabi frequencies have the same phase. Then, the atomic occupation is limited to the $|a\rangle$ and $|b\rangle$ ground states, and their ratio is proportional to Ω_2/Ω_1. For the case of all Rabi frequencies equal and also equal damping rates from $|c\rangle$ and $|d\rangle$ states, the occupations of these states are proportional to $(1 - \cos(\Phi_{\rm FL}))/2$.

For the diamond configuration investigated in Ref. [31] the dependence on the $\Phi_{\rm FL}$ parameter is different. The interference of the excitation amplitudes starting from the $|c\rangle$ ground state and reaching the double-excited $|d\rangle$ state takes place only for saturation of the lower optical transitions. In the case of saturation with Rabi frequencies larger than all damping rates, the $|d\rangle$ state occupation depends on the phase as $(1 + \cos(\Phi_{\rm FL}))/2$. Therefore, for $\Phi_{\rm FL} = n2\pi$ the interference leads to a maximum occupation and for $\Phi_{\rm FL} = (2n+1)\pi$ to a zero occupation, a result shifted by π in respect of the triangle and double-Λ schemes.

3 Closed-Loop Experiments

3.1 Triangle

The first triangle closed-loop experimental investigation was reported in Ref. [32] within the Λ configuration of Fig. 1a and was based on a Raman process transferring population from $|a\rangle \rightarrow |b\rangle$. The experiment was performed of sodium atoms making use of the optical transition to the first excited state, with the $|a\rangle, |b\rangle$ states corresponding to the $F = 2$ and $F = 1$ ground hyperfine states. The loop was closed applying a microwave radiation resonant with the hyperfine $F = 1 \rightarrow F = 2$ transition. The Φ_T phase dependence was tested by measuring the Raman transfer efficiency for 0 and $\pi/2$ phase values, obtaining different values, similar to the Fig. 2 prediction for the excited state occupation.

A similar and more complete investigation was performed in Ref. [33] within an EIT configuration always based on a triangle-Λ scheme in rubidium with energy levels equivalent to the sodium ones of Ref. [32]. As typical of EIT the authors measured the transmission of a weak optical probe laser in the presence of a strong optical laser and different microwave parameters of the $|a\rangle \rightarrow |c\rangle$ driving. The transmission was measured as function of the Φ_T phase, and the plot of Fig. 2 was totally verified. The experiment analysis took into account the sign of Clebsch–Gordan coefficients in calculating the dipole moments, i.e. the Rabi frequencies. Notice that a negative Rabi frequency corresponds to a $\Phi_T = \pi$ phase within the overall phase loop.

The periodic Φ_T phase dependence above was tested in Ref. [34] on the same rubidium triangle-Λ system by monitoring the optical absorption from one lower state to a fourth level, not belonging to the triangle scheme.

3.2 Double-Λ

This four-level configuration has played an important role for the generation of photon pairs, correlated photon states, and entangled imaging for quantum communication with atomic ensembles [35–38]. However, in all those experiments the photon generation process automatically controls the loop phase.

The preparation of the total dark state was investigated in experiments by Windholz group [39–41] based on sodium optical transitions close to the triangle one reported above. The $|a\rangle, |b\rangle$ ground states of Fig. 3 correspond to the $F = 1$ and $F = 2$ hyperfine ground states, and the $|c\rangle$ and $|d\rangle$ states correspond to the first and second excited states. The phase of one laser beam was modified either by changing the optical path or by inserting an electro-optics modulator. The Φ_{FL} periodic modulation of the excited state occupation was verified. The excited state occupations were monitored in the presence of Δ_{FL} detunings different from zero, with values around 10 Hz. An occupation modulation was observed corresponding to the first component of the Fourier expansion in Eq. 6. The authors applied a different, but equivalent, interpretation: the population temporal variation is produced by the $\Delta_{\mathrm{FL}}t$ temporal phase modulation associated with the evolving t observation time. The authors measured how the modulation amplitude was modified by the laser intensities, in agreement with the contribution of different multiphoton processes discussed for the triangle configuration by [20] and mentioned within the “Appendix”.

In an investigation based on the equivalent Rb double-Λ system in Ref. [42], a phase-controlled light switching was observed.

3.3 Diamond

The experimental investigations of the four-level diamond configuration, mainly triggered by success with the double-Λ experiments, were concentrated on the rubidium atom, the excitation starting from one hyperfine level of the $5S_{1/2}$ ground state, via the $5P_{1/2}$ and $5P_{3/2}$ intermediate states and reaching as final state either the $6S_{1/2}$ state or the $5D_{1/2}$ one, using a different wavelength for each upper transition. This configuration allowed the generation of photon pairs [43], but once again the photon generation process controlled the loop phase. A clear test of the Φ_{FL} phase dependence was performed in the experiment of Ref. [44] based on a frequency comb excitation. All the comb components contribute to the excitation process, with interference between their transition amplitudes. However, within the broad-band comb spectrum the components nearly resonant with a given optical transition produce the largest contributions to the excitation process. The Φ_{FL} dependence of this comb investigation was performed by applying a selective phase shift of the frequency comb components resonant with the left half-part of the Fig. 3b diamond, i.e. the $|c\rangle \rightarrow |b\rangle$ and $|b\rangle \rightarrow |d\rangle$ transitions. The experiment verified the Φ_{FL} phase periodic dependence. Because a phase shift was simultaneously applied to two transition of the closed-loop, the phase dependence periodicity was two times larger than that reported in Fig. 2. A phase offset was also detected and associated with the near-resonant excitation of additional intermediate hyperfine levels with different dipole moments and detunings. The visibility of the phase oscillations was measured, but a comparison with theoretical predictions was not performed.

4 Conclusions and Perspectives

Several theoretical treatments examined the different issues related to the phase-sensitive closed-loop spectroscopy. Few experiments tested the main features of the closed systems, observing (1) the optical sidebands created for a closed-loop detuning different from zero, and (2) the periodic dependence of the atomic response on the overall phase. The additional theoretically predicted features may lead to a larger spectrum of spectroscopic applications.

The bichromatic excitation on a three-level open system leads to quantum control, as subluminal or superluminal light propagation [11]. In connection with this target, the theoretical analysis of [19] demonstrated that trichromatic excitation on a triangle system leads to a greater flexibility in sub/super-luminal light generation, with a commutation between the two regimes produced by changing Φ_T. This result was obtained for the $\Delta_T = 0$ case. The refractive index modification leading sub/super-luminal light was not explored for $\Delta_T \neq 0$, and this case may lead to a more efficient control of light propagation. It should be noted that owing to Fourier expansion of the density matrix elements the atomic response will generate transmitted light sidebands at the Δ_T frequency. Their presence could modify the temporal shape of the laser pulse and also the sub/superluminal pulse speed. It may be important to explore how the Δ_T choice will preserve the slow/fast light propagation in combination with a limited deformation of the laser pulse. Given the narrow frequency bandwidth reached today within the dark state preparation in several systems, Δ_T sidebands in the ten MHz range should be a good choice for the required compromise between light speed and pulse deformation. In order to plan for an experimental investigation, precise simulations should be performed. In addition also the four-level systems should be explored on the topic of slow/fast light generation.

As a different exploration direction, the requirement on the closed-loop Φ phase may be used to explore the properties of the atomic/molecular states composing the level configuration. The Φ phase is determined by the driving laser phase, but also by the phases associated with the Rabi frequencies. From the definition of Eq. 4 the Rabi frequency is a real number, except for the $\pm$ sign, the minus sign corresponding to a π phase shift. In fact the authors of Ref. [33, 44] verified the π phase shift presence on the measured periodic excitation. Any level mixing produced by configuration interaction, off-diagonal fine and hyperfine couplings leads to a modification of the Rabi frequency amplitude or $\pm$ sign without adding an extra phase corresponding to a complex Rabi frequency. Instead a wavefunction phase appears into the evolution of a discrete autoionized state, i.e. a state interacting with a near-resonant continuum [14, 15]. That phase appears also into the Rabi frequency for a transition between the autoionized state and any discrete state. If that state belongs to a closed-loop system, an experimental investigation for the loop phase will provide a direct measurement of the discrete–continuum coupling, with an accuracy similar to the one in Eq. 7.

Control of the relative phase between the carrier wave and the pulse envelope is a critical issue for stabilizing mode-locked femtosecond laser systems. Such carrier-envelope phase stabilization was recently realized through (1) interference in the fluorescence light of a ZnO crystal excited by the carrier and its second-harmonic [45], (2) quantum-interference semiconductor current sensitive to the relative phase between two coherent laser pulse trains [46] and (3) phase-sensitive photoelectron emission from a metal surface [47] following a proposition in [48]. The closed-loop spectroscopy could represent an alternative tool for the same target of phase control. Additional loop-phase spectroscopy applications can be found in similar solid state system.

Acknowledgements This work is in honour of Ted Hänsch, a friend and a pioneer in the revolution of the atomic spectroscopy. We met for the first time at the Carberry Tower (Scotland) in July 1969 for the Summer School on Quantum Optics organized by the Scottish Universities, both of us looking for post-docs opportunity, finally Ted in USA and myself in England. Ten years later at Stanford, I was fascinated by Ted playing with a toy train along the hall of the Varian Physics building and setting up for a simple wavemeter. An Alexander von Humboldt Award generated my long visit to Munich in 2002 with a special admittance to Ted personal laboratory in Schelling Straße.The author acknowledges the collaboration with Hema Ramachandran that stimulated the interest into closed-loop systems, and Thomas Udem for discussion on the phase determination.

Appendix: Triangle Density Matrix Equations and Solution

We write the ρ density matrix equations including the decay rates depending on the level configuration. For the case of triangle-Λ configuration the only nonzero decay elements are γ_g the ground state rate equilibrating the $|a\rangle$ and $|c\rangle$ populations, and γ_b the $|b\rangle$ state decay rate into $|a\rangle$ and $|c\rangle$ ground states, for simplicity with equal branching ratios. Applying the rotating-wave approximation, the Hamiltonian of Eq. 1 leads to the following equations for populations and coherences:

$$\begin{aligned}
\dot{\rho}_{aa} &= \gamma_g(\rho_{cc}-\rho_{aa})+\frac{\gamma_b}{2}\rho_{bb} \\
&\quad -\left[i\frac{\Omega_3}{2}e^{-i\omega_3 t}\rho_{ab}+i\frac{\Omega_1}{2}e^{-i(\omega_1 t-\Phi)}\rho_{ac}+\text{H.c.}\right], \\
\dot{\rho}_{bb} &= -\gamma_b\rho_{bb}+\left[i\frac{\Omega_3}{2}e^{-i\omega_3 t}\rho_{ab}-i\frac{\Omega_2}{2}e^{-i\omega_2 t}\rho_{bc}+\text{H.c.}\right], \\
\dot{\rho}_{cc} &= \gamma_g(\rho_{aa}-\rho_{cc})+\frac{\gamma_b}{2}\rho_{bb} \\
&\quad +\left[i\frac{\Omega_2}{2}e^{-i\omega_2 t}\rho_{bc}+i\frac{\Omega_1}{2}e^{-i(\omega_1 t-\Phi)}\rho_{ac}+\text{H.c.}\right], \\
\dot{\rho}_{ab} &= -\left(\frac{\gamma_b+\gamma_g}{2}-i\omega_b\right)\rho_{ab}+i\frac{\Omega_1}{2}e^{i\omega_1 t}(\rho_{bb}-\rho_{aa}) \\
&\quad +i\frac{\Omega_3}{2}e^{i(\omega_3 t-\Phi_T)}\rho_{cb}-i\frac{\Omega_2}{2}e^{-i\omega_2 t}\rho_{ac}, \\
\dot{\rho_{bc}} &= -\left[\frac{\gamma_b+\gamma_g}{2}-i(\omega_c-\omega_b)\right]\rho_{bc}+i\frac{\Omega_2}{2}e^{i\omega_2 t}(\rho_{cc}-\rho_{bb}) \\
&\quad +i\frac{\Omega_3}{2}e^{-i(\omega_3 t-\Phi_T)}\rho_{ac}-i\frac{\Omega_1}{2}e^{-i\omega_1 t}\rho_{ba}, \\
\dot{\rho}_{ac} &= -(\gamma_g-i\omega_c)\rho_{ca}+i\frac{\Omega_3}{2}e^{i(\omega_3 t-\Phi_T)}(\rho_{cc}-\rho_{aa}) \\
&\quad +i\frac{\Omega_1}{2}e^{i\omega_1 t}\rho_{bc}-i\frac{\Omega_3}{2}e^{i(\omega_3 t-\Phi_T)}\rho_{ab}.
\end{aligned} \tag{10}$$

Introducing the standard transformation to a rotating frame [16]

$$\begin{aligned}
\rho_{ba} &= \sigma_{ba}e^{-i\omega_1 t}, \rho_{cb} = \sigma_{cb}e^{-i\omega_2 t}, \\
\rho_{ca} &= \sigma_{ca}e^{-i(\omega_3 t-\Phi_T)}, \rho_{ii} = \sigma_{ii},
\end{aligned} \tag{11}$$

with $(i=a,b,c)$, we obtain a new set of equations

$$\begin{aligned}
\dot{\sigma}_{aa} &= \gamma_g(\sigma_{cc}-\sigma_{aa}) + \frac{\gamma_b}{2}\sigma_{bb} - i\frac{\Omega_1}{2}(\sigma_{ab}-\sigma_{ba}) - i\frac{\Omega_3}{2}(\sigma_{ac}-\sigma_{ca}),\\
\dot{\sigma}_{bb} &= -\gamma_b\sigma_{bb} + i\frac{\Omega_1}{2}(\sigma_{ab}-\sigma_{ba}) - i\frac{\Omega_2}{2}(\sigma_{bc}-\sigma_{cb}),\\
\dot{\sigma}_{cc} &= \gamma_g(\sigma_{aa}-\sigma_{cc}) + \frac{\gamma_b}{2}\sigma_{bb} + i\frac{\Omega_2}{2}(\sigma_{bc}-\sigma_{cb}) + i\frac{\Omega_3}{2}(\sigma_{ac}-\sigma_{ca}),\\
\dot{\sigma}_{ab} &= -(\frac{\gamma_b+\gamma_g}{2} + i\delta_1)\sigma_{ab} + i\frac{\Omega_1}{2}(\sigma_{bb}-\sigma_{aa})\\
&\quad + i\frac{\Omega_3}{2}e^{i\Delta_T t}e^{-i\Phi_T}\sigma_{cb} - i\frac{\Omega_2}{2}e^{i\Delta_T t}e^{-i\Phi_T}\sigma_{ac},\\
\dot{\sigma}_{bc} &= -\left(\frac{\gamma_b+\gamma_g}{2} + i\delta_2\right)\sigma_{bc} + i\frac{\Omega_2}{2}(\sigma_{cc}-\sigma_{bb})\\
&\quad + i\frac{\Omega_1}{2}e^{i\Delta_T t}e^{-i\Phi_T}\sigma_{ac} - i\frac{\Omega_3}{2}e^{i\Delta_T t}e^{-i\Phi_T}\sigma_{ba},\\
\dot{\sigma}_{ac} &= -(\gamma_g + i\delta_3)\sigma_{ac} + i\frac{\Omega_3}{2}(\sigma_{cc}-\sigma_{aa})\\
&\quad + i\frac{\Omega_1}{2}e^{-i\Delta_T t}e^{i\Phi_T}\sigma_{bc} - i\frac{\Omega_2}{2}e^{-i\Delta_T t}e^{i\Phi_T}\sigma_{ab}.
\end{aligned} \tag{12}$$

These equations become time-independent only in the case of $\Delta_T = 0$, leading to time-independent populations ρ_{ii} and to ρ_{ij} coherences with oscillations at the corresponding laser driving frequency given by Eq. 11.

For $\Delta_T \neq 0$ the time-dependent density matrix equations were solved in Ref. [20] through a series expansion in terms of the Rabi frequencies, where the connection between Rabi frequency power dependence and the high-order multiphoton processes was pointed out.

Within the approach of Eq. 6, each density matrix element is expanded into a Fourier expansion with all the harmonics of the Δ_T angular frequency. The Fourier harmonics are proportional to the Rabi frequencies, higher-order terms containing higher powers of the Rabi frequencies, and therefore produced by high-order multiphoton processes. The Fourier components satisfy recurrent equations that can be solved through a continued fraction solution as in refs. [23, 49]. The triangle configuration with a 9×9 density matrix leads to a continued fraction of matrices with the same order. Instead the four-level system is solved on the basis of 16×16 continued fraction matrices.

References

1. T.W. Hänsch, I.S. Shahin, A.L. Schawlow, Phys. Rev. Lett. **27**, 707 (1971)
2. W. Demtröder, *Laser Spectroscopy*, vol. I, II (Springer, Berlin, 2008)
3. J.L. Hall, Rev. Mod. Phys. **78**, 1279 (2006)
4. T.W. Hänsch, Rev. Mod. Phys. **78**, 1297 (2006)
5. W.E. Bell, A.L. Bloom, Phys. Rev. Lett. **6**, 280 (1961)
6. E.B. Aleksandrov, Opt. Spectrosc. **14**, 233 (1963)
7. A. Corney, G.W. Series, Proc. Phys. Soc. **83**, 207–213 (1964)
8. G. Alzetta, A. Gozzini, L. Moi, G. Orriols, Nuovo Cim. B **36**, 5 (1976)
9. H.R. Gray, R.M. Whitley, C.R. Stroud Jr., Opt. Lett. **3**, 218 (1978)

10. T.W. Hänsch, P. Toschek, Z. Phys. **236**, 213 (1970)
11. M. Fleischhauer, A. Imamoglu, J.P. Marangos, Rev. Mod. Phys. **77**, 633 (2005)
12. M. Artoni, A. Zavatta, Phys. Rev. Lett. **115**, 113005 (2015)
13. Y.R. Shen, Ann. Rev. Phys. Chem. **64**, 129 (2013)
14. M. Kotur, D. Guénot, A. Jiménez-Galán, D. Kroon, E.W. Larsen, M. Louisy, S. Bengtsson, M. Miranda, J. Mauritsson, C.L. Arnold, S.E. Canton, M. Gisselbrecht, T. Carette, J.M. Dahlstrlóm, E. Lindroth, A. Maquet, L. Argenti, F. Martín, A. L'Huillier, Nat. Commun. **7**, 10566 (2016)
15. U. Fano, Phys. Rev. **124**, 1866 (1961)
16. S.J. Buckle, S.M. Barnett, P.L. Knight, M.A. Lauder, D.T. Pegg, Opt. Acta **33**, 1129 (1986)
17. D.V. Kosachiov, G.G. Matisov, Y.U. Rozhdestvensky, J. Phys. At. Mol. Phys. B **25**, 2473 (1992)
18. R.G. Unanyan, L.P. Yatsenko, K. Bergmann, B.W. Shore, Opt. Commun. **139**, 48 (1997)
19. M. Sahrai, H. Tajalli, K.T. Kapale, M.S. Zubairy, Phys. Rev. A **70**, 023813 (2004)
20. R. Fleischhaker, J. Evers, Phys. Rev. A **77**, 043805 (2008)
21. J. Korociński, A. Raczyński, J. Zaremba, S. Zielińska-Kaniasty, J. Opt. Soc. Am. B **30**, 1517 (2013)
22. J. Korociński, Phys. Scr. T **160**, 014022 (2014)
23. A. Raczyński, J.J. Zaremba, S. Zielińska-Raczyńska, J. Opt. Soc. Am. B **32**, 1229 (2015)
24. M. Manjappa, S. Sainadh Undurti, A. Karigowda, A. Narayanan, B.C. Sanders, Phys. Rev. A **90**, 043859 (2014)
25. A.A. Rangelov, N.V. Vitanov, B.W. Shore, Phys. Rev. A **77**, 033404 (2008)
26. E. Cerboneschi, E. Arimondo, Phys. Rev. A **52**, R1823 (1995)
27. M.D. Lukin, P.R. Hemmer, M. Löffler, M.O. Scully, Phys. Rev. Lett. **81**, 2675 (1998)
28. A.A. Korsunsky, D.V. Kosachiov, Phys. Rev. A **60**, 4996 (1999)
29. H. Shpaisman, A.D. Wilson-Gordon, H. Friedmann, Phys. Rev. A **71**, 043812 (2005)
30. A. Raczyński, J. Zaremba, S. Zieliska-Kaniasty, Phys. Rev. A **69**, 043801 (2004)
31. G. Morigi, S. Franke-Arnold, G.-L. Oppo, Phys. Rev. A **66**, 053409 (2002)
32. M.S. Shahriar, P.R. Hemmer, Phys. Rev. Lett. **65**, 1865 (1990)
33. H. Li, V.A. Sautenkov, Y.V. Rostovtsev, G.R. Welch, P.R. Hemmer, M.O. Scully, Phys. Rev. A **80**, 023820 (2009)
34. T.M. Preethi, M. Manukumara, K. Asha, J. Vijay, D.A. Roshi, A. Narayanan, Eur. Phys. Lett. **95**, 34005 (2011)
35. M. Jain, H. Xia, G.Y. Yin, A.J. Merriam, S.E. Harris, Phys. Rev. Lett. **77**, 4326 (1996)
36. A. Kuzmich, W.P. Bouwen, A.D. Boozer, A. Boca, C.W. Chou, L.-M. Duan, H.J. Kimble, Nature **423**, 731 (2003)
37. C.H. van der Wal, M.D. Eisaman, A. André, R.L. Walsworth, D.F. Phillips, A.S. Zibrov, M.D. Lukin, Science **301**, 196 (2003)
38. V. Boyer, A.M. Marino, R.C. Pooser, P.D. Lett, Science **321**, 544–547 (2008)
39. W. Maichen, R. Gaggl, E.A. Korsunsky, L. Windholz, Europhys. Lett. **31**, 189 (1995)
40. E.A. Korsunsky, N. Leinfellner, A. Huss, S. Baluschev, L. Windholz, Phys. Rev. A **59**, 2302 (1999)
41. A.F. Huss, E.A. Korsunsky, L. Windholz, J. Mod. Opt. **49**, 141 (2002)
42. H. Kang, G. Hernandez, J. Zhang, Y. Zhu, Phys. Rev. A **73**, 011802R (2006)
43. R.T. Willis, F.E. Becerra, L.A. Orozco, S.L. Rolston, Phys. Rev. A **82**, 053842 (2010)
44. M.C. Stowe, A. Pe'er, J. Ye, Phys. Rev. Lett. **100**, 203001 (2008)
45. O.D. Mücke, T. Tritschler, M. Wegener, U. Morgner, F.X. Kärtner, Opt. Lett. **27**, 2127 (2002)
46. P.A. Roos, Q. Quraishi, S.T. Cundiff, R.D.R. Bhat, J.E. Sipe, Opt. Express **11**, 2081 (2003)
47. A. Apolonski, P. Dombi, G.G. Paulus, M. Kakehata, R. Holzwarth, Ch. Th Udem, K.Torizuka Lemell, J. Burgdörfer, T.W. Hänsch, F. Krausz, Phys. Rev. Lett. **92**, 073902 (2004)
48. C. Lemell, X.-M. Tong, F. Krausz, J. Burgdörfer, Phys. Rev. Lett. **90**, 076403 (2003)
49. V. Wong, R.S. Bennink, A.M. Marino, R.W. Boyd, C.R. Stroud Jr., F.A. Narducci, Phys. Rev. A **70**, 053811 (2004)

10. F.W. Hänsch, P. Toschek, Z. Phys. 236, 213 (1970)
11. M. Fleischhauer, A. Imamoglu, J.P. Marangos, Rev. Mod. Phys. 77, 633 (2005)
12. M. Artoni, A. Zavatta, Phys. Rev. Lett. 115, 113005 (2015)
13. Y.R. Shen, Ann. Rev. Phys. Chem. 64, 129 (2013)
14. M. Kotur, D. Guénot, Á. Jiménez-Galán, D. Kroon, E.W. Larsen, M. Louisy, S. Bengtsson, [illegible]
[illegible]

Precision Measurements and Test of Molecular Theory in Highly Excited Vibrational States of H_2 ($v = 11$)

T. Madhu Trivikram, M.L. Niu, P. Wcisło, W. Ubachs, and E.J. Salumbides

Abstract Accurate $EF^1\Sigma_g^+ - X^1\Sigma_g^+$ transition energies in molecular hydrogen were determined for transitions originating from levels with highly excited vibrational quantum number, $v = 11$, in the ground electronic state. Doppler-free two-photon spectroscopy was applied on vibrationally excited H_2^*, produced via the photodissociation of H_2S, yielding transition frequencies with accuracies of 45 MHz or 0.0015 cm^{-1}. An important improvement is the enhanced detection efficiency by resonant excitation to autoionizing $7p\pi$ electronic Rydberg states, resulting in narrow transitions due to reduced ac-Stark effects. Using known EF level energies, the level energies of $X(v = 11, J = 1, 3–5)$ states are derived with accuracies of typically 0.002 cm^{-1}. These experimental values are in excellent agreement with and are more accurate than the results obtained from the most advanced ab initio molecular theory calculations including relativistic and QED contributions.

This article is part of the topical collection "Enlightening the World with the Laser" - Honoring T. W. Hänsch guest edited by Tilman Esslinger, Nathalie Picqué, and Thomas Udem.

T.M. Trivikram • M.L. Niu • W. Ubachs
Department of Physics and Astronomy, LaserLaB, Vrije Universiteit Amsterdam, De Boelelaan 1081, 1081 HV Amsterdam, The Netherlands

P. Wcisło
Institute of Physics, Faculty of Physics, Astronomy and Informatics, Nicolaus Copernicus University, Grudziądzka 5, 87-100 Toruń, Poland

E.J. Salumbides (✉)
Department of Physics and Astronomy, LaserLaB, Vrije Universiteit Amsterdam, De Boelelaan 1081, 1081 HV Amsterdam, The Netherlands

Department of Physics, University of San Carlos, 6000 Cebu City, Philippines
e-mail: e.j.salumbides@vu.nl

D. Meschede et al. (eds.), *Exploring the World with the Laser*,
https://doi.org/10.1007/978-3-319-64346-5_37

1 Introduction

The advance of precision laser spectroscopy of atomic and molecular systems has, over the past decades, been closely connected to the development of experimental techniques such as tunable laser technology [1], saturation spectroscopy [2], two-photon Doppler-free spectroscopy [3], cavity-locking techniques [4], and ultimately, the invention of the frequency comb laser [5], developments to which Prof. Theodor Hänsch has greatly contributed. These inventions are being exploited to further investigate at ever-increasing precision the benchmark atomic system—the hydrogen atom, resulting in the advance of spectroscopic accuracy in atomic hydrogen measurements by more than seven orders of magnitude since the invention of the laser [6]. The spectroscopy of the 1S–2S transition in atomic hydrogen, at 4×10^{-15} relative accuracy [7], provides a stringent test of fundamental physical theories, in particular quantum electrodynamics (QED). Currently, the theoretical comparison with precision measurements on atomic hydrogen is limited by uncertainties in the proton charge radius r_p. The finding that the r_p-value obtained from muonic hydrogen spectroscopy is in disagreement by some $7 - \sigma$ [8] is now commonly referred to as the *proton-size puzzle*.

Molecular hydrogen, both the neutral and ionic varieties, is benchmark systems in molecular physics, in analogy to its atomic counterpart. Present developments in the ab initio theory of the two-electron neutral H_2 molecule and the one-electron ionic H_2^+ molecule, as well as the respective isotopologues have advanced in accuracy approaching that of its atomic counterpart despite the increased complexity. The most accurate level energies of the entire set of rotational and vibrational states in the ground electronic state of H_2 were calculated by Komasa et al. [9]. An important breakthrough in these theoretical studies was the inclusion of higher-order relativistic and QED contributions, along with a systematic assessment of the uncertainties in the calculation. Recently, further improved calculations of the adiabatic [10] as well as non-adiabatic [11] corrections have been performed, marking the steady progress in this field.

In the same spirit as in atomic hydrogen spectroscopy, the high-resolution experimental investigations in molecular hydrogen are aimed toward confronting the most accurate ab initio molecular theory. For the H_2^+ and HD^+ ions, extensive efforts by Korobov and co-workers over the years, have recently led to the theoretical determination of ground electronic state level energies at 0.1 ppb accuracies [12]. The latter accuracy enables the extraction of the proton/electron mass ratio, m_p/m_e, when combined with the recent HD^+ spectroscopy using a laser-cooled ion trap [13]. These are currently at lower precision than other methods but prospects exist that competitive values can be derived from molecular spectroscopy. Similarly Karr et al. [14] recently discussed the possibility of determining R_∞ using H_2^+ (or HD^+) transitions as an alternative to atomic hydrogen spectroscopy. Even for the neutral system of molecular hydrogen, the determination of r_p from spectroscopy is projected to be achievable, from the ongoing efforts in both calculation [15] and experiments [16]. Molecular spectroscopy might thus be posed to contribute toward the resolution of the proton-size puzzle.

In contrast to atomic structure, the added molecular complexity due to the vibrational and rotational nuclear degrees of freedom could constitute an important feature, with a multitude of transitions (in the ground electronic state) that can be conscripted toward the confrontation of theory and experiments. From both experimental and theoretical perspectives, this multiplicity allows for consistency checks and assessment of systematic effects. In recent years, we have tested the most accurate H_2 quantum chemical calculations using various transitions, for example, the dissociation limit D_0 or binding energy of the ground electronic state [17]; the rotational sequence in the vibrational ground state [18]; and the determination of the ground tone frequency $(v = 0 \rightarrow 1)$ [19]. The comparisons exhibit excellent agreement thus far and have in turn been interpreted to provide constraints of new physics, such as fifth forces [20] or extra dimensions [21].

Recently, we reported a precision measurement on highly excited vibrational states in H_2 [22]. The experimental investigation of such highly excited vibrational states probes the region where the calculations of Komasa et al. [9] are the least accurate, specifically in the $v = 6$–12 range. The production of excited H_2^* offers a unique possibility on populating the high-lying vibrational states that would otherwise be practically inaccessible in thermal equilibrium (corresponding temperature of $T \sim 47{,}000$ K for $v = 11$).

Here, we present measurements of level energies of $v = 11$ rovibrational quantum states that extend the spectroscopy in Ref. [22] and that implement improvements, leading to a narrowing of the resonances. This is achieved by the use of a resonant ionization step to molecular Rydberg states, thereby enhancing the detection efficiency significantly. The enhancement allows for the use of a low-intensity spectroscopy laser minimizing the effect of ac-Stark-induced broadening and shifting of lines. The ac-Stark effect is identified as the major source of systematic uncertainty in the measurements, and a detailed treatment of the line shape models used to describe the asymmetric Stark-broadened profiles is also included in this contribution.

2 Experiment

The production of excited H_2^* from the photodissociation of hydrogen sulfide was first demonstrated by Steadman and Baer [23], who observed that the nascent H_2^* molecules were populated at predominantly high vibrational quanta in the two-photon dissociation of H_2S at UV wavelengths. That study used a single powerful laser for dissociation, for subsequent H_2 spectroscopy, and to induce dissociative ionization for signal detection. Niu et al. [22] utilized up to three separate laser sources to address the production, probe, and detection steps in a better controlled fashion. The present study, targeting H_2 $(v = 11)$ levels, is

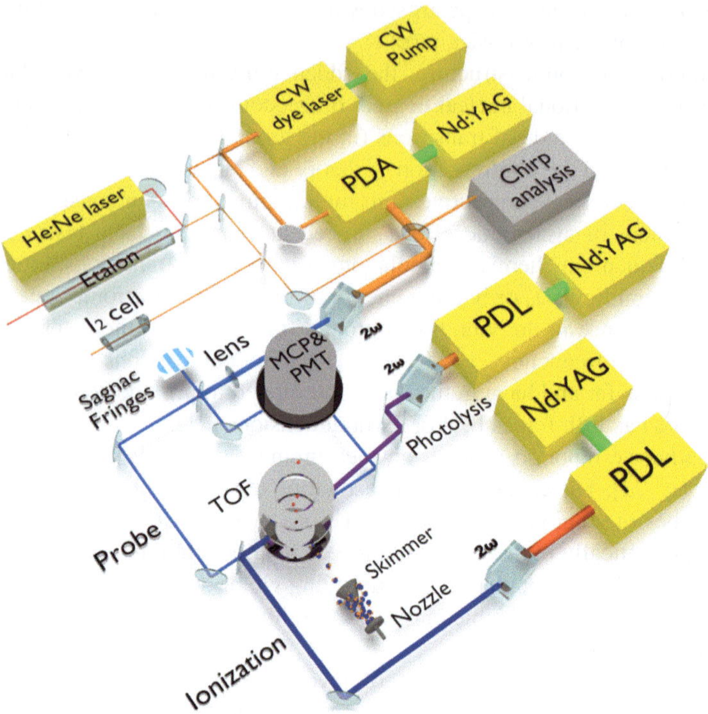

Fig. 1 Schematic of experimental setup indicating the three main radiation sources: the *photolysis* laser at 293 nm for H_2^* production from H_2S dissociation; the *probe* laser from a narrowband dye amplifier (PDA) for the *EF–X* spectroscopy transition; and the *ionization* laser for resonant H_2^+ ion production, subsequently detected as signal. Doppler-free two-photon excitation is facilitated by the Sagnac interferometric alignment of the counter-propagating probe beams. Absolute frequency calibration is performed with respect to I_2 hyperfine reference lines, aided by the relative frequency markers from the transmission fringes of a length-stabilized Fabry–Pérot etalon. The frequency offset between the cw-seed and PDA pulse output, induced by chirp effects in the dye amplifier, is measured and corrected for post-measurement. *cw* Continuous wave, *PDL* pulsed dye laser, 2ω frequency-doubling stage, *TOF* time-of-flight region, *MCP* multichannel plates, *PMT* photomultiplier tube

performed using the same experimental setup as in Ref. [22], depicted schematically in Fig. 1. The photolysis laser at 293 nm, generated from the second harmonic of the output of a commercial pulsed dye laser (PDL) with rhodamine B dye, serves in the production of H_2^* by photolyzing H_2S. The narrowband spectroscopy laser radiation at around 300–304 nm is generated by frequency upconversion of the output of a continuous-wave (cw)-seeded pulsed dye amplifier system (PDA) running on rhodamine 640 dye. The ionization laser source at 302–305 nm, from the frequency-doubled output of another PDL (also with rhodamine 640 dye), is used to resonantly excite from the *EF* to autoionizing Rydberg states to eventually form H_2^+ ions. This $2 + 1'$ resonance-enhanced multiphoton ionization (REMPI) scheme results in much improved sensitivities compared to our previous study [22].

The H_2S molecular beam, produced by a pulsed solenoid valve in a source vacuum chamber, passes through a skimmer toward a differentially pumped interaction chamber, where it intersects the laser beams perpendicularly. The probe or spectroscopy laser beam is split into two equidistant paths and subsequently steered in a counter-propagating orientation, making use of a Sagnac interferometer alignment for near-perfect cancelation of the residual first-order Doppler shifts [24]. Moreover, the probe laser beams pass through respective lenses, of $f = 50$-cm focal length, to focus and enhance the probe intensity at the interaction volume. Finally, the ionization beam is aligned in almost co-linear fashion with the other laser beams to ensure maximum spatial overlap. To avoid ac-Stark shifts during the spectroscopic interrogation, induced by the photolysis laser (~6-mJ typical pulse energy; ~10-ns pulse duration), a 15-ns delay between the photolysis and probe pulses is established with a delay line. For a similar reason, the ionization pulse (~1-mJ typical pulse energy; ~10-ns pulse duration) is also delayed by 30 ns with respect to the probe pulse. The 1-mJ ionization pulse energy is sufficient for saturating the ionization step.

The ions produced in the interaction volume are accelerated by ion lenses, further propagating through a field-free time-of-flight (TOF) mass separation region before impinging on a multichannel plate (MCP) detection system. Scintillations in a phosphor screen behind the MCP are monitored by a photomultiplier tube (PMT) and a camera, culminating in the recording of the mass-resolved signals. In the non-resonant ionization step as in Ref. [22], predominantly H^+ ions were produced and were thus used as the signal channel for the *EF–X* excitation. In contrast, the resonant ionization scheme employed here predominantly produces H_2^+ ions. In addition to the enhancement of sensitivity, the H_2^+ channel offers another important advantage as it is a background-free channel, whereas the H^+ channel includes significant contributions from H_2S, as well as SH, dissociative ionization products. To avoid dc-Stark effects on the transition frequencies, the acceleration voltages of the ion lens system are pulsed and time-delayed with respect to the probe laser excitation.

Niu et al. [22] confirmed the observation of H_2 two-photon transitions in various *EF–X* ($v', v'' = 10 - 12$) bands, first identified by Steadman and Baer [23], but only for transitions to the outer F-well of the *EF* electronic potential in H_2. Franck–Condon factor (FCF) calculations, to assess the transition strengths of the photolysis-prepared levels of $X(v'')$ to levels in the combined inner (E) and outer (F) wells of the *EF* double well potential, were performed by Fantz and Wünderlich [25, 26]. While Niu et al. [22] performed precision measurements probing the $X(v'' = 12)$ levels, presently $X(v'' = 11)$ levels are probed. Note that for the excited state two different numberings of vibrational levels exist: one counting the levels in the combined *EF* well and the other counting the levels in the E and F wells separately. Then $EF(v' = 5)$ corresponds to $F(v' = 3)$, while $EF(v' = 1)$ corresponds to $F(v' = 0)$. In the following, we will refer to the vibrational assignments using the F-well notation. The FCF for the *F–X* (3, 2) band, used in Ref. [22], amounts to 0.047 [26] and that of the presently used F–X (0, 11) band amounts to 0.17, making the latter band's transitions three times stronger.

2.1 Resonant Ionization

The non-resonant ionization step was the major limitation in [22], since this prohibited the spectroscopy to be carried out at sufficiently low probe laser intensities. Due to ac-Stark effects, the lines were broadened to more than 1 GHz, while the expected instrumental linewidth is less than 200 MHz. Moreover, at higher probe intensities asymmetric line profiles are observed, reducing the accuracy of the line position determination and ultimately limiting the ac-Stark extrapolation to the unperturbed line position.

While signal improvement was observed when employing a detection laser in the range between 202 and 206 nm in Ref. [22], the enhancement was limited since no sharp resonances were found, indicating excitation to some continuum. For the present study, a thorough search for resonances from the F state was undertaken. The $np\pi$ and $np\sigma$ Rydberg series, with principal quantum number $n = 5$–7, were identified as potential candidates based on the FCFs for the outer F-well, as can be inferred from Fig. 2. The search was based on reported FCFs for the $D^1\Pi_u - F^1\Sigma_g^+$ $(v', 1)$ bands [26]. It was further assumed that the FCFs for the $np\pi\,^1\Pi_u - F^1\Sigma_g^+$ electronic systems are comparable to that of $3p\pi\, D^1\Pi_u - F^1\Sigma_g^+$, since the potential energy curves for the $np\pi\,^1\Pi_u$ Rydberg states are similar as they all converge to the H_2^+ ionic potential. Note the particular characteristic of the $n = 5$–7 $np\,^1\Pi_u$ Rydberg states, which dissociate to a ground state atom and another with a principal quantum number H(n − 1), i.e., $5p\pi \rightarrow$ H(1s) + H(4f); $6p\pi \rightarrow$ H(1s) + H(5f); $7p\pi \rightarrow$ H (1s) + H(6d) [27–29]. The electron configuration changes as a function of the internuclear distance R, e.g., the low vibrational levels of the $5p\,^1\Pi_u$ follow a diabatic potential that extrapolates to the $n = 5$ limit, and not the $n = 4$ dissociation limit at $R \rightarrow \infty$.

The $np\pi\,^1\Pi_u$ Rydberg states decay via three competing channels: by fluorescence to lower n electronic configurations; by predissociation, where the nascent H atom is further photoionized to yield H^+ ions; and lastly, by autoionization to yield H_2^+ ions [30]. Recent analysis of one-photon absorption measurements using XUV synchrotron radiation in the range of 74–81 nm demonstrated that autoionization completely dominates over the other two competing channels in the case of $n > 5$ [31]. Using the $n = 5$–7 level energies of the Rydberg levels reported in [27–29], the detection laser was scanned in the vicinity of the expected transition energies, where it turned out that transitions to $7p\pi$, $v = 5$, resulted in sufficient H_2^+ signal enhancement. The maximum FCF overlap for the $D^1\Pi_u - F^1\Sigma_g^+$ system is 0.34 for the (8, 0) band, while the corresponding $D - F$ (5, 0) band only has an FCF of 0.022 [26]. Although the (8, 0) band with better FCF could be used, the ionization step is already saturated using the weaker (5, 0) band. Autoionization resonances

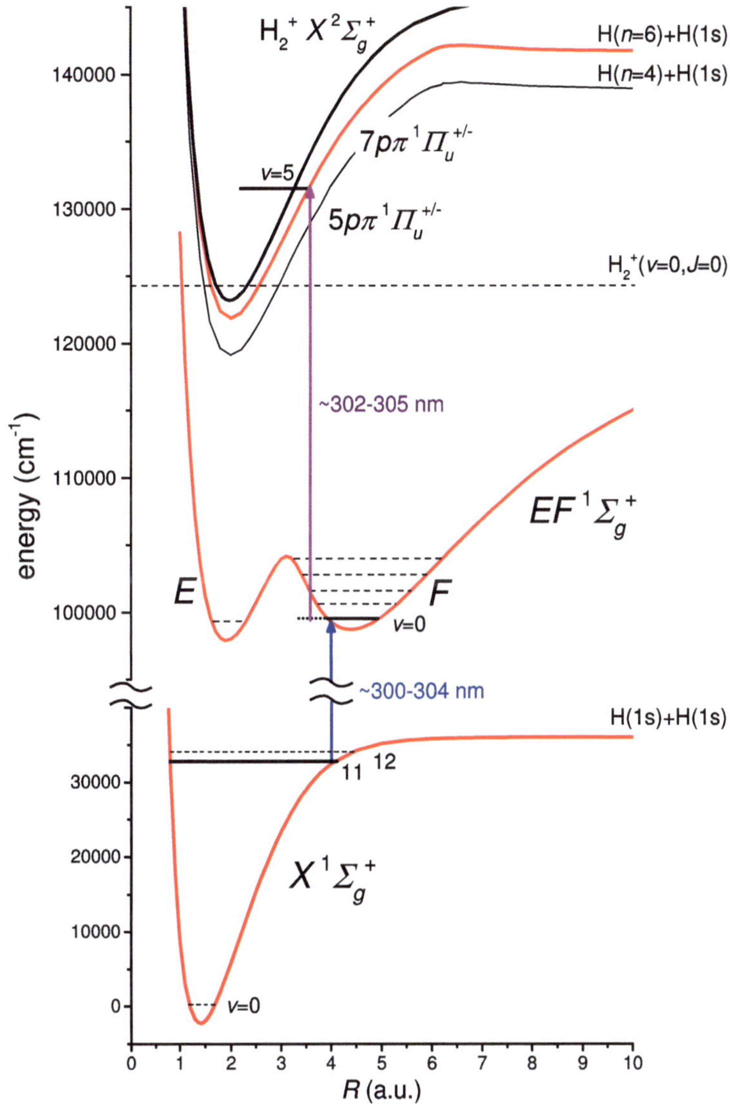

Fig. 2 Potential energy diagram showing the relevant H_2 electronic states in the $2 + 1'$ REMPI study. Two-photon Doppler-free spectroscopy is applied on the *F*–*X* (0, 11) band. Resonant excitation to the $7p\pi$ state by the detection laser follows the spectroscopic excitation, leading to subsequent autoionization yielding enhanced H_2^+ ion signal. The $6p\pi$ potential converging to H ($n = 5$) in between the $5p\pi$ and $7p\pi$ is not shown to reduce congestion

are shown in Fig. 3 and assigned to $R(3)$ and $P(4)$ lines in the $7p\pi\,^1\Pi_u - F\,^1\Sigma_g^+$ (5, 0) bands, whose widths are in good agreement with the synchrotron data [27]. The neighboring resonances of the $R(3)$ line in Fig. 3 are not yet assigned, but this is not relevant to the *F*–*X* investigation presented here. Appropriate $7p\pi - F$ transitions are used for the ionization of particular *F*–*X* two-photon $Q(J)$ transitions.

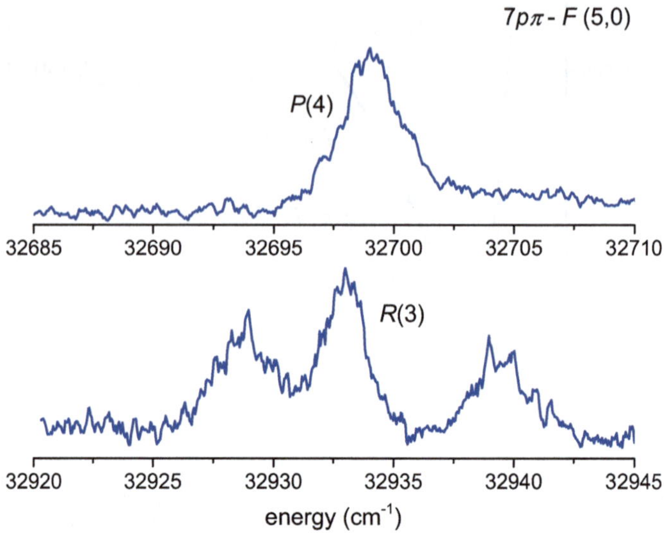

Fig. 3 Autoionizing $R(3)$ and $P(4)$ resonances of the $7p\pi\,^1\Pi_u - F\,^1\Sigma_g^+$ (5, 0) band which are crucial in enhancing the H_2^+ signal strength of the F–X spectroscopy

2.2 Frequency Calibration

A representative high-resolution spectrum of the $F\,^1\Sigma_g^+$–$X\,^1\Sigma_g^+$ (0, 11) $Q(5)$ transition taken at low probe intensity is displayed in Fig. 4. Simultaneously with the H_2 spectroscopy, the transmission fringes of the PDA cw-seed radiation through a Fabry–Pérot interferometer were also recorded to serve as relative frequency markers, with the free spectral range of FSR = 148.96(1) MHz. The etalon is temperature-stabilized, and its length is actively locked to a frequency-stabilized HeNe laser. The absolute frequency calibration is obtained from the I_2 hyperfine-resolved saturation spectra using part of the cw-seed radiation. For the $Q(5)$ line in Fig. 4, the I_2 $B-X(11,2)\,P(94)$ transition is used, where the line position of the hyperfine feature marked with an * is 16,482.833 12(1) cm^{-1} [32, 33]. The accuracy of the frequency calibration for the narrow H_2 transitions is estimated to be 1 MHz in the fundamental or 4 MHz (1.3×10^{-4} cm^{-1}) in the transition frequency, after accounting for a factor of 4 for the harmonic upconversion and two-photon excitation.

For sufficiently strong transitions probed at the lowest laser intensities, linewidths as narrow as 150 MHz were obtained. This approaches the Fourier-transform limited instrumental bandwidth of 110 MHz, for the 8-ns pulsewidths at the fundamental, approximated to be Gaussian, that also includes a factor two to account for the frequency upconversion. The narrow linewidth obtained demonstrates that despite the photodissociation process imparting considerable kinetic energy on the produced H_2^*, additional Doppler broadening is not observed.

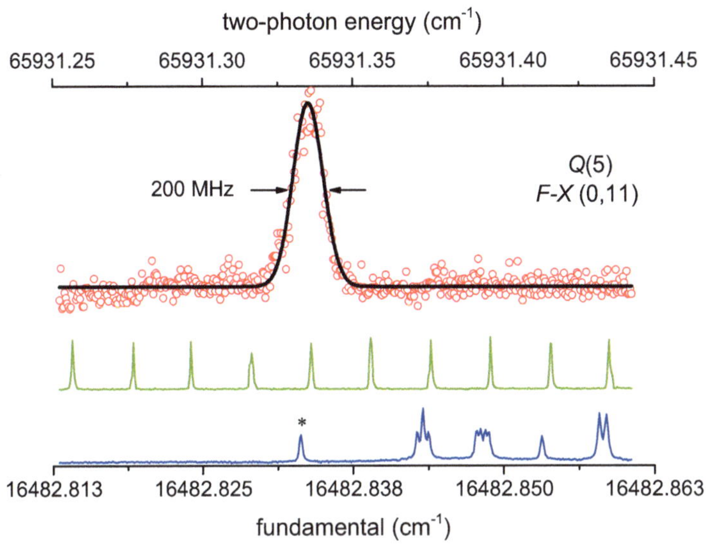

Fig. 4 Recording of the $F\,^1\Sigma_g^+ - X\,^1\Sigma_g^+$ (0, 11) $Q(5)$ transition is shown at a probe laser intensity of 280 MW/cm^2 and detection in the H_2^+ signal channel. The transmission markers of a length-stabilized Fabry–Pérot etalon (FSR = 148.96 MHz) are used in the relative frequency calibration, while the hyperfine feature marked with * of the I_2 $B - X(11,2)\,P(94)$ transition serves as an absolute frequency reference (* at 16,482.83312 cm^{-1}). Note that the spectroscopy wave number is the second harmonic of the *fundamental*, and there is an additional factor of 2 for the *two-photon* process

Although not unexpected due to the Doppler-free experimental scheme implemented, this strengthens the claim that residual Doppler shifts are negligible.

Since the frequency calibration is performed using the cw-seed, while the spectroscopy is performed using the PDA output pulses, any cw–pulse frequency offset needs to be measured and corrected for [34]. A typical recording of the chirp-induced frequency offset for a fixed PDA wavelength is shown in Fig. 5. While the measurements can be done online for each pulse, this comes at the expense of a slower data acquisition speed and was only implemented for a few recordings in order to assess any systematic effects. A flat profile of the cw–pulse offset, when the wavelength was tuned over the measurement range accessed in this study, justifies this offline correction. Typical cw–pulse frequency offset values were measured to be −8.7(1.2) MHz in the fundamental, which translates to −35(5) MHz [0.0012 (2) cm^{-1}] in the transition frequency.

2.3 Uncertainty Estimates

The sources of uncertainties and the respective contributions are shown in Table 1. The contributions of each source are summed in quadrature in order to obtain the

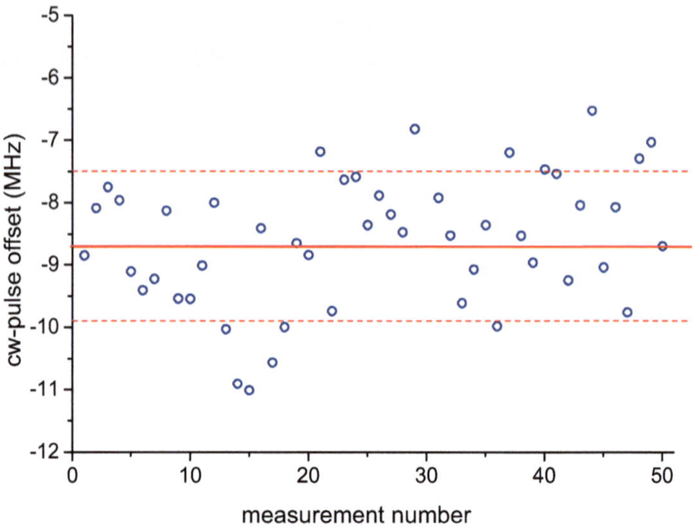

Fig. 5 Chirp-induced cw–pulse frequency offset of the fundamental radiation for a fixed PDA wavelength. The *solid line* indicates the average, and the *dashed lines* indicate the standard deviation

Table 1 Uncertainty contributions in units of 10^{-3} cm^{-1}

Source	Correction	Uncertainty
Line fitting	–	0.5
ac-Stark[a]	–	1.0
Frequency calibration	–	0.3
cw–pulse offset	−1.2	0.2
Residual Doppler	0	<0.1
dc-Stark	0	<0.1
Total		1.5

[a]Correction depends on transition

final uncertainty for each transition. Data sets from which separate ac-Stark extrapolations to zero power were performed on different days and were verified to exhibit consistency within the statistical uncertainty of 0.0014 cm^{-1}. Note that the estimates shown in Table 1 are only for the low probe intensity measurements used to obtain the highest resolutions. The uncertainties of the present investigation constitute more than a factor of two improvements over our previous study in [22]. The dominant source of systematic uncertainty is the ac-Stark shift and is discussed in more detail in the following section.

3 ac-Stark Shift and Broadening

In the perturbative regime, the leading-order energy level shift ΔE_n of a state $|n\rangle$ induced by a linearly polarized optical field with an amplitude $\mathcal{E}_0$ and frequency ν can be described as

$$\Delta E_n = \frac{1}{2}\sum_m \left\{ \frac{\left\langle n|\vec{\mu}\cdot\vec{\mathcal{E}}_0|m\right\rangle\left\langle m|\vec{\mu}\cdot\vec{\mathcal{E}}_0|n\right\rangle}{E_n - E_m - h\nu} + \frac{\left\langle n|\vec{\mu}\cdot\vec{\mathcal{E}}_0|m\right\rangle\left\langle m|\vec{\mu}\cdot\vec{\mathcal{E}}_0|n\right\rangle}{E_n - E_m + h\nu} \right\}, \quad (1)$$

where $\langle n|\mu|m\rangle$ is the transition dipole moment matrix element between states n and m, with an energy E_m for the latter [35]. Thus ΔE_n has a quadratic dependence on the field or a linear dependence on intensity for this frequency-dependent ac-Stark level shift. In a simple case, when there is one near-resonant coupling to state m whose contribution dominates ΔE_n, the sign of the detuning with respect to transition frequency $\nu_{mn} = |E_n - E_m|/h$ determines the direction of the light shifts with intensity. [Note the sign difference in the denominators of the two terms in Eq. (1).] When the probing radiation is blue-detuned, i.e., $\nu > \nu_{mn}$, the two levels $|n\rangle$ and $|m\rangle$ shift toward each other, while for red-detuning, $\nu < \nu_{mn}$, the levels repel each other. In the case when all accessible states are far off resonant, both terms in Eq. (1) contribute for each state m, and numerous m-states need to be included in the calculations to explain the magnitude and sign of ΔE_n. The energy shifts of the upper (ΔE_u) and lower (ΔE_l) levels in turn translate into an ac-Stark shift,

$$h\delta_S = \alpha I, \quad (2)$$

where α is the ac-Stark coefficient. The measured *transition* energy is $h\nu = h\nu_0 + \alpha I$, where $\nu_0 = |E_u - E_l|/h$ is the unperturbed (zero-field) transition frequency. The ac-Stark coefficient α depends on the coupling strengths of the $|u\rangle$ and $|l\rangle$ levels to the dipole-accessible $|m\rangle$ states, as well as the magnitude and sign of the detuning. We note that in a so-called *magic wavelength* configuration, the frequency ν is selected so that the level shifts of the upper and lower states cancel out, leading to ac-Stark free transition frequencies [36].

The first experimental study of ac-Stark effects in molecules associated with REMPI processes was performed by Otis and Johnson [37] on NO. The broad ac-Stark-induced features in NO were later explained in the extensive models by Huo et al. [38]. An investigation of ac-Stark effects on two-photon transitions in CO was performed by Girard et al. [39]. For molecular hydrogen, various studies have been performed on the two-photon excitation in the *EF*–*X* system over the years [19, 40–43]. In the following, we present our evaluation of the ac-Stark effect in *F*–*X* (0, 11) transitions where we first discuss line shape effects. This is followed by a discussion on the ac-Stark coefficients extracted from the analysis and comparisons with previous determinations on the *EF–X* system.

3.1 Line Shape Model

The line profiles of the *F–X* (0, 11) *Q*(3) transition, recorded at different probe laser intensities, are displayed in Fig. 6. The ac-Stark broadening and asymmetry is readily apparent at higher intensities, and only at low intensities can the profile be fitted by a simple Gaussian line shape. Note that the shift in peak position at the highest probe intensity amounts to several linewidths of the lowest intensity recording.

The asymmetry at high intensities is highly problematic with regard to the extraction of the line positions. In our previous study [22], a skewed Gaussian function $g(f)$ was used to fit the spectra,

$$g(f) = \frac{A}{\Gamma_G}\exp\left(\frac{-(f-f_c)^2}{2\Gamma_G^2}\right) \times \left\{1+\mathrm{erf}\left(\xi\frac{f-f_c}{\sqrt{2}\Gamma_G}\right)\right\}, \tag{3}$$

where f_c is the Gaussian peak position in the absence of asymmetry, Γ_G is the linewidth, A is an amplitude scaling parameter and ξ is the asymmetry parameter. The center f_c of the error function, erf(f), is arbitrarily chosen to coincide with the

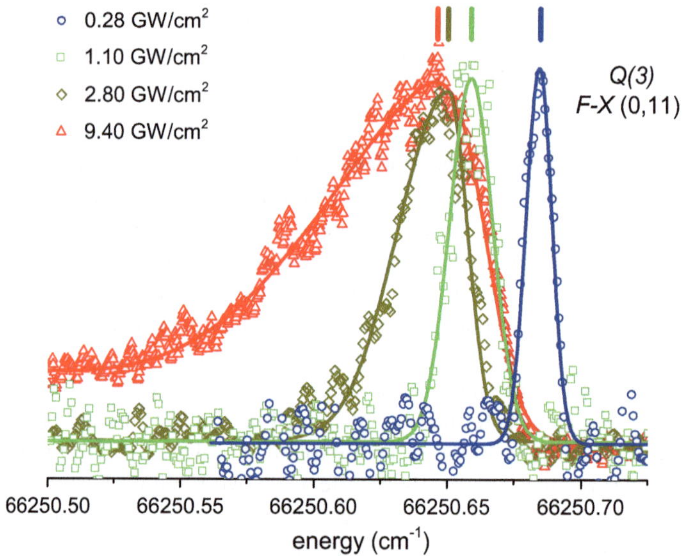

Fig. 6 Line profiles of the $F-X\,(0,11)\,Q(3)$ transition recorded at different probe laser intensities. The *vertical lines* above each profile denote the peak position used in the subsequent extrapolation

Gaussian center. For sufficiently high intensities, a satisfactory fit is only possible if a linear background $B(f)$ is added, and thus a revised fitting function, $g'(f) = g(f) + B(f)$, is used. This phenomenological fit function resulted in better fits than symmetric Lorentzian, Gaussian or Voigt profiles. However, since it does not include any consideration of the underlying physics, the interpretation of the extracted f_c and ξ parameters, as well as the background $B(f)$, is not straightforward. The energy position of the skewed profile maximum, instead of f_c, is used as the ac-Stark-shifted frequency in the subsequent extrapolations.

The ambiguity in the attribution of the ac-Stark-shifted transition frequency from the skewed Gaussian fitting prompted us to pursue a model that takes into account the physical origins of the asymmetry. A more physically motivated asymmetric line shape function was derived by Li et al. [44] for the analysis of multiphoton resonances in the NO $A\,^2\Sigma^+ - X\,^2\Pi\,(0,0)$ band. Their closed-form line shape model accounted for effects of the spatial and temporal distributions of the light intensity. Here, we reproduce their line shape as a function of $\delta_\mathrm{L} = \nu - \nu_0$, the laser frequency shifts from the zero-field line position

$$S(\delta_0, \Gamma, \delta_\mathrm{L}) = \kappa \int_0^{\delta_0} \mathrm{d}\delta' \frac{K\left(\ln\left(\frac{\delta_0}{\delta'}\right)\right)}{\delta'} G(\Gamma, \delta' - \delta_\mathrm{L}), \tag{4}$$

where δ_0 is the maximum ac-Stark shift induced at the peak intensity I_0. K contains the dependence on the temporal profile, as well as the transverse (Gaussian beam profile) and longitudinal intensity (focused) distribution, parameterized in [44] as

$$\begin{aligned} K(x) &= \left[0.6366/x + 2.087\mathrm{e}^{x/2} \right. \\ &\quad \left. - \mathrm{e}^{-x/2}(1.087 + 0.90x + 0.45x^2 + 0.3x^3)\right]^{-1}. \end{aligned} \tag{5}$$

G is a Gaussian distribution with full width at half maximum (FWHM) Γ,

$$\begin{aligned} G(\Gamma, \delta' - \delta_\mathrm{L}) = &\ \frac{1}{\sqrt{\pi}} \frac{2\sqrt{\ln 2}}{\Gamma} \\ &\times \exp\left(-\left(\frac{2\sqrt{\ln 2}}{\Gamma}\right)^2 (\delta' - \delta_\mathrm{L})^2\right), \end{aligned} \tag{6}$$

that approximates all other sources of line broadening, such as the spectral width of the laser. The parameter $\kappa = 1.189$ is a normalization factor which ensures that

$$\int_{-\infty}^{\infty} S(\delta_0, \Gamma, \delta_\mathrm{L}) \mathrm{d}\delta_\mathrm{L} = 1 \tag{7}$$

for any Γ and δ_0. It appears that the spatial and temporal intensity distribution was also treated in the investigations of Huo et al. [38] and Girard et al. [39], but the expressions were not explicitly given. When comparing different probe intensity recordings, the normalized profile given by Eq. (4) should be multiplied by a factor that scales with light intensity as $\sim I_0^2$, with the exact form given in Ref. [44]. Note that the error function in the skewed Gaussian model of Eq. (3) effectively captures the result of the integration in Eq. (4). However, the physical interpretation of the f_c and ξ parameters from the skewed Gaussian model is ambiguous, while the background, $B(f)$, is an ad hoc addition.

Li et al. [44] presented intuitive explanations of qualitative behavior of the line profile at two extreme cases: (1) of a perfectly collimated probe beam and (2) of a conically focused beam. In case (1) the laser intensity is spatially homogeneous, so that the temporal intensity distribution is the dominant effect. Almost all contribution to the resonant excitation comes from the peak of the pulse, causing the line peak position to be shifted by almost δ_0. In case (2) the strongly inhomogeneous spatial intensity plays the dominant role, and molecules located at the focus, having the highest Stark shift δ_0, have a smaller contribution relative to those from the entire interaction volume. The majority of the excited molecules come from a region of low intensities outside the focus; therefore, the integrated line profile is only slightly shifted from the field-free resonance. Our experimental conditions lie in between these two cases, where a loose focus is implemented and a molecular beam that overlaps within a few Rayleigh ranges of the laser beam is employed.

The asymmetric line profiles can be fitted very well using the line shape model expressed in Eq. (4) with appropriate experimental parameters. For a recording at a particular intensity I_0, the maximum ac-Stark shift from the zero-field resonance, δ_0, is obtained from the fit. The line shape asymmetry, in particular the skew handedness, is consistent with the direction of the light shift observed at different intensities, validating the expected behavior from Eq. (4). In Fig. 7, fits using the physical line shape model and skewed Gaussian profile are shown for the Q (3) transition recorded at ~9.4 GW/cm^2. The linear background $B(f)$ (dashed line) was necessary for a satisfactory fit with the skewed Gaussian, while no additional background functions were used for the line shape model. The extracted line positions are indicated in Fig. 7 by vertical lines above the profiles, where the difference in the line positions of the two fit functions amounts to about 0.006 cm^{-1}. For reference, the f_c position obtained by using Eq. (3) is also indicated by a dotted line, although this is not used further in the analysis. That the phenomenological skewed Gaussian model does not yield a reliable value for f_c, is what motivated us to adopt a *physically based* line shape model for the asymmetric profiles.

Spectra are recorded at a fixed intensity setting, where each datapoint in a particular laser frequency scan is averaged ten times. The probe laser power is set by adjusting the waveplate before the second-harmonic generation stage, so that the beam focusing condition is not altered. The average laser power (and intensity) varies by about ~10% as monitored by a photodetector. To complete an ac-Stark set for a certain transition, several intensity values are used with the maximum value that is about ten times the minimum intensity value. Several complete ac-Stark sets are measured for each transition on different days to check for consistency and reproducibility.

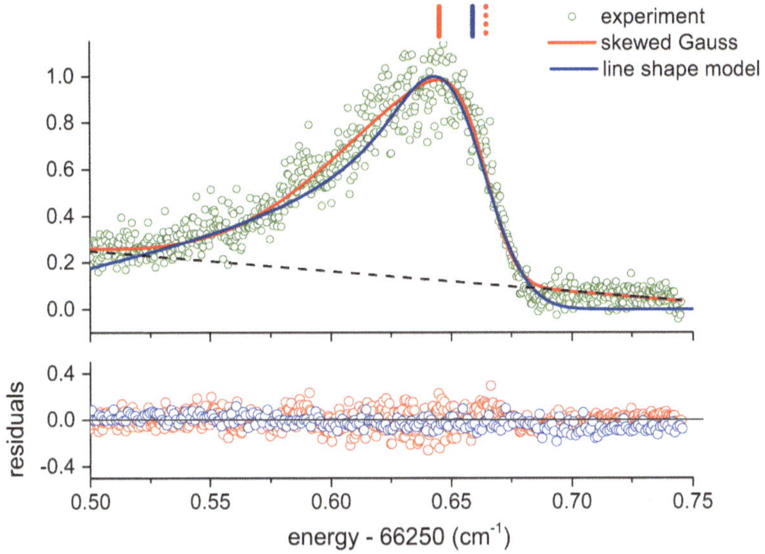

Fig. 7 $F-X\,(0,11)\,Q(3)$ transition taken at high probe intensity (9.4 GW/cm^2) fitted with a skewed Gaussian (*red solid line*), with the fitted linear background indicated by the *black dashed line*. The fitted curve for line shape model is plotted as the *blue solid line*, which includes the effects of spatial and temporal intensity distribution. The *red vertical line* above the profile indicates the line position for the skewed Gaussian fit, while the *blue vertical line* indicates position shifted by δ_0 obtained from the line shape model. For comparison, the f_c parameter obtained from skewed Gaussian fitting is also indicated (*dotted red line*)

A comparison of the *F*–*X* (0, 11) *Q*(3) line positions extracted from the skewed Gaussian and from the line shape model based on Li et al. [44] is shown in Fig. 8. The line positions show different trends that separate into low-intensity, with symmetric line profiles, and high-intensity subsets, with asymmetric ones. The line profile models are in agreement at low intensities as expected, but show a discrepancy at higher intensities as explained above (see Fig. 7). The extrapolated zero-intensity positions for the low-intensity measurements converge to within 0.0001 cm^{-1} for both line profile models.

When using only the high-intensity data to extrapolate the zero-intensity frequency, a shift of ∼0.02 cm^{-1} with respect to the low-intensity subset is found. The latter difference is a concern when only high-intensity data are available as in Ref. [22] for the *F*–*X* (3, 12) band. The ac-Stark coefficients, however, are an order of magnitude lower for the *F*–*X* (3, 12) band compared with the present (0, 11) band; thus, any systematic offset is still expected to be within the uncertainty estimates in that study [22]. It is comforting to note that a second-order polynomial fit, also shown in Fig. 8 as dash-dotted curve, results in an extrapolated zero-field frequency that is within 0.0002 cm^{-1} of the low-intensity linear fits.

The ac-Stark shifts of the different *F*–*X* (0, 11) transitions exhibit a nonlinear dependence on intensity, contrary to the expected behavior in Eq. (2). A correlation

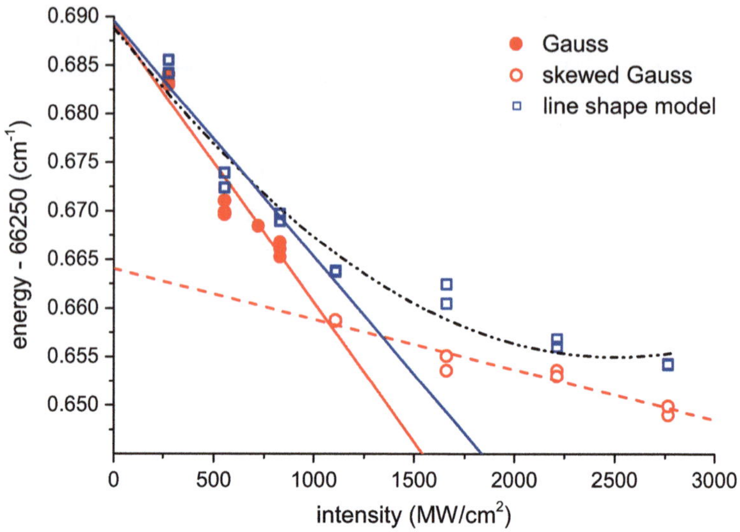

Fig. 8 Extracted line positions of the $F - X\,(0, 11)\,Q(3)$ transition from measurements at different probe laser intensities. *Squares* (*blue*) are obtained from fits using the line shape model of Eq. (4), while *circles* (*red*) are obtained from Gaussian fits. The *solid circles* (*red*) are symmetric profiles fitted with a simple Gaussian, while *unfilled circles* are fitted with skewed Gaussian profiles. The *solid lines* are linear fits to low-intensity data points, while the *dashed lines* are a linear fit using the high-intensity points only. The *dash-dotted line* is a second-order polynomial fit for the whole intensity range using the data points obtained from the line shape model (*squares*)

is also observed between the nonlinearity and the ac-Stark coefficient α, where the onset of nonlinearity occurs at a higher intensity for transitions with smaller α. The nonlinearity may indicate close proximity to a near-resonant state, which may signal the breakdown of the perturbative approximation in Eq. (1). Possibly, this requires contributions beyond the second-order correction [39] that lead to a higher-power dependence on intensity. A similar behavior was observed by Liao and Bjorkholm [45] in their study of the ac-Stark effect in the two-photon excitation of sodium. In that investigation, they observed the nonlinear dependence of the ac-Stark shift at some probe detuning that is sufficiently close to resonance with an intermediate state.

3.2 ac-Stark Coefficients

The ac-Stark coefficient α, as defined in Eq. (2), was obtained by Hannemann et al. [43] for the H_2 $EF^1\Sigma_g^+ - X^1\Sigma_g^+$ (0, 0) band, where they reported +13(7) and +6 (4) MHz per MW/cm^2, respectively, for the $Q(0)$ and $Q(1)$ lines. Investigations on the E–X (0, 1) band in Ref. [19], and more extensively in Ref. [46], also resulted in positive ac-Stark coefficients for $Q(J = 0–3)$ transitions that are about an order of magnitude lower than for the E–X (0, 0) band. Eyler and coworkers [42] found that the ac-Stark slopes vary considerably, typically at a few tens of MHz per (MW/cm^2), for different transitions in the H_2 E–X system. The Rhodes group in Chicago has performed a number of excitation studies with high-power lasers on the EF–X system in hydrogen, where they also investigated optical Stark shifts (e.g., [40]). Following excitation of molecular hydrogen by 193-nm radiation, intense stimulated emission on both the Lyman and Werner bands is observed. Using excitation intensities of ~600 GW/cm^2, they obtained shifts in the order of 2 MHz per MW/cm ^{2} for the EF–X (2, 0) band. While the work of Vrakking et al. [41] was primarily on the detection sensitivity of REMPI on the H_2 EF–X system, they also obtained ac-Stark coefficients of ~15 MHz per MW/cm^2 similar to the value found in Ref. [43]. Vrakking et al. [41] also made reference to a private communication with Hessler on the ac-Stark effect (shift) amounting to 3–6 MHz per MW/cm^2, presumably obtained in the study by Glab and Hessler [47].

The ac-Stark coefficients α obtained in this study for the F–X (0, 11) transitions are plotted in Fig. 9. The error bars in the figure comprise two contributions, with the larger one dominated by the accuracy in the absolute determination of the probe intensity. The difficulties include estimating the effective laser beam cross sections in the interaction volume that should take into account the overlap of the counterpropagating probe beams and also the overlap of the photodissociation

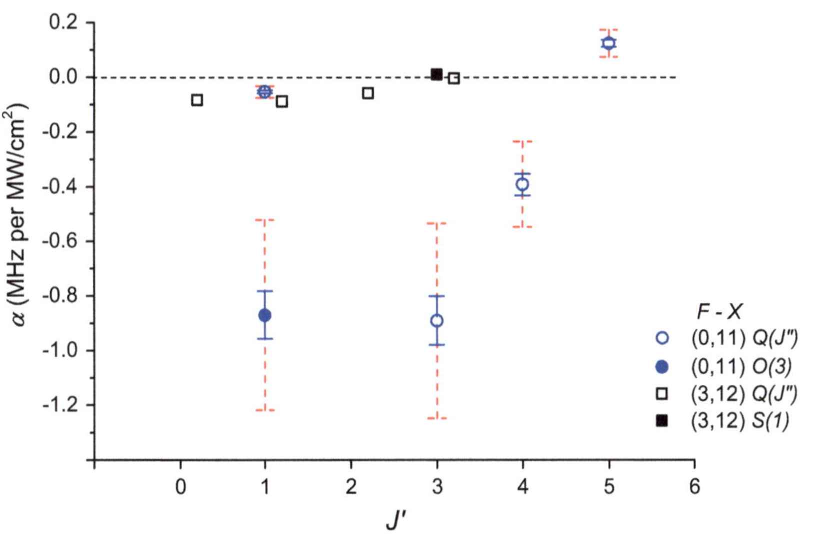

Fig. 9 ac-Stark coefficients plotted against the upper J'-quantum number for the different transitions of the $F - X\,(0,11)$ band (*blue circles*). The α-values of the $F - X\,(3,12)$ transitions (*black squares*) from Ref. [34] are also plotted and displaced for clarity. The datapoint indicated by a *filled circle* is obtained from the F–X (0, 11) $O(3)$ transition, while the *filled* (*black*) *square* is from the F–X (3, 12) $S(1)$ transition

and ionization beams. The smaller error bars are obtained from fits using relative intensities, shown here to emphasize that the differences in α-coefficients are significant. Also included in Fig. 9 are the Stark coefficients for *F–X* (3, 12) obtained in [34] that differ by about an order of magnitude with respect to the values obtained for *F–X* (0, 11). The sign of the *F–X* α-coefficients is mostly negative for both bands except for the *F–X* (0, 11) *Q*(5) line.

The different signs in the ac-Stark coefficients, which are positive for the *E–X* transitions and mostly negative for the *F–X* transitions, seem to be a feature of the ac-Stark effect in H_2. For the *E–X* (0, 0) transitions probed at around 202 nm and *E–X* (0, 1) transitions probed at around 210 nm, all intermediate states are far off resonant with respect to transitions from the *E* or *X* levels. Using Eq. (1), the ac-Stark shift is estimated based on the approximation of Rhodes and coworkers [48, 49] for the *E–X* (2, 0) band. In those studies, they assumed that the intermediate states are predominantly the Rydberg series at principal quantum number $n > 2$, clustered at an average energy of $E_m = 14.7$ eV. The present estimates of ac-Stark shifts via Eq. (1), using the probe frequency and intensity in Ref. [43], are in agreement with within an order of magnitude of the *E–X* (0, 0) observations. A similar estimate for *E–X* (0, 1) transitions is also within an order-of-magnitude agreement of the measurements in [19, 46]. The blue-detuned probe in the aforementioned *E–X* transitions explains the observed light shift direction as expected from Eq. (1).

In *F–X* excitation from *X*, $v = 11$ and $v = 12$, the fundamental probe wavelengths are around 300 nm. Rydberg *np* levels can be close to resonance at the probe wavelengths, a fact that we have exploited in the resonant ionization (using an additional laser source) for the REMPI detection. This could explain the difference in sign of α between the *E–X* and *F–X* transitions. Furthermore, Fig. 9 displays a trend of the *F–X* (0, 11) transitions, where a small negative coefficient is observed for *Q*(1), increasing in magnitude at *Q*(3), decreasing in magnitude again, and eventually becoming positive at *Q*(5). Interestingly, the *F–X* (0, 11) *O*(3) transition displays a relatively large α with respect to that of the *Q*(1) transition, despite having the same upper *F* level $J' = 1$. The *Q*(1) and *O*(3) fundamental probe energies differ only by some 250 cm^{-1}, but this results in significant change in α. These phenomena hint toward a scenario where some near-resonant levels are accessed, so that the summation (and cancelation) of the contributions in Eq. (1) depend more sensitively on the detunings of the probe laser frequency with respect to ν_{mn}. The variation in α-magnitudes as well as the change from negative to positive sign could be explained by a change from red to blue detuning when traversing across a dominant near-resonance intermediate level. The latter behavior was expected in the NO investigation of Huo et al. [38], where the α-coefficients for individual *M* sublevels varied from −4.1 to 53 MHz per MW/cm^2. The order-of-magnitude difference between α-values of the *F–X* (0, 11) and (3, 12) transitions may be due to less-favorable FCF overlaps of the near-resonant intermediate levels for the latter band. This is correlated with the similar value of the *F–X* (3, 12) *Q* (3) and *S*(1) transition, displaying a different trend to that of the *Q*(1) and *O*(3) α-coefficients in the *F–X* (0, 11) band. The nonlinear intensity dependence of the transitions discussed in the previous subsection may be consistently explained by the same argument on the importance of near-resonant intermediate levels. For a quantitative explanation of the *F–X* ac-Stark shift, a more extensive theoretical study is necessary to account for the dense intermediate Rydberg levels involved that should also include considerations of FCF overlaps and dipole coupling strengths at the appropriate internuclear distance.

4 Results and Discussion

The resulting two-photon transition energies for the $F\,^1\Sigma_g^+ - X\,^1\Sigma_g^+$ (0, 11) lines are listed in Table 2. The results presented are principally based on high-resolution measurements using transitions with symmetric linewidths narrower than 1 GHz. This ensures that the ac-Stark shift extrapolation can be considered robust and reliable as discussed above.

Combination differences between appropriate transition pairs allow for the confirmation of transition assignments as well as consistency checks of the measurements, where most of the systematic uncertainty contributions cancel. The Q (1) and O(3) transitions share a common upper *EF* level, and the energy difference of 248.7329(21) cm^{-1} gives the ground state splitting $X, v'' = 11, J'' = 1 \rightarrow 3$. This can be compared to the theoretical splitting derived from Komasa et al. [9] of 248.731(7) cm^{-1}. In analogous fashion, the Q(3) and O(3) share the same lower X level, which enables the extraction of the $EF, v' = 1, J' = 1 \rightarrow 3$ energy splitting of 61.1194(21) cm^{-1}. This is in good agreement with the derived experimental splitting of 61.1191(10) cm^{-1} from Bailly et al. [50].

To extract the ground electronic $X\,^1\Sigma_g^+$, $v = 11, J$ level energies, from the *EF–X* transition energy measurements, we use the level energy values of the $F(v' = 0)$ states determined by Bailly et al. [50]. The derived experimental level energies are listed in Table 3, where the uncertainty is limited by the present *F–X* determination

Table 2 Measured two-photon transition energies of *F–X* (0, 11) band

Line	Experiment
Q(1)	66,438.2920(15)
Q(3)	66,250.6874(15)
Q(4)	66,105.8695(15)
Q(5)	65,931.3315(15)
O(3)	66,189.5591(15)

All values in cm^{-1}

Table 3 Experimental and theoretical level energies of the $X\,^1\Sigma_g^+$, $v = 11$, J levels

J	Experiment	Theory	Exp–theo
1	32,937.7554(16)	32,937.7494(53)	0.0060(55)
3	33,186.4791(16)	33,186.4802(52)	−0.0011(54)
4	33,380.1025(33)	33,380.1019(52)	0.0006(62)
5	33,615.5371(18)	33,615.5293(51)	0.0078(54)

except for the $J = 4$. The calculated values obtained by Komasa et al. [9] are also listed in Table 3. The experimental and theoretical values are in good agreement, except for $J = 5$ that deviate by $1.5 - \sigma$. The combined uncertainty of the difference is dominated by the theoretical uncertainty. However, improvements in the calculations of the non-relativistic energies, limited by the accuracy of fundamental constants m_p/m_e and R_∞, have recently been reported [15], and improved calculations of QED corrections up to the $m\alpha^6$-order are anticipated.

As has been pointed out previously in Ref. [22], the uncertainties in the calculations [9] are five times worse for the $v = 8$–11 in comparison with the $v = 0$ level energies. Along with the previous measurements on the X, $v = 12$ levels in Ref. [22], the measurements presented here probe the highest uncertainty region of the most advanced first-principle quantum chemical calculations.

5 Conclusion

H_2 transition energies of $F\,^1\Sigma_g^+(v' = 0) - X\,^1\Sigma_g^+(v'' = 11)$ rovibrational states were determined at 0.0015 cm^{-1} absolute accuracies. Enhanced detection efficiency was achieved by resonant excitation to autoionizing $7p\pi$ electronic Rydberg states, permitting excitation with low probe laser intensity that led to much narrower transitions due to reduced ac-Stark effects. The asymmetric line broadening, induced by the ac-Stark effect, at high probe intensities was found to be well explained by taking into account the spatial and temporal intensity beam profile of the probe laser. The extracted ac-Stark coefficients for the different transitions F–X, as well as previously determined E–X transitions, are consistent with qualitative expectations. However, a quantitative explanation awaits detailed calculations of the ac-Stark effect that account for molecular structure, i.e., including a proper treatment of relevant intermediate states.

Using the F level energies obtained by Bailly et al. [50], the level energies of X ($v = 11$, $J = 1$, 3–5) states are derived with accuracies better than 0.002 cm^{-1} except for $J = 3$, limited by F level energy accuracy. The derived experimental values are in excellent agreement with, thereby confirming, the results obtained from the most advanced and accurate molecular theory calculations. The experimental binding energies reported here are about thrice more accurate than the present theoretical values and may provide further stimulus toward advancements in the already impressive state-of-the-art ab initio calculations.

Acknowledgements This work is part of the research program of the Foundation for Fundamental Research on Matter (FOM), which is part of the Netherlands Organisation for Scientific Research (NWO). P. W. received support from LASERLAB-EUROPE within the EC's Seventh Framework Programme (Grant No. 284464). W. U. received funding from the European Research Council (ERC) under the European Unions Horizon 2020 research and innovation program (Grant No. 670168).

References

1. T.W. Hänsch, Appl. Opt. **11**, 895 (1972)
2. T.W. Hänsch, M.D. Levenson, A.L. Schawlow, Phys. Rev. Lett. **26**, 946 (1971)
3. T.W. Hänsch, S.A. Lee, R. Wallenstein, C. Wieman, Phys. Rev. Lett. **34**, 307 (1975)
4. T.W. Hänsch, B. Couillaud, Opt. Commun. **35**, 441 (1980)
5. R. Holzwarth, T. Udem, T.W. Hänsch, J.C. Knight, W.J. Wadsworth, P.S.J. Russell, Phys. Rev. Lett. **85**, 2264 (2000)
6. T.W. Hänsch, Rev. Mod. Phys. **78**, 1297 (2006)
7. C.G. Parthey, A. Matveev, J. Alnis, B. Bernhardt, A. Beyer, R. Holzwarth, A. Maistrou, R. Pohl, K. Predehl, T. Udem et al., Phys. Rev. Lett. **107**, 203001 (2011)
8. A. Antognini, F. Nez, K. Schuhmann, F.D. Amaro, F. Biraben, J.M.R. Cardoso, D.S. Covita, A. Dax, S. Dhawan, M. Diepold et al., Science **339**, 417 (2013)
9. J. Komasa, K. Piszczatowski, G. Łach, M. Przybytek, B. Jeziorski, K. Pachucki, J. Chem. Theory Comput. **7**, 3105 (2011)
10. K. Pachucki, J. Komasa, J. Chem. Phys. **141**, 224103 (2014)
11. K. Pachucki, J. Komasa, J. Chem. Phys. **143**, 034111 (2015)
12. V.I. Korobov, L. Hilico, J.-P. Karr, Phys. Rev. A **89**, 032511 (2014)
13. J. Biesheuvel, J.-P. Karr, L. Hilico, K.S.E. Eikema, W. Ubachs, J.C.J. Koelemeij, Nat. Commun. **7**, 10385 (2016)
14. J.-P. Karr, L. Hilico, J. Koelemeij, V. Korobov, Phys. Rev. A **94**, 050501(R) (2016)
15. K. Pachucki, J. Komasa, J. Chem. Phys. **144**, 164306 (2016)
16. W. Ubachs, J. Koelemeij, K. Eikema, E. Salumbides, J. Mol. Spectrosc. **320**, 1 (2016)
17. J. Liu, E.J. Salumbides, U. Hollenstein, J.C.J. Koelemeij, K.S.E. Eikema, W. Ubachs, F. Merkt, J. Chem. Phys. **130**, 174306 (2009)
18. E.J. Salumbides, G.D. Dickenson, T.I. Ivanov, W. Ubachs, Phys. Rev. Lett. **107**, 043005 (2011)
19. G.D. Dickenson, M.L. Niu, E.J. Salumbides, J. Komasa, K.S.E. Eikema, K. Pachucki, W. Ubachs, Phys. Rev. Lett. **110**, 193601 (2013)
20. E.J. Salumbides, J.C.J. Koelemeij, J. Komasa, K. Pachucki, K.S.E. Eikema, W. Ubachs, Phys. Rev. D **87**, 112008 (2013)
21. E.J. Salumbides, A.N. Schellekens, B. Gato-Rivera, W. Ubachs, New J. Phys. **17**, 033015 (2015)
22. M.L. Niu, E.J. Salumbides, W. Ubachs, J. Chem. Phys. **143**, 081102 (2015a)
23. J. Steadman, T. Baer, J. Chem. Phys. **91**, 6113 (1989)
24. S. Hannemann, E.J. Salumbides, W. Ubachs, Opt. Lett. **32**, 1381 (2007)
25. U. Fantz, D. Wünderlich, At. Data. Nucl. Data Tables **92**, 853 (2006)
26. U. Fantz, D. Wünderlich, *Monograph INDC(NDS)-457* (University of Augsburg, Augsburg, 2004)
27. M. Glass-Maujean, C. Jungen, H. Schmoranzer, I. Tulin, A. Knie, P. Reiss, A. Ehresmann, J. Mol. Spectrosc. **293–294**, 11 (2013a)
28. M. Glass-Maujean, C. Jungen, A. Spielfiedel, H. Schmoranzer, I. Tulin, A. Knie, P. Reiss, A. Ehresmann, J. Mol. Spectrosc. **293–294**, 1 (2013b)
29. M. Glass-Maujean, C. Jungen, H. Schmoranzer, I. Tulin, A. Knie, P. Reiss, A. Ehresmann, J. Mol. Spectrosc. **293–294**, 19 (2013c)
30. M. Sommavilla, F. Merkt, J.Z. Mezei, C. Jungen, J. Chem. Phys. **144**, 084303 (2016)
31. M. Glass-Maujean, C. Jungen, H. Schmoranzer, A. Knie, I. Haar, R. Hentges, W. Kielich, K. Jänkälä, A. Ehresmann, Phys. Rev. Lett. **104**, 183002 (2010)
32. S. Xu, R. van Dierendonck, W. Hogervorst, W. Ubachs, J. Mol. Spectrosc. **201**, 256 (2000)
33. B. Bodermann, H. Knöckel, E. Tiemann, Eur. J. Phys. D **19**, 31 (2002)
34. M.L. Niu, F. Ramirez, E.J. Salumbides, W. Ubachs, J. Chem. Phys. **142**, 044302 (2015b)
35. J. Bakos, Phys. Rep. **31**, 209 (1977)
36. J. Ye, H.J. Kimble, H. Katori, Science **320**, 1734 (2008)

37. C.E. Otis, P.M. Johnson, Chem. Phys. Lett. **83**, 73 (1981)
38. W.M. Huo, K.P. Gross, R.L. McKenzie, Phys. Rev. Lett. **54**, 1012 (1985)
39. B. Girard, N. Billy, J. Vigue, J. Lehmann, Chem. Phys. Lett. **102**, 168 (1983)
40. T. Srinivasan, H. Egger, T. Luk, H. Pummer, C. Rhodes, IEEE J. Quantum Electron. **19**, 1874 (1983a)
41. M. Vrakking, A. Bracker, T. Suzuki, Y. Lee, Rev. Sci. Instrum. **64**, 3 (1993)
42. A. Yiannopoulou, N. Melikechi, S. Gangopadhyay, J.C. Meiners, C.H. Cheng, E.E. Eyler, Phys. Rev. A **73**, 022506 (2006)
43. S. Hannemann, E.J. Salumbides, S. Witte, R.T. Zinkstok, E.J. van Duijn, K.S.E. Eikema, W. Ubachs, Phys. Rev. A **74**, 062514 (2006)
44. L. Li, B.-X. Yang, P.M. Johnson, J. Opt. Soc. Am. B **2**, 748 (1985)
45. P.F. Liao, J.E. Bjorkholm, Phys. Rev. Lett. **34**, 1 (1975)
46. M.L. Niu, E.J. Salumbides, G.D. Dickenson, K.S.E. Eikema, W. Ubachs, J. Mol. Spectrosc. **300**, 44 (2014)
47. W.L. Glab, J.P. Hessler, Phys. Rev. A **35**, 2102 (1987)
48. H. Pummer, H. Egger, T.S. Luk, T. Srinivasan, C.K. Rhodes, Phys. Rev. A **28**, 795 (1983)
49. T. Srinivasan, H. Egger, H. Pummer, C. Rhodes, IEEE J. Quantum Electron. **19**, 1270 (1983b)
50. D. Bailly, E. Salumbides, M. Vervloet, W. Ubachs, Mol. Phys. **108**, 827 (2010)

Ranging with Frequency-Shifted Feedback Lasers: From μm-Range Accuracy to MHz-Range Measurement Rate

J.I. Kim, V.V. Ogurtsov, G. Bonnet, L.P. Yatsenko, and K. Bergmann

Abstract We report results on ranging based on frequency-shifted feedback (FSF) lasers with two different implementations: (1) An Ytterbium-fiber system for measurements in an industrial environment with accuracy of the order of 1 μm, achievable over a distance of the order of meters with potential to reach an accuracy of better than 100 nm; (2) A semiconductor laser system for a high rate of measurements with an accuracy of 2 mm @ 1 MHz or 75 μm @ 1 kHz and a limit of the accuracy of ≥10 μm. In both implementations, the distances information is derived from a frequency measurement. The method is therefore insensitive to detrimental influence of ambient light. For the Ytterbium-fiber system, a key feature is the injection of a single-frequency laser, phase modulated at variable frequency Ω, into the FSF-laser cavity. The frequency $\Omega_{\max}$ at which the detector signal is maximal yields the distance. The semiconductor FSF-laser system operates without external injection seeding. In this case, the key feature is frequency

This article is part of the topical collection "Enlightening the World with the Laser" - Honoring T. W. Hänsch guest edited by Tilman Esslinger, Nathalie Picqué, and Thomas Udem.

J.I. Kim
Department of Physics, Technical University Kaiserslautern, 67653 Kaiserslautern, Germany

Present Address: Defense R&D Center, Hanwha Corporation, 99 Oesam-ro, Yuseong-gu, Daejeon 305-156, Korea

V.V. Ogurtsov
Department of Physics, Technical University Kaiserslautern, 67653 Kaiserslautern, Germany

Present Address: Zollamstr. 45, 67663 Kaiserslautern, Germany

G. Bonnet
SPHERON-VR AG, Hauptstrasse 186, 67714 Waldfischbach-Burgalben, Germany

L.P. Yatsenko (✉)
Institute of Physics, National Academy of Sciences of Ukraine, Prospect Nauki 46, 03028 Kyiv, Ukraine
e-mail: leonid.yatsenko@gmail.com

K. Bergmann
Department of Physics, Technical University Kaiserslautern, 67653 Kaiserslautern, Germany

OPTIMAS Research Center, Technical University Kaiserslautern, 67653 Kaiserslautern, Germany

D. Meschede et al. (eds.), *Exploring the World with the Laser*,
https://doi.org/10.1007/978-3-319-64346-5_38

counting that allows convenient choice of either accuracy or speed of measurements simply by changing the duration of the interval during which the frequency is measured by counting.

1 Introduction

Following up on earlier work in theory [1, 2] and experiment [3, 4], we report new results using the concept of ranging based on a frequency-shifted feedback laser (FSF laser). Central to these concepts is the use of a frequency comb. Therefore, our work appears fitting for this volume honoring T. Hänsch at the occasion of his 75th birthday. However, the frequency comb used in our work is created by a cavity-internal acousto-optic frequency shifter (AOFS) and thus differs from the type of optical comb for which T. Hänsch received the Nobel Prize [5]. Our work is directed toward the development of systems which are suitable for industrial applications, i.e., which are accurate, robust, and economic.

In Sect. 2, we present work based on a fiber-laser system aiming at high accuracy of the measurement over distances of the order of meter. An accuracy $\delta z \approx 1$ μm is documented ($\delta z/z \approx 10^{-6}$), with an expected limit of the achievable accuracy <100 nm ($\delta z/z < 10^{-7}$).

In Sect. 5, we present work based on a semiconductor laser aiming at a high rate of measurements. We document the potential to achieve a measurement rate up to 1 MHz with mm accuracy. At a lower measurement rate, the accuracy of measurements with that system approaches 10 μm.

1.1 The FSF-Laser Concept

The history of work on lasers with cavity-internal frequency shift goes back many decades, at least to the early 1970, when a Stanford group [6] later including also T. Hänsch [7] showed that insertion of an AOFS into the laser cavity allowed fast wavelength tuning. That work led to the development of a new type of modelocking: Kowalski et. al. [8] showed that robust modelocking occurs when the AOFS frequency shift equals the free-spectral range of the cavity. This was demonstrated first in a passive cavity [8] and later with an active cavity [9, 10]. The group at Golden [11, 12] was also the first to start characterizing the optical output spectrum of a FSF laser.

The latter work triggered similar experimental and theoretical efforts in Kaiserslautern [13–16] and later in Sendai [17, 18]. In parallel, the group of T. Hänsch continued to further develop the scheme as a means for tuning laser wavelength [19] or bandwidth control [20]. Aspects of the nonlinear dynamics of a FSF laser were analyzed via simulation studies in [21–23] or in experiments [24]. More recent work looking at the properties of FSF lasers can be found, e.g., for fiber lasers doped with Erbium [25, 26] or Thulium [27], or for semiconductor lasers in [28, 29].

A comprehensive general theory of the FSF-laser operation was first presented in [30] followed by work with special emphasis on the coherence properties of the optical output [31]. Further detailed analysis of FSF-laser properties and some new applications was presented by the Grenoble group [32–38].

1.2 Other Applications of FSF Lasers

Although our present work treats *ranging* based on FSF lasers, we briefly mention a few other applications of such lasers. Early work using specific properties of the FSF laser concerned the mechanical action of light on atoms, such as slowing and cooling of atoms in a beam [39] following up on a proposal by [40] or light-induced drift separation of Rb isotopes [41]. Further applications include accurate frequency-interval measurements [42]; efficient optical pumping of atoms [43, 44]; efficient excitation of atoms in the higher atmosphere for preparation of a guide star [45–47]; observation of dark resonances in optical excitation of atoms [48]; realization of a FSF laser with high [49] or ultrahigh and variable [50] pulse rate; control of pulse duration [51] or new schemes for frequency stabilization [52]; rapid tuning of frequency [53]; and generation of a frequency comb for specific tasks in underwater sensing [54].

1.3 Laser-Based Distance Measurements

Numerous methods for laser-based distance measurement are known. Several of them are implemented in commercial devices. Here, we only mention a few reviews that cover many of the well-known techniques at various levels of detail [55–59]. An interesting line of more recent development is based on Hänsch-type frequency combs, often with sub-mm accuracy, some of them developed for applications over very long distances in space (e.g., [60–67]). One common feature of most of the techniques discussed in those reviews is the need to accurately measure, in one way or another, intensities returned from the object. Ambient light reaching the detector may therefore be detrimental to the achievement of high sensitivity or accuracy. To conclude this brief overview of past work, we mention early work [68] suggesting ranging based on interferometry with phase-modulated light. Because this approach relies on quantitative measurement of light intensity-modulation amplitude, it has not found widespread application for ranging.

1.4 Ranging Based on FSF Lasers

An early application of a FSF laser for distance measurement was reported by the Sendai group [69] using a Nd:YAG laser followed by later work [70, 71] including the application to the monitoring of bridge movements over 100 m range distances [72]. A powerful modification of the FSF-ranging scheme was introduced by the Kaiserslautern group through theoretical work [1] followed by experimental demonstrations using lasers with an Ytterbium-doped fiber [3, 73] or Erbium-doped fiber [4] as gain medium. The theoretical frame of that work is discussed in some detail in Sect. 3. Further contributions to the scheme came from [74] and in particular from the Grenoble group [36]. Applications using a variety of gain media, e.g., Titanium-Sapphire [75] and distributed feedback systems [76, 77] have also been reported.

A significant feature that sets FSF laser-based ranging technique apart from more traditional methods is the fact that the distance of an object is deduced from a frequency measurement.

2 The Fiber-Laser System: Overview

2.1 The Ytterbium System

The characteristic elements of the FSF laser used for this experiment are shown in Fig. 1. The essential components are a broadband gain medium (here a Yb^{3+}-doped optical fiber), an intra-cavity acousto-optic frequency shifting element (AOFS), a pump laser (diode laser), and an output port. The AOFS is inserted into the ring via fiber pigtails adjusted such that only the frequency-shifted first diffraction order of the AOFS is allowed to continue circulating in the ring cavity. A specific and crucial feature of this setup is a single-frequency CW laser, the radiation of which is injected into the ring cavity after passing through an electro-optical phase modulator [80].

The radiation reflected from the reference surface and the time-delayed comb scattered from the object are superimposed on the detector leading to a multi-component signal, see Eqs. (8) and (9) below. The follow-up electronics looks at the amplitude of the ac-component of the signal at frequency Ω. The modulation frequency Ω_0 at which this signal is maximal yields—according to Eq. (13) below—directly the distance to the object. Thus, a measurement proceeds by tuning the modulation frequency Ω over a suitable frequency range. During the scan, the frequency Ω_0 is determined.

Ranging with an accuracy of the order of 1 μm requires a broad optical spectrum which can be realized by, e.g., a fiber-laser system as used in this experiment. Yb^{3+}-doped optical fibers have a large gain bandwidth due to strongly broadened laser transitions in glasses. Ytterbium ions Yb^{3+} have a simple electronic level structure,

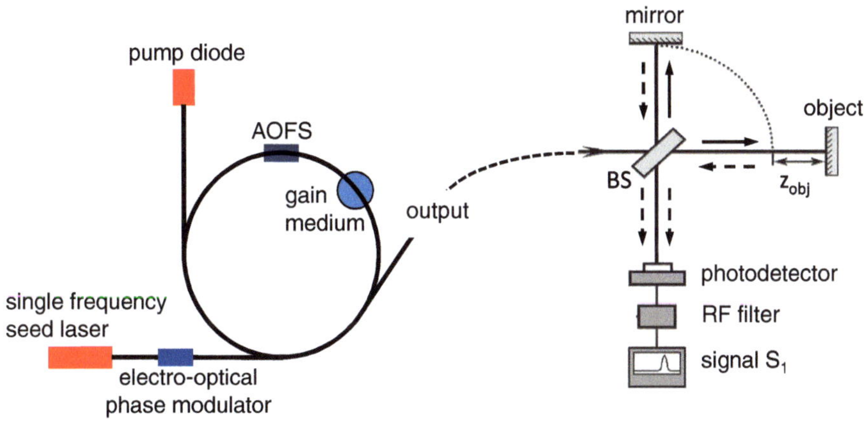

Fig. 1 On the *left side*, the schematics of the FSF ring laser are shown, with a Ytterbium-doped fiber as gain medium, an acousto-optic frequency-shifter AOFS, a pump diode, and a single-frequency (≈1 MHz bandwidth) seed laser injected into the fiber ring after passing an electro-optical phase modulator. The laser output is launched into the ranging setup, shown at the *right side*, with reference mirror, object, beam splitter (BS), photodetector, RF-filter, and evaluation electronics. The central frequency of the RF-filter coincides with the variable frequency $\Omega(t)$ of phase modulation

with only one excited state manifold ($^2F_{5/2}$) which can be exited from the ground-state manifold ($^2F_{7/2}$) with near-infrared radiation. The level scheme has a quasi-three-level character. Compared to the semiconductor gain medium (see Sect. 5) Yb $^{3+}$-doped fiber gain has a much larger saturation intensity and longer upper-state lifetime (typically of the order of 1–2 ms).

All components are spliced together leaving no adjustable elements. Figure 1 shows also the schematics of the Michelson arrangement implemented for detecting the ranging signal. An extra mirror for the reference path may not be needed if the reflection from the end facet of the fiber, which delivers the radiation to the object, serves as reference beam.

2.2 *Comparison with FMCW-Ladar*

At first glance, the FSF-laser ranging concept appears similar to the frequency-modulated scheme for laser detection and ranging (FMCW-Ladar, see e.g., [58, 78, 79]), a technique which allows measuring a beat frequency resulting from single-frequency laser radiation, the frequency of which is linearly changed in time with a chirp rate γ_c periodically repeated in a saw-tooth pattern. That radiation is sent via a beam splitter to a reference surface and the object before being recombined at the detector. Any difference z_{obj} in path length to the reference mirror and the object results in a beat frequency Ω_{beat} from which z_{obj} is deduced according to

$$z_{\mathrm{obj}} = \frac{c}{2}\frac{\Omega_{\mathrm{beat}}}{\gamma_{\mathrm{c}}}. \tag{1}$$

Obviously for high accuracy a large Ω_{beat} and thus a large γ_{c} (a steep chirp) is wanted. Furthermore, high accuracy requires a strictly linear variation of frequency with time.

However, the physics underlying the generation of a ranging signal by a FSF laser differs from the physics of FMCW-Ladar, see Sect. 3.

3 Theoretical Background

3.1 The FSF-Laser Field

Following [30], the output field $E(t)$ of the FSF laser, seeded by laser radiation of amplitude $\mathcal{E}_0$, frequency ω_s and phase $\varphi_s(t)$, is a sum of many discrete components

$$E(t) = \sum_{n=0}^{\infty} \mathcal{E}_n \exp[-i(\omega_s + n\Delta)t - i\Phi_n - i\varphi_s(t - n\tau)], \tag{2}$$

because the seed laser is strong enough to dominate over any seeding by spontaneous emission within the gain medium. The frequency of component n is shifted from ω_s by $n\Delta$ where Δ is the AOFS-induced frequency shift per round trip. The phase of this component is given by the phase φ_s of the seed laser taken at the time $t_0 = t - n\tau$ (which is the time when the radiation relevant for this component entered the cavity), augmented by the additional phase shift Φ_n due to the propagation within the cavity,

$$\Phi_n = -n\tau[\omega_s + (n+1)\Delta/2], \tag{3}$$

where τ is the cavity round-trip of the FSF laser. Saturated gain from the amplifying medium and loss from any spectral filter that may act in the cavity alter the original seed amplitude $\mathcal{E}_0$ to the values $\mathcal{E}_n$. The distribution of amplitudes $\mathcal{E}_n$ of the discrete spectral components is well described [30] by a Gaussian

$$\mathcal{E}_n = \mathcal{E}_0 \exp\left[-\frac{(n - n_{\mathrm{max}})^2}{n_W^2}\right], \tag{4}$$

centered at the component n_{max} where the intensity reaches its maximum. The parameter Δn_{max} is the shift of the maximum of the optical spectrum of the FSF laser from the frequency of the seed laser. The parameter n_W is the width of the Gaussian distribution of amplitudes $\mathcal{E}_n$. Accordingly, the width of the optical spectrum is determined by the distribution of intensities $\propto |\mathcal{E}_n|^2$ and is equal to $\Delta n_W/\sqrt{2}$. Typically we have $n_{\mathrm{max}} \sim n_W$ and $10^3 < n_W < 10^4$.

The phase of the seed laser is periodically changed by an electro-optical modulator (EOM) according to

$$\varphi_s(t) = \beta \sin(\Omega t + \vartheta), \tag{5}$$

where Ω is the modulation frequency, β the modulation index, and ϑ an (irrelevant) initial phase. The frequency Ω of this phase modulation is a crucial parameter in the experiment.

3.2 The Structure of the Ranging Signal

The fields returned from the reference surface and from the object are combined at the detector. The signal delivered by the detector depends on t, Ω and T, where T is the delay of the time of arrival of radiation coming from the reference surface and from the object. The signal is evaluated from

$$S(t;\Omega,T) = \left\langle |E(t;\Omega) + E(t-T;\Omega)|^2 \right\rangle, \tag{6}$$

where $\langle\,\rangle$ means averaging over a period long compared to the optical cycle ω_s^{-1} but short compared with the modulation cycle Ω^{-1}.

The signal $S(t;\Omega,T)$ is a superposition of many harmonics of the modulation frequency [1]

$$S(t;\Omega,T) = \sum_l S_l(\Omega,T)\exp(il\Omega t). \tag{7}$$

We use an electronic filter to restrict detection of the signal at the modulation frequency Ω yielding $S_1(\Omega,T)$. Using Eqs. (2), (3), (5), and (7), we obtain

$$S_1(\Omega,T) = J_1(X)\sum_n |\mathcal{E}_n|^2 \left[P_n^- - P_n^+\right] \tag{8}$$

with

$$P_n^{(\pm)} = \exp\{-i[n(\Omega\tau \pm T\Delta) - \psi]\}. \tag{9}$$

Here, $J_1(X)$ is a Bessel function and $X = 2\beta\sin(\Omega T/2)$. The phase

$$\psi = \omega_s T - (\Omega T + \pi)/2 + \theta \tag{10}$$

does not depend on n. The term $n\Omega\tau$ originates from the propagation within the FSF cavity and $nT\Delta$ results from the accumulated AOFS-induced shift of the optical frequency of the nth component of the field, Eq. (2).

In the experiment, the signal $|S_1(\Omega,T)|$ is measured as a function of Ω. For each n, we have two contributions to $S_1(\Omega,T)$ with amplitude according to Eq. (4). The phase factors $P_n^{(\pm)}$, Eq. (9), depend on n, the system parameters τ and Δ and the variables Ω and T. As n increases from $n=1$ to $n>n_{\max}$ ($n_{\max}\approx 10^4$), the phases $n(\Omega\tau \pm T\Delta)-\psi$ of $P_n^{(\pm)}$ will go, for given Ω and T, through very many cycles of 2π, and thus, their contribution to $|S_1(\Omega,T)|$ will be small unless we have either $\Omega\tau - T\Delta = 2\pi m$ or $\Omega\tau + T\Delta = 2\pi k$ with integer m and k. In that case, the phases of either the first term, associated with P_n^-, or the second term, associated with P_n^+, will be equal to $2\pi mn-\psi$ or $2\pi kn-\psi$, respectively, meaning that P_n^- or P_n^+ is in effect independent of n.

Assuming $T>0$ (as in our experiment), we realize that for the modulation frequency $\Omega=\Omega_q^+$ given by

$$\Omega_q^+ = \gamma_c T + q\Delta_{\text{FSR}}; \quad q=-M_0,\, -M_0+1,\, \ldots \tag{11}$$

the terms of Eq. (8) associated with P_n^+ tend to sum up to a very small value, if not zero. However, the coefficients P_n^- are independent of n and the sum reaches a possibly large absolute value $S_{\max}$. In Eq. (11) $\gamma_c=\Delta/\tau$ is the chirp rate in the moving comb model of FSF laser [30, 81], $\Delta_{\text{FSR}}=2\pi/\tau$, and M_0 is the integer part of $T\Delta/2\pi$. Similarly, for the modulation frequency $\Omega=\Omega_p^-$ given by

$$\Omega_p^- = -\gamma_c T + p\Delta_{\text{FSR}}; \quad p=M_0+1, M_0+2,\, \ldots \tag{12}$$

the terms of Eq. (8) associated with P_n^- tend to sum up to a very small value. In this case, the coefficients P_n^+ are independent of n and the absolute value of the sum reaches the same large value $S_{\max}$.

The electronics for the data analysis tracks the detector signal $|S_1(\Omega,T)|$, for a given $T=2z_{\text{obj}}/c$. Here, $z_{\text{obj}}=0$ marks the location of the end of the fiber, where the reference laser beam and the beam sent to the object are separated. In the present experiment, this is the end facet of the optical fiber which launches the laser beam toward the object. The modulation frequency $\Omega_{\max}$ at which $|S_1(\Omega,T)|$ reaches its maximum is recorded. For a short distance $T<2\pi/\Delta$ (meaning $M_0=0$), as is typical for our experiment, the first two frequency $\Omega_{\max}$ are $\Omega_0^+=\gamma_c T$ and $\Omega_1^- = -\gamma_c T + \Delta_{\text{FSR}}$. We can use both of these frequencies to find the distance z_{obj}. In the present experiment, we use the resonance at $\Omega_0^+=\gamma_c T$ and find the distance z_{obj} from

$$z_{\text{obj}} = \frac{c}{2}\frac{\Omega_0^+}{\gamma_c}. \tag{13}$$

When the measurement aims at an accuracy of 1 μm the order q or p will usually be known. When this is not the case, the ambiguity can be resolved experimentally, see Sect. 5.

In case of a smooth dependence of the amplitudes $\mathcal{E}_n$ on n, the function $S_1(\Omega,T)$, Eq. (8), represents the Fourier transform of the optical spectrum of the FSF laser.

For a Gaussian optical spectrum, Eq. (4), the variation of the $|S_1(\Omega, T)|$ with $\delta\Omega = \Omega - \Omega_{\max}$ is also Gaussian:

$$|S_1(\Omega, T)| = S_{\max} \exp\left\{-\left[\frac{\delta\Omega^2}{\delta\Omega_0^2}\right]\right\}, \tag{14}$$

where

$$\delta\Omega_0 = \frac{2\sqrt{2}}{n_W \tau}. \tag{15}$$

The method is not sensitive to detrimental influence of ambient light or to a variation of the intensity of the light returned from the object, as long as $|S_1(\Omega, T)|$ is large enough to reliably determine $\Omega_{\max}$.

Furthermore, the accuracy δz of a measurement is determined by the width $\delta\Omega_0$ of $|S_1(\Omega, T)|$. The accuracy $\delta\Omega_{\max}$ of a measurement of $\Omega_{\max}$ is $\delta\Omega_{\max} = \epsilon\delta\Omega_0$ where ϵ is determined by the electronic recording system and laser noise. In our experiment, we typically have $10^{-3} < \epsilon < 10^{-2}$. According to Eqs. (4), (13), and (15), the uncertainty of $\Omega_{\max}$ leads to uncertainty δz of distance

$$\delta z = \epsilon \frac{2c}{\Gamma_{\text{opt}}}, \tag{16}$$

where $\Gamma_{\text{opt}} = n_w \Delta / \sqrt{2}$ is the width of the optical spectrum of the FSF laser. Thus, the attainable accuracy of the distance measurements is given exclusively by the width of the optical spectrum and does *not* depend on distance. Furthermore, because the signal $S_1(\Omega, T)$ is caused by beating the radiation which propagates along the reference path (I_{ref}) and the path to the object (I_{obj}), its amplitude is $\propto \sqrt{I_{\text{ref}} I_{\text{obj}}}$ and thus varies with z_{obj} as $1/z_{\text{obj}}$ rather than as $1/z_{\text{obj}}^2$.

4 Experiment

4.1 The Fiber FSF Laser

The all-fiber Yb^{3+}-doped FSF ring laser (schematically shown in Fig. 1) with a free-spectral range $\Delta_{\text{FSR}} = 113.6$ MHz ($\tau = 8.8$ ns) is a compact transportable system similar to the one used in refs. [3, 4] and [73]. The gain medium of the laser is a 40-cm-long Yb^{3+}-doped active fiber pumped by a diode laser ($\lambda = 976$ nm, $P = 300$ mW) which is coupled into the ring via a wavelength division multiplexer (WDM). The intra-cavity fiber-pigtail acousto-optic frequency shifter increases the frequency of the radiation circulating in the cavity by 200 MHz per round trip, leading to a "chirp rate" of $\gamma_c = \Delta/\tau = 2.3 \cdot 10^{16}$ Hz/s. The single-frequency seed laser is a diode laser in Littmann-Metcalf configuration [82, 83]. The voltage driving the EOM phase modulator is generated by a digital synthesizer.

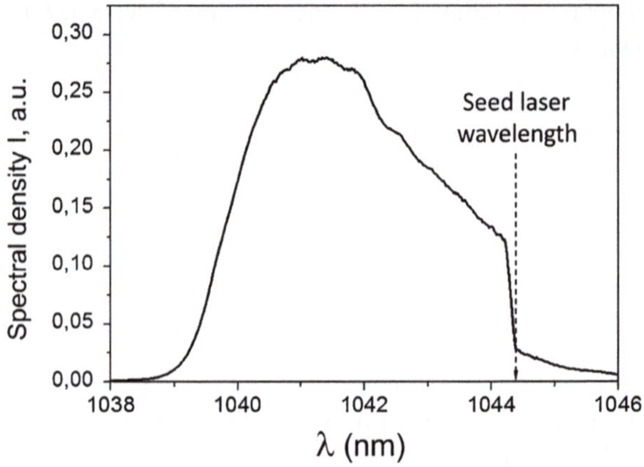

Fig. 2 Spectrum of the FSF-laser radiation recorded with low resolution. The comb structure is not resolved. The seed-laser wavelength is marked. The AOFS shift is to higher frequencies

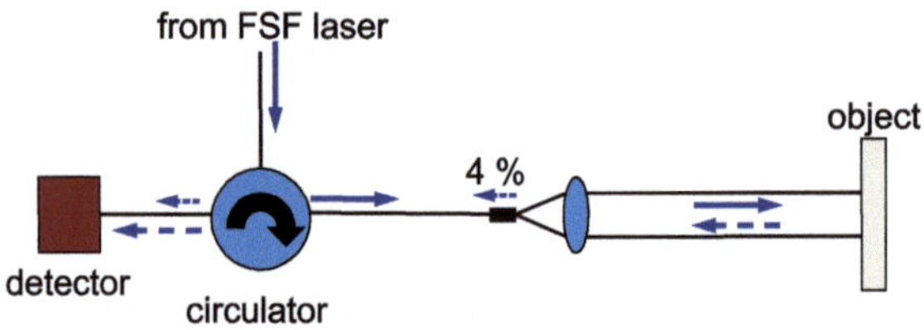

Fig. 3 Details of the interferometer setup. The circulator directs the radiation from the FSF laser via a fiber to the object. The light reflected from the end facet of the fiber serves as the reference beam. The reflected beams are directed by the circulator to the detector

Figure 2 shows a typical spectrum of the FSF-laser radiation with its central wavelength near $\lambda_c = 1041$ nm and width of 5 nm. Experiments are typically done with an output power of $P = 10\,\mathrm{mW}$. Fiber couplers are used to inject the seed laser into the cavity and to couple radiation out of the cavity. All fiber components are connected by fusion splices.

4.2 The Detection System

The modulation frequency $\Omega(t)$ is linearly scanned across an interval $\Delta\Omega$ which includes the resonance frequency according to Eq. (11) or Eq. (12). Typically, $\Delta\Omega$ is set to cover a range of changes of z_{obj} less than 2 mm. The FSF-laser radiation is sent via fibers through a circulator before being launched toward the object. The reference beam is either generated by the reflection of the end facet of the fiber, see Fig. 3, or in a Michelson interferometer setup, see Fig. 1.

The returning radiation is coupled into the same fiber, reaches the circulator again and is directed by it to the fast InGaAs photodetector of area 0.2 mm^2 with fixed gain, a bandwidth of 150 MHz, and a sensitivity range of 800–1700 nm. The detector signal passes an electronically controlled bandpass filter with a high transmission in the vicinity of the modulation frequency Ω before being sent to the evaluation electronics which records the modulation frequency Ω_{max} at which the signal $|S_1(\Omega, T)|$ reaches its maximum. An automatic level-control assures that the evaluation electronics processes a signal of about the same magnitude, within limits independent of the intensity of the light returned from the object.

4.3 *Results*

4.3.1 Accuracy

The data shown in Fig. 4 are taken with the FSF-laser radiation launched toward the object via a telescope (Fig. 4a) which expands the beam diameter to 8 mm before it is focused to a 15 μm spot onto the surface. For testing the accuracy of the system, the object and the focusing lens are both mounted on a platform which is moved to

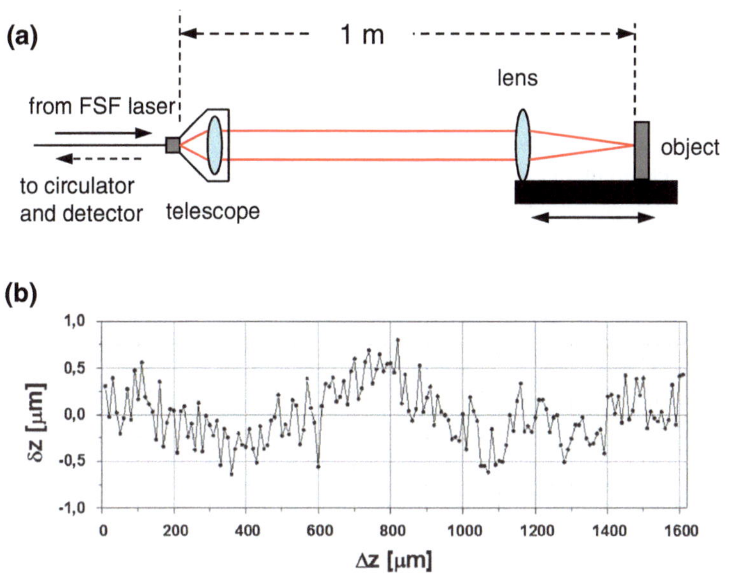

Fig. 4 Accuracy of the FSF laser-based measurement. **a** Optical setup. The object and the focusing lens are jointly mounted on a platform. The position of the platform is varied under the control of a linear encoder with better than 100 nm accuracy. **b** Deviation δz of the result provided by the linear encoder and by the FSF laser-based measurement, while the position of the platform is changed over a range of $\Delta z_{max} = 1.6$ mm. The position $\Delta z = 0$ corresponds to $z_{obj} = 1002, 139$ mm, known from the frequency measurement to an accuracy of 1 μm

change the distance z_{obj}. Thus, the size of the laser spot on the object does not change when z_{obj} is varied. The variation of the position z_{obj} of the platform, and thus the position of the object, is independently monitored with a Heidenhain linear encoder with an accuracy of better than 100 nm. An Abbe error may arise from an angular deviation θ of the direction along which the platform motion is measured from the direction of propagation and the laser beam. Therefore, the angle θ was kept below 0.1°. In this case, the Abbe error does not exceed 10 nm.

The accuracy of the laser-ranging result is documented in Fig. 4b with the object placed at $z_{obj} \approx 1$ m (the exact reading from the frequency measurement for $\Delta z = 0$ is $z_{obj} = 1002, 129$ mm accurate to 1 μm) and moved in steps of 10 μm over the distance $\Delta z = 1.6$ mm, while the variation of the position is measured through both the FSF-laser system and the linear decoder. Figure 4b shows the deviation δz of these two readings. Each point is an average over one hundred measurements, each of 2 ms duration. The deviation δz of the results of the laser measurement from the result of the linear encoder is $\delta z \leq 0.8$ μm ($\delta z/z < 10^{-6}$). The apparently periodic deviation from $\delta z = 0$ by no more than 0.6 μm is caused by diffraction on the focusing lens which can shift Ω_{max} from the value given by Eq. (11). An analysis [Yatsenko and Bergmann, to be published] shows that under uncontrolled conditions (non-ideal focusing, titled object, non-Gaussian FSF-laser spectrum), this deviation may reach the order of μm.

Further characterization of the system is documented in Fig. 5. Here, the distribution of results from 10^3 sweeps of Ω across the bandwidth $\Delta\Omega$ is shown for a fixed distance $z_{obj} = 1002, 756$ mm. The distribution is characterized by a standard deviation of $\sigma = 1.1$ μm. We emphasize that the width of the distribution is independent on the distance z_{obj} to the object. The accuracy of a single measurement, taken during a 2-ms sweep of Ω, is $3\sigma \approx 3$ μm. If time is available for, e.g., a thousand measurements (in a total of a few seconds), the maximum of the distribution, like the one shown in Fig. 5, can easily be determined to less than 10% of its width corresponding to an uncertainty of 100 nm or less. The accuracy of the measurements is determined by the width of the resonance (see Eq. (16)). Like most

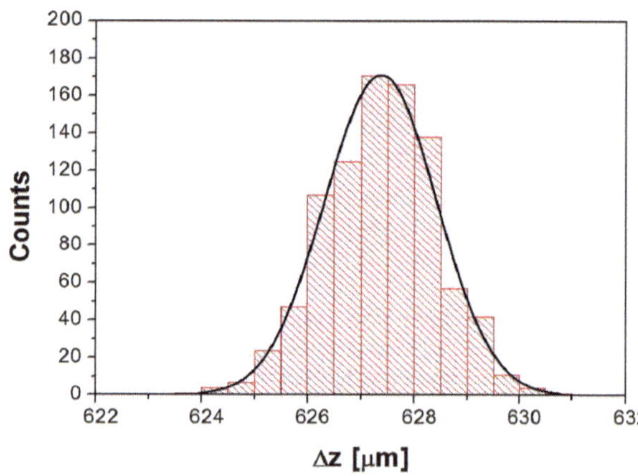

Fig. 5 Distribution of the result of 10^3 FSF laser-based measurements taken at a fixed position near $\Delta z = 627$ μm, see Fig. 4. A Gaussian fit leads for this distribution to the standard deviation of $\sigma = 1.1$ μm

other absolute ranging methods, the FSF-laser ranging technique determines the optical path length between the laser and the object. Therefore, for larger distances, the variation of the refractive index along the path during the measurement could be important. Due to air turbulence, variation of the temperature profile, etc., the air refractive index will deviate from the value $(n_{\rm air} - 1)10^8 = 27{,}403.6$ [84]. According to [67], the relative uncertainty of the distance can be estimated to be about $\delta z_{\rm obj}/z_{\rm obj} = 2 \times 10^{-8}$. For our experimental conditions, the influence of the uncertainty of $n_{\rm air}$ needs to be considered only for $z_{\rm obj} > 5$ m.

4.3.2 Ambiguity

Like in any modulation-based ranging system, FSF-ranging exhibits ambiguity regarding the results. A measured distance is known only modulo the addition of a multiple of the unambiguity range, which—for our Ytterbium system—is of the order of 1 m. The latter range is determined exclusively by the size of the AOFS frequency shift. However, being interested in micron-range accuracy, the prior uncertainty is usually much less than the ambiguity range. Should that not be the case, the procedure discussed in Sect. 5.2 is applicable.

4.3.3 Scanning Object Surfaces

A series of experiments was conducted to demonstrate the high quality of surface topography recording. In this case, the focusing lens, shown in Fig. 4a, is not attached to the table. The scanning of the object in the $x - y$-plane is done with stepper motors in increments of 10 μm. The variation of the position is monitored by linear encoders of the type mentioned above. The coordinate z is measured through the FSF-laser ranging method. The surface topography is reconstructed from the measured coordinates (x, y, z).

Figure 6 shows a step of 110 μm in an otherwise flat blackened metal surface positioned at a distance of $z_{\rm obj} \simeq 1$ m. The surface normal is not oriented exactly parallel to the direction of the incident laser beam axis. The average amplitude of the wiggles marking the boundary between colors is consistent with the accuracy of the order of 1 μm, as documented in Figs. 4 and 5.

Figure 7 shows the surface topography of the letter “E” of a Euro 2-cent coin (see Fig. 8) also taken from a distance of $z_{\rm obj} \simeq 1$ m. Figure 8 shows the same letter “E”, now using the full information content of the data set (see Fig. 7), after rendering, together with a photograph of the coin. The high potential of our approach may be best appreciated by comparison with [85], which shows the topography of an US 1-cent coin recorded with a 4×4 array of detectors, using a time-domain frequency-modulated continuous wave ranging scheme.

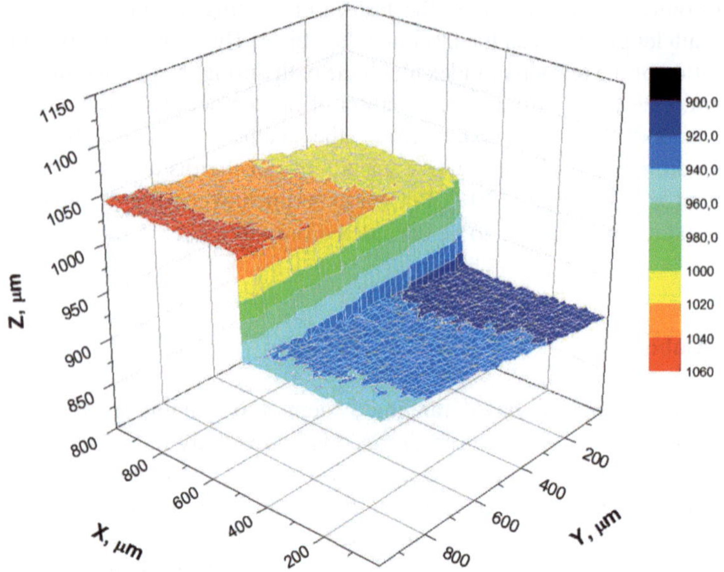

Fig. 6 False *color* representation of the topography of a flat black unpolished metal surface with a 110 µm step, taken from a distance of 1 m. The surface normal was not exactly parallel to the axis of the laser beam. The step size in the *x*- and *y*-direction was 10 µm. The diameter of the laser spot on the surface was 15 µm

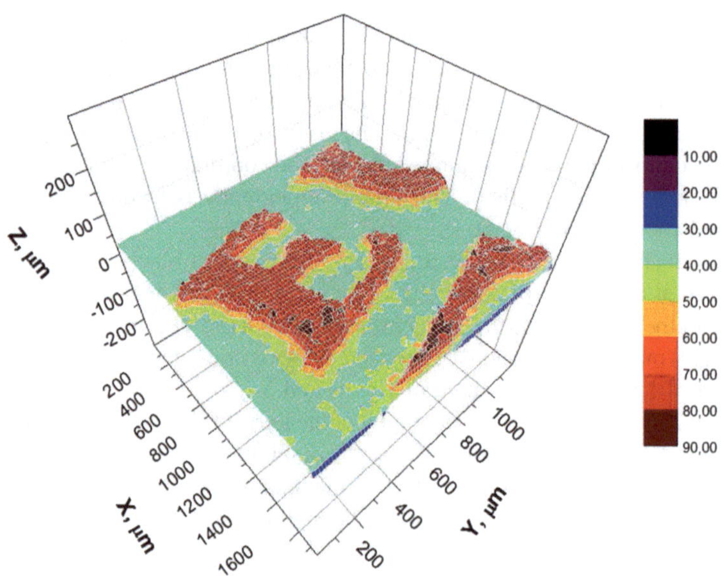

Fig. 7 False color representation of the surface topography of a segment from a Euro 2-cent coin taken from a distance of 1 m. The experimental conditions are the same as for Fig. 6

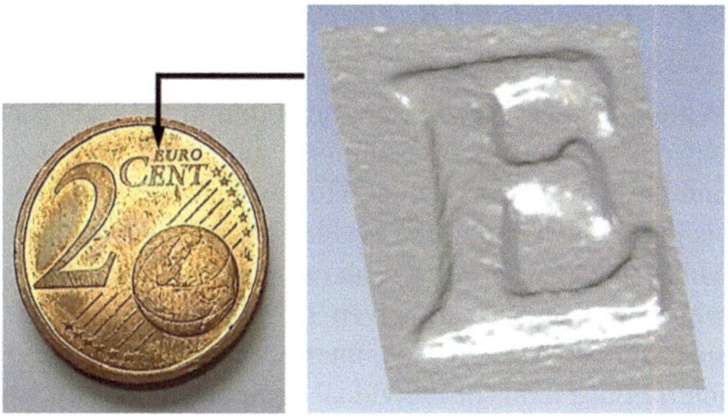

Fig. 8 The *right frame* shows the region of a Euro 2-cent coin (see *left frame*) near the *letter E* as recorded by FSF-laser ranging at a distance of 1 m after rendering, based on the same set of data used for Fig. 7

5 The Semiconductor-Laser System

Many applications are anticipated (e.g., documentation of objects of meter-size or larger) where speed of measurement, rather than the highest possible accuracy is important. The minimum duration $\Delta t_{\min}$ of a measurement of the kind described in the previous section for a fiber-laser system is set by the width of the RF resonance. Recalling (see Eq. (4)) that $n_W = \sqrt{2}\Gamma_{\text{opt}}/\Delta$, we find from Eq. (15) with $\Delta t_{\min} = 1/\delta\Omega_0$

$$\Delta t_{\min} = \frac{\Gamma_{\text{opt}}}{2\Delta}\tau. \tag{17}$$

Thus, for high-speed measurements, it is preferable to work with a smaller optical bandwidth, a large Δ and in particular a short cavity. The lower limit for the length of the cavity is given by the size of the AOFS and the dimensions of the gain medium. When a semiconductor gain medium is used, the size of the cavity can be much smaller compared to a system based on a doped fiber. In fact, we have $\tau = 0.8$ ns ($\Delta_{\text{FSR}} = 1.3$ GHz) for the system discussed below.

However, some properties of a semiconductor gain medium differ distinctly from those of doped fibers. For instance, the upper-state lifetime in a semiconductor system is orders of magnitudes shorter than in a fiber system and the rate of spontaneous emission is accordingly larger. It is therefore difficult, if not impossible, to control the emission spectrum of the FSF laser toward yielding a single-frequency comb by injecting a narrow band laser as successfully done for the fiber system. As a consequence, we do not expect for a semiconductor gain medium the optical spectrum of the FSF laser to consist of a well-defined frequency comb with isolated frequency components separated by Δ. Rather, each photon spontaneously emitted into a cavity mode will be amplified in successive round trips and create a comb with frequency components separated by Δ. Because the spontaneous emission processes occur throughout the spectral profile, the superposition of all the frequency combs started by individual emission processes will lead to a continuum spectrum with no immediately obvious comb structure.

5.1 Theoretical Background

Without external seed laser, the output of the FSF laser is dominated by spontaneous emission processes within the gain medium. These processes are uncorrelated and intrinsically incoherent. However, the fourth-order correlation-function analysis [31], which allows the evaluation of the RF spectrum of the output intensity of a Michelson interferometer, shows that this spectrum comprises a set of doublets, whose frequencies are given by Eqs. (11) and (12). This structure in the RF spectrum of a Michelson interferometer output results from the correlation of interference terms of individual components of a cyclostationary stochastic process [31]. The spectrum carries information about the path length difference of the interferometer arms (the distance to be measured) and was used in early work for ranging [69]. Of interest is the frequency for which the spectral density of the fluctuation reaches a maximum. This approach is similar to the one described for the fiber system in Sect. 2 but is less accurate because the origin of the resonance is noise [30].

Another option, used here, for accessing the information about distance is the direct frequency counting of the quasi-monochromatic output of the RF filter. This approach requires the isolation of only a single resonance in the RF spectrum of the interferometer output (see. Sect. 5.2). This approach allows a convenient choice of either speed or precision of the measurement by simply choosing an appropriate duration of the gate during which the frequency is determined by counting oscillation periods.

The interferometer output intensity filtered near one of the frequencies given by Eq. (11) or Eq. (12) is the quasi-monochromatic signal

$$I(t) = \delta I(t) \cos\left[\Omega_{max} t + \phi(t)\right]. \tag{18}$$

Here, $\delta I(t)$ and $\phi(t)$ are slow random function of time with autocorrelation time $1/\delta\Omega_0$ where $\delta\Omega_0$ is the RF resonance width given by Eq. (15). Theoretical analysis [Yatsenko and Bergmann, to be published] shows that the mean value $\delta I_0 = \sqrt{\langle \delta I^2(t) \rangle}$ of the amplitude of the oscillating part of the interferometer output depends on the width Γ_{opt} of the optical spectrum of the FSF laser and is given by

$$\delta I_0 = 2\sqrt{(1-R)R}|r|\frac{(2\pi)^{1/4}}{\sqrt{\Gamma_{\text{opt}}\tau}} I_0. \tag{19}$$

Here, R is the ratio of the power sent to the reference mirror and the object surface, the coefficient r describes the efficiency of the collection system efficiency, I_0 is the mean output intensity of the FSF laser. The amplitude δI_0 can be quite large especially for the semiconductor gain medium used for our experiment in combination with a short cavity length. For example, for $R = 1/2, r = 1$ (which is the case for a highly reflecting surface), $\Gamma_{\text{opt}} = 200$ GHz, $\tau = 1$ ns, we have $\delta I_0 \simeq 0.045 I_0$ which can be conveniently recorded and evaluated.

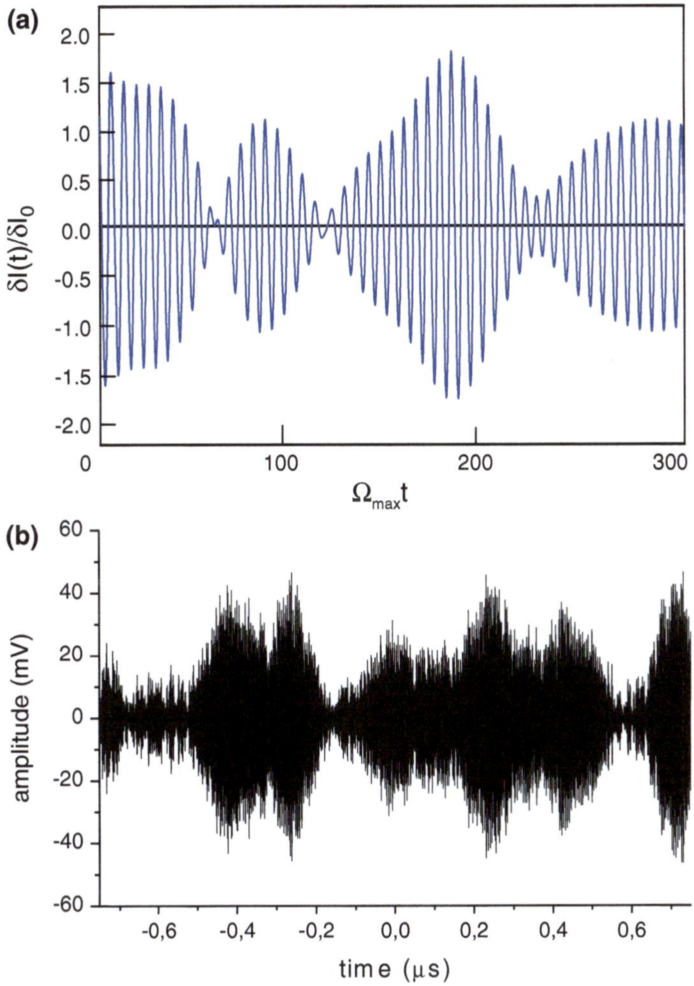

Fig. 9 **a** A simulated realization of the noise process, Eq. (18), for a Gaussian spectrum of width $\Gamma = 0.1\Omega_{\max}$. **b** Experimental trace of the oscillating component of the linear-cavity FSF-laser output (see Sect. 5.3)

However, the noise origin of the oscillatory signal leads to a challenge in the practical implementation of the scheme. Figure 9 shows a simulated realization of the process, Eq. (18), for a Gaussian spectrum with width $\Gamma = 0.1\Omega_{\max}$ and an experimental trace of the oscillating component of the linear-cavity FSF-laser output described in Sect. 5.3. The fluctuation of the amplitude does not prevent the frequency counting except when the signal amplitude is close to zero, as occurs at random times. The short period, during which the amplitude falls below a threshold needed for the counting, must be excluded by electronic means to obtain the highest possible accuracy.

5.2 Elimination of Ambiguity

Interferometric measurements determine the distance only to within an integer multiple of an unambiguity distance z_{ua} [4]. Procedures which allow resolving the ambiguity of the detector reading are well known. The measures required in our case needs a discussion of the RF spectrum of the detector signal.

Because any fluctuation of the laser intensity is likely to experience some recurrence after one round trip of the radiation in the cavity, the RF spectrum of the output of the FSF laser may show a set of spectral components which are separated in frequency by the free-spectral range Δ_{FSR} of the cavity. Such components separated by Δ_{FSR} are schematically shown as long thick vertical lines in Fig. 10.

When the distance of the object, and thus $\gamma_c T$, increases the spectral components shown as short vertical dashed lines in Fig. 10 decrease in frequency, while the frequency of the other components shown as short vertical solid lines in Fig. 10 increase. Each one of these components carries the information about $\gamma_c T$. The structure shown in Fig. 10 is repeated within neighboring frequency intervals of bandwidth Δ_{FSR}. The following discussion deals only with one of the spectral regions, for instance, with the one including the RF components in the range $0 < \Omega/\Delta_{FRS} < 1$. Electronic filters assure that spectral components outside this spectral range are not detected. In the case shown, it happens that $\Omega^+_{q-1} < \Omega^-_p$. However, depending on the value of $\gamma_c T$, we could also have $\Omega^+_{q-1} > \Omega^-_p$. The frequencies of these two components are symmetric about the middle of the given spectral range. For frequency counting to successfully determine T (and thus the distance z_{obj}), only *one* single spectral component is allowed in the spectral range in which the detector plus subsequent electronics has nonzero sensitivity. In order to avoid low-frequency noise, it is best to use an electronic filter which restricts detection to the higher-frequency part of width $\Delta_{FSR}/2$, shown in Fig. 10 as shaded area.

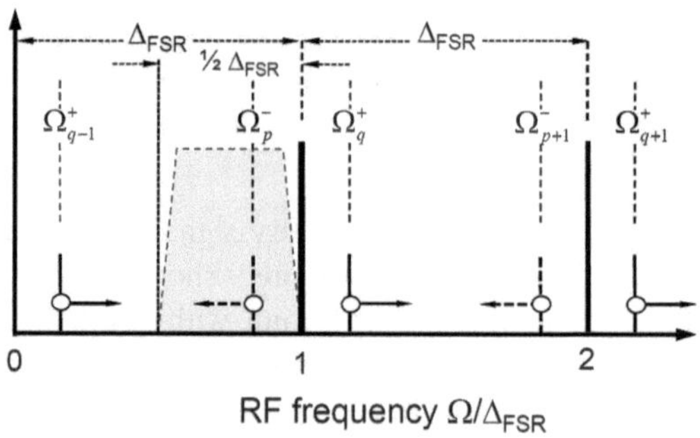

Fig. 10 RF spectrum recorded by the detector. The frequency range to be selected for detection and frequency counting is marked by the *trapezoidal shaded area*. The *arrows* attached to the *short dashed* or *solid lines* indicate the direction of change of their frequency when z_{obj} and thus $\gamma_c T$ increase

We will now discuss the straight forward procedure for (i) the determination of the order p or q, resolving the ambiguity issue and (ii) how to deal with the situation when $\gamma T \approx M\Delta_{\mathrm{FSR}}/2$ (M is an integer number), i.e., when the spectral components Ω^+_{q-1} or Ω^-_p fall near the edge of the sensitive range set by the electronic filter. The latter (problematic) situation is easily overcome by installing a second reference path. The light returned from this additional reference surface is superimposed with a fraction of the light returned from the object on a second detector. The length of the second reference path is adjusted such that the spectral components are shifted by about an odd integer of $\Delta_{FSR}/2$. With this arrangement, there is always one detector for which the relevant spectral feature is not close the edge of the filtered bandwidth.

The determination of the order p or q is easily done by repeating a given measurement with a slightly different AOFS frequency Δ. From Eq. (11) and recalling that $\gamma_c = \Delta/\tau$, we determine the variation $\delta\Omega^\pm$ of $\Omega^\pm$ when the frequency shift Δ is changed by $\delta\Delta$ and arrive for Ω^+ at

$$q = \frac{\Omega^+}{\Delta_{\mathrm{FSR}}}\left(1 + \frac{\Delta}{\Omega^+}\frac{\delta\Omega^+}{\delta\Delta}\right), \tag{20}$$

and accordingly for Ω^- and the order p, starting from Eq. (12). Because Δ and Δ_{FSR} are known, measuring $\Omega^\pm$ and $\delta\Omega^\pm/\delta\Delta$ will lead to the value of the wanted order. Experiments have confirmed that a variation of $\delta\Delta/\Delta \le 10^{-3}$ suffices to reliably determine the order p or q. The sign of $\delta\Omega^\pm/\delta\Delta$ will reveal whether we detect a component of type Ω^+ or Ω^-.

We conclude this discussion by determining the unambiguity range z_{ua} for our experimental conditions. As discussed above, the determination of z_{obj} is unambiguous when $\Omega = \gamma_c T \le \Delta_{\mathrm{FSR}}/2$, what leads with $T = 2z_{\mathrm{obj}}/c$ to

$$z_{\mathrm{ua}} = \frac{c}{4\Delta}. \tag{21}$$

We also note an alternative approach for resolving the ambiguity issue used in [36], an approach which is specific to the FSF-laser system developed for that work. Because that laser radiation involves pulses, the distance is measured in a first step sufficiently accurate by the time-of-flight method to determine the integer number p or q.

5.3 Experiments

The experimental setup is shown in Fig. 11. The semiconductor gain chip (COVEGA TFP780A) working at 780 nm has one tilted facet with antireflection coating and one normal facet coated for high reflection. This gain medium and the

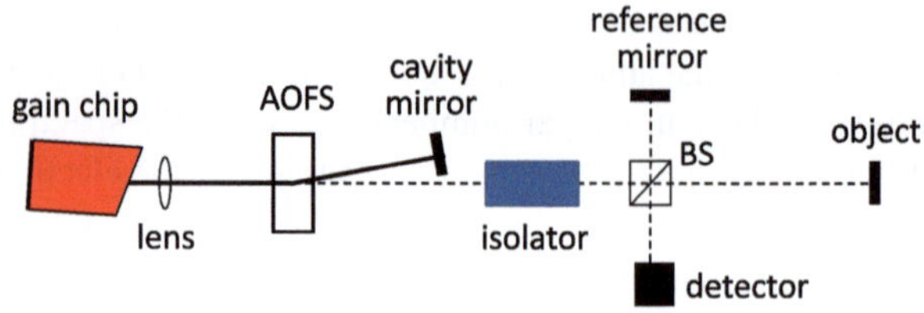

Fig. 11 Schematics of the linear-cavity FSF-laser setup, with a gain chip, a collimating lens, an acousto-optic frequency shifter (AOFS) and an interferometer with 50:50 beam splitter (BS), a reference mirror, the object and the detector. An optical isolator prevents feedback into the laser. For recording the data shown in Figs. 13 and 14, the cavity mirror was replaced by a diffraction grating and the zero-order reflection of that grating used as output beam

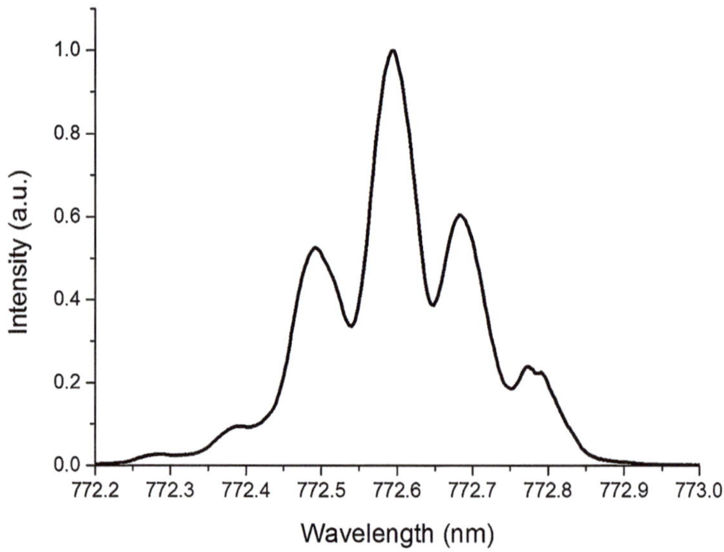

Fig. 12 Optical spectrum of the linear-cavity semiconductor FSF-laser radiation

AOFS with $\Delta_{AOFS} = 80$ MHz frequency shift are placed in a short linear cavity with free-spectral range $\Delta_{FSR} = 1.3$ GHz.

The cavity is closed via the frequency-shifted first diffraction order of the AOFS. Thus, the net frequency-shift per round trip is $\Delta = 2\Delta_{AOFS} = 160$ MHz, leading to a "chirp rate" of $\gamma_c = 2.1 \cdot 10^{17}$ Hz/s. Typically, the undiffracted beam of the latter with a power of a few mW is used for the measurements. A 50:50 beam-splitter cube (BS) sends this beam to both, the reference surface and the object, some 8 cm away. The light, returning from these two surfaces, is recombined at the same BS and sent to the detector. The frequency of the oscillating component of the detector output is measured by frequency counting (Tektronix FCA3003).

Figure 12 shows a spectrum of the linear-cavity semiconductor FSF-laser radiation with its central wavelength near $\lambda_c = 772.6$ nm and an overall width of 0.4 nm. The residual reflection on the output facet of the gain chip leads to the mode structure as shown in Fig. 12, with the separation Δ_{chip} of the peaks given by the dimension of the gain chip. The individual spectral components of width ≤ 0.08 nm are mutually incoherent. Therefore, the bandwidth which determines the accuracy in a ranging experiment is relatively small.

Experiments are typically done with an output power of $P = 10$ mW. Because of the low output power and small effective bandwidth of the optical spectrum, the data shown below serve as proof of principle rather than as convincing documentation of the full potential of this approach. For practical ranging application, power amplification is needed, e.g., by a tapered amplifier. Amplification does not modify the essential spectral properties of the radiation emitted by the FSF laser. Furthermore, suitable modification of the gain chips is likely to lead to a larger coherence bandwidth and thus to a higher accuracy of ranging results.

5.4 Results

The beat frequency, Ω_{beat}, is measured over a range of about 30 mm as shown in Fig. 13a. The variation of the accuracy of the measurement for the fixed distance of $\Delta z = 83$ mm with the gate time during which the frequency is measured by counting oscillation periods is shown in Figs. 13b and 14. Each entry is based on 10^3 individual runs. Figure 13b shows the mean frequency and 1σ-width of the distribution of recorded frequencies, while Fig. 14 shows the data of 13b converted to the 1σ-width of the related distribution of readings for the distance.

An important observation from Fig. 13b is the fact that the mean value of the frequency measurement (for fixed distance), does not change when the gate time is varied over nearly six orders of magnitude. The data shown in Fig. 14 confirm the typical trade-off between accuracy and data collection time. When the gate time is 1 μs (allowing a measurement rate of the order of 1 MHz) the uncertainty of a measurement is 2 mm. That uncertainty shrinks to about 75 μm when the gate time is 1 ms (allowing a measurement rate of the order of 1 kHz). The accuracy continues to improve with further increase of the gate duration and approaches the level of $\sigma = 10$ μm. The influence of systematic errors becomes relevant for gate times longer than 100 ms.

6 Summary and Outlook

We have presented results based on powerful modifications of schemes for ranging using FSF-laser radiation, characterized through (i) the implementation of a novel scheme with injection of phase-modulated single-mode radiation into the cavity for accurate (order of μm) measurements or (ii) by using frequency counting for fast measurements (order of MHz rate). For (i) we used doped fibers as gain medium

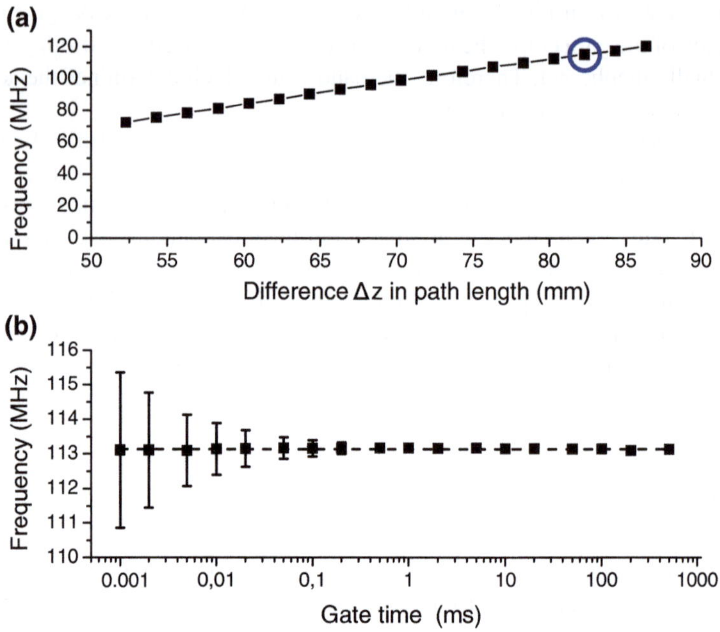

Fig. 13 **a** The variation of the center frequency of the beat signals as a function of the distance Δz. This variation is consistent with the "chirp rate" $\gamma_c = 2.1 \cdot 10^{17}$ Hz/s, given by the AOFS frequency and the cavity roundtrip time. **b** The variation of the 1σ-value of the distribution of results from 10^3 frequency measurements (taken at the distance marked by the *circle* in frame (**a**)) with gate time during which oscillation periods are counted. It is a most important feature that the mean of the measurement does not change when the gate time is varied over nearly six orders of magnitude

exploiting the large gain bandwidth of the laser transitions leading to the large bandwidth (about 5 nm) of the generated frequency comb with smooth envelope. Such a laser is robust and compact and thus suitable for operation in an industrial environment. For (ii) we used a semiconductor gain medium with the high linear gain and saturation power. Such a FSF laser can be realized with a short cavity providing short round-trip time (≤ 1 ns) necessary for high-speed ranging.

Equations (1) and (13) reveal a striking similarity of the FMCW-ladar technique with the FSF-approach. Although the physics behind the signal generation is distinctly different, it is in both cases a parameter γ_c which maps "distance" onto "frequency". In FMCW-ladar, γ_c characterizes the chirp of the radiation with a periodic, saw-tooth like, change of the frequency. For precision of the measurements, γ_c must be constant, meaning that efforts are needed to assure that the chirp is strictly linear across the entire bandwidth which is covered. In FSF-ranging, γ_c is typically (much) larger, meaning that a given distance z_{obj} is mapped onto a (much) larger frequency. Furthermore, because in FSF-ranging γ_c is given by well-controlled parameters (the frequency shift Δ induced by the an acousto-optic frequency shifter and the cavity round-trip time τ), it is straight forward to keep it constant to very high precision. Finally, because the seeding (either by the injected

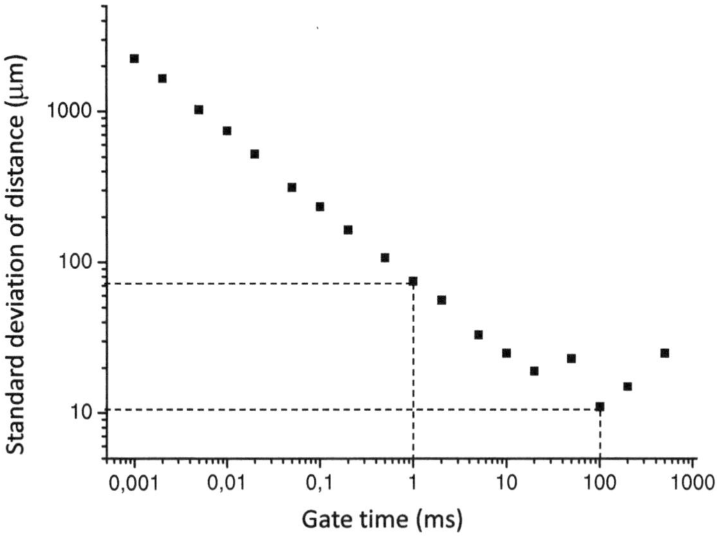

Fig. 14 Variation of the standard deviation σ of distance measurement. We find $\sigma = 2$ mm for a gate time of 1 μs and $\sigma = 80\,\mu\text{m}$ for a gate time of 1 ms. For a gate time > 10 ms the limit of $\sigma \geq 10\,\mu\text{m}$ is approached

radiation or by spontaneous emission within the gain medium) occurs continuously followed by frequency shift across the gain profile, the radiation is truly continuous.

An obvious next step is to go beyond the current arrangement with circular focus and a single detector and use, for even higher speed of scanning surfaces of objects, a cylindrical focus and a linear array of detectors or even a planar array of detectors when the entire surface of the object surface is illuminated by the FSF radiation.

Acknowledgements We acknowledge support from the German "Bundesministerium für Bildung und Forschung" (BMBF) under the projects numbered 13-N-9345 and 13-N-9346. K.B. acknowledges additional support from the research center OPTIMAS of the state of Rhineland-Palatinate. We also thank B.W.Shore for carefully reading the manuscript.

References

1. L.P. Yatsenko, B.W. Shore, K. Bergmann, Ranging and interferometry with a frequency shifted feedback laser. Opt. Commun. **242**, 581–598 (2004)
2. L.P. Yatsenko, B.W. Shore, K. Bergmann, An intuitive picture of optical ranging using frequency shifted feedback lasers seeded by a phase modulated field. Opt. Commun. **282**, 2212–2216 (2009)
3. V.V. Ogurtsov, L.P. Yatsenko, V.M. Khodakovskyy, B.W. Shore, G. Bonnet, K. Bergmann, High accuracy ranging with Yb^{3+}-doped fiber-ring frequency-shifted feedback laser with phase-modulated seed. Opt. Commun. **266**, 266–273 (2006)

4. V.V. Ogurtsov, V.M. Khodakovsky, L.P. Yatsenko, G. Bonnet, B.W. Shore, K. Bergmann, An all-fiber frequency-shifted feedback laser for optical ranging; signal variation with distance. Opt. Commun. **281**, 1679–1685 (2008)
5. T.W. Hänsch, Nobel lecture: passion for precision. Rev. Mod. Phys. **78**, 1297–1309 (2006)
6. W. Streifer, J.R. Whinnery, Analysis of a dye laser using the acousto-optic filter. Appl. Phys. Lett. **17**, 335–337 (1970)
7. D.J. Taylor, S.E. Harris, S.T.K. Nieh, T.W. Hänsch, Electronic tuning of a dye laser using the acousto-optic filter. Appl. Phys. Lett. **19**, 269–271 (1971)
8. F.V. Kowalski, J.A. Squier, J.T. Pinckney, Pulse generation with an acousto-optic frequency shifter in a passive cavity. Appl. Phys. Lett. **50**, 711–713 (1987)
9. F.V. Kowalski, S.J. Shattil, P.D. Hale, Optical pulse generation with a frequency shifted feedback laser. Appl. Phys. Lett. **53**, 734–736 (1988)
10. G. Bonnet, S. Balle, T. Kraft, K. Bergmann, Dynamics and self-modelocking of a titanium-sapphire laser with intracavity frequency shifted feedback. Opt. Commun. **123**, 790–800 (1996)
11. F.V. Kowalski, P.D. Hale, S.J. Shattil, Broadband continuous-wave laser. Opt. Lett. **13**, 622–624 (1988)
12. P.D. Hale, F.V. Kowalski, Output characterization of a frequency shifted feedback laser: theory and experiment. IEEE J. Quant. Electron. **26**, 1845–1851 (1990)
13. I.C.M. Littler, S. Balle, K. Bergmann, Continuous-wave laser without frequency-domain-mode structure: investigation of emission properties and build-up dynamics. J. Opt. Soc. Am. B **8**, 1412–1420 (1991)
14. I.C.M. Littler, S. Balle, K. Bergmann, The cw modeless laser: spectral control, performance data and build-up dynamics. Opt. Commun. **88**, 514–522 (1992)
15. S. Balle, F.V. Kowalski, K. Bergmann, Frequency shifted feedback dye laser operating at a small shift frequency. Opt. Commun. **102**, 166–174 (1993)
16. F.V. Kowalski, S. Balle, I.C.M. Littler, K. Bergmann, Lasers with internal frequency-shifted feedback. Opt. Eng. **33**, 1146–1151 (1994)
17. K. Nakamura, F. Abe, K. Kasahara, T. Hara, M. Sato, H. Ito, Spectral characteristics of an all solid-state frequency-shifted feedback laser. IEEE J. Quant. Electron. **33**, 103–111 (1997)
18. K. Nakamura, T. Miyahara, H. Ito, Observation of a highly phase-correlated chirped frequency comb output from a frequency-shifted feedback laser. Appl. Phys. Lett. **72**, 2631–2633 (1998)
19. P.I. Richter, L. Jakab, T.W. Hänsch, N. Adoph, A cw dye-laser tuned by an acoustooptic filter. Opt. Commun. **84**, 159–161 (1991)
20. P.I. Richter, T.W. Hänsch, Diode lasers in external cavities with frequency-shifted feedback. Opt. Commun. **85**, 414–418 (1991)
21. H. Sabert, E. Brinkmeyer, Pulse generation in fiber lasers with frequency shifted feedback. J. Lightwave Technol. **12**, 1360–1368 (1994)
22. M. Stellpflug, G. Bonnet, B.W. Shore, K. Bergmann, Dynamics of frequency shifted feedback lasers: simulation studies. Opt. Express **11**, 2060–2080 (2003)
23. J.-N. Maran, P. Besnard, S. LaRochelle, Theoretical analysis of a pulsed regime observed with a frequency-shifted-feedback fiber laser. J. Opt. Soc. Am. B **23**, 1302–1311 (2006)
24. J. Martin, Y. Zhao, S. Balle, M.P. Fewell, K. Bergmann, Visible-wavelength diode laser with weak frequency-shifted optical feedback. Opt. Commun. **112**, 109–121 (1994)
25. L.A. Vazquez-Zuniga, Y. Jeong, Study of a mode-locked erbium-doped frequency-shifted-feedback fiber laser incorporating a broad bandpass filter: Experimental results. Opt. Commun. **306**, 1–8 (2013)
26. L.A. Vazquez-Zuniga, Y. Jeong, Study of a mode-locked erbium-doped frequency-shifted-feedback fiber laser incorporating a broad bandpass filter: numerical results. Opt. Commun. **322**, 54–60 (2014)
27. H. Chen, S.P. Chen, Z.F. Jiang, J. Hou, Diversified pulse generation from frequency shifted feedback Tm-doped fibre lasers. Sci. Rep. **6**, 26431 (2016)

28. C. Ndiaye, T. Hara, F.V. Kowalski, H. Ito, Spectral characteristics of a frequency-shifted feedback ring laser using a semiconductor optical amplifier. Jpn. J. Appl. Phys. **47**, 3483–3485 (2008)
29. M. Sellahi, M. Myara, G. Beaudoin, I. Sagnes, A. Garnache, Highly coherent modeless broadband semiconductor laser. Opt. Lett. **40**, 4301–4304 (2015)
30. L.P. Yatsenko, B.W. Shore, K. Bergmann, Theory of a frequency shifted feedback laser. Opt. Commun. **236**, 183–202 (2004)
31. L.P. Yatsenko, B.W. Shore, K. Bergmann, Coherence in the output spectrum of frequency shifted feedback lasers. Opt. Commun. **282**, 300–309 (2009)
32. H.G. de Chatellusand, J.P. Pique, Coherence properties of modeless lasers. PoS **9**, 1 (2009)
33. H.G. de Chatellusand, J.P. Pique, Statistical properties of frequency shifted feedback lasers. Opt. Commun. **283**, 71–77 (2010)
34. H.G. de Chatellus, E. Lacot, W. Glastre, O. Jacquin, O. Hugon, The hypothesis of the moving comb in frequency shifted feedback lasers. Opt. Commun. **284**, 4965–4970 (2011)
35. H.G. de Chatellus, E. Lacot, W. Glastre, O. Jacquin, O. Hugon, Theory of Talbot lasers. Phys. Rev. A **88**, 033828 (2013)
36. J.P. Pique, Pulsed frequency shifted feedback laser for accurate long distance measurements: beat order determination. Opt. Commun. **286**, 233–238 (2013)
37. H.G. de Chatellus, O. Jacquin, O. Hugon, E. Lacot, Quiet broadband light. Phys. Rev. A **90**, 033810 (2014)
38. H.G. de Chatellus, L.R. Cortés, J. Azaña, Optical real-time Fourier transformation with kilohertz resolutions. Optica **3**(1), 1–8 (2016)
39. I.C.M. Littler, H.-M. Keller, U. Gaubatz, K. Bergmann, Velocity control and cooling of an atomic-beam using a modeless laser. Z. Phys. **D18**, 307–308 (1991)
40. J. Hoffnagle, Proposal for continuous white-light cooling of an atomic beam. Opt. Lett. **13**, 102–104 (1988)
41. D.T. Mugglin, A.D. Streater, S. Balle, K. Bergmann, Observation of white light-induced drift separation of Rb isotopes. Opt. Commun. **104**, 165–174 (1993)
42. J.R.M. Barr, G.Y. Young, M.W. Phillipe, Accurate optical frequency-interval measurement by use of nonresonant frequency comb generation. Opt. Lett. **18**, 1010–1012 (1993)
43. M.J. Lim, C.I. Sukenik, T.H. Stiefvater, P.H. Bucksbaum, R.S. Conti, Improved design of a frequency-shifted feedback diode laser for optical pumping at high magentic field optics. Opt. Commun. **147**, 99–102 (1998)
44. M. Cashen, V. Bretin, H. Metcalf, Optical pumping in ^{4} He * with frequency-shifted feedback amplification of light. JOSA **B17**, 530–533 (2000)
45. J.-P. Pique, S. Farinotti, Efficient modeless laser for a mesospheric sodium laser guide star. JOSA **B20**, 2093–2101 (2003)
46. F. Marc, I.C. Moldovan, H.G. de Chatellus, J.P. Pique, High power modeless lasers for sodium laser guide stars. Ann. Phys. **32**, 83–85 (2007)
47. J.-P. Pique, V. Fesquet, S. Jacob, Pulsed frequency-shifted feedback laser for laser guide stars: intracavity preamplifier. Appl. Opt. **50**, 6294–6301 (2011)
48. V.I. Romanenko, A.V. Romanenko, L.P. Yatsenko, G.A. Kazakov, A.N. Litvinov, B.G. Matisov, Y.V. Rozhdestvensky, Dark resonances in the field of frequency-shifted feedback laser radiation. J. Phys. B **43**, 215402 (2010)
49. S. Balle, K. Bergmann, Self-pulsing and instabilities in a unidirectional ring dye laser with intracavity frequency shift. Opt. Commun. **116**, 136–142 (1995)
50. H.G. de Chatellus, O. Jacquin, O. Hugon, W. Glastre, E. Lacot, J. Marklof, Generation of ultrahigh and tunable repetition rates in CW injection-seeded frequency-shifted feedback lasers. Opt. Express **21**, 15065–15074 (2013)
51. M.P. Nikodem, E. Kluzniak, K. Abramski, Wavelength tunability and pulse duration control in frequency shifted feedback Er-doped fiber laser. Opt. Express **17**, 3299–3304 (2009)
52. M.P. Nikodem, K. Abramski, Controlling the frequency of the frequency-shifted feedback fiber laser using injection-seeding technique. Opt. Commun. **283**, 2202–2205 (2010)

53. A. Lyakh, R. Barron-Jimenez, I. Dunayevskiy, R. Go, E. Tsvid, C.K.N. Patel, Progress in rapidly-tunable external cavity quantum cascade lasers with a frequency-shifted feedback. Photonics **3**, 19 (2016)
54. H. Zhang, M. Brunel, M. Romanelli, M. Vallet, Green pulsed lidar-radar emitter based on a multipass frequency-shifting external cavity. Appl. Opt. **55**, 2467–2473 (2016)
55. M.-C. Amann, T. Bosch, M. Lescure, R. Myllyla, M. Rioux, Laser ranging: a critical review of usual techniques for distance measurement. Opt. Eng. **40**, 10–19 (2001)
56. P. de Groot, Unusual techniques for absolute distance measurement. Opt. Eng. **40**, 28–32 (2001)
57. J. Geng, Structered-light 3D surface imaging. Adv. Opt. Photon. **3**, 128–160 (2011)
58. G. Berkovic, E. Shafir, Optical methods for distance and displacement measurements. Adv. Opt. Photon. **4**, 441–471 (2012)
59. P.J. Delfyett, D. Mandridis, M.U. Piracha, D. Nguyen, K. Kim, S. Lee, Chirped pulse laser sources and applications. Prog. Quant. Electron. **36**, 475–540 (2012)
60. I. Coddington, W.C. Swann, L. Nenadovic, N.R. Newbury, Rapid and precise absolute distance measurements at long range. Nat. Photon. **3**, 352–356 (2009)
61. J. Lee, Y.-J. Kim, K. Lee, S. Lee, S.-W. Kim, Time-of-flight measurement with femtosecond light pulses. Nat. Photon. **4**, 716–720 (2010)
62. S.A. van den Berg, S.T. Persijn, G.J.P. Kok, M.G. Zeiouny, N. Bhattacharya, Many-wavelength interferometry with thousands of lasers for absolute distancemeasurement. Phys. Rev. Lett. **108**, 183901 (2012)
63. E. Baumann, F.R. Giorgetta, J.D. Deschênes, W.C. Swann, I. Coddington, N.R. Newbury, Comb-calibrated laser ranging for three-dimensional surface profiling with micrometer-level precision at a distance. Opt. Express **22**, 24914–24928 (2014)
64. Y. Liang, J. Huang, M. Ren, B. Feng, X. Chen, E. Wu, G. Wu, H. Zeng, 1550-nm time-of-flight ranging system employing laser with multiple repetition rates for reducing the range ambiguity. Opt. Express **22**, 4662–4670 (2014)
65. R. Yang, F. Pollinger, K. Meiners-Hagen, M. Krystek, J. Tan, H. Bosse, Absolute distance measurement by dual-comb interferometry with multi-channel digital lock-in phase detection. Meas. Sci. Technol. **26**, 084001 (2015)
66. S.A. van den Berg, S. van Eldik, N. Bhattacharya, Mode-resolved frequency comb interferometry for high-accuracy long distance measurement. Sci. Rep. **5**, 14661 (2015). doi:10.1038/srep14661
67. Y.S. Jang, G. Wang, S. Hyun, H.J. Kang, B.J. Chun, Y.J. Kim, S.W. Kim, Comb-referenced laser distance interferometer for industrial nanotechnology. Sci. Rep. **6**, 31770 (2016). doi:10.1038/srep31770
68. D.J. Webb, R.M. Taylor, J.D.C. Jones, D.A. Jackson, Interferometric optical path difference measurement using sinusoidal frequency modulation of a diode laser. Opt. Commun. **66**, 245–247 (1988)
69. K. Nakamura, T. Hara, M. Yoshida, T. Miyahara, H. Ito, Optical frequency domain ranging by a frequency-shifted feedback laser. IEEE J. Quant. Electron. **36**, 305–316 (2000)
70. H. Ito, T. Hara, C. Ndiaye, Frequency-shifted-feedback laser for precise remote 3D measurement for industry applications. Rev. Laser Eng. Suppl. **2008**(36), 1038–1041 (2008)
71. C. Ndiaye, T. Hara, H. Ito, Performance of a solid-state frequency-shifted feedback laser in optical ranging. JEOS: RP **4**, 09010 (2009)
72. S. Umemoto, M. Fujii, N. Miyamoto, T. Okamoto, T. Hara, H. Ito, Y. Fujino, Deflection measurement for bridges with frequency-shifted feedback laser. in *Proceedings of the Bridge Maintenance, Safety, Management and Life-cycle Optimization*, pp. 2570–2574 (2010)
73. V.V. Ogurtsov, L.P. Yatensko, V.M. Khodakovskyy, B.W. Shore, G. Bonnet, K. Bergmann, Experimental characterization of an Yb^{3+}-doped fiber ring laser with frequency-shifted feedback. Opt. Commun. **266**, 627–637 (2006)
74. K.A. Shore, D.M. Kane, Optimum modulation frequency for FM seeded FSF laser ranging. IEEE Proc.-Optoelectron. **153**, 284–286 (2006)

75. M.F. Brandl, O.D. Mücke, Narrow-linewidth chirped frequency comb from a frequency-shifted feedback Ti: sapphire laser seeded by a phase-modulated single-frequency fiber laser. Opt. Lett. **35**, 4223–4225 (2010)
76. J. Paul, Y. Hong, P.S. Spencer, I. Pierce, K.A. Shore, Simple and accurate optical frequency domain ranging using off-the-shelf dfb lasers subject to frequency-shifted optical feedback. IEEE Photon. Technol. Lett. **19**, 1708–1710 (2007)
77. J. Paul, Y. Hong, P.S. Spencer, I. Pierce, K.A. Shore, Optical frequency-domain ranging using a frequency-shifted feedback distributed-feedback laser. IET Optoelectron **1**, 277–279 (2007)
78. A. Dieckmann, M. Amann, Frequency modulated continous-wave lidar. SPIE Proc. **2271**, 134 (1994)
79. A.B. Mateo, Z.W. Barber, Precision and accuracy testing of FMCW ladar-based length metrology. Appl. Opt. **54**, 6019–6024 (2015)
80. K. Bergmann, L.P. Yatsenko, G. Bonnet, B.W. Shore: Method and device for measuring distance. European patent EP1470621 (2003) and United states patent US 7684019 (2010)
81. K. Kasahara, K. Nakamura, M. Sato, H. Ito, Dynamic properties of an all solid-state frequency-shifted feedback laser. IEEE J. Quantum Electron. **34**, 190–203 (1998)
82. M.G. Littmann, H.J. Metcalf, Spectrally resolved narrow pulsed dye laser without beam expander. Appl. Opt. **17**, 2224–2227 (1978)
83. K. Liu, M.G. Littman, Novel geometry for single-mode scanning of tunable lasers. Opt. Lett. **6**, 117–18 (1981)
84. B. Edlen, The refractive index of air. Metrologia **2**, 71–80 (1966)
85. F. Aflatouni, B. Abiri, A. Rekhi, A. Hajimiri, Nanophotonic coherent imager. Opt. Express **23**, 5117–5125 (2015)
86. J.V. Moloney, J. Hader, S.W. Koch, Quantum design of semiconductor active materials. Laser & Photon. Rev. **1**, 24–43 (2007)

Absorption Spectroscopy of Xenon and Ethylene–Noble Gas Mixtures at High Pressure: Towards Bose–Einstein Condensation of Vacuum Ultraviolet Photons

Christian Wahl, Rudolf Brausemann, Julian Schmitt, Frank Vewinger, Stavros Christopoulos, and Martin Weitz

Abstract Bose–Einstein condensation is a phenomenon well known for material particles as cold atomic gases, and this concept has in recent years been extended to photons confined in microscopic optical cavities. Essential for the operation of such a photon condensate is a thermalization mechanism that conserves the average particle number, as in the visible spectral regime can be realized by subsequent absorption re-emission processes in dye molecules. Here we report on the status of an experimental effort aiming at the extension of the concept of Bose–Einstein condensation of photons towards the vacuum ultraviolet spectral regime, with gases at high-pressure conditions serving as a thermalization medium for the photon gas. We have recorded absorption spectra of xenon gas at up to 30 bar gas pressure of the $5p^6$–$5p^5 6s$ transition with a wavelength close to 147 nm. Moreover, spectra of ethylene noble gas mixtures between 158 and 180 nm wavelength are reported.

1 Introduction

For now approaching 50 years, Theodor Hänsch has revolutionized spectroscopy by the use of lasers. One of the authors (M.W.) had the opportunity to carry out both his doctorate and his habilitation degree work in Theodor Hänsch's laboratory at the Max-Planck-Institute of Quantum Optics near Munich. It is a pleasure to devote this article to Theodor Hänsch on the occasion of his 75th birthday.

For bosonic particles with a nonvanishing rest mass Bose–Einstein condensation to a macroscopically occupied ground state minimizes the free energy when cooled to very low temperature given a sufficient particle density [1–4]. In distinct contrast,

---

This article is part of the topical collection "Enlightening the World with the Laser" - Honoring T. W. Hänsch guest edited by Tilman Esslinger, Nathalie Picqué, and Thomas Udem.

C. Wahl • R. Brausemann • J. Schmitt • F. Vewinger • S. Christopoulos • M. Weitz (✉)
Institut für Angewandte Physik, Universität Bonn, Wegelerstraße 8, 53115 Bonn, Germany
e-mail: wahl@iap.uni-bonn.de; weitz@uni-bonn.de

D. Meschede et al. (eds.), *Exploring the World with the Laser*,
https://doi.org/10.1007/978-3-319-64346-5_39

Bose–Einstein condensation usually does not occur for photons [5]. For the perhaps most widely discussed photon gas, blackbody radiation [6, 7], photons disappear in the system walls when cooled to low temperature instead of exhibiting condensation to the ground mode. This is what is expressed by the common statement of the chemical potential vanishing for the photon case. Early theory work has proposed photon Bose–Einstein condensation in Compton scattering of X-rays [8], and Chiao et al. proposed a photon fluid in a nonlinear cavity [9, 10]. More recently, Bose–Einstein condensation of exciton–polaritons, which are mixed states of matter and light in the strongly coupled regime, has been observed experimentally in several groups [11–13]. Collisions of the material part of the polaritons here drive the system into or near thermal equilibrium.

Our group has in 2010 observed Bose–Einstein condensation of photons in a dye-solution-filled microcavity [14–16], a result confirmed more recently by Marelic and Nyman [17]. A cavity with very small mirror spacing here imprints a low-frequency cut-off for photons in the visible spectral range, yielding a range of allowed photon frequencies well above the thermal energy in frequency units. The confinement results in the photon gas becoming two-dimensional, with the longitudinal modal quantum number being frozen.

Thermalization of the cavity photons is achieved by repeated absorption re-emission processes by the dye molecules, equilibrating the photon gas to the rovibrational temperature of the dye, which is at room temperature. The dye molecules here fulfil the Kennard–Stepanov relation, a thermodynamic Boltzmann-type frequency scaling between absorption and emission spectra well known to apply for systems with both upper and lower electronic states rovibrational manifolds in thermal equilibrium [18–20], as here achieved from frequent collisions with solvent molecules in the dye solution. When the thermalization of the photon gas is faster than photon loss through, e.g. the cavity mirrors, photons condense into the macroscopically occupied mode at the position of the low-frequency cut-off on top of a spectrally broad thermal photon cloud. The opposite limit of photons leaving the cavity before they have a chance to thermalize, corresponds to the usual laser case [21, 22].

In contrast to a usual laser in a photon Bose–Einstein condensation set-up, no inverted active medium is required to generate coherent optical emission, and spontaneous emission is retrapped [14]. This makes the system particular attractive for ultraviolet or vacuum ultraviolet coherent optical sources, a spectral regime at which inversion, and correspondingly laser operation, is hard to achieve. We are aware of ongoing other work aiming at high optical frequency coherent emission using quantum interference effects, forming "lasers without inversion" [23, 24]. A possible suitable medium for photon thermalization in the vacuum ultraviolet regime could be high-pressure noble gas samples, which exhibit electronic transitions starting from the electronic ground state in this spectral regime, or both mixtures between atoms or molecules and noble gases. Alkali–noble gas collisions are long known to be extremely elastic [25, 26], and for the case of rubidium-argon gas mixtures at a few 100 bar of argon pressure the applicability of the thermodynamic Kennard–Stepanov scaling between absorption and emission has been validated for the visible rubidium D-lines [27]. The latter results can be well described by a simple model that assumes equilibrium between external degrees of freedom and the internal structure of alkali–noble gas quasi-molecules in the electronic ground and excited state manifolds, as due to the frequent collisions in the dense buffer gas system. For the case of xenon noble gas, vacuum ultraviolet absorption spectra have in earlier work been carried out up to 20 bar pressure [28]. For simple molecules, a large collection of both absorption and partly also emission data exists in the vacuum ultraviolet spectral regime, but mostly only for pure samples or at moderate buffer gas pressures [29–31].

As a first step towards a photon condensate in the vacuum ultraviolet spectral regime, we have started experiments to test for the applicability of the Kennard–Stepanov relation of gas samples, to provide a spectroscopic test of possible photon gas thermalization. Using the emission of a VUV deuterium lamp, we have recorded absorption spectra of xenon gas in the 142–160 nm wavelength range for up to 30 bar of pressure. Moreover, spectra of ethylene molecules have been recorded in the 155–180 nm wavelength range for up to 130 bar of argon buffer gas pressure.

A further motivation for the here described work stems from the possibility to carry out collisional redistribution laser cooling of the dense gas samples. In earlier work of our group, collisional laser cooling has been carried out with rubidium gases subject to near 200 bar argon buffer gas pressure, where frequent rubidium–argon collisions shift the alkali atomic transition into resonance with a far-red-detuned laser beam, while spontaneous decay occurs near the unperturbed resonance frequency [32, 33]. The reaching of a sufficient alkali vapour pressure near 1 mbar here requires for a preheating of the sample cells, while gases that are gaseous at room temperatures, as ethylene, CO_2 or pure noble gases, could allow for redistribution laser cooling starting from room temperature conditions. The electronic transitions of such gases frequently are in the ultraviolet or vacuum ultraviolet spectral regime, as investigated in the here described work.

In the following, Sect. 2 describes the used experimental set-up, and Sect. 3 gives obtained spectroscopic data for the case of xenon gas. Further, in Sect. 4 we present results obtained with molecular ethylene gas both for the case without and with noble gas buffer gas. Finally, Sect. 5 gives conclusions.

2 Experimental Set-Up

A schematic of the used experimental apparatus for VUV absorption measurements is shown in Fig. 1a. The optical set-up including the high-pressure cell containing the investigated gas sample is placed in a vacuum chamber evacuated to 5×10^{-5} mbar. As a light source for the spectroscopy set-up we use a 200 W deuterium lamp (model: Heraeus D200VUV), whose optical emission spans the spectral range from 120 to 400 nm. Its emission is not uniform within this spectral range, but rather exhibits variations by two orders of magnitude, see Fig. 1b. Therefore, a two monochromator set-up is used, which allows for a large dynamic range of the spectroscopy signal. At first, a home-built grating monochromator (see Fig. 1a)

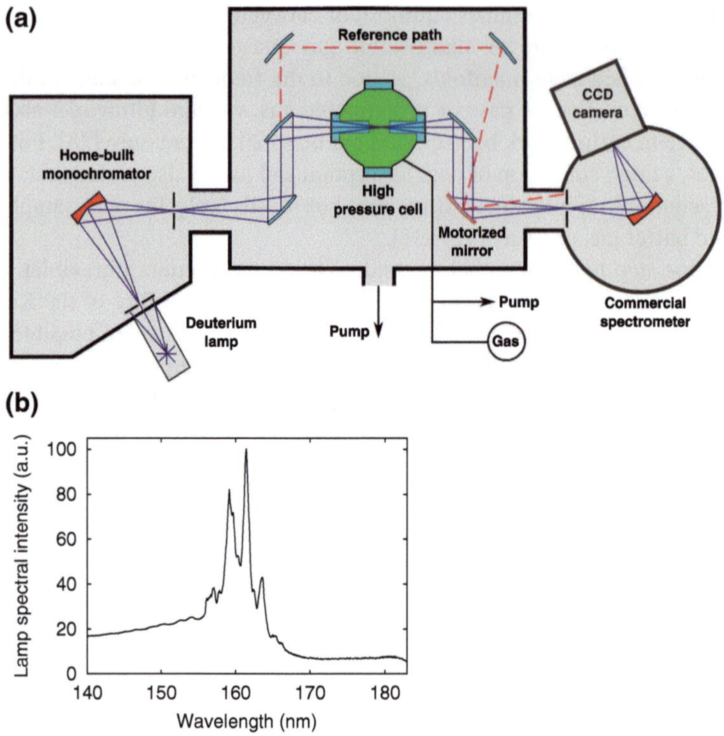

Fig. 1 **a** Schematic representation of the experimental set-up. Light emitted from the deuterium lamp is spectrally filtered by the first monochromator and send through a high-pressure cell containing the gas to be analysed. A fraction of the light passes around the cell and is used to monitor the source output. A flip mirror is used to guide light from one of the two paths into the detection spectrometer. All optical elements in this set-up are coated with aluminium and a protective magnesium difluoride (MgF_2) layer. **b** Deuterium source spectral intensity between 140 and 185 nm

provides initial wavelength selection of the source, which besides the enhancement of the possible dynamic range of the spectroscopy set-up also suppresses transitions in the gas sample driven by radiation with wavelength outside of the detection window. The home-built monochromator is equipped with a grating with 1200 groves/mm with a blaze angle optimized for 200 nm radiation. The concave grating focuses the emission to a first exit slit of 400 μm width. The width of spectral selection of this first monochromator is 2.2 nm. The spectrally filtered radiation is split into a reference and a measurement beam. The latter is focused into a high-pressure cell containing the sample and then directed onto a commercial grating spectrometer (model: McPherson 234/302) consisting of a 10-μm-wide entrance slit and a concave grating with 2400 lines/mm, which images the entrance slit onto an open nose VUV sensitive CCD camera. The spectral resolution of the commercial spectrometer is 0.05 nm. The reference beam is guided around the sample cell and by means of a motorized mirror can be, alternatively to the measurement beam, sent onto the commercial spectrometer. This allows for a normalization of the detected spectrum. This was done for each data point individually since we found the intensity of the detected signal to fluctuate over time by nearly 30% on a 30-min timescale, an issue that is most likely due to polarization variations of the source which lead to intensity changes by the used optical elements, e.g. from a polarization dependent reflectivity of the used mirrors.

The samples investigated in this work include high-purity ethylene, argon, helium, and xenon gases. High gas pressures, above 40 bar, are measured using a 200 bar gauge with a 1 bar resolution. Intermediate pressures in the range between 150 mbar and 40 bar are determined using a digital gauge with a 10 mbar resolution. For pressures below 150 mbar, a fine needle gauge is utilized, giving a resolution of 0.05 mbar. The pressure cell is constructed to house pressures of up to 200 bar. To ensure a largely contamination free environment, the high-pressure cell is repeatedly flushed with argon and evacuated to 10^{-5} mbar. Before each absorption measurement, a reference is taken to ensure that no residual absorption is observed. In order to seal the cell MgF_2 step windows of different step lengths are used, which further allow to vary the absorption length in the cell.

To measure broadband absorption spectra both monochromators are first coupled, set to the same wavelength and subsequently scanned over the whole region of interest. Since the width of the spectral selection of the first monochromator is 2.2 nm, wavelength steps of 1 nm are performed while measuring the absorption and lamp output for each wavelength setting. The resulting signals are then, after being corrected for the current power of the deuterium lamp, used to create a spectrum spanning the region of interest. Broadband absorption spectra are then determined from the Lambert–Beer law.

3 Xenon VUV Absorption Measurements

To begin with, we have recorded xenon absorption spectra around the transition from the $5p^6$ ground state to the $5p^5 6s$ electronically excited state near 147 nm. At the used high xenon gas pressures frequent collisions between atoms occur and the quasi-molecular character in a regime beyond the impact limit determines the line shapes. Figure 2 shows calculated potential curves for the binary xenon system including the relevant levels. The lines in Fig. 3 show experimentally observed absorption data for various xenon pressures in the range of 0.1–30 bar. We observe significant pressure broadening, with the red wing being strongly enhanced with respect to the blue wing. This is well understood from the potential curves shown in Fig. 2, with the ground state for not to small internuclear distances being mostly flat and the excited state decreasing in energy. As the average internuclear distance decreases with increasing pressure lower photon energies are sufficient to excite the transition, leading to an enhanced pressure broadening of the red wing with respect to that of the blue wing. Above a pressure of 0.4 bar, the absorption coefficient at the line centre cannot be resolved due to the here very high optical density.

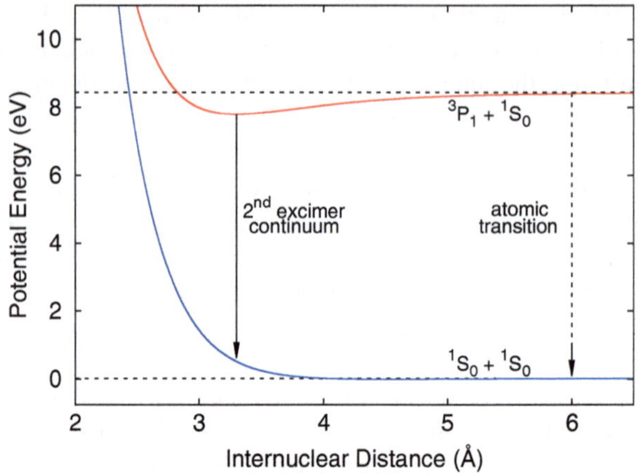

Fig. 2 Calculated energy in the xenon dimer system versus the internuclear distance [34, 35]. The *horizontal dashed lines* give the asymptotic energy levels for infinitely large internuclear distance

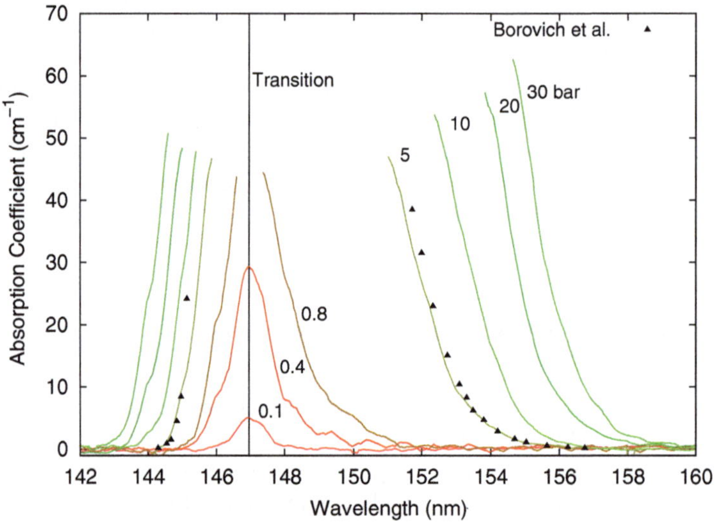

Fig. 3 Absorption coefficients of pure xenon in the wavelength range of 142–160 nm for pressures between 0.1 and 30 bar. Xenon pressures are indicated next to the respective absorption spectrum. Saturated regions are omitted for clarity, and the *black triangles* indicate comparable measurements from Borovich et al. for the 5 bar dataset [28]

Correspondingly, we also cannot determine the linewidth for the high-pressure data, but already at the quoted pressure value of 0.4 bar the linewidth exceeds the thermal energy k_BT in wavelength units of 0.5 nm at room temperature for a wavelength of 147 nm. Our results are in good agreement with earlier measurements by Borovich et al. [28], see the exemplary data points for a pressure of 5 bar in Fig. 3.

The spontaneous emission of the xenon system at pressures above 50 mbar, where the collision rate is sufficient that excimer molecules form faster than spontaneous decay, is known to be dominated by the second continuum around 172 nm, see the corresponding potential curve of Fig. 2. Though a linear extrapolation of our presented data with pressure does not indicate a significant overlap with emission at the second continuum at pressures below values reachable with our cell (200 bar), upon reaching of the critical pressure of 58.4 bar in the xenon system a highly nonlinear increase in the density with pressure does occur, which will strongly influence the spectra. This has been indicated in earlier narrowband optical measurements [36]. We are currently working on corresponding broadband measurements in this regime.

4 Absorption Spectroscopy of Ethylene

An alternative promising approach for a thermalization medium for a photon Bose–Einstein condensate in the VUV spectral regime are simple molecules in high-pressure noble buffer gas environments. Ethylene molecules show electronic transitions in the relevant wavelength range, and the low-pressure regime (below 0.3 mbar) of pure ethylene has previously been studied [29–31]. To begin with, we have investigated corresponding spectra of pure ethylene up to 20 mbar pressure in the wavelength range between 158 and 180 nm, see Fig. 4a. The absorption coefficient scales proportionally with increasing ethylene concentration as expected from Beer's law, while no evident pressure-broadening effect can be observed. Earlier experiments are in very good agreement with the present work, presenting comparatively negligible discrepancies (on the order of 0.1 nm) in the observed peak positions. Figure 4b gives corresponding ethylene spectra both without (bottom) and with argon buffer gas, for buffer gas pressures of up to 130 bar. The ethylene molecular partial pressure here was 1 mbar. One observes that the addition of the buffer gas broadens the individual resonances, while the overall spectral structure is not significantly modified. Additionally, we observe for the three visible doublets with minima at 165.6, 168.4, and 173.5 nm, respectively, that the long wavelength peak is dominant at low buffer gas pressures compared to the short wavelength peak, which is inverted at high buffer gas pressures. When using helium instead of argon as a buffer gas no significant modification of the spectra is observed, see the exemplary curve in Fig. 4b.

To quantify the pressure broadening by the buffer gas we focus on the spectral peak near 174.1 nm. As we cannot unambiguously determine the width of this line due to the many nearby resonances and the asymmetry of the line shapes, we instead use a normalized doublet visibility $V = (\kappa_{\max} - \kappa_{\min})/(\kappa_{\max} + \kappa_{\min})$, where $\kappa_{\max}$ and $\kappa_{\min}$ denote the absorption coefficients on resonance and at the minimum between the resonances of the doublet structure at 172.8 and 174.1 nm, respectively, to quantify the pressure broadening, see also Fig. 4b. The variation of the doublet visibility for both argon and helium buffer gas with pressure is shown in

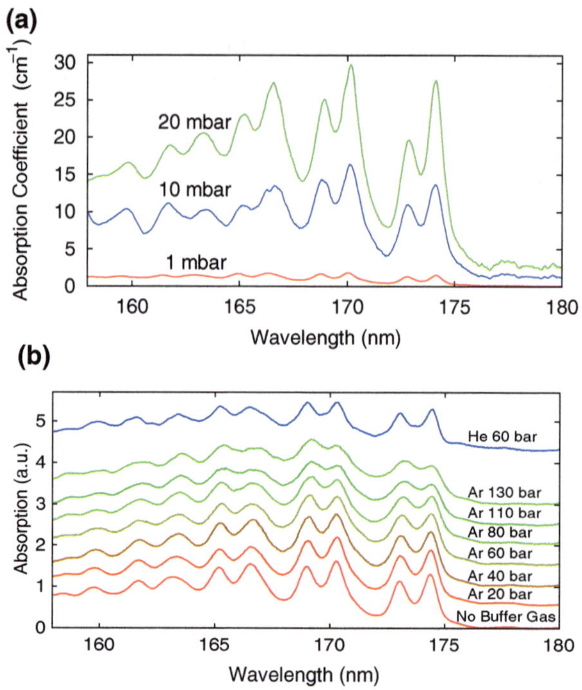

Fig. 4 **a** Absorption coefficients in the wavelength range of 158–180 nm for three different pressures of pure ethylene, 1 mbar (*red*), 10 mbar (*blue*), and 20 mbar (*green*). **b** Absorption spectra for 1 mbar of ethylene with the addition of 20–130 bar argon. Spectra are vertically shifted by 0.5 for clarity. It is observed that the peaks in the ethylene spectrum experience a *red*-shift dependent on argon pressure. Further, a strong decrease in contrast between peak and underlying continuum is apparent. The *top* spectrum (*blue line*) gives an exemplary spectrum for helium buffer gas at 60 bar pressure

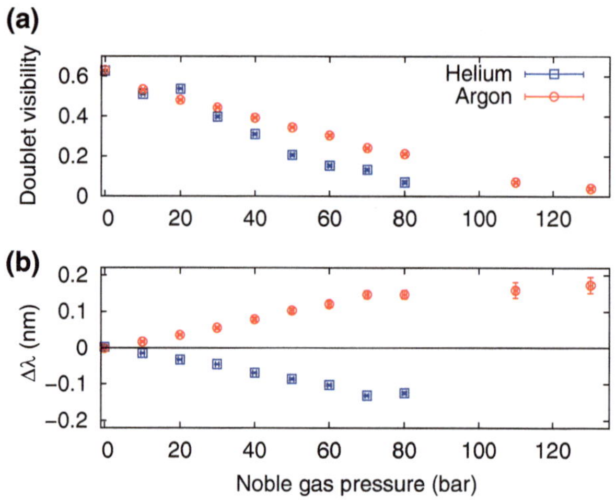

Fig. 5 **a** Extracted visibility doublet and **b** shift $\Delta\lambda$ of the position for the peak found at 174.1 nm in 1 mbar of ethylene with added argon (*red cirlces*) and helium (*blue squares*) buffer gas versus the noble gas pressure. While both noble gases lead to a similar change in visibility, an opposing effect can be seen with respect to the wavelength shift

Fig. 5a. The data show that the loss of doublet visibility with pressure and correspondingly also the pressure broadening is somewhat larger for the case of helium than for argon buffer gas. Figure 5b shows the pressure-induced shift $\Delta\lambda$ of the 174.1 nm resonance for both of the buffer gases, which has been obtained from a fit to the data of the resonance line. We observe a pressure shift of similar magnitude, but of opposite sign for the two noble gases. Similar effects have been reported in earlier works for other systems when using different noble gas perturbers and were attributed to the differing collisional kinetics of the gas mixtures [37, 38]. Here we find that for pressures of up to 60 bar the shift is linear to good approximation, with a slope of 0.1 nm per 50 bar, while for higher pressures a saturation of the shift is observed. At 130 bar argon buffer gas pressure, the observed pressure shift is 0.18 nm, corresponding to $0.29 k_{\mathrm{B}} T$ in wavelength units at room temperature for a wavelength of 174 nm. This shift is roughly a factor of eight smaller than the pressure broadening responsible for the above-discussed reduction in the visibility of the doublet spaced by 1.4 nm.

5 Conclusions and Outlook

To conclude, we have presented absorption spectra of both xenon gas and binary mixtures of ethylene with noble gases at high pressure. Our measurements show a broadening of resonances of order of the thermal energy $k_{\mathrm{B}} T$. For the case of the ethylene data, both the data with and without noble buffer gas can provide valuable input for future calculations of molecular potentials.

In the future, we plan to test for the thermodynamic Kennard–Stepanov scaling of the dense gas samples, for which additionally fluorescence spectra at the corresponding parameters are required. Besides the here presented xenon system, the lighter noble gases offer even shorter wavelength closed electronic transitions, as candidate systems for photon condensates in the deeper ultraviolet spectral regime. Provided that the Kennard–Stepanov relation in the dense system holds, this implies a high quantum efficiency of the involved transitions [39]. Besides photon Bose–Einstein condensation, this would make the system also a promising candidate for collisional redistribution laser cooling starting from room temperature [32].

Acknowledgements We acknowledge support of the Deutsche Forschungsgemeinschaft (within SFB/TRR 185 and We1748-17) and the European Research Council (INPEC). We thank G. Wallstabe for his contributions in the early phase of this experiment.

References

1. A. Einstein, Zur Quantentheorie des idealen Gases. Sitz. ber. Preuss. Akad. der Wiss. **1**, 1–3 (1925)
2. E.A. Cornell, C.E. Wieman, Nobel lecture: Bose–Einstein condensation in a dilute gas, the first 70 years and some recent experiments. Rev. Mod. Phys. **74**, 875 (2002)
3. W. Ketterle, Nobel lecture: when atoms behave as waves: Bose–Einstein condensation and the atom laser. Rev. Mod. Phys. **74**, 1131 (2002)

4. M.H. Anderson, J.R. Ensher, M.R. Matthews, C.E. Wieman, E.A. Cornell, Observation of Bose–Einstein condensation in a dilute atomic vapor. Science **269**, 198 (1995)
5. K. Huang, *Statistical Mechanics*, 2nd edn. (Wiley, Hoboken, 1987)
6. M. Planck, On the law of distribution of energy in the normal spectrum. Ann. Phys. **4**, 1 (1901)
7. S.N. Bose, Plancks Gesetz und Lichtquantenhypothese. Z. Phys. **26**, 178 (1924)
8. Y.B. Zel'Dovich, E. Levich, Bose condensation and shock waves in photon spectra. Sov. Phys. JETP **28**, 11 (1969)
9. R.Y. Chiao, Bogoliubov dispersion relation for a "photon fluid": is this a superfluid? Opt. Commun. **179**, 157 (2000)
10. E. Bolda, R. Chiao, W. Zurek, Dissipative optical flow in a nonlinear Fabry–Pérot cavity. Phys. Rev. Lett. **86**, 416 (2001)
11. J. Kasprzak et al., Bose–Einstein condensation of exciton polaritons. Nature **443**, 409 (2006)
12. H. Deng, H. Haug, Y. Yamamoto, Exciton–polariton Bose–Einstein condensation. Rev. Mod. Phys. **82**, 1489 (2010)
13. R. Balili, V. Hartwell, D. Snoke, L. Pfeiffer, K. West, Bose–Einstein condensation of microcavity polaritons in a trap. Science **316**, 1007 (2007)
14. J. Klaers, J. Schmitt, F. Vewinger, M. Weitz, Bose–Einstein condensation of photons in an optical microcavity. Nature **468**, 545 (2010)
15. J. Klaers, J. Schmitt, T. Damm, F. Vewinger, M. Weitz, Bose–Einstein condensation of paraxial light. Appl. Phys. B **105**, 17 (2011)
16. J. Schmitt, T. Damm, D. Dung, F. Vewinger, J. Klaers, M. Weitz, Observation of grand-canonical number statistics in a photon Bose–Einstein condensate. Phys. Rev. Lett. **112**, 030401 (2014)
17. J. Marelic, R. Nyman, Experimental evidence for inhomogeneous pumping and energy-dependent effects in photon Bose–Einstein condensation. Phys. Rev. A **91**, 033813 (2015)
18. E.H. Kennard, Phys. Rev. **11**, 29 (1918)
19. B. Stepanov, Dokl. Akad. Nauk SSSR **112**, 839 (1957)
20. D.A. Sawicki, R.S. Knox, Universal relationship between optical emission and absorption of complex systems: an alternative approach. Phys. Rev. A **54**, 4837 (1996)
21. P. Kirton, J. Keeling, Nonequilibrium model of photon condensation. Phys. Rev. Lett. **111**, 100404 (2013)
22. J. Schmitt, T. Damm, D. Dung, F. Vewinger, J. Klaers, M. Weitz, Thermalization kinetics of light: from laser dynamics to equilibrium condensation of photons. Phys. Rev. A **92**, 011602 (2015)
23. A. Zibrov, M. Lukin, D. Nikonov, L. Hollberg, M. Scully, V. Velichansky, H. Robinson, Experimental demonstration of laser oscillation without population inversion via quantum interference in Rb. Phys. Rev. Lett. **75**, 1499 (1995)
24. O. Kocharovskaya, Amplification and lasing without inversion. Phys. Rep. **219**, 175 (1992)
25. E. Speller, B. Staudenmayer, V. Kempter, Z. Phys, Quenching cross sections for alkali–inert gas collisions. Z. Phys. A At. Nucl. **291**, 311 (1979)
26. U. Vogl, M. Weitz, Spectroscopy of atomic rubidium at 500-bar buffer gas pressure: approaching the thermal equilibrium of dressed atom-light states. Phys. Rev. A **78**, 011401 (2008)
27. P. Moroshkin, L. Weller, A. Saß, J. Klaers, M. Weitz, Kennard–Stepanov relation connecting absorption and emission spectra in an atomic gas. Phys. Rev. Lett. **113**, 063002 (2014)
28. B. Borovich, V. Zuev, D. Stavrovsky, Pressure-induced ultraviolet absorption in rare gases: absorption coefficients for mixtures of Xe and Ar at pressures up to 40 atm in the vicinity of 147 nm. J. Quant. Spectr. Radiat. Transf. **13**, 1241 (1973)
29. H.C. Lu, H.K. Chen, B.M. Cheng, Analysis of C_2H_4 in C_2H_6 and C_2H_5D with VUV absorption spectroscopy and a method to remove C_2H_4 from C_2H_6 and C_2H_5D. Anal. Chem. **76**, 5965 (2004)
30. M. Zelikoff, K. Watanabe, Absorption coefficients of ethylene in the vacuum ultraviolet. JOSA **43**, 756 (1953)

31. V.L. Orkin, R.E. Huie, M.J. Kurylo, Rate constants for the reactions of OH with HFC-245cb ($CH_3CF_2CF_3$) and some fluoroalkenes (CH_2CHCF_3, CH_2CFCF_3, CF_2CFCF_3, and CF_2CF_2). J. Phys. Chem. A **101**, 9118 (1997)
32. U. Vogl, M. Weitz, Laser cooling by collisional redistribution of radiation. Nature **461**, 70 (2009)
33. S. Yeh, P. Berman, Theory of collisionally aided radiative excitation. Phys. Rev. A **19**, 1106 (1979)
34. O. Dutuit, M. Castex, J. Le Calve, M. Lavollee, Synchrotron radiation study of the molecular xenon fluorescence around 2000 Å. J. Chem. Phys. **73**, 3107 (1980)
35. K.K. Docken, T.P. Schafer, Spectroscopic information on ground-state Ar 2, Kr 2, and Xe 2 from interatomic potentials. J. Mol. Spectrosc. **46**, 454 (1973)
36. H.A. Koehler, L.J. Ferderber, D.L. Redhead, P.J. Ebert, Vacuum-ultraviolet emission from high-pressure xenon and argon excited by high-current relativistic electron beams. Phys. Rev. A **9**, 768 (1974)
37. C. Beer, R. Bernheim, Hyperfine pressure shift of Cs 133 atoms in noble and molecular buffer gases. Phys. Rev. A **13**, 1052 (1976)
38. N. Allard, J. Kielkopf, The effect of neutral nonresonant collisions on atomic spectral lines. Rev. Mod. Phys. **54**, 1103 (1982)
39. J. Klaers, The thermalization, condensation and flickering of photons. J. Phys. B At. Mol. Opt. **47**, 243001 (2014)

Extraction of Enhanced, Ultrashort Laser Pulses from a Passive 10-MHz Stack-and-Dump Cavity

Sven Breitkopf, Stefano Wunderlich, Tino Eidam, Evgeny Shestaev, Simon Holzberger, Thomas Gottschall, Henning Carstens, Andreas Tünnermann, Ioachim Pupeza, and Jens Limpert

Abstract Periodic dumping of ultrashort laser pulses from a passive multi-MHz repetition-rate enhancement cavity is a promising route towards multi-kHz repetition-rate pulses with Joule-level energies at an unparalleled average power. Here, we demonstrate this so-called stack-and-dump scheme with a 30-m-long cavity. Using an acousto-optic modulator, we extract pulses of 0.16 mJ at 30-kHz repetition rate, corresponding to 65 stacked input pulses, representing an improvement in three orders of magnitude over previously extracted pulse energies. The ten times

This article is part of the topical collection "Enlightening the World with the Laser" - Honoring T. W. Hänsch guest edited by Tilman Esslinger, Nathalie Picqué, and Thomas Udem.

S. Breitkopf (✉) • T. Gottschall
Institute of Applied Physics, Abbe Center of Photonics, Friedrich-Schiller-Universität Jena, Albert-Einstein-Str. 15, 07745 Jena, Germany
e-mail: sven.breitkopf@uni-jena.de; thomas.gottschall@uni-jena.de

S. Wunderlich
Institute of Applied Physics, Abbe Center of Photonics, Friedrich-Schiller-Universität Jena, Albert-Einstein-Str. 15, 07745 Jena, Germany

Active Fiber Systems GmbH, Wildenbruchstr. 15, 07745 Jena, Germany
e-mail: stefano.wunderlich@uni-jena.de

T. Eidam
Active Fiber Systems GmbH, Wildenbruchstr. 15, 07745 Jena, Germany
e-mail: eidam@afs-jena.de

E. Shestaev
Institute of Applied Physics, Abbe Center of Photonics, Friedrich-Schiller-Universität Jena, Albert-Einstein-Str. 15, 07745 Jena, Germany

Helmholtz-Institute Jena, Fröbelstieg 3, 07743 Jena, Germany
e-mail: evgeny.shestaev@uni-jena.de

S. Holzberger
Max-Planck-Institute of Quantum Optics, Hans-Kopfermann-Str. 1, 85748 Garching, Germany

Department of Physics, Ludwig-Maximilians-Universität München, Am Coulombwall 1, 85748 Garching, Germany

Present Address: Menlo Systems GmbH, Am Klopferspitz 19a, 82152 Martinsried, Germany
e-mail: s.holzberger@menlosystems.com

D. Meschede et al. (eds.), *Exploring the World with the Laser*,
https://doi.org/10.1007/978-3-319-64346-5_40

longer cavity affords three essential benefits over former approaches. First, the time between subsequent pulses is increased to 100 ns, relaxing the requirements on the switch. Second, it allows for the stacking of strongly stretched pulses (here from 800 fs to 1.5 ns), thus mitigating nonlinear effects in the cavity optics. Third, the choice of a long cavity offers increased design flexibility with regard to thermal robustness, which will be crucial for future power scaling. The herein presented results constitute a necessary step towards stack-and-dump systems providing access to unprecedented laser parameter regimes.

1 Introduction

A number of visionary applications like laser wake-field acceleration of elementary particles [1] or space debris removal [2] ask for a dramatically improved performance of femtosecond laser systems with high repetition rates [3]. In particular, Joule-level pulse energies at average powers in the multi-kilowatt regime with diffraction-limited beam quality are required. This combination of parameters greatly exceeds the capabilities of today's laser systems, and the scalability of the average and of the pulse peak power of single-aperture amplifier solutions does not suffice these demands [4–7]. Current limitations which need to be overcome are

H. Carstens
Max-Planck-Institute of Quantum Optics, Hans-Kopfermann-Str. 1, 85748 Garching, Germany

Department of Physics, Ludwig-Maximilians-Universität München, Am Coulombwall 1, 85748 Garching, Germany
e-mail: henning.carstens@mpq.mpg.de

A. Tünnermann
Institute of Applied Physics, Abbe Center of Photonics, Friedrich-Schiller-Universität Jena, Albert-Einstein-Str. 15, 07745 Jena, Germany

Helmholtz-Institute Jena, Fröbelstieg 3, 07743 Jena, Germany

Fraunhofer Institute for Applied Optics and Precision Engineering, Albert-Einstein-Str. 7, 07745 Jena, Germany
e-mail: andreas.tuennermann@iof.fraunhofer.de

I. Pupeza
Max-Planck-Institute of Quantum Optics, Hans-Kopfermann-Str. 1, 85748 Garching, Germany
e-mail: ioachim.pupeza@mpq.mpg.de

J. Limpert
Institute of Applied Physics, Abbe Center of Photonics, Friedrich-Schiller-Universität Jena, Albert-Einstein-Str. 15, 07745 Jena, Germany

Active Fiber Systems GmbH, Wildenbruchstr. 15, 07745 Jena, Germany

Helmholtz-Institute Jena, Fröbelstieg 3, 07743 Jena, Germany

Fraunhofer Institute for Applied Optics and Precision Engineering, Albert-Einstein-Str. 7, 07745 Jena, Germany
e-mail: jens.limpert@uni-jena.de

mainly caused by thermal or nonlinear effects in the amplifier media [8, 9]. Recently, multi-aperture spatial combining approaches have emerged as one possibility to circumvent these limitations [10, 11]. Additionally, temporal combining techniques aimed at artificially extending the stretched pulse duration and, thus, overcoming pulse peak power limitations have been successfully demonstrated. Among those, the most straightforward approach is the so-called divided-pulse amplification (DPA) [12]. Here, in order to reduce the peak power-related limitations, each pulse is split into several temporally separated replicas before the final amplification stage and recombined afterwards. Alternatively, the creation of temporal replicas can be avoided, if a pulse train with a much higher repetition rate is amplified and subsequently temporally combined to achieve the repetition rate demanded by the application. Here, the general idea is to increase the pulse peak power at the cost of a reduced repetition rate by temporally stacking successive pulses after their amplification. One implementation of this approach, which we refer to as stack and dump (SND), is to superpose amplified pulses in an enhancement cavity (EC) and periodically extract them using a fast and efficient switch [13, 14].

Passive ECs have been subject to intensive research and development for several decades [15–17]. They are employed for a multitude of intracavity optical conversion processes such as high-harmonic generation [18, 19] or inverse Compton scattering [20]. Due to the energy enhancement in such a cavity, average powers in the MW range [21] and multi-GW peak power levels [22] are achievable within the cavity at multi-MHz repetition rates. In 2002 and 2003, the extraction of pulses from such an enhancement cavity was proposed [23] and demonstrated at around 80-MHz with nJ-level, picosecond pulses by the Ye and Hänsch groups [24, 25]. In 2004, slightly stretched femtosecond pulses were first enhanced and then extracted from a 100-MHz cavity [26]. Recently, concepts making use of the vast potential of ECs as stacking devices for stretched ultrashort pulses were published [13, 27].

In this paper, we demonstrate the SND scheme in a 30-m-long EC, corresponding to a length increase of a factor of 10 over the state of the art. Towards tapping the full potential of ECs as stacking devices for ultrashort pulses, this constitutes a crucial design criterion relaxing the thermal stress in the switch and in the cavity optics [28] and allowing for longer times between successive pulses. The EC supported a steady-state power enhancement factor exceeding 200 and was seeded with a 10-MHz repetition-rate train of 3-μJ pulses. The cavity enabled the enhancement of strongly stretched pulses (∼1.5 ns). A systematic investigation of different dumping rates was performed with an intracavity acousto-optic modulator (AOM). Pulses with the accumulated energy of up to 65 input pulses, i.e. 0.2 mJ, were extracted at 30 kHz. These pulses were recompressed to the initial duration of 800 fs, demonstrating the feasibility of SND with strongly stretched pulses and energies surpassing previous results by three orders of magnitude. These results, even if not stating new laser parameter records on their own, constitute the first milestone towards a power-scalable device and, thus, are a necessary step towards the first stack-and-dump system providing truly unprecedented laser parameters. Peak power-related and thermal limitations of this technique are discussed.

2 Cavity Set-Up and Steady-State Enhancement

In a first experiment, the steady-state behaviour of the EC without the AOM was investigated. The 30-m-long ring cavity (Fig. 1) was seeded with an average power of $P_{in} = 30$ W at 10-MHz repetition rate. Hence, the energy of the incoming pulses was $E_{in} = 3$ µJ. The pulses, spectrally centred around $\lambda = 1038$ nm were stretched to ~1.5-ns duration (measured at the −5-dB level of the maximum pulse intensity). A telescope was used to match the spatial mode of the incoming beam to the TEM_{00} mode of the EC, providing a measured overlap of $U = 80\%$ (which includes the spatial and the spectral overlap).

This set-up allowed for an energy enhancement factor

$$V = E_{circ}/E_{in}, \tag{1}$$

of 213, where E_{circ} is the energy of the pulse circulating in the EC. During steady-state operation, the intracavity average power was measured to be 6.4 kW, corresponding to an energy of the circulating pulses of 0.64 mJ. The round-trip losses L within the cavity were estimated to be 0.22%.

The power enhancement was mainly limited by the reflectivity of the input-coupling (IC) mirror $R = 99\%$, which did not fulfil the impedance matching condition ($R = 1 - L$) corresponding to optimum steady-state enhancement. However, this IC was purposely chosen to allow for a comparison to the non-steady-state experiment, in which a higher reflectivity would have been disadvantageous due to the additional losses induced by the switching device. The calculated caustic and the measured steady-state beam profile are shown in Fig. 2.

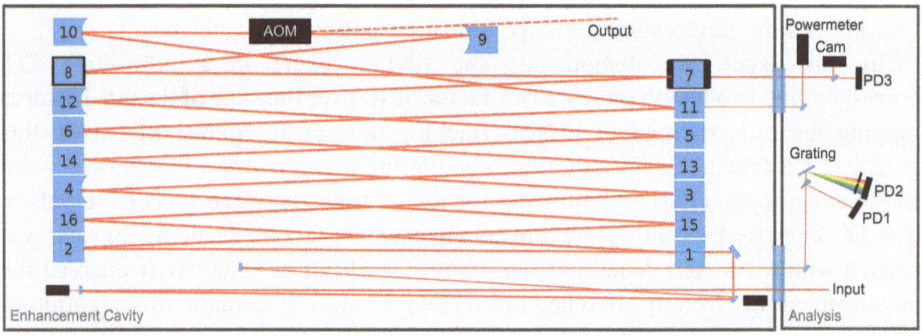

Fig. 1 Schematic of the EC. The 30-m-long EC consists of one input-coupling mirror (1, $R = 99\%$) and 15 highly reflective (HR) mirrors (2–16). Two of the HR mirrors are curved (4, 12) in order to form a stable resonator (see Fig. 2a). The mirrors 8 and 9, which are plane for the steady-state experiment, are replaced by curved ones once the AOM is inserted for the non-steady-state experiment (see Fig. 2a). The beams transmitted through mirrors 7 and 11 are sent to diagnostics such as a camera (Cam) and photodiodes (PD). The photodiode behind the grating is used for the Pound–Drever–Hall stabilization scheme [29]

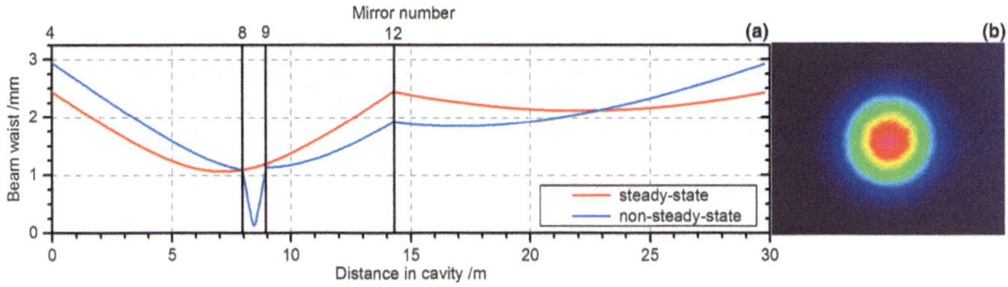

Fig. 2 **a** Cavity caustic for the steady-state experiment (*red*) and the non-steady-state experiment (*blue*). The positions of the curved mirrors 4, 12 and 8, 9 (only for the non-steady-state experiment) are indicated by *vertical lines*. **b** Transmitted beam profile of the intracavity beam at an enhancement factor of 213

3 Non-steady-State Operation: Pulse Extraction

For the dumping of pulses from the EC, an AOM was employed because of its simple implementation and fast switching times. The commercially available AOM (MQ80-A0.7-L1030.1064 from AA Opto-Electronic) used here offered an active aperture with a diameter of 0.7 mm and had a thickness of 23.5 mm. The rise time of the acoustic waves in this AOM is 110 ns/mm and depends on the diameter of the beam d. Using a TTL trigger signal with the experimentally optimized duration of ~50 ns, single-pulse extraction was enabled. The facets were antireflection coated for wavelengths between 1030 and 1060 nm. The AOM introduced additional transmission losses of about 0.6% and hence increased the overall round-trip losses L to 0.9% and the achieved steady-state enhancement to about 90. The cavity caustic was modified with respect to the steady-state experiment (see Fig. 2) in order to achieve an appropriate spot size in the AOM. The plane cavity mirrors 8 and 9 were replaced by concave mirrors with $R_{8,9} = 1000$ mm to obtain a focus with a $1/e^2$-diameter of $d = 0.3$ mm within the AOM, leading to negligible clipping losses while still providing a diffraction efficiency η_{diff} of around 72%. Hence, the circulating pulse could only be partially extracted and the energy enhancement of the extracted pulses compared to the input pulses (short: extracted enhancement) can be defined as

$$V_{\text{extr}} = \eta_{\text{diff}} \times V. \tag{2}$$

The energy that remains in the cavity after the extraction (see Fig. 3) changes the subsequent build-up cycle. After adjusting the mode-matching telescope, the beam overlap U was similar to the steady-state case. Dumping via the AOM was triggered synchronously to the laser repetition frequency after an integer number of pulses, employing a gate function just wide enough for a single pulse. The intracavity and output signals are shown exemplarily in Fig. 3 for 100 stacked pulses.

The switching rate (the repetition rate of the extracted pulses) can be calculated as

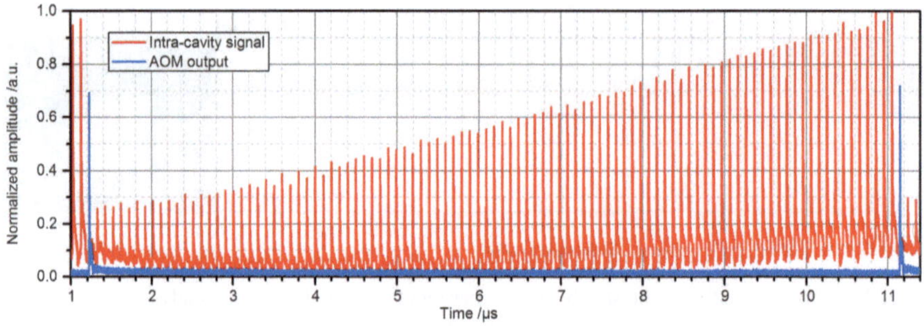

Fig. 3 Exemplary measurement of the entire pulse build-up within the EC using an AOM with 72% diffraction efficiency and a switching rate of 100 kHz (100 stacked pulses)

$$f_{\mathrm{switch}} = f_{\mathrm{rep}}/N, \tag{3}$$

were N represents the number of stacked pulses. As mentioned before, 28% of the enhanced pulse energy remained inside the cavity at the end of each build-up cycle, affecting the subsequent build-up. However, after a certain number of round trips an equilibrium is reached and the extracted enhancement after a large number of build-up cycles can therefore be described analytically with the following equation (derived similarly to [30]):

$$V_{\mathrm{extr}} = \eta_{\mathrm{diff}} \frac{(1-R)\left(\sqrt{AR}^{N} - 1\right)^2}{\left(\sqrt{AR} - 1\right)^2 \left[1 - \sqrt{AR}^{N}\sqrt{1-\eta_{\mathrm{diff}}}\right]^2}, \tag{4}$$

where $A = 1 - L$ is the round-trip attenuation. This equation was used for the simulations of the cavity behaviour. The power at the output port of the AOM was measured for different switching rates and used to determine the cavity efficiency

$$\eta = \eta_{\mathrm{stack}} \times \eta_{\mathrm{diff}} = P_{\mathrm{out}}/P_{\mathrm{in}}, \tag{5}$$

which describes how much of the input average power is conserved during the increase in the pulse energy of a laser system via stack and dump. The energy of the extracted pulses can be derived as

$$E_{\mathrm{out}} = P_{\mathrm{out}}/f_{\mathrm{switch}}. \tag{6}$$

The extracted enhancement can therefore also be written as

$$V_{\mathrm{extr}} = \eta_{\mathrm{diff}} \times V = E_{\mathrm{out}}/E_{\mathrm{in}}. \tag{7}$$

Figure 4 shows a plot of the measured extracted enhancement and cavity efficiency over the number of stacked pulses and the switching rate, compared to theoretical predictions. In agreement with the theory, a smaller number of stacked

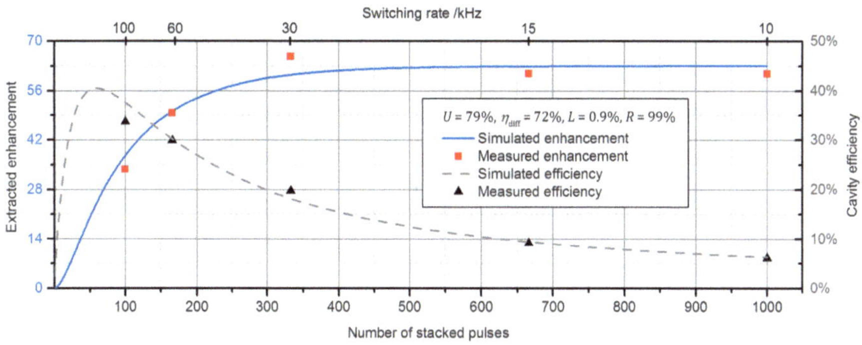

Fig. 4 Extracted enhancement and cavity efficiency for various switching rates along with the theoretical predictions calculated from Eq. 4

pulses lead to a smaller extracted enhancement at a given input coupler reflectivity. The measured efficiency shows a clear maximum for 100 stacked pulses, reaching 34%. When the number of stacked pulses is increased, a saturation of the enhancement sets in and the efficiency therefore drops continuously.

The efficiency as well as the extracted enhancement can be further optimized by adapting the input coupler reflectivity R for each switching rate as discussed in [13]. For a given input coupler reflectivity, the optimum working point in terms of the switching rate depends on whether the highest pulse energy or the highest efficiency is desired. The small deviations of the measured values from the ones predicted by theory (Fig. 4) are caused by variations in the alignment, slightly changing the overlap between the incoming beam and the cavity mode. It is noteworthy that the stabilization of the oscillator to the cavity was barely affected by the dumping process. Only at the highest investigated switching rates, the partial dumping occasionally leads to a collapse of the lock.

A switching rate of 30 kHz offered the highest extracted enhancement (~65), corresponding to an output pulse energy of 197 μJ. Figure 5 shows the photodiode signal of the intracavity pulses and the output of the system during this measurement. The dumped 30-kHz pulses were recompressed using a grating compressor with an efficiency of ~80% resulting in an energy of 0.16 mJ.

Figure 6 depicts a measurement of the autocorrelation and of the spectrum of the pulses. Additionally, a reference autocorrelation (Fig. 6a) was acquired in a single cavity-pass set-up without any input-coupling mirror and hence without any enhancement. Using the measured output spectrum and the spectral phase, the duration of the enhanced pulses was estimated to be around 800 fs. This is only slightly longer than the input pulses, which is also confirmed by the almost identical spectrum of the input and of the extracted pulses (see Fig. 6b).

In order to estimate the limitations induced by the nonlinear phase associated with self-phase-modulation (SPM) in the AOM, the pulse build-up in the cavity was numerically simulated step by step under consideration of this effect. At a certain threshold, the reduced overlap of the electric field of the intracavity pulse with the

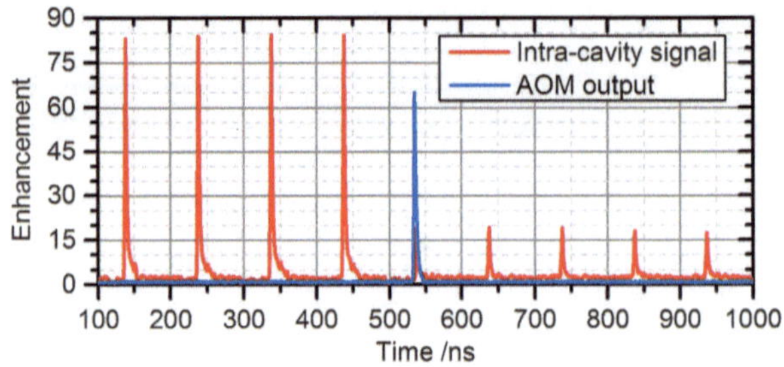

Fig. 5 Photodiode signals of the intracavity pulse (*red*) and of the output pulse (*blue*) for a switching rate of 30 kHz, revealing the extraction of a single pulse. A fraction of the pulse remains in the cavity due to the limited diffraction efficiency of the AOM

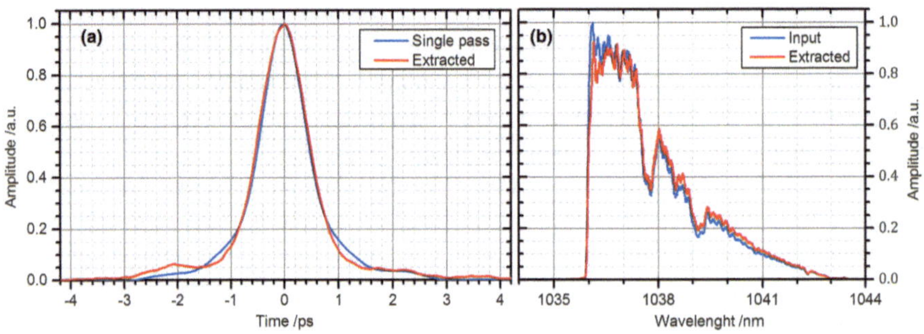

Fig. 6 **a** Autocorrelation traces (AC) of a diffracted pulse in the single-pass set-up (zeroth order of the AOM blocked) and of an extracted pulse during cavity operation. In the latter case, the shape is slightly different and the AC duration increased from 1.00 to 1.08 ps. The duration of the extracted pulse was estimated to be around 800 fs. **b** Spectrum of the signal before the EC (*blue*) and of the extracted pulse (*red*), both clearly showing the hard-cut of the stretcher at 1036 nm

seeding pulse will lead to a drop in the stacked pulse energy and therefore in the extracted energy. The simulation was carried out for different switching rates, using the input spectrum (Fig. 6b), a length of the AOM of 23.5 mm, a spot diameter in the AOM of 0.4 mm (as required in a 10-MHz cavity), a stretched pulse duration of 1.5 ns and a diffraction efficiency of the AOM of 72%. As shown in Fig. 7, for the experimental setting presented here ($E_{\text{in}} = 3$ μJ, black dotted line), the occurrence of SPM in the AOM was already starting to limit the extracted pulse energy. For 3 μJ of input energy and between 100 and 1000 stacked pulses, the possible extracted enhancement was slightly reduced compared to the SPM-free case. Due to the larger number of round trips and hence a greater acquired nonlinear phase, the effect becomes more critical when N is increased.

The spot size in the AOM is crucial for the peak intensity, but on the other hand it is determined by the required switching time for a given cavity length. To achieve an efficient enhancement at an increased input energy, using an AOM as a switching device, it is hence necessary to extend the cavity length. This measure

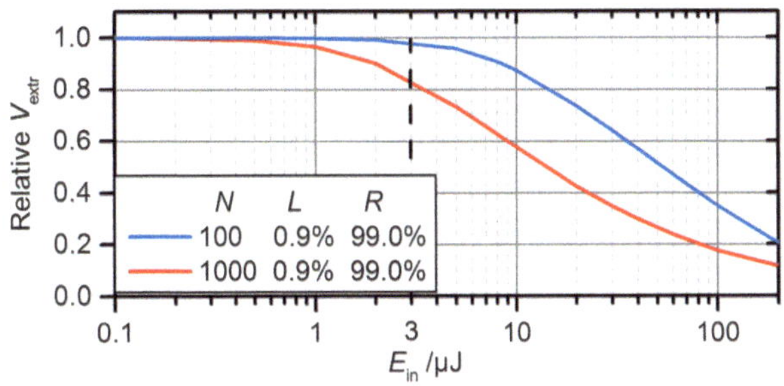

Fig. 7 Relative decrease in the extracted enhancement due to the nonlinear phase acquired via SPM in the AOM. Simulated for the pulse build-up in a 10-MHz cavity for the smallest and largest number of stacked pulses as a function of the input pulse energy (diameter in AOM of 0.4 mm, stretched pulse durations of 1.5 ns)

offers the additional benefit that for a certain desired switching rate (i.e. output repetition rate) the required number of stacked pulses is lower for longer cavities (see Eq. 3) which decreases the magnitude of the acquired nonlinear phase due to lower average round trips of the pulses. Naturally a sweet spot has to be found in the trade-off between a high enhancement and a minimized nonlinear phase.

4 Conclusion and Outlook

The experiments presented here constitute a first demonstration that pulse stacking of stretched fs pulses in a 10-MHz cavity is a promising route towards increasing the pulse peak power of high-average-power ultrafast laser systems. An extracted enhancement of 65, delivering pulse energies of 0.16 mJ at 30 kHz with a pulse duration of 800 fs, demonstrates a significant improvement over previous results [24–27]. To achieve this progress, it was necessary to lengthen the cavity in order to allow for a longer switching time with the additional benefit of a cavity mode that is less sensitive to thermal effects [28]. Other temporal pulse combining techniques, like divided-pulse amplification [31, 32], are limited in the number of combined pulses due to an increasing complexity. Stack and dump currently constitutes the most promising way for the superposition of a large number of pulses.

In the next experimental step, the AR-coated AOM will be exchanged for a Brewster-cut AOM and a state-of-the-art seed system [33] will be used to deliver up to 1 mJ at 2-MHz repetition rate with a spectrum enabling pulse durations below 300 fs. Our simulations under consideration of SPM-related effects in the AOM show that at the corresponding cavity length of 150 m an extracted enhancement of around 50 should be feasible by stacking 100 pulses. Together with a good management of the thermal lenses [34] that may occur in the cavity, these modifications will enable the extraction of 50-mJ pulses at 20-kHz repetition rate with a cavity efficiency of 50%, conserving 1 kW average power. Finally, a purely reflective switch [13, 14] overcoming AOM-related limitations is highly desirable to further increase the performance and efficiency of the system to Joule-class pulse energy and multi-kW average powers.

Acknowledgements This work has been partly supported by the European Research Council under the ERC Grant Agreement No. [617173] "ACOPS", by the German Federal Ministry of Education and Research (BMBF) under contract 13N13167 "MEDUSA" and by the Fraunhofer and Max Planck cooperation program "MEGAS".

References

1. T. Tajima, J. Dawson, Phys. Rev. Lett. **43**, 267–270 (1979)
2. R. Soulard, M.N. Quinn, T. Tajima, G. Mourou, Acta Astronaut. **105**, 192–200 (2014)
3. G. Mourou, B. Brocklesby, T. Tajima, J. Limpert, Nat. Photonics **7**, 258–261 (2013)
4. T. Eidam, S. Hanf, E. Seise, T.V. Andersen, T. Gabler, C. Wirth, T. Schreiber, J. Limpert, A. Tünnermann, Opt. Lett. **35**, 94–96 (2010)
5. P. Russbueldt, T. Mans, J. Weitenberg, H.D. Hoffmann, R. Poprawe, Opt. Lett. **35**, 4169–4171 (2010)
6. W.P. Leemans, R. Duarte, E. Esarey, S. Fournier, C.G.R. Geddes, D. Lockhart, C.B. Schroeder, C. Toth, S. Zimmermann, AIP Conf. Proc. **1299**(3), 3–11 (2010)
7. C.J. Saraceno, F. Emaury, C. Schriber, A. Diebold, M. Hoffmann, M. Golling, S. Thomas, U. Keller, IEEE J. Sel. Top. Quantum Electron. **21**, 1–18 (2015)
8. H.-J. Otto, F. Stutzki, F. Jansen, T. Eidam, C. Jauregui, J. Limpert, A. Tünnermann, Opt. Express **20**, 15710 (2012)
9. G. P. Agrawal: Nonlinear Fiber Optics (Academic, 2001)
10. S.J. Augst, J.K. Ranka, T.Y. Fan, A. Sanchez, J. Opt. Soc. Am. B **24**, 1707 (2007)
11. J. Limpert, A. Klenke, M. Kienel, S. Breitkopf, T. Eidam, S. Hadrich, C. Jauregui, A. Tunnermann, IEEE J. Sel. Top. Quantum Electron. **20**, 1–10 (2014)
12. S. Zhou, F.W. Wise, D.G. Ouzounov, Opt. Lett. **32**, 871–873 (2007)
13. S. Breitkopf, T. Eidam, A. Klenke, L. von Grafenstein, H. Carstens, S. Holzberger, E. Fill, T. Schreiber, F. Krausz, A. Tünnermann, I. Pupeza, J. Limpert, Light Sci. Appl. **3**, 1–7 (2014)
14. N. Lilienfein, S. Holzberger, I. Pupeza, Applied Physics B (2016) (**to appear**)
15. A. Ashkin, G.D. Boyd, J.M. Dziedzic, IEEE J. Quantum Electron. **2**, 109 (1966)
16. I. Hartl, T.R. Schibli, A. Marcinkevicius, D.C. Yost, D.D. Hudson, M.E. Fermann, J. Ye, Opt. Lett. **32**, 2870–2872 (2007)
17. I. Pupeza, T. Eidam, J. Rauschenberger, B. Bernhardt, A. Ozawa, E. Fill, A. Apolonski, T. Udem, J. Limpert, Z.A. Alahmed, A.M. Azzeer, A. Tünnermann, T.W. Hänsch, F. Krausz, Opt. Lett. **35**, 2052–2054 (2010)
18. I. Pupeza, S. Holzberger, T. Eidam, H. Carstens, D. Esser, J. Weitenberg, P. Rußbüldt, J. Rauschenberger, J. Limpert, T. Udem, A. Tünnermann, T.W. Hänsch, A. Apolonski, F. Krausz, E. Fill, Nat. Photonics **7**, 608–612 (2013)
19. C. Gohle, T. Udem, M. Herrmann, J. Rauschenberger, R. Holzwarth, H.A. Schuessler, F. Krausz, T.W. Hänsch, Nature **436**, 234–237 (2005)
20. H. Shimizu, A. Aryshev, Y. Higashi, Y. Honda, J. Urakawa, Nucl. Instrum. Methods Phys. Res. Sect A Accel. Spectrom. Detect. Assoc. Equip. **745**, 63–72 (2014)
21. H. Carstens, N. Lilienfein, S. Holzberger, C. Jocher, T. Eidam, J. Limpert, A. Tünnermann, J. Weitenberg, D.C. Yost, A. Alghamdi, Z. Alahmed, A. Azzeer, A. Apolonski, E. Fill, F. Krausz, I. Pupeza, Opt. Lett. **39**, 2595–2598 (2014)

22. S. Holzberger, N. Lilienfein, H. Carstens, T. Saule, M. Högner, F. Lücking, M. Trubetskov, V. Pervak, T. Eidam, J. Limpert, A. Tünnermann, E. Fill, F. Krausz, I. Pupeza, Phys. Rev. Lett. **115**, 023902 (2015)
23. R.J. Jones, J. Ye, Opt. Lett. **27**, 1848–1850 (2002)
24. E.O. Potma, C. Evans, X.S. Xie, R.J. Jones, J. Ye, Opt. Lett. **28**, 1835–1837 (2003)
25. Y. Vidne, M. Rosenbluh, T.W. Hänsch, Opt. Lett. **28**, 2396–2398 (2003)
26. R.J. Jones, J. Ye, Opt. Lett. **29**, 2812–2814 (2004)
27. T. Zhou, J. Ruppe, C. Zhu, I.-N. Hu, J. Nees, A. Galvanauskas, Opt. Express **23**, 7442 (2015)
28. H. Carstens, S. Holzberger, J. Kaster, J. Weitenberg, V. Pervak, A. Apolonski, E. Fill, F. Krausz, I. Pupeza, Opt. Express **21**, 11606 (2013)
29. R. W. P. Drever, J. L. Hall, F. V. Kowalski, J. Hough, G. M. Ford, a. J. Munley, and H. Ward: Appl. Phys. B Photophysics Laser Chem. 31, 97–105 (1983)
30. W. Nagourney, Quantum Electronics for Atomic Physics (Oxford University Press, 2010)
31. M. Kienel, A. Klenke, T. Eidam, S. Hädrich, J. Limpert, A. Tünnermann, Opt. Lett. **39**, 1049–1052 (2014)
32. Y. Zaouter, F. Guichard, L. Daniault, M. Hanna, F. Morin, C. Hönninger, E. Mottay, F. Druon, P. Georges, Opt. Lett. **38**, 106 (2013)
33. M. Kienel, M. Müller, A. Klenke, J. Limpert, A. Tünnermann, Opt. Lett. **41**, 3343–3346 (2016)
34. N. Lilienfein, H. Carstens, S. Holzberger, C. Jocher, T. Eidam, J. Limpert, A. Tünnermann, A. Apolonski, F. Krausz, I. Pupeza, Opt. Lett. **40**, 843–846 (2015)

External Cavity Diode Laser Setup with Two Interference Filters

Alexander Martin, Patrick Baus, and Gerhard Birkl

Abstract We present an external cavity diode laser setup using two identical, commercially available interference filters operated in the blue wavelength range around 450 nm. The combination of the two filters decreases the transmission width, while increasing the edge steepness without a significant reduction in peak transmittance. Due to the broad spectral transmission of these interference filters compared to the internal mode spacing of blue laser diodes, an additional locking scheme, based on Hänsch–Couillaud locking to a cavity, has been added to improve the stability. The laser is stabilized to a line in the tellurium spectrum via saturation spectroscopy, and single-frequency operation for a duration of two days is demonstrated by monitoring the error signal of the lock and the piezo drive compensating the length change of the external resonator due to air pressure variations. Additionally, transmission curves of the filters and the spectra of a sample of diodes are given.

1 Introduction

Modern optical spectroscopy as well as the manipulation of the internal and external degrees of freedom of atoms and molecules (neutral and charged) requires stable laser sources in a wide range of wavelengths. External cavity diode lasers (ECDLs) are an excellent choice in many cases [1] and are specifically used because of their low cost and the broad wavelength availability, which covers almost the entire spectrum from the near-infrared to near-ultraviolet. The first compact ECDL setup for single-frequency operation using a grating for the external cavity was described by Ted Hänsch in 1995 [2] and has remained an important configuration ever since. Twenty years later, ECDLs with the external feedback

This article is part of the topical collection "Enlightening the World with the Laser" - Honoring T. W. Hänsch guest edited by Tilman Esslinger, Nathalie Picqué, and Thomas Udem.

A. Martin • P. Baus (✉) • G. Birkl
Institut für Angewandte Physik, Technische Universität Darmstadt, Schlossgartenstraße 7, 64289 Darmstadt, Germany
e-mail: patrick.baus@physik.tu-darmstadt.de

D. Meschede et al. (eds.), *Exploring the World with the Laser*,
https://doi.org/10.1007/978-3-319-64346-5_41

realized by a narrow-bandwidth interference filter and a semitransparent mirror [3] improved the stability against external perturbations. In recent years, the development of blue laser diodes based on gallium nitride and indium gallium nitride [4] extended the range of ECDLs to shorter wavelength. Nevertheless, due to the specific characteristics of these laser diodes, stable single-frequency operation in the blue wavelengths region remains a challenge.

In this work, we present an ECDL setup for stable single-frequency operation in the wavelength range around 450 nm (Fig. 1). We combine the design of an interference filter-based ECDL [3] in the specific realization of Ref. [5], adding a

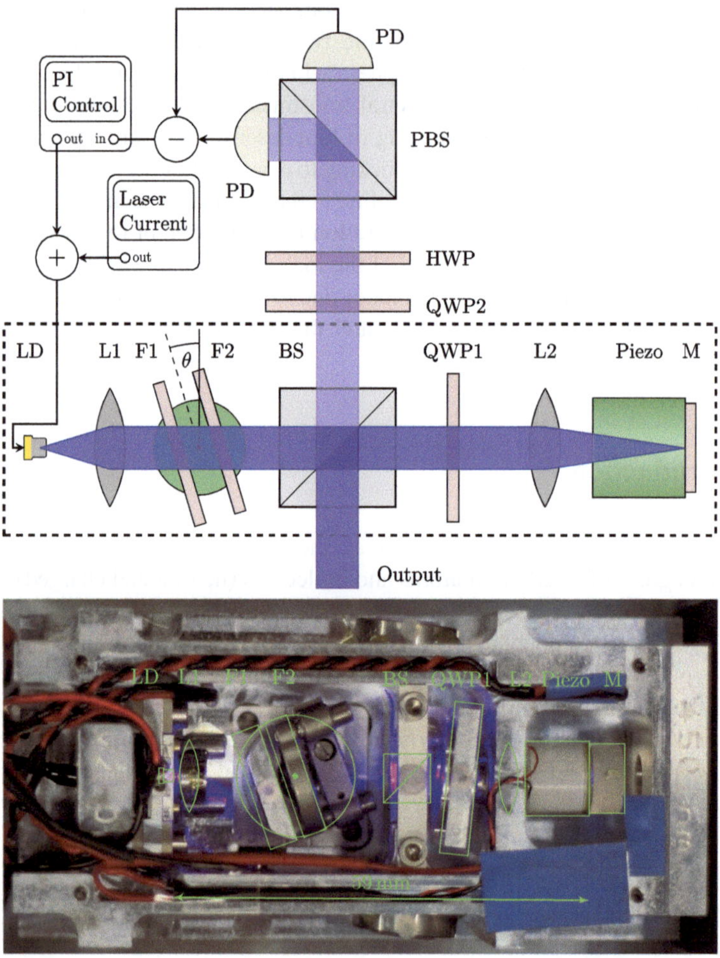

Fig. 1 (*Top*) schematic setup of the external cavity diode laser (*dashed box*) with two interference filters and the additional elements for polarization-stabilization of the cavity length. The two filters are mounted on a rotation stage for easy wavelength tuning. The external resonator has a total geometrical length of 59 mm from laser diode LD to mirror M. (*bottom*) photograph of the laser setup

second filter and the intra-cavity stabilization technique presented by Führer et al. [6, 7]. The latter is based on the polarization-dependent locking scheme introduced by Hänsch and Couillaud [8] for locking lasers to an external cavity. We use a PL-450B indium gallium nitride (InGaN) laser diode manufactured by OSRAM [9] as active element. This single transverse mode diode has an optical output power of up to 80 mW in free-running mode. The high availability in volume and the low price make it an interesting candidate for ECDL developments. On the other hand, the spectral characteristics are less favorable for single-frequency operation than for more expensive diode variants as used in grating stabilized ECDLs in the blue so far [10, 11].

In the next sections, we first discuss the realization of the ECDL setup. Then, we will present the characterization of a set of free-running PL-450B laser diodes and the interference filter used. This is followed by the characterization of the complete diode laser and by the demonstration of the long-term stability of the laser locked to a molecular reference.

2 Laser Setup

The laser setup depicted in Fig. 1 is a variation of the design of Ref. [3] in the form introduced in Ref. [5]. The light from the laser diode (LD) is collimated by a lens (L1) with 3.1 mm focal length. The external resonator consists of a silver coated mirror (M) (reflectivity $R_{\mathrm{M}} = 99\%$) and lens L2 ($f = 18.4\,\mathrm{mm}$) in a cateye configuration. The resonator has a total length of 59 mm. For adjustment of the resonator length, the mirror is glued to a piezoelectric ring actuator with a length change coefficient of approximately 100 nm/V. For outcoupling of the laser light, we use a polarization-dependent beam splitter cube (BS) combined with a quarter wave plate (QWP1). The BS has a transmission of $T_{\mathrm{p}} \approx 90\%$ (10%) for p-polarized (s-polarized) light. Depending on the rotation angle of QWP1, the back-reflectivity of the combination of BS, QWP1, and M can be varied between $T_{\mathrm{p}} \cdot R_{\mathrm{M}} \cdot T_{\mathrm{s}} \approx 9\%$ and $T_{\mathrm{p}} \cdot R_{\mathrm{M}} \cdot T_{\mathrm{p}} \approx 80\%$. The laser diode must therefore be mounted with its polarization axis in the p-plane of BS. Two band-pass filters (F1 and F2) provide the wavelength selection. We use interference filters (Semrock Laser-Line LL01-458) with a center wavelength of 457.9 nm and a measured full width at half maximum (FWHM) of 1.6 nm at normal incidence. The filters are mounted on a rotary stage for easy tuning of the transmission wavelength. If required, a nonzero relative angle ($\theta_1 \neq \theta_2$) allows the modification of the effective spectral transmission characteristics.

The central element for improved long-term stability of this laser setup is the additional intra-cavity lock based on the Hänsch–Couillaud locking scheme [8] using the fact that a mismatch between the internal modes of the laser diode gain medium and the external resonator modes changes the state of polarization of the light in the external resonator. Following the implementation of Führer et al. [6], the second output beam produced by the beam splitter BS is used for generation of the respective error signal: The ratio between p- and s-polarization in the external resonator is measured by subtracting the photodiode signals after the extra-cavity polarizing beam splitter (PBS) as shown in Fig. 1. With QWP2 and a half-wave plate (HWP), the signal-to-noise ratio of the error signal can be maximized. A proportional-integral (PI) controller locks the internal to the external resonator modes by regulating the laser current.

3 Experimental Results

3.1 Laser Diode and Filter

According to the datasheet [9], the center wavelengths of the PL-450B laser diodes spread over a range of 20 nm. We measured the emission spectra of twelve diodes from different lots and retailers with an optical spectrum analyzer (OSA). It has a wavelength accuracy of 0.05 nm and a resolution of 0.01 nm. The lasing spectra of a representative subset of seven diodes are shown in Fig. 2a. All spectra were measured with 60 mA current without an external resonator. The central lasing wavelengths cover a range of about 8 nm between 445.15 and 452.95 nm with a mean value of 447.66 nm. The measurement also revealed an internal mode separation of 0.05 nm corresponding to 80 GHz. Figure 2b shows the spectrum of a diode below the laser threshold. The spontaneous emission profile has a FWHM of 1.82 nm, which is much smaller than the 20 nm width of typical infrared diodes. In addition, the wavelength shift with temperature of 0.06 nm/K for InGaN [4] is about five times smaller than for InGaAs. This requires the preselection of a diode for the targeted wavelength.

The spectral transmission of the interference filter (Semrock Laser-Line LL01-458) was also measured with the OSA using a blue LED as broadband light source. Figure 3a depicts the spectral filter transmission for different angles of incidence. As expected, the transmission peak shifts to shorter wavelengths with increasing angle. The maximum transmittance decreases from above 95% at 0° to below 80% at 35°. At normal incidence, the filter has a flattop profile, but at incidence angles of $\theta \geq 20°$ the profile is nearly Gaussian.

The FWHM of the transmission stays below 1.8 nm for incident angles $\theta < 25°$. As shown in Fig. 3b, this width is larger than the full emission spectrum of the laser diode and corresponds to the combined spectral width of more than 30 internal cavity modes. These are significantly more modes than in the ECDL setup of Ref. [5] in the infrared. Their setup uses an interference filter and laser diode combination, in which only five internal cavity modes fit into the transmission band of the filter. In that case, utilizing the edges of the gain profile of the diode allowed for a stabilization of the laser wavelength. The smaller mode spacing and the fact that the

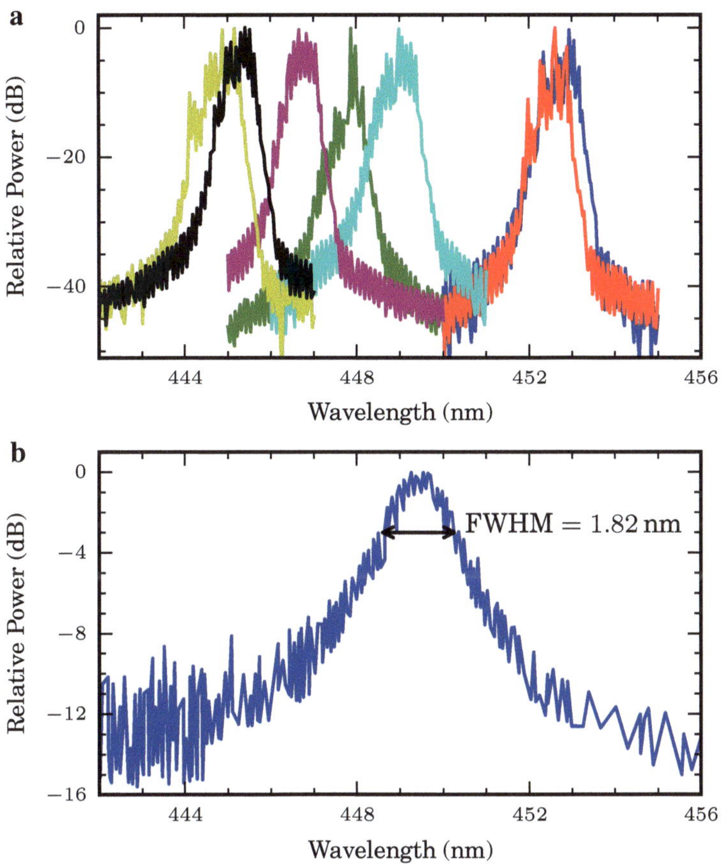

Fig. 2 **a** Normalized spectra of different free-running laser diodes at 60 mA. For a better overview, only 7 from the 12 measured diode spectra are shown. **b** Spectrum of a laser diode below laser threshold. The FWHM of 1.82 nm is significantly smaller than the one of infrared laser diodes

edges of the filters are not as steep as in the infrared contribute significantly to the difficulty of achieving stable single-frequency operation of blue diode lasers. At our targeted wavelength of operation of 452.756 nm, the filter ($\theta \approx 20^\circ$) has a FWHM = 1.64 nm, an edge steepness of 1.01 nm (10–90% value), and a peak transmittance of ≈98%. To improve the filter performance, we mounted two filter in series. This decreased the FWHM to 1.23 nm and improved the edge steepness to 0.80 nm without a significant loss in transmittance. Under small angles of incidence ($\theta < 20^\circ$), a slight variation of the relative angle between the two filters ($\theta_1 \neq \theta_2$) changes the flattop profile of the filter combination to a more Gaussian shape. This reduces the flattop contribution and decreases the FWHM. At higher incident angles ($\theta \geq 20^\circ$), the transmission characteristics is already Gaussian. In this case, a

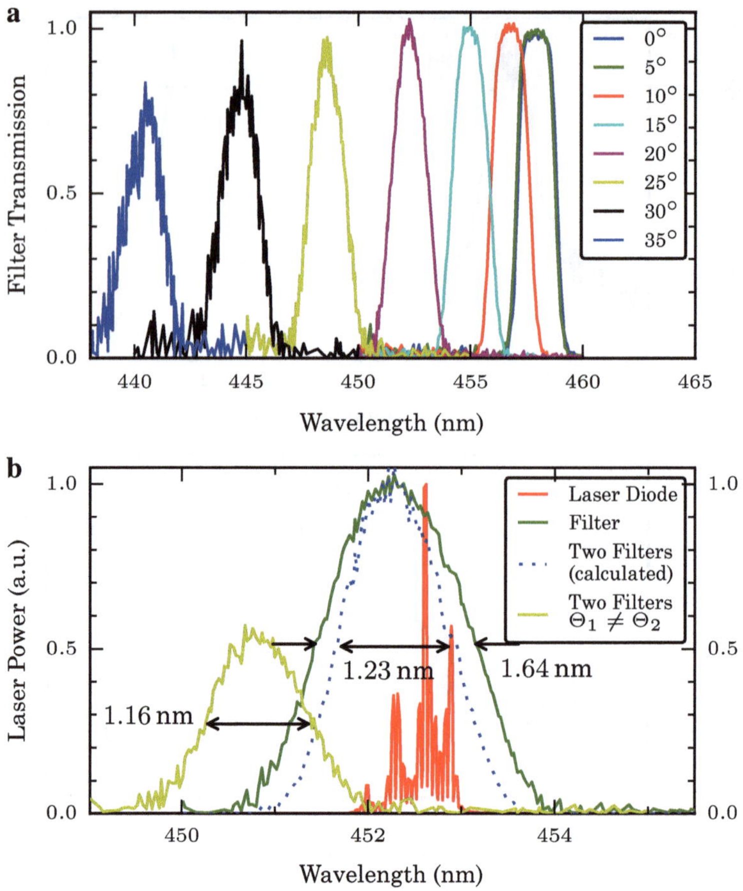

Fig. 3 **a** Spectral transmission of the interference filter for incidence angles between 0° and 35° in steps of 5°. The peak wavelength decreases with increasing angle, and the filter profile changes from a flattop to a Gaussian profile. **b** Comparison of the emission spectrum of a free-running laser diode (*red*) with filter transmission curves: (*green*) measured transmission of one interference filter at 20°, (*dotted blue*) calculated transmission of two parallel filters at 20° in series, and (*yellow*) two filters in series at ≈23° with a slight relative tilt ($\theta_1 \neq \theta_2$)

nonzero relative angle between the filters leads to a smaller total transmittance without further improvement in comparison with two parallel filters, as shown in Fig. 3b. For this reason, we have chosen to align the two filters in parallel.

3.2 ECDL Characterization

For the determining of the optimum feedback of the external resonator of the ECDL, the intra-cavity beam splitter (BS) was replaced by a polarization beam splitter (PBS) resulting in an effective cavity mirror with adjustable reflectivity and constant polarization: By rotating QWP1, the overall reflectivity of the combination of PBS, QWP1, and mirror M could be changed. Above 45% reflectivity, the laser showed increasing instability. The best stability could be achieved between 10 and 20% reflectivity, depending on the targeted internal laser mode. A maximum output power of 17 mW could be reached at a wavelength of 454 nm. Since this configuration does not allow for the implementation of the intra-cavity polarization lock, all further measurements were performed with the ECDL in the original setup described above with beam splitter BS. To receive a complete picture, we also tested the ECDL setup from Baillard et al. [3] with one filter and a 15% back reflecting mirror used for feedback and outcoupling. Here, an output power of up to 22 mW was achieved, but single-frequency operation was limited to a duration of one hour typically.

To further characterize the setup, the wavelength and the output power of the double-filter ECDL ($\theta \approx 20^\circ$) were measured as a function of the diode injection current (see Fig. 4). The wavelength was measured with a commercial wavemeter and the optical output power with a calibrated photodiode. With increasing current, the wavelength and the output power increase. The laser diode current was changed in steps of 21 μA. Data were only recorded when the wavemeter confirmed single-mode operation of the laser. Below 38 mA and above 86 mA, no stable operation was achieved. Following the internal modes (indicated by linear fits with specific colors), the output power rises linearly with the applied current, with different slopes for different modes. More than 10 mW of output power could be achieved at

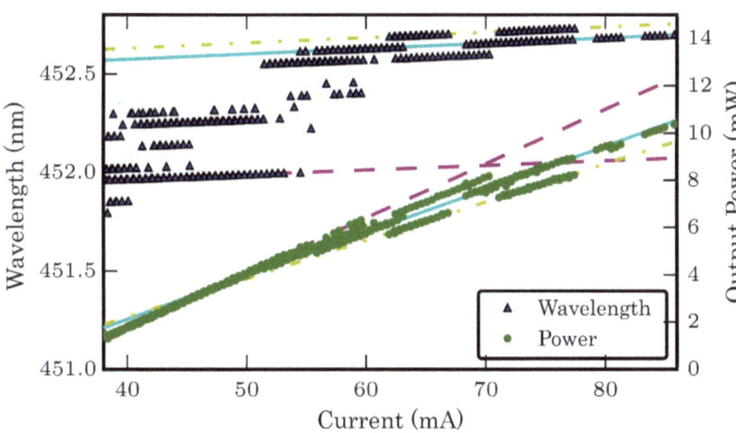

Fig. 4 Output power and wavelength characteristics of the two-filter ECDL. The wavelength jumps in discrete steps of 0.05 nm between the internal cavity modes of the laser diode. A specific mode shows a shift of 2.59(4) pm/mA. The slope of the corresponding output power is dependent on the specific internal mode. This correspondence is indicated by three pairs of linear fits with matching *line style* and *color* (*solid cyan, dashed magenta, dash-dotted yellow*) for power and wavelength

86 mW injection current. The spatial profile of the output mode resembles an elliptical Gaussian TEM_{00} mode with a 2.5:1 ratio of the spatial widths. We achieve a coupling efficiency of about 60% of the output mode into a single-mode optical fiber. The slope of the wavelength shift for all laser modes is similar at about 2.59(4) pm/ mA.

With rising current, additional jumps in the wavelength of 0.05 nm toward longer wavelengths occur. The step size agrees with the internal mode spacing of the laser diode. Due to the broad filter transmission compared to the internal mode spacing, the wavelength can be tuned over a wide range of 0.8 nm without changing the angle θ of the filters. The stability is highest at high currents: Here, the laser medium favors emission at longer wavelengths, whereas the edge of the filters suppresses internal modes with too large wavelengths. The measurement also shows a current range of several tens of μA for stable operation at a given laser mode before the next mode hop occurs. Because of this small range, we use a digital current driver which allows us to vary the current in steps of 2 μA.

Due to the broad transmission spectrum of the utilized interference filters compared to the laser diode mode spacing, the frequency stability of the ECDL is strongly dependent on the external cavity configuration. With the original setup of Ref. [3] and one filter, single-frequency operation could be sustained for up to 1 h. Switching to the two-filter configuration described in Sect. 2 without intra-cavity locking applied, extended the duration of single-frequency operation to several hours. The laser remained sensitive to small disturbances. These can lead to a misalignment between internal and external resonator, which causes mode jumps. A significant improvement in the long-term stability could be achieved by adding the intra-cavity polarization locking scheme described above. For a characterization of the long-term stability, the laser was frequency-stabilized via Doppler-free saturation spectroscopy to $^{130}Te_2$ which has a dense line spectrum in the blue wavelength region [12]. The stabilization on a Lamb-dip of tellurium uses a second lock-in servo loop, which generates the derivative of the Lamb-dip via a 40-kHz modulation of the laser current. This modulation is faster than the feedback bandwidths of both lock-in servo loops. Figure 5a shows a scan of the dispersion signal of the targeted tellurium Lamb-dip at 452.756 nm which is an unlabeled line about halfway between the lines labeled 521 ($22085.9292\,\mathrm{cm}^{-1}$) and 522 ($22087.6258\,\mathrm{cm}^{-1}$) in Ref. [12]. With a second PI controller, this signal is fed to the piezo of the ECDL cavity and to the laser diode current.

For a long-term measurement of the frequency stability, the tellurium lock error signal and the piezo voltage for stabilizing the ECDL cavity length were recorded in 5-s intervals. For each interval, the signals were averaged over 100 samples to reduce the noise. In order to determine the influence of the laboratory conditions, room temperature, ambient air pressure, and humidity were recorded every five minutes. Figure 5b, c shows a respective measurement over duration of two days. The laser stays locked over the full duration and thus demonstrates excellent long-term stability. The piezo voltage shows a strong correlation with air pressure, superimposed by a small negative piezo voltage drift, starting after the first half of the measurement time. The room temperature of 21.5 °C was stable to better than

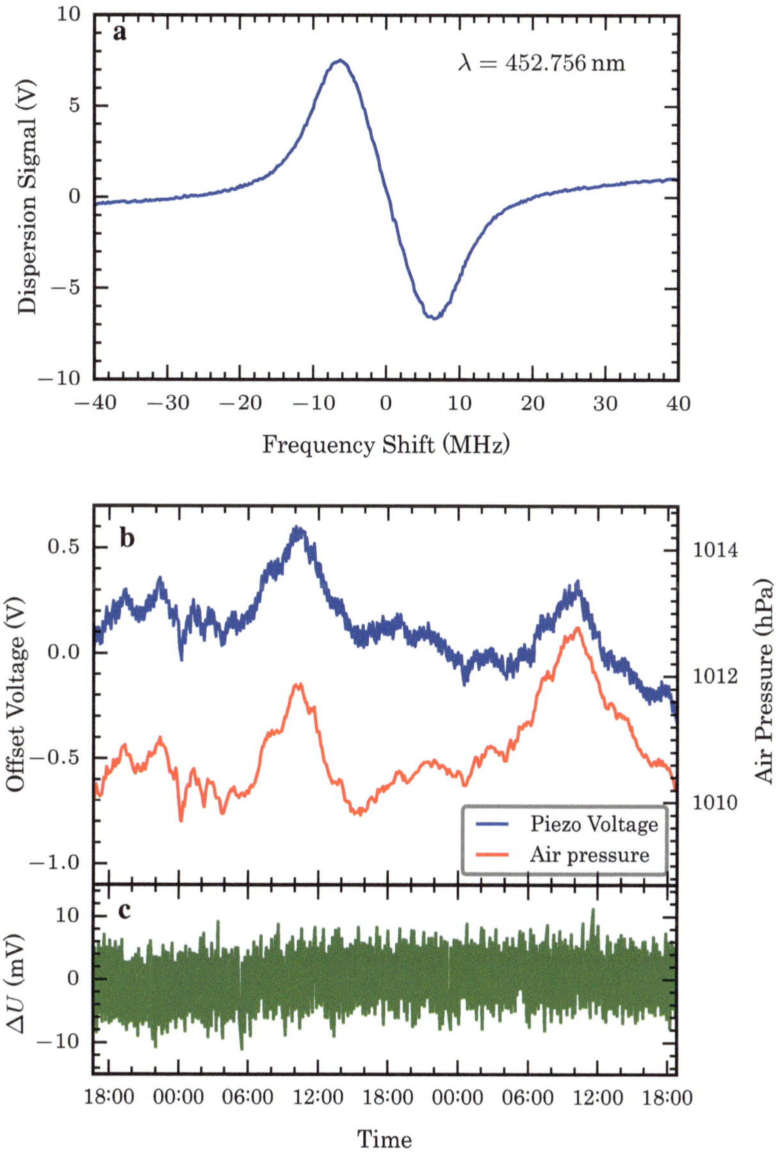

Fig. 5 **a** Measured Doppler-free dispersion spectrum of an absorption line of molecular tellurium at 452.756 nm. **b** Long-term stability of the laser recorded over a period of two days. The laser is stabilized to the Doppler-free tellurium line shown in **a**. The applied offset voltage of the ring piezo compensates the length change of the external resonator predominantly caused by air pressure variations. **c** The corresponding error signal of the tellurium stabilization loop demonstrates the stable laser operation over the full measurement period

0.5 K and the relative humidity stayed between 40 and 50%. The absolute variation of the piezo voltage does not exceed 1 V which corresponds to an upper bound for the change of the cavity length of 100 nm. Figure 5c confirms that the laser stayed in lock during the full measurement time. The in-loop deviation of the error signal is symmetric around the setpoint of 0 V with a mean excursion of only a few mV. An out-of-lock event would show a large discontinuous jump of several tens of mV.

4 Conclusion

We built and characterized an external cavity diode laser setup with two inexpensive, commercially available interference filters giving an optical output power of more than 10 mW at about 450 nm. The use of two filters instead of one for wavelength selection in the external resonator increased the edge steepness and thus enhanced the stability of single-frequency operation significantly. In addition, the configuration of the external resonator with a beam splitter cube, a quarter wave plate, and a highly reflective mirror allowed for fine-adjustment of the back-reflectivity and for the implementation of an intra-cavity locking scheme for matching the diode laser modes to the ECDL modes. This scheme increased the duration of stable single-frequency operation from a few hours to more than two days. This could be demonstrated by stabilizing the laser to a tellurium absorption line at 452.756 nm.

To further optimize the stability of the ECDL, the length of the external resonator can be decreased, reducing the effects of air pressure and temperature variations. If a higher output power is needed, the measured optimal feedback condition can be implemented by building an ECDL with two filters and a reflectivity optimized semitransparent output mirror in cateye configuration with less intra-cavity losses. The intra-cavity locking scheme still can be applied. For a reflectivity between 10 and 20%, an output power above 20 mW is achievable.

Acknowledgements We dedicate this contribution to Ted Hänsch in honor of his 75th birthday. Without his seminal contributions to physics in general, laser science and technology, atomic physics, and quantum optics in specific, modern physics would not be as fascinating as it is, and the work reported here would not have been possible. We wish to thank Holger John and the group of Thomas Walther for fruitful discussions and advise on the laser design and the support with the OSA measurements. We appreciate the contributions of Florian Vollrath to the stabilization system. This work received partial financial support from the Deutsche Forschungsgemeinschaft (DFG) under the Grant No. BI 647/4-1. A. Martin and P. Baus acknowledge support from HGS-HIRe. We thank the ARTEMIS collaboration for continuing support. The experiments have been performed within the framework of the HITRAP facility at the Helmholtz Center for Heavy Ion Research (GSI) and the Facility for Antiproton and Ion Research (FAIR) at Darmstadt.

References

1. C.E. Wieman, L. Hollberg, Using diode lasers for atomic physics. Rev. Sci. Instrum. **62**(1), 1–20 (1991). http://scitation.aip.org/content/aip/journal/rsi/62/1/10.1063/1.1142305
2. L. Ricci, M. Weidemüller, T. Esslinger, A. Hemmerich, C. Zimmermann, V. Vuletic, W. König, T. Hänsch, A compact grating-stabilized diode laser system for atomic physics. Opt. Commun. **117**(5), 541–549 (1995). http://www.sciencedirect.com/science/article/pii/003040189500146Y
3. X. Baillard, A. Gauguet, S. Bize, P. Lemonde, P. Laurent, A. Clairon, P. Rosenbusch, Interference-filter-stabilized external-cavity diode lasers. Opt. Commun. **266**(2), 609–613 (2006). http://www.sciencedirect.com/science/article/pii/S0030401806004561
4. S. Nakamura, S.F. Chichibu, *Introduction to Nitride Semiconductor Blue Lasers and Light Emitting Diodes* (Taylor & Francis, Boca Raton, 2000)
5. D.J. Thompson, R.E. Scholten, Narrow linewidth tunable external cavity diode laser using wide bandwidth filter. Rev. Sci. Instrum. **83**(2), 023107 (2012). http://scitation.aip.org/content/aip/journal/rsi/83/2/10.1063/1.3687441
6. T. Führer, D. Stang, T. Walther, Actively controlled tuning of an external cavity diode laser by polarization spectroscopy. Opt. Express **17**, 4991–4996 (2009). http://www.opticsexpress.org/abstract.cfm?URI=oe-17-7-4991
7. T. Führer, S. Euler, T. Walther, Model for tuning an external-cavity diode laser by polarization locking. J. Opt. Soc. Am. B **28**(3), 508–514 (2011). http://josab.osa.org/abstract.cfm?URI=josab-28-3-508
8. T.Hänsch, B. Couillaud, Laser frequency stabilization by polarization spectroscopy of a reflecting reference cavity. Opt. Commun. **35**(3), 441–444 (1980). http://www.sciencedirect.com/science/article/pii/0030401880900693
9. OSRAM Opto Semiconductors, *Blue Laser Diode in TO38 ICut Package* (data sheet, 2013). http://www.osram-os.com/Graphics/XPic5/00193831_0.pdf/PL%20450B.pdf
10. Y. Shimada, Y. Chida, N. Ohtsubo, T. Aoki, M. Takeuchi, T. Kuga, Y. Torii, A simplified 461-nm laser system using blue laser diodes and a hollow cathode lamp for laser cooling of Sr. Rev. Sci. Instrum. **84**(6), 063101 (2013). http://www.ncbi.nlm.nih.gov/pmc/articles/PMC3689831/
11. T. Dutta, D.D. Munshi, M. Mukherjee, Absolute Te_2 reference for barium ion at 455.4 nm. J. Opt. Soc. Am. B **33**(6), 1177–1181 (2016). http://josab.osa.org/abstract.cfm?URI=josab-33-6-1177
12. J. Cariou, P. Luc, *Atlas du Spectre d'Absorption de la Molécule de Tellure*. 21100 cm^{-1} – 23800 cm^{-1} (Laboratoire Aime-Cotton, 1980). http://www.lac.u-psud.fr/spip.php?article497

References

1. C.E. Wieman, L. Hollberg, Using diode lasers for atomic physics. Rev. Sci. Instrum. 62(1), [illegible]

[illegible]

Bose–Einstein Condensates in an Optical Cavity with Sub-recoil Bandwidth

J. Klinder, H. Keßler, Ch. Georges, J. Vargas, and A. Hemmerich

Abstract This article provides a brief synopsis of our recent work on the interaction of Bose–Einstein condensates with the light field inside an optical cavity exhibiting a bandwidth on the order of the recoil frequency. Three different coupling scenarios are discussed giving rise to different physical phenomena at the borderline between the fields of quantum optics and many-body physics. This includes sub-recoil opto-mechanical cooling, cavity-controlled matter wave superradiance and the emergence of a superradiant superfluid or a superradiant Mott insulating many-body phase in a self-organized intra-cavity optical lattice with retarded infinite range interactions.

1 Introduction

Experimental progress in coupling single atoms to single-mode electromagnetic radiation in a cavity has prepared the ground for the field of cavity quantum electrodynamics during the 1980s and 1990s [1, 2]. The focus has since changed toward macroscopic objects, e.g., submicron mechanical oscillators like cantilevers or membranes [3, 4], superconducting qubits [5], quantum dots [6], or droplets of quantum degenerate atomic gases [7], as in the work described in this article. The fragile nature of ultracold quantum ensembles limits one to dispersive light–matter interactions excluding near-resonant excitations followed by spontaneous emission. Hence, the cavity should be pumped by radiation, which is far detuned from any atomic resonance, such that Rayleigh scattering prevails. The first experiments, exploring this regime with optical cavities and laser-cooled thermal atomic samples [8–12] or Bose–Einstein condensates (BECs) [13], have used ring geometries. The use of extreme values of the cavity finesse exceeding several 10^5 constrains scattering of photons to the modes resonant with the cavity, owing to the Purcell

This article is part of the topical collection "Enlightening the World with the Laser" - Honoring T. W. Hänsch guest edited by Tilman Esslinger, Nathalie Picqué, and Thomas Udem.

J. Klinder • H. Keßler • C. Georges • J. Vargas • A. Hemmerich (✉)
Institut für Laserphysik, Universität Hamburg, 22761 Hamburg, Germany
e-mail: hemmerich@physnet.uni-hamburg.de

D. Meschede et al. (eds.), *Exploring the World with the Laser*,
https://doi.org/10.1007/978-3-319-64346-5_42

effect [14, 15]. At the same time, this allows one to access the most interesting regime of strong cooperative coupling, when the atom sample acts to shift the cavity transmission resonance by more than its linewidth, and hence, the back-action of the atoms on the intra-cavity light field is significant [10, 12, 16–20].

In addition, an important aspect of the work discussed here is the combination of a high finesse with a large cavity length, yielding the regime of sub-recoil resolution, for which the cavity linewidth is comparable to or smaller than the single-photon recoil frequency, which corresponds to the kinetic energy transferred to a resting atom by backscattering of a single photon. This circumstance has two important consequences. On a single atom level, a photon scattering blockade scenario arises: A first single-photon scattering process drives an atom out of resonance for a subsequent second cavity-assisted scattering process. On a many-body level, the time scale for the evolution of the matter degrees of freedom, i.e., the recoil time, is comparable to the time scale for the evolution of the cavity field, i.e., the photon life time in the cavity. For short cavities with MHz bandwidth (for example, as in Refs. [17–20]), the cavity light field can be adiabatically eliminated such that its net effect on the atoms is described in terms of an instantaneous long-range interaction, while for long cavities, as considered here, the induced long-range interaction has a finite propagation time.

The perhaps most elementary atom–cavity configuration is a BEC interacting with a single longitudinal mode of a standing wave resonator, which is coupled by an external pump field using one of the three elementary coupling geometries depicted in Fig. 1: (a) The pump field arises from a laser beam axially coupled to the cavity through one of its mirrors; (b) the pump field is due to a laser beam irradiating the BEC perpendicularly with respect to the cavity axis, and (c) the pump field has a standing wave geometry resulting from a retro-reflected laser beam intersecting the BEC perpendicularly with respect to the cavity axis. These coupling geometries give rise to surprisingly different physics. Case (a) is characterized by optical bistability, persistent oscillations and superradiant backscattering instabilities, case (b) leads to matter wave superradiance, and case (c) opens a new arena

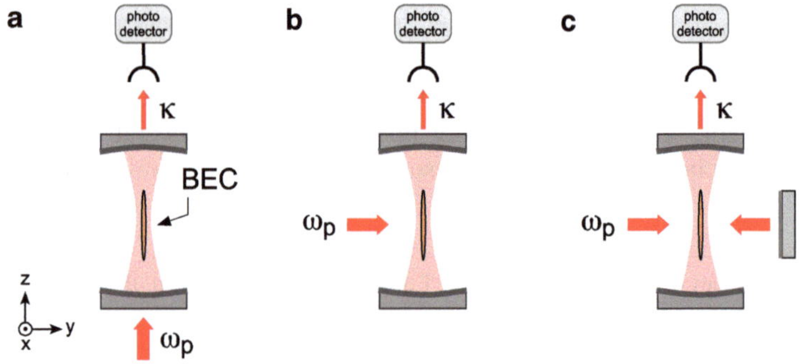

Fig. 1 Alternative coupling geometries for the pump light at frequency ω_p. The BEC and the cavity mode are not drawn to scale. **a** Axial coupling, **b** transverse traveling wave coupling, **c** transverse standing wave coupling

to investigate physics related to the Dicke model [21, 22] and the famous Dicke–Hepp–Lieb superradiant phase transition [23, 24]. In this brief review, a number of characteristic examples of the physics emerging for the three couplings schemes are revisited. Our synopsis of several studies that spread over a number of original articles is complemented by various unpublished results.

2 BEC Preparation and Cavity Setup

A cigar-shaped BEC with Thomas–Fermi radii $(3.1, 3.3, 26.8)$ μm and $N \approx 10^5$ ^{87}Rb -atoms, prepared in the upper hyperfine component of the ground state $|F = 2, m_F = 2\rangle$, is confined by 3-cm-sized solenoids arranged to provide a magnetic trap with a nonzero bias field with trap frequencies $\omega/2\pi = (215.6 \times 202.2 \times 25.2)$ Hz [25, 26]. Two oppositely mounted mirrors form a standing wave high finesse cavity integrated into the coil setup with the cavity axis along the direction of gravity. For the best possible mode match between the atomic ensemble and the cavity mode, the weekly confining z-axis of the magnetic trap is aligned with the cavity axis. The cavity has a free spectral range $\Delta\nu_{\mathrm{FSR}} = 3$ GHz, a finesse $\mathcal{F} = 3.44 \times 10^5$, a Purcell factor $\eta_c = 44$ [14] and a field decay rate $\kappa = 2\pi \times 4.45$ kHz. The BEC is exposed to pump fields arranged as shown in Fig. 1) operating at the wavelength $\lambda = 803$ nm, i.e., at large detuning to the negative side of the principle fluorescence lines of rubidium at 780 and 795 nm. For a uniform atomic sample and left circularly polarized light at 803 nm, the cavity resonance frequency is dispersively shifted with respect to the case of an empty cavity by an amount $\delta_- = \frac{1}{2} N_a \Delta_-$ with an experimentally determined light shift per photon $\Delta_- \approx -2\pi \times 0.36$ Hz. With $N_a = 5 \times 10^4$ atoms $\delta_- = -2\pi \times 9$ kHz, which amounts to -2κ, i.e., the cavity operates in the regime of strong cooperative coupling. For σ_+-light $\Delta_+ \approx -2\pi \times 0.16$ Hz.

3 Axial Coupling

The unique feature of the cavity in the present work is its field decay rate κ on the order of the recoil frequency $\omega_{\mathrm{rec}} \equiv \hbar k^2/2m$ (where $k \equiv 2\pi/\lambda$ and $m =$ atomic mass). As a consequence, cavity-assisted backscattering can only occur in a narrow resonance window such that only very few motional states are dynamically coupled. This is sketched in Fig. 2 for the simplified case when the possible formation of an intra-cavity optical lattice is neglected, and hence, the atoms are considered as freely moving rather than populating Bloch states. For an atom in the zero-momentum state $|0\rangle$ [see (a)], backscattering of a photon requires that the atom receives $2\hbar k$ momentum and hence a kinetic energy of $4\hbar\omega_{\mathrm{rec}}$. Because of the narrow cavity linewidth [indicated by the blue bar in (a)], this process only becomes possible, if the frequency of the pump photon ω_{p} is detuned to the blue side of the

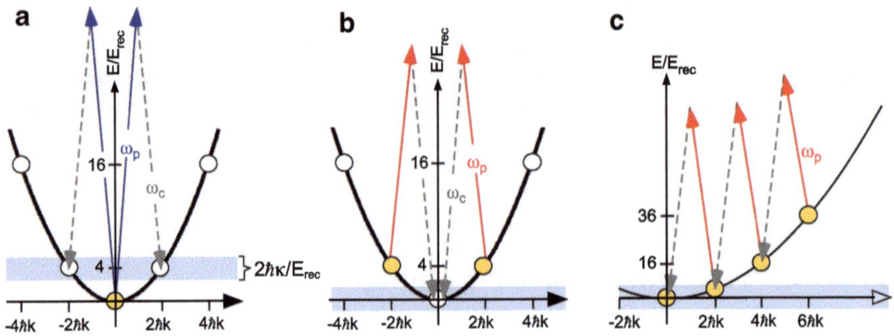

Fig. 2 Basic backscattering processes. The momentum states accessible by backscattering (at multiples of $2\hbar k$) are indicated by disks (*orange* if initially populated, *white* if initially unpopulated). **a** Heating process: The blue detuned ($\omega_p > \omega_c$) pump photons (*solid blue arrows*) are resonantly backscattered into the empty cavity (*dashed gray arrows*). The *light blue horizontal bar* represents the cavity linewidth. **b** Cooling process: Red detuned ($\omega_p < \omega_c$) pump photons (*solid red arrows*) are resonantly backscattered into the cavity mode by atoms initially prepared at $\pm 2\hbar k$. **c** Succession of cooling pulses with decreasing negative detuning $\omega_p - \omega_c$

cavity resonance ω_c by $4\omega_{rec}$. After the photon emitted into the cavity has left, a second backscattering process involving the same atom is not supported by the cavity. Hence, backscattering irreversibly populates the momentum states $|\pm 2\,\hbar k\rangle$. Similarly, if initially atoms in the $|\pm 2\,\hbar k\rangle$ states are prepared, upon offering photons detuned to the red side of the cavity resonance by $4\,\hbar\omega_{rec}$ these atoms are transferred back to the zero-momentum state [see (b)]. In (c) it is indicated how by applying a suitable sequence of light pulses, with frequencies approaching the cavity resonance from the red side, it should be possible to cool thermal atoms to less than the recoil temperature, in analogy to conventional Raman cooling, albeit at largely increased atomic densities and without the need of spontaneous photons, such that even quantum degeneracy could be reached [27].

3.1 *Recoil Selective Cavity-Assisted Scattering*

In Ref. [28], the elementary processes, sketched in Fig. 2a, b, are experimentally studied. Examples are shown in Fig. 3. In (a) and (b), after preparation of the BEC in the zero-momentum state $|0\rangle$, a (heating) pump pulse of 200-μs duration is applied with positive effective detuning $\delta_{eff}/\kappa = 4$ (with $\delta_{eff} \equiv \omega_p - \omega_c - \delta_-$). The power transmitted through the cavity (blue trace) recorded in (a), which monitors the intra-cavity lattice depth, shows oscillatory dynamics. Via backscattering by the atoms, pump photons at frequency ω_p are converted into photons resonant with the cavity, which yields a beat in the transmitted light at the difference frequency. After few cycles, most of the atoms are transferred to an excited motional state composed of the momentum states $|\pm 2\,\hbar k\rangle$, which can be seen in (b), where the momentum distribution at the end of the pulse is recorded. This is accomplished by mapping the population of the nth band onto the nth Brillouin zone. In the experiment, the lattice

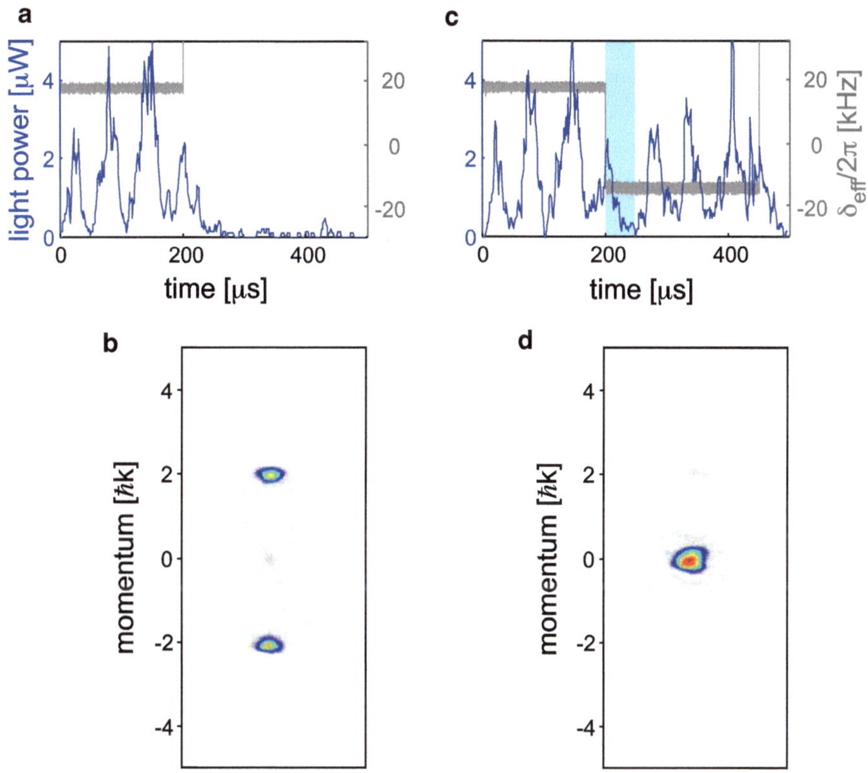

Fig. 3 Resonant momentum transfer. **a** Intra-cavity power for a 200-μs-long heating pulse with positive detuning $\delta_{\mathrm{eff}}/\kappa = 4$. **b** Population of momentum classes produced by the pulse in **a**. **c**, **d** correspond to **a**, **b**, however, after an additional 200-μs-long cooling pulse with negative detuning $\delta_{\mathrm{eff}}/\kappa = -3.1$. The *light blue bar* indicates a 40-μs-long period with the pump beam blocked

is switched off with moderate speed determined by κ and, after a 25-ms ballistic flight, the atomic density distribution is recorded. The plot in (b) shows that after the excitation pulse, both momenta $\pm 2\hbar k$ are equally populated, while the zero-momentum class is entirely depleted. The grayish background arises because atoms in the different momentum states $(0, \pm 2\hbar k)$ elastically scatter into a continuum of scattering states. While these atoms (up to 60%) remain in the cavity and contribute to the atom–cavity coupling, their dynamics is not captured by a dual-mode description. Note that no atoms with momenta $\pm 2n\hbar k$ with $n > 1$ are observed, which shows that excitations into higher bands by multiple backscattering are in fact suppressed due to the sub-recoil energy resolution of the cavity. In (c) and (d) about 40 μs after termination of the blue detuned excitation pulse discussed in (a) and (b), a red detuned (cooling) pulse of the same duration is applied ($\delta_{\mathrm{eff}}/\kappa = -3.1$), which transfers the atoms back to zero momentum. A detailed study of the axial coupling scenario in Ref. [29] further investigates the superradiant dissipative character of the processes shown in Fig. 3 and discusses the occurrence of limit cycle dynamics and the observation of dispersive optical bistability in cavity transmission spectra.

3.2 Opto-Mechanical Bloch Oscillations

A useful application of the axial coupling scenario is the in-situ monitoring of Bloch oscillations [30, 31]. In optical lattices with their perfect periodic structure, Bloch oscillations can be used as a quantum mechanical force sensor, for example, for precise measurements of the gravitational acceleration. However, the destructive character of the measurement process in previous experimental implementations typically poses notable limitations for the precision of such measurements [32–34]. In Ref. [35], we have explored a nonlinear opto-mechanical scenario consisting of a Bose–Einstein condensate (BEC) performing Bloch oscillations in the intra-cavity lattice formed in an optical cavity strongly coupled to the atoms. We show that back-action of the oscillating atoms on the cavity mode leads to a concomitant oscillation of the intra-cavity lattice depth, which can be non-destructively monitored in the light leaking out of the cavity. Hence, with a single atomic sample, the Bloch oscillation dynamics can be mapped out, while in the case of Bloch oscillations in a conventional optical lattice, each data point requires the preparation of a new atomic cloud. According to theoretical works [36–40], the frequency of our coupled system agrees with the Bloch frequency in a conventional lattice with the same lattice constant. Hence, the use of a cavity-based monitor should help to improve the precision of Bloch oscillation measurements for metrological purposes.

The mechanism leading to an oscillation at the Bloch frequency in the cavity transmission is as follows: In a simplified picture of Bloch oscillation dynamics, the atomic wave packet may be approximated by a Bloch function with a quasi-momentum $q(t) = q_0 + d^{-1}\Omega_B\, t$ linearly increasing in time, which therefore periodically crosses through the first Brillouin zone (FBZ). Here, $\Omega_B \equiv Fd/\hbar$ is the Bloch frequency exclusively depending on the size of the force F and the lattice constant d. The associated atomic density oscillates between a nearly constant shape for q values near to the center of the FBZ and a grating commensurate with the intra-cavity light intensity for q values near the edge of the FBZ. As a consequence, the resulting atom–cavity coupling strength oscillates with the frequency Ω_B. This leads to an oscillating shift of the cavity resonance frequency and hence to a periodic change of the in-coupled light intensity, which gives rise to a phase and amplitude modulation of the transmitted light intensity at the frequency Ω_B.

An example of a typical data set, taken from Ref. [35], is shown in Fig. 4 with the details deferred to the figure caption. In this implementation, the number of observable Bloch cycles is limited by an extensive binary collision rate that leads to decoherence and heating. This can be avoided in experiments optimized for a

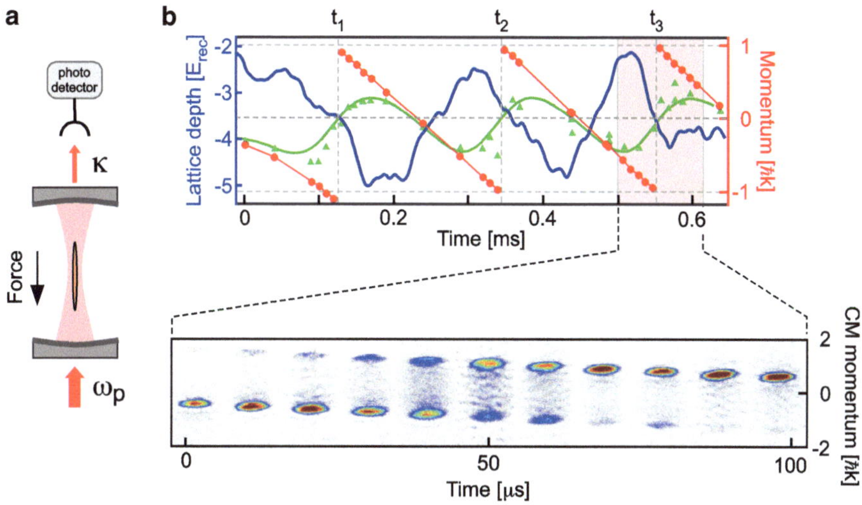

Fig. 4 **a** Sketch (not drawn to scale) of a BEC placed inside a standing wave cavity, which is axially pumped at frequency ω_p. A magnetic force is applied along the cavity axis. **b** Typical data set: The *blue thick line* shows the intra-cavity intensity [parameterized in terms of the associated (negative) intra-cavity light shift], observed during ca. 600 μs by the photodetector in **a** for the same BEC in a single experimental run. The *red disks* (connected by *solid red straight line* segments for eye guiding) show the positions of the density maxima of the atomic cloud versus the holding time in the lattices. The *green triangles* show the corresponding center of mass of the atomic cloud. The *green solid lines* show a calculation for $\delta_c = -5.8\,\kappa$. Each *data point* requires preparation of a new BEC. For the section of the time axis highlighted by a *reddish background*, images of the BECs after ballistic expansion are shown associated with the *data points*

precise determination of the Bloch frequency by adjusting zero scattering length exploiting a Feshbach resonance [37, 41] or by using atoms with a naturally small scattering length like strontium [32].

4 Transverse Traveling Wave Coupling

4.1 Matter Wave Superradiance

In this section, we discuss a study of the collective nature of Rayleigh scattering by a BEC, when the scattered photons are guided by a high finesse cavity combining a Purcell factor far above unity with extreme energy resolution beyond the single-photon recoil energy [42]. A schematic of the setup is sketched in Fig. 5a. This scenario extends a previous experimental study at MIT of Rayleigh scattering from a BEC in free space [43]. In the MIT work, an elongated BEC was irradiated by a monochromatic off-resonant pump laser beam perpendicularly with respect to its long axis. Above a critical pump intensity, the emission of superradiant light pulses along the long axis of the BEC was observed, while the atoms were collectively

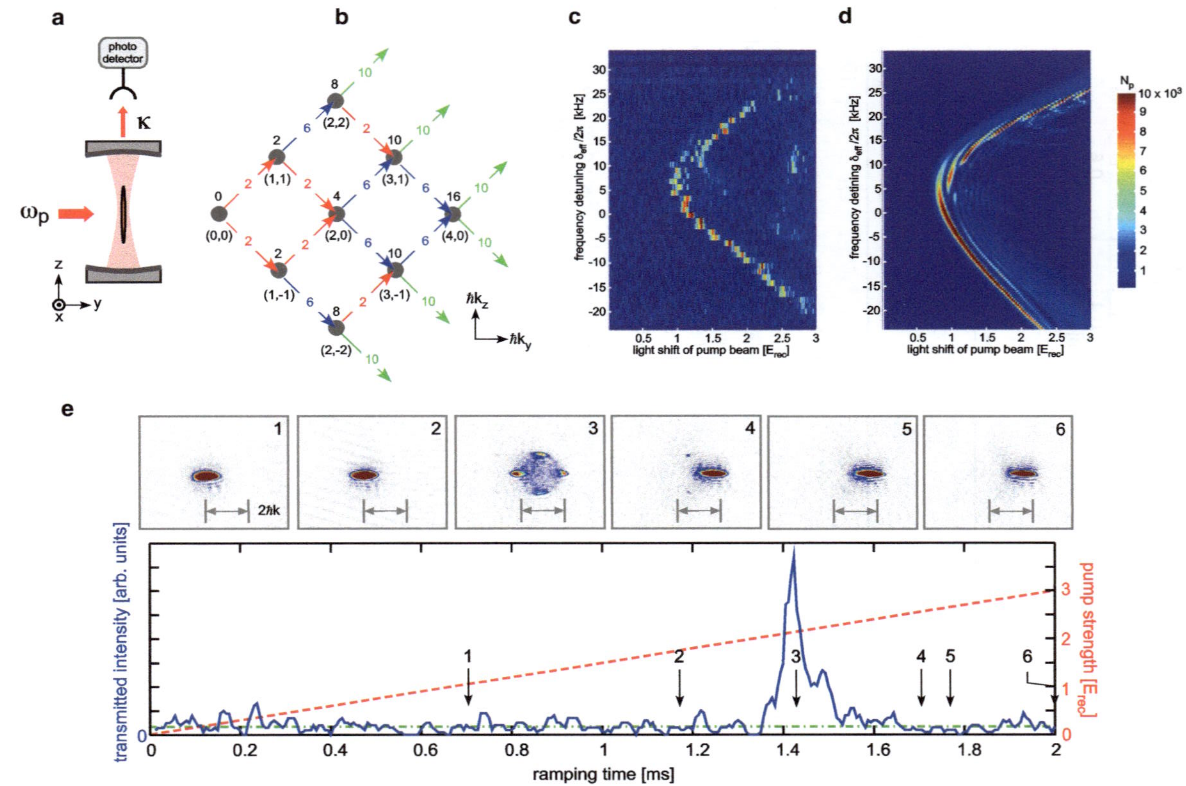

Fig. 5 **a** Sketch of the scattering geometry. The pump is linearly polarized along the x-axis. **b** Relevant momentum states coupled to the BEC via scattering of pump photons. The tuples (n, m) below the filled *gray disks* denote the momenta along the y and z directions in units of $\hbar k$. The *numbers* above the disks

scattered into a fan of discrete free-space momentum states. If the BEC is enclosed by a recoil-resolving cavity as in our recent study in Ref. [42], the scattering of atoms into higher momentum states can be precisely controlled via adjustment of the cavity resonance frequency. Via control of the available scattering channels, complex spatially periodic matter states can be synthesized, which preserve the full coherence of the initial condensate. With appropriate pulse sequences, the entire atomic sample can be collectively accelerated or decelerated by multiples of two recoil momenta.

As a consequence of the sub-recoil bandwidth, similarly as in the case of axial coupling, only a few selected motional states are coupled. This is sketched in Fig. 5b for the simplified case when the transient formation of an intra-cavity optical lattice and the external trap are neglected, and hence, the atoms are considered as freely moving. Scattering of a pump photon by a BEC atom into the cavity corresponds to a transition from the $(0,0)\,\hbar k$ to the $(1,\pm 1)\,\hbar k$-momentum states. Energy conservation requires $\omega_p - \omega_{\text{scat}} = 2\,\omega_{\text{rec}}$ with ω_p and ω_{scat} denoting the pump frequency and the frequency of the scattered photon, respectively. The scattering process is best supported by the cavity, if ω_{scat} coincides with the effective cavity resonance frequency $\omega_{c,\text{eff}} \equiv \omega_c - \delta_-$ with δ_- according to Sec. 2 and ω_c denoting the resonance frequency of the empty cavity, i.e., the effective detuning $\delta_{\text{eff}} \equiv \omega_p - \omega_{\text{c,eff}}$ should satisfy $\delta_{\text{eff}} = 2\,\omega_{\text{rec}}$. The same detuning allows to resonantly scatter a second photon bringing the atom to the $(2,0)\,\hbar k$ state. Further scattering, which would transfer the entire atomic sample via the $(3,\pm 1)$ $\hbar k$ states to the $(4,0)\,\hbar k$ state, is not supported by the cavity unless δ_{eff} is modified to account for the significantly larger energy costs of 6 recoil energies per atom. Due to the back-action of the scattered photons on the atomic sample, the scattering mechanism is expected to acquire collective character leading to the emission of a superradiant light pulse along the cavity axis: If the initial sample, a BEC in the $(0,0)\,\hbar k$ state, was perfectly homogeneous, scattering would be prevented by destructive interference from contributions from different locations within the BEC. Hence, quantum or thermal fluctuations are required to start the scattering process. Once a few photons are scattered into the cavity, the light field building up inside the cavity oscillating at frequency $\omega_p - 2\omega_{\text{rec}}$ interferes with the pump wave with frequency ω_p. The result is a moving optical potential and a commensurate moving density grating formed by the atoms transferred into the

Fig. 5 (continued) indicate the kinetic energy in units of the recoil energy. The *colored numbers* on the *arrows* indicate the kinetic energy transfer associated with the respective scattering process. **c** The intra-cavity photon number N_p is recorded for variable fixed values of δ_{eff}, while ε_p is linearly increased from 0 to 3 in 2 ms. **d** Mean field calculation of the intra-cavity photon number for the pump strength ramp applied in **c**. **e** The *blue solid trace* shows the intensity leaking out of the cavity, while the pump strength is ramped from zero to $3\,E_{\text{rec}}$ in 2 ms (as indicated by the *red dashed trace*) with negative detuning $\delta_{\text{eff}}/2\pi = -12$ kHz. The *insets* numbered 1–6 show (single shot) momentum spectra taken at times marked by the *black arrows*. The *green dashed-dotted line* indicates the noise floor of the light detection

$(1, \pm 1)\hbar k$-momentum states. This grating acts as a Bragg grating, which enhances the scattering efficiency such that the optical standing wave and the corresponding matter grating grow in an exponential process reaching maximal values, when most atoms populate $(1, \pm 1)\hbar k$. Their further transfer to $(2,0)\hbar k$ suppresses superradiance again, since in this state, no density grating along the cavity axis can be formed. Hence, scattering terminates and the atoms remain in the state $(2,0)\hbar k$ with no photons in the cavity.

Figure 5c, d shows a measurement (c) and a calculation (d) of the instability boundary in the plane spanned by the effective pump detuning δ_{eff} and the pump strength ε_p. In both plots, the pump strength is ramped up from zero to $3\,\omega_{\text{rec}}$ in 2 ms for different settings of δ_{eff}, and the intra-cavity intensity, is plotted according to the specified color code. The observed and calculated instability boundaries depend on the duration of the ε_p-ramp, since the system is unstable in the entire $(\varepsilon_p, \delta_{\text{eff}})$-plane, as a linearization of the dynamical equations around the solution of zero intra-cavity intensity and no condensate depletion shows [42]. As expected from Fig. 5b, the most efficient scattering (i.e., scattering for minimal pump strength) arises at about 7 kHz corresponding to $\delta_{\text{eff}} \approx 2\,\omega_{\text{rec}}$, when the $2\,\omega_{\text{rec}}$-process is resonant. In (e) a horizontal section through (c) is analyzed in more detail with momentum spectra recorded at several instances of time. It is seen that a rapid sudden transfer of the BEC to the $(2,0)\hbar k$ momentum state is observed with a superradiant light pulse emitted into the cavity.

4.2 Superradiant Atom Accelerator

The cavity-assisted steering of the scattering channels may be used for controlled acceleration of a BEC. In order to successively transfer the BEC through the momentum states $(2n,0)\,\hbar k$ from $n = 0$ to $n = 5$ (see Fig. 6a), a series of five pump pulses were applied with frequencies adjusted to successively tune the scattering by atoms populating the $(2n,0)\,\hbar k$ momentum state to resonance. The variation of the pump detuning is recorded as the green solid line in Fig. 6b. The blue solid line shows the superradiant pulses emitted into the cavity enabling the momentum transfer. In Fig. 6c, a series of momentum spectra demonstrate that the entire BEC is in fact accelerated. The total momentum after five scattering pulses, each transferring $2\,\hbar k$ momentum, turns out to be less that $10\,\hbar k$, because the atoms are also decelerated by climbing up the harmonic trap potential.

5 Transverse Standing Wave Coupling

If the traveling pump wave in Fig. 5a is replaced by a standing wave, surprisingly different physics is encountered. At negative pump detuning δ_{eff}, beyond a critical pump strength, a stationary self-organized intra-cavity optical lattice can build

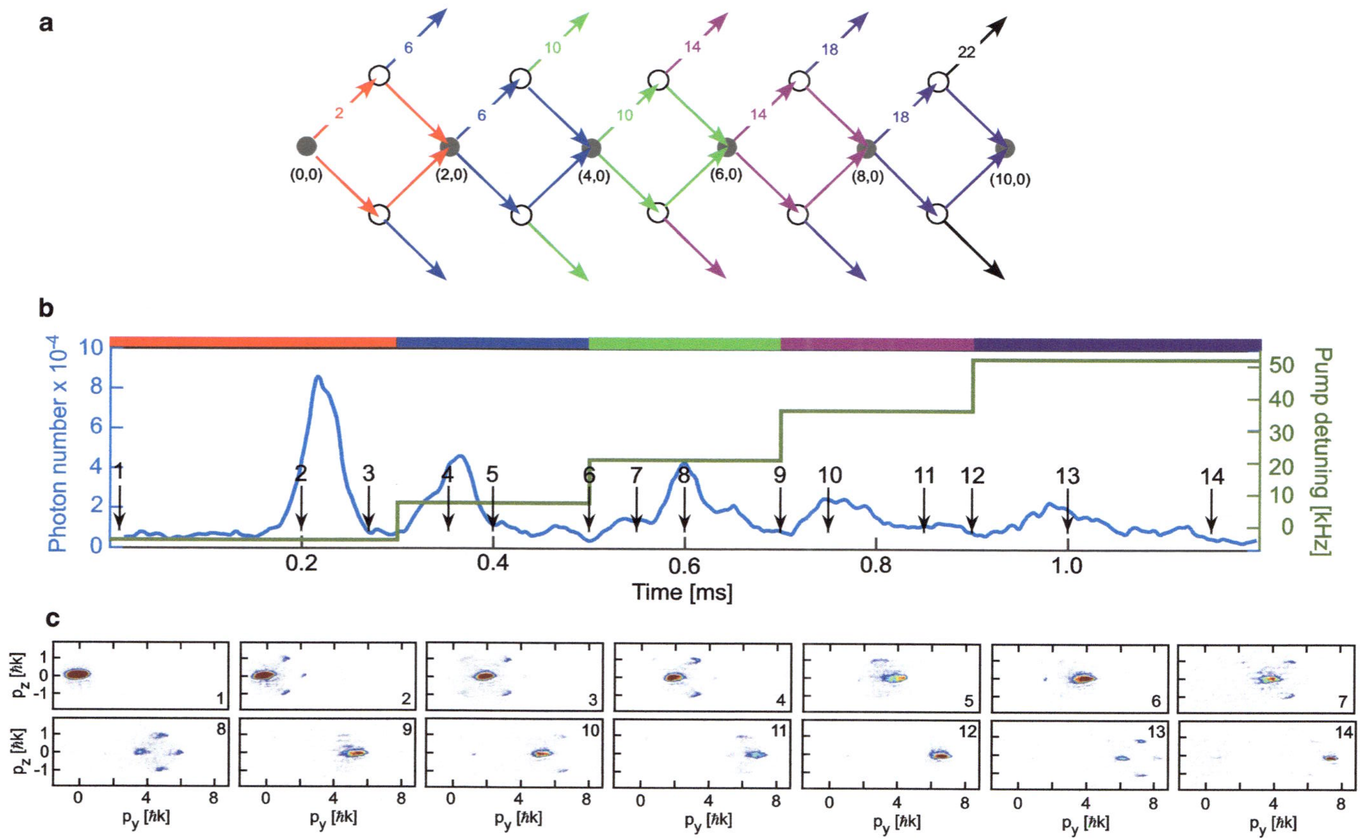

Fig. 6 **a** The *white* and *gray disks* symbolize different momentum states with momenta $(n, m)\,\hbar k$ specified below the *gray disks*. The *colored arrows* symbolize different scattering processes in momentum space. The *numbers* on *top of the arrows* specify the associated number of recoil energies transferred. **b** The *green solid line* shows the detuning of the pump beam. The *solid blue line* shows the observed intra-cavity intensity. **c** Momentum spectra recorded at the instances of time indicated by *numbers* and *arrows* in **b**

up. The momentum and energy budget for this process is illustrated in Fig. 7a. Since we operate at negative detuning with respect to the relevant atomic resonances (the case of normal dispersion), the intra-cavity lattice comes with a negative light shift, such that scattering of a pump photon negatively detuned (red arrow in Fig. 7a) with respect to the cavity resonance allows one to reach a bound state in the intra-cavity lattice (indicated by a green bar in the lower detail). The associated self-organization phase transition has been predicted and experimentally demonstrated for thermal atomic samples in Refs. [44, 45], respectively. Later, a similar transition was demonstrated using a BEC [46] and it was emphasized that close to the transition boundary, the system can be approximated by the driven open Dicke model [46–49]. For positive pump detuning (blue arrow in Fig. 7a), bound states in an intra-cavity lattice cannot be reached and hence a stationary intra-cavity lattice is not formed in this regime. Rather, for arbitrarily small pump strength, the system is excited to free-space momentum states in a cascade of pulsed superradiant scattering events, similarly as for any value of δ_{eff} in the case of a traveling wave pump, considered in the previous section.

5.1 Superradiant Self-organization Transition

The self-organization transition at negative δ_{eff} may be understood according to Fig. 7b, c. For small pump intensity (b), the atoms are merely trapped in the pancake-like high-intensity regions of the quasi 1D pump lattice along the y-axis, which should lead to a momentum spectrum showing, aside from the dominating zero-momentum condensate peak, further higher-order contributions at multiples of $p_y = \pm 2\,\hbar k$ [as sketched in the lower detail of (b)]. Along the cavity axis (z-axis), a nearly homogeneous density distribution prevails such that, for every atom ready to scatter a photon into the cavity, a second atom can be found, which would scatter with opposite phase, such that destructive interference prevents scattering altogether. As a consequence, no intra-cavity light field arises. This phase is henceforth addressed as subradiant, superfluid (SubSF). Above a critical strength of the pump field, initiated by quantum or thermal fluctuations, an exponential instability occurs and a 2D light field is formed by interference of the pump field and the intra-cavity field, which are both linearly polarized along the x-axis. This is illustrated in Fig. 7c. In the relevant case of normal dispersion, if $|\delta_{\text{eff}}| \gg \kappa$, the intra-cavity field is created in phase with the pump field at the positions of all contributing scatterers. Two distinct position offsets of the resulting optical potential and the associated density grating are possible (separated by $\lambda/2$ along the z-axis). Which of them is actually realized depends on the geometry of the initial density fluctuation. The density grating acts as a Bragg grating, thus giving rise to superradiant scattering and hence a rapid increase of the optical potential, which in turn yields increasing contrast of the density grating. The emergence of this self-organized lattice is indicated by additional Bragg resonances in the momentum spectrum at

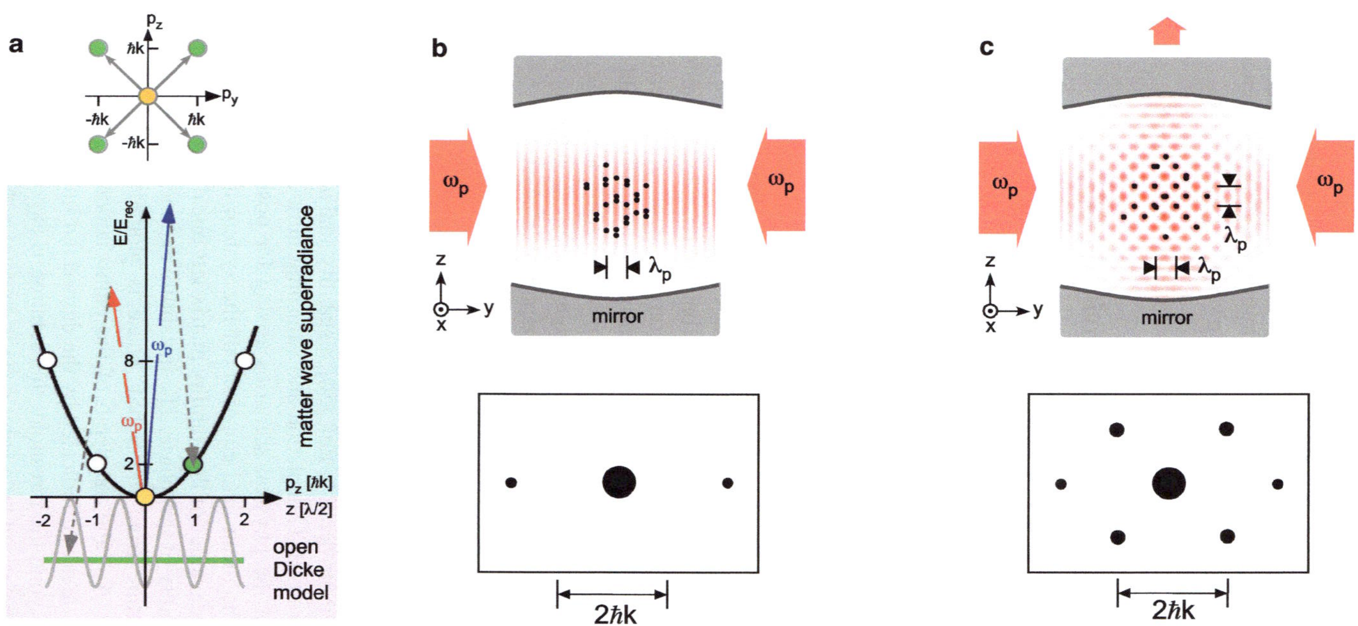

Fig. 7 **a** Momentum (upper detail) and energy budget (lower detail). **b** Subradiant superfluid phase (SubSF): The BEC is trapped in the ground state of the quasi 1D pump lattice, and no light is scattered into the cavity. **c** Superradiant superfluid phase (SupSF): A 2D optical lattice is formed; the atoms are spatially structured to form a Bragg grating, which optimizes scattering into the cavity. The lower details in **b** and **c** show schematics of the expected atomic momentum distributions

multiples of $(p_x, p_y) = (\pm\hbar k, \pm\hbar k)$. Once the optical lattice is formed, the atomic density grating is commensurate with the optical trapping potential, i.e., every available lattice site carries the same average population. For this reason, it appears misleading to call this phase a supersolid or a density wave although its position offset arises via spontaneous breaking of an initial Z_2 symmetry. Henceforth, it is denoted as superradiant, superfluid phase (SupSF).

5.2 Validity of Dicke Model and Cavity Bandwidth

Close to the critical pump strength, the self-organization phase transition simulates the second-order phase transition expected to occur in the driven open Dicke model [46–49]. This applies in the regime where it is justified to assume a two-mode scenario for the motional states of the atomic matter, i.e., the BEC mode $|0,0\rangle$ and the superposition $\frac{1}{4}\sum_{\nu,\mu\in\{-1,1\}}|\nu\hbar k, \mu\hbar k\rangle$. In atom–cavity systems with a large cavity bandwidth, the momentum states $|\pm\hbar k, \pm\hbar k\rangle$ are coupled to a multitude of higher-order motional states such that a description in terms of the Dicke Hamiltonian is only justified in the limit of vanishing condensate depletion, i.e., very close to the self-organization phase transition boundary. In case of a narrow cavity bandwidth, due to the scattering blockade effect, the coupling to higher-order momentum states is significantly decreased (but not completely eliminated), and hence, the regime of validity of the Dicke picture is extended. Furthermore, due to the comparable time scales for the evolution of the matter and light degrees of freedom, the system is easily driven away from equilibrium. Hence, the observation of the equilibrium phase transition requires an extremely slow tuning of the pump strength across the critical value in order to remain adiabatic.

5.3 Stability Analysis and Phase Diagram

As is discussed in more detail in Ref. [50], a stability analysis for the SubSF phase, characterized by zero BEC depletion and vanishing intra-cavity field, reveals the different nature of the physical scenarios arising for positive and negative δ_{eff}. Linearization of the mean field equations of the system in the SubSF phase and analysis of the eigenvalues of the stability matrix lets one determine the rate for an exponential instability of this phase. The instability exponent, which is the rate for exponential departure from the SubSF phase is plotted in Fig. 8a versus the effective detuning δ_{eff} and the depth of the pump lattice ε_p. For negative δ_{eff}, below the critical threshold indicated by the dashed red line, the SubSF phase is a stable solution. Beyond the threshold, this solution is exponentially unstable with an exponent parameterized by a color code in units of κ. Note the interesting fact that when δ_{eff} approaches zero, the SubSF phase remains stable for arbitrarily strong

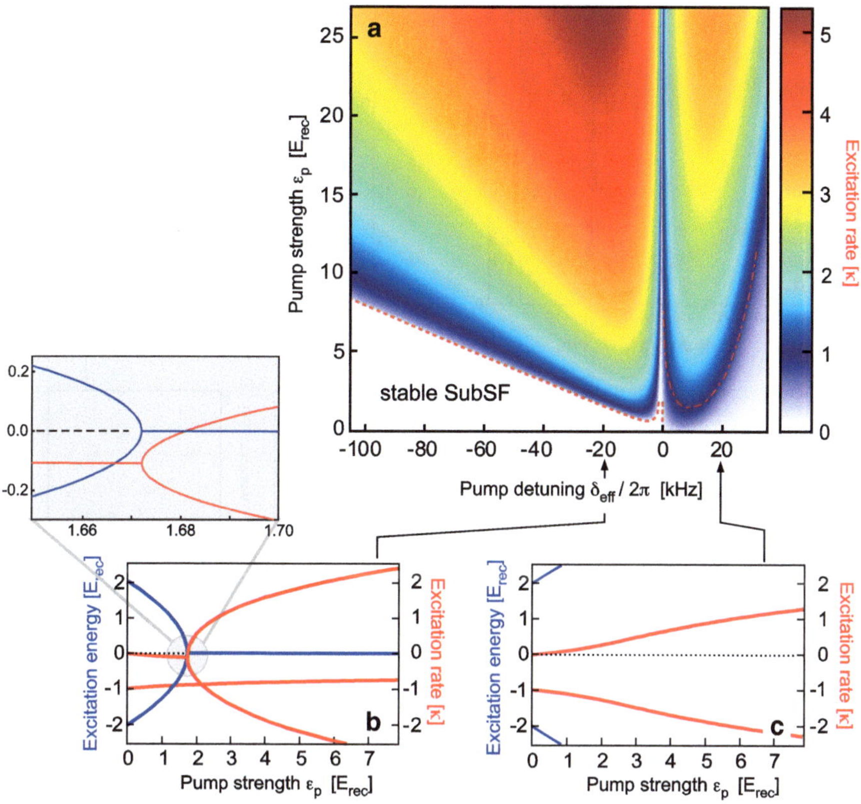

Fig. 8 **a** Instability exponent plotted versus the effective detuning δ_{eff} and the depth of the pump lattice ε_p. **b**, **c** show sections through **a** for $\delta_{\text{eff}}/2\pi = \pm 20$ kHZ. In addition to the instability exponent (*red traces*), also the excitation energies are plotted (*blue traces*)

pumping. The reason is as follows: When $|\delta_{\text{eff}}|$ becomes smaller than κ, the relative phase between the intra-cavity field and the pump field approaches $\pi/2$ as $|\delta_{\text{eff}}| \to 0$. This suppresses the interference between both fields, i.e., their intensities simply add. As a result, the unit cell develops a second minimum, which approaches equal depth for $\delta_{\text{eff}} \to 0$. The associated density grating, now composed of atoms trapped in both classes of minima, no longer supports Bragg scattering of pump photons into the cavity, and hence, the intra-cavity field and the density grating collapse. In Fig. 8b, a vertical section through (a) at $\delta_{\text{eff}}/2\pi = -20$ kHz is shown. In addition to the instability exponent (red traces), also the excitation energies are plotted (blue traces). The plot shows that the lowest excitation (of the $2\,E_{\text{rec}}$-mode) undergoes softening when the critical pump strength is approached. An expansion of the vicinity of the critical pump strength shows that there is a small interval of ε_p, where this mode has acquired zero frequency, while the SubSF phase yet remains stable. In contrast, in the case of positive δ_{eff}, the SubSF phase is everywhere

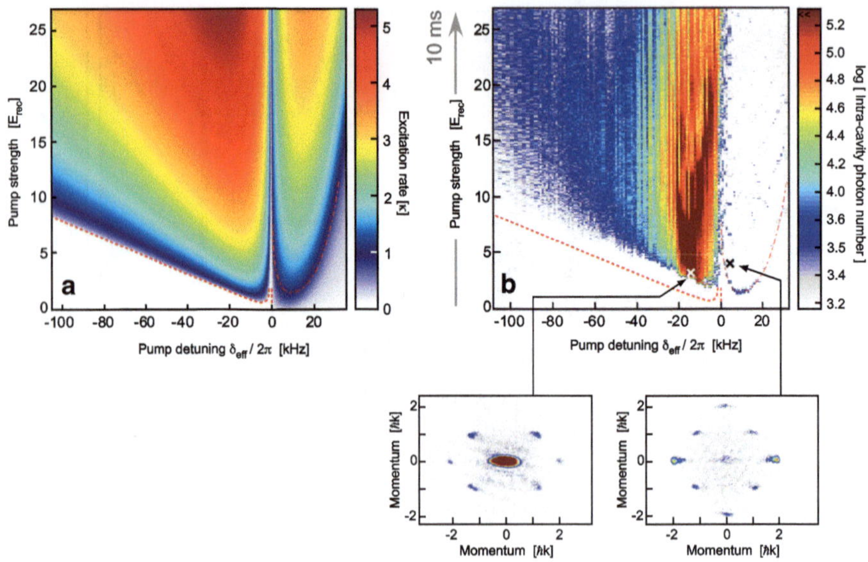

Fig. 9 The stability diagram of Fig. 8 is repeated in **a** and compared to an observation of the intra-cavity lattice depth in **b**. *Insets* show momentum spectra recorded at the positions indicated by a *white* and a *black cross*, respectively

unstable. This is emphasized by the positive red trace in Fig. 8c showing a vertical section through (a) for $\delta_{\mathrm{eff}}/2\pi = 20$ kHz.

In Fig. 9 the stability diagram of Fig. 8, repeated in (a), is compared to observations for the same section of the $(\delta_{\mathrm{eff}}, \varepsilon_p)$-plane in (b). In (b) the intra-cavity lattice depth (parameterized by the indicated color scheme) is recorded while ramping up ε_p in 10 ms from zero to $26\,E_{\mathrm{rec}}$ for different settings of δ_{eff}. In the half-plane of negative δ_{eff} a stationary intra-cavity lattice is observed above a critical value of ε_p, slightly larger than the predicted second-order equilibrium phase boundary indicated by the dashed red line in (a). A momentum spectrum recorded at the position indicated by the white cross shows distinct Bragg peaks as expected for a coherent stationary 2D optical square lattice. In the positive δ_{eff} half-plane, no stationary lattice arises. Rather, when the contour of the instability exponent indicated by the red dashed line in (a) is crossed, a short intense spike of light is emitted into the cavity, and the system becomes irreversibly excited. This is exemplified by the momentum spectrum recorded at the position indicated by the black cross, which shows a nearly complete depletion of the condensate peak. The emission of superradiant light pulses observed for positive δ_{eff} reflects the same physics of matter wave superradiance found to occur at any value of δ_{eff} in the case of traveling wave coupling, discussed in Sec. 4.

5.4 Quench Across Phase Boundary, Hysteresis

A more complete picture of the physics encountered for negative δ_{eff} is provided in Fig. 10, where the transition through the phase boundary is studied for $\delta_{\mathrm{eff}} = -2\pi \times 17.5$ kHz in more detail. In (a) ε_p is ramped up from 0 to 4 in 1.5 ms and back to 0 again in 1.5 ms. The solid blue and red lines show the observed intra-cavity intensity for the increasing and decreasing sections of the ramp, respectively. Note that this quantity measures the depth of the intra-cavity lattice emerging in the SupSF phase and hence corresponds to the square of the order parameter for the Dicke phase transition. A significant hysteresis is observed. For increasing ε_p, a sudden jump of the intra-cavity intensity arises on a time scale corresponding to the cavity decay rate. On the way back, the intra-cavity intensity is smoothly tuned to zero. In the center row of the figure [below (a) and (b)], a series of consecutively numbered momentum spectra are shown, recorded at different instances of time during the ε_p-ramp, indicated by the correspondingly numbered arrows in (a). As the intra-cavity intensity assumes finite values, a coherent optical lattice is formed (arrow 2), as is seen from the occurrence of higher-order Bragg peaks. As the lattice depth grows (arrows 3 and 4), tunneling amplitudes decrease, and the relatively increased collisional interaction acts to reduce particle number fluctuations resulting in a partial loss of coherence. When ramping back to small values of ε_p, the BEC is recovered with no notable atom loss and only few low-energy Bogoliubov excitations (arrow 5).

In Fig. 10b a mean field calculation is shown for a homogeneous, infinite system without collisional interaction, which shows the same signatures as observed in (a) including dynamical details as the oscillation of the red trace around $\varepsilon_p \approx 2.5$ and the overshooting of the blue trace around $\varepsilon_p \approx 3.5$. The observed hysteresis appears fundamentally different from that known to occur in conventional bistable systems, where discontinuities arise for both critical values, where the system becomes unstable. We do not find a discontinuity at the lower critical value $\varepsilon_{p,1}$ in Fig. 10a, b; however, the system always follows the blue curve, when this point is passed with increasing ε_p, irrespective of the duration τ_Q of the applied ε_p-ramp. For increasing cavity bandwidths, our mean field calculations predict that the area enclosed by the hysteresis decreases and finally is obscured by increasing opto-mechanical oscillations at the phase boundary.

5.5 Power Law Behavior

The dependence of the threshold values $\varepsilon_{p,1}$ and $\varepsilon_{p,2}$ for the dynamical transitions in Fig. 10a, b upon the quench time τ_Q is studied in (c)–(e). These quantities are determined as those values of ε_p, where the intra-cavity intensity assumes 5% of its maximal value reached for $\varepsilon_p = 4$. In (c) the values of $\Delta\varepsilon_{p,\mu}(\tau_Q) \equiv \varepsilon_{p,\mu}(\tau_Q) - \varepsilon_{p,\mu}$ $(\tau_Q = \infty)$ $(\mu \in \{1,2\})$, calculated from curves as that shown in (b), are

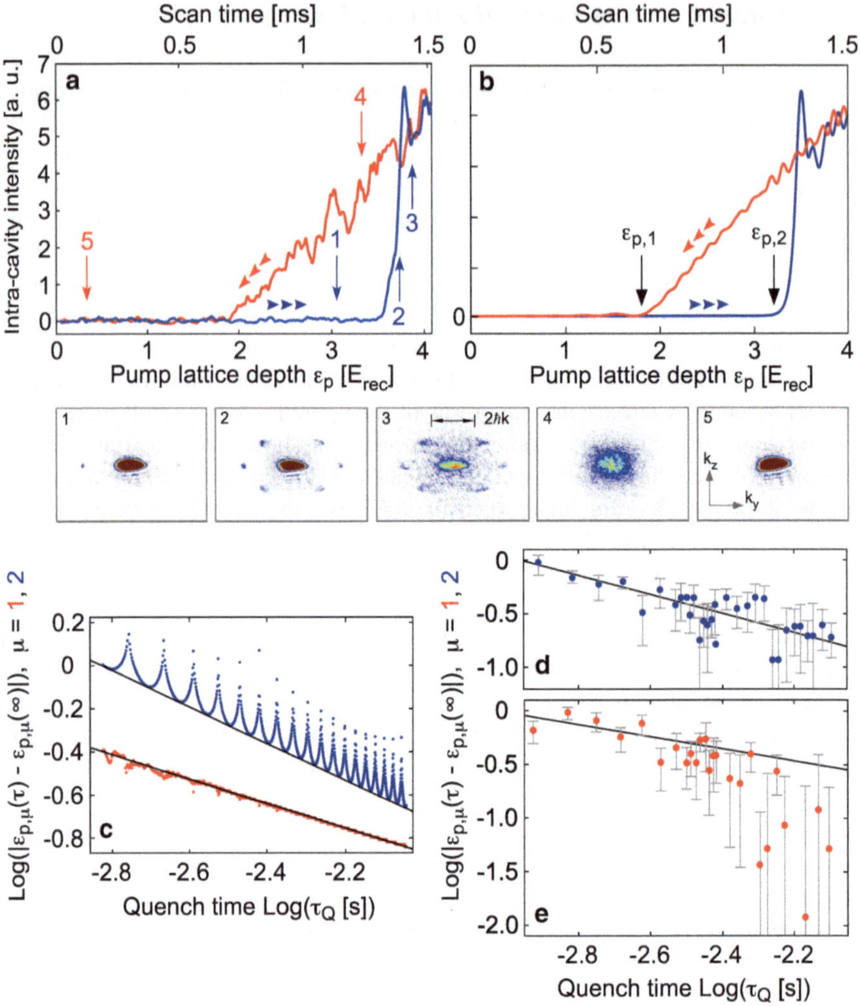

Fig. 10 **a** For fixed $\delta_{\text{eff}} = -2\pi \times 17.5$ kHz the intra-cavity intensity is plotted with the pump strength ramped from 0 to $4\,E_{\text{rec}}$ in 1.5 ms (*blue line*) and back (*red line*). Below **a**, a series of consecutively numbered momentum spectra are shown (*1*–*5*), recorded at increasing times during the ε_p-ramp, indicated by the correspondingly numbered arrows in **a**. **b** A mean field calculation according to **a** for a homogeneous, infinite system without collisional interaction. In **c** mean field calculations of $\varepsilon_{p,1}$ (*lower graph*, *red dots*) and $\varepsilon_{p,2}$ (*upper graph*, *blue dots*) are shown. The *solid lines* show power laws with exponents $n_1 = -0.57$ and $n_2 = -0.85$ in the *lower* and *upper graph*, respectively. The measured dependence of the critical values $\varepsilon_{p,\mu}, \mu \in \{1, 2\}$ upon the quench time τ_Q is shown for $\mu = 2$ in **d** and for $\mu = 1$ in **e**. The *solid lines* repeat the power laws found in the mean field calculations in **c**. The *error bars* reflect the standard deviations for 10 measurements

plotted versus τ_Q. As shown by the solid lines, the τ_Q-dependences follow power laws $\Delta\varepsilon_{p,\mu}(\tau_Q) \propto \tau_Q{}^{n_\mu}$ with $n_1 = -0.57$ and $n_2 = -0.85$. The phase offset of the sharp resonances occurring periodically at a frequency $\Omega = 0.682\,\omega_{\mathrm{rec}}$ in the upper graph depends on the specific choice of a small initial excitation, necessary to drive the system out of the SubSF phase, which is provided by quantum and thermal fluctuations in the experiment. The exponents n_1, n_2 turn out independent of the exact initial conditions (for details see Ref. [50]). In (d) and (e) we plot the experimentally observed values of $\Delta\varepsilon_{p,\mu}(\tau_Q)$ with $\mu = 2$ and $\mu = 1$, respectively. The solid lines repeat the power laws found in the calculations in (c) with $n_2 = -0.85$ in (d) and $n_1 = -0.57$ in (e). While in (d) the data nicely agree with the power law behavior, in (e) this is only the case for the first half of the plot. At later times, the data points assume an exponential rather than a power law decay, which is in accordance with the observation that for long ramp times at the end of the descending ramp, notable particle loss sets in.

5.6 *Kibble–Zurek Picture*

The fact that the observed intra-cavity intensity is a direct measure of the order parameter together with the observations of power law behavior of $\Delta\varepsilon_{p,\mu}(\tau_Q)$ suggests an interpretation within the universal model introduced by Kibble and Zurek [51–53], which applies for second-order phase transitions in isolated many-body systems. According to this model, a quench between two phases is approximated by a succession of an adiabatic approach toward and a departure from the equilibrium critical point $\varepsilon_{p,c}$ conjoined by a diabatic passage through the critical point, where the dynamics is completely frozen (c.f. Fig. 11). The time $t(\varepsilon_p)$ elapsing during the ε_p ramp after passing $\varepsilon_{p,c}$ may be parameterized as $t(\varepsilon_p) = \frac{\varepsilon_p - \varepsilon_{p,c}}{\varepsilon_{p,2} - \varepsilon_{p,1}}\,\tau_Q$ with the quench time τ_Q. Furthermore, a power law dependence for the relaxation time is assumed, i.e., $\tau_\mu(\varepsilon_p) \propto |\varepsilon_p - \varepsilon_{p,c}|^{-z_\mu \nu_\mu}$ with $\mu \in \{1, 2\}$ if $\varepsilon_p < \varepsilon_{p,c}$ and $\varepsilon_p > \varepsilon_{p,c}$, respectively. The identification of $\Delta\varepsilon_{p,\mu}$ with the lower and upper bounds of the diabatic region around $\varepsilon_{p,c}$ (i.e., $|t(\varepsilon_{p,\mu})| = \tau_\mu(\varepsilon_{p,\mu})$) then leads

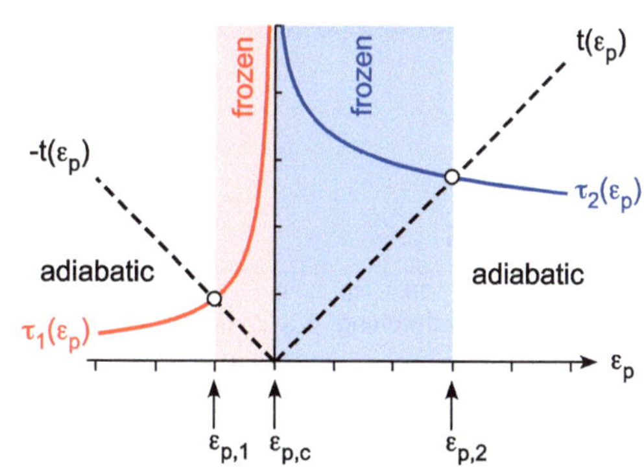

Fig. 11 Kibble–Zurek scenario: $t(\varepsilon_p)$ denotes the time elapsed during ramping from $\varepsilon_{p,c}$ to ε_p and $\tau_\mu(\varepsilon_p), \mu \in \{1, 2\}$ denote the relaxation times of the system

to the prediction that $z_\mu \nu_\mu = -(1 + \frac{1}{n_\mu})$, i.e., in our system: $z_1\nu_1 = 0.75, z_2\nu_2 = 0.18$ (for details, see the supplement of Ref. [50]). A physical understanding of these values would require a comprehensive extension of the concept of universality to the case of driven open systems [54], which is a topic of intensive ongoing research.

5.7 Crossover Between Standing Wave and Traveling Wave Coupling

The striking difference of the physics observed for transverse traveling wave and transverse standing wave coupling gives rise to the question, how these two regimes are connected. Experimentally, it is easy to gradually decrease the intensity in the retro-reflected pump wave in Fig. 1c, such that the two coupling scenarios are smoothly transferred into one another. In Fig. 12, this crossover is illustrated by showing the phase diagram, recorded similarly as in Fig. 9b, for different fractions of the retro-reflected intensity $\zeta = 1, 0.25, 0.15, 0$. The plots clearly show how the

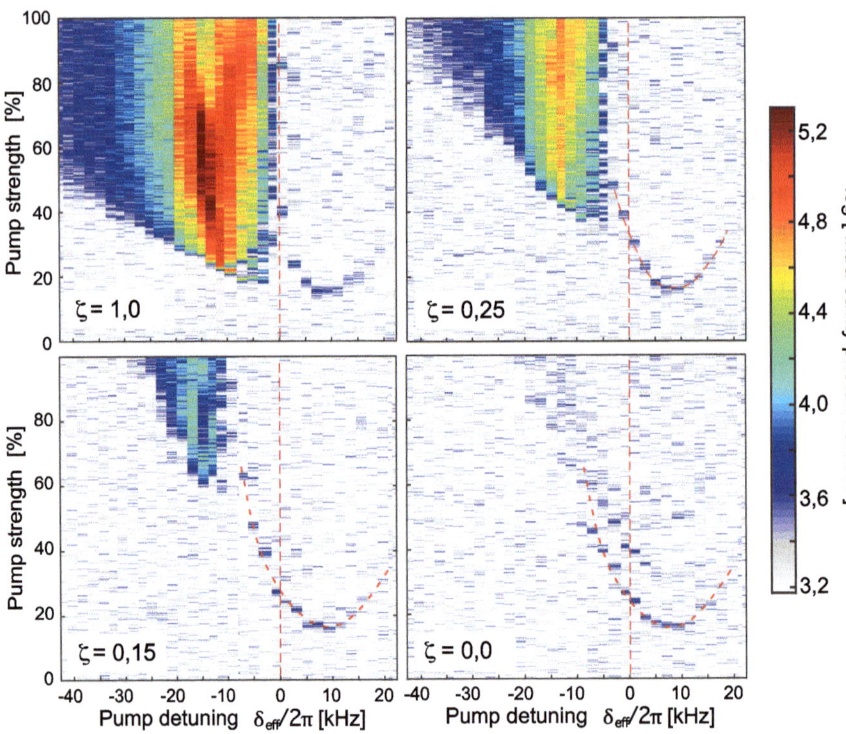

Fig. 12 Plots analogous to Fig. 9b, however, with the intensity of the retro-reflected pump wave (cf. Fig. 1c) reduced to a fraction ζ. The *vertical dashed red lines* emphasize $\delta_{\mathrm{eff}} = 0$. The *curved dashed red lines* emphasize the instability boundary for matter wave superradiance

SupSF phase pulls back toward larger values of ε_p and increasingly negative values of $\delta_{\rm eff}$ as ζ approaches zero, while the regime of matter wave superradiance progressively extends into the half-lane of negative detuning.

5.8 Superradiant Mott Insulator

Theoretical work [55, 56] has suggested that the SupSF phase discussed above should have a Mott insulating counterpart, if contact interaction dominates tunneling, similarly as found in conventional optical lattices simulating the bosonic Hubbard model [57]. To enter this regime, tight confinement is required along all spatial directions. In this subsection, we consider the setup of Fig. 1c extended by an additional standing wave oriented perpendicularly with respect to the drawing plan in Fig. 1c (i.e., parallel to the x-axis), which has been studied in Ref. [58]. An even further extended scenario with a second external lattice along the cavity axis (z-axis in Fig. 1c) has been reported in Ref. [59]. The polarization of the additional light field applied in Ref. [58] points along the cavity axis (z-axis) such that it cannot give rise to light scattering into the cavity and hence its only effect is to provide an additional periodic potential. The depth of the resulting lattice is adjusted to $14\,E_{\rm rec}$ such that tunneling is suppressed along the x-axis. As a consequence, in addition to the SupSF phase entered upon tuning the pump strength above a critical value, a further increase of the pump strength above a second critical value leads to a sudden complete loss of coherence in time-of-flight spectra, indicating the emergence of a self-organized superradiant Mott insulating phase (SupMI). By successively traversing the phase boundary in both directions, it is found that coherence is restored as the SupSF phase is reentered. This shows that the loss of coherence observed in the SupMI region cannot be attributed to irreversible heating.

By tuning $\delta_{\rm eff}$ and ε_p we drive the system along the three trajectories $\overline{\rm ABCD}$, $\overline{\rm DCD}$ and $\overline{\rm EFE}$ that intersect the SubSF, SupSF and SubMI phases in different regions of the $(\delta_{\rm eff}, \varepsilon_p)$-plane, connecting locations labeled by the capital letters A–F in Fig. 13a. Along these trajectories, in addition to the intra-cavity photon number, momentum spectra are recorded. Evaluating the widths of the Bragg resonances in these spectra, we obtain information on the degree of phase coherence of the sample along the trajectory. The exemplary case of the path $\overline{\rm ABCD}$, which is traversed in 12 ms, is detailed in Fig. 13b. The upper panel shows how $\delta_{\rm eff}$ and ε_p are changed versus time, thus successively passing the points "A–D." In the lower panel, the intra-cavity photon number N_p (red disks) and the width (FWHM) $\mathcal{W}$ of the observed zero-momentum Bragg resonance (blue diamonds) with regard to the z-direction are plotted. A reference momentum spectrum for $\varepsilon_p = 0$ is shown in Fig. 13c, labeled "1." Due to technical reasons detailed in Ref. [58], as the pump strength is raised to $5\,E_{\rm rec}$ without a notable intra-cavity photon number yet arising, $\mathcal{W}$ slightly decreases. We thus have chosen "B" as the reference point for $\mathcal{W}$, setting

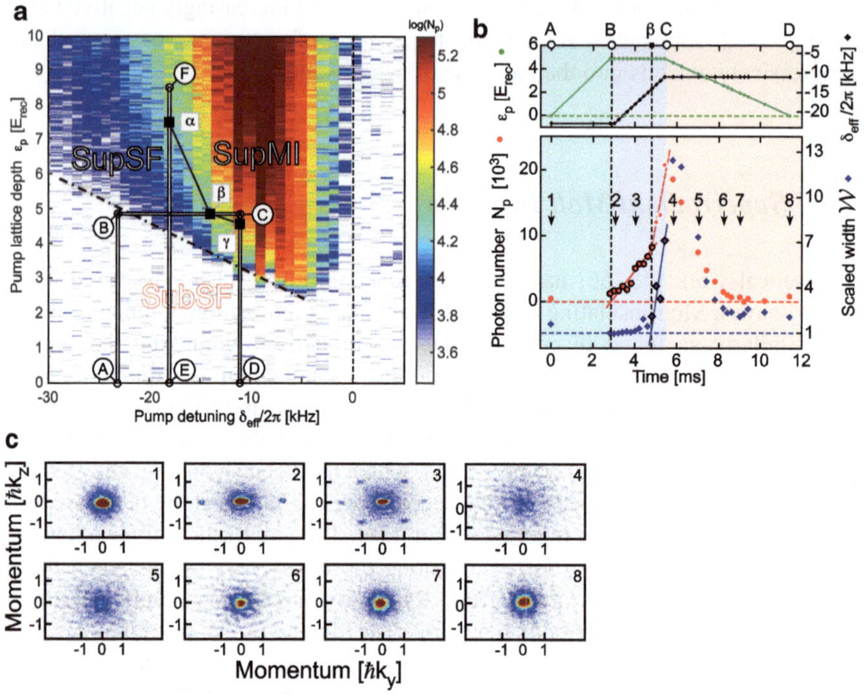

Fig. 13 **a** The intra-cavity photon number is plotted versus δ_{eff} and ε_p. The SubSF–SupSF phase boundary is highlighted by a *thick dashed–dotted black line*. Six locations in the phase diagram are highlighted by capital letters *A–F*. Three trajectories $\overline{\text{ABCD}}$, $\overline{\text{DCD}}$ and $\overline{\text{EFE}}$ are indicated along which the degree of coherence of the atomic sample is observed. The *black squares* labeled by Greek letters α, β, γ show the observed boundary between the SupSF and the SupMI phase. **b** The *upper panel* quantifies the tuning of δ_{eff} (*black squares*) and ε_p (*green disks*) along the path $\overline{\text{ABCD}}$ in **a**. In the *lower panel*, the intra-cavity photon number N_p (*red disks*) and the width (FWHM) $\mathcal{W}$ of the observed zero-momentum Bragg resonance (*blue diamonds*) with regard to the *z*-direction are plotted. The *red* and *blue solid lines* are linear fits including the data points highlighted by *black margins*. The *red dashed line* corresponds to the white delimited data points. The SubSF–SupSF and the SupSF–SupMI phase boundaries are defined by the intersections of the *red* and *blue lines* with the $(N_p = 0)$- and $(\mathcal{W} = 1)$-lines, respectively. In **c** momentum spectra (in the (k_y, k_z)-plane) are shown recorded at the locations on the path $\overline{\text{ABCD}}$ indicated by the *black arrows* in **b**

$\mathcal{W}(B) = 1$. Shortly past point "B," the transition into the SupSF phase is observed indicated by the sudden increase of the intra-cavity photon number. The width of the zero-momentum Bragg peak practically maintains its reference value of unity, which indicates complete coherence, until point "β" is reached. This is supported by the two exemplary momentum spectra shown for this section in Fig. 13c, labeled "2" and "3." These spectra show increasing population of higher-order Bragg peaks. In "2" the $(\pm 2, 0)\,\hbar k$ Bragg peaks become visible, which result from the 5 E_{rec} deep pump lattice. The intra-cavity contribution to the overall lattice potential is yet negligible. In "3" the visibility of the $(\pm 1, \pm 1)\,\hbar k$ peaks results from the presence of a notable intra-cavity photon number of 5.6×10^3, which yields an overall lattice potential with a well depth of 5.8 E_{rec} with respect to the plane spanned by the cavity and the pump wave. Only when the point "β" is passed, which corresponds to an intra-cavity photon number 8.2×10^3 and 8.0 E_{rec} well depth, a sudden more than tenfold increase of $\mathcal{W}$ is encountered. At this point, a kink in the dependence of N_p upon δ_{eff} appears, indicating an increase of the superradiant scattering efficiency due to reduced particle number fluctuations [60]. At point "C" coherence is completely lost as is also directly seen in the momentum spectrum "4" in Fig. 13c. In the subsequent section $\overline{\mathrm{CD}}$ the pump strength is reduced again to zero and $\mathcal{W}$ is observed to decrease again, finally reaching nearly the value initially prepared at point "A." The significant recovery of coherence indicates that its loss at large lattice depths is not a consequence of excessive heating, but rather indicates the emergence of the SupMI state. The particle number N_a at point "D" is reduced to 60% of the initial number at point "A." As is detailed in Ref. [58] we attribute this to three-body loss associated with the large peak density in the initial BEC. Without the additional external lattice, the observed particle loss is only on the 1% level [58].

6 Conclusion

Quantum gases subjected to high finesse optical cavities provide a rich experimental platform to explore phenomena at the boundary between many-body physics and quantum optics. Atom–cavity setups are particularly suited to study driven open systems far from equilibrium. Here, we have discussed examples such as recoil-resolved cavity-assisted scattering, opto-mechanical Bloch oscillations, cavity-controlled matter wave superradiance, the emergence of a superradiant superfluid and a superradiant Mott insulating many-body phase in a self-organized intra-cavity optical lattice with retarded infinite range interactions. The present article takes a very limited perspective, exclusively focussing on recent studies in Hamburg. Readers with an interest in a broader viewpoint may wish to consider a more extended review article, e.g., Ref. [7]. However, the field of atom–cavity physics is still rapidly growing such that a complete overview of the present state of the art requires to consult the original literature.

Acknowledgements It is a pleasure to acknowledge invaluable theory support by Reza Bakthiari, Michael Thorwart, Ludwig Mathey and Prasanna Venkatesh. This work was partially supported by DFG-SFB 925 and the Hamburg Centre of Ultrafast Imaging (CUI).

References

1. J.M. Raimond, M. Brune, S. Haroche, Rev. Mod. Phys. **73**, 565 (2001)
2. H. Walther, B.T.H. Varcoe, B.-G. Englert, T. Becker, Rep. Prog. Phys. **69**, 1325 (2006)
3. T. Kippenberg, K. Vahala, Science **321**, 1172 (2008)
4. M. Aspelmayer, T.J. Kippenberg, F. Marquardt, arXiv:1303.0733v1 (2013)
5. M.H. Devoret, R.J. Schoelkopf, Science **339**, 1169 (2013)
6. P. Lodahl, S. Mahmoodian, S. Stobbe, Rev. Mod. Phys. **87**, 347 (2015)
7. H. Ritsch, P. Domokos, F. Brennecke, T. Esslinger, Rev. Mod. Phys. **85**, 553 (2013)
8. D. Kruse et al., Phys. Rev. A **67**, 051802(R) (2003)
9. B. Nagorny et al., Phys. Rev. A **67**, 031401(R) (2003)
10. B. Nagorny et al., Phys. Rev. Lett. **91**, 153003 (2003)
11. Th Elsässer, B. Nagorny, A. Hemmerich, Phys. Rev. A **67**, 051401(R) (2003)
12. Th Elsässer, B. Nagorny, A. Hemmerich, Phys. Rev. A **69**, 033403 (2004)
13. S. Slama, S. Bux, G. Krenz, C. Zimmermann, Ph.W. Courteille, Phys. Rev. Lett. **98**, 053603 (2007)
14. E.M. Purcell, Phys. Rev. **69**, 681 (1946)
15. D. Kleppner, Phys. Rev. Lett. **47**, 233–236 (1981)
16. J. Klinner, M. Lindholdt, B. Nagorny, A. Hemmerich, Phys. Rev. Lett. **96**, 023002 (2006)
17. F. Brennecke, T. Donner, S. Ritter, T. Bourdel, M. Köhl, T. Esslinger, Nature **450**, 268 (2007)
18. Y. Colombe, T. Steinmetz, G. Dubois, F. Linke, D. Hunger, J. Reichel, Nature **450**, 272 (2007)
19. F. Brennecke, S. Ritter, T. Donner, T. Esslinger, Science **322**, 235 (2008)
20. K.W. Murch, K.L. Moore, S. Gupta, D.M. Stamper-Kurn, Nat. Phys. **4**, 561 (2008)
21. R.H. Dicke, Phys. Rev. **93**, 99 (1954)
22. R.H. Dicke, in *Proceedings of the Third International Congress on Quantum Electronics*, ed. by P. Grivet, N. Bloembergen (Columbia Univ. Press, New York, 1964), pp. 35–54
23. K. Hepp, E.H. Lieb, Ann. Phys. **76**, 360 (1973)
24. R. Gilmore, L.M. Narducci, Phys. Rev. A **17**, 1747 (1978)
25. D. Hansen, A. Hemmerich, Phys. Rev. Lett. **96**, 073003 (2006)
26. J. Klinner, M. Wolke, A. Hemmerich, Phys. Rev. A **81**, 043414 (2010)
27. R.M. Sandner, W. Niedenzu, H. Ritsch, EPL **104**, 43001 (2013)
28. M. Wolke, J. Klinner, H. Keßler, A. Hemmerich, Science **337**, 85 (2012)
29. H. Keßler, J. Klinder, M. Wolke, A. Hemmerich, New J. Phys. **16**, 053008 (2014)
30. F. Bloch, Z. Phys. **52**, 555–600 (1929)
31. C. Zener, Proc. R. Soc. Lond. A **145**, 523–529 (1934)
32. G. Ferrari, N. Poli, F. Sorrentino, G.M. Tino, Phys. Rev. Lett. **97**, 060402 (2006)
33. N. Poli, F.-Y. Wang, M.G. Tarallo, A. Alberti, M. Prevedelli, G.M. Tino, Phys. Rev. Lett. **106**, 038501 (2011)
34. M.G. Tarallo, A. Alberti, N. Poli, M.L. Chiofalo, F.-Y. Wang, G.M. Tino, Phys. Rev. A **86**, 033615 (2012)
35. H. Keßler, J. Klinder, B.P. Venkatesh, Ch. Georges, A. Hemmerich, New J. Phys. **18**, 102001 (2016)
36. D. Witthaut, M. Werder, S. Mossmann, H.J. Korsch, Phys. Rev. E **71**, 036625 (2005)
37. M. Gustavsson, E. Haller, M.J. Mark, J.G. Danzl, G. Rojas-Kopeinig, H.-C. Nägerl, Phys. Rev. Lett. **100**, 080404 (2008)
38. A.R. Kolovsky, H.J. Korsch, E.-M. Graefe, Phys. Rev. A **80**, 023617 (2009)
39. F. Meinert, M.J. Mark, E. Kirilov, K. Lauber, P. Weinmann, M. Gröbner, H.-C. Nägerl, Phys. Rev. Lett. **112**, 193003 (2014)
40. B.P. Venkatesh, D.H.J. O'Dell, J. Goldwin, Atoms **4**, 2 (2016)
41. T. Köhler, K. Góral, P.S. Julienne, Rev. Mod. Phys. **78**, 1311–61 (2006)
42. H. Keßler, J. Klinder, M. Wolke, A. Hemmerich, Phys. Rev. Lett. **113**, 070404 (2014)
43. S. Inouye et al., Science **285**, 571–574 (1999)
44. P. Domokos, H. Ritsch, Phys. Rev. Lett. **89**, 253003 (2002)

45. A.T. Black, H.W. Chan, V. Vuletic, Phys. Rev. Lett. **91**, 203001 (2003)
46. K. Baumann, C. Guerlin, F. Brennecke, T. Esslinger, Nature **464**, 1301 (2010)
47. D. Nagy, G. Konya, G. Szirmai, P. Domokos, Phys. Rev. Lett. **104**, 130401 (2010)
48. M.J. Bhaseen, J. Mayoh, B.D. Simons, J. Keeling, Phys. Rev. A **85**, 013817 (2012)
49. V.M. Bastidas, C. Emary, B. Regler, T. Brandes, Phys. Rev. Lett. **108**, 043003 (2012)
50. J. Klinder, H. Keßler, M. Wolke, L. Mathey, A. Hemmerich, Proc. Natl. Acad. Sci. USA **112**, 3290 (2015)
51. T.W.B. Kibble, J. Phys. A **9**, 1387–1398 (1976)
52. W.H. Zurek, Nature **317**, 505–508 (1985)
53. A. del Campo, W. Zurek, Int. J. Mod. Phys. A **29**, 1430018 (2014)
54. L.M. Sieberer, S.D. Huber, E. Altman, S. Diehl, Phys. Rev. Lett. **110**, 195301 (2013)
55. Y. Li, L. He, W. Hofstetter, Phys. Rev. A **87**, 051604 (2013)
56. M.R. Bakhtiari, A. Hemmerich, H. Ritsch, M. Thorwart, Phys. Rev. Lett. **114**, 123601 (2015)
57. M. Lewenstein et al., Adv. Phys. **56**, 243–379 (2007)
58. J. Klinder, H. Keßler, M. Reza Bakhtiari, M. Thorwart, A. Hemmerich, Phys. Rev. Lett. **115**, 230403 (2015)
59. R. Landig, L. Hruby, N. Dogra, M. Landini, R. Mottl, T. Donner, T. Esslinger, Nature **532**, 476–479 (2016)
60. I.B. Mekhov, Ch. Maschler, H. Ritsch, Phys. Rev. Lett. **98**, 100402 (2007)

Ted's Stanford Days

David H. McIntyre and Allister I. Ferguson

1 Stanford Arrival

In 1969 one of the first ever summer schools in Quantum Optics was held at the historic Carberry Towers [1], just outside Edinburgh. This was one of a long line of Scottish Universities Summer Schools in Physics that have attracted leading physicists and promising students over many generations. A newly graduated Theo Hänsch was a student at the summer school and lectures were given by such luminaries as Roy Glauber, Dan Walls, Nicolaas Bloembergen [2] and George Series. Seminars were given by Alfred Kastler, Art Schawlow and Leonard Mandel. Art Schawlow's lecture on "Lasers-Present and Future" inspired the young Hänsch who was "*immediately captivated by his warmth, his keen mind, and his contagious sense of humour*." Ted was seeking to broaden his experience and he persuaded Art to take him on as a NATO postdoctoral fellow and in March 1970 Ted left Germany for Stanford University. Over the next 16 years Ted entered one of the most innovative phases of his career and built one of the most successful scientific partnerships of the twentieth century with his friend and colleague, Art Schawlow.

On his way to Stanford Ted stopped off to visit Bell Labs at Holmdel where he saw the nitrogen laser pumped dye laser developed by Chuck Shank. This laser operated at 100 Hz and looked almost continuous. Ted was fascinated by the possibilities that this could open up for spectroscopy. Ted realised that if could make the linewidth of this laser sufficiently narrow he could perform Doppler-free saturation spectroscopy. When he got to Stanford he persuaded Art to purchase an AVCO nitrogen laser and proceeded to make his own dye laser. The laser that Ted

D.H. McIntyre (✉)
Department of Physics, Oregon State University, Corvallis, OR, USA
e-mail: david.mcintyre@oregonstate.edu

A.I. Ferguson
Department of Physics, University of Strathclyde, Glasgow, Scotland, UK

D. Meschede et al. (eds.), *Exploring the World with the Laser*,
https://doi.org/10.1007/978-3-319-64346-5_43

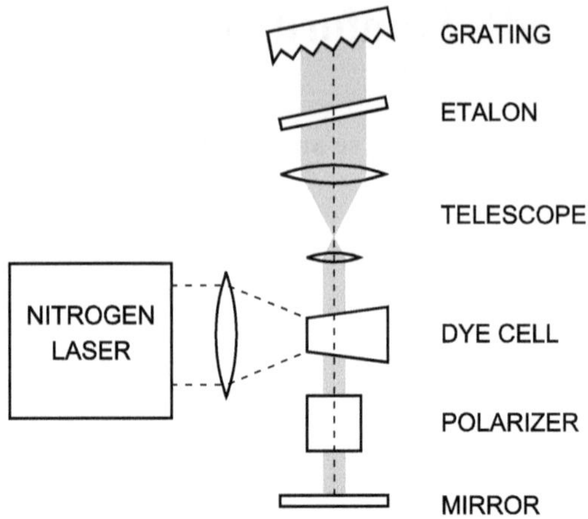

Fig. 1 Hänsch dye laser. This simple laser revolutionized the techniques of laser spectroscopy. The telescope in the resonator increased the number of lines illuminated on the grating and produced a very narrow linewidth. The addition of a low finesse etalon permitted single frequency tunable laser operation making this the workhorse for new spectroscopic techniques

had seen in Bell Labs was tuned by a diffraction grating but the linewidth was still quite broad. Ted argued that by increasing the number of lines illuminated on the grating he could improve the resolution of the grating and thereby reduce the linewidth of the laser. Ted introduced a pocket telescope into the laser cavity to broaden the illumination of the grating and immediately achieved a much narrower linewidth. Thus was born the Hänsch dye laser shown in Fig. 1 [3]. This new tool permitted the observation of the spectra of atoms of many different types, including sodium and hydrogen, for the first time and essentially led to a revolution in spectroscopy. Once the news of the Hänsch design of dye laser became public it was emulated and employed by researchers across the globe. Indeed this laser and the spectroscopic techniques that it enabled had such an impact that Art and Ted were named "California Scientists of the Year" in 1973.

Ted's experience in developing the dye laser taught him some lessons that have persisted throughout his distinguished career. Firstly, a new tool will open up a myriad of new possibilities and secondly that "*a simple and imperfect proof-of-principle experiment can sometimes find much wider resonance than a complex experiment of intimidating perfection.*"

Together Art and Ted pioneered a range of spectroscopic techniques that enabled the spectra of atoms and molecules to be investigated with unprecedented resolution by eliminating the blurring effects of Doppler broadening. This permitted tests of atomic theory, particularly in the hydrogen atom, and the measurement of fundamental constants leading to Art being awarded the Nobel Prize in Physics in 1981. Their seminal paper on cooling of gases by laser radiation underpinned a revolution in laser cooling of atoms that led to the subsequent award of several Nobel Prizes in Physics in later years. Ted's work on the use of mode-locked lasers to perform high resolution spectroscopy and precision measurement heralded the use of laser generated spectral combs for optical frequency metrology and led to him being awarded the Nobel Prize in Physics in 2005 together with John Hall.

Ted's work at Stanford was characterised by experiments that could be done with a single student or a student and a visitor or postdoctoral fellow. This permitted

Fig. 2 Ted Hänsch in his office at Stanford in 1978

great agility and a "let's do it" attitude. This was the perfect environment for Ted's insight and imagination to flourish and gave rise to many new spectroscopic techniques, new types of lasers, new stabilisation techniques and many scientific insights. It also led to a large number of students graduating who have gone on themselves to make very significant contributions to the field. Figure 2 shows Ted in his Stanford office in 1978.

The atmosphere created by Ted and Art was one of great excitement and of a great deal of fun. They shared a joint passion for the emerging microcomputer devices that were being developed in the Bay Area. Ted was a regular attendee at the Stanford Homebrew Computer Club where Bill Gates would sell early versions of his BASIC program on rolls of punched paper tape. Steve Jobs took Ted's classes in electricity and magnetism before going on to found Apple Corporation. Ted even formed a company that he called Menlo Systems Inc. in order to get more attention at personal computer shows. Exhibitors were less interested in talking to a professor from Stanford than they were talking to the CEO of a Bay Area start-up even though the start-up consisted only of some business cards.

2 Saturated Absorption Spectroscopy

In the early 1970s Hänsch [4, 5] and Borde [6] independently developed the new high-resolution technique of saturated absorption spectroscopy. In traditional absorption spectroscopy of a gaseous sample, the thermal motion of the atoms causes varying Doppler shifts that broaden the observed spectral line. Saturated absorption spectroscopy uses the nonlinear response of an atom to resonant laser

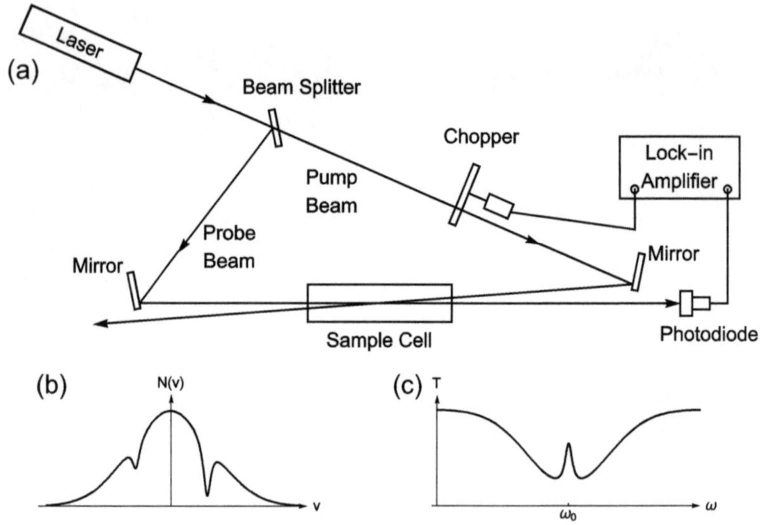

Fig. 3 (**a**) The experiment uses counter-propagating pump and probe beams that, (**b**) burn holes in the velocity distribution and (**c**) yield a Doppler-free signal when the beams interact with atoms near zero velocity

light as a tool to access the small group of atoms in a sample that do not have a first-order Doppler shift. In the typical experimental realization, two counter-propagating laser beams overlap within a gas cell, as shown in Fig. 3(a). The beam splitter is used to create a strong "pump" beam and a weaker "probe" beam. The transmission of the probe beam is monitored with a photodetector as the laser frequency is scanned over the desired optical resonance transition. Moving atoms absorb laser light that is Doppler shifted from the resonance frequency ω_0 for an atom at rest. An atom with velocity $\vec{\mathbf{v}}$ observes a laser beam with wave vector $\vec{\mathbf{k}}$ to have a frequency $\omega = \omega_0 - \vec{\mathbf{k}} \cdot \vec{\mathbf{v}}$. The opposite directions of the pump and probe beams result in opposite Doppler shifts, so that each beam is absorbed by classes of atoms with different velocities and hence creates a hole in the ground state velocity distribution, as shown in Fig. 3(b). As the laser frequency is scanned across the resonance, the probe beam experiences absorption with the traditional Doppler profile except near the center of the Doppler profile where the pump beam has greatly depleted the number of atoms, causing a decrease in the absorption of the probe beam as shown in Fig. 3(c). This Doppler-free signal is created by only those atoms with nearly zero velocity component along the direction defined by the laser beams. In the ideal case, this saturated absorption dip has a linewidth approaching the natural linewidth of the resonance.

In 1971, Ted and graduate student Marc Levenson used saturated absorption spectroscopy to observe the hyperfine structure of a molecular iodine line that was within the narrow tunable range of a krypton ion laser [5]. They observed spectral components with linewidths of about 10 MHz, representing a resolution exceeding 10^8. The advent of tunable dye lasers allowed this technique to be applied to any visible transition. In the same year, Ted and Issa Shahin used a pulsed tunable dye laser to observe Doppler-free spectra of the atomic sodium D transitions at 589 nm [7]. They recorded spectra with linewidths of 40 MHz, twice the 20 MHz natural linewidth. The difference was attributed to the laser linewidth and to residual Doppler broadening caused by the finite crossing angle of the pump and probe beams. This experiment opened the door to the wide spread use of this powerful technique, which Ted noted at the end of Ref. [7]: "*Moreover it is demonstrated with the experiment described here that it is now possible, by using narrow-band tunable dye lasers, to apply the powerful techniques of laser saturation spectroscopy in principle to any atomic or molecular transition even between highly excited states throughout the visible spectrum.*"

Given these new powerful tools, Art Schawlow suggested that Ted apply saturated absorption spectroscopy to the Balmer-α line of Hydrogen at 656 nm. The key role of the Balmer spectrum in the history of atomic physics made this idea very attractive to Ted, who often referred to hydrogen as the "*Rosetta stone that allowed us to decipher the laws of quantum mechanics.*" In 1972, Ted and Issa Shahin measured Doppler-free spectra of Hydrogen that revealed fine-structure features that had previously been hidden by the large Dopper broadening of the light hydrogen atom [8]. Their high-resolution spectra (300 MHz linewidths) also demonstrated the first optical observation of the Lamb shift. This experiment was the beginning of Ted's long passion for precision. At the end of Ref. [8], Ted was already foreseeing the importance of high-resolution laser spectroscopy of hydrogen in measurements of the Rydberg constant, H, D, and T isotope shifts, and QED corrections.

3 Hydrogen Energy Levels

Ted's desire to study the hydrogen atom stems from its pivotal role in the development of both quantum mechanics and the theory of quantum electrodynamics. The visible spectrum of the hydrogen atom was first described quantitatively in 1885 by Balmer, who was able to account for the positions of the spectral lines with a simple formula

$$\frac{1}{\lambda} = R\left(\frac{1}{n_1^2} - \frac{1}{n_2^2}\right)$$

where λ is the wavelength of a line, R is a constant, $n_1 = 2$, and n_2 takes on the integer values 3, 4, 5, etc. This series of spectral lines has since come to be called the Balmer series. A more general version of this empirical formula was deduced independently in 1889 by Rydberg, whose formula accounted for the spectra of many elements and included Balmer's hydrogen formula as a special case. The constant R is now called the Rydberg constant and can be determined by measuring the wavelength of a spectral line of hydrogen. Some of the key spectral lines are shown in the energy level diagram in Fig. 4.

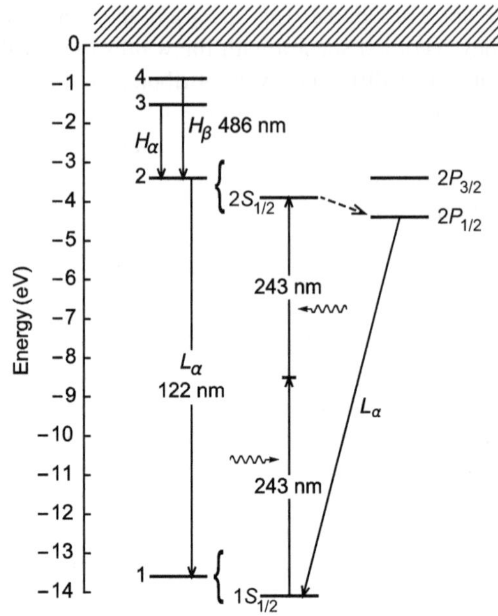

Fig. 4 Hydrogen energy levels. This shows the coincidence between the wavelengths of the Balmer β at 486 nm and two-photon excitation of the 1*S* to 2*S* transition at 243 nm achieved by frequency doubling a tunable laser at 486 nm

The empirical formula of Balmer inspired Bohr to derive the quantum theory of the atom in 1912. Bohr showed that the energy states in the hydrogen atom could be expressed as

$$E_n = -\frac{1}{n^2}\left(\frac{m_e e^4}{2\hbar^2}\right).$$

Bohr thus provided a theoretical prediction for the value of the Rydberg constant,

$$R = \frac{m_e e^4}{4\pi\hbar^3 c}$$

that was in excellent agreement with the experimental value determined previously by Rydberg. Corrections to Bohr's energy levels caused by special relativistic effects and the spin of the electron were incorporated in Dirac's relativistic formulation of quantum mechanics in 1928. These fine structure corrections were better able to explain experimental observations, but there remained some lingering discrepancies that were not resolved until after World War II.

In 1947, Lamb and Retherford found that the hydrogen $2S_{1/2}$ and $2P_{1/2}$ levels were not degenerate, as Dirac theory suggested, but were separated by approximately 1000 MHz. The need to explain this "Lamb shift" was instrumental in the development of the theory of quantum electrodynamics (QED), which now stands as one of the most successful theories in physics.

In 1974, Ted and his group realized their first measurement of the Rydberg constant [9]. They recorded Doppler-free saturated absorption spectra of the Balmer-α lines of hydrogen and deuterium using a pulsed dye laser. To measure absolute wavelengths, they used a Fabry–Perot interferometer that was calibrated with a He–Ne laser that was stabilized to iodine reference lines. This first Doppler-free Rydberg measurement had a precision of four parts in 10^9 and was ten times more precise than previous measurements.

4 Polarization Spectroscopy

In 1976, Ted and his graduate student Carl Wieman demonstrated the new technique of polarization spectroscopy, which is similar to Doppler-free saturated absorption spectroscopy but has improved signal-to-noise ratio [10]. This technique relies on the fact that small changes in laser polarization are more easily detected than small changes in intensity. Hence, the standard saturated absorption experiment is changed to have the probe beam pass through two nearly crossed polarizers with the gas sample placed between the polarizers. The probe beam detector measures negligible signal unless the gas sample causes the probe beam polarization to be changed. If the pump beam is circularly polarized, then it produces different levels of saturation in different Zeeman sublevels of the absorbing gas. The linearly polarized probe beam may be considered as a superposition of left- and right-circularly polarized beams, which then probe the differently saturated Zeeman levels of the gas. As in standard saturated absorption spectroscopy, this effect is only present when the pump and probe beams both interact with the same group of atoms at near zero velocity, ensuring a Doppler-free signal. The measured probe signal includes effects from both the saturated absorption and the saturated dispersion of the gas, i.e. both the real and imaginary components of the atomic response. The relative weights depend on the small crossing angle of the polarizers. At non-zero angles, the signal is larger than at zero angle because the unmodified probe field acts a local oscillator to magnify the effect of the field modified by the pump beam.

In 1978, Ted's group used polarization spectroscopy to measure the Rydberg constant again, with several improvements [11]. They improved the instrument resolution by utilizing a continuous-wave (cw) dye laser with a narrower linewidth than the previous pulsed lasers (1 MHz vs. 30 MHz). They used the $2S$-$3P$ transition rather than the $2P$-$3D$ transition to take advantage of the metastability of the $2S$ state and hence improved the atomic resolution from 110 to 29 MHz. They were able to use the weaker $2S$-$3P$ transition because of the better signal-to-noise ratio of polarization spectroscopy. They achieved a three-fold improvement of the Rydberg precision compared to the 1974 measurement. In both the 1974 and 1978 papers, Ted alluded to much better resolution in the future through use of the narrow two-photon $1S$-$2S$ transition, whose linewidth is limited only by the 1/7-sec lifetime of the $2S$ state.

5 Two-Photon Spectroscopy

Two-photon spectroscopy was first proposed in 1970 by Vasilenko, Chebotaev, and Shishaev [12, 13]. Two-photon spectroscopy is attractive for several reasons. (1) The sum energy of the two photons adds up to the total energy separation of the two atomic states, so one can use visible light to access energy states that might normally require ultraviolet light (as indicated in Fig. 4 for the case of hydrogen). (2) It allows access to states of the same parity as the ground state, which often means narrower transition linewidths. (3) By arranging for the two photons to come from opposite directions, the Doppler shifts of the two photons cancel to give a Doppler-free signal. Moreover, this Doppler cancellation occurs for all atoms in the sample, not just for the zero-velocity atoms as in saturation spectroscopy. Using all the atoms is important because the probability for a two-photon transition is typically much less than for a one-photon transition.

The first two-photon experiment at Stanford occurred in 1974, when Ted and his group observed Doppler-free signals from the sodium $3s$-$4d$ transition [14]. The sodium signal was quite strong because the two-photon transition probability is enhanced by the near resonance of the strongly allowed single-photon Na $3s$-$3p$ resonance line. In hydrogen however, there are no energy levels near the half-way point of the $1S$-$2S$ transition and so the transition is very weak. But the very narrow natural linewidth (1 Hz) makes this transition very attractive to study. In 1975 Ted's group used 243-nm light generated by frequency doubling of a 486-nm pulsed dye laser to observe the $1S$-$2S$ two-photon transition in hydrogen [15, 16], beginning a series of experiments that continues to this day. The 486-nm light is nearly coincident with the hydrogen Balmer-β transition from the $n = 2$ states to the $n = 4$ states. The coincidence would be exact if not for Dirac fine-structure and QED corrections to the Bohr energy levels so a comparison of the 486-nm laser frequencies required to excite the $1S$-$2S$ and Balmer-β transitions determines the QED corrections. This experiment thus provided the first measurement of the Lamb shift in the hydrogen $1S$ state.

The Stanford $1S$-$2S$ experiment in hydrogen underwent a series of improvements until Ted's departure for Munich in 1986. In 1980 Carl Weiman and Ted used a cw dye laser at 486 nm to perform the Balmer-β polarization spectroscopy and to seed the pulsed dye laser amplifier for second harmonic generation of 243-nm light [17]. In 1986 Ted's group added an atomic beam to the experiment to avoid pressure shifts and also made an absolute frequency measurement by comparing the 486-nm light with calibrated molecular tellurium reference lines [18, 19]. In the early 1980s Ted's group expended considerable effort to develop a cw source of the 243-nm light needed for the two-photon excitation [20, 21]. This culminated in the first cw-excitation of the hydrogen $1S$-$2S$ transition and a Rydberg constant measurement with a precision of seven parts in 10^{10} [22–24].

6 Laser Spectroscopy with Multiple Pulses

While working to improve the resolution of the Stanford hydrogen 1*S*-2*S* experiment, Ted came across an idea that laid the ground work for his later frequency comb work in Germany. The challenge of frequency doubling 486-nm light to produce the 243-nm ultraviolet light to excite the 1*S*-2*S* transition in hydrogen led Ted to the notion of using a sequence of light pulses to enhance the resolution, analogous to the Ramsey separated fields approach but in this case the fields were separated in time. Ted and his students Rich Teets and Jim Eckstein initially used a pulsed dye laser in a passive cavity to produce the train of coherent pulses needed for the technique [25]. This technique showed promise but was clearly not going to be the technique of choice for really high resolution. In a paper by Baklanov and Chebotayev the use of a train of pulses from a mode-locked laser to access the 1*S*-2*S* transition was proposed [26]. The idea is illustrated in Fig. 5. A mode-locked laser producing a train of pulses equally spaced in time appears as a comb of modes spaced by the repetition rate of the laser in the frequency domain. In a two-photon Doppler-free configuration with counter-propagating laser pulses if one imagines the nth mode being in two-photon resonance then the $n + m$ and the $n - m$ modes will also be in resonance thereby ensuring that all the modes contribute to the two-photon signal.

This idea was tested using a synchronously-pumped dye laser operating around 580 nm and a two-photon resonance in sodium [27]. For the first time it was experimentally demonstrated that pulses which individually were broadband could contribute to a high resolution Doppler-free two-photon spectrum. In this sense the continuously mode-locked pulse train could be considered as a large number of single mode lasers all equally spaced and locked in phase such as to constructively interfere to produce an ultrashort pulse once per cavity round trip. This simple notion led to the idea that the pulse train itself could be used to calibrate the spectrum and relate the optical frequency to a multiple of the repetition rate. The big challenge in using this technique for absolute calibration of the optical frequency was the phase slips relative to the pulse envelope caused by dispersion in the cavity. This was explained in the appendix to Ted's student Jim Eckstein's thesis

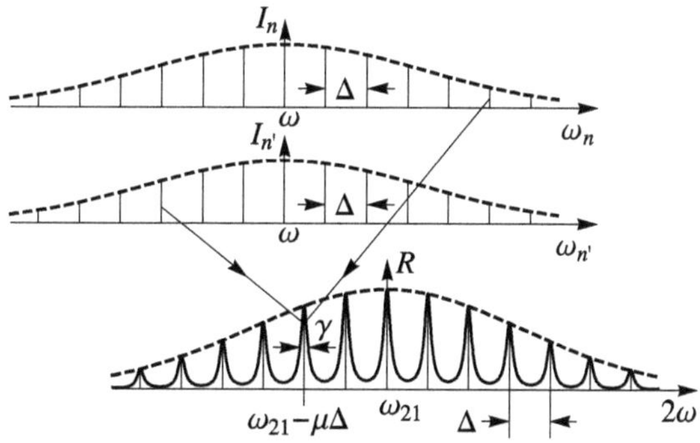

Fig. 5 Two-photon spectrum with a mode-locked laser. The laser excitation is a comb of equally spaced modes and so when resonance is achieved between a pair of counter-propagating modes the other modes also contribute to the resonance

[28]. The solution to this challenge took several decades and advances in the mode-locking of lasers and nonlinear optics of fibers.

Ted later returned to the use of mode-locked lasers to calibrate optical frequencies and to relate radio frequencies to optical frequencies when he returned to Germany and was able to devote many more resources to the problem. The ultimate solution required the development of the Kerr lens mode-locked Ti:sapphire laser and photonic crystal fibers. These developments, at last, permitted the measurement and control of the offset frequency and ultimately led to the award of the Nobel Prize to Ted and John Hall in 2005. These fascinating developments leading up to the measurement of absolute optical frequencies are well documented and are especially well described in the Nobel lectures of 2005.

7 Laser Cooling

The spectroscopic techniques that Ted developed were Doppler-free to first order, but second-order Doppler effects and transit time broadening were still present. As a possible solution to this problem, Ted and Art Schawlow proposed a method in 1975 to use laser light to cool an ensemble of atoms in a gas [29]. The method relies on the cycle of absorption and emission shown in Fig. 6. The three steps illustrated are: (1) A resonant laser beam is incident on an atom in the ground state of the two-level system. (2) The atom absorbs a photon, which promotes the electron to the excited state and causes the atom to recoil in the direction of the incident laser with momentum change $\Delta\mathbf{p} = \hbar\mathbf{k}$. (3) The excited atom decays back down to the ground state via spontaneous emission of another photon. The spontaneous photon is emitted in a random direction, so the recoil kicks due to the spontaneously emitted photons average to zero over many absorption–emission cycles and the average momentum change per complete absorption emission cycle is $\langle\Delta\mathbf{p}\rangle_{cycle} = \hbar\mathbf{k}$, due only to the momenta of the absorbed photons. Once the atom returns to the ground state, it is ready to absorb another photon and begin the cycle anew. This radiation force is directed along the incident laser beam. To achieve

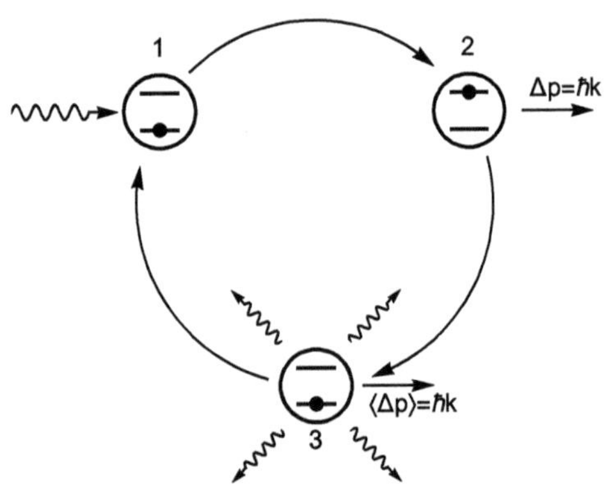

Fig. 6 Laser cooling cycle. Repeated absorption and emission events of resonant photons allow for slowing and cooling of atoms

cooling of the atoms, the force must be velocity dependent and have components along all directions. The Doppler shift provides the velocity dependence and six laser beams aligned along both directions of the three Cartesian axes ensure that all directions are impacted. The laser frequency must be tuned below the resonance line so that the laser beam that opposes an atom's motion is Doppler shifted closer to resonance to enhance the radiation force while the laser beam that follows an atom's motion is Doppler shifted farther from resonance to decrease its radiation force. While Ted and Art did propose that this method would be useful in their quest to improve the resolution of the hydrogen $1S$-$2S$ transition, they fortunately left this field for others to explore. This configuration was later dubbed optical molasses [30].

8 Stanford Departure

In the mid-1980s Ted agonised for several years over whether to remain at Stanford or to be lured back to Germany. In March 1986, almost exactly 16 years after first arriving at Stanford, Ted was lured back to Germany to take up the position of Professor of Experimental Physics at Ludwig-Maximilians University in Munich and Director of the Max Planck Institute for Quantum Optics, which had recently moved into a new building in Garching. The move to Germany allowed Ted to build larger groups, he was able to run laboratories in central Munich and at Garching, and he could take on longer term challenging research projects.

Ted's time at Stanford also exposed him to "*the liberating climate of entrepreneurship that was omnipresent in the heart of Silicon Valley*." Together with former students Ronald Holzwarth and Michael Mei, Ted formed Menlo Systems GmBH to develop the commercial potential for frequency comb synthesizer and frequency comb metrology thereby bringing Silicon Valley thinking to Germany and permitting a much wider group of researchers to have access to the world of precision optical frequency measurement.

Ted's Stanford days were characterized by fun, excitement, innovation and discovery, something that he has continued to enjoy throughout his long career and something that he has brought to all of us who have had the pleasure of working with him and being influenced by him.

References

1. See Peter Higgs' Nobel lecture where he attended a previous Summer School at Carberry Towers and was the wine steward
2. One of the authors (AIF) was privileged to be a student when Bloembergen gave a set of lectures at a subsequent Scottish Universities Summer School in Physics held in Edinburgh in 1975

3. T.W. Hänsch, Repetitively pulsed tunable dye laser for high-resolution spectroscopy. Appl. Opt. **11**, 895–898 (1972)
4. P.W. Smith, T. Hänsch, Cross-relaxation effects in the saturation of the 6328-Å neon-laser line. Phys. Rev. Lett. **26**, 740–743 (1971)
5. T.W. Hänsch, M.D. Levenson, A.L. Schawlow, Complete hyperfine structure of a molecular iodine line. Phys. Rev. Lett. **26**, 946–949 (1971)
6. C. Bordé, Saturated absorption spectroscopy of various molecules by means of carbon dioxide and nitrous oxide lasers. C. R. Acad. Sci. Paris **271**, 371–374 (1970)
7. T.W. Hänsch, I.S. Shahin, A.L. Schawlow, High-resolution saturation spectroscopy of sodium D lines with a pulsed tunable dye laser. Phys. Rev. Lett. **27**, 707–710 (1971)
8. T.W. Hänsch, I.S. Shahin, A.L. Schawlow, Optical resolution of the Lamb shift in atomic-hydrogen by laser saturation spectroscopy. Nature **235**, 63–65 (1972)
9. T.W. Hänsch, M.H. Nayfeh, S.A. Lee, S.M. Curry, I.S. Shahin, Precision measurement of the Rydberg constant by laser saturation spectroscopy of the Balmer α line in hydrogen and deuterium. Phys. Rev. Lett. **32**, 1336–1340 (1974)
10. C. Wieman, T.W. Hänsch, Doppler-free laser polarization spectroscopy. Phys. Rev. Lett. **36**, 1170–1173 (1976)
11. J.E.M. Goldsmith, E.W. Weber, T.W. Hänsch, New measurement of the Rydberg constant using polarization spectroscopy of H_{α}. Phys. Rev. Lett. **41**, 1525–1528 (1978)
12. L.S. Vasilenko, V.P. Chebotaev, A.V. Shishaev, Line shape of two-photon absorption in a standing-wave field in a gas. JETP Lett. **12**, 113–116 (1970)
13. E.V. Baklanov, V.P. Chebotaev, On the precise measurement of the frequency transition 1*S*–2*S* of the hydrogen atom. Opt. Commun. **12**, 312–314 (1974)
14. T.W. Hänsch, K.C. Harvey, G. Meisel, A.L. Schawlow, Two-photon spectroscopy of Na 3*s*-4*d* without Doppler broadening using a CW dye laser. Opt. Commun. **11**, 50–53 (1974)
15. T.W. Hänsch, S.A. Lee, R. Wallenstein, C. Wieman, Doppler-free two-photon spectroscopy of hydrogen 1*S*-2*S*. Phys. Rev. Lett. **34**, 307–309 (1975)
16. S.A. Lee, R. Wallenstein, T.W. Hänsch, Hydrogen 1*S*-2*S* isotope-shift and 1*S* Lamb shift measured by laser spectroscopy. Phys. Rev. Lett. **35**, 1262–1266 (1975)
17. C. Wieman, T.W. Hänsch, Precision-measurement of the 1*S* Lamb shift and of the 1*S*-2*S* isotope shift of hydrogen and deuterium. Phys. Rev. A **22**, 192–205 (1980)
18. E.A. Hildum, U. Boesl, D.H. McIntyre, R.G. Beausoleil, T.W. Hänsch, Measurement of the 1*S*-2*S* frequency in atomic-hydrogen. Phys. Rev. Lett. **56**, 576–579 (1986)
19. J.R.M. Barr, J.M. Girkin, A.I. Ferguson, G.P. Barwood, P. Gill, W.R.C. Rowley, R.C. Thompson, Interferometric frequency measurements of $^{130}Te_2$ transitions at 486 nm. Opt. Commun. **54**, 217–221 (1985)
20. B. Couillaud, L.A. Bloomfield, J.E. Lawler, A. Siegel, T.W. Hänsch, Saturation spectroscopy of ultraviolet transitions in mercury with a frequency-doubled cw ring dye-laser. Opt. Commun. **35**, 359–362 (1980)
21. B. Couillaud, L.A. Bloomfield, T.W. Hänsch, Generation of continuous-wave radiation near 243 nm by sum-frequency mixing in an external ring cavity. Opt. Lett. **8**, 259–261 (1983)
22. C.J. Foot, B. Couillaud, R.G. Beausoleil, T.W. Hänsch, Continuous-wave 2-photon spectroscopy of the 1*S*-2*S* transition in hydrogen. Phys. Rev. Lett. **54**, 1913–1916 (1985)
23. R.G. Beausoleil, D.H. McIntyre, C.J. Foot, E.A. Hildum, B. Couillaud, T.W. Hänsch, Continuous-wave measurement of the 1*S*-Lamb shift in atomic-hydrogen. Phys. Rev. A **35**, 4878–4881 (1987)
24. D.H. McIntyre, R.G. Beausoleil, C.J. Foot, E.A. Hildum, B. Couillaud, T.W. Hänsch, Continuous-wave measurement of the hydrogen 1*S*-2*S* transition frequency. Phys. Rev. A **39**, 4591–4598 (1989)
25. R. Teets, J. Eckstein, T.W. Hänsch, Coherent 2-photon excitation by multiple light-pulses. Phys. Rev. Lett. **38**, 760–764 (1977)
26. Y.V. Baklanov, V.P. Chebotayev, Narrow resonances of two-photon absorption of super-narrow pulses in a gas. Appl. Phys. **12**, 97–99 (1977)

27. J.N. Eckstein, A.I. Ferguson, T.W. Hänsch, High-resolution 2-photon spectroscopy with picosecond light-pulses. Phys. Rev. Lett. **40**, 847–850 (1978)
28. J. N. Eckskein, Ph.D. Thesis, Stanford University, 1978
29. T.W. Hänsch, A.L. Schawlow, Cooling of gases by laser radiation. Opt. Commun. **13**, 68–69 (1975)
30. S. Chu, L. Hollberg, J.E. Bjorkholm, A. Cable, A. Ashkin, Three-dimensional viscous confinement and cooling of atoms by resonance radiation pressure. Phys. Rev. Lett. **55**, 48–51 (1985)

From Laser Cooling to the Superfluid Mott Insulator

Tilman Esslinger

His arguably most succinct paper Ted Hänsch wrote together with Art Schawlow, proposing laser cooling of an atomic gas on a mere two pages of *Optics Communications* [1]. Once asked why he did not further pursue their idea of laser cooling at Stanford, he gave two reasons. First, he was not so keen on building a complicated ultra-high vacuum apparatus. Second, and perhaps more importantly, he had heard of Chebotayev's proposal for two-photon spectroscopy, which he thought would be a much more effective approach to circumvent the Doppler shift in laser spectroscopy. Fortunately for many researchers in the field of cold atoms who had learned their profession in Ted's laboratories in Munich and Garching, he eventually decided to give the field of laser cooling a second chance. This decision was characteristic for his style of running a large research group since coming back to Germany. He supports his PhD students and postdocs in following their own ideas and initiatives, even if the topics have fairly little to do with laser spectroscopy.

1 Laser Cooling and Diode Lasers

The first laser-cooling set-up in Ted Hänsch's lab was designed by Andreas Hemmerich, one of his first PhD students in Garching, and by David H. McIntyre, who stayed on as a postdoc when Ted returned to Germany in 1986. Their experiment was based on an atomic rubidium beam traversing a region where it interacted with laser light of different configurations, to be finally detected in a hot-wire detector [2]. Rubidium was chosen as its transitions are readily accessible with laser diodes used in CD players, which were cheaply available. Already the

T. Esslinger (✉)
ETH Zurich, Institute for Quantum Electronics, Otto-Stern-Weg-1, HPF D4, 8093 Zurich, Switzerland
e-mail: esslinger@phys.ethz.ch

D. Meschede et al. (eds.), *Exploring the World with the Laser*,
https://doi.org/10.1007/978-3-319-64346-5_44

second round of experiments with this apparatus introduced rather innovative and influential elements. A two-dimensional standing laser wave with time–phase control was added perpendicular to the atomic beam axis [3]. Further downstream, the atoms deflected by the light potential then passed through a resonant sheet of light and their fluorescence was imaged in real-time using an intensified CCD camera [4], my diploma project.

These first laser-cooling activities had started in Ted's labs at the Max-Planck Institute for Quantum Optics in Garching, but soon more lab space became available on the second and third floor of a building owned by the Ludwig-Maximilian University in the center of Munich's Schwabing borough. There, at Schellingstrasse 4, several cooling and trapping experiments were designed and set-up by Ted's team. The initially used cumbersome technology of stabilizing a diode laser to an external optical cavity [2] was abandoned. Instead, a simple, robust and elegantly designed grating-stabilized diode laser in Littrow configuration [5] was employed (see Fig. 1), which helped accelerating progress. It was Ted himself who designed and demonstrated the first prototypes of grating-stabilized diode lasers in his famous, well-equipped and carpeted optics lab on the third floor of Schellingstrasse 4. Mastering this diode-laser technology at an early stage was crucial for the success of all subsequent cold-atom experiments in Munich. Refined designs would later become Toptica's DL100 laser. It should not remain unmentioned that his carpeted lab is his private lab and it is where Ted cooks up his many ideas, plays with the latest gadgets and also produced a by now rather well-known whimsical movie showing a variety of Paul Traps [6]. The traps operated in air and the particles were charged powdery spores.

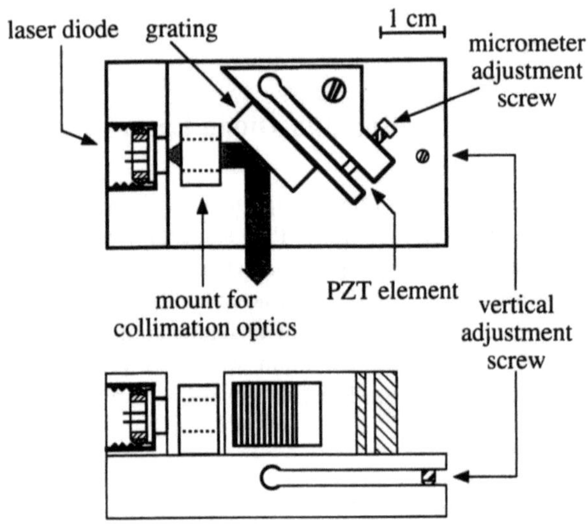

Fig. 1 Schematics of the mechanical setup of the grating-stabilized diode laser system [5]

2 Magneto-Optical Traps

In one of Ted's other labs, Andreas Hemmerich—by then a postdoc—created one of the early vapor-cell magneto-optical traps. An amazingly simple vacuum-chamber design sporting anti-reflection-coated high-quality glass windows was destined for housing the first generation of two- and three-dimensional optical lattices [7]. In these pioneering experiments, the lattices were operated close to the atomic resonance, providing first insights into the quantized motion of atoms in the band structure of such artificial crystals [8]. Ted immediately realized the huge potential of this field; Fig. 2 shows a list of his ideas, as of 1993 [courtesy of Wilhelm Zwerger]. A truly memorable lattice experiment was performed by Matthias Weidemüller. He aligned a blue laser beam such that it would undergo Bragg reflection on the sparsely populated optical lattice [9]. The reflected beam could be seen with the naked eye.

The second activity was magnetic trapping of neutral atoms, partly motivated by potential applications for spectroscopy. Leonardo Ricci, a PhD student working with postdoc Claus Zimmermann, built a magnetic trap for lithium atoms using permanent magnets. This was followed by seminal experiments using magnetic microtraps for lithium atoms, featuring super-steep potentials [10]. The experiment was made possible by the heroic effort of Vladan Vuletic, who managed to operate a Ti:Saphire laser at 671 nm, necessary to excite lithium atoms.

The third experiment—part of my PhD project—used a magneto-optical trap as a source for creating a slow atomic beam. Combining the trap with a spatially resolving fluorescence detector, we studied velocity-selective dark states and pioneered sub-recoil laser cooling in rubidium, reaching temperatures as low as 100 nanokelvin [11]. The heydays of laser cooling came to an end, but the experimental set-up with differentially pumped vacuum chambers would turn out to be a perfect starting point for what was to come. In laser cooling, the mysteries were solved, the limits known and the differences between various schemes were on the level of producing different polarization patterns with the laser fields. In hindsight, one should have of course combined the magnetic trapping and laser-cooling know-how to go a step further towards the unknown territory of quantum gases. Yet the goal of a dense atomic gas was probably too far away from our mindset, formed by spectroscopy and the beauty of laser-cooling mechanisms.

In June 1995 at the International Conference on Laser Spectroscopy, organized by Massimo Inguscio on the beautiful island of Capri in Italy, a new age for atomic physics started. During the Sunday-evening welcome reception rumors were spreading that Eric Cornell had seen strange features in the absorption images of evaporatively cooled clouds of rubidium atoms: rings around the expanded cloud, as well as bimodal and anisotropic velocity distributions. Probably aware of the historic moment, Ted, sitting in the audience, had his brand new little video camera ready to record when Eric Cornell started his talk with the slide "Evidence for Bose–Einstein condensation".

Some Possible Experiments with 2D- or 3D- "Crystalline Molasses"

T.W.Hänsch, April 30, 1993

Explore Probe Geometries
Vary probe angle, change focussing. Evidence for holographic gratings? Use spatially scanning probe? Study diffusion, tunelling,...

Explore Trap Geometries
Effect of deliberate misalignements? Introduce vortex forces.

Zone Melting
Move shadow through trapping light beams. Study "healing". Produce more highly ordered crystals?

Bragg Scattering
Study Bragg scattering of short wavelength radiation. Obtain information on short-range and long-range order. Example: interaction of atoms via light-induced dipoles should lead to pearl-string-like regular spacing of atoms in linear De Broglie waveguides.

Fluorescence Spectroscopy
Analyze spectrum of resonance fluorescence on trapping transition and on separate probe transition. Find Raman sidebands. Search for signals from atom pairs in same trap or in neighboring traps.

Nonlinear Optics
Try to observe harmonic generation, self-focussing, bistability, stimulated light scattering, photorefractive effects, phase conjugation...

Noise Analysis
Observe amplitude or frequency fluctuations of transmitted probe light.
Study atom dynamics?

"Saturation Spectroscopy"
Use 2 beams in addition to trapping light field (pump and probe), observe cross-talk
Vary relative frequency, polarization, angle, parallel displacement, timing.
Investigate spatial diffusion, tunelling, spin exchange.
Apply additional static field gradients - analogy to quantum wells

Towards a Better Harmonic Oscillator
Use more intense light to provide a larger number of almost evenly spaced levels.

Micro-Molasses
Obtain more intense light by tighter focussing (lenses, near-spherical build-up cavities)

Fig. 2 A list of possible experiments for two-dimensional and three-dimensional optical lattices, proposed by Ted in 1993 [courtesy of Wilhelm Zwerger]

New Geometries
Use more beams. Make different crystal geometries (quasi-crystals?).
Use coarser interference patterns from nearly co-propagating beams as trapping fields.
Vary shape of potential wells by spatial Fourier synthesis or by imaging of mask patterns. Example: more widely separated potential wells, larger region of purely circular polarization near center of wells, approximate square wells, Coulomb-wells, etc. Trapping in non-diffracting "Bessel beams"?

Taking Pictures
Observe atom crystal with microscope, fast intensified CCD camera. Use sheet-like geometry for well defined distance lens - object. Thin planar vacuum cell?

Single Atom
Try to observe (image?) single trapped atom. Study diffusion, tunelling

Dark Traps
Feasibility of atom crystals with blue detuned light?

Adiabatic Cooling
Try to change trap potentials continuously. Cool below recoil limit?

Collective Vibrations
Excite synchronous vibrations by "shaking the egg-tray".
Apply phase modulation to trapping light beams.
Apply oscillating external electric or magnetic field gradients.
Stroboscopic Observation?

Nonclassical Atomic Motion
Try to produce and observe squeezing, Fock-states, Glauber states, Schrödinger cat states...

Image Vibrational Wavefunctions
Use local scanning probe. Scanning electron microscope? Ion beam? Thin wire or tip? VUV or soft X-ray ionizing probe? Using ion-optic magnification by photoionization + acceleration in electric field of fine metal tip. Trap atoms with evanescent standing wave, observe deposits on surface. Mode selection?

Negative Mass
Try to observe change in effective mass of atoms in periodic potentials. Response to external force (electric, magnetic, optical, gravity). Atoms with negative mass should fall upwards.

Two Species
Simultaneous trapping of 2 different species (isotopes?).
Study interactions. Trap separately, move crystals through each other?

Molecular Crystals
Explore trapping of molecules. Use CO_2 laser? Sympathetic cooling? Buckey balls?

Fig. 2 (continued)

Back in Munich we all looked very carefully at the recorded video of Eric's talk and were wondering about the mysterious aluminum foil which was scattered around the TOP-trap experiment. Probably the picture was taken soon after a bake out of the vacuum apparatus. A few weeks later, Ted supported my plan to convert the set-up for slow atomic rubidium beams into a Bose–Einstein–condensation (BEC) experiment. A year later, when the double magneto-optical trap was running, I went to his office and asked whether we could hire a PhD student to join the project. Ted picked up a folder, which happened to lie on his desk, and said that in a week's time an applicant who had done his Diploma thesis with Dieter Meschede and was now working with Mark Kasevich would come and visit. Immanuel Bloch joined the new experiment in September 1997 to do his PhD. We did not quite make it to producing the first Bose–Einstein condensate outside the US, but the second—Gerhard Rempe's team, then at the University of Constance, was a bit faster. Our experiment featured a simple magnetic trap and μ-metal shield [12], so we could still make our mark with the continuous-wave atom laser [13] (see Fig. 3) and the direct measurement of off-diagonal long-range order [14].

Ted generously let me talk at Varenna Summer School on BEC at Lake Como in 1998, where I had a very inspirational conversation with Eric Cornell on how to design future BEC experiments. This resulted in two similar, but distinctly different, concepts for magnetic-transport experiments, using either overlapping coils or a moving pair of coils. Markus Greiner, who had already joined us for an internship to produce phase-contrast images of our Bose–Einstein condensates, worked out how to use overlapping magnetic coils for magnetic transport [15] and set up this second-generation BEC experiment during his Diploma (Masters) project. It was designed for the use of both rubidium isotopes [16] and optical lattices [17].

In the summer of 2001, we obtained our first results of the superfluid-to-Mott-insulator transition. When we presented our first momentum pictures (see Fig. 4) to Ted, he commented with a cheeky smile on his face: "Oh, I've got many research proposals on my desk that have the long-term goal of what you have done just now". The Mott insulator was a new beast and turned out to be rather robust.

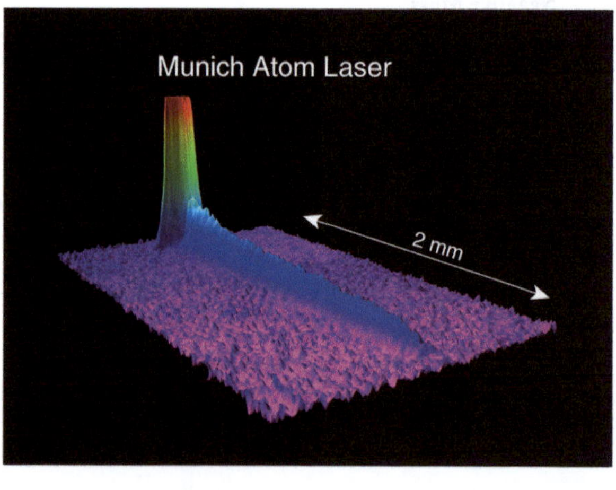

Fig. 3 The absorption image of an atom-laser beam continuously extracted from a Bose–Einstein condensate [13]

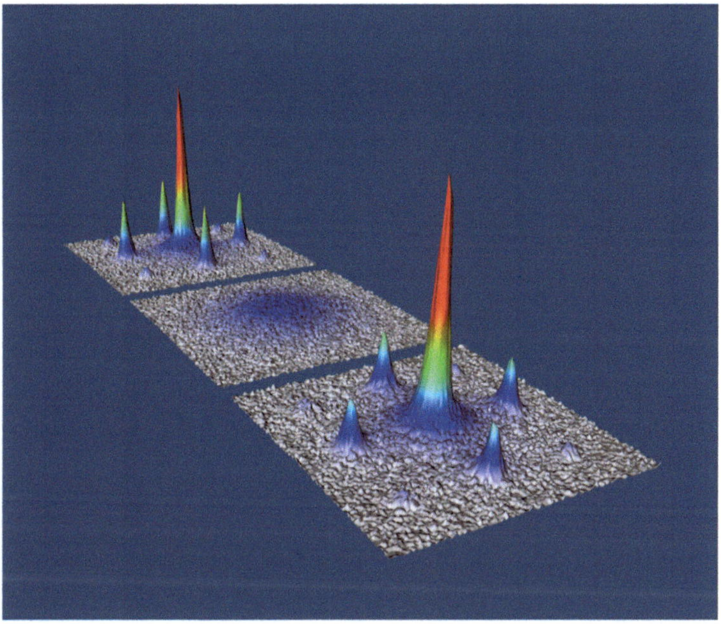

Fig. 4 Quantum phase transition from a superfluid to a Mott insulator and back. The images show the presence and absence of matter-wave interference pattern [18]

It could not easily be destroyed; only when we tilted the lattice rather vigorously by applying a magnetic-field gradient, we managed to excite the system and measure the gap spectrum [18]. In Schellingstrasse 4, we benefitted a lot from Wilhelm Zwerger being dean of the Physics department in the same building, two floors up. After his administrative duties he would often pass by and help us understand the physics of Bose–Einstein condensates and Hubbard models.

Ted created a unique research-oriented yet playful atmosphere in Schellingstrasse 4. And he was, and still is, always keen to play with the latest gadgets. I will always remember the day when I entered Ted's carpeted private optics labs to show him some of our latest data. He was sat on the floor next to John Hall, who had come for a visit, both playing with a magnetic levitron that John had brought with him from Boulder as a present. Unfortunately phones with built-on cameras were not ubiquitous back then and I could not take a picture of this scene of the two future Nobel laureates.

Ted Hänsch gave us the freedom to spread out into many new directions. Failed attempts would usually remain uncommented. He would always welcome a good science discussion, in particular on Sundays, when his office door was usually ajar, which we interpreted as a sign that he has time for a discussion. When we were cooking up something new in the labs, there was from time to time a very gentle knock on the lab door and you could be sure that it was Ted. He was curious, but at the same time he did not want to disturb that one activity he values the most: Doing science.

References

1. T.W. Hänsch, A.L. Schawlow, Cooling of gases by laser radiation. Opt. Commun. **13**, 68–69 (1975)
2. A. Hemmerich, D. McIntyre, D. Schropp, D. Meschede, T. Hänsch, Optically stabilized narrow linewidth semiconductor laser for high resolution spectroscopy. Opt. Commun. **75**, 118–122 (1990)
3. A. Hemmerich, D. Schropp, T. Esslinger, T. Hänsch, Elastic scattering of rubidium atoms by two crossed standing waves. Europhys. Lett. **18**, 391–395 (1992)
4. T. Esslinger, A. Hemmerich, T. Hänsch, Imaging an atomic beam in two dimensions. Opt. Commun. **93**, 49–53 (1992)
5. L. Ricci, M. Weidemüller, T. Esslinger, A. Hemmerich, C. Zimmermann, V. Vuletic, W. König, T.W. Hänsch, A compact grating-stabilized diode laser system for atomic physics. Opt. Commun. **117**, 541–549 (1995)
6. Paul Traps, A TWH production (1991), https://www.youtube.com/watch?v=bkYXNeJ8IP0
7. A. Hemmerich, M. Weidemüller, T. Esslinger, T.W. Hänsch, Collective atomic dynamics in a magneto-optical trap. Europhys. Lett. **21**, 445–450 (1993)
8. A. Hemmerich, T. Hänsch, Two-dimensional atomic crystal bound by light. Phys. Rev. Lett. **70**, 410–413 (1993)
9. M. Weidemüller, A. Hemmerich, A. Görlitz, T. Esslinger, T. Hänsch, Bragg scattering in an atomic lattice bound by light. Phys. Rev. Lett. **75**, 4583–4586 (1995)
10. V. Vuletic, T.W. Hänsch, C. Zimmermann, Steep magnetic trap for ultra cold atoms. EPL **36**, 349–354 (1996)
11. T. Esslinger, F. Sander, M. Weidemüller, A. Hemmerich, T. Hänsch, Subrecoil laser cooling with adiabatic transfer. Phys. Rev. Lett. **76**, 2432–2435 (1996)
12. T. Esslinger, I. Bloch, T.W. Hänsch, Bose-Einstein condensation in a quadrupole-ioffe-configuration trap. Phys. Rev. A **58**, 2664–2667 (1998)
13. I. Bloch, T.W. Hänsch, T. Esslinger, Atom laser with a cw output coupler. Phys. Rev. Lett. **82**, 3008–3011 (1999)
14. I. Bloch, T.W. Hänsch, T. Esslinger, Measurement of the spatial coherence of a trapped Bose gas at the phase transition. Nature **403**, 166–170 (2000)
15. M. Greiner, I. Bloch, T.W. Hänsch, T. Esslinger, Magnetic transport of trapped cold atoms over a large distance. Phys. Rev. A **63**, 031401 (2001)
16. I. Bloch, M. Greiner, O. Mandel, T.W. Hänsch, T. Esslinger, Sympathetic cooling of 85Rb and 87Rb. Phys. Rev. A **64**, 021402 (2001)
17. M. Greiner, I. Bloch, O. Mandel, T.W. Hänsch, T. Esslinger, Exploring phase coherence in a 2D lattice of Bose-Einstein condensates. Phys. Rev. Lett. **87**, 160405 (2001)
18. M. Greiner, O. Mandel, T. Esslinger, T.W. Hänsch, I. Bloch, Quantum phase transition from a superfluid to a Mott insulator in a gas of ultracold atoms. Nature **415**, 39–44 (2002)

FSC
www.fsc.org
MIX
Papier aus verantwortungsvollen Quellen
Paper from responsible sources
FSC® C105338

If you have any concerns about our products,
you can contact us on
ProductSafety@springernature.com

In case Publisher is established outside the EU,
the EU authorized representative is:
Springer Nature Customer Service Center GmbH
Europaplatz 3, 69115 Heidelberg, Germany

Printed by Libri Plureos GmbH
in Hamburg, Germany